Lecture Notes in Computer Science 5126

Commenced Publication in 1973
Founding and Former Series Editors:
Gerhard Goos, Juris Hartmanis, and Jan van Leeuwen

Editorial Board

David Hutchison
   Lancaster University, UK
Takeo Kanade
   Carnegie Mellon University, Pittsburgh, PA, USA
Josef Kittler
   University of Surrey, Guildford, UK
Jon M. Kleinberg
   Cornell University, Ithaca, NY, USA
Alfred Kobsa
   University of California, Irvine, CA, USA
Friedemann Mattern
   ETH Zurich, Switzerland
John C. Mitchell
   Stanford University, CA, USA
Moni Naor
   Weizmann Institute of Science, Rehovot, Israel
Oscar Nierstrasz
   University of Bern, Switzerland
C. Pandu Rangan
   Indian Institute of Technology, Madras, India
Bernhard Steffen
   University of Dortmund, Germany
Madhu Sudan
   Massachusetts Institute of Technology, MA, USA
Demetri Terzopoulos
   University of California, Los Angeles, CA, USA
Doug Tygar
   University of California, Berkeley, CA, USA
Gerhard Weikum
   Max-Planck Institute of Computer Science, Saarbruecken, Germany

Luca Aceto   Ivan Damgård
Leslie Ann Goldberg
Magnús M. Halldórsson
Anna Ingólfsdóttir   Igor Walukiewicz (Eds.)

Automata, Languages and Programming

35th International Colloquium, ICALP 2008
Reykjavik, Iceland, July 7-11, 2008
Proceedings, Part II

 Springer

Volume Editors

Luca Aceto
Magnús M. Halldórsson
Anna Ingólfsdóttir
Reykjavik University, School of Computer Science
Kringlan 1, 103 Reykjavík, Iceland
E-mail: {luca, mmh, annai}@ru.is

Ivan Damgård
University of Aarhus, Department of Computer Science, IT-Parken
Åbogade 34, 8200 Århus N, Denmark
E-mail: ivan@daimi.au.dk

Leslie Ann Goldberg
University of Liverpool, Department of Computer Science
Ashton Building, Liverpool L69 3BX, UK
E-mail: l.a.goldberg@liverpool.ac.uk

Igor Walukiewicz
Université de Bordeaux-1, LaBRI
351, Cours de la Libération, 33405 Talence cedex, France
E-mail: igw@labri.fr

Library of Congress Control Number: 2008930136

CR Subject Classification (1998): F, D, C.2-3, G.1-2, I.3, E.1-2

LNCS Sublibrary: SL 1 – Theoretical Computer Science and General Issues

ISSN 0302-9743
ISBN 978-3-540-70582-6 Springer Berlin Heidelberg New York

Springer is a part of Springer Science+Business Media

springer.com

© Springer-Verlag Berlin Heidelberg 2008

Typesetting: Camera-ready by author, data conversion by Scientific Publishing Services, Chennai, India
Printed on acid-free paper   SPIN: 12322992   06/3180   5 4 3 2 1 0

Preface

ICALP 2008, the 35th edition of the International Colloquium on Automata, Languages and Programming, was held in Reykjavik, Iceland, July 7–11, 2008. ICALP is a series of annual conferences of the European Association for Theoretical Computer Science (EATCS) which first took place in 1972. This year, the ICALP program consisted of the established Track A (focusing on algorithms, automata, complexity and games) and Track B (focusing on logic, semantics and theory of programming), and of the recently introduced Track C (focusing on security and cryptography foundations).

In response to the call for papers, the Program Committees received 477 submissions, the highest ever: 269 for Track A, 122 for Track B and 86 for Track C. Out of these, 126 papers were selected for inclusion in the scientific program: 70 papers for Track A, 32 for Track B and 24 for Track C. The selection was made by the Program Committees based on originality, quality, and relevance to theoretical computer science. The quality of the manuscripts was very high indeed, and many deserving papers could not be selected.

ICALP 2008 consisted of five invited lectures and the contributed papers. This volume of the proceedings contains all contributed papers presented in Track B and Track C together with the papers by the invited speakers Ran Canetti (IBM T.J. Watson Research Center and MIT, USA) and Javier Esparza (Technische Universität München, Germany). A companion volume includes all contributed papers presented at the conference in Track A, together with the papers by the invited speakers S. Muthukrishnan (Google, USA) and Bruno Courcelle (Labri, Universitè Bordeaux, France). The program had an additional invited lecture by Peter Winkler (Dartmouth, USA), which does not appear in the proceedings.

The following workshops were held as satellite events of ICALP 2008:

ALGOSENSORS 2008 – 4th International Workshop on Algorithmic Aspects of Wireless Sensor Networks
CL&C 2008 – Second International Workshop on Classical Logic and Computation
FOCLASA 2008 – 7th International Workshop on the Foundations of Coordination Languages and Software Architectures
FIMN 2008 – Foundations of Information Management in Networks
FBTC 2008 – From Biology To Concurrency and Back
ICE 2008 – Interaction and Concurrency Experience
MatchUP 2008 – Matching Under Preferences - Algorithms and Complexity
MSFP 2008 – Second Workshop on Mathematically Structured Functional Programming
PAuL 2008 – Third International Workshop on Probabilistic Automata and Logics
QPL/DCM 2008 – 5th Workshop on Quantum Physics and Logic and 4th Workshop on Development of Computational Models

SOS 2008 – 5th Workshop on Structural Operational Semantics
IMAGINE 2008 – Second International Workshop on Mobility, Algorithms and
Graph Theory in Dynamic Networks
DYNAMO 2008 – Second Training School on Algorithmic Aspects of Dynamic
Networks

We wish to thank all authors who submitted extended abstracts for consideration, the Program Committees for their scholarly effort, and all referees who assisted the Program Committees in the evaluation process.

Thanks to the sponsors (CCP Games, Icelandair, IFIP TC1, Teymi) for their support, and to Reykjavik University for hosting ICALP 2008. We are also grateful to all members of the Organizing Committee in the School of Computer Science and to the Facilities and Technical staff of Reykjavik University. Thanks to Andrei Voronkov and Shai Halevi for writing the conference-management systems EasyChair and Web-Submission-and-Review software, which were used in handling the submissions and the electronic PC meeting as well as in assisting in the assembly of the proceedings.

May 2008   Luca Aceto
                                               Ivan Damgård
                                          Leslie Ann Goldberg
                                        Magnús M. Halldórsson
                                           Anna Ingólfsdóttir
                                            Igor Walukiewicz

Organization

Program Committee

Track A

Michael Bender, State University of New York at Stony Brook, USA
Magnus Bordewich, Durham University, UK
Lenore Cowen, Tufts University, USA
Pierluigi Crescenzi, Università di Firenze, Italy
Artur Czumaj, University of Warwick, UK
Edith Elkind, University of Southampton, UK
David Eppstein, University of California at Irvine, USA
Leslie Ann Goldberg, University of Liverpool, UK (Chair)
Martin Grohe, Humboldt-Universität zu Berlin, Germany
Giuseppe F. Italiano, Università di Roma "Tor Vergata", Italy
Christos Kaklamanis, University of Patras, Greece
Peter Bro Miltersen, University of Aarhus, Denmark
Michael Mitzenmacher, Harvard University, USA
Ian Munro, University of Waterloo, Canada
Ryan O'Donnell, Carnegie Mellon University, USA
Dana Ron, Tel-Aviv University, Israel
Tim Roughgarden, Stanford University, USA
Christian Scheideler, Technische Universität München, Germany
Christian Sohler, University of Paderborn, Germany
Luca Trevisan, University of California at Berkeley, USA
Berthold Voecking, RWTH Aachen University, Germany
Gerhard Woeginger, Eindhoven University of Technology, The Netherlands

Track B

Parosh Abdulla, Uppsala University, Sweden
Luca de Alfaro, University of California, Santa Cruz, USA
Christel Baier, Technische Universität Dresden, Germany
Giuseppe Castagna, Université Paris 7, France
Rocco de Nicola, Università di Firenze, Italy
Javier Esparza, Technische Universität München, Germany
Marcelo Fiore, University of Cambridge, UK
Erich Grädel, RWTH Aachen, Germany
Jason Hickey, California Institute of Technology, USA
Martin Hofmann, Ludwig-Maximilians-Universität München, Germany
Hendrik Jan Hoogeboom, Leiden University, The Netherlands

Radha Jagadeesen, DePaul University, USA
Madhavan Mukund, Chennai Mathematical Institute, India
Luke Ong, Oxford University, UK
Dave Schmidt, Kansas State University, USA
Philippe Schnoebelen, ENS Cachan, France
Igor Walukiewicz, Labri, Université Bordeaux, France (Chair)
Mihalis Yannakakis, Columbia University, USA
Wieslaw Zielonka, Université Paris 7, France

Track C

Christian Cachin, IBM Research Zürich, Switzerland
Jan Camenisch, IBM Research Zürich, Switzerland
Ivan Damgård, University of Aarhus, Denmark (Chair)
Stefan Dziembowski, Università di Roma "La Sapienza", Italy
Dennis Hofheinz, CWI Amsterdam, The Netherlands
Susan Hohenberger, Johns Hopkins University, USA
Yuval Ishai, Technion Haifa, Israel
Lars Knudsen, DTU Copenhagen, Denmark
Arjen Lenstra, EPFL Lausanne, Switzerland
Anna Lysyanskaya, Brown University, USA
Rafael Pass, Cornell University, USA
David Pointcheval, ENS Paris, France
Dominique Unruh, Saarland University, Germany
Serge Vaudenay, EPFL Lausanne, Switzerland
Bogdan Warinschi, Bristol University, UK
Douglas Wikström, KTH Stockholm, Sweden
Stefan Wolf, ETH Zürich, Switzerland

Organizing Committee

Luca Aceto, Reykjavik University (Conference Chair)
Bjarni V. Halldórsson, Reykjavik University (Workshop Co-chair)
Magnús M. Halldórsson, Reykjavik University (Conference Chair)
Anna Ingólfsdóttir, Reykjavik University (Conference Chair)
MohammadReza Mousavi, Eindhoven University of Technology (Workshop
    Co-chair)

Sponsoring Institutions

CCP Games
Icelandair
IFIP TC1
Reykjavik University
Teymi

Referees

Michel Abdalla
Andreas Abel
Jiri Adamek
Ben Adida
Foto Afrati
Benjamin Aminof
Torben Amtoft
Vikraman Arvind
Eugene Asarin
Giuseppe Ateniese
Albert Atserias
Jean-Philippe Aumasson
Thomas Baignres
Steffen van Bakel
Vince Barany
Joerg Bauer
Nick Benton
Còme Berbain
Martin Berger
Lennart Beringer
Nathalie Bertrand
Puneet Bhateja
Henrik Björklund
Bruno Blanchet
Manuel Bodirsky
Mikolaj Bojanczyk
Benedikt Bollig
Michele Boreale
Joppe Bos
Charles Bouillaguet
Patricia Bouyer
Tomas Brazdil
Thomas Brihaye
Andrei Bulatov
Manuela Burojani
Thomas Bäck
Cristian S. Calude
Olivier Carton
Frank Cassez
Dario Catalano
Balder ten Cate
Rafik Chaabouni
Amine Chaieb
Supratik Chakraborty

Prakash Chandrasekaran
Krishnendu Chatterjee
Jan Chomicki
Evelyne Contejean
Scott Contini
Veronique Cortier
Arnaud Da Costa
Deepak D'Souza
Victor Dalmau
Mads Dam
Pierpaolo Degano
Giorgio Delzanno
Stephane Demri
Yuxin Deng
Alex Dent
Josee Desharnais
Dan Dougherty
Ross Duncan
Irène Durand
Stephen A. Edwards
Joost Engelfriet
Javier Esparza
John Fearnley
Serge Fehr
Jerome Feret
Anna Lisa Ferrara
Maribel Fernandez
Marc Fischlin
Matthias Fitzi
Riccardo Focardi
Wan Fokkink
Georg Fuchsbauer
Murdoch Gabbay
Fabio Gadducci
Tobias Ganzow
Juan Garay
Stephane Gaubert
Blaise Genest
Silvia Ghilezan
Giuseppe De Giacomo
Hugo Gimbert
Cinzia Di Giusto
Rob van Glabeek
Stefania Gnesi

Emmanuel Godard
Rodolfo Gomez
Michaela Goetz
Mikael Goldmann
Daniele Gorla
Nathaniel Gray
Gianluigi Greco
Matthew Green
Alain Griffault
Gary Griffing
Serge Grigorieff
Colas Le Guernic
Stefano Guerrini
Peter Habermehl
Serge Haddad
Esfandiar Haghverdi
Matthew Hague
Noomene Ben Henda
Monika Henzinger
Holger Hermanns
Thomas Hildebrandt
Jane Hillston
Peter Hines
Markus Holzer
Andreas Holzer
Haruo Hosoya
Nick Howgrave-Graham
Cătălin Hriţcu
Juraj Hromkovic
Emeline Hufschmitt
Radu Iosif
Ellen Jochemsz
Jan Johannsen
Lisa Kaati
Joost-Pieter Katoen
Stefan Kiefer
Eike Kiltz
Joachim Klein
Joost Kok
Pavel Krcal
Gunnar Kreitz
Manfred Kufleitner
Stefan Kugele
K. Narayan Kumar

Michal Kunc
Alp Kupcu
Anna Labella
Yassine Lakhnech
Matthew R. Lakin
Martin Lange
François Laroussinie
Sławomir Lasota
Axel Legay
Stephane Lengrand
Jerome Leroux
Martin Leucker
Ming Li
Leonid Libkin
Huijia Lin
Moses Liskov
Kamal Lodaya
Hans-Wolfgang Loidl
Sylvain Lombardy
Michele Loreti
Michael Luttenberger
Vadim Lyubashevsky
Sergio Maffeis
Mila Majster-
  Cederbaum
Rupak Majumdar
Nicolas Markey
Paulo Mateus
Ralph Matthes
Krystian Matusiewicz
Alex May
Richard Mayr
Massimo Merro
Antoine Meyer
Christian Michaux
Dale Miller
Paul Morrissey
Francesco Zappa Nardelli
Frank Neven
Joachim Niehren
Damian Niwinski
David Noblet
Jakob Nordström
Aditya Nori
Alexander Okhotin
Vincent van Oostrom

Khaled Ouafi
Raphael Overbeck
Prakash Panangaden
Paritosh Pandya
Matthew Parkinson
Sylvain Pasini
Rafael Pass
Michael Østergaard
  Pedersen
Paul Pettersson
Benjamin Pierce
Krzysztof Pietrzak
Jean-Eric Pin
Libor Polak
François Pottier
Bartosz Przydatek
Rosario Pugliese
Christophe Raffalli
George Rahonis
R. Ramanujam
Jean-François Raskin
Dominik Raub
Jason Reed
Renato Renner
Arend Rensink
Gwénaël Richomme
Tom Roeder
Sabina Rossi
Abhik Roychoudhury
Albert Rubio
Jacques Sakarovitch
Sylvain Salvati
Juraj Sarinay
Christian Schaffner
Christian Schallhart
Alan Schmitt
Gerardo Schneider
Stefan Schwoon
Ulrich Schöpp
Roberto Segala
Luc Segoufin
Helmut Seidl
Peter Sewell
Andrey Sidorenko
Jeremy Sproston
Srikanth Srinivasan

Ludwig Staiger
Martijn Stam
Sam Staton
Benjamin Steinberg
Thomas Streicher
S.P. Suresh
Gregoire Sutre
Paulo Tabuada
Alain Tapp
Stefano Tessaro
Denis Therien
Hayo Thielecke
Soren S. Thomsen
Tayssir Touili
Jan Tretmans
Dustin Tseng
Emilio Tuosto
Irek Ulidowski
Paweł Urzyczyn
Muthu
  Venkitasubramaniam
Damien Vergnaud
Björn Victor
Aymeric Vincent
Walter Vogler
Martin Vuagnoux
Dirk Walther
Yongge Wang
Benne de Weger
Pascal Weil
Philipp Weis
Thomas Wilke
Erik Winfree
Christopher Wolf
Verena Wolf
James Worrell
Jörg Wullschleger
Shaofa Yang
Nina Yevtushenko
Nobuko Yoshida
Dae Hyun Yum
Gianluigi Zavattaro
Lisa Zhang
Vassilis Zikas

Table of Contents – Part II

Invited Lectures

Track B: Logic, Semantics, and Theory of Programming

Bounds

Distributed Computation

Real-Time and Probabilistic Systems

Logic and Complexity

Words and Trees

Nonstandard Models of Computation

Reasoning about Computation

Verification

Track C: Security and Cryptography Foundations

Theory

Secure Computation

Two-Party Protocols and Zero-Knowledge

Encryption with Special Properties/Quantum Cryptography

Various Types of Hashing

Public-Key Cryptography/Authentication

Table of Contents – Part I

Random Walks and Random Structures

Design and Analysis of Algorithms

Scheduling

Codes and Coding

Coloring

Randomness in Computation

Online and Dynamic Algorithms

Approximation Algorithms

Property Testing

Parameterized Algorithms and Complexity

Graph Algorithms

Computational Complexity

Games and Automata

Group Testing, Streaming, and Quantum

Algorithmic Game Theory

Quantum

Composable Formal Security Analysis: Juggling Soundness, Simplicity and Efficiency

Ran Canetti*

IBM Research
canetti@csail.mit.edu

Abstract. A security property of a protocol is *composable* if it remains intact even when the protocol runs alongside other protocols in the same system. We describe a method for asserting composable security properties, and demonstrate its usefulness. In particular, we show how this method can be used to provide security analysis that is formal, relatively simple, and still does not make unjustified abstractions of the underlying cryptographic algorithms in use. It can also greatly enhance the feasibility of *automated* security analysis of systems of realistic size.

1 Introduction

Security analysis of protocols is a slippery business. On the one hand, we want to capture all "feasible attacks". On the other hand, we want to allow those protocols that do not succumb to attacks. Indeed, time and again attacks are found against protocols that were thoroughly analyzed and sometime even deployed and standardized (see e.g. [Ble98, Low96]). The situation is particularly tricky when the analyzed protocol uses "cryptographic primitives", namely algorithms that guarantee certain behaviors only when the adversarial components of system are computationally bounded.

A crucial first step in any rigorous security analysis is to devise an appropriate mathematical model for representing protocols and formulating the desired security properties. Indeed, the analysis can only be meaningful to the degree that the devised model and the formulated security requirements are meaningful.

Many models for analyzing security of protocols have been proposed over the past few decades, each with its own advantages and drawbacks. Roughly, there are two main analytical approaches, which differ in the way the cryptographic primitives used by the protocol and their security properties are modeled. In symbolic models, devised mainly within the formal analysis community, cryptographic primitives are treated as abstract, or symbolic operations with rigid interfaces that restrict the way in which the primitives can be used - and, more importantly, the ways in which the primitives can be *mis*used by adversarial components. In a way, this models the cryptographic primitives in use as "ideal

* IBM T.J. Watson Research Center. Supported by NSF grant CFF-0635297 and US-Israel Binational Science Foundation Grant 2006317.

L. Aceto et al. (Eds.): ICALP 2008, Part II, LNCS 5126, pp. 1–13, 2008.
© Springer-Verlag Berlin Heidelberg 2008

boxes" that provide "absolute security". Quintessential examples of such models include the Dolev-Yao model [DY83], the BAN logic [BAN89], Spi-calculus [AG97] and their many derivatives.

In contrast, computational models (such as those of [GM84, GMR89, BR93] and many others) explicitly treat cryptographic constructs as algorithms, and consider adversaries that have full access to the actual input and output strings of these algorithms. In that respect, these models directly reflect the actual capabilities of adversaries in realistic systems. Here, meaningful formalizations of security requirements have to be probabilistic. Furthermore, they have to incorporate computational bounds on the adversarial entities involved. In addition, given the current state of the art in complexity theory, such analysis has to rely on computational hardness assumptions.

These two analytical approaches provide a clear tradeoff: The symbolic approach is much simpler and easier to work with than the computational approach. Also, it is conceptually attractive since it allows for clear separation of the analysis at the "protocol level" from the analysis of the underlying primitives. However, at least a priori, the computational approach is the only one that is sound; that is, it is the only approach that can actually provide security guarantees for protocols in realistic settings.

A recent research program, initiated in [AR02] and followed in many works since, is aimed at combining these two analytical approaches in a single model that provides the best of both: Soundness together with the ability to argue about protocols in a symbolic and mechanical way. A number of approaches have been proposed to carry out this combination. This paper reviews one such approach, that builds on security models that provide a general *security-preserving composition* guarantee. Specifically, the approach uses the universal composition theorem [PW00, BPW04, Can01], which guarantees that a protocol that uses an abstractly specified primitive can be securely "composed" with a protocol that realizes this specification, "without bad side effects". Here the symbolic model would correspond to the protocol that uses the abstract primitive, and the soundness would follow from the security preserving composition theorem. See more details within.[1]

While the initial thrust of the above work is to argue the soundness of the symbolic approach, there is an additional aspect here that we wish to highlight. (Indeed, this aspect seems to have been overlooked by most works in that area.) The symbolic approach, being dramatically simpler than the computational one, lends naturally to mechanization and automation of the analysis (see e.g. [Mea96, MMS97, Bla03]). Still, traditional automated symbolic analysis is feasible only for relatively small systems: the complexity of analysis is typically exponential in the number of variables, parties, and protocol instances in the analyzed system (see e.g. [MS01]). In fact, when the protocol description and number

---

[1] Our notion of "secure composition" differs from other notions of compositionality, such as the one in, say, [DMP01], which is a more fine-grained approach for synthesizing protocols from elementary instructions, and does not carry composition theorems akin to the ones here.

of instances is taken to be part of the input, the question whether a symbolic protocol satisfies a certain property is NP-hard. When the number of protocol instances is unbounded the question becomes undecidable [EG83, DLMS99].

Composable security can help overcome this complexity barrier in certain systems of interest. Indeed, when asserting composable security properties, it suffices to apply the symbolic analysis to small, single-instance systems. Security of large composite systems would then follow from the composition theorem.

Organization. This paper is organized as follows. Section 2 briefly reviews symbolic protocol analysis. Section 3 briefly reviews the universally composable (UC) security framework. Section 4 reviews ways in which UC security has been used to assert the soundness of symbolic analysis and to enable its efficient automation. Section 5 concludes with some directions for further research.

Throughout, we do not attempt to give a broad survey of all relevant works. Rather, the intention is to present the main ideas, concerns and challenges, as seen by the author, in a way that is accessible to the non-expert. We apologize for any misrepresentations and omissions.

2 Symbolic Analysis in a Nutshell

There are a number of approaches for formulating models for protocol analysis where the cryptographic primitives are represented in an "idealized", or abstract way. Examples include the Dolev-Yao model [DY83], which essentially amounts to formulating a protocol-dependent abstract algebra where security properties translate to representability questions in the algebra (see, e.g. [Pau98, FHG98]); the BAN logic [BAN89] where security properties are translated to assertions in a protocol-dependent logic; or the spi-calculus [AG97] where security properties are translated to observational equivalence assertions in an extension of the π-calculus [Mil89]. Here we briefly sketch one of these approaches, namely the Dolev-Yao model, which is relatively simple and self contained. (On the down side, this model tends to be specific for a given class of tasks and protocols, and has to be reformulated whenever the task or class of protocols changes.)

The Dolev-Yao model has several components. First, the model defines a symbolic algebra. The atomic elements of the algebra represent primitive structures such as party identifiers, public and secret keys for the cryptographic algorithms in use, and random challenges (nonces).

Operations in the algebra represent the allowed usage of the cryptographic primitives — both by the legitimate protocol parties and by adversarial entities. For instance, in the case of public key encryption the symbolic encryption operation Enc takes a public key symbol ek and arbitrary symbol m (say, an identifier or a nonce) to return a compound "ciphertext" symbol $Enc_{ek}(m)$. The symbolic decryption operation Dec takes a private key symbol dk and a symbol of the from $Enc_{ek}(m)$ where ek and dk are paired, and returns the symbol m.

Inputs, outputs, and protocol messages are represented as compound elements in the algebra. That is, each message (compound element) represents a "parse tree", or the sequence of operations needed to obtain the compound element from elementary ones. The algebra is free: it admits no equalities other than the identity. That is, each message has exactly one representation. In the above example, for instance, this means that there is no way to retrieve the symbol m (or gain any information on it) from $Enc_{ek}(m)$ without explicitly using the special symbol dk.

Symbolic protocols are defined via a function from the sequence of messages received so far to the next move, when a move consists of a message to be transmitted or alternatively some local output. All inputs, outputs, and messages are compound elements from the algebra.

The symbolic adversary is defined in two parts: its initial knowledge (a set of symbolic messages), and the adversary operations it can use to deduce new messages from known ones. (These known messages consist of the initial knowledge and the messages sent during the protocol execution.) The adversary operations are bound by the operations specified in the algebra. Typically, these operations are limited to the operations that represent the cryptographic primitives in use, plus simple operations such as concatenation and de-concatenation.

The closure of a message (or a set of messages) is the set of all messages that the adversary can potentially derive from the given message (or set). That is, the closure operation defines the messages which the adversary can create and transmit at any point.

A protocol execution in this model consists of a sequence of events where each event consists of the delivery of an adversarially generated message to some party, followed by the generation of new outgoing message, or a new local input, by that party.

The trace of an execution is the sequence of these events. The security properties of protocols are typically (but not always) predicates on sets of traces: A protocol satisfies such a security property if the predicate is satisfied by the set of that protocol's possible (or valid) traces.

As discussed in the introduction, this model has two substantial limitations: First, it does not provide any guarantees regarding the security of protocols that use concrete algorithms to implement the abstract cryptographic primitives postulated by the algebra. Second, mechanic verification of security properties of protocols is intractable in general. We'll see that both of these limitations can be overcome by taking a compositional approach to security analysis.

3   Universally Composable Security

We turn to a brief review of the universally composable (UC) security framework. (The first variant of the framework appears in [Can01]; some context and related work are briefly discussed below). The framework takes the cryptographic approach to protocol analysis; namely, the adversarial entities are given unrestricted access to the actual bits of the communication between parties. Also, adversaries are taken to be computationally bounded and the security properties are stated in probabilistic and asymptotic terms.

In this setting, the framework provides a general way for specifying the security requirements of cryptographic tasks, and asserting whether a given protocol realizes the specification. A salient property of this framework is that it provides strong composability guarantees: A protocol that meets a specification in isolation continues to meet the specification regardless of the activity in the rest of the network. We give here a very high level sketch of the framework, as well as some motivation. See [Can01, Can06] for a more thorough treatment.

The trusted party paradigm. The underlying definitional idea (which originates in [GMW87], albeit very informally) proceeds as follows. To determine whether a given protocol is secure for some cryptographic task, first envision an ideal process for carrying out the task in a secure way. In the ideal process all parties secretly hand their inputs to an external trusted party who locally computes the outputs according to the specification, and secretly hands each party its prescribed outputs. This ideal process can be regarded as a "formal specification" of the security requirements of the task. (For instance, when the task is to compute a joint function f of the local inputs of the parties, the trusted party simply evaluates f on the inputs provided by the parties, and hands the outputs back to the parties. If the function is probabilistic then the trusted party also makes the necessary random choices.) The protocol is said to securely realize a task if running the protocol amounts to "emulating" the ideal process for the task, in the sense that any damage that can be caused by an adversary interacting with the protocol can also happen in the ideal process for the task.

An attractive property of this approach is its generality: It seems possible to capture the requirements of very different tasks by considering different sets of instructions for the external trusted party. Another attractive property is potential compositionality: It seems almost "built into the definitional approach" that if a protocol successfully mimics the behavior of some trusted party then any protocol that uses the protocol should continue to behave the same when the protocol is replaced by the trusted party.

Still, substantiating this approach in a way that maintains its intuitive appeal and materializes the potential generality and composability turns out to be nontrivial. Indeed, several general frameworks for representing cryptographic protocols and specifying the security requirements of tasks were developed over the years, e.g. [GL90, MR91, Bea91, Can00, HM00, DM00, PW00, Can01, PMS03, K06]. While all of these frameworks follow the above paradigm in one way or another, they differ greatly in their expressibility (i.e., the range of security concerns and tasks that can be captured), in the computational models addressed, and in many significant technical details. They also support different types of security-preserving composition theorems.

The basic formalism. Defining what it means for a protocol π to "securely realize" a certain task is done in three steps, as follows. First, we formulate a model for executing the protocol. This model consists of the parties running π, plus two adversarial entities: the environment $\mathcal{Z}$, which generates the inputs for the parties and reads their outputs, and the adversary $\mathcal{A}$, which reads the

R. Canetti

outgoing messages generated by the parties and delivers incoming messages to the parties. The adversary and the environment can interact freely during the protocol execution.

The adversary represents attacks against a single instance of the analyzed protocol. The environment represents "everything else that happens in the system," including both the the immediate users of the protocol, and other parties and protocols. Letting $\mathcal{A}$ and $\mathcal{Z}$ interact freely during the computation represents the continual information flow between an execution of a protocol and the rest of the system. Indeed, this provision turns out to be critical for the universal composition theorem to hold.

Next, we formulate the ideal process, in a straightforward way. Here the protocol participants simply pass their inputs to an additional, incorruptible *trusted party*, who locally computes the desired outputs and hands them back to the parties. The program run by the trusted party is called an **ideal functionality** and is intended to capture the security and correctness specifications of the task. For convenience, the ideal process with ideal functionality $\mathcal{F}$ is formulated as the process of running a special protocol $I_{\mathcal{F}}$ called the **ideal protocol** for $\mathcal{F}$. That is, in protocol $I_{\mathcal{F}}$ the parties simply pass all inputs to the trusted party, and output whatever information they obtain from the trusted party. Here the adversary does not interact directly with the parties; instead, it interacts with $\mathcal{F}$ in a way specified by $\mathcal{F}$. The communication between the adversary and the environment remains arbitrary.

Finally, we say that protocol π **UC-emulates** protocol ϕ if for *any* polytime adversary $\mathcal{A}$ there *exists* a polytime adversary $\mathcal{S}$ such that no polytime environment $\mathcal{Z}$ can tell with non-negligible probability whether it is interacting with an execution of π and adversary $\mathcal{A}$, or alternatively with protocol ϕ and adversary $\mathcal{S}$. We say that π **UC-realizes** an ideal functionality $\mathcal{F}$ if it UC-emulates the ideal protocol $I_{\mathcal{F}}$. Somewhat more formally, let $\text{EXEC}_{\mathcal{Z},\mathcal{A},\pi} = \{\text{EXEC}_{\mathcal{Z},\mathcal{A},\pi}(n,z)\}_{n\in N, z\in\{0,1\}^n}$ denote the probability ensemble describing the output of environment $\mathcal{Z}$ in an interaction with adversary $\mathcal{A}$ and protocol π with security parameter n and external input z for $\mathcal{Z}$. Then:

Definition 1. *Protocol* π UC-emulates *protocol* ϕ *if for any polytime adversary* $\mathcal{A}$ *there exists a polytime adversary* $\mathcal{S}$ *such that for any polytime environment* $\mathcal{Z}$ *we have* $\text{Prob}(\text{EXEC}_{\mathcal{Z},\mathcal{A},\pi} = 1) - \text{Prob}(\text{EXEC}_{\mathcal{Z},\mathcal{A},\pi} = 1)| < \nu(n)$, *where* ν *is a negligible function.*

π UC-realizes *an ideal functionality* $\mathcal{F}$ *if it UC-emulates the ideal protocol* $I_{\mathcal{F}}$.

Very informally, the goal of the above requirement is to guarantee that any information gathered by the adversary $\mathcal{A}$ when interacting with π, as well as any "damage" caused by $\mathcal{A}$, could have also been gathered or caused by an adversary $\mathcal{S}$ in the ideal process for $\mathcal{F}$. Now, since the ideal process is designed so that *no* $\mathcal{S}$ can gather information or cause damage more than what is explicitly permitted in the ideal process for $\mathcal{F}$, we can conclude that $\mathcal{A}$ too, when interacting with π, cannot gather information or cause damage more than what is explicitly permitted by $\mathcal{F}$. In particular, the I/O behavior of the good parties in the protocol execution is essentially the same as that of the ideal functionality; similarly, the information that $\mathcal{Z}$ learns from $\mathcal{A}$ can be generated (or, "simulated") by $\mathcal{S}$, who is given only the information that it can learn legally from interacting with $\mathcal{F}$.

We remark that the notion of UC emulation can be viewed as a relaxation of the notion of *observational equivalence* of processes (see, e.g., [Mil89]); indeed, observational equivalence essentially fixes the entire system outside the protocol instances, whereas emulation allows the analyst to choose an appropriate simulator that will make the two systems look observationally equivalent. In a way, this relaxation allows the analyst to specify which properties of the analyzed protocol are "salient" and which are "unimportant", and thereby allow for many proofs of security of cryptographic protocols to go through.

Universal composition. The following universal composition theorem holds in this framework. Let π be a protocol that UC-emulates protocol ϕ, and let ρ be a protocol that has access to (multiple instances of) ϕ. Let $\rho^{\pi/\phi}$ be the "composed protocol" which is identical to ρ except that inputs to ϕ are replaced by inputs to π, and outputs from π are treated as outputs from ϕ. Then, protocol $\rho^{\pi/\phi}$ behaves in an indistinguishable way from the original ρ:

Theorem 1. *Let* ρ, π, ϕ *be protocols such that* π *UC-emulates* ϕ. *Then* $\rho^{\pi/\phi}$ *UC-emulates* ρ.

4   Composable Formal Security Analysis

Providing soundness. The idea underlying the use of security-preserving protocol composition for asserting soundness of formal analysis is simple: Intuitively, the formal (or, symbolic) model appears to naturally correspond to a model where protocols have access to a "trusted party") that embodies the abstract properties of the cryptographic primitives in use, just as in the definition of UC realization. Thus, the universal composition theorem should imply that any security property enjoyed by the symbolic protocol continues to be enjoyed by the protocol even when the symbolic cryptographic primitive is replaced by a concrete protocol that realizes the corresponding ideal functionality. This idea was mentioned already in [PW00] with respect to their formalism (which bears some significant similarities with the UC framework) and also in [Can01].

Substantiating this idea involves a number of steps. Specifically, one has to carry out the following:

1. Formulate ideal functionalities, within the UC framework, that capture in an abstract way the functionality and security properties of the cryptographic primitives in use.
2. Devise concrete protocols that UC-realize the formulated functionalities.
3. Formulate a class (or, rather, a "programming language") of concrete protocols that make use of (i.e., subroutine calls to) the formulated ideal functionalities. (We call these protocol **hybrid** protocols, since they are a hybrid of a concrete protocol with an abstract ideal functionality.)
4. Formulate a symbolic model that models the cryptographic primitives in use in an abstract way, akin to the formulated ideal functionalities.

5. Formulate a security property (goal) for the concrete protocols.
6. Formulate a translation of this property in the symbolic model, and a method for asserting this property in the symbolic model.
7. Demonstrate that if a symbolic protocol satisfies the symbolic property (within the symbolic model) then the corresponding hybrid protocol, within the devised language, satisfies the corresponding concrete property.

Now, we can translate hybrid protocols to fully concrete ones by replacing the ideal functionalities with the protocols that UC-realize them, and use the universal composition theorem to deduce that the fully concrete protocols enjoy the same security properties enjoyed by the hybrid protocols.

A substantiation of these ideas, along the lines of the above sketch, is given in [BPW03]. That work concentrates on protocols where the cryptographic primitives in use are public-key encryption, digital signatures, and secure communication channels. They also provide symbolic constructs that correspond to the use of random challenges, or nonces. (Indeed, these primitives are the ones addressed by traditional symbolic models.)

Specifically, an ideal functionality is formulated, that provides the interface expected from the above primitives, along with absolute security properties. For instance, to model public-key encryption, the [BPW03] ideal functionality provides an encryption interface, that takes a public key symbol and a message and returns an abstract handle, and a decryption interface that takes a handle and a decryption key symbol, and returns the message associated with the handle and the corresponding encryption key - in case these are defined. (Else the decryption interface returns an error symbol.) Digital signatures and secure channels are modeled via handles in a similar way.

Next, [BPW03] show that their ideal functionality can be realized using known cryptographic protocols. Specifically, any combination of an encryption scheme that's semantically secure against chosen ciphertext attacks [RS91, DDN00] with a signature scheme that's existentially unforgeable against chosen message attacks, along with appropriate symmetric encryption and authentication schemes (for obtaining secure communication channels), suffice.

This work opens the door for abstract security analysis of protocols that use the above primitives. All there is to do is to write the protocol in a way that uses the [BPW03] ideal functionality for all its cryptographic operations. Now, the protocol becomes considerably simpler; in fact, in many cases it becomes deterministic, akin to the symbolic ("Dolev-Yao") model. Security of the corresponding concrete protocol follows from the universal composition theorem, as discussed above.

We note that the [BPW03] modeling does not formulate a dedicated abstract model along the lines of the original Dolev-Yao analysis. Instead, even the abstract protocols are defined and analyzed in the same cryptographic model in which the full-fledged cryptographic protocols are. This forces the analyst to either analyze the abstract protocol in a relatively complex model, or alternatively simplify the model at the price of reduced generality in terms of expressing realistic concerns and situations.

Still, this approach has proven to be very useful, serving as a basis for analyzing a number of protocols, e.g. [Bac04, BP04, BD05, Bac06, BP06, BCJ$^+$06]. The security properties asserted in these works are mainly mutual authentication and generation of a common secret key ("key exchange"), as well as other properties such as transactional integrity in payment systems. Also, this work has been the basis for semi-automated security analysis of protocols, using the Isabelle theorem prover [SBB$^+$06, Pau88].

Feasible mechanization and automation. So far, we have seen how to perform symbolic (abstract) security analysis that provides security guarantees even for fully concrete protocols. However, in spite of its apparent simplicity, traditional symbolic analysis still has a serious shortcoming: As argued in the introduction, performing such analysis in a fully mechanical (or, automated) way is intractable, even for systems of moderate size. Consequently, we can feasibly analyze in a fully mechanical way only systems of relatively small size. In particular, we cannot directly analyze systems which consist of unboundedly many concurrent protocol instances, even when all these instances are instances of the same protocol.

Also here, composable security offers a natural solution: When coming to analyze security of a complex system, first de-compose the system to relatively small components; then, use symbolic analysis to mechanically analyze the security of each component; finally, use the composition theorem to re-compose the components and deduce security properties of the whole system. In particular, when coming to analyze a system which consists of an unbounded number of sessions of the same protocol, it suffices to analyze a single session of this protocol, in isolation.

Two main issues need to be addressed in order to make good of this approach, when carrying out the steps described above: First, in order to be able to perform the symbolic analysis separately in each component, independently of all other components, the ideal functionality in Step 1 above needs to be "de-composable" into multiple independent, simpler ideal functionalities, where each such simpler functionality is used only within a single component.

Second, in order to be able to deduce a security property of the re-composed system from the security of the individual components, the security properties asserted by the symbolic analysis (see Step 6 above) needs to be phrased as *composable* security properties. (In the UC framework, this means that the symbolic security properties need to be translatable to assertions of UC-realizing some ideal functionalities.)

A first attempt for coming up with a formalism that addresses the above two concerns would be to try to use the [BPW03] formalism described above. However, it turns out that this formalism does not address the first concern. (Indeed, here it seems essential that all instances of all cryptographic algorithms will reside within a single ideal functionality. See more discussion in [Can04]). Furthermore, the security properties asserted within this formalism (see above literature) are not composable; thus the second concern is not addressed either.

We are thus motivated to look for alternative ways to substantiate the composable approach to symbolic analysis, that will allow us to materialize the prospective efficiency gains. Such an alternative approach is given in [CH04]. That work concentrates on protocols that use a single cryptographic primitive, namely public-key encryption. As in [BPW03], an ideal functionality is presented that captures the behavior of ideal encryption. Here, however, the formalism is such that multiple instances of the ideal functionality can co-exist in the same system where each instance represents encryption via a different set of keys. (On a technical level, this change requires, among other things, abandoning the convenient abstraction of "handles;" instead, the ideal functionality returns "dummy ciphertexts", which are strings generated by an adversarial computational entity without knowledge of the plaintext.) Still, it is shown in [CH04] that any public-key encryption scheme that's semantically secure against chosen ciphertext attacks can be used to UC-realize the devised ideal functionality. This addresses the first concern mentioned above.

The security properties asserted in [CH04] are the traditional ones: Mutual Authentication and Key Exchange. However, in order to address the second concern, these properties are formulated as composable security properties. Specifically, in [CH04] a special-purpose symbolic algebra is devised for representing the class of protocols considered. Next, symbolic Mutual Authentication and Key Exchange properties are formulated. It it then shown that a symbolic protocol satisfies the symbolic Mutual Authentication (resp., symbolic Key Exchange) property *if and only if* the corresponding concrete protocol UC-realizes an ideal Mutual Authentication (resp., Key Exchange) functionality.

To demonstrate the validity of their approach, [CH04] encode the devised symbolic properties in the language of the ProVerif verification tool [Bla03], and use it to automatically assert security of a systems consisting of an unbounded number of concurrent instances of some variants of the Needham-Schroeder-Lowe protocol. The analysis takes less than a second on a standard commodity laptop. We remind the reader that directly analyzing such a system using traditional means is undecidable.

We remark that [CH04] is strongly influenced by [MW04]. In fact, for the case of mutual authentication [CH04] follows the approach of [MW04] quite closely. However, [MW04] is not formulated within a composable framework and thus it cannot provide the efficiency gains provided by [CH04].

5   Future Research

We are at the early stages of capitalizing on the potential of composable notions of security in enabling sound automated analysis of complex systems. Directions for further research include:

1. Widen the range of cryptographic primitives that can be modeled in an abstract, symbolic, and composable way. In the same vein, widen the range of security properties and tasks that can be asserted symbolically.

2. Construct new tools (or, improve existing ones) to allow for efficient automated security analysis, capitalizing on the composable approach to analysis, with the end goal being to perform fully automated security analysis of real-life systems. An interesting challenge here is to mechanize the process of de-composing a system to small components.
3. In a slightly different vein, it might be interesting to formulate and assert the composability of security properties directly in a symbolic model, without having to rely on the composability properties of the underlying computational framework.

Acknowledgements. I thank the program committee of ICALP 2008 for inviting me to talk at the conference and for soliciting this paper. Special thanks is also due to Oded Goldreich for his invaluable conceptual advice and direction.

References

[AG97]   Abadi, M., Gordon, A.D.: A calculus for cryptographic protocols: The spi calculus. In: 4th ACM Conference on Computer and Communications Security, pp. 36–47 (1997), http://www.research.digital.com/SRC/abadi

[AR02]   Abadi, M., Rogaway, P.: Reconciling two views of cryptography (the computational soundness of formal encryption). J. Cryptology 15(2), 103–127 (2002)

[Bac04]   Backes, M.: A cryptographically sound Dolev-Yao style security proof of the Otway-Rees protocol. In: Samarati, P., Ryan, P.Y.A., Gollmann, D., Molva, R. (eds.) ESORICS 2004. LNCS, vol. 3193, pp. 89–108. Springer, Heidelberg (2004)

[Bac06]   Backes, M.: Real-or-random key secrecy of the Otway-Rees protocol via a symbolic security proof. Electr. Notes Theor. Comput. Sci. 155, 111–145 (2006)

[BAN89]   Burrows, M., Abadi, M., Needham, R.: A logic for authentication. DEC Systems Research Center Technical Report 39 (February 1990); Earlier versions in the Second Conference on Theoretical Aspects of Reasoning about Knowledge, 1988, and the Twelfth ACM Symposium on Operating Systems Principles (1989)

[BCJ+06] Backes, M., Cervesato, I., Jaggard, A.D., Scedrov, A., Tsay, J.-K.: Cryptographically sound security proofs for basic and public-key Kerberos. In: Gollmann, D., Meier, J., Sabelfeld, A. (eds.) ESORICS 2006. LNCS, vol. 4189, pp. 362–383. Springer, Heidelberg (2006)

[BD05]   Backes, M., Dürmuth, M.: A cryptographically sound Dolev-Yao style security proof of an electronic payment system. CSFW, 78–93 (2005)

[Bea91]   Beaver, D.: Foundations of secure interactive computing. In: Feigenbaum, J. (ed.) CRYPTO 1991. LNCS, vol. 576. Springer, Heidelberg (1992)

[Bla03]   Blanchet, B.: Automatic proof of strong secrecy for security protocols. In: IEEE Security and Privacy Conference, pp. 86–102 (2003)

[Ble98]   Bleichenbacher, D.: Chosen ciphertext attacks against protocols based on the RSA encryption standard PKCS #1. In: Krawczyk, H. (ed.) CRYPTO 1998. LNCS, vol. 1462, pp. 1–12. Springer, Heidelberg (1998)

[BP04]   Backes, M., Pfitzmann, B.: A cryptographically sound security proof of the Needham-Schroeder-Lowe public-key protocol. IEEE Journal on Selected Areas in Communications 22(10), 2075–2086 (2004)

[BP06]   Backes, M., Pfitzmann, B.: On the cryptographic key secrecy of the strengthened Yahalom protocol. SEC, 233–245 (2006)

[BPW03]   Backes, M., Pfitzmann, B., Waidner, M.: A composable cryptographic library with nested operations. In: 10th ACM CCS (2003), http://eprint.iacr.org/2003/015/

[BPW04]   Backes, M., Pfitzmann, B., Waidner, M.: A general composition theorem for secure reactive systems. In: Naor, M. (ed.) TCC 2004. LNCS, vol. 2951, pp. 336–354. Springer, Heidelberg (2004)

[BR93]   Bellare, M., Rogaway, P.: Entity authentication and key distribution. In: Stinson, D.R. (ed.) CRYPTO 1993. LNCS, vol. 773, pp. 232–249. Springer, Heidelberg (1994), http://www-cse.ucsd.edu/users/mihir/

[Can00]   Canetti, R.: Security and composition of multi-party cryptographic protocols. J. Cryptology 13(1) (2000)

[Can01]   Canetti, R.: Universally composable security: A new paradigm for cryptographic protocols. In: FOCS, pp. 136–145 (2001); Long version at IACR Eprint Archive entry 2000/067

[Can04]   Canetti, R.: Universally composable signature, certification, and authentication. CSFW. Long version at eprint.iacr.org/2003/239 (2004)

[Can06]   Canetti, R.: Security and composition of cryptographic protocols: A tutorial. SIGACT News 37(3&4) (2006); Available also at the Cryptology ePrint Archive, Report 2006/465

[CH04]   Canetti, R., Herzog, J.: Universally composable symbolic analysis of cryptographic protocols (the case of encryption-based mutual authentication and key-exchange). In: 3rd TCC, 2006. Full version at Cryptology ePrint Archive, Report 2004/334 (2004)

[DDN00]   Dolev, D., Dwork, C., Naor, M.: Non-malleable cryptography. SIAM Journal on Computing 30(2), 391–437 (2000)

[DLMS99] Durgin, N.A., Lincoln, P.D., Mitchell, J.C., Scedrov, A.: Undecidability of bounded security protocols. In: Workshop on Formal Methods and Security Protocols (FMSP) (1999)

[DM00]   Dodis, Y., Micali, S.: Secure computation. In: CRYPTO 2000 (2000)

[DMP01]   Durgin, N.A., Mitchell, J.C., Pavlovic, D.: A compositional logic for protocol correctness. SCFW (2001)

[DY83]   Dolev, D., Yao, A.: On the security of public-key protocols. IEEE Transactions on Information Theory 2(29) (1983)

[EG83]   Even, S., Goldreich, O.: On the security of multi-party ping-pong protocols. In: 24th FOCS, pp. 34–39 (1983)

[FHG98]   Fabrega, F.J.T., Herzog, J.C., Guttman, J.D.: Strand spaces: Why is a security protocol correct? In: IEEE Symposium on Security and Privacy (1998)

[GL90]   Goldwasser, S., Levin, L.: Fair computation of general functions in presence of immoral majority. In: Menezes, A., Vanstone, S.A. (eds.) CRYPTO 1990. LNCS, vol. 537, pp. 77–93. Springer, Heidelberg (1991)

[GM84]   Goldwasser, S., Micali, S.: Probabilistic encryption. JCSS 28(2), 270–299 (1984)

[GMR89]   Goldwasser, S., Micali, S., Rackoff, C.: The knowledge complexity of interactive proof systems. SIAM Journal on Comput. 18(1), 186–208 (1989)

[GMW87]   Goldreich, O., Micali, S., Wigderson, A.: How to play any mental game. In: 19th Symposium on Theory of Computing (STOC), pp. 218–229 (1987)

[HM00]   Hirt, M., Maurer, U.: Complete characterization of adversaries tolerable in secure multi-party computation. J. Cryptology 13(1), 31–60 (2000)

[K06]   Küsters, R.: Simulation based security with inexhaustible interactive Turing machines. In: 19th CSFW (2006)

[Low96]   Lowe, G.: Breaking and fixing the Needham-Schröder public-key protocol using CSP and FDR. In: 2nd International Workshop on Tools and Algorithms for the construction and analysis of systems (1996)

[Mea96]   Meadows, C.: The NRL protocol analyzer: An overview. J. Log. Program. 26(2), 113–131 (1996)

[Mil89]   Milner, R.: Communication and concurrency. Prentice Hall, Englewood Cliffs (1989)

[MMS97]   Mitchell, J.C., Mitchell, M., Stern, U.: Automated analysis of cryptographic protocols using Murφ. In: Proceedings, 1997 IEEE Symposium on Security and Privacy, pp. 141–153 (1997)

[MR91]   Micali, S., Rogaway, P.: Secure computation (abstract). In: McCurley, K.S., Ziegler, C.D. (eds.) Advances in Cryptology 1981 - 1997. LNCS, vol. 1440, pp. 392–404. Springer, Heidelberg (1999)

[MS01]   Millen, J.K., Shmatikov, V.: Constraint solving for bounded-process cryptographic protocol analysis. In: ACM Conference on Computer and Communications Security (CCS) (2001)

[MW04]   Micciancio, D., Warinschi, B.: Soundness of formal encryption in the presence of active adversaries. In: 1st TCC, pp. 133–151 (2004)

[Pau88]   Paulson, L.C.: Isabelle: the next seven hundred theorem provers (system abstract). In: 9th International Conf. on Automated Deduction. LNCS, vol. 310, pp. 772–773. Springer, Heidelberg (1988),
          http://www.cl.cam.ac.uk/Research/HVG/Isabelle/

[Pau98]   Paulson, L.C.: The inductive approach to verifying cryptographic protocols. Journal of Computer Security 6, 85–128 (1998)

[PMS03]   Mitchell, J.C., Mateus, P., Scedrov, A.: Composition of cryptographic protocols in a probabilistic polynomial-time process calculus. In: 14th CONCUR, pp. 323–345 (2003)

[PW00]   Pfitzmann, B., Waidner, M.: Composition and integrity preservation of secure reactive systems. In: 7th ACM Conf. on Computer and Communication Security (CCS), pp. 245–254 (2000)

[RS91]   Rackoff, C., Simon, D.: Non-interactive zero-knowledge proof of knowledge and chosen ciphertext attack. In: Feigenbaum, J. (ed.) CRYPTO 1991. LNCS, vol. 576. Springer, Heidelberg (1992)

[SBB+06] Sprenger, C., Backes, M., Basin, D.A., Pfitzmann, B., Waidner, M.: Cryptographically sound theorem proving. CSFW, 153–166 (2006)

Newton's Method for ω-Continuous Semirings*

Javier Esparza, Stefan Kiefer, and Michael Luttenberger

Institut für Informatik, Technische Universität München, 85748 Garching, Germany
{esparza,kiefer,luttenbe}@model.in.tum.de

Abstract. Fixed point equations $X = f(X)$ over ω-continuous semirings are a natural mathematical foundation of interprocedural program analysis. Generic algorithms for solving these equations are based on Kleene's theorem, which states that the sequence $0, f(0), f(f(0)), \ldots$ converges to the least fixed point. However, this approach is often inefficient. We report on recent work in which we extend Newton's method, the well-known technique from numerical mathematics, to arbitrary ω-continuous semirings, and analyze its convergence speed in the real semiring.

1 Introduction

In the last two years we have investigated generic algorithms for solving systems of fixed point equations over ω-*continuous semirings* [15]. These semirings provide a nice mathematical foundation for program analysis. A program can be translated (in a syntax-driven way) into a system of $O(n)$ equations over an abstract semiring, where n is the number of program points. Depending on the information about the program one wants to compute, the carrier of the semiring and its abstract sum and product operations can be instantiated so that the desired information is the least solution of the equations. Roughly speaking, the translation maps choice and sequential composition at program level into the sum and product operators of the semiring. Procedures, even recursive ones, are first order citizens and can be easily translated. The translation is very similar to the one that maps a program into a monotone framework [16].

Kleene's fixed point theorem applies to ω-continuous semirings. It shows that the least solution μf of a system of equations $X = f(X)$ is equal to the supremum of the sequence $(\kappa^{(i)})_{i\in\mathbb{N}}$ of *Kleene approximants* given by $\kappa^{(0)} = 0$ and $\kappa^{(i+1)} = f(\kappa^{(i)})$. This yields a procedure (let's call it *Kleene's method*) to compute or at least approximate μf. If the domain satisfies what is usually known as the *ascending chain condition*, then the procedure terminates, because there exists an i such that $\kappa^{(i)} = \kappa^{(i+1)} = \mu f$.

Kleene's method is generic and robust: it always converges when started at 0, for any ω-continuous semiring, and whatever the shape of f is. On the other hand, its efficiency can be very unsatisfactory. If the ascending chain condition fails, then the sequence of Kleene approximants hardly ever reaches the solution after a finite number of steps. Another problem of the Kleene sequence arises in the area of quantitative program analysis. Quantitative information, like average runtime and probability of termination

---

* This work was in part supported by the DFG project *Algorithms for Software Model Checking*.

(for programs with a stochastic component) can also be computed as the least solution of a system of equations, in this case over the semiring of the non-negative real numbers plus infinity. While in these analyses one cannot expect to compute the exact solution by any iterative method (it may be irrational and not even representable by radicals), it is very important to find approximation techniques that converge fast to the solution. However, the convergence of the Kleene approximants can be extremely slow. An artificial but illustrative example is the case of a procedure that can either terminate or call itself twice, both with probability $1/2$. The probability of termination of this program is given by the least solution of the equation $X = 1/2 + 1/2X^2$. It is easy to see that the least solution is equal to 1, but we have $\kappa^{(i)} \leq 1 - \frac{1}{i+1}$ for every $i \geq 0$, i.e., in order to approximate the solution within i bits of precision we have to compute about 2^i Kleene approximants. For instance, we have $\kappa^{(200)} = 0.9990$.

Faster approximation techniques for equations over the reals have been known for a long time. In particular, Newton's method, suggested by Isaac Newton more than 300 years ago, is a standard efficient technique to approximate a zero of a differentiable function. Since the least solution of $X = 1/2 + 1/2X^2$ is a zero of $1/2 + 1/2X^2 - X$, the method can be applied, and it yields $\nu^{(i)} = 1 - 2^{-i}$ for the i-th *Newton approximant*. So i bits of precision require to compute only i approximants, i.e., Newton's method converges exponentially faster than Kleene's in this case. However, Newton's method on the real field is by far not as robust and well behaved as Kleene's method on semirings. The method may converge very slowly, converge only when started at a point very close to the zero, or even not converge at all [17].

So there is a puzzling mismatch between the current states of semantics and program analysis on the one side, and numerical mathematics on the other. On ω-continuous semirings, the natural domain of semantics and program analysis, Kleene's method is robust and generally applicable, but inefficient in many cases, in particular for quantitative analyses. On the real field, the natural domain of numerical mathematics, Newton's method can be very efficient, but it is not robust.

We became aware of this mismatch two years ago through the the the work of Etessami and Yannakakis on Recursive Markov Chains and our work on Probabilistic Pushdown Automata. Both are abstract models of probabilistic programs with procedures, and their analysis reduces to or at least involves solving systems of fixed point equations. The mismatch led us to investigate the following questions:

- Can Newton's method be generalized to arbitrary ω-continuous semirings?
  I.e., could it be the case that Newton's method is in fact as generally applicable as Kleene's, but nobody has noticed yet?
- Is Newton's method robust when restricted to the real semiring?
  I.e., could it be the case that the difficult behaviour of Newton's method disappears when we restrict its application to the non-negative reals, but nobody has noticed yet?

The answer to both questions is essentially affirmative, and has led to a number of papers [5, 4, 14, 6]. In this note we present the results, devoting some attention to those examples and intuitions that hardly ever reach the final version of a conference paper due to the page limit.

2 From Programs to Fixed Point Equations on Semirings

Recall that a semiring is a set of *values* together with two binary operations, usually called sum and product. Sum and product are associative and have neutral elements 0 and 1, respectively. Moreover, sum is commutative, and product distributes over sum. The *natural order* relation $\sqsubseteq$ on a semiring is defined by setting $a \sqsubseteq a + d$ for every d. A semiring is *naturally ordered* if $\sqsubseteq$ is a partial order.

An ω-*continuous* semiring is a naturally ordered semiring extended by an infinite summation-operator $\sum$ that satisfies some natural properties. In particular, for every sequence $(a_i)_{i \geq 0}$ the supremum $\sup\{\sum_{0 \leq i \leq k} a_i \mid k \in \mathbb{N}\}$ w.r.t. $\sqsubseteq$ exists, and is equal to $\sum_{i \in \mathbb{N}} a_i$ [15].

We show how to assign to a procedural program a set of abstract equations by means of an example. Consider the (very abstractly defined) program consisting of three procedures X_1, X_2, X_3, and associate to it a system of equations. For our discussion it is not relevant which is the main procedure. The flow graphs of the procedures are shown in Figure 1. For instance, procedure X_1 can either execute the abstract action b and terminate, or execute a, call itself recursively, and, after the call has terminated, call procedure X_2.

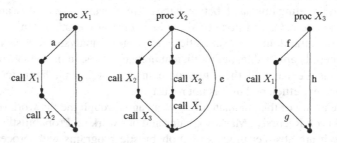

Fig. 1. Flowgraphs of three procedures

We associate to the program the following three abstract equations [1]

$$X_1 = a \cdot X_1 \cdot X_2 + b$$
$$X_2 = c \cdot X_2 \cdot X_3 + d \cdot X_2 \cdot X_1 + e \tag{1}$$
$$X_3 = f \cdot X_1 \cdot g + h$$

where $+$ and $\cdot$ are the abstract semiring operations, and $\{a, b, \ldots, h\}$ are semiring values. Notice that we slightly abuse language and use the same symbol for a program action and its associated value.

---

[1] One for each procedure. A systematic translation from programs to equations yields one variable and one equation for each program point. We have not done it in order to keep the number of equations small.

2.1 Some Semiring Interpretations

Many interesting pieces of information about our program correspond to the least solution of the system of equations over different semirings.[2] For the rest of the section
let $\Sigma = \{a, b, \ldots, h\}$ be the set of actions in the program, and let σ denote an arbitrary
element of Σ.

Language interpretation. Consider the following semiring. The carrier is 2^{Σ^*} (i.e., the
set of languages over Σ). A program action $\sigma \in \Sigma$ is interpreted as the singleton language $\{\sigma\}$. The sum and product operations are union and concatenation of languages,
respectively. We call it *language semiring* over Σ. Under this interpretation, the system
of equations (1) is nothing but the following context-free grammar:

$$X_1 \rightarrow aX_1X_2 \mid b$$
$$X_2 \rightarrow cX_2X_3 \mid dX_2X_1 \mid e$$
$$X_3 \rightarrow fX_1g \mid h$$

The least solution of (1) is the triple $(L(X_1), L(X_2), L(X_3))$, where $L(X_i)$ denotes
the set of terminating executions of the program with X_i as main procedure, or, in
language-theoretic terms, the language of the associated grammar with X_i as axiom.

Relational interpretation. Assume that an action σ corresponds to a program instruction
whose semantics is described by means of a relation $R_\sigma(V, V')$ over a set V of program
variables (as usual, primed and unprimed variables correspond to the values before and
after executing the instruction). Consider now the following semiring. The carrier is
the set of all relations over V, V'. The semiring element σ is interpreted as the relation R_σ. The sum and product operations are union and join of relations, respectively,
i.e., $(R_1 \cdot R_2)(V, V') = \exists V'' R_1(V, V'') \wedge R_2(V'', V')$. Under this interpretation, the
i-th component of the least solution of (1) is the *summary* relation $R_i(V, V')$ containing
the pairs V, V' such that if procedure X_i starts at valuation V, then it may terminate at
valuation V'.

Counting interpretation. Assume we wish to know how many as, bs, etc. we can
observe in a (terminating) execution of the program, but we are not interested in the
order in which they occur. In the terminology of abstract interpretation [2], we abstract
an execution $w \in \Sigma^*$ by the vector $(n_a, \ldots, n_h) \in \mathbb{N}^{|\Sigma|}$, where $n_a, \ldots, n_h$ are the
number of occurrences of $a, \ldots, h$ in w. We call this vector the *Parikh image* of w. We
wish to compute the vector $(P(X_1), P(X_2), P(X_3))$ where $P(X_i)$ contains the Parikh
images of the words of $L(X_i)$. It is easy to see that this is the least solution of (1) for the
following semiring. The carrier is $2^{\mathbb{N}^{|\Sigma|}}$. The i-th action of Σ is interpreted as the sin-
gleton set $\{(0, \ldots, 0, 1, 0 \ldots, 0)\}$ with the "1" at the i-th position. The sum operation
is set union, and the product operation is given[3] by

$$U \cdot V = \{(u_a + v_a, \ldots, u_h + v_h) \mid (u_a, \ldots, u_h) \in U, (v_a, \ldots, v_h) \in V\}.$$

---

[2] This will be no surprise for the reader acquainted with monotone frameworks or abstract interpretation, but the examples will be used throughout the paper.

[3] Abstract interpretation provides a general recipe to define these operators.

Probabilistic interpretations. Assume that the choices between actions are stochastic. For instance, actions a and b are chosen with probability p and $(1 - p)$, respectively. The probability of termination is given by the least solution of (1) when interpreted over the following semiring (the *real semiring*) [8, 9]. The carrier is the set of non-negative real numbers, enriched with an additional element ∞. The semiring element σ is interpreted as the probability of choosing σ among all enabled actions. Sum and product are the standard operations on real numbers, suitably extended to ∞ – if we are instead interested in the probability of the most likely execution, we just have to reinterpret the sum operator as maximum.

As a last example, assume that actions are assigned not only a probability, but also a *duration*. Let d_σ denote the duration of σ. We are interested in the expected termination time of the program, under the condition that the program terminates (the *conditional expected time*). For this we consider the following semiring. The elements are the set of pairs (r_1, r_2), where r_1, r_2 are non-negative reals or ∞. We interpret σ as the pair (p_σ, d_σ), i.e., the probability and the duration of σ. The sum operation is defined as follows (where to simplify the notation we use $+_e$ and $\cdot_e$ for the operations of the semiring, and $+$ and $\cdot$ for sum and product of reals)

$$(p_1, d_1) +_e (p_2, d_2) = \left(p_1 + p_2, \frac{p_1 \cdot d_1 + p_2 \cdot d_2}{p_1 + p_2} \right)$$

$$(p_1, d_1) \cdot_e (p_2, d_2) = (p_1 \cdot p_2, d_1 + d_2)$$

One can easily check that this definition satisfies the semiring axioms. The i-th component of the least solution of (1) is now the pair (t_i, e_i), where t_i is the probability that procedure X_i terminates, and e_i is its conditional expected time.

3   Fixed Point Equations

Fix an arbitrary ω-continuous semiring with a set S of values. We define systems of fixed point equations and present Kleene's fixed point theorem.

Given a finite set $\mathcal{X}$ of variables, a *monomial* is a finite expression

$$a_1 X_1 a_2 \cdots a_k X_k a_{k+1}$$

where $k \geq 0$, $a_1, \ldots, a_{k+1} \in S$ and $X_1, \ldots, X_k \in \mathcal{X}$. A *polynomial* is an expression of the form $m_1 + \ldots + m_k$ where $k \geq 0$ and $m_1, \ldots, m_k$ are monomials.

A *vector* is a mapping v that assigns to every variable $X \in \mathcal{X}$ a value denoted by v_X or v_X, called the X-component of v. The value of a monomial $m = a_1 X_1 a_2 \cdots a_k X_k a_{k+1}$ at v is $m(v) = a_1 v_{X_1} a_2 \cdots a_k v_{X_k} a_{k+1}$. The value of a polynomial at v is the sum of the values of its monomials at v. A polynomial induces a mapping from vectors to values that assigns to v the vector $f(v)$. A vector of polynomials is a mapping f that assigns a polynomial f_X to each variable $X \in \mathcal{X}$; it induces a mapping from vectors to vectors that assigns to a vector v the vector $f(v)$ whose X-component is $f_X(v)$. A *fixed point of* f is a solution of the equation $X = f(X)$.

It is easy to see that polynomials are monotone and continuous mappings w.r.t. $\sqsubseteq$. Kleene's theorem can then be applied (see e.g. [15]), which leads to this proposition:

Proposition 3.1. *A vector $\boldsymbol{f}$ of polynomials has a unique least fixed point $\mu\boldsymbol{f}$ which is the $\sqsubseteq$-supremum of the* Kleene *sequence given by $\kappa^{(0)} = \boldsymbol{0}$, and $\kappa^{(i+1)} = \boldsymbol{f}(\kappa^{(i)})$.*

4 Newton's Method for ω-Continuous Semirings

We recall Newton's method for approximating a zero of a differentiable function, and apply it to find the least solution of a system of fixed point equations over the reals. Then, we present the generalization of Newton's method to arbitrary ω-continuous semirings we obtained in [5]. We focus on the univariate case (one single equation in one variable), because it already introduces all the basic ideas of the general case.

Given a differentiable function $g\colon \mathbb{R} \to \mathbb{R}$, Newton's method computes a zero of g, i.e., a solution of the equation $g(X) = 0$. The method starts at some value $\nu^{(0)}$ "close enough" to the zero, and proceeds iteratively: given $\nu^{(i)}$, it computes a value $\nu^{(i+1)}$ closer to the zero than $\nu^{(i)}$. For that, the method *linearizes* g at $\nu^{(i)}$, i.e., computes the tangent to g passing through the point $(\nu^{(i)}, g(\nu^{(i)}))$, and takes $\nu^{(i+1)}$ as the zero of the tangent (i.e., the x-coordinate of the point at which the tangent cuts the x-axis).

We formulate the method in terms of the *differential* of g at a given point v. This is is the mapping $Dg|_v \colon \mathbb{R} \to \mathbb{R}$ that assigns to each $x \in \mathbb{R}$ a linear function, namely the one corresponding to the tangent of g at v, but represented in the coordinate system having the point $(v, g(v))$ as origin. If we denote the differential of g at v by $Dg|_v$, then we have $Dg|_v(X) = g'(v) \cdot X$ (for example, if $g(X) = X^2 + 3X + 1$, then $Dg|_3(X) = 9X$). In terms of differentials, Newton's method starts at some $\nu^{(0)}$, and computes iteratively $\nu^{(i+1)} = \nu^{(i)} + \Delta^{(i)}$, where $\Delta^{(i)}$ is the solution of the linear equation $Dg|_{\nu^{(i)}}(X) + g(\nu^{(i)}) = 0$ (assume for simplicity that the solution of the linear system is unique).

Computing a solution of a fixed point equation $f(X) = X$ amounts to computing a zero of $g(X) = f(X) - X$, and so we can apply Newton's method. Since for every real number v we have $Dg|_v(X) = Df|_v(X) - X$, the method for computing the least solution of $f(X) = X$ looks as follows:

Starting at some $\nu^{(0)}$, compute iteratively

$$\nu^{(i+1)} = \nu^{(i)} + \Delta^{(i)} \tag{2}$$

where $\Delta^{(i)}$ is the solution of the linear equation

$$Df|_{\nu^{(i)}}(X) + f(\nu^{(i)}) - \nu^{(i)} = X . \tag{3}$$

So Newton's method "breaks down" the problem of solving a non-linear system $f(X) = X$ into solving the sequence (3) of linear systems.

4.1 Generalizing Newton's Method

In order to generalize Newton's method to arbitrary ω-continuous semirings we have to overcome two obstacles. First, differentials are defined in terms of derivatives, which are the limit of a quotient of differences. This requires both the sum and product operations to have inverses, which is not the case in general semirings. Second, Equation (3) contains the term $f(\nu^{(i)}) - \nu^{(i)}$, which again seems to be defined only if the sum operation has an inverse.

The first obstacle. Differentiable functions satisfy well-known algebraic rules with respect to sums and products of functions. We take these rules as the *definition* of the differential of a polynomial f over an ω-continuous semiring.

Definition 4.1. *Let f be a polynomial in one variable X over an ω-continuous semiring with carrier S. The* differential *of f at the point v is the mapping $Df|_v : S \to S$ inductively defined as follows for every $a \in S$:*

$$Df|_v(a) = \begin{cases} 0 & \text{if } f \in S \\ a & \text{if } f = X \\ Dg|_v(a) \cdot h(v) + g(v) \cdot Dh|_v(a) & \text{if } f = g \cdot h \\ \sum_{i \in I} Df_i|_v(a) & \text{if } f = \sum_{i \in I} f_i(a) \,. \end{cases}$$

On commutative semirings, like the real semiring, we have $Df|_v(a) = f'(v) \cdot a$ for all $v, a \in S$, where $f'(v)$ is the derivative of f. This no longer holds when product is not commutative. For a function $f(X) = a_0 X a_1 X a_2$ we have

$$Df|_v(a) = a_0 \cdot a \cdot a_1 \cdot v \cdot a_2 + a_0 \cdot v \cdot a_1 \cdot a \cdot a_2.$$

The second obstacle. It turns out that the Newton sequence is well-defined if we choose $\nu^{(0)} = f(0)$. More precisely, in [5] we *guess* that this choice will solve the problem, define the Newton sequence, and then *prove* that the guess is correct. The precise guess is that this choice implies $\nu^{(i)} \sqsubseteq f(\nu^{(i)})$ for every $i \geq 0$. By the definition of $\sqsubseteq$, the semiring then contains a value $\delta^{(i)}$ such that $f(\nu^{(i)}) = \nu^{(i)} + \delta^{(i)}$. We can replace $f(\nu^{(i)}) - \nu^{(i)}$ by any such $\delta^{(i)}$. This leads to the following definition:

Definition 4.2. *Let f be a polynomial in one variable over an ω-continuous semiring. The* Newton sequence *$(\nu^{(i)})_{i \in \mathbb{N}}$ is given by:*

$$\nu^{(0)} = f(0) \quad \text{and} \quad \nu^{(i+1)} = \nu^{(i)} + \Delta^{(i)} \tag{4}$$

where $\Delta^{(i)}$ is the least solution of

$$Df|_{\nu^{(i)}}(X) + \delta^{(i)} = X \tag{5}$$

and $\delta^{(i)}$ is any element satisfying $f(\nu^{(i)}) = \nu^{(i)} + \delta^{(i)}$.

Notice that for arbitrary semirings the Newton sequence is not unique, since we may have different choices for $\delta^{(i)}$.

The definition can be easily generalized to the multivariate case. Fix a set $\mathcal{X} = \{X_1, \ldots, X_n\}$ of variables. Given a multivariate polynomial f, we define the differential of f at the vector v *with respect to the variable X* by almost the same equations as above:

$$D_X f|_v(a) = \begin{cases} 0 & \text{if } f \in S \text{ or } f \in \mathcal{X} \setminus \{X\} \\ a_X & \text{if } f = X \\ D_X g|_v(a) \cdot h(v) + g(v) \cdot D_X h|_v(a) & \text{if } f = g \cdot h \\ \sum_{i \in I} D_X f_i|_v(a) & \text{if } f = \sum_{i \in I} f_i \,. \end{cases}$$

Then the differential of f at the vector v is defined as $Df|_v = D_{X_1} f|_v + \cdots + D_{X_n} f|_v$. Finally, for a vector of polynomials f we set $Df|_v = (Df_{X_1}|_v, \ldots, Df_{X_n}|_v)$.

Definition 4.3. *Let* $f\colon V \rightarrow V$ *be a vector of polynomials. The* Newton *sequence* $(\nu^{(i)})_{i \in \mathbb{N}}$ *is given by:*

$$\nu^{(0)} = f(0) \quad and \quad \nu^{(i+1)} = \nu^{(i)} + \Delta^{(i)}, \tag{6}$$

where $\Delta^{(i)}$ *is the solution of*

$$Df|_{\nu^{(i)}}(X) + \delta^{(i)} = X . \tag{7}$$

and $\delta^{(i)}$ *is any vector satisfying* $f(\nu^{(i)}) = \nu^{(i)} + \delta^{(i)}$.

Theorem 4.4. *Let* $f\colon V \rightarrow V$ *be a vector of polynomials. For every* ω-*continuous semiring and every* $i \in \mathbb{N}$:

- *There exists at least one Newton sequence, i.e., there exists a vector* $\delta^{(i)}$ *such that* $f(\nu^{(i)}) = \nu^{(i)} + \delta^{(i)}$;
- $\kappa^{(i)} \sqsubseteq \nu^{(i)} \sqsubseteq f(\nu^{(i)}) \sqsubseteq \mu f = \sup_j \kappa^{(j)}$.

5 Newton's Method on Different Semirings

In this section we Introduce the main results of our study of Newton's method [5, 4, 14, 6] by focusing on three representative semirings: the language, the counting, and the real semiring. We first show that the Newton approximants of the language semiring are the context-free languages of *finite index*, a notion extensively studied in the 60s [20, 11, 19, 12]. We then explain how the algebraic technique for solving fixed point equations over the counting semiring presented by Hopkins and Kozen in [13] is again nothing but a special case of Newton's method. Finally, we show that in the real semiring Newton's method is just as robust as Kleene's.

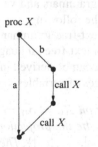

Fig. 2. Flowgraph of a recursive program with one procedure

We present the results for the three semirings with the help of an example. Consider the recursive program from the introduction that can execute action a and terminate, or action b, after which it recursively calls itself twice, see Figure 2. Its corresponding abstract equation is

$$X = a + b \cdot X \cdot X \tag{8}$$

We solve this equation in the three semirings, point out some of its peculiarities, and then introduce the general results.

5.1 The Language Semiring

Consider the language semiring with $\Sigma = \{a, b\}$. Recall that the product operation is concatenation of languages, and hence non-commutative. So we have

$Df|_v(X) = bvX + bXv$. It is easy to show that when sum is idempotent the definition of the Newton sequence can be simplified to

$$\nu^{(0)} = f(0) \quad \text{and} \quad \nu^{(i+1)} = \Delta^{(i)}, \tag{9}$$

where $\Delta^{(i)}$ is the least solution of

$$Df|_{\nu^{(i)}}(X) + f(\nu^{(i)}) = X . \tag{10}$$

For the program of Figure 2 Equation (10) becomes

$$\underbrace{b\nu^{(i)}X + bX\nu^{(i)}}_{Df|_{\nu^{(i)}}(X)} + \underbrace{a + b\nu^{(i)}\nu^{(i)}}_{f(\nu^{(i)})} = X . \tag{11}$$

Its least solution, and by (9) the $i+1$-th Newton approximant, is a context-free language. Let $G^{(i)}$ be a grammar with axiom $S^{(i)}$ such that $\nu^{(i)} = L(G^{(i)})$. Since $\nu^{(0)} = f(0)$, the grammar $G^{(0)}$ contains one single production, namely $S^{(0)} \to a$. Equation (11) allows us to define $G^{(i+1)}$ in terms of $G^{(i)}$, and we get:

$$G^{(0)} = \{S^{(0)} \to a\}$$
$$G^{(i+1)} = G^{(i)} \cup \{S^{(i+1)} \to a \mid bXS^{(i)} \mid bS^{(i)}X \mid bS^{(i)}S^{(i)}\}$$

and it is easy to see that in this case $L(G^{(i)}) \neq L(G^{(i+1)})$ for every $i \geq 0$.

It is well known that in a language semiring, context-free grammars and vectors of polynomials are essentially the same, so we identify them in the following.

We can characterize the Newton approximants of a context-free grammar by the notion of *index*, a well-known concept from the theory of context-free languages [20, 11, 19, 12]. Loosely speaking, a word of $L(G)$ has index i if it can be derived in such a way that no intermediate word contains more than i occurrences of variables.

Definition 5.1. *Let G be a grammar, and let D be a derivation $X_0 = \alpha_0 \Rightarrow \cdots \Rightarrow \alpha_r = w$ of $w \in L(G)$, and for every $i \in \{0, \ldots, r\}$ let β_r be the projection of α_r onto the variables of G. The* index *of D is the maximum of $\{|\beta_0|, \ldots, |\beta_r|\}$. The index-$i$ approximation of $L(G)$, denoted by $L_i(G)$, contains the words derivable by some derivation of G of index at most i.*

Finite-index languages have been extensively investigated under different names by Salomaa, Gruska, Yntema, Ginsburg and Spanier, among others [19,12,20,11](see [10] for historical background). In [4] we show that for a context-free grammar in Chomsky normal form, the Newton approximants coincide with the finite-index approximations:

Theorem 5.2. *Let G be a context-free grammar in CNF with axiom S and let $(\nu^{(i)})_{i \in \mathbb{N}}$ be the Newton sequence associated with G. Then $(\nu^{(i)})_S = L_{i+1}(G)$ for every $i \geq 0$.*

In particular, it follows from Theorem 5.2 that the (S-component of the) Newton sequence for a context-free grammar G converges in finitely many steps if and only if $L(G) = L_i(G)$ for some $i \in \mathbb{N}$.

5.2   The Counting Semiring

Consider the counting semiring with $a = \{(1,0)\}$ and $b = \{(0,1)\}$. Since the sum operation is union of sets of vectors, it is idempotent and Equations (9) and (10) hold. Since the product operation is now commutative, Equation (10) becomes

$$b \cdot \nu^{(i)} \cdot X + a + b \cdot \nu^{(i)} \cdot \nu^{(i)} = X \,. \tag{12}$$

By virtue of Kleene's fixed point theorem the least solution of a *linear* equation $X = u \cdot X + v$ over an ω-continuous semiring is given by the supremum of the sequence

$$v, \quad v + uv, \quad v + uv + uuv, \ldots$$

i.e. by $(\sum_{i \in \mathbb{N}} u^i) \cdot v = u^* \cdot v$, where $*$ is Kleene's iteration operator. The least solution $\Delta^{(i)}$ of Equation (12) is then given by

$$\Delta^{(i)} = (b \cdot \nu^{(i)})^* \cdot (a + b \cdot \nu^{(i)} \cdot \nu^{(i)})$$

and we obtain:

$$\begin{aligned}
\nu^{(0)} &= a = \{(1,0)\} \\
\nu^{(1)} &= (b \cdot a)^* \cdot (a + b \cdot a \cdot a) \\
&= \{(n,n) \mid n \geq 0\} \cdot \{(1,0),(2,1)\} \\
&= \{(n+1,n) \mid n \geq 0\} \\
\nu^{(2)} &= (\{(n,n) \mid n \geq 1\})^* \cdot (\{(1,0)\} \cup \{(2n+2,2n+1) \mid n \geq 0\}) \\
&= (\{(n,n) \mid n \geq 0\})^* \cdot (\{(1,0)\} \cup \{(2n+2,2n+1) \mid n \geq 0\}) \\
&= \{(n+1,n) \mid n \geq 0\}
\end{aligned}$$

So the Newton sequence reaches a fixed point after only one iteration.

It turns out that the Newton sequence *always* reaches a fixed point in the counting semiring. This immediately generalizes to any finitely generated commutative idempotent ω-continuous semiring as we simply can opt not to evaluate the products and sums. More surprisingly, this is even the case for all semirings where sum is idempotent and product is commutative. This was first shown by Hopkins and Kozen in [13], who introduced the sequence without knowing that it was Newton's sequence (see [5] for the details). Hopkins and Kozen also gave an $O(3^n)$ upper bound for the number of iterations needed to reach the fixed point of a system of n equations. In [5] we reduced this upper bound from $O(3^n)$ to n, which is easily shown to be tight.

Theorem 5.3. *Let f be a vector of n polynomials over a commutative idempotent ω-continuous semiring. Then $\mu f = \nu^{(n)}$, i.e., Newton's method reaches the least fixed point after n iterations.*

We have mentioned above that the least solution of $X = u \cdot X + v$ is $u^* \cdot v$. Using this fact it is easy to show that the Newton approximants of equations over commutative semirings can be described by regular expressions. A corollary of this result is Parikh's theorem, stating that the Parikh image of a context-free language is equal to the Parikh image of some regular language [18]. To see why this is the case, notice that a context-free language is the least solution of a system of fixed point equations over the language semiring. Its Parikh image is the least solution of the same system over the counting semiring. Since Newton's method terminates over the counting semiring, and Newton approximants can be described by regular expressions, the result follows.

Notice that we are by no means the first to provide an algebraic proof of Parikh's theorem. A first proof was obtained by Aceto et al. in [1], and in fact the motivation of Hopkins and Kozen for the results of [13] was again to give a proof of the theorem. Our results in [5] make two contributions: first, the aesthetically appealing connection between Newton and Parikh, and, second, an algebraic algorithm for computing the Parikh image with a tight bound on the the the number of iterations.

We conclude the section with a final remark. The counting semiring is a simple example of a semiring that does not satisfy the ascending chain condition. Kleene's method does not terminate for any program containing at least one loop. However, Newton's method always terminates!

5.3   The Real Semiring

Consider again Equation (8), but this time over the real semiring (non-negative real numbers enriched with ∞) and with $a = b = 1/2$. We get the equation

$$X = 1/2 + 1/2 \cdot X^2 \tag{13}$$

which was already briefly discussed in the introduction. We have $Df|_v(X) = v \cdot X$, and a single possible choice for $\delta^{(i)}$, namely $\delta^{(i)} = f(\nu^{(i)}) - \nu^{(i)} = 1/2 + 1/2\,(\nu^{(i)})^2 - \nu^{(i)}$. Equation (5) becomes

$$\nu^{(i)}\,X + 1/2 + 1/2\,(\nu^{(i)})^2 - \nu^{(i)} = X$$

with $\Delta^{(i)} = (1 - \nu^{(i)})/2$ as its only solution. So we get

$$\nu^{(0)} = 1/2 \qquad \nu^{(i+1)} = (1 + \nu^{(i)})/2$$

and therefore $\nu^{(i)} = 1 - 2^{(i+1)}$. The Newton sequence converges to 1, and gains one bit of accuracy per iteration, i.e., the relative error is halved at each iteration.

In [14,6] we have analyzed in detail the convergence behaviour of Newton's method. Loosely speaking, our results say that Equation (13) is an example of the worst-case behaviour of the method.

To characterize it, we use the term *linear convergence*, a notion from numerical analysis that states that the number of bits obtained after i iterations depends linearly on i. If $\|\mu f - v\| / \|\mu f\| \le 2^{-i}$ (in the maximum-norm), we say that the approximation v of μf has (at least) i bits of accuracy . Newton's method converges linearly provided that f has a finite least fixed point and is in an easily achievable normal form (the polynomials have degree at most 2, and μf is nonzero in all components). More precisely [6]:

Theorem 5.4. *Let f be a vector of n polynomials over the real semiring in the above mentioned normal form. Then Newton's method converges linearly: there exists a $t_f \in \mathbb{N}$ such that the Newton approximant $\nu^{(t_f + i \cdot (n+1) \cdot 2^n)}$ has at least i bits of accuracy.*

Theorem 5.4 is essentially tight. Consider the following family of equation systems.

$$X_1 = 1/2 + 1/2 \cdot X_1^2$$
$$X_2 = 1/4 \cdot X_1^2 + 1/2 \cdot X_1 X_2 + 1/4 \cdot X_2^2$$
$$\vdots \qquad\qquad\qquad\qquad\qquad (14)$$
$$X_n = 1/4 \cdot X_{n-1}^2 + 1/2 \cdot X_{n-1} X_n + 1/4 \cdot X_n^2$$

Its least solution is $(1, \ldots, 1)$. We show in [14,6] that at least $i \cdot 2^{n-1}$ iterations of Newton's method are needed to obtain i bits. More precisely, we show that after $i \cdot 2^{n-1}$ iterations no more than $i \cdot 2^{n-1}$ bits of accuracy have been obtained for the first component (cf. the convergence behaviour of (13) above) and that the number of accurate bits of the $(k+1)$-th component is at most one half of the number of accurate bits of the k-th component, for all $k < n$. This implies that for the n-th component we have obtained at most i bits of accuracy.

This example exploits the fact that X_k depends only on the X_l for $l \leq k \leq n$. In fact, Theorem 5.4 can be substantially strengthened if f is *strongly connected*. More formally, let a variable X *depend* on Y if Y appears in f_X. Then, f is said to be strongly connected if every variable depends transitively on every variable. For those systems we show that Newton's method gains 1 bit of accuracy per iteration after the "threshold" t_f has been reached. In addition (and even more importantly from a computational point of view) we can give bounds on t_f [6]:

Theorem 5.5. *Let f be as in Theorem 5.4, and, additionally, strongly connected. Further, let m be the size of f (coefficients in binary). Then i bits of accuracy are attained by $\nu^{(n2^{n+2}m+i)}$. This improves to $\nu^{(5n^2m+i)}$, if $f(0)$ is positive in all components.*

In [7], a recent invited paper, we discuss equation systems over the real semiring, the motivation and complexity of computing their least solutions, and our results [14, 6] on Newton's method for the real semiring in more detail. We present an extension of Newton's method on polynomials with min and max operators in [3].

6 Conclusion

We have shown that the two questions we asked in the introduction have an affirmative answer. Newton's method, a 300 years old technique for approximating the solution of a system of equations over the reals, can be extended to arbitrary ω-continuous semirings. And, when restricted to the real semiring, the pathologies of Newton's method—no convergence, or only local and slow convergence—disappear: the method always exhibits at least linear convergence.

We like to look at our results as bridges between numerical mathematics and the foundations of program semantics and program analysis. On the one hand, while numerical mathematics has studied Newton's method in large detail, it has not payed much attention to its restriction to the real semiring. Our results indicate that this is an interesting case certainly deserving further research.

On the other hand, program analysis relies on computational engines for solving systems of equations over a large variety of domains, and these engines are based, in one way or another, on Kleene's iterative technique. This technique is very slow when working on the reals, and numerical mathematics has developed much faster ones, Newton's method being one of the most basic. The generalization of these techniques to the more general domains of semantics and program analysis is an exciting research program.

References

1. Aceto, L., Ésik, Z., Ingólfsdóttir, A.: A fully equational proof of Parikh's theorem. Informatique Théorique et Applications 36(2), 129–153 (2002)
2. Cousot, P., Cousot, R.: Abstract interpretation: A unified lattice model for static analysis of programs by construction or approximation of fixpoints. In: POPL, pp. 238–252 (1977)
3. Esparza, J., Gawlitza, T., Kiefer, S., Seidl, H.: Approximative methods for monotone systems of min-max-polynomial equations. In: Aceto, L., et al. (eds.) ICALP 2008, Part II. 5126, vol. 5126, Springer, Heidelberg (2008)
4. Esparza, J., Kiefer, S., Luttenberger, M.: An extension of Newton's method to ω-continuous semirings. In: Harju, T., Karhumäki, J., Lepistö, A. (eds.) DLT 2007. LNCS, vol. 4588, pp. 157–168. Springer, Heidelberg (2007)
5. Esparza, J., Kiefer, S., Luttenberger, M.: On fixed point equations over commutative semirings. In: Thomas, W., Weil, P. (eds.) STACS 2007. LNCS, vol. 4393, pp. 296–307. Springer, Heidelberg (2007)
6. Esparza, J., Kiefer, S., Luttenberger, M.: Convergence thresholds of Newton's method for monotone polynomial equations. In: Proceedings of STACS, pp. 289–300 (2008)
7. Esparza, J., Kiefer, S., Luttenberger, M.: Solving monotone polynomial equations. In: Proceedings of IFIP TCS 2008. Springer, Heidelberg (to appear, 2008)
8. Esparza, J., Kučera, A., Mayr, R.: Model checking probabilistic pushdown automata. In: LICS 2004. IEEE Computer Society, Los Alamitos (2004)
9. Etessami, K., Yannakakis, M.: Recursive Markov chains, stochastic grammars, and monotone systems of nonlinear equations. In: Diekert, V., Durand, B. (eds.) STACS 2005. LNCS, vol. 3404, pp. 340–352. Springer, Heidelberg (2005)
10. Fernau, H., Holzer, M.: Conditional context-free languages of finite index. In: New Trends in Formal Languages, pp. 10–26 (1997)
11. Ginsburg, S., Spanier, E.: Derivation-bounded languages. Journal of Computer and System Sciences 2, 228–250 (1968)
12. Gruska, J.: A few remarks on the index of context-free grammars and languages. Information and Control 19, 216–223 (1971)
13. Hopkins, M.W., Kozen, D.: Parikh's theorem in commutative Kleene algebra. In: Logic in Computer Science, pp. 394–401 (1999)
14. Kiefer, S., Luttenberger, M., Esparza, J.: On the convergence of Newton's method for monotone systems of polynomial equations. In: Proceedings of STOC, pp. 217–226. ACM, New York (2007)
15. Kuich, W.: Their Relevance to Formal Languages and Automata. In: Handbook of Formal Languages, ch.9. Semirings and Formal Power Series, vol. 1, pp. 609–677. Springer, Heidelberg (1997)
16. Nielson, F., Nielson, H.R., Hankin, C.: Principles of Program Analysis. Springer, Heidelberg (1999)
17. Ortega, J.M., Rheinboldt, W.C.: Iterative solution of nonlinear equations in several variables. Academic Press, London (1970)
18. Parikh, R.J.: On context-free languages. J. Assoc. Comput. Mach. 13(4), 570–581 (1966)
19. Salomaa, A.: On the index of a context-free grammar and language. Information and Control 14, 474–477 (1969)
20. Yntema, M.K.: Inclusion relations among families of context-free languages. Information and Control 10, 572–597 (1967)

The Tractability Frontier for NFA Minimization[*]

Henrik Björklund and Wim Martens

TU Dortmund
henrik.bjoerklund@udo.edu,
wim.martens@udo.edu

Abstract. We essentially show that minimizing finite automata is NP-hard as soon as one deviates from the class of deterministic finite automata. More specifically, we show that minimization is NP-hard for all finite automata classes that subsume the class that is unambiguous, allows at most one state q with a non-deterministic transition for at most one alphabet symbol a, and is allowed to visit state q at most once in a run. Furthermore, this result holds even for automata that only accept finite languages.

1 Introduction

The regular languages are immensely important, not only in theoretical computer science, but also in practical applications. When using regular languages in practice, the developer is often faced with a trade-off between the descriptive complexity and the complexity of optimization. Concretely, it has been known for a long time that there are regular languages for which non-deterministic automata (NFAs) can provide an exponentially more succinct description than deterministic finite automata (DFAs) [13]. On the other hand, many decision problems that are solvable in polynomial time for DFAs, i.e., equivalence, inclusion, and universality, are computationally hard for NFAs.

The choice of a representation mechanism can therefore be crucial. If the set of regular languages used in an application is relatively constant, membership tests are the main language operations, and economy of space is an issue, NFAs are probably the right choice. If, on the other hand, the languages change frequently, and inclusion or equivalence tests are frequent, DFAs may be more attractive.

Since both NFAs and DFAs have their disadvantages, a lot of effort has been spent on trying to find intermediate models, i.e., finite automata that have some *limited* form of non-determinism. A rather successful intermediate model is the class of *unambiguous* finite automata (UFAs).[1] While in general still being exponentially more succinct than an equivalent DFA for the same language, static analysis questions such as inclusion and equivalence can be solved in PTIME on UFAs [17]. However, UFAs do not allow for tractable state minimization [11]. Therefore, the question whether there are good intermediate models between

---

[*] This work was supported by the DFG Grant SCHW678/3-1.

[1] An automaton is unambiguous if it has at most one accepting run for each word.

L. Aceto et al. (Eds.): ICALP 2008, Part II, LNCS 5126, pp. 27–38, 2008.
© Springer-Verlag Berlin Heidelberg 2008

DFAs and NFAs needs to be revisited for state minimization. The main result of this paper is that, probably, no such good models exist. Even the tiniest bit of non-determinism makes the minimization problem NP-hard.

Minimizing unrestricted NFAs is PSPACE-complete [18] and every undergraduate computer science curriculum teaches its students how to minimize a DFA in polynomial time. The minimization problem for automata with varying degrees of non-determinism was studied in a seminal paper by Jiang and Ravikumar in 1993 [11]. Among other results, they thoroughly investigated the minimization problem for UFAs. They showed the following.

- Given a UFA, finding the minimal equivalent UFA is NP-complete.
- Given a DFA, finding the minimal equivalent UFA is NP-complete.
- Given a DFA, finding the minimal equivalent NFA is PSPACE-complete.

Minimization problems have even been studied for automata with unary alphabets; see, e.g., [10,5].

Recently, Malcher [12] improved on the results of Jiang and Ravikumar in the sense that he showed that finite automata with quite a small amount of non-determinism are hard to minimize. More precisely, he showed the following:

(a) Minimization is NP-complete for automata that can non-deterministically choose between a fixed number of initial states, but are otherwise deterministic.
(b) Minimization is NP-complete for non-deterministic automata with a constant number of computations for each string.[2]

Whereas Malcher made significant progress in showing that minimization is hard for non-deterministic automata, he was not yet able to solve the entire question. Therefore, he states the question whether there are relaxations of the deterministic automata model *at all* for which minimization is tractable as an important open problem. In this respect, he mentions the class of automata with at most two computations for each string and the two classes (a) and (b) above with the added restriction of unambiguousness as important remaining cases. In the present paper, we settle these open questions and provide a uniform NP-hardness proof for all classes of automata mentioned above.

In brief, we define a class δNFA of automata that are unambiguous, have at most two computations per string, and have at most one state q with two outgoing a-transitions, for at most one symbol a. Then, we show that minimization is NP-hard for all classes of finite automata that include δNFAs, and show that these hardness results can also be adapted to the setting of unambiguous automata that can non-deterministically choose between two start states, but are deterministic everywhere else. This solves the open cases mentioned by Malcher. On the other hand, there *are* relaxations of the deterministic automaton model that allow tractable minimization. We show that, if we add to the definition of

---

[2] Actually, he showed this for automata with constant *branching*, which is slightly different from the number of computations; see Section 2.1.

δNFAs that each word should have at most one *rejecting* computation, minimization becomes tractable again. However, the minimal automata in this class are the DFAs, so this class is not likely to be very useful in practice.

Other related work. An overview of state and transition complexity of NFAs can be found in [15]. Known results about the trade-off between amount of non-determinism and descriptional complexity are surveyed in [2]. The problems of producing small NFAs from regular expressions has been considered in [9,16]. There has also been some work on approximating minimal NFAs [3,4,6], basically showing that approximation is very hard. The work of Hromkovic et al. [8] about measuring non-determinism in finite automata will be relevant to us in Section 2. Since the minimization problem for NFAs is hard, the bisimulation minimization problem has also been considered; see, e.g., [14,1].

Due to space restrictions, some proofs and parts of proofs have been omitted, and will appear in the full version of the paper.

2   Preliminaries

By Σ we always denote a finite alphabet. A *(non-deterministic) finite automaton (NFA)* over Σ is a tuple $A = (\mathrm{States}(A), \mathrm{Alpha}(A), \mathrm{Rules}(A), \mathrm{init}(A), \mathrm{Final}(A))$, where $\mathrm{States}(A)$ is its finite set of states, $\mathrm{Alpha}(A) = \Sigma$, $\mathrm{init}(A) \in \mathrm{States}(A)$ is its initial state, $\mathrm{Final}(A) \subseteq \mathrm{States}(A)$ is its set of final states, and $\mathrm{Rules}(A)$ is a set of transition rules of the form $q_1 \xrightarrow{a} q_2$, where $q_1, q_2 \in \mathrm{States}(A)$ and $a \in \Sigma$. The *size* of an automaton is $|\mathrm{States}(A)|$, i.e., its number of states. A finite automaton is *deterministic* if, for each $q_1 \in \mathrm{States}(A)$ and $a \in \mathrm{Alpha}(A)$, there is at most one $q_2 \in \mathrm{States}(A)$ such that $q_1 \xrightarrow{a} q_2 \in \mathrm{Rules}(A)$. By DFA we denote the class of deterministic finite automata.

A *run*, or *computation*, r of A on a word $w = a_1 \cdots a_n \in \Sigma^*$ is a string $q_1 \cdots q_n \in \mathrm{States}(A)^*$ such that $\mathrm{init}(A) \xrightarrow{a_1} q_1 \in \mathrm{Rules}(A)$ and, for each $i = 1, \ldots, n-1$, $q_i \xrightarrow{a_{i+1}} q_{i+1} \in \mathrm{Rules}(A)$. The run is *accepting* if $q_n \in \mathrm{Final}(A)$. The *language of A* is the set of words w such that there exists an accepting run of A on w. A finite automaton A is *unambiguous* if, for each string w, there exists at most one accepting run of A on w.

Let $\mathcal{N}_1$ and $\mathcal{N}_2$ be two classes of NFAs. We say that $\mathcal{N}_1 \subseteq \mathcal{N}_2$ if each automaton in $\mathcal{N}_1$ also belongs to $\mathcal{N}_2$. For example, DFA $\subseteq$ NFA.

2.1   Notions of Non-determinism

We recall some standard measures of non-determinism in a finite automaton. For a state q and an alphabet symbol a, the *degree of non-determinism* of a pair (q, a), denoted by $\mathrm{degree}(q, a)$ is the number k of different states $q_1, \ldots, q_k$ such that, for all $1 \leq i \leq k$, $q \xrightarrow{a} q_i \in \mathrm{Rules}(A)$. We say that A has *degree of non-determinism k*, denoted by $\mathrm{degree}(A) = k$, if $\mathrm{degree}(q, a) \leq k$ for every $(q, a) \in \mathrm{States}(A) \times \mathrm{Alpha}(A)$, and there is at least one pair (q, a) such that $\mathrm{degree}(q, a) = k$.

The *branching* of an automaton is intuitively defined as the maximum product of the degrees of non-determinism over states in a possible run. Formally, the branching of A on a word $w = a_1 \cdots a_n$ is $\text{branch}_A(w) = \max\{\Pi_{i=1}^n \text{degree}(q_{i-1}, a_i) \mid q_1 \cdots q_n$ is a run of A on $a_1 \cdots a_n$, and $\text{init}(A) = q_0\}$. The *branching of* A, denoted $\text{branch}(A)$, is $\max\{\text{branch}_A(w) \mid w \in L(A)\}$ if this quantity is defined, and otherwise ∞.

Hromkovic et al. [8] define three measures non-determinism for a finite automaton A: $advice(A)$, $computations(A)$, and $ambig(A)$. These measures are defined as follows: $advice(A)$ is the maximum number of non-deterministic choices during any computation of A, $computations(A)$ is the maximum number of different computations of A on any word,[3] and $ambig(A)$ is the maximum number of different *accepting* computation of A on any word. For the formal definitions of these concepts, we refer to [8].

2.2 A Notion of Very Little Non-determinism

Next we define the notion of a δNFA. The intuition is that such an automaton should allow only a very small amount of non-determinism.

Definition 1. A δ*NFA* is an NFA A with the following properties

- A is unambiguous;
- $\text{branch}(A) \leq 2$; and
- there is at most one pair (q, a) such that $\text{degree}(q, a) = 2$.

For δNFAs, we have that $\text{degree}(A) \leq 2$, $advice(A) \leq 1$, $computations(A) \leq 2$, and $ambig(A) = 1$. Notice that any of $\text{degree}(A) = 1$, $advice(A) = 0$, or $computations(A) = 1$ implies that A is deterministic. Also, $ambig(A) = 1$ is the minimum value possible for any automaton that accepts at least one string.

2.3 The Minimization Problem

We define the minimization problem in two flavors. For two classes of finite automata $\mathcal{N}_1$ and $\mathcal{N}_2$ the $\mathcal{N}_1 \to \mathcal{N}_2$ *minimization problem* is the following problem. Given a finite automaton A in $\mathcal{N}_1$ and an integer k, does there exist a finite automaton B in $\mathcal{N}_2$ of size at most k such that $L(A) = L(B)$? For a class $\mathcal{N}$ of finite automata, the *minimization problem for* $\mathcal{N}$ is then simply the $\mathcal{N} \to \mathcal{N}$ minimization problem.

Suppose that the $\mathcal{N}_1 \to \mathcal{N}_2$ minimization problem is hard for a complexity class C, and let $\mathcal{N}_3$ be a class of automata such that $\mathcal{N}_1 \subseteq \mathcal{N}_3$. Then the $\mathcal{N}_3 \to \mathcal{N}_2$ minimization problem is also trivially hard for C. However, assuming that $\mathcal{N}_1 \to \mathcal{N}_2$ is hard for C and that $\mathcal{N}_2 \subseteq \mathcal{N}_3$, there is, as far as we know, no general argument that also makes the $\mathcal{N}_1 \to \mathcal{N}_3$ minimization problem hard for C, as finding a small $\mathcal{N}_3$ automaton might be easier than finding a small $\mathcal{N}_2$ automaton in general.[4] Therefore, we will prove directly that minimization is NP-hard for *all* classes of automata between δNFAs and NFAs.

---

[3] Hromkovic et al. wrote $\text{leaf}(A)$ instead of $computations(A)$.

[4] This is also why, e.g., Malcher explicitly proves NP-hardness for minimizing various classes of automata that are included in one another (Lemmas 3 and 11 in [12]).

2.4   Are δNFAs the Closest Possible to Determinism?

Before we give more intuition about this question, we first note that there are in fact *two* incomparable notions of determinism for finite automata: determinism and *reverse* determinism.[5] Both classes can be efficiently minimized by the same algorithm, modulo a simple pre- and post-processing step for reverse deterministic automata. We view these two classes as the two possible "optima" in the spectrum of determinism, as they arise very naturally from the fact that one can either read strings from left to right or from right to left. From now on, we only consider the proximity of δNFAs to (left-to-right deterministic) DFAs.

We believe that one can always think of more and more exotic notions of non-determinism that come closer and closer to DFAs.[6] We provide an example here. Define the class $\mathcal{C}$ to be the class of δNFAs with the additional condition that, for each word w, there can be at most one *rejecting* computation of A on w. (Thus, for each w, there can be at most two runs — one accepting and one rejecting.) This notion of non-determinism lies strictly between DFAs and δNFAs (DFA $\subseteq \mathcal{C} \subseteq$ δNFA).

Consider the minimization problem for $\mathcal{C}$ and let A be an arbitrary automaton in $\mathcal{C}$. We will argue that the minimal $\mathcal{C}$-automaton for $L(A)$ is a DFA. Suppose that A is not a DFA. Let q and a be the unique state and label so that degree$(q, a) = 2$. Let q_1 and q_2 be the two states such that $q \xrightarrow{a} q_1$ and $q \xrightarrow{a} q_2$ are in Rules(A). Let w be an arbitrary string that leads A to state q. By definition of $\mathcal{C}$, A must accept every string of the form waw', where w' is an arbitrary word in Alpha$(A)^*$. (If waw' would be rejected, then there would be two rejecting runs, one over q_1 and one over q_2.) Therefore, we can make A strictly smaller by merging the two states q_1, q_2 to q_3, making q_3 a final state, and adding loop transitions to q_3 for each alphabet symbol. Moreover, by this operation, A becomes deterministic. Hence, every automaton A in $\mathcal{C}$ that is not a DFA can be rewritten as a smaller DFA. This means that, in the class $\mathcal{C}$, the minimal automata are the DFAs. In particular, this also puts the minimization problem for $\mathcal{C}$ into PTIME.

From the above it is clear that δNFAs are certainly not the closest possible to determinism that one can get. Rather, it is the closest class to DFAs we were able to find that takes advantage of the succinctness of nondeterminism in a nontrivial way.

Our NP-hardness result for the minimization of δNFAs therefore puts the tractability frontier precisely between δNFAs and the above mentioned class $\mathcal{C}$; two classes that are extremely close to one another.

3   Minimizing Non-deterministic Automata is Hard

The main result of this section is the following.

---

[5] The latter is the class for which the inverted transitions are deterministic.
[6] One could, for instance, take the class of DFAs and add a single NFA.

Theorem 2. *Let $\mathcal{N}$ be a class of finite automata. If $\delta NFA \subseteq \mathcal{N}$ then DFA $\to \mathcal{N}$ minimization is NP-hard.*

Corollary 3. *For each class $\mathcal{N}$ of finite automata such that $\delta NFA \subseteq \mathcal{N}$, the minimization problem for $\mathcal{N}$ is NP-hard.*

We start by formally defining the decision problems that are of interest to us, and then sketch an intuitive overview of our proof. Given an undirected graph $G = (V, E)$ such that V is its set of vertices and $E \subseteq V \times V$ is its set of edges, we say that a set of vertices $VC \subseteq V$ is a *vertex cover* of G if, for every edge $(v_1, v_2) \in E$, VC contains v_1, v_2, or both.

If B and C are finite collections of finite sets, we say that B is a *set basis* for C if, for each $c \in C$, there is a subcollection B_c of B whose union is c. We say that B is a *normal set basis* for C if, for each $c \in C$, there is a *pairwise disjoint* subcollection B_c of B whose union is c. We say that B is a *separable normal set basis* for C if B is a normal set basis for C and B can be written as a disjoint union $B_1 \uplus B_2$ such that, for each $c \in C$, the subcollection B_c of B contains at most one element from B_1 and at most one from B_2.

The following decision problems are considered in this paper. *Vertex Cover* asks, given a pair (G, k) where G is a graph and k is an integer, whether there exists a vertex cover of G of size at most k. *Set Basis*, *Normal Set Basis*, and *Separable Normal Set Basis* ask, given a pair (C, s) where C is a finite collection of finite sets and s is an integer, whether there exists a set basis, resp., normal set basis, resp., separable normal set basis for C containing at most s sets.

The proof of Theorem 2 proceeds in several steps. First, we provide a slightly modified version of a known reduction from Vertex Cover to Normal Set Basis (Lemma 4 in [11]), showing that the latter problem is NP-hard. Second, we proceed to show that the set **I** of instances of Normal Set Basis obtained through this reduction has a number of interesting properties (Lemma 5). In particular, we show that if such an instance has a set basis of a certain size s, then it also has a *normal* set basis of size s. Third, we show that the the Normal Set Basis problem, for instances in **I** reduces to minimization for δNFAs (Lemma 6).

The statement of Theorem 2 says that given a DFA, finding the minimal equivalent automaton in class $\mathcal{N}$ is NP-hard, for any class of finite automata that contains the δNFAs. As argued in Section 2.3, using a DFA instead of a δNFA as input of the problem strengthens the statement. Also, showing that DFA $\to \delta$NFA is NP-hard doesn't immediately imply that DFA $\to \mathcal{N}$ is hard for every $\mathcal{N}$ that contains all δNFAs. To show that this is actually the case, we prove that for the languages obtained in our reduction, the minimal NFAs are precisely one state smaller than the minimal δNFAs (Lemma 6). For these languages, the minimization problem for δNFAs and for NFAs is essentially the same problem.

We revisit a slightly modified reduction which is due to Jiang and Ravikumar [11], as our further results rely on a construction in their proof.

Lemma 4 (Jiang and Ravikumar [11]). *Normal Set Basis is NP-complete.*

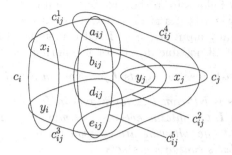

Fig. 1. The constructed sets $c_i, c_j, c_{ij}^1, \ldots, c_{ij}^5$ in the proof of Lemma 4

Proof (Sketch). Obviously, Normal Set Basis is in NP. Indeed, given an input (C, s) for Normal Set Basis, the NP algorithm simply guesses a collection B containing at most s sets, guesses the subcollections B_c for each $c \in C$, and verifies whether the sets B_c satisfy the necessary conditions.

We give a reduction from Vertex Cover to Normal Set Basis but omit the correctness proof. Given an input (G, k) of Vertex Cover, where $G = (V, E)$ is a graph and k is an integer, we construct in LOGSPACE an input (C, s) of Normal Set Basis, where C is a finite collection of finite sets and s is an integer. In particular, (C, s) is constructed such that G has a vertex cover of size at most k if and only if C has a normal set basis containing at most s sets.

For a technical reason which will become clear in later proofs, we assume without loss of generality that $k < |E| - 3$. Notice that, under this restriction, Vertex Cover is still NP-complete under LOGSPACE reductions. Indeed, if $k \geq |E| - 3$, Vertex Cover can be solved in LOGSPACE by testing all possibilities of the at most 3 vertices which are not in the vertex cover, and verifying that there does not exist an edge between 2 of these 3 vertices.

Formally, let $V = \{v_1, \ldots, v_n\}$. For each $i = 1, \ldots, n$, define c_i to be the set $\{x_i, y_i\}$ which intuitively corresponds to the node v_i. Let (v_i, v_j) be in E with $i < j$. To each such edge we associate five sets as follows:

$$c_{ij}^1 := \{x_i, a_{ij}, b_{ij}\}, \quad c_{ij}^4 := \{x_j, e_{ij}, a_{ij}\}, \text{ and}$$
$$c_{ij}^2 := \{y_j, b_{ij}, d_{ij}\}, \quad c_{ij}^5 := \{a_{ij}, b_{ij}, d_{ij}, e_{ij}\}.$$
$$c_{ij}^3 := \{y_i, d_{ij}, e_{ij}\},$$

Figure 1 contains a graphical representation of the constructed sets $c_i, c_j, c_{ij}^1, \ldots, c_{ij}^5$ for some $(v_i, v_j) \in E$. Then, define

$$C := \{c_i \mid 1 \leq i \leq n\} \cup \{c_{ij}^t \mid (v_i, v_j) \in E, i < j, \text{ and } 1 \leq t \leq 5\}$$

and $s := n + 4|E| + k$. Notice that the collection C contains $n + 5|E|$ sets. Obviously, C and s can be constructed from G and k in polynomial time.

It can be shown that G has a vertex cover of size at most k if and only if C has a (separable) normal set basis containing at most s sets. □

The next lemma now follows from the proof of Lemma 4. It shows that C has a set basis containing s sets if and only if C has a separable normal set basis containing s sets for any input (C, s) in $\mathbf{I}$. Of course, the latter property does not hold for the set of all possible inputs for the normal set basis problem.

Lemma 5. *There exists a set of inputs $\mathbf{I}$ for Normal Set Basis, such that*

(1) Normal Set Basis is NP-complete for inputs in $\mathbf{I}$;
(2) for each (C, s) in $\mathbf{I}$, C contains every set at most once and $s < |C| - 3$;
(3) for each $(C, s) \in \mathbf{I}$, the following are equivalent:
   (a) C has a set basis containing s sets.
   (b) C has a separable normal set basis containing s sets.
(4) for each (C, s) in $\mathbf{I}$, each solution B for (C, s) writes at least two sets of C as a union of at least two sets in B.

Proof. The set $\mathbf{I}$ is obtained by applying the reduction in Lemma 4 to inputs (G, k) of Vertex Cover such that $k \leq |E| - 3$. This immediately shows (1) and (2). We continue by proving the other cases.

(3) The direction from (b) to (a) is trivial. For the other direction the full proof of Lemma 4 actually shows that if G has a vertex cover of size k, then C has a *separable* normal set basis containing s sets. Conversely, if C has a *normal* set basis containing at most s sets, then G has a vertex cover of size k. This would imply that C also has a separable normal set basis containing s sets.

Hence, we still need to prove that, if C has a set basis of at most s sets, then C also has a normal set basis containing at most s sets. Let (C, s) be an instance in $\mathbf{I}$, i.e., there is an $n \in \mathbb{N}$ and $E \subseteq \{(i, j) \mid 1 \leq i < j \leq n\}$ such that $C = \{c_i \mid 1 \leq i \leq n\} \cup \{c_{ij}^r \mid (i, j) \in E \wedge 1 \leq r \leq 5\}$, and suppose C has a set basis $B = \{b_1, \ldots, b_s\}$ of size s. We construct a *normal* set basis for C of size s.

Suppose that there is an i such that B contains both $\{x_i\}$ and $\{x_i, y_i\}$. Then we can replace $\{x_i, y_i\}$ with $\{y_i\}$ and still have a set basis for C, since c_i is the only set in C of which $\{x_i, y_i\}$ is a subset. Therefore, we can assume without loss of generality that B either contains $\{x_i, y_i\}$ or $\{x_i\}$ and $\{y_i\}$, but never both $\{x_i, y_i\}$ and $\{x_i\}$ (or, symmetrically, $\{x_i, y_i\}$ and $\{y_i\}$).

Suppose there are $1 \leq i < j \leq n$ and $1 \leq r \leq 4$ such that c_{ij}^r cannot be formed as a disjoint union of sets from B. Without loss of generality, we may assume that $r = 1$, i.e., $c_{ij}^r = c_{ij}^1 = \{x_i, a_{ij}, b_{ij}\}$, since all other cases follow by symmetry. Since there are no disjoint sets from B whose union is c_{ij}^1, there must be two different sets b^1 and b^2 in B that are subsets of c_{ij} and contain precisely two elements each. At least one of these subsets must contain x_i. Assume w.l.o.g. that this set is b^1. No subset of size two of c_{ij}^1 that contains x_i is a subset of any set of C other than c_{ij}^1. This means that we can replace b^1 with $b^1 - b^2$ in B and still have a set basis of size at most s. Thus we can assume that for any i, j and any $r \in \{1, \ldots, 4\}$, the set c_{ij}^r can be formed as a union of disjoint sets from B.

The only remaining case is if there are $1 \leq i < j \leq n$ such that c_{ij}^5 cannot be formed as a disjoint union of sets from B. Let B_{ij} be a subset of B such that the union of the sets in B_{ij} is c_{ij}^5. We can assume that B_{ij} is inclusion free, i.e., there are no two sets in B_{ij} such that one is a subset of the other. If B_{ij} has four members, then we can replace B_{ij} with the four singletons and still have a set basis of size s, so we can assume that B_{ij} has at most three members. Suppose there is a set b in B_{ij} such that b is not a subset of any set in C other than c_{ij}^5. Then we can replace b with $b - (\bigcup_{b' \in B_{ij} - \{b\}} b')$ in B and still have a set basis of size at most s. Thus we can assume that each member of B_{ij} is a subset of some set from C other that c_{ij}^5. In particular, this means that each member of B_{ij} has at most two elements. If we take three different subsets of c_{ij}^5 with at most two elements, that are also subsets of other sets from C than c_{ij}^5, then at least two of them are disjoint; see Figure 1. Let these two disjoint sets be b^1 and b^2. If b^1 and b^2 both contain two elements, we can replace B_{ij} with $\{b^1, b^2\}$. Thus we can assume that there are at most two sets in B_{ij} with two elements. Furthermore, these two sets must overlap. This means that we can assume that B_{ij} has exactly three members, two with two elements and one singleton.

Without loss of generality, we may assume that $B_{ij} = \{\{a_{ij}, b_{ij}\}, \{b_{ij}, d_{ij}\}, \{e_{ij}\}\}$. All other cases are symmetrical. We may also assume that neither $\{a_{ij}\}$ nor $\{b_{ij}\}$ belong to B. (If $\{a_{ij}\}$ belongs to B then we can replace $\{a_{ij}, b_{ij}\}$ by $\{b_{ij}\}$ in B_{ij}, and if $\{b_{ij}\}$ belongs to B we can replace $\{b_{ij}, d_{ij}\}$ by $\{d_{ij}\}$ in B_{ij}.) This means that in order to form c_{ij}^1 either $\{x_i\}$, $\{x_i, a_{ij}\}$, $\{x_i, b_{ij}\}$, or $\{x_i, a_{ij}, b_{ij}\}$ must be a member of B. If it is not $\{x_i\}$, we can replace it with $\{x_i\}$, since none of the other sets is a subset of any other set in C than c_{ij}^1. But if $\{x_i\} \in B$ we can also assume that $\{y_i\} \in B$. To form c_{ij}^3, B must, apart from $\{y_i\}$ and $\{e_{ij}\}$, contain some subset of c_{ij}^3 that contains d_{ij}. Since we have both $\{y_i\}$ and $\{e_{ij}\}$ in B, we may replace this subset with $\{d_{ij}\}$. Once we have $\{d_{ij}\}$ in B, we can replace $\{b_{ij}, d_{ij}\}$ by $\{b_{ij}\}$ in B and still have a set basis for C of size at most s, one that can form c_{ij}^5 as a union of disjoint members.

In summary, we have shown that from any set basis for C of size s we can form a *normal* set basis for C of size at most s.

(4) One simply has to observe that a normal set basis writing at most one set of C as a union of at least two sets must contain at least $|C|$ sets, and hence cannot be a solution for (C, s). ☐

The proof of the following lemma is partly inspired by the proof of Theorem 3.1 of [11], but we significantly strengthen it for our purposes.

Lemma 6. *There exists a set of regular languages $\mathcal{L}$ such that*

(1) DFA $\rightarrow$ δNFA minimization is NP-complete for DFAs accepting $\mathcal{L}$ and

(2) for each $L \in \mathcal{L}$, the size of the minimal NFA for L is equal to the size of the minimal δNFA for L, minus 1.

Proof (Sketch). The NP upper bound is immediate, as equivalence testing for unambiguous finite automata is in PTIME [17]. We can guess a δNFA of sufficiently small size and test in PTIME whether it is equivalent to the given δNFA.

For the lower bound, we reduce from Separable Normal Set Basis. To this end, let (C, s) be an input of Separable Normal Set Basis. Hence, C is a collection of n sets and s is an integer. According to Lemma 5, we can assume without loss of generality that $(C, s) \in \mathbf{I}$, that is, C has a separable normal set basis containing

s sets if and only if C has a normal set basis of size s. Moreover, we can assume that $s < n - 3$.

We construct in LOGSPACE a δNFA A and an integer ℓ such that the following are equivalent:

– C has a separable normal set basis of size at most s.
– There exists a δNFA N_δ for $L(A)$ of size at most ℓ.
– There exists an NFA N for $L(A)$ of size at most $\ell - 1$.

The δNFA A accepts the language $\{acb \mid c \in C \text{ and } b \in c\}$, which is a finite language of strings of length three.

Formally, let $C = \{c_1, \ldots, c_n\}$ and $c_i = \{b_{i,1}, \ldots, b_{i,n_i}\}$. Then, A is defined over $\mathrm{Alpha}(A) = \{a\} \cup \bigcup_{1 \le i \le n} \{c_i, b_{i,1}, \ldots, b_{i,n_i}\}$. The state set of A is $\mathrm{States}(A) = \{q_0, q_0', q_1, \ldots, q_n, q_f\}$, and the initial and final state sets of A are q_0 and q_f, respectively. The transitions $\mathrm{Rules}(A)$ are formally defined as follows:

– $q_0 \xrightarrow{a} q_0'$;
– for every $i = 1, \ldots, n$, $q_0' \xrightarrow{c_i} q_i$; and
– for every $i = 1, \ldots, n$ and $j = 1, \ldots, n_i$, $q_i \xrightarrow{b_{i,j}} q_f$.

Finally, define $\ell := s + 4$.

Obviously, A and ℓ can be constructed from C and s using logarithmic space. Observe that due to Lemma 5, C contains every set at most once, and hence does not contain $c_i = c_j$ with $i \ne j$. Hence, A is a minimal DFA for $L(A)$.

We now show that,

(a) if C has a separable normal set basis containing at most s sets, then there exists a δNFA N_δ for $L(A)$ of size at most ℓ and an NFA N for $L(A)$ of size at most $\ell - 1$;
(b) if there exists a δNFA N_δ for $L(A)$ of size at most ℓ then C has a separable normal set basis containing at most s sets; and
(c) if there exists an NFA N for $L(A)$ of size at most $\ell - 1$ then C has a separable normal set basis containing at most s sets.

(a) Assume that C has a separable normal set basis containing s sets. We construct a δNFA N_δ for $L(A)$ of size $\ell = s + 4$.

Let $B = \{r_1, \ldots, r_s\}$ be the separable normal set basis for C containing s sets. Also, let B_1 and B_2 be disjoint subcollections of B such that each element of C is either an element of B_1, an element of B_2, or a disjoint union of an element of B_1 and an element of B_2.

To describe N_δ, we first fix the representation of each set c in C as a disjoint union of at most one set in B_1 and at most one set in B_2. Say that each basic member of B in this representation *belongs to* c.

We define the state set of N_δ as $\mathrm{States}(N_\delta) = \{q_0, q_1, q_2, q_f\} \cup \{r_i \in B_1\} \cup \{r_i \in B_2\}$. The transition rules of N_δ are defined as follows. First, $\mathrm{Rules}(N_\delta)$ contains the non-deterministic transitions $q_0 \xrightarrow{a} q_1$ and $q_0 \xrightarrow{a} q_2$. Furthermore, for every $i = 1, \ldots, n$, $j = 1, \ldots, s$, and $m = 1, 2$, $\mathrm{Rules}(N_\delta)$ contains the rule

- $q_m \xrightarrow{c_i} r_j$, if $r_j \in B_m$ and r_j belongs to c_i; and
- $r_j \xrightarrow{b} q_f$, if $r_j \in B_m$ and $b \in r_j$.

Notice that the size of N_δ is $|B| + 4 = s + 4 = \ell$. By construction, we have that $L(N_\delta) = L(A)$. It can be shown that N_δ is a δNFA.

(b) This part of the proof is omitted.

(c) It can be shown that an NFA for $L(A)$ of size at most $\ell - 1$ gives rise to a set basis of size s. From Lemma 5, it now follows that C also has a separable normal set basis containing at most s sets. □

Theorem 2 now follows from the proof of Lemma 6.

Until now, our results focused on classes of finite automata that can accept all regular languages. Our proof shows that this is not even necessary, as the NP-hard instances we construct only accept strings of length tree. Therefore, we also have the following Corollary.

Corollary 7. *Let δNFA_{finite} be the class of δNFAs that accept only finite languages. Let $\mathcal{N}$ be class of finite automata. If $\delta NFA_{finite} \subseteq \mathcal{N}$ then the DFA $\rightarrow \mathcal{N}$ minimization problem is NP-hard.*

4   Succinctness and Uniqueness

As mentioned in the introduction, when a developer selects a description mechanism for regular languages, she faces a trade-off between succinctness and complexity of minimization. The following proposition shows that in the case of δNFAs, the succinctness bought at the price of NP-completeness is limited.

Proposition 8. *For every δNFA of size n, there is an equivalent DFA of size $O(n^2)$.*

On the other hand, if we were to remove the branch$(A) \leq 2$ condition in the definition of δNFAs, then there would be an exponential gain in succinctness. This is witnessed by the standard family of languages $(a+b)^* a(a+b)^n$ for $n \geq 0$ that shows that NFAs are exponentially more succinct than DFAs in general. The canonical NFA for this language is unambiguous and has only one pair (q, a) for which degree$(q, a) = 2$.

Proposition 9. *The minimal δNFA for a regular language is not unique.*

5   Automata with Multiple Initial States

Throughout the paper, to simplify definitions, we have assumed that finite automata have a unique start state. As we mentioned in the Introduction, the minimization problem for finite automata that can non-deterministically choose between multiple initial states, but are otherwise deterministic, has also been studied [7,12].

Proposition 10. *Minimization is NP-hard for unambiguous finite automata that have at most two initial states but are otherwise deterministic.*

Together with Theorem 2 this solves all the open cases mentioned by Malcher [12].

References

1. Abdulla, P., Deneux, J., Kaati, L., Nilsson, M.: Minimization of non-deterministic automata with large alphabets. In: CIAA, pp. 31–42 (2006)
2. Goldstine, J., Kappes, M., Kintala, C., Leung, H., Malcher, A., Wotschke, D.: Descriptional complexity of machines with limited resources. J. Univ. Comp. Science 8(2), 193–234 (2002)
3. Gramlich, G., Schnitger, G.: Minimizing NFAs and regular expressions. JCSS 73(6), 908–923 (2007)
4. Gruber, H., Holzer, M.: Finding lower bounds for nondeterministic state complexity is hard. In: H. Ibarra, O., Dang, Z. (eds.) DLT 2006. LNCS, vol. 4036, pp. 363–374. Springer, Heidelberg (2006)
5. Gruber, H., Holzer, M.: Computational complexity of NFA minimization for finite and unary languages. In: LATA, pp. 261–272 (2007)
6. Gruber, H., Holzer, M.: Inapproximability of nondeterministic state and transition complexity assuming P ≠ NP. In: Harju, T., Karhumäki, J., Lepistö, A. (eds.) DLT 2007. LNCS, vol. 4588, pp. 205–216. Springer, Heidelberg (2007)
7. Holzer, M., Salomaa, K., Yu, S.: On the state complexity of k-entry deterministic finite automata. J. Automata, Languages, and Combinatorics 6(4), 453–466 (2001)
8. Hromkovic, J., Karhumäki, J., Klauck, H., Schnitger, G., Seibert, S.: Measures of nondeterminism in finite automata. In: Welzl, E., Montanari, U., Rolim, J.D.P. (eds.) ICALP 2000. LNCS, vol. 1853, pp. 199–210. Springer, Heidelberg (2000)
9. Hromkovic, J., Schnitger, G.: Comparing the size of NFAs with and without epsilon-transitions. TCS 380(2), 100–114 (2007)
10. Jiang, T., McDowell, E., Ravikumar, B.: The structure and complexity of minimal NFAs over unary alphabet. Int. J. Found. Comp. Science 2, 163–182 (1991)
11. Jiang, T., Ravikumar, B.: Minimal NFA problems are hard. Siam J. Comp. 22(6), 1117–1141 (1993)
12. Malcher, A.: Minimizing finite automata is computationally hard. TCS 327(3), 375–390 (2004)
13. Meyer, A., Fischer, M.J.: Economy of descriptions by automata, grammars, and formal systems. In: FOCS, pp. 188–191. IEEE, Los Alamitos (1971)
14. Paige, R., Tarjan, R.: Three parition refinement algorithms. Siam J. Comp. 16, 973–989 (1987)
15. Salomaa, K.: Descriptional complexity of nondeterministic finite automata. In: Harju, T., Karhumäki, J., Lepistö, A. (eds.) DLT 2007. LNCS, vol. 4588, pp. 31–35. Springer, Heidelberg (2007)
16. Schnitger, G.: Regular expressions and NFAs without epsilon-transitions. In: Durand, B., Thomas, W. (eds.) STACS 2006. LNCS, vol. 3884, pp. 432–443. Springer, Heidelberg (2006)
17. Stearns, R.E., Hunt III., H.B.: On the equivalence and containment problems for unambiguous regular expressions, regular grammars and finite automata. Siam J. Comp. 14(3), 598–611 (1985)
18. Stockmeyer, L., Meyer, A.: Word problems requiring exponential time: Preliminary report. In: STOC, pp. 1–9. ACM, New York (1973)

Finite Automata, Digraph Connectivity, and Regular Expression Size

(Extended Abstract)

Hermann Gruber[1] and Markus Holzer[2]

[1] Institut für Informatik, Ludwig-Maximilians-Universität München,
Oettingenstraße 67, D-80538 München, Germany
`gruberh@tcs.ifi.lmu.de`
[2] Institut für Informatik, Technische Universität München,
Boltzmannstraße 3, D-85748 Garching bei München, Germany
`holzer@in.tum.de`

Abstract. Recently lower bounds on the minimum required size for the conversion of deterministic finite automata into regular expressions and on the required size of regular expressions resulting from applying some basic language operations on them, were given by Gelade and Neven [8]. We strengthen and extend these results, obtaining lower bounds that are in part optimal, and, notably, the presented examples are over a binary alphabet, which is best possible. To this end, we develop a different, more versatile lower bound technique that is based on the star height of regular languages. It is known that for a restricted class of regular languages, the star height can be determined from the digraph underlying the transition structure of the minimal finite automaton accepting that language. In this way, star height is tied to cycle rank, a structural complexity measure for digraphs proposed by Eggan and Büchi, which measures the degree of connectivity of directed graphs.

1 Introduction

One of the most basic theorems in formal language theory is that every regular expression can be effectively converted into an equivalent finite automaton, and *vice versa* [16]. While algorithms accomplishing these tasks have been known for a long time, there has been a renewed interest in these classical problems during the last few years. For instance, new algorithms for converting regular expressions into finite automata outperforming classical algorithms have been found only recently, as well as a matching lower bound of $\Omega(n \cdot \log^2 n)$ on the number of transitions required by any equivalent nondeterministic finite automaton (NFA). The lower bound is, however, only attained for growing alphabet size, and a better algorithm is known for constant alphabet size, see [26] for the current state of the art.

In contrast, much less is known about the converse direction, namely of converting finite automata into regular expressions. Apart from the fundamental

L. Aceto et al. (Eds.): ICALP 2008, Part II, LNCS 5126, pp. 39–50, 2008.

Table 1. Comparing the lower bound results for conversion problems of deterministic finite automata (DFA) and regular expressions (RE), where $\cap$ denotes intersection, $\neg$ complementation, and $\amalg$ the shuffle operation on formal languages. Entries with a bound in $\Theta(\cdot)$ indicate that the result is best possible, i.e., refers to a lower bound matching a known upper bound.

Conversion	Gelade and Neven [8]	this paper with $|\Sigma| = 2$	
planar DFA to RE^1	—	$2^{\Theta(\sqrt{n})}$	[Thm. 11]
DFA to RE	$2^{\Omega(\sqrt{n/\log n})}$ for $|\Sigma| = 4$	$2^{\Theta(n)}$	[Thm. 16]
$RE \cap RE$ to RE	$2^{\Omega(\sqrt{n})}$ for $|\Sigma| = O(n)$	$2^{\Omega(n)}$	[Cor. 8]
$RE \amalg RE$ to RE		$2^{\Omega(n)}$	[Cor. 9]
$\neg RE$ to RE	$2^{2^{\Omega(n)}}$ for $|\Sigma| = 4$	$2^{2^{\Omega(\sqrt{n \log n})}}$	[Thm. 10]

nature of the problem, some applications lie in control flow normalization, including uses in software engineering such as automatic translation of legacy code [20]. All known algorithms covering the general case of infinite languages are based on the classical ones, which are compared in the survey [25]. The drawback is that all of these (structurally similar) algorithms return expressions of size $2^{O(n)}$ in the worst case, and Ehrenfeucht and Zeiger exhibit a family of languages over an alphabet of size n^2 for which this exponential blow-up is inevitable [6]. These examples naturally raise the question whether a size blow-up of $2^{\Omega(n)}$ can also occur for constant alphabet size, a question posed in [7]. One of the main results in this paper is a positive answer to this question, even in the case of a binary alphabet; note that the conversion problem becomes polynomial for unary languages [7]. Currently, there are not many lower bound techniques for regular expression size. A notable exception is the technique used in the above mentioned work [6], which however requires, in its original version, a largely growing alphabet. Recently, a variation of Ehrenfeucht and Zeiger's method was used in [8] to get similar but weaker lower bounds on the conversion problem for small alphabets. The above mentioned question, however, was left open. A technique based on communication complexity that applies only for finite languages, is proposed in [10]. They give an optimal bound of $n^{\Theta(\log n)}$ for the conversion problem in the case of finite languages.

Independently of [8], we take a different direction, by relating the descriptional complexity of regular languages (alphabetic width) to their structural complexity (star height). The star height is a structural complexity measure of regular languages that has been intensively studied in the literature for more than 40 years, see [11,15] for a recent treatment. Determining the star height can be in some cases reduced to the easier task of determining the cycle rank of a certain digraph. The latter concept is related to the cycle rank of digraphs, a digraph connectivity measure defined by Eggan and Büchi [5] in the 1960s. Since measuring the connectivity of digraphs is a very active research area, see,

---

[1] The lower bound result on the conversion of planar deterministic finite automata to regular expressions holds for $|\Sigma| = 4$.

e.g., [1,2,14,22], and as we feel that cycle rank is a interesting concept in its own right, we summarize and further develop the theory of cycle rank. For a more thorough treatment, including all proofs and comparison to some other recently proposed measures we refer to [9]. These connections turn out to be fruitful, allowing not only for proving a tight lower bound on the problem of converting finite automata into regular expressions, but also for giving reasonably good lower bounds for the alphabetic width of some basic regular language operations, namely intersection, complement, and shuffle. In this way, we independently improve on and extend the recently obtained results in [8]—we summarize and compare the obtained results in Table 1.

2   Basic Definitions

We introduce some basic notions in formal language and automata theory—for a thorough treatment, the reader might want to consult a textbook such as [12]. In particular, let Σ be a finite alphabet and Σ^* the set of all words over the alphabet Σ, including the empty word ε. The length of a word w is denoted by $|w|$, where $|\varepsilon| = 0$. A *(formal) language* over the alphabet Σ is a subset of Σ^*.

The *regular expressions* over an alphabet Σ are defined recursively in the usual way:[2] $\emptyset$, ε, and every letter a with $a \in \Sigma$ is a regular expression; and when r_1 and r_2 are regular expressions, then $(r_1 + r_2)$, $(r_1 \cdot r_2)$, and $(r_1)^*$ are also regular expressions. The language defined by a regular expression r, denoted by $L(r)$, is defined as follows: $L(\emptyset) = \emptyset$, $L(\varepsilon) = \{\varepsilon\}$, $L(a) = \{a\}$, $L(r_1 + r_2) = L(r_1) \cup L(r_2)$, $L(r_1 \cdot r_2) = L(r_1) \cdot L(r_2)$, and $L(r_1^*) = L(r_1)^*$. The *size* or *alphabetic width* of a regular expression r over the alphabet Σ, denoted by $\mathrm{alph}(r)$, is defined as the total number of occurrences of letters of Σ in r. For a regular language L, we define its alphabetic width, $\mathrm{alph}(L)$, as the minimum alphabetic width among all regular expressions describing L.

It is well known that regular expressions and finite automata are equally powerful, i.e., for every regular expression one can construct an equivalent (deterministic) finite automaton and *vice versa*. Finite automata are defined as follows: A *nondeterministic finite automaton* (NFA) is a 5-tuple $A = (Q, \Sigma, \delta, q_0, F)$, where Q is a finite set of states, Σ is a finite set of input symbols, $\delta : Q \times \Sigma \to 2^Q$ is the transition function, $q_0 \in Q$ is the initial state, and $F \subseteq Q$ is the set of accepting states. The *language accepted* by the finite automaton A is defined as $L(A) = \{ w \in \Sigma^* \mid \delta(q_0, w) \cap F \neq \emptyset \}$, where δ is naturally extended to a function $Q \times \Sigma^* \to 2^Q$. A nondeterministic finite automaton $A = (Q, \Sigma, \delta, Q_0, F)$ is *deterministic*, for short a DFA, if $|\delta(q, a)| \leq 1$, for every $q \in Q$ and $a \in \Sigma$. In this case we simply write $\delta(q, a) = p$ instead of $\delta(q, a) = \{p\}$. Two (deterministic or nondeterministic) finite automata are *equivalent* if they accept the same language.

---

[2] For convenience, parentheses in regular expressions are sometimes omitted and the concatenation is simply written as juxtaposition. The priority of operators is specified in the usual fashion: concatenation is performed before union, and star before both product and union.

In the remainder of this section we fix some basic notions from graph theory. A *directed graph*, or *digraph*, $G = (V, E)$ consists of a finite set of vertices V with an associated set of edges $E \subseteq V \times V$. An edge whose start and end vertex are identical is called a *loop*. If G has no loops, then G is called *loop-free*. If the edge relation of G is symmetric, then G is an *undirected graph*, or simply *graph*. It is often convenient to view the set of edges of an undirected graph as a set of unordered pairs $\{u, v\}$, with u and v in V. Only if there is no risk of confusion, for an undirected graph G, we refer to the set $\{\{u, v\} \mid (u, v) \in E\}$ as the set of edges of G, and, abusing notation, denote it by E. A digraph $H = (U, F)$ is a *subdigraph*, or simply *subgraph*, of a digraph $G = (V, E)$, if $U \subseteq V$ and for each edge $(u, v) \in F$ with $u, v \in U$, the pair (u, v) is an edge in E. A subgraph H is called *induced*, if furthermore for each edge $(u, v) \in E$ with $u, v \in U$, the pair (u, v) is also an edge in F. In the latter case, H is referred to as the subgraph of G induced by U, and denoted by $G[U]$. When removing a set of vertices U, or a single vertex u, from G, it is often handy to write $G - U$ and $G - u$ to denote the induced subgraphs $G[V \setminus U]$ and $G[V \setminus \{u\}]$, respectively.

We recall the definitions of some other important concepts related to walks and reachability. A subgraph $H = (U, F)$ of G is *strongly connected*, if for every vertices u and v, both u is reachable from v and v is reachable from u. A strongly connected subgraph H is called *nontrivial* if H has at least one edge, otherwise it is called trivial. Note that every trivial strongly connected subgraph has at most one vertex, but if G is not loop-free, it also has nontrivial strongly connected subgraphs with only one vertex. A set of vertices $\emptyset \subset C \subseteq V$ is a *strongly connected component* if $G[C]$ is strongly connected, but for every proper superset $C' \supset C$, the induced subgraph $G[C']$ is not strongly connected.

3   Star Height of Regular Languages and Cycle Rank of Digraphs

3.1   Definitions and Early Results

For a regular expression r over Σ, the star height, denoted by $h(r)$, is a structural complexity measure inductively defined by: $h(\emptyset) = h(\varepsilon) = h(a) = 0$, $h(r_1 \cdot r_2) = h(r_1 + r_2) = \max(h(r_1), h(r_2))$, and $h(r_1^*) = 1 + h(r_1)$. The star height of a regular language L, denoted by $h(L)$, is then defined as the minimum star height among all regular expressions describing L. We will later establish a relation between star height and alphabetic width of regular languages. This relation will allow us to reduce the task of proving lower bounds on alphabetic width to the one of proving lower bounds on star height.

First, we call to attention a structural complexity measure for digraphs intimately related to the star height of regular languages, called the cycle rank, suggested by Eggan and Büchi in the course of investigating the star height of regular languages [5].

Definition 1. *The* cycle rank *of a directed graph* $G = (V, E)$, *denoted by* $cr(G)$, *is inductively defined as follows: (1) If* G *is acyclic, then* $cr(G) = 0$. *(2) If* G *is* *strongly connected, then* $cr(G) = 1 + \min_{v \in V}\{cr(G - v)\}$, *where* $G - v$ *denotes the graph with the vertex set* $V \setminus \{v\}$ *and appropriately defined edge set. (3) If* G *is not strongly connected, then* $cr(G)$ *equals the maximum cycle rank among all strongly connected components of* G.

In the following, we will be sometimes concerned with the cycle rank of the digraph underlying the transition structure of finite automata, so for a given finite automaton A, let its cycle rank, denoted by $cr(A)$, be defined as the cycle rank of the underlying graph. The following relation between cycle rank of automata and star height of regular languages became known as Eggan's Theorem [5]:

Theorem 2 (Eggan's Theorem). *The star height of a regular language* L *equals the minimum cycle rank among all nondeterministic finite automata accepting* L.

The star height of a regular language appears to be a more difficult concept than alphabetic width, see, e.g., [11,15]. In light of this consideration, proving lower bounds on alphabetic width *via* lower bounds on star height appears to be trading a hard problem for an even harder one. But early research on the star height problem established a subclass of regular languages for which the star height is determined more easily, namely the family of bideterministic regular languages, which are defined as follows: A regular language L is *bideterministic* if there exists a deterministic finite automaton A with a single final state such that a deterministic finite automaton accepting the reversed language L^R is obtained from A by reverting the direction of each transition and exchanging the roles of the initial and final state. The star height of bideterministic languages was shown to be computable in [18], building on earlier work which was, however, published only later in [19]:

Theorem 3 (McNaughton's Theorem). *Let* L *be a bideterministic language, and let* A *be the minimal trim, i.e., without a dead state, deterministic finite automaton accepting* L. *Then* $h(L) = cr(A)$.

In fact, the minimality requirement in the above theorem is not needed, since every bideterministic finite automaton in which all states are useful is already a trim minimal deterministic finite automaton. Here, a state is useful if it is both reachable from the start state, and from which the final state is reachable from it.

In order to relate star height to alphabetic width, and to find lower bound techniques for the cycle rank, we study the latter concept in more detail. First, we establish a basic fact about cycle rank, which is used throughout the following sections. The second part of the following statement is found in [19, Theorem 2.4.], and the other part is established by an easy induction:

Lemma 4. *Let* $G = (V, E)$ *be a digraph and let* $U \subseteq V$. *Then we have the inequalities* $cr(G) - |U| \leq cr(G - U) \leq cr(G)$, *where* $G - U$ *denotes the graph with vertex set* $V \setminus U$ *and appropriately defined edge set.*   □

3.2 Cycle Rank *Via* Cops and Robbers

The characterization of cycle rank in terms of some "game against the graph" was already suggested in [19]. We give a modern formulation in terms of a cops and robber game. This characterization provides a useful tool in proving lower bounds on the cycle rank of specific families of digraphs. Moreover, many other digraph connectivity measures proposed recently admit a characterization in terms of some cops and robber game; this allows to compare the cycle rank with these other measures.

The *cops and strong visible robber game*, defined in [14], is given as follows: Let $G = (V, E)$ be a digraph. Initially, the cops occupy some set of $X \subseteq V$ vertices, with $|X| \leq k$, and the robber is placed on some vertex $v \in V \setminus X$. At any time, some of the cops can reside outside the graph, say, in a helicopter. In each round, the cop player chooses the next location $X' \subseteq V$ for the cops. The stationary cops in $X \cap X'$ remain in their positions, while the others go to the helicopter and fly to their new position. During this, the robber player, knowing the cops' next position X' from wire-tapping the police radio, can run at great speed to any new position v', provided there is both a (possibly empty) directed path from v to v', and a (possibly empty) directed path back from v' to v in $G - (X \cap X')$, i.e., he has to avoid to run into a stationary cop, and to run along a path in and to stay in the same strongly connected component of the remaining graph induced by the non-blocked vertices. Afterwards, the helicopter lands the cops at their new positions, and the next round starts, with X' and v' taking over the roles of X and v, respectively. The cop player wins the game if the robber cannot move any more, and the robber player wins if the robber can escape indefinitely.

The *immutable cops* variant of the above game restricts the movements of the cops in the following way: Once a cop has been placed on some vertex of the graph, he has to stay there forever. The *hot-plate* variant of the game restricts the movements of the robber in that he has to move along a nontrivial path in each move—even if the path consists only of a self-loop. These games are robust in the sense that small variations of rules, such as letting the robber player begin, or allowing only the placement of one cop at a time, do not alter the number of required cops. Also note that at most one additional cop is needed if we drop the hot-plate restriction. The following theorem gives a characterization of the cycle rank in terms of such a game. Due to space constraints, the proof is omitted.

Theorem 5. *Let G be a digraph and $k \geq 0$. Then k cops have a winning strategy for the immutable cops and hot-plate strong visible robber game if and only if the cycle rank of G is at most k, i.e., $cr(G) \leq k$.* □

4 Lower Bounds on Regular Expression Size

Now we have developed enough tools to derive lower bounds on alphabetic width in terms of star height.

Theorem 6. *Let $L \subseteq \Sigma^*$ be a regular language. Then $\mathrm{alph}(L) \geq 2^{\frac{1}{3}(h(L)-1)} - 1$.*

Proof. Let r be a regular expression over Σ of alphabetic width $n = \text{alph}(L)$. Then the construction given in [13] shows how to transform this expression into an equivalent nondeterministic finite automaton A with ε-transitions having at most $n+1$ states. It is not hard to see that the digraph underlying the transition structure of the constructed automaton has undirected treewidth at most 2. With a graph separator technique, we show the following claim:

> Let G be a digraph with n vertices and undirected treewidth at most k. Then $cr(G) \leq 1 + (k+1) \cdot \log n$.

We argue as follows: First, we lift some notions and results concerning graph separators known for undirected graphs (see, e.g., [21]), to the case of digraphs: Let $G = (V, E)$ be a digraph and let $U \subseteq V$ be a set of vertices. A set of vertices S is a *weak separator* for U if every strongly connected component of $G[U \setminus S]$ contains at most $\frac{1}{2}|U|$ vertices. For real numbers $0 \leq k \leq |V|$, let $s(G, k)$ denote the maximum of the size of the smallest weak separator for U, where the maximum is taken over all subsets U of size at most k of V. The *weak separator number* of G, denoted by $s(G)$, is defined as $s(G, |V|)$.

Next we prove the following relation: Let $G = (V, E)$ be a digraph with $n \geq 1$ vertices. Then

$$cr(G) \leq 1 + \sum_{0 \leq k \leq \log n - 1} s\left(G, \frac{n}{2^k}\right). \tag{1}$$

The proof proceeds by induction on n. In the case $n = 1$, we have $s(G) = 0$, and the sum in the statement of the lemma is empty, as desired. The induction step is as follows: By definition of weak separator number, G has a weak separator S of size at most $s(G, n)$. Let $C_1, C_2, \ldots, C_p$ be the strongly connected components of $G - S$. Each of these has cardinality at most $\frac{n}{2}$. With Lemma 4, we obtain

$$cr(G) \leq |S| + \max_{1 \leq i \leq p} cr(C_i) \leq s\left(G, \frac{n}{2^0}\right) + \max_{1 \leq i \leq p} cr(C_i).$$

Since for each $k \leq n$ and for each strongly connected component C_i obviously holds $s(C_i, k) \leq s(G, k)$, we have by induction hypothesis

$$\max_{1 \leq i \leq p} cr(C_i) \leq 1 + \sum_{0 \leq k \leq \log(n/2) - 1} s\left(G, \frac{n/2}{2^k}\right) = 1 + \sum_{1 \leq k \leq \log n - 1} s\left(G, \frac{n}{2^k}\right),$$

where the right hand side is obtained by simply shifting the summation index. By putting the two inequalities together, the proof of Inequality (1) is completed.

This establishes a relation between cycle rank and weak separator number, namely $cr(G) \leq 1 + s(G) \cdot \log n$, if G is a digraph with n vertices. Moreover, it is known from [24] that digraphs with undirected treewidth at most k have weak separator number at most $k + 1$, thus establishing our claim. Thus, we obtain $cr(A) \leq 1 + 3\log(n+1)$. Finally, the proof is completed by using Theorem 2.   □

This bound is almost tight: Define the language L_n inductively by $L_0 = \varepsilon$ and $L_i = (a \cdot L_{i-1} \cdot b)^*$, for $i > 0$. Then $\text{alph}(L_n)$ is clearly at most $2n$, but it is known from [19] that $h(L_{2^k}) = k$, for each $k \geq 1$. In contrast, there cannot exist an upper bound on the alphabetic width in terms of star height, since all finite languages have star height 0, but there are only finitely many languages of bounded alphabetic width.

4.1   Lower Bounds on Alphabetic Width of Language Operations

As a first application of Theorem 6, we exhibit a family of languages over a binary alphabet that shows that several natural operations on regular languages such as complement, intersection and shuffle cannot be supported efficiently by regular expressions; most notably, complementation can require an almost doubly-exponential blow-up in regular expression size. These languages have an appealingly simple structure, and their star height was already studied, although not completely determined, in the very first paper on star height of regular languages [5].

Theorem 7. *For $m, n \in \mathbb{N}$, define $K_m = \{\, w \in \{a,b\}^* \mid |w|_a \equiv 0 \mod m \,\}$ and $L_n = \{\, w \in \{a,b\}^* \mid |w|_b \equiv 0 \mod n \,\}$. Then we have $h(K_m \cap L_n) = m$, if $m = n$, and $h(K_m \cap L_n) = \min(m,n) + 1$, otherwise.*

Proof. The stated upper bound on the star height is proved already in [5, Corollary 2, pp. 394f.], so it remains to show a matching lower bound. It is straightforward to construct deterministic finite automata with m (n, respectively) states describing the languages K_m and L_n, respectively. By applying the standard product construction on these automata, we obtain a deterministic finite automaton A accepting the language $K_m \cap L_n$. It is not hard to see that this automaton is a minimal trim deterministic finite automaton, and furthermore that it is bideterministic. Therefore Theorem 3 shows $h(K_m \cap L_n) = cr(A)$.

The digraph underlying automaton A is the directed discrete $(m \times n)$-torus arising from the Cartesian graph product of two directed cycles, whose entanglement was determined by similar means in [2]. We give a lower bound on the cycle rank of this digraph using the game characterization given in Theorem 5. By symmetry, assume the torus has m rows and n columns, with $m \leq n$. At any stage of the game, we call a row (column, respectively) *free*, if each of the vertices in the row (column, respectively) is neither yet occupied, nor announced to be occupied in the current move of the cops. In the kth move of the cops, there are at least $m - k$ free rows and $n - k$ free columns. As long as $k < m$, the robbers' strategy is to reside on the subgraph induced by the rows and columns that are currently free. For $k < m$, each free row or column is strongly connected itself, and each pair of free columns is strongly connected to each other *via* the (nonempty) set of free rows. The strategy always yields a valid game position, and this already shows the desired lower bound in the case $m = n$. In the case $m > n$, as soon as the last free row is threatening to be occupied, the robber can still flee to one of the remaining free columns. Thus an additional cop is needed, since each free column itself forms a nontrivial strongly connected subgraph, even though the columns are no longer strongly connected to each other.   □

Together with Theorem 6, we immediately obtain some results about the alphabetic width of operations on regular languages. The classical way to extend the syntax of regular expressions is to allow intersection, thus obtaining the semi-extended regular expressions, or to allow also complement, resulting in extended regular expressions. It is known that semi-extended regular expressions can be exponentially more succinct even than nondeterministic finite automata, and hence than ordinary regular expressions. The former fact no longer holds if the number of occurrences of the intersection operator is bounded. But for regular expressions, already a single intersection operation can infer a huge blow-up in the needed description size:

Corollary 8. *For every $m \geq n$, there exist languages K_m and L_n over a binary alphabet with $\mathrm{alph}(K_m) \leq m$ and $\mathrm{alph}(L_n) \leq n$, such that $\mathrm{alph}(K_m \cap L_n) = 2^{\Omega(n)}$.* $\square$

This improves a lower bound independently obtained in [8]. Another language operation is the shuffle of two languages, which naturally arises in modeling the interleaving of the action traces of two processes. The shuffle of two languages L_1 and L_2 over alphabet Σ is $\{\, w \in \Sigma^* \mid w \in x \amalg y \text{ for some } x \in L_1 \text{ and } y \in L_2 \,\}$, where the shuffle of two words x and y is defined as the set of all words of the form $x_1 y_1 x_2 y_2 \ldots x_n y_n$, where $x = x_1 \ldots x_n$, $y = y_1 \ldots y_n$ with $x_i, y_i \in \Sigma^*$, for $1 \leq i \leq n$ and $n \geq 1$, and is denoted by $x \amalg y$. While the shuffle operation preserves regularity, it is known that regular expressions extended with the shuffle operator can be exponentially more succinct than regular expressions—in fact, the same holds for nondeterministic finite automata [17]. As with intersection, a similar blow-up can be caused already by a single application of the shuffle operator (which cannot be deduced from an argument solely based on automaton size). Namely, the language from Theorem 7 can be written as $(a^m)^* \amalg (b^n)^*$.

Corollary 9. *For every $m \geq n$, there exist languages L_m and L_n over a binary alphabet with $\mathrm{alph}(K_m) \leq m$ and $\mathrm{alph}(L_n) \leq n$, such that $\mathrm{alph}(K_m \amalg L_n) = 2^{\Omega(n)}$.* $\square$

For numbers n that have many distinct prime factors, the language $\{a, b\}^* \setminus (K_n \cap L_n)$, where K_n and L_n are defined as in Theorem 7, can be expressed very succinctly by a regular expression using a kind of Chinese Remainder Representation. In this way, we obtain for the complementation operation a lower bound that is roughly doubly exponential, even for binary alphabets, thus complementing a result given in [8] for 4-symbol alphabets—the proof is omitted due to lack of space:

Theorem 10. *There exists an infinite family of languages L_n over a binary alphabet Σ with $\mathrm{alph}(L_n) \leq n$, such that $\mathrm{alph}(\Sigma^* \setminus L_n) = 2^{2^{\Omega(\sqrt{n \log n})}}$.* $\square$

4.2   A Lower Bound for Converting DFAs into Regular Expressions

From the results in the previous chapter, it can be deduced that there are very simple examples of languages over a binary alphabet for which a blow-up in size of $2^{\Omega(\sqrt{n})}$ is inevitable when converting from an n-state deterministic finite automaton to an equivalent regular expression. Next, we can show that this bound can even be reached for *planar* deterministic finite automata, first studied in [4], thus complementing a corresponding algorithmic result from [7] with an optimal lower bound—again, the proof has to be omitted, but we note that the transition structure of the witness DFA are undirected grid graphs.

Theorem 11. *For alphabet size $|\Sigma| \geq 4$, there is an infinite family of languages L_n over alphabet Σ acceptable by n-state planar deterministic finite automata, such that* $\mathrm{alph}(L_n) \geq 2^{\Omega(\sqrt{n})}$. □

The obvious question is now if a lower bound of $2^{\Omega(n)}$ can be reached over a constant alphabet, when starting with non-planar deterministic finite automata. The rest of this section is devoted to a proof of this fact.

By Theorem 5, the cycle rank of an undirected graph G, i.e., a symmetric digraph, can be described in terms of the immutable cops and strong visible robber game. Note that in this case every connected component of size at least two is also a nontrivial strongly connected component. The *greedy* strategy for the robber player is to choose in each step the largest connected component he can reach in the remaining graph. We will identify a class of graphs in which the greedy strategy is particularly successful, namely expander graphs.

Definition 12. *Let $G = (V, E)$ be an undirected graph. For a subset $U \subset V$, the boundary of U, denoted by δU, is defined as $\delta U = \{ v \in V \setminus U \mid \{u, v\} \in E$ for some $u \in U \}$. An (undirected) d-regular graph $G = (V, E)$ with n vertices is called a (n, d, c)-expander, for $c > 0$, if each subset $U \subset V$ of vertices satisfies $|\delta U| > c \cdot |U|$, if $|U| < n/2$ and $|\delta U| \geq c \cdot (n - |U|)$, if $|U| \geq n/2$.*

A now standard probabilistic argument, originally from [23], shows that expander graphs are the rule rather than the exception among d-regular graphs, for all $d \geq 3$.

Theorem 13 (Pinsker). *There exists a fixed $c > 0$ such that for any $d \geq 3$ and even integer n, there is an (n, d, c)-expander, which is furthermore d-edge-colorable.*[3]

The proof of the following theorem is similar to that of [3, Theorem 4], where it was shown that each directed expander graph contains a long directed path.

Theorem 14. *Let G be a (n, d, c)-expander with $n \geq 3$. Then the cycle rank of G is at least $\frac{c}{d+1}(n - 1)$, i.e., $cr(G) \geq \frac{c}{d+1}(n - 1)$.* □

The next lemma shows that such a graph, equipped with an edge coloring, can be easily converted into a bideterministic finite automaton that accepts a language of large star height and uses only the edge colors as input alphabet.

---

[3] That is, one can assign to its edges d colors such that no pair of incident edges receives the same color.

Lemma 15. *For every d-edge colorable, connected undirected graph G with n vertices of cycle rank k, there exists an n-state deterministic finite automaton A over a d-symbol alphabet such that the star height of $L(A)$ is k.*

Proof. Let $G = (V, E)$ be such a graph, with $V = \{1, 2, \ldots, n\}$. and maximum degree d, equipped with an edge coloring $c : E \to \{0, 1, \ldots, d\}$ such that no pair of incident edges receives the same color. Given this colored graph, we construct a deterministic finite automaton over the alphabet $\Sigma = \{a_1, a_2, \ldots, a_d\}$ with state set V, start and single final state $v_0 \in V$ (arbitrary), and whose transition relation is defined as follows: $\delta(p, a_i) = q$ if the colored graph G has an i-colored edge $\{p, q\}$. It is not hard to see that this automaton is a trim bideterministic automaton, and therefore minimal. Furthermore, its underlying digraph is symmetric, and its undirected version is isomorphic to G. By Theorem 3, the star height of $L(A)$ equals k. $\square$

For the main theorem of this section we need the existence of a suitable homomorphism that preserves star height. The existence of reasonably economic binary encodings with this property have been already conjectured in [5], and their existence was proved constructively in [19]: Let $\Sigma = \{a_1, a_2, \ldots, a_d\}$ be a finite alphabet, $d \geq 1$, and let $\varphi : \Sigma^* \to \{a, b\}^*$ be the homomorphism defined by $\varphi(a_i) = a^i b^{d-i+1}$, for $i = 1, 2, \ldots, d$. Then for every regular language $L \subseteq \Sigma^*$ the star height of L equals the star height of $\varphi(L)$. Then Lemma 15 and Theorems 6, 13, and 14 can be combined with the above presented star height preserving homomorphism to give the following theorem.

Theorem 16. *For alphabet size $|\Sigma| \geq 2$, there is an infinite family of languages L_n over alphabet Σ acceptable by deterministic finite automata with at most n states, such that $\mathrm{alph}(L_n) = 2^{\Omega(n)}$.* $\square$

This gives an affirmative answer to "Open Problem 3" in [7], which asked whether such a family of languages exists, over some constant alphabet.

References

1. Berwanger, D., Dawar, A., Hunter, P., Kreutzer, S.: Dag-width and parity games. In: Durand, B., Thomas, W. (eds.) STACS 2006. LNCS, vol. 3884, pp. 524–536. Springer, Heidelberg (2006)
2. Berwanger, D., Grädel, E.: Entanglement—A measure for the complexity of directed graphs with applications to logic and games. In: Baader, F., Voronkov, A. (eds.) LPAR 2004. LNCS (LNAI), vol. 3452, pp. 209–223. Springer, Heidelberg (2005)
3. Björklund, A., Husfeldt, T., Khanna, S.: Approximating longest directed paths and cycles. In: Díaz, J., Karhumäki, J., Lepistö, A., Sannella, D. (eds.) ICALP 2004. LNCS, vol. 3142, pp. 222–233. Springer, Heidelberg (2004)
4. Book, R.V., Chandra, A.K.: Inherently nonplanar automata. Acta Informatica 6, 89–94 (1976)
5. Eggan, L.C.: Transition graphs and the star height of regular events. Michigan Mathematical Journal 10, 385–397 (1963)

6. Ehrenfeucht, A., Zeiger, H.P.: Complexity measures for regular expressions. Journal of Computer and System Sciences 12(2), 134–146 (1976)
7. Ellul, K., Krawetz, B., Shallit, J., Wang, M.: Regular expressions: New results and open problems. Journal of Automata, Languages and Combinatorics 10(4), 407–437 (2005)
8. Gelade, W., Neven, F.: Succinctness of the complement and intersection of regular expressions. In: Albers, S., Weil, P. (eds.) Symposium on Theoretical Aspects of Computer Science. Dagstuhl Seminar Proceedings, vol. 08001, pp. 325–336. IBFI (2008)
9. Gruber, H., Holzer, M.: Finite automata, digraph connectivity and regular expression size. Technical report, Technische Universität München (December 2007)
10. Gruber, H., Johannsen, J.: Optimal lower bounds on regular expression size using communication complexity. In: Amadio, R. (ed.) Foundations of Software Science and Computation Structures. LNCS, vol. 4962, pp. 273–286. Springer, Heidelberg (2008)
11. Hashiguchi, K.: Algorithms for determining relative star height and star height. Information and Computation 78(2), 124–169 (1988)
12. Hopcroft, J.E., Ullman, J.D.: Introduction to Automata Theory, Languages and Computation. Addison-Wesley, Reading (1979)
13. Ilie, L., Yu, S.: Follow automata. Information and Computation 186(1), 140–162 (2003)
14. Johnson, T., Robertson, N., Seymour, P.D., Thomas, R.: Directed tree-width. Journal of Combinatorial Theory, Series B 82(1), 138–154 (2001)
15. Kirsten, D.: Distance desert automata and the star height problem. RAIRO – Theoretical Informatics and Applications 39(3), 455–509 (2005)
16. Kleene, S.C.: Representation of events in nerve nets and finite automata. In: Shannon, C.E., McCarthy, J. (eds.) Automata Studies. Annals of Mathematics Studies, pp. 3–42. Princeton University Press, Princeton (1956)
17. Mayer, A.J., Stockmeyer, L.J.: Word problems – This time with interleaving. Information and Computation 115(2), 293–311 (1994)
18. McNaughton, R.: The loop complexity of pure-group events. Information and Control 11(1/2), 167–176 (1967)
19. McNaughton, R.: The loop complexity of regular events. Information Sciences 1, 305–328 (1969)
20. Morris, P.H., Gray, R.A., Filman, R.E.: Goto removal based on regular expressions. Journal of Software Maintenance 9(1), 47–66 (1997)
21. Nešetřil, J., de Mendez, P.O.: Tree-depth, subgraph coloring and homomorphism bounds. European Journal of Combinatorics 27(6), 1022–1041 (2006)
22. Obdržálek, J.: Dag-width: Connectivity measure for directed graphs. In: ACM-SIAM Symposium on Discrete Algorithms, pp. 814–821. ACM Press, New York (2006)
23. Pinsker, M.S.: On the complexity of a concentrator. In: Annual Teletraffic Conference, pp. 318/1–318/4 (1973)
24. Robertson, N., Seymour, P.D.: Graph minors. II. Algorithmic aspects of tree-width. Journal of Algorithms 7(3), 309–322 (1986)
25. Sakarovitch, J.: The language, the expression, and the (small) automaton. In: Farré, J., Litovsky, I., Schmitz, S. (eds.) CIAA 2005. LNCS, vol. 3845, pp. 15–30. Springer, Heidelberg (2006)
26. Schnitger, G.: Regular expressions and NFAs without ε-transitions. In: Durand, B., Thomas, W. (eds.) STACS 2006. LNCS, vol. 3884, pp. 432–443. Springer, Heidelberg (2006)

Leftist Grammars Are Non-primitive Recursive*

Tomasz Jurdziński

Institute of Computer Science, University of Wrocław, Poland
tju@ii.uni.wroc.pl

Abstract. Leftist grammars were introduced by Motwani et. al. [7], as
a tool to show decidability of the accessibility problem in certain general protection systems. It is shown that the membership problem for
languages defined by leftist grammars is non-primitive recursive. Therefore, by the reduction of Motwani et. al., the accessibility problem in the
appropriate protection systems is non-primitive recursive as well.

1 Introduction

A protection system is a set of policies that prescribes the ways in which *objects* interact with each other in a computer system. By objects we mean users,
processes or other entities. Interactions can include access rights, information
sharing privileges and other mechanisms. The accessibility problem for a protection system is formulated in the form "Can object p gain (illegal) access to
object q by a series of legal moves (as prescribed by the policy)?". A formal
treatment of accessibility was first presented by Harrison, et. al. [3] who showed
that the accessibility problem is undecidable for a general access-matrix model.
This result prompted a broad research on trade-offs between expressibility and
verifiability in protection systems. We consider the model proposed in [2,8] in
the context of Java virtual worlds, called here the *Saraswat's model*. The accessibility problem is decidable for the Saraswat's model, which was obtained
by relating this problem to the intersection problem for leftist grammars [7].
Further refinement and applications of this model are presented in [9].

Leftist grammars can be characterized in terms of rules of the form $ab \rightarrow b$
and $c \rightarrow dc$, where a, b, c, d belong to the finite alphabet Σ. A fixed symbol
$x \in \Sigma$ is called the final symbol and a word $w \in \Sigma^*$ belongs to the language
defined by the grammar G iff there exists a derivation which starts with wx and
ends with x. Observe that each leftist grammar is actually a semi-Thue system
with leftist rewriting rules.

It is known that the membership problem problem for leftist grammars is decidable [7]. The result from [7] implies also that any lower bound for complexity of
the membership problem for leftist grammars induces the same lower bound for
the accessibility problem in the restricted variant of the Saraswat's model (up to
a polynomial factor). Simplicity of leftist grammars led to the conjecture that the
actual complexity of the membership problem is small [7]. As shown in [6,1], quite

---

* Partially supported by MNiSW grant number N206 024 31/3826, 2006-2008.

L. Aceto et al. (Eds.): ICALP 2008, Part II, LNCS 5126, pp. 51–62, 2008.

natural restrictions imply context-freenes or even regularity of languages defined by leftist grammars. However, it has been shown that the membership problem for general leftist grammars is PSPACE-hard [4]. In this paper, we strengthen this result in the following way. We prove that, if a grammar and an input word form the input for the problem, the membership problem is non-primitive recursive. In the case that a grammar is fixed, we show that, for each primitive recursive function f, there exists a grammar $\mathcal{G}_f$ such that the membership problem for $\mathcal{G}_f$ is not in space $O(f(n))$, where n is the length of an input word.

In Section 2 we provide basic definitions and notations. In Sections 3 and 4 we show how leftist grammars can "compute" certain families of functions related to the Ackermann's function. In Section 5, we present the main result of the paper. Finally, concluding remarks are presented in Section 6. Because of the page limit not all proofs are given in the paper. For a complete and detailed presentation we refer to the technical report [5].

2   Definitions

Throughout the paper λ denotes the empty word, $\mathbb{N}$ denotes the set of non-negative integers. For a word x, $|x|$, $x[i]$ and $x[i,j]$ denote the length of x, the ith symbol of x and the factor $x[i] \cdots x[j]$ respectively, where $0 < i \leq j \leq |x|$. Moreover, let $[i,j] = \{l \in \mathbb{N} \mid i \leq l \leq j\}$, let $i\%j$ be equal the remainder of the division i/j and let $\bar{i} = 1 - i$ for $i \in [0,1]$. Furthermore, we identify regular expressions with languages defined by them.

We say that sets $A_0, \ldots, A_s$ form the **partition** $\mathcal{A} = (A_0, \ldots, A_s)$ of the set $A = \bigcup_{i=0}^{s} A_i$ iff $A_i \neq \emptyset$ and $A_i \cap A_j = \emptyset$ for each $i, j \in [0, s]$, $i \neq j$. A word $w \in A^*$ is an **alternating word** with respect to the partition $\mathcal{A}$ if w is a subword of a word from the language $(A_0^+ A_1^+ \cdots A_s^+)^*$. Note that, if $s = 1$ then each word over A is an alternating word wrt $\mathcal{A}$. Let w be an alternating word wrt to $\mathcal{A}$. Then, the "alternation measure" $\|w\|_{\mathcal{A}}$ is equal to 0 if $w = \lambda$, $\|w\|_{\mathcal{A}} = 1$ if $w \in A_i^+$ for $i \in [0, s]$, and $\|w\|_{\mathcal{A}} = \|w_2\|_{\mathcal{A}} + 1$ if $w = w_1 w_2$ for $w_1 \in A_i^+$, $i \in [0, s]$, and $w_2 \in A_{(i+1)\%2}^+ A^*$. If w is not an alternating word wrt $\mathcal{A}$, $\|w\|_{\mathcal{A}}$ is not defined.

2.1   Leftist Grammars

A *leftist grammar* $\mathcal{G} = (\Sigma, P, x)$ consists of a finite alphabet Σ, a final symbol $x \in \Sigma$, and a set of production rules P of the following two types,

$$ab \to b \;\; (\text{Delete Rule}), \; c \to dc \;\; (\text{Insert Rule})$$

where $a, b, c, d \in \Sigma$. We denote the above productions as $b \xrightarrow{\text{del}} a$ and $c \xrightarrow{\text{ins}} d$. We say that $u \Rightarrow_{\mathcal{G}} v$ (or shortly $u \Rightarrow v$) is a **derivation step** for $u, v \in \Sigma^*$, if $u = u_1 y u_2$ and $v = u_1 z u_2$ such that $y \to z$ is a production rule in P. A sequence of derivation steps $u_1 \Rightarrow u_2 \Rightarrow \ldots \Rightarrow u_m$ is called a **derivation**. Finally, the language of $\mathcal{G}$ is defined to be $L(\mathcal{G}) = \{w \in \Sigma^* \mid wx \Rightarrow^* x\}$. The **membership problem** for a fixed leftist grammar $\mathcal{G} = (\Sigma, P, x)$ is, given a word $w \in \Sigma^*$ as an input, to decide whether $w \in L(\mathcal{G})$. The **variable membership problem** is, given a word $w \in \Sigma^*$ and a leftist grammar $\mathcal{G} = (\Sigma, P, x)$ as an input, to decide whether $w \in L(\mathcal{G})$.

We think of symbols as objects which can insert/delete other symbols and can be inserted/deleted in a derivation. In order to simplify notations, we identify the particular occurrence of a symbol a with its value a.

We say that the symbol b in a delete rule $ab \rightarrow b$ is **active**. Similarly, the symbol c is **active** in an insert rule $c \rightarrow dc$. Let $u \Rightarrow v$, where $u = u_1 y u_2$ and $v = v_1 z v_2$ such that $y \rightarrow z$ is a production rule in P. Then, we say that the rightmost symbol of the prefix $u_1 y$ of $u_1 y u_2$ is **active** in the derivation step $u \Rightarrow v$. Though it might be the case that the choice of the active symbol for a step $u \Rightarrow v$ is not unique, one can avoid this ambiguity [6]. So, we assume that the active symbol can be determined uniquely for each derivation step.

We say that d is a **descendant** of b with respect to a derivation U if (b, d) belongs to the reflexive and transitive closure of the relation

$$\{(e, f) \mid v\underline{e}w \Rightarrow v f \underline{e} w \text{ is a derivation step in } U \text{ for } v, w \in \Sigma^*, \ e, f \in \Sigma\}.$$

Let $\mathcal{G} = (\Sigma, P, x)$ be a leftist grammar, where $\Sigma = \{a_i\}_{i=1}^r$. The *insert graph* of $\mathcal{G}$ is $G(\Sigma, E)$, where $E = \{(a_i, a_j) \mid (a_i \rightarrow a_j a_i) \in P\}$. Similarly, the *delete graph* of $\mathcal{G}$ is $G(\Sigma, E)$, where $E = \{(a_i, a_j) \mid (a_j a_i \rightarrow a_i) \in P\}$.

Let $\mathsf{Del}(a) = \{b \mid b \overset{\mathsf{del}}{\rightarrow} a\}$ for $a \in \Sigma$ and $\mathcal{G} = (\Sigma, P, x)$, and let $\mathsf{Del}(A)$ for $A \subseteq \Sigma$ be $\bigcup_{a \in A} \mathsf{Del}(a)$. That is, $\mathsf{Del}(a)$ is the set of in-neighbors of a in the delete graph of $\mathcal{G}$. A set $A \subseteq \Sigma$ is **homogeneous** if a is not active in any production rule of $\mathcal{G}$ for each $a \in A$, and $\mathsf{Del}(a) = \mathsf{Del}(b)$ for each $a, b \in A$.

A derivation $u_1 \Rightarrow u_2 \Rightarrow \ldots \Rightarrow u_m$ is a **leftmost derivation** if each symbol from u_i located to the left of the symbol active in $u_i \Rightarrow u_{i+1}$ is not active in $u_i \Rightarrow^* u_m$ for every $i \in [1, m-1]$. For each $u, v \in \Sigma^*$ such that $u \Rightarrow_{\mathcal{G}}^* v$, there exists a leftmost derivation which starts with u and ends with v [6].

We say that a word u **eliminates** a word $w \in \Sigma^+$ in a derivation $z_1 w z_2 \Rightarrow^*$ z', if all elements of w are not active in this derivation, and all elements of w are deleted by the elements of u and their descendants. Moreover, we say that $u \in \Sigma^+$ is **able to eliminate** $v \in \Sigma^+$ if there exists a derivation $vu \Rightarrow^* v'$, where u eliminates v. Observe that if u is able to eliminate v, then u is able to eliminate v' for each v' obtained from v by deleting some symbols of v.

Definition 1 (Greedy derivation). *A derivation* $U \equiv (u_1 \Rightarrow u_2 \Rightarrow \cdots \Rightarrow u_m)$ *is greedy if the following conditions are satisfied:*

(a) U is a leftmost derivation;

(b) if $u_j = zbav$ for $a, b \in \Sigma$, $j \in [1, m-1]$ and the grammar contains the production $a \overset{\mathsf{del}}{\rightarrow} b$, then a is active in at least one step of $u_j \Rightarrow^* u_m$;

(c) there is no derivation step $u\underline{a}v \Rightarrow u\underline{ba}v$ in U such that b does not eliminate any element of u during U.

It is known that there exists a greedy derivation $wx \Rightarrow_{\mathcal{G}}^* x$ for each leftist grammar $\mathcal{G}$ and $w \in L(\mathcal{G})$ [4]. So, we consider only greedy derivations.

For a grammar $\mathcal{G}$, let size of $\mathcal{G}$, $|\mathcal{G}|$, be equal to the size of its alphabet.

2.2 Ackermann's Function

For a function $f : \mathbb{N} \to \mathbb{N}$, let $f^{(0)}(n) = n$ and $f^{(l)}(n) = f(f^{(l-1)}(n))$ for each $n \in \mathbb{N}$ and $l \geq 1$. Let the functions $\phi_p : \mathbb{N} \to \mathbb{N}$ be defined as follows:

$$\phi_2(n) = 2 \cdot n \qquad \text{for } n > 0,$$
$$\phi_p(n) = \phi_{p-1}^{(n)}(1) \quad \text{for } p > 2 \text{ and } n > 0.$$

Moreover, let $\phi_p(0) = 1$ for each $p \geq 2$. One of possible definitions of Ackermann's function is $Ack(n) = \phi_n(3)$, see [10]. We define inverse functions $(\phi_p^{-1})_{p \geq 2}$ as

$$\phi_p^{-1}(m) := \min\{n \mid \phi_p(n) \geq m\}.$$

Notice that $\phi_p^{-1}(m) = n$ if and only if $(\phi_{p-1}^{-1})^{(n)}(m) = 1$ for $p > 2$. Moreover, observe that ϕ_ps are monotone, i.e., $\phi_p(n+1) \geq \phi_p(n)$ for $n \geq 0$.

It is well-known that $Ack(n)$ dominates any primitive recursive function of n.

3 Expansion

Our first goal is to build a grammar which somehow "computes" the function ϕ_p for $p \in \mathbb{N}$. The idea is as follows. Let I (the "input alphabet") and O (the "output alphabet") be disjoint finite sets and let $g \notin I \cup O$. Given a word $w \in I^*$, g is able to eliminate w only in a derivation $wg \Rightarrow^* w'g$, such that $w' \in O^*$ and $|w'| = \phi_p(|w|)$. Due to limitations of leftist grammars, the following definition is more complicated than the above scenario. In particular, the result of a "computation" is approximate in such a way that $|w'|$ might be larger than $\phi_p(|w|)$, but not smaller. Moreover, numbers are identified with the alternation measures of words, not with their lengths.

Definition 2. *Let I, B, O, F and $\{g\}$ be disjoint, finite and nonempty sets of symbols and let $\mathcal{I} = (I_0, I_1)$, $\mathcal{O} = (O_0, O_1)$ be partitions of I and O, respectively. Moreover, let the following conditions be satisfied in a grammar $\mathcal{G}$:*

1. *For each $v \in B^+$ and $w \in I^*$, there exists a derivation $vwg \Rightarrow^* v'w'g$ in which g eliminates vw, such that $v' \in F^+$, $w' \in O^+$ and $\|w'\|_O = \phi_p(\|w\|_\mathcal{I})$.*
2. *Let $v \in B^*$, $w \in I^*$ such that $|vw| > 0$ and let $U \equiv (vwg \Rightarrow^* zg)$ be a greedy derivation in which g eliminates vw. If $z \in (F \cup O)^*$, then $z = v'w'g$, where $v' \in F^+$, $w' \in O^+$ and $\|w'\|_O \geq \phi_p(\|w\|_\mathcal{I})$.*
3. *$\mathrm{Del}(B) \subseteq F$, the elements of $\mathrm{Del}(B)$ are able to eliminate only the elements of B and g is not able to eliminate the elements of $\mathrm{Del}(B)$.*
4. *The sets I_0, I_1 and B are homogeneous in $\mathcal{G}$.*

Then, we say that $\mathcal{G}$ satisfies p-expansion wrt $\mathcal{I}$, B, $\mathcal{O}$, F and $\{g\}$, we denote this fact by $\mathsf{ex}_p(\mathcal{I}, B, \mathcal{O}, F, g)$.

The statement 1. of the above definition corresponds to the property that it is possible to "compute" the function ϕ_p precisely, while the second statement denotes the property that each "computation" is such that the result is not smaller than the actual value of the function ϕ_p. The subalphabets B (the "border alphabet") and F (the "frontier alphabet") are introduced for technical reasons. Their roles are essential in the inductive construction in Lemma 2. The third statement of Def. 2 guarantees that g is not able to eliminate the elements of $I \cup O$ which appear to the left of a symbol from B. This property helps in the proof that the inductive construction from Lemma 2 works.

For $\mathcal{G}$ satisfying $\mathrm{ex}_p(\mathcal{I}, B, \mathcal{O}, F, g)$, we use the notation $\mathcal{I}(\mathcal{G}) = \mathcal{I}$, $B(\mathcal{G}) = B$, $\mathcal{O}(\mathcal{G}) = \mathcal{O}$, and $F(\mathcal{G}) = F$. According to Def. 2 (statement 4), the construction of a grammar satisfying p-expansion is independent of $\mathcal{I}$ and B. Therefore, if $\mathcal{I}, B$ do not matter, we denote p-expansion by $\mathrm{EX}_p(\mathcal{O}, F, g)$. Given I, B and a partition $\mathcal{I} = (I_0, I_1)$ of I, we build a grammar $\mathcal{G}_{\phi_2}$ with the following production rules:

$(i)\ \ g \xrightarrow{\text{ins}} a_i$ $\quad$ for $i \in \{0, 2\}$ $\quad (vi)\ a_3 \xrightarrow{\text{ins}} f_1$

$(ii)\ \ g \xrightarrow{\text{ins}} a_0'$ $\qquad\qquad\qquad\quad (vii)\ f_j \xrightarrow{\text{ins}} f$ $\quad$ for $j \in [0, 1]$

$(iii)\ a_0' \xrightarrow{\text{ins}} f$ $\qquad\qquad\qquad (viii)\ f_j \xrightarrow{\text{del}} f_{1-j}$ for $j \in [0, 1]$

$(iv)\ a_i \xrightarrow{\text{ins}} a_{(i+1)\%4}$ for $i \in [0, 3]$ $\quad (ix)\ f_j \xrightarrow{\text{del}} \iota$ $\quad$ for $j \in [0, 1], \iota \in I_j$

$(v)\ \ a_1 \xrightarrow{\text{ins}} f_0$ $\qquad\qquad\qquad\quad (x)\ \ f \xrightarrow{\text{del}} b$ $\quad$ for $b \in B$

where a_0', a_j, f_i, $g \notin I \cup B$ for each $i \in [0, 1]$ and $j \in [0, 3]$.

Lemma 1. *The grammar $\mathcal{G}_{\phi_2}$ satisfies $\mathrm{ex}_2(\mathcal{I}, B, \mathcal{O}, F, g)$, where $\mathcal{O} = (O_0, O_1)$, $F = \{f_0, f_1, f\}$, $O_0 = \{a_0, a_0', a_2\}$, and $O_1 = \{a_1, a_3\}$.*

The idea of the proof of Lemma 1 is based on the following observations:

- each greedy derivation in which g and its descendants are active corresponds to a path in the insert graph of $\mathcal{G}_{\phi_2}$ which starts with g, and goes either through a_0' or through one of the cycles $a_0a_1a_2a_3$, $a_2a_3a_0a_1$ (possibly many times);
- the cycle $a_0a_1a_2a_3$ ($a_2a_3a_0a_1$) describes the fact that the factor $a_3a_2a_1a_0$ ($a_1a_0a_3a_2$) is added to the left of g and each subword from $I_1^*I_0^*$ ($I_0^*I_1^*$, respectively) can be deleted (see the productions (v), (vi), (ix), and $(viii)$).

Lemma 2. *Assume that a grammar $\mathcal{G}_{\phi_p}$ satisfies $\mathrm{EX}_p(\widehat{\mathcal{O}}, \widehat{F}, \widehat{g})$. Then, one can build a grammar $\mathcal{G}_{\phi_{p+1}}$ of size at most $5|\mathcal{G}_{\phi_p}|$ which satisfies $\mathrm{EX}_{p+1}(\mathcal{O}, F, g)$ for some sets of symbols $O, F, \{g\}$ and a partition $\mathcal{O} = (O_0, O_1)$ of O.*

Proof. Let $\mathcal{G}_0, \dots, \mathcal{G}_3$ be copies of the grammar $\mathcal{G}_{\phi_p}$ such that:

- $\mathcal{G}_j$ satisfies $\mathrm{EX}_p(\widehat{\mathcal{O}}_j, \widehat{F}_j, \widehat{g}_j)$, where $\widehat{\mathcal{O}}_j = (\widehat{O}_{j,0}, \widehat{O}_{j,1})$ is a partition of $\widehat{O}_j$ for $j \in [0, 3]$,
- $\Sigma_j \cap \Sigma_l = \emptyset$ for $j \neq l$, where Σ_k is the set of symbols accessible from $\widehat{g}_k$ in the insert graph of $\mathcal{G}_k$.

The partition $\mathcal{I}(\mathcal{G}_j)$ and the set $B(\mathcal{G}_j)$ for $j \in [0, 3]$ are as follows:

$$\mathcal{I}(\mathcal{G}_j) = \widehat{\mathcal{I}}_j := \widehat{\mathcal{O}}_{(j+1)\%2}, \text{ and } B(\mathcal{G}_j) = \widehat{B}_j := \widehat{F}_{(j+1)\%2}.$$

That is,

$$\mathcal{G}_j \text{ satisfies } \mathsf{ex}_p(\widehat{\mathcal{O}}_{(j+1)\%2}, \widehat{F}_{(j+1)\%2}, \widehat{\mathcal{O}}_j, \widehat{F}_j, \widehat{g}_j) \qquad\qquad (\mathbf{e}(j))$$

for $j \in [0,3]$. So, the "output" subalphabet of the subgrammar $\mathcal{G}_j$ for $j \in [0,1]$ is the "input" subalphabet of $\mathcal{G}_{(j+1)\%2}$ and $\mathcal{G}_{2+(j+1)\%2}$ (see Figure 1). Given disjoint nonempty sets I, B and a partition $\mathcal{I} = (I_0, I_1)$ of I, the grammar $\mathcal{G}_{\phi_{p+1}}$ is obtained by combining $\mathcal{G}_0, \dots, \mathcal{G}_3$, adding new symbols g, f_0, f_1, f_2, and f_3 to the alphabet and the following production rules (see Figure 1):

$$
\begin{aligned}
&(i) && e \xrightarrow{\text{ins}} f_j && \text{for each } e \in \mathsf{Del}(\widehat{B}_j) \cap \widehat{F}_j, j \in [0,3]\\
&(ii) && f_j \xrightarrow{\text{del}} \imath && \text{for each } \imath \in I_j, j \in [0,1]\\
&(iii) && f_j \xrightarrow{\text{del}} \imath && \text{for each } \imath \in B, j \in [2,3]\\
&(iv) && f_j \xrightarrow{\text{del}} f_l && \text{for each } j \in [0,3], l = (j+1)\%2\\
&(v) && g \xrightarrow{\text{ins}} \widehat{g}_j && \text{for each } j \in [0,3]\\
&(vi) && g \xrightarrow{\text{del}} \widehat{g}_j && \text{for each } j \in [0,3]
\end{aligned}
$$

Our goal is to show that $\mathcal{G}_{\phi_{p+1}}$ defined in this way satisfies

$$\mathsf{ex}_{p+1}(\mathcal{I}, \ B, \ \mathcal{O}, \ \widehat{F}_2 \cup \widehat{F}_3 \cup \{f_2, f_3\}, \ g),$$

where $\mathcal{O} = (O_0, O_1)$ is a partition of $O := \widehat{O}_2 \cup \widehat{O}_3$, $O_k = \widehat{O}_{2,k} \cup \widehat{O}_{3,k}$ for $k \in [0,1]$.

The correctness of the above construction is based on the following observations:

- since $\widehat{g}_j$ is able to eliminate elements of I_j and $\widehat{g}_j$ is not able to eliminate elements of I_{1-j} for $j \in [0,1]$, it is necessary to apply $\mathcal{G}_0$ and $\mathcal{G}_1$ alternately;
- each application of $\mathcal{G}_j$ deleting the element of I_j requires that the sequence over $\widehat{O}_{1-j}$ following the element of I_j is deleted as well; however, since $\mathcal{I}(\mathcal{G}_j) = \mathcal{O}(\mathcal{G}_{1-j})$, this subderivation replaces a sequence $y \in \widehat{O}_{1-j}^*$ with $y' \in \widehat{O}_j^*$ such that $\|y'\|_{\widehat{O}_j} \geq \phi_p(\|y\|_{\widehat{O}_{1-j}})$.

Below, we prove the lemma formally by checking whether $\mathcal{G}_{\phi_{p+1}}$ satisfies the statements of Def. 2.

First, observe that the statements 3 and 4 of Def. 2 hold for $\mathcal{G}_{\phi_{p+1}}$ by the fact that $\mathsf{Del}(B) = \{f_2, f_3\}$ and by the constraints of (iii).

Next, we show that the stat. 1 of Def. 2 holds for $\mathcal{G}_{\phi_{p+1}}$. Let $v \in B^+$, $w \in I^*$ and let $w = w_1 \cdots w_n$, where $w_l \in I_{l\%2}^+$ for each $l \in [1,n]$, i.e. $\|w\|_{\mathcal{I}} = n$.

Below, we describe a derivation $U \equiv (vwg \Rightarrow^* v'w'g)$ which consists of the stages $U_n, U_{n-1}, \dots, U_1, U_0$ such that the stage U_k for $k > 0$ applies the productions (i)-(vi) and the productions of the grammar $\mathcal{G}_{k\%2}$. Moreover, $\mathcal{G}_0$ applies $\mathcal{G}_2$ or $\mathcal{G}_3$, w_k is deleted in U_k for $k > 0$ and v is deleted in U_0. W.l.o.g., assume that we are going to eliminate a supersequence vwv'_n of vw in U, where $v'_n \in f_{\overline{n\%2}}\widehat{F}_{\overline{n\%2}}$. The following algorithm describes the derivation U:

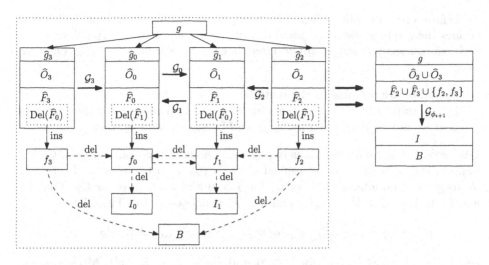

Fig. 1. An illustration for the construction of $\mathcal{G}_{\phi_{i+1}}$. An edge labeled by a grammar $\mathcal{G}$, going from the box $[g_1, \mathcal{O}, F]$ to the box $[g_2, \mathcal{I}, B]$ denotes that $\mathsf{ex}_i(g_1, \mathcal{I}, B, \mathcal{O}, F)$.

1. $w'_n := \lambda$, $v'_n := f_{\overline{n\%2}}a$ for some $a \in \widehat{F}_{\overline{n\%2}}$;
2. For $k = n, n-1, \ldots, 2, 1$ do

$$w_1 \cdots w_k v'_k w'_k g \Rightarrow^*_{(v),\text{ind.ass.},(i),(iv),(ii),(vi)} w_1 \cdots w_{k-1} v'_{k-1} w'_{k-1} g,$$

where $v'_j \in f_{\overline{j\%2}} \widehat{F}^+_{\overline{j\%2}}$, $w'_j \in \widehat{O}^*_{\overline{j\%2}}$ for $j \in [0,n]$, and $\|w'_{k-1}\|_{\widehat{O}_{k\%2}} =$
$\phi_p(\|w'_k\|_{\widehat{O}_{\overline{k\%2}}}) = \phi_p^{(n-k+1)}(0)$, by the assumption $e(k\%2)$ on page 56.
3. Apply $\mathcal{G}_2$ in order to eliminate v:

$$vv'_0 w'_0 g \Rightarrow^*_{\text{ind.ass.}} vf_1 v'' w'' g \Rightarrow^*_{(i,iv)} vf_2 v'' w'' g \Rightarrow^*_{(iii)} f_2 v'' w'' g = v' w' g$$

where $f_2 v'' = v' \in f_2 \widehat{F}^+_2 \subset F^+$, $w' = w'' \in \widehat{O}^+_2$ and

$$\|w'\|_{\widehat{O}_2} = \phi_p(\|w'_0\|_{\widehat{O}_1}) = \phi_p(\phi_p^{(n)}(0)) = \phi_p^{(n)}(1).$$

Finally, we have obtained a derivation $vwg \Rightarrow^* vv'_0 w'_0 g \Rightarrow^* v'w'g$, where $v' \in F^+$, $w' \in O^+$ and $\|w'\|_O = \phi_{p+1}(\|w\|_{\mathcal{I}})$. This shows that the grammar $\mathcal{G}_{\phi_{p+1}}$ satisfies the statement 1 of Definition 2. (It is easy to build a similar derivation for the case that $w_j \in I^+_{\overline{j\%2}}$ for each j; then, we use $\mathcal{G}_3$ instead of $\mathcal{G}_2$ in step 3.)

The final step is to show that the statement 2 of Def. 2 holds for $\mathcal{G}_{\phi_{p+1}}$. Let

$$U \equiv (vwg \Rightarrow^* z'g),$$

where $v \in B^*$, $w \in I^*$, $|vw| > 0$, and g eliminates vw in U. Moreover, assume that

$$z' \in (O \cup F)^* = (\widehat{O}_2 \cup \widehat{O}_3 \cup \widehat{F}_2 \cup \widehat{F}_3 \cup \{f_2, f_3\})^* \tag{1}$$

$$w = w_1 \cdots w_n, \text{ where } w_k \in I^+_{k\%2} \text{ for each } k \in [1,n] \text{ and } n \text{ is even.} \tag{2}$$

As before, one can split U into *stages* $U_m, U_{m-1}, \ldots, U_0$ such that each stage applies the productions (i)-(vi) and one of the grammars $\mathcal{G}_0, \ldots, \mathcal{G}_3$. Moreover, each two consecutive stages apply different grammars from $\mathcal{G}_0, \ldots, \mathcal{G}_3$. Observe that

- $\widehat{g}_2$ and $\widehat{g}_3$ are not able to eliminate the elements of I;
- $\widehat{g}_2$ ($\widehat{g}_3$, resp.) and its descendants insert a word which contains the elements that cannot be eliminated by $\widehat{g}_0, \widehat{g}_1$ and $\widehat{g}_3$ ($\widehat{g}_0, \widehat{g}_1$, and $\widehat{g}_2$, resp.).

The second of these observations implies that none of the grammars $\mathcal{G}_0, \mathcal{G}_1, \mathcal{G}_3$ ($\mathcal{G}_0, \mathcal{G}_1$, and $\mathcal{G}_2$, resp.) can be applied after $\mathcal{G}_2$ ($\mathcal{G}_3$, resp.) is applied. Therefore, U_j uses the productions of $\mathcal{G}_0$ or $\mathcal{G}_1$ for $j > 0$ and U_0 uses $\mathcal{G}_2$ or $\mathcal{G}_3$. That is, $w \in I^+$ is deleted in $U_m, \ldots, U_1$ and $v \in B^*$ is deleted in U_0. Thus, let

$$vwg = z_m g \Rightarrow^*_{U_m} z_{m-1} g \Rightarrow^*_{U_{m-1}} \cdots \Rightarrow^*_{U_1} z_0 g \Rightarrow^*_{U_0} z' g,$$

i.e., $U_j \equiv (z_j g \Rightarrow^* z_{j-1} g)$ for $j > 0$ and $U_0 \equiv (z_0 g \Rightarrow^* z')$. Moreover, let $\widehat{F}_k := \widehat{F}_k \cup \{f_k\}$ for $k \in [0, 3]$. We make use of the following claim.

Claim 3. *Let* $U \equiv (vwg = z_m g \Rightarrow^*_{U_m} z_{m-1} g \Rightarrow^*_{U_{m-1}} \cdots \Rightarrow^*_{U_1} z_0 g \Rightarrow^*_{U_0} z' g)$ *be a derivation in which* g *eliminates* vw, *and the conditions* (1), (2) *are satisfied. Then, for each* $j \in [0, m-1]$, $z_j g$ *is equal to* $vw_1 \cdots w_l v'_j w'_j g$ *such that*

(a) $v'_j \in (\widehat{F}_0 \cup \widehat{F}_1)^* \widehat{F}^+_{\gamma(j)}$, $w'_j \in \widehat{O}^+_{\gamma(j)}$;
(b) $j \geq l$ *(i.e., at most one of* w_l*'s is deleted in each stage);*
(c) $\|w'_j\|_{\widehat{O}_{\gamma(j)}} \geq \phi_p^{(m-j)}(0)$;

where $\gamma(j) = (m - j + 1)\%2$ *and* $U_j \equiv (z_j g \Rightarrow^* z_{j-1} g)$.

The proof of Claim 3 is based on the observation that, if the scenario described in the claim does not hold, the word to the left of g contains a subsequence $u \in (\widehat{O}_0 \cup \widehat{O}_1)(\widehat{F}_0 \cup \widehat{F}_1)$ at the end of some stage. Then, by the fact that $\mathcal{G}_j$'s satisfy the statement 3 of Def. 2, the prefix of u belonging to $\widehat{O}_0 \cup \widehat{O}_1$ cannot be deleted, which contradicts the assumption (1) on page 57.

Claim 3 implies that $U_m, \ldots, U_1$ form a subderivation $vwg \Rightarrow^* z_0 g = vv'_0 w'_0 g$ of U, where $v'_0 \in (\widehat{F}_0 \cup \widehat{F}_1)^+ \widehat{F}^+_{\gamma(0)}$, $w'_0 \in \widehat{O}^+_{\gamma(0)}$, and $\|w'_0\|_{\widehat{O}_{\gamma(0)}} \geq \phi_p^{(m)}(0) = \phi_p^{(m-1)}(1)$. Finally, $z_0 = vv'_0 w'_0$ has to be eliminated in the stage U_0 using $\mathcal{G}_2$ or $\mathcal{G}_3$, since U finishes with $z'g$ and z' satisfies the assumption (1) on page 57. That is, by the assumption $e(3 - \gamma(0))$ on page 56, $U_0 \equiv (vv'_0 w'_0 g \Rightarrow^* v'w'g)$, where $v' \in \widehat{F}_{3-\gamma(0)}$, $w' \in \widehat{O}^+_{3-\gamma(0)}$ and

$$\|w'\|_{\widehat{O}_{3-\gamma(0)}} \geq \phi_p(\|w'_0\|_{\widehat{O}_{\gamma(0)}}) \geq_{\text{(Claim 3(c))}} \phi_p^{(m)}(1) = \phi_{p+1}(m) \geq \phi_{p+1}(n),$$

since $m \geq n$, which follows from Claim 3(b). Thus, we obtain the statement 2 of Definition 2 for the case $w \neq \lambda$. One can easily check that this statement holds for $w = \lambda$ as well. $\qquad\square$

4   Shrinking

Similarly to the expansion property, we define the shrinking property which describes a way in which the functions ϕ_p^{-1} are "computed" by leftist grammars.

Definition 4. *Let I, O, F and $\{g\}$ be disjoint, finite and nonempty sets of symbols and let $\mathcal{I} = (I_0, \ldots, I_3)$, $\mathcal{O} = (O_0, \ldots, O_3)$ be partitions of I and O, respectively. Moreover, let the following conditions be satisfied in a grammar $\mathcal{G}$:*

1. *Let $w \in I^+$ be an alternating word wrt $\mathcal{I}$. Then, there exists a derivation $wg \Rightarrow^* v'w'g$ in which g eliminates w, such that $v' \in F^+$, $w' \in O^*$ is an alternating word wrt $\mathcal{O}$ and $\|w'\|_{\mathcal{O}} = \phi_p^{-1}(\|w\|_{\mathcal{I}})$.*
2. *Let $w \in I^+$ and let $U \equiv (wg \Rightarrow^* zg)$ be a derivation in which g eliminates w. If $z \in (F \cup O)^*$, then $z = v'w'$ such that $v' \in F^+$, $w' \in O^*$, and $\|w'\|_{\mathcal{O}} \geq \phi_p^{-1}(\|w\|_{\mathcal{I}})$.*
3. *The sets $I_0, \ldots, I_3$ are homogeneous in $\mathcal{G}$.*

Then, we say that $\mathcal{G}$ satisfies p-shrinking wrt $\mathcal{I}$, $\mathcal{O}$, F and $\{g\}$, we denote this fact by $\mathsf{sh}_p(\mathcal{I}, \mathcal{O}, F, g)$.

For a grammar $\mathcal{G}$ satisfying $\mathsf{sh}_p(\mathcal{I}, \mathcal{O}, F, g)$, we use the notation $\mathcal{I}(\mathcal{G}) := \mathcal{I}$, $\mathcal{O}(\mathcal{G}) := \mathcal{O}$, and $F(\mathcal{G}) := F$. Since the construction of a grammar satisfying p-shrinking is independent of $\mathcal{I}$, we denote p-shrinking by $\mathsf{SH}_p(\mathcal{O}, F, g)$.

Let $\mathcal{G}_{\phi_2^{-1}}$ be a grammar which contains the following productions:

$$(i) \quad g \xrightarrow{\text{ins}} a_l \qquad \text{for each } l \in [0, 1]$$
$$(ii) \quad a_i \xrightarrow{\text{ins}} a_{(i+1)\%4}$$
$$(iii) \quad a_i \xrightarrow{\text{ins}} f_i$$
$$(iv) \quad a_i \xrightarrow{\text{ins}} f_i'$$
$$(v) \quad g \xrightarrow{\text{ins}} f_i''$$
$$(vi) \quad f_i \xrightarrow{\text{del}} f_{(i-1)\%4}$$
$$(vii) \quad f_i' \xrightarrow{\text{del}} f_{(i-1)\%4}'$$
$$(viii) \quad f_i \xrightarrow{\text{del}} \imath \qquad \text{for each } \imath \in I_j \text{ such that } i\%2 = \lfloor j/2 \rfloor$$
$$(ix) \quad f_i' \xrightarrow{\text{del}} \imath \qquad \text{for each } \imath \in I_j \text{ such that } i\%2 = |\lfloor (j-1)/2 \rfloor|$$
$$(x) \quad f_i'' \xrightarrow{\text{del}} \imath \qquad \text{for each } \imath \in I_i.$$

for each $i \in [0, 3]$.

Lemma 3. *The grammar $\mathcal{G}_{\phi_2^{-1}}$ satisfies $\mathsf{sh}_2(\mathcal{I}, \mathcal{O}, F, g)$, where $\mathcal{O} = (O_0, \ldots, O_3)$ is a partition of $O = \bigcup_{j=0}^3 O_j$, $O_i = \{a_i\}$ for $i \in [0, 3]$ and $F = \bigcup_{i=0}^3 \{f_i, f_i', f_i''\}$.*

Lemma 4. *Assume that a grammar $\mathcal{G}_{\phi_p^{-1}}$ satisfies $\mathsf{SH}_p(\widehat{\mathcal{O}}, \widehat{F}, \widehat{g})$. Then, one can build a grammar $\mathcal{G}_{\phi_{p+1}^{-1}}$ of size at most $6|\mathcal{G}_{\phi_p^{-1}}|$ which satisfies $\mathsf{SH}_{p+1}(\mathcal{O}, F, g)$ for some sets O, F, $\{g\}$ and a partition $\mathcal{O} = (O_0, \ldots, O_3)$ of O.*

Proof. (Sketch) Let $\mathcal{G}_0, \ldots, \mathcal{G}_4$ be copies of the grammar $\mathcal{G}_{\phi_p^{-1}}$ such that $\mathcal{G}_j$ satisfies $\mathsf{SH}_{\phi_p^{-1}}(\widehat{\mathcal{O}}_j, \widehat{F}_j, \widehat{g}_j)$.

Let $\mathcal{I}(\mathcal{G}_4) = \widehat{\mathcal{I}}_4 := \mathcal{I}$ and $\mathcal{I}(\mathcal{G}_j) = \widehat{\mathcal{I}}_j := \widehat{\mathcal{O}}_{(j-1)\%4} \cup \widehat{\mathcal{O}}_4$ for $j \in [0,3]$. That is,

$$\mathcal{G}_j \text{ satisfies } \mathsf{sh}_p(\widehat{\mathcal{O}}_{(j-1)\%4} \cup \widehat{\mathcal{O}}_4, \widehat{\mathcal{O}}_j, \widehat{F}_j, \widehat{g}_j) \text{ for } j \in [0,3]$$
$$\mathcal{G}_4 \text{ satisfies } \mathsf{sh}_p(\mathcal{I}, \widehat{\mathcal{O}}_4, \widehat{F}_4, \widehat{g}_4)$$

Given a set I and a partition $\mathcal{I} = (I_0, \ldots, I_3)$ of I, the grammar $\mathcal{G}_{\phi_{p+1}^{-1}}$ is obtained by combining the alphabets and the rules of $\mathcal{G}_0, \ldots, \mathcal{G}_4$, adding a new symbol g to the alphabet and the rules $g \xrightarrow{\text{ins}} \widehat{g}_j$, $g \xrightarrow{\text{del}} \widehat{g}_j$ for $j \in [0,4]$. The grammar $\mathcal{G}_{\phi_{p+1}^{-1}}$ defined in this way satisfies $\mathsf{sh}_{p+1}(\mathcal{I}, \mathcal{O}, F, g)$, where $\mathcal{O} = (\widehat{F}_0, \ldots, \widehat{F}_3)$ is a partition of $O = \bigcup_{j=0}^{3} \widehat{F}_j$ and $F = \widehat{F}_4$. □

By combining Lemma 3 and Lemma 4, one can build a grammar of size exponential with p, which "computes" ϕ_p^{-1}.

5 Reduction

One can show by standard methods of computability theory that the problem whether a one-tape Turing machine M halts on an empty input in $Ack(|M|)$ space is non-primitive recursive. The question whether a one-tape TM halts on empty input in $f(n)$ space can be reduced to the question if a linear-bounded automaton (LBA) M' simulating M step by step accepts a word $\flat^{f(n)}$ where $\flat$ is a blank symbol. In [4], a reduction from the membership problem for LBAs to the membership problem for leftist grammars is presented. We combine a grammar obtained as a result of this reduction with grammars which "compute" the functions ϕ_p and ϕ_p^{-1}. As a result, we reduce (in exponential time) the problem whether a Turing Machine M accepts an empty word in space $Ack(|M|)$ to the membership problem for a leftist grammar.

Theorem 1. *The variable membership problem for leftist grammars is non-primitive recursive.*

Proof. First, we recall some results concerning leftist grammars.

Theorem 2. *[4][1] Let M be a linear-bounded automaton (LBA) with an input alphabet Σ_M, and let $\mathcal{I} = (\bigcup_{a \in \Sigma_M} I(a,0), \bigcup_{a \in \Sigma_M} I(a,1))$ be a partition of the set $I = \bigcup_{a \in \Sigma_M, i \in [0,1]} I(a,i)$, where $I(a,i) \neq \emptyset$ for each $a \in \Sigma_M$ and $i \in [0,1]$. For a word $w \in \Sigma_M^*$ of length n, let*
$$A(w) = \{v_1 \cdots v_n \mid v_i \in I(w[i], i\%2) \text{ for each } i \in [1,n]\}.$$
Then, there exists a grammar $\mathcal{G}_M$ of size $poly(|M|)$ and a set O with a partition $\mathcal{O} = (O_0, \ldots, O_3)$, such that $I \cap O = \emptyset$ and

– $I(a,j)$ is homogenous in $\mathcal{G}_M$ for each $a \in \Sigma_M$, $j \in [0,1]$,

---

[1] This result corresponds to a part of the main construction from [4].

– if v is an alternating word wrt $\mathcal{I}$, g eliminates v in a derivation $vg \Rightarrow^*_{\mathcal{G}_M} w'g$ and $w' \in O^*$, then w' is an alternating word wrt $\mathcal{O}$, and $\|w'\|_{\mathcal{O}} \geq \|v\|_{\mathcal{I}}$.

– $w \in L(M)$ iff there exists a derivation $vg \Rightarrow^*_{\mathcal{G}_M} w'g$ for each $v \in A(w)$ such that g eliminates v, $\|w'\|_{\mathcal{O}} = \|v\|_{\mathcal{I}}$ and w' is an alternating word wrt $\mathcal{O}$.

(Note that, due to the fact that $I(a,j)$'s are homogeneous, the above two statements hold for each $v \in A(w)$ iff they hold for any $v \in A(w)$.)

If $\mathcal{G}_M$ satisfies the above conditions, we denote this fact by $\mathsf{cmp}_M(\mathcal{I}, \mathcal{O}, g)$.   □

Theorem 3. [6][2] *Let I be a set of symbols and let $\mathcal{I} = (I_0, I_1)$ be a partition of I. Then, there exists a grammar $\mathcal{G}$ and a set O with a partition $\mathcal{O} = (O_0, O_1)$ such that elements of O cannot insert other symbols and $z \in (O_0 \cup O_1)^* O_j$ is able to eliminate $w \in (I_0 \cup I_1)^* I_j$ iff $\|z\|_{\mathcal{O}} \geq \|w\|_{\mathcal{I}}$. This property is denoted by* $\mathsf{eq}(\mathcal{I}, \mathcal{O})$.   □

For a Turing machine M and natural numbers p, n, we build the grammar $\mathcal{H}_M$ verifying if M halts on an empty input word in space $\phi_p(n)$. The grammar $\mathcal{H}_M$ consists of:

$$\begin{aligned}
\mathcal{G}_1 \text{ which satisfies } &\quad \mathsf{ex}_p\,(\mathcal{I}_1, B_1, \mathcal{O}_1, F_1, g_1) \\
\mathcal{G}_2 \text{ which satisfies } &\quad \mathsf{cmp}_{M'}(\mathcal{I}_2, \mathcal{O}_2, g_2) \\
\mathcal{G}_3 \text{ which satisfies } &\quad \mathsf{sh}_p\,(\mathcal{I}_3, \mathcal{O}_3, F_3, g_3) \\
\mathcal{G}_4 \text{ which satisfies } &\quad \mathsf{eq}\,(\mathcal{I}_4, \mathcal{O}_4).
\end{aligned}$$

where M' is an **LBA** simulating M, $\mathcal{I}_2 = (\bigcup_{a \in \Sigma_{M'}} I_2(a,0), \bigcup_{a \in \Sigma_{M'}} I_2(a,1))$, $I_2(b,i) := O_{1,i}$, $\mathcal{I}_3 := \mathcal{O}_2$, $\mathcal{I}_4 := (O_{3,0} \cup O_{3,2}, O_{3,1} \cup O_{3,3})$, $\mathcal{O}_i = (O_{i,0}, \ldots, O_{i,j_i})$, $O_i = O_{i,0} \cup \cdots \cup O_{i,j_i}$, $j_1 = j_4 = 1$, and $j_2 = j_3 = 3$. The grammar $\mathcal{H}_M$ is obtained by combining alphabets and productions of $\mathcal{G}_1, \ldots, \mathcal{G}_4$, adding symbols $\imath_0, \imath_1, b$ such that $\mathcal{I}_1 = (\{\imath_0\}, \{\imath_1\})$, $B_1 = \{b\}$, adding a final symbol x and new production rules

$$\begin{aligned}
(i) &\quad x \xrightarrow{\text{del}} a &&\text{for each } a \in O_4 \cup F_1 \cup F_3 \\
(ii) &\quad g_i \xrightarrow{\text{del}} g_{i-1} &&\text{for } i \in [2,3] \\
(iii) &\quad a \xrightarrow{\text{del}} g_3 &&\text{for each } a \in O_4.
\end{aligned}$$

The question whether M accepts with empty input in space $\phi_p(n)$ is reduced to the question whether the word

$$R(M, p, n) \equiv \psi_1(n) g_1 g_2 g_3 \psi_4(n) x$$

belongs to the language $L(\mathcal{H}_M)$, where $\psi_1(n) = b \imath_0^{n\%2} (\imath_1 \imath_0)^{\lfloor n/2 \rfloor}$, $\psi_4(n) = a_0^{n\%2} (a_1 a_0)^{\lfloor n/2 \rfloor}$ and $a_j \in O_{4,j}$ for $j \in [0,1]$. The correctness of this reduction is based on the observation that the only possible greedy derivation $R(M, p, n) x \Rightarrow^* x$ works according to the following scenario:

1. Expansion: $\psi_1(n) g_1 \Rightarrow^*_{\mathcal{G}_1} v_1 w_1 g_1$, where $v_1 \in F_1^+$, $w_1 \in [O_{1,0}](O_{1,1} O_{1,0})^*$ and $\|w_1\|_{\mathcal{O}_1} = \phi_p(n)$.

---

[2] This theorem describes a simple generalization of a grammar presented in the proof of Theorem 2 in [6].

2. Computation: $v_1 w_1 g_1 g_2 \Rightarrow^* v_1 w_2 g_2$, where $w_2 \in O_{2,0}(O_{2,1}O_{2,0})^*$ and $\|w_2\|_{O_2} = \|w_1\|_{O_1}$. Here, w_1 is considered as an element of $A(w)$, where $w = \flat^{\phi_p(n)}$.

3. Shrinking: $v_1 w_2 g_2 g_3 \Rightarrow^* v_1 v_3 w_3 g_3$, where
$\|w_3\|_{O_3} = \phi_p^{-1}(\|w_2\|_{O_2}) = \phi_p^{-1}(\phi_p(n)) = n$.

4. Verification: $v_1 v_3 w_3 g_3 \psi_4(n) \Rightarrow^* v_1 v_3 w_4$, where $w_4 \in O_4^*$.

5. Final steps: $v_1 v_3 w_4 x \Rightarrow^* x$.

However, such a derivation is possible only if $w \in L(M)$ (see Theorem 2).

By applying the reduction $R(M, p, n)$ for $p := |M|$ and $n := 3$, we see that the variable membership problem for leftist grammars is non-primitive recursive, since $Ack(m) = \phi_m(3)$. □

Using a similar reduction, one can prove the following result.

Theorem 4. *There is no primitive recursive upper time bound for the ("static") membership problem for leftist grammars.*

6 Conclusions and Open Problems

We have shown that the variable membership problem for leftist grammars is non-primitive recursive. An interesting research direction is to find a restricted variant of leftist grammars with feasible complexity and large expressive power.

Interestingly, large complexity of the membership problem for leftist grammars can be obtained by a technique which is similar to the method applied in the proof that verifying lossy channels has non-primitive recursive complexity [10]. In both models, one can "weakly compute" recursive functions.

References

1. Bandyopadhyay, S., Mahajan, M., Narayan Kumar, K.: A non-regular leftist language (manuscript, 2005)
2. Cheiner, O., Saraswat, V.: Security Analysis of Matrix. Technical report, AT&T Shannon Laboratory (1999)
3. Harrison, M., Ruzzo, W., Ullman, J.: Protection in operating systems. Communications of the ACM 19(8), 461–470 (1976)
4. Jurdziński, T.: On Complexity of Grammars Related to the Safety Problem. Theoretical Computer Science 389(1-2), 56–72 (2007); An extended abstract appeared in ICALP 2006 Proceedings
5. Jurdziński, T.: Leftist Grammars are Non-primitive Recursive, Technical Report of the Institute of Computer Science, University of Wroclaw (2008/01)
6. Jurdziński, T., Loryś, K.: Leftist Grammars and the Chomsky Hierarchy. Theory of Computing Systems 41(2), 233–256 (2007)
7. Motwani, R., Panigrahy, R., Saraswat, V.A., Venkatasubramanian, S.: In: STOC 2000, pp. 306–315 (2000)
8. Saraswat, V.: The Matrix Design. Technical report, AT&T Laboratory (April 1997)
9. Saraswat, V., Jagadeesan, R.: Static support for capability-based programming in Java (manuscript)
10. Schnoebelen, P.: Verifying Lossy Channel Systems has Nonprimitive Recursive Complexity. Information Processing Letters 83(5), 251–261 (2002)

On the Computational Completeness of Equations over Sets of Natural Numbers

Artur Jeż[1,*] and Alexander Okhotin[2,3,**]

[1] Institute of Computer Science, University of Wrocław, Poland
aje@ii.uni.wroc.pl
[2] Academy of Finland
[3] Department of Mathematics, University of Turku, Finland
alexander.okhotin@utu.fi

Abstract. Systems of equations of the form $\varphi_j(X_1, \ldots, X_n) = \psi_j(X_1, \ldots, X_n)$ with $1 \leqslant j \leqslant m$ are considered, in which the unknowns X_i are sets of natural numbers, while the expressions φ_j, ψ_j may contain singleton constants and the operations of union (possibly replaced by intersection) and pairwise addition $S + T = \{m + n \mid m \in S, \ n \in T\}$. It is shown that the family of sets representable by unique (least, greatest) solutions of such systems is exactly the family of recursive (r.e., co-r.e., respectively) sets of numbers. Basic decision problems for these systems are located in the arithmetical hierarchy.

1 Introduction

Consider equations, in which the variables assume values of sets of natural numbers, and the left- and right-hand sides use Boolean operations and pairwise addition of sets defined as $S + T = \{m + n \mid m \in S, \ n \in T\}$. The simplest example of such an equation is $X = (X + X) \cup \{2\}$, with the set of all even numbers as the least solution. On one hand, such equations constitute a basic mathematical object, which is closely related to *integer expressions* introduced in the seminal paper by Stockmeyer and Meyer [18] and later systematically studied by McKenzie and Wagner [11]. On the other hand, they can be regarded as *language equations* over a one-letter alphabet, with the sum of sets representing concatenation of such languages.

Language equations are equations with formal languages as unknowns, which recently became an active area of research, with unexpected connections to computability established. Undecidability of the solution existence problem for language equations with concatenation and Boolean operations was shown by Charatonik [1]. Later it was determined by Okhotin [13,15,16] that the family of sets representable by unique (least, greatest) solutions of such equations is exactly the family of recursive languages (recursively enumerable, co-recursively enumerable, respectively). Kunc [8] constructed an equation of

---

* Supported by MNiSW grant number N206 024 31/3826, 2006–2008.
** Supported by the Academy of Finland under grant 118540.

L. Aceto et al. (Eds.): ICALP 2008, Part II, LNCS 5126, pp. 63–74, 2008.

the form $XL = LX$, where L is a finite constant language, with a computationally universal greatest solution. See Kunc [9] for a recent survey of the area.

The cited results essentially use languages over alphabets containing at least two symbols, and, until recently, language equations over a unary alphabet received fairly little attention. Systems of the form

$$X_i = \varphi_i(X_1, \ldots, X_n) \quad (1 \leqslant i \leqslant n) \tag{*}$$

with union and concatenation represent context-free grammars and their solutions over a unary alphabet are well-known to be regular. Constructing any equation with a non-regular unique solution is already not a trivial task; the first example of such an equation using the operations of concatenation and complementation was presented by Leiss [10]. Recently Jeż [5] constructed a system (*) using concatenation, union and intersection with a non-regular solution. This result was extended to a large class of unary languages by Jeż and Okhotin [6,7], who showed that these equations can simulate *trellis automata* [2] (which are the simplest type of cellular automata) recognizing positional notation of numbers.

These recent advances suggest the question of understanding the exact limits of the expressive power of equations over sets of numbers. Unexpectedly, this paper establishes computational completeness of systems of equations of the form $\varphi_j(X_1, \ldots, X_n) = \psi_j(X_1, \ldots, X_n)$, in which X_i are sets of natural numbers and φ_j, ψ_j contain sum and either union or intersection. To be precise, it is proved that a set is representable as a component of a unique solution of such a system if and only if this set is recursive. Similar characterizations are obtained for least and greatest solutions. The results are established by re-creating the existing computational completeness results for language equations using a much more restricted object, equations over sets of numbers. Before proceeding with the arguments, let us review the key result on language equations.

2   Language Equations and Their Computational Completeness

Let Σ be a finite alphabet and consider systems of equations of the form

$$\varphi_j(X_1, \ldots, X_n) = \psi_j(X_1, \ldots, X_n), \tag{**}$$

where the unknowns X_i are languages over Σ, while φ_j and ψ_j are expressions using union, intersection and concatenation, as well as singleton constants.

Theorem 1 (Okhotin [13,15]). *Let (**) be a system that has a unique (least, greatest) solution $(L_1, \ldots, L_n)$. Then each component L_i is recursive (r.e., co-r.e., respectively). Conversely, for every recursive (r.e., co-r.e.) language $L \subseteq \Sigma^*$ (with $|\Sigma| \geqslant 2$) there exists a system (**) with the unique (least, greatest, respectively) solution $(L, \ldots)$.*

As this paper considers a much more restricted family of equations, the first part of Theorem 1 will apply as it is, while the lower bound proofs will have to be entirely remade. Let us summarize the proof of the second part of Theorem 1, which will serve as a model for the arguments presented later.

The main technical device used in the construction of such a system is the language of computation histories of a Turing machine, defined and used by Hartmanis [4]. In short, for every TM T over an input alphabet Σ one can construct an alphabet Γ and an encoding of computations $C_T : \Sigma^* \to \Gamma^*$, so that for every $w \in L(T)$ the string $C_T(w)$ lists the configurations of T on each step of its accepting computation on w, and the language

$$\mathrm{VALC}(T) = \{w\natural C_T(w) \mid C_T(w) \text{ is an accepting computation}\},$$

where $\natural \notin \Sigma \cup \Gamma$, is an intersection of two linear context-free languages. Since equations (**) can directly simulate context-free grammars and are equipped with intersection, for every Turing machine it is easy to construct a system in variables $(X_1, \ldots, X_n)$ with a unique solution $(L_1, \ldots, L_n)$, so that $L_1 = \mathrm{VALC}(T)$.

It remains to "extract" $L(T)$ out of $\mathrm{VALC}(T)$ using a language equation. Let Y be a new variable and consider the inequality

$$\mathrm{VALC}(T) \subseteq Y\natural\Gamma^*,$$

which can be formally rewritten as an equation $X_1 \cup Y\natural\Gamma^* = Y\natural\Gamma^*$. This inequality states that for every $w \in L(T)$, the string $w\natural C_T(w)$ should be in $Y\natural\Gamma^*$, that is, w should be in Y. This makes $L(T)$ the least solution of this inequality and proves the second part of Theorem 1 with respect to r.e. sets and least solutions. The construction for a co-r.e. set and a greatest solution is established by a dual argument, and these two constructions can be then combined to represent every recursive set [15].

At the first glance, the idea that the same result could hold if the alphabet consists of a single letter sounds odd. However, this is what will be proved in this paper, and, moreover, the general plan of the argument remains essentially the same.

3   Resolved Systems with $\{\cup, \cap, +\}$

A formal language L over the alphabet $\Sigma = \{a\}$ can be regarded as a set of numbers $\{n \mid a^n \in L\}$, and so equations over sets of numbers represent a very special subclass of language equations. Let us first review the recent results on *resolved systems* over sets of natural numbers of the form

$$X_i = \varphi(X_1, \ldots, X_n) \quad (1 \leqslant i \leqslant n)$$

Here the right-hand sides φ_i may contain union, intersection and addition, as well as singleton constants. To minimize the number of brackets, assume that the addition has the highest precedence, followed by intersection, while the precedence of union is the least.

If intersection is disallowed, such systems are basically context-free grammars over a one-letter alphabet, and hence their solutions are ultimately periodic. Equations with both union and intersection are equivalent to an extension of context-free grammars, the *conjunctive grammars* [12], and the question whether any non-periodic set can be specified by such a system of equations has been open for some years, until answered by the following example:

Example 1 (Jeż [5]). The least solution of the system

$$X_1 = (X_2 + X_2 \cap X_1 + X_3) \cup \{1\} \qquad X_3 = (X_6 + X_6 \cap X_1 + X_2) \cup \{3\}$$
$$X_2 = (X_6 + X_2 \cap X_1 + X_1) \cup \{2\} \qquad X_6 = X_3 + X_3 \cap X_1 + X_2$$

is $(\{4^n \mid n \geqslant 0\}, \{2 \cdot 4^n \mid n \geqslant 0\}, \{3 \cdot 4^n \mid n \geqslant 0\}, \{6 \cdot 4^n \mid n \geqslant 0\})$.

To understand this construction, it is useful to consider positional notation of numbers. Let $\Sigma_k = \{0, 1, \ldots, k-1\}$ be digits in base-k notation. For every $w \in \Sigma_k^*$, let $(w)_k$ be the number defined by this string of digits. Define $(L)_k = \{(w)_k \mid w \in L\}$. Now the solution of the above system can be conveniently represented in base-4 notation as $((10^*)_4, (20^*)_4, (30^*)_4, (120^*)_4)$.

The following generalization of this example has been obtained:

Theorem 2 (Jeż [5]). *For every $k \geqslant 2$ and for every regular language $L \subseteq \Sigma_k^+$ there exists a resolved system over sets of natural numbers in variables X, $Y_2, \ldots, Y_n$ with the least solution $X = (L)_k$ and $Y_i = K_i$ for some $K_i \subseteq \mathbb{N}$.*

A further extension of this result allows one to take a *trellis automaton* (one-way real-time cellular automaton) recognizing a positional notation of a set of numbers, and construct a system of equations representing this set of numbers.

A trellis automaton [2,14], defined as a quintuple $(\Sigma, Q, I, \delta, F)$, processes an input string of length $n \geqslant 1$ using a uniform array of $\frac{n(n+1)}{2}$ nodes, as presented in the figure below. Each node computes a value from a fixed finite set Q. The nodes in the bottom row obtain their values directly from the input symbols using a function $I : \Sigma \to Q$. The rest of the nodes compute the function $\delta : Q \times Q \to Q$ of the values in their predecessors. The string is accepted if and only if the value computed by the topmost node belongs to the set of accepting states $F \subseteq Q$.

Definition 1. *A trellis automaton is a quintuple $M = (\Sigma, Q, I, \delta, F)$, in which:*
- *Σ is the input alphabet,*
- *Q is a finite non-empty set of states,*
- *$I : \Sigma \to Q$ is a function that sets the initial states,*
- *$\delta : Q \times Q \to Q$ is the transition function, and*
- *$F \subseteq Q$ is the set of final states.*

Extend δ to a function $\delta : Q^+ \to Q$ by $\delta(q) = q$ and

$$\delta(q_1, \ldots, q_n) = \delta(\delta(q_1, \ldots, q_{n-1}), \delta(q_2, \ldots, q_n)),$$

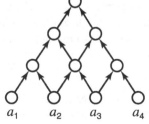

while I is extended to a homomorphism $I : \Sigma^* \to Q^*$.
Let $L_M(q) = \{w \mid \delta(I(w)) = q\}$ and define $L(M) = \bigcup_{q \in F} L_M(q)$.

Theorem 3 (Jeż, Okhotin [6]). *For every $k \geqslant 2$ and for every trellis automaton M over Σ_k with $L(M) \cap 0\Sigma_k^* = \varnothing$ there exists a resolved system over sets of natural numbers in variables $X, Y_2, \ldots, Y_n$ with the least solution $X = (L(M))_k$ and $Y_i = K_i$ for some $K_i \subseteq \mathbb{N}$.*

An important example of a set representable according to this theorem is the numeric version of the set of computational histories of a given Turing machine. The symbols needed to represent the standard language of computations of a Turing machine are interpreted as digits, and then every string from this language is represented by a number. Since the standard language of computations can be recognized by a trellis automaton, by Theorem 3 there is a system of equations representing the corresponding set of numbers.

In the next section, such a set of numbers will be used for the same purpose as the standard language VALC in the computational completeness proofs for language equations [13,15,16].

4   Unresolved Systems with $\{\cup, \cap, +\}$

Consider systems of equations of the form

$$\varphi_j(X_1, \ldots, X_n) = \psi_j(X_1, \ldots, X_n) \quad (1 \leqslant j \leqslant m),$$

where the unknowns X_i are sets of natural numbers and φ_j, ψ_j may use union, intersection and addition, as well as singleton constants.

The ultimate result of this paper is the computational completeness of such systems using *either* union *or* intersection. However, let us start with the case of systems that use both Boolean operations. The case of only one Boolean operation presents additional challenges, since Theorem 3 as it is requires both union and intersection; these issues will be discussed later in Section 6.

Theorem 4. *The family of sets of natural numbers representable by unique (least, greatest) solutions of systems of equations of the form $\varphi_i(X_1, \ldots, X_n) = \psi_i(X_1, \ldots, X_n)$ with union, intersection and addition, is exactly the family of recursive (r.e., co-r.e., respectively) sets.*

These solutions are recursive (r.e., co-r.e., respectively) because so are the solutions of language equations with union, intersection and concatenation, see Theorem 1. So the task is to take any recursive (r.e., co-r.e.) set of numbers and to construct a system of equations representing this set by a solution of the corresponding kind. The construction is based upon a rather complicated arithmetization of Turing machines, which proceeds in several stages.

First, valid accepting computations of a Turing machine are represented as numbers, so that these numbers could be recognized by a trellis automaton working on base-6 positional notations of these numbers. While trellis automata are rather flexible and could accept many different encodings of such computations, the subsequent constructions require a set of numbers of a very specific form. This form will now be defined.

First, computations are expressed as strings in the standard way:

Definition 2. *Let T be a Turing machine recognizing numbers given to it in base-6 notation. Let $V \supset \Sigma_6$ be its tape alphabet, let Q be its set of states, and define $\Gamma = V \cup Q \cup \{\sharp\}$.*

For every number $n \in L(T)$, denote the instantaneous description of T after i steps of computation on n as a string $ID_i = \alpha q a \beta \subseteq V^* Q V V^*$, where T is in state q scanning $a \in \Gamma$ and the tape contains $\alpha a \beta$. Define

$$\widetilde{C}_T(n) = ID_0 \cdot \sharp \cdot ID_1 \cdot \sharp \cdot \ldots \cdot \sharp \cdot ID_{\ell-1} \cdot \sharp\sharp \cdot ID_\ell \cdot \sharp \cdot (ID_\ell)^R \cdot \sharp \cdot \ldots \cdot \sharp \cdot (ID_1)^R \cdot \sharp \cdot (ID_0)^R$$

Next, consider any code $h : \Gamma^* \to \Sigma_6^*$, under which every codeword is in $\{30, 300\}^+$. Define $C_T(n) = h(\widetilde{C}_T(n))300$.

The language $\{\widetilde{C}_T(n) \mid n \in L(T)\} \subseteq \Gamma^*$ is an intersection of two linear context-free languages and hence is recognized by a trellis automaton [2,14]. By the known closure of trellis automata under codes, the language $\{C_T(n) \mid n \in L(T)\} \subseteq \Sigma_6^+$ is recognized by a trellis automaton as well.

Now the set of accepting computations of a Turing machine is represented as the following six sets of numbers:

Definition 3. *Let T be a Turing machine recognizing numbers given in base-6 notation. For every $i \in \{1, 2, 3, 4, 5\}$, the valid accepting computations of T on numbers $n \geqslant 6$ with their base-6 notation beginning with the digit i is*

$$\mathrm{VALC}_i(T) = \{(C_T(n)1w)_6 \mid n = (iw)_6, \, n \in L(T)\},$$

The computations of T on numbers $n \in \{0, 1, 2, 3, 4, 5\}$, provided that they are accepting, are represented by the following finite set of numbers:

$$\mathrm{VALC}_0(T) = \{(C_T(n))_6 + n \mid n \in \{0, 1, 2, 3, 4, 5\} \text{ and } n \in L(T)\}$$

For example, under this encoding, the accepting computation on a number $n = (543210)_6$ will be represented by a number $(30300300\ldots30300143210)_6 \in \mathrm{VALC}_5(T)$, where the whole computation is encoded by blocks of digits 30 and 300, the digit 1 acts as a separator and the lowest digits 43210 represent n with its leading digit cut. A crucial property of this encoding is that the digits representing n can be separated from the digits representing the computation:

Lemma 1. *Let $L \subseteq (1\Sigma_6^+)_6$. Then for every $m \in (\{30, 300\}^*3000^\ell)_6$ and for every $n \in (1\Sigma_6^{\leqslant \ell})_6$, if $m + n \in (\{30, 300\}^*3000^*)_6 + L$, then $n \in L$.*

Trellis automata recognizing the base-6 notation of numbers in $\mathrm{VALC}_i(T)$, by Theorem 3, give us the following system of equations:

Lemma 2. *For every Turing machine T recognizing numbers there exists a system of equations $X_i = \varphi_i(X_1, \ldots, X_n)$ over sets of natural numbers using union, intersection and addition, such that its least solution is $(L_0, L_1, \ldots, L_5, L_6, \ldots, L_n)$ with $L_i = \mathrm{VALC}_i(T)$ for $0 \leqslant i \leqslant 5$.*

Using these sets as constants, the required equations can be constructed. The first case to be established is the case of least solutions and r.e. sets.

Lemma 3. *For every recursively enumerable set of numbers $L_0 \subseteq \mathbb{N}$ there exists a system of equations of the form*

$$\varphi_j(Y, X_1, \ldots, X_m) = \psi_j(Y, X_1, \ldots, X_m)$$

with union, intersection and addition, which has the set of solutions

$$\{ (L, f_1(L), \ldots, f_m(L)) \mid L_0 \subseteq L \}$$

where $f_1, \ldots, f_m : 2^{\mathbb{N}} \to 2^{\mathbb{N}}$ are some monotone functions on sets of numbers defined with respect to L_0. In particular, there is a least solution with $Y = L_0$.

Proof. Consider any Turing machine T recognizing L_0. A system in variables $(Y, Y_1, \ldots, Y_5, Y_0, X_7, \ldots, X_m)$ will be constructed, where the number m will be determined below, and the set of solutions of this system will be defined by the following conditions, which ensure that the statement of the lemma is fulfilled:

$$L(T) \cap \{0, 1, 2, 3, 4, 5\} \subseteq Y_0 \subseteq \{0, 1, 2, 3, 4, 5\}, \tag{1a}$$

$$\{(1w)_6 \mid w \in \Sigma_6^+, \, (iw)_6 \in L(T)\} \subseteq Y_i \subseteq (1\Sigma_6^+)_6 \quad (1 \leqslant i \leqslant 5), \tag{1b}$$

$$Y = Y_0 \cup \bigcup_{i=1}^{5} \{(iw)_6 \mid (1w)_6 \in Y_i\}, \tag{1c}$$

$$X_j = K_j \quad (7 \leqslant j \leqslant m). \tag{1d}$$

The sets $K_7, \ldots, K_m$ are some constants needed for the construction to work. These constants and the equations needed to specify them will be implicitly obtained in the proof. The constructed system will use inequalities of the form $\varphi \subseteq \psi$, which can be equivalently rewritten as equations $\varphi \cup \psi = \psi$ or $\varphi \cap \psi = \varphi$.

For each $i \in \{1, 2, 3, 4, 5\}$, consider the above definition of $\mathrm{VALC}_i(T)$ and define a variable Y_i with the equations

$$Y_i \subseteq (1\Sigma_6^+)_6, \tag{2a}$$

$$\mathrm{VALC}_i(T) \subseteq (\{30, 300\}^* 3000^*)_6 + Y_i. \tag{2b}$$

Both constants are given by regular languages of base-6 representations, and therefore can be specified by equations according to Theorem 2. It is claimed that this system is equivalent to (1b).

Suppose (1b) holds for Y_i. Then (2a) immediately follows. To check (2b), consider any $(C_T^i(iw)1w)_6 \in \mathrm{VALC}_i(T)$. Since this number represents the computation of T on $(iw)_6$, this implies $(iw)_6 \in L(T)$, and hence $(1w)_6 \in Y_i$ by (1b). Then $(C_T^i(iw)1w)_6 \in (\{30, 300\}^* 3000^{|1w|})_6 + (1w)_6 \subseteq (\{30, 300\}^* 3000^*)_6 + Y_i$, which proves the inclusion (2b).

Conversely, assuming (2), it has to be proved that for every $(iw)_6 \in L(T)$, where $w \in \Sigma_6^+$, the number $(1w)_6$ must be in Y_i. Since $(iw)_6 \in L(T)$, there exists an accepting computation of T: $(C_T^i(iw)1w)_6 \in \mathrm{VALC}_i(T)$. Hence, $(C_T^i(iw)1w)_6 \in (\{30, 300\}^* 3000^*)_6 + Y_i$ due to the inclusion (2b), and therefore $(1w)_6 \in Y_i$ by Lemma 1.

Define one more variable Y_0 with the equations

$$Y_0 \subseteq \{0, 1, 2, 3, 4, 5\}, \tag{3a}$$
$$\text{VALC}_0(T) \subseteq (\{30, 300\}^* 300)_6 + Y_0. \tag{3b}$$

The claim is that (3) holds if and only if (1a).

Assume (1a) and consider any number $(C_T(n))_6 + n \in \text{VALC}_0(T)$, where $n \in \{0, 1, 2, 3, 4, 5\}$ by definition. Then n is accepted by T, and, by (1a), $n \in Y_0$. Since $(C_T(n))_6 \in (\{30, 300\}^* 300)_6$, the addition of n affects only the last digit, and $(C_T(w))_6 + n \in (\{30, 300\}^* 300)_6 + n \subseteq (\{30, 300\}^* 300)_6 + Y_0$, which proves (3b).

The converse claim is that (3) implies that every $n \in L(T) \cap \{0, 1, 2, 3, 4, 5\}$ must be in Y_0, The corresponding $(C_T(n))_6 + n \in \text{VALC}_0(T)$ is in $(\{30, 300\}^* 300)_6 + n$ by (3b). Since n is represented by a single digit, the number $(C_T(n))_6 + n$ ends with this digit. The set $(\{30, 300\}^* 300)_6 + Y_0$ contains a number of such a form only if $n \in Y_0$.

Next, combine the above six systems together and add a new variable Y with the following equation:

$$Y = Y_0 \cup Y_1 \cup \bigcup_{\substack{i \in \{2,3,4,5\} \\ i' \in \Sigma_k}} \left((Y_i \cap (1i' \Sigma_k^*)_6) + ((i-1)0^*)_6 \cap (ii' \Sigma_k^*)_6 \right). \tag{4}$$

This equation has been borrowed from the authors' previous paper [6, Th. 3], where it was proved equivalent to $Y = Y_0 \cup \{iw \mid 1w \in Y_i\}$, that is, to (1c).

The final step of the construction is to express constants used in the above systems through singleton constants, which can be done by Theorem 2 and Lemma 2. The variables needed to specify these languages are denoted $(X_7, \ldots, X_n)$, and the equations for these variables have a unique solution $X_j = K_j$ for all j.

This completes the description of the set of solutions of the system. It is easy to see that there is a least solution in this set, with $Y = L(T)$, $Y_0 = L(T) \cap \{0, 1, 2, 3, 4, 5\}$, $Y_i = \{(1w)_6 \mid w \in \Sigma_6^+, (iw)_6 \in L(T)\}$ and $X_j = K_j$. □

The representation of co-recursively enumerable sets by greatest solutions is dual to the case of least solutions and is established by an analogous argument.

Lemma 4. *For every co-recursively enumerable set of numbers $L_0 \subseteq \mathbb{N}$ there exists a system of equations of the form*

$$\varphi_j(Z, X_1, \ldots, X_m) = \psi_j(Z, X_1, \ldots, X_m)$$

with union, intersection and addition, which has the set of solutions

$$\{ (L, f_1(L), \ldots, f_m(L)) \mid L \subseteq L_0 \},$$

where $f_1, \ldots, f_m : 2^{\mathbb{N}} \to 2^{\mathbb{N}}$ are some monotone functions on sets of numbers defined with respect to L_0. In particular, there is a greatest solution with $Z = L_0$.

Finally, the case of recursive languages and unique solutions can be established by combining the constructions of Lemmata 3 and 4 as follows:

Lemma 5. *For every recursive set of numbers $L \subseteq \mathbb{N}$ there exists a system of equations of the form $\varphi_i(Y, Z, X_1, \ldots, X_n) = \psi_i(Y, Z, X_1, \ldots, X_n)$ with union, intersection and addition, such that its unique solution is $Y = Z = L$, $X_i = K_i$, where $(K_1, \ldots, K_n)$ is some vector of sets.*

Proof. As a recursive language, L is both recursively enumerable and co-recursively enumerable, hence both Lemmata 3 and 4 apply. Consider both systems of language equations given by these lemmata, let Y be the variable from Lemma 3 let Z be the variable from Lemma 4, and let $X_1, \ldots, X_n$ be the rest of the variables in these systems combined. The set of solutions of the systems obtained is

$$\{ (Y, Z, f_1(Y, Z), \ldots, f_n(Y, Z)) \mid Z \subseteq L \subseteq Y \}.$$

Add one more equation to the system:

$$Y = Z.$$

This condition collapses the bounds $Z \subseteq L \subseteq Y$ to $Z = L = Y$, and the resulting system has the unique solution

$$\{(L, L, f_1(L, L), \ldots, f_n(L, L))\},$$

which completes the proof.   □

5 Decision Problems

Consider basic properties of equations, such as the existence and the uniqueness of solutions. For the more general case of language equations it is known that these and a few other properties are undecidable [13,15,16], and their exact position in the arithmetical hierarchy has been determined. These results will now be re-created for equations over sets of numbers, based upon the constructions from the previous section.

Theorem 5. *The problem of whether a system of equations $\varphi_i(X_1, \ldots, X_n) = \psi_i(X_1, \ldots, X_n)$ over sets of natural numbers has a solution is Π_1-complete.*

Proof. The problem is in Π_1 in the more general case of language equations [13]. Its Π_1-hardness is proved by a reduction from the emptiness problem for Turing machines. Let T be a TM and construct a system of equations in variables $(Y_0, \ldots, Y_5, X_1, \ldots, X_m)$ with the unique solution $Y_i = \mathrm{VALC}_i(T)$, $X_j = K_j \subseteq \mathbb{N}$. Since $L(T) = \varnothing$ if and only if $\bigcup_{i=0}^{5} \mathrm{VALC}_i(T) = \varnothing$, it is sufficient to add a new equation $\bigcup_{i=0}^{5} Y_i = \varnothing$ so that the resulting system has a solution if and only if $L(T) = \varnothing$.   □

Theorem 6. *Testing whether a system $\varphi_i(X_1, \ldots, X_n) = \psi_i(X_1, \ldots, X_n)$ over sets of natural numbers has a unique solution is a Π_2-complete problem.*

Proof. The Π_2 upper bound is known from the case of language equations [13].

Π_2-hardness is proved by a reduction from the known Π_2-complete Turing machine universality problem, which can be stated as follows: "Given a TM M working on natural numbers, determine whether it accepts every $n \in \mathbb{N}_0$". Given M, construct the system of equations as in Lemma 3. It has a unique solution if and only if the bounds $L(T) \subseteq L \subseteq \mathbb{N}$ are tight, that is, if and only if the TM accepts every number. This completes the reduction. $\square$

Theorem 7. *The problem whether a system $\varphi_i(X_1, \ldots, X_n) = \psi_i(X_1, \ldots, X_n)$ over sets of natural numbers has finitely many solutions is Σ_3-complete.*

Proof. The problem is in Σ_3 for language equations [16].

To prove Σ_3-hardness, consider the co-finiteness problem for Turing machines, which is stated as "Given a TM T working on natural numbers, determine whether $\mathbb{N} \setminus L(T)$ is finite", which is known to be Σ_3-complete [17, Cor. 14-XVI]. Given M, use Lemma 3 to construct the system of equations with the set of solutions $\{(L, f_1(L), \ldots, f_k(L)) \mid L(T) \subseteq L\}$. This set is finite if and only if $\mathbb{N} \setminus L(T)$ is finite, which completes the reduction. $\square$

6   Unresolved Systems with $\{\cup, +\}$ and $\{\cap, +\}$

All results so far have been established using equations with addition, union and intersection. In fact, the same results hold for equations using addition and *either* union *or* intersection. Establishing all results in this stronger form, in particular, requires rewriting the basic constructions of Theorems 2 and 3 [5,6]. The proof of the new Theorem 4 also has to undergo some changes.

An outline of these modifications is explained in this section. The first basic result is a simulation of a resolved system of a specific form using union, intersection and addition by an unresolved system that does not use intersection.

Lemma 6. *Let $X_i = \varphi_i(X_1, \ldots, X_n)$ be a resolved system of equations with union, intersection and addition and with constants from a set C containing only positive integers. Let $(L_1, \ldots, L_n)$ be its least solution. Assume that for every variable X_{i_0} there exists a subset of variables $\{X_i\}_{i \in I}$ containing X_{i_0}, such that*

- *the sets $\{L_i\}_{i \in I}$ are pairwise disjoint and their union is in C,*
- *the equations for all $\{X_i\}_{i \in I}$ are either all of the form $X_i = \bigcup_j \alpha_{ij}$, or all of the form $X_i = \bigcap_j \alpha_{ij} \cup C$, where C is a constant and $\alpha_{ij} = A_1 + \ldots + A_k$, with $k \geq 1$ and with each A_t being a constant or a variable,*

In addition, assume that there are no cyclic chain dependencies in the system. Then there exists an unresolved system with union and addition, with constants from C, which has the unique solution $(L_1, \ldots, L_n)$.

The construction is by replacing each equation $X_i = \bigcap_j \alpha_{ij} \cup C$ with equations $X_i \subseteq \alpha_{ij} \cup C$ for all j. In addition, for each subset of variables $\{X_i\}_{i \in I}$ with $\bigcup_{i \in I} L_i = C_I$, the equation $\bigcup_{i \in I} X_i = C_I$ is added.

A similar construction produces equations with intersection instead of union:

Lemma 7. *Under the assumptions of Lemma 6, there exists an unresolved system with intersection and addition and with constants from C, which has a unique solution that coincides with the least solution of the given system.*

Here every equation $X_j = \bigcup_j \alpha_{ij}$ is replaced with $\alpha_i \subseteq X$ for each i. For every subset of variables $\{X_i\}_{i \in I}$ with union C_I, an equation $X_i \subseteq C_I$ is added for each $i \in I$, as well as an equation $X_i \cap X_j = \varnothing$ for all $i, j \in I$ with $i \neq j$.

The next task is to apply Lemmata 6 and 7 to resolved systems constructed in the proofs of Theorems 2 and 3. For the lemmata to be applicable, the existing equations (see Jeż [5] and Jeż and Okhotin [6]) need to be decomposed into smaller parts and slightly changed. Then the variables can be grouped into subsets, as required by the lemmata. In the end, the following variant of Theorem 3 is obtained:

Lemma 8. *For every $k \geqslant 2$ and for every trellis automaton M over Σ_k, with $L(M) \cap 0\Sigma_k^* = \varnothing$, there exists and can be effectively constructed an unresolved system of equations over sets of natural numbers using union and addition (intersection and addition, respectively) and singleton constants, such that its unique solution contains a component $(L(M))_k$.*

It remains to modify the proofs of Lemmata 3–5. The equations in there use mostly inclusions, which can be simulated using either union or intersection. The only exception is the equation (4). It can be equivalently replaced by the following unresolved equations using intersection and addition:

$$Y \cap (ii'\Sigma_6^*)_6 = (Y_i \cap (1i'\Sigma_6^*)_6) + ((i-1)0^*)_6 \cap (ii'\Sigma_6^*)_6 \quad (i, j \in \Sigma_6, \, i \geqslant 2)$$

$$Y \cap \{0, 1, \ldots, 5\} = Y_0$$

$$Y \cap (1\Sigma_6^+)_6 = Y_1$$

The construction of equivalent unresolved equations using union and addition is slightly more complicated and is omitted due to space constraints. Altogether, the following improved statement of Theorem 4 is established:

Theorem 8. *The sets of natural numbers representable by unique (least, greatest) solutions of systems of equations of the form $\varphi_i(X_1, \ldots, X_n) = \psi_i(X_1, \ldots, X_n)$ with addition, and union or intersection are exactly the recursive (r.e., co-r.e., respectively) sets.*

Theorems 5–7 on the undecidability level of decision problems for such equations hold in this case as well.

7 Conclusion

The equations considered in this paper are a pure mathematical object. Unexpectedly, it turned out to be equivalent to the notion of effective computability.

This can be compared to Diophantine equations, which have been proved to be computationally complete by Matiyasevich. Due to this result, it is known, for instance, that there is a Diophantine equation for which the range of admissible values of a certain variable x is exactly the set of primes. Similarly, our Lemma 3 allows one to construct a system of equations over sets of natural numbers, which has a unique solution with one of its components being exactly the set of primes.

Among the applications of this result, it settles the expressive power of a generalization of integer circuits [11], as well as shows that language equations are computationally complete even in the seemingly trivial case of a unary alphabet.

References

1. Charatonik, W.: Set constraints in some equational theories. Information and Computation 142(1), 40–75 (1998)
2. Culik II, K., Gruska, J., Salomaa, A.: Systolic trellis automata, I and II. International Journal of Computer Mathematics 15, 16, 195–212, 3–22 (1984)
3. Ginsburg, S., Rice, H.G.: Two families of languages related to ALGOL. Journal of the ACM 9, 350–371 (1962)
4. Hartmanis, J.: Context-free languages and Turing machine computations. In: Proceedings of Symposia in Applied Mathematics, vol. 19, pp. 42–51. AMS (1967)
5. Jeż, A.: Conjunctive grammars can generate non-regular unary languages. In: Harju, T., Karhumäki, J., Lepistö, A. (eds.) DLT 2007. LNCS, vol. 4588, pp. 1242–1253. Springer, Heidelberg (2007)
6. Jeż, A., Okhotin, A.: Conjunctive grammars over a unary alphabet: undecidability and unbounded growth. In: Diekert, V., Volkov, M.V., Voronkov, A. (eds.) CSR 2007. LNCS, vol. 4649, pp. 168–181. Springer, Heidelberg (2007)
7. Jeż, A., Okhotin, A.: Complexity of equations over sets of numbers. In: STACS 2008 (2008)
8. Kunc, M.: The power of commuting with finite sets of words. Theory of Computing Systems 40(4), 521–551 (2007)
9. Kunc, M.: What do we know about language equations? In: Harju, T., Karhumäki, J., Lepistö, A. (eds.) DLT 2007. LNCS, vol. 4588, pp. 23–27. Springer, Heidelberg (2007)
10. Leiss, E.L.: Unrestricted complementation in language equations over a one-letter alphabet. Theoretical Computer Science 132, 71–93 (1994)
11. McKenzie, P., Wagner, K.W.: The complexity of membership problems for circuits over sets of natural numbers. Computational Complexity 16, 211–244 (2007)
12. Okhotin, A.: Conjunctive grammars. Journal of Automata, Languages and Combinatorics 6(4), 519–535 (2001)
13. Okhotin, A.: Decision problems for language equations with Boolean operations. In: Baeten, J.C.M., Lenstra, J.K., Parrow, J., Woeginger, G.J. (eds.) ICALP 2003. LNCS, vol. 2719, pp. 239–251. Springer, Heidelberg (2003)
14. Okhotin, A.: On the equivalence of linear conjunctive grammars to trellis automata. Informatique Théorique et Applications 38(1), 69–88 (2004)
15. Okhotin, A.: Unresolved systems of language equations: expressive power and decision problems. Theoretical Computer Science 349(3), 283–308 (2005)
16. Okhotin, A.: Strict language inequalities and their decision problems. In: Jedrzejowicz, J., Szepietowski, A. (eds.) MFCS 2005. LNCS, vol. 3618. Springer, Heidelberg (2005)
17. Rogers Jr., H.: Theory of Recursive Functions and Effective Computability (1967)
18. Stockmeyer, L.J., Meyer, A.R.: Word problems requiring exponential time. In: STOC 1973, pp. 1–9 (1973)

Placement Inference for a Client-Server Calculus

Matthias Neubauer and Peter Thiemann

Institut für Informatik, Universität Freiburg, Georges-Köhler-Allee 079
79110 Freiburg, Germany
{neubauer,thiemann}@informatik.uni-freiburg.de

Abstract. Placement inference assigns locations to operations in a distributed program under the constraints that some operations can only execute on particular locations and that values may not be transferred arbitrarily between locations. An optimal choice of locations additionally minimizes the run time of the program, given that operations take different time on different locations and that a cost is associated to transferring a value from one location to another.

We define a language with a time- and location-aware semantics, formalize placement inference in terms of constraints, and show that solving these constraints is an NP-complete problem. We then show that optimal placements are computable via a reformulation of the semantics in terms of matrices and an application of the max-plus spectral theory. A prototype implementation validates our results.

1   Introduction

A multi-tier architecture is the best practice approach to the construction of a distributed client-server system. Each tier corresponds to a component with a well-defined set of interfaces that can be developed independently. Moreover, it is possible to upgrade a component, to later change the underlying technology of a component, or to individually maintain and test a component without affecting other parts of the system. Multi-tier applications typically scale well because the overhead of running several incarnations of a component simultaneously is manageable and is easily outweighed by the performance gain.

Structuring such a system is no easy task, in particular if the physical design does not quite match up with the logical design. The requirements dictated by the characteristics of distribution must be kept in mind by designers and programmers alike. Local and remote versions of interfaces must be provided and must be kept synchronized, middleware technology has to be selected and integrated into the system, and all sorts of idiosyncratic restrictions have to be catered for to obtain a working system. In short, distribution leads to many new sources of error.

Recent work [21, 19, 9, 22] indicates that distributed client-server applications may also be developed in one piece. This way, an application may be structured entirely following logical design. Distribution is introduced later by a program transformation which—guided by annotations—splits the program into components that run on different tiers and introduces the necessary communications.

L. Aceto et al. (Eds.): ICALP 2008, Part II, LNCS 5126, pp. 75–86, 2008.

Expr $\ni e ::= \textbf{halt}^s \mid f^s\,\overline{x} \mid \textbf{let}\ x = op^s(\overline{x})\ \textbf{in}\ e$	$x_1, \ldots, x_{n_v} \in \mathsf{Var}$
$\mid \textbf{if}^s\ x\ \textbf{then}\ e\ \textbf{else}\ e$	$f_1, \ldots, f_{n_f} \in \mathsf{FVar}\quad \mathsf{pfun} \in \mathsf{POp}$
$\mid \textbf{trans}^s\ \overline{x}\ \textbf{in}\ e$	$s_1, \ldots, s_{n_s} \in \mathsf{SVar}\quad op \in \mathsf{EOp}$
Def $\ni d ::= f(\overline{x}) = e$	$op \in \mathsf{Op} = \mathsf{POp} \cup \mathsf{EOp}$
Prog $\ni p ::= \overline{d}\,e$	$= \{\textbf{if}\} \uplus \{op_2, \ldots, op_{n_{op}}\}$

Fig. 1. Syntax of IL_S

The transformation preserves the semantics, hence reasoning about the system and its functional test can happen in a non-distributed setting.

In previous work [21], we have proposed a theoretical framework that consists of a sequential calculus (to model the application), a placement specification (to formalize the annotation of operations with the locations where they run), and a program transformation for introducing processes and communication.

Contribution. We extend our previous work on placement specification [21] to placement inference, a static analysis that computes a placement that obeys the placement restrictions of the basic operations and exhibits the shortest worst-case run time.

Sec. 2 presents a language and its time- and location-aware semantics. Sec. 3 specifies a constraint-based analysis that infers all valid placements. The problem of finding a valid placement is shown to be NP-complete.

Sec. 4 applies the theory of (max, +) algebras [6] to compute the expected worst-case run time for a valid placement. The challenge is to define such a notion also for nonterminating programs. This step requires to reformulate the timed semantics in terms of matrix multiplication. Finally, we exhibit an algorithm that finds a placement with minimal worst-case run time.

The results in this paper are excerpted from Neubauer's PhD thesis [20], which contains all proofs.

2   The Calculus IL_S

2.1   Syntax and Informal Standard Semantics

Figure 1 defines the syntax of the IL_S calculus, *i.e.*, first-order recursive program schemes with primitive operations, conditionals and explicit transfer operations. There are five disjoint denumerable sets, Var, FVar, SVar, POp, and EOp, that represent variable identifiers, function names, placement labels, names of pure and effectful operations, respectively. An operation name, op, can stand for a pure or an effectful operation.

To simplify the formalization of the static analysis, we assume that all placement labels in a program are unique, each variable is bound at most once, and that variables, function names, placement labels, and operations are numbered consecutively. In some contexts, if serves as operation op_1.

$$\ell \in \mathcal{L} = \{1, \dots, n_l\} \qquad \text{locations}$$
$$\mathbf{c} \in \mathsf{Op} \hookrightarrow \mathcal{D}^{n_l} \qquad \text{calculation cost } \mathbf{c}^j \text{ for operation } j$$
$$\mathbf{T} \in \mathcal{D}^{n_l \times n_l} \qquad \text{transfer cost matrix}$$
$$\mathbf{o} \in \mathsf{Op} \times \mathsf{SVar} \hookrightarrow \mathcal{D}^{n_l} \qquad \text{operation placement } \mathbf{o}^{j,k} \text{ for operation } j \text{ annotated by } s_k$$
$$\mathbf{S} \in \mathsf{SVar} \times \mathsf{Var} \hookrightarrow \mathcal{D}^{n_l \times n_l} \qquad \text{transfer placement matrix } \mathbf{S}^{s,x} \text{ for var. } x \text{ annotated by } s$$

Fig. 2. Components of a timed configuration

Neubauer's PhD dissertation [20] defines the standard semantics of the language along with a simple type system and a type soundness proof. Here, we summarize only the most important points.

A program is a sequence of mutually recursive function definitions, d, followed by a single main expression. An expression of IL_S performs a single operation on multiple locations at the same time. Each expression carries a placement label annotation, s, to specify the locations on which the expression acts. The variable bindings are location dependent, that is, a variable's value may be available on one location, but unavailable on another. Expressions do not produce a final value. They either end with a $\mathtt{halt}^s$ expression or with a control transfer to another top level function, $f^s \, \overline{x}$. Arguments to operators, conditionals, and jumps are restricted to variables.

Compound expressions sequence the evaluation of expressions. The expression $\mathtt{let} \; x = op^s(\overline{x}) \; \mathtt{in} \; e$ executes a primitive operation (with or without side effect) before continuing with e. It presumes that its arguments are available on the locations indicated with s. The conditional expression $\mathtt{if}^s \; x \; \mathtt{then} \; e_1 \; \mathtt{else} \; e_2$ branches on the condition x and continues with either e_1 or e_2. The expression $\mathtt{trans}^s \; \overline{x} \; \mathtt{in} \; e$ transfers the values bound to the variables $\overline{x}$ between locations as indicated through the placement label annotation s before continuing with e. After a transfer, the values are available on more locations as before.

For this work, our prime objective is a nonstandard semantics that computes the time taken by executing a program according to the standard semantics.

2.2 Timed Semantics with Localities

The nonstandard semantics specifies the (maximal) duration that a distributed computation takes for a certain location assignment. It is a small-step operational semantics which models durations as elements of the $(\max, +)$ algebra $\mathcal{M} = (\mathbb{R} \cup \{-\infty\}, \otimes, \oplus, \mathbf{1}, \mathbf{0})$ [6], where $\otimes = +$ combines the durations of sequential operations, $\oplus = \max$ combinues durations of parallel operations, $\mathbf{1} = 0$, and $\mathbf{0} = -\infty$. We simplify the timing behavior of a distributed system by ignoring the actual values of the operands as well as any external factors on which the duration of a transfer or an operation may depend.

A *timed configuration* $\chi = (\mathcal{L}, \mathbf{c}, \mathbf{T}, \mathbf{o}, \mathbf{S})$ forms the context of evaluation for an IL_S program. Fig.2 describes its components. The set of *locations*, $\mathcal{L}$, models

$$\text{T-S-OP}$$
$$\vartheta_c'(\ell) = \text{if } \mathbf{o}_\ell^{s,k} = \mathbf{1} \text{ then } \mathbf{c}_\ell^k \otimes \left(\bigoplus_j \vartheta_v(y_j)(\ell) \oplus \vartheta_c(\ell)\right) \text{ else } \vartheta_c(\ell)$$
$$\vartheta_v' = \vartheta_v[z \mapsto \lambda\ell.\mathbf{o}_\ell^{s,k} \otimes \vartheta_c'(\ell)]$$
$$\rule{9cm}{0.4pt}$$
$$\mathcal{L}, \mathbf{c}, \mathbf{T}, \mathbf{o}, \mathbf{S} \mid \vartheta_c, \vartheta_v, \overline{d} \text{ let } z = op_k^s(\overline{y_j}) \text{ in } e, \longrightarrow_f \vartheta_c', \vartheta_v', \overline{d} \; e$$

$$\text{T-S-IF}$$
$$\frac{\vartheta_c'(\ell) = \text{if } \mathbf{o}_\ell^{s,1} = \mathbf{1} \text{ then } \mathbf{c}_\ell^1 \otimes (\vartheta_v(x)(\ell) \oplus \vartheta_c(\ell)) \text{ else } \vartheta_c(\ell)}{\mathcal{L}, \mathbf{c}, \mathbf{T}, \mathbf{o}, \mathbf{S} \mid \vartheta_c, \vartheta_v, \overline{d} \text{ if}^s \; x \text{ then } e_1 \text{ else } e_2 \longrightarrow_f \vartheta_c', \vartheta_v, \overline{d} \; e_i} \quad i \in \{1,2\}$$

$$\text{T-S-TRANS}$$
$$\frac{\vartheta_v'(x)(\ell) = \text{if } x \in \overline{x} \land \mathbf{S}_{\ell',\ell}^{s,x} = \mathbf{1} \text{ then } \mathbf{T}_{\ell',\ell} \otimes \vartheta_v(x)(\ell') \text{ else } \vartheta_v(x)(\ell)}{\mathcal{L}, \mathbf{c}, \mathbf{T}, \mathbf{o}, \mathbf{S} \mid \vartheta_c, \vartheta_v, \overline{d} \text{ trans}^s \; \overline{x} \text{ in } e \longrightarrow_f \vartheta_c, \vartheta_v', \overline{d} \; e}$$

$$\text{T-S-JUMP}$$
$$\frac{d_j \equiv f_j(\overline{x}) = e_j \qquad (\forall i) \; \vartheta_v'(x_i) = \vartheta_v(z_i)}{\chi \mid \vartheta_c, \vartheta_v, \overline{d} \; f_j^s \, \overline{z} \longrightarrow_f \vartheta_c, \vartheta_v', \overline{d} \; e_j}$$

Fig. 3. Timed reduction rules for IL_S programs

entities where computations can take place simultaneously and independently. The *calculation cost* $\mathbf{c}_\ell^j$ is the time to execute op_j on location ℓ. The *transfer cost* $\mathbf{T}_{\ell_1,\ell_2}$ is the time to send a value from location ℓ_1 to ℓ_2. They constitute a cost declaration for operations and transfers. The cost $\mathbf{0}$ declares unavailability.

The remaining components select operations and transfers for execution. The *operation placement* $\mathbf{o} = \mathbf{o}^{j,k}$ has $\mathbf{o}_\ell = \mathbf{1}$ iff operation op_j annotated by placement label s_k is executed at location ℓ. Otherwise $\mathbf{o}_\ell = \mathbf{0}$. The *transfer placement matrix* $\mathbf{S}_{\ell',\ell}^{s,x} \in \{\mathbf{0}, \mathbf{1}\}$ contains $\mathbf{1}$ to indicate a possible transfer operation labeled with s to transfer x from location ℓ' to location ℓ. Otherwise it is $\mathbf{0}$. In addition, each column of a *transfer placement matrix* is allowed to only have at most one $\mathbf{1}$ entry, because each target location must has at most one source location.

An *evaluation state* $\vartheta_c, \vartheta_v, \overline{d} \, e$ comprises a program and two timed environments that indicate when values are available at different locations and for specific arguments. The *clock environment*, $\vartheta_c \in \mathcal{L} \to \mathcal{D}$, expresses the time elapsed on each location. The *arrival environment*, $\vartheta_v \in \text{Var} \to \mathcal{L} \to \mathcal{D}$, indicates the time when the value of a variable will be available at a given location.

Fig. 3 specifies the dynamic semantics as a reduction relation $\chi \mid \vartheta_c, \vartheta_v, \overline{d} \, e \longrightarrow_f \vartheta_c', \vartheta_v', \overline{d} \, e'$, which relates an evaluation state to one of its successor states in the context of a timed configuration, χ.

The rule (T-S-OP) specifies the timing behavior of operation op_k. For each location ℓ on which op_k is performed, evaluation begins after the current time on ℓ and after all arguments of op_k are available. Then op_k takes $\mathbf{c}_\ell^k$ time, which yields the new time at ℓ. If op_k is not performed, then the time at ℓ does not advance. The arrival times of the newly bound variable z are set accordingly.

The rule (T-S-IF) has two possible outcomes, one for each branch, because it cannot decide the condition. Execution of the conditional starts after the current time on the selected locations and after the condition is available. Then either of the branches starts after the time taken for executing the conditional. The arrival environment does not change.

A transfer expression (rule (T-S-TRANS)) does not change the clock environment because transfers happen asynchronously. Sending from a location ℓ' starts right after the current time and after the variable's value is available. It arrives at location ℓ with a delay specified by the transfer cost $\mathbf{T}_{\ell',\ell}$.

The rule (T-S-JUMP) models a function call by setting the arrival times of the formal parameters according to the actual parameters. Evaluation proceeds with the body of the called function.

3   Placement Analysis

A timed configuration includes placements of operations and transfers. These placements must be *valid*, *i.e.*, satisfy restrictions to ensure that all arguments of an operation scheduled to run at location ℓ are eventually present at ℓ.

This section develops a constraint-based placement analysis for IL_S that identifies all valid placements for a program.[1] Then we prove that finding a solution for the resulting constraints is an NP-complete problem.

Availabilities. A name, $M, N, \cdots \in \mathcal{N}$, represents a fixed *set of locations*, $\mathcal{L}(M) \in \mathcal{P}(\mathcal{L})$. A *location set variable* $\alpha, \beta, \cdots \in$ LSVar ranges over a set of locations. An *availability*, Avail $\ni A ::= \alpha \mid N$, specifies possible locations for a value of a base type. A *function availability*, FAvail $\ni \overline{A} \xrightarrow{A',A''} 0$, states that the arguments of a function must be available on locations $\overline{A}$, at least locations A' are needed to run the function, and effects may be visible on locations A''.

An *availability operation placement*, $\Xi_{op} \in$ Op $\hookrightarrow$ SVar $\hookrightarrow$ Avail, associates each occurrence of an operator with a availability, which may be prescriptive by specifying a set of locations or which may just associate a location set variable with the occurrence. An *availability transfer placement*, $\Xi_{tr} \in$ SVar $\hookrightarrow$ Var $\hookrightarrow$ Avail $\times$ Avail, similarly associates each variable occurrence with a pair of availabilities. A *function availability assumption*, $\Delta \in$ FVar $\hookrightarrow$ FAvail, connects a function declaration with its uses. An *availability assumption*, $\Gamma \in$ Var $\hookrightarrow$ Avail, associates a variable with its availability.

Constraints. Constraints, $C ::= A \leqslant A \mid A \dashrightarrow A \mid op \triangleright A \mid \mathsf{single}\, A \mid A \overset{?}{=} A$, express demands on availabilities: A is a subtype of A' ($A \leqslant A'$), a transfer from any location in A to any location in A' is possible ($A \dashrightarrow A'$), operation op is available on A ($op \triangleright A$), A represents a single location ($\mathsf{single}\, A$), and A and A' are equal ($A \overset{?}{=} A'$). We write $\mathcal{C}$ for a set of constraints.

---

[1] The first author's dissertation [20] includes a specification of the analysis and proves its equivalence with the constraint-based variant.

$$\Xi_{op}; A \vdash op^s \mid \mathcal{C}, \alpha \qquad\qquad \text{operator placement access}$$
$$\Gamma \vdash x : \alpha \mid \mathcal{C} \qquad\qquad\qquad \text{variable availability access}$$
$$\Xi_{op}; \Xi_{tr}; \Delta; \Gamma; A \vdash e \,!\, \alpha' \mid \mathcal{C} \;\; \text{expression availabilities}$$

C-S-POP
$$\frac{\Xi_{op}(\mathbf{pfun})(s) = \alpha \qquad \mathcal{C} \supseteq \{A_c \leqslant \alpha, \mathbf{pfun} \triangleright \alpha\}}{\Xi_{op}; A_c \vdash \mathbf{pfun}^s \mid \mathcal{C}, \alpha}$$

C-S-EOP
$$\frac{\Xi_{op}(op)(s) = \alpha \qquad \mathcal{C} \supseteq \{A_c \leqslant \alpha, op \triangleright \alpha, \mathsf{single}\,\alpha\}}{\Xi_{op}; A_c \vdash op^s \mid \mathcal{C}, \alpha}$$

C-S-VAR
$$\frac{\Gamma(x) = A' \qquad \mathcal{C} \supseteq \{A' \leqslant \alpha\}}{\Gamma \vdash x : \alpha \mid \mathcal{C}}$$

C-S-HALT
$$\frac{\alpha' \text{ fresh}}{\Xi_{op}; \Xi_{tr}; \Delta; \Gamma; A \vdash \mathbf{halt}^s \,!\, \alpha' \mid \emptyset}$$

C-S-LET
$$\frac{\Xi_{op}; A \vdash op^s \mid \mathcal{C}, \alpha_0 \qquad \mathcal{C} \supseteq \{\alpha' \leqslant \alpha_0\} \quad (\forall i \in [m]) \; \Gamma \vdash x_i : \alpha_0 \mid \mathcal{C} \quad \Xi_{op}; \Xi_{tr}; \Delta; \Gamma, x_0 : \alpha_0; A \vdash e \,!\, \alpha' \mid \mathcal{C}}{\Xi_{op}; \Xi_{tr}; \Delta; \Gamma; A \vdash \mathbf{let}\; x_0 = op(\overline{x}) \;\mathbf{in}\; e \,!\, \alpha' \mid \mathcal{C}}$$

C-S-IF
$$\frac{\begin{array}{c}\Xi_{op}; A \vdash \mathbf{if}^s \mid \mathcal{C}\Gamma \vdash x : \alpha_0 \mid \mathcal{C} \\ (\forall i \in [2]) \; \Xi_{op}; \Xi_{tr}; \Delta; \Gamma, x : \alpha_0; A \vdash e_i \,!\, \alpha_i \mid \mathcal{C} \\ \mathcal{C} \supseteq \{\alpha_0 \leqslant \alpha_1, \alpha_0 \leqslant \alpha_2, \alpha' \leqslant \alpha_0\} \qquad \alpha', \alpha_1, \alpha_2 \text{ fresh}\end{array}}{\Xi_{op}; \Xi_{tr}; \Delta; \Gamma; A \vdash \mathbf{if}\; x \;\mathbf{then}\; e_1 \;\mathbf{else}\; e_2 \,!\, \alpha' \mid \mathcal{C}}$$

C-S-TRANS
$$\frac{\begin{array}{c}\Gamma \vdash x_i : \alpha_i \mid \mathcal{C} \\ \Xi_{op}; \Xi_{tr}; \Delta; \Gamma, \overline{x_i : \alpha_i'}; A \vdash e \,!\, \alpha' \mid \mathcal{C} \\ (\forall i \in [m]) \; (\Xi_{tr}(s)(x_i) = (\alpha_i, \alpha_i') \\ \Rightarrow \mathcal{C} \supseteq \{\alpha_i \dashrightarrow \alpha_i'\})\end{array}}{\Xi_{op}; \Xi_{tr}; \Delta; \Gamma; A \vdash \mathbf{trans}^s\; \overline{x} \;\mathbf{in}\; e \,!\, \alpha' \mid \mathcal{C}}$$

C-S-JUMP
$$\frac{\begin{array}{c}\Delta(f) = \overline{\alpha_i} \xrightarrow{\alpha'', \alpha'''} 0 \\ (\forall i \in [m]) \; \Gamma \vdash x_i : \alpha_i \mid \mathcal{C} \\ \mathcal{C} \supseteq \{A \leqslant \alpha'', \alpha''' \overset{?}{=} \alpha'\} \qquad \alpha' \text{ fresh}\end{array}}{\Xi_{op}; \Xi_{tr}; \Delta; \Gamma; A \vdash f^s\; \overline{x} \,!\, \alpha' \mid \mathcal{C}}$$

Fig. 4. Availability rules for IL_S

An *availability substitution* $\sigma \in \mathsf{LSVar} \hookrightarrow \mathcal{P}(\mathcal{L})$ substitutes a set of locations for each location set variable. It extends pointwise to availabilities, function availabilities, placements, and assumptions. It applies $\mathcal{L}(\cdot)$ to each name.

Definition 1. *Given a location mapping for operations $\Sigma \in \mathsf{Op} \hookrightarrow \mathcal{P}(\mathcal{L})$ that specifies possible locations for operations, and a transfer matrix $\Theta \subseteq \mathcal{L} \times \mathcal{L}$ that represents all possible transfer paths between locations, an availability substitution σ is a (Σ, Θ)-solution of a constraint set $\mathcal{C}$, written $\sigma \models_{\Sigma, \Theta} \mathcal{C}$, if for all $C \in \mathcal{C}$, it holds that*

   (i) if $C \equiv A \leqslant A'$, then $\sigma(A) \supseteq \sigma(A')$,
   (ii) if $C \equiv A \dashrightarrow A'$, then $(\forall \ell \in \sigma(A), \ell' \in \sigma(A')) \; \ell \rightsquigarrow \ell' \in \Theta$,
   (iii) if $C \equiv op \triangleright A$, then $\Sigma(op) \supseteq \sigma(A)$,
   (iv) if $C \equiv \mathsf{single}\,A$, then $|\sigma(A)| = 1$, and
   (v) if $C \equiv A \overset{?}{=} A'$, then $\sigma(A) = \sigma(A')$.

Constraint Availability Rules. Fig. 4 specifies the judgments and rules for constraint generation. Given a program, p, and preassigned availabilities for functions, variables and static placements, Δ, Γ, Ξ_{op}, and Ξ_{tr}, the system generates a set of constraints, which expresses validity conditions.

The (C-S-POP) rule considers a pure operation **pfun** at placement label s with available locations A_c. The (C-S-EOP) rule considers an effectful operation **op** at placement label s with available locations A_c. It extends the rule (C-S-POP) by assuring that operation **op** is placed on exactly one location. The (C-S-VAR) rule generates a subtyping constraint following the variable lookup.

Rule (C-S-HALT) places no restriction. Rule (C-S-LET) ensures that all arguments for an operation placed according to α_0 are present at all locations in α_0 and that the body of the let can also execute at least on α_0. Rule (C-S-IF) ensures that the condition is available at α_0 where the conditional is supposed to execute and that these locations also support execution of the branches of the conditional. Rule (C-S-TRANS) states that the availability of variable values must match the transfer facilities specified in Ξ_{tr}. Rule (C-S-JUMP) matches the argument placement with the placement required by the function's type. It furthermore threads the currently running locations A through the function's effect. Neubauer [20] presents the remaining judgments and rules.

Solving Constraints and Complexity. Neubauer's PhD thesis [20, chapter 4.2] presents and proves sound and complete a constraint simplification procedure that transforms a set of constraints C to a substitution that solves C. In general, there may be more than one solution.

It remains to assess the complexity of constraint solving. The execution of a program without effectful operations cannot be observed, thus the trivial placement that allows no transfers and no computation is valid and even optimal. For a program with effectful operations, determining the set of valid placements is an NP-complete problem if there are more than two distinct locations.

Theorem 1. *If $n_l > 2$ then the problem of determining the valid placements of a program, p, with effectful operations is NP-complete.*

Proof. The problem is obviously in NP. It remains to show that it is NP-hard by giving a polynomial-time Karp-reduction [4] from the problem of checking the n_l-colorability of a graph (an NP-complete problem for $n_l > 2$ [12]) to the computation of valid placements for an IL_S program.

Let $G = (V, E)$ be a graph with $|V| = n$ and $E = \{(v_{i_1}, v_{j_1}), \ldots, (v_{i_e}, v_{j_e})\}$. From this graph, we construct a program as prescribed in Fig. 5. It can be shown that each valid placement for p corresponds to a n_l-coloring of G, thus proving our claim. Details may be found in Neubauer's thesis [20].

Informally, the parameters of the top-level function f simulate the vertices of the graph. A coloring of a vertex by one of the n_l colors is emulated by searching a location assignment for the corresponding parameter. For each edge between two vertices, we "connect" the parameters by introducing a data transfer between them. The requirement that two adjacent vertices have to be colored differently is

$$p \equiv f(x_0, x_1, \ldots, x_n) = D_{i_1, j_1}[D_{i_2, j_2}[\ldots [D_{i_e, j_e}[\mathbf{halt}]]]]$$

$$\mathbf{let}\ y_0 = \mathbf{pfun}^{s_0}()\ \mathbf{in}$$
$$\mathbf{let}\ y_1 = \mathbf{pfun}^{s_1}()\ \mathbf{in}$$
$$\vdots$$
$$\mathbf{let}\ y_n = \mathbf{pfun}^{s_n}()\ \mathbf{in}$$
$$f^{s_{n+1}}(y_0, y_1, \ldots, y_n)$$

where for each edge $(v_i, v_j) \in E$ the contexts $C_{i,j}$ and $D_{i,j}$ are defined by

$$C_{i,j}[\cdot] \equiv \mathbf{let}\ x_j' = \mathbf{op}^{s_{i,j}}(x_i)\ \mathbf{in}$$
$$\mathbf{trans}^{s_{i,j}'}\ x_j'\ \mathbf{in}$$
$$\mathbf{if}^{s_{i,j}''}\ x_0\ \mathbf{then}\ f^{s_{i,j}'''}(x_0, x_1, \ldots, x_i, \ldots, x_j', \ldots, x_n)\ \mathbf{else}\ [\cdot]$$

$$D_{i,j}[\cdot] \equiv C_{i,j}[C_{j,i}[\cdot]]$$

Fig. 5. Program constructed from a graph

M-S-Trans

$$\mathbf{S} = (\mathbf{S}^{k,l})_{k \in [n_s], l \in [n_v]}$$
$$\mathbf{D} = \mathrm{diagB}(\mathbf{I}, \mathbf{B}_1, \ldots, \mathbf{B}_{n_v})$$
$$\mathbf{B}_l = \begin{cases} \mathbf{S}^{k,l} \bullet \mathbf{T} & \text{if } x_l \in \bar{x} \\ \mathbf{I} & \text{otherwise} \end{cases}$$

$$\overline{c; \mathbf{T}; o; \mathbf{S} \mid \mathbf{d}, \bar{d}\ \mathbf{trans}^k\ \bar{x}\ \mathbf{in}\ e \longrightarrow_m \mathbf{D} \otimes \mathbf{d}, \bar{d}\ e}$$

Fig. 6. Matrix-based reduction rule for the transfer expression

enforced by choosing a transfer cost function which only allows transfers between *different* locations.

4   Properties of Placement

The calculation of an optimal placement amounts to finding the placement which minimizes the *worst-case run time* of an IL_S program. This section presents a method to calculate the worst-case run time by reformulating the timed semantics in terms of matrices over a dioid and then applying results of the spectral theory of such matrices. The difficulty here is to assign a meaningful notion of worst-case run time also to potentially non-terminating programs, which is where the spectral theory kicks in.

The first step is the reformulation of the timed semantics. It is now defined by a judgment $c; \mathbf{T}; o; \mathbf{S} \mid \mathbf{d}, \bar{d}\ e \longrightarrow_m \mathbf{d}', \bar{d}\ e'$ which relates two evaluation states of the form $\mathbf{d}, \bar{d}\ e$. The *environment vector* $\mathbf{d} \in \mathcal{D}^{n_l + n_v n_l}$ combines the clock and arrival environments of the previous formulation in a vector where the first n_l entries correspond to ϑ_c and the remaining entries to ϑ_v.

Fig. 6 shows the rule (M-S-TRANS) for transfers as an example. It corresponds to rule (T-S-TRANS) from Fig. 3. The rule obtains the new environment vector by multiplying $\mathbf{d}$ with a block diagonal matrix $\mathbf{D}$ where the upper left block is $\mathbf{I} \in \mathcal{D}^{n_l \times n_l}$ to copy the ϑ_c-part unchanged and the remaining blocks $\mathbf{B}_l$ compute the new arrival times of all variables by adding the transfer times from $\mathbf{T}$ to the variable availabilities as determined by the placement through $\mathbf{S}^{k,l}$. The operation $\bullet$ stands for Hadamard (pointwise) multiplication of matrices.

In fact, each transition of the timed semantics can be expressed as a multiplication of the environment vector with a suitable transition matrix $\mathbf{D}$ that depends only on $\mathbf{c}$, $\mathbf{T}$, $\mathbf{o}$, and $\mathbf{S}$ [20, Lemma 4.33 and 4.34]. These matrices completely describe the transition behavior of the program:

Lemma 1. *There exists a finite family of matrices,* $(\mathbf{D}_{i,j})_{i,j}$, *such that*

$$\mathbf{c}; \mathbf{T}; \mathbf{o}; \mathbf{S} \mid \mathbf{d}, \overline{d}\ ^{s_j} e' \longrightarrow_m \mathbf{d}', \overline{d}\ ^{s_i} e'' \qquad \textit{iff} \qquad \mathbf{D}_{i,j} \neq \mathbf{Z} \wedge \mathbf{d}' = \mathbf{D}_{i,j} \otimes \mathbf{d}$$

where $i, j \in [n_s]$ *range over the indices of program expressions.*

We set $\mathbf{D}_{i,j} = \mathbf{Z}$ if there is no transition from expression j to expression i. The *transition matrix* $\mathbf{M} = \mathbf{M}(p, \mathbf{o}, \mathbf{S}) \in \mathcal{D}^{n_l n_s \times n_l n_s}$ combines this family of matrices into one big block matrix where the blocks are defined by $\mathbf{M}_{i,j} = \mathbf{D}_{i,j}$. This matrix encompasses the behavior of the entire program and it depends on $\mathbf{o}$ and $\mathbf{S}$.

Theorem 2. *Let* $\mathbf{M} = \mathbf{M}(\overline{d}\ e, \mathbf{o}, \mathbf{S})$ *be the transition matrix. Then*

$$\mathbf{c}; \mathbf{T}; \mathbf{o}; \mathbf{S} \mid \mathbf{d}, \overline{d}\ ^{s_j} e' \longrightarrow_m \mathbf{d}', \overline{d}\ ^{s_i} e'' \qquad \textit{iff} \qquad \mathbf{D} \neq \mathbf{Z} \wedge \mathbf{d}' = \mathbf{D} \otimes \mathbf{d}$$

where $\mathbf{A}_j = \begin{bmatrix} \mathbf{Z} \cdots \underset{j}{\mathbf{I}} \cdots \mathbf{Z} \end{bmatrix}^T$, $\mathbf{B}_i = \begin{bmatrix} \mathbf{Z} \cdots \underset{i}{\mathbf{I}} \cdots \mathbf{Z} \end{bmatrix}$ *and* $\mathbf{D} = \mathbf{B}_i \otimes \mathbf{M} \otimes \mathbf{A}_j$.

Hence, n-fold multiplication with M corresponds to n reduction steps. Thus, we may compute the n-step worst-case duration as the maximum of the clocks and availabilities at any location after n steps starting from expression s_j as follows:

Definition 2. *The* n-step worst-case duration *of program* p *from* s_j *is*

$$d_n^{s_j, p, \mathbf{o}, \mathbf{S}} = [1, \ldots, 1] \otimes \mathbf{M}(p, \mathbf{o}, \mathbf{S})^n \otimes \begin{bmatrix} \mathbf{Z} \cdots \underset{j}{\mathbf{I}} \cdots \mathbf{Z} \end{bmatrix}^T.$$

Worst-Case Mean Duration. Most programs do not terminate under the timed semantics, so that the n-step worst-case duration grows without bound as $n \to \infty$. However, the *mean duration* of a single step taken over all finite path segments is an indicator for the expected performance of a program that makes sense both for terminating and for nonterminating programs. The worst-case behavior under this indicator is the *maximal mean duration* of all paths. In keeping with the (max, +)-notation, we write $d^{1/n}$ for the division of a real number d by a natural number n.

Definition 3. *The* worst-case mean duration *of* p *from* s_i *with respect to* $\mathbf{M}$ *is defined as* $\rho^{s_i, p, \mathbf{o}, \mathbf{S}} = \lim \sup_n (d_n^{s_i, p, \mathbf{o}, \mathbf{S}})^{1/n}$.

This definition happens to coincide with the definition of the *maximum cycle mean* for a transition matrix. According to the $(\max, +)$ spectral theory [6] [13, Lemma 1], the following quantities equally describe the maximum cycle mean, $\rho(\mathbf{A})$, of a matrix, $\mathbf{A}$:

Lemma 2. *For all* $\mathbf{A} \in \mathbb{R}^{n \times n}_{(\max,+)}$: $\limsup_k ||\mathbf{A}^k||^{1/k} = \bigoplus_{1 \leq k \leq n} (trace(\mathbf{A}^k))^{1/k}$.

As the maximum cycle mean of such a matrix is computable, we can calculate the worst-case mean duration of a program with a fixed placement.

Theorem 3. *If p is a program with transition matrix $\mathbf{M}$, then its worst-case mean duration from s_i is either* $\rho^{s_i, p, \mathbf{o}, \mathbf{S}} = \mathbf{0}$ *or* $= \bigoplus_{1 \leq k \leq n_l n_s} (trace(\mathbf{M}^k))^{1/k}$.

Optimal Placements. The constraint-based placement analysis of Section 3 gives us a way to find *all* valid placements for a certain program. The last step is to statically determine one valid placement that minimizes the worst-case mean duration among all valid placements.

Definition 4 (Optimal Placements). *Let p be a program and $\mathbf{c}, \mathbf{T}$ the associated cost matrices. Let V be the set of valid placements with respect to $\mathbf{c}, \mathbf{T}$:*[2]

$$V = \{(\mathbf{o}, \mathbf{S}) \mid \Xi_{op}; \Xi_{\mathbf{tr}}; \Delta; \Gamma; \alpha \vdash p \,!\, \alpha' \mid \mathcal{C}, (\exists \sigma) \; \sigma \models_{\Sigma, \Theta} \mathcal{C}, \mathbf{o} \sim \sigma \Xi_{op}, \mathbf{S} \sim \sigma \Xi_{\mathbf{tr}}\}$$

$(\mathbf{o}, \mathbf{S}) \in V$ *is an* optimal placement *if* $\rho^{s_1, p, \mathbf{o}, \mathbf{S}} \leq \rho^{s_1, p, \mathbf{o}', \mathbf{S}'}$, *for all* $(\mathbf{o}', \mathbf{S}') \in V$.

The set of valid placements, V, depends on calculation costs, $\mathbf{c}$, transfer costs, $\mathbf{T}$, and a program, p. A placement, $(\mathbf{o}, \mathbf{S})$, is valid if there exist availability assumptions and constraints generated by the constrained-based availability analysis and the constraints are solvable by an availability substitution.

Corollary 1. *For each program, p, operation costs, $\mathbf{c}$, and transfer costs, $\mathbf{T}$, the set of optimal placements of p is computable.*

Implementation. There is a prototype implementation of the placement analysis for IL_S, which determines an optimal placement from calculation costs and transfer costs. Our algorithm is a variation of the *backtracking integer optimizer* of Marriott and Stuckey [18, Chapter 3, Figure 3.19].

We encode placement inference as a boolean satisfaction problem and determine solutions (valid placements) with a backtracking algorithm. The valuation of a valid placement computes the maximum cycle mean using an adaption of Howard's algorithm for solving Markov decision processes. In practice [11], this algorithm is reported to run in quadratic time in the length of the input program.

Still, enumerating all solutions of a placement problem may take time exponential in the length of the input program. Our implementation checks consistency at each choice point to abort the search as soon as the constraint set is no longer solvable. The implementation avoids multiple traversals of the search tree by interleaving the search for an optimal placement with the search for the next solution. We additionally minimize the need to calculate the maximum cycle mean by using the best intermediate solution as additional search bound. For sample program, our implementation finds valid placements instantly.

---

[2] $\mathbf{o} \sim \sigma \Xi_{op}$ and $\mathbf{S} \sim \sigma \Xi_{\mathbf{tr}}$ denote the correspondence between matrices and relations.

5 Related Work and Conclusion

There are extensions of process algebras and automata with timed semantics, which use absolute time values or time intervals (*e.g.* Chen's thesis on Timed CCS [8]). Hybrid system, typically modeled by hybrid automata, also allow to model dynamic behavior, in particular timing, by considering time constraints on transitions [14]. Other approaches add time constraints to automata using real-valued clocks: Alur and Dill's timed automata [2], Lynch's MMT timed automata [17] [16, Chapter 23], and others.

Our model is similar to weighted automata [15]. We rely on the $(\max, +)$ algebra as a time domain, because we are only interested in worst-case run time. The $(\max, +)$ algebra and related systems are studied in the context of discrete event systems [6]. Originally, this theory models concurrent computations with timed Petri nets. Our model is mainly influenced by Gaubert's work on non-deterministic finite automata with duration functions over $(\max, +)$ [13]. Buchholz and Kemper [7] have developed a notion of weak bisimulation for $(\max, +)$ automata.

Many static analysis problems can be solved by reduction to constraint satisfaction problems, *e.g.*, type inference for simply typed languages [5] or for calculi with subtyping [23] and many others.

The constraint language of our location analysis is a restricted form of set constraints [1]. In general, constraint problems are instances of constraint programming [18,3]. General methodologies to characterize and examine the tractability of constraint-satisfaction problems are considered, among others, by Cooper et al. [10].

To conclude, we have shown that placement inference is feasible and that it yields useful (though not optimal) results instantly, despite the NP-completeness of the underlying problem. These results encourage further experimentation with the analysis. We plan to integrate it in a compiler that implements the splitting transformation of our previous work [21] and evaluate its performance with realistic programs. It also remains to see if the predicted timing behavior corresponds with actual measurements.

References

1. Aiken, A., Wimmers, E.L.: Solving systems of set constraints. In: Proc. 1992 IEEE Symposium on Logic in Computer Science. IEEE Computer Society Press, Los Alamitos (1992)
2. Alur, R., Dill, D.L.: A theory of timed automata. Theoretical Computer Science 126(2), 183–235 (1994)
3. Apt, K.R.: Explaining constraint programming. In: Middeldorp, A., van Oostrom, V., van Raamsdonk, F., de Vrijer, R.C. (eds.) Processes, Terms and Cycles. LNCS, vol. 3838, pp. 55–69. Springer, Heidelberg (2005)
4. Ausiello, G., Crescenzi, P., Gambosi, G., Kann, V., Marchetti-Spaccamela, A., Protasi, M.: Complexity and Approximation: Combinatorial Optimization Problems and Their Approximability Properties. Springer, New York (1999)
5. Baader, F., Snyder, W.: Unification theory. In: Robinson, A., Voronkov, A. (eds.) Handbook of Automated Reasoning, ch.8, vol. I, pp. 445–534. Elsevier Science, Amsterdam (2001)

6. Baccelli, F., Cohen, G., Olsder, G.J., Quadrat, J.-P.: Synchronization and Linearity, An Algebra for Discrete Event Systems. John Wiley and Sons, Chichester (1992)
7. Buchholz, P., Kemper, P.: Weak bisimulation for (max/+) automata and related models. Journal of Automata, Languages and Combinatorics 8(2), 187–218 (2003)
8. Chen, L.: Timed Processes: Models, Axioms and Decidabilty. PhD thesis, University of Edinburgh (1992)
9. Cooper, E., Lindley, S., Wadler, P., Yallop, J.: Links: Web programming without tiers. In: Post-proceedings of FMCO 2006. LNCS, vol. 4709. Springer, Heidelberg (2006)
10. Cooper, M.C., Cohen, D.A., Jeavons, P.G.: Characterising tractable constraints. Artificial Intelligence 65(2), 347–361 (1994)
11. Dasdan, A.: Experimental analysis of the fastest optimum cycle ratio and mean algorithms. ACM Transactions on Design Automation of Electronic Systems 9(4), 385–418 (2004)
12. Garey, M.R., Johnson, D.S.: Computers and Intractability; A Guide to the Theory of NP-Completeness. W. H. Freeman & Co., New York (1990)
13. Gaubert, S.: Performance evaluation of (max,+) automata. IEEE Transactions On Automatic Control 40(12) (December 1995)
14. Henzinger, T.A.: The theory of hybrid automata. In: Proceedings of the 11th Annual Symposium on Logic in Computer Science, pp. 278–292. IEEE Computer Society Press, Los Alamitos (1996)
15. Kuich, W., Salomaa, A.: Semirings, automata, languages. Springer, London (1986)
16. Lynch, N.A.: Distributed Algorithms. Morgan Kaufmann, San Francisco (1996)
17. Lynch, N.A., Vaandrager, F.W.: Forward and backward simulations for timing-based systems. In: de Bakker, J.W., Huizing, K., de Roever, W.P., Rozenberg, G. (eds.) Proceedings REX Workshop on Real-Time: Theory in Practice, Mook, The Netherlands, June 1991. LNCS, vol. 600, pp. 397–446. Springer, Berlin (1992)
18. Marriott, K., Stuckey, P.J.: Programming with constraints: an introduction. MIT Press, Cambridge (1998)
19. Murphy VII, T., Crary, K., Harper, R.: Type-safe distributed programming with ML5. In: Trustworthy Global Computing 2007 pre-proceedings (November 2007)
20. Neubauer, M.: Multi-Tier Programming. PhD thesis, Universität Freiburg (April 2007), http://www.freidok.uni-freiburg.de/volltexte/3104/
21. Neubauer, M., Thiemann, P.: From sequential programs to multi-tier applications by program transformation. In: Abadi, M. (ed.) Proc. 32nd ACM Symp. POPL, Long Beach, CA, USA, January 2005, pp. 221–232. ACM Press, New York (2005)
22. Serrano, M., Gallesio, E., Loitsch, F.: HOP, a language for programming the Web 2.0. In: Proceedings of the First Dynamic Languages Symposium, Portland, OR, USA (October 2006)
23. Su, Z., Aiken, A., Niehren, J., Priesnitz, T., Treinen, R.: The first-order theory of subtyping constraints. In: Mitchell, J. (ed.) Proc. 29th ACM Symp. POPL, Portland, OR, USA, January 2002, pp. 203–216. ACM Press, New York (2002)

Extended pi-Calculi

Magnus Johansson, Joachim Parrow, Björn Victor, and Jesper Bengtson

Department of Information Technology, Uppsala University

Abstract. We demonstrate a general framework for extending the pi-calculus with data terms. In this we generalise and improve on several related efforts such as the spi calculus and the applied pi-calculus, also including pattern matching and polyadic channels. Our framework uses a single untyped notion of agent, name and scope, an operational semantics without structural equivalence and a simple definition of bisimilarity. We provide general criteria on the semantic equivalence of data terms; with these we prove algebraic laws and that bisimulation is preserved by the operators in the usual way. The definitions are simple enough that an implementation in an automated proof assistant is feasible.

1 Introduction

The pi-calculus [1] is a foundational calculus for describing communicating systems with dynamic connectivity. Allowing only names of communication channels to be transmitted between processes, it is still expressive enough to encode other types of data such as booleans, integers, lists etc. Such data structures are convenient when modelling protocols, programming languages, and other complicated applications. However, having to work with encodings muddles the models and complicates their analysis – and constructing correct encodings can be quite difficult. To overcome these difficulties, extensions of the pi-calculus have been introduced, where higher-level data structures (and operations on them) are given as primitive. Our contribution in this paper is to establish a unifying untyped framework where a range of such calculi can be formulated, using a lean and symmetric semantics that is well suited for an automated proof assistant.

Perhaps the simplest and oldest of data structures for the pi-calculus is to allow tuples of names to be transmitted, leading to the polyadic pi-calculus [2] and its typed variants. This is a quite mild extension where for example the agent (or process) $\overline{a}\langle b, c \rangle . P$ can send the two names b and c to a receiving $a(x, y) . Q$ in one go. This needs to be faithful to the pi-calculus principle of scope extrusion, which says that if a name is transmitted outside its scope then that scope is extended to include the receiver. If for example both b and c are scoped, as in $(\nu b, c)\overline{a}\langle b, c \rangle . P$, then the scope of both b and c are extended to encompass Q. In this way the effect is the same as sending b and c individually in subsequent transmissions.

With subsequent advances of calculi for cryptographic applications another view has emerged, where a fresh key is represented as a scoped name, and a data

L. Aceto et al. (Eds.): ICALP 2008, Part II, LNCS 5126, pp. 87–98, 2008.

structure such as enc(m, k) represents the encryption of m by the key k. Two main examples are the spi calculus [3] and the applied pi-calculus [4].

The spi calculus is equipped with encryption/decryption primitives. An example process

$$P = (\nu k, m)\overline{a}\langle \text{enc}(m, k)\rangle . P' \qquad \text{where} \quad P' = b(x) . \text{if } x = m \text{ then } \overline{c}\langle m\rangle$$

sends out, over a, the fresh name m encrypted with the fresh key k, and then receives a value x over b. If the received value is m, an output is sent on c. Intuitively, since the environment does not know k (or m), P' can never receive m, and the output on c can never happen.

However, a labelled transition for the process is $P \xrightarrow{(\nu k, m)\overline{a}\langle \text{enc}(m,k)\rangle} P'$. Although the label contains the names m and k, and therefore their scopes are opened by the transition, the names should not become known to the environment since it cannot decrypt the message and find them. Thus, any reasonable equivalence must explicitly keep track of which names are known. There are several bisimulation equivalences of varying complexity for the spi calculus – see [5] for an overview.

The applied pi-calculus [4] improves the situation in several ways. Firstly, it is more general since the calculus is parameterised by a signature and an equation system for data structures. Secondly, and more importantly, the transition labels expose only what should be revealed. This is achieved by introducing variables with so called *active substitutions* $\{^M/_x\}$ of data terms for variables, the structural rule $\{^M/_x\} \mid P \equiv \{^M/_x\} \mid P[x := M]$, and by disallowing the sending of complex data directly – it must be sent using an *alias* variable such as x. Let us return to the process P above; we have $P \equiv Q$ where

$$Q = (\nu k, m, z)(\{^{\text{enc}(m,k)}/_z\} \mid \overline{a}\langle z\rangle . P') \xrightarrow{(\nu z)\overline{a}\langle z\rangle} (\nu k, m)(\{^{\text{enc}(m,k)}/_z\} \mid P')$$

Here, the transition label only reveals that a (fresh) value is sent, but not how the value is constructed; the scope of k and m is not opened, and their confidentiality to the environment is clear; the bisimulation definition is correspondingly simple. However, sending out a tuple of channel names as in the polyadic pi-calculus is not possible in the labelled semantics of the applied pi-calculus – see Section 3.

In this paper we define a generalisation of the (applied) pi-calculus where we allow *both* possibilities: data terms can be sent using an alias, not revealing the construction of the value, or as the text or syntax tree for the term, exposing all details. The process needs to explicitly use one form or the other, unlike the case in the applied pi-calculus where a term in an object position is always implicitly rewritten using an alias. Which form to use is up to the intended purpose of the term: to keep secrets the alias form must be used, while to extrude the scope of e.g. communication channels, the explicit form should be used. P and Q are not at all equivalent, and the structural rule from the applied pi-calculus mentioned above is neither needed nor valid. It is even possible to mix the different modes by including an alias in an explicit data structure; the parts represented by the alias remain hidden.

Our new framework allows pattern matching inputs: e.g., $a(f(x,y)) \cdot P'$ can only communicate with an output $\bar{a}\langle f(M,N)\rangle$ for some data terms M, N. This can be seen as a generalisation of the polyadic pi-calculus where inputs of a certain arity can only communicate with outputs of the same arity. Another significant extension is that we allow arbitrary data terms also as communication channels. Thus it is possible to include functions that create channels and also polyadic synchronisation, cf. [6], where a channel may be composed of several names.

A reader might fear that to achieve all this the framework must be a complicated union of all work cited above, with a plethora of primitives. This is not the case. We use a single basic framework for extended pi-calculi, with a single untyped notion of agent and a single-level structural operational semantics that does not rely on any structural equivalence to rewrite agents, a single notion of name and scope, and a simple definition of bisimilarity. We provide general criteria on the semantic equivalence of data terms; with these we prove algebraic laws and that bisimulation is preserved by the operators in the usual way. Our framework facilitates comparisons between different approaches, and proofs about the calculi can be conducted using straightforward inductions over transitions. In this way we improve on existing definitions of the applied pi-calculus. The framework is lean enough that it is profitable to use an automated proof assistant and our preliminary efforts in this direction (using Isabelle [7]) hold promise.

Related Work. The different bisimulation definitions for the spi calculus are presented in [8,9,10,5] and an overview can be found in [5]. Another example of an environment sensitive equivalence can be found in [11]. In [12], a spi calculus parameterised by a data signature and evaluation function is defined. The relation between the applied pi-calculus and extended pi-calculi is explored in Section 3.

The applied pi-calculus has been used for analysis of several security protocols, e.g. [13,14,15]. Ad-hoc variants of the applied pi-calculus have also been used to prove non-security properties, e.g. in [16] correctness of network based storage is shown using a variant of an applied pi-calculus with polyadicity (but without labelled semantics).

Polyadic synchronisation in the pi-calculus [6] is a subcalculus of an extended pi-calculi as discussed in Section 3. In [17] a Spi calculus with pattern matching is presented.

Disposition. In Section 2 we present the syntax and operational semantics of extended pi-calculi. In Section 3, we illustrate by examples how these can be put to use, and relate informally to other variants of the pi-calculus. Section 4 presents the labelled bisimulation relation, and our main results about it. Section 5 concludes with suggestions for further work.

2 Definitions

Assume a signature with a set of *symbols* f, g, each with a nonnegative arity. Among the symbols there is a countably infinite set of *names* $a, b, \ldots$, all with arity 0. (There may also be other symbols with arity 0.)

Definition 1 (Terms). *The* data terms, *or for short just* terms, *ranged over by* $M, N, \ldots$, *are defined by* $M ::= f(M_1, \ldots, M_n)$ *where f has arity n.*

If f has arity 0 we write just f for $f()$; in particular a name is a term. The *substitution* of a term N for a name a in a term M, written $M[a := N]$ is defined by $a[a := N] = N$ and $f \neq a \implies f[a := N] = f$, extending homomorphically to all terms.

In the following we shall define agents, actions, transitions etc. in the style of nominal datatypes [18] building on our previous experience of nominal types for the π-calculus [19]. It is not necessary to understand nominal datatypes to appreciate the work presented here; suffice it to say that all alpha-equivalent agents, actions etc. are considered equal.

The set of free names (i.e. the names with an non-bound occurrence) of an object X is written as $\mathrm{fn}(X)$, and we abbreviate $\mathrm{fn}(X) \cup \mathrm{fn}(Y)$ to $\mathrm{fn}(X, Y)$ etc. We write $\mathrm{n}(X)$ for the names (bound and free) in X. We also write $a \# X$ ("a is fresh in X") to mean $a \notin \mathrm{fn}(X)$. Name permutation of a and b (exchanging all a for b and vice versa) in P is written $(a\,b) \bullet P$. In the following $\tilde{a}$ means a finite (possibly empty) sequence of distinct names, $a_1, \ldots, a_n$. When occurring as an operand of a set operator, $\tilde{a}$ means the corresponding set of names $\{a_1, \ldots, a_n\}$. Concatenation of sequences $\tilde{a}$ and $\tilde{b}$ is written $\tilde{a} \cdot \tilde{b}$ and the empty sequence is written ϵ.

Definition 2 (Agents and aliased names). *The* agents, *ranged over by* P, Q, *are the following:*

0	Nil	$(\nu a)P$	Restriction
$\overline{M}\,N.P$	Output	$P \vert Q$	Parallel
$M(N).P$	Input	$!P$	Replication
if $M = N$ **then** P **else** Q	Conditional	$\{^M\!/_a\}$	Alias

We say that a is the aliased name *of $\{^M\!/_a\}$, and define the* aliased names *$\mathrm{an}(P)$ of an agent P to be the set of free names aliased in any subagent in P. In the input $M(N).P$ it is required that $\mathrm{n}(N) \cap \mathrm{an}(P) = \emptyset$. Input and restriction are binding occurrences of $\mathrm{n}(N)$ and a.*

All operators except Alias are familiar from the pi-calculus (we omit the sum operator but inclusion of guarded sum would not greatly affect our results). The Alias is similar in intention to what in the applied pi-calculus is called an "active substitution", though semantically it is a bit different in that our Alias will not always enforce a substitution. An aliased name cannot be bound by input. The reason is that with an input bound alias we can express aliasing of terms for terms: $M(a).\{^N\!/_a\}$ can receive a term K to replace a, becoming $\{^N\!/_K\}$, which is not a syntactically correct agent.

Note that an aliased name is not a binder. Suppose we would restrict to a sub-calculus with a construct like "**alias** $a = M$ **in** P" to represent $(\nu a)(\{^M\!/_a\} \vert P)$, i.e., the construct binds a. In this sub-calculus it is impossible to directly express an agent such as $(\nu a)(((\nu k)(\{^{\mathrm{enc}(M,k)}\!/_a\} \vert P)) \vert Q)$. This agent could represent that a term M encrypted with key k has been sent to P and Q, and that only P has the key. So Q cannot use M until it receives k in a communication (opening the scope of k). In order to express aliases that become usable only when their "keys" are received, it seems that a binding "**alias...**" construct is not enough.

The Input construct contains an implicit pattern matching. For example $M(f(a, b)).P$ can input objects only of shape $f(K, L)$, thereby substituting K for a and L for b in P. We shall use the silent prefix $\tau.P$, which can be thought of as a shorthand for $(\nu a)(\overline{a}\, a.\mathbf{0} \mid a(c).P)$ for some $a, c \# P$, and let γ range over prefixes. A generalised restriction $(\nu\tilde{a})P$ means $(\nu a_1)\cdots(\nu a_n)P$, or just P if $n = 0$.

The *substitution* of M for a in the agent P is written $P[a := M]$, and is defined if $a \notin \mathrm{an}(P)$; in other words it is not possible to substitute something for an aliased name. Substitution is defined in the usual way, homomorphic on all operators and renaming bound names to avoid captures. When Z is a term or agent, $\tilde{a} = a_1, \ldots a_n$ are pairwise distinct and $\tilde{L} = L_1, \ldots, L_n$ we write $Z[\tilde{a} := \tilde{L}]$ to mean the simultaneous substitution of each L_i for a_i in Z.

Definition 3 (Frames). *A frame F is of kind $(\nu\tilde{a}_F)R_F$, where $\tilde{a}_F$ is a sequence of names and R_F is a finite relation between names and terms. The names in $\tilde{a}_F$ are binding occurrences in the frame. The* domain $\mathrm{dom}(F)$ *is defined to be* $\mathrm{dom}(R_F) - \tilde{a}_F$, *i.e. the domain of the relation but not including the bound names.*

The intuition is that a frame contains information about what terms should be considered equal, by relating aliases to terms. The bound names in a frame represent local placeholders, and cannot be contained in the terms to be compared for equality. Frames may be nondeterministic in that two different terms have the same alias (in that case the alias can represent either term) and even circular — we have found no reason to forbid such aliases even though in some contexts they would not make sense.

We abbreviate the empty frame $(\nu\epsilon)\emptyset$ to $\emptyset$ when no confusion can arise. We also write $(\nu a)F$ for $\nu(a \cdot \tilde{a}_F)R_F$, and $\{^T/_a\}$ for $(\nu c)\{(a, T)\}$, and $F \cup G$ for $\nu(\tilde{a}_F \cdot \tilde{a}_G)(R_F \cup R_G)$, as always alpha-converting to avoid clashes.

Definition 4 (Equal-in-Frame relation). *An* Equal-in-Frame relation *(EF-relation) is a ternary relation between a frame F and two terms M and N, written $F \vdash M = N$.*

The intuition is that $F \vdash M = N$ means that given the knowledge of the aliases in F, it is possible to infer $M = N$. We shall not now define exactly how such inferences are done; rather, we shall give a set of conditions (in Section 4 below) on the EF-relation for our results to hold.

Definition 5 (Static equivalence). *Two frames F and G are* statically equivalent, *written $F \simeq G$, when $\mathrm{dom}(F) = \mathrm{dom}(G)$ and when for all terms M and N we have $F \vdash M = N$ iff $G \vdash M = N$.*

Definition 6 (Actions). *The actions ranged over by α, β are of the following three kinds: An* output *of kind $(\nu\tilde{a})\overline{M}\, N$, where $\tilde{a}$ are binding occurrences, an* input *of kind $M\, N$, and the* silent action τ.

Table 1. Structural operational semantics. Symmetric versions are elided.

$$\text{IN} \ \frac{F \vdash M = K}{F \triangleright M(N).P \ \xrightarrow{K\,N[\tilde{a}:=\tilde{L}]} \ P[\tilde{a} := \tilde{L}]} \ \tilde{a} = n(N) \qquad \text{OUT} \ \frac{F \vdash M = K}{F \triangleright \overline{M}\,N.P \ \xrightarrow{\overline{K}\,N} \ P}$$

$$\text{THEN} \ \frac{F \vdash M = N}{F \triangleright \text{if } M = N \text{ then } P \text{ else } Q \ \xrightarrow{\tau} \ P}$$

$$\text{ELSE} \ \frac{\neg(F \vdash M = N)}{F \triangleright \text{if } M = N \text{ then } P \text{ else } Q \ \xrightarrow{\tau} \ Q}$$

$$\text{COM} \ \frac{F \cup \mathcal{F}(Q) \triangleright P \ \xrightarrow{(\nu\tilde{a})\overline{M}\,N} \ P' \qquad F \cup \mathcal{F}(P) \triangleright Q \ \xrightarrow{M\,N} \ Q'}{F \triangleright P \,|\, Q \ \xrightarrow{\tau} \ (\nu\tilde{a})(P' \,|\, Q')} \ \tilde{a}\#Q$$

$$\text{PAR} \ \frac{F \cup \mathcal{F}(Q) \triangleright P \ \xrightarrow{\alpha} \ P'}{F \triangleright P \,|\, Q \ \xrightarrow{\alpha} \ P' \,|\, Q} \ \text{bn}(\alpha) \cap \text{fn}(Q) = \emptyset$$

$$\text{SCOPE} \ \frac{F \triangleright P \ \xrightarrow{\alpha} \ P'}{F \triangleright (\nu a)P \ \xrightarrow{\alpha} \ (\nu a)P'} \ a\#\alpha, F$$

$$\text{OPEN} \ \frac{F \triangleright P \ \xrightarrow{(\nu\widetilde{b_1}\cdot\widetilde{b_2})\overline{M}\,N} \ P'}{F \triangleright (\nu a)P \ \xrightarrow{(\nu\widetilde{b_1}\cdot a\cdot\widetilde{b_2})\overline{M}\,N} \ P'} \ \begin{matrix} a \in n(N), \\ a\#M, F, \tilde{b_1}, \tilde{b_2} \end{matrix} \qquad \text{REP} \ \frac{F \triangleright P \,|\, !P \ \xrightarrow{\alpha} \ P'}{F \triangleright !P \ \xrightarrow{\alpha} \ P'}$$

For an output action $(\nu\tilde{a})\overline{M}\,N$ it will hold that $n(M) \cap \tilde{a} = \emptyset$ and $\tilde{a} \subseteq n(N)$. If $\tilde{a}$ is empty we write the action as just $\overline{M}\,N$; this corresponds to a free output in standard pi. In the actions above we will refer to M as the *subject*, corresponding to the channel over which communication takes places, and N as the *object* transferred in the communication. Note that the subject is a term and not necessarily a name. This admits functions in the signature that construct channels. When an output object N is syntactically complicated we sometimes write $\overline{M}\langle N\rangle$ for $\overline{M}\,N$, to facilitate reading, in both actions and prefixes.

The frame of an agent is, intuitively, what the agent contributes to its environment for resolution of aliases. It contains all the unguarded aliases, preserving the scope of names.

Definition 7 (Frame of an agent). *The function $\mathcal{F}$ from agents to frames is defined inductively as follows.*
$\mathcal{F}(0) = \mathcal{F}(\gamma.P) = \mathcal{F}(\text{if } M = N \text{ then } P \text{ else } Q) = \emptyset, \quad \mathcal{F}(\{^M/_a\}) = \{^M/_a\},$
$\mathcal{F}((\nu a)P) = (\nu a)\mathcal{F}(P), \quad \mathcal{F}(P \,|\, Q) = \mathcal{F}(P) \cup \mathcal{F}(Q), \quad \mathcal{F}(!P) = \mathcal{F}(P)$

Definition 8 (Transitions). *A transition is of kind $F \triangleright P \ \xrightarrow{\alpha} \ Q$, meaning that when the environment contains the frame F the agent P can do an α to become Q. The transitions are defined inductively in Table 1.*

The frame is used to determine the action subjects. So if the EF-relation equates the different terms M and N in every frame then the agents $\overline{M}\,K.P$ and $\overline{N}\,K,P$ are bisimilar. In contrast, the frame does not affect objects, so the agents $\overline{K}\,M.P$ and $\overline{K}\,N,P$ are *not* bisimilar, and can indeed be distinguished by the agent $K(a).(\overline{b}\,\langle t_2(a,M)\rangle.\mathbf{0}\mid b\,(t_2(c,c)).Q)$, where the non-linear pattern matching along b succeeds only if the corresponding output is a tuple of two identical terms. We have experimented with versions where the prefix rules, or the COM-rule, or the definition of bisimilarity (and combinations thereof) use the frame to generate or compare objects. In all cases we have encountered technical problems in that the scope extension law $(\nu a)(P\mid Q)\sim P\mid(\nu a)Q$ if $a\#P$ fails, or restriction fails to preserve bisimilarity.

3   Examples

We will now show how the extended pi-calculi relate to a few other calculi and give some examples.

The pi-calculus and the Polyadic pi-calculus. Any monadic pi-calculus agent [1] is also an extended pi-calculus agent. Using only names as symbols, and with an EF-relation with only identity on names, the extended pi semantics directly corresponds to an early operational semantics for the pi-calculus (without sum). The frame of a pi agent will always be empty since there is no aliasing in the pi-calculus.

Adding tupling symbols t_n for tuples of arity n, we can also easily encode the polyadic pi-calculus [2] in the extended pi-calculus. The encoding $[\![\cdot]\!]_{\mathrm{P2E}}$ uses pattern matching in the input rule:

$$[\![a(b_1,\ldots,b_n).P]\!]_{\mathrm{P2E}} = a(t_n(b_1,\ldots,b_n)).[\![P]\!]_{\mathrm{P2E}}$$
$$[\![\overline{a}\langle c_1,\ldots,c_n\rangle.P]\!]_{\mathrm{P2E}} = \overline{a}\langle t_n(c_1,\ldots,c_n)\rangle.[\![P]\!]_{\mathrm{P2E}}$$

Pi-calculus with Polyadic Synchronisation. In [6] the subject of an action can be a vector of names. This is useful if we want a couple of processes to atomically communicate only if they share a set of parameters. Example applications of polyadic synchronisation given in [6] are modelling e-services [20] where a client and a server can only communicate if they agree on a set of service parameters, and simple representations of localities and cryptography. Using polyadic synchronisation we can gradually enable a communication by opening the scope of names in a subject.

Using tuples of names as subjects in an extended pi-calculus we gain the expressiveness of polyadic synchronisation (strictly greater than standard pi-calculus). With the tupling symbols above, we can, e.g., encode **if** $a = b$ **then** P **else 0** as $(\nu c)(\overline{t_2(c,a)}\langle c\rangle\mid \underline{t_2(c,b)}(d).P$ where $c,d\#P$, and the underlining is intended to clarify the input subject.

The Applied pi-Calculus. Our work is inspired by the applied pi-calculus [4]. For the most part the applied pi-calculus is a subset and can be translated directly into an extended pi-calculus, but there are a few noteworthy issues. Firstly, in the applied pi-calculus there is a distinction between variables and names (collectively called *atoms*) where only variables can be substituted. Secondly, as seen in Section 1, in the applied pi-calculus there are limitations on what can be output in an output action. For example the agent $\overline{a}\,M.P$, where M is not an atom, has no transitions since the only allowed output actions are of the form $\overline{a}\,u$ or $(\nu u)\overline{a}\,u$ where u is an atom. In order to derive a transition from $\overline{a}\,M.P$ it must first be rewritten using rules for structural equivalence: $\overline{a}\,M.P \equiv (\nu x)(\{M/x\} \mid \overline{a}\,x.P) \xrightarrow{(\nu x)\overline{a}\,x} \{M/x\} \mid P$. In a similar fashion names are the only subjects in the applied pi-calculus. Any other term in subject position in a prefix must be rewritten to a name using structural equivalence before a transition can be derived.

Assuming an EF-relation which, as in the applied pi-calculus, is based on a set of user supplied equations, given an agent A in the applied pi-calculus, let $[\![A]\!]_{\text{A2E}}$ be a translation into an extended pi-calculus. $[\![\cdot]\!]_{\text{A2E}}$ maps variables to names and is a homomorphism except for the output prefix which is translated as follows:

$$[\![\overline{u}\,M.P]\!]_{\text{A2E}} = (\nu a)(\{M/a\} \mid \overline{u}\,a.[\![P]\!]_{\text{A2E}}),\ a \notin atoms(\overline{u}\,M.P)\ (M \text{ not an atom})$$
$$[\![\overline{u}\,M.P]\!]_{\text{A2E}} = \overline{u}\,M.[\![P]\!]_{\text{A2E}} \quad (\text{when } M \text{ is an atom})$$

In the applied pi-calculus we have that $\overline{u}\,M.P \equiv (\nu x)(\{M/x\} \mid \overline{u}\,x.P)$ where $x \notin atoms(\overline{u}\,M.P)$, which exactly matches our translation. Assuming a well sorted original process and well sorted inputs, this translation gives rise to a process that has the same transitions as the original, except for outputs of names that are not channels, which in applied pi-calculus can be output both literally and as aliases. To handle this in an extended pi-calculus we would need a choice operator.

As indicated in the introduction, we can express cryptographic operations, given an appropriate signature and EF-relation. Using aliases for cryptographic terms, we can protect the secrecy of keys and encrypted messages. In contrast to the applied pi-calculus, we also have the possibility to send data terms *as is*, without using aliases. This is important to be able to directly represent the polyadic pi-calculus. Consider an example based on our encoding of polyadic pi above, where z is a variable in the applied pi-calculus represented as a name in our framework:

$$P = [\![(\nu c, d)\overline{a}\langle c, d\rangle.c(z).P']\!]_{\text{P2E}} = (\nu c, d)\overline{a}\langle \mathsf{t}_2(c, d)\rangle.c(z).P''$$

In an extended pi-calculus, $P \xrightarrow{(\nu c,d)\overline{a}\langle \mathsf{t}_2(c,d)\rangle} c(z).P''$ and $c(z).P'' \xrightarrow{c\,M} P''[z := M]$ for some M. In the applied pi-calculus, to have a transition, P must first be rewritten to $Q = (\nu c, d, x)(\{\mathsf{t}_2(c,d)/x\} \mid \overline{a}\,x\,.\,c(z)\,.\,P'')$. This rewritten process can, however, not open the scopes of c and d, but has the only transition $Q \xrightarrow{(\nu x)\overline{a}\,x}$ $(\nu c, d)(\{\mathsf{t}_2(c,d)/x\} \mid c(z)\,.\,P'')$. This process is deadlocked.

The spi calculus. If we assume an EF-relation based on an equation system and a signature with primitives for encryption and decryption, similarly to what we did for the applied pi-calculus above, we can encode the spi calculus [3] using the translation used above for the applied pi-calculus. Note in particular that the encoding ensures that encryptions are always sent as aliases.

4   Theory

This section gives our main results: our labelled bisimulation is preserved by the operators in the usual way, and although we do not use structural congruence in our operational semantics, the usual structural rules are sound.

Definition 9 (Bisimulation). *A bisimulation $\mathcal{R}$ is a ternary relation between frames and pairs of agents such that $\mathcal{R}(F, P, Q)$ implies all of*

1. *Static equivalence: $\mathcal{F}(P) \simeq \mathcal{F}(Q)$*
2. *Symmetry: $\mathcal{R}(F, Q, P)$*
3. *Extension of arbitrary frame: $\forall G. \; \mathcal{R}(F \cup G, P, Q)$*
4. *Simulation: for all α, P' such that the bound names in α are disjoint from $n(Q, F)$ there exists Q' such that*

$$F \triangleright P \xrightarrow{\alpha} P' \quad \Longrightarrow \quad F \triangleright Q \xrightarrow{\alpha} Q' \;\; \wedge \;\; \mathcal{R}(F, P', Q')$$

We define $P \sim Q$ to mean that there exists a bisimulation $\mathcal{R}$ such that $\mathcal{R}(\emptyset, P, Q)$.

Because of the universal quantification of G in clause 3, bisimilarity corresponds to equal behaviour in all frames.

In order to establish properties of $\sim$ we will need to make assumptions about the EF-relation.

Equivariance: $F \vdash M = N \implies (a\,b) \bullet F \vdash (a\,b) \bullet M = (a\,b) \bullet N.$
Equivalence: $\{(M, N) : F \vdash M = N\}$ is an equivalence relation.
Strengthening: $a \# (F, M, N) \wedge F \cup \{^T/_a\} \vdash M = N \implies F \vdash M = N.$
Weakening: $F \vdash M = N \implies F \cup \{^T/_a\} \vdash M = N.$
Scope Introduction: $F \vdash M = N \wedge a \# (M, N) \implies (\nu a) F \vdash M = N.$
Scope Elimination: $(\nu a) F \vdash M = N \implies F \vdash M = N.$
Idempotence: $F \simeq F \cup F$
Union: $F \simeq G \implies F \cup H \simeq G \cup H.$

Weakening and strengthening say that no fewer equalities can be proved by adding to the frame or by subtracting an alias that is never used. Scope introduction and elimination similarly say that by removing a scope no less can be proved (here the bound name a is guaranteed to be distinct from anything in M or N), and likewise for adding a scope that does not capture anything in M and N. Note that a property for restriction, $F \simeq G \implies (\nu a)F \simeq (\nu a)G$, follows from scope introduction and elimination.

Table 2. Structural rules

$$P \equiv P \,|\, \mathbf{0} \qquad\qquad (\nu a)\mathbf{0} \equiv \mathbf{0}$$
$$P \,|\, (Q \,|\, R) \equiv (P \,|\, Q) \,|\, R \qquad P \,|\, (\nu a)Q \equiv (\nu a)(P \,|\, Q) \quad \text{if } a\#P$$
$$P \,|\, Q \equiv Q \,|\, P \qquad\qquad \overline{M} \, N.(\nu a)P \equiv (\nu a)\overline{M} \, N.P \quad \text{if } a\#M,N$$
$$!P \equiv P \,|\, !P \qquad\qquad M(N).(\nu a)P \equiv (\nu a)M(N).P \quad \text{if } a\#M,N$$

$$\textbf{if } M = N \textbf{ then } P \textbf{ else } (\nu a)Q \equiv (\nu a)(\textbf{if } M = N \textbf{ then } P \textbf{ else } Q) \quad \text{if } a\#M,N,P$$
$$\textbf{if } M = N \textbf{ then } (\nu a)P \textbf{ else } Q \equiv (\nu a)(\textbf{if } M = N \textbf{ then } P \textbf{ else } Q) \quad \text{if } a\#M,N,Q$$

$$(\nu a)(\nu b)P \equiv (\nu b)(\nu a)P \qquad (\nu a)\{^M\!/_a\} \equiv \mathbf{0} \qquad \{^M\!/_a\} \equiv \{^N\!/_a\} \text{ if } \emptyset \vdash M = N$$

With these conditions on EF we can prove the following:

Theorem 1. *If $P \equiv Q$ from the structural laws in Table 2 then $P \sim Q$.*

Bisimilarity is preserved by all operators except input. Interestingly, the reason that it is not preserved by input is not the same as in the ordinary pi-calculus. Let $a \neq b$ and consider the agent $P = \textbf{if } a = b \textbf{ then } P' \textbf{ else } \mathbf{0}$; in the ordinary pi-calculus (where this **if** is represented by a match) we have $P \sim \mathbf{0}$ and $c\,(a).P \not\sim c\,(a).\mathbf{0}$, since the input may instantiate a to b. But in our framework we do *not* have that $P \sim \mathbf{0}$. The reason is that a frame can identify a and b, as in $\{^a\!/_b\} \rhd P \xrightarrow{\tau} P'$. Instead, the counterexample is

$$P = (\nu a)(\overline{a}\, b\,.\,\mathbf{0} \mid a\,(f)\,.\,\mathbf{0})$$

where f is a symbol that is not a name. Here P has no transitions because the pattern matching will never succeed (remember that the frame is not used when comparing the objects in a communication). So $P \sim \mathbf{0}$. But $d(b)\,.\,P$ can instantiate b in P to f, so $d(b)\,.\,P \not\sim d(b)\,.\,\mathbf{0}$.

Theorem 2

1. $P \sim Q \Longrightarrow P \,|\, R \sim Q \,|\, R$.
2. $P \sim Q \Longrightarrow (\nu a)P \sim (\nu a)Q$.
3. $P \sim Q \Longrightarrow\, !P \sim\, !Q$.
4. $P \sim R \wedge Q \sim S \Rightarrow \textbf{if } M = N \textbf{ then } P \textbf{ else } Q \sim \textbf{if } M = N \textbf{ then } R \textbf{ else } S$.
5. $P \sim Q \Longrightarrow \overline{M} \, N.P \sim \overline{M} \, N.Q$.
6. $(\forall \widetilde{L}.\ P[\widetilde{a} := \widetilde{L}] \sim Q[\widetilde{a} := \widetilde{L}]) \Longrightarrow M\,(N).P \sim M\,(N).Q$, *where* $\widetilde{a} = \mathrm{n}(N)$.

5   Further Work

Our framework for extended pi-calculi has been designed to facilitate formalisation with a theorem prover. We intend to extend our previous work [19] where we formalised a significant part of the pi-calculus using the nominal datatype package [21] of the interactive theorem prover Isabelle [7].

A few generalisations of the current framework will be considered. In particular, we would like to generalise the notion of a frame so that it can contain more than just aliases. In a typed version of the applied pi-calculus [22], processes are extended with first order logical predicates which are used to prove correspondence assertions of cryptographic protocols. This mechanism could be modeled by allowing the frame to include such predicates.

There are several versions of applied pi-calculi focusing on cryptographic properties of agents, where type checking rather than bisimilarity proves a process secure [23,24,22]. By typing extended pi-calculi we will include this kind of reasoning.

We intend to add weak bisimulation to our framework. Moving from strong early to weak early bisimilarity is usually not difficult and we do not anticipate any problems here. We also intend to explore barbed bisimilarity.

We have given some preliminary encodings of other calculi into our framework but we have not formally proved any properties about them. It would be interesting to investigate in a formal fashion how they relate and see what equivalences are preserved by the encodings.

Finally, we aim to give a symbolic semantics for our framework. Non-symbolic semantics suffer from problems with exploding state spaces making it difficult for automatic tools such as the Mobility or Concurrency Workbenches [25,26] to reason about them. A symbolic semantics could make use of the generalisations mentioned above to let the frame contain conditions on symbolic values.

References

1. Milner, R., Parrow, J., Walker, D.: A calculus of mobile processes, part I/II. Journal of Information and Computation 100, 1–77 (1992)
2. Milner, R.: The polyadic π-calculus: A tutorial. In: Bauer, F.L., Brauer, W., Schwichtenberg, H. (eds.) Logic and Algebra of Specification. Series F., NATO ASI, vol. 94. Springer, Heidelberg (1993)
3. Abadi, M., Gordon, A.D.: A calculus for cryptographic protocols: The Spi calculus. Journal of Information and Computation 148, 1–70 (1999)
4. Abadi, M., Fournet, C.: Mobile values, new names, and secure communication. In: Proceedings of POPL 2001, pp. 104–115. ACM, New York (2001)
5. Borgström, J., Nestmann, U.: On bisimulations for the spi calculus. In: Kirchner, H., Ringeissen, C. (eds.) AMAST 2002. LNCS, vol. 2422, pp. 287–303. Springer, Heidelberg (2002)
6. Carbone, M., Maffeis, S.: On the expressive power of polyadic synchronisation in π-calculus. Nordic Journal of Computing 10(2), 70–98 (2003)
7. Nipkow, T., Paulson, L.C., Wenzel, M.: Isabelle/HOL. LNCS, vol. 2283. Springer, Heidelberg (2002)
8. Abadi, M., Gordon, A.D.: A bisimulation method for cryptographic protocols. Nordic Journal of Computing 5(4), 267–303 (1998)
9. Elkjær, A.S., Höhle, M., Hüttel, H., Overgård, K.: Towards automatic bisimilarity checking in the spi calculus. In: Calude, C.S., Dinneen, M.J. (eds.) Combinatorics, Computation & Logic. Australian Computer Science Communications, vol. 21(3), pp. 175–189. Springer, Heidelberg (1999)

10. Boreale, M., De Nicola, R., Pugliese, R.: Proof techniques for cryptographic processes. SIAM Journal on Computing 31(3), 947–986 (2002)
11. Durante, L., Sisto, R., Valenzano, A.: Automatic testing equivalence verification of spi calculus specifications. ACM Trans. Softw. Eng. Methodol. 12(2), 222–284 (2003)
12. Borgström, J.: Equivalences and Calculi for Formal Verifiation of Cryptographic Protocols. PhD thesis, EPFL, Lausanne (to appear, 2008)
13. Fournet, C., Abadi, M.: Hiding names: Private authentication in the applied pi calculus. In: Okada, M., Pierce, B.C., Scedrov, A., Tokuda, H., Yonezawa, A. (eds.) ISSS 2002. LNCS, vol. 2609, pp. 317–338. Springer, Heidelberg (2003)
14. Abadi, M., Blanchet, B., Fournet, C.: Just fast keying in the pi calculus. ACM Trans. Inf. Syst. Secur. 10(3) (2007)
15. Bhargavan, K., Fournet, C., Gordon, A.D.: A semantics for web services authentication. Theor. Comput. Sci. 340(1), 102–153 (2005)
16. Chaudhuri, A., Abadi, M.: Formal security analysis of basic network-attached storage. In: FMSE 2005: Proceedings of the 2005 ACM workshop on Formal methods in security engineering, pp. 43–52. ACM, New York (2005)
17. Haack, C., Jeffrey, A.: Pattern-matching spi-calculus. Information and Computation 204(8), 1195–1263 (2006)
18. Pitts, A.M.: Nominal logic, a first order theory of names and binding. Information and Computation 186, 165–193 (2003)
19. Bengtson, J., Parrow, J.: Formalising the pi-calculus using nominal logic. In: Seidl, H. (ed.) FOSSACS 2007. LNCS, vol. 4423, pp. 63–77. Springer, Heidelberg (2007)
20. Carbone, M., Coccia, M., Ferrari, G., Maffeis, S.: Process algebra-guided design of java mobile network applications. In: Informal Proceedings of the FMTJP 2001 Workshop, Budapest (2001)
21. Urban, C.: Nominal techniques in Isabelle/HOL. Journal of Automatic Reasoning (to appear, 2007)
22. Fournet, C., Gordon, A.D., Maffeis, S.: A type discipline for authorization in distributed systems. In: Proc. of CSF 2007 (to appear, 2007)
23. Gordon, A.D., Jeffrey, A.: Secrecy despite compromise: Types, cryptography, and the pi-calculus. In: Abadi, M., de Alfaro, L. (eds.) CONCUR 2005. LNCS, vol. 3653, pp. 186–201. Springer, Heidelberg (2005)
24. Fournet, C., Gordon, A.D., Maffeis, S.: A type discipline for authorization policies. In: Sagiv, M. (ed.) ESOP 2005. LNCS, vol. 3444, pp. 141–156. Springer, Heidelberg (2005)
25. Victor, B., Moller, F.: The Mobility Workbench — a tool for the π-calculus. In: Dill, D.L. (ed.) CAV 1994. LNCS, vol. 818, pp. 428–440. Springer, Heidelberg (1994)
26. Cleaveland, R., Parrow, J., Steffen, B.: The Concurrency Workbench: a semantics-based tool for the verification of concurrent systems. ACM Trans. Program. Lang. Syst. 15(1), 36–72 (1993)

Completeness and Logical Full Abstraction in Modal Logics for Typed Mobile Processes

Martin Berger[1], Kohei Honda[2], and Nobuko Yoshida[1]

[1] Imperial College London
[2] Queen Mary, University of London

Abstract. We study an extension of Hennessy-Milner logic for the π-calculus which gives a sound and complete characterisation of representative behavioural preorders and equivalences over typed processes. New connectives are introduced representing actual and hypothetical typed parallel composition and hiding. We study three compositional proof systems, characterising the May/Must testing preorders and bisimilarity. The proof systems are uniformly applicable to different type disciplines. Logical axioms distill proof rules for parallel composition studied by Amadio and Dam. We demonstrate the expressiveness of our logic through verification of state transfer in multiparty interactions and fully abstract embeddings of program logics for higher-order functions.

1 Introduction

Communication is becoming a foremost element of computing, from web services to sensor networks to multicore programming. The diversity of behaviour these communicating systems exhibit is staggering, including functional and stateful, sequential and concurrent, and deterministic and non-deterministic. A useful way of understanding this diversity is to classify behaviour into *types*. A compositional universe of types has fundamental merit in engineering, helping distilled understanding of the semantics of behaviour and guaranteeing basic safety such as the absence of communication errors.

The π-calculus [17] is an expressive formalism for concurrency, representing a vast array of communication behaviours with its small syntax. Starting from Milner's sorting [16], many different notions of types have been studied to classify different universes of interactions. For example, one linear type discipline turns the π-calculus into a semantic universe which exactly captures call-by-name and call-by-value higher-order sequential computation [4].

Built on the preceding studies of modal logics for the untyped π-calculus [2,8,18] and CCS [20,21], as well as on our own works on program logics [3,10,25], the present work introduces a sound and complete modal logic for typed π-calculi which is uniformly applicable to diverse type disciplines. Its adaptability comes from three logical operators, representing actual and hypothetical parallel composition and hiding. The introduction of these operators is less about sheer expressiveness than about the organisation of proof rules. Compositional reasoning is now confined to the proof rules of the logic, which precisely follow the syntactic structures of processes; whereas extracting the modal content of composition is relegated to the axioms of the assertion language.

L. Aceto et al. (Eds.): ICALP 2008, Part II, LNCS 5126, pp. 99–111, 2008.

This organisation helps us uniformly treat multiple type disciplines and their mixture in logic: different type disciplines induce different axioms for these operators, reflecting their distinct semantic effects, while keeping the identical proof rules.

Typed composition in the π-calculus often yields locally deterministic interactions, which allows us to abstract away silent actions semantically. This is often essential for reasoning about embeddings of data structures and programming languages. To capture this effect, the present study considers modal assertions and proof systems for weak typed transitions. Suggested by our study on logics for higher-order functions [10], we construct three proof systems, the first one based on the May modality, the second one on Must, and the third one which mixes these modalities. By deriving characteristic formulae, we prove completeness of these proof systems with respect to the May/Must testing preorders and bisimilarity. These results are established for the integration of three channel type disciplines widely found in the literature, *non-deterministic*, *linear* and *replicated*. These results extend to other linear and non-linear disciplines.

The combination of types and logics offers a powerful reasoning framework. We show two case studies. First we reason about a practical business protocol, using a new axiom for fixed point formulae for merging states in synchronised interactions. Second we show our logic can fully abstractly embed the total and partial program logics for call-by-value higher-order functions studied in [10]. The result extends to other program logics, offering a unifying view on logics for sequential and concurrent programs.

Related Work. Hennessy-Milner logic of the untyped π-calculus is first studied in [18] where early and late bisimilarities are characterised. Amadio and Dam [2] study model checking and proof systems of Hennessy-Milner logic of the untyped π-calculus with minimal and maximal fixed points. Dam [8] presents a proof system with ordinal-indexed fixed point formulae with a powerful discharge rule and presents specifications on Milner's encoding of data structures. Our logic is built on these works. One of the key contributions of the present work is the introduction of axioms for parallel composition based on typed synchronisation algebra, through which we can logically capture the semantics of typed processes. As far as we know, ours is the first modal logic for mobile processes which fully characterises typed semantics.

Other process logics for the untyped π-calculus include [15,23], which study efficient proof search using a freshness quantifier ∇; [6], which presents a logic for spatial properties using a hiding operator and a freshness operator; and [5], which extends Abramsky's logical characterisation of a class of CPOs to obtain a negation-less logic which corresponds to a power domain constructed by Fiore and others and which characterises a strong late bisimilarity.

The logical operators for actual and hypothetical parallel composition appeared in Stirling's early work [20,21]. Their usage in the present work originates in [3]. The operator for hypothetical composition allows rely-guarantee-based reasoning [12], whose analogue in the sequent format is studied by Simpson [19] as well as in [2,7,8]. Logical full abstraction of PCF is studied in [14] in the context of CPOs. A derivation of a program logic from a typed process logic is studied in [9]. A fully abstract embedding of a program logic in a modal process logic may not be found in the literature.

The full version of the present paper [1] lists detailed proofs and further examples.

2   Processes and Types

Processes. We use a typed π-calculus with three kinds of channel types: *linear, non-deterministic* and *replicated*. Linear types are based on *session types* [11,22] which allow legible description of structured communication. For simplicity, we omit the delegation primitive. The grammar of processes $(P, Q, \ldots)$ is given by:

$$P ::= \mathbf{0} \mid a(k).P \mid !a(k).P \mid \overline{a}(k).P \mid k(x).P \mid \overline{k}\langle e\rangle.P \mid k \triangleleft l.P \mid k \triangleright [l_i : P_i]_{i \in I}$$
$$\mid \ \text{if } e \text{ then } P \text{ else } Q \mid P|Q \mid (\nu u)P \mid (\mathbf{rec}\,X(\tilde{x}).P)\langle \tilde{e}\rangle \mid X\langle \tilde{e}\rangle$$

$k, k', ..$ are *linear channels*; $a, b, c, ..$ *shared channels*; $u, u', ..$ their union; $v, w, ..$ *values*, which are constants (numbers and booleans) and channels; $x, y, ..$ variables; $X, Y, ..$ process variables; and $l, l_i, ..$ labels for branching. Expressions $(e, e', ..)$ are variables, constants, arithmetic/boolean operations (such as $e + e'$) and linear/shared channels.

The process $a(k).P$ receives a request to establish a session from $\overline{a}(k).Q$. $!a(k).P$ is the replicated version of $a(k).P$. In all of these three prefixes, k is bound in the body. $k(x).P$ receives a value from $\overline{k}\langle e\rangle.Q$ via k; and $k \triangleright [l_i : P_i]_{i \in I}$ (with I finite) waits with $\{l_i\}_{i \in I}$-labelled branches from which $k \triangleleft l.P$ selects one. $P \mid Q$ is a parallel composition and $(\nu u)P$ is a hiding. A recursive process $(\mathbf{rec}\,X(\tilde{x}).P)\langle \tilde{e}\rangle$ consists of a recursive definition $(\mathbf{rec}\,X(\tilde{x}).P)$ and actual parameters $\tilde{e}$. In $\mathbf{rec}\,X(\tilde{x}).P$, a process variable X and formal parameters $\tilde{x}$ are binders. $\mathsf{fn}(P)$ denotes the free channels in P. We often omit $\mathbf{0}$ and the empty vector. For example we write $\overline{k}$ for $\overline{k}\langle\rangle.\mathbf{0}$ and $\mathbf{rec}\,X.P$ for $(\mathbf{rec}\,X().P)\langle\rangle$.

The structural congruence $\equiv$ is standard [22,11], in which we include the unfolding rule for recursion: $(\mathbf{rec}\,X(\tilde{x}).P)\langle \tilde{e}\rangle \equiv P[\tilde{v}/\tilde{x}][\mathbf{rec}\,X(\tilde{x}).P/X]$ with $e_i \downarrow v_i$, where $e \downarrow v$ means e evaluates to v. The reduction rules are generated by:

$$a(k).P \mid \overline{a}(k).Q \longrightarrow (\nu k)(P \mid Q) \qquad !a(k).P \mid \overline{a}(k).Q \longrightarrow !a(k).P \mid (\nu k)(P \mid Q)$$
$$k(x).P \mid \overline{k}\langle e\rangle.Q \longrightarrow P[v/x] \mid Q \quad (e \downarrow v) \qquad k \triangleright [l_i : P_i]_{i \in I} \mid k \triangleleft l_j.Q \longrightarrow P_j \mid Q \quad (j \in I)$$

with the standard if-then-else rules, closing under the evaluation contexts and structure rules. The first rule carries out *session initiation* via bound name passing. The second rule is for value passing and the third for branching.

As an example, a simple ATM process with an initial value 300 is given below.

$$\mathbf{rec}\,X(x).(a(k).(\mathbf{rec}\,Y(yk).k \triangleright [\ \mathsf{balance} : \overline{k}\langle y\rangle.Y\langle yk\rangle,$$
$$\mathsf{deposit} : k(w).\overline{k}\langle y + w\rangle.Y\langle y + wk\rangle,$$
$$\mathsf{quit} : X\langle y\rangle] \ )\langle xk\rangle \qquad\qquad)\langle 300\rangle$$

This ATM first establishes a session identified by k; and offers three options, balance, deposit and quit. If balance is selected, then it shows the balance of the account, and recurs with the same amount (y). If deposit is selected, then it receives a deposited amount w, and recurs with the new state $(y + w)$. If quit is chosen, it exits the loop and terminates the conversation. The actual parameter 300 indicates the initial balance.

Types and Typing. The grammar of types follows [11], augmented with replicated types, $(\tau)^!$ and $(\tau)^?$, from [4].

$$\alpha ::= \mathtt{nat} \mid \mathtt{bool} \mid (\tau) \mid (\tau)^! \mid (\tau)^? \mid \mathbf{rec}\,t.\alpha \mid t$$
$$\tau ::= \downarrow\alpha; \tau \mid \uparrow\alpha; \tau \mid \&\{l_i : \tau_i\}_{i \in I} \mid \oplus\{l_i : \tau_i\}_{i \in I} \mid \mathbf{rec}\,t.\tau \mid \mathtt{end} \mid t \mid \bot$$

We call α a *shared type*, which consists of *non-deterministic type* (τ); *server type* $(\tau)^!$ and *client types* $(\tau)^?$ together called *replicated types*; atomic type nat and bool; recursive type rec$t.\tau$; and a type variable. We take an *equi-recursive* view of types, not distinguishing between a type rec$t.\alpha$ and its unfolding $\alpha[\text{rec}t.\alpha/t]$. τ is a *linear type*. Type $\downarrow\alpha;\tau$ represents first inputting a value of type α, then performing the actions typed by τ; type $\uparrow\alpha;\tau$ is its dual. Type $\&\{l_i : \tau_i\}_{i\in I}$ represents waiting with n options, and behaves as τ_i if the i-th action is selected; type $\oplus\{l_i : \tau_i\}_{i\in I}$ is its dual. Type end represents inaction and is often omitted. $\bot$ indicates that no further connection is possible at a given channel. The *dual type* of α is defined by exchanging ! and ?, $\uparrow$ and $\downarrow$, and $\&$ and $\oplus$. Atomic types, (τ), end, and t are self-dual.

The partial commutative and associative operator $\odot$ [22,4], which controls a parallel composition, is defined by: (1) $\tau \odot \overline{\tau} = \bot$; (2) $\alpha \odot \alpha = \alpha$ if $\overline{\alpha} = \alpha$; and (3) $(\tau)^! \odot (\overline{\tau})^? = (\tau)^!$ and $(\tau)^? \odot (\tau)^? = (\tau)^?$. (1) says that once we compose two processes at a linear channel, the channel is no longer composable. (3) says a server should be unique, while an arbitrary number of clients can request interactions. Δ_0 and Δ_1 are *compatible*, written $\Delta_0 \asymp \Delta_1$, if $\Delta_0(u) \odot \Delta_1(u)$ is defined for each $u \in \text{dom}(\Delta_0) \cap \text{dom}(\Delta_1)$; $\Delta_i(u) = \alpha$, then $u \in \text{dom}(\Delta_j)$; and process variables are disjoint. If $\Delta_0 \asymp \Delta_1$, we set $\Delta_0 \odot \Delta_1 = \{(\Delta_0 \odot \Delta_1)(u) \mid u \in \text{dom}(\Delta_0) \cap \text{dom}(\Delta_1)\} \cup \Delta_0 \setminus \text{dom}(\Delta_1) \cup \Delta_1 \setminus \text{dom}(\Delta_0)$.

Typing environments $\Gamma, \Delta, \dots$ are given by $\Gamma ::= \emptyset \mid \Gamma, a : \alpha \mid \Gamma, X : \tilde{\alpha}\tilde{\tau} \mid \Gamma, k : \tau$. The typing judgement for process P is given as $\Gamma \vdash P$. The typing rules are identical with [22,11] for linear/non-deterministic types, augmented with the typing for replicated types from [4] (allowing only client typed channels to be free under a replicated prefix). We only list the following rule for parallel composition.

$$\Gamma_i \vdash P_i \text{ with } i = 1,2 \text{ and } \Gamma_1 \asymp \Gamma_2, \text{ then } \Gamma_1 \odot \Gamma_2 \vdash P_1 \mid P_2$$

As an example, session channel k in ATM is typed by:

$$\tau = \text{rec}t.\&\{\text{balance} : \uparrow\text{nat};t, \text{ deposit} : \downarrow\text{nat};\uparrow\text{nat};t, \text{ quit} : \text{end}\}$$

The same session from the user's viewpoint is typed dually as $\overline{\tau} = \text{rec}t. \oplus \{\text{balance} : \downarrow\text{nat};t, \text{ deposit} : \uparrow\text{nat};\downarrow\text{nat};t, \text{ quit} : \text{end}\}$, composable with τ by $\odot$.

Bisimilarity and Testing. *Transition labels* $(\ell, \ell', ..)$ are given by the grammar:

$$\ell ::= \tau \mid a(k) \mid \overline{a}(k) \mid kv \mid \overline{k}v \mid k(a) \mid \overline{k}(a) \mid k \triangleright l \mid k \triangleleft l$$

where k and a in (k) and (a) introduce binding. ℓ is *shared* if it has shape $a(k)$ or $\overline{a}(k)$; *linear* if it is neither shared nor τ. We write $\overline{\ell}$ for the dual of ℓ, defined by exchanging the input and output (for example $\overline{a(k)} = \overline{a}(k)$). $\overline{\tau}$ is undefined. We use the standard early transition relation augmented with $k \triangleleft l.P \xrightarrow{k \triangleleft l} P$ and $k \triangleright [l_i : P_i]_{i \in I} \xrightarrow{k \triangleright l_j} P_j$ $(j \in I)$. The *typed early transition* is defined by setting $\Gamma \vdash P \xrightarrow{\ell} \Gamma \setminus \ell \vdash Q$ if $P \xrightarrow{l} Q$ and if the operation $\Gamma \setminus \ell$ is defined, where $\Gamma \setminus \ell$ is defined if ℓ conforms to Γ, in which case $\Gamma \setminus \ell$ denotes the resulting environment. For example, assuming $\Gamma = \Delta, k : \&\{l_i : \tau_i\}_{i \in I}$, if $l = l_j$ $(j \in I)$ then $\Gamma \setminus k \triangleright l = \Delta, k : \tau_j$; otherwise $\Gamma \setminus k \triangleright l_j$ is undefined. We often leave Γ and Δ implicit. $\Longrightarrow$ stands for a reflexive and transitive closure of $\xrightarrow{\tau}$. We define the *early weak bisimilarity*, the *weak May preorder* and the *(divergence-insensitive) weak Must preorder* in the standard way, written $\approx$, $\sqsubseteq_{may}$ and $\sqsubseteq_{must}$, respectively.

3   Assertions

A Logical Language. Our logical language is Hennessy-Milner logic with equality, value/name passing modality and fixed point formulae [2,8], augmented with new operators. The grammar of assertions $(A,B,C,\ldots)$ follows.

$$A ::= e_1 = e_2 \mid A \wedge B \mid \neg A \mid \forall x^\rho .A \mid \langle\!\langle\,\rangle\!\rangle A \mid \langle \ell \rangle A \mid (\mu X(\tilde{x}).A)\langle \tilde{e}\rangle \mid X\langle \tilde{e}\rangle$$
$$\mid \; \mathsf{v} x^\rho .A \mid A \circ B \mid A \triangleright B$$

Above ℓ ranges over $a(k)$, $\bar{a}(k)$, $k\langle e\rangle$, $\bar{k}\langle e\rangle$, $k \triangleright l_i$ and $k \triangleleft l_i$. ρ stands for either α or τ. We define $A \vee B$, $A \supset B$, $\exists x^\rho .A$, $[\ell]A$, $[\![\,]\!]A$, and $(\mathsf{v} X(\tilde{x}).A)\langle \tilde{e}\rangle$, by dualisation.

$\langle \ell \rangle A$ says that the process has some immediate, or strong, ℓ action, satisfying A as the result. $\langle\!\langle\,\rangle\!\rangle A$ says that after some sequence of zero or more silent actions, the process will satisfy A (dually, in $[\![\,]\!]A$, after whatever zero or more silent actions, the process will satisfy A). We write $\langle\!\langle \ell \rangle\!\rangle A$ for $\langle\!\langle\,\rangle\!\rangle\langle \ell \rangle\langle\!\langle\,\rangle\!\rangle A$, saying that some weak ℓ-transition leads to A. Dually $[\![\ell]\!]A$ says that any weak ℓ-transition ends up satisfying A. The combination of strong and weak modalities is important for proof systems and axioms.

The minimal and maximal fixed points use parameters following [2,8], which are essential *for describing* state-changing *loops*, as in the ATM example. We assume that $X\langle \tilde{e}\rangle$ never occurs in A negatively (the assumption part of $\triangleright$ is contravariant) [8].

$A \circ B$ (read as "A *par* B") is understood as A, B in [21]. Informally, a process $\Gamma \vdash P$ satisfies $A \circ B$ when $\Gamma \vdash P$ has the same observable behaviour as $Q|R$, together typed under Γ, such that Q satisfies A and R satisfies B. This puts typing constraints on A and B: if A and B have minimal typings Δ and Δ', we demand $\Delta \asymp \Delta'$ and $\Delta \odot \Delta' \subset \Gamma$.

$A \triangleright B$ (read as "*rely A then B*") is a typed version of the consequence relation studied in [20]. A process $\Gamma \vdash P$ satisfies $A \triangleright B$ if, for each appropriately typed Q satisfying A, $P|Q$ satisfies B. Again this constrains the typing of A and B: if A has the minimal typing Δ, we demand $\Gamma \asymp \Delta$ and that B is typed under $\Gamma \odot \Delta$. For example, for $\Gamma \vdash P$ with $\Gamma(k) = \bar{\tau}$ to satisfy $B \triangleright C$, k can be typed as τ in B, and, if so, k is typed $\perp$ in C.

$\mathsf{v} x^\rho .A$ is the quantifier for name hiding. A process, say P, satisfies $\mathsf{v} x^\rho .A$ if there is a fresh name u of type ρ and P' such that $(\mathsf{v} u)P' \approx P$ and P' satisfies A. Its logical nature differs substantially from $\exists$, as studied in [25].

We often omit type annotations for quantifiers. T denotes $1 = 1$, F its negation. The standard association of operators is assumed, e.g. $\forall x.A \wedge B \supset C$ is parsed as $((\forall x.A) \wedge B) \supset C$ ($\circ$, $\triangleright$, $\mathsf{v} x.A$ associate as $\wedge$, $\supset$, $\exists x.A$). We use the following notation:

Definition 1 (mixed modality). $(\!\langle \ell \rangle\!)A \; = \; [\![\,]\!](\langle \ell \rangle \mathsf{T} \wedge [\ell]A)$.

The modal formula $(\!\langle \ell \rangle\!)A$ (read: "surely ℓ then A") says that now or after any silent actions the process may have, it can do a strong ℓ-action, and then it satisfies A.

Examples of Assertions. We illustrate $\circ$ and $\triangleright$ using a simple example.

$$P \equiv \,!b(k).k(x).\bar{k}\langle x+1\rangle.\mathbf{0} \qquad Q \equiv \bar{b}(k).\bar{k}\langle 2\rangle.k(y).\bar{h}\langle y\rangle.\mathbf{0}$$

P accepts a session request, receives a number and returns its increment: Q requests a session, sends 2 and receives and forwards the result to h. P and Q are typed under $b : (\downarrow\mathtt{nat};\uparrow\mathtt{nat};\mathtt{end})^!$ and $b : (\uparrow\mathtt{nat};\downarrow\mathtt{nat};\mathtt{end})^?, h : \uparrow\mathtt{nat};\mathtt{end}$, respectively.

We now assert for P and Q and their composition. First for P and Q individually:

$$A \;=\; \forall x^{\mathtt{nat}}.\langle\!\langle b(k)\rangle\!\rangle\,\langle\!\langle kx\rangle\!\rangle\,\langle\!\langle \overline{k}x+1\rangle\!\rangle\mathsf{T} \qquad B \;=\; \forall y^{\mathtt{nat}}.\langle\!\langle \overline{b}(k)\rangle\!\rangle\,\langle\!\langle \overline{k}2\rangle\!\rangle\,\langle\!\langle ky\rangle\!\rangle\,\langle\!\langle \overline{h}y\rangle\!\rangle\mathsf{T}$$

From this we assert $A \circ B$ for $P|Q$. Since $A \circ B \supset \langle\!\langle \overline{h}3\rangle\!\rangle\mathsf{T}$ (by the axioms in Section 4 later), we know $P|Q$ can emit 3 via h. From this entailment we also know Q satisfies $A \rhd \langle\!\langle \overline{h}3\rangle\!\rangle\mathsf{T}$, i.e. when composed with any behaviour satisfying A, it can emit 3 via h.

Above we only used the May modality. In fact, we can strengthen A and B using the mixed modality (cf. Definition 1) as follows.

$$A' \;=\; \forall x^{\mathtt{nat}}.(\!|b(k)|\!)\,(\!|kx|\!)\,(\!|\overline{k}x+1|\!)\mathsf{T} \qquad\qquad B' \;=\; \forall y^{\mathtt{nat}}.(\!|\overline{b}(k)|\!)\,(\!|\overline{k}2|\!)\,(\!|ky|\!)\,(\!|\overline{h}y|\!)\mathsf{T}$$

We can then show that $A' \circ B'$ entails $(\!|\overline{h}3|\!)\mathsf{T}$, hence $P|Q$ surely emits 3 via h. This entailment depends on the type of b: if b's type is non-deterministic, e.g. $b : (\downarrow\mathtt{nat};\uparrow\mathtt{nat};\mathtt{end})$, then this assertion can*not* be derived (as discussed in Proposition 4 later).

Next we consider a specification of the simple ATM, given as:

$$(\!|a(k)|\!)\,(\nu Y(yk).(k \rhd \mathsf{balance})\,(\!|\overline{k}y|\!)Y\langle yk\rangle)\langle 300k\rangle \tag{3.1}$$

The assertion says the process is ready to receive a session request via a: then it enters a loop, and, if asked to show a balance, it shows y, and recurs. The initial balance is 300. Now a user of ATM may satisfy: $\forall x.(\!|\overline{a}(k)|\!)\,(k \lhd \mathsf{balance})\,(\!|kx|\!)\,(\!|\overline{h}x|\!)\mathsf{T}$. which, when combined with (3.1) by $\circ$, gives us $\langle\!\langle \overline{h}300\rangle\!\rangle\mathsf{T}$. In contrast to the previous example, we can*not* derive $(\!|\overline{h}300|\!)\mathsf{T}$ since another user may interfere at the shared channel a before this user. This distinction will be formally underpinned in Proposition 4 later.

Semantics of Assertions. The interpretation of assertions follows [8], extended to the typed setting. We list the key points. First, a *property* (written $p, q, ..$) is a set of typed processes under an identical typing which are without free value/process variables and which are closed under $\approx$. We define operators on properties as:

$$p|q = \bigcup_{P \in p, Q \in q}[P|Q]_{\approx} \qquad (\nu u)p = \bigcup_{P \in p}[(\nu u)P]_{\approx}$$
$$\langle\!\langle\rangle\!\rangle p' = \{P \mid P \Longrightarrow P' \in p'\} \qquad \langle\ell\rangle p' = \{P \mid P \approx P_0 \xrightarrow{\ell} P' \in p'\}$$

A *parametrised property of type* $\tilde{\rho}$ (written $f, g, \dots$) is a function which maps a vector of values typed $\tilde{\rho}$ to a property. An interpretation of variables ($\xi, \xi', ..$) follows [8], mapping a variable to a value and an assertion variable to a parametrised property. Given $\Gamma \vdash A$ where Γ types the free channels in A, the *interpretation of* $\Gamma \vdash A$ *under* ξ, written $\langle\!\langle \Gamma \vdash A\rangle\!\rangle\xi$, or $\langle\!\langle A\rangle\!\rangle\xi$ if Γ is known from the context, is given by the standard clauses for equality, conjunction, universal quantifier, negation and assertion variable, augmented with the following clauses. For modality, we set:

$$\langle\!\langle \Gamma \vdash \langle\!\langle\rangle\!\rangle A\rangle\!\rangle\xi = \langle\!\langle\rangle\!\rangle\langle\!\langle \Gamma \vdash A\rangle\!\rangle\xi, \qquad \langle\!\langle \Gamma \vdash \langle\ell\rangle A\rangle\!\rangle\xi = \langle\ell\rangle\langle\!\langle \Gamma\setminus\ell \vdash A\rangle\!\rangle\xi$$

where $\Gamma\setminus\ell$ adds a mapping w.r.t. ℓ. For $\circ$, $\rhd$ and ν we set:

$$\langle\!\langle \Gamma \vdash A \circ B\rangle\!\rangle\xi = \bigcup_{\Delta\odot\Theta = \Gamma}\langle\!\langle \Delta \vdash A\rangle\!\rangle\xi\,|\,\langle\!\langle \Theta \vdash B\rangle\!\rangle\xi \qquad \langle\!\langle \Gamma \vdash \nu x^\rho.A\rangle\!\rangle\xi = (\nu u)\langle\!\langle \Gamma, u : \rho \vdash A\rangle\!\rangle(\xi\cdot x \mapsto u)$$
$$\langle\!\langle \Gamma \vdash A \rhd B\rangle\!\rangle\xi = \max p^\Gamma.\,((p\,|\,\langle\!\langle \Delta \vdash A\rangle\!\rangle\xi) \subset \langle\!\langle \Delta\odot\Gamma \vdash B\rangle\!\rangle\xi)$$

Above $\max p^\Gamma.\mathcal{P}$ denotes the maximum property (by set inclusion) typed under Γ which satisfies $\mathcal{P}$. The following clause for μ-recursion is from [8].

$$\langle\!\langle \Gamma \vdash (\mu X(\tilde{x}).A)\langle\tilde{e}\rangle\rangle\!\rangle\xi = (\mathsf{fix}\,\lambda f.\lambda\tilde{v}.(\langle\!\langle A\rangle\!\rangle(\xi\cdot X \mapsto f\cdot\tilde{x}\mapsto\tilde{v})))(\xi(\tilde{e}))$$

where fix is the least fixed point and $\xi(e)$ is the interpretation of e under ξ.

$$\frac{E \vdash P \blacktriangleright A}{E \vdash a(k).P \blacktriangleright \langle\!\langle a(k)\rangle\!\rangle A} \qquad \frac{E \vdash P \blacktriangleright A}{E \vdash \bar{a}(k).P \blacktriangleright \langle\!\langle \bar{a}(k)\rangle\!\rangle A} \qquad \frac{E \vdash P \blacktriangleright A}{E \vdash\, !a(k).P \blacktriangleright \langle\!\langle a(k)\rangle\!\rangle A} \qquad \frac{E \vdash P \blacktriangleright A}{E \vdash \bar{a}(k).P \blacktriangleright \langle\!\langle \bar{a}(k)\rangle\!\rangle A}$$
Acc,Req,Ser,CReq

$$\frac{E \vdash P \blacktriangleright A}{E \vdash k(x).P \blacktriangleright \forall x.\langle\!\langle kx\rangle\!\rangle A} \qquad \frac{E \vdash P \blacktriangleright A}{E \vdash \bar{k}\langle e\rangle.P \blacktriangleright \langle\!\langle \bar{k}e\rangle\!\rangle A} \qquad \frac{E \vdash P_i \blacktriangleright A_i \quad i=1,2}{E \vdash P_1 \,|\, P_2 \blacktriangleright A_1 \circ A_2} \qquad \frac{E \vdash P \blacktriangleright A \quad x \text{ fresh}}{E \vdash (\nu u)P \blacktriangleright \nu x.A[x/u]}$$
Rcv,Send,Conc, Res

$$\frac{E \vdash P_i \blacktriangleright A_i \quad \forall i \in I}{E \vdash k \triangleright [l_i : P_i]_{i \in I} \blacktriangleright \bigwedge_{i \in I} \langle\!\langle k \triangleright l_i\rangle\!\rangle A_i} \qquad \frac{E \vdash P \blacktriangleright A_j}{E \vdash k \triangleleft l_j.P \blacktriangleright \langle\!\langle k \triangleleft l_j\rangle\!\rangle A_j} \qquad \frac{-}{E \vdash 0 \blacktriangleright \mathsf{T}}$$
Bra,Sel,Inact

$$\frac{-}{E, X : (\tilde{x})A \vdash X\langle \tilde{e}\rangle \blacktriangleright A[\tilde{e}/\tilde{x}]} \qquad \frac{E, X : (\tilde{x})(\forall j \leq i.A(j)) \vdash P \blacktriangleright A(i)}{E \vdash (\mathbf{rec}X.(\tilde{x}).P)\langle \tilde{e}\rangle \blacktriangleright \forall i.A(i)[\tilde{e}/\tilde{x}]}$$
Var,Rec-ind

$$\frac{E \vdash P_1 \blacktriangleright e \supset A \quad E \vdash P_2 \blacktriangleright \neg e \supset A}{E \vdash \mathtt{if}\ e\ \mathtt{then}\ P_1\ \mathtt{else}\ P_2 \blacktriangleright A} \qquad \frac{E \vdash P \blacktriangleright A \quad A \supset B}{E \vdash P \blacktriangleright B}$$
If, Conseq

Fig. 1. Proof System (the May Modality)

4 Proof Rules, Axioms and Completeness

Rules for the May Modality. Write $\Gamma; E \vdash P \blacktriangleright A$ for the provability judgement where Γ types P and A (except auxiliary variables in A) and E contains assignments of the form $X : (\tilde{x})A$, mapping a process variable to a parametrised formula ($\tilde{x}$ are binders). We often write $E \vdash P \blacktriangleright A$, leaving Γ implicit. We consider three systems, one for the May modality, one for Must, and one for their combination. They soundly and completely characterise the May/Must preorders and bisimilarity, respectively.

The proof rules for the May modality are given in Figure 1. There is a single rule for each typing rule, except that Conseq has no corresponding rules. The typing is not mentioned, assuming it follows the typing rules. The first eight rules are standard (Ser does not use a fixed point, which suffices due to the semantics of replication, cf. Proposition 4 (6) later). Conc and Res hide complexity of process composition under $\circ$ and ν, which is to be unfolded by the axioms for these operators.

Inact and Var are standard. In Rec-ind, we assume i, j are in some well-ordered set [10]. We make this rule applicable to fixed point operators by introducing the notation $(\mu/\nu X^{\kappa}(\tilde{x}).A)\langle \tilde{e}\rangle$ from [8], with κ ranging over ordinals. The notation stands for the standard approximant to the least fixed point, given as: $(\mu X^0(\tilde{x}).A)\langle \tilde{e}\rangle \equiv \mathsf{F}$, $(\mu X^{\kappa+1}(\tilde{x}).A)\langle \tilde{e}\rangle \equiv A[(\mu X^{\kappa}(\tilde{x}).A)/X][\tilde{e}/\tilde{x}]$, and $(\mu X^{\lambda}(\tilde{x}).A)\langle \tilde{e}\rangle \equiv \exists_{i \leq \lambda}(\mu X^i(\tilde{x}).A)\langle \tilde{e}\rangle$ with λ a limit ordinal. Dually, via ν-recursion. For example, via this notation, an inference for $(\mathbf{rec}X(k).\bar{k}1.X\langle k\rangle)\langle k\rangle$ is given as follows, setting $A(i) = \nu Y^i(k).\langle\!\langle \bar{k}1\rangle\!\rangle Y\langle k\rangle$.

$$\frac{X : (k)\forall j \leq i.A(j) \vdash \bar{k}1.X\langle k\rangle \blacktriangleright A(i)}{\vdash (\mathbf{rec}X(k).\bar{k}1.X\langle k\rangle)\langle k\rangle \blacktriangleright (\nu Y(k).\langle\!\langle \bar{k}1\rangle\!\rangle Y\langle k\rangle)\langle k\rangle}$$

Using higher ordinals becomes necessary when we have a lexicographic ordering, as with the behaviour with nested recursions.

The conditional rule is standard. The final proof rule is the consequence rule as found in Hoare logic.

Rules for the Must and Mixed Modalities. The May proof rules ensure that a process *can* reach a certain state: in contrast, the Must rules ensure that a process *cannot* reach a certain state. We first define the abbreviation $\mathsf{noact}(\Gamma)$, which says: "no actions at $\mathsf{dom}(\Gamma)$ are possible". Let $\mathsf{noact}(k:\downarrow\alpha;\tau) = \forall x^\alpha.[\![kx]\!]\mathsf{F}$, $\mathsf{noact}(k:\&\{l_i:\tau_i\}) = \wedge_i[\![k\triangleright l_i]\!]\mathsf{F}$, $\mathsf{noact}(k:\mathsf{end}) = \mathsf{noact}(x:\mathtt{nat}) = \mathsf{noact}(x:\mathtt{bool}) = \mathsf{T}$ and similarly for outputs and shared names. Set $\mathsf{noact}(\tilde{u}:\tilde{\rho}) = \wedge_i\mathsf{noact}(u_i:\rho_i)$. We then write $[\![\ell,\Gamma]\!]A$ for $[\![\,]\!]([\ell]A \wedge \mathsf{noact}(\Gamma))$ with $\ell \neq \tau$, which says: "ℓ is the only action possible and if it ever happens then A follows". Using this predicate, the proof system for the Must modality is given by replacing $\langle\!\langle\ell\rangle\!\rangle$ in each prefix rule in Figure 1 with $[\![\ell,\Delta]\!]$, where Δ is the typing of a process minus that of ℓ; and for Inact, replacing T with $\mathsf{noact}(\Gamma)$, assuming Γ is the implicit typing. Other rules stay unchanged, except for adding:

$$\frac{E, X:(\tilde{x})A \vdash P \blacktriangleright A \quad A \text{ admissible}}{E \vdash (\mathbf{rec}\,X(\tilde{x}).P)\langle\tilde{e}\rangle \blacktriangleright A[\tilde{e}/\tilde{x}]} \quad \text{Rec-adm}$$

where admissibility is defined via syntactic unfoldings [10]. Given $R \equiv (\mathbf{rec}\,X(\tilde{x}).P)\langle\tilde{e}\rangle$, let $P^0 \equiv \mathbf{0}$ and $P^{n+1} \equiv P[(\tilde{x})P^n/X]$ (where we set $((\tilde{x})Q)\langle\tilde{e}\rangle = Q[\tilde{e}/\tilde{x}]$). Then a closed formula A is *admissible* if: (1) P^0 satisfies A; and (2) If P_i satisfies A for each $i \geq 0$, then $(\mathbf{rec}\,X(\tilde{x}).P)\langle\tilde{x}\rangle$ also satisfies A. This is extended to open formulae closing under admissible properties. In practice, we may use a tractable variant of admissibility: for example, if we restrict P to be sequential (i.e. without parallel composition), there is a simple syntactic characterisation of admissibility.

To capture both modalities in a single proof system, we strengthen the Must prefix rules through the use of the combined modality $(\!|\ell,\Delta|\!)A$, which stands for $(\!|\ell|\!)A \wedge \mathsf{noact}(\Delta)$ (cf. Definition 1). The proof system is given by replacing $\langle\!\langle\ell\rangle\!\rangle$ in each prefix rule in Figure 1 with $(\!|\ell,\Delta|\!)$, fully capturing the semantics of prefix. Other rules remain identical except for adding the following recursion rule, due to Larsen [13].

$$\frac{E, X:(\tilde{x})X'\langle\tilde{x}\rangle \vdash P \blacktriangleright A}{E \vdash (\mathbf{rec}\,X(\tilde{x}).P)\langle\tilde{e}\rangle \blacktriangleright (\mu X'(\tilde{x}).A)\langle\tilde{e}\rangle} \quad \text{Rec-mix}$$

Soundness and Relative Completeness. Let us write $\Gamma; E \vdash_{may} P \blacktriangleright A$, $\Gamma; E \vdash_{must} P \blacktriangleright A$ and $\Gamma; E \vdash_{mix} P \blacktriangleright A$, for provability in the May/Must/Mixed proof systems, respectively. We also write $\Gamma; E \models P \blacktriangleright A$ (read: $\Gamma \vdash P$ *satisfies* A *under* E), when we have $P \in \langle\!\langle\Gamma \vdash A\rangle\!\rangle(\xi \cdot \langle\!\langle E\rangle\!\rangle\xi)$ for each ξ, where $\langle\!\langle E\rangle\!\rangle\xi$ is the obvious interpretation of process variables under E and ξ. We first observe:

Theorem 2 (soundness). $\Gamma; E \vdash_{may} P \blacktriangleright A$ *implies* $\Gamma; E \models P \blacktriangleright A$, *similarly for* $\Gamma; E \vdash_{must} P \blacktriangleright A$ *and* $\Gamma; E \vdash_{mix} P \blacktriangleright A$.

Thus the three proof systems are all sound under the same satisfaction relation, allowing the mixed use of their proof rules in reasoning. Further each system precisely captures a distinct process semantics, as shown by the following completeness result. The proof is by syntactically deriving characteristic formulae, which also entails observational and descriptive completeness in the sense of [10].

Theorem 3 (completeness). *Let* $\Gamma \vdash P$ *and A be closed. Then* $\models P \blacktriangleright A$ *with A being an upper-closed property w.r.t.* $\sqsubseteq_{may}$ *(resp. a downward-closed property w.r.t.* $\sqsubseteq_{must}$*) implies* $\vdash_{may} P \blacktriangleright A$ *(resp.* $\vdash_{must} P \blacktriangleright A$*). Further for any A, if* $\models P \blacktriangleright A$ *then* $\vdash_{mix} P \blacktriangleright A$.

Basic Axioms. The operators $\circ$ and ν, used in the proof rules, do not directly describe the communication behaviour of a process: It is through the axioms of the assertion language that modal behaviours are extracted. Some of the basic axioms follow.

Proposition 4. *Below we assume well-typedness of formulae.*

1. $B \supset (A \rhd (A \circ B))$, $A \rhd (B \rhd C) \equiv (A \circ B) \rhd C$ *and* $(A \circ (A \rhd B)) \supset B$.
2. $A \circ B \equiv A \wedge B$ *if* $\mathsf{fn}(A) \cap \mathsf{fn}(B) = \emptyset$ *and all free channels are server typed.*
3. $(\!(\ell)\!)A) \circ B \equiv (\!(\ell)\!)(A \circ B)$ *and* $(\!(\ell)\!)A \circ (\!(\bar{\ell})\!)B \equiv \nu\mathsf{bn}(\ell).(A \circ B)$, *with ℓ linear.*
4. $(\!(\ell)\!)A \circ (\!(\bar{\ell})\!)B \supset (\langle\!\langle \ell \rangle\!\rangle (A \circ (\!(\bar{\ell})\!)B) \wedge \langle\!\langle \rangle\!\rangle (A \circ B) \wedge \langle\!\langle \bar{\ell} \rangle\!\rangle ((\!(\ell)\!)A \circ B))$
5. $(\!(a(k))\!)A \circ (\!(\bar{a}(k))\!)B \equiv (\!(a(k))\!)A \circ \nu k.(A \circ B)$, *where a has a server type.*
6. $(\nu X(\tilde{x}).A)\langle \tilde{e} \rangle \circ (\nu Y(\tilde{y}).B)\langle \tilde{g} \rangle \supset (\nu Z(\tilde{x}\tilde{y}).C[Z\langle \tilde{e}\tilde{g}\rangle]_i)\langle \tilde{e}\tilde{g} \rangle$ *where* $(A \circ B \supset C[X\langle \tilde{e} \rangle \circ Y\langle \tilde{g} \rangle]_i)$ *is valid and* $C[X\langle \tilde{e} \rangle_i \circ Y\langle \tilde{g} \rangle_i]_{i \in I}$ *denotes a formula with multiple holes indexed by I, assuming all occurrences of X and Y are thus exhausted.*

The three axioms in (1) relate $\rhd$ and $\circ$. In (2), $\mathsf{fn}(A)$ is the set of names and variables of channel types. In (3), the second axiom eliminates dual actions. In (4) the prefixing $\langle\!\langle \rangle\!\rangle$ cannot be removed due to state change, unlike (2). In (5), the axiom relies on a having a server type, corresponding to the replication law in [16,24]. In the Server-Client example in Section 3, if we type b with a non-deterministic type, we cannot apply this axiom, hence cannot derive $(\!(\bar{h}3)\!)\top$. In (6), A and B indicate well-synchronised recursive interactions, in which case we can merge their states under recursions.

Elimination of $\circ$ and ν. Through these and other axioms, we can transform formulae into those without $\circ$ and ν. We discuss a basic result for such elimination, using deterministic type disciplines from [4,24] (the typing in [4] ensures determinacy, to which [24] adds a causality constraint to ensure strong normalisation: essentially the same result holds for processes in Section 2 without non-deterministic types). We extend $\langle a(k) \rangle$ to $\langle a\tilde{b}(k) \rangle$ (dually for output) since, in [4,24], a server channel (a) carries not only a linear channel (k) but client-typed channels ($\tilde{b}$). We also replace the use of bisimilarity in Section 3 with the standard reduction-based congruence [4,24], denoted $\cong$, which adds semantic precision. In correspondence, we refine the interpretation of equality and quantification over server-typed names. Below let α be server-typed.

$$\langle\!\langle \Gamma \vdash e_1^\alpha = e_2^\alpha \rangle\!\rangle \xi = \{ \Gamma \vdash P \mid P[\xi(e_1)\xi(e_2)/\xi(e_2)\xi(e_1)] \cong P \}$$

$$\langle\!\langle \Gamma\xi \vdash \forall x^\alpha.A \rangle\!\rangle \xi = \max p^{\Gamma\xi}.(\forall q^{u:\alpha}.p \mid q \subset \langle\!\langle \Gamma, x : \alpha \vdash A \rangle\!\rangle (\xi \cdot x \mapsto u))$$

The first clause says that two replicated channels are equal if the corresponding behaviours are. Together these clauses treat replicated channels as the behaviours they represent, while maintaining the standard axioms for equality and quantifiers. Their significance will become clear when we discuss logical full abstraction in Section 5. The same proof systems satisfy completeness for $\cong$ and the corresponding precongruences.

Now let us say A is $\circ$-*free* (resp. ν-*free*) if $\circ$ (resp. ν) does not occur in A. A is *approximately* $\circ$-*free* if $\circ$ occurs only in fixed point formulae whose finite unfoldings are $\circ$-free up to logical equivalence. We also say A *characterises* P when $\Gamma \models P \blacktriangleright A$ and, moreover, whenever $\Gamma \models Q \blacktriangleright A$ we have $P \cong Q$.

Theorem 5 (elimination of $\circ$ and ν under determinism). *Let P be typable by the type discipline in [4] (resp. [24]). Then there is an algorithm to find a ν-free and approximately $\circ$-free formula (resp. a ν-free and $\circ$-free formula) which characterises P.*

5   Applications

State Transfer: Synchronising Stateful Interactions. As the first reasoning example, we extend the previous ATM in Sections 2 and 3 to three-party interactions among User, ATM and Bank. Our purpose is to demonstrate how we can reason about the transfer of state induced by synchronised actions among multiple parties. ATM is extended with withdraw option, in which ATM asks Bank each time it receives a request from User, and forwards the answer to User. The π-calculus term representing this behaviour, which we call *ATM*, is given as:

$$a(k).\overline{b}(k').\mathbf{rec}\, Y.(\, k \triangleright [\, \mathsf{balance}: \overline{k'} \triangleleft \mathsf{balance}.k'(z).\overline{k}\langle z\rangle.Y,$$
$$\mathsf{withdraw}: k(n).\overline{k'} \triangleleft \mathsf{withdraw}.\overline{k'}\langle n\rangle.k' \triangleright [\mathsf{ok}: k \triangleleft \mathsf{ok}.Y, \mathsf{no}: k \triangleleft \mathsf{no}.Y\,],$$
$$\mathsf{quit}: \overline{k'} \triangleleft \mathsf{quit}\,]\,)$$

The new ATM no longer has its own state, dispensing with parameters in its recursion. At the same time, the state change in Bank is reflected onto ATM through interactions, so that ATM will *behave to User as if it were stateful*. In turn, User would demand the following invariance: if User withdraws money several times *within a single session*, the withdrawal of an amount n succeeds if n is within the immediately preceding balance, say z, with the resulting balance $z - n$. Below we give a specification of ATM, as seen from User, asserting this invariance. The specification $\mathsf{ATMSpec}(a,x)$, where x is an initial balance, is given as the formula $\langle\!| a(k) |\!\rangle \langle\!\langle \,\rangle\!\rangle (\nu Z(z).A)\langle x\rangle$ where we set A to be:

$$\langle\!| k \triangleright \mathsf{withdraw} |\!\rangle \,\forall n.\langle\!| kn |\!\rangle \,(\, z \geq n \supset \langle\!| k \triangleleft \mathsf{ok} |\!\rangle Z\langle z - n\rangle \,\wedge\, z < n \supset \langle\!| k \triangleleft \mathsf{no} |\!\rangle Z\langle z\rangle\,)$$

Let $\mathsf{BankSpec}(b,x)$ be a specification for Bank given as $\langle\!| b(k') |\!\rangle (\nu Z(z).B)\langle x\rangle$ where $B = A[k'/k]$ with k' fresh in A. We now show:

$$ATM \models \mathsf{BankSpec}(b,300) \,\triangleright\, \mathsf{ATMSpec}(a,300)$$

To reach this judgement, we start from a formula directly derived by the proof rules, which we call $\mathsf{ATMSpec}_0(a,b)$, defined as $\langle\!| a(k) |\!\rangle\,(\!| \overline{b}(k') |\!\rangle \nu Y.A_0$ where we set A_0 to be:

$$\langle\!| k \triangleright \mathsf{withdraw} |\!\rangle (\!| k' \triangleleft \mathsf{withdraw} |\!\rangle \,\forall n.\langle\!| kn |\!\rangle (\!| \overline{k}'n |\!\rangle (\,\langle\!| k' \triangleright \mathsf{ok} |\!\rangle (\!| k \triangleleft \mathsf{ok} |\!\rangle.Y \,\wedge\, \langle\!| k' \triangleright \mathsf{no} |\!\rangle (\!| k \triangleleft \mathsf{no} |\!\rangle.Y\,)$$

It thus suffices to show $\mathsf{ATMSpec}_0(a,b) \circ \mathsf{BankSpec}(b,300) \supset \mathsf{ATMSpec}(a,300)$. We first calculate $A_0 \circ B \supset A$ by compensating all dual strong linear actions by Axiom (2) in Proposition 4. This and Axiom (6) of the same proposition give us:

$$(\nu Y.A_0) \circ (\nu Z(z).B)\langle x\rangle \quad \supset \quad (\nu Z(z).A)\langle x\rangle$$

Thus we have successfully transferred Bank's state to the specification for ATM. Finally by Axiom (4) in Proposition 4 we calculate:

$$(b(k')).(\nu Z(z).B)\langle 300 \rangle \circ (a(k))(\overline{b}(k'))(\nu Y.A_0)$$
$$\supset (a(k))((b(k'))(\nu Z(z).B)\langle 300 \rangle \circ (\overline{b}(k'))(\nu Y.A_0))$$
$$\supset (a(k)) \langle\!\langle \rangle\!\rangle (\nu Z(z).A)\langle 300 \rangle$$

Above the logical calculation of interaction at b induces $\langle\!\langle \rangle\!\rangle$ in the final line, indicating a shared, hence possibly nondeterministic, interaction: in contrast, all actions *within* a session have strong modality. In this way the present framework allows specifications and reasoning about the fine-grained mixture of determinism and non-determinism.

Logical Full Abstraction of PCFv. One of the notable effects of types in the π-calculus is to enhance the semantic precision of the embedding of diverse calculi and programming languages in this calculus. When a type discipline is sufficiently strong, the embedding even enjoys full abstraction [4]. In the following we demonstrate that the proposed logic inherits this feature at a logical level. We use the complete program logic for call-by-value PCF (henceforth PCFv) from [10] and the process logic under the type discipline of [4] based on the reduction-based equality $\cong$, discussed in Section 4.

We first review PCFv and its logic. PCFv-types are either atomic types (nat and bool) or arrow types ($\alpha \Rightarrow \beta$). PCFv-terms ($M, N, \ldots$) and formulae ($A, B, \ldots$) are given by the following grammar.

$$M ::= x \mid op(\tilde{M}) \mid \lambda x^\alpha.M \mid MN \mid \mu x^{\alpha \Rightarrow \beta}.\lambda y^\alpha.M \mid \text{if } M \text{ then } N_1 \text{ else } N_2$$
$$A ::= e_1 = e_2 \mid A \wedge B \mid \forall x^\alpha.A \mid \neg A \mid x \bullet y \searrow z$$

In the first line (terms), $op(\tilde{M})$ denotes the standard first-order operations (including constants). In the second line (formulae), $x \bullet y \searrow z$, called *evaluation formula*, specifies that a function x, when applied to an argument y, converges and results in a value z. The semantics of these formulae exactly follows [10]. The judgement $\models [A]M :_u [B]$ intuitively says that if the free variables in M satisfy A, the program M terminates and whose result, named u, satisfies B. For its formal definition, see [10].

We use Milner's encoding of call-by-value λ-calculus [16]. Below we only show primary ones.

$$\langle\!\langle x \rangle\!\rangle_k = \overline{k}\langle x \rangle \qquad \langle\!\langle \lambda x.M \rangle\!\rangle_k = (\nu a)(\overline{k}\langle a \rangle \mid !a(xk').\langle\!\langle M \rangle\!\rangle_{k'})$$
$$\langle\!\langle MN \rangle\!\rangle_k = (\nu k_1)(\langle\!\langle M \rangle\!\rangle_{k_1} \mid k_1(m).(\nu k_2)(\langle\!\langle N \rangle\!\rangle_{k_2} \mid k_2(n).\overline{m}\langle nk \rangle))$$

The last line uses free name passing unlike [4], following [24, §6]. The embedding of types is given accordingly [4]. For formulae, the standard constructs are mapped directly: $\langle\!\langle e_1 = e_2 \rangle\!\rangle \equiv e_1 = e_2$, $\langle\!\langle A \wedge B \rangle\!\rangle \equiv \langle\!\langle A \rangle\!\rangle \wedge \langle\!\langle B \rangle\!\rangle$, $\langle\!\langle \neg A \rangle\!\rangle \equiv \neg \langle\!\langle A \rangle\!\rangle$ and $\langle\!\langle \forall x^\alpha.A \rangle\!\rangle \equiv \forall x.\langle\!\langle A \rangle\!\rangle$. In the first map, equality of two names in the PCFv-logic denotes equality of their denotations: to embed this notion in the process logic, we need the refinement of semantics of equality in Section 4. For evaluation formulae we set:

$$\langle\!\langle x \bullet y \searrow z \rangle\!\rangle \equiv \langle\!\langle xy(k) \rangle\!\rangle \langle\!\langle \overline{k}z \rangle\!\rangle \mathsf{T},$$

which decomposes an evaluation formula to a modal formula with the May modality (which corresponds to total correctness under determinism).

Below we say a formula A of PCFv-logic with $\mathsf{fv}(A) = \{u\}$ is *upper-closed with respect to u* [10] if, whenever V named u satisfies A, and if W is greater than V in the standard observational precongruence of PCFv, then W named u also satisfies A.

Theorem 6 (logical full abstraction of PCFv). *Let V be a well-typed closed PCFv-term and A be upper-closed with respect to u and, moreover, $\mathsf{fv}(A) = \{u\}$. Then we have* $\models [\mathsf{T}]V :_u [A]$ *if and only if* $\langle\!\langle V \rangle\!\rangle_k \models \exists u.(\langle\!\langle \overline{k}x \rangle\!\rangle \mathsf{T} \wedge \langle\!\langle A \rangle\!\rangle [x/u])$.

The proof uses the correspondence of characteristic formulae on both sides, observing that the May preorder and the contextual preorder coincide via the encoding of terms, and that validity in upper-closed formulae is preserved and reflected via the encoding of assertions. By translating partial correctness formulae using the Must modality, we obtain logical full abstraction for the PCFv-logic for partial correctness in [10].

References

1. Full version of this paper as a DoC technical report, Imperial College London (to appear, 2008) www.dcs.qmul.ac.uk/~kohei/processlogic
2. Amadio, R., Dam, M.: A modal theory of types for the π-calculus. In: Jonsson, B., Parrow, J. (eds.) FTRTFT 1996. LNCS, vol. 1135, pp. 347–365. Springer, Heidelberg (1996)
3. Berger, M.: A program logic for sequential higher-order control (1): stateless case. Typescript, 36 pages (October 2007)
4. Berger, M., Honda, K., Yoshida, N.: Sequentiality and the π-calculus. In: Abramsky, S. (ed.) TLCA 2001. LNCS, vol. 2044, pp. 29–45. Springer, Heidelberg (2001)
5. Bonsangue, M., Kurz, A.: Pi-calculus in logical form. In: LICS 2007, pp. 303–312. IEEE, Los Alamitos (2007)
6. Caires, L., Cardelli, L.: A spatial logic for concurrency. I& C 186(2), 194–235 (2003)
7. Cardelli, L., Gordon, A.D.: Anytime, anywhere: Modal logics for mobile ambients. In: POPL, pp. 365–377 (2000)
8. Dam, M.: Proof systems for pi-calculus logics. In: Logic for Concurrency and Synchronisation. Trends in Logic, Studia Logica Library, pp. 145–212. Kluwer, Dordrecht (2003)
9. Honda, K.: From process logic to program logic. In: ICFP 2004, pp. 163–174. ACM, New York (2004)
10. Honda, K., Berger, M., Yoshida, N.: Descriptive and relative completeness for logics for higher-order functions. In: Bugliesi, M., Preneel, B., Sassone, V., Wegener, I. (eds.) ICALP 2006. LNCS, vol. 4052, pp. 360–371. Springer, Heidelberg (2006)
11. Honda, K., Vasconcelos, V.T., Kubo, M.: Language Primitives and Type Disciplines for Structured Communication-based Programming. In: Hankin, C. (ed.) ESOP 1998 and ETAPS 1998. LNCS, vol. 1381, pp. 22–138. Springer, Heidelberg (1998)
12. Jones, C.B.: Specification and design of (parallel) programs. In: IFIP Congress, pp. 321–332 (1983)
13. Larsen, K.G.: Proof systems for satisfiability in Hennessy-Milner logic with recursion. Theor. Comput. Sci. 72(2&3), 265–288 (1990)
14. Longley, J., Plotkin, G.: Logical full abstraction and PCF. In: Tbilisi Symposium on Logic, Language and Information, CLSI (1998)
15. Miller, D., Tiu, A.: A proof theory for generic judgments. ACM Transactions on Computational Logic 6(4), 749–783 (2005)

16. Milner, R.: The polyadic π-calculus: A tutorial. In: Proceedings of the International Summer School on Logic Algebra of Specification, Marktoberdorf (1992)
17. Milner, R., Parrow, J., Walker, D.: A Calculus of Mobile Processes, Parts I and II. Info.& Comp. 100(1) (1992)
18. Milner, R., Parrow, J., Walker, D.: Modal logics for mobile processes. TCS 114, 149–171 (1993)
19. Simpson, A.: Sequent calculi for process verification: Hennessy-Milner logic for an arbitrary GSOS. J. Log. Algebr. Program. 60-61, 287–322 (2004)
20. Stirling, C.: A complete compositional model proof system for a subset of CCS. In: Brauer, W. (ed.) ICALP 1985. LNCS, vol. 194, pp. 475–486. Springer, Heidelberg (1985)
21. Stirling, C.: Modal logics for communicating systems. TCS 49, 311–347 (1987)
22. Takeuchi, K., Honda, K., Kubo, M.: An Interaction-based Language and its Typing System. In: Halatsis, C., Philokyprou, G., Maritsas, D., Theodoridis, S. (eds.) PARLE 1994. LNCS, vol. 817, pp. 398–413. Springer, Heidelberg (1994)
23. Tiu, A.F.: Model checking for pi-calculus using proof search. In: Abadi, M., de Alfaro, L. (eds.) CONCUR 2005. LNCS, vol. 3653, pp. 36–50. Springer, Heidelberg (2005)
24. Yoshida, N., Berger, M., Honda, K.: Strong Normalisation in the π-Calculus. Information and Computation 191, 145–202 (2004)
25. Yoshida, N., Honda, K., Berger, M.: Logical reasoning for higher-order functions with local state. In: Seidl, H. (ed.) FOSSACS 2007. LNCS, vol. 4423, pp. 361–377. Springer, Heidelberg (2007)

On the Sets of Real Numbers Recognized by Finite Automata in Multiple Bases*

Bernard Boigelot[1], Julien Brusten[1,**], and Véronique Bruyère[2]

[1] Institut Montefiore, B28
Université de Liège
B-4000 Liège, Belgium
{boigelot,brusten}@montefiore.ulg.ac.be
[2] Université de Mons-Hainaut
Avenue du Champ de Mars, 6
B-7000 Mons, Belgium
veronique.bruyere@umh.ac.be

Abstract. This paper studies the expressive power of finite automata recognizing sets of real numbers encoded in positional notation. We consider Muller automata as well as the restricted class of *weak deterministic automata*, used as symbolic set representations in actual applications. In previous work, it has been established that the sets of numbers that are recognizable by weak deterministic automata in two bases that do not share the same set of prime factors are exactly those that are definable in the first order additive theory of real and integer numbers $\langle \mathbb{R}, \mathbb{Z}, +, < \rangle$. This result extends *Cobham's theorem*, which characterizes the sets of integer numbers that are recognizable by finite automata in multiple bases.

In this paper, we first generalize this result to *multiplicatively independent* bases, which brings it closer to the original statement of Cobham's theorem. Then, we study the sets of reals recognizable by Muller automata in two bases. We show with a counterexample that, in this setting, Cobham's theorem does not generalize to multiplicatively independent bases. Finally, we prove that the sets of reals that are recognizable by Muller automata in two bases that do not share the same set of prime factors are exactly those definable in $\langle \mathbb{R}, \mathbb{Z}, +, < \rangle$. These sets are thus also recognizable by weak deterministic automata. This result leads to a precise characterization of the sets of real numbers that are recognizable in multiple bases, and provides a theoretical justification to the use of weak automata as symbolic representations of sets.

1 Introduction

By using the positional notation, real numbers can be encoded as infinite words over an alphabet composed of a fixed number of digits, with an additional symbol

* This work is supported by the *Interuniversity Attraction Poles* program *MoVES* of the Belgian Federal Science Policy Office, and by the grant 2.4530.02 of the Belgian Fund for Scientific Research (F.R.S.-FNRS).

** Research fellow ("Aspirant") of the Belgian Fund for Scientific Research (F.R.S.-FNRS).

L. Aceto et al. (Eds.): ICALP 2008, Part II, LNCS 5126, pp. 112–123, 2008.

for separating their integer and fractional parts. This encoding scheme maps sets of numbers onto languages that describe precisely those sets.

This paper studies the sets of real numbers whose encodings can be accepted by finite automata. The motivation is twofold. First, since regular languages enjoy good closure properties under a large range of operators, automata provide powerful theoretical tools for establishing the decidability of arithmetic theories. In particular, it is known that the sets of numbers that are definable in the first-order additive theory of integers $\langle \mathbb{Z}, +, < \rangle$, also called *Presburger arithmetic*, are encoded by regular finite-word languages [Büc62, BHMV94]. This result translates into a simple procedure for deciding the satisfiability of Presburger formulas. Moving to infinite-word encodings and ω-regular languages, it can be extended to sets of real numbers definable in $\langle \mathbb{R}, \mathbb{Z}, +, < \rangle$, i.e., the first-order additive theory of real and integer variables. [BBR97, BJW05].

The second motivation is practical. Since finite automata are objects that are easily manipulated algorithmically, they can be used as actual data structures for representing symbolically sets of values. This idea has successfully been exploited in the context of computer-aided verification, leading to representations suited for the sets of real and integer vectors handled during symbolic state-space exploration [WB95, BJW05, EK06]. A practical limitation of this approach is the high computational cost of some operations involving infinite-word automata, in particular language complementation [Saf88, Var07]. However, it has been shown that a restricted form of automata, *weak deterministic* ones, actually suffices for handling the sets definable in $\langle \mathbb{R}, \mathbb{Z}, +, < \rangle$ [BJW05]. Weak automata can be manipulated with essentially the same cost as finite-word ones [Wil93], which alleviates the problem and leads to an effective representation system.

Whether a set of numbers can be recognized by an automaton generally depends on the chosen encoding base. For integer numbers, it is known that a set $S \subseteq \mathbb{Z}$ is recognizable in a base $r > 1$ iff it is definable in the theory $\langle \mathbb{Z}, +, <, V_r \rangle$, where V_r is a base-dependent function [BHMV94]. Furthermore, the well known *Cobham's theorem* states that if a set $S \subseteq \mathbb{Z}$ is simultaneously recognizable in two bases $r > 1$ and $s > 1$ that are *multiplicatively independent*, i.e., such that $r^p \neq s^q$ for all $p, q \in \mathbb{N}_{>0}$, then S is *ultimately periodic*, i.e., it differs from a periodic subset of $\mathbb{Z}$ only by a finite set [Cob69]. Equivalently, such a set is definable in $\langle \mathbb{Z}, +, < \rangle$ [BHMV94]. It follows that such a set S is recognizable in every base. Our aim is to generalize as completely as possible Cobham's theorem to automata recognizing real numbers, by precisely characterizing the sets that are recognizable in multiple bases. We first consider the case, relevant for practical applications, of weak deterministic automata. In previous work, it has been established that a set of real numbers is simultaneously recognizable by weak deterministic automata in two bases that do not share the same set of prime factors iff this set is definable in $\langle \mathbb{R}, \mathbb{Z}, +, < \rangle$ [BB07]. As a first contribution, we extend this result to pairs of multiplicatively independent bases. Since recognizability in two multiplicatively dependent bases is equivalent to recognizability in only one of them [BRW98], this result provides a complete characterization of the sets that are recognizable in multiple bases by weak deterministic automata.

Then, we move to sets recognized by Muller automata. We prove that there exists a set of real numbers recognizable in two multiplicatively independent bases that share the same set of prime factors, but that is not definable in $\langle \mathbb{R}, \mathbb{Z}, +, < \rangle$. This shows that Cobham's theorem does not directly generalize to Muller automata recognizing sets of real numbers. Finally, we establish that a set $S \subseteq \mathbb{R}$ is simultaneously recognizable in two bases that do not share the same set of prime factors iff S is definable in $\langle \mathbb{R}, \mathbb{Z}, +, < \rangle$. As a corollary, such a set must then be recognizable by a weak deterministic automaton. Our result thus provides a theoretical justification to the use of weak automata, by showing that their expressive power corresponds precisely to the sets of reals recognizable by infinite-word automata in every encoding base.

2   Representing Sets of Real Numbers with Finite Automata

Let $r > 1$ be an integer numeration *base* and let $\Sigma_r = \{0, \ldots, r - 1\}$ be the corresponding set of *digits*. We encode a real number x in base r, most significant digit first, by words of the form $w_I \star w_F$, where $w_I \in \{0, r-1\}\Sigma_r^*$ encodes the integer part x_I of x and $w_F \in \Sigma_r^\omega$ encodes its fractional part x_F. Negative numbers are represented by their r's-complement. The length p of w_I is not fixed but has to be large enough for $-r^{p-1} \leq x_I < r^{p-1}$ to hold; thus, the most significant digit of an encoding of a real number is equal to 0 for positive numbers and to $r-1$ for negative ones [BBR97]. Some numbers have two distinct encodings with the same integer-part length, e.g., in base 10, the number $11/2$ admits the encodings $0^+5 \star 50^\omega$ and $0^+5 \star 49^\omega$. For a word $w = b_{p-1}^I b_{p-2}^I \ldots b_1^I b_0^I \star b_1^F b_2^F b_3^F \ldots \in \{0, r-1\}\Sigma_r^* \star \Sigma_r^\omega$, we denote by $[w]_r$ the real number encoded by w in base r, i.e.,

$$[w]_r = \sum_{i=0}^{p-2} b_i^I r^i + \sum_{i>0} b_i^F r^{-i} + \begin{cases} 0 & \text{if } b_{p-1}^I = 0, \\ -r^{p-1} & \text{if } b_{p-1}^I = r - 1. \end{cases}$$

For finite words $w \in \Sigma_r^*$, we denote by $[w]_r$ the natural number encoded by w, i.e., $[w]_r = [0w \star 0^\omega]_r$.

If the language formed by all the base-r encodings of the elements of a set $S \subseteq \mathbb{R}$ is ω-regular, then it can be accepted by a (non-unique) infinite-word automaton, called a *Real Number Automaton (RNA)* recognizing S. Such a set S is then said to be r-*recognizable*. RNA can be generalized into *Real Vector Automata (RVA)*, suited for subsets of $\mathbb{R}^n$, with $n > 0$ [BBR97].

RNA and RVA have originally been defined as Büchi automata [BBR97]. In this paper, we will instead consider them to be *deterministic Muller automata*. This adaptation can be made without loss of generality, since both classes of automata share the same expressive power [McN66, PP04]. The fact that RNA have a deterministic transition relation will simplify technical developments.

The subsets of $\mathbb{R}$ that are r-recognizable are exactly those that are definable in the first-order theory $\langle \mathbb{R}, \mathbb{Z}, +, <, X_r \rangle$, where $X_r(x, u, k)$ is a base-dependent predicate that holds whenever u is an integer power of r and there exists an encoding of x in which the digit at the position specified by u is equal to k [BRW98].

It is known that the full expressive power of infinite-word automata is not needed for representing the subsets of $\mathbb{R}$ that are definable in the first-order theory $\langle \mathbb{R}, \mathbb{Z}, +, < \rangle$ [BJW05]. Indeed, such sets can be recognized by *weak deterministic automata*, i.e., deterministic Büchi automata such that their set of states can be partitioned into disjoint subsets $Q_1, \ldots, Q_m$, where each Q_i contains only either accepting or non-accepting states, and there exists a partial order $\leq$ on the sets $Q_1, \ldots, Q_m$ such that for every transition (q, a, q') of the automaton, with $q \in Q_i$ and $q' \in Q_j$, we have $Q_j \leq Q_i$.

A set recognized by a weak deterministic automaton in base r is said to be *weakly r-recognizable* and such an automaton is called a *weak RNA*.

It has been established [BJW05] that the r-recognizable sets $S \subseteq \mathbb{R}$ that are not weakly r-recognizable are exactly those that satisfy the *dense oscillating property*: One has $\exists x_1 \forall \varepsilon_1 \exists x_2 \forall \varepsilon_2 \exists x_3 \forall \varepsilon_3 \cdots$ such that $|x_{i+1} - x_i| < \varepsilon_i$ for all $i \geq 1$, $x_i \in S$ for all odd i, and $x_i \notin S$ for all even i.

In the technical sections of this paper, we will need to apply transformations to sets represented by RNA (or weak RNA), or to the chosen encoding base. The following results are immediate corollaries of [BRW98] and [BJW05].

Theorem 1. *Let* $S \subseteq \mathbb{R}$, $r \in \mathbb{N}_{>1}$ *and* $a, b \in \mathbb{Q}$. *If* S *is (resp. weakly) r-recognizable then the sets* $aS + b$ *and* $S \cap [a, b]$ *are (resp. weakly) r-recognizable as well.*

Theorem 2. *Let* $S \subseteq \mathbb{R}$, $r \in \mathbb{N}_{>1}$, *and* $k \in \mathbb{N}_{>0}$. *The set* S *is (resp. weakly) r-recognizable iff it is (resp. weakly) r^k-recognizable.*

3   Problem Reductions

In the next sections, we will consider sets $S \subseteq \mathbb{R}$ that are simultaneously recognizable, either by RNA or by weak RNA, in two bases r and s that satisfy some conditions. We will then tackle the problem of proving that such sets are definable in $\langle \mathbb{R}, \mathbb{Z}, +, < \rangle$. In this section, we reduce this problem, by restricting the domain to the interval $[0, 1]$, and introducing the notion of boundary point.

3.1   Reduction to [0, 1]

Let $S \subseteq \mathbb{R}$ be a set recognized by a (resp. weak) RNA $\mathcal{A}$. Each accepting path of $\mathcal{A}$ reads exactly one occurrence of the symbol $\star$. Since $\mathcal{A}$ is finite-state, its accepted language $L(\mathcal{A})$ has the form $\bigcup_i L_i^I \star L_i^F$, where the union is finite, and the languages L_i^I and L_i^F contain, respectively, integer and fractional parts of the encodings of the elements of S. This induces a decomposition of S into a finite union $\bigcup_i (S_i^I + S_i^F)$, where for each i, we have $S_i^I \subseteq \mathbb{Z}$ and $S_i^F \subseteq [0, 1]$. It has been shown [BB07] that this decomposition is independent from the encoding base. Besides, every set S_i^I and S_i^F is recognizable by the same type of automaton as $\mathcal{A}$.

Assume now that $S \subseteq \mathbb{R}$ is simultaneously (resp. weakly) r- and s-recognizable, with respect to bases r and s that are multiplicatively independent. By Cobham's theorem [Cob69], each set S_i^I is thus definable in $\langle \mathbb{Z}, +, < \rangle$. This reduces the problem of establishing that S is definable in $\langle \mathbb{R}, \mathbb{Z}, +, < \rangle$ to the same problem for each set S_i^F. Since we have $S_i^F \subseteq [0, 1]$ for all i, the problem has thus been reduced from the domain $\mathbb{R}$ to the interval $[0, 1]$.

3.2   Boundary Points

The following notions are adapted from [BB07]. Given a point $x \in \mathbb{R}$ and a value $\varepsilon > 0$, a *neighborhood* of x is the set $N_\varepsilon(x) = \{y \in \mathbb{R} \mid |x - y| < \varepsilon\}$. A point $x \in \mathbb{R}$ is a *boundary point* of a set $S \subseteq \mathbb{R}$ iff all its neighborhoods contain at least one point from S as well as from its complement $\overline{S} = \mathbb{R} \setminus S$.

Lemma 1. *If a (resp. weakly) r-recognizable set $S \subseteq \mathbb{R}$ has only finitely many boundary points, then it is definable in $\langle \mathbb{R}, \mathbb{Z}, +, < \rangle$.*

Proof sketch. If $S \subseteq \mathbb{R}$ has only finitely many boundary points, then it can be decomposed into a finite union of intervals. In order to prove that S is definable in $\langle \mathbb{R}, \mathbb{Z}, +, < \rangle$, it is sufficient to show that the extremities of these intervals are rational numbers. Since S is (resp. weakly) r-recognizable, it is definable in $\langle \mathbb{R}, \mathbb{Z}, +, <, X_r \rangle$, and so is the set S' containing only those interval extremities. The set S' is thus finite and r-recognizable, and its elements are encoded by words sharing a finite number of fractional parts. These are necessarily ultimately periodic, from which the elements of S' are rational. $\square$

4   Multiplicatively Independent Bases

Let $r, s \in \mathbb{N}_{>1}$ be two *multiplicatively independent* bases, i.e., such that $r^p \neq s^q$ for all $p, q \in \mathbb{N}_{>0}$. We consider a set $S \subseteq [0, 1]$ that is both (resp. weakly) r- and s-recognizable. In the next section, we derive some properties under the assumption that S has infinitely many boundary points. We will see that this assumption leads to a contradiction in the case of weak RNA, showing that S is definable in $\langle \mathbb{R}, \mathbb{Z}, +, < \rangle$ by Lemma 1. This will be no longer true for RNA.

4.1   Product Stability

Let $\mathcal{A}_r$ be a (resp. weak) RNA recognizing S in base r. We assume w.l.o.g. that the transition relation of $\mathcal{A}_r$ is complete.

Since S is (resp. weakly) r-recognizable, it is definable in $\langle \mathbb{R}, \mathbb{Z}, +, <, X_r \rangle$, and so is the set B_S of all boundary points of S, which is thus r-recognizable. Let $\mathcal{A}_r^B$ be a RNA recognizing B_S.

By hypothesis, S has infinitely many boundary points, hence there exist infinitely many distinct paths of $\mathcal{A}_r^B$ that end up cycling in the same set of accepting states. One can thus extract from $\mathcal{A}_r^B$ an infinite language $L = 0 \star uv^* tw^\omega$, where $t, u, v, w \in \Sigma_r^*$, $|v| > 0$, $|w| > 0$, and L encodes an infinite subset of the boundary points of S. We then define $y = [0 \star uv^\omega]_r$ and, for each $k \in \mathbb{N}_{>0}$, $y_k = [0 \star uv^k tw^\omega]_r$. The sequence $y_1, y_2, y_3, \ldots \in \mathbb{Q}^\omega$ forms an infinite sequence of distinct boundary points of S, converging towards $y \in \mathbb{Q}$. If we have $y_k > y$ for infinitely many k, then we define $S^1 = (S - y) \cap [0, 1]$. Otherwise, we define $S^1 = (-S + y) \cap [0, 1]$. From Theorem 1, the set S^1 is both (resp. weakly) r- and s-recognizable. Moreover, this set admits an infinite sequence of distinct boundary points that converges to 0.

Let $\mathcal{A}_r^1$ and $\mathcal{A}_s^1$ be (resp. weak) RNA recognizing S^1 in the respective bases r and s. The path π_0 of $\mathcal{A}_r^1$ that reads $0 \star 0^\omega$ is composed of a prefix labeled by $0\star$, followed by an acyclic path of length $p \geq 0$, and finally by a cycle of length $q > 0$. It follows that a word of the form $0 \star 0^p t$, with $t \in \Sigma_r^\omega$, is accepted by $\mathcal{A}_r^1$ iff the word $0 \star 0^{p+q} t$ is accepted as well. Remark that the set S^1 admits infinitely many boundary points with a base-r encoding beginning with $0 \star 0^p$. Similar properties hold for $\mathcal{A}_s^1$. In this automaton, the path π_0' recognizing $0 \star 0^\omega$ reads the symbols 0 and $\star$, and then follows an acyclic sequence of length p' before reaching a cycle of length q'.

We now define $S^2 = r^p S^1 \cap [0, 1]$. Like S^1, the set S^2 admits an infinite sequence of boundary points that converges to 0. Moreover, by Theorem 1, S^2 is both (resp. weakly) r- and s-recognizable. Let $\mathcal{A}_r^2$ be a (resp. weak) RNA recognizing S^2 in base r. For every $t \in \Sigma_r^\omega$, the word $0 \star t$ is accepted by $\mathcal{A}_r^2$ iff the word $0 \star 0^q t$ is accepted as well. In other words, the fact that a number $x \in [0, 1]$ belongs or not to S^2 is not influenced by the insertion of q zero digits in its encodings, immediately after the symbol $\star$. This amounts to dividing the value of x by r^q, which leads to the following definition.

Definition 1. *Let $D \subseteq \mathbb{R}$ be a domain, and let $f \in \mathbb{R}_{>0}$. A set $S \subseteq D$ is f-product-stable in the domain D iff for all $x \in D$ such that $fx \in D$, we have $x \in S \Leftrightarrow fx \in S$.*

From the previous discussion, we have that S^2 is r^q-product-stable in $[0, 1]$. We then define $S^3 = s^{p'} S^2 \cap [0, 1]$. The set S^3 is r^q-product-stable in $[0, 1]$ as well. By Theorem 1, S^3 is also both (resp. weakly) r- and s-recognizable. Besides, since $S^3 = r^p s^{p'} S^1 \cap [0, 1]$, the set S^3 can alternatively be obtained by first defining $S^4 = s^{p'} S^1 \cap [0, 1]$, which is both (resp. weakly) r- and s-recognizable by Theorem 1. Then, one has $S^3 = r^p S^4 \cap [0, 1]$. By a similar reasoning in base s, we get that S^3 is $s^{q'}$-product-stable in $[0, 1]$. Like S^2, the set S^3 admits an infinite sequence of distinct boundary points that converges to 0.

Finally, we replace the bases r and s by $r' = r^q$ and $s' = s^{q'}$, thanks to Theorem 2. The results of this section are then summarized by the following lemma.

Lemma 2. *Let r and s be two multiplicatively independent bases, and let $S \subseteq [0, 1]$ be a set that is both (resp. weakly) r- and s-recognizable, and that admits infinitely many boundary points. There exist powers $r' = r^i$ and $s' = s^j$ of r and s, with $i, j \in \mathbb{N}_{>0}$, and a set $S' \subseteq [0, 1]$ that is both (resp. weakly) r'- and*

s'-recognizable, both r'- and s'-product-stable in $[0,1]$, and that admits infinitely many boundary points.

4.2   Recognizability by Weak RNA

We are now ready to prove that the sets $S \subseteq [0,1]$ that are recognizable by weak RNA in two multiplicatively independent bases r and s can only have finitely many boundary points.

By contradiction, suppose that such a set S has infinitely many boundary points. By Lemma 2, we can assume w.l.o.g. that S is r- and s-product-stable in $[0,1]$.

Hence, there exist $\alpha, \beta \in (0,1]$ such that $\alpha \in S$ and $\beta \notin S$. For every $i,j \in \mathbb{Z}$ such that $r^i s^j \alpha \in (0,1]$, we thus have $r^i s^j \alpha \in S$. Similarly, for every $i,j \in \mathbb{Z}$ such that $r^i s^j \beta \in (0,1]$, we have $r^i s^j \beta \notin S$.

Let γ be an arbitrary point in the open interval $(0,1)$. Since r and l are multiplicatively independent, it follows from Kronecker's approximation theorem [HW85] that any open interval of $\mathbb{R}_{>0}$ contains some number of the form r^i/s^j with $i,j \in \mathbb{N}_{>0}$ [Per90]. Hence, for every sufficiently small $\varepsilon > 0$ and $\delta \in \{\alpha, \beta\}$, there exist $i,j \in \mathbb{N}_{>0}$ such that

$$0 < \gamma - \varepsilon < (r^i/s^j)\delta < \gamma + \varepsilon < 1$$

showing that every sufficiently small neighborhood $N_\varepsilon(\gamma)$ of γ contains one point from S as well as from $\overline{S}$. The latter property leads to a contradiction, since it implies that S satisfies the dense oscillating property, and therefore cannot be recognized by a weak RNA.

Taking into account the problem reductions introduced in Sections 3.1 and 3.2, we thus have established the following result, that fully generalizes Cobham's theorem to weak RNA.

Theorem 3. *Let r and s be two multiplicatively independent bases. If a set $S \subseteq \mathbb{R}$ is weakly r- and s-recognizable, then it is definable in $\langle \mathbb{R}, \mathbb{Z}, +, < \rangle$.*

Thanks to the above mentioned reductions, we can rephrase this theorem as follows. If a set $S \subseteq \mathbb{R}$ is weakly r- and s-recognizable in two multiplicatively independent bases, then it is a finite union $\bigcup_i (S_i^I + S_i^F)$, where each $S_i^I \subseteq \mathbb{Z}$ is ultimately periodic and each $S_i^F \subseteq [0,1]$ is a finite union of intervals with rational extremities. It is worth mentioning that, as observed in [Wei99], such a structural description of subsets S of $\mathbb{R}$ is equivalent to the definability of S in $\langle \mathbb{R}, \mathbb{Z}, +, < \rangle$.

4.3   Recognizability by RNA

Theorem 3 cannot be directly generalized to automata that are not restricted to be weak and deterministic. Indeed, with RNA, a set can be recognizable in two multiplicatively independent bases without being definable in $\langle \mathbb{R}, \mathbb{Z}, +, < \rangle$. This property is established by the following theorem.

Theorem 4. *For every pair of bases r and s that share the same set of prime factors, there exists a set S that is both r- and s-recognizable, and that is not definable in $\langle \mathbb{R}, \mathbb{Z}, +, < \rangle$.*

Proof sketch. A counterexample is provided by the set $S = \{n/(f_1^{i_1} f_2^{i_2} \cdots f_k^{i_k}) \mid n \in \mathbb{Z}, i_1, i_2, \ldots, i_k \in \mathbb{N}\}$, where $f_1, f_2, \ldots f_k$ are the prime factors of r and s. Indeed, in either base $t \in \{r, s\}$, this set is encoded by the language $L_t = \{0, t-1\}\Sigma_t^* \star \Sigma_t^* (0^\omega \cup (t-1)^\omega)$. This language is clearly ω-regular, hence S is both r- and s-recognizable. However, S satisfies the dense oscillating property, which prevents it from being recognized by a weak RNA. It follows that S is not definable in $\langle \mathbb{R}, \mathbb{Z}, +, < \rangle$.   □

The case of bases that do not share the same set of prime factors is investigated in the next section.

5   Bases with Different Sets of Prime Factors

We now consider a subset of $[0,1]$ that is recognizable by RNA in two bases that have different sets of prime factors. Recall that according to Lemma 1, in order to prove that the set is definable in $\langle \mathbb{R}, \mathbb{Z}, +, < \rangle$, it is sufficient to show that this set has only finitely many boundary points. Like in Section 4, we proceed by contradiction, and assume that the set has infinitely many boundary points. By Lemma 2, there exist bases r and s with different sets of prime factors, and a set $S \subseteq [0,1]$ that is both r- and s-recognizable, both r- and s-product-stable in $[0,1]$, and that has infinitely many boundary points. Without loss of generality, we assume that there is a prime factor of s that does not divide r.

5.1   Sum Stability

Our strategy consists in exploiting Cobham's theorem so as to derive additional properties of S. The first step is to build from S a set $S' \subseteq \mathbb{R}_{\geq 0}$ that coincides with S over $[0,1]$, shares the same recognizability and product-stability properties, and contains numbers with non-trivial integer parts.

Lemma 3. *Let $r, s \in \mathbb{N}_{>1}$ be two bases with different sets of prime factors, and let $S \subseteq [0,1]$ be a set that is r- and s-recognizable, r- and s-product-stable in $[0,1]$, and that has infinitely many boundary points. There exists a set $S' \subseteq \mathbb{R}_{\geq 0}$ that is r- and s-recognizable, r- and s-product-stable in $\mathbb{R}_{\geq 0}$, and that has infinitely many boundary points.*

Proof sketch. Let $S' = \{r^k x \mid x \in S \wedge k \in \mathbb{N}\}$. This set is clearly r-product-stable in $\mathbb{R}_{\geq 0}$. Since S is r-product-stable in $[0,1]$, we have $S' \cap [0,1] = S$ showing that S' has infinitely many boundary points. We build a RNA $\mathcal{A}'_r$ recognizing S' in base r from a RNA $\mathcal{A}_r$ recognizing S as follows. The automaton $\mathcal{A}'_r$ is similar to $\mathcal{A}_r$, except that it delays arbitrarily the reading of the symbol $\star$.

In order to prove that S' is s-recognizable, notice that, since S is both r- and s-product-stable in $[0,1]$, we have $S' = \{r^i s^j x \mid x \in S \wedge i, j \in \mathbb{Z}\}$. The set S' can therefore be expressed as $S' = \{s^k x \mid x \in S \wedge k \in \mathbb{N}\}$. By the same reasoning as in base r, this set is s-recognizable, and s-product-stable in $\mathbb{R}_{\geq 0}$. $\qquad\square$

Consider now a set S' obtained from S by Lemma 3. As discussed in Section 3.1, this set can be expressed as a finite union $S' = \bigcup_i (S_i^I + S_i^F)$, where for each i, we have $S_i^I \subseteq \mathbb{N}$ and $S_i^F \subseteq [0, 1]$. Moreover, for each i, the set S_i^I is both r- and s-recognizable, and it follows from Cobham's theorem that this set is definable in $\langle \mathbb{N}, +, < \rangle$. Since such a set is ultimately periodic, there exists $n_i \in \mathbb{N}_{>0}$ for which $\forall x \in \mathbb{N}, x \geq n_i : x \in S_i^I \Leftrightarrow x + n_i \in S_i^I$. By defining $n = \mathrm{lcm}_i(n_i)$, we have $\forall x \in \mathbb{R}_{\geq 0}, x \geq n : x \in S' \Leftrightarrow x + n \in S'$. This prompts the following definition.

Definition 2. *Let $D \subseteq \mathbb{R}$ be a domain, and let $t \in \mathbb{R}$. A set $S \subseteq D$ is t-sum-stable in D iff for all $x \in D$ such that $x + t \in D$, we have $x \in S \Leftrightarrow x + t \in S$.*

Let us show that the set $S'' = (1/n)S'$ is 1-sum-stable in $\mathbb{R}_{>0}$. For every $x \geq 1$, we have $x \in S'' \Leftrightarrow x + 1 \in S''$. For $x < 1$, we choose $k \in \mathbb{N}$ such that $r^k x \geq 1$. Exploiting the properties of S' (transposed to S''), we get $x \in S'' \Leftrightarrow r^k x \in S'' \Leftrightarrow r^k x + r^k \in S'' \Leftrightarrow x + 1 \in S''$. Lemma 3 can thus be refined as follows.

Lemma 4. *Let $r, s \in \mathbb{N}_{>1}$ be two bases with different sets of prime factors, and let $S \subseteq [0, 1]$ be a set that is r- and s-recognizable, r- and s-product-stable in $[0, 1]$, and that has infinitely many boundary points. There exists a set $S' \subseteq \mathbb{R}_{>0}$ that is r- and s-recognizable, has infinitely many boundary points, and is r-product-, s-product- and 1-sum-stable in $\mathbb{R}_{>0}$.*

Note that Lemmas 3 and 4 still hold if the bases r and s are multiplicatively independent.

5.2 Exploiting Sum-Stability Properties

Consider a set $S' \subseteq \mathbb{R}_{>0}$ that satisfies the properties expressed by Lemma 4. It remains to show that these properties lead to a contradiction. The hypothesis on the prime factors of r and s is explicitly used in this section.

We proceed by characterizing the numbers $t \in \mathbb{R}$ for which S' is t-sum-stable in $\mathbb{R}_{>0}$. These form the set $T_{S'} = \{t \in \mathbb{R} \mid \forall x \in \mathbb{R}_{>0} : x + t \in \mathbb{R}_{>0} \Rightarrow (x \in S' \Leftrightarrow x + t \in S')\}$. Since S' is r-recognizable, it is definable in $\langle \mathbb{R}, \mathbb{Z}, +, <, X_r \rangle$, and so is $T_{S'}$, that is therefore r-recognizable as well.

The set $T_{S'}$ enjoys interesting closure properties:

Property 1. *For every $t, u \in T_{S'}$ and $a, b \in \mathbb{Z}$, we have $at + bu \in T_{S'}$.*

The set $T_{S'}$ is also r- and s-product stable in $\mathbb{R}$. Since $1 \in T_{S'}$, this yields the following property.

Property 2. *For every $k \in \mathbb{Z}$, we have $r^k \in T_{S'}$ and $s^k \in T_{S'}$.*

Intuitively, being able to add or subtract r^k from a number, for any k, makes it possible to change in an arbitrary way finitely many digits in its base-r encodings, without influencing the fact that this number belongs or not to S'. Our next step will be to show that this property can be extended to all digits of base-r encodings, implying either $S' = \emptyset$ or $S' = \mathbb{R}_{>0}$. This would then contradict our assumption that S' has infinitely many boundary points.

Property 3. *There exist $l, m \in \mathbb{N}_{>0}$ such that, for every $k \in \mathbb{N}_{>0}$, we have*

$$m/(r^{lk} - 1) \in T_{S'}.$$

Proof. By Property 2, we have $1/s^k \in T_{S'}$ for all $k \in \mathbb{N}$. The base-r encodings of $1/s^k$ are of the form $0^+ \star v_k u_k^\omega$, where u_k is their *period*. Hence, $1/s^k = a_k/(r^{|v_k|}(r^{|u_k|} - 1))$, with $a_k \in \mathbb{N}_{>0}$. Recall that, by hypothesis, there exists a prime factor f of s that does not divide r. Thus f^k must divide $r^{|u_k|} - 1$. It follows that the length of the periods u_k must be unbounded w.r.t. k.

Consider a RNA $\mathcal{A}_r^T$ recognizing $T_{S'}$ in base r. We study the rational numbers accepted by $\mathcal{A}_r^T$, which have base-r encodings of the form $v \star w u^\omega$. We assume w.l.o.g. that the considered periods u are the shortest possible ones. It follows from the unboundedness of u_k that $T_{S'}$ contains rational numbers with infinitely many distinct periods. As a consequence, there exist u, u', v, v', w, w' such that u^ω is not a suffix of $(u')^\omega$, the words $v \star w u^\omega$ and $v' \star w'(u')^\omega$ are both accepted by $\mathcal{A}_r^T$, and the paths π and π' of $\mathcal{A}_r^T$ reading them end up cycling in exactly the same subset of accepting states. (Recall that RNA are deterministic Muller automata.)

Let q be one of these states, and $u_1, u_2 \in \Sigma_r^+$ be periods of the (respective) words read by π and π' after reaching q in their final cycle. These periods can be repeated arbitrarily, hence we can assume w.l.o.g. that $|u_1| = |u_2|$. Moreover we can assume w.l.o.g. that $[u_2]_r > [u_1]_r$, otherwise u^ω would be a suffix of $(u')^\omega$. Besides, there exist $v, w \in \Sigma_r^*$ such that $v \star w$ reaches q. From the structure of $\mathcal{A}_r^T$, it follows that for every $k \geq 0$, the word $v \star w(u_1^k u_2)^\omega$ is accepted by $\mathcal{A}_r^T$.

For each $k \geq 0$, we thus have $[v \star w(u_1^k u_2)^\omega]_r \in T_{S'}$. Developing, we get $d_k/r^{|w|} + [vw \star 0^\omega]_r/r^{|w|} \in T_{S'}$, with $d_k = [\star(u_1^k u_2)^\omega]_r$. Thanks to Properties 1 and 2, and the r-product-stability property of $T_{S'}$, this implies $d_k \in T_{S'}$. We now express d_k in terms of $[u_1]_r$, $[u_2]_r$, and k:

$$d_k = \frac{[u_1^k u_2]_r}{r^{l(k+1)} - 1} = \frac{[u_2]_r - [u_1]_r}{r^{l(k+1)} - 1} + \frac{[u_1]_r}{r^l - 1}, \text{ where } l = |u_1| = |u_2|.$$

The next step will consist in getting rid of the second term of this expression. By Properties 1 and 2, we have for all $k \in \mathbb{N}$,

$$(r^l - 1)d_k - [u_1]_r = \frac{m}{r^{l(k+1)} - 1} \in T_{S'},$$

where $m = (r^l - 1)([u_2]_r - [u_1]_r)$ is such that $m \in \mathbb{N}_{>0}$. For all $k > 0$, we thus have $m/(r^{lk} - 1) \in T_{S'}$. $\qquad\square$

We are now ready to conclude. Given l and m by Property 3, we define $S'' = (1/m)S'$. Like S', this set has infinitely many boundary points. The set $T_{S''}$ of the values t for which S'' is t-sum-stable in $\mathbb{R}_{>0}$ is given by $T_{S''} = (1/m)T_{S'}$. This set is thus r-recognizable. From Properties 1 and 2, we have for every $k \in \mathbb{N}$, $1/r^k \in T_{S''}$. Finally, from Property 3, we have for every $k > 0$, $1/(r^{lk}-1) \in T_{S''}$.

Property 4. *The set $T_{S''}$ is equal to $\mathbb{R}$.*

Proof. Since $T_{S''}$ and $\mathbb{R}$ are both r-recognizable, and two ω-regular languages are equal iff they share the same subset of ultimately periodic words [PP04], it is actually sufficient to show $T_{S''} \cap \mathbb{Q} = \mathbb{Q}$. Every rational t admits a base-r encoding of the form $v \star wu^\omega$, where $|u| = lk$ for some $k \in \mathbb{N}_{>0}$. We have

$$t = \frac{[vw \star 0^\omega]_r}{r^{|w|}} + \frac{[u]_r}{r^{|w|}(r^{lk} - 1)}.$$

Since $1/r^{|w|} \in T_{S''}$ and $1/(r^{lk} - 1) \in T_{S''}$, the closure and product-stability properties of $T_{S''}$ imply $t \in T_{S''}$.   □

As a consequence, we either have $S'' = \emptyset$ or $S'' = \mathbb{R}_{>0}$, which contradicts the hypothesis that this set has infinitely many boundary points. We thus finally have the following theorem.

Theorem 5. *Let r and s be two bases that do not share the same set of prime factors. If a set $S \subseteq \mathbb{R}$ is r- and s-recognizable, then it is definable in $\langle \mathbb{R}, \mathbb{Z}, +, < \rangle$.*

6   Conclusions

In this paper, we have established that the sets of real numbers that can be recognized by finite automata in two sufficiently different bases are exactly those that are definable in the first-order additive theory of real and integer variables $\langle \mathbb{R}, \mathbb{Z}, +, < \rangle$. In the case of weak deterministic automata, used in actual implementations of symbolic representation systems, the condition on the bases turns out to be multiplicative independence. It is worth mentioning that recognizability in multiplicatively dependent bases is equivalent to recognizability in one of them, and that definability in $\langle \mathbb{R}, \mathbb{Z}, +, < \rangle$ implies recognizability in every base. We have thus obtained a complete characterization of the sets of numbers recognizable in multiple bases, similar to the one known for the integer domain [Cob69].

For Muller automata, we have demonstrated that multiplicative independence of the bases is not a strong enough condition, and that the bases must have different sets of prime factors in order to force definability of the represented sets in $\langle \mathbb{R}, \mathbb{Z}, +, < \rangle$. Recall that the sets definable in that theory can all be recognized by weak deterministic automata. We have thus established that the sets of real numbers that can be recognized by infinite-word automata in all encoding bases are exactly those that are recognizable by weak deterministic automata. This result provides a theoretical justification to the use of weak automata as symbolic data structures for representing sets of real and integer numbers.

References

[BB07]   Boigelot, B., Brusten, J.: A generalization of Cobham's theorem to au-
           tomata over real numbers. In: Arge, L., Cachin, C., Jurdziński, T., Tar-
           lecki, A. (eds.) ICALP 2007. LNCS, vol. 4596, pp. 813–824. Springer,
           Heidelberg (2007)
[BBR97]   Boigelot, B., Bronne, L., Rassart, S.: An improved reachability analysis
           method for strongly linear hybrid systems. In: Proc. 9th CAV, Haifa, June
           1997. LNCS, vol. 1254, pp. 167–177. Springer, Heidelberg (1997)
[BHMV94]   Bruyère, V., Hansel, G., Michaux, C., Villemaire, R.: Logic and p-
           recognizable sets of integers. Bulletin of the Belgian Mathematical So-
           ciety 1(2), 191–238 (1994)
[BJW05]   Boigelot, B., Jodogne, S., Wolper, P.: An effective decision procedure for
           linear arithmetic over the integers and reals. ACM Transactions on Com-
           putational Logic 6(3), 614–633 (2005)
[BRW98]   Boigelot, B., Rassart, S., Wolper, P.: On the expressiveness of real and
           integer arithmetic automata. In: Larsen, K.G., Skyum, S., Winskel, G.
           (eds.) ICALP 1998. LNCS, vol. 1443, pp. 152–163. Springer, Heidelberg
           (1998)
[Büc62]   Büchi, J.R.: On a decision method in restricted second order arithmetic.
           In: Proc. International Congress on Logic, Methodoloy and Philosophy of
           Science, pp. 1–12. Stanford University Press, Stanford (1962)
[Cob69]   Cobham, A.: On the base-dependence of sets of numbers recognizable by
           finite automata. Mathematical Systems Theory 3, 186–192 (1969)
[EK06]   Eisinger, J., Klaedtke, F.: Don't care words with an application to the
           automata-based approach for real addition. In: Ball, T., Jones, R.B. (eds.)
           CAV 2006. LNCS, vol. 4144, pp. 67–80. Springer, Heidelberg (2006)
[HW85]   Hardy, G.H., Wright, E.M.: An introduction to the theory of numbers,
           5th edn. Oxford University Press, Oxford (1985)
[McN66]   McNaughton, R.: Testing and generating infinite sequences by a finite
           automaton. Information and Control 9(5), 521 530 (1966)
[Per90]   Perrin, D.: Finite automata. In: van Leeuwen, J. (ed.) Handbook of Theo-
           retical Computer Science. Formal Models and Semantics, vol. B. Elsevier
           and MIT Press (1990)
[PP04]   Perrin, D., Pin, J.E.: Infinite words. Pure and Applied Mathematics,
           vol. 141. Elsevier, Amsterdam (2004)
[Saf88]   Safra, S.: On the complexity of ω-automata. In: Proc. 29th Symposium on
           Foundations of Computer Science, pp. 319–327. IEEE Computer Society,
           Los Alamitos (1988)
[Var07]   Vardi, M.: The Büchi complementation saga. In: Thomas, W., Weil, P.
           (eds.) STACS 2007. LNCS, vol. 4393, pp. 12–22. Springer, Heidelberg
           (2007)
[WB95]   Wolper, P., Boigelot, B.: An automata-theoretic approach to Presburger
           arithmetic constraints. In: Mycroft, A. (ed.) SAS 1995. LNCS, vol. 983,
           pp. 21–32. Springer, Heidelberg (1995)
[Wei99]   Weispfenning, V.: Mixed real-integer linear quantifier elimination. In:
           Proc. ACM SIGSAM ISSAC, Vancouver, pp. 129–136. ACM Press, New
           York (1999)
[Wil93]   Wilke, T.: Locally threshold testable languages of infinite words. In: Proc.
           10th STACS, Würzburg. LNCS, vol. 665, pp. 607–616. Springer, Heidel-
           berg (1993)

On Expressiveness and Complexity
in Real-Time Model Checking

Patricia Bouyer[1,2], Nicolas Markey[2], Joël Ouaknine[1], and James Worrell[1]

[1] Oxford University Computing Laboratory
{First.Last}@comlab.ox.ac.uk
[2] Laboratoire Spécification & Vérification
{First.Last}@lsv.ens-cachan.fr

Abstract. Metric Interval Temporal Logic (MITL) is a popular formalism for expressing real-time specifications. This logic achieves decidability by restricting the precision of timing constraints, in particular, by banning so-called *punctual* specifications. In this paper we introduce a significantly more expressive logic that can express a wide variety of punctual specifications, but whose model-checking problem has the same complexity as that of MITL. We conclude that for model checking the most commonly occurring specifications, such as invariance and bounded response, punctuality can be accommodated at no cost.

1 Introduction

One of the most successful approaches to verification is *model checking*: given a representation S of a system together with a specification φ, determine whether S satisfies φ. In the world of real time, a prominent modelling framework is to use timed automata to represent systems and Metric Temporal Logic (MTL) as the specification formalism.

MTL was proposed nearly twenty years ago by Koymans [12] and has since been extensively studied. MTL is an extension of Linear Temporal Logic (LTL) which allows one to specify a wide range of timed behaviours. The formula $\Box(p \rightarrow \Diamond_{\{1\}}q)$, for example, asserts that whenever the system finds itself in a p-state, then it will be in a q-state precisely one time unit later.

Unfortunately, the model-checking and satisfiability problems for MTL over dense time are undecidable [3,16]. In fact, it was widely held until quite recently that any formalism in which 'punctual' (exact) timing constraints could be expressed would automatically be undecidable—see [3,4,9], among others. The formula given in the previous paragraph is a typical example of a punctual specification.

Many researchers were thus led to consider relaxations and variations of the original MTL formalism in search of decidability and tractability. The identification of *Metric Interval Temporal Logic* (MITL) as a decidable fragment of MTL is a classic result in real-time verification. MITL consists of those formulas in which every constraining interval is non-singular. This syntactic restriction directly removes the problem of punctuality, but correspondingly loses considerable expressiveness. Satisfiability and model checking for MITL were shown to be EXPSPACE-complete in [2] via a translation of formulas into equivalent timed automata; see also [11,15].

L. Aceto et al. (Eds.): ICALP 2008, Part II, LNCS 5126, pp. 124–135, 2008.
© Springer-Verlag Berlin Heidelberg 2008

The starting point of this paper is to identify a new decidable fragment of MTL, which we call Bounded-MTL. This is the subset of MTL in which the constraining intervals appearing in any formula have finite length. For instance $\square_{[0,25)}(p \rightarrow \lozenge_{\{1\}} q)$ is a Bounded-MTL formula. Note that, unlike in MITL, punctual formulas are permitted. We show that Bounded-MTL is decidable in EXPSPACE if the time constraints in formulas are encoded in binary, and in PSPACE if time constraints are encoded in unary. Notwithstanding these bounds, we provide examples of Bounded-MTL formulas that can only be satisfied by signals whose variability is doubly exponential in the size of the formula. Moreover we observe that there exist Bounded-MTL formulas for which there is no equivalent timed automata, unlike the situation for MITL formulas.

Bounded-MTL shows that, at least in the time-bounded setting, punctuality need not be fatal for the complexity of model checking. However the restriction to time-bounded modalities in Bounded-MTL is severe, for example prohibiting the expression of basic safety properties such as invariance. This leads us to isolate the notion of *flatness*, which generalises boundedness. We introduce coFlat-MTL, a natural extension of both MITL and Bounded-MTL, which is closed under the *always* operator $\square$ and the bounded until operator U_I. In particular, if φ is a Bounded-MTL formula, expressing some time-bounded property, then the invariance specification $\square\varphi$ is in coFlat-MTL.

Our main result is that the model checking problem for coFlat-MTL on timed automata is EXPSPACE-complete, that is, in the same complexity class as MITL model checking. This substantiates the main thesis of this paper—that in model checking the most common specifications, including invariance and bounded response, punctuality can be accommodated for free. However we note that coFlat-MTL is not closed under negation, and its satisfiability problem is undecidable. In this respect coFlat-MTL is similar to the branching-time logic TCTL for which model checking is PSPACE-complete but satisfiability is undecidable (again due to the problem of punctuality).

This paper adopts the standard semantics for MTL in which a model of a formula is a *signal*: a function from the positive reals into a finite set, indicating which propositions hold at every instant in time. An alternative approach, used in our earlier work [6], is the so-called *point-based* semantics, which represents models as countable sequences of timestamped snapshots. The signal semantics can be shown to generalise the pointwise semantics. To accommodate this extra generality we had to move from the automata-based proof techniques used in [6] to model-theoretic ones. As a side benefit, this shift has allowed us to lift our previous restriction to finitely-variable models. Finally, the logic which we term coFlat-MTL in the present paper strictly generalises the logic by the same name in [6]; in particular, MITL is now a fragment of coFlat-MTL, so that our results also extend the original EXPSPACE model checking of MITL [2].

2   Metric Temporal Logic

Given a set P of atomic propositions, the formulas of MTL are built from P using Boolean connectives, and time-constrained versions of the *until* operator U as follows:

$$\varphi ::= p \mid \neg\varphi \mid \varphi \wedge \varphi \mid \varphi\, U_I\, \varphi,$$

where $I \subseteq (0, \infty)$ is an interval of reals with endpoints in $\mathbb{N} \cup \{\infty\}$. We sometimes abbreviate $U_{(0,\infty)}$ to U, calling this the *unconstrained* until operator. We assume a dag representation of formulas, and define the size of a formula φ, denoted $|\varphi|$, to be the number of distinct subformulas of φ. We also write M_φ for the maximum finite integer occurring as an endpoint of a constraining interval in φ.

We denote by $\mathbb{R}_+$ the set of nonnegative real numbers. Given a set X, a *signal* is a function $f : \mathbb{R}_+ \to X$. We say that f has *finite variability* if its set of discontinuities has no limit points. We say that f has *variability* $n \in \mathbb{N}$ if it has at most n discontinuities in any open unit-length subinterval $(k, k + 1)$ of its domain, where $k \in \mathbb{N}$. Given an MTL formula φ over the set of propositional variables P, and a signal $f : \mathbb{R}_+ \to 2^P$, the satisfaction relation $f \models \varphi$ is defined inductively, with the classical rules for atomic propositions and Boolean operators, and with the following rule for the "until" modality, where f^t denotes the signal $f^t(s) = f(t + s)$:

$$f \models \varphi_1 \ U_I \ \varphi_2 \text{ iff for some } t \in I, \ f^t \models \varphi_2 \text{ and } f^u \models \varphi_1 \text{ for all } u \in (0, t).$$

Note that we adopt a *strict* semantics for U_I, in which the judgement $f \models \varphi_1 \ U_I \ \varphi_2$ is independent of $f(0)$ (recall that $0 \notin I$ by assumption). In the following we write $\varphi_1 \ U \ \varphi_2$ for $\varphi_1 \ U_{(0,\infty)} \ \varphi_2$.

In general we do not assume that signals are finitely variable. Indeed there are formulas that are satisfiable only by *infinitely* variable signals: *e.g.* $\neg(p \ U \ p) \wedge \neg(\neg p \ U \ \neg p)$.

Further connectives can be defined following the usual conventions. In addition to propositions $\top$ (true) and $\bot$ (false), and to disjunction $\vee$, we have the *constrained eventually* operator $\Diamond_I \varphi \equiv \top \ U_I \ \varphi$, the *constrained always* operator $\Box_I \varphi \equiv \neg \Diamond_I \neg \varphi$, and the *constrained dual until* operator $\varphi_1 \ \widetilde{U}_I \ \varphi_2 \equiv \neg((\neg \varphi_1) \ U_I \ (\neg \varphi_2))$.

Admitting only $\widetilde{U}_I$ as an extra connective one can transform any MTL formula into an equivalent *negation normal form*, in which negation is only applied to propositional variables.

3 Decidable Sublogics

It is well known that both model checking and satisfiability for MTL are highly undecidable (Σ_1^1-complete) [2]. Here we consider syntactic restrictions yielding sublogics with decidable model checking problem.

One approach, due to Alur, Feder and Henzinger [2], involves placing restrictions on *punctuality*. We say that a formula φ is *punctual* if its outermost connective is a temporal modality with a singular constraining interval, e.g., $\Diamond_{\{1\}} p$. Intuitively, a punctual formula specifies an exact timing constraint. *Metric Interval Temporal Logic* (MITL) is the subset of MTL in which all constraining intervals are non-singular, that is, in which punctual formulas are banned. The satisfiability and model checking problems for MITL are EXPSPACE-complete.

In this paper our starting point is, in some sense, dual to that of [2]. Rather than ban constraining intervals that are *too small*, we ban constraining intervals that are *too big*. We define Bounded-MTL to be the subset of MTL in which all constraining intervals have finite length, and we show that the satisfiability and model checking problems for Bounded-MTL are EXPSPACE-complete (or PSPACE-complete if timing constraints are encoded in unary), matching the complexity of MITL. However the following example illustrates the fundamentally different character of MITL and Bounded-MTL.

Example 1. Consider the **Bounded-MTL** formula $\varphi \equiv \Box_{(0,1)}(p \leftrightarrow \Diamond_{\{1\}}p)$. A variation on a well-known result tells us that the set of signals satisfying φ is not realisable as the language of a timed automaton [1,5]. Therefore φ defines a property that is not expressible in **MITL** since **MITL** formulas can be transformed into equivalent timed automata [2].

MITL and **Bounded-MTL** represent two different approaches to defining decidable metric temporal logics, and they have incomparable expressive power. In particular, **Bounded-MTL** is not capable of expressing invariance—one of the most basic safety specifications. To repair this deficiency we introduce *flatness* as a generalisation of boundedness. Our use of this term is motivated by similarities with logics introduced in [7,8].

We say that an **MTL** formula in negation normal form is *flat* if *(i)* in any subformula of the form $\varphi_1 \, U_I \, \varphi_2$, either I is bounded or φ_1 is in **MITL**, and *(ii)* in any subformula of the form $\varphi_1 \, \tilde{U}_I \, \varphi_2$, either I is bounded or φ_2 is in **MITL**. For example $\Box\varphi \equiv \bot \, \tilde{U} \, \varphi$ is flat if φ is in **MITL**. The intuition behind flatness is that *potentially persistent* subformulas must be in **MITL**. We write **Flat-MTL** for the fragment of **MTL** composed of all flat formulas.

Flatness is a key technical notion in this paper, however our main results are most naturally understood in terms of the dual notion, *coflatness*. A formula is coflat if it is the negation of a flat formula. More explicitly we say that a formula is *coflat* if *(i)* in any subformula of the form $\varphi_1 \, U_I \, \varphi_2$, either I is bounded or φ_2 is in **MITL**, and *(ii)* in any subformula of the form $\varphi_1 \, \tilde{U}_I \, \varphi_2$, either I is bounded or φ_1 is in **MITL**. If we write **coFlat-MTL** for the sublogic of coflat formulas then **coFlat-MTL** includes both **Bounded-MTL** and **MITL**, is closed under $\Box_I$ for arbitrary I (invariance), and is closed under U_I for bounded I (bounded liveness). Thus, for specifications, coflatness is a much less restrictive property than flatness. While this generality renders the satisfiability problem for **coFlat-MTL** undecidable (the undecidability proof of [2] for the satisfiability of **MTL** makes use only of formulas in **coFlat-MTL**), we show that the model checking problem is no harder than for **MITL**.

Example 2. The formula $(\Box\Diamond_{(0,1)} \bigvee_{m \in M} in_m) \wedge (\Box \bigwedge_{m \in M}(in_m \rightarrow \Diamond_{\{1\}} out_m))$ specifies the behaviour of a perfect buffer which processes each message in one time unit, operating in an environment where at least one message arrives every time unit. This formula is in **coFlat-MTL**, but is not in **Bounded-MTL** (due to the unconstrained $\Box$) and is not in **MITL** (due to the punctual $\Diamond_{\{1\}}$).

Model Checking. The model checking problem for **coFlat-MTL** asks, given a timed automaton $\mathcal{A}$ and a **coFlat-MTL** formula φ, whether all (finitely variable) signals accepted by $\mathcal{A}$ also satisfy φ. Rather than formally introducing timed automata we rely on a result of [10,17] that for each timed automaton $\mathcal{A}$ there is an **MITL** formula $\varphi_{\mathcal{A}}$, of size polynomial in $\mathcal{A}$, such that the language of $\mathcal{A}$ is a projection of the language of $\varphi_{\mathcal{A}}$. Since **Flat-MTL** subsumes **MITL**, using this result we can reduce the model checking problem for **coFlat-MTL** to the satisfiability problem for the dual logic **Flat-MTL**. The main result of this paper is that the latter problem is **EXPSPACE**-complete, as is the same problem for **MITL** [2].

Theorem 1. *The model-checking problem for* coFlat-MTL *is* EXPSPACE-*complete.*

The proof of Theorem 1 occupies Sections 5 and 6. The decision procedure involves a satisfiability-respecting translation of Flat-MTL into Linear Temporal Logic over the reals. In this translation the non-punctual connectives in Flat-MTL are handled using similar techniques to [11]. Dealing with the punctual connectives, however, requires completely new ideas.

4 Hardness

Proposition 1. *The satisfiability problem for* Bounded-MTL *is* EXPSPACE-*hard.*

Proof. Given a 2^n-space-bounded Turing machine $\mathcal{M}$ with input X, we construct in logarithmic space a Bounded-MTL formula $\varphi_{\mathcal{M},X}$ that is satisfiable if and only if $\mathcal{M}$ accepts X. This reduction bears some similarities with the undecidability proof for MTL [2], but it also differs in important respects. Indeed, directly applying the latter proof to Bounded-MTL would only yield EXPTIME-hardness.

We now sketch the main ideas behind the definition of $\varphi_{\mathcal{M},X}$. Suppose that $\mathcal{M}$ has set of control states S and tape alphabet Σ. The set of atomic propositions used by $\varphi_{\mathcal{M},X}$ is $P \cup \dot{P}$, where $P = \{p_\sigma, p_{\sigma,s} : \sigma \in \Sigma, s \in S\}$ and $\dot{P} = \{\dot{p} : p \in P\}$. Intuitively, proposition p_σ represents a tape cell that currently contains σ, whereas $p_{\sigma,s}$ represents a tape cell that currently contains σ and is pointed to by the head of $\mathcal{M}$, while $\mathcal{M}$ is in control state s. The dot is used as a pointer to aid in simulating $\mathcal{M}$: an entire computation of $\mathcal{M}$ is encoded in each time unit, and each step of the computation is checked using the distinguished dotted propositions in two consecutive unit intervals.

$\varphi_{\mathcal{M},X}$ is written as the conjunction of three components

$$\varphi_{\mathcal{M},X} \equiv \varphi_{UNIQUE} \wedge \varphi_{COPY} \wedge \varphi_{CHECK}.$$

The formula φ_{UNIQUE}, which is straightforward to formalise, ensures that any signal satisfying $\varphi_{\mathcal{M},X}$ defines a left-continuous function $f : [0, 2^n] \to P \cup \dot{P}$, that is, only one proposition holds at each moment, and propositions do not hold instantaneously.

The purpose of φ_{COPY} and φ_{CHECK} is to ensure that in any signal satisfying $\varphi_{\mathcal{M},X}$ the sequence of propositions holding in the time interval $[0, 1)$ encodes the computation history of $\mathcal{M}$ on X. Within this, the job of φ_{COPY} is to copy the sequence of propositions holding in each unit-duration time interval into the subsequent time interval, at the same time moving the dot superscript 'one place to the right'. Formally we have

$$\varphi_{COPY} = \bigwedge_{p \in P} \Box_{[0,2^n]}(p \to \Diamond_{\{1\}}(p \vee \dot{p}))$$

$$\wedge \bigwedge_{p,q \in P} \Box_{[0,2^n]}((\dot{p}\, U_{(0,1)}\, q) \leftrightarrow \Diamond_{\{1\}}(p\, U_{(0,1)}\, \dot{q})),$$

where $\Box_{[0,2^n]}\psi$ is a shorthand for $\psi \wedge \Box_{(0,2^n]}\psi$.

Thus the sequence of propositions holding in each subsequent time interval $[k, k+1)$, $k = 1, \ldots, n-1$, should also represent the computation history of $\mathcal{M}$ on X. The only difference is that in the interval $[k, k+1)$ the dot should decorate exactly those propositions encoding the contents of the k-th tape cell in each configuration in the computation history.

The role of φ_{CHECK} is to verify that the sequence of propositions holding in each subsequent unit-length interval does indeed encode the computation history of $\mathcal{M}$ on X. As it 'reads' the segment of the input signal defined over the time interval $[k, k+1)$, φ_{CHECK} uses the dots as pointers to check the correctness of the k-th tape cell in each configuration. Thus, in 2^n time units the whole computation is checked. We omit the details of φ_{CHECK}, but point out that it is equivalent to an LTL formula. (In fact each modality is decorated with the constraining interval $(0, 2^n)$ merely to ensure that φ_{CHECK} is in **Bounded-MTL**). $\square$

The proof of Proposition 1 assumes that constants are encoded in binary (in order to concisely write $\square_{[0,2^n]}$). It can be proved that model checking **Bounded-MTL** drops to **PSPACE** when constants are encoded in unary. However, for the more expressive logic **Flat-MTL** we can adapt the above encoding to show **EXPSPACE**-hardness assuming only unary encoding of constants. Thus Theorem 1 holds irrespective of whether constants are encoded in unary or binary.

5 Closure Labellings

It is well-known that the constrained until and dual-until operators U_I and $\widetilde{U}_I$ can be expressed in terms of the unconstrained operators U and $\widetilde{U}$ and the unary operators $\square_I$ and $\diamondsuit_I$ [11,15]. Unfortunately, adopting this simplification makes it impossible to express flatness as a syntactic property, hence we prefer to retain a bit more flexibility in our basic syntax. To this end we say that an **MTL** formula is in *constraint normal form* if it is generated by the grammar

$$\varphi ::= p \mid \neg p \mid \varphi_1 \wedge \varphi_2 \mid \varphi_1 \vee \varphi_2 \mid \varphi_1 \, U_I \, \varphi_2 \mid \varphi_1 \, \widetilde{U}_I \, \varphi_2 \mid \diamondsuit_J \varphi \mid \square_J \varphi,$$

where I is a left-open, initial (*i.e.*, with left end-point 0) interval, while J is arbitrary.

Any **MTL** formula can be transformed into an equivalent constraint normal form using equivalences such as

$$\varphi_1 \, U_{(\ell,r]} \, \varphi_2 \quad \leftrightarrow \quad \square_{(0,\ell]}(\varphi_1 \, U_{(0,r]} \, \varphi_2) \wedge \diamondsuit_{(\ell,r]}\varphi_2.$$

This transformation is linear with respect to the DAG-size of formulas and it preserves both **MITL** and **Flat-MTL**. Henceforth, without loss of generality, we assume that all formulas are in constraint normal form.

Given $I \subseteq \mathbb{R}_+$ and $n \in \mathbb{N}$, write $I - n = \{x \in (0, \infty) : x + n \in I\}$. Define the closure $cl(\varphi)$ of a formula φ to be the smallest set such that the following hold (where we adopt the identifications $\square_\emptyset \varphi \equiv \top$ and $\diamondsuit_\emptyset \varphi \equiv \bot$).

C1. $cl(\varphi)$ contains all subformulas of φ
C2. $\varphi_1 \, U_I \, \varphi_2 \in cl(\varphi)$ implies $\varphi_1 \, U \, \varphi_2, \diamondsuit_I \varphi_2 \in cl(\varphi)$
C3. $\varphi_1 \, \widetilde{U}_I \, \varphi_2 \in cl(\varphi)$ implies $\varphi_1 \, \widetilde{U} \, \varphi_2, \square_I \varphi_2 \in cl(\varphi)$

C4. $\Box_J \varphi_1 \in cl(\varphi)$ implies $\Box_{J-1} \varphi_1 \in cl(\varphi)$
C5. $\Diamond_J \varphi_1 \in cl(\varphi)$ implies $\Diamond_{J-1} \varphi_1 \in cl(\varphi)$.

For example, $cl(\Box_{(1,\infty)} p \wedge \Diamond_{\{1\}} q) = \{\bot, p, q, \Diamond_{\{1\}} q, \Box_{(1,\infty)} p, \Box_{(0,\infty)} p, \Box_{(1,\infty)} p \wedge \Diamond_{\{1\}} q\}$.

It is straightforward to verify that $cl(\varphi)$ has cardinality $O(|\varphi| \cdot M_\varphi)$. We note also that if $\varphi \in$ MITL, then $cl(\varphi) \subseteq$ MITL; in particular, the interval $I - 1$ is a singleton only if I is a singleton.

Given an MTL formula φ in constraint normal form, we define a *closure labelling* to be a signal $f: \mathbb{R}_+ \to 2^{cl(\varphi)}$ such that Rules CL1–CL10 below are satisfied for all $s \in \mathbb{R}_+$. Closure labellings are continuous-time counterparts of Hintikka sequences [19]. Here we denote by P the set of propositions mentioned in φ. We also assume that Rules CL5 and CL6 apply to $\Diamond \varphi_1$ and $\Box \varphi_1$, respectively, under the identifications $\Diamond \varphi_1 \equiv \top \, U \, \varphi_1$ and $\Box \varphi_1 \equiv \bot \, \tilde{U} \, \varphi_1$.

CL1. $\bot \notin f(s)$;
CL2. exactly one of p and $\neg p$ lies in $f(s)$ for any $p \in P$;
CL3. $\varphi_1 \wedge \varphi_2 \in f(s)$ implies $\varphi_1 \in f(s)$ and $\varphi_2 \in f(s)$;
CL4. $\varphi_1 \vee \varphi_2 \in f(s)$ implies $\varphi_1 \in f(s)$ or $\varphi_2 \in f(s)$;
CL5. $\varphi_1 \, U \, \varphi_2 \in f(s)$ implies there exists $t > s$ such that $\varphi_2 \in f(t)$ and $\varphi_1 \, U \, \varphi_2, \varphi_1 \in f(u)$ for all $u \in (s, t)$;
CL6. $\varphi_1 \, \tilde{U} \, \varphi_2 \in f(s)$ implies for all $t > s$, if $\varphi_2 \notin f(t)$ then there exists $u \in (s, t)$ with $\varphi_1 \in f(u)$, and if $\varphi_1 \, \tilde{U} \, \varphi_2 \notin f(t)$ then there exists $u \in (s, t]$ with $\varphi_1 \in f(u)$;
CL7. $\varphi_1 \, U_I \, \varphi_2 \in f(s)$ implies $\varphi_1 \, U \, \varphi_2 \in f(s)$ and $\Diamond_I \varphi_2 \in f(s)$;
CL8. $\varphi_1 \, \tilde{U}_I \, \varphi_2 \in f(s)$ implies $\varphi_1 \, \tilde{U} \, \varphi_2 \in f(s)$ or $\Box_I \varphi_2 \in f(s)$;
CL9. $\Box_J \varphi_1 \in f(s)$ implies $\Box_{J-1} \varphi_1 \in f(s+1)$ and $\varphi_1 \in f(s+\delta)$ for all $\delta \in (0, 1] \cap J$;
CL10. $\Diamond_J \varphi_1 \in f(s)$ implies $\Diamond_{J-1} \varphi_1 \in f(s + 1)$ unless $\varphi_1 \in f(s + \delta)$ for some $\delta \in (0, 1] \cap J$.

Rules CL1–CL10 encode the semantics of MTL in a natural way. However it is worth noting though that constrained until U_I and dual until $\tilde{U}_I$ are handled indirectly, via Rules CL7 and CL8. Note also that the correctness of CL7 and CL8 depends on the assumption that the interval I appearing in these rules is initial, which holds because φ is in constraint normal form.

The following straightforward proposition expresses the expected property of closure labellings.

Proposition 2. *A* Flat-MTL *formula φ is satisfiable iff there is a closure labelling $g: \mathbb{R}_+ \to 2^{cl(\varphi)}$ with $\varphi \in g(0)$.*

5.1 The Partition Lemma

Next we identify some structure on the closure labellings of Flat-MTL formulas. To this end, say that $E \subseteq \mathbb{R}_+$ is a *basic set* if it can be written as a finite union of compact intervals with integer end-points: $E = E_1 \cup E_2 \cup \cdots \cup E_n$. We define $length(E)$ in the obvious manner as the sum of the lengths of the E_i. Given a basic set $E \subseteq \mathbb{R}_+$, we say that a closure labelling g is *E-rigid* if $g(t)$ contains punctual formulas[1] only when $t \in E$. The term *rigid* anticipates the development in Section 6.2.

The following result crucially relies on the flatness of φ:

Lemma 1 (Partition Lemma). *Let φ be a* Flat-MTL *formula and $g \colon \mathbb{R}_+ \to 2^{cl(\varphi)}$ a closure labelling with $\varphi \in g(0)$. Then there is a basic set E with $length(E) \leqslant M_\varphi \cdot 2^{|\varphi|}$ and an E-rigid closure labelling h with $\varphi \in h(0)$.*

Remark 1. In case $\varphi \in$ Bounded-MTL, the Partition Lemma can be strengthened by requiring that $length(E) \leqslant M_\varphi \cdot |\varphi|$. This makes our algorithm be in PSPACE for Bounded-MTL with unary-encoded integers.

Given a closure labelling $f \colon \mathbb{R}_+ \to 2^{cl(\varphi)}$, the *non-punctual part* of f is the function $f_{np} \subseteq f$, where $f_{np}(t)$ consists of the set of formulas in $f(t)$ of the form $\Box_I \varphi_1$ or $\Diamond_I \varphi_1$ with I non-singular.

Consider a signal $f \colon \mathbb{R}_+ \to 2^P$, and assume that $t_1 < t_2 < t_3 < t_1 + 1$, and that f^{t_1} and f^{t_3} both satisfy $\Box_I \psi$ with I non-singular; then it is easily shown that f^{t_2} also satisfies $\Box_I \psi$. Thus $\Box_I \psi$ changes its truth value at most 3 times in any unit interval. By duality, it also holds that $\Diamond_I \psi$ also changes its truth value at most 3 times in any unit interval. Following this line of reasoning we can assume in Proposition 2 that g_{np} has variability at most $3 \cdot M_\varphi \cdot |\varphi|$. Moreover the construction underlying the proof of the Partition Lemma is such that the only discontinuities in h, other than those in g, are integer-valued. In summary we have:

Proposition 3. *A* Flat-MTL *formula φ is satisfiable if, and only if, there is a basic set E with $length(E) \leq M_\varphi \cdot 2^{|\varphi|}$ and an E-rigid closure labelling $h \colon \mathbb{R}_+ \to 2^{cl(\varphi)}$ such that $\varphi \in h(0)$ and h_{np} has variability at most $3 \cdot M_\varphi \cdot |\varphi|$.*

6 The Decision Procedure

In this section we describe an EXPSPACE decision procedure for the Flat-MTL satisfiability problem. As explained in Section 3, this implies that the model checking problem for coFlat-MTL is also in EXPSPACE. To achieve this we utilise a technique, inspired by [11], to give a translation of Flat-MTL into LTL+Past[2] that respects the satisfiability of formulas.

6.1 Tableaux

The rules CL9 and CL10 in Section 5 treat punctual and non-punctual connectives alike. We now introduce a modified notion of closure labelling, called a *tableau*, in which punctual and non-punctual connectives are handled differently. To motivate the definition of a tableau, consider the following 'stacking' construction on a closure labelling $g \colon \mathbb{R}_+ \to 2^{cl(\varphi)}$. Given an integer $k \geq 1$, define $T \colon \mathbb{R}_+ \to \left(2^{cl(\varphi)} \right)^k$

---

[1] We recall (see Section 3) that a formula is *punctual* if its outermost connective is a temporal modality with a singular constraining interval.

[2] LTL+Past is the classical extension of LTL with past-time modalities [13].

by $T(t) = \langle g(t), g(t+1), \ldots, g(t+k) \rangle$. We can think of T as a multi-track closure labelling in which the i-th track is the function $T_i \colon \mathbb{R}_+ \to 2^{cl(\varphi)}$ defined by $T_i(t) = T(t)_i$. Notice that the $(i+1)$-th track is one time unit ahead of the i-th track. Motivated by this construction, we axiomatise the notion of a *tableau* below.

Given an integer $k \geq 1$, we say that a signal $T \colon \mathbb{R}_+ \to \left(2^{cl(\varphi)}\right)^k$ is a tableau if the following rules are satisfied for each $0 \leq i \leq k-1$ and $s \in \mathbb{R}_+$:

TH1. T_i satisfies the closure labelling axioms CL1–CL8;

TH2. T_i satisfies the versions of CL9 and CL10 in which the constraining interval J is non-singular;

TV1. If $0 \leq i < k-1$ then $\Box_J \psi \in T(s)_i$ implies $\Box_{J-1} \psi \in T(s)_{i+1}$, $\psi \in T_i(s+\delta)$ for all $\delta \in (0,1] \cap J$ such that $s + \delta \leq \lceil s \rceil$, and $\psi \in T_{i+1}(s+\delta-1)$ for all $\delta \in (0,1] \cap J$ such that $\lfloor s \rfloor \leq s + \delta - 1$;

TV2. If $0 \leq i < k-1$ then $\Diamond_J \psi \in T(s)_i$ implies that either $\Diamond_{J-1} \psi \in T(s)_{i+1}$, or there exists $\delta \in (0,1] \cap J$ such that either $s + \delta \leq \lceil s \rceil$ and $\psi \in T(s+\delta)_i$, or $s + \delta - 1 \geq \lfloor s \rfloor$ and $\psi \in T(s+\delta-1)_{i+1}$;

TV3. if $0 \leq i < k-1$ then for each $n \in \mathbb{N}$ such that $n > 0$ we have $T(n)_i = T(n-1)_{i+1}$.

We think of TH1 and TH2 as *horizontal* rules, since they concern individual tracks of T. They say that each track T_i would be a closure labelling but for the fact that rules CL9 and CL10 need only hold for non-punctual connectives.

Next come the *vertical* rules TV1–TV3, which relate points on different tracks of T. TV1 and TV2 are vertical counterparts of CL9 and CL10 for punctual and non-punctual formulas. Note how TV3 reflects the intuition that $T_{i+1}(s) = T_i(s+1)$.

Since TH2 does not apply to punctual connectives, the tableau axioms do not accurately capture the semantics of arbitrary **MTL** formulas. However, for **Flat-MTL** formulas the notion of rigidity, defined in Section 5, comes to the rescue. Intuitively in a tableau of an **Flat-MTL** formula we rely on the vertical rules to handle punctual subformulas and we rely on the horizontal rules otherwise. Since the number of tracks of a tableau is finite, the correctness of this idea depends on the existence of bounds on the parts of the tableau containing punctual subformulas. To this end we first extend the notion of rigidity to tableaux by saying that a tableau T is E-rigid iff each track T_i is E_i-rigid, where $E_i = \{t : t + i \in E\}$. Then we have the following result:

Proposition 4. *Given a basic set E and $k \geq length(E)$, there is an E-rigid closure labelling $g \colon \mathbb{R}_+ \to 2^{cl(\varphi)}$ with $\varphi \in g(0)$ if, and only if, there is an E-rigid tableau $T \colon \mathbb{R}_+ \to \left(2^{cl(\varphi)}\right)^k$ with $\varphi \in T_0(0)$.*

Proof (sketch). If $g \colon \mathbb{R}_+ \to 2^{cl(\varphi)}$ is an E-rigid closure labelling with $\varphi \in g(0)$, then $T \colon \mathbb{R}_+ \to \left(2^{cl(\varphi)}\right)^k$ defined by $T(t) = \langle g(t), g(t+1), \ldots, g(t+k) \rangle$ is an E-rigid tableau with $\varphi \in T_0(0)$. Indeed, only rules TV1–TV3 need to be checked, and TV1 (resp. TV2) is simply a consequence of CL9 (resp. CL10). The rule TV3 is satisfied by construction.

Conversely, given an E-rigid tableau T we construct a closure labelling g by splicing together unit-length segments from different tracks of T. The idea is that if a given segment contains a punctual formula then it is concatenated with the segment immediately below on the next track; otherwise it is concatenated with its right neighbour on the same track. More precisely, define $\sigma \colon \mathbb{N} \to \{0, \ldots, k-1\}$ by $\sigma(0) = 0$ and

$$\sigma(n+1) = \begin{cases} \sigma(n) + 1 & \text{if } [n, n+1] \subseteq E \\ \sigma(n) & \text{otherwise.} \end{cases}$$

Then $g(t) = T(t - \sigma(\lfloor t \rfloor))_{\sigma(\lfloor t \rfloor)}$ is an E-rigid closure labelling ($k \geq length(E)$). □

Combining Lemma 1 and Proposition 4, we obtain the following result.

Corollary 1. *A* Flat-MTL *formula φ is satisfiable if, and only if, there is a basic set E with $length(E) \leq M_\varphi \cdot 2^{|\varphi|}$ and an E-rigid tableau $T \colon \mathbb{R}_+ \to \left(2^{cl(\varphi)}\right)^k$ with $\varphi \in T_0(0)$ and $k = length(E)$. Moreover we can assume that T_{np} has variability $3 \cdot M_\varphi^2 \cdot |\varphi| \cdot 2^{|\varphi|}$.*

6.2 The Stretching Lemma

Say that two signals $f, g \colon \mathbb{R}_+ \to X$ are *stretching equivalent*, denoted $f \sim g$, if there is a monotone bijection $h \colon \mathbb{R}_+ \to \mathbb{R}_+$ such that $g = f \circ h$. In this case it is easy to see that f and g satisfy the same LTL+Past properties. Our translation from Flat-MTL to LTL+Past relies on an observation of [11] that simple metric properties can be specified in LTL+Past up to stretching equivalence.

   Given an integer N, and set of atomic propositions $\Delta_N = \{d_j, d'_j : 1 \leq j \leq N\} \cup \{p_\checkmark\}$, let θ_N be an LTL formula enforcing the following properties: *(i)* the propositions d_j, d'_j and $p_\checkmark$ all hold punctually; *(ii)* the propositions d_j are mutually exclusive and the d'_j are also mutually exclusive; *(iii)* $p_\checkmark$ holds at time 0 and thereafter holds infinitely often; *(iv)* in between each occurrence of $p_\checkmark$ each d_j holds exactly once, and the d_j hold in the order $d_1, d_2, \ldots, d_N$ (and similarly for the d'_j). We omit the formal definition of θ_N, which is straightforward. Lemma 2, below, states that a signal f that satisfies *(i)–(iv)* can be stretched into one in which $p_\checkmark$ holds precisely at integer time-points and every time d_j holds then d'_j holds one time unit later. The proof uses a construction from [11, Lemma 10].

Lemma 2 (Stretching Lemma). *If $f \models \theta_N$ then there exists a signal $g \sim f$ such that $g^t \models p_\checkmark$ iff $t \in \mathbb{N}$, and $g^{t+1} \models d'_j$ whenever $g^t \models d_j$.*

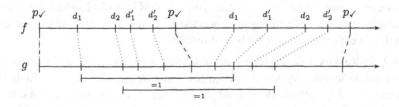

Fig. 1. The stretching lemma

6.3   Translation to LTL+Past

Given a Flat-MTL formula φ we define an LTL+Past formula $\varphi°$ such that φ is satisfiable iff $\varphi°$ is satisfiable. The idea is that $\varphi°$ encodes the tableau rules for φ and the E-rigidity condition. To this end, $\varphi°$ uses the set of propositions $Q = \{p_{\psi,i} : \psi \in cl(\varphi), 0 \le i \le k-1\}$, where k, which represents the height of the tableau, will be chosen later. Then given a signal $f: \mathbb{R}_+ \to 2^Q$, the stretching equivalent signal g given by Lemma 2 naturally encodes a function $T: \mathbb{R}_+ \to 2^{(cl(\varphi))^k}$ by $T_i(t) = \{\psi : p_{\psi,i} \in g(t)\}$. The definition of $\varphi°$ is such that $f \models \varphi°$ iff $g \models \varphi°$ iff T is a tableau for φ.

Most of the tableau rules can be straightforwardly encoded in $\varphi°$. For example, TH1 is captured by formulas such as $\Box(p_{\varphi_1 \wedge \varphi_2,i} \to (p_{\varphi_1,i} \wedge p_{\varphi_2,i}))$ and $\Box(p_{\varphi_1 U \varphi_2,i} \to p_{\varphi_1,i} \ U \ p_{\varphi_2,i})$; corresponding to Rule TV1 we have formulas such as $\Box(p_{\Box_I \psi,i} \to p_{\Box_{I-1}\psi,i+1})$.

The most interesting part of the translation concerns the rule TH2: the horizontal rule dealing with the constrained and non-punctual connectives $\Box_J$ and $\Diamond_J$. To help encode this, we choose a suitable constant N (more on this later) and include formula θ_N, from Section 6.2, as a conjunct of $\varphi°$. We furthermore specify in $\varphi°$ that propositions of the form $p_{\Diamond_I \psi,i}$ and $p_{\Box_I \psi,i}$ with I non-singular only change truth value when one of the d_j holds (think of the d_j as marking discontinuities in T_{np}). Now consider some signal that satisfies $\varphi°$; let s_j^n denote the time-point of the n-th occurrence of d_j and let t_j^n denote the time-point of the n-th occurrence of d'_j. By Lemma 2 we can assume without loss of generality that $t_j^n = s_j^n + 1$. But now it is easy to encode TH2. For instance, by referring to the propositions d_j and d'_j we can specify in $\varphi°$ that if $p_{\Box_I \psi,i}$ holds in an interval (s_j^n, s_{j+1}^n) then $p_{\Box_{I-1}\psi,i}$ holds in the interval $(t_j^{n+1}, t_{j+1}^{n+1})$.

It only remains to choose the constants k and N. The choice should be such that if φ is satisfiable then there should exist a tableau T with k tracks such that T_{np} has variability at most N. But then Corollary 1 shows that we can take $k = M_\varphi \cdot 2^{|\varphi|}$ and $N = 3 \cdot M_\varphi^2 \cdot |\varphi| \cdot 2^{|\varphi|}$. Note that since k and N are both exponential in the size of the description of φ, formula $\varphi°$ may be exponentially bigger than φ. The correctness of the construction is stated below.

Theorem 2. *Let φ be a Flat-MTL formula, and $\varphi°$ be the LTL+Past formula defined above. Then, φ is satisfiable iff $\varphi°$ is satisfiable.*

In summary, we have a satisfiability-respecting exponential translation from Flat-MTL to LTL+Past. Now it is known that the satisfiability problem for LTL+Past over $\mathbb{R}_+$ is PSPACE-complete [18,14] and we conclude that the satisfiability problem for Flat-MTL and the model checking problem for coFlat-MTL are both in EXPSPACE. As a final remark we observe that, due to the factor $2^{|\varphi|}$ in the expressions for N and k, the exponential blow-up in the translation from Flat-MTL to LTL+Past arises even if the timing constraints in formulas are encoded in unary (as mentioned at the end of Section 4, this exponential blow-up is unavoidable).

Remark 2. In Remark 1 we noticed that the length of the basic set for Bounded-MTL formulas can be bounded by $M_\varphi \cdot |\varphi|$ when constants are encoded in unary. In that case, we can take $k = M_\varphi \cdot |\varphi|$ and $N = 3 \cdot M_\varphi^2 \cdot |\varphi|^2$, and the size of the LTL+Past formula is now polynomial. Hence, under the hypothesis that constants are encoded in unary, the satisfiability and model checking problems for Bounded-MTL are in PSPACE.

References

1. Alur, R., Dill, D.: A theory of timed automata. TCS 126(2), 183–235 (1994)
2. Alur, R., Feder, T., Henzinger, T.A.: The benefits of relaxing punctuality. J. of the ACM 43(1), 116–146 (1996)
3. Alur, R., Henzinger, T.A.: Logics and models of real time: A survey. In: Real-Time: Theory in Practice, Proc. REX Workshop 1991. LNCS, vol. 600, pp. 74–106. Springer, Heidelberg (1992)
4. Alur, R., Henzinger, T.A.: Real-time logics: Complexity and expressiveness. Inf. & Comp. 104(1), 35–77 (1993)
5. Alur, R., Madhusudan, P.: Decision problems for timed automata: A survey. In: Bernardo, M., Corradini, F. (eds.) SFM-RT 2004. LNCS, vol. 3185, pp. 1–24. Springer, Heidelberg (2004)
6. Bouyer, P., Markey, N., Ouaknine, J., Worrell, J.: The cost of punctuality. In: Proc. 22nd Ann. IEEE Symp. Logic in Computer Science (LICS 2007), pp. 109–118. IEEE, Los Alamitos (2007)
7. Comon, H., Cortier, V.: Flatness is not a weakness. In: Clote, P.G., Schwichtenberg, H. (eds.) CSL 2000. LNCS, vol. 1862, pp. 262–276. Springer, Heidelberg (2000)
8. Demri, S., Lazić, R., Nowak, D.: On the freeze quantifier in constraint LTL: Decidability and complexity. Inf. & Comp. 205(1), 2–24 (2007)
9. Henzinger, T.A.: It's about time: Real-time logics reviewed. In: Sangiorgi, D., de Simone, R. (eds.) CONCUR 1998. LNCS, vol. 1466, pp. 439–454. Springer, Heidelberg (1998)
10. Henzinger, T.A., Raskin, J.-F., Schobbens, P.-Y.: The regular real-time languages. In: Larsen, K.G., Skyum, S., Winskel, G. (eds.) ICALP 1998. LNCS, vol. 1443, pp. 580–591. Springer, Heidelberg (1998)
11. Hirshfeld, Y., Rabinovich, A.: Timer formulas and decidable metric temporal logic. Inf. & Comp. 198(2), 148–178 (2005)
12. Koymans, R.: Specifying real-time properties with metric temporal logic. Real-Time Systems 2(4), 255–299 (1990)
13. Lichtenstein, O., Pnueli, A., Zuck, L.D.: The glory of the past. In: Proc. Conference on Logics of Programs. LNCS, vol. 193, pp. 413–424. Springer, Heidelberg (1985)
14. Lutz, C., Walther, D., Wolter, F.: Quantitative temporal logics over the reals: PSPACE and below. Inf. & Comp. 205(1), 99–123 (2007)
15. Maler, O., Nickovic, D., Pnueli, A.: From MITL to timed automata. In: Asarin, E., Bouyer, P. (eds.) FORMATS 2006. LNCS, vol. 4202, pp. 274–289. Springer, Heidelberg (2006)
16. Ouaknine, J., Worrell, J.: On metric temporal logic and faulty Turing machines. In: Aceto, L., Ingólfsdóttir, A. (eds.) FOSSACS 2006 and ETAPS 2006. LNCS, vol. 3921, pp. 217–230. Springer, Heidelberg (2006)
17. Raskin, J.-F.: Logics, Automata and Classical Theories for Deciding Real-Time. PhD thesis, Université de Namur, Belgium (1999)
18. Reynolds, M.: The complexity of the temporal logic over the reals (submitted 2004)
19. Wolper, P.: Constructing automata from temporal logic formulas: A tutorial. In: European Educational Forum: School on Formal Methods and Performance Analysis. LNCS, vol. 2090, pp. 261–277. Springer, Heidelberg (2000)

STORMED Hybrid Systems

Vladimeros Vladimerou, Pavithra Prabhakar, Mahesh Viswanathan,
and Geir Dullerud

University of Illinois at Urbana-Champaign
Champaign, Illinois, USA

Abstract. We introduce STORMED hybrid systems, a decidable class
of hybrid systems which is similar to o-minimal hybrid automata in that
the continuous dynamics and constraints are described in an o-minimal
theory. However, unlike o-minimal hybrid automata, the variables are not
initialized in a memoryless fashion at discrete steps. STORMED hybrid
systems require flows which are monotonic with respect to some vector
in the continuous space and can be characterised as bounded-horizon
systems in terms of their discrete transitions. We demonstrate that such
systems admit a finite bisimulation, which can be effectively constructed
provided the o-minimal theory used to describe the system is decidable.
As a consequence, many verification problems for such systems have
effective decision algorithms.

1 Introduction

Embedded processors and electronic controllers are seeing increasingly ubiquitous
use, and in critical cases require extremely accurate and predictable functionality.
Such devices compute discrete steps while interacting with an environment that
has continuous dynamics and meeting real-time constraints. *Hybrid automata* [1]
are a popular formal model used to describe such systems. They have (finitely
many) discrete states, and continuous states evolving with time. The discrete and
continuous states dictate when discrete transitions take place as well as what the
effect of the transition is on the continuous part. Once such a system is modeled,
the verification problem asks whether the formal model meets certain correctness
requirements.

While the problem of verifying a general hybrid automaton against even simple properties (like invariants) is known to be undecidable, important decidable
classes have been identified. *Timed automata* [2], certain special kinds of *rectangular hybrid automata* [10], and *o-minimal hybrid automata* [11] are important
classes of general hybrid automata for which many verification problems are
decidable. The decidability in all these cases is proved by demonstrating the
existence of a finite, computable partition of the state space that is *bisimilar*
to the original system. However, all these classes of decidable automata suffer
from serious drawbacks — timed and rectangular hybrid automata have very
simple dynamics for the way the continuous variables evolve, while o-minimal
systems have strong reset conditions on discrete transitions, that decouples the

L. Aceto et al. (Eds.): ICALP 2008, Part II, LNCS 5126, pp. 136–147, 2008.

discrete dynamics from the continuous one, leaving the continuous state largely unaffected by the discrete transitions. The many undecidability results in the area [1,10,3,4,13] have reinforced the folklore belief that one must either restrict the continuous dynamics or the discrete dynamics to something simple, in order to achieve decidability. Notable exceptions like dynamical systems with piecewise constant derivatives [3] and polygonal hybrid systems [8] are however restricted to very low dimensions (only 2 variables are allowed to obtain decidability).

In this paper we introduce a new class of hybrid automata that we call *STORMED* hybrid systems (STORMED h.s.). These adhere to the following constraints. First the guards of any two transitions are separable in space by some minimum, non-zero distance. Next, all the constraints (i.e, the guards, invariants, and flows) must be definable in a *order-minimal* (or **o**-minimal) theory. Further we require the existence of a vector ϕ such that the flows in all the control states have positive projections on ϕ, and the projections of the guards onto ϕ have delimited-ends. These automata also have monotonic resets, which either leave the continuous state unchanged or advance its projection along ϕ. A form of monotonicity was also captured in [5].

Our main result in this paper is that STORMED h.s. can be shown to be *bisimilar* to a finite state transition system. Moreover the finite transition system can be effectively constructed provided the o-minimal theory in which the automaton is defined is decidable. Thus, STORMED h.s. can be verified against rich branching time properties expressed in logics such as CTL and μ-calculus [7].

STORMED h.s. are both more general in some respects, and more restrictive in other ways, when compared with other subclasses of hybrid automata investigated in previous publications. They allow for a richer continuous dynamics than timed automata and rectangular hybrid automata, and the discrete transitions can affect the continuous dynamics in non-trivial ways unlike o-minimal systems. However, they are required to have separable guards, monotonic flows/resets and delimited ends on guard constraints. In spite of these restrictions, we believe STORMED h.s. can be conveniently used to model interesting systems. For example, monotonicity is implicitly present in terms of a depleting resource, like fuel or time, while separability of guards translates to infrequency of discrete steps.

Finally we look at some relaxations of the STORMED model, and prove that removal of any single constraint cannot be tolerated. Such an investigation demonstrates that our model is reasonably tight; most relaxations of the constraints yield undecidable models.

2   Preliminaries

Equivalence Relations and Partitions. A binary relation R on a set A is a subset of $A \times A$. We will say aRb to denote $(a, b) \in R$. An *equivalence relation* on a set A is a binary relation R that is reflexive, symmetric and transitive. An equivalence relation partitions the set A into *equivalence classes*: $[a]_R = \{b \in A \mid aRb\}$. A partition Π of the set A defines a natural equivalence relation $\equiv_\Pi$, where

$a \equiv_{\Pi} b$ iff a and b belong to the same partition in Π. In this paper, we will use the partition Π to mean both the partition, as well as the equivalence relation associated with it. Finally, we will say an equivalence relation R_1 *refines* another equivalence relation R_2 iff $R_1 \subseteq R_2$.

Transition Systems and Bisimulation. A *transition system* is given by $\mathcal{S} = (Q, Q^0, \rightarrow)$, where Q is a set of states, $Q^0 \subseteq Q$ is the set of initial states, and $\rightarrow \subseteq Q \times Q$ is the transition relation. For a transition system $\mathcal{S} = (Q, Q^0, \rightarrow)$, a *simulation relation* is a binary relation $R \subseteq Q \times Q$ such that if $(q_1, q_1') \in R$ and $q_1 \rightarrow q_2$ then there is q_2' such that $q_1' \rightarrow q_2'$ and $(q_2, q_2') \in R$. A binary relation R is said to be a *bisimulation* iff both R and R^{-1} are simulation relations. q_1 is said to be *bisimilar* to q_2 when there is a bisimulation R such that $(q_1, q_2) \in R$, and we denote this by $q_1 \cong q_2$. Bisimilarity $\cong$ is an equivalence relation and a bisimulation [12]. It is said to be of *finite* index if it has finitely many equivalence classes. A bisimulation R is said to *respect* a partition Π iff R refines the equivalence relation defined by Π.

Definability. Recall that a k-ary relation $S \subseteq A^k$, where A is the domain of $\mathcal{A}$, is said to be *definable* in the structure $\mathcal{A}$ if there is a formula $\varphi(x_1, x_2, \ldots x_k)$, with free variables $x_1, \ldots x_k$, such that $S = \{(a_1, \ldots, a_k) \mid \mathcal{A} \models \varphi[x_i \mapsto a_i]_{i=1}^k\}$. A k-ary function f will be said to be definable if its graph, i.e., the set of all $(x_1, \ldots, x_k, f(x_1, \ldots x_k))$, is definable. A *theory* $T(\mathcal{A})$ of a structure $\mathcal{A}$ is the set of all sentences that hold in $\mathcal{A}$. $T(\mathcal{A})$ (or sometimes simply $\mathcal{A}$) is said to be *decidable* if there is an effective procedure to decide membership in the set $T(\mathcal{A})$.

O-minimality. A binary relation $\leq$ on a set A is said to be a *total ordering* if it is reflexive, transitive, antisymmetric $((a \leq b \ \wedge \ b \leq a) \Rightarrow a = b)$, and total $(a \leq b \vee b \leq a)$. The set A is said to be totally ordered if there is a total order on it. An *interval* is a subset of a totally order set defined, using one or two bounds, as follows: $\{x : a \leq x \leq b\}$, $\{x : x \leq a\}$, and $\{x : a \leq x\}$. Trivially, $\{x : a \leq x \leq b\}$ with $a = b$, is an interval consisting of a single point. We write $\mathcal{A} = (A, \leq, \ldots)$ to convey that the τ-structure $\mathcal{A}$ has an ordering relation $\leq$ and other elements in its structure. A totally ordered first-order structure $\mathcal{A} = (A, \leq, \ldots)$ is *o-minimal* (order-minimal) if every definable set is a finite union of intervals [17]. The theory of this structure is also called o-minimal. Examples of o-minimal structures include $(\mathbb{R}, <, +, -, \cdot, \exp)$ and $(\mathbb{R}, <, +, -, \cdot)$, where $+, -, \cdot, \exp$ are the addition, subtraction, multiplication and exponentiation operations on reals, respectively. Additional examples can be found in [16,17]. The theory of $(\mathbb{R}, <, +, -, \cdot)$ is known to be decidable [15].

3   Hybrid Systems and Special Subclasses

Hybrid systems mix discrete events with continuous dynamics. One formal representation that has been found to conveniently model the behavior of such systems is *hybrid automata* [10]. In this section, we recall the basic definition and introduce special classes of such systems, as a prelude to STORMED hybrid systems that we define in the next section and is the main object of study in this paper.

Definition 1. *A hybrid automaton $\mathcal{H}$ is a tuple $(Q, \Delta, X, X_0, q_0, \mathcal{I}, \mathcal{F}, \mathcal{R}, \mathcal{G})$ where*

- Q *is a finite set of* (discrete) *control states,*
- $\Delta \subseteq Q \times Q$ *is the set of edges between control states,*
- $X = \mathbb{R}^n$, *is the domain of the* continuous (part of the) *state,*
- $X_0 \subseteq X$ *is the set of* initial continuous states,
- $q_0 \in Q$ *is* initial control state,
- $\mathcal{I} : Q \rightarrow 2^X$, *associates an* invariant *with every control state*
- $\mathcal{F} : Q \times X \rightarrow (\mathbb{R}_+ \rightarrow X)$ *associates a flow function with each $(q, x) \in Q \times X$, describing how the continuous state evolves with time,*
- $\mathcal{G} : \Delta \rightarrow 2^X$ *assigns a guard to each edge, which is a condition on the continuous state that must hold in order to take the discrete transition,*
- $\mathcal{R} : \Delta \rightarrow 2^{X \times X}$ *associates a reset with each edge, which is a binary relation that describes how the continuous state changes when a discrete transition is taken.*

In the above hybrid automaton, we call n the *dimension* of $\mathcal{H}$.

Notation: To make the text more readable, we will often write the argument of a function as a subscript. In particular, $\mathcal{I}_q$ will be used to denote the invariant associated with control state q instead of $\mathcal{I}(q)$, and similarly $\mathcal{G}_{(p,q)}$ and $\mathcal{R}_{(p,q)}$ to denote the guard and reset conditions associated with an edge (p, q) instead of $\mathcal{G}(p, q)$ and $\mathcal{R}(p, q)$. We will use $\mathcal{F}_{(q,x)}$ for the flow associated with (q, x) instead of $\mathcal{F}(q, x)$. Also, we call members of $Q \times X$ locations.

Before defining the semantics of the hybrid automata, we observe some conditions that the flow function must satisfy for it to define "reasonable continuous dynamics"; we call this *time-independent spatially consistent*.

Definition 2. *The flow function $\mathcal{F} : Q \times X \rightarrow (\mathbb{R}_+ \rightarrow X)$ is said to be time-independent spatially-consistent (TISC) if for every $q \in Q$ and $x \in X$, $\mathcal{F}_{(q,x)}$ satisfies the following conditions:*

1. *$\mathcal{F}_{(q,x)}$ is continuous and $\mathcal{F}_{(q,x)}(0) = x$.*
2. *It satisfies the following "semi-group" property: for every $t \geq 0$ and $x' \in X$, if $\mathcal{F}_{(q,x)}(t) = x'$ then for every $t' \geq 0$, $\mathcal{F}_{(q,x)}(t + t') = \mathcal{F}_{(q,x')}(t')$.*

Henceforth, we will assume all flows in the hybrid automata to be TISC flows.

Remark 3. TISC flows are a very basic requirement on the continuous dynamics satisfied by most definitions of hybrid automata in the literature (except in [6]). Typically the requirement is ensured by specifying the continuous dynamics in a control state by a differential equation which gives the derivative with respect to time of the continuous state evolution. The flow itself is then the solution of this differential equation. In this paper, we find it convenient to instead directly talk about the flows themselves, rather than the differentials. Notice that a TISC flow is not required to be differentiable and therefore it allows for more general dynamics than is typically considered.

The semantics of a hybrid automaton $\mathcal{H}$ is defined in terms of a transition system $[\![\mathcal{H}]\!] = (C, C_0, \rightarrow)$, where

- $C = Q \times X$ is the set of states,
- $C_0 = q_0 \times X_0$ is the set of initial states, and
- the transition relation $\rightarrow$ is the union of *time transitions* $\rightarrow_t$ and discrete transitions $\rightarrow_d$ given by:
  - $(q_1, x_1) \rightarrow_t (q_2, x_2)$ iff $q_1 = q_2$ and there is a $t \in \mathbb{R}_+$ such that $x_2 = \mathcal{F}_{(q, x_1)}(t)$ and for all $t' \in [0, t]$, $\mathcal{F}_{(q, x_1)}(t') \in \mathcal{I}_{q_1}$.
  - $(q_1, x_1) \rightarrow_d (q_2, x_2)$ iff there is an edge $(q_1, q_2) \in \Delta$ such that $x_1 \in \mathcal{I}_{q_1}$, $x_2 \in \mathcal{I}_{q_2}$, $x_1 \in \mathcal{G}_{(q_1, q_2)}$, and $(x_1, x_2) \in \mathcal{R}_{(q_1, q_2)}$.

In a time transition, the discrete part q_1 of the state does not change but the continuous part changes according to the flow $\mathcal{F}_{q_1}$ while remaining within the invariant $\mathcal{I}_{q_1}$. On the other hand, in a discrete transition, control state changes according to an edge in the automaton, the continuous part of the state before the transition is required to satisfy the guard associated with the edge, and the result of taking the transition changes the continuous state according to the reset conditions associated with the edge.

An *execution* starting from state (q, x) is a sequence of states $(q_1, x_1), (q_2, x_2)$, $\ldots, (q_k, x_k)$ such that $(q_1, x_1) = (q, x)$, and for all $i < k$, $(q_i, x_i) \rightarrow (q_{i+1}, x_{i+1})$. (q_k, x_k) is said to be *reachable* from (q, x). For a hybrid automaton $\mathcal{H}$, we say a control state q is *reachable*, if for some $x \in X$, $x_0 \in X_0$, (q, x) is reachable from an initial state (q_0, x_0). For a hybrid automaton $\mathcal{H}$, the *reachability problem* is to determine if a given control state is reachable.

3.1 Special Definitions

Here we look at some special restrictions on hybrid automata that will be relevant for defining STORMED hybrid systems that we consider in this paper.

3.2 Separable Guards

A hybrid system $\mathcal{H} = (Q, \Delta, X, X_0, q_0, \mathcal{I}, \mathcal{F}, \mathcal{R}, \mathcal{G})$ is said to have *separable guards* if there exists $d_{min} > 0$ such that for every pair of distinct edges $(p_1, q_1), (p_2, q_2) \in \Delta$, $\min\{\|x_1 - x_2\| \mid x_1 \in \mathcal{G}_{(p_1, q_1)} \text{ and } x_2 \in \mathcal{G}_{(p_2, q_2)}\} \geq d_{min}$. The guards of $\mathcal{H}$ are said to be d_{min}-separable.

Here $\| \cdot \|$ denotes euclidean distance. Also, we will be using the dot product $x \cdot y$, where $x, y \in X$, to denote the real value of the length of the projection of y onto x as it is commonly used.

Guard separability can help remove the so-called Zeno behavior, i.e. it helps avoid unbounded number of discrete steps in finite time.

Thus far, our discussion on hybrid automata did not address the issue of how the automaton is formally presented. The general definition presented does not give an effective presentation. We will consider automata where all the conditions, guards, invariants, etc. are described in a logical theory, and even more specifically in an o-minimal theory.

3.3 O-Minimal Definability

A hybrid system $\mathcal{H}$ is said to be *definable in an o-minimal structure* $\mathcal{A} = \{A, \leq , \dots\}$ (or simply called o-minimal), if all its initial conditions, invariants, flows, resets and guards are definable in $\mathcal{A}$.

Remark 4. In the literature, o-minimal hybrid automata [11] refer to hybrid automata as defined above with the additional restriction that all resets are *strong*. In other words, for any edge (p, q) the reset $\mathcal{R}_{(p,q)}$ is of the form $\mathcal{G}_{(p,q)} \times X'$ for some $X' \subseteq X$. This allows one to decouple the system into separate dynamical systems, with the discrete transitions "resetting" the continuous state on each discrete step. Even in Extended O-minimal Hybrid Automata [9], strong resets (at each control graph cycle) seems inevitable. We do not need this decoupling in STORMED, but we do make use of o-minimality.

The subclass of hybrid automata that we will consider in this paper will have monotonicity requirements on the flow. We define these next.

3.4 Monotonic Flows

The set of flows $\mathcal{F}$ of $\mathcal{H}$ is *monotonic* with respect to a vector $\phi \in X$, if there exists an $\epsilon > 0$ such that for every $q \in Q, x \in X$, and $t, \tau \geq 0$,

$$\phi \cdot (\mathcal{F}_{(q,x)}(t + \tau) - \mathcal{F}_{(q,x)}(t)) \geq \epsilon \|\mathcal{F}_{(q,x)}(t + \tau) - \mathcal{F}_{(q,x)}(t)\|,$$

where $a \cdot b$ refers to the dot-product between the vectors. We call such a set of flows (ϵ, ϕ)-monotonic.

The above monotonicity requirement says that as the continuous state evolves with time according to any flow, the projection on the vector ϕ increases at a minimum rate ϵ. This guarantees that the projection on ϕ will never decrease.

Some obvious examples of monotonic flows are:

1. Linear flows of the form $\mathcal{F}_{(q,x)}(t) = x + \alpha_q(t)$, where $x \in \mathbb{R}^n$, and $\alpha_q \in (\mathbb{R}_+ - \{0\})^n$.
2. Analytic flows s.t. for some ϕ, for all $q \in Q$ and $x \in X$, satisfy $\nabla_t \mathcal{F}_{(q,x)}(t) \cdot \phi > \epsilon \|\nabla_t \mathcal{F}_{(q,x)}(t)\|$.

3.5 Monotonic Resets

The collection of reset sets $\mathcal{R}$ of $\mathcal{H}$ is said to be *monotonic* with respect to some $\phi \in X$, if there exist $\epsilon, \zeta > 0$ such that for every $(p, q) \in \Delta$ and $x_1, x_2 \in X$ s.t. $(x_1, x_2) \in \mathcal{R}_{(p,q)}$, we have:

(i) if $p = q$, then either $x_1 = x_2$ or $\phi \cdot (x_2 - x_1) \geq \zeta$, and
(ii) if $p \neq q$, then $\phi \cdot (x_2 - x_1) \geq \epsilon \|x_2 - x_1\|$.

We call such a collection of resets (ϵ, ζ, ϕ)-monotonic.

Remark 5. Notice that in the case when the discrete state changes, we do not require the reset to move the continuous state along ϕ by a minimum value. It only requires the change in the continuous state along ϕ is lower bounded by the actual change in the continuous state. In particular, it forbids resets that take the continuous state back along ϕ. Also our definition allows for identity resets.

Our definition guarantees that a minimum distance of $\min\{\zeta, \epsilon d_{min}\}$ is traveled along ϕ between two successive discrete transitions when the flow of the hybrid systems is (ϵ, ϕ)-monotonic and its guards are d_{min}-separable. The only exception is the trivial[1] identity map. In all other cases, condition (i) avoids Zeno behaviors in a discrete self-loop, and condition (ii) ensures that we cannot have infinitely fast switching while moving only a finite distance along ϕ when the guards are separable. To see the last remark, if the reset itself changes the value of the continuous state enough to move it to another guard, then $\|x_2 - x_1\|$ will be at least d_{min}. Hence the distance traveled along ϕ would be at least ϵd_{min}. Otherwise, suppose $\|x_2 - x_1\| < d_{min}$, it moves at least $\phi \cdot (x_2 - x_1)$ along ϕ which is at least $\epsilon \|x_2 - x_1\|$, and it needs to travel a minimum of $(d_{min} - \|x_2 - x_1\|)$ before taking the next transition. But since the flow is (ϵ, ϕ)-monotonic, it moves another $\epsilon(d_{min} - \|x_2 - x_1\|)$ at least along ϕ. Hence it moves at least ϵd_{min} in total.

4   STORMED Hybrid Systems

In this section we formally introduce the special class of hybrid systems that we study in this paper, and show that they admit a finite bisimulation.

Definition 6 (STORMED Hybrid Systems). *A STORMED hybrid system is a tuple* $(\mathcal{H}, \mathcal{A}, \phi, b_-, b_+, d_{min}, \epsilon, \zeta)$ *where* $\mathcal{H} = (Q, \Delta, X, X_0, q_0, \mathcal{I}, \mathcal{F}, \mathcal{R}, \mathcal{G})$ *is a hybrid automaton,* $\mathcal{A}$ *is an o-minimal structure,* $b_-, b_+, d_{min} \in \mathbb{R}$, *and* $\phi \in X$ *is a vector such that the following conditions are satisfied:*

(S) *The guards of* $\mathcal{H}$ *are* d_{min}-**S***eparable.*

(T) *The flows of* $\mathcal{H}$ *are* **T***ISC.*

(O) $\mathcal{H}$ *is definable in the* **O***-minimal structure* $\mathcal{A}$.

(RM) **R***esets and flows* $\mathcal{F}_{(\cdot,\cdot)}(\cdot)$ *are* **M***onotonic:* (ϵ, ζ, ϕ)-*monotonic and* (ϵ, ϕ)-*monotonic respectively.*

(ED) **E***nds are* **D***elimited: for all* $(p, q) \in \Delta$ *we have* $\{\phi \cdot x : x \in G_{(p,q)} \in (b-, b+)$ *meaning that the projection of each of the guard sets on* ϕ *is bounded below by (or is greater than)* b_- *and bounded above by (or is less than)* b_+.

Before we turn to proving our main result on the existence of a bisimulation for the STORMED systems, we will introduce a few definitions and a lemma to aid the proof.

---

[1] We ignore the trivial (identity) discrete transitions, i.e. $(q, x) \rightarrow_d (q, x)$, which are allowed by monotonic resets because we can all the same consider they do not happen.

Definition 7. *Given a partition $\mathcal{V}$ of $Q \times X$, define $F_t^\star(\mathcal{V})$ to be the coarsest bisimulation[2] with respect to only $\rightarrow_t$ that respects $\mathcal{V}$. Further, define $F_d(\mathcal{V}) :=$ $\{(s_1, s_2)|(\exists s_1' \ . \ s_1 \rightarrow_d s_1') \Rightarrow (\exists s_2' \ . \ s_2 \rightarrow_d s_2' \wedge s_1' \mathcal{V} s_2')\} \cap \mathcal{V}$.*

It can be easily observed that (a) The functionals $F_t^\star(\cdot)$ and $F_d(\cdot)$ are monotonic; (b) $F_t^\star(\mathcal{V})$ is a refinement of $\mathcal{V}$ and so is $F_d(\mathcal{V})$, i.e. $F_t^\star(\mathcal{V}) \subseteq \mathcal{V}$ and $F_d(\mathcal{V}) \subseteq \mathcal{V}$; (c) $F_t^\star(\cdot)$ is idempotent, i.e. $F_t^\star(F_t^\star(\mathcal{V})) = F_t^\star(\mathcal{V})$

Definition 8. *For a hybrid system, we define the i-th neighborhood $N_i \in Q \times X$ to be the set of all locations starting from which there is no execution that can have more than i non-trivial discrete transitions. Note that $N_{i+1} \supseteq N_i$.*

Lemma 9. *Given a STORMED Hybrid System $(\mathcal{H}, \mathcal{A}, \phi, b_-, b_+, d_{min}, \epsilon, \zeta)$ and a partition $\mathcal{P} = \{P_1, P_2, ..., P_k\}$ of its state space $Q \times X$, let $\cong$ to be a bisimulation relation on $\mathcal{H}$ refining $\mathcal{P}$. Define a sequence of partitions $\{W_0, W_1, ...\}$ inductively by setting $W_0 = F_t^\star(\mathcal{P})$ and $W_{i+1} = F_t^\star(F_d(W_i))$. The following hold for all $i \geq 0$:*

(a) W_i is a finite partition definable in the o-minimal theory,
(b) $\cong \subseteq W_i$, and
(c) W_i is a bisimulation on locations in the i-th neighborhood N_i that respects $\mathcal{P}$.

Proof: We use o-minimality to prove (a), then (b) and (c) follow for any hybrid transition system. Details of the induction can be found in [18].

Lemma 10. *Given a STORMED Hybrid System $(\mathcal{H}, \mathcal{A}, \phi, b_-, b_+, d_{min}, \epsilon, \zeta)$, any execution of the system can have at most $i^\star = \lceil \frac{b_+ - b_-}{\eta} \rceil$ non-trivial discrete transitions, where $\eta := min\{\zeta, \epsilon d_{min}\}$.*

Proof: Detailed proof using remark 5 in [18].

Theorem 11. (Finite Bisimulation) *The transition system of a STORMED hybrid system $(\mathcal{H}, \mathcal{A}, \phi, b_-, b_+, d_{min}, \epsilon, \zeta)$ has a finite bisimulation that respects any $\mathcal{A}$-definable partition $\mathcal{P}$. Moreover, if $\mathcal{A}$ is decidable, then there is an effective algorithm for constructing that bisimulation.*

Proof: Again, let $\eta := min\{\zeta, \epsilon d_{min}\}$ and $i^\star := \lceil \frac{b_+ - b_-}{\eta} \rceil$. We can simply observe that since, by Lemma 10, any execution in a STORMED system can go through at most $i^\star$ discrete transitions, all reachable states belong to $N_{i^\star}$. Therefore, by Lemma 9, $W_{i^\star}$ is a bisimulation for all reachable states in $Q \times X$, it respects $\mathcal{P}$ and it is definable in $\mathcal{A}$. Therefore, if $\mathcal{A}$ is decidable, there exists an effective algorithm for constructing $W_{i^\star}$. ∎

Corollary 12. (Reachability) *Given a STORMED hybrid system $(\mathcal{H}, \mathcal{A}, \phi, b_-, b_+, d_{min}, \epsilon, \zeta)$,*

---

[2] The coarsest bisimulation with respect to a subset of the transition relation $\rightarrow' \subseteq \rightarrow$ is the coarsest partition $\mathcal{P} = \{P_i\}$ of the state space $Q \times X$ such that $\mathcal{P}$ is a bisimulation relation of the transition system given by $(Q \times X, q_0 \times X_0, \rightarrow')$.

1. *the set-to-set reachability problem (i.e. given two sets $S_1, S_2 \subseteq Q \times X$, if there is a point in S_1 that can reach some point in S_2) is decidable, if $\mathcal{A}$ is.*
2. *Claim 1 is true even if the guards are not delimited, as long as the initial conditions satisfy $\{\phi \cdot x : \exists q \in Q \ . (q, x) \in S_1\} \in [b_-, \infty]$ and the final set satisfies $\{\phi \cdot x : \exists q \in Q \ . (q, x) \in S_2\} \in [-\infty, b_+]$.*

Proof: First note that Claim 2 reduces to Claim 1 since there can be no discrete transitions outside the set of states $\{(q, x) : x \in [b_-, b_+], q \in Q\}$ that can reach the set S_2. Therefore we can restrict all guards along ϕ to $[b_-, b_+]$ and be able to answer the same question. To check reachability of a set $S_2 \subseteq Q \times X$ from a non-intersecting set S_1, we can partition the state space to $\mathcal{P} = \{S_1, S_2, Q \times X \setminus (S_1 \cup S_2)\}$ and get a finite bisimulation that respects $\mathcal{P}$. This is possible because of Theorem 11. The reachability problem then reduces to the reachability problem of a finite automaton which is constructible if $\mathcal{A}$ is decidable, and hence the reachability problem for STORMED h.s. is decidable.   ∎

5   Examples of STORMED Hybrid Systems

We believe that STORMED h.s. model will be useful in modeling many systems. The STORMED h.s. constraints are realized in some physical systems as follows.

- Monotonicity can be associated with energy or time depletion, or in vehicle control problems, with non-decreasing trajectories.
- The Ends-Delimited property can be present as a deadline on the monotonic direction or a spatial confinement.
- Separability of guards represents infrequency in making control decisions, also based on location or time.
- TISC flows arise naturally, whereas o-minimality is not necessarily a common property, but can be used as an approximation most of the time. Linearization and other model reductions may also result to o-minimal realizations.

In [18] we give a toy example illustrating how the characteristics of a physical system map to the constraints imposed by a STORMED h.s.

6   Relaxations of the STORMED Model

In this section we show that relaxing the various constraints of the STORMED model makes the reachability problem undecidable, and thus justify the tightness of our definition of STORMED model. We consider TISC property of the flows and o-minimal definablility of the system as intrinsic to our model. The theorem below identifies relaxations which render the model undecidable.

Theorem 13.   *1. The reachability problem of the STORMED model with the constraint on the monotonicity of resets removed is undecidable.*
2. *The reachability problem of the STORMED model with the constraint on the ends being limited removed is undecidable.*

Proof: We first present a proof of the undecidability of the reachability problem of multi-rate timed automata along the lines of [1] , and then describe how it can be modified to serve our purpose. Multi-rate timed automata can simulate two counter-machines thus reducing the reachability problem for two counter-machines to that of multi-rate automata. Consider a 2 counter-machine M with counters C and D. In the multi-rate automaton A simulating it, there are two variables x and y which store the values corresponding to the values of the counters. A counter value of n is stored in the corresponding variable as $1/2^n$. Hence an increment will halve the value of the variable and similarly a decrement will double the value. The execution of A will synchronize with that of M every two time units in the sense that if the i-th configuration of M points to location p with the two counter values m and n, then A at time instant $2i$ will be in state p with values of counters $1/2^m$ and $1/2^n$. The parts of the automaton corresponding to the operations increment, decrement and test for 0 is given in Figure 1. Here g is a variable which keeps track of the global time. All variables not shown are assumed to have a flow of 0.

Observe that automaton A satisfies all the STORMED constraints except monotonic resets and separable guards. In order to prove part 1 of the above theorem we modify A to obtain A_1 such that A_1 simulates M but has separable guards. With every state q we associate a distinct even number h_q. We introduce a new variable v, and include in the transition going out of p a constraint $v \in (h_p, h_p + 1]$. If there is only one transition going out of p' we add to its guard the constraint $v \in (h_{p'}, h_{p'} + 1]$, otherwise we add to the transition going from p' to q the constraint $v = h_{p'} + 1$, and to the transition going from p' to r the constraint $v \in (h_{p'}, h_{p'} + 1/2]$. We have three more variables g', x' and y' whose values equal that of g, x and y, respectively, while entering any state. However the values of x' and y' do not change while in state p and the value of g' does not change in state p'. It is easy to see that this can be ensured by treating the variables x', y' and g' similar to x, y and g respectively everywhere, except that in state p, $\dot{x}' = 0$ and $\dot{y}' = 0$ and in state p', $\dot{g}' = 0$. Finally we set $\dot{v} = h_p/(2 - x') + x'/(2 - x')$ in state p corresponding to an operation on C. In state p' we set $\dot{v} = h_{p'}/(2 - g') + g'/(2 - g')$. Hence the value of v upon exiting p would be $h_p + v_1$ and that upon exiting p' would be $h_{p'} + v_1$ where v_1 is the value of x when entering p. At any point of time the transitions that are enabled in A_1 is the same as that of A.

Now returning to part 2 of the theorem, we show how we can construct the automaton A_2 which restores the monotonicity of resets. However the ends will no more be delimited. A_2 is obtained from A_1 by adding a new variable n which increases monotonically at rate 1. The monotonicity is now along the flow of n. This proves the above theorem.   ∎

Relaxing combinations of the STORMED constraints causes undecidability at very low dimensions. Without separability of guards and ends-delimited we have undecidability in 4 dimensions. This follows from the results of [3] where piecewise constant derivatives (PCD) with delimited ends in 3 dimensions is shown undecidable. PCD flows are not monotonic but they can be made monotonic by

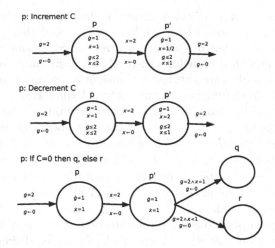

Fig. 1. The parts of the multi-rate automaton A corresponding to the operations increment, decrement and test for zero of the 2-counter machine M

introducing a fourth dimension along which the flows are monotonic. The results in [3] also imply that the reachability problem for STORMED h.s. without guard separability or monotonicity is undecidable in 3 dimensions. By just relaxing separability of guards, it follows from the results in [14] that finite bisimulation does not exist even in two dimensions.

7   Conclusions

We introduced STORMED h.s., a new class of hybrid automata and showed that they admit a finite bisimulation. Further, the bisimulation is constructible if the o-minimal theory in which the elements of the system are defined is decidable. STORMED h.s. allow the continuous variables to have rich dynamics, while at the same time not decoupling the discrete states. However, they require monotonic flows/resets and separable guards. But such constraints are often present in real systems, for example, monotonicity appears in the form of a depleting resource. We also demonstrated that the relaxation of certain constraints from the STORMED h.s. model results in a model that is undecidable. In the future it would be useful to build a tool to algorithmically analyze systems described as STORMED h.s., and evaluate its performance on models of embedded systems.

References

1. Alur, R., Courcoubetis, C., Halbwachs, N., Henzinger, T.A., Ho, P.-H., Nicollin, X., Olivero, A., Sifakis, J., Yovine, S.: The algorithmic analysis of hybrid systems. Theoretical Computer Science 138(1), 3–34 (1995)
2. Alur, R., Dill, D.L.: A theory of timed automata. Theoretical Computer Science 126(2), 183–235 (1994)

3. Asarin, E., Maler, O., Pnueli, A.: Reachability analysis of dynamical systems having piecewise-constant derivatives. Theoretical Computer Science 138(1), 35–65 (1995)
4. Blondel, V.D., Bournez, O., Koiran, P., Papadimitriou, C.H., Tsitsiklis, J.N.: Deciding stability and mortality of piecewise affine dynamical systems. Theoretical Computer Science 255(1–2), 687–696 (2001)
5. Bouyer, P., Brihaye, T., Chevalier, F.: Weighted o-minimal hybrid systems are more decidable than weighted timed automata? In: Artemov, S.N., Nerode, A. (eds.) LFCS 2007. LNCS, vol. 4514, pp. 69–83. Springer, Heidelberg (2007)
6. Brihaye, T.: Verification and control of o-minimal hybrid systems and weighted timed automata. PhD thesis, Academie Universitaire Wallonie-Bruxelles (2006)
7. Clarke, E.M., Grumberg, O., Peled, D.A.: Model Checking. MIT Press, Cambridge (2000)
8. Asarin, E., Schneider, G., Yovine, S.: Algorithmic analysis of polygonal hybrid systems, part i: Reachability. Theor. Comput. Sci. 379(1-2), 231–265 (2007)
9. Gentilini, R.: Reachability Problems on Extended O-minimal Hybrid Automata. Lecture Notes in Compute Sceience, vol. 3829/2006, pp. 162–176. Springer, Heidelberg (2005)
10. Henzinger, T.A., Kopke, P.W., Puri, A., Varaiya, P.: What's decidable about hybrid automata? In: Proc. 27th Annual ACM Symp. on Theory of Computing (STOC), pp. 373–382 (1995)
11. Lafferriere, G., Pappas, G., Sastry, S.: O-minimal hybrid systems (1998)
12. Milner, R.: Communication and Concurrency. Prentice-Hall, Inc., Englewood Cliffs (1989)
13. Mysore, V., Pnueli, A.: Refining the undecidability frontier of hybrid automata. In: Proceedings of the International Conference on the Foundations of Software Technology and Theoretical Computer Science, pp. 261–272 (2005)
14. Prabhakar, P., Vladimerou, V., Viswanathan, M., Dullerud, G.: A decidable class of planar linear hybrid systems. Technical report, University of Illinois at Urbana-Champaign, UIUCDCS-R-2008-2927 (2008)
15. Tarski, A.: A Decision Method for Elementary Algebra and Geometry, 2nd edn. University of California Press (1951)
16. van den Dries, L., Mille, C.: On the real exponential field with restricted analytic functions. Israel Journal of Mathematics 85, 19–56 (1994)
17. van den Dries, L.: Tame Topology and O-minimal Structures. Cambridge Univesity Press, Cambridge (1998)
18. Vladimerou, V., Prabhakar, P., Viswanathan, M., Dullerud, G.: STORMED Hybrid Systems. Technical report, University of Illinois at Urbana-Champaign (2008)

Controller Synthesis and Verification for Markov Decision Processes with Qualitative Branching Time Objectives*

Tomáš Brázdil, Vojtěch Forejt, and Antonín Kučera

Faculty of Informatics, Masaryk University, Botanická 68a, 60200 Brno, Czech Republic
{brazdil,forejt,kucera}@fi.muni.cz

Abstract. We show that the controller synthesis and verification problems for Markov decision processes with qualitative PECTL* objectives are 2-**EXPTIME** complete. More precisely, the algorithms are *polynomial* in the size of a given Markov decision process and doubly exponential in the size of a given qualitative PECTL* formula. Moreover, we show that if a given qualitative PECTL* objective is achievable by *some* strategy, then it is also achievable by an effectively constructible *one-counter* strategy, where the associated complexity bounds are essentially the same as above. For the fragment of qualitative PCTL objectives, we obtain **EXPTIME** completeness and the algorithms are only singly exponential in the size of the formula.

1 Introduction

A *Markov decision process (MDP)* [16,11] is a finite directed graph $G = (V, E, (V_\square, V_\bigcirc), Prob)$ where the vertices of V are partitioned into *non-deterministic* and *stochastic* subsets (denoted $V_\square$ and $V_\bigcirc$, resp.), $E \subseteq V \times V$ is a set of edges, and *Prob* assigns a fixed probability to every edge $(s, s') \in E$ where $s \in V_\bigcirc$ so that $\sum_{(s,s') \in E} Prob(s, s') = 1$ for every fixed $s \in V_\bigcirc$. Without restrictions, we assume that each vertex has at least one and at most two outgoing edges.

MDPs are used as a generic model for discrete systems where one can make *decisions* (by selecting successors in non-deterministic vertices) whose outcomes are *uncertain* (this is modeled by stochastic vertices). The application area of MDPs includes such diverse fields as ecology, chemistry, or economics. In this paper, we focus on more recent applications of MDPs in the area of computer systems (see, e.g., [18]). Here, non-deterministic vertices are used to model the environment, unpredictable users, process scheduler, etc. Stochastic vertices model stochastic features such as coin-tossing in randomized algorithms, bit-flips and other hardware errors whose probability is known empirically, probability distribution on input events, etc. There are two main problems studied in this area:

- *Controller synthesis.* The task is to construct a "controller" which selects appropriate successors at non-deterministic vertices so that a certain objective is achieved.

* Supported by the research center Institute for Theoretical Computer Science (ITI), project No. 1M0545.

- *Verification.* Here, we wonder whether a given objective is achieved for all "adversaries" that control the non-deterministic vertices. In other words, we want to know whether a given system behaves correctly in all environments, under all interleavings produced by a scheduler, etc.

Both "controller" and "adversary" are mathematically captured by the notion of *strategy*, i.e., a function which to every computational history $vs \in V^*V_\Box$ ending in a non-deterministic vertex assigns a probability distribution over the set of outgoing edges of s. General strategies are also referred to as HR strategies because the decision depends on the *history* of the current computation (H) and it is *randomized* (R). Strategies that always return a Dirac distribution are *deterministic* (D), and strategies which depend just on the currently visited vertex are *memoryless* (M). Thus, one can distinguish among HR, HD, MD, and MR strategies.

Since the original application field of MDPs was mainly economics and performance evaluation, there is a rich and mature mathematical theory of MDPs with *discounted* and *limit-average* objectives [16,11]. In the context of computer systems, one is usually interested in objectives related to safety, liveness, fairness, etc. , and these can be naturally formalized as *temporal properties*. In particular, the subclass of *linear-time* properties (such as Büchi, parity, Rabin, Street, or Muller properties) is relatively well understood even in a more general framework of simple stochastic games [12,19,8,6]. Another class of temporal objectives studied in the literature are *linear-time multi-objectives* [10,7], which are Boolean combinations of linear-time objectives.

In this paper, we deal with a more general class of temporal properties that are specified as formulae of probabilistic branching-time logics PCTL, PCTL*, and PECTL* [13]. These logics are obtained from their non-probabilistic counterparts CTL, CTL*, and ECTL* (see, e.g., [9,17]) by replacing the universal and existential path quantifiers with the probabilistic operator $\mathcal{P}^{\bowtie \varrho}$, where ϱ is a rational constant and $\bowtie$ is a comparison such as $\leq$ or $>$. Intuitively, the formula $\mathcal{P}^{\bowtie \varrho}\varphi$ says "the probability of all runs that satisfy φ is $\bowtie$-related to ϱ". If the bound ϱ is restricted just to 0 and 1, we obtain the *qualitative fragment* of a given logic. Controller synthesis for MDPs with branching-time objectives has been considered in [1] where it is shown that strategies for fairly simple qualitative PCTL objectives may require memory and/or randomization. Hence, the classes of MD, MR, HD, and HR strategies (see above) form a strict hierarchy. Moreover, in the same paper it is also proved that the controller synthesis problem for PCTL objectives is **NP**-complete for the subclass of MD strategies. A trivial consequence of this result is **coNP**-completeness of the verification problem for PCTL objectives and MD strategies. In [15], the subclass of MR strategies is examined, and it is proved that the controller synthesis problem for PCTL objectives and MR strategies is in **PSPACE** (the same holds for the verification problem). Some results about history-dependent strategies are presented in [3], where it is shown that controller synthesis for PCTL objectives and HD (and also HR) strategies is *highly undecidable* (in fact, this problem is complete for the Σ_1^1 level of the analytical hierarchy). In [3], it is also demonstrated that the controller synthesis and verification problems are **EXPTIME**-complete for HD/HR strategies and the fragment of PCTL that contains only the qualitative connectives $\mathcal{P}^{=1}\mathcal{F}$, $\mathcal{P}^{=1}\mathcal{G}$, and $\mathcal{P}^{>0}\mathcal{F}$. Moreover, it is shown that strategies for this type of objectives require only *finite memory*, and can be effectively constructed in exponential time. This study is continued in [4] where the memory requirements for objectives of various fragments of qualitative PCTL are classified in a systematic way.

Our contribution. In this paper we solve the controller synthesis and verification problems for all qualitative PCTL and qualitative PECTL* objectives and history-dependent (i.e., HR and HD) strategies. For the sake of simplicity, we first unify HR and HD strategies into a single notion of *history-dependent combined (HC)* strategy. Let $G = (V, E, (V_\Box, V_\bigcirc), Prob)$ be a MDP and let (V_D, V_R) be a partitioning of $V_\Box$ into the subsets of *Dirac* and *randomizing* vertices. A *HC strategy* is a HR strategy σ such that $\sigma(vs)$ is a Dirac distribution for every $vs \in V^*V_D$. Hence, HC strategies coincide with HR and HD strategies when $V_D = \emptyset$ and $V_D = V_\Box$, respectively. Nevertheless, our solution covers also the cases when $\emptyset \neq V_D \neq V_\Box$. Now we can formulate the main result of this paper.

Theorem 1. *Let $G = (V, E, (V_\Box, V_\bigcirc), Prob)$ be a MDP, (V_D, V_R) a partitioning of $V_\Box$, and φ a qualitative PECTL* formula.*

- *The problem whether there is a HC strategy that achieves the objective φ is 2-**EXPTIME**-complete. More precisely, the problem is solvable in time which is polynomial in $|G|$ and doubly exponential in $|\varphi|$. Since qualitative PECTL* objectives are closed under negation, the same complexity results hold for the verification problem.*
- *If the objective φ is achievable by some HC strategy, then it is also achievable by a one-counter strategy (see Definition 3). Moreover, the corresponding one-counter automaton can effectively be constructed in time which is polynomial in $|V|$, doubly exponential in $|\varphi|$, and singly exponential in bp, where bp is the number of bits of precision for the constants employed by Prob.*
- *In the special case when φ is a qualitative PCTL formula, the controller synthesis problem is **EXPTIME**-complete and the algorithms are only singly exponential in the size of the formula.*

This result gives a substantial generalization of the partial results discussed above and solves some of the major open questions formulated in these papers. In some sense, it complements the undecidability result for quantitative PCTL objectives given in [3].

The principal difficulty which requires new ideas and insights is that strategies for qualitative branching-time objectives need infinite memory in general. In Section 3 we give examples demonstrating this fact. Another difference from the previous work is that the precise values of probabilities that are employed by a given strategy *do influence* the (in)validity of qualitative PECTL* objectives. This is very different from qualitative linear-time (multi-)objectives whose (in)validity depends just on the information what edges have zero/positive probability.

Due to space constraints, we could not include all technical definitions and proofs. These can be found in the full version of this paper [5].

2   Definitions

In this section we recall basic definitions that are needed for understanding key results of this paper. For reader's convenience, we also repeat the definitions that appeared already in Section 1.

In the rest of this paper, $\mathbb{N}$, $\mathbb{N}_0$, $\mathbb{Q}$, and $\mathbb{R}$ denote the set of positive integers, non-negative integers, rational numbers, and real numbers, respectively. We also use the standard notation for intervals of real numbers, writing, e.g., $(0, 1]$ to denote the set $\{x \in \mathbb{R} \mid 0 < x \leq 1\}$.

The set of all finite words over a given alphabet Σ is denoted Σ^*, and the set of all infinite words over Σ is denoted Σ^ω. Given two sets $K \subseteq \Sigma^*$ and $L \subseteq \Sigma^* \cup \Sigma^\omega$, we use $K \cdot L$ (or just KL) to denote the concatenation of K and L, i.e., $KL = \{ww' \mid w \in K, w' \in L\}$. We also use Σ^+ to denote the set $\Sigma^* \setminus \{\varepsilon\}$ where ε is the empty word. The length of a given $w \in \Sigma^* \cup \Sigma^\omega$ is denoted $length(w)$, where the length of an infinite word is ω. Given a word (finite or infinite) over Σ, the individual letters of w are denoted $w(0), w(1), \ldots$.

A *probability distribution* over a finite or countably infinite set X is a function $f : X \to [0, 1]$ such that $\sum_{x \in X} f(x) = 1$. A probability distribution is *Dirac* if it assigns 1 to exactly one element. A σ-*field* over a set Ω is a set $\mathcal{F} \subseteq 2^\Omega$ that includes Ω and is closed under complement and countable union. A *probability space* is a triple $(\Omega, \mathcal{F}, \mathcal{P})$ where Ω is a set called *sample space*, $\mathcal{F}$ is a σ-field over Ω whose elements are called *events*, and $\mathcal{P} : \mathcal{F} \to [0, 1]$ is a *probability measure* such that, for each countable collection $\{X_i\}_{i \in I}$ of pairwise disjoint elements of $\mathcal{F}$, $\mathcal{P}(\bigcup_{i \in I} X_i) = \sum_{i \in I} \mathcal{P}(X_i)$, and moreover $\mathcal{P}(\Omega) = 1$.

Definition 1 (Markov Chain). *A Markov chain is a triple $M = (S, \to, Prob)$ where S is a finite or countably infinite set of states, $\to \subseteq S \times S$ is a transition relation, and Prob is a function which to each transition $s \to t$ of M assigns its probability $Prob(s \to t) \in (0, 1]$ so that for every $s \in S$ we have $\sum_{s \to t} Prob(s \to t) = 1$ (as usual, we write $s \xrightarrow{x} t$ instead of $Prob(s \to t) = x$).*

A *path* in M is a finite or infinite word $w \in S^+ \cup S^\omega$ such that $w(i-1) \to w(i)$ for every $1 \leq i < length(w)$. A *run* in M is an infinite path in M. The set of all runs that start with a given finite path w is denoted $Run[M](w)$. When M is clear from the context, we write $Run(w)$ instead of $Run[M](w)$.

When defining the semantics of probabilistic logics (see below), we need to measure the probability of certain sets of runs. Formally, to every $s \in S$ we associate the probability space $(Run(s), \mathcal{F}, \mathcal{P})$ where $\mathcal{F}$ is the σ-field generated by all *basic cylinders* $Run(w)$ where w is a finite path starting with s, and $\mathcal{P} : \mathcal{F} \to [0, 1]$ is the unique probability measure such that $\mathcal{P}(Run(w)) = \Pi_{i=1}^{length(w)-1} x_i$ where $w(i-1) \xrightarrow{x_i} w(i)$ for every $1 \leq i < length(w)$. If $length(w) = 1$, we put $\mathcal{P}(Run(w)) = 1$. Hence, only certain subsets of $Run(s)$ are $\mathcal{P}$-measurable, but in this paper we only deal with "safe" subsets that are guaranteed to be in $\mathcal{F}$.

Definition 2 (Markov Decision Process). *A Markov decision process (MDP) is a finite directed graph $G = (V, E, (V_\square, V_\bigcirc), Prob)$ where the vertices of V are partitioned into non-deterministic and stochastic subsets (denoted $V_\square$ and $V_\bigcirc$, resp.), $E \subseteq V \times V$ is a set of edges, and Prob assigns a fixed positive probability to every edge $(s, s') \in E$ where $s \in V_\bigcirc$ so that $\sum_{(s,s') \in E} Prob(s, s') = 1$ for every fixed $s \in V_\bigcirc$. For technical convenience, we require that each vertex has at least one and at most two outgoing edges.*

Let $G = (V, E, (V_\square, V_\bigcirc), Prob)$ be a MDP. A *strategy* is a function which to every $vs \in V^* V_\square$ assigns a probability distribution over the set of outgoing edges of s. Each strategy σ determines a unique Markov chain G_σ where states are finite paths in G and $vs \xrightarrow{x} vss'$ iff either s is stochastic, $(s, s') \in E$, and $Prob((s, s')) = x$, or s is non-deterministic, $(s, s') \in E$, and x is the probability of (s, s') chosen by $\sigma(vs)$. General strategies are also called HR strategies, because they are *history-dependent (H)* and *randomized (R)*. We say that σ is *memoryless (M)* if $\sigma(vs)$ depends just on the last vertex s, and *deterministic* if $\sigma(vs)$ is a Dirac distribution. Thus, we obtain the classes of HR, HD, MR, and MD strategies. For the sake of clarity and uniformity of our presentation, we also introduce the notion of *history-dependent combined (HC)* strategy. Here we assume that the non-deterministic vertices of $V_\square$ are split into two disjoint subsets V_D and V_R of *Dirac* and *randomizing* vertices. A HC strategy is a HR strategy σ such that $\sigma(vs)$ is a Dirac distribution for every $vs \in V^*V_D$. Hence, in the special case when $V_D = \emptyset$ (or $V_D = V_\square$), every HC strategy is a HD strategy (or a HR strategy). A special type of history-dependent strategies are *finite-memory (F)* strategies. A finite-memory strategy σ is specified by a deterministic finite-state automaton $\mathcal{A}$ over the input alphabet V (see, e.g., [14]), where $\sigma(vs)$ depends just on the control state entered by $\mathcal{A}$ after reading the word vs. In this paper we also consider *one-counter* strategies which are specified by *one-counter automata*.

Definition 3 (One counter automaton). *A one counter automaton is a tuple $C = (Q, \Sigma, q_{in}, \delta^{=0}, \delta^{>0})$ where Q is a finite set of control states, Σ is a finite input alphabet, $q_{in} \in Q$ is the initial state, and $\delta^{=0} : Q \times \Sigma \to Q \times \{0, 1\}$, $\delta^{>0} : Q \times \Sigma \to Q \times \{0, 1, -1\}$ are transition functions. The set of configurations of C is $Q \times \mathbb{N}_0$. For every $u \in \Sigma^+$ we define a binary relation $\xrightarrow{u}$ over configurations inductively as follows:*

- *for all $a \in \Sigma$ we put $(q, c) \xrightarrow{a} (q', c + i)$ iff either $c = 0$ and $\delta^{=0}(q, a) = (q', i)$, or $c > 0$ and $\delta^{>0}(q, a) = (q', i)$;*
- *$(q, c) \xrightarrow{au} (q', c')$ iff there is (q'', c'') such that $(q, c) \xrightarrow{a} (q'', c'')$ and $(q'', c'') \xrightarrow{u} (q', c')$.*

For every $u \in \Sigma^+$, let $q_u \in Q$ and $c_u \in \mathbb{N}_0$ be the unique elements such that $(q_{in}, 0) \xrightarrow{u} (q_u, c_u)$.

Let $G = (V, E, (V_\square, V_\bigcirc), Prob)$ be a MDP and (V_D, V_R) a partitioning of $V_\square$. A *one-counter strategy* is a HC strategy σ for which there is a one-counter automaton $C = (Q, V, q_{in}, \delta^{=0}, \delta^{>0})$ and a constant $k \in \mathbb{N}$ such that

- for every $vs \in V^*V_D$, $\sigma(vs)$ is a Dirac distribution that depends only on q_{vs} and the information whether c_{vs} is zero or not;
- for every $vs \in V^*V_R$ such that s has two outgoing edges, $\sigma(vs)$ is either a Dirac distribution or a distribution that assigns $k^{-c_{vs}}$ to one edge, and $1 - k^{-c_{vs}}$ to the other edge. The choice depends solely on q_{vs}.

Before presenting the definition of the logic PECTL*, we need to recall the notion of Büchi automaton. Our definition of Büchi automaton is somewhat nonstandard in the sense that we consider only special alphabets of the form $2^{\{1,...,n\}}$ and the symbols assigned to transitions in the automaton are interpreted in a special way. These differences are not fundamental but technically convenient.

Definition 4 (Büchi automaton). *A Büchi automaton of arity $n \in \mathbb{N}$ is a tuple $\mathcal{B} = (Q, q_{in}, \delta, A)$, where Q is a finite set of control states, $q_{in} \in Q$ is the initial state, $\delta : Q \times 2^{\{1,...,n\}} \to 2^Q$ is a transition function, and $A \subseteq Q$ is a set of accepting states. A* *given infinite word w over the alphabet* $2^{\{1,...,n\}}$ *is* accepted by $\mathcal{B}$ *if there is an* accepting computation *for w, i.e., an infinite sequence of states* $q_0, q_1, \ldots$ *such that* $q_0 = q_{in}$, $q_j \in A$ *for infinitely many* $j \in \mathbb{N}_0$, *and for all* $i \in \mathbb{N}_0$ *there is* $\alpha_i \in 2^{\{1,...,n\}}$ *such that* $q_{i+1} \in \delta(q_i, \alpha_i)$ *and* $\alpha_i \subseteq w(i)$. *The set of all infinite words accepted by* $\mathcal{B}$ *is denoted* $L(\mathcal{B})$.

Let $Ap = \{a, b, c, \ldots\}$ be a countably infinite set of *atomic propositions*. The syntax of PECTL* formulae is defined by the following abstract syntax equation:

$$\varphi \quad ::= \quad a \mid \neg a \mid \mathcal{P}^{\bowtie \varrho}\mathcal{B}(\varphi_1, \ldots, \varphi_n)$$

Here a ranges over Ap, $\bowtie$ is a comparison (i.e., $\bowtie \in \{<, >, \leq, \geq, =\}$), ϱ is a rational constant, $n \in \mathbb{N}$, and the $\mathcal{B}$ in $\mathcal{B}(\varphi_1, \ldots, \varphi_n)$ is a Büchi automaton of arity n. The *qualitative fragment* of PECTL* is obtained by restricting ϱ to 0 and 1. For simplicity, from now on we write $\mathcal{B}^{\bowtie \varrho}(\varphi_1, \ldots, \varphi_n)$ instead of $\mathcal{P}^{\bowtie \varrho}\mathcal{B}(\varphi_1, \ldots, \varphi_n)$.

Let $M = (S, \rightarrow, Prob)$ be a Markov chain, and let $\eta : S \rightarrow 2^{Ap}$ be a *valuation*. The validity of PECTL* formulae in the states of M is defined inductively as follows: $s \models^\eta a$ iff $a \in \eta(s)$, $s \models^\eta \neg a$ iff $a \notin \eta(s)$, and

$$s \models^\eta \mathcal{B}^{\bowtie \varrho}(\varphi_1, \ldots, \varphi_n) \text{ iff } \mathcal{P}(\{w \in Run(s) \mid w[\varphi_1, \ldots, \varphi_n] \in L(\mathcal{B})\}) \bowtie \varrho$$

Here $w[\varphi_1, \ldots, \varphi_n]$ is the infinite word over the alphabet $2^{\{1,...,n\}}$ where $w[\varphi_1, \ldots, \varphi_n](i)$ is the set of all $1 \leq j \leq n$ such that $w(i) \models^\eta \varphi_j$. Let us note that the set of runs $\{w \in Run(s) \mid w[\varphi_1, \ldots, \varphi_n] \in L(\mathcal{B})\}$ is indeed $\mathcal{P}$-measurable in the above introduced probability space $(Run(s), \mathcal{F}, \mathcal{P})$, and hence the definition of PECTL* semantics makes sense for all PECTL* formulae. In the rest of this paper, we often write $s \models \varphi$ instead of $s \models^\eta \varphi$ when η is clear from the context.

The syntax of PECTL* is rather terse and does not include conventional temporal operators such as $\mathcal{G}$ and $\mathcal{F}$. This is convenient for our purposes (proofs become simpler), but the intuition about the actual expressiveness of PECTL* and its sublogics is lost. As a little compensation, we show how to encode conjunction, disjunction, and temporal connectives $\mathcal{G}$, $\mathcal{F}$, $\mathcal{U}$ and $\mathcal{X}$ (the negation of φ corresponds to $\mathcal{B}_\wedge^{=0}(\varphi, \varphi)$).

For example, the formula $\varphi_1 \wedge \mathcal{F}^{=1}\varphi_2$ is then a shortcut for $\mathcal{B}_\wedge^{=1}(\varphi_1, \mathcal{B}_\mathcal{F}^{=1}(\varphi_2))$, and in our examples we stick to this simpler notation. The PCTL fragment of PECTL* is obtained by restricting the syntax to $\varphi ::= a \mid \neg a \mid \varphi_1 \wedge \varphi_2 \mid \varphi_1 \vee \varphi_2 \mid \mathcal{X}^{\bowtie \varrho}\varphi \mid \varphi_1 \mathcal{U}^{\bowtie \varrho} \varphi_2$. We also write $a \Rightarrow \varphi$ instead of $\neg a \vee \varphi$.

3   The Result

As we have already noted, qualitative PECTL* formulae are closed under negation, and hence it suffices to consider only the controller synthesis problem (a solution for the verification problem is then obtained as a trivial corollary). Formally, the controller

synthesis problem for qualitative PECTL* objectives and HC strategies is specified as
follows:

Problem: Controller synthesis for qualitative PECTL* objectives and HC strategies.

Instance: A MDP $G = (V, E, (V_\square, V_\bigcirc), Prob)$, a partition (V_D, V_R) of $V_\square$, $s_{in} \in V$, $\eta :$
$V \to 2^{Ap}$, and a qualitative PECTL* formula φ. (The η is extended to all
$vs \in V^*V$ by stipulating $\eta(vs) = \eta(s)$.)

Question: Is there a HC strategy σ such that $s_{in} \models^\eta \varphi$ in G_σ ?

Our solution of the problem (see Theorem 1) is based on one central idea underpinned
by many technically involved observations which "make it work". Roughly speaking,
a given objective φ is first split into finitely many "sub-objectives" $\varphi_1, \ldots, \varphi_n$ that are
achievable by effectively constructible *finite-memory* strategies $\sigma_1, \ldots, \sigma_n$. Then, the
finite-memory strategies $\sigma_1, \ldots, \sigma_n$ are combined into a single one-counter strategy σ
that achieves the original objective φ.

Let us illustrate this idea on a concrete example. Consider the MDP G of the following figure, where s_{in} is Dirac.

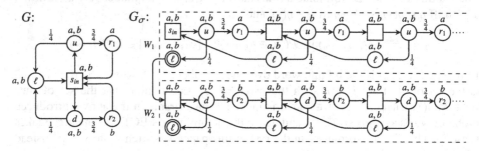

The winning objective is the formula $\varphi \equiv \varphi_a \wedge \varphi_b$, where $\varphi_a \equiv G^{=1}(a \Rightarrow G^{>0}a)$ and
$\varphi_b \equiv G^{=1}(b \Rightarrow G^{>0}b)$. The validity of a, b in the vertices of G is also indicated in the
figure. In this case, the "sub-objectives" are the formulae φ_a and φ_b, that are achievable by memoryless strategies σ_u and σ_d that always select the transitions $s_{in} \to u$ and
$s_{in} \to d$, respectively. Obviously, $s_{in} \models \varphi_a$, $s_{in} \not\models \varphi_b$ in G_{σ_u}, and similarly $s_{in} \models \varphi_b$,
$s_{in} \not\models \varphi_a$ in G_{σ_d}. Hence, none of these two strategies achieves the objective φ (in fact,
one can easily show that φ is not achievable by any *finite-memory* strategy). Now we
show how to combine the strategies σ_u and σ_d into a single one-counter strategy σ such
that $s_{in} \models \varphi$ in G_σ.

Let us start with an informal description of the strategy σ. During the whole play,
the *mode* of σ is either σ_u or σ_d, which means that σ makes the same decision as σ_u
or σ_d, respectively. Initially, the mode of σ is σ_u, and the counter is initialized to 1. If
(and only if) the counter reaches zero, the current mode is switched to the other mode,
and the counter is set to 1 again. This keeps happening ad infinitum. During the play,
the counter is modified as follows: each visit to ℓ decrements the counter, and each visit
to r_1 or r_2 increments the counter.

Obviously, σ is a one-counter strategy. However, it is not so obvious why it works.
The structure of the play G_σ is indicated in the figure above, where the initial state is
labeled s_{in} (the actual graph of G_σ is an infinite tree obtained by *unfolding* the graph
shown in the figure). The play G_σ closely resembles an "infinite sequence" $W_1, W_2, \ldots$

of one-dimensional random walks. In each W_i, the probability of going right is $\frac{3}{4}$, the probability of going left is $\frac{1}{4}$, and whenever the "left end" is entered (i.e., the counter becomes zero), the next random walk W_{i+1} in the sequence is started. All W_i, where i is odd/even, correspond to the σ_u/σ_d mode. In the above figure, only W_1 and W_2 are shown, and their "left ends" are indicated by double circles. By applying standard results about one-dimensional random walks, we can conclude that for every state s of every W_i that is not a "left end", the probability of reaching the "left end" of W_i from s is strictly less than one. Now it suffices to realize the following:

- Let s be a state of W_i, where i is odd. Then $s \models G^{>0}a$ in G_σ. This is because all states of W_i satisfy a, and the probability of reaching the "left end" of W_i from s is strictly less than one. For the same reason, all states of W_i, where i is even, satisfy the formula $G^{>0}b$.
- Let s be a state of W_i, where i is odd, such that $s \models b$. Then $s \models G^{>0}b$. This is because there is a *finite* path to a state s' in W_{i+1} along which b holds (this path leads through the "left end" of W_i). Since $s' \models G^{>0}b$ (as justified in the previous item), we obtain that $s \models G^{>0}b$. For the same reason, for every state s of every W_i such that i is even and $s \models a$ we have that $s \models G^{>0}a$.

Both claims can easily be verified by inspecting the figure on the previous page. Hence, $s_{in} \models \varphi$ in G_σ as needed.

The main idea of "combining" the constructed finite-memory strategies $\sigma_1, \ldots, \sigma_n$ into a single one-counter strategy σ is illustrated quite well by the above example. One basically "rotates" among the strategies $\sigma_1, \ldots, \sigma_n$ ad infinitum. Of course, some issues are (over)simplified in this example. In particular,

- in general, the "sub-objectives" do *not* correspond to subformulae of φ. They depend both on a given φ and a given G;
- the events counted in the counter are not just individual visits to selected vertices;
- the individual random walks obtained by "rotating" the modes $\sigma_1, \ldots, \sigma_n$ do not form an infinite sequence but an infinite tree;
- in the previous example, the only way how to leave a given W_i is to pass through its "left end". In general, each state of a given W_i can have a transition which "leaves" W_i. However, these transitions have progressively smaller and smaller probabilities so that the probability of "staying within" W_i remains positive.

Note that the last item explains why the definition of one-counter strategy admits the use of "exponentially small" probabilities that depend on the current counter value (the one-counter strategy defined in the above example only tested the counter for zero). To demonstrate that the use of "exponentially small" probabilities is unavoidable, consider the MDP $\hat{G}$ of the following figure, where $\hat{s}_{in}$ is randomizing.

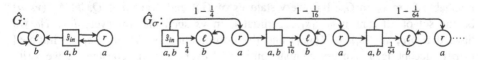

Let $\hat{\varphi} \equiv \mathcal{G}^{>0}(a \wedge (b \Rightarrow \mathcal{G}^{>0}b))$. We claim that every HC strategy κ which achieves the objective $\hat{\varphi}$ must satisfy the following: Let K be the set of all probabilities that are assigned to the edge $\hat{s}_{in} \to \ell$ in the play $\hat{G}_\kappa$. Then all elements of K are positive and $\inf(K) = 0$, otherwise the formula $\hat{\varphi} \equiv \mathcal{G}^{>0}(a \wedge (b \Rightarrow \mathcal{G}^{>0}b))$ would not hold. Hence, κ must inevitably assign "smaller and smaller" positive probability to the edge $\hat{s}_{in} \to \ell$. This is achievable by a one-counter strategy $\hat{\sigma}$ where $\hat{\sigma}(v\hat{s}_{in})$ assigns $4^{-c(v\hat{s}_{in})}$ to $\hat{s}_{in} \to \ell$ and $1 - 4^{-c(v\hat{s}_{in})}$ to $\hat{s}_{in} \to r$, where $c(v\hat{s}_{in})$ is the number of occurrences of $\hat{s}_{in}$ in $v\hat{s}_{in}$. The play $\hat{G}_{\hat{\sigma}}$ is also shown in the above figure. It is easy to see that $\hat{s}_{in} \models \mathcal{G}^{>0}(a \wedge (b \Rightarrow \mathcal{G}^{>0}b))$ in $\hat{G}_{\hat{\sigma}}$.

A Formal Proof of the Result. Due to space constraints, we cannot give a full proof of Theorem 1 (it can be found in [5]). Here we only outline the structure of our proof, identify the milestones, and try to "map" the vague notions introduced earlier to precise technical definitions. Roughly speaking, our proof has two major phases.

(1) The controller synthesis problem for qualitative PECTL* objectives and HC strategies is reduced to the controller synthesis problem for "consistency objectives" and HC strategies. The "consistency objectives" are technically simpler than PECTL* objectives, and they in fact represent the very core of the whole problem.
(2) The controller synthesis problem for consistency objectives and HC strategies is solved.

The most important insights are concentrated in Phase (2). Our complexity results are based on a careful analysis of the individual steps which constitute Phase (1) and (2). Since all of our constructions are effective, one can also effectively construct the strategy for the original PECTL* objective by taking the strategy for the constructed consistency objective and modifying it accordingly.

We start by a formal definition of consistency objectives. First, we need to recall the notion of a *deterministic Muller automaton*, which is a tuple $\mathcal{M} = (Q, \Sigma, \delta, A)$ where Q is a finite set of control states, Σ is a finite alphabet, $\delta : Q \times \Sigma \to Q$ is a transition function (which is extended to the elements of $Q \times \Sigma^*$ in the standard way), and $A \subseteq 2^Q$ is a set of accepting sets. A *computation* of $\mathcal{M}$ on $w \in \Sigma^\omega$ initiated in $q \in Q$ is the (unique) infinite sequence of control states $\gamma = q_0, q_1, \ldots$ such that $q_0 = q$ and $\delta(q_i, w(i)) = q_{i+1}$ for all $i \in \mathbb{N}_0$. A computation γ is *accepting* if $\inf(\gamma) \in A$, where $\inf(\gamma)$ is the set of all control states that occur infinitely often in γ.

Definition 5 (Consistency objective). *Let* $G = (V, E, (V_\square, V_\bigcirc), \text{Prob})$ *be a MDP,* $s_{in} \in V$ *an initial vertex, and* (V_D, V_R) *a partition of* $V_\square$. *A consistency objective is a triple* $(\mathcal{M}, (Q_{>0}, Q_{=1}), L)$, *where* $\mathcal{M} = (Q, V, \delta, A)$ *is a deterministic Muller automaton over the alphabet* V, $(Q_{>0}, Q_{=1})$ *is a partition of* Q *s.t. for all* $q \in Q_{>0}$, $q' \in Q_{=1}$ *and* $w \in V^*$ *we have that* $\delta(q, w) \in Q_{>0}$ *and* $\delta(q', w) \in Q_{=1}$, *and* $L : V \to 2^Q$ *is a labeling.*

Let σ be a HC strategy, and let $G_\sigma^{s_{in}}$ be the play G_σ restricted to states that are reachable from s_{in} in G_σ. For every state vs of $G_\sigma^{s_{in}}$ and every $q \in Q$, let $Acc(vs, q)$ be the set of all runs $v_0 s_0, v_1 s_1, \ldots$ initiated in vs such that for every $i \in \mathbb{N}_0$ we have that $\delta(q, s_0 \cdots s_i) \in L(s_{i+1})$ and the computation of $\mathcal{M}$ on $s_0 s_1 \cdots$ initiated in q is accepting. For every comparison $\bowtie$ and every rational constant ϱ, we write $vs \models_\sigma Acc^{\bowtie \varrho}(q)$ if $\mathcal{P}(Acc(vs, q)) \bowtie \varrho$ in G_σ. A HC strategy σ achieves the consistency objective $(\mathcal{M}, (Q_{>0}, Q_{=1}), L)$ if for every state $vs \in V^*V$ of the play $G_\sigma^{s_{in}}$, every $q \in Q$, and every $\bowtie\varrho \in \{=1, >0\}$ we have that if $q \in Q_{\bowtie\varrho} \cap L(s)$, then $vs \models Acc^{\bowtie\varrho}(q)$.

Phase (1). Let $G = (V, E, (V_\square, V_\bigcirc), Prob)$ be a MDP, (V_D, V_R) a partition of $V_\square$, $s_{in} \in V$, $\eta : V \to 2^{Ap}$ a valuation, and φ a qualitative PECTL* formula. We construct a MDP $G' = (V', E', (V'_\square, V'_\bigcirc), Prob')$, a partitioning (V'_D, V'_R), a vertex $s'_{in} \in V$, and a consistency objective $(\mathcal{M}, (Q_{>0}, Q_{=1}), L)$ such that the existence of a HC strategy σ where $s_{in} \models^\eta \varphi$ in G_σ implies the existence of a HC strategy π that achieves the objective $(\mathcal{M}, (Q_{>0}, Q_{=1}), L)$ in $G'^{s'_{in}}_\pi$, and vice versa. The size of G' is polynomial in $|G|$ and exponential in $|\varphi|$.

The construction is partly based on ideas of [4] and proceeds as follows. First, all Büchi automata that occur in φ are replaced with equivalent deterministic Muller automata. The resulting formula is further modified so that all probability bounds take the form ">0" or "=1" (to achieve that, some of the deterministic Muller automata may be complemented). Thus, we obtain a formula φ'. Let $M^{>0}$ and $M^{=1}$ be the sets of all deterministic Muller automata that appear in φ' with the probability bound >0 and =1, respectively. The automaton $\mathcal{M}$ is essentially the disjoint union of all automata in $M^{>0}$ and $M^{=1}$. The sets $Q_{>0}$ and $Q_{=1}$ are unions of sets of control states of all Muller automata in $M^{>0}$ and $M^{=1}$, respectively. The tricky part is the construction of G'. Intuitively, the MDP G' is the same as G, but several instances of Muller automata from $M^{>0} \cup M^{=1}$ are simulated "on the fly". Moreover, some "guessing" vertices are added so that a strategy can decide what "subformulae of φ'" are to be satisfied in a given vertex. The structure of G' itself does not guarantee that the commitments chosen by the strategy are fulfilled. This is done by the automaton $\mathcal{M}$ and the condition that $vs \models Acc^{\bowtie\varrho}(q)$ for all $q \in Q_{\bowtie\varrho} \cap L(s)$. (Intuitively, this condition says that the play $G'^{s'_{in}}_\pi$ is "consistent" with the commitments chosen in the guessing vertices.)

Phase (2). The controller synthesis problem for consistency objectives and HC strategies is solved in three steps:

(a) We solve the special case when the set $Q_{>0}$ (see Definition 5) is empty.
(b) We solve the special case when the strategy is *strictly randomizing* (see below), using the result of (a).
(c) We reduce the general (unrestricted) case to the special case of (b).

Now we describe the three steps in more detail. Let $G = (V, E, (V_\square, V_\bigcirc), Prob)$ be a MDP, $s_{in} \in V$ an initial vertex, (V_D, V_R) a partition of $V_\square$, and $(\mathcal{M}, (Q_{>0}, Q_{=1}), L)$ a consistency objective, where $\mathcal{M} = (Q, V, \delta, A)$.

As for step (a), the key insight is the following observation (the proposition holds under the non-restrictive assumption that for all $s, t \in V$ such that $(s, t) \in E$ and for all $p \in Q_{=1}$ such that $p \in L(s)$ we have $\delta(p, s) \in L(t)$):

Proposition 1. *Let us assume that $Q_{>0} = \emptyset$. Then the objective $(\mathcal{M}, (Q_{>0}, Q_{=1}), L)$ is achievable by some HC strategy iff there is a HC strategy σ such that for every state vs of $G_\sigma^{s_{in}}$, every $p \in L(s) \cap Q_{=1}$, and almost all runs $v_0 s_0, v_1 s_1, \ldots$ initiated in vs there are $k \in \mathbb{N}_0$, $q \in Q$, and $X \in A$ such that $\delta(p, s_0 \cdots s_{k-1}) = q$ and almost all runs $\hat{v}_0 \hat{s}_0, \hat{v}_1 \hat{s}_1, \ldots$ initiated in $v_k s_k$ satisfy the following conditions: $\delta(q, \hat{s}_0 \cdots \hat{s}_j) \in X$ for every $j \in \mathbb{N}_0$, and for every $r \in X$ there are infinitely many $j \in \mathbb{N}_0$ such that $\delta(q, \hat{s}_0 \cdots \hat{s}_j) = r$.*

In other words, if $Q_{>0} = \emptyset$, then the objective is achievable by a strategy which simply "guesses" an appropriate moment and an appropriate $X \in A$, and then it suffices to verify that the guess was correct, i.e., almost all simulated computations of M visit only the states of X and each of them is visited infinitely often. This can be effectively implemented by a qualitative Büchi objective, and hence we can rely on the existing algorithms (see Section 1). At this point, there is no need for infinite memory.

In step (b), we concentrate on another special case where both $Q_{>0}$ and $Q_{=1}$ may be non-empty, but the set of strategies is restricted to *strictly randomizing HC (srHC)* strategies. A srHC strategy is a HC strategy σ such that $\sigma(vs)$ assigns a positive probability to *all* outgoing edges whenever $s \in V_R$. This is perhaps the most demanding part of the whole construction, where we formalize the notion of "sub-objective" mentioned earlier, invent the technique of "rotating" the finite-memory strategies for the individual "sub-objectives", etc. The main technical ingredient is the notion of *entry point*.

Definition 6. *A set $X \subseteq V$ is closed if each $s \in X$ has at least one immediate successor in X, and every $s \in X$ which is stochastic or randomizing has all immediate successors in X. Each closed X determines a* sub-MDP $G|X$ *which is obtained from G by restricting the set of vertices to X.*

Let X be a closed set. An entry point *for X is a pair $(s, q) \in X \times Q_{>0}$ for which there is a HD strategy ξ in $G|X$ satisfying the following conditions:*

1. $s \models_\xi Acc^{=1}(q)$;
2. *for every state vt of $(G|X)_\xi^s$ and every $p \in L(t) \cap Q_{=1}$ we have that $vt \models_\xi Acc^{=1}(p)$;*
3. *for all states vt of $(G|X)_\xi^s$ and all $p \in L(t) \cap Q_{>0}$ we have the following: if there is no state of $V^*V_\bigcirc$ reachable from vt in $(G|X)_\xi^s$, then either $wt \models_\xi Acc^{=1}(p)$, or there is a finite path $v_0 t_0, \ldots, v_k t_k$ initiated in vt such that $t_k \in V_R$ and t_k has two outgoing edges $(t_k, r_1), (t_k, r_2) \in E$ such that $\xi(v_k t_k)$ selects the edge (t_k, r_1) and $\delta(p, t_0 \cdots t_k) \in L(r_2) \cap Q_{>0}$.*

Intuitively, entry points correspond to the finitely many "sub-objectives" discussed earlier. The next step is to show that the set of all entry points for a given closed set X can be effectively computed in time which is polynomial in $|G|$ and exponential in $|Q|$. Further, we show that for each entry point (s, q) one can effectively construct a *finite-memory deterministic* strategy $\xi(s, q)$ which has the same properties as the HD strategy ξ of Definition 6 (this is what we meant by "achieving a sub-objective"). Here we use the results of step (a). Technically, the key observation of step (b) is the following proposition (this proposition holds under some technical assumptions that are not listed explicitly here).

Proposition 2. *The consistency objective $(M, (Q_{>0}, Q_{=1}), L)$ is achievable by a srHC strategy σ iff there is a closed $X \subseteq V$ such that $s_{in} \in X$ and for all $s_0 \in X$ and $q_0 \in L(s_0) \cap Q_{>0}$ there is finite sequence $(s_0, q_0), \ldots, (s_n, q_n)$ such that $(s_i, s_{i+1}) \in E$, $q_i \in L(s_i)$ and $\delta(q_i, s_i) = q_{i+1}$ for all $0 \leq i < n$, and (s_n, q_n) is an entry point for X.*

Both directions of the proof require effort, and the "if" part can safely be declared as difficult. This is where we introduce the counter and "rotate" the $\xi(s, q)$ strategies for the individual entry points to obtain a srHC strategy that achieves the objective $(M, (Q_{>0}, Q_{=1}), L)$. This part is highly non-trivial and relies on many subtle observations. Nevertheless, the whole construction is effective and admits a detailed complexity analysis.

Step (c) is relatively simple (compared to step (a) and particularly step (b)). The 2-**EXPTIME** lower bound for qualitative PECTL* objectives also requires a proof (the bound does not follow from the previous work). Here we use a standard technique for simulating an exponentially bounded alternating Turing machine, employing some ideas presented in [2]. The **EXPTIME** lower bound for qualitative PCTL has been established already in [3].

References

1. Baier, C., Größer, M., Leucker, M., Bollig, B., Ciesinski, F.: Controller synthesis for probabilistic systems. In: Proceedings of IFIP TCS 2004, pp. 493–506. Kluwer, Dordrecht (2004)
2. Brázdil, T., Brožek, V., Forejt, V.: Branching-time model-checking of probabilistic pushdown automata. In: Proceedings of INFINITY 2007, pp. 24–33 (2007)
3. Brázdil, T., Brožek, V., Forejt, V., Kučera, A.: Stochastic games with branching-time winning objectives. In: Proceedings of LICS 2006, pp. 349–358. IEEE, Los Alamitos (2006)
4. Brázdil, T., Forejt, V.: Strategy synthesis for Markov decision processes and branching-time logics. In: Caires, L., Vasconcelos, V.T. (eds.) CONCUR. LNCS, vol. 4703, pp. 428–444. Springer, Heidelberg (2007)
5. Brázdil, T., Forejt, V., Kučera, A.: Controller synthesis and verification for Markov decision processes with qualitative branching time objectives. Technical report FIMU-RS-2008-05, Faculty of Informatics, Masaryk University (2008)
6. Chatterjee, K., de Alfaro, L., Henzinger, T.: Trading memory for randomness. In: Proceedings of 2nd Int. Conf. on Quantitative Evaluation of Systems (QEST 2004), pp. 206–217. IEEE, Los Alamitos (2004)
7. Chatterjee, K., Majumdar, R., Henzinger, T.: Markov decision processes with multiple objectives. In: Durand, B., Thomas, W. (eds.) STACS 2006. LNCS, vol. 3884, pp. 325–336. Springer, Heidelberg (2006)
8. de Alfaro, L.: Quantitative verification and control via the mu-calculus. In: Amadio, R.M., Lugiez, D. (eds.) CONCUR 2003. LNCS, vol. 2761, pp. 102–126. Springer, Heidelberg (2003)
9. Emerson, E.A.: Temporal and modal logic. Handbook of TCS B, 995–1072 (1991)
10. Etessami, K., Kwiatkowska, M., Vardi, M., Yannakakis, M.: Multi-objective model checking of Markov decision processes. In: Grumberg, O., Huth, M. (eds.) TACAS 2007. LNCS, vol. 4424, pp. 50–65. Springer, Heidelberg (2007)
11. Filar, J., Vrieze, K.: Competitive Markov Decision Processes. Springer, Heidelberg (1996)
12. Grädel, E.: Positional determinacy of infinite games. In: Diekert, V., Habib, M. (eds.) STACS 2004. LNCS, vol. 2996, pp. 4–18. Springer, Heidelberg (2004)
13. Hansson, H., Jonsson, B.: A logic for reasoning about time and reliability. Formal Aspects of Computing 6, 512–535 (1994)
14. Hopcroft, J.E., Ullman, J.D.: Introduction to Automata Theory, Languages, and Computation. Addison-Wesley, Reading (1979)
15. Kučera, A., Stražovský, O.: On the controller synthesis for finite-state Markov decision processes. In: Proceedings of FST&TCS 2005. LNCS, vol. 3821, pp. 541–552. Springer, Heidelberg (2005)
16. Puterman, M.L.: Markov Decision Processes. Wiley, Chichester (1994)
17. Stirling, C.: Modal and temporal logics. Handbook of Logic in Comp. Sci. 2, 477–563 (1992)
18. Vardi, M.: Automatic verification of probabilistic concurrent finite-state programs. In: Proceedings of FOCS 1985, pp. 327–338. IEEE, Los Alamitos (1985)
19. Walukiewicz, I.: A landscape with games in the background. In: Proceedings of LICS 2004, pp. 356–366. IEEE, Los Alamitos (2004)

On Datalog vs. LFP

Anuj Dawar and Stephan Kreutzer

[1] University of Cambridge Computer Lab
anuj.dawar@cl.cam.ac.uk
[2] Oxford University Computing Laboratory
kreutzer@comlab.ox.ac.uk

Abstract. We show that the homomorphism preservation theorem fails for LFP, both in general and in restriction to finite structures. That is, there is a formula of LFP that is preserved under homomorphisms (in the finite) but is not equivalent (in the finite) to a Datalog program. This resolves a question posed by Atserias. The results are established by two different methods: (1) a method of diagonalisation that works only in the presence of infinite structures, but establishes a stronger result showing a hierarchy of homomorphism-preserved problems in LFP; and (2) a method based on a pumping lemma for Datalog due to Afrati, Cosmadakis and Yannakakis which establishes the result in restriction to finite structures. We refine the pumping lemma of Afrati et al. and relate it to the power of Monadic Second-Order Logic on tree decompositions of structures.

1 Introduction

Among the important classical results of model theory, relating syntactic to semantic properties of first-order logic, are the preservation theorems. For instance, the Łoś-Tarski theorem tells us that a sentence of first-order logic is equivalent to an *existential sentence* if, and only if, the class of its models is closed under extensions and Lyndon's theorem states that a sentence is monotone in a relation R if, and only if, it is equivalent to one that is positive in R (see [12]). The study of preservation theorems has played an important role in the development of finite model theory, with many early results demonstrating that such results fail when we restrict consideration to finite structures (see, for instance, [8]).

One important exception to the general failure of preservation theorems in the finite is Rossman's proof of the homomorphism preservation theorem [17]. This shows that on the class of finite structures, just as on the class of all structures (finite or infinite) a sentence of first-order logic is equivalent to an *existential positive* sentence if, and only if, it is preserved under homomorphisms. The homomorphism preservation property in finite structures has aroused much interest in theoretical computer science through its connections with questions in database theory and the study of constraint satisfaction problems (CSPs).

Each of the preservation theorems mentioned has two directions, one of which is generally quite easy to establish: namely that the syntactic restriction (such as the restriction to existential positive sentences) implies the semantic restriction (being preserved under homomorphisms). Moreover, this direction holds generally on any class of structures C.

L. Aceto et al. (Eds.): ICALP 2008, Part II, LNCS 5126, pp. 160–171, 2008.

The other direction, sometimes known as *expressive completeness*, states that any sentence that satisfies the semantic restriction is equivalent to one of the simple syntactic form. When we restrict this statement to a class C, we weaken both the hypothesis and the conclusion of the statement. Thus, even for classes C and C' where $C \subseteq C'$, it is impossible to deduce either the validity or the failure of a preservation theorem on C from the statement for C'. In particular, the statements for the class of all structures and for the class of finite structures alone are quite independent statements. Recently, there has been a growing interest in investigating the status of preservation theorems for classes C more restrictive than the class of all finite structures [5,6].

Atserias [3] (see [1, Question 4.3]) asked whether the homomorphism preservation theorem holds for LFP—the extension of first-order logic with an operator for defining least fixed points of monotone formulas. Fixed-point logics have arguably played a more important role in finite model theory than first-order logic. In particular, it is known that LFP expresses all polynomial time computable properties of finite ordered structures [13,18]. Thus, the question of whether a homomorphism preservation theorem can be established for this logic arises naturally. The language formed by extending existential positive formulas by means of a least fixed-point operator is Datalog and it has been extensively studied as a database query language. It has also received attention in the study of constraint satisfaction problems as it provides a general means of classifying many CSPs as tractable. It is easily seen that any query defined in Datalog is preserved under homomorphisms. Thus, Atserias' question asks whether it is the case that every sentence of LFP that is preserved under homomorphisms is equivalent to a Datalog program. We show in this paper that this is not the case, either on the class of all (finite or infinite) structures or in restriction to the class of finite structures.

The homomorphism preservation question for extensions of first-order logic was also studied by Feder and Vardi [10]. They showed that on finite structures, the homomorphism preservation property holds for a number of *existential* infinitary and fixed-point logics. In particular, they established that any query definable in Datalog($\neg, \neq$) that is closed under homomorphisms is already definable in Datalog. The former language is the extension of Datalog with inequality and negation on EDB predicates. Just as Datalog can be seen as the existential positive fragment of LFP, Datalog($\neg, \neq$) is its existential fragment. Thus, our results show that the theorem of Feder and Vardi cannot be extended from Datalog($\neg, \neq$) to LFP.

The two examples we construct separating LFP from Datalog bear some similarity to each other in that they are defined in terms of graphs having path lengths in some set S. In addition, to guarantee that the classes we consider are closed under homomorphisms, we take the union with the class of all graphs containing a cycle. The main differences in the two results are in the choice of the set S and in the method used to prove that the resulting class of graphs is not definable in Datalog. In the case where we allow infinite structures, the proof is somewhat simpler as we can construct a set S that is undefinable in Datalog (over the natural numbers) using standard diagonalisation arguments and then obtain the result by means of a reduction of the graph problem to this set. This actually establishes something stronger. It shows that for every k, there are formulas of LFP that are preserved under homomorphisms but not definable by a formula with only k nested alternations of the fixed-point operator with negation. These results are established in Section 4.

When we restrict ourselves to finite structures, such diagonalisation methods are unavailable and we adapt a pumping lemma due to Afrati et al. [2] for our purpose. Afrati et al. use their pumping lemma to demonstrate polynomial-time monotone properties that are not definable in Datalog. In order to adapt it to the LFP-definable properties we are interested in, we need to show that it works on a class of acyclic graphs. What we establish is that if π is a Datalog program which accepts a directed acyclic graph $(\mathbf{G}, s, t)$ if, and only if, $\mathbf{G}$ contains a path from s to t of length p for some p in a given set S, then S cannot grow too fast (the precise statement is given in Lemma 5.2). This suffices to establish the result we seek. An apparently stronger pumping lemma (saying that S cannot grow faster than linearly) is stated in [2], but without the restriction to acyclic graphs. In the absence of this restriction, we cannot use their lemma directly and it is not clear from their description of the proof that it can be adapted. This is explained in more detail in Section 5. One virtue of our proof of this pumping lemma is that it connects it with other recent innovations in the analysis of Datalog queries, namely their relationship with tree decompositions and with the power of monadic second-order logic over these. This new insight into Datalog may be of independent interest.

One source of interest in the relationship between LFP and Datalog is research on the classification of tractable constraint satisfaction problems. We can associate with any structure $\mathbf{B}$, the decision problem CSP($\mathbf{B}$) of determining for a given structure $\mathbf{A}$ whether there is a homomorphism $\mathbf{A} \to \mathbf{B}$. This is the *constraint satisfaction problem* associated with $\mathbf{B}$ (see [9]). Much research work has been devoted to classifying those structures $\mathbf{B}$ for which this problem is decidable in polynomial time. It is immediate from the above definition that the complement of CSP($\mathbf{B}$) is closed under homomorphisms. If we could find a *finite* structure $\mathbf{B}$ for which the complement of CSP($\mathbf{B}$) is definable in LFP but not in Datalog, this would resolve certain conjectures on the classification of tractable CSPs (see [4] for a discussion). We note here that our example of a homomorphism closed class separating LFP from Datalog on finite structures is not the complement of CSP($\mathbf{B}$) for any finite $\mathbf{B}$, but is of this form for an *infinite* structure $\mathbf{B}$.

We begin in Section 2 with definitions, including those of LFP and Datalog as well as first-order and monadic second-order logic. We also recall the definitions of tree decompositions of structures and relate them to Datalog programs.

2   Preliminaries

We briefly introduce the fundamental concepts and notation we need in later sections.

Homomorphisms and Preservation. Let σ be a finite signature. We use boldface letters for structures $\mathbf{A}, \mathbf{B}, \ldots$ and corresponding Roman letters $A, B, \ldots$ to denote their universe. We also write $\boldsymbol{a}$ for a tuple $a_1, \ldots, a_k$.

Definition 2.1. Let σ be a relational signature possibly with constant symbols and let $\mathbf{A}, \mathbf{B}$ be σ-structures. A *homomorphism* from $\mathbf{A}$ to $\mathbf{B}$ is a function $h : A \to B$ such that for every k-ary relation symbol $R \in \sigma$ and every k-tuple $\boldsymbol{a} \in A^k$ if $\boldsymbol{a} \in R^{\mathbf{A}}$ then $(h(a_1), \ldots, h(a_k)) \in R^{\mathbf{B}}$ and for every constant symbol $c \in \sigma$, $h(c^{\mathbf{A}}) = c^{\mathbf{B}}$. We write $\mathbf{A} \to \mathbf{B}$ to denote that there is a homomorphism from $\mathbf{A}$ to $\mathbf{B}$.

Definition 2.2. Let $\mathcal{C}$ be a class of structures. A subclass $\mathcal{D} \subseteq \mathcal{C}$ is *closed under homomorphisms* if whenever $\mathbf{A}, \mathbf{B} \in \mathcal{C}$ so that $\mathbf{A} \in \mathcal{D}$ and there is a homomorphism $h : \mathbf{A} \to \mathbf{B}$ then $\mathbf{B} \in \mathcal{D}$. We are particularly interested in model classes of sentences. If φ is a sentence of a logic, we say φ *is preserved under homomorphisms on* $\mathcal{C}$ if the class $Mod_{\mathcal{C}}(\varphi) := \{\mathbf{A} \in \mathcal{C} : \mathbf{A} \models \varphi\}$ is closed under homomorphisms.

First-Order Logic, Monadic Second-Order Logic and Types. Let σ be a signature. We assume that the reader is familiar with first-order logic. We write $\mathrm{FO}(\sigma)$ for the class of all first-order formulas over the signature σ. *Monadic Second-Order Logic* (MSO) is the extension of first-order logic by quantification over sets of elements, i.e. there are quantifiers $\exists X, \forall X$, where X is a unary relation variable, and a formula $\exists X \varphi$ is true in a structure $\mathbf{A}$, written $\mathbf{A} \models \exists X \varphi$, if there is a set $X \subseteq A$ such that $(\mathbf{A}, X) \models \varphi$. The semantics of $\forall X \varphi$ is defined analogously. See e.g. [8] for more on MSO.

The *quantifier rank* $\mathrm{qr}(\varphi)$ of a formula φ (of FO or MSO) is the maximal depth of nesting of quantifiers in φ. Note that up to logical equivalence there are only finitely many MSO-formulas of quantifier rank at most q in a finite signature σ. We write MSO_q for the class of MSO-formulas of quantifier rank at most q. We write $\mathbf{A} \equiv^q \mathbf{B}$ to denote that two structures $\mathbf{A}$ and $\mathbf{B}$ cannot be distinguished in MSO_q.

A *type* is a maximally consistent class of formulas. For a structure $\mathbf{A}$ and $q \in \mathbb{N}$, the MSO_q-type of $\mathbf{A}$ is the class of MSO-sentences of quantifier rank at most q which are true in $\mathbf{A}$ and if $a \in A^k$, then the MSO_q-type of a in $\mathbf{A}$ is the class of all MSO_q-formulas $\varphi(x)$ such that $\mathbf{A} \models \varphi(a)$. As, for each $q \in \mathbb{N}$, MSO_q only contains finitely many formulas up to equivalence, the MSO_q-type of a tuple or a structure can completely be described by a single formula in MSO_q (see [8]). We will use the following decomposition theorem for MSO. See e.g. [15] or [11].

Lemma 2.3. *Let A and B be structures and let u be a tuple listing the vertices in the intersection of A and B. The MSO_q-type of u in $\mathbf{A} \cup \mathbf{B}$ is uniquely determined by the MSO_q-types of u in $\mathbf{A}$ and in $\mathbf{B}$.*

In particular, if $\mathbf{A}, \mathbf{B}_1$ and $\mathbf{B}_2$ are structures such that $A \cap B_1 = A \cap B_2 =: u$ and the MSO_q-types of u in $\mathbf{B}_1$ and $\mathbf{B}_2$ are the same, then $\mathbf{A} \cup \mathbf{B}_1 \equiv^q \mathbf{A} \cup \mathbf{B}_2$.

Least Fixed-Point Logic. We first present a brief introduction to least fixed-point logic. For a detailed exposition see [8]. Let σ be a signature and let $\varphi(R, x)$ be a formula of signature σ which is *positive* in the k-ary relation variable R, i.e. every atom of the form Rt in φ occurs within the scope of an *even* number of negation symbols. For every σ-structure $\mathbf{A}$, φ defines a monotone operator[1] $F_{\mathbf{A},\varphi} : \mathrm{Pow}(A^k) \to \mathrm{Pow}(A^k)$ via $F_{\mathbf{A},\varphi}(P) := \{a \in A^k : (\mathbf{A}, P) \models \varphi[a]\}$, for every $P \subseteq A^k$. A theorem due to Knaster and Tarski shows that on every structure $\mathbf{A}$ every monotone operator $F_{\mathbf{A},\varphi}$ has a least fixed point which we denote by $\mathrm{lfp}(F_{\mathbf{A},\varphi})$.

The logic $\mathrm{LFP}(\sigma)$ is the extension of $\mathrm{FO}(\sigma)$ by least fixed-point operators. To be precise: $\mathrm{LFP}(\sigma)$ contains $\mathrm{FO}(\sigma)$ and is closed under Boolean connectives and first-order quantification; and if $\varphi(R, x, z, Q)$ is an $\mathrm{LFP}(\sigma)$-formula which is positive in the k-ary relation variable R then for every k-tuple t of terms $[\mathrm{lfp}_{R,x} \varphi](t)$ is an $\mathrm{LFP}(\sigma)$-formula such that for every $(\sigma \,\dot\cup\, \{z, Q\})$-structure $\mathbf{A}$ and every tuple $a \in A^k$ we have $\mathbf{A} \models [\mathrm{lfp}_{R,x} \varphi](a)$ if, and only if, $a \in \mathrm{lfp}(F_{\mathbf{A},\varphi})$.

---

[1] An operator $F : \mathrm{Pow}(M) \to \mathrm{Pow}(M)$ is *monotone* iff $F(A) \subseteq F(B)$ for all $A \subseteq B \subseteq M$.

The *alternation depth* of an LFP formula φ is defined as the maximal number of alternations between fixed-point operators and negations inside φ. We write LFP^k for the class of LFP formulas of alternation depth at most k.

Datalog. Datalog is a database query language which could be defined as the collection of formulas of LFP which do not use negation or universal quantification. However, the usual presentation of the language is in terms of function-free Horn clauses, and we follow this presentation below as the structure of the program in terms of rules is useful for our proof of the pumping lemma in Section 5.

A *Datalog program* is a finite set of rules of the form $T_0 \leftarrow T_1, \ldots, T_m$, where each T_i is an atomic formula. T_0 is called the *head* of the rule, while the right-hand side is called the *body*. The relation symbols that occur in the heads are the *intensional* database predicates (IDBs), while all others are the *extensional* database predicates (EDBs). Note that IDBs may occur in the bodies too, thus, a Datalog program is a recursive specification of the IDBs with semantics obtained via least fixed-points of monotone operators. The collection of EDB predicates occurring in π constitute its *signature* σ, and a Datalog program of signature σ is interpreted in σ-structures. One IDB predicate is distinguished as the *goal predicate*. In general, we will assume that the goal predicate is a 0-ary predicate, so that the program defines a Boolean query. In the interests of space, we will not give a formal definition of the semantics of the program, which can be found in standard textbooks such as [8]. A key parameter in analysing Datalog programs is the number of variables used. We write k-Datalog for the collection of all Datalog programs with at most k distinct variables in total.

A formula of first-order logic is said to be a *conjunctive query* if it is obtained from atomic formulas using only conjunctions and existential quantification. Every finite structure $\mathbf{A}$ with n elements gives rise to a *canonical conjunctive query* $\varphi_{\mathbf{A}}$, which is obtained by first associating a different variable x_i with every element a_i of $\mathbf{A}$, $1 \leq i \leq n$, then forming the conjunction of all atomic facts true in $\mathbf{A}$, and finally existentially quantifying all variables x_i, $1 \leq i \leq n$. In other words, the formula $\varphi_{\mathbf{A}}$ is the existential closure of the *positive diagram* of $\mathbf{A}$ (see [12]). The significance of these queries lies in the fact (first noted by Chandra and Merlin [7]) that for any structure $\mathbf{B}$, $\mathbf{B} \models \varphi_{\mathbf{A}}$ if, and only if, there is a homomorphism from $\mathbf{A}$ to $\mathbf{B}$.

For every positive integer k, let CQ^k be the collection of conjunctive queries that have at most k distinct variables. Note that each variable may be reused, so its number of occurrences may be arbitrarily large. The significance of CQ^k lies in that the number of variables required to express $\varphi_{\mathbf{A}}$ is closely related to the *tree width* of $\mathbf{A}$. We first review the definition of tree width and then state its relationship with CQ^k.

Let $\mathbf{A}$ be a σ-structure. A *tree-decomposition* of $\mathbf{A}$ is a pair $(\mathbf{T}, B)$ where $\mathbf{T}$ is a directed tree oriented from the root to the leaves and B is a labelling that associates to each node t of $\mathbf{T}$ a non-empty set of elements $B_t \subseteq A$ such that
1. for every tuple a in some relation R of $\mathbf{A}$, there is a node $t \in T$ such that a is contained in B_t; and
2. for every $a \in A$, the set $\{t \in T : a \in B_t\}$ forms a connected subtree of $\mathbf{T}$.

The *width* of a tree-decomposition is the maximum cardinality of a set B_t minus one. The *treewidth* of $\mathbf{A}$ is the smallest k for which $\mathbf{A}$ has a tree-decomposition of width k.

The connection between the number of variables in $\varphi_{\mathbf{A}}$ and the tree width of $\mathbf{A}$ can now be summarised as follows (see [14,6]).

Lemma 2.4. *If* **A** *has tree width less than* k, *then* $\varphi_{\mathbf{A}}$ *is equivalent to a formula of* CQ^k. *For any satisfiable formula* φ *in* CQ^k, *there is a structure* **A** *with tree width less than* k, *such that* $\varphi_{\mathbf{A}}$ *is logically equivalent to* φ.

A Datalog program π can be *unfolded* into a conjunctive query, by repeatedly expanding the rules. There are infinitely many such unfoldings for a recursive program. We are interested in the structures, called *expansions* of π, for which these unfoldings are the canonical conjunctive queries.

Definition 2.5. Given a Datalog program π, a *partial unfolding* of π is any conjunctive query obtained using the following rules:
- The goal predicate G of π is a partial unfolding of π;
- If ϑ is a partial unfolding of π; R is an IDB predicate of π; $R(x)$ is an atomic formula occurring in ϑ; and $R(y) \leftarrow T_1(z_1), \ldots, T_m(z_m)$ is a rule of π, let $\varphi(x)$ be the formula obtained from $\exists z(T_1(z_1) \wedge \cdots \wedge T_m(z_m))$ (where z includes all variables occurring in the rule except for those in y) by replacing the variables in y by x. Then, the formula ϑ' obtained from ϑ by replacing the occurrence $R(x)$ by $\varphi(x)$ is also a partial unfolding of π.

An *unfolding* of π is a partial unfolding in which no IDB predicate occurs.

It is not difficult to see that any unfolding of a Datalog program is a conjunctive query, and more particularly, if π is a k-Datalog program, then any unfolding of π is in CQ^k. It is also easily established that a structure **A** is in the query defined by π if, and only if, there is some unfolding ϑ of π such that $\mathbf{A} \models \vartheta$.

Definition 2.6. An *expansion* of a k-Datalog program π is a structure **A** of tree width less than k such that the canonical conjunctive query $\varphi_{\mathbf{A}}$ is logically equivalent to an unfolding of π.

Now, it is clear, by Lemma 2.4, that $\mathbf{B} \models \pi$ if, and only if, $\mathbf{A} \to \mathbf{B}$ for some expansion **A** of π. Indeed, the models of π are generated from expansions whose tree decompositions are given by the unfolding of π.

Definition 2.7. A *decorated expansion* of the k-Datalog program π is a tree decomposition $(\mathbf{T}, B)$ of an expansion **A** of π along with a labelling L that associates to each node t of **T** a pair (r, ρ) where r is either a rule of π or an atomic formula $R(x)$ (for an EDB predicate R); and ρ is an injective mapping from the variables of r to B_t.

The labelling L must satisfy the following conditions:

1. If $L(t) = (r, \rho)$ and r is an atomic formula, then t is a leaf of **T**.
2. If $L(t) = (r, \rho)$ and r is a rule $R(x) \leftarrow T_1(z_1), \ldots, T_m(z_m)$, then t has exactly m children $t_1, \ldots, t_m$ where for each i, if $L(t_i) = (r_i, \rho_i)$ then r_i is either an atomic formula $T_i(y)$ or a rule whose head is $T_i(y)$. Further $\rho_i(y) = \rho(z_i)$.

3   LFP Definable Classes Closed Under Homomorphisms

In this section we introduce the classes of structures which we will use to separate LFP from Datalog, and show that they are LFP definable, though proofs are omitted for lack of space.

A *source-target graph* is a (finite or infinite) directed graph $\mathbf{G}$ with two distinguished vertices s and t, i.e. a structure over the signature $\{E, s, t\}$ where E is a binary relation symbol and s, t are constant symbols. For a source-target graph $\mathbf{A} = (\mathbf{G}, s, t)$, let $n^{\mathbf{A}}$ denote $\sup\{p : \mathbf{G}$ contains a simple path of length p starting at $s\}$. Note that $n^{\mathbf{A}}$ is either a finite ordinal or ω. In the sequel, when we speak about a graph, we mean a source-target graph.

Fix a set $S \subseteq \omega$ of natural numbers. We define the following classes of graphs.

- Cyc – the class of graphs that contain a cycle.
- Unb – the class of graphs $\mathbf{A}$ for which $n^{\mathbf{A}} = \omega$.
- P_S – the class of graphs $\mathbf{A}$ that contain a path from s to t of length p for some $p \in S$.
- $C_S := P_S \cup \mathrm{Cyc}$.
- $C_S^{\infty} = (P_S \cap \mathrm{Unb}) \cup \mathrm{Cyc}$.

It is the classes C_S and C_S^{∞} (for suitable choices of the set S) which we show separate LFP from Datalog. Note that all acyclic graphs in C_S^{∞} are infinite, while C_S may contain finite as well as infinite acyclic graphs. We begin first by noting that these classes are closed under homomorphisms.

Lemma 3.1. *The classes C_S and C_S^{∞} are closed under homomorphisms.*

It can be shown that even the classes P_S are closed under homomorphisms. The reason we work with the classes C_S and C_S^{∞} is for the sake of definability in LFP. It is difficult to use LFP to determine the lengths of paths in the presence of cycles. In fact, the longest path problem is NP-complete and hence unlikely to be definable in LFP. By including all graphs with cycles, we make the problem easier, as then we only have to consider the longest path in acyclic digraphs. We now aim to show that if the set S is definable in LFP, in some sense, then the classes C_S and C_S^{∞} are also definable.

For an ordinal $\alpha \in [0, \omega]$, we write $(\alpha, succ)$ to denote the structure whose universe is $\{\beta : \beta < \alpha\}$ and where $succ$ is interpreted as the binary successor relation.

Lemma 3.2. *There is a uniform LFP interpretation of $(n^{\mathbf{A}}, succ)$ in acyclic source-target graphs $\mathbf{A}$.*

The proof of Lemma 3.2 relies on the use of *stage comparison relations*, see [16]. We remark that the interpretation in Lemma 3.2 is already definable in LFP[1], the alternation free fragment of LFP.

Lemma 3.3. *There is a formula φ_{unb} of LFP that defines Unb on acyclic graphs.*

This is used to show the definability of the classes C_S and C_S^{∞}.

Lemma 3.4. *If $S \subseteq \omega$ is definable in the structure $(\omega, succ)$ by a formula of LFP^k, then the class C_S^{∞} is defined by a sentence of LFP^{k+1}.*

Note that the class of sets S that are definable by LFP formulas in $(\omega, succ)$ is very rich. In particular, it includes all Π_1^1-definable sets of numbers.

Lemma 3.5. *If the class of finite structures $S = \{(n, succ) : n \in S\}$ is definable in LFP, then C_S is defined by a sentence of LFP.*

Note that $\{(n, succ) : n \in S\}$ is definable in LFP if, and only if, the set S, represented in unary, is decidable in polynomial time.

4   The Diagonalisation Method

The main result of this section is the following theorem.

Theorem 4.1. *There is a sentence of* LFP *that is preserved under homomorphisms on the class of all structures but which is not equivalent to any* Datalog *program.*

Since Datalog is in the negation-free fragment of LFP, it is clear that every Datalog program is equivalent to a formula in LFP^1. Using diagonalisation methods, one can show that for each k there is a subset $S_k \subset \omega$ such that S_k can be defined in the structure $(\omega, succ)$ by an LFP^{k+1} formula $\varphi_k(x)$ but not by any formula in LFP^k, where $succ$ denotes the successor relation on ω. See e.g. [16]. Thus, we can choose a set S of natural numbers which is definable in LFP on the structure $(\omega, succ)$ but not in Datalog. Our aim is to show that the class C_S^∞ is not definable in Datalog.

Lemma 4.2. *For any set S, if there is a* Datalog *program defining C_S^∞, then S is definable in $(\omega, succ)$ by a* Datalog *program.*

This allows us to prove Theorem 4.1, as we can choose a set S that is definable in LFP but not in LFP^1. Then, Lemma 3.4, 3.1 and 4.2 together imply the theorem. The proof actually implies a somewhat stronger result.

Corollary 4.3. *For every k, there is an* LFP^{k+2}*-definable class of structures which is closed under homomorphisms but which cannot be defined in* LFP^k.

5   The Pumping Method

The result in the previous section relies crucially on infinite structures. In particular, the class C_S^∞ restricted to finite structures is just the class of all graphs containing a cycle, and this is definable in Datalog. Moreover, the stronger Corollary 4.3 cannot hold on finite structures since it is known that every formula of LFP is equivalent, in the finite, to a formula of LFP^1 (see [13]). Still, in this section we establish that the homomorphism preservation property fails even when we restrict ourselves to finite structures.

Theorem 5.1. *There is an* LFP *sentence φ which is preserved under homomorphisms on the class of all finite structures such that there is no* Datalog *program equivalent to φ on finite structures.*

Specifically, we show that there are sets of numbers S, which are polynomial-time decidable when written in unary, such that there is no Datalog program whose finite models are exactly the ones in C_S. This is established by showing the following pumping lemma.

Lemma 5.2. *Let $S \subseteq \omega$ be an infinite set of numbers and π a* Datalog *program which accepts a directed acyclic graph $(\mathbf{G}, s, t)$ if, and only if, $\mathbf{G}$ contains a path from s to t of length p for some $p \in S$. Then, there is a constant c and an increasing sequence $(a_i)_{i \in \omega}$ of numbers such that:*

1. $a_{i+1} < a_i^c$ *for all i; and*
2. $S \cap [a_i, a_{i+1}] \neq \varnothing$ *for all i.*

Before we give a proof, a few remarks are in order. Recall that a Datalog program π determines a collection $\mathcal{C}$ of expansions of bounded tree width such that a structure $\mathbf{B}$ is accepted by π if, and only if, $\mathbf{A} \to \mathbf{B}$, for some $\mathbf{A} \in \mathcal{C}$. If π is as in Lemma 5.2, then it accepts a structure $(\mathbf{G}, s, t)$ where $\mathbf{G}$ is a simple path of length $p \in S$. The expansion $\mathbf{A}$ that maps to this structure must be an acyclic graph in which all paths from s to t are of length p. To prove the lemma, we proceed from a decorated expansion for $\mathbf{A}$ to "pump" a portion of the tree decomposition and obtain a sequence of expansions $\mathbf{A}_i$ which are all acyclic and such that the lengths of all paths in $\mathbf{A}_i$ from s to t are in the interval $[a_i, a_{i+1}]$ for a suitably defined sequence $(a_i)_{i \in \omega}$. This establishes the result.

It should be noted that a similar pumping lemma is stated by Afrati et al. [2], and proved by similar means. Indeed, their statement is apparently stronger in that condition (1) can be replaced by $a_{i+1} < c + a_i$, which is to say that the sequence $(a_i)_{i \in \omega}$ can be chosen to grow linearly in i rather than exponentially. However, their statement is not confined to acyclic graphs, which is an essential restriction for our result. It would suffice for our purposes if, in the proof of the pumping lemma of Afrati et al., it could be shown that when an acyclic expansion is pumped, we always obtain an acyclic expansion, but we are unable to recover this fact from their proof. To be precise, they present the proof in detail only for the case when the expansion $\mathbf{A}$ is itself a simple path. In this case, the proof below can also be used to yield a linear sequence $(a_i)_{i \in \omega}$. They then state that the general case can be handled similarly, by choosing in the decorated expansion of $\mathbf{A}$ a collection of pairs of points to pump such that each simple path crosses exactly one such pair. We are unable to determine how such a collection could be chosen and, if the points at which we pump an expansion are crossed by more than one path, it is quite possible that pumping may create shortcuts. This is the reason why, in the proof below, we have to pump each pair of points multiple times, forcing an exponential growth in the sequence $(a_i)_{i \in \omega}$. However, this is still sufficient to establish Theorem 5.1, which is our aim here. We now proceed to a proof of Lemma 5.2.

Proof of Lemma 5.2. Let π be a Datalog-program that accepts a directed acyclic graph $(\mathbf{G}, s, t)$ if, and only if, $\mathbf{G}$ contains a path from s to t of length p for some $p \in S$ and let k be the number of variables in π. Then, for any such $(\mathbf{G}, s, t)$, there is an expansion $\mathbf{A}$ of π such that $\mathbf{A} \to (\mathbf{G}, s, t)$, and there is a corresponding decorated expansion $(\mathbf{T}, B, L)$ where $(\mathbf{T}, B)$ is a tree decomposition of $\mathbf{A}$ of width $k - 1$. We can assume, without loss of generality that each B_u, $u \in T$, has exactly k elements. It will be clear how to adapt the construction to the case where this is not so. Since $\mathbf{A}$ is acyclic (otherwise there would be no homomorphism $\mathbf{A} \to \mathbf{G}$), we let $<$ be the (partial) order on vertices of $\mathbf{A}$ induced by distance from $s^{\mathbf{A}}$ (where vertices that are not reachable from s have distance ∞).

We now represent the decorated expansion $(\mathbf{T}, B, L)$ as a relational structure $\mathbf{D}$ as follows:

- the universe of $\mathbf{D}$ is $D := T \dot{\cup} A$, the disjoint union of T and A;
- the constants s and t are interpreted in $\mathbf{D}$ by $s^{\mathbf{A}}$ and $t^{\mathbf{A}}$;
- $\mathbf{D}$ has a $k + 1$-ary relation B such that for each $u \in T$ there is exactly one tuple $(u, a_1, \ldots, a_k) \in B$, and it satisfies: $B_u = \{a_1, \ldots, a_k\}$ and $a_1 \leq a_2 \cdots \leq a_k$; and

- for every rule r of π and every mapping ρ from the variables of π to $\{1, \ldots, k\}$, there is a unary relation $L_{r,\rho}$ interpreted in $\mathbf{D}$ by $\{u \in T : L(u) = (r, \rho')\}$, where ρ' is the map that takes x to $a_{\rho(x)}$ where $(u, a_1, \ldots, a_k) \in B$.

We will not distinguish notationally between (T, B, L) and $\mathbf{D}$ in the sequel, as it will always be clear from the context in which presentation we formally work. It is easily seen that we can write a formula φ of MSO such that $\mathbf{D} \models \varphi$ if, and only if, $\mathbf{D}$ is a decorated expansion of π and the underlying expansion $\mathbf{A}$ is acyclic. Let q be the quantifier rank of φ and let Q be the number of distinct MSO-types of quantifier rank at most q. Note that the values of q and Q are determined by π and do not depend on the choice of the expansion $\mathbf{A}$.

For $x \in T$, we write $\mathbf{D}_x$ for the substructure of $\mathbf{D}$ induced by the subtree of T rooted at x, and the elements related to nodes of this subtree by B. Note that the only elements that $\mathbf{D}_x$ shares with the rest of $\mathbf{D}$ are in B_x. We write $\mathbf{D}[x/\mathbf{D}']$ for the structure obtained from $\mathbf{D}$ by replacing $\mathbf{D}_x$ by $\mathbf{D}'$. That is, it is the disjoint union of the structure $\mathbf{D} \setminus \mathbf{D}_x$, obtained by removing $\mathbf{D}_x$ from $\mathbf{D}$, with the structure $\mathbf{D}'$ while identifying the elements in $B_r^{\mathbf{D}'}$ (where r is the root of $\mathbf{D}'$) with $B_x^{\mathbf{D}}$. It is then an easy consequence of Lemma 2.3 that $\mathbf{D} \equiv^q \mathbf{D}[x/\mathbf{D}']$ if $\mathbf{D}_x \equiv^q \mathbf{D}'$. In particular this implies that if $\mathbf{D}$ is an acyclic decorated expansion then $\mathbf{D}[x/\mathbf{D}']$ is also an acyclic expansion. For $x, y \in T$, we write $y \prec x$ to denote that y is an ancestor of x in T.

We begin with an informal account of the proof of Lemma 5.2. The idea is to start with an acyclic expansion $\mathbf{D}$ that maps homomorphically to a simple path of length N, for some large enough N. This enables us to find a pair $x, y \in T$ such that $\mathbf{D}_x \equiv^q \mathbf{D}_y$ and $y \prec x$. We can then *pump*, i.e. consider the expansions $\mathbf{D}' := \mathbf{D}[x/\mathbf{D}_y]$ and $\mathbf{D}'' := \mathbf{D}[x/\mathbf{D}'_y]$, etc. in order to obtain larger acyclic expansions with longer s-t-paths. If $\mathbf{D}$ itself consisted of a single path, x and y could be chosen so that the pumped expansions themselves consisted of simple paths and we would obtain a set of such paths growing linearly in length. However, if $\mathbf{D}$ contains multiple intersecting paths, the process of pumping may create new paths, including ones shorter than N. Moreover, in order to ensure that *all* paths in the new expansion are affected by pumping, it is not sufficient to choose one pumping pair (x, y), rather we need pairs intersecting (in a suitable way) all s-t-paths in $\mathbf{D}$. Unfortunately, these pairs may overlap and we need to define the process of pumping carefully.

The difficult part of the proof is therefore to choose the set of pairs (x, y) we want to use, and to define the process of pumping carefully. In the construction outlined below, we show how such a set of pairs can be found such that after repeating the pumping process n times, every s-t-path has length at least n and at most n^c, for some $c \in \mathbb{N}$ that depends on $\mathbf{D}$ but not on n. This is enough to prove the lemma. We begin by giving a definition of pumping for a set C of pairs (x, y) which form an *anti-chain* in $\mathbf{D}$ (in a sense we make precise below). We then use this to inductively define the pumped expansions for more general sets C.

Pumping at an antichain: Let $\mathbf{D} = (T, B, L)$ be a decorated expansion and $C \subseteq T^2$ a set of pairs (x, y) such that $y \prec x$ and if $(x, y), (x', y') \in C$ then $x \neq x'$ and $y \not\prec y'$. We define the expansions $\mathbf{D}_n^C$ by induction on n: $\mathbf{D}_0^C := \mathbf{D}$ and $\mathbf{D}_{n+1}^C := \mathbf{D}[x/(\mathbf{D}_n^C)_y : (x, y) \in C]$.

In other words, $\mathbf{D}_n^C$ is obtained from $\mathbf{D}$ by pumping each pair (x,y) in C simultaneously n times. Since, for distinct pairs (x,y) and (x',y'), y and y' (and hence also x and x') are incomparable, this is well-defined. We now use this to define pumping for sets of pairs C which are not necessarily incomparable. To be specific, suppose that $C \subseteq T^2$ is a set of pairs (x,y) with $y \prec x$ and $x \neq x'$ for distinct pairs (x,y) and (x',y'). We define a partial order on C by letting $(x,y) \sqsubset (x',y')$ just in case $y' \prec y$ and let $\mathrm{ht}(x,y)$ denote the length of the maximal $\sqsubset$-chain below the pair (x,y). Let m be the maximal value of $\mathrm{ht}(x,y)$ among all pairs in C. Write C^p for the set $\{(x,y) \in C : \mathrm{ht}(x,y) = p\}$.

Pumping: We define the pumped expansions by induction on p: $\mathbf{D}_n^0 = \mathbf{D}_n^{C^0}$ and $\mathbf{D}_n^{p+1} = (\mathbf{D}_n^p)_n^{C^{p+1}}$. Finally, let $\mathbf{D}_n^C$ denote $\mathbf{D}_n^m$.

Intuitively, given $\mathbf{D}$ and C we pump $\mathbf{D}$ by working bottom-up in $\mathbf{D}$ and replacing recursively for each pair $(x,y) \in C$ the tree rooted at x by the tree rooted at y and repeat n times. Note that if C is chosen so that for each $(x,y) \in C$, $\mathbf{D}_x \equiv^q \mathbf{D}_y$, then we also have $\mathbf{D}_n^C \equiv^q \mathbf{D}$. In particular, $\mathbf{D}_n^C$ is an acyclic expansion of π. The following claim is easily established by induction on p.

Claim. Every s-t-path in $\mathbf{D}_n^C$ is of length at most $n^m \cdot N$.

Let b be the maximal branching degree in any decorated expansion of π (note that this depends only on π) and choose $N \in S$ with $N > b^{Q \cdot K}$, where $K := 2^{k^2} \cdot k^4$. Let $\mathbf{A}$ be an expansion witnessing that a simple path of length N is accepted by π and $\mathbf{D} = (\mathbf{T}, B, L)$ be the corresponding decorated expansion. By the choice of N, every s-t-path P in $\mathbf{A}$ must contain K distinct internal vertices $v_1, \ldots v_K$ such that there is a chain $\boldsymbol{x}_P := x_1 \prec \cdots \prec x_K$ in $\mathbf{T}$ with $\mathbf{D}_{x_i} \equiv^q \mathbf{D}_{x_j}$ for all i,j and $v_i \in B_{x_i}$. Choose for each s-t-path such a chain and let $\Gamma := \{\boldsymbol{x}_P : P \text{ is an } s\text{-}t\text{-path in } \mathbf{A}\}$.

If B_x consists of the elements $a_1, \ldots, a_k$ in order, we say that a path P crosses x at (α, β) (for $1 \leq \alpha < \beta \leq k$) if P contains a_α and a_β and no intermediate element of B_x. For a fixed $\boldsymbol{x}_P$, by the choice of K, we can find a pair (α, β) and a subsequence $\boldsymbol{x}_P'$ of $\boldsymbol{x}_P$ of length at least $2^{k^2} \cdot k^2$ such that for each x in $\boldsymbol{x}_P'$, P crosses x at (α, β). Let Γ' be the collection of the pairs $(\boldsymbol{x}_P', (\alpha, \beta))$ for $\boldsymbol{x}_P \in \Gamma$.

Distant: Say a pair $(u,v) \subseteq B_y$, for some $y \in T$, is *distant* if for every path P from u to v in $\mathbf{A}$ there is some $(\boldsymbol{x}, (\alpha, \beta)) \in \Gamma'$ such that P crosses each $x \in \boldsymbol{x}$ at (α, β).

By construction, (s,t) is distant. For each $(\boldsymbol{x}, (\alpha, \beta)) \in \Gamma'$ we can choose a pair $x, y \in \boldsymbol{x}$ with $(x, a_1, \ldots, a_k) \in B$ and $(y, b_1, \ldots, b_k) \in B$ such that $y \prec x$ and (a_i, a_j) is distant if, and only if, (b_i, b_j) is distant for all i,j. Indeed, as $\boldsymbol{x}$ has at least $2^{k^2} \cdot k^2$ elements, we have at least k^2 distinct choices for x. This ensures that we can choose C to be a collection of such pairs (x,y), including one from each $(\boldsymbol{x}, (\alpha, \beta)) \in \Gamma'$ such that no two pairs in C share the same first component.

For $u, v \in B_x$, for some $x \in \mathbf{T}$, define the *pumping height* of (u,v) to be the length of the maximal chain (with respect to the order $\sqsubset$) in C below x. The following claim is the key to the pumping argument.

Claim. For all p, n, if the pumping height of (u,v) is at most p and (u,v) is distant then the distance of u and v in $\mathbf{D}_n^p$ is at least n.

In particular, the claim implies that for $n \in \mathbb{N}$, every s-t-path in $\mathbf{D}_n^C$ is of length at least n. As C, and hence m, only depend on the initial choice of $\mathbf{D}$ and not on n, we have that every s-t-path in $\mathbf{D}_n^C$ is of length at most $n^m \cdot N$. To complete the proof of Lemma 5.2, take $a_1 = N + 1$ and $a_{i+1} = a_i^{m+1}$. $\square$

To complete the proof of Theorem 5.1, consider the class C_S where $S = \{2^{2^{n^2}} : n \in \mathbb{N}\}$ which is clearly decidable in polynomial time. It is easily verified that there is no sequence $(a_i)_{i \in \omega}$ that satisfies the conditions of Lemma 5.2 for this set. Finally, we note also that the restriction of the class C_S to finite structures can be characterised as $\{\mathbf{A} : \mathbf{A}$ finite and $\mathbf{A} \not\rightarrow \mathbf{B}\}$ for a fixed infinite structure $\mathbf{B}$. Simply take $\mathbf{B}$ to be the structure formed from the disjoint union of all finite $\mathbf{A} \notin C_S$ by identifying all copies of s and t.

References

1. Open Problems List for the MathsCSP Workshop, Oxford (2006), http://www.cs.rhul.ac.uk/home/green/mathscsp/
2. Afrati, F., Cosmadakis, S., Yannakakis, M.: On Datalog vs. Polynomial Time. Journal of Computer and System Sciences 51, 177–196 (1995)
3. Atserias, A.: The homomorphism preservation property. In: Talk at International Workshop on Mathematics of Constraint Satisfaction, Oxford (2006)
4. Atserias, A., Bulatov, A., Dawar, A.: Affine systems of equations and counting infinitary logic. In: Proc. 34th International Colloquium on Automata, Languages and Programming. LNCS, vol. 4596, pp. 558–570. Springer, Heidelberg (2007)
5. Atserias, A., Dawar, A., Grohe, M.: Preservation under extensions on well-behaved finite structures. In: Caires, L., Italiano, G.F., Monteiro, L., Palamidessi, C., Yung, M. (eds.) ICALP 2005. LNCS, vol. 3580, pp. 1437–1449. Springer, Heidelberg (2005)
6. Atserias, A., Dawar, A., Kolaitis, P.G.: On preservation under homomorphisms and unions of conjunctive queries. Journal of the ACM 53, 208–237 (2006)
7. Chandra, A.K., Merlin, P.M.: Optimal implementation of conjunctive queries in relational databases. In: Proc. 9th ACM Symp. on Theory of Computing, pp. 77–90 (1977)
8. Ebbinghaus, H.-D., Flum, J.: Finite Model Theory, 2nd edn. Springer, Heidelberg (1999)
9. Feder, T., Vardi, M.Y.: Computational structure of monotone monadic SNP and constraint satisfaction: A study through Datalog and group theory. SIAM Journal of Computing 28, 57–104 (1998)
10. Feder, T., Vardi, M.Y.: Homomorphism closed vs existential positive. In: Proc. of the 18th IEEE Symp. on Logic in Computer Science, pp. 311–320 (2003)
11. Grohe, M.: Logic, graphs, and algorithms. In: Flum, J., Grädel, E., Wilke, T. (eds.) Logic and Automata History and Perspectives, Amsterdam University Press (2007)
12. Hodges, W.: Model Theory. Cambridge University Press, Cambridge (1993)
13. Immerman, N.: Relational queries computable in polynomial time. Information and Control 68, 86–104 (1986)
14. Kolaitis, P.G., Vardi, M.Y.: Conjunctive query containment and constraint satisfaction. Journal of Computer and System Sciences 61, 302–332 (2000)
15. Makowsky, J.A.: Algorithmic uses of the Feferman-Vaught theorem. Annals of Pure and Applied Logic 126, 159–213 (2004)
16. Moschovakis, Y.N.: Elementary Induction on Abstract Structures. North Holland, Amsterdam (1974)
17. Rossman, B.: Existential positive types and preservation under homomorphisisms. In: 20th IEEE Symposium on Logic in Computer Science, pp. 467–476 (2005)
18. Vardi, M.Y.: The complexity of relational query languages. In: Proc. of the 14th ACM Symp. on the Theory of Computing, pp. 137–146 (1982)

Directed st-Connectivity Is Not Expressible in Symmetric Datalog*

László Egri[1], Benoît Larose[2], and Pascal Tesson[3]

[1] School of Computer Science, McGill University
legri1@cs.mcgill.ca
[2] Département d'Informatique et de Génie Logiciel, Université Laval
pascal.tesson@ift.ulaval.ca
[3] Department of Mathematics and Statistics, Concordia University
larose@mathstat.concordia.ca

Abstract. We show that the directed st-connectivity problem cannot be expressed in symmetric Datalog, a fragment of Datalog introduced in [5]. It was shown there that symmetric Datalog programs can be evaluated in logarithmic space and that this fragment of Datalog captures logspace when augmented with negation, and an auxiliary successor relation S together with two constant symbols for the smallest and largest elements with respect to S. In contrast, undirected st-connectivity is expressible in symmetric Datalog and is in fact one of the simplest examples of the expressive power of this logic. It follows that undirected non-st-connectivity can be expressed in restricted symmetric monotone Krom SNP, whereas directed non-st-connectivity is only definable in the more expressive restricted monotone Krom SNP. By results of [8], the inexpressibility result for directed st-connectivity extends to a wide class of homomorphism problems that fail to meet a certain algebraic condition.

1 Introduction

Separating deterministic logspace from non-deterministic logspace remains an outstanding challenge of computational complexity. Because undirected and directed st-connectivity are respectively complete for L [10] and NL, the question is tied to the distinction between the hardness of these two problems. Ajtai and Fagin gave the first proof that st-connectivity is "harder" for directed graphs than for undirected graphs in a precise technical sense [2]. They showed that unlike undirected st-connectivity, directed st-connectivity is not definable in monadic Σ_1^1. The result presented here is similar in spirit: we prove that unlike undirected st-connectivity, directed st-connectivity is not definable in *symmetric Datalog*.

This result is part of a research program investigating the descriptive complexity of constraint satisfaction problems or, equivalently, of the problem Hom($\mathbf{B}$) of determining whether a homomorphism exists between an input relational structure $\mathbf{A}$ and the fixed template $\mathbf{B}$. Feder and Vardi [6] showed that in a number

---

* Research supported in part by NSERC, FQRNT and CRM.

L. Aceto et al. (Eds.): ICALP 2008, Part II, LNCS 5126, pp. 172–183, 2008.
© Springer-Verlag Berlin Heidelberg 2008

of important cases, a polynomial time algorithm for the problem Hom(**B**) could be obtained by showing that the set of structures which are *not* homomorphic to **B** is definable in Datalog (in the sequel, we abuse terminology and say simply that Hom(**B**) is definable in Datalog). Pursuing that line of research, Dalmau proved that cases of the homomorphism problem that are known to lie in NL are all related to the linear fragment of Datalog. Finally, symmetric Datalog (itself a restriction of linear Datalog), was introduced in [5]. Symmetric Datalog programs can be evaluated in logspace and all cases of the homomorphism problem currently known to lie in L are in fact expressible in this fragment.

Over the last ten years, the complexity of Hom(**B**) has also been studied through a powerful algebraic approach [3], whose description is beyond the scope of this paper. The algebraic angle of attack was initially developed in parallel to the aforementioned logical one, but the links between them have been increasingly apparent. In particular, it is conjectured in [8] that Hom(**B**) (1) belongs to logspace iff (2) it is definable in symmetric Datalog iff (3) the algebra associated to **B** satisfies a technical universal-algebraic condition. The implication (2) $\rightarrow$ (1) is a consequence of Reingold's theorem [10]. The inexpressibility of directed st-connectivity in symmetric Datalog is the missing piece of a puzzle establishing the implication (2) $\rightarrow$ (3). Note that (1) $\leftrightarrow$ (3) (or (1) $\leftrightarrow$ (2)) can only hold if L $\neq$ NL, whereas the equivalence of (2) and (3) may still hold if L = NL.

The second order logic fragments restricted monotone Krom SNP and restricted symmetric monotone Krom SNP [4, 5, 7] were shown to be equivalent to linear Datalog and symmetric Datalog respectively. From our main theorem, it follows that undirected non-st-connectivity can be expressed in both second order logic fragments while directed non-st-connectivity is only definable in restricted monotone Krom SNP. Grädel showed that a logic closely related to symmetric monotone Krom SNP captures Logspace[1] in the presence of an auxiliary successor relation S and constant symbols for the smallest and largest element with respect to S. Similarly, symmetric Datalog captures logarithmic space if it is augmented with negation, an auxiliary successor relation S and constant symbols for the smallest and largest elements with respect to S [5]. Separating the complexity classes L and NL is equivalent to extending our inexpressibility result and showing that directed st-connectivity is inexpressible in symmetric Datalog even in the presence of negation and an auxiliary successor relation.

In the remainder of this section, we review the basic notions required for the exposition of our results. Section 2 introduces the key technical ingredients of our arguments and Section 3 presents a sketch of the proof of our main result. Because of space restrictions, most technical arguments are omitted in this extended abstract.

1.1 Relational Structures and Homomorphisms

A *vocabulary* is a finite set of relation symbols. In the following, τ denotes a vocabulary. Every relation symbol R in τ has an associated *arity* r. A *relational*

---

[1] Grädel's result uses the class co-SL now known to coincide with L.

structure **A** over the vocabulary τ, or simply a τ-*structure* consists of a set A called the *universe* or *domain* of **A**, and a relation $R^{\mathbf{A}} \subseteq A^r$ for every relation symbol $R \in \tau$, where r is the arity of R. We use boldface letters to denote relational structures.

Definition 1. *Let* $\mathbf{A} = \langle A; R_1^{\mathbf{A}}, \ldots, R_q^{\mathbf{A}} \rangle$ *and* $\mathbf{B} = \langle B; R_1^{\mathbf{B}}, \ldots, R_q^{\mathbf{B}} \rangle$ *be relational structures over the same vocabulary. A function* $h : A \to B$ *is called a homomorphism from* **A** *to* **B** *if* $\langle h(a_1), \ldots, h(a_{r_i}) \rangle \in R_i^{\mathbf{B}}$ *whenever* $\langle a_1, \ldots, a_{r_i} \rangle \in R_i^{\mathbf{A}}$, $1 \leq i \leq q$. *We write* $\mathbf{A} \xrightarrow{h} \mathbf{B}$ *if* h *is a homomorphism from* **A** *to* **B** *and simply* $\mathbf{A} \to \mathbf{B}$ *if such an* h *exists.*

For a fixed τ-structure **B**, the *homomorphism problem* for **B**, denoted Hom(**B**), consists of determining whether a given τ-structure **A** is homomorphic to **B**. Alternatively, one can think of Hom(**B**) as the set of such structures, i.e. Hom(**B**) = $\{\mathbf{A} : \mathbf{A} \xrightarrow{h} \mathbf{B}\}$. For a graph H with distinguished sets S and T of start and target vertices, the directed ST-connectivity problem consists of determining if there is a directed path from some $s \in S$ to some $t \in T$. We view the triple (H, S, T) as a relational structure **H** over the set of vertices with a binary edge-relation E and unary relations S and T. One can easily verify that a directed path from some $s \in S$ to some $t \in T$ exists in H iff there is no homomorphism from **H** to the two-element structure **B** defined by $E^{\mathbf{B}} = \{(0,0), (0,1), (1,1)\}$, $S^{\mathbf{B}} = \{1\}$ and $T^{\mathbf{B}} = \{0\}$. Similarly, there is an undirected ST-path in H iff **H** is not homomorphic to the two-element structure **C** defined as $E^{\mathbf{C}} = \{(0,0), (1,1)\}$, $S^{\mathbf{C}} = \{1\}$ and $T^{\mathbf{C}} = \{0\}$. In the sequel, we therefore regard (un)directed ST-connectivity as a homomorphism problem.

1.2 Datalog

Datalog is a query and rule language for deductive databases. A Datalog program $\mathcal{D}$ over a vocabulary τ is a finite set of rules of the form $h \leftarrow b_1; \ldots; b_m$ where h and each b_i are atomic formulas $R_j(v_1, \ldots, v_k)$. We say that h is the *head* of the rule and that $b_1; \ldots; b_m$ is its *body*. Relational predicates R_j which appear in the head of some rule of $\mathcal{D}$ are called *intensional database predicates* (*IDBs*) and are not part of the vocabulary τ. All other relational predicates are called *extensional database predicates* (*EDBs*) and are in τ.

A rule of $\mathcal{D}$ is *linear* if its body contains at most one IDB and is *non-recursive* if its body contains only EDBs. A linear but recursive rule is of the form $I_1(\bar{x}) \leftarrow I_2(\bar{y}); E_1(\bar{z}_1); \ldots; E_k(\bar{z}_k)$ where I_1, I_2 are IDBs and the E_i are EDBs[2]. Each such rule has a *symmetric* $I_2(\bar{y}) \leftarrow I_1(\bar{x}); E_1(\bar{z}_1); \ldots; E_k(\bar{z}_k)$. A Datalog program is *non-recursive* if all its rules are non-recursive, *linear* if all its rules are linear and *symmetric* if it is linear and if the symmetric of each recursive rule of $\mathcal{D}$ is also a rule of $\mathcal{D}$. We further say that $\mathcal{D}$ has *width* (j, k) if each rule of $\mathcal{D}$ has at most k variables and at most j variables in the head.

A Datalog program $\mathcal{D}$ takes a τ-structure **A** as input and returns a structure $\mathcal{D}(\mathbf{A})$ over the vocabulary $\tau' = \tau \cup \{I : I \text{ is an IDB in } \mathcal{D}\}$. We also want to

---

[2] Note that the variables occurring in $\bar{x}, \bar{y}, \bar{z}_i$ are not necessarily distinct.

view a Datalog program as being able to accept or reject an input τ-structure and this is achieved by choosing one of the IDB's of $\mathcal{D}$ as the *goal predicate*: the τ-structure $\mathbf{A}$ is *accepted by* $\mathcal{D}$ if the goal predicate G is non-empty in $\mathcal{D}(\mathbf{A})$.

The semantics of Datalog are very intuitive and we only illustrate them through an example. A formal definition can be found, for example, in [4, 9]. Consider the problem of two-coloring. An undirected graph is two-colorable if and only if it contains no cycles of odd length. The following Datalog program $\mathcal{D}$ defines two-coloring because the goal predicate becomes non-empty if and only if the input graph contains an odd cycle.

$$O(x, y) \leftarrow E(x, y)$$
$$O(x, y) \leftarrow O(x, w); E(w, z); E(z, y)$$
$$O(x, w) \leftarrow O(x, y); E(w, z); E(z, y)$$
$$G \leftarrow O(x, x)$$

Here E is the binary EDB representing the adjacency relation in the input graph, O is a binary IDB whose intended meaning is "there exists an odd-length path from x to y" and G is the 0-ary goal predicate. Intuitively, the program first finds a path of length one using the only non-recursive rule and then iteratively finds paths of higher odd lengths using the middle two rules. Whenever the path begins and ends at the same vertex x, the goal predicate becomes non-empty indicating the presence of a cycle of odd length.

Note that the two middle rules form a symmetric pair. In the above description, we have not included the symmetric of the last rule. In fact, the fairly counterintuitive rule $O(x, x) \leftarrow G$ can be added to the program without changing the class of structures accepted by the program since the rule only becomes relevant if an odd cycle has already been detected in the graph.

Assume that a program $\mathcal{D}$ accepts a structure $\mathbf{A}$. Intuitively, a *derivation tree* over $\mathbf{A}$ is a representation of the "proof" that $\mathcal{D}$ accepts $\mathbf{A}$. Consider, for example, the following (linear) Datalog program $\mathcal{D}$ over the vocabulary consisting of a binary relation symbol E and two unary relation symbols S and T:

$$I(y) \leftarrow S(y)$$
$$I(y) \leftarrow I(x); E(x, y)$$
$$G \leftarrow I(y); T(y)$$

We choose G as the goal predicate. One can verify that an input structure $\mathbf{A}$ is accepted if and only if there is a path in the graph $\mathbf{A}$ from a vertex in S to a vertex in T. Let $\mathbf{A}$ be the input structure in Figure 1. Notice that $\mathcal{D}$ accepts $\mathbf{A}$ because it contains a path v_5, v_6, v_3, v_4 from a vertex in S to a vertex in T. Therefore one possible derivation tree for $\mathcal{D}$ over $\mathbf{A}$ is shown in Figure 1. Intuitively, the derivation tree "follows" the path from v_5 to v_4.

Formally, a *derivation tree* T for a Datalog program $\mathcal{D}$ over a structure $\mathbf{A}$ is a tree such that (1) the root of T is the goal predicate; (2) an internal node (including the root) together with its children correspond to a rule $\mathcal{R}$ of $\mathcal{D}$ in the following sense. The internal node is the head IDB of $\mathcal{R}$ and the children are the

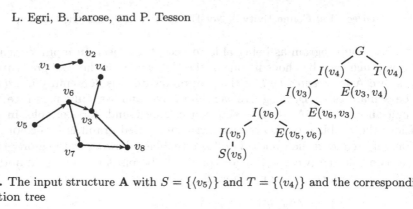

Fig. 1. The input structure **A** with $S = \{\langle v_5\rangle\}$ and $T = \{\langle v_4\rangle\}$ and the corresponding derivation tree

predicates in the body of $\mathcal{R}$; (3) an internal node of $\mathcal{T}$ is an IDB predicate $I(\bar{a})$ together with an instantiation of variables where if I has arity r then $\bar{a} \in A^r$; (4) the predicate children of a parent node inherit an instantiation to elements of A from their parents; (5) the leaf nodes of $\mathcal{T}$ are EDB predicates $E(\bar{b})$ such that if E has arity s then $\bar{b} \in A^s$ and $\bar{b} \in E^\mathbf{A}$. By design, a derivation tree for a Datalog program $\mathcal{D}$ and a structure $\mathbf{A}$ exist if and only if $\mathcal{D}$ accepts $\mathbf{A}$. For linear and symmetric Datalog we use the term *derivation path* instead of derivation tree because the IDBs in a derivation tree form a path.

Let τ consist of a single binary relation E. We say that a symmetric Datalog program $\mathcal{D}$ *computes reflexive transitive closure* if $\mathcal{D}$ has a binary IDB G such that for any τ-structure $\mathbf{A}$ (i.e. a digraph) we have that $G^{\mathcal{D}(\mathbf{A})}$ is the reflexive transitive closure of $E^\mathbf{A}$. For this case, we slightly abuse terminology and call G the goal predicate. We show (Lemma 11) that there exist a symmetric $\mathcal{D}$ computing reflexive transitive closure iff directed ST-connectivity is expressible in symmetric Datalog. Although our main objective is the second statement, our proof establishes the former statement for technical convenience.

In order to show that no symmetric Datalog program $\mathcal{D}$ computes reflexive transitive closure, we use a form of pumping argument reminiscent of [1]. Roughly speaking, we prove that if $\mathcal{D}$ is complete (i.e. if for all $\mathbf{H}$ we have $\langle u, v\rangle \in G^{\mathcal{D}(\mathbf{H})}$ whenever $\langle u, v\rangle$ is in the reflexive transitive closure of $E^\mathbf{H}$) then it must be unsound (i.e. there exists $\mathbf{H}'$ such that $G^{\mathcal{D}(\mathbf{H}')}$ contains a pair $\langle u, v\rangle$ which is *not* in the closure of $E^{\mathbf{H}'}$). We consider the behavior of $\mathcal{D}$ on an input $\mathbf{H}$ which is a "sufficiently long" simple path $s \to h_1 \to \ldots \to h_r \to t$. If $\mathcal{D}$ is complete then the pair $\langle s, t\rangle$ must be in $G^{\mathcal{D}(\mathbf{H})}$ and there must exist a derivation path $\mathcal{P}$ witnessing this fact. Now, consider a subpath of $\mathcal{P}$, say $I_3 - I_2 - I_1$. For example, suppose that by successively using rules $\mathcal{R}_1$ and $\mathcal{R}_2$, the program derives $I_3(a, b)$ from $I_1(c)$ as follows: $I_2(c, a) \xleftarrow{\mathcal{R}_1} I_1(c); E(c, a)$ and $I_3(a, b) \xleftarrow{\mathcal{R}_2} I_2(a, c); E(a, b)$. Notice that since the symmetric rules $\mathcal{R}'_1, \mathcal{R}'_2$ are also in $\mathcal{D}$ we can use $\mathcal{R}'_2$ followed by $\mathcal{R}'_1$ to re-obtain $I_1(c)$ from $I_3(a, b)$ and, from there, re-derive $I_3(a, b)$ from $I_1(c)$. In other words, we can artificially lengthen $\mathcal{P}$ by replacing the subpath $I_3 - I_2 - I_1$ of $\mathcal{P}$ by $I_3 - I_2 - I_1 - I_2 - I_3 - I_2 - I_1$. By using this "pumping" trick on carefully chosen subpaths, we obtain from $\mathcal{P}$ a new derivation path $\mathcal{P}'$ and then construct from $\mathcal{P}'$ a new digraph $\mathbf{H}'$ such that $\mathcal{P}'$ witnesses the membership of a pair $\langle u, v\rangle$ in $G^{\mathcal{D}(\mathbf{H}')}$ even though no path from u to v exists in $\mathbf{H}'$. In the next section, we formalize the above intuition by introducing mirror operators and zig-zags: mirroring corresponds to the lengthening process just described and zig-zags allow us to describe the form of the derivation path obtained through a sequence of mirroring operations.

2   Zig-Zags, Mirror Operators and Free Derivation Paths

A *min-max pair*, denoted by $[a, b]$, is a pair of integers a, b such that $a < b$. Similarly, an *index pair* $\langle i, j \rangle$ is a pair of integers such that $i < j$. A *zig-zag* is a sequence of integers $Z = t_1, \ldots, t_p$ such that $|t_i - t_{i+1}| = 1$ for each $1 \le i \le p - 1$. Given a min-max pair $[a, b]$ and a zig-zag Z, $\mathrm{MaxP}_{[a,b]}(Z)$ denotes the set of all index pairs $\langle k, \ell \rangle$ such that (i) if $t \in \{t_k, t_{k+1}, ..., t_\ell\}$ then $a \le t \le b$, (ii) $a, b \in \{t_k, t_{k+1}, ..., t_\ell\}$ and (iii) neither $a \le t_{k-1} \le b$ nor $a \le t_{\ell+1} \le b$ holds. Let $\mathcal{Z}$ denote the set of all zig-zags. A *mirror operator* $\mu_{[a,b],r} : \mathcal{Z} \to \mathcal{Z}$ is a function with a min-max pair parameter and $r \in \mathbb{Z}^+$. Let Z be a zig-zag. Then $\mu_{[a,b],r}(Z)$ is the zig-zag such that for each index pair $\langle i, j \rangle \in \mathrm{MaxP}_{[a,b]}(Z)$ we insert r consecutive copies of the sequence $t_{j-1}, t_{j-2}, \ldots t_i, t_{i+1}, \ldots, t_j$ after t_j in Z.

Our main theorem relies on corollaries 6 and 8 which, in turn, rely on Lemma 2. The proof of Lemma 2 is not difficult but laborious and is omitted.

Lemma 2. *Let* $\mu(Z) = \mu_{[a,b],r_1}(Z)$ *and* $\nu(Z) = \mu_{[c,d],r_2}(Z)$ *be mirror operators, and let* Z *be a zig-zag. Then*

$$\nu(\mu(Z)) = \mu(\nu(Z)).$$

2.1   The Free Derivation Path

A *free derivation path* is obtained from a derivation path by replacing the domain elements with the underlying variables and renaming all quantified variables to different names. For example, let $\mathcal{D}$ be the symmetric Datalog program in Figure 2a. Let the input structure **A** be the graph in Figure 2b together with the unary relations $S = \{s\}$ and $T = \{t\}$. Then a derivation path $\mathcal{P}$ obtained from $\mathcal{D}$ over the input **A** is shown in Figure 2c. In Figure 2d we obtain the corresponding free derivation path by renaming each variable of $\mathcal{P}$ such that the variables of an IDB and EDBs in the body of a rule inherit the variables of the head IDB and all other variables are renamed to new elements.

Let τ be the input vocabulary of a symmetric Datalog program $\mathcal{D}$ and let $\mathcal{F}$ be a free derivation path. We can associate a τ-structure **F** with $\mathcal{F}$ as follows. First, the domain F of **F** consists of all the variables appearing in $\mathcal{F}$. Second, let $R \in \tau$ have arity r and put a tuple $\langle x_1, \ldots, x_r \rangle \in F^r$ into the relation $R^{\mathbf{F}}$ if $R(x_1, \ldots, x_r)$ is present in $\mathcal{F}$. Observe that $\mathcal{F}$ is a derivation path for **F**. **F** is called the *free structure* associated with $\mathcal{F}$.

Given a zig-zag $Z = t_1, \ldots, t_q$ and a free derivation path $\mathcal{F}$ having q occurrences of IDBs, we construct a corresponding *labeled free derivation path* $\mathcal{F}^Z$ as follows. Label the i-th IDB of $\mathcal{F}$ starting from the goal predicate with t_i, $1 \le i \le q$. If an IDB I is labeled with t_i we denote it by I_{t_i}. Let $\mathcal{F}^Z$ be a

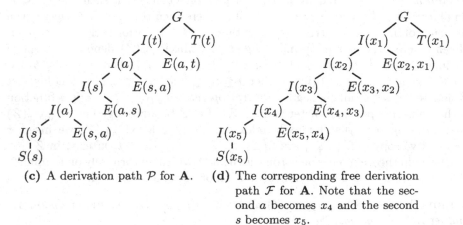

$$I(y) \leftarrow S(y)$$
$$I(y) \leftarrow I(x); E(x, y)$$
$$I(x) \leftarrow I(y); E(x, y)$$
$$G \leftarrow I(y); T(y)$$

(a) The symmetric Datalog program $\mathcal{D}$.

(b) The edge relation of the input structure **A**.

(c) A derivation path $\mathcal{P}$ for **A**.

(d) The corresponding free derivation path $\mathcal{F}$ for **A**. Note that the second a becomes x_4 and the second s becomes x_5.

Fig. 2. The construction of a free derivation path

labeled free derivation path and let $\mu = \mu_{[a,b],r}$ be a mirror operator. We extend the action of μ to labeled free derivation paths in a natural way. First, for each index pair $\langle i, j \rangle \in \mathrm{MaxP}_{[a,b]}(Z)$ we insert r consecutive copies of the sequence $I_{t_{j-1}}, I_{t_{j-2}}, \ldots I_{t_i}, I_{t_{i+1}}, \ldots, I_{t_j}$ after I_{t_j} in $\mathcal{F}^Z$. Second, let $I_{t_k} - I_{t_{k+1}}$ be a segment of $\mathcal{F}^Z$ and let $\mathcal{R}$ be the corresponding rule. Whenever $I_{t_k} - I_{t_{k+1}}$ is inserted in the new derivation path the corresponding rule is $\mathcal{R}$ and the parent of the EDBs of $\mathcal{R}$ is I_{t_k}. On the other hand, whenever we insert $I_{t_{k+1}} - I_{t_k}$ the rule corresponding to $I_{t_{k+1}} - I_{t_k}$ is the symmetric rule $\mathcal{R}'$ of $\mathcal{R}$ and accordingly, the parent of the EDBs of $\mathcal{R}'$ is $I_{t_{k+1}}$. Third, $\mu(\mathcal{F}^Z)$ is labeled with $\mu(Z)$. Finally, starting at the goal predicate, traverse the variables of $\mu(\mathcal{F}^Z)$ and rename them to a new name whenever possible. This ensures that $\mu(\mathcal{F}^Z)$ is free. Clearly, if $\mathcal{F}^Z$ is a labeled free derivation path constructed from the rules of a symmetric program $\mathcal{D}$ then $\mu(\mathcal{F}^Z)$ can also be constructed from the rules of $\mathcal{D}$. The definition of the free structure associated with a labeled free derivation path is analogous to the definition of the free structure associated with a free derivation path.

For instance, consider again the program $\mathcal{D}$ in Figure 2a; a free derivation path $\mathcal{F}$ for this program is shown in Figure 3a. Let $Z = 1, 2, 3$ be a zig-zag. Then $\mathcal{F}^Z$ is shown in Figure 3b. Figure 3c shows the intermediate step when the variables are not yet renamed to new elements and Figure 3d shows $\mu_{[2,3],1}(\mathcal{F}^Z)$.

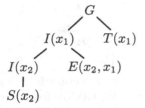

(a) The free derivation path $\mathcal{F}$.

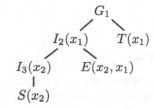

(b) The labeled free derivation path $\mathcal{F}^Z$, where $Z = 1, 2, 3$.

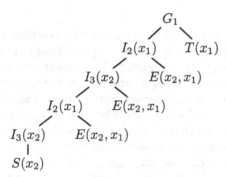

(c) Intermediate step in the construction of $\mu_{[2,3],1}(\mathcal{F}^Z)$.

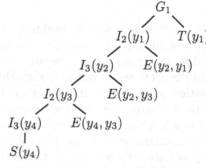

(d) By renaming the variables we get the new labeled free derivation path $\mu_{[2,3],1}(\mathcal{F}^Z)$.

Fig. 3. Constructing $\mu_{[2,3],1}(\mathcal{F}^Z)$

Before we present an outline of the general proof, we demonstrate the proof idea by showing that the symmetric $(1,2)$-Datalog program $\mathcal{D}$ in Figure 2a cannot decide the directed ST-connectivity problem. Program $\mathcal{D}$ is complete in the sense that it accepts any structure $\mathbf{A}$ in which there exist vertices $s \in S^{\mathbf{A}}$ and $t \in T^{\mathbf{A}}$ such that there is a path from s to t in $E^{\mathbf{A}}$. In fact, $\mathcal{D}$ decides the undirected ST-connectivity problem.

Observe that $\mathcal{D}$ accepts the directed ST-connectivity instance $\langle U; S, T, E \rangle$, where $U = \{u, v\}, S = \{u\}, T = \{v\}, E = \{\langle u, v \rangle\}$. There could be many corresponding derivation paths but pick the one that contains only a single application of a recursive rule. Then the corresponding free derivation path is the one in Figure 3a and the corresponding free structure is $\mathbf{F}_1$ in Figure 4a. Applying $\mu_{[2,3],1}$ and constructing the free structure yields $\mathbf{F}_2$ in Figure 4b which is also accepted by $\mathcal{D}$ even though there is no path from y_4 to y_1.

This trick can also be applied to any $(1, k)$-Datalog program. The difficulty is to generalize this argument to (j, k)-programs for $j > 1$. To see this challenge, assume that we input a directed path from a vertex u to a vertex v together with $S = \{u\}$ and $T = \{v\}$ to a symmetric (j, k)-Datalog program with $j > 1$. Then the free structure $\mathbf{F}$ can be rather more complicated than before. First, $\mathbf{F}$ could contain many different paths from u to v. For each path p we can

$$x_2 \quad x_1$$

$$S = \{x_2\} \qquad T = \{x_1\}$$

$$y_2 \quad y_1$$

$$y_4 \quad y_3$$

$$S = \{y_4\} \qquad T = \{y_1\}$$

(a) The free structure $\mathbf{F}_1$ associated with the free derivation path in Figure 3a.

(b) The free structure $\mathbf{F}_2$ associated with the labeled free derivation path $\mathcal{F}^Z$ in Figure 3d.

Fig. 4. Fooling a simple symmetric program by "pumping" the input structure

find a mirror operator that produces a labeled free derivation path such that in the corresponding free structure p "disappears". After we compose these mirror operators to obtain a new operator. Using the commutativity property stated in Corollaries 6 and 8 we show that this new operator produces a labeled free derivation path that is still accepted but in the corresponding free structure each path "disappears". Second, notice that if $j = 1$ then when we traverse the edges of a path in a free structure we must either move in the free derivation path monotonically towards the goal predicate or monotonically away from the goal predicate. When $j > 1$ this is not the case any more. The location of the edges of the path in the free structure can be much more "disordered" in the free derivation path. This is why we need to mirror a segment of a labeled free derivation path more than once, i.e. to set r greater than 1 in a mirror operator $\mu_{[a,b],r}$.

3 Proof of the Main Theorems: Outline

We provide in this section an overview of the proof of the main theorems. Due to space restrictions, we simply describe the main technical results and show how they can be assembled to obtain Theorem 3 below and its alternative formulation Theorem 12.

Theorem 3. *No symmetric Datalog program $\mathcal{D}$ computes reflexive transitive closure.*

As noted at the end of the previous section, our proof revolves around a generalization of the basic argument described through Figure 4.

Definition 4. *Let $\mathcal{F}^Z$ and $\mathcal{F}^{Z'}$ be labeled free derivation paths and $\mathbf{F}$ and $\mathbf{F}'$ be the corresponding relational structures, respectively. We say that there is a homomorphism from $\mathcal{F}^Z$ to $\mathcal{F}^{Z'}$ if there is a homomorphism h from $\mathbf{F}$ to $\mathbf{F}'$, and we denote this by $\mathcal{F}^Z \xrightarrow{h} \mathcal{F}^{Z'}$.*

Lemma 5. *Let $\mathcal{F}^Z$ be a labeled free derivation path and $\mu_1, \mu_2, \ldots, \mu_n$ be a sequence of mirror operators. Let $\mathcal{F}_0^Z, \mathcal{F}_1^Z, \ldots, \mathcal{F}_n^Z$ be a sequence of labeled free derivation paths defined by $\mathcal{F}_0^Z = \mathcal{F}^Z$ and $\mathcal{F}_{i+1}^Z = \mu_{i+1}(\mathcal{F}_i^Z)$, $0 \le i \le n-1$. Then $\mathcal{F}_n^Z \xrightarrow{h} \mathcal{F}^Z$.*

Proof. It clearly suffices to show that $\mathcal{F}_{i+1}^Z \xrightarrow{h} \mathcal{F}_i^Z$. Notice that when we apply μ_i to $\mathcal{F}_i^Z$ to obtain $\mathcal{F}_{i+1}^Z$ we insert additional sequences into $\mathcal{F}_i^Z$ and we create new variables by renaming the original variables to new names whenever possible. Let h be the function that maps each new variable in $\mathcal{F}_{i+1}^Z$ to the original in $\mathcal{F}_i^Z$. Clearly, h is a homomorphism.

For example, we obtained the labeled free derivation path in Figure 3d by applying $\mu_{[2,3],1}$ to the labeled free derivation path in Figure 3b. Define h to be $y_1 \mapsto x_1, y_2 \mapsto x_2, y_3 \mapsto x_1, y_4 \mapsto x_2$. Observe that h is a homomorphism from $\mu_{[2,3],1}(\mathcal{F}^Z)$ to $\mathcal{F}^Z$.

One of our main tools is Corollary 6, which follows directly from Lemma 2.

Corollary 6. *Let $\mu(Z) = \mu_{[a,b],r_1}(Z)$ and $\nu(Z) = \mu_{[c,d],r_2}(Z)$ be mirror operators, let Z be a zig-zag, and let $\mathcal{F}$ be a free derivation path. Then, up to renaming variables,*

$$\nu(\mu(\mathcal{F}^Z)) = \mu(\nu(\mathcal{F}^Z)).$$

Given a labeled free derivation path $\mathcal{F}^Z$ over τ we define $\mathcal{E}(\mathcal{F}^Z)$ as follows. Index the EDBs in $\mathcal{F}^Z$ by their distance from the root $G(u,v)$ and let $\mathcal{E}(\mathcal{F}^Z)$ be the set of all indexed EDBs that appear in $\mathcal{F}^Z$. Furthermore, we say that a set $X \subseteq \mathcal{E}(\mathcal{F}^Z)$ *contains a path from* x_1 *to* x_n if there are indexed EDBs in X such that $E_{i_1}(x_1, x_2), E_{i_2}(x_2, x_3), \ldots, E_{i_k}(x_{n-1}, x_n)$ for some variables $x_2, x_3, \ldots, x_{n-1}$. The indices of the EDBs are used to differentiate between paths which would be the same if we removed the indices. For example, a labeled free derivation path $\mathcal{F}^Z$ could contain two paths p_1 and p_2 whose EDBs are exactly the same but an EDB $E(x,y)$ of p_1 appears at a different level in $\mathcal{F}^Z$ than the same EDB of p_2. If we had no indices this difference would be lost.

Definition 7. *Let $\mu_1, \mu_2, \ldots, \mu_n$ be mirror operators, let M_i be the operator $M_i = \mu_i \circ \cdots \circ \mu_2 \circ \mu_1$ and let $M = M_n$. Let $X \subseteq \mathcal{E}(\mathcal{F}^Z)$. We define the M-expansion $X' \subseteq \mathcal{E}(M(\mathcal{F}^Z))$ of X inductively as follows. The M_0-expansion of X is X. Assume that the M_i-expansion of X is X_i. Consider the construction of $M_{i+1}(\mathcal{F}^Z)$ from $M_i(\mathcal{F}^Z)$. For each indexed EDB E_ℓ in X_i add the corresponding indexed EDB in $M_{i+1}(\mathcal{F}^Z)$ to X_{i+1} and any new copies of E_ℓ. (Note that indices of the EDBs in $M_{i+1}(\mathcal{F}^Z)$ are recomputed.) The M-expansion of X is X_n.*

For example, the $\mu_{[2,3],1}$-expansion of $\{E_2(x_2, x_1)\}$ in Figure 3b (after indexing the EDBs) is $\{E_2(y_2, y_1), E_3(y_2, y_3), E_4(y_4, y_3)\}$ in Figure 3d. We need the following corollary of Lemma 2.

Corollary 8. *Let Z be a zig-zag, let μ, ν be mirror operators, and let $\mathcal{F}$ be a free derivation path. Let $\mathcal{F}' = (\nu \circ \mu)(\mathcal{F}^Z) = (\mu \circ \nu)(\mathcal{F}^Z)$. Let $X \subseteq \mathcal{E}(\mathcal{F}^Z)$. Then in $\mathcal{F}'$, the $(\nu \circ \mu)$-expansion of X is the same as the $(\mu \circ \nu)$-expansion of X.*

Finally, we also need the following two lemmas.

Lemma 9. *Let $\mathcal{F}^Z$ be a labeled free derivation path and $M = \mu_n \circ \cdots \circ \mu_2 \circ \mu_1$ where $\mu_1, \mu_2, \ldots, \mu_n$ are mirror operators. Let u, v be the variables appearing in* the goal predicate $G(u, v)$, and let $E_{u \not\to v} \subseteq \mathcal{E}(\mathcal{F}^Z)$ be a set that contains no path from u to v. Let $E'_{u \not\to v} \subseteq \mathcal{E}(M(\mathcal{F}^Z_n))$ be the M-expansion of $E_{u \not\to v}$. Then $E'_{u \not\to v}$ contains no path from u to v.

Proof. Assume that $E'_{u \not\to v}$ contains a path from u to v. Notice that if $\mu_1(\mathcal{F}^Z)$ contains a path from u to v then so does $\mathcal{F}^Z$. Repeating this argument for all $i \geq 2$, we obtain a path in $E_{u \not\to v}$ and this leads to a contradiction.

Lemma 10. *Let $\mathcal{F}$ be a labeled free derivation path originating from a symmetric (j, k)-Datalog program which has a binary goal predicate G. Assume that the top IDB of $\mathcal{F}$ is $G(u, v)$. Let q be the number of IDBs in $\mathcal{F}$ and consider the zig-zag $Z = 1, 2, \ldots, q$. Define a function $L_y(x)$ recursively by setting $L_y(1) = 3(y - 1)$ and $L_y(x) \geq 4L_y(x - 1) + 6$. Assume that $\mathcal{E}(\mathcal{F}^Z)$ contains a path p from u to v of length at least ℓ where $\ell = L_k(j)$. Then there exists a mirror operator μ such that the μ-expansion of p in $\mathcal{E}(\mu(\mathcal{F}^Z))$ does not contain any path from u to v.*

We now have the intermediate results required to prove Theorem 3 which we restate for convenience.

Theorem 3. *No symmetric Datalog program $\mathcal{D}$ computes reflexive transitive closure.*

Proof. Suppose for contradiction that $\mathcal{D}$ is a symmetric Datalog program that computes transitive closure. Let $\mathbf{B}$ be a structure with an edge relation $E^{\mathbf{B}}$ such that $E^{\mathbf{B}}$ is a simple path from a to b. Let the length ℓ of this path satisfy the length condition of Lemma 10. Obtain the free derivation path $\mathcal{F}$ from the derivation path that witnesses the fact that $\mathcal{D}$ derives $\langle a, b \rangle$. Assume that the variables in the binary goal G at the top of $\mathcal{F}$ are u and v. Let q be the number of IDBs in $\mathcal{F}$ and let Z be the zig-zag $1, 2, \ldots, q$ be a zig-zag. Observe that any path from u to v in $\mathcal{E}(\mathcal{F}^Z)$ must have length exactly ℓ. Let $P = \{p_1, \ldots, p_n\}$ be the set of all paths from u to v in $\mathcal{E}(\mathcal{F}^Z)$. For each $1 \leq i \leq n$, use Lemma 10 to find a mirror operator μ_i such that the μ_i-expansion of p_i does not contain a path from u to v. Let $\mathcal{F}^{\not\to} = (\mu_n \circ \mu_{n-1} \circ \cdots \circ \mu_1)(\mathcal{F}^Z)$. Observe that by Corollary 6, $\mathcal{F}^{\not\to} = (\mu_n \circ \cdots \circ \mu_{i+1} \circ \mu_{i-1} \circ \cdots \circ \mu_1 \circ \mu_i)(\mathcal{F})$ for each $1 \leq i \leq n$. We claim that $\mathcal{E}(\mathcal{F}^{\not\to})$ does not contain any path from u to v.

For the sake of contradiction assume that $\mathcal{E}(\mathcal{F}^{\not\to})$ contains a path w from u to v. Let h be the homomorphism defined in Lemma 5 from $\mathcal{F}^{\not\to}$ to $\mathcal{F}^Z$. Then $h(w)$ is a path in $\mathcal{E}(\mathcal{F}^Z)$ and therefore $h(w) = p_i$ for some i. By construction, the μ_i-expansion of p_i in $\mathcal{E}(\mu_i(\mathcal{F}^Z))$ does not contain a path from u to v. Using Lemma 9 we have that the $(\mu_n \circ \ldots \circ \mu_{i+1} \circ \mu_{i-1} \circ \ldots \circ \mu_1 \circ \mu_i)$-expansion of p_i in $\mathcal{E}(\mathcal{F}^{\not\to})$ does not contain a path from u to v. By Corollary 8 the $(\mu_n \circ \ldots \circ \mu_{i+1} \circ \mu_{i-1} \circ \ldots \circ \mu_1 \circ \mu_i)$-expansion and the $(\mu_n \circ \mu_{n-1} \circ \ldots \circ \mu_1)$-expansion of p_i are the same, hence the $(\mu_n \circ \mu_{n-1} \circ \ldots \circ \mu_1)$-expansion of p_i in $\mathcal{F}^{\not\to}$ does not contain a path from u to v. This leads to a contradiction since w is such a path.

Lemma 11. *The following statements are equivalent:*

(1) Let $\tau = \{E\}$ where E is a binary relation symbol. There exists a symmetric Datalog program with binary goal predicate G such that for any τ-structure $\mathbf{A}$, after running the program on input $\mathbf{A}$, G is the reflexive transitive closure of $E^{\mathbf{A}}$.
(2) Let $\sigma = \{S, T, E\}$ where E is a binary and S and T are unary relation symbols. Then there exists a symmetric Datalog program $\mathcal{D}$ such that for any σ-structure $\mathbf{B}$, $\mathcal{D}$ accepts if and only if there exist two vertices $s \in S^{\mathbf{B}}$ and $t \in T^{\mathbf{B}}$ such that there is a path from s to t in $E^{\mathbf{B}}$.

We have thus proved the following theorem, the second half of which relies on results of [4] and [5] linking fragments of Datalog and monotone Krom SNP.

Theorem 12
(a) Directed ST-connectivity is not definable in symmetric Datalog. More formally, let σ be a vocabulary as in Lemma 11. There is no symmetric Datalog program $\mathcal{D}$ such that $\mathcal{D}$ accepts an input σ-structure $\mathbf{H}$ iff $E^{\mathbf{H}}$ contains a directed path from some $s \in S^{\mathbf{H}}$ to some $t \in T^{\mathbf{H}}$.
(b) The complement of directed ST-connectivity cannot be defined in restricted symmetric monotone Krom SNP.

Minor modifications to our arguments show that the above theorem still holds if we allow negation of EDBs in the Datalog programs.

Acknowledgment. We thank the anonymous referees for their helpful comments on an earlier version of the paper.

References

1. Afrati, F., Cosmadakis, S.S., Yannakakis, M.: On Datalog vs. polynomial time. J. Comput. Syst. Sci. 51(2), 177–196 (1995)
2. Ajtai, M., Fagin, R.: Reachability is harder for directed than for undirected finite graphs. J. Symb. Log. 55(1), 113–150 (1990)
3. Cohen, D., Jeavons, P.: The complexity of constaint languages. In: Handbook of Constraint Programming, pp. 245–280 (2006)
4. Dalmau, V.: Linear Datalog and bounded path duality of relational structures. Logical Methods in Computer Science 1(1) (2005)
5. Egri, L., Larose, B., Tesson, P.: Symmetric Datalog and constraint satisfaction problems in logspace. In: LICS 2007: Proceedings of the 22nd Annual IEEE Symposium on Logic in Computer Science, pp. 193–202 (2007)
6. Feder, T., Vardi, M.Y.: The computational structure of monotone monadic SNP and constraint satisfaction: A study through Datalog and group theory. SIAM J. Comput. 28(1), 57–104 (1999)
7. Grädel, E.: Capturing complexity classes by fragments of second-order logic. Theor. Comput. Sci. 101(1), 35–57 (1992)
8. Larose, B., Tesson, P.: Universal algebra and hardness results for constraint satisfaction problems[3]. In: ICALP, pp. 267–278 (2007)
9. Libkin, L.: Elements of finite model theory. Springer, Heidelberg (2004)
10. Reingold, O.: Undirected st-connectivity in log-space, pp. 376–385 (2005)

---

[3] Extended version is to appear in Theoretical Computer Science.

Non-dichotomies in
Constraint Satisfaction Complexity

Manuel Bodirsky[1] and Martin Grohe[2]

[1] École Polytechnique (CNRS), France
[2] Humboldt-Universität zu Berlin, Germany

Abstract. We show that every computational decision problem is polynomial-time equivalent to a constraint satisfaction problem (CSP) with an infinite template. We also construct for every decision problem L an ω-*categorical* template Γ such that L reduces to $\mathrm{CSP}(\Gamma)$ and $\mathrm{CSP}(\Gamma)$ is in coNP^L (i.e., the class coNP with an oracle for L). CSPs with ω-categorical templates are of special interest, because the universal-algebraic approach can be applied to study their computational complexity.

Furthermore, we prove that there are ω-categorical templates with coNP-complete CSPs and ω-categorical templates with coNP-intermediate CSPs, i.e., problems in coNP that are neither coNP-complete nor in P (unless P=coNP). To construct the coNP-intermediate CSP with ω-categorical template we modify the proof of Ladner's theorem. A similar modification allows us to also prove a non-dichotomy result for a class of left-hand side restricted CSPs, which was left open in [10]. We finally show that if the so-called *local-global conjecture* for *infinite constraint languages* (over a finite domain) is false, then there is no dichotomy for the constraint satisfaction problem for infinite constraint languages.

1 Introduction

Let Γ be a relational structure over a finite signature τ, also called *template* or *constraint language* in the following. The *constraint satisfaction problem (CSP)* of Γ, denoted by $\mathrm{CSP}(\Gamma)$, is the computational problem to decide whether there exists a homomorphism from a given finite τ-structure to Γ. For finite relational structures Γ it has been conjectured that the computational complexity of $\mathrm{CSP}(\Gamma)$ exhibits a *dichotomy*: it is in P or NP-complete [9,6].

Ladner's theorem [14] states that there are computational problems in NP that are neither in P nor NP-complete (unless P=NP); we also say that these problems are *NP-intermediate*. However, the problems constructed by Ladner are highly artificial, and the lack of natural computational problems that could be NP-intermediate is one of the phenomena in complexity theory that is not well understood. Even for well-studied problems in NP that are neither known to be in P nor known to be NP-hard, such as the Graph Isomorphism problem, there is usually no strong evidence that they are not in P. From the computational

L. Aceto et al. (Eds.): ICALP 2008, Part II, LNCS 5126, pp. 184–196, 2008.

complexity perspective, it would therefore be very interesting if there were NP-intermediate CSPs with finite domains (since CSPs for finite templates are, arguably, relatively natural computational problems).

The CSP dichotomy conjecture is wide open. Bulatov, Krokhin, and Jeavons [6] give a sufficient condition for a finite domain CSP to be NP-hard, and they conjecture that all other problems are in P. This conjecture is based on the so-called *universal-algebraic approach* to the CSP, and the mentioned condition is formulated in terms of an algebra that can be associated to the template Γ. Most work in the area goes into finding larger and larger tractable classes of CSPs [13, 3, 4]. But not all researchers in the area believe in the conjecture. On the negative side, Feder and Vardi [9] introduced several heavily restricted subclasses of NP –for example monadic strict NP (MSNP)– that do *not* have a dichotomy. In fact, every problem in NP has a polynomial-time equivalent problem in MSNP.

In this paper we are mainly interested in infinite templates Γ. A particularly interesting class of infinite templates is the class of ω-*categorical*[1] structures, because ω-categorical templates Γ generalize finite templates in such a way that the above mentioned universal-algebraic approach still applies. For example, it holds that the complexity of the $\mathrm{CSP}(\Gamma)$ is fully captured by the polymorphism clone of Γ (even by the pseudo-variety generated by the algebra of the polymorphism clone [2]).

Results

We show that for every computational decision problem L there exists a polynomial-time equivalent constraint satisfaction problem with an infinite template Γ. This improves previous complexity results about infinite domain constraint satisfaction problems obtained by Bauslaugh [1] and Schwandtner [17]. We also construct an ω-categorical template Γ such that L reduces to $\mathrm{CSP}(\Gamma)$ and $\mathrm{CSP}(\Gamma)$ is in coNP^L (i.e., the class coNP with an oracle for L).

Furthermore, we prove that there are ω-categorical templates with coNP-complete CSPs and ω-categorical templates with *coNP-intermediate* CSPs, i.e., problems in coNP that are neither coNP-complete nor in P (unless P=coNP). To show this we use templates that are *countable homogeneous directed graphs*. These graphs are ω-categorical, and even though there are uncountably many non-isomorphic countable homogeneous directed graphs, they are model-theoretically well-understood [8]. In our construction we apply a modification of Ladner's proof technique. It remains open whether there are ω-categorical structures Γ such that $\mathrm{CSP}(\Gamma)$ is NP-intermediate.

In another line of research, the complexity of the constraint satisfaction problem has been studied for restricted classes of input structures. Let τ be a finite relational signature, and let $\mathcal{C}$ and $\mathcal{D}$ be two classes of finite τ-structures. Then $\mathrm{CSP}(\mathcal{C}, \mathcal{D})$ is the computational problem where the input consists of *two*

---

[1] A structure is ω-categorical if it has for all k only a finite number of k-*types* (in the model-theoretic sense [12]).

structures $C \in \mathcal{C}$ and $D \in \mathcal{D}$, and the question is whether there exists a homomorphism from C to D. It has been shown that when $\mathcal{D}_0$ is the set of all finite τ-structures, then for any recursively enumerable class of finite τ-structures $\mathcal{C}$ the problem $\mathrm{CSP}(\mathcal{C}, \mathcal{D}_0)$ is tractable if and only if $\mathcal{C}$ has bounded tree-width modulo homomorphic equivalence [10]. Note that this is not a dichotomy theorem.

With a similar modification of Ladner's proof as in our previous result, we can show that there is an efficiently decidable class $\mathcal{C}$ of finite (undirected) graphs such that $\mathrm{CSP}(\mathcal{C}, \mathcal{D}_0)$ is NP-intermediate (here, D_0 is the class of all finite undirected graphs); this was left open in [10]. In particular, this shows that there is no dichotomy for problems of the form $\mathrm{CSP}(\mathcal{C}, \mathcal{D})$. The same result was independently obtained in a recent paper by Chen, Thurley, and Weyer [7].

Finally, we show a connection between the dichotomy conjecture for *infinite constraint languages over a finite domain* (implied by Conjecture 4.12 of [5]) and the so-called *local-global conjecture*. If Γ is a relational structure with an infinite signature and a finite domain, then $\mathrm{CSP}(\Gamma)$ is called *locally tractable* if $\mathrm{CSP}(\Gamma')$ is tractable for all reducts Γ' of Γ with a finite signature. Note that for the notion of local tractability we can fix an arbitrary representation of the relations in the input instances. $\mathrm{CSP}(\Gamma)$ is called *globally tractable* if there is a polynomial-time algorithm that solves $\mathrm{CSP}(\Gamma)$ where the relations in the input instances are represented by fully specifying the relation $R \subset D(\Gamma)^k$. Obviously, global tractability implies local tractability. The *local-global conjecture* says that $\mathrm{CSP}(\Gamma)$ is locally tractable if and only if it is globally tractable.

If the local-global conjecture is true, then the dichotomy for finite constraint languages implies the dichotomy for infinite constraint languages. We show that in a certain sense the converse is true as well: if the global tractability conjecture is false, then there is no dichotomy for infinite constraint languages. In other words, if the dichotomy conjecture for finite constraint languages is true, then local and global tractability are equivalent.

Preliminaries

A *relational signature* τ is a set of *relation symbols* R_i, each of which has an associated finite *arity* k_i. The signature will be finite unless stated otherwise. A *relational structure* Γ over the signature τ (also called τ-*structure*) consists of a set D_Γ (the *domain*) together with a relation $R \subseteq D_\Gamma^{k_i}$ for each relation symbol of arity k_i from τ. If the reference to the relational structure is clear from the context, we use for simplicity the same symbol for a relation symbol and the corresponding relation. If necessary, we write R^Γ for the relation R belonging to the structure Γ. We call the elements of D_Γ the *vertices* of Γ.

Let Γ and Γ' be τ-structures. A *homomorphism* from Γ to Γ' is a function f from D_Γ to $D_{\Gamma'}$ such that for each n-ary relation symbol R in τ and each n-tuple $(a_1, \ldots, a_n)$, if $(a_1, \ldots, a_n) \in R^\Gamma$, then $(f(a_1), \ldots, f(a_n)) \in R^{\Gamma'}$. In this case we say that the map f *preserves* the relation R. Injective homomorphisms that also preserve the complement of each relation are called *embeddings*. Surjective embeddings are called *isomorphisms*, and isomorphisms between Γ and Γ are called *automorphisms*.

The *order* of a structure Γ, denoted by $|\Gamma|$, is the cardinality of D_Γ. An *induced substructure* of a structure Γ is a structure Γ' with $D_{\Gamma'} \subseteq D_\Gamma$ and $R^{\Gamma'} = R^\Gamma \cap D_{\Gamma'}^n$, for each n-ary $R \in \tau$. The *union* of two τ-structures Γ, Γ' is the structure $\Gamma \cup \Gamma'$ with domain $D_\Gamma \cup D_{\Gamma'}$ and relations $R^{\Gamma \cup \Gamma'} = R^\Gamma \cup R^{\Gamma'}$ for all $R \in \tau$. The intersection $\Gamma \cap \Gamma'$ is defined similarly. A *disjoint union* of Γ and Γ' is the union of isomorphic copies of Γ and Γ' with disjoint domains. As disjoint unions are unique up to isomorphism, we usually speak of *the* disjoint union of Γ and Γ'. A structure is called *connected* if it is not the disjoint union of two nonempty structures. For a mapping f defined on the domain of a τ-structure Γ, we let $f(\Gamma)$ be the τ-structure with domain $f(D_\Gamma)$ and relations $R^{f(\Gamma)} = \{(f(a_1), \ldots, f(a_n)) \mid (a_1, \ldots, a_n) \in R^\Gamma\}$ for every n-ary $R \in \tau$.

Classes of structures are always assumed to be closed under isomorphism.

2 Templates of All Complexities

In this section we show that for every computational decision problem there exists a polynomial-time equivalent constraint satisfaction problem with an infinite template Γ. Previously, it was known that for every recursive funtion f there exists an infinite structure Γ such that CSP(Γ) is decidable, but has time complexity at least f (a result due to Bauslaugh [1]). Recently, Schwandtner gave upper and lower bounds in the exponential time hierarchy for some infinite domain CSPs [17]; but these bounds leave an exponential gap.

Theorem 1. *Let L be a language over a finite alphabet Σ. Then there is an infinite relational structure Γ such that L is polynomial-time Turing equivalent to CSP(Γ).*

Proof (Idea). We construct an infinite relational structure Γ over the signature τ that contains pairwise distinct unary relation symbols P_a for all elements of $a \in \Sigma$; moreover, τ contains a binary relation symbol N and unary relation symbols S and T.

For each word $w \in \Sigma^*$ let W be the following τ-structure with vertices $1, \ldots, |w|$. The relation N^W is $\{(i, i + 1) \mid 1 \leq i < |w|\}$. The unary relation symbol P_a^W holds on $j \in \{1, \ldots, |w|\}$ iff the jth symbol in w is a. Finally, $S^W = \{1\}$ and $T^W = \{|w|\}$. Let $\mathcal{A}$ be $\{W \mid w \in L\}$.

Let $\mathcal{X}$ be the set of τ-structures with domain $\{1, \ldots, n\}$, for some n, where the symbol N is interpreted by the relation $\{(i, i + 1) \mid 1 \leq i < n\}$, each vertex i is contained in at most one relation P_a, for $a \in \Sigma$, and at least one of the following conditions is satisfied:

- S holds for none of the elements $1, \ldots, n$.
- T holds for none of the elements $1, \ldots, n$.
- There is an element from $1, \ldots, n$ such that for all $a \in \Sigma$ the relation P_a does not hold.

The structure Γ is the infinite disjoint union over all structures in $\mathcal{A} \cup \mathcal{X}$. The proof that Γ has the required properties has been omitted due to space restrictions. □

3 ω-Categorical Templates of Various Complexities

A relational structure is called ω-*categorical* if for all $k \geq 1$ there are at most finitely many inequivalent first-order formulas with k free variables. This definition of ω-categoricity is by the theorem of Ryll-Nardzewski equivalent to the standard definition of ω-categoricity [12]. With the definition given here it is easy to see that the structure Γ constructed in Section 2 is *not* ω-categorical, since there are infinitely many inequivalent first-order formulas with two free variables x and y that express that x and y are at distance k with respect to the relation N.

In this section we study the computational complexity of $CSP(\Gamma)$ if Γ is an ω-categorical structure, and prove the following.

Theorem 2. *For every language L over a finite signature Σ there exists an ω-categorical structure Γ such that L reduces to $CSP(\Gamma)$ and $CSP(\Gamma)$ is in $coNP^L$.*

For constructing ω-categorical templates, we need a few preliminaries from model theory. Let B_1, B_2 be τ-structures such that $A = B_1 \cap B_2$ is an induced substructure of both B_1 and B_2. Then we call $B_1 \cup B_2$ the *free amalgam* of B_1, B_2 over A. More generally, a τ-structure C is an *amalgam of B_1 and B_2 over A* if for $i = 1, 2$ there are embeddings f_i of B_i to C such that $f_1(a) = f_2(a)$ for all $a \in D_A$. Recall that classes of structures are always assumed to be closed under isomorphism. A class $\mathcal{A}$ of τ-structures has the *amalgamation property* if for all $A, B_1, B_2 \in \mathcal{A}$ with $A = B_1 \cap B_2$ there is a $C \in \mathcal{A}$ that is an amalgam of B_1 and B_2 over A. A class $\mathcal{K}$ of finite τ-structures that has the amalgamation property and is closed under taking induced substructures is called an *amalgamation class*.

The following basic result is known as Fraïssé's Theorem in model theory (see Theorem 6.1.2 in [12]):

Fact 3. *Let $\mathcal{K}$ be an amalgamation class. Then there is an ω-categorical τ-structure Γ such that $\mathcal{K}$ is the class of finite induced substructures of Γ.*

The structure Γ, which is unique up to isomorphism, is called the Fraïssé limit *of $\mathcal{K}$.*

A few remarks are necessary to relate our version of the theorem to the more general version stated, for example, in [12]: Firstly, the amalgamation property is usually defined in a slightly more complicated way where the structure A is not necessarily an induced substructure of the B_i, but embedded into B_i. As we assume classes of structures to be closed under isomorphism, this makes no difference. Secondly, the class $\mathcal{A}$ in Fraïssé's Theorem usually has to have another property known as the *joint embedding property*. However, for relational structures the joint embedding property is subsumed by the amalgmation property. And thirdly, in general the Fraïssé limit Γ is *homogeneous*, but not necessarily ω-categorical. But for finite relational vocabularies, homogeneity implies ω-categoricity.

Let us now turn to the proof of Theorem 2. We encode words over the alphabet Σ by structures similarly, but not exactly as in the proof of Theorem 1. Let τ be a signature that contains the binary relation symbols $N, \neq$, the unary relation symbols S, T, and a unary relation symbol P_a for each $a \in \Sigma$. With each word $w = a_1 \ldots a_n \in \Sigma^*$ we associate the τ-structure W with universe $\{0, 1, \ldots, n+1\}$, $N^W = \{(i, i+1) \mid 0 \leq i < n+1\}$, $\neq^W = \{(i,j) \mid 0 \leq i, j \leq n+1, i \neq j\}$, $S^W = \{0\}$, $T^W = \{n+1\}$, and $P_a = \{i \mid 1 \leq i \leq n, a_i = a\}$.

For every τ-structure A, we define an undirected graph $G_N(A)$ to be the graph with vertex set D_A and an edge between $a, b \in D_A$ if and only if $a \neq b$ and $(a, b) \in N^A$ or $(b, a) \in N^A$. We say that A is *connected by* N if the graph $G_N(A)$ is connected.

Let $\mathcal{C}$ be the class of all τ-structures isomorphic to a structure W for some $w \in \Sigma^* \setminus L$. Let $\mathcal{K}$ be the class of all τ-structures A with the following properties.

(1) The binary relation $\neq^A$ is *anti-reflexive*, i.e., it does not contain pairs of the form (x, x).

(2) The relations S^A, T^A, and P_a for $a \in \Sigma$ are pairwise disjoint.

(3) N^A is anti-reflexive and anti-symmetric. Furthermore, if $\neq^A$ has cardinality $|A|(|A| - 1)$, then the graph $G_N(A)$ is acyclic.

(4) A does not contain a structure from $\mathcal{C}$ as an induced substructure.

Lemma 1. *The class $\mathcal{K}$ is an amalgamation class.*

Proof. It is straightforward to verify that $\mathcal{K}$ is closed under isomorphisms and induced substructures. To show that $\mathcal{K}$ has the amalgamation property, let $B_1, B_2 \in \mathcal{K}$ such that $A = B_1 \cap B_2$ is an induced substructure of both B_1 and B_2. We claim that the free amalgam $C = B_1 \cup B_2$ is contained in $\mathcal{K}$.

We have to prove that C has properties (1)–(4) from page 189. This is obvious for properties (1) and (2), because both B_1 and B_2 have the property and their intersection A is an induced substructure of both structures. To see that C has property (3), note that either $C = B_1$ or $C = B_2$, and C inherits the property from the respective structure, or $D_C \setminus D_{B_1} \neq \emptyset$ and $D_C \setminus D_{B_2} \neq \emptyset$, and $\neq^C = \neq^{B_1} \cup \neq^{B_2}$ has cardinality less than $|C|(|C|-1)$. It remains to prove that C has property (4). Suppose for contradiction that C has an induced substructure $W \in \mathcal{C}$. As $\neq^W$ connects all pairs of distinct vertices of W, the structure W must be an induced substructure of B_1 or of B_2, which is a contradiction.

Let Γ be the Fraïssé-limit of $\mathcal{K}$.

Lemma 2. *A finite structure A has a homomorphism to Γ if and only if A has properties (1)–(4).*

Proof (of Theorem 2). There is the following reduction from L to $\mathrm{CSP}(\Gamma)$. Given a word $w \in \Sigma^*$, let W be the corresponding τ-structure. If $w \notin L$, then W is in $\mathcal{C}$, and hence W does not satisfy 4. Lemma 2 then implies that there is no homomorphism from W to Γ. If $w \in L$, then $W \notin \mathcal{C}$. The structure W has no induced substructure from $\mathcal{C}$, because every proper induced substructure of W either does not have an element in S, or does not have an element in T, or is not connected by N. Therefore, W homomorphically maps to Γ. This shows that the function that returns for a given word w its word-structure W is a polynomial-time many-one reduction from L to CSP(Γ).

We finally show that CSP(Γ) can be decided by a universal-nondeterministic polynomial-time algorithm with an oracle for L, by the following algorithm.

Input: A

If A does not have properties (1)–(3) then reject.
For all induced substructures W of A:
    // (can be implemented non-deterministically)
    If W is the τ-structure of a word $w \in \Sigma^*$ then
        If $w \notin L$ then reject.
Accept.

Lemma 2 shows that A homomorphically maps to Γ if and only if A satisfies (1)–(4), and this is what the algorithm tests. The algorithm can be implemented on a non-deterministic polynomial-time Turing machine such that there exists a run on input A where the algorithm rejects if and only if A homomorphically maps to Γ.

4   coNP-Intermediate ω-Categorical Templates

In this section we construct an ω-categorical directed graph Γ such that CSP(Γ) is in coNP, but neither coNP-complete nor in P (unless coNP=P). As in the previous section, the infinite structures studied here are all defined as Fraïssé limits. All structures in this section will be directed graphs.

Henson [11] used Fraïssé limits to construct 2^ω many ω-categorical directed graphs. If $\mathcal{N}$ is a class of τ-structures, $Forb(\mathcal{N})$ denotes the class of all finite τ-structures A such that no structure from $\mathcal{N}$ embeds into A. A *tournament* is a directed graph G (without self-loops) such that for all pairs x, y of distinct vertices exactly one of the pairs (x, y), (y, x) is an arc in G. Note that for all classes $\mathcal{N}$ of finite tournaments, $Forb(\mathcal{N})$ is an amalgamation class, because if G_1 and G_2 are directed graphs in $Forb(\mathcal{N})$ such that $H = G_1 \cap G_2$ is an induced substructure of both G_1 and G_2, then the free amalgam $G_1 \cup G_2$ is also in $Forb(\mathcal{N})$. We write $\Gamma_\mathcal{N}$ for the Fraïssé-limit of $Forb(\mathcal{N})$. Observe that for finite $\mathcal{N}$ the problem CSP$(\Gamma_\mathcal{N})$ can be solved in deterministic polynomial time, because for a given instance S of this problem an algorithm simply has to check whether there is a homomorphism from one of the structures in $\mathcal{N}$ to S, which is the case if and only if there is a homomorphism from S to $\Gamma_\mathcal{N}$.

Henson in his proof specified an infinite set $\mathcal{T}$ of tournaments $T_1, T_2, \ldots$ with the property that T_i does not embed into T_j if $i \neq j$. Note that this implies that for two distinct subsets $\mathcal{T}_1$ and $\mathcal{T}_2$ of $\mathcal{T}$ the two sets $Forb(\mathcal{T}_1)$ and $Forb(\mathcal{T}_2)$ are distinct as well. Since there are 2^ω many subsets of the infinite set $\mathcal{T}$, there are also that many distinct ω-categorical directed graphs. The tournament T_n in Henson's set $\mathcal{T}$ has vertices $0, \ldots, n + 1$, and the following edges:

- (i, j) for $j = i + 1$ and $0 \le i \le n$;
- $(0, n + 1)$;
- (j, i) for $j > i + 1$ and $(i, j) \ne (0, n + 1)$.

Proposition 1. *The problem $CSP(\Gamma_{\mathcal{J}})$ is coNP-complete.*

Proof. The problem is contained in coNP, because we can efficiently test whether a sequence $v_1, \ldots, v_k$ of distinct vertices of a given directed graph G induces T_k in G, i.e., whether (v_i, v_j) is an arc in G if and only if (i, j) is an arc in T_k, for all $i, j \in \{1, \ldots, k\}$. If for all such sequences of vertices this test is negative, we can be sure that G is from $Forb(\mathcal{J})$, and hence homomorphically maps to $\Gamma_{\mathcal{J}}$. Otherwise, G embeds a structure from $\mathcal{J}$, and hence does not homomorphically map to $\Gamma_{\mathcal{J}}$.

The proof of coNP-hardness goes by reduction from the complement of the NP-complete 3-SAT problem, and is inspired by a classical reduction from 3-SAT to Clique. For a given 3-SAT instance, we create an instance G of $CSP(\Gamma_{\mathcal{J}})$ as follows: If

$$\{x_0^1, x_0^2, x_0^3\}, \ldots, \{x_{k+1}^1, x_{k+1}^2, x_{k+1}^3\}$$

are the clauses of the 3-SAT formula (we assume without loss of generality that the 3-SAT instance has at least three clauses), then the vertex set of G is $\{(0, 1), (0, 2), (0, 3), \ldots, (k + 1, 1), (k + 1, 2), (k + 1, 3)\}$, and the arc set of G consists of all pairs $((i, j), (p, q))$ of vertices such that $x_i^j \ne \neg x_p^q$ and such that (i, p) is an arc in T_k.

We claim that a 3-SAT instance is unsatisfiable if and only if the created instance G homomorphically maps to $\Gamma_{\mathcal{J}}$. The 3-SAT instance is satisfiable iff there is a mapping from the variables to true and false such that in each clause at least one literal, say $x_0^{j_0}, \ldots, x_{k+1}^{j_{k+1}}$, is true. This is the case if and only if the vertices $(0, j_1), \ldots, (k + 1, j_{k+1})$ induce T_k in G, i.e., $((i, j_i), (p, j_p))$ is an edge if and only if (i, p) is an edge in T_k. This is the case if and only if T_k embeds into G, and if and only if G does not homomorphically map to $\Gamma_{\mathcal{J}}$. $\qquad\square$

We now modify the proof of Ladner's Theorem given in [16] (which is basically Ladner's original proof) to create a subset $\mathcal{J}_0$ of $\mathcal{J}$ such that $CSP(\Gamma_{\mathcal{J}_0})$ is in coNP, but neither in P nor coNP-complete (unless coNP=P). One of the ideas in Ladner's proof is to 'blow holes into SAT', such that the resulting problem is too sparse to be NP-complete and to dense to be in P. Our modification is that we do not blow holes into a computational problem itself, but that we 'blow holes into the obstruction set $\mathcal{J}$ of $CSP(\Gamma_{\mathcal{J}})$'.

In the following, we fix one of the standard encodings of graphs as strings over the alphabet $\{0, 1\}$. Let $M_1, M_2, \ldots$ be an enumeration of all polynomial-time bounded Turing machines, and let $R_1, R_2, \ldots$ be an enumeration of all polynomial time bounded reductions. We assume that these enumerations are effective; it is well-known that such enumerations exist.

The definition of $\mathcal{J}_0$ uses a Turing machine F that computes a function $f : \mathbb{N} \to \mathbb{N}$, which is defined below. The set $\mathcal{J}_0$ is then defined as follows.

$$\mathcal{J}_0 = \{T_n \mid f(n) \text{ is even }\}$$

The input number n is given to the machine F in unary representation. The computation of F proceeds in two phases. In the first phase, F simulates itself[2] on input 1, then on input 2, 3, and so on, until the number of computation steps of F in this phase exceeds n (we can always maintain a counter during the simulation to recognize when to stop). Let k be the value $f(i)$ for the last input i for which the simulation was completely performed by F.

In the second phase, the machine stops if phase two takes more than n computation steps, and F returns k. We distinguish whether k is even or odd. If k is even, all directed graphs G on $s = 1, 2, 3, \ldots$ vertices are enumerated. For each directed graph G in the enumeration the machine F simulates $M_{k/2}$ on the encoding of G. Moreover, F computes whether G homomorphically maps to $\Gamma_{\mathcal{J}_0}$. This is the case if for all structures $T_l \in \mathcal{J}$ that embed into G the value of $f(l)$ is even. So F tests for $l = 1, 2, \ldots, s$ whether T_l embeds to G (F uses any straightforward exponential time algorithm for this purpose), and if it does, simulates itself on input l to find out whether $f(l)$ is even. If

(1) $M_{k/2}$ rejects and G homomorphically maps to $\Gamma_{\mathcal{J}_0}$, or
(2) $M_{k/2}$ accepts and G does not homomorphically map to $\Gamma_{\mathcal{J}_0}$,

then F returns $k + 1$ (and $f(n) = k + 1$).

The other case of the second phase is that k is odd. Again F enumerates all directed graphs G on $s = 1, 2, 3, \ldots$ vertices, and simulates the computation of $R_{\lfloor k/2 \rfloor}$ on the encoding of G. Then F computes whether the output of $R_{\lfloor k/2 \rfloor}$ encodes a directed graph G' that homomorphically maps to $\Gamma_{\mathcal{J}_0}$. The graph G' homomorphically maps to $\Gamma_{\mathcal{J}_0}$ iff for all tournaments T_l that embed into G' the value $f(l)$ is even. Whether T_l embeds into G' is tested with a straightforward exponential-time algorithm. To test whether $f(l)$ is even, F simulates itself on input l. Finally, F tests with a straightforward exponential-time algorithm whether G homomorphically maps to $\Gamma_{\mathcal{J}}$. If

(3) G homomorphically maps to $\Gamma_{\mathcal{J}}$ and G' does not homomorphically map to $\Gamma_{\mathcal{J}_0}$, or
(4) G does not homomorphically map to $\Gamma_{\mathcal{J}}$ and G' homomorphically maps to $\Gamma_{\mathcal{J}_0}$,

then F returns $k + 1$.

Lemma 3. *The function f is a non-decreasing function, that is, for all n we have $f(n) \leq f(n + 1)$.*

Lemma 4. *For all n_0 there is $n > n_0$ such that $f(n) > f(n_0)$ (unless $coNP \neq P$).*

Proof. Assume for contradiction that there exists an n_0 such that $f(n)$ equals a constant k_0 for all $n \geq n_0$. Then there also exists an n_1 such that for all $n \geq n_1$ the value of k computed by the first phase of F on input n is k_0.

---

[2] Note that by the fixpoint theorem of recursion theory we can assume that F has access to its own description.

If k_0 is even, then on all inputs $n \geq n_1$ the second phase of F simulates $M_{k_0/2}$ on encodings of an enumeration of graphs. Since the output of F must be k_0, for all graphs neither (1) nor (2) can apply. Since this holds for all $n \geq n_1$, the polynomial-time bounded machine $M_{k_0/2}$ correctly decides $\mathrm{CSP}(\Gamma_{\mathcal{T}_0})$, and hence $\mathrm{CSP}(\Gamma_{\mathcal{T}_0})$ is in P. But then there is the following polynomial-time algorithm that solves $\mathrm{CSP}(\Gamma_{\mathcal{T}})$, a contradiction to coNP-completeness of $\mathrm{CSP}(\Gamma_{\mathcal{T}})$ (Proposition 1) and our assumption that coNP $\neq$ P.

Input: A directed graph G.

Test whether G homomorphically maps to $\Gamma_{\mathcal{T}_0}$.
If yes, accept.
If no, test whether one of the finitely many graphs in $\mathcal{T} \setminus \mathcal{T}_0$ embeds into G.
Accept if none of them embeds into G.
Reject otherwise.

If k_0 is odd, then on all inputs $n \geq n_1$ the second phase of F does not find a graph G for which (3) or (4) applies, because the output of F must be k_0. Hence, $R_{\lfloor k_0/2 \rfloor}$ is a polynomial-time reduction from $\mathrm{CSP}(\Gamma_{\mathcal{T}})$ to $\mathrm{CSP}(\Gamma_{\mathcal{T}_0})$, and by Proposition 1 the problem $\mathrm{CSP}(\Gamma_{\mathcal{T}_0})$ is coNP-hard. But note that because $f(n)$ equals the odd number k_0 for all but finitely many n, the set $\mathcal{T}_0$ is finite. Therefore, $\mathrm{CSP}(\Gamma_{\mathcal{T}_0})$ can be solved in polynomial time, contradicting our assumption that coNP $\neq$ P.   □

Theorem 4. *CSP($\Gamma_{\mathcal{T}_0}$) is in coNP, but neither in P nor coNP-complete (unless coNP=P).*

Proof. It is easy to see that $\mathrm{CSP}(\Gamma_{\mathcal{T}_0})$ is in coNP. On input G the algorithm non-deterministically chooses a sequence of l vertices, and checks in polynomial time whether this sequence induces a copy of T_l. If yes, the algorithm computes $f(l)$, which can be done in linear time by executing F on the unary representation of l. If $f(l)$ is even, the algorithm accepts. Recall that G does not homomorphically map to $\Gamma_{\mathcal{T}_0}$ iff a tournament $T_l \in \mathcal{T}_0$ embeds into G, which is the case iff there is an accepting computation path for the above non-deterministic algorithm.

Suppose that $\mathrm{CSP}(\Gamma_{\mathcal{T}_0})$ is in P. Then for some i the machine M_i decides $\mathrm{CSP}(\Gamma_{\mathcal{T}_0})$. By Lemma 3 and Lemma 4 there exists an n_0 such that $f(n_0) = 2i$. Then there must also be an $n_1 > n_2$ such that the value k computed during the first phase of F on input n_1 equals $2i$. Since M_i correctly decides $\mathrm{CSP}(\Gamma_{\mathcal{T}_0})$, the machine F returns $2i$ on input n_1. By Lemma 3, the machine F also returns $2i$ for all inputs from n_1 to n_2, and by induction it follows that it F returns $2i$ for *all* inputs larger than $n \geq n_0$, in contradiction to Lemma 4.

Finally, suppose that $\mathrm{CSP}(\Gamma_{\mathcal{T}_0})$ is coNP-complete. Then for some i the machine R_i is a valid reduction from $CSP(\Gamma_{\mathcal{T}})$ to $\mathrm{CSP}(\Gamma_{\mathcal{T}_0})$. Again, by Lemma 3 and Lemma 4 there exists an n_1 such that the value k computed during the first phase of F on input n_1 equals $2i$. Since the reduction R_i is correct, the machine F returns $2i$ on input n_1, and in fact returns $2i$ on all inputs greater than n_1. This contradicts Lemma 4.   □

5   Left-Hand Side Restrictions

Let S be a class of finite τ-structures. Then $CSP(S, \_)$ is the computational problem to decide whether for a given pair (A, B) of finite τ-structures with $A \in S$ there is a homomorphism from A to B.

As an example, let τ be the signature that consists of a single binary relation. In this case, τ-structures can be considered as directed graphs (potentially with loops). If C is the set of all complete graphs (without loops!), then $CSP(C, \_)$ is essentially a formulation of the Clique problem.

The following question was left open in [10]: *Are there classes of structures S such that $CSP(S, \_)$ is in NP, but neither in P nor NP-complete?*

We answer this question positively, and construct such a class S, which can even be decided in polynomial time. Again we use a modification of Ladner's theorem. The modification is similar to the modification presented in Section 4. This time, we *'blow holes into the possible clique sizes for the clique problem'* and obtain a class $C_0 \subseteq C$ such that $CSP(C_0, \_)$ is in NP $\setminus$ P and not NP-complete (unless P=NP).

The idea is to define C_0 in such a way that the $C \setminus C_0$ becomes finite when $CSP(C_0, \_)$ is in P; hence, $CSP(C_0, \_)$ is polynomial-time equivalent to the Clique problem, a contradiction unless P=NP. Moreover, the construction of C_0 is such that C_0 is finite if $CSP(C_0, \_)$ is NP-hard. But for finite C_0, the problem $CSP(C_0, \_)$ is in P, again contradicting the assumption that P $\neq$ NP. We also take extra care to make C_0 polynomial-time decidable.

Theorem 5. *$CSP(C_0, \_)$ is in NP, but neither in P nor NP-complete (unless P=NP). Moreover, the set C_0 can be decided in deterministic polynomial time.*

In our proof we do not use any specific properties of the class C and the clique problem, but in fact we can construct classes $S_0 \subseteq S$ with NP-intermediate $CSP(S_0, \_)$ for any class S where $CSP(S, \_)$ is NP-complete.

6   The Local-Global Conjecture

The complexity of the constraint satisfaction problem has also been studied for templates Γ with an *infinite signature*. Several well-known computational problems can be modeled as CSPs only if we allow (countably) infinite constraint languages: examples are boolean Horn-satisfiability, Ord-Horn constraints in temporal reasoning [15], or solving linear equation systems over a finite field.

If the local-global conjecture as stated in Section 1 is true, then the dichotomy for finite constraint languages implies the dichotomy for infinite constraint languages: if an infinite constraint language Γ has a finite reduct Γ' such that $CSP(\Gamma')$ is NP-hard, then $CSP(\Gamma)$ is clearly NP-hard as well. On the other hand, if $CSP(\Gamma)$ is locally tractable, then the conjecture implies that $CSP(\Gamma)$ is globally tractable.

We show that if the local-global conjecture is false, then there is no dichotomy for infinite constraint languages.

Theorem 6. *If the local-global conjecture is false, then there exists a template Γ_0 with finite domain D and infinite signature such that $CSP(\Gamma_0)$ is neither globally tractable nor NP-complete (unless P=NP); moreover, the meta-problem for Γ_0 is efficiently decidable, i.e., given a relation R over D, we can decide in polynomial-time whether R is in Γ_0.*

The proof of Theorem 6 is essentially again a modification of Ladner's theorem, but we have to overcome a complication: for a straightforward application of Ladner's theorem, we need that if Γ is an expansion of Γ' by finitely many relation symbols, and $CSP(\Gamma)$ is NP-hard, then $CSP(\Gamma')$ is NP-hard as well: but this is not true in general. We only sketch the basic setting of Ladner's construction, which we have already seen twice in this paper, and focus on the complication.

Proof (Sketch). Assume that there is an infinite constraint language Γ with a finite domain such that $CSP(\Gamma)$ is locally tractable, but not globally tractable. If Γ is not NP-complete, we are already done with $\Gamma_0 = \Gamma$, so assume that Γ is NP-complete. We claim that we can assume without loss of generality that Γ contains all primitive positive definable relations (for an introduction to this basic concept in model theory and its applications in constraint satisfaction theory, see e.g. [6]). To see this, first observe that the expansion Γ' of Γ by all those relations trivially still has an NP-complete CSP. Moreover, all reducts of Γ' with a finite signature have a CSP that is in P, because all relations in this reduct can be defined by finitely many relations from Γ. So, Γ' can be obtained from a finite reduct of Γ by expansion with finitely many primitive positive relations, and hence $CSP(\Gamma')$ is in P.

We construct a reduct Γ_0 of Γ such that $CSP(\Gamma_0)$ is neither in P nor NP-complete (unless P=NP). Again, the definition of Γ_0 is by a Turing machine F that computes a function $f : \mathbb{N} \to \mathbb{N}$, and the n-th relation of Γ (according to some fixed enumeration of the relations of Γ) is in Γ_0 if $f(n)$ is even.

As in the proofs before, n is given to the machine F in unary representation, and we can define F in such a way that

- it runs in polynomial time in n, and
- Γ_0 becomes finite if $CSP(\Gamma_0)$ is NP-hard, and
- Γ is an expansion of Γ_0 by finitely many relations if $CSP(\Gamma_0)$ is in P.

If Γ_0 is finite, then $CSP(\Gamma_0)$ is tractable, because every reduct of Γ with a finite signature is by assumption tractable, and we obtain that P=NP. Now, suppose that we are in the other case, and that Γ is an expansion of Γ_0 by finitely many relations. We want to show that $CSP(\Gamma_0)$ is NP-hard by reducing $CSP(\Gamma)$ to $CSP(\Gamma_0)$.

Let S be an instance of $CSP(\Gamma)$. Note that S might contain constraints for the relations from Γ that are not in Γ_0, but there is a k such that all those constraints have arity less than k. Because Γ contains all primitive positive definable relations, Γ contains in particular for every l-ary relation R the k-ary relation $R' := R \times D \times \cdots \times D$. We now replace each constraint in S with a relation

R from Γ that is not in Γ_0 by a constraint for R', introducing $k - l$ new dummy variables for the last $k - l$ arguments of R'. Even though the representation of R' is much larger than the representation of R, this can only lead to a linear increase in the size of the instance S, because both D and k are fixed. The resulting structure S' is an instance of $\mathrm{CSP}(\Gamma_0)$, and S' homomorphically maps to Γ_0 if and only if S homomorphically maps to Γ. Therefore $\mathrm{CSP}(\Gamma_0)$ is NP-complete, which again implies that P=NP.   □

Acknowledgements. We would like to thank Hubie Chen for his helpful comments on an earlier version of this paper.

References

1. Bauslaugh, B.L.: The complexity of infinite h-coloring. J. Comb. Theory, Ser. B 61(2), 141–154 (1994)
2. Bodirsky, M.: Constraint satisfaction problems with infinite templates. Survey (to appear, 2007)
3. Bulatov, A.: Tractable conservative constraint satisfaction problems. In: Proceedings of LICS 2003, pp. 321–330 (2003)
4. Bulatov, A.: A graph of a relational structure and constraint satisfaction problems. In: Proceedings of LICS 2004, Turku, Finland (2004)
5. Bulatov, A., Jeavons, P.: Algebraic structures in combinatorial problems. Technical report MATH-AL-4-2001, Technische Universitat Dresden, submitted to International Journal of Algebra and Computing (2001)
6. Bulatov, A., Krokhin, A., Jeavons, P.G.: Classifying the complexity of constraints using finite algebras. SIAM Journal on Computing 34, 720–742 (2005)
7. Chen, Y., Thurley, M., Weyer, M.: Understanding the complexity of induced substructure isomorphisms. In: Aceto, L., Damgaard, I., Goldberg, L.A., Halldorsson, M.M., Ingolfsdottir, A., Walukiewicz, I. (eds.) ICALP 2008. LNCS, vol. 5125. Springer, Heidelberg (2008)
8. Cherlin, G.: The classification of countable homogeneous directed graphs and countable homogeneous n-tournaments. AMS Memoir. 131(621) (January 1998)
9. Feder, T., Vardi, M.: The computational structure of monotone monadic SNP and constraint satisfaction: A study through Datalog and group theory. SIAM Journal on Computing 28, 57–104 (1999)
10. Grohe, M.: The complexity of homomorphism and constraint satisfaction problems seen from the other side. Journal of the ACM 54(1) (2007)
11. Henson, C.W.: Countable homogeneous relational systems and categorical theories. Journal of Symbolic Logic 37, 494–500 (1972)
12. Hodges, W.: A shorter model theory. Cambridge University Press, Cambridge (1997)
13. Idziak, P.M., Markovic, P., McKenzie, R., Valeriote, M., Willard, R.: Tractability and learnability arising from algebras with few subpowers. In: LICS 2007, pp. 213–224 (2007)
14. Ladner, R.E.: On the structure of polynomial time reducibility. J. ACM 22(1), 155–171 (1975)
15. Nebel, B., Bürckert, H.-J.: Reasoning about temporal relations: A maximal tractable subclass of Allen's interval algebra. J. ACM 42(1), 43–66 (1995)
16. Papadimitriou, C.H.: Computational Complexity. Addison-Wesley, Reading (1994)
17. Schwandtner, G.: Datalog on infinite structures. Dissertation, Humboldt-Universität zu Berlin (submitted, 2008)

Quantified Constraint Satisfaction and the Polynomially Generated Powers Property

(Extended Abstract)

Hubie Chen

Departament de Tecnologies de la Informació i les Comunicacions
Universitat Pompeu Fabra
Barcelona, Spain
hubie.chen@upf.edu

Abstract. The quantified constraint satisfaction probem (QCSP) is the problem of deciding, given a relational structure and a sentence consisting of a quantifier prefix followed by a conjunction of atomic formulas, whether or not the sentence is true in the structure. The general intractability of the QCSP has led to the study of restricted versions of this problem. In this article, we study restricted versions of the QCSP that arise from prespecifying the relations that may occur via a set of relations called a *constraint language*. A basic tool used is a correspondence that associates an algebra to each constraint language; this algebra can be used to derive information on the behavior of the constraint language.

We identify a new combinatorial property on algebras, the *polynomially generated powers* (PGP) property, which we show is tightly connected to QCSP complexity. We also introduce another new property on algebras, *switchability*, which both implies the PGP property and implies positive complexity results on the QCSP. Our main result is a classification theorem on a class of three-element algebras: each algebra is either switchable and hence has the PGP, or provably lacks the PGP. The description of non-PGP algebras is remarkably simple and robust.

1 Introduction

Background. The *constraint satisfaction problem* (CSP) is the problem of deciding, given a relational structure and a *primitive positive sentence*

$$\exists x_1 \ldots \exists x_m (R(x_{i_1}, \ldots, x_{i_k}) \wedge \ldots)$$

that is, a conjunction of atomic formulas in front of which all variables are existentially quantified, whether or not the sentence is true in the structure. The *quantified constraint satisfaction problem* (QCSP) is the generalization of the CSP where universal quantification is permitted in addition to existential quantification. Each of these problems constitutes a natural syntactic restriction of model checking in first-order logic.

The general intractability of the CSP and the QCSP–they are NP-complete and PSPACE-complete, respectively–has prompted the study of restricted versions of these problems. In this paper, we study restricted versions of the QCSP that are obtained by prespecifying the relations that may occur using a set of relations called a *constraint*

L. Aceto et al. (Eds.): ICALP 2008, Part II, LNCS 5126, pp. 197–208, 2008.
© Springer-Verlag Berlin Heidelberg 2008

language.[1] This form of restriction has its origins in a 1978 paper of Schaefer [24], who gave a classification theorem showing that all constraint languages over a two-element domain give rise to a case of the CSP that is either polynomial-time tractable or NP-complete. The non-trivial tractable cases identified by this result are *2-SAT*, where each constraint is equivalent to a length 2 clause; *Horn-SAT*, where each constraint is equivalent to a propositional Horn clause, and *Affine-SAT*, where each constraint is an equation over the two-element field. The quantified generalizations of these three problems are known to be tractable [1,21,7,14], and a classification theorem on two-element QCSP complexity shows that all other constraint languages over a two-element domain give rise to a PSPACE-complete case of the QCSP [14,11].

An approach to studying the complexity of constraint languages based directly on concepts and tools from universal algebra was introduced by Bulatov, Jeavons and Krokhin [5]. The cornerstone of this approach is to associate, to each constraint language Γ, an algebra $\mathbb{A}_\Gamma$ whose operations are the *polymorphisms* of Γ. (Roughly speaking, an operation f is a polymorphism of a constraint language Γ if each relation of Γ is closed under the coordinate-wise action of f.) This algebra is used to derive information about the constraint language. This approach has provided new and promising vistas on the complexity of constraint languages; one celebrated achievement that it has thus far yielded is the CSP complexity classification of constraint languages over a three-element domain by Bulatov [6]. One can further name [3,16,4,22,20] as a sampling of recent work using this viewpoint.

Contributions. In this paper, we develop the algebraic theory of the QCSP and present both structural algebraic results as well as new complexity results on the QCSP.

Our starting point is collapsibility, a previously studied property on algebras [12] which provides a sufficient condition on $\mathbb{A}_\Gamma$ for the reducibility of the QCSP over Γ to the CSP over Γ (which in turn immediately gives an NP upper bound on the complexity of the QCSP). This condition was shown to hold on all algebras $\mathbb{A}_\Gamma$ whose corresponding QCSP is polynomial-time tractable, in the two-element case. At its heart, the QCSP-to-CSP reduction provided by collapsibility exploits a property on algebras that we define here and call the *polynomially generated powers* (PGP) property: an algebra $\mathbb{A}$ has the PGP property if its nth power $\mathbb{A}^n$ has a polynomial-size generating set. Indeed, it can be directly shown that collapsibility of an algebra $\mathbb{A}$ implies that $\mathbb{A}$ has the PGP property via generating sets of a specific form.[2] Although the PGP property is, in

---

[1] It is worth mentioning that the complexity of the QCSP under *structural restrictions*– restrictions based on variable interaction–have also recently attracted attention: see for example the papers [9,13] by the present author (the latter with Dalmau), the paper [18] by Gottlob, Greco and Scarcello, and the paper [23] by Pan and Vardi.

[2] In fact, as discussed in this paper (Section 2), $\mathbb{A}_\Gamma$ having the PGP property, along with a mild computational assumption, yields a simple algorithm for the QCSP-to-CSP reduction on Π_2 formulas. This relationship can be used to present simple and self-contained proofs of this reduction on Π_2 formulas; examples of such proofs are given in [11, Section 1]. The essential proof technique applies across various constraint languages, and gives a uniform derivation of previously existing results in the literature, namely, the collapse results of Grädel [19] in descriptive complexity, as well as a theorem proved by Karpinski et al. [21] on the Π_2 fragment of *Quantified Horn-SAT*.

our view, a natural combinatorial property on algebras, to the best of our knowledge, it has not yet been systematically studied.

As we observe early on in this paper, in the two-element case, collapsibility "tells the full story" for the PGP property and also for polynomial-time tractability: for a constraint language Γ over a two-element set, either $\mathbb{A}_\Gamma$ is collapsible, has the PGP, and the QCSP on Γ is polynomial-time tractable; or, $\mathbb{A}_\Gamma$ does not have the PGP and the QCSP on Γ is PSPACE-complete. However, the three-element case is not yet understood, neither with respect to the PGP property nor with respect to QCSP complexity. A partial description of three-element non-collapsible idempotent algebras was achieved [12], but this result does not readily yield a characterization of the three-element idempotent algebras with respect to the PGP.[3] The present investigation was initiated with the hope of understanding three-element idempotent algebras from the mentioned perspectives. In particular, two specific questions that we set out to answer were: Are there three-element idempotent algebras having the PGP property other than those that are collapsible, or does collapsibility give a full characterization of the PGP property for such algebras? And, for all such algebras $\mathbb{A}_\Gamma$ having the PGP property, is this QCSP on Γ polynomial-time tractable, or at least reducible to the CSP?

This paper presents a classification of three-element idempotent algebras with respect to the PGP property, which identifies further, non-collapsible algebras having this property. In particular, we define a new property on algebras, *switchability*, and demonstrate that, even though switchability strictly generalizes collapsibility, it still implies both the PGP as well as a QCSP-to-CSP reduction. We then show our classification: for any three-element idempotent algebra $\mathbb{A}$ lacking an algebraic sufficient condition for CSP/QCSP intractability (the "G-set condition"), either $\mathbb{A}$ has switchability and indeed is polynomial-time tractable, or $\mathbb{A}$ has a particular structure that readily implies absence of the PGP property. For the described class of algebras, we therefore answer both of the aforementioned research questions.[4]

On the "computer science" side, our introduction and study of switchability provide new polynomial-time tractable cases of the QCSP. Additionally, they give a sufficient condition for a QCSP-to-CSP reduction that is general in that it can be applied to universes of all sizes. We would like to emphasize that the description of non-switchable/non-PGP algebras given by our classification is robust in the sense that the terms of any such algebra may be viewed as a subclone of a single clone satisfying the description. In addition to being appealing in its own right, we believe that this description's mathematical robustness will facilitate study of these algebras and their QCSP complexity.

Our classification in fact shows that the non-switchable/non-PGP algebras have the *exponentially generated powers* (EGP) property: for any sequence of generating sets $X_1, X_2, \ldots$ for the powers $\mathbb{A}^1, \mathbb{A}^2, \ldots$ of such an algebra $\mathbb{A}$, these sets must have exponential size in that $|X_n|$ must be $\Omega(b^n)$ for some $b > 1$. We thus obtain a dichotomy

---

[3] This paper focuses on the the QCSP where constants are permitted. It is known that to study this problem relative to a constraint language Γ, it suffices to study the algebra whose operations are the idempotent polymorphisms of Γ, hence our focus on idempotent algebras.

[4] The first is answered by identifying new PGP algebras; the second, by showing these algebras to be polynomial-time tractable.

result with respect to a new combinatorial measure on algebras–the requisite size of generating sets for powers. Note that a dichotomy between the PGP property and EGP property is by no means entailed by the definitions of these two properties, as there are "intermediate" growth rates such as $n^{\log n}$ that are neither polynomial nor exponential, according to our definitions. Our dichotomy thus exhibits a chasm in the growth rates that occur naturally in this context, and it is curious that such intermediate growth rates do not occur.

We conclude this introduction with a brief methodological/philosophical discussion of this article's approach and context. On the one hand, an algebraic notion was introduced by complexity considerations: the definition of the PGP property was inspired by considering the Π_2 fragment of the QCSP. On the other hand, algebraic investigation led to insight on complexity: during the course of obtaining the results in this article, the author identified some non-collapsible algebras to have the PGP property prior to establishing any complexity-theoretic result on any of them (in particular, before proving that any of them possessed a QCSP-to-CSP reduction). It is our understanding that the recent investigations on the so-called *few subalgebras property* followed a similar storyline: a purely combinatorial property on algebras was defined from computational considerations [15,10], a classification of such algebras with respect to the property was established [2], and then the algebras possessing the property were shown to entail a desirable computational property–in this case, CSP tractability [20].[5] We believe and hope that the CSP and its variants will continue to effect this mutual cross-pollination between algebra and complexity, and certainly look forward to further work along these lines.

Preliminaries. Our notations and definitions are fairly standard, and similar to those used in [12]. Here, we confine ourselves to a few remarks. We use $[n]$ to denote the set containing the first n positive integers, $\{1, \ldots, n\}$. We use $\text{QCSP}_c(\Gamma)$ to denote the QCSP over constraint language Γ where constants may appear in constraints; $\text{CSP}_c(\Gamma)$ is the restriction of $\text{QCSP}_c(\Gamma)$ to formulas having only existential quantifiers. We use $\text{QCSP}(\Gamma)$ and $\text{CSP}(\Gamma)$ to denote the same problems, but where constants may not appear in constraints.

2   Properties

In this section, we introduce the two combinatorial properties on algebras–the *polynomially generated powers* and the *exponentially generated powers* properties–that will be studied.

For an algebra $\mathbb{A}$ and a subset $X \subseteq A$, we use $\langle X \rangle$ to denote the intersection of all $\mathbb{A}$-subalgebras containing X (that is, the smallest subalgebra of $\mathbb{A}$ containing X). We call $\langle X \rangle$ the subalgebra *generated* by X. We say that a function $f : \mathbb{N} \to \mathbb{N}$ is a *polynomial* if there exists $k \geq 1$ such that $f(n)$ is $O(n^k)$.

---

[5] A remark: M. Valeriote has communicated to us that in studying the few subalgebras property, it was observed by the authors of [2] that the few subalgebras property implies the PGP property. The converse does not hold: the algebra $(\{0, 1\}, \{\wedge\})$ is an example of an algebra having the PGP property but not having the few subalgebras property.

Definition 1. *An algebra* $\mathbb{A}$ *has the* polynomially generated powers (PGP) property *if there exists a polynomial* $p(n)$ *such that for all* $n \geq 1$, *there exists a subset* $X_n \subseteq A^n$ *of size* $|X_n| \leq p(n)$ *that generates the algebra* $\mathbb{A}^n$.

We now show that the PGP property, along with a polynomial-time algorithm that computes generating sets, implies a reduction from the Π_2 fragment of the QCSP to the CSP.

Proposition 1. *Let* Γ *be a constraint language. If* $\mathbb{A}_\Gamma$ *has the PGP property and there exists an algorithm that outputs a generating set* X_n *for* $\mathbb{A}_\Gamma^n$ *in polynomial time (in* n), *then* Π_2-$\mathrm{QCSP_c}(\Gamma)$ *reduces to* $\mathrm{CSP_c}(\Gamma)$.

This proposition is implicit in [11]. As noted there, it can be readily used to derive the collapse results of Grädel [19] in descriptive complexity, as well as a theorem proved by Karpinski et al. [21] on the Π_2 fragment of *Quantified Horn-SAT*. We refer the reader to [11] for further discussion.

Definition 2. *An algebra* $\mathbb{A}$ *has the* exponentially generated powers (EGP) property *if for any sequence of subsets* $X_1 \subseteq A^1, X_2 \subseteq A^2, X_3 \subseteq A^3, \ldots$ *where for all* $n \geq 1$ *the subset* X_n *generates* $\mathbb{A}^n$, *there exists* $b > 1$ *such that the size function* $|X_n|$ *is* $\Omega(b^n)$.

The following proposition furnishes examples of algebras having the EGP property. Say that an operation $f : A^k \rightarrow A$ is *essentially unary* if there exists a coordinate i and a unary operation $g : A \rightarrow A$ such that $f(a_1, \ldots, a_k) = g(a_i)$ for all $(a_1, \ldots, a_k) \in A^k$. Say that an algebra is *essentially unary* if all of its operations are essentially unary.

Proposition 2. *An essentially unary algebra* $\mathbb{A}$ *with finite universe of non-trivial size has the EGP property.*

3 Collapsibility

In this section, we review the notion of collapsibility as well as the results on this notion that will be relevant to the present work. This section should be taken as a presentation of previous work; other than Propositions 4, 5, and 7, the results and concepts are either explicit or implicit in [12].

Definition 3. *Let* $n \geq 1$ *and* $\mathbb{A}$ *be an algebra. A* rectangular adversary *(of length* n) *is a set of tuples having the form* $B_1 \times \cdots \times B_n$, *where* $B_i \subseteq A$ *for all* $i \in [n]$.[6]

Definition 4. *Let* $n, w \geq 1$ *and* $\mathbb{A}$ *be an algebra. Say that a rectangular adversary* $B_1 \times \cdots \times B_n$ *is* w-bounded *if there exists a value* $a \in A$ *and a subset* $S \subseteq [n]$ *with* $|S| \leq w$ *such that* $B_s = A$ *for all* $s \in S$ *and* $B_i = \{a\}$ *for all* $i \in [n] \setminus S$. *That is, a rectangular adversary is* w-bounded *if* w *or fewer of its sets are equal to* A, *and the rest are equal to* $\{a\}$ *for the same constant* $a \in A$.

---

[6] We remark that what we call a *rectangular adversary* here is simply called an *adversary* in [12].

Definition 5. *An algebra $\mathbb{A}$ is collapsible if there exists $w \geq 1$ such that for all $n \geq 1$, there exist an $\mathbb{A}$-term operation $f : A^k \rightarrow A$ and w-bounded rectangular adversaries (of length n) $B_{11} \times \cdots \times B_{1n}, \ldots, B_{k1} \times \cdots \times B_{kn}$ such that for all $j \in [n]$, $A = f(B_{1j}, \ldots, B_{kj})$.*

The following is the primary computational property of collapsibility.

Theorem 3. *[12] Let Γ be a constraint language. If $\mathbb{A}_\Gamma$ is collapsible, then $\mathrm{QCSP_c}(\Gamma)$ reduces to $\mathrm{CSP_c}(\Gamma)$.*

We show that collapsibility directly implies the PGP.

Proposition 4. *If an algebra $\mathbb{A}$ is collapsible, then it has the polynomially generated powers property.*

We may now observe that collapsibility characterizes the PGP property in the two-element case. We have the following dichotomy result.

Proposition 5. *Let $\mathbb{A}$ be an algebra having a 2-element universe. Either $\mathbb{A}$ is collapsible and has the PGP property, or $\mathbb{A}$ has the EGP property.*

We can remark that, in the two-element case, the boundary line between the PGP property and the EGP property matches the boundary between the tractability and in-tractability of $\mathrm{QCSP}(\Gamma)$: for a constraint language Γ over a two-element set, when $\mathbb{A}_\Gamma$ has the PGP property, $\mathrm{QCSP}(\Gamma)$ is polynomial-time tractable; and, when $\mathbb{A}_\Gamma$ has the EGP property, $\mathrm{QCSP}(\Gamma)$ is PSPACE-complete. (This is readily derived from the dichotomy on two-element $\mathrm{QCSP}(\Gamma)$ complexity [14,11] and the proof of Proposition 5.)

The following theorem was the result of attempts to understand those algebras which are not collapsible, nor have a G-set.

Theorem 6. *[12] Let $\mathbb{A}$ be an idempotent algebra having three-element universe A. Either:*

1. *$\mathbb{A}$ is collapsible and for any constraint language Γ with $\mathbb{A}_\Gamma = \mathbb{A}$, $\mathrm{QCSP_c}(\Gamma)$ is in P;*
2. *$\mathbb{A}$ has a G-set as factor and for any constraint language Γ with $\mathbb{A}_\Gamma = \mathbb{A}$, $\mathrm{QCSP_c}(\Gamma)$ is NP-hard; or,*
3. *There is a way to label the elements of A as $\{a, b, c\}$ such that:*
   - *the size 2 subalgebras of $\mathbb{A}$ are $\{a, c\}$ and $\{b, c\}$, which we denote by α and β respectively,*
   - *$\mathbb{A}$ has as a term operation the semilattice operation $s_{abc} : A \times A \rightarrow A$ defined by $s_{abc}(x, y) = c$ if $x \neq y$, and $s_{abc}(x, y) = x$ if $x = y$, and*
   - *for every term operation $f : A^k \rightarrow A$ of $\mathbb{A}$ and subalgebras $S_1, \ldots, S_k \in \{\alpha, \beta\}$, there exists $T \in \{\alpha, \beta\}$ such that $f(S_1, \ldots, S_k) \subseteq T$.[7]*

---

[7] Regarding the statement of this theorem, we remark that it is not the case that for every three-element idempotent algebra $\mathbb{A}$ there exists a constraint language Γ such that $\mathbb{A} = \mathbb{A}_\Gamma$; the operations of an algebra of the form $\mathbb{A}_\Gamma$ are closed under taking term operations, which is not required of an algebra in general.

The algebras $\mathbb{A}_\Gamma$ for which the tractability/intractability of $\text{QCSP}_c(\Gamma)$ is not yet fully understood are those of type (3). These fall into a "gap": on the one hand, we cannot derive tractability using the condition of collapsibility; on the other hand, we cannot derive intractability using the G-set condition. We hence refer to them as *gap algebras*.

Definition 6. *A gap algebra is a three-element idempotent algebra that is not collapsible and has no G-set as factor.*

We now introduce some terminology that can be used to discuss gap algebras, but also apply more generally to three-element algebras.

Definition 7. *Let A be a three-element set, and let α and β be distinct two-element subsets of A.*

- *We say that an operation $f : A^k \to A$ can be* realized *as an operation $g :$ $\{\alpha, \beta\}^k \to \{\alpha, \beta\}$ if for all $S_1, \ldots, S_k \in \{\alpha, \beta\}$, it holds that $f(S_1, \ldots, S_k) \subseteq g(S_1, \ldots, S_k)$.*
- *We say that an operation $f : A^k \to A$ is $\alpha\beta$-projective if it can be realized by an operation g on $\{\alpha, \beta\}$ that is a projection.*

We remark that any term operation f of a gap algebra can be realized as an operation g, by the third condition given on such algebras by Theorem 6, with respect to the α and β described in that theorem.

We now define the notion of an $\alpha\beta$-projective *algebra*.

Definition 8. *Let $\mathbb{A}$ be an algebra with three-element universe A. We say that $\mathbb{A}$ is $\alpha\beta$-projective if there exist distinct two-element subsets α and β of A with respect to which all operations of $\mathbb{A}$ are $\alpha\beta$-projective.*

Note that the notion of $\alpha\beta$-projective operation is robust: the composition of $\alpha\beta$-projective operations is also an $\alpha\beta$-projective operation. As all projections are $\alpha\beta$-projective, all term operations of an $\alpha\beta$-projective algebra are $\alpha\beta$-projective.

Proposition 7. *An algebra that is $\alpha\beta$-projective has the EGP property.*

4   A Curious Operation

In this section, we focus on a particular algebra that is defined by a single operation. Define $r : \{a, b, c\}^4 \to \{a, b, c\}$ to be the operation where

$$r(a, b, b, b) = r(b, a, b, b) = r(b, b, b, b) = b$$

and

$$r(a, a, b, a) = r(a, a, a, b) = r(a, a, a, a) = a$$

and is equal to c otherwise. Define $\mathbb{A}_r$ to be the algebra $(\{a, b, c\}, \{r\})$. Note that $\mathbb{A}_r$ has s_{abc} as a term operation: $s_{abc}(x, y) = r(x, x, y, y)$; hence, $\mathbb{A}_r$ has no G-set factor.

Does $\mathbb{A}_r$ have the PGP property? We may observe that $\mathbb{A}_r$ is not $\alpha\beta$-projective, rendering this sufficient condition for the EGP property unusable here. We do this as follows. An algebra cannot be $\alpha\beta$-projective with respect to a two-element subset that is not a subalgebra; hence, if $\mathbb{A}_r$ is $\alpha\beta$-projective at all, it must be $\alpha\beta$-projective with respect to the two two-element subalgebras $\alpha = \{a, c\}$ and $\beta = \{b, c\}$ of $\mathbb{A}_r$. However, from the relationships

$$b = r(a, b, b, b) \in r(\alpha, \beta, \beta, \beta) \nsubseteq \alpha$$

$$b = r(b, a, b, b) \in r(\beta, \alpha, \beta, \beta) \nsubseteq \alpha$$

$$a = r(a, a, b, a) \in r(\alpha, \alpha, \beta, \alpha) \nsubseteq \beta$$

$$a = r(a, a, a, b) \in r(\alpha, \alpha, \alpha, \beta) \nsubseteq \beta$$

we may see that r cannot be realized as any of the arity 4 projections (on $\{\alpha, \beta\}$), and hence is not $\alpha\beta$-projective. Indeed, one might view r as being "diagonalized away" from being $\alpha\beta$-projective.

So, we were unable to apply our sufficient condition for the EGP property to $\mathbb{A}_r$. What about trying to show that $\mathbb{A}_r$ satisfies collapsibility, our sufficient condition for the PGP property? This fails as well.

Proposition 8. *The algebra $\mathbb{A}_r$ is not collapsible.*

As we have now obtained that the algebra $\mathbb{A}_r$ has no G-set, is not $\alpha\beta$-projective, and is not collapsible, we have that $\mathbb{A}_r$ is a non-$\alpha\beta$-projective gap algebra. We now show that this algebra has the PGP property.

Proposition 9. *The algebra $\mathbb{A}_r$ has the PGP property.*

The proof of this proposition uses the following definition. For a tuple $(t_1, \ldots, t_n) \in T^n$ over a set T, say that a coordinate $i \in [n-1]$ is a *switch* (of the tuple) if $t_i \neq t_{i+1}$.

In the next sections, we will see that the algebra $\mathbb{A}_r$ is a member of a class of algebras that we introduce called *switchable* algebras–algebras which, as with collapsible algebras, both have the PGP property and imply a QCSP-to-CSP reduction. The condition of switchability generalizes the condition of collapsibility; the algebra $\mathbb{A}_r$ will witness that switchability is a strict generalization of collapsibility.[8]

5   Switchability

The goal of this section is to present the notion of *switchability* as well as some of its basic properties. We review and introduce some basic concepts concerning quantified formulas to be used (Section 5.1), present a notion of composition for sets of tuples that is used in the definition of switchability (Section 5.2), and then proceed to develop the notion of switchability (Section 5.3).

---

[8] In the present section, we showed that the algebra $\mathbb{A}_r$ is not collapsible (Proposition 8); it follows from Theorem 17 that the algebra $\mathbb{A}_r$ is switchable.

5.1   Truth and Adversaries

We now review a characterization of truth for quantified constraint formulas. This characterization comes from the concept of Skolemization [17]. When Φ is a quantified constraint formula, let V^Φ denote the variables of Φ, let E^Φ denote the existentially quantified variables of Φ, let U^Φ denote the universally quantified variables of Φ, and for each $x \in E^\Phi$, let U^Φ_x denote the variables in U^Φ that come before x in the quantifier prefix of Φ. (We may drop the Φ superscript if the formula is clear from the context.) Let $[B \to A]$ denote the set of functions mapping from B to A.

Definition 9. *A strategy for a quantified constraint formula Φ is a sequence of partial functions*

$$\sigma = \{\sigma_x : [U^\Phi_x \to A] \to A\}_{x \in E^\Phi}.$$

That is, a strategy has a mapping σ_x for each existentially quantified variable $x \in E^\Phi$, which tells how to set the variable x in response to an assignment to the universal variables coming before x. Let $\tau : U^\Phi \to A$ be an assignment to the universal variables. We define $\langle \sigma, \tau \rangle$ to be the mapping from V^Φ to A such that $\langle \sigma, \tau \rangle(v) = \tau(v)$ for all $v \in U^\Phi$, and $\langle \sigma, \tau \rangle(x) = \sigma_x(\tau|_{U^\Phi_x})$ for all $x \in E^\Phi$. The mapping $\langle \sigma, \tau \rangle$ is undefined if $\sigma_x(\tau|_{U^\Phi_x})$ is not defined for all $x \in E^\Phi$. The intuitive point here is that a strategy σ along with an assignment τ to the universally quantified variables naturally yields an assignment $\langle \sigma, \tau \rangle$ to all of the variables, so long as the mappings σ_x are defined at the relevant points.

We have the following characterization of truth for quantified constraint formulas.

Fact 10. *A quantified constraint formula Φ is true if and only if there exists a strategy σ for Φ such that for all mappings $\tau : U^\Phi \to A$, the assignment $\langle \sigma, \tau \rangle$ is defined and satisfies the constraints of Φ.*

Note that a strategy satisfying the condition of Fact 10 must consist only of total functions. We have defined a strategy to be a sequence of *partial* functions as we will be interested in strategies σ that need not yield an assignment $\langle \sigma, \tau \rangle$ for all τ.

Definition 10. *An adversary is a set of tuples on a set A, all of the same length; the length of an adversary is considered to be the length of one (any) of its tuples.*

Let us say that an adversary $\mathcal{A}$ is an adversary *for* a quantified constraint formula Φ if the length of $\mathcal{A}$ matches the number of universally quantified variables in Φ. When this is the case, the adversary $\mathcal{A}$ naturally induces the set of assignments $\mathcal{A}[\Phi] = \{\tau \in [U^\Phi \to A] \mid (\exists(a_1, \dots, a_n) \in \mathcal{A})(\forall i \in [n])(\tau(y_i) = a_i)\}$. Here, we assume that $y_1, \dots, y_n$ are the universally quantified variables of Φ, ordered according to quantifier prefix, from outside to inside.

We say that an adversary is Φ-*winnable* if in the modified game, the existential player can win: that is, if there is a strategy that can handle all assignments that the adversary gives rise to, as formalized in the following definition.

Definition 11. *Let Φ be a quantified constraint formula, and let $\mathcal{A}$ be an adversary for Φ. We say that $\mathcal{A}$ is Φ-winnable if there exists a strategy σ for Φ such that for all assignments $\tau \in \mathcal{A}[\Phi]$, the assignment $\langle \sigma, \tau \rangle$ is defined and satisfies the constraints of Φ.*

We have previously given a characterization of truth for quantified constraint formulas (Fact 10). This characterization can be formulated in the terminology just introduced.

Fact 11. *The adversary A^n is Φ-winnable if and only if Φ is true.*

5.2 Reactive Composition

Let $f : A^k \to A$ be an operation and let $\mathcal{A}, \mathcal{B}_1, \ldots, \mathcal{B}_k$ be adversaries of length n. We say that $\mathcal{A}$ is f-*reactively composable* from $\mathcal{B}_1, \ldots, \mathcal{B}_k$, denoted $\mathcal{A} \trianglelefteq f(\mathcal{B}_1, \ldots, \mathcal{B}_k)$, if there exist partial functions $g_i^j : A^i \to A$ for $i \in [n], j \in [k]$ such that, for every tuple $(a_1, \ldots, a_n) \in \mathcal{A}$:

- for every $j \in [k]$, the values $g_1^j(a_1), g_2^j(a_1, a_2), \ldots, g_n^j(a_1, \ldots, a_n)$ are defined,
- for every $j \in [k]$, the tuple $(g_1^j(a_1), g_2^j(a_1, a_2), \ldots, g_n^j(a_1, \ldots, a_n))$ is contained in $\mathcal{B}_j$, and
- for each $i \in [n]$, $a_i = f(g_i^1(a_1, \ldots, a_i), \ldots, g_i^k(a_1, \ldots, a_i))$.

When $\mathbb{A}$ is an algebra and $\mathcal{A}, \mathcal{B}_1, \ldots, \mathcal{B}_k$ are adversaries of the same length, we say that $\mathcal{A}$ is $\mathbb{A}$-*reactively composable* from $\mathcal{B}_1, \ldots, \mathcal{B}_k$ if there exists a term operation f of $\mathbb{A}$ such that $\mathcal{A} \trianglelefteq f(\mathcal{B}_1, \ldots, \mathcal{B}_k)$.

Theorem 12. *Let Φ be a quantified constraint formula, assume that $f : A^k \to A$ is an idempotent polymorphism of all relations of Φ, and let $\mathcal{A}, \mathcal{B}_1, \ldots, \mathcal{B}_k$ be adversaries for Φ. If each of the adversaries $\mathcal{B}_1, \ldots, \mathcal{B}_k$ is Φ-winnable and $\mathcal{A} \trianglelefteq f(\mathcal{B}_1, \ldots, \mathcal{B}_k)$, then the adversary $\mathcal{A}$ is Φ-winnable.*

The notion of reactive composition as well as this theorem appeared in [8], although in a slightly different formulation.

Proposition 13. *Let $\mathbb{A}$ be an algebra, and let S and S' be sets of adversaries, all of the same length. If an adversary $\mathcal{A}$ is $\mathbb{A}$-reactively composable from adversaries in S', and all adversaries in S' are $\mathbb{A}$-reactively composable from adversaries in S, then $\mathcal{A}$ is $\mathbb{A}$-reactively composable from adversaries in S.*

5.3 Definition and Basic Properties

Recall that we define the *switches* of a tuple $\bar{s} = (s_1, \ldots, s_n) \in S^n$ over a set S to be the coordinates $\{i \in [n-1] \mid s_i \neq s_{i+1}\}$; the number of switches of $\bar{s}$ is the cardinality of this set.

Definition 12. *Let $T \subseteq A^n$ be a set of tuples, and let $w \geq 1$. Define $\mathcal{S}(T, w)$ to be the set $\{\bar{t} \in T \mid \bar{t}$ has w or fewer switches $\}$.*

Definition 13. *An algebra $\mathbb{A}$ is switchable if there exists $w \geq 1$ such that for all $n \geq 1$, there exists an $\mathbb{A}$-term operation $f : A^k \to A$ such that*

$$A^n \trianglelefteq f(\mathcal{S}(A^n, w), \ldots, \mathcal{S}(A^n, w)).$$

Observe that, for a fixed $w \geq 1$, the set $S(A^n, w)$ has polynomial size in n: a tuple in this set is determined by the location of its switches and a tuple over A of length up to $w + 1$ specifying, in order, the values it takes on. We may thus upper bound the size of $S(A^n, w)$ by $\left(\binom{n}{1} + \cdots + \binom{n}{w} \right) \cdot (|A|^{w+1})$ which is $O(n^w)$.

Proposition 14. *Let $\mathbb{A}$ be an algebra. If $\mathbb{A}$ is collapsible, then $\mathbb{A}$ is switchable.*

Theorem 15. *Let Γ be a constraint language. If $\mathbb{A}_\Gamma$ is switchable, then $\mathrm{QCSP}_c(\Gamma)$ reduces to $\mathrm{CSP}_c(\Gamma)$.*

Theorem 16. *Let $\mathbb{A}$ be an algebra. If $\mathbb{A}$ is switchable, then $\mathbb{A}$ has the PGP property.*

6   Classification Theorem

Theorem 17. *(Classification theorem) A three-element idempotent algebra not having a G-set is either switchable or is $\alpha\beta$-projective.*

Notice that the terms of an $\alpha\beta$-projective algebra may be viewed as a subclone of the clone containing all $\alpha\beta$-projective operations.

6.1   Corollaries

Corollary 1. *A three-element idempotent algebra not having a G-set either has the PGP property, or has the EGP property.*

Corollary 2. *For every 3-element constraint language Γ, either $\mathrm{QCSP}_c(\Gamma)$ is in P, $\mathrm{QCSP}_c(\Gamma)$ is NP-hard, or the algebra $\mathbb{A}_\Gamma$ is $\alpha\beta$-projective.*

Acknowledgements. The author thanks Víctor Dalmau for useful discussions, Manuel Bodirsky for suggestions on a draft of this paper, and Matt Valeriote for helpful remarks and aid with terminological decisions. Some of the results and ideas in this paper were presented at the Workshop on Universal Algebra and the Constraint Satisfaction Problem held at Vanderbilt University in June 2007; the author thanks the participants of this workshop for their interest.

References

1. Aspvall, B., Plass, M.F., Tarjan, R.E.: A linear-time algorithm for testing the truth of certain quantified boolean formulas. Information Processing Letters 8(3), 121–123 (1979)
2. Berman, J., Idziak, P., Markovic, P., McKenzie, R., Valeriote, M., Willard, R.: Varieties with few subalgebras of powers (submitted for publication)
3. Bulatov, A.: Tractable conservative constraint satisfaction problems. In: Proceedings of 18th IEEE Symposium on Logic in Computer Science (LICS 2003), pp. 321–330 (2003)
4. Bulatov, A., Dalmau, V.: A simple algorithm for mal'tsev constraints. SIAM Journal of Computing 36(1), 16–27 (2006)
5. Bulatov, A., Jeavons, P., Krokhin, A.: Classifying the complexity of constraints using finite algebras. SIAM J. Computing 34(3), 720–742 (2005)

6. Bulatov, A.A.: A dichotomy theorem for constraint satisfaction problems on a 3-element set. Journal of the ACM (J. ACM) 53 (2006)
7. Büning, H.K., Karpinski, M., Flögel, A.: Resolution for quantified boolean formulas. Information and Computation 117(1), 12–18 (1995)
8. Chen, H.: The computational complexity of quantiifed constraint satisfaction. Ph.D. thesis, Cornell University (August 2004)
9. Chen, H.: Quantified constraint satisfaction and bounded treewidth. In: ECAI (2004)
10. Chen, H.: The expressive rate of constraints. Annals of Mathematics and Artificial Intelligence 44(4), 341–352 (2005)
11. Chen, H.: A rendezvous of logic, complexity, and algebra. SIGACT News Logic Column (December 2006)
12. Chen, H.: The complexity of quantified constraint satisfaction: Collapsibility, sink algebras, and the three-element case. SIAM Journal on Computing 37(5), 1674–1701 (2008)
13. Chen, H., Dalmau, V.: From pebble games to tractability: An ambidextrous consistency algorithm for quantified constraint satisfaction. In: Ong, L. (ed.) CSL 2005. LNCS, vol. 3634, Springer, Heidelberg (2005)
14. Creignou, N., Khanna, S., Sudan, M.: Complexity Classification of Boolean Constraint Satisfaction Problems. SIAM Monographs on Discrete Mathematics and Applications. Society for Industrial and Applied Mathematics (2001)
15. Dalmau, V.: Computational complexity of problems over generalized formulas. Ph.D. Thesis, UPC
16. Dalmau, V.: Generalized majority-minority operations are tractable. In: LICS (2005)
17. Ebbinghaus, H.D., Flum, J., Thomas, W.: Mathematical Logic. Springer, Heidelberg (1984)
18. Gottlob, G., Greco, G., Scarcello, F.: The complexity of quantified constraint satisfaction problems under structural restrictions. In: IJCAI (2005)
19. Grädel, E.: Capturing complexity classes by fragments of second order logic. Theoretical Computer Science 101, 35–57 (1992)
20. Idziak, P., Markovic, P., McKenzie, R., Valeriote, M., Willard, R.: Tractability and learnability arising from algebras with few subpowers (extended abstract). In: LICS (2007)
21. Karpinski, M., Büning, H.K., Schmitt, P.H.: On the computational complexity of quantified horn clauses. In: CSL 1987, pp. 129–137 (1987)
22. Kiss, E., Valeriote, M.: On tractability and congruence distributivity. In: LICS (2006)
23. Pan, G., Vardi, M.Y.: Fixed-parameter hierarchies inside pspace. In: LICS (2006)
24. Schaefer, T.J.: The complexity of satisfiability problems. In: Proceedings of the ACM Symposium on Theory of Computing (STOC), pp. 216–226 (1978)

When Does Partial Commutative Closure Preserve Regularity?

Antonio Cano Gómez[1], Giovanna Guaiana[2], and Jean-Éric Pin[3],[*]

[1] Departamento de Sistemas Informáticos y Computación, Universidad Politécnica de Valencia, Camino de Vera s/n, P.O. Box: 22012, E-46020 - Valencia
[2] LITIS EA 4108, Université de Rouen, BP12, 76801 Saint Etienne du Rouvray, France
[3] LIAFA, Université Paris-Diderot and CNRS, Case 7014, 75205 Paris Cedex 13, France

The closure of a regular language under commutation or partial commutation has been extensively studied [1,11,12,13], notably in connection with regular model checking [2,3,7] or in the study of Mazurkiewicz traces, one of the models of parallelism [14,15,16,22]. We refer the reader to the survey [10,9] or to the recent articles of Ochmański [17,18,19] for further references.

In this paper, we present new advances on two problems of this area. The first problem is well-known and has a very precise statement. The second problem is more elusive, since it relies on the somewhat imprecise notion of robust class. By a *robust class*, we mean a class of **regular languages** closed under some of the usual operations on languages, such as Boolean operations, product, star, shuffle, morphisms, inverses of morphisms, residuals, etc. For instance, regular languages form a very robust class, *commutative languages* (languages whose syntactic monoid is commutative) also form a robust class. Finally, *group languages* (languages whose syntactic monoid is a finite group) form a semi-robust class: they are closed under Boolean operation, residuals and inverses of morphisms, but not under product, shuffle, morphisms or star.

Here are the two problems:

Problem 1. *When is the closure of a regular language under [partial] commutation still regular?*

Problem 2. *Are there any robust classes of languages closed under [partial] commutation?*

The classes considered in this paper are all closed under polynomial operations. Recall that, given a class $\mathcal{L}$ of regular languages, the *polynomial languages* of $\mathcal{L}$ are the finite unions of languages of the form $L_0 a_1 L_1 \cdots a_k L_k$ where $a_1, \ldots, a_k$ are letters and $L_0, \ldots, L_k$ are languages of $\mathcal{L}$. Taking the polynomial closure usually increase robustness. For instance, the class $\mathrm{Pol}(\mathcal{G})$ of polynomials of group languages is closed under union, intersection, quotients, product, shuffle and inverses of morphisms.

---

[*] The authors acknowledge support from the AutoMathA programme of the European Science Foundation.

L. Aceto et al. (Eds.): ICALP 2008, Part II, LNCS 5126, pp. 209–220, 2008.
© Springer-Verlag Berlin Heidelberg 2008

Let I be a partial commutation and let D be its complement in $A \times A$. Our main results on Problems 1 and 2 can be summarized as follows:

(1) The class $\mathrm{Pol}(\mathcal{G})$ is closed under commutation. If D is transitive, it is also closed under I-commutation.

(2) Under some simple conditions on the graph of I, the closure of a language of $\mathrm{Pol}(\mathcal{G})$ under I is regular.

(3) There is a very robust class of languages $\mathcal{W}$ which is closed under commutation. This class, which contains $\mathrm{Pol}(\mathcal{G})$, is closed under intersection, union, shuffle, concatenation, residual, length preserving morphisms and inverses of morphisms. Further, it is decidable and can be defined as the largest positive variety of languages not containing $(ab)^*$.

(4) If I is transitive, the closure of a language of $\mathcal{W}$ under I is regular.

Result (3) is probably the most important of the four results. It is, in a sense, optimal since $(ab)^*$ is the canonical example of a regular language whose commutative closure is not regular.

The proofs are nontrivial and combine several advanced techniques, including combinatorial Ramsey type arguments, algebraic properties of the syntactic monoid [5,6], finiteness conditions on semigroups [8] and properties of insertion systems [4].

Our paper is organised as follows. We first survey the known results in Section 1. Then we establish some combinatorial properties of group languages in Section 2. Our results on commutative closure are established in Section 3 and those on closure under partial commutation in Section 4.

1   Known Results

Let A be an alphabet and let I be a symmetric and irreflexive relation on A (often called the *independence relation*). We denote by $\sim_I$ the congruence on A^* generated by the set $\{ab = ba \mid (a, b) \in I\}$. If L is a language on A^*, we also denote by $[L]_I$ the closure of L under $\sim_I$. When I is the relation $\{(a, b) \in A \times A \mid a \neq b\}$, we simplify the notation to $\sim$ and $[L]$, respectively. Thus $\sim$ is the commutation relation and $[L]$ is the *commutative closure* of L. The *dependence relation* associated with I is the relation $D = \{(a, b) \in A \times A \mid (a, b) \notin I\}$. The relations I and D define two graphs (A, I) and (A, D) with A as set of vertices. A class $\mathcal{C}$ of languages is *closed under I-commutation* if $L \in \mathcal{C}$ implies $[L]_I \in \mathcal{C}$.

1.1   The First Problem

For the commutative closure, the problem is solved [11,12,13]: the commutative closure of the language $(ab)^*$ is not regular, but one can effectively decide whether the commutative closure of a regular language is regular or not.

For partial commutations, the result of Sakarovitch [22] concluded a series of previous partial results.

Theorem 1.1. *One can decide whether the closure $[L]_I$ of a regular language L is regular if and only if I is a transitive relation.*

1.2   The Second Problem

Only a few results are known for the second problem. They concern the following classes of languages:

(1) the class $Pol(\mathcal{I})$ of finite unions of languages of the form $A^* a_1 A^* \cdots a_k A^*$, with $a_1, \ldots, a_k \in A$,

(2) the class $\mathcal{J}$ of piecewise testable languages (the Boolean closure of $Pol(\mathcal{I})$),

(3) the class $Pol(\mathcal{J})$, which consists of finite unions of languages of the form $A_0^* a_1 A_1^* \cdots a_k A_k^*$ with $A_i \subseteq A$ and $a_1, \ldots, a_k \in A$. These languages are also called *APC (Alphabetic Pattern Constraints)* in [2],

(4) the class $Pol(\mathcal{C}om)$ of polynomials of commutative languages.

The following theorem summarises the results of Guaiana, Restivo and Salemi [14,15], Bouajjani, Muscholl and Touili [2,3] and Cécé, Héam and Mainier [7].

Theorem 1.2. *Let I be any independence relation. Then*

(1) *the class $Pol(\mathcal{I})$ is closed under commutation,*

(2) *the class $\mathcal{J}$ is closed under commutation,*

(3) *the class $Pol(\mathcal{J})$ is closed under I-commutation,*

(4) *the class $Pol(\mathcal{C}om)$ is closed under I-commutation.*

1.3   Star-Free Languages

Two nice results on star-free languages were proved by Muscholl and Petersen [16]. The first one is the counterpart of Theorem 1.1 for star-free languages.

Theorem 1.3. *One can decide whether the closure $[L]_I$ of a star-free language L is star-free if and only if I is a transitive relation.*

The second result is related to our second problem.

Theorem 1.4. *Let L be a star-free language. If D is transitive, then $[L]_I$ is either star-free or non regular. If D is not transitive, then there exist star-free languages such that $[L]_I$ is regular but not star-free.*

2   Properties of Group Languages

Recall that a *group language* is a regular language whose syntactic monoid is a group, or, equivalently, is recognized by a finite deterministic automaton in which each letter defines a permutation of the set of states. We gather in this section the properties of these languages needed in this paper.

An *insertion system* is a special type of rewriting system whose rules are of the form $1 \to r$ for all r in a given language R. We write $u \to_R v$ if $u = u'u''$ and $v = u'ru''$ for some $r \in R$. We denote by $\overset{*}{\to}_R$ the reflexive transitive closure of the relation $\to_R$. Given a language L of A^*, its closure under $\overset{*}{\to}_R$ is the language

$$[L]_{\overset{*}{\to}_R} = \{v \in A^* \mid \text{there exists } u \in L \text{ such that } u \overset{*}{\to}_R v\}$$

We are especially interested in the case $R = \pi^{-1}(1)$, where π is a morphism from A^* onto a finite group G. Let F be the set of words of R of length $\leqslant |G|$. It is not difficult to see that the relations $\xrightarrow{*}_F$ and $\xrightarrow{*}_R$ coincide.

The next lemma is a well-known consequence of Ramsey's theorem [20].

Lemma 2.1. *For any $n > 0$, there exists $N > 0$ such that, for any u_0, u_1, ..., $u_N \in A^*$ there exists a sequence $0 \leqslant i_0 < i_1 < \ldots < i_n \leqslant N$ such that*
$$\pi(u_{i_0} u_{i_0+1} \cdots u_{i_1-1}) = \pi(u_{i_1} u_{i_1+1} \cdots u_{i_2-1}) = \ldots = \pi(u_{i_{n-1}} \cdots u_{i_n-1}) = 1.$$

Define a relation $\leqslant_\pi$ on A^* by setting $u \leqslant_\pi v$ if and only if $u = a_1 \cdots a_n$, with $a_0, \ldots, a_n \in A$ and $v = v_0 a_1 v_1 \cdots v_{n-1} a_n v_n$ for some words $v_0, \ldots, v_n$ such that $\pi(v_0) = \ldots = \pi(v_n) = 1$. The next proposition follows from the results of Bucher, Ehrenfeucht and Haussler [4].

Proposition 2.2. *The relation $\leqslant_\pi$ is a well preorder on A^* and for any language L, the language $[L]_{\xrightarrow{*}_R}$ is regular.*

We prove a slightly more precise result.

Proposition 2.3. *For any language L, the language $[L]_{\xrightarrow{*}_R}$ is a polynomial of group languages.*

Proof. By construction, R is a group language. If $u = a_1 \cdots a_n$, the language $[u]_{\xrightarrow{*}_R}$ is equal to $R a_1 R \cdots R a_n R$, a polynomial of group languages. Now since $\leqslant_\pi$ is a well preorder, every language of the form $[L]_{\xrightarrow{*}_R}$ is equal to a language of the form $[E]_{\xrightarrow{*}_R}$ with E finite and thus is a finite union of languages of the form $R a_1 R \cdots R a_n R$. It is therefore a polynomial of group languages. $\square$

3   Commutative Closure

This section contains three new results. The first one concerns group languages, the second one polynomials of group languages and the third one a robust class introduced in [5,6] and denoted by $\mathcal{W}$.

Recall that if L is a language of A^*, the *syntactic preorder of L* is the relation $\leqslant_L$ defined on A^* by $u \leqslant_L v$ if and only if, for every $x, y \in A^*$, $xvy \in L$ implies $xuy \in L$. The syntactic congruence $\sim_L$ is defined by $u \sim_L v$ if and only if $u \leqslant_L v$ and $v \leqslant_L u$.

3.1   Group Languages

Theorem 3.1. *The commutative closure of a group language is regular.*

Proof. Let $L \subseteq A^*$ be a group language and let $\pi : A^* \to G$ be its syntactic morphism. Let $n = |G|$ and let N be the integer given by Lemma 2.1. We claim that for any letter $a \in A$, $a^N \sim_{[L]} a^{N+n}$. Let $g = \pi(a)$.

Suppose that $xa^N y \in [L]$. Then there exists a word w of L commutatively equivalent to $xa^N y$. It follows that wa^n is commutatively equivalent to $xa^{N+n}y$.

Further, since G is a finite group, one has $g^n = 1$ by Lagrange's theorem, whence $\pi(wa^n) = \pi(w)\pi(a^n) = \pi(w)$. Thus the words w and wa^n have the same syntactic image by π and hence $wa^n \in L$. Therefore $xa^{N+n}y \in [L]$.

Conversely, assume that $xa^{N+n}y \in [L]$. Then $xa^{N+n}y$ is commutatively equivalent to some word of L, say $w = u_0au_1a\cdots u_{N-1}au_Nau_{N+1}$. By applying Lemma 2.1 to the sequence of words u_0a, u_1a, $\ldots$, u_Na, we obtain a sequence $0 \leqslant i_0 < i_1 < \ldots < i_n \leqslant N$ such that

$$\pi(u_{i_0}a\cdots au_{i_1-1}a) = \pi(u_{i_1}a\cdots au_{i_2-1}a) = \ldots = \pi(u_{i_{n-1}}a\cdots au_{i_n-1}a) = 1 \quad (1)$$

This implies in particular

$$\pi(u_{i_0}a\cdots au_{i_1-1}) = \pi(u_{i_1}a\cdots au_{i_2-1}) = \ldots = \pi(u_{i_{n-1}}a\cdots au_{i_n-1}) = g^{-1} \quad (2)$$

Let r and s be the words defined by

$$w = r(u_{i_0}a\cdots au_{i_1-1}a)(u_{i_1}a\cdots au_{i_2-1}a)(u_{i_{n-1}}a\cdots au_{i_n-1}a)s$$

Since w is commutatively equivalent to $xa^{N+n}y$, the word

$$w' = r(u_{i_0}a\cdots au_{i_1-1})(u_{i_1}a\cdots au_{i_2-1})\cdots(u_{i_{n-1}}a\cdots au_{i_n-1})s$$

is commutatively equivalent to xa^Ny. Furthermore, Formulas (1) and (2) show that $\pi(w) = \pi(r)\pi(s)$ and $\pi(w') = \pi(r)(g^{-1})^n\pi(s)$. Since $(g^{-1})^n = 1$ by Lagrange's theorem, $\pi(w) = \pi(w')$ and thus $w' \in L$. It follows that $xa^Ny \in [L]$, which proves the claim.

Now, the syntactic monoid of $[L]$ is a commutative monoid in which each generator has a finite index. Since the alphabet is finite, this monoid is finite and thus $[L]$ is regular. $\qquad\square$

Theorem 3.1 indicates that the commutative closure of a group language is a commutative regular language. One may wonder whether, in turn, any commutative regular language is the commutative closure of a group language. The answer is no, but requires an improved version of Theorem 3.1.

Theorem 3.2. *The commutative closure of a group language is a polynomial of group languages.*

Proof. Let L be a group language and let $\pi : A^* \to G$ be its syntactic morphism. We claim that $[L]$ is a filter for $\leqslant_\pi$, which will give the result by Proposition 2.3. Let us show that if $a_1\cdots a_n \in [L]$ and $v_0, v_1, \ldots, v_n \in \pi^{-1}(1)$, then $v_0a_1v_1\cdots a_nv_n \in [L]$. Since $a_1\cdots a_n \in [L]$, there exists a word $w \in L$ which is commutatively equivalent to $a_1\cdots a_n$. Thus the word $wv_0v_1\cdots v_n$ is commutatively equivalent to $v_0a_1v_1\cdots a_nv_n$. Now $\pi(wv_0v_1\cdots v_n) = \pi(w)\pi(v_0)\cdots\pi(v_n) = \pi(w)$. Therefore $wv_0v_1\cdots v_n \in L$, proving the claim. $\qquad\square$

Note that the commutative closure of a group language is not necessarily a group language. Indeed, consider the set of all words of $\{a, b\}^*$ having an even number of (scattered) subwords equal to ab. Its commutative closure, $A^*aA^*bA^* \cup A^*bA^*aA^*$ is not a group language. However, Theorem 3.2 can be extended to polynomials of group languages.

Theorem 3.3. *The commutative closure of a polynomial of group languages is also a polynomial of group languages.*

Proof. It is shown in [21] that for any polynomial of group languages L, there exists a morphism $\pi : A^* \to G$ from A^* onto a finite group G such that L is a finite union of languages of the form $Ra_1R \cdots Ra_nR$, with $R = \pi^{-1}(1)$. Thus it suffices to show that if $K = Ra_1R \cdots Ra_nR$ for some letters $a_1, \ldots, a_n$, then $[K]$ is a polynomial of group languages.

We claim that $[K]$ is a filter for $\leqslant_\pi$, which will give the result by Proposition 2.3. Let us show that if $b_1 \cdots b_m \in [K]$ and $v_0, v_1, \ldots, v_m \in R$, then $v_0b_1v_1 \cdots b_mv_m \in [K]$. Let w be a word of K commutatively equivalent to $b_1 \cdots b_m$. As an element of K, w can be written as $r_0a_1r_1 \cdots a_nr_n$ for some words $r_0, \ldots, r_n \in R$. Since the words $v_0, \ldots, v_m$ are in R, the word $wv_0v_1 \cdots v_m$ also belongs to K and is commutatively equivalent to $v_0b_1v_1 \cdots b_mv_m$. This proves the claim and concludes the proof.   □

3.2   Languages of $\mathcal{W}$

We now define the class of regular languages $\mathcal{W}$ first introduced and studied in [5,6]. Recall that a *positive variety* of languages is a class of regular languages closed under union, intersection, residuals and inverses of morphisms.

The class $\mathcal{W}$ is the unique maximal positive variety of languages which does not contain the language $(ab)^*$, for all letters $a \neq b$. It is also the unique maximal positive variety satisfying the two following conditions: it is *proper*, that is, strictly included in the variety of regular languages, and it is closed under the shuffle operation. It is also the largest proper positive variety closed under length preserving morphisms. Being closed under intersection, union, shuffle, concatenation, length preserving morphisms and inverses of morphisms, $\mathcal{W}$ is a quite robust class, which strictly contains the classes APC, Pol($\mathcal{C}om$) and Pol($\mathcal{G}$) introduced previously.

The class $\mathcal{W}$ has an algebraic characterization [5,6]. Let a and b be two elements of a monoid. Recall that b is an *inverse of* a if $aba = a$ and $bab = b$. Now, a regular language belongs to $\mathcal{W}$ if and only if its syntactic ordered monoid belongs to the variety of finite ordered monoids **W** defined as follows: an ordered monoid $(M, \leqslant)$ belongs to **W** if and only if, for any pair (a, b) of mutually inverse elements of M, and any element z of the minimal ideal of the submonoid generated by a and b, $(abzab)^\omega \leqslant ab$ (see [6, p.435–436] for a precise definition of the semigroup notions used in this characterization). This description might appear quite involved, but has an important consequence: the variety $\mathcal{W}$ is decidable. That is, given a regular language L, one can decide whether or not L belongs to $\mathcal{W}$. We also mention for the specialists that **W** contains the variety of finite monoids **DS**.

The main result of this section states that $\mathcal{W}$ is closed under commutative closure. In fact, we prove a stronger result, which relies on the notion of a period that we now introduce.

Let M be a finite monoid. The *exponent* of M is the least integer ω such that for all $x \in M$, x^ω is idempotent. Its *period* is the least integer p such that for all $x \in M$, $x^{\omega+p} = x^{\omega}$. By extension, the *period* (respectively *exponent*) of a regular language is the period (respectively exponent) of its syntactic monoid.

Proposition 3.4. *Let L be a commutative language of A^* and let d be a positive integer. If, for each letter c of A, there exists $N > 0$ such that $c^{N+d} \leqslant_L c^N$, then L is regular and its period divides d.*

Proof. It follows from [8, Theorem 6.6.2, page 215] that, under these conditions, L is a regular language. Let ω be the exponent of L. The relation $c^{N+d} \leqslant_L c^N$ gives $c^{N(\omega-1)}c^{N+d} \leqslant_L c^{N(\omega-1)}c^N$, whence $c^{N\omega+d} \leqslant_L c^{N\omega}$ and since $c^{\omega} \sim_L c^{2\omega} \sim_L c^{N\omega}$, one gets finally $c^{\omega+d} \leqslant_L c^{\omega}$. It follows that

$$c^{\omega} \sim_L c^{\omega+\omega d} \leqslant_L \cdots \leqslant_L c^{\omega+2d} \leqslant_L c^{\omega+d} \leqslant_L c^{\omega}$$

and hence $c^{\omega} \sim_L c^{\omega+d}$. Since L is commutative, its syntactic monoid is commutative and therefore $u^{\omega} \sim_L u^{\omega+d}$ for all $u \in A^*$. It follows that the period of L divides d.   □

We can now state:

Theorem 3.5. *Let L be a language of $\mathcal{W}(A^*)$. Then $[L]$ is regular and commutative (and hence in $\mathcal{W}(A^*)$) and its period divides that of L.*

Proof. Let L be a language of $\mathcal{W}(A^*)$ and let $[L]$ be its commutative closure. Then there exist an ordered monoid $(M, \leqslant) \in \mathbf{W}$, a surjective monoid morphism $\varphi : A^* \to M$ and an order ideal P of $(M, \leqslant)$ such that $\varphi^{-1}(P) = L$. Let ω be the exponent of M and let p be its period. Let also d be any number such that, for all $t \in M$, t^d is idempotent. In particular, d can be either ω or $\omega + p$. We claim that, for every such d, there exists an integer N such that, for every letter $c \in A$, $c^{N+d} \leqslant_{[L]} c^N$. If the claim holds, then Proposition 3.4 shows that $[L]$ is regular and that its period divides d. Taking $d = \omega$ and $d = \omega + p$ then proves that this period also divides p.

The rest of the proof consists in proving the claim. We need two combinatorial results. The first one is a slight variation of Lemma 2.1.

Lemma 3.6. *Let c be a letter of A. For any $n \geqslant 0$, there exists an integer N such that, for every word u of A^* containing at least $N + 1$ occurrences of c, there exist an idempotent e of M and a factorization $u = v_0 v_1 c v_2 c \cdots v_n c v_{n+1}$ such that, for $1 \leqslant i \leqslant n$, $\varphi(v_i c) = e$.*

The second one requires an auxiliary definition. A word u of $\{a, b\}^*$ is said to be *balanced* if $|u|_a = |u|_b$.

Proposition 3.7. *Let $B = \{a, b\}$. There exists a balanced word $z \in B^*$ such that, for any morphism $\gamma : B^* \to M$, $\gamma(z)$ belongs to the minimal ideal of the monoid $\gamma(B^*)$.*

Proof. Let $n = |M|$ and let z be a balanced word of B^* containing all words of length $\leqslant n$ as a factor. Let $\gamma : B^* \to M$ be a morphism and let m be an element of the minimal ideal J of $\gamma(B^*)$. Then one can show there is a word u of length $\leqslant n$ such that $\gamma(u) = m$. Since $|u| \leqslant n$, u is a factor of z and $\gamma(z)$ belongs to $M\gamma(u)M$. Now since $m \in J$, $M\gamma(u)M = MmM = J$ and hence $\gamma(z) \in J$.   □

Let us continue the proof of Theorem 3.5. Let $n = |M|$ and let z be the balanced word given by Proposition 3.7. Let $r = |z|_a = |z|_b$, $n_3 = d(1+r)$, $n_2 = nn_3$ and $n_1 = 3n_2$. Finally let $N = N(n_1)$ be the constant given by Lemma 3.6.

Let $x, y \in A^*$. If $xc^N y \in [L]$, there exists a word u of L commutatively equivalent to $xc^N y$ and hence containing at least N occurrences of c. By Lemma 3.6, there exist an idempotent e of M and a factorization $u = v_0 v_1 c \cdots v_{n_1} c v_{n_1+1}$ such that, for $1 \leqslant i \leqslant n_1$, $\varphi(v_i c) = e$.

Now, since $n_1 = 3n_2$, one can also write u as $u = v_0(f_1 g_1) \cdots (f_{n_2} g_{n_2}) v_{n_1+1}$ where, for $1 \leqslant i \leqslant n_2$, $f_i = v_{3i-2} c v_{3i-1}$ and $g_i = c v_{3i} c$.

Lemma 3.8. *For $1 \leqslant i \leqslant n_2$, the elements $\varphi(f_i)$ and $\varphi(g_i)$ are mutually inverse.*

Proof. We omit this proof, but it is a straightforward verification.   □

Setting $\bar{s} = \varphi(c)e$, one gets $\varphi(g_i) = \bar{s}$ for $1 \leqslant i \leqslant n_2$. Further, by the choice of n_2 and by the pigeonhole principle, one can find n_3 indices $i_1 < \ldots < i_{n_3}$ and an element $s \in M$ such that $\varphi(f_{i_1}) = \ldots = \varphi(f_{i_{n_3}}) = s$. Setting

$$w_0 = v_0 f_1 g_1 \cdots f_{i_1-1} g_{i_1-1} \qquad x_1 = f_{i_1} \qquad y_1 = g_{i_1}$$
$$w_1 = f_{i_1+1} g_{i_1+1} \cdots f_{i_2-1} g_{i_2-1} \qquad x_2 = f_{i_2} \qquad y_2 = g_{i_2}$$

$$\vdots \qquad\qquad\qquad\qquad \vdots$$

$$w_{n_3-1} = f_{i_{n_3-1}+1} g_{i_{n_3-1}+1} \cdots f_{i_{n_3}-1} g_{i_{n_3}-1} \qquad x_{n_3} = f_{i_{n_3}} \qquad y_{n_3} = g_{i_{n_3}}$$
$$w_{n_3} = f_{i_{n_3}+1} g_{i_{n_3}+1} \cdots f_{n_2} g_{n_2} v_{n_1+1}$$

we obtain a factorization

$$u = w_0 x_1 y_1 w_1 x_2 y_2 w_2 \cdots w_{n_3-1} x_{n_3} y_{n_3} w_{n_3} \qquad (3)$$

such that $\varphi(w_1) = \ldots = \varphi(w_{n_3-1}) = e$, $\varphi(x_1) = \ldots = \varphi(x_{n_3}) = s$ and $\varphi(y_1) = \ldots = \varphi(y_{n_3}) = \bar{s}$.

Recall that $n_3 = d(1+r)$ where $r = |z|_a = |z|_b$. We now define words $z_1, \ldots, z_d$ as follows: the word z_j is obtained by replacing in z the first occurrence of a by $x_{d+(j-1)r+1}$, the second occurrence of a by $x_{d+(j-1)r+2}$, $\ldots$, the r's occurrence of a by x_{d+jr} and, similarly, the first occurrence of b by $y_{d+(j-1)r+1}$, the second occurrence of b by $y_{d+(j-1)r+2}$, $\ldots$, the r's occurrence of b by y_{d+jr}. Finally, set

$$u' = w_0(v_{3i_1-2} cc v_{3i_1-1} c z_1 v_{3i_1} c)(v_{3i_2-2} cc v_{3i_2-1} c z_2 v_{3i_2} c) \cdots$$
$$(v_{3i_d-2} cc v_{3i_d-1} c z_d v_{3i_d} c) w_1 \cdots w_{n_3} \qquad (4)$$

We are now ready for the three final steps.

Lemma 3.9. *The word u' is commutatively equivalent to $xc^{N+d}y$.*

Proof. It is clear that u' is commutatively equivalent to

$$c^d w_0 (v_{3i_1-2} c v_{3i_1-1} c v_{3i_1} c) \cdots (v_{3i_d-2} c v_{3i_d-1} c v_{3i_d} c)(z_1 \cdots z_d)(w_1 \cdots w_{n_3})$$

Now, $v_{3i_1-2} c v_{3i_1-1} c v_{3i_1} c = f_{i_1} g_{i_1} = x_1 y_1, \ldots, v_{3i_d-2} c v_{3i_d-1} c z_d v_{3i_d} c = f_{i_d} g_{i_d} = x_d y_d$. Further, by construction, $z_1 \cdots z_d \sim x_{d+1} y_{d+1} \cdots x_{n_3} y_{n_3}$. Therefore

$$u' \sim c^d w_0 x_1 y_1 w_1 x_2 y_2 w_2 \cdots w_{n_3-1} x_{n_3} y_{n_3} w_{n_3}$$

and finally $u' \sim u c^d \sim x c^{N+d} y$. $\qquad\square$

Let T be the submonoid of M generated by s and $\bar{s}$ and let $\gamma : \{a, b\}^* \to T$ be the morphism defined by $\gamma(a) = s$ and $\gamma(b) = \bar{s}$. By Proposition 3.7, $\gamma(z)$ belongs to the minimal ideal of T and since $e = s\bar{s}$, the definition of **W** shows that in M, $(e\gamma(z)e)^d \leqslant e$.

Lemma 3.10. *One has* $\varphi(z_1) = \ldots = \varphi(z_d) = \gamma(z)$.

Proof. Each of the words z_j is obtained by replacing in z the occurrences of a by some x_k and each occurrence of b by some y_k. Since all the x_k (resp. y_k) have the same image by φ, namely s (resp. $\bar{s}$), $\varphi(z_j)$ is equal to $\gamma(z)$. $\qquad\square$

Lemma 3.11. *The word u' belongs to L.*

Proof. It follows from (3) that $\varphi(u) = \varphi(w_0) e \varphi(w_{n_3})$, and hence, since $P = \varphi(L)$, $\varphi(w_0) e \varphi(w_{n_3}) \in P$. Now, observe that

$$\varphi(v_{3i_1-2} c c v_{3i_1-1} c z_1 v_{3i_1} c) = \varphi(v_{3i_1-2} c) \varphi(c) \varphi(v_{3i_1-1} c) \varphi(z_1) \varphi(v_{3i_1} c)$$
$$= e\varphi(c) e \varphi(z_1) e = e\bar{s}\gamma(z)e \quad \text{by Lemma 3.10}$$

By a similar argument, one has

$$\varphi(v_{3i_1-2} c c v_{3i_1-1} c z_1 v_{3i_1} c) = \ldots = \varphi(v_{3i_d-2} c c v_{3i_d-1} c z_d v_{3i_d} c) = e\bar{s}\gamma(z)e$$

Finally, since $\varphi(w_1) = \ldots = \varphi(w_{n_3-1}) = e$, it follows from (4) that

$$\varphi(u') = \varphi(w_0)(e\bar{s}\gamma(z)e)^d \varphi(w_{n_3})$$

Furthermore, since $\bar{s} \in T$, $\bar{s}\gamma(z)$ belongs to the minimal ideal of T and since M is in **W**, one has $(e\bar{s}\gamma(z)e)^d \leqslant e$. Since $\varphi(L)$ is an order ideal, the element $\varphi(w_0)(e\bar{s}\gamma(z)e)^d \varphi(w_{n_3})$ is also in $\varphi(L)$ and hence $u' \in L$. $\qquad\square$

Putting Lemmas 3.9 and 3.11 together, we conclude that $x c^{N+d} y \in [L]$, which proves the claim and the theorem. $\qquad\square$

4   Closure Under Partial Commutation

Some of the results of Section 3 can be extended to partial commutations, under some restrictions on the set I.

4.1 The Case Where D Is Transitive

The condition that D is transitive is equivalent to requiring that $A^*/{\sim_I}$ is isomorphic to a direct product of free monoids $A_1^* \times \cdots \times A_k^*$. Denote by π_j the projection from A^* onto A_j^* and let π_I be the morphism from A^* onto $A_1^* \times \cdots \times A_k^*$ defined by $\pi_I(u) = (\pi_1(u), \ldots, \pi_k(u))$. This morphism is intimately connected to our problem, since $u \sim_I v$ if and only if $\pi_I(u) = \pi_I(v)$. Let us denote by Ш the shuffle product. The (easy) proof of the next result is omitted. The second part of the statement relies on Mezei's theorem characterizing the recognizable subsets of a direct product of monoids.

Proposition 4.1. *Let L be a language of A^*. If*

$$\pi_I(L) = \bigcup_{1 \leqslant i \leqslant n} L_{i,1} \times \cdots \times L_{i,k} \tag{5}$$

where for $1 \leqslant j \leqslant k$, the languages $L_{1,j}, \ldots, L_{n,j}$ are languages of A_j^*, then $[L]_I = \bigcup_{1 \leqslant i \leqslant n} L_{i,1}$ Ш $\cdots$ Ш $L_{i,k}$. In particular, if $\pi_I(L)$ is a recognizable subset of $A_1^* \times \cdots \times A_k^*$, then $[L]_I$ is regular.

If L is a group language, one can adapt an argument from [5, Proposition 9.6] to show that $\pi_I(L)$ can be decomposed as in (5), where each $L_{i,j}$ belongs to Pol($\mathcal{G}$). Therefore, since Pol($\mathcal{G}$) is closed under shuffle, we get:

Theorem 4.2. *Suppose that D is transitive. If L is a group language, then $[L]_I$ is a polynomial of group languages.*

Still some work is needed to obtain the following result.

Theorem 4.3. *Suppose that D is transitive. If L is a polynomial of group languages, then $[L]_I$ is also a polynomial of group languages.*

This result cannot be extended to $\mathcal{W}$. Indeed, let $A = \{a, b, c, d\}$ and $I = \{(a, b), (b, c), (c, d), (d, a)\}$. Then the language $(abcd)^* + A^*aaA^* + A^*bbA^* + A^*ccA^* + A^*ddA^* + A^*ababA^* + A^*bcbcA^* + A^*cdcdA^* + A^*dadaA^*$ belongs to $\mathcal{W}$ but $[L]_I$ is not regular, although D is transitive in this case.

4.2 The Case Where I Is Transitive

We now consider the case where I is transitive. In this case, $A^*/{\sim_I}$ is a free product of free commutative monoids.

Theorem 4.4. *Let L be a language of $\mathcal{W}(A^*)$ and let I be a transitive independence relation. Then $[L]_I$ is a regular language.*

Proof. (Sketch) Let $\mathcal{P} = \{A_1, \ldots, A_k\}$ be the partition of A such that $A^*/{\sim_I}$ is isomorphic to the free product $\mathbb{N}^{A_1} * \cdots * \mathbb{N}^{A_k}$.

Let $\mathcal{A} = (Q, A, \cdot, q_0, F)$ be the minimal automaton of L. Recall that the states of Q are partially ordered by the relation $\leqslant$ defined by $p \leqslant q$ if and only if, for all $u \in A^*$, $q \cdot u \in F$ implies $p \cdot u \in F$.

We now construct a generalized automaton $\mathcal{B}$, over the same set of states Q, in which transitions are labelled by regular languages. More precisely, for each pair of states (p, q), we create a transition from p to q labelled by

$$R_{p,q} = \bigcup_{1 \leqslant i \leqslant k} [\{u \in A_i^* \mid p \cdot u \leqslant q\}]$$

Each language $\{u \in A_i^* \mid p \cdot u \leqslant q\}$ can be written as the intersection of A_i^* and of the language $K_{p,q} = \{u \in A^* \mid p \cdot u \leqslant q\}$. Since $L \in \mathcal{W}(A^*)$, one also has $K_{p,q} \in \mathcal{W}(A^*)$ and since $\mathcal{W}$ is closed under commutation by Theorem 3.5, so does $R_{p,q}$. The remainder of the proof (omitted for lack of space) consists in proving that $\mathcal{B}$ recognizes $[L]_I$. It follows that $[L]_I$ is regular.   □

Note that we don't know whether $[L]_I$ also belongs to $\mathcal{W}(A^*)$. However, the proof of Theorem 4.4 can be adapted to prove another result.

Let $I_1, \ldots, I_k$ be the connected components of the graph (A, I). Then $A^*/\!\!\sim_I$ is isomorphic to the free product $A^*/\!\!\sim_{I_1} * \cdots * A^*/\!\!\sim_{I_k}$. Let us modify the construction of the automaton $\mathcal{B}$ by taking

$$R_{p,q} = \bigcup_{1 \leqslant j \leqslant k} [\{u \in A_j^* \mid p \cdot u \leqslant q\}]_{I_j}$$

Then one can prove that if each language $[\{u \in A_j^* \mid p \cdot u \leqslant q\}]_{I_j}$ is regular, then $[L]_I$ is regular. Putting $D_j = \{(a, b) \in A_j \times A_j \mid (a, b) \notin I_j\}$ for $1 \leqslant j \leqslant k$, one can show, thanks to Theorem 4.3, that $R_{p,q}$ is regular if L is a polynomial of group languages and each relation D_j is transitive.

There is a simple graph theoretic interpretation of this latter condition. One can show that I satisfies it if and only if the restriction of the graph (A, I) to any four letter subalphabet is not one of the graphs P_4 and Paw represented below:

We can now state our last result.

Theorem 4.5. *Let L be a polynomial of group languages. If the graph (A, I) is (P_4, Paw)-free, then $[L]_I$ is regular.*

References

1. Achache, A.: Opérateurs de fermeture semi-commutatifs. Novi Sad J. Math. 34(1), 79–87 (2004)
2. Bouajjani, A., Muscholl, A., Touili, T.: Permutation Rewriting and Algorithmic Verification. In: Proc. 16th Symp. on Logic in Computer Science (LICS 2001), Boston (MA), USA, pp. 399–409. IEEE Pub., Los Alamitos (2001)

3. Bouajjani, A., Muscholl, A., Touili, T.: Permutation Rewriting and Algorithmic Verification. Information and Computation 205(2), 199–224 (2007)
4. Bucher, W., Ehrenfeucht, A., Haussler, D.: On total regulators generated by derivation relations. Theor. Comput. Sci. 40(2-3), 131–148 (1985)
5. Cano Gómez, A., Pin, J.-E.: On a conjecture of Schnoebelen. In: Ésik, Z., Fülöp, Z. (eds.) DLT 2003. LNCS, vol. 2710, pp. 35–54. Springer, Heidelberg (2003)
6. Cano Gómez, A., Pin, J.-É.: Shuffle on positive varieties of languages. Theoret. Comput. Sci. 312, 433–461 (2004)
7. Cécé, G., Héam, P.-C., Mainier, Y.: Efficiency of Automata in Semi-Commutation Verification Techniques. Theoret. Informatics Appl. 42, 197–215 (2008)
8. De Luca, A., Varricchio, S.: Finiteness and regularity in semigroups and formal languages. Monographs in Theoretical Computer Science. An EATCS Series. Springer, Berlin (1999)
9. Diekert, V., Métivier, Y.: Partial commutation and traces. In: Handbook of formal languages. Beyond words, vol. 3, pp. 457–533. Springer, New York (1997)
10. Diekert, V., Rozenberg, G. (eds.): The book of traces. World Scientific Publishing Co. Inc., River Edge (1995)
11. Ginsburg, S., Spanier, E.H.: Bounded regular sets. Proc. Amer. Math. Soc. 17, 1043–1049 (1966)
12. Ginsburg, S., Spanier, E.H.: Semigroups, Presburger formulas, and languages. Pacific J. Math. 16, 285–296 (1966)
13. Gohon, P.: An algorithm to decide whether a rational subset of $\mathbb{N}^k$ is recognizable. Theor. Comput. Sci. 41, 51–59 (1985)
14. Guaiana, G., Restivo, A., Salemi, S.: On the product of trace languages. In: Bertoni, S.C.R.A., Goldwurm, M. (eds.) Proc. of the Workshop Trace theory and code parallelization, pp. 54–67. University of Milan, Italy (2000)
15. Guaiana, G., Restivo, A., Salemi, S.: On the trace product and some families of languages closed under partial commutations. J. Autom. Lang. Comb. 9(1), 61–79 (2004)
16. Muscholl, A., Petersen, H.: A note on the commutative closure of star-free languages. Inform. Process. Lett. 57(2), 71–74 (1996)
17. Ochmański, E., Stawikowska, K.: On closures of lexicographic star-free languages. In: Automata and formal languages, pp. 227–234. Univ. Szeged. Inst. Inform., Szeged (2005)
18. Ochmański, E., Stawikowska, K.: Star-free star and trace languages. Fund. Inform. 72(1-3), 323–331 (2006)
19. Ochmański, E., Stawikowska, K.: A Star Operation for Star-Free Trace Languages. In: Harju, T., Karhumäki, J., Lepistö, A. (eds.) DLT 2007. LNCS, vol. 4588, pp. 337–345. Springer, Heidelberg (2007)
20. Pin, J.-E.: Varieties of formal languages. North Oxford, London (1986); (Traduction de Variétés de langages formels)
21. Pin, J.-E.: Polynomial closure of group languages and open sets of the Hall topology. Theoret. Comput. Sci. 169, 185–200 (1996)
22. Sakarovitch, J.: The "last" decision problem for rational trace languages. In: Simon, I. (ed.) LATIN 1992. LNCS, vol. 583, pp. 460–473. Springer, Heidelberg (1992)

Weighted Logics for Nested Words and Algebraic Formal Power Series

Christian Mathissen*

Institut für Informatik, Universität Leipzig
D-04009 Leipzig, Germany
mathissen@informatik.uni-leipzig.de

Abstract. Nested words, a model for recursive programs proposed by Alur and Madhusudan, have recently gained much interest. In this paper we introduce quantitative extensions and study nested word series which assign to nested words elements of a semiring. We show that regular nested word series coincide with series definable in weighted logics as introduced by Droste and Gastin. For this, we establish a connection between nested words and series-parallel-biposets. Applying our result, we obtain a characterization of algebraic formal power series in terms of weighted logics. This generalizes a result of Lautemann, Schwentick and Thérien on context-free languages.

1 Introduction

Model checking of finite state systems has become an established method for automatic hardware and software verification and led to numerous verification programs used in industrial application. In order to verify recursive programs, it is necessary to model them as pushdown systems rather than finite automata. This has motivated Alur and Madhusudan [2,3] to define the classes of nested word languages and visibly pushdown languages which is a proper subclass of the class of context-free languages and exceeds the regular languages. These classes gained much interest and set a starting point for a new research field (see e.g. [1,4] among many others).

The goal of this paper will be: 1. to introduce a quantitative automaton model and a quantitative logic for nested words that are equally expressive, 2. to establish a connection between nested words and series-parallel-biposets which were studied by Ésik and Németh [11] and others, 3. to give a characterization of the important class of algebraic formal power series by means of weighted logics.

In order to be able to model quantitative aspects, extensions of existing models were investigated, such as weighted automata or probabilistic pushdown automata. In this paper we introduce and investigate weighted nested word automata which we propose as a quantitative model for sequential programs with recursive procedure calls. Due to the fact that we define them over arbitrary semirings they are very flexible and can model, for example, probabilistic or stochastic systems. As the first main result of this paper we characterize their expressiveness using weighted logics, generalizing a result of Alur and Madhusudan. Weighted logics were introduced by Droste and Gastin [7].

---

* Supported by the GK 446/3 of the German Research Foundation.

L. Aceto et al. (Eds.): ICALP 2008, Part II, LNCS 5126, pp. 221–232, 2008.

They enriched the classical language of monadic second-order logic with values from a semiring in order to add quantitative expressiveness. For example one may now express how often a certain property holds, how much execution time a process needs or how reliable it is. The result of Droste and Gastin has been extended to infinite words, trees, texts, pictures and traces [9,10,17,16,18]. Moreover, a restriction of Łukasiewicz multi-valued logic coincides with this weighted logics [19].

For our result we establish a new connection between series-parallel-biposets and nested words. The class of sp-biposets forms the free bisemigroup which was investigated by Hashiguchi et al. (e.g. [12]). Moreover, a language theory for series-parallel-biposets was developed by Ésik and Németh [11]. We anticipate that the connection between nested words and sp-biposets can be utilized to obtain further results. We give an indication in the conclusions at the end of the paper.

Using projections of nested word series and applying the above mentioned result, we obtain the second main result, a characterization of algebraic formal power series. These form an important generalization of context-free languages. Algebraic formal power series were already considered initially by Chomsky and Schützenberger [5] and have since been intensively studied by Kuich and others. For a survey see [13] or [14]. Here we are able to give a characterization of algebraic formal power series in terms of weighted logics, generalizing a result of Lautemann, Schwentick and Thérien [15] on context-free languages.

2   Weighted Automata on Nested Words

Definition 2.1 (Alur & Madhusudan [3]). *A* nested word *(over a finite alphabet Δ) is a pair (w, ν) such that $w \in \Delta^+$ and ν is a binary* nesting relation *on $\{1, \ldots, |w|\}$ that satisfies (a) if $\nu(i,j)$, then $i < j$, (b) if $\nu(i,j)$ and $\nu(i,j')$, then $j = j'$, (c) if $\nu(i,j)$ and $\nu(i',j)$, then $i = i'$ and (d) if $\nu(i,j)$ and $\nu(i',j')$ and $i < i'$, then $j < i'$ or $j' < j$.*
If $\nu(i,j)$ we say i is a call position *and j is a* return position.

We collect all nested words over Δ in $\mathrm{NW}(\Delta)$. Let $nw = (w, \nu) \in \mathrm{NW}(\Delta)$ where $w = a_1 \ldots a_n$. The *factor* $nw[i,j]$ for $i \leq j$ is the restriction of nw to positions between i and j; more formally $nw[i,j] = (a_i \ldots a_j, \nu[i,j])$ where $\nu[i,j] = \{(k,l) \mid (k+i-1, l+i-1) \in \nu, 1 \leq k, l \leq j-i+1\}$.

Nested words have been introduced in order to model executions of recursive programs, as well as nested data structures such as XML documents. Here we model quantitative behavior of systems such as runtime or the probability of an execution of a randomized program. That is we assign to a nested word a quantity expressing, for example, runtime or probability.

Example 2.2.   1. Probabilistic automata have been used to model fault-tolerant systems or to model randomized programs. Consider the randomized recursive pseudo-procedure `bar` (see next page) where `flip(Y)` means flipping a fair coin Y. Consider furthermore the alphabet $\Delta = \{r, w, b, call, ret\}$ of atomic events which stand for read, write, beep, call and return. Then the nested word $nw = (w, \nu)$ defined by $w = r.call.r.b.ret.w.w.ret$ and $\nu = \{(2,5)\}$ models an execution of `bar`. We calculate the probability of the execution by multiplying the probability of each atomic action, i.e. $1 \cdot 1/2 \cdot 1 \cdot 1/2 \cdot 1/2 \cdot 1/2 \cdot 1/2 \cdot 1/2 = 1/64$.

2. As Alur and Madhusudan point out XML documents or bibtex databases can naturally be modeled as nested words where the nesting relation captures open and close tags [3]. Suppose we model bibtex databases as nested words. Then we may assign to a nested word e.g. the number of technical reports it stores.

To be as flexible as possible we take the quantities we assign to a nested word from a commutative semiring. A *commutative semiring* $\mathbb{K}$ is an algebraic structure $(\mathbb{K}, +, \cdot, 0, 1)$ such that $(\mathbb{K}, +, 0)$ and $(\mathbb{K}, \cdot, 1)$ are commutative monoids, multiplication distributes over addition and 0 is absorbing. For example the natural numbers $(\mathbb{N}, +, \cdot, 0, 1)$ form a commutative semiring. An important example is also the max-plus semiring $(\mathbb{Z} \cup \{-\infty\}, \max, +, -\infty, 0)$ which has been used to model real-time systems or discrete event systems.

```
proc bar(){
  read(x);
  flip(Y);if(Y==head)
    beep;
  else
    bar();
  flip(Y);while(Y==head)
    write(x);
    flip(Y);
  exit;}
```

This semiring possesses the property that any finitely generated submonoid of $(\mathbb{K}, +, 0)$ is finite. Such semirings are called *additively locally finite*. Another important example of an additively locally finite semiring is the probabilistic semiring $([0, 1], \max, \cdot, 0, 1)$. We call a semiring *locally finite* if any finitely generated subsemiring is finite. Examples include any Boolean algebra such as the trivial Boolean algebra $\mathbb{B} = (\{0, 1\}, \vee, \wedge, 0, 1)$ as well as $(\mathbb{R}_+ \cup \{\infty\}, \max, \min, 0, \infty)$ and the fuzzy semiring $([0, 1], \max, \min, 0, 1)$.

Definition 2.3. *A weighted nested word automaton (WNWA for short) is a quadruple* $\mathcal{A} = (Q, \iota, \delta, \kappa)$ *where* $\delta = (\delta_c, \delta_i, \delta_r)$ *such that*

1. *Q is a finite set of states,*
2. *$\delta_c, \delta_i : Q \times \Delta \times Q \to \mathbb{K}$ are the* call *and* internal *transition functions,*
3. *$\delta_r : Q \times Q \times \Delta \times Q \to \mathbb{K}$ is the* return *transition function,*
4. *$\iota, \kappa : Q \to \mathbb{K}$ are the* initial *and* final *distribution.*

A *run* of $\mathcal{A}$ on $nw = (a_1 \ldots a_n, \nu)$ is a *sequence of states* $r = (q_0, \ldots, q_n)$; we also write $r : q_0 \overset{nw}{\to} q_n$. The *weight* of r *at position* $1 \leq j \leq n$ is given by

$$\text{wgt}_{\mathcal{A}}(r, j) = \begin{cases} \delta_c(q_{j-1}, a_j, q_j) & \text{if } \nu(j, i) \text{ for some } j < i \leq n \\ \delta_r(q_{i-1}, q_{j-1}, a_j, q_j) & \text{if } \nu(i, j) \text{ for some } 1 \leq i < j \\ \delta_i(q_{j-1}, a_j, q_j) & \text{otherwise.} \end{cases}$$

The *weight* $\text{wgt}_{\mathcal{A}}(r)$ of r is defined by $\text{wgt}_{\mathcal{A}}(r) = \prod_{1 \leq j \leq n} \text{wgt}_{\mathcal{A}}(r, j)$ and the *behavior* $\|\mathcal{A}\| : \text{NW}(\Delta) \to \mathbb{K}$ of $\mathcal{A}$ is given by

$$\|\mathcal{A}\| (nw) = \sum_{q_0, q_n \in Q} \iota(q_0) \cdot \sum_{r: q_0 \overset{nw}{\to} q_n} \text{wgt}_{\mathcal{A}}(r) \cdot \kappa(q_n).$$

A function $S : \text{NW}(\Delta) \to \mathbb{K}$ is called a *nested word series*. As for formal power series we write (S, nw) for $S(nw)$. We define the *scalar multiplication* $\cdot$ and the *sum* $+$ pointwise, i.e. for $k \in \mathbb{K}$ and nested word series S_1, S_2 we let $(k \cdot S_1, nw) = k \cdot (S_1, nw)$

and $(S_1+S_2, nw) = (S_1, nw)+(S_2, nw)$ for all $nw \in \mathrm{NW}(\Delta)$. A series S is *regular* if there is a WNWA $\mathcal{A}$ with $\|\mathcal{A}\| = S$. For the Boolean semiring $\mathbb{B}$ WNWA are equivalent to (unweighted) nested word automata [3]. A language $L \subseteq \mathrm{NW}(\Delta)$ is recognized by a nested word automaton iff its characteristic function $\mathbb{1}_L : \mathrm{NW}(\Delta) \to \mathbb{B}$ is regular.

Example 2.4. The procedure `bar` of Example 2.2 can be modeled by a WNWA over $\mathbb{K} = ([0,1], \max, \cdot, 0, 1)$ with four states $\{q_1, \ldots, q_4\}$. The transitions (only the ones with weight $\neq 0$) are given as follows. We let $\iota(q_1) = 1$ and $\kappa(q_4) = 1$. Moreover,

$$\delta_i(q_1, r, q_2) = 1, \qquad \delta_i(q_2, b, q_3) = \delta_i(q_3, w, q_3) = \delta_i(q_3, ret, q_4) = 1/2,$$
$$\delta_c(q_2, call, q_1) = 1/2, \qquad \delta_r(q_2, q_3, ret, q_3) = 1/2.$$

Intuitively, each of the states corresponds to a line in the procedure `bar` which is the next to be executed: q_1 to line 2, q_2 to line 3, q_3 to line 7 and q_4 is only reached at the end of an execution. Consider the nested word nw of Example 2.2(a). There is exactly one run $r : q_1 \overset{nw}{\to} q_4$ with $\mathrm{wgt}(r) \neq 0$. Observe that the automaton assigns $1/64$ to nw.

3   Weighted Logics for Nested Words

We now introduce a formalism for specifying nested word series. For this we interpret a nested word $nw = (a_1 \ldots a_n, \nu)$ as a relational structure consisting of the domain $\mathrm{dom}(nw) = \{1, \ldots, n\}$ together with the unary relations $\mathrm{Lab}_a = \{i \in \mathrm{dom}(w) \mid a_i = a\}$ for all $a \in \Delta$, the binary relation ν and the usual $\leq$ relation on $\mathrm{dom}(nw)$.

First, we recall classical MSO logic. Formulae of MSO are inductively built from the atomic formulae $x = y$, $\mathrm{Lab}_a(x)$, $x \leq y$, $\nu(x,y)$, $x \in X$ using negation $\neg$, the connective $\vee$ and the quantifications $\exists x.$ and $\exists X.$ (where x, y range over individuals and X over sets). Let $\varphi \in \mathrm{MSO}$, let $\mathrm{Free}(\varphi)$ be the set of free variables, let $\mathcal{V} \supseteq \mathrm{Free}(\varphi)$ be a finite set of variables and let γ be a $(\mathcal{V}, nw)$-*assignment* (assigning variables of $\mathcal{V}$ an element or a set of $\mathrm{dom}(nw)$, resp.). For $i \in \mathrm{dom}(nw)$ and $T \subseteq \mathrm{dom}(nw)$ we denote by $\gamma[x \to i]$ and $\gamma[X \to T]$ the $(\mathcal{V} \cup \{x\}, nw)$-assignment (resp. $(\mathcal{V} \cup \{X\}, nw)$-assignment) which equals γ on $\mathcal{V} \setminus \{x\}$ (resp. $\mathcal{V} \setminus \{X\}$) and assumes i for x (resp. T for X). We let $\mathscr{L}_{\mathcal{V}}(\varphi) = \{(nw, \gamma) \mid (nw, \gamma) \models \varphi\}$ and $\mathscr{L}(\varphi) = \mathscr{L}_{\mathrm{Free}(\varphi)}(\varphi)$.

Let $Z \subseteq \mathrm{MSO}$. A language $L \subseteq \mathrm{NW}(\Delta)$ is Z-*definable* if $L = \mathscr{L}(\varphi)$ for a sentence $\varphi \in Z$. First-order formulae, that is formulae containing only quantification over individuals, are collected in FO. Monadic second-order logic and (unweighted) nested word automata turned out to be equally expressive (Alur and Madhusudan [3, 2]).

We now turn to weighted MSO logics as introduced in [7]. Formulae of $\mathrm{MSO}(\mathbb{K})$ are built from the atomic formulae k (for $k \in \mathbb{K}$), $x = y$, $\mathrm{Lab}_a(x)$, $x \leq y$, $\nu(x,y)$, $x \in X$, $\neg(x = y)$, $\neg \mathrm{Lab}_a(x)$, $\neg(x \leq y)$, $\neg(\nu(x,y))$, $\neg(x \in X)$ using the connectives $\vee, \wedge$ and the quantifications $\exists x., \exists X., \forall x., \forall X.$. Let $\varphi \in \mathrm{MSO}(\mathbb{K})$ and $\mathrm{Free}(\varphi) \subseteq \mathcal{V}$. The weighted semantics $\llbracket \varphi \rrbracket_{\mathcal{V}}$ of φ is a function which assigns to each pair (nw, γ) an element of $\mathbb{K}$. For $k \in \mathbb{K}$ we put $\llbracket k \rrbracket_{\mathcal{V}}(nw, \gamma) = k$. For all other atomic formulae φ we let $\llbracket \varphi \rrbracket_{\mathcal{V}}$ be the characteristic function $\mathbb{1}_{\mathscr{L}_{\mathcal{V}}(\varphi)}$. Moreover:

$$\llbracket \varphi \vee \psi \rrbracket_{\mathcal{V}}(nw, \gamma) \quad = \quad \llbracket \varphi \rrbracket_{\mathcal{V}}(nw, \gamma) + \llbracket \psi \rrbracket_{\mathcal{V}}(nw, \gamma),$$
$$\llbracket \varphi \wedge \psi \rrbracket_{\mathcal{V}}(nw, \gamma) \quad = \quad \llbracket \varphi \rrbracket_{\mathcal{V}}(nw, \gamma) \cdot \llbracket \psi \rrbracket_{\mathcal{V}}(nw, \gamma),$$

$$[\![\exists x.\varphi]\!]_\mathcal{V}(nw,\gamma) \quad = \quad \sum\nolimits_{i\in\text{dom}(nw)}[\![\varphi]\!]_{\mathcal{V}\cup\{x\}}(nw,\gamma[x\to i]),$$

$$[\![\exists X.\varphi]\!]_\mathcal{V}(nw,\gamma) \quad = \quad \sum\nolimits_{T\subseteq\text{dom}(nw)}[\![\varphi]\!]_{\mathcal{V}\cup\{X\}}(nw,\gamma[X\to T]),$$

$$[\![\forall x.\varphi]\!]_\mathcal{V}(nw,\gamma) \quad = \quad \prod\nolimits_{i\in\text{dom}(nw)}[\![\varphi]\!]_{\mathcal{V}\cup\{x\}}(nw,\gamma[x\to i]),$$

$$[\![\forall X.\varphi]\!]_\mathcal{V}(nw,\gamma) \quad = \quad \prod\nolimits_{T\subseteq\text{dom}(nw)}[\![\varphi]\!]_{\mathcal{V}\cup\{X\}}(nw,\gamma[X\to T]).$$

In the following, we shortly write $[\![\varphi]\!]$ for $[\![\varphi]\!]_{\text{Free}(\varphi)}$.

Remark. A formula $\varphi \in \text{MSO}(\mathbb{K})$ which does not contain a subformula $k \in \mathbb{K}$ can also be interpreted as an unweighted formula. Moreover, if $\mathbb{K}$ is the Boolean semiring $\mathbb{B}$, then it is easy to see that weighted logics and classical MSO logic coincide. In this case k is either 0 (false) or 1 (true).

Example 3.1. 1. As in Example 2.2 suppose we model bibtex databases as nested words. Moreover, assume that tecrep $\in \Delta$ marks the beginning of an entry containing a technical report. Now, let $\mathbb{K} = \mathbb{N}$ be the semiring of the natural numbers. Then $[\![\exists x. \text{Lab}_{\text{tecrep}}(x)]\!](nw)$ counts the number of technical reports of the bibtex database modeled by nw.

2. The *nesting depth* of a position of a nested word is the number of open call positions (i.e. where the corresponding return position has not occurred yet). The nesting depth of a nested word is the maximum over all positions. Let $\mathbb{K} = (\mathbb{Z} \cup \{-\infty\}, \max, +, -\infty, 0)$. Define

$$\text{open}(x) := \forall y.(y \leq x \wedge \text{call}(y)) \to 1 \vee (y \leq x \wedge \text{return}(y)) \to -1$$

where $\text{call}(x) := \exists y.\nu(x,y)$ and $\text{return}(x) := \exists y.\nu(y,x)$ (the precise definition of $\to$ is given below). Then $[\![\exists x.\text{open}(x)]\!]$ assigns to a nested word its nesting depth.

Let $Z \subseteq \text{MSO}(\mathbb{K})$. A series $S : \text{NW}(\Delta) \to \mathbb{K}$ is *Z-definable* if $S - [\![\varphi]\!]$ for a sentence $\varphi \in Z$. Already for words, examples [7] show that unrestricted application of universal quantification does not preserve regularity as the resulting series may grow to fast. Therefore we now define different fragments of $\text{MSO}(\mathbb{K})$. For the fragment RMSO for words, which we do not define here, Droste and Gastin [7] showed that a formal power series is regular iff it is RMSO-definable. Unfortunately, RMSO is a semantic restriction and it is not clear if it can be decided. In order to have a decidable fragment, we syntactically define the fragment sRMSO. For this we follow the approach of [8]. Given a classical MSO formula φ we assign to it formulae φ^+ and φ^- such that $[\![\varphi^+]\!] = \mathbb{1}_{\mathscr{L}(\varphi)}$ and $[\![\varphi^-]\!] = \mathbb{1}_{\mathscr{L}(\neg\varphi)}$. The problem that arises is that by definition, e.g. $\vee$ is interpreted as $+$. Hence, for a formula $\varphi \vee \psi$ one might not end up with a sum which equals 0 or 1. One possible solution is to evaluate φ only if ψ evaluates to 0. Similar for $\exists x.$ and $\exists X.$. This leads to the following definition:

1. If φ is of the form $x = y$, $\text{Lab}_a(x)$, $x \leq y$, $\nu(x,y)$, then $\varphi^+ = \varphi$ and $\varphi^- = \neg\varphi$.
2. If $\varphi = \neg\psi$ then $\varphi^+ = \psi^-$ and $\varphi^- = \psi^+$.
3. If $\varphi = \psi \vee \psi'$, then $\varphi^+ = \psi^+ \vee (\psi^- \wedge \psi'^+)$ and $\varphi^- = \psi^- \wedge \psi'^-$.
4. If $\varphi = \exists x.\psi$, then $\varphi^+ = \exists x.\psi^+ \wedge \forall y.(y < x \wedge \psi(y))^-$ and $\varphi^- = \forall x.\psi^-$.

In order to disambiguate set quantification, we have to define a linear order on the subsets of the domain of a nested word or equivalently on nested words (of fixed length) over the alphabet $\{0, 1\}$. We take the lexicographic order $<$ which is given by the following formula: $X < Y := \exists y.y \in Y \wedge \neg y \in X \wedge \forall z.[z < y \rightarrow (z \in X \leftrightarrow z \in Y)]^+$. We proceed:

5. If $\varphi = \exists X.\psi$, then $\varphi^+ = \exists X.\psi^+ \wedge \forall Y.(Y < X \wedge \psi(Y))^-$ and $\varphi^- = \forall X.\psi^-$.

Formulae of the form φ^+ or φ^- for some $\varphi \in \mathrm{MSO}$ are called *syntactically unambiguous*. In the following, we shortly write $\varphi \rightarrow \psi$ for $\varphi^- \vee (\varphi^+ \wedge \psi)$ for any two weighted formulae φ, ψ where φ does not contain subformulae of form k ($k \in \mathbb{K}$).

We define aUMSO, the collection of *almost unambiguous* formulae, to be the smallest subset of $\mathrm{MSO}(\mathbb{K})$ containing all constants k ($k \in \mathbb{K}$) and all syntactically unambiguous formulae which is closed under conjunction and disjunction.

Definition 3.2. *A weighted formula φ is in* sRMSO *(syntactically restricted MSO) if:*
1. Whenever it contains a subformula $\forall X.\psi$, then ψ is syntactically unambiguous.
2. Whenever it contains a subformula $\forall x.\psi$, then $\psi \in$ aUMSO.

Let now wUMSO, the collection of *weakly unambiguous* formulae, be the smallest subset of $\mathrm{MSO}(\mathbb{K})$ containing all constants k ($k \in \mathbb{K}$) and all syntactically unambiguous formulae which is closed under conjunction, disjunction and existential quantification.

Definition 3.3. *A weighted formula φ is in* swRMSO *(syntactically weakly restricted MSO) if: 1. Whenever it contains a subformula $\forall X.\psi$, then ψ is syntactically unambiguous. 2. Whenever it contains a subformula $\forall x.\psi$, then $\psi \in$ wUMSO.*

Clearly, sRMSO $\subset$ swRMSO $\subset$ MSO($\mathbb{K}$). The first main result of this paper is the characterization of regular nested word series using weighted logics.

Theorem 3.4. *Let $\mathbb{K}$ be a commutative semiring and let $S : \mathrm{NW}(\Delta) \rightarrow \mathbb{K}$. Then:*

(a) The series S is regular iff S is sRMSO*-definable.*
(b) If $\mathbb{K}$ is additively locally finite, then S is regular iff S is swRMSO*-definable.*
(c) If $\mathbb{K}$ is locally finite, then S is regular iff S is MSO*-definable.*

We prove the result in the next section by interpreting nested words in sp-biposets, which we first investigate. The results are interesting in their own rights.

4 Nested Words and SP-Biposets

A bisemigroup is a set together with two associative operations. Several authors investigated the free bisemigroup as a fundamental, two-dimensional extension of classical automaton theory, see e.g. Ésik and Németh [11] and Hashiguchi et al. (e.g. [12]). Ésik and Németh considered as a representation for the free bisemigroup the so-called *sp-biposets*, a certain class of biposets. A Δ-*labeled biposet* is a finite nonempty set V of vertices equipped with two partial orders $\leq_h$ and $\leq_v$ and a labeling function $\lambda : V \rightarrow \Delta$. Let $p_1 = (V_1, \lambda_1, \leq_h^1, \leq_v^1), p_2 = (V_2, \lambda_2, \leq_h^2, \leq_v^2)$ be biposets. We define $p_1 \circ_h p_2 = (V_1 \uplus V_2, \lambda_1 \cup \lambda_2, \leq_h, \leq_v)$ by letting $\leq_h = \leq_h^1 \cup \leq_h^2 \cup (V_1 \times V_2)$

and $\leq_v = \leq_v^1 \cup \leq_v^2$. The operation $\circ_v$ is defined dually. Clearly, both products are associative. The set of biposets generated from the singletons by $\circ_h$ and $\circ_v$ is denoted $\mathrm{SPB}(\Delta)$. Its elements are called *sp-biposets*.

Weighted parenthesizing automata operating on sp-biposets generalizing the automata of Ésik and Németh [11] were defined in [16].

Definition 4.1. *A weighted parenthesizing automaton (WPA for short) over Δ is a tuple $\mathcal{P} = (\mathcal{H}, \mathcal{V}, \Omega, \mu, \mu_{op}, \mu_{cl}, \lambda, \gamma)$ where*

1. *$\mathcal{H}, \mathcal{V}$ are finite disjoint sets of horizontal and vertical states, respectively,*
2. *Ω is a finite set of parentheses (to help the intuition we write $(_s$ or $)_s$ for $s \in \Omega$),*
3. *$\mu : (\mathcal{H} \times \Delta \times \mathcal{H}) \cup (\mathcal{V} \times \Delta \times \mathcal{V}) \to \mathbb{K}$ is the transition function,*
4. *$\mu_{op}, \mu_{cl} : (\mathcal{H} \times \Omega \times \mathcal{V}) \cup (\mathcal{V} \times \Omega \times \mathcal{H}) \to \mathbb{K}$ are the opening and closing parenthesizing functions and*
5. *$\lambda, \gamma : \mathcal{H} \cup \mathcal{V} \to \mathbb{K}$ are the initial and final weight functions.*

A *run* r of $\mathcal{P}$ is a word over the alphabet $(\mathcal{H} \cup \mathcal{V}) \times (\Delta \cup \Omega) \times (\mathcal{H} \cup \mathcal{V})$ defined inductively as follows. We also define its *label* $\mathrm{lab}(r)$, its *weight* $\mathrm{wgt}_{\mathcal{P}}(r)$, its *initial state* $\mathrm{init}(r)$ and its *final state* $\mathrm{fin}(r)$.

1. (q_1, a, q_2) is a run for all $(q_1, q_2) \in (\mathcal{H} \times \mathcal{H}) \cup (\mathcal{V} \times \mathcal{V})$ and $a \in \Delta$. We set

$$\mathrm{lab}((q_1, a, q_2)) = a \in \mathrm{SPB}(\Delta), \qquad \mathrm{wgt}_{\mathcal{P}}((q_1, a, q_2)) = \mu(q_1, a, q_2),$$

$$\mathrm{init}((q_1, a, q_2)) = q_1 \text{ and } \mathrm{fin}((q_1, a, q_2)) = q_2.$$

2. Let r_1 and r_2 be runs such that $\mathrm{fin}(r_1) = \mathrm{init}(r_2) \in \mathcal{H}$ (resp. $\mathcal{V}$). Then $r = r_1 r_2$ is a run having

$$\mathrm{lab}(r) = \mathrm{lab}(r_1) \circ_h \mathrm{lab}(r_2) \quad (\text{resp. } \mathrm{lab}(r) = \mathrm{lab}(r_1) \circ_v \mathrm{lab}(r_2)),$$

$$\mathrm{wgt}_{\mathcal{P}}(r) = \mathrm{wgt}_{\mathcal{P}}(r_1) \cdot \mathrm{wgt}_{\mathcal{P}}(r_2), \mathrm{init}(r) = \mathrm{init}(r_1) \text{ and } \mathrm{fin}(r) = \mathrm{fin}(r_2).$$

3. Let r be a run resulting from 2. such that $\mathrm{fin}(r), \mathrm{init}(r) \in \mathcal{H}$ (resp. $\mathcal{V}$). Let $q_1, q_2 \in \mathcal{V}$ (resp. $\mathcal{H}$) and $s \in \Omega$. Then $r' = (q_1, (_s, \mathrm{init}(r)) \, r \, (\mathrm{fin}(r),)_s, q_2)$ is a run. We set

$$\mathrm{lab}(r') = \mathrm{lab}(r), \mathrm{wgt}_{\mathcal{P}}(r') = \mu_{op}((q_1, (_s, \mathrm{init}(r))) \cdot \mathrm{wgt}_{\mathcal{P}}(r) \cdot \mu_{cl}((\mathrm{fin}(r),)_s, q_2)),$$

$$\mathrm{init}(r') = q_1 \text{ and } \mathrm{fin}(r') = q_2.$$

Let $p \in \mathrm{SPB}(\Delta)$. If r is a run of $\mathcal{P}$ with $\mathrm{lab}(r) = p$, $\mathrm{init}(r) = q_1$, $\mathrm{fin}(r) = q_2$, we write $r : q_1 \xrightarrow{p} q_2$. Since we do not allow repeated application of rule 3, there are only finitely many runs with label p. The behavior of $\mathcal{P}$ is a function $\|\mathcal{P}\| : \mathrm{SPB}(\Delta) \to \mathbb{K}$ with

$$(\|\mathcal{P}\|, p) = \sum_{q_1, q_2 \in \mathcal{H} \cup \mathcal{V}} \lambda(q_1) \cdot \sum_{r : q_1 \xrightarrow{p} q_2} \mathrm{wgt}_{\mathcal{P}}(r) \cdot \gamma(q_2).$$

An sp-biposet series, i.e. a function $S : \mathrm{SPB}(\Delta) \to \mathbb{K}$, is *regular* if there is a WPA $\mathcal{P}$ such that $\|\mathcal{P}\| = S$. For sp-biposets, $\mathrm{MSO}(\mathbb{K})$ can be defined similar as for nested words. Moreover, for any sp-biposet $p = (V, \lambda, \leq_h, \leq_v)$ the union $\leq := \leq_h \cup \leq_v$ gives a linear order [11]. Using this linear order, we can define syntactically unambiguous formulae as for nested words and then sRMSO and swRMSO. From Theorems 6.3 and 5.6 of [16] we obtain (here without proof due to space constraints) the following result which generalizes the Büchi-type result of Ésik and Németh on the coincidence of MSO-definable and regular languages of sp-biposets [11].

Theorem 4.2. *Let* $\mathbb{K}$ *be a commutative semiring and let* $S : \mathrm{SPB}(\Delta) \to \mathbb{K}$. *Then:*

(a) *The series* S *is regular iff* S *is* sRMSO-*definable.*
(b) *If* $\mathbb{K}$ *is additively locally finite, then* S *is regular iff* S *is* swRMSO-*definable.*
(c) *If* $\mathbb{K}$ *is locally finite, then* S *is regular iff* S *is* MSO-*definable.*

We note that under the assumptions of Theorem 4.2, given an sRMSO (resp. swRMSO, resp. MSO) formula φ we can effectively construct a WPA $\mathcal{P}$ such that $\llbracket \varphi \rrbracket = \|\mathcal{P}\|$ (and vice versa).

We will now derive similar results for nested words as for sp-biposets by interpreting the different structures within each other. For this, we utilize definable transductions as introduced by Courcelle [6]. Let σ_1 and $\sigma_2 = ((R_i)_{i \in I}, \rho)$ be two relational signatures where $\rho : I \to \mathbb{N}^+$ assigns to each relation symbol R_i a positive arity and let $\mathcal{C}_1$ and $\mathcal{C}_2$ be classes of finite σ_1- and σ_2-structures, respectively.

Definition 4.3. *A* (σ_1, σ_2)-*1-copying definition scheme* (*without parameter*) *is a tuple* $\mathcal{D} = (\vartheta, \delta, (\varphi_i)_{i \in I})$ *of formulae in* $\mathrm{MSO}(\sigma_1)$ *such that* $\mathrm{Free}(\vartheta) = \emptyset$, $\mathrm{Free}(\delta) = \{x_1\}$ *and* $\mathrm{Free}(\varphi_i) = \{x_1, \ldots, x_{\rho(i)}\}$.

Let $\mathcal{D}$ be a (σ_1, σ_2)-1-copying definition scheme. For each $s_1 \in \mathcal{C}_1$ such that $s_1 \models \vartheta$ define the σ_2-structure $\mathbf{def}_{\mathcal{D}}(s_1) = s_2 = (\mathrm{dom}(s_2), (R_i^{s_2})_{i \in I})$ where $\mathrm{dom}(s_2) = \{v \in \mathrm{dom}(s_1) \mid (s_1, v) \models \varphi\}$ and $R_i^{s_2} = \{(v_1, \ldots, v_r) \in \mathrm{dom}(s_2)^r \mid (s_1, v_1, \ldots, v_r) \models \varphi_i\}$ with $r = \rho(i)$. Now a partial function $\Phi : \mathcal{C}_1 \to \mathcal{C}_2$ is *definable* if there is a definition scheme $\mathcal{D}$ such that $\Phi = \mathbf{def}_{\mathcal{D}}$.

Clearly, $\mathrm{MSO}(\mathbb{K})$ can be defined for $\mathcal{C}_1$ and $\mathcal{C}_2$ along the lines as for nested words. In order to disambiguate a formula, we need a linear order on each $s \in \mathcal{C}_1$ (resp. $\mathcal{C}_2$). For the next proposition we therefore assume that there are binary relation symbols $\leq_1 \in \sigma_1$ and $\leq_2 \in \sigma_2$ such that the interpretation of $\leq_i$ in s is a linear order for any $s \in \mathcal{C}_i$ ($i = 1, 2$). Using these linear orders, we can define syntactically unambiguous formulae and then sRMSO($\mathbb{K}$) and swRMSO($\mathbb{K}$) over σ_1 and σ_2. Now, let $\Phi : \mathcal{C}_1 \to \mathcal{C}_2$ be a partial function with domain $\mathrm{dom}(\Phi)$ and let $S : \mathcal{C}_2 \to \mathbb{K}$. Define $\Phi^{-1}(S)$ by letting $(\Phi^{-1}(S), s_1) = (S, \Phi(s_1))$ for all $s_1 \in \mathrm{dom}(\Phi)$ and $(\Phi^{-1}(S), s_1) = 0$ otherwise.

Proposition 4.4. *Let* $\Phi : \mathcal{C}_1 \to \mathcal{C}_2$ *be a definable function. If* $S : \mathcal{C}_2 \to \mathbb{K}$ *is* MSO-*definable* (*resp.* sRMSO-*definable,* swRMSO-*definable*), *then so is* $\Phi^{-1}(S)$.

Remark. It suffices that $\leq_i$ can be defined by a formula $\varphi_i(x, y)$ such that $\llbracket \varphi_i \rrbracket = \mathbb{1}_{\mathscr{L}(\varphi_i)}$. This is the case for sp-biposets where we let $\varphi(x, y) = x \leq_v y \vee x \leq_h y$.

Now we define two embeddings of nested words into sp-biposets $\Phi_v, \Phi_h : \mathrm{NW}(\Delta) \to \mathrm{SPB}(\Delta)$ as follows. Let $nw = (w, \nu) \in \mathrm{NW}(\Delta)$ where $w = a_1 \ldots a_n$. If $\nu = \emptyset$, then let $\Phi_h(nw) = a_1 \circ_h \ldots \circ_h a_n$ and $\Phi_v(nw) = a_1 \circ_v \ldots \circ_v a_n$. If $\nu \neq \emptyset$, let i be the minimal call position and j the corresponding return position. Let $nw' = nw[i+1, j-1]$ and $nw'' = nw[j+1, n]$. Suppose for the moment that $i + 1 \leq j - 1$ and $j + 1 \leq n$. Then

$$\Phi_h(nw) = a_1 \circ_h \ldots \circ_h a_{i-1} \circ_h (a_i \circ_v \Phi_v(nw') \circ_v a_j) \circ_h \Phi_h(nw''),$$
$$\Phi_v(nw) = a_1 \circ_v \ldots \circ_v a_{i-1} \circ_v (a_i \circ_h \Phi_h(nw') \circ_h a_j) \circ_v \Phi_v(nw'').$$

If $i + 1 > j - 1$ or $j + 1 > n$, we just ignore the terms $\Phi_h(nw')$, $\Phi_h(nw'')$, $\Phi_v(nw')$ and $\Phi_v(nw'')$, respectively, in the definition above. We identify the domain of $\Phi(nw)$ with $\{1, \ldots, n\}$ in the obvious way. Observe that Φ_h and Φ_v are injective.

Lemma 4.5. *Let* $nw = (a_1 \ldots a_n, \nu) \in \mathrm{NW}(\Delta)$, $\Phi_h(nw) = (\{1, \ldots, n\}, \lambda, \leq_h, \leq_v)$. *Moreover, let* $1 \leq k \leq i \leq j \leq l \leq n$ *with* $(k, l) \in \nu$ *such that there is no* $(k', l') \in \nu$ *with* $k < k' \leq i \leq j \leq l' < l$. *Then* $i \leq_h j$ *iff* k *has even nesting depth (cf. Ex. 3.1).*

Recall that for an sp-biposet $p = (V, \lambda, \leq_h, \leq_v)$ we let $\leq := \leq_h \cup \leq_v$. A *clan* of p is an interval $[i, j] = \{k \in V \mid i \leq k \leq j\}$ which can not be distinguished from outside, i.e. if for all $i \leq k, k' \leq j$ and $l < i$ or $j < l$ we have $k \leq_v l$ iff $k' \leq_v l$ and $k \leq_h l$ iff $k' \leq_h l$ and $l \leq_v k$ iff $l \leq_v k'$. A *prime clan* is a clan that does not overlap with any other, i.e. there is no clan $[k, l]$ such that $k < i < l < j$ or $i < k < j < l$.

Lemma 4.6. *Let* $nw = (a_1 \ldots a_n, \nu) \in \mathrm{NW}(\Delta)$, $\Phi_h(nw) = (\{1, \ldots, n\}, \lambda, \leq_h, \leq_v)$. *Then* $(i, j) \in \nu$ *iff* $i < j$, $[i, j]$ *is a prime clan and not* $i = 1$, $j = n$ *and* $1 \leq_h n$.

Since the conditions of Lemma 4.6 and Lemma 4.5 can be expressed in MSO (actually the latter can be expressed in FO), we obtain:

Corollary 4.7. *The (partial) functions* $\Phi_h, \Phi_v, \Phi_h^{-1}, \Phi_v^{-1}$ *are definable.*

We will now show that not only the formulae can be translated, but that WPA can simulate WNWA and vice versa. More precisely:

Proposition 4.8. *A series* $S : \mathrm{NW}(\Delta) \to \mathbb{K}$ *is regular iff* $(\Phi_h^{-1})^{-1}(S)$ *is regular.*

Proof (Sketch). *(If).* Let $\mathcal{P} = (\mathcal{H}, \mathcal{V}, \Omega, \mu, \mu_{\mathrm{op}}, \mu_{\mathrm{cl}}, \lambda, \gamma)$ be a WPA. There is a WNWA $\mathcal{A} = (Q, \iota, \delta, \kappa)$ with state space $Q = (\mathcal{H} \uplus \mathcal{V}) \times (\Omega \uplus \{i\})$ such that $\|\mathcal{A}\| = \Phi_h^{-1}(\|\mathcal{P}\|)$. Intuitively, in the first component one simulates the states of the WPA and in the second component one stores the most recent open bracket. This has to be updated when reading a return position using the look-back ability of the nested word automaton.

(Only if). Let $\mathcal{A} = (Q, \iota, \delta, \kappa)$ be a WNWA. There is a WPA where $\mathcal{H}, \mathcal{V}$ are disjoint copies of $Q \times (\{c, i\} \uplus \Delta)$ and $\Omega = Q$ such that $(\|\mathcal{P}\|, \Phi_h(nw)) = (\|\mathcal{A}\|, nw)$ for all $nw \in \mathrm{NW}(\Delta)$. Intuitively, in the first component one simulates the states of the WNWA, in the second component one either selects if the next transition is a call or an internal transition or one stores the letter to simulate a return transition in the next bracket. Look-back behavior is simulated storing a state in the opening bracket and closing it at the appropriate return position. $\square$

Theorem 3.4 now follows from Cor. 4.7, Prop. 4.4 and Thm. 4.2 together with Prop. 4.8.

We note that Proposition 4.8 also holds for non-commutative semirings. Moreover, we note that under the assumptions of Theorem 3.4, given an sRMSO (resp. swRMSO, resp. MSO) formula φ we can effectively construct a WNWA $\mathcal{A}$ such that $[\![\varphi]\!] = \|\mathcal{A}\|$ (and vice versa). Furthermore from Corollaries 5.7 and 5.8 of [16] we obtain

Corollary 4.9. *Let* $\mathbb{K}$ *be a computable field (resp. computable locally finite semiring). It is decidable whether two sentences* $\varphi, \psi \in$ *sRMSO (resp. MSO) satisfy* $[\![\varphi]\!] = [\![\psi]\!]$.

5 Algebraic Formal Power Series

In this section we consider algebraic formal power series and show that they arise as the projections of regular nested word series. Algebraic formal power series were already considered initially by Chomsky and Schützenberger [5] and have since been intensively studied by Kuich and others. For a survey see [13] or [14].

A formal power series is a mapping $S : \Delta^+ \to \mathbb{K}$. Given two formal power series S_1, S_2, their *Cauchy product*, denoted $S_1 \cdot S_2$, is given by $(S_1 \cdot S_2, w) = \sum_{w_1 w_2 = w} (S_1, w_1)(S_2, w_2)$. Let $\mathbb{1}_u$ denote the characteristic series of a word $u \in \Delta^+$.

Let $\mathcal{X} = \{X_1, \ldots, X_n\}$ be a set of variables. A polynomial P over $(\Delta \cup \mathcal{X})$ with values in $\mathbb{K}$ is a mapping $P : (\Delta \cup \mathcal{X})^+ \to \mathbb{K}$ such that its support is finite, i.e. the set $\mathrm{supp}(P) = \{w \in (\Delta \cup \mathcal{X})^+ \mid (P, w) \neq 0\}$ is finite. A collection of polynomials P_i for $i = 1, \ldots, n$ is called an *algebraic system* with variables in $\mathcal{X}$. The supports of the P_i's are, thus, finite sets consisting of words of the form $u_1 X_{i_1} \ldots u_k X_{i_k} u_{k+1}$ where $u_j \in \Delta^*$ and $X_{i_j} \in \mathcal{X}$. A collection $(S_i)_{1 \leq i \leq n}$ of formal power series $S_i : \Delta^+ \to \mathbb{K}$ is the solution of the algebraic system $(P_i)_{1 \leq i \leq n}$ if

$$S_i = \sum_{u_1 X_{i_1} \ldots u_k X_{i_k} u_{k+1} \in \mathrm{supp}(P_i)} P_i(u_1 X_{i_1} \ldots u_k X_{i_k} u_{k+1}) \mathbb{1}_{u_1} \cdot S_{i_1} \cdots \mathbb{1}_{u_k} \cdot S_{i_k} \cdot \mathbb{1}_{u_{k+1}}.$$

for $1 \leq i \leq n$. An algebraic system $(P_i)_{1 \leq i \leq n}$ is *proper* if $(P_i, X_j) = 0$ for all $1 \leq i, j \leq n$. Proper algebraic systems have a unique solution [14]. If a formal power series S is the (component of) a solution of a proper algebraic system, then S is called an *algebraic formal power series*. Over the trivial Boolean algebra $\mathbb{B}$ these series correspond exactly to the ε-free context-free languages (the bijection is given by supp).

We now consider the projections of regular nested word series and show that they give rise exactly to the algebraic series. The projection $\pi(nw)$ of a nested word $nw = (w, \nu)$ is simply the word w, i.e. we forget the nesting relation. This projection is canonically generalized to nested word series S by letting $\pi(S) : w \mapsto \sum_{w = \pi(nw)} (S, nw)$.

Proposition 5.1. *Let $S : \mathrm{NW}(\Delta) \to \mathbb{K}$ be a regular nested word series. Then $\pi(S)$ is an algebraic formal power series.*

Proof (Sketch). Let $\mathcal{A} = (Q, \iota, \delta, \kappa)$ be a WNWA such that $\|\mathcal{A}\| = S$ and let $q_1, q_2 \in Q$. We define polynomials $P_{q_1, q_2} : (\Delta \cup Q^2)^* \to \mathbb{K}$ as follows (in order to obtain a more compact presentation we consider here the empty word ε): For any $a, b, c, d \in \Delta$ and $q_1, \ldots, q_8 \in Q$ let $(P_{q_1, q_1}, \varepsilon) = 1$, $(P_{q_1, q_2}, a) = \delta_i(q_1, a, q_2)$, $(P_{q_1, q_2}, a(q_3, q_4)b) = \delta_i(q_1, a, q_3) \cdot \delta_i(q_4, b, q_2) + \delta_c(q_1, a, q_3) \cdot \delta_r(q_1, q_4, b, q_2)$, $(P_{q_1, q_2}, a(q_3, q_4)b(q_5, q_6)c) = \delta_c(q_1, a, q_3) \cdot \delta_r(q_1, q_4, b, q_5) \cdot \delta_i(q_6, c, q_2) + \delta_i(q_1, a, q_3) \cdot \delta_c(q_4, b, q_5) \cdot \delta_r(q_4, q_6, c, q_2)$ and $(P_{q_1, q_2}, a(q_3, q_4)b(q_5, q_6)c(q_7, q_8)d) = \delta_c(q_1, a, q_3) \cdot \delta_r(q_1, q_4, b, q_5) \cdot \delta_c(q_6, c, q_7) \cdot \delta_r(q_6, q_8, d, q_2)$. Finally, let $(P_{q_1, q_2}, w) = 0$ in any other case. This gives a so-called strict algebraic system with variables in Q^2 having a necessarily unique solution $(S_{q_1, q_2})_{q_1, q_2 \in Q}$ which is algebraic [14]. By induction on the length of w one gets

$$(S_{q_1, q_2}, w) = \sum_{\pi(nw) = w} \sum_{r : q_1 \overset{nw}{\to} q_2} \mathrm{wgt}_{\mathcal{A}}(r).$$

Now use that algebraic series are closed under sum and scalar multiplication. $\quad\square$

Our aim is to give a logical characterization for algebraic formal power series in the spirit of Lautemann, Schwentick and Thérien [15]. They showed that the context-free languages are precisely the languages which can be defined by sentences of the form $\exists \nu. \varphi$ where φ is a first-order formula and ν a binary predicate ranging over nesting relations. Let φ be a weighted MSO formula over nested words, $\mathrm{Free}(\varphi) \subseteq \mathcal{V}$, $w \in \Delta^+$ and γ a $(\mathcal{V}, w)$-assignment. We define the semantics $[\![\exists \nu. \varphi]\!]^{\mathrm{nest}} : \Delta^+ \to \mathbb{K}$ by letting

$$[\![\exists\nu.\varphi]\!]^{\text{nest}}(w,\gamma) = \sum_{\text{nesting rel. } \nu} [\![\varphi]\!]((w,\nu),\gamma).$$

Using Theorem 3.4, we may now reformulate Proposition 5.1.

Corollary 5.2. *Let $\varphi \in$ sRMSO be a sentence, then $[\![\exists\nu.\varphi]\!]^{\text{nest}}$ is an algebraic formal power series.*

Conversely, we can construct a nested word automaton $\mathcal{A}$ such that $\pi(\|\mathcal{A}\|)$ is the solution of a given algebraic system in Greibach normal form [14], i.e. we require $\text{supp}(P_i) \subseteq \Delta \cup \Delta\mathcal{X} \cup \Delta\mathcal{X}\mathcal{X}$. Elements of $\Delta\mathcal{X}\mathcal{X}$ produce call transitions, elements in $\Delta\mathcal{X}$ internal transitions and elements in Δ return transitions. Therefore we conclude:

Proposition 5.3. *Let $R : \Delta^+ \to \mathbb{K}$ be an algebraic formal power series. Then there is a regular nested word series $S : \text{NW}(\Delta) \to \mathbb{K}$ such that $\pi(S) = R$.*

Even stronger, we can restrict the series to first-order definable ones:

Proposition 5.4. *Let $S : \text{NW}(\Delta) \to \mathbb{K}$ be a regular nested word series. Then there is a first-order sentence $\varphi \in$ sRMSO such that $[\![\exists\nu.\varphi]\!]^{\text{nest}} = \pi(S)$.*

Proof (Hint). Let $\mathcal{A}$ be a weighted nested word automaton such that $\|\mathcal{A}\| = S$. Construct an algebraic system as in the proof of Proposition 5.1. This system has a form as required in the proof of [15]. The rest of the proof follows [15]; one has to ensure not to count weights twice and to obtain $\varphi \in$ sRMSO. $\square$

In the following theorem we summarize what we obtained in this section.

Theorem 5.5. *Let $\mathbb{K}$ be a commutative semiring and let $S : \Delta^+ \to \mathbb{K}$. Then the following are equivalent:*

1. *S is an algebraic formal power series.*
2. *$S = \pi(R)$ for some regular nested words series $R : \text{NW}(\Delta) \to \mathbb{K}$.*
3. *There is a first-order sentence $\varphi \in$ sRMSO such that $[\![\exists\nu.\varphi]\!]^{\text{nest}} = S$.*

Let $S : \{a\}^+ \to \mathbb{N}$ be an algebraic formal power series. One can show that $(S, a^n) \leq c^n$ for some constant c and all $n \in \mathbb{N}$. Thus, in item 3 of the last result we may not replace sRMSO by MSO since $([\![\forall x.\exists y.1]\!], a^n) = n^n$.

Again we note that given an algebraic $S : \Delta^+ \to \mathbb{K}$ we can effectively construct a sentence $\varphi \in \text{FO} \cap \text{sRMSO}$ such that $S = [\![\exists\nu.\varphi]\!]^{\text{nest}}$ and vice versa.

Conclusion and Consequences. We introduced a quantitative automaton model and a quantitative logic for nested words and showed that they are equally expressive. This generalizes the logical characterization in the unweighted case as given in [3]. Moreover, we established a new connection between nested words and sp-biposets. Presumably, the logical characterization of regular nested word series could also be obtained by structural induction. However, the connection between sp-biposets and nested words enables us to also obtain a generalization of the second main result of [15]. In this paper another logical characterization of the algebraic formal power series is given where quantification over nesting relations is now replaced by quantification over tree-definable orders. It is easy to see that there is a definable bijection between sp-biposets and the class of words together with a tree-definable order [11]. Thus, using the connection between nested words and sp-biposets we can conclude that every algebraic formal power series can be defined by a formula $\exists \nu. \varphi$ where $\varphi \in FO \cap sRMSO$ and ν ranges now over tree-definable orders. The converse can be shown by simulating weighted parenthesizing automata by weighted pushdown automata (as defined in [14]).

Acknowledgments. Thanks to M. Droste and A. Maletti for helpful comments, D. Kuske for pointing to [15], as well as to three anonymous referees for insightful remarks.

References

1. Alur, R., Kumar, V., Madhusudan, P., Viswanathan, M.: Congruences for visibly pushdown languages. In: Caires, L., Italiano, G.F., Monteiro, L., Palamidessi, C., Yung, M. (eds.) ICALP 2005. LNCS, vol. 3580, pp. 1102–1114. Springer, Heidelberg (2005)
2. Alur, R., Madhusudan, P.: Visibly pushdown languages. In: Proc. of the 36th STOC, Chicago, pp. 202–211. ACM, New York (2004)
3. Alur, R., Madhusudan, P.: Adding nesting structure to words. In: H. Ibarra, O., Dang, Z. (eds.) DLT 2006. LNCS, vol. 4036, pp. 1–13. Springer, Heidelberg (2006)
4. Arenas, M., Barceló, P., Libkin, L.: Regular languages of nested words: Fixed points, automata, and synchronization. In: Arge, L., Cachin, C., Jurdziński, T., Tarlecki, A. (eds.) ICALP 2007. LNCS, vol. 4596, pp. 888–900. Springer, Heidelberg (2007)
5. Chomsky, N., Schützenberger, M.P.: The algebraic theory of context-free languages. In: Computer Programming and Formal Systems, pp. 118–161. North-Holland, Amsterdam (1963)
6. Courcelle, B.: Monadic second-order definable graph transductions: a survey. Theoretical Computer Science 126, 53–75 (1994)
7. Droste, M., Gastin, P.: Weighted automata and weighted logics. Theoretical Computer Science 380, 69–86 (2007)
8. Droste, M., Gastin, P.: Weighted automata and weighted logics. In: Droste, M., Kuich, W., Vogler, H. (eds.) Handbook of Weighted Automata, ch.5. Springer, Heidelberg (to appear, 2008)
9. Droste, M., Rahonis, G.: Weighted automata and weighted logics on infinite words. In: H. Ibarra, O., Dang, Z. (eds.) DLT 2006. LNCS, vol. 4036, pp. 49–58. Springer, Heidelberg (2006)
10. Droste, M., Vogler, H.: Weighted tree automata and weighted logics. Theoretical Computer Science 366, 228–247 (2006)
11. Ésik, Z., Németh, Z.L.: Higher dimensional automata. Journal of Automata, Languages and Combinatorics 9(1), 3–29 (2004)
12. Hashiguchi, K., Ichihara, S., Jimbo, S.: Formal languages over free binoids. Journal of Automata, Languages and Combinatorics 5(3), 219–234 (2000)
13. Kuich, W.: Word, Language, Grammar. In: Handbook of Formal Languages, ch.9. Semirings and formal power series, vol. 1, pp. 609–677. Springer, Heidelberg (1997)
14. Kuich, W., Salomaa, A.: Semirings, Automata, Languages. Springer, Heidelberg (1986)
15. Lautemann, C., Schwentick, T., Thérien, D.: Logics for context-free languages. In: CSL 1994. LNCS, vol. 933, pp. 205–216. Springer, Heidelberg (1994)
16. Mathissen, C.: Definable transductions and weighted logics for texts. In: Harju, T., Karhumäki, J., Lepistö, A. (eds.) DLT 2007. LNCS, vol. 4588, pp. 324–336. Springer, Heidelberg (2007)
17. Mäurer, I.: Weighted picture automata and weighted logics. In: Durand, B., Thomas, W. (eds.) STACS 2006. LNCS, vol. 3884, pp. 313–324. Springer, Heidelberg (2006)
18. Meinecke, I.: Weighted logics for traces. In: Grigoriev, D., Harrison, J., Hirsch, E.A. (eds.) CSR 2006. LNCS, vol. 3967, pp. 235–246. Springer, Heidelberg (2006)
19. Schwarz, S.: Łukasiewicz logics and weighted logics over MV-semirings. Journal of Automata, Languages and Combinatorics 12(4), 485–499 (2007)

Tree Languages Defined in First-Order Logic with One Quantifier Alternation[*]

Mikołaj Bojańczyk[1] and Luc Segoufin[2]

[1] Warsaw University
[2] INRIA - LSV

Abstract. We study tree languages that can be defined in Δ_2. These are tree languages definable by a first-order formula whose quantifier prefix is $\exists^*\forall^*$, and simultaneously by a first-order formula whose quantifier prefix is $\forall^*\exists^*$, both formulas over the signature with the descendant relation. We provide an effective characterization of tree languages definable in Δ_2. This characterization is in terms of algebraic equations. Over words, the class of word languages definable in Δ_2 forms a robust class, which was given an effective algebraic characterization by Pin and Weil [11].

1 Introduction

We say a logic has a decidable characterization if the following decision problem is decidable: "given as input a finite automaton, decide if the recognized language can be defined using a formula of the logic". Representing the input language by a finite automaton is a reasonable choice, since many known logics (over words or trees) are captured by finite automata.

This type of problem has been successfully studied for word languages. Arguably best known is the result of McNaughton, Papert and Schützenberger [12, 9], which says that the following two conditions on a regular word language L are equivalent: a) L can be defined in first-order logic with order and label tests; b) the syntactic semigroup of L does not contain a non-trivial group. Since condition b) can be effectively tested, the above theorem gives a decidable characterization of first-order logic. This result demonstrates the importance of this type of work: a decidable characterization not only gives a better understanding of the logic in question, but it often reveals unexpected connections with algebraic concepts. During several decades of research, decidable characterizations have been found for fragments of first-order logic with restricted quantification and a large group of temporal logics, see [10] and [16] for references.

An important part of this research has been devoted to the quantifier alternation hierarchy, where each level counts the alterations between $\forall$ and $\exists$ quantifiers in a first-order formula in prenex normal form. Formulas that have $n-1$ alternations are called Σ_n if they begin with $\exists$, and Π_n if they begin with $\forall$. For instance, the word property "some position has label a" can be defined by a

---

[*] Work partially funded by the AutoMathA programme of the ESF, the PHC programme Polonium, and by the Polish government grant no. N206 008 32/0810.

Σ_1 formula $\exists x.\ a(x)$, while the language a^*ba^* can be defined by the Σ_2 formula $\exists x\forall y.\ b(x) \wedge (y \neq x \Rightarrow a(y))$.

A lot of attention has been devoted to analyzing the low levels of the quantifier alternation hierarchy. The two lowest levels are easy: a word language is definable in Σ_1 (resp. Π_1) if and only if it is closed under inserting (removing) letters. Both properties can be tested in polynomial time based on a recognizing automaton, or semigroup. However, just above Σ_1, Π_1, and even before we get to Σ_2, Π_2, we already find two important classes of languages. A fundamental result, due to Simon [14], says that a language is defined by a boolean combination of Σ_1 formulas if and only if its syntactic monoid is $\mathcal{J}$-trivial. Above the boolean combination of Σ_1, we find Δ_2, i.e. languages that can be defined simultaneously in Σ_2 and Π_2. As we will describe later on, this class turns out to be surprisingly robust and it is the focus of this paper. Another fundamental result, due to Pin and Weil [11], says that a regular language is in Δ_2 if and only if its syntactic monoid is in DA. The limit of our knowledge is level Σ_2: it is decidable if a language can be defined on level Σ_2 [1, 11], but there are no known decidable characterization for boolean combinations of Σ_2, for Δ_3, for Σ_3, and upwards.

For trees even less is known. No decidable characterization has been found for what is arguably the most important proper subclass of regular tree languages, first-order logic with the descendant relation, despite several attempts. Similarly open are chain logic and the temporal logics CTL, CTL* and PDL. However, there has been some recent progress. In [5], decidable characterizations were presented for some temporal logics, while Benedikt and Segoufin [2] characterized tree languages definable in first-order logic with the successor relation (but without the descendant relation).

This paper is part of a program to understand the expressive power of first-order logic on trees, and the quantifier alternation hierarchy in particular. The idea is to try to understand the low levels of the quantifier alternation hierarchy before taking on full first-order logic (which is contrary to the order in which word languages were analyzed). We focus on a signature that contains the ancestor order on nodes and label tests. In particular, there is no order between siblings. As shown in [3], there is a reasonable notion of concatenation hierarchy for tree languages that corresponds to the quantifier alternation hierarchy. Levels Σ_1 and Π_1 are as simple for trees as they are for words. A recent unpublished result [8] extends Simon's theorem to trees, by giving a decidable characterization of tree languages definable by a Boolean combination of Σ_1 formulas. There is no known characterization of tree languages definable in Σ_n for $n \geq 2$.

The contribution of this paper is a decidable characterization of tree languages definable in Δ_2, i.e. definable both in Σ_2 and Π_2. As we signaled above, for word languages the class Δ_2 is well studied and important, with numerous equivalent characterizations. Among them one can find [11, 15, 13, 7]: a) word languages that can be defined in the temporal logic with operators F and F^{-1}; b) word languages that can be defined by a first-order formula with two variables, but with unlimited quantifier alternations; c) word languages whose syntactic semigroup belongs to the semigroup variety DA; d) word languages recognized by two-way ordered deterministic automata; e) a certain form of "unambiguous" regular expressions.

It is not clear how to extend some of these concepts to trees. Even when natural tree counterparts exist, they are not equivalent. For instance, the temporal logic in a) can be defined for trees—by using operators "in some descendant" and "in some ancestor". This temporal logic was studied in [4], however it was shown to have an expressive power incomparable with that of Δ_2. A characterization of Δ_2 was left as an open problem, one which is solved here.

We provide an algebraic characterization of tree languages definable in Δ_2. This characterization is effectively verifiable if the language is given by a tree automaton. It is easy to see that the word setting can be treated as a special case of the tree setting. Hence our characterization builds on the one over words. However the added complexity of the tree setting makes both formulating the correct condition and generalizing the proof quite nontrivial.

2   Notation

Trees, forests and contexts. In this paper we work with finite unranked ordered trees and forests over a finite alphabet A. Formally, these are expressions defined inductively as follows: If s is a forest and $a \in A$, then as is a tree. If $t_1, \ldots, t_n$ is a finite sequence of trees, then $t_1 + \cdots + t_n$ is a forest. This applies as well to the empty sequence of trees, which is called the *empty forest,* and denoted 0 (and which provides a place for the induction to start). Forests and trees alike will be denoted by the letters $s, t, u, \ldots$ When necessary, we will remark on which forests are trees, i.e. contain only one tree in the sequence.

A set L of forests over A is called a *forest language.*

The notions of node, descendant and ancestor relations between nodes are defined in the usual way. We write $x < y$ to say that x is an ancestor or y or, equivalently, that y is a descendant of x.

If we take a forest and replace one of the leaves by a special symbol $\square$, we obtain a *context.* Contexts will be denoted using letters p, q, r. A forest s can be substituted in place of the hole of a context p, the resulting forest is denoted by ps. There is a natural composition operation on contexts: the context qp is formed by replacing the hole of q with p. This operation is associative, and satisfies $(pq)s = p(qs)$ for all forests s and contexts p and q.

When a is a letter, we will sometimes also write a for the context that has one root with label a and a hole below. For instance, any tree with label a in the root can be written as at, for some forest t.

We say a forest s is an *immediate piece* of a forest s' if s, t can be decomposed as $s = pt$ and $s' = pat$ for some contexts p, some label a, and some forest t. The reflexive transitive closure of the immediate piece relation is called the *piece* relation. We write $s \preceq t$ to say that s is a piece of t. In other words, a piece of t is obtained by removing nodes from t. We extend the notion of piece to contexts. In this case, the hole must be preserved while removing the nodes. The notions of piece for forests and contexts are related, of course. For instance, if p, q are contexts with $p \preceq q$, then $p0 \preceq q0$. Also, conversely, if $s \preceq t$, then there are contexts $p \preceq q$ with $s = p0$ and $t = q0$. The figure below depicts two contexts, the left one being a piece of the right one, as can be seen by removing the white nodes.

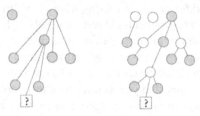

We will be considering three types of languages in the paper: *forest languages* i.e. sets of forests, denoted L; *context languages*, i.e. sets of contexts, denoted K, and *tree languages*, i.e. sets of trees, denoted M.

Logic. The focus of this paper is the expressive power of first-order logic on trees. A forest can be seen as a logical relational structure. The domain of the structure is the set of nodes. The signature contains a unary predicate P_a for each symbol a of A plus the binary predicate $<$ for the ancestor relation. A formula without free variables over this signature defines a set of forests, these are the forests where it is true. We are particularly interested in formulas of low quantifier complexity. A Σ_2 formula is a formula of the form

$$\exists x_1 \cdots x_n \; \forall y_1 \cdots y_m \; \gamma \,,$$

where γ is quantifier free. Properties defined in Σ_2 are closed under disjunction and conjunction, but not necessarily negation. The negation of a Σ_2 formula is called a Π_2 formula, equivalently this is a formula whose quantifier prefix is $\forall^* \exists^*$. A forest property is called Δ_2 if it can be expressed both by a Σ_2 and a Π_2 formula.

The problem. We want an algorithm deciding whether a given regular forest language is definable in Δ_2.

Notice that the forest property of "being a tree" is definable in Δ_2. The Σ_2 formula says there exists a node that is an ancestor of all other nodes, while the Π_2 says that for every two nodes, there exists a common ancestor. Hence a solution of the problem for forest languages also gives a solution for tree languages.

As noted earlier, the corresponding problem for words was solved by Pin and Weil: a word language L is definable in Δ_2 if and only if its syntactic monoid $M(L)$ belongs to the variety DA, i.e. it satisfies the identity

$$(mn)^\omega = (mn)^\omega m (mn)^\omega$$

for all $m, n \in M(L)$. The power ω means that the identity holds for sufficiently large powers (in different settings, ω is defined in terms of idempotent powers, but the condition on sufficiently large powers is good enough here). Since one can effectively test if a finite monoid satisfies the above property (it is sufficient to verify the power $|M(L)|$), it is decidable whether a given regular word language is definable in Δ_2. We assume that the language L is given by its syntactic monoid and syntactic morphism, or by some other representation, such as a finite automaton, from which these can be effectively computed.

We will show that a similar characterization can be found for forests; although the identities will be more involved. For decidability, it is not important how the input language is represented. In this paper, we will represent a forest language by a forest algebra that recognizes it. Forest algebras are described in the next section.

Basic properties of Σ_2. Most of the proofs in the paper will work with Σ_2 formulas. We present some simple properties of such formulas in this section.

Apart from defining forest languages, we will also be using Σ_2 formulas to define languages of contexts. To define a context language we use Σ_2 formulas with a free variable; such a formula is said to hold in a context if it is true when the free variable is mapped to the hole of the context.

Fact 1. *Let K be a context language, L a forest language, and M a tree language. If these languages are all definable in Σ_2, then so are:*

1. *For any letter a, the forest language KaL.*
2. *The forest language $M \oplus L$. This is the set of forests $t_1 + t + t_2$ such t is a tree in M, and the concatenation of forests $t_1 + t_2$ is in L.*

Proof

We only do the proof for KaL. The formula places an existentially quantified variable x on the node a, and then relativizes the formulas for languages K and L to nodes that are, respectively, not descendants of x and descendants of x. ∎

3   Forest Algebras

Forest algebras were introduced by Bojańczyk and Walukiewicz as an algebraic formalism for studying regular tree languages [6]. Here we give a brief summary of the definition of these algebras and their important properties. A forest algebra consists of a pair (H, V) of finite monoids, subject to some additional requirements, which we describe below. We write the operation in V multiplicatively and the operation in H additively, although H is not assumed to be commutative. We accordingly denote the identity of V by $\square$ and that of H by 0. We require that V act on the left of H. That is, there is a map $(h, v) \mapsto vh \in H$ such that $w(vh) = (wv)h$ for all $h \in H$ and $v, w \in V$. We further require that this action be *monoidal*, that is, $h \cdot \square = h$ for all $h \in H$, and that it be *faithful*, that is, if $vh = wh$ for all $h \in H$, then $v = w$. Finally we require that for every $g \in H$, V contains elements $(\square + g)$ and $(g + \square)$ defined by $(\square + g)h = h + g$, $(g + \square)h = g + h$ for all $h \in H$.

A morphism $\alpha : (H_1, V_1) \to (H_2, V_2)$ of forest algebras is actually a pair (γ, δ) of monoid morphisms such that $\gamma(vh) = \delta(v)\gamma(h)$ for all $h \in H$, $v \in V$. However, we will abuse notation slightly and denote both component maps by α.

Let A be a finite alphabet, and let us denote by H_A the set of forests over A, and by V_A the set of contexts over A. Clearly (H_A, V_A) with forest substitution as action, forms a forest algebra which we denote A^Δ.

We say that a forest algebra (H, V) *recognizes* a forest language $L \subseteq H_A$ if there is a morphism $\alpha : A^\Delta \to (H, V)$ and a subset X of H such that $L = \alpha^{-1}(X)$. It is easy to show that a forest language is regular if and only if it is recognized by a finite forest algebra [6].

Given any finite monoid M, there is a number $\omega(M)$ (denoted by ω when M is understood from the context) such that for all element x of M, x^ω is an idempotent: $x^\omega = x^\omega x^\omega$. Therefore for any forest algebra (H, V) and any element u of V and g of H we will write u^ω and $\omega(g)$ for the corresponding idempotents.

Given $L \subseteq H_A$ we define an equivalence relation $\sim_L$ on H_A by setting $s \sim_L s'$ if and only if for every context $x \in V_A$, hx and $h'x$ are either both in L or both outside of L. We further define an equivalence relation on V_A, also denoted $\sim_L$, by $x \sim_L x'$ if for all $h \in H_A$, $xh \sim_L x'h$. This pair of equivalence relations defines a congruence of forest algebras on A^Δ, and the quotient (H_L, V_L) is called the *syntactic forest algebra* of L. Each equivalence class of $\sim_L$ is called a *type*.

We now extend the notion of piece to elements of a forest algebra (H, V). The general idea is that a context $v \in V$ is a piece of a context $w \in V$ if one can construct a term (using elements of H and V) which evaluates to w, and then take out some parts of this term to get v.

Definition 2. *Let (H, V) be a forest algebra. We say $v \in V$ is a piece of $w \in V$, denoted by $v \preceq w$, if $\alpha(p) = v$ and $\alpha(q) = w$ hold for some morphism*

$$\alpha : A^\Delta \to (H, V)$$

and some contexts $p \preceq q$ over A. The relation $\preceq$ is extended to H by setting $g \preceq h$ if $g = v0$ and $h = w0$ for some contexts $v \preceq w$.

4 Characterization of Δ_2

In this section we present the main result of the paper: a characterization of Δ_2 in terms of two identities.

Theorem 3
A forest language is definable in Δ_2 if and only if its syntactic forest algebra satisfies the following identities:

$$h + g = g + h \tag{1}$$

$$v^\omega w v^\omega = v^\omega \qquad for\ w \preceq v \tag{2}$$

Corollary 4. *It is decidable whether a forest language can be defined in Δ_2.*

Proof

We assume that the language is represented as a forest algebra. This representation can be computed based on other representations, such as automata or monadic second-order logic.

Once the forest algebra is given, both conditions (1) and (2) can be tested in polynomial time by searching through all elements of the algebra. The relation $\preceq$ can be computed in polynomial time, using a fixpoint algorithm as in [4]. ∎

Theorem 3 is stated in terms of forest languages, but as mentioned earlier, the same result works for trees.

We begin with the easier implication in Theorem 3, that the syntactic forest algebra of a language definable in Δ_2 must satisfy the identities (1) and (2). The first identity must clearly be satisfied since the signature only contains the descendant relation. The other identity follows from the following claim, whose proof is standard.

Lemma 5. *Let φ be a formula of Σ_2 and let $p \preceq q$ be two contexts. For $n \in \mathbb{N}$ sufficiently large, forests satisfying φ are closed under replacing $p^n p^n$ with $p^n q p^n$.*

The rest of the paper contains the more difficult implication of Theorem 3. We will show that if a language is recognized by a forest algebra satisfying identities (1) and (2), then it is definable in Δ_2.

Proposition 6. *Fix a morphism $\alpha : A^\Delta \to (H, V)$, with (H, V) satisfying (1) and (2). For every $h \in H$, the forest language $\alpha^{-1}(h)$ is definable in Σ_2.*

Before proving this Proposition, we show how it concludes the proof of Theorem 3. The nontrivial part is showing that every forest language $\alpha^{-1}(h)$ is also definable in Π_2, and not just Σ_2, as the proposition says (the rest follows by closure of Δ_2 under boolean operations). But this is a consequence of finiteness of H:

$$t \in \alpha^{-1}(h) \qquad \Leftrightarrow \qquad t \notin \bigcap_{g \neq h} \alpha^{-1}(g) \,,$$

since the intersection on the right-hand side is Σ_2, and therefore non-membership is a Π_2 condition.

The rest of this section is devoted to showing Proposition 6. The proof is by induction on two parameters: the first is the size of the algebra, and the second is the position of h in a certain pre-order defined below. The second parameter corresponds to a bottom-up pass through the forest, as the types h that are small in the pre-order correspond to forests that are close to the leaves. Moreover, for some types h in the bottom-up pass, we will need a nested induction, involving a top-down pass.

5   Bottom-Up Phase

We now define the pre-order on H, which is used in the induction proof of Proposition 6. We say that a type h is *reachable from* a type g, and denote this by $g \sqsubseteq h$, if there is a context $v \in V$ such that $h = vg$. If h and g are mutually reachable from each other, then we write $h \sim g$. Note that $\sim$ is an equivalence relation. A type h is said to be *maximal* if h can be reached from all types reachable from h.

The proof of Proposition 6 is by induction on the size of the algebra (H, V) and then on the position of h in the reachability pre-order. The two parameters are ordered lexicographically, the most important parameter being the size of the algebra. As far as h is concerned, the induction corresponds to a bottom-up pass, where types close to the leaves are treated first.

Let then $h \in H$ be a type. By induction, using Proposition 6, for each $g \sqsubseteq h$ with $g \not\sim h$, we have a Σ_2 formula defining the language of forests of type g. (The case when there are no such types g corresponds to the induction base, which is treated the same way as the induction step.) In this section we will use these formulas to produce a Σ_2 formula defining those forests s such that $\alpha(s) = h$.

In the following, we will be using two sets:

$$stab_V(h) = \{v : vh \sim h\} \subseteq V \qquad stab_H(h) = \{g : g + h \sim h\} \subseteq H .$$

The main motivation for introducing this notation is that equation (2) implies that they are both submonoids of V and H, respectively.

Lemma 7. *The sets $stab_V(h), stab_H(h)$ only depend on the $\sim$-class of h. In particular, both sets are submonoids (of V and H, respectively).*

Proof

We prove the Lemma for $stab_V(h)$, the case of $stab_H(h)$ being similar. We need to show that if $h \sim h'$ then $stab_V(h) = stab_V(h')$. Assume $v \in stab_V(h)$. Then $vh \sim h$. Hence we have u_1, u_2, u_3 such that $h = u_1 vh, h = u_2 h'$ and $h' = u_3 h$. This implies that $h' = u_3 u_1 vu_2 h'$ and therefore $h' = (u_3 u_1 vu_2)^\omega h'$. From (2) we have that

$$h' = (u_3 u_1 vu_2)^\omega h' = (u_3 u_1 vu_2)^\omega v(u_3 u_1 vu_2)^\omega h' = (u_3 u_1 vu_2)^\omega vh' .$$

Hence h' is reachable from vh'. Since vh' is clearly reachable from h', we get $h \sim h'$ and $v \in stab_V(h')$.   ∎

Recall now the piece order $\preceq$ on H from Definition 2, which corresponds to removing nodes from a forest. We say a set $F \subseteq H$ of forest types is *closed under pieces* if $h \preceq g \in F$ implies $h \in F$. A similar definition is also given for contexts. Another consequence of equation (2) is:

Lemma 8. *Both $stab_V(h), stab_H(h)$ are closed under pieces.*

Proof

We consider only the case of $stab_V(h)$, the case of $stab_H(h)$ being similar. From the definition of piece we need to show that if $u \in stab_V(h)$ and $u' \preceq u$ then $u' \in stab_V(h)$. By definition we have a context v such that $h = vuh$. We are looking for a context w such that $wu'h = h$. From $h = vuh$ we get $h = (vu)^\omega h$. Hence by (2) we have $h = (vu)^\omega u'(vu)^\omega h = (vu)^\omega u'h$ as desired.   ∎

We now consider two possible cases: either h belongs to $stab_H(h)$, or it does not. In the first case we will conclude by induction on the size of the algebra while in the second case we will conclude by induction on the partial order $\sqsubseteq$. These are treated separately in Sections 5.1 and 5.2, respectively.

5.1 $h \notin stab_H(h)$

For $v \in V$, we write K_v for the set of contexts of type v. For $g \in H$, we write L_g for the set of forests of type h. For $g \in H$ and $F \subseteq H$, we write L_g^F for the set of forests t of type h that can be decomposed as $t = t_1 + \ldots + t_n$, with each t_i a tree with of type in F.

Let G be the set of forest types g such that h is reachable from g but not vice-versa. By induction assumption, each language L_g is definable in Σ_2, for $g \in G$. Our goal is to give a formula for L_h.

Lemma 9. *A forest has type h if and only if it belongs to L_h^G or a language $K_u a L_g^G$, with $u\alpha(a)g = h$ and $u \in stab_V(h)$.*

Proof
Let t be a forest of type h, and choose s a subtree of t that has type equivalent to h, but no subtree with a type equivalent to h. If such s does not exist, then t belongs to L_h^G as a concatenation of trees with type in G. By minimality, s must belong to some set aL_g^G. Let p be the context such that $t = ps$. Since the type of s is equivalent to h, and the type of t is h, then the type u of p belongs to $stab_V(\alpha(s))$ which is the same as $stab_V(h)$ by Lemma 7.   ■

In Lemmas 10 and 11, we will show that the languages K_u and L_g^G above can be defined in Σ_2. To be more precise, we only give an over-approximation φ_g^G of the language L_g^G, however all forests in the over-approximation have type g, which is all we need. Proposition 6 then follows by closure of Σ_2 under finite union and Fact 1.

We begin by giving the over-approximation of L_g^G.

Lemma 10. *For any type $g \in H$, there is a formula φ_g^G of Σ_2 such that:*

- *Any forest L_g^G satisfies φ_g^G; and*
- *Any forest satisfying φ_g^G has type g.*

Proof
The proof of the lemma is in two steps. In the first step, we introduce a condition (*) on a forest t, and show that: a) any forest in L_g^G satisfies (*); and b) any forest satisfying (*) has type g. Then we will show that condition (*) can be expressed in Σ_2.

  (*) For some $m \leq n$, the forest t can be decomposed, modulo commu-
      tativity, as the concatenation $t = t_1 + \cdots + t_n$ of trees $t_1, \ldots, t_n$, with
      types $g_1, \ldots, g_n$, such that
      1. $g_1 + \cdots + g_m = g$.

2. Each type from G is represented at most ω times in $g_1, \ldots, g_m$.
3. If a tree s is a piece of $t_{m+1} + \cdots + t_n$, then $\alpha(s) \preceq g_i$ holds for some type g_i that occurs ω times in the sequence $g_1, \ldots, g_m$.

We first show that condition (*) is necessary. Let $t_1, \ldots, t_n$ be all the trees in a forest t, and let $g_1, \ldots, g_n \in G$ be the types of these trees. Without loss of generality, we may assume that trees are ordered so that for some m, each type of g_i with $i > m$ already appears ω times in $g_1, \ldots, g_m$. It is not hard to see that identity (2) implies aperiodicity of the monoid H, i.e.

$$\omega \cdot f = \omega \cdot f + f \qquad \text{for all } f \in H . \tag{3}$$

In particular, it follows that $g = g_1 + \cdots + g_m$ since all of $g_{m+1}, \ldots, g_n$ are swallowed by the above. It remains to show item 3 of condition (*). Let then s be the piece of a tree t_i with $i > m$. We get the desired result since the type of t_i already appears in $g_1, \ldots, g_m$.

We now show that condition (*) implies $\alpha(t) = g$. Let then $m \leq n$ and $t = t_1 + \cdots + t_n$ be as in (*). We will show that for any $j > m$, we have $g + g_j = g$, which shows that the type of t is g. By item 3, $g_j \preceq g_i$ holds for some some type g_i that occurs ω times in the sequence $g_1, \ldots, g_m$. By (3), we have $g = g + g_i = g + \omega \cdot g_i$. It therefore remains to show that $\omega \cdot g_i + g_j = \omega \cdot g_i$:

$$\omega \cdot g_i + g_j = \omega \cdot g_i + g_j + \omega \cdot g_i =$$
$$(\square + g_i)^\omega (\square + g_j)(\square + g_i)^\omega 0 = (\square + g_i)^\omega 0 = \omega \cdot g_i$$

In the above we have used identity (2). Note that the requirement in (2) was satisfied, since $g_j \preceq g_i$ implies $\square + g_j \preceq \square + g_i$.

It now remains to show that forests satisfying condition (*) can be defined in Σ_2. Note that m cannot exceed $|G| \cdot \omega$, and therefore there is a finite number of cases to consider for $g_1, \ldots, g_m$. Fix some sequence $g_1, \ldots, g_m$. The only nontrivial part is to provide a Σ_2 formula that describes the set of forests $t_{m+1} + \ldots + t_n$ that satisfy item 3 of condition (*). From this construction, the formula for (*) follows by closure of Σ_2 under finite union and $\oplus$ (recall Fact 1), as well as the assumption that each type in G can be defined in Σ_2.

In order to define forests as in item 3 we use a Π_1 formula that forbids the appearance of certain pieces of bounded size inside $t_{m+1} + \cdots + t_n$. Let F be the types in $g_1, \ldots, g_m$ that appear at least ω times. We claim that a sequence of trees $t_{m+1} + \cdots + t_n$ satisfies item 3 if and only if it satisfies item 3 with respect to pieces s that have at most $|H|^{|H|}$ nodes. The latter property can be expressed by a Π_1 formula. The reason for this is that, thanks to a pumping argument, any tree has a piece that has the same type, but at most $|H|^{|H|}$ nodes.   ∎

Lemma 11. *For any $u \in stab_V(h)$, the context language K_u is definable in Σ_2.*

To prove this lemma, we will use a more general result, Proposition 12, stated below. The proof of this Proposition will appear in the journal version of this paper. We say a tree t is a subtree of a context p if t is the subtree of some node in p that is not an ancestor of the hole.

Proposition 12. *Let $F \subseteq H$ be a set of forest types definable in Σ_2 that is closed under pieces. For any $u \in V$, there is a Σ_2 formula that defines the set of contexts with type u that have no subtree of type outside F.*

Proof (of Lemma 11)

Let $F = stab_H(h)$. The result will follow from Proposition 12 once we show that a context in K_u cannot have a subtree outside F, and that F satisfies the conditions in the proposition.

By Lemma 8, the set $F = stab_H(h)$ is closed under pieces. We now show that $F \subseteq G$, and therefore each type in F is definable in Σ_2. To the contrary, if F would contain a type outside G, i.e. a type reachable from h, then by closure under pieces it would also contain h, contradicting our assumption on $h \notin stab_H(h)$. Finally, each subtree t of a context in $stab_V(h)$ is a subtree—and therefore also a piece—of a tree in $stab_H(h) = F$.   ∎

5.2 $h \in stab_H(h)$

Lemma 13. *If $h \in stab_H(h)$ then $(stab_H(h), stab_V(h))$ is a forest algebra.*

Proof

We need to show that the two sets are closed under all operations.

$$stab_V(h)stab_V(h) \subseteq stab_V(h)$$
$$stab_H(h) + stab_H(h) \subseteq stab_H(h)$$
$$\square + stab_H(h) \subseteq stab_V(h)$$
$$stab_V(h)stab_H(h) \subseteq stab_H(h)$$

The first two of the above inclusions follow from Lemma 7. The third follows straight from the definition of *stab*. For the last inclusion, consider $v \in stab_V(h)$ and $g \in stab_H(h)$. We need to show that $vg \in stab_H(h)$. This means showing that $vg + h \sim h$. Since we have $g + h \sim h$ and $vh \sim h$ we have $u, u' \in V$ such that $h = u(g + h)$ and $h = u'vh$. Hence $h = u'vu(g + h)$ and $vg \preceq h$. We conclude using Lemma 8 and the fact that $h \in stab_H(h)$.   ∎

We have two subcases depending whether $(stab_H(h), stab_V(h))$ is a proper subalgebra of (H, V) or not.

Assume first that h is not maximal. Hence there exists a type g reachable from h but not vice-versa. Let u be a context such that $g = uh$. It is clear that u is not in $stab_V(h)$. Therefore $(stab_H(h), stab_V(h))$ must be a proper subalgebra of (H, V), as we have that $stab_V(h) \subsetneq V$. Furthermore, this algebra contains all pieces of h; so it still recognizes the language $\alpha^{-1}(h)$; at least as long as the alphabet is reduced to include only letters that can appear in h. We can then use the induction assumption on the smaller algebra to get the Σ_2 formula required in Proposition 6.

If h is maximal then the algebra is not proper and we need to do more work. The Σ_2 formula required in Proposition 6 is obtained by taking $v = \square$ in the proposition below. The proof of this proposition introduces a pre-order on V and is done by induction using that pre-order simulating a top-down process. The details are omitted here and will appear in the journal version of this paper.

Proposition 14. *Fix a morphism* $\alpha : A^\Delta \to (H, V)$, *a context type* $v \in V$ *and a maximal forest type* h. *The following forest language is definable in* Σ_2:

$$\{t : v\alpha(t) = h\}$$

6   Discussion

Apart from label tests, the signature we have used contains only the descendant relation. What about other predicates? For instance, if we add the lexicographic order on nodes, we lose commutativity $g + h = h + g$, although the remaining identity (2) remains valid. Is the converse implication true, i.e. can every language whose algebra satisfies (2) be defined by a Δ_2 formula with the lexicographic and descendant order? What is the expressive power of Δ_2 in the other signatures, with predicates such as the closest common ancestor, next sibling or child?

Probably the most natural continuation would be an effective characterization of Σ_2. Note that this would strenghten our result: a language L is definable in Δ_2 if and only if both L and its complement are definable in Σ_2. We conjecture that, as in the case for words [1], the characterization of Σ_2 requires replacing the equivalence in (2) by a one-sided implication, which says that a language definable in Σ_2 is closed under replacing v^ω by $v^\omega w v^\omega$, for $w \preceq v$.

References

1. Arfi, M.: Opérations polynomiales et hiérarchies de concaténation. Theor. Comput. Sci. 91(1), 71–84 (1991)
2. Benedikt, M., Segoufin, L.: Regular tree languages definable in FO and in FO+mod (preliminary version in STACS 2005) (manuscript, 2008)
3. Bojańczyk, M.: Forest expressions. In: Duparc, J., Henzinger, T.A. (eds.) CSL 2007. LNCS, vol. 4646, pp. 146–160. Springer, Heidelberg (2007)
4. Bojańczyk, M.: Two-way unary temporal logic over trees. In: Logic in Computer Science, pp. 121–130 (2007)
5. Bojańczyk, M., Walukiewicz, I.: Characterizing EF and EX tree logics. Theoretical Computer Science 358(2-3), 255–273 (2006)
6. Bojańczyk, M., Walukiewicz, I.: Forest algebras. In: Automata and Logic: History and Perspectives, pp. 107–132. Amsterdam University Press (2007)
7. Etessami, K., Vardi, M.Y., Wilke, T.: First-order logic with two variables and unary temporal logic. Inf. Comput. 179(2), 279–295 (2002)
8. Straubing, H., Bojańczyk, M., Segoufin, L.: Piecewise testable tree languages. In: Logic in Computer Science (2008)
9. McNaughton, R., Papert, S.: Counter-Free Automata. MIT Press, Cambridge (1971)

10. Pin, J.-É.: Logic, semigroups and automata on words. Annals of Mathematics and Artificial Intelligence 16, 343–384 (1996)
11. Pin, J.-É., Weil, P.: Polynomial closure and unambiguous product. Theory Comput. Systems 30, 1–30 (1997)
12. Schützenberger, M.P.: On finite monoids having only trivial subgroups. Information and Control 8, 190–194 (1965)
13. Schwentick, T., Thérien, D., Vollmer, H.: Partially-ordered two-way automata: A new characterization of DA. In: Devel. in Language Theory, pp. 239–250 (2001)
14. Simon, I.: Piecewise testable events. In: Automata Theory and Formal Languages, pp. 214–222 (1975)
15. Thérien, D., Wilke, T.: Over words, two variables are as powerful as one quantifier alternation. In: STOC, pp. 256–263 (1998)
16. Wilke, T.: Classifying discrete temporal properties. In: Meinel, C., Tison, S. (eds.) STACS 1999. LNCS, vol. 1563, pp. 32–46. Springer, Heidelberg (1999)

Duality and Equational Theory
of Regular Languages

Mai Gehrke[1], Serge Grigorieff[2], and Jean-Éric Pin[2,*]

[1] Radboud University Nijmegen, The Netherlands
[2] LIAFA, University Paris-Diderot and CNRS, France

This paper presents a new result in the equational theory of regular languages, which emerged from lively discussions between the authors about Stone and Priestley duality. Let us call *lattice of languages* a class of regular languages closed under finite intersection and finite union. The main results of this paper (Theorems 5.2 and 6.1) can be summarized in a nutshell as follows:

> *A set of regular languages is a lattice of languages if and only if it can be defined by a set of profinite equations.*

> *The product on profinite words is the dual of the residuation operations on regular languages.*

In their more general form, our equations are of the form $u \to v$, where u and v are profinite words. The first result not only subsumes Eilenberg-Reiterman's theory of varieties and their subsequent extensions, but it shows for instance that any class of regular languages defined by a fragment of logic closed under conjunctions and disjunctions (first order, monadic second order, temporal, etc.) admits an equational description. In particular, the celebrated McNaughton-Schützenberger characterisation of first order definable languages by the aperiodicity condition $x^\omega = x^{\omega+1}$, far from being an isolated statement, now appears as an elegant instance of a very general result.

How is this equational theory related to duality? The connection between profinite words and Stone spaces was already discovered by Almeida [2], [3, Theorem 3.6.1], but Pippenger [14] was the first to formulate it in terms of Stone duality. Almeida (implicitly) and Pippenger (explicitly) both observed that the Boolean algebra of regular languages over A^* is dual to the Stone space $\widehat{A^*}$, the set of profinite words. Pippenger actually came very close to our first result, since he mentioned that this duality extends to a one-to-one correspondence between Boolean algebras of regular languages and quotients of $\widehat{A^*}$. Our first result is the full-fledged consequence of the similar one-to-one correspondence for all lattices of languages provided by Priestley duality.

However, this link to duality theory is in fact much stronger and encompasses not only the underlying lattices and spaces involved but also the algebraic operations including the product of profinite words. That is the content of our

* The authors acknowledge support from the AutoMathA programme of the European Science Foundation.

L. Aceto et al. (Eds.): ICALP 2008, Part II, LNCS 5126, pp. 246–257, 2008.

second result. It means that the profinite semigroup structure, in its entirety, is a dual structure and thus the entire theory is a special case of duality theory. In particular, the deep and highly evolved theory of duality and relational semantics from modal logic applies, and, in the other direction, the wealth of knowledge and examples from semigroup theory enriches our understanding of general duality theory. In this sense, the results described here are just the tip of an iceberg yet to be explored.

Due to the lack of space, most of the proofs are omitted.

1 Historical Background

Our starting point was Eilenberg's variety theorem [7]. Recall that a *variety of languages* is a class of regular languages closed under Boolean operations, inverses of morphisms and left and right quotients by words. Eilenberg's theorem states that varieties of languages are in one-to-one correspondence with *varieties of finite monoids*, that is, classes of finite monoids closed under taking submonoids, quotient monoids and finite direct products.

The notion of a variety of finite monoids is similar to that of variety of monoids introduced by Birkhoff: a *variety of monoids* is a class of monoids closed under taking submonoids, quotient monoids and direct products. Birkhoff proved in [6] that his varieties can be characterized by sets of identities: for instance the identity $xy = yx$ characterizes the variety of commutative monoids. Almost fifty years later, Reiterman [18] extended Birkhoff's theorem to varieties of finite monoids: any variety of finite monoids can be characterized by a set of profinite identities. A *profinite identity* is an identity between two profinite words. The precise definition of profinite words will be given in Section 2, but they can be viewed as limits of sequences of words for a certain metric, the profinite metric. For instance, one can show that the sequence $x^{n!}$ converges to a profinite word denoted by x^ω and the variety of finite aperiodic monoids can be defined by the identity $x^\omega = x^{\omega+1}$.

Eilenberg's and Reiterman's theorems have been extended several times over the last twenty years by relaxing the definition of a variety of languages. In [11], the third author considered *positive varieties*, for which the closure under complement is not required and showed they correspond to varieties of finite ordered monoids. The counterpart of Reiterman's theorem, obtained by Pin-Weil [13], makes use of identities of the form $u \leqslant v$, where u and v are profinite words.

Pippenger [14] proposed to relax another condition by introducing *strains of languages*, which share the same properties as varieties of languages except for the closure under quotients by words, which is not required. Finally, Straubing [21] and independently, Esik [8], relaxed the closure under inverses of morphisms. Esik just required the closure under inverses of length-preserving morphisms. Straubing considered a class $\mathcal{C}$ of morphisms between free monoids containing the length-preserving morphisms and closed under composition and called $\mathcal{C}$-variety a class of regular languages closed under Boolean operations, quotients and inverses of morphisms from the class $\mathcal{C}$. The counterpart of Reiterman's theorem for this case was given by Kunc [10] (see also [12]).

2 Profinite Topology

In this paper, A denotes a *finite* alphabet. A morphism $\varphi : A^* \to M$ *separates* two words u and v of A^* if $\varphi(u) \neq \varphi(v)$. By extension, we say that a monoid M *separates* two words if there is a morphism from A^* onto M that separates them. One can show that two distinct words can always be separated by a finite monoid. Given two words $u, v \in A^*$, we set

$$r(u, v) = \min \{|M| \mid M \text{ is a monoid that separates } u \text{ and } v\}$$
$$d(u, v) = 2^{-r(u,v)}$$

with the usual conventions $\min \emptyset = +\infty$ and $2^{-\infty} = 0$. One can show that d is an *ultrametric*, that is, satisfies the following properties, for all $u, v, w \in A^*$,

(1) $d(u, v) = 0$ if and only if $u = v$,

(2) $d(u, v) = d(v, u)$,

(3) $d(u, w) \leqslant \max\{d(u, v), d(v, w)\}$.

Moreover, the relations $d(uv, u'v') \leqslant \max\{d(u, u'), d(v, v')\}$ hold for all $u, u', v, v' \in A^*$, so that the concatenation product on A^* is uniformly continuous.

Thus (A^*, d) is a metric space. Its completion, denoted by $\widehat{A^*}$, is called the *free profinite monoid* on A and its elements are called *profinite words*.

We now briefly review the main properties of $\widehat{A^*}$. The reader is referred to [22, 4] for more details. First, $\widehat{A^*}$ is compact. Second, the topology defined by d is the *profinite topology*, that is, the least topology which makes continuous every morphism from A^* onto a finite monoid (considered as a discrete metric space). It follows that every morphism φ from A^* onto a finite monoid F extends uniquely to a (uniformly) continuous morphism $\hat{\varphi} : \widehat{A^*} \to F$. Thirdly, since the product on A^* is uniformly continuous, it can be extended in a unique way to a uniformly continuous product on $\widehat{A^*}$. This product makes $\widehat{A^*}$ a monoid.

Recall that a set is *clopen* if it is both open and closed. There is a strong connection between clopen sets of $\widehat{A^*}$ and regular languages of A^*. Indeed, a language L is regular if and only if $\overline{L}$ is clopen in $\widehat{A^*}$ and $L = \overline{L} \cap A^*$ [4]. The languages of the form $\overline{L}$, where L is a regular language, actually form a basis for the topology and hence $\widehat{A^*}$ is *zero-dimensional*. It is also *totally disconnected* since its connected components are singletons.

What about sequences? First, every profinite word is the limit of some converging sequence of words. Next, a sequence of profinite words $(u_n)_{n \geqslant 0}$ is converging to a profinite word u if and only if, for every morphism φ from A^* onto a finite monoid, $\hat{\varphi}(u_n)$ is ultimately equal to $\hat{\varphi}(u)$.

For instance, if u is a word (or even a profinite word), one can prove that the sequence $u^{n!}$ is converging. Its limit is denoted by u^ω for the following reason: if φ is a morphism from A^* onto a finite monoid M, the sequence $\hat{\varphi}(u)^{n!}$ is ultimately equal to the unique idempotent power of $\hat{\varphi}(u)$, which is traditionally denoted by $\hat{\varphi}(u)^\omega$ in semigroup theory. Thus the notation u^ω is justified by the formula $\hat{\varphi}(u^\omega) = \hat{\varphi}(u)^\omega$.

The closure in $\widehat{A^*}$ of a regular language of A^* can be characterized as follows.

Proposition 2.1. *Let L be a regular language of A^* and let $u \in \widehat{A^*}$. The following conditions are equivalent:*

(1) $u \in \overline{L}$,

(2) $\hat{\varphi}(u) \in \varphi(L)$, *for all morphisms φ from A^* onto a finite monoid,*

(3) $\hat{\varphi}(u) \in \varphi(L)$, *for some morphism φ from A^* onto a finite monoid that recognizes L,*

(4) $\hat{\eta}(u) \in \eta(L)$, *where η is the syntactic morphism of L.*

3 Duality for Distributive Lattices

In Stone duality, the dual space of a bounded distributive lattice D is based on the set S_D of prime filters of D. As identified already by Birkhoff, there is a lattice embedding e of D into $\mathcal{P}(S_D)$, defined by:

$$e(d) \text{ is the set of prime filters containing } d.$$

A description of the range of e, both for Boolean algebras and then for distributive lattices was first provided by Stone [19, 20]. He showed that if one generates a topology on the space of prime filters with the sets in the image of the embedding e, then the resulting space is, in the Boolean case, a compact 0-dimensional space, and in the distributive lattice case a *spectral space*, i.e. a compact (not necessarily Hausdorff), sober space with a ring of compact-open sets as a basis. An answer in complete lattice theoretic terms is the result by Jónsson and Tarski on canonical extensions. This is the most advantageous point of view when considering additional structure on lattices and spaces such as the semigroup operation.

For distributive lattices, Priestley [16] gave a slightly different topological characterization of the range of e than Stone. If one generates a topology τ, not just with the sets in the range of e, but also with their complements, one obtains the dual space of the free Boolean extension of the lattice and, crucially, one may reconstruct the original lattice if one remembers, in addition to the dual space of the free Boolean extension of the lattice, also the inclusion order on the space of prime filters. Thus in Priestley duality the dual of a distributive lattice is the ordered topological space $(S_D, \subseteq, \tau)$. It is characterized by the property that it is compact and totally order disconnected. An ordered topological space is *totally order disconnected* provided the points of the space are separated by the upwards saturated clopen subsets. This is the duality we will use here.

One of the most powerful facts about dualities is that we get a complete correspondence between subobjects on one side and quotients on the other. Here we are interested in sublattices of regular languages, and these will of course correspond, under Priestley duality, to Priestley space quotients or equivalently, certain compatible preorders on the dual space of the lattice of all regular languages. Working out this correspondence dates back to work by M. E. Adams [1]. If D is a subalgebra of B, we obtain a dual quotient $S_B \twoheadrightarrow S_D$ by mapping a prime filter p of B to $p \cap D$. The topological condition that is needed is that the quotient is a continuous (and, in the DL case, order preserving) map. An equivalence relation (preorder for DL subalgebras) on the space S_B corresponds to a subalgebra provided the clopen subsets that are saturated with respect to the equivalence relation (preorder for DL subalgebras) separate the equivalence classes (of the equivalence relation corresponding to the preorder in the DL case).

4   Duality Applied to $\mathrm{Reg}(A^*)$

The proof that the dual space of $\mathrm{Reg}(A^*)$ is none other than the space $\widehat{A^*}$ of profinite words can be found in Pippenger's paper [14]. It relies on two facts. First, given a prime filter p of $\mathrm{Reg}(A^*)$, there is a unique profinite word u such that, for every morphism from A^* onto a finite monoid, $\varphi(u)$ is the unique element m of M such that $\varphi^{-1}(m) \in p$. In the opposite direction, if u is a profinite word, the set

$$p_u = \{L \in \mathrm{Reg}(A^*) \mid \varphi^{-1}(\hat{\varphi}(u)) \subseteq L$$
$$\text{for some morphism } \varphi \text{ from } A^* \text{ onto a finite monoid } \} \tag{1}$$

is a prime filter of $\mathrm{Reg}(A^*)$.

Theorem 4.1 (See [14]). *The topological space underlying the profinite completion* $\widehat{A^*}$ *is equal to the dual space of the Boolean algebra* $\mathrm{Reg}(A^*)$. *Furthermore, the canonical embedding is given by the topologial closure:* $e(L) = \overline{L}$.

5   Equational Characterization of Lattices

Formally, a *profinite equation* is a pair (u, v) of profinite words of $\widehat{A^*}$. We also use the term *explicit equation* when both u and v are words of A^*. We say that a regular language L of A^* satisfies the profinite equation $u \to v$ (or $v \leftarrow u$) if the condition $u \in \overline{L}$ implies $v \in \overline{L}$. Proposition 2.1 immediately gives some equivalent definitions:

Corollary 5.1. *Let L be a regular language of A^*, let η be its syntactic morphism and let φ be any morphism onto a finite monoid recognizing L. The following conditions are equivalent:*

(1) *L satisfies the equation $u \to v$,*

(2) *$\hat{\eta}(u) \in \eta(L)$ implies $\hat{\eta}(v) \in \eta(L)$,*

(3) *$\hat{\varphi}(u) \in \varphi(L)$ implies $\hat{\varphi}(v) \in \varphi(L)$.*

Given a set E of equations of the form $u \to v$, the set of all regular languages of A^* satisfying all the equations of E is called the set of languages *defined by* E. It is not hard to see that the set of languages defined by a set E of equations is a lattice. Our first result states that the converse is true as well.

Theorem 5.2. *A set of regular languages of A^* is a lattice of languages if and only if it can be defined by a set of equations of the form $u \to v$, where $u, v \in \widehat{A^*}$.*

Proof. The proof is an instantiation of the duality between sublattices of $\mathrm{Reg}(A^*)$ and preorders on its dual space $\widehat{A^*}$. Given a lattice D of regular languages, we get dually a quotient map $q_D : \widehat{A^*} \twoheadrightarrow S_D$ given by $p_u \mapsto p_u \cap D$, where p_u is defined by Formula (1). Equivalently, we may describe this quotient map by the preorder Q_D on $\widehat{A^*}$ given by $u\, Q_D\, v$ if and only if $q_D(p_u) \subseteq q_D(p_v)$. But this latter condition is equivalent to requiring that, for all $L \in D$, $u \in \overline{L}$ implies $v \in \overline{L}$. That is, in our terminology, the preorder on $\widehat{A^*}$ determining the quotient dual to D is exactly the equational theory of D:

$$Q_D = \{(u,v) \mid \text{for all } L \in D \ (L \text{ satisfies } u \to v)\}.$$

On the other hand, in the duality, given a preorder Q on $\widehat{A^*}$ giving rise to a Priestley quotient $\widehat{A^*}/Q$, the corresponding lattice is the set of all $L \in \mathrm{Reg}(A^*)$ so that their representation $\overline{L}$ is saturated with respect to the preorder. That is, $u \in \overline{L}$ implies $v \in \overline{L}$ for all $(u, v) \in Q$. But, by our earlier definition, this is exactly what we call the set of languages defined by Q if we identify each pair (u, v) in Q with the corresponding equation $u \to v$.

Since, coming from D, going to the preorder Q_D, and then going back to the set of languages defined by Q_D under duality gives us back D, we see that D is the set of languages defined by Q_D. $\qquad\square$

Writing $u \leftrightarrow v$ for ($u \to v$ and $v \to u$), we get an equational description of the Boolean algebras of languages.

Corollary 5.3. *A set of regular languages of A^* is a Boolean algebra of languages if and only if it can be defined by a set of equations of the form $u \leftrightarrow v$, where $u, v \in \widehat{A^*}$.*

6 Duality for Quotienting Operations

As announced in the introduction, our second main result is that the product on $\widehat{A^*}$ itself is dual to operations on $\mathrm{Reg}(A^*)$. The pertinent operations are the *residuals* of the product of languages, $\backslash$ and $/$, defined, for all $L, M, N \in \mathrm{Reg}(A^*)$, by the conditions

$$LM \subseteq N \iff M \subseteq L\backslash N \iff L \subseteq N/M.$$

More explicitly, the *right* and *left residuals* of N by M are given by:

$$M\backslash N = \{u \in A^* \mid Mu \subseteq N\} = \{u \in A^* \mid \text{for all } v \in M,\ vu \in N\}$$
$$N/M = \{u \in A^* \mid uM \subseteq N\} = \{u \in A^* \mid \text{for all } v \in M,\ uv \in N\}.$$

In extended Priestley duality [9], the additional operations are captured by additional relational structure on the dual space. A well-known case of this is the capture of a modality on the dual frame by its binary Kripke relation. More generally, n-ary relations on lattices are captured by $(n + 1)$-ary relations on their dual spaces. Remarkably, in the case of the algebra $(\operatorname{Reg}(A^*), \backslash, /)$, the dual relation common to the two additional operations is functional and turns out to be the product on profinite words.

Theorem 6.1. *The dual space of the algebra* $(\operatorname{Reg}(A^*), \backslash, /)$ *under extended duality is the topological monoid of profinite words* $(\widehat{A^*}, \tau, \cdot)$. *The relational dual of the operations* $\backslash$ *and* $/$ *is the product of profinite words. The closure of* $\operatorname{Reg}(A^*)$ *under* $\backslash$ *and* $/$ *accounts for the right and left continuity of the product, respectively, and the equational property* $(H \backslash K)/L = H \backslash (K/L)$ *of* $(\operatorname{Reg}(A^*), \backslash, /)$ *corresponds to the associativity of the product.*

The proof of Theorem 6.1 requires advanced machinery from duality theory and space does not allow us to give even a sketch of the proof here.

This theorem has far-reaching consequences. To mention just two, the syntactic ordered monoid of a regular language is none other than the dual space of the subalgebra of $(\operatorname{Reg}(A^*), \backslash, /)$ generated by the singleton set $\{L\}$ under the lattice operations and the residuation operations with arbitrary denominators, and closure of $\operatorname{Reg}(A^*)$ under product of languages corresponds to the fact that product for profinite words is an open mapping. In the next section we use Theorem 6.1 to give an important specialisation of Theorem 5.2.

The following observations will come in handy in the next section: for each $a \in A$ the residuals with denominator $\{a\}$ are central in language theory. We denote them by $a^{-1}(\ )$ and $(\ )a^{-1}$ instead of $\{a\} \backslash (\ )$ and $(\ )/\{a\}$, respectively, and call them *quotienting operations*.

We call a lattice of languages a *quotienting algebra of languages* provided it is closed under the quotienting operations. For instance, the lattice $\operatorname{Reg}(A^*)$ is a quotienting algebra. It is easy to prove that, for sets of regular languages closed under finite intersections, closure under the residuals with arbitrary denominators amounts to the same as closure under the quotienting operators.

7   Lattices of Languages Closed Under Quotienting

In this section we characterise those lattices of languages for which the dual quotient is not only a topological quotient but also an ordered monoid quotient. Recall that an *ordered monoid* is a partially ordered monoid in which the monoid operation is order preserving in each coordinate. Note that the map $\widehat{A^*} \twoheadrightarrow S_D$ defined in the proof of Theorem 5.2 is an ordered monoid quotient if and only if the relation Q_D is a congruence of ordered monoid.

Let u and v be two profinite words of $\widehat{A^*}$. We say that L *satisfies the semigroup equation* $u \leqslant v$ if, for all $x, y \in \widehat{A^*}$, it satisfies the equation $xvy \to xuy$. Since A^* is dense in $\widehat{A^*}$, it is equivalent to state that L satisfies these equations only for all $x, y \in A^*$. But there is a much more convenient characterization using the syntactic ordered monoid of L.

Proposition 7.1. *Let L be a regular language of A^*, let $(M, \leqslant_L)$ be its syntactic ordered monoid and let $\eta : A^* \to M$ be its syntactic morphism. Then L satisfies the equation $u \leqslant v$ if and only if $\hat{\eta}(u) \leqslant_L \hat{\eta}(v)$.*

Proof. Corollary 5.1 shows that L satisfies the equation $u \leqslant v$ if and only if, for every $x, y \in A^*$, $\hat{\eta}(xvy) \in \eta(L)$ implies $\hat{\eta}(xuy) \in \eta(L)$. Since $\hat{\eta}(xvy) = \hat{\eta}(x)\hat{\eta}(v)\hat{\eta}(y) = \eta(x)\hat{\eta}(v)\eta(y)$ and since η is surjective, this is equivalent to saying that, for all $s, t \in M$, $s\hat{\eta}(v)t \in \eta(L)$ implies $s\hat{\eta}(u)t \in \eta(L)$, which exactly means that $\hat{\eta}(u) \leqslant_L \hat{\eta}(v)$. $\qquad\square$

Using the fact that in the extended duality, preservation of operations on the algebraic side corresponds to bounded morphisms [9] on the other, one can now prove the following specialisation of Theorem 5.2.

Theorem 7.2. *Let D be a lattice of languages of A^*. The following conditions are equivalent:*

(1) *D is a quotienting algebra of languages,*

(2) *D can be defined by a set of semigroup equations $u \leqslant v$, where $u, v \in \widehat{A^*}$,*

(3) *the corresponding dual quotient $\widehat{A^*} \twoheadrightarrow S_D$ is an ordered quotient monoid.*

Theorem 7.2 can be readily extended to Boolean algebras. Let u and v be two profinite words. We say that a regular language L *satisfies the equation $u = v$* if it satisfies the equations $u \leqslant v$ and $v \leqslant u$. Proposition 7.1 now gives immediately:

Proposition 7.3. *Let L be a regular language of A^* and let η be its syntactic morphism. Then L satisfies the equation $u = v$ if and only if $\hat{\eta}(u) = \hat{\eta}(v)$.*

This leads to the following equational description of the Boolean algebras of languages closed under quotients.

Proposition 7.4. *A set of regular languages of A^* is a Boolean quotienting algebra if and only if it can be defined by a set of semigroup equations of the form $u = v$, where $u, v \in \widehat{A^*}$.*

8 Classes of Languages Closed Under Inverses of Morphisms

The results of this section and the previous section permit in particular to recover the equational characterization of Eilenberg's varieties and Straubing's C-varieties.

Denote by C a class of morphisms between free monoids containing the length-preserving morphisms and closed under composition. These morphisms will be called C-*morphisms*. Examples include the classes of all *length-preserving* morphisms (morphisms for which the image of each letter is a letter), all *length-multiplying* morphisms (morphisms such that, for some integer k, the length of the image of a word is k times the length of the word), all *non-erasing* morphisms (morphisms for which the image of each letter is a nonempty word), all *length-decreasing* morphisms (morphisms for which the image of each letter is either a letter of the empty word) and all morphisms.

A *class of language lattices* $\mathcal{L}$ associates with every finite alphabet A a lattice of languages $\mathcal{L}(A^*)$. Theorem 5.2 gives an equational description for each of these lattices, but these equations depend on the alphabet A. We now show that if $\mathcal{L}$ is closed under inverses of $\mathcal{C}$-morphisms, a single set of equations suffices to characterize the whole class $\mathcal{L}$.

Indeed, if $u \to v$ is an equation of $\mathcal{L}(A^*)$ and $\varphi : A^* \to B^*$ is a $\mathcal{C}$-morphism, then $\hat{\varphi}(u) \to \hat{\varphi}(v)$ is an equation of $\mathcal{L}(B^*)$. This leads naturally to the following definition. Let Σ be a countable alphabet. A regular language L of A^* satisfies the $\mathcal{C}$-*identity* $u \leqslant v$, where $u, v \in \widehat{\Sigma^*}$ if, for each $\mathcal{C}$-morphism $\varphi : \Sigma^* \to A^*$, L satisfies the equation $\hat{\varphi}(v) \to \hat{\varphi}(u)$. Then one gets the following result:

Theorem 8.1. *A class of language lattices is closed under quotienting and under inverses of $\mathcal{C}$-morphisms if and only if it can be defined by a set of $\mathcal{C}$-identities of the form $u \leqslant v$, where $u, v \in \widehat{\Sigma^*}$.*

In practice, one may consider a $\mathcal{C}$-identity as an equation in which each letter represents a variable. If $\mathcal{C}$ is the class of length-preserving morphisms, these variables can be replaced by letters, if it is the class of length-multiplying morphisms, they can be replaced by words of the same fixed length, etc.

Of course, similar results hold for identities of the form $u \leftrightarrow v$, $u \leqslant v$ or $u = v$. Our main result thus offers multifarious aspects, which are summarized in the following table. Reiterman's theorem corresponds to the strongest assumptions.

Closed under	Equations	Definition
$\cup, \cap$	$u \to v$	$\hat{\eta}(u) \in \hat{\eta}(L) \Rightarrow \hat{\eta}(v) \in \hat{\eta}(L)$
quotienting	$u \leqslant v$	for all x, y, $xuy \to xvy$
complement	$u \leftrightarrow v$	$u \to v$ and $v \to u$
quotienting and complement	$u = v$	for all x, y, $xuy \leftrightarrow xvy$
Closed under inverses of morphisms		**Interpretation of variables**
all morphisms		words
nonerasing morphisms		nonempty words
length multiplying morphisms		words of equal length
length preserving morphisms		letters

9   Examples of Equational Definitions

In this section, we give a few examples of equational characterizations for classes of languages that are not closed under inverses of morphisms and hence do not form a variety of languages. The language A^* is called the *full language*.

9.1   Languages with Zero and Nondense Languages

A *language with zero* is a language whose syntactic monoid has a zero. The class of regular languages with zero is closed under Boolean operations and residuals. According to Proposition 7.4, it has an equational definition, but finding one explicitly requires a little bit of work.

Let us fix a total order on the alphabet A. Let $u_0, u_1, \ldots$ be the ordered sequence of all words of A^+ in the induced shortlex order. For instance, if $A = \{a, b\}$ with $a < b$, the first elements of this sequence would be $1, a, b, aa, ab,$ $ba, bb, aaa, aab, aba, abb, baa, bab, bba, bbb, aaaa, \ldots$ It is proved in [17,5] that the sequence of words $(v_n)_{n \geqslant 0}$ defined by $v_0 = u_0$, $v_{n+1} = (v_n u_{n+1} v_n)^{(n+1)!}$ converges to an idempotent ρ_A of the minimal ideal of $\widehat{A^*}$. We can now state:

Proposition 9.1. *A regular language has a zero if and only if it satisfies the equation $x\rho_A = \rho_A = \rho_A x$ for all $x \in A^*$.*

Proof. Let L be a regular language and let $\eta : A^* \to M$ be its syntactic monoid. Since ρ_A belongs to the minimal ideal of $\widehat{A^*}$, $\hat{\eta}(\rho_A)$ is an element of the minimal ideal of M. In particular, if M has a zero, $\hat{\eta}(\rho_A) = 0$ and L satisfies the equations $x\rho_A = \rho_A = \rho_A x$ for all $x \in A^*$.

Conversely, assume that L satisfies these equations. Let $m \in M$ and let $x \in A^*$ be such that $\eta(x) = m$. Then the equations $\hat{\eta}(x\rho_A) = \hat{\eta}(\rho_A) = \hat{\eta}(\rho_A x)$ give $m\hat{\eta}(\rho_A) = \hat{\eta}(\rho_A) = \hat{\eta}(\rho_A)m$, showing that $\hat{\eta}(\rho_A)$ is a zero of M. Thus L has a zero. $\qquad\square$

In the sequel, we shall use freely the symbol 0 in equations to mean that a monoid has a zero. For instance the equation $x \leqslant 0$ of Theorem 9.2 below should be formally replaced by the three equations $x\rho_A = \rho_A = \rho_A x$ and $x \leqslant \rho_A$.

A language L of A^* is *dense* if, for every word $u \in A^*$, $L \cap A^* u A^* \neq \emptyset$. Note that dense languages are not closed under intersection: $(A^2)^*$ and $(A^2)^* A \cup \{1\}$ are dense, but their intersection is not dense. However, one can show that regular nondense or full languages form a lattice of languages closed under quotients.

We now give an equational description of the form foretold by Theorem 7.2.

Theorem 9.2. *A language of A^* is nondense or full if and only if it satisfies the equations $x \leqslant 0$ for all $x \in A^*$.*

9.2   Languages Defined by Density

The *density* of a language $L \subseteq A^*$ is the function which counts the number of words of length n in L. More formally, it is the function $d_L : \mathbb{N} \to \mathbb{N}$ defined by $d_L(n) = |L \cap A^n|$. See [23] for a general reference.

If $d_L(n) = O(1)$, then L is called a *slender language*. It is well known that a regular language is slender if and only if it is a finite union of languages of the form xu^*y, where $x, u, y \in A^*$. Regular slender languages form a lattice of languages closed under residuals and morphisms.

Note that if $|A| \leqslant 1$, all regular languages are slender. For $|A| \geqslant 2$, slender or full languages admit a simple equational characterization. Let us denote by $i(u)$ the first letter (or *initial*) of a word u.

Theorem 9.3. *Suppose that* $|A| \geqslant 2$. *A regular language of* A^* *is slender or full if and only if it satisfies the equations* $x \leqslant 0$ *for all* $x \in A^*$ *and the equation* $x^\omega u y^\omega = 0$ *for each* $x, y \in A^+$, $u \in A^*$ *such that* $i(uy) \neq i(x)$.

We now also consider the Boolean closure of slender languages. A language is called *coslender* if its complement is slender.

Theorem 9.4. *Suppose that* $|A| \geqslant 2$. *A regular language of* A^* *is slender or coslender if and only if its syntactic monoid has a zero and satisfies the equations* $x^\omega u y^\omega = 0$ *for each* $x, y \in A^+$, $u \in A^*$ *such that* $i(uy) \neq i(x)$.

Note that if $A = \{a\}$, the language $(a^2)^*$ is slender but its syntactic monoid, the cyclic group of order 2, has no zero. Therefore the condition $|A| \geqslant 2$ in Theorem 9.4 is mandatory.

A language is *sparse* if it has polynomial density, that is, if $d_L(n) = O(n^k)$ for some $k > 0$. It is well known that a regular language is sparse if and only if it is a finite union of languages of the form $u_0 v_1^* u_1 \cdots v_n^* u_n$, where $u_0, v_1, \ldots,$ v_n, u_n are words. Regular sparse languages from a lattice of languages and are closed under concatenation product, morphisms and residuals.

Theorem 9.5. *Suppose that* $|A| \geqslant 2$. *A regular language of* A^* *is sparse or full if and only if it satisfies the equations* $x \leqslant 0$ *for all* $x \in A^*$ *and the equations* $(x^\omega y^\omega)^\omega = 0$ *for each* $x, y \in A^+$ *such that* $i(x) \neq i(y)$.

Pursuing the analogy with slender languages, we consider now the Boolean closure of sparse languages. A language is *cosparse* if its complement is sparse.

Theorem 9.6. *Suppose that* $|A| \geqslant 2$. *A regular language of* A^* *is sparse or cosparse if and only if its syntactic monoid has a zero and satisfies the equations* $(x^\omega y^\omega)^\omega = 0$ *for each* $x, y \in A^+$ *such that* $i(x) \neq i(y)$.

10 Conclusion

We proved that every lattice of regular languages is given by an equational theory, a result that subsumes Eilenberg's variety theorem and its extensions to positive varieties and $\mathcal{C}$-varieties. One could further extend this result to classes of regular languages only closed under finite intersection by using the syntactic semiring introduced by Polák [15]. Our result could also be adapted to languages of infinite words, words over ordinals or linear orders, and even perhaps to tree languages.

Our second main result does not in itself give a new result in the theory of automata and semigroups, but it reveals a very strong link between two theories pertaining to the foundations of computer science: the theory of relational semantics for non-classical (modal, intuitionistic, many-valued, etc.) logics on the one side and the algebraic theory of automata on the other. We have indicated how the fundamental tools of semigroup theory fit into the duality perspective, obtaining an extensive repertoire of equational theories as a modular family of results so typical of modal correspondence theory. Further duality results will be presented in the full version of this paper.

References

1. Adams, M.E.: The Frattini sublattice of a distributive lattice. Alg. Univ. 3, 216–228 (1973)
2. Almeida, J.: Residually finite congruences and quasi-regular subsets in uniform algebras. Partugaliæ Mathematica 46, 313–328 (1989)
3. Almeida, J.: Finite semigroups and universal algebra. World Scientific Publishing Co. Inc., River Edge (1994)
4. Almeida, J.: Profinite semigroups and applications. In: Structural theory of automata, semigroups, and universal algebra. NATO Sci. Ser. II Math. Phys. Chem., vol. 207, pp. 1–45. Springer, Dordrecht (2005); Notes taken by Alfredo Costa
5. Almeida, J., Volkov, M.V.: Profinite identities for finite semigroups whose subgroups belong to a given pseudovariety. J. Algebra Appl. 2(2), 137–163 (2003)
6. Birkhoff, G.: On the structure of abstract algebras. Proc. Cambridge Phil. Soc. 31, 433–454 (1935)
7. Eilenberg, S.: Automata, languages, and machines, vol. B. Academic Press [Harcourt Brace Jovanovich Publishers], New York (1976)
8. Ésik, Z.: Extended temporal logic on finite words and wreath products of monoids with distinguished generators. In: Ito, M., Toyama, M. (eds.) DLT 2002. LNCS, vol. 2450, pp. 43–58. Springer, Heidelberg (2003)
9. Goldblatt, R.: Varieties of complex algebras. Ann. Pure App. Logic 44, 173–242 (1989)
10. Kunc, M.: Equational description of pseudovarieties of homomorphisms. Theoretical Informatics and Applications 37, 243–254 (2003)
11. Pin, J.-E.: A variety theorem without complementation. Russian Mathematics (Iz. VUZ) 39, 80–90 (1995)
12. Pin, J.-É., Straubing, H.: Some results on $\mathcal{C}$-varieties. Theoret. Informatics Appl. 39, 239–262 (2005)
13. Pin, J.-É., Weil, P.: A Reiterman theorem for pseudovarieties of finite first-order structures. Algebra Universalis 35, 577–595 (1996)
14. Pippenger, N.: Regular languages and Stone duality. Theory Comput. Syst. 30(2), 121–134 (1997)
15. Polák, L.: Syntactic semiring of a language. In: Sgall, J., Pultr, A., Kolman, P. (eds.) MFCS 2001. LNCS, vol. 2136, pp. 611–620. Springer, Heidelberg (2001)
16. Priestley, H.A.: Representation of distributive lattices by means of ordered Stone spaces. Bull. London Math. Soc. 2, 186–190 (1970)
17. Reilly, N.R., Zhang, S.: Decomposition of the lattice of pseudovarieties of finite semigroups induced by bands. Algebra Universalis 44(3-4), 217–239 (2000)
18. Reiterman, J.: The Birkhoff theorem for finite algebras. Algebra Universalis 14(1), 1–10 (1982)
19. Stone, M.: The theory of representations for Boolean algebras. Trans. Amer. Math. Soc. 40, 37–111 (1936)
20. Stone, M.H.: Applications of the theory of Boolean rings to general topology. Trans. Amer. Math. Soc. 41(3), 375–481 (1937)
21. Straubing, H.: On logical descriptions of regular languages. In: Rajsbaum, S. (ed.) LATIN 2002. LNCS, vol. 2286, pp. 528–538. Springer, Heidelberg (2002)
22. Weil, P.: Profinite methods in semigroup theory. Int. J. Alg. Comput. 12, 137–178 (2002)
23. Yu, S.: Regular languages. In: Rozenberg, G., Salomaa, A. (eds.) Handbook of language theory, ch. 2, vol. 1, pp. 679–746. Springer, Heidelberg (1997)

Reversible Flowchart Languages and the Structured Reversible Program Theorem

Tetsuo Yokoyama[1], Holger Bock Axelsen[2], and Robert Glück[2]

[1] NCES, Graduate School of Information Science, Nagoya University
[2] DIKU, Department of Computer Science, University of Copenhagen
yokoyama@nagoya-u.jp, funkstar@diku.dk, glueck@acm.org

Abstract. Many irreversible computation models have reversible counterparts, but these are poorly understood at present. We introduce reversible flowcharts with an assertion operator and show that any
reversible flowchart can be simulated by a structured reversible flowchart
using only three control flow operators. Reversible flowcharts are *r-
Turing-complete*, meaning that they can simuluate reversible Turing machines without garbage data. We also demonstrate the *injectivization* of
classical flowcharts into reversible flowcharts. The reversible flowchart
computation model provides a theoretical justification for low-level machine code for reversible microprocessors as well as high-level block-
structured reversible languages. We give examples for both such
languages and illustrate them with a lossless encoder for permutations
given by Dijkstra.

1 Introduction

In the microprocessor industry, the *circuit model*, based on well-known logical
connectives such as OR and AND, reigns supreme. In recent years, however,
energy efficiency has become an increasing concern, since standard desktop processors dissipate on the order of 100W of power, which must be removed as heat.
Lowering power consumption while increasing computing power is a non-trivial
obstacle for the microprocessor industry, and efforts to do this have involved
computer science, physics and engineering.

Non-standard models of computing have therefore received increased attention [17]. One such model is *reversible computing*, which is the only approach
known to date that can circumvent the hard, physical barrier to the energy
efficiency of irreversible computations (such as the ubiquitous NAND-gate).
This physical barrier, the *von Neumann-Landauer* limit, provides a strict lower
boundary to the energy dissipated as heat with every bit of information destroyed, whence *irreversibility*. Reversible computing, as well as *reversible programming*, are poorly understood at present. This is unfortunate, since a good
understanding of reversible computing is also essential for *quantum computing*,
in that every operation on a quantum state must be *unitary*, and therefore invertible and reversible. Low-power CMOS and quantum computing are two of
the possible applications for the reversible computing model.

L. Aceto et al. (Eds.): ICALP 2008, Part II, LNCS 5126, pp. 258–270, 2008.

A reversible computing model allows deterministic time-invertible computations, in which not only the next computation state, but also the previous computation state is determined uniquely by the current state. All computations are forward and backward deterministic. Store updates are non-destructive. Although there are several reversible computation models, such as reversible Turing machines [2] and invertible cellular automata [18], they are not sufficiently program-oriented to relate theoretical considerations and recent practical developments [7,9].

Most modern programming languages are *imperative*, with block-structured control flow operators (CFOs) such as if and while. Structured programs are more readable and maintainable [6]. The theoretical foundation for structured programming is the classic *Structured Program Theorem* from the 1960s [4,5], which guarantees that any unstructured program can be written using only three structured CFOs: *sequence*, *selection* and *loop*. The same property is desirable for reversible programming languages, but it is not obvious that it should carry over from classical computing models.

The main goal of the present paper is to provide the theoretical justification for the design, translation and computational strength of high-level imperative reversible languages, such as Janus [14,19,20], and low-level machine code for reversible architectures, such as the Pendulum microprocessor [9,1]. The flowchart model is well suited for this purpose, as it accommodates both low-level aspects such as jumps and high-level aspects such as structured control flow operators.

We identify three reversible CFOs that are sufficient for the definition of a structured reversible flowchart language. We show that reversible flowcharts are *r-Turing-complete*, in that they can simulate *reversible Turing machines* without garbage data. We show the *injectivization* of classical flowcharts into reversible flowcharts, indicating that the latter are Turing-complete, if garbage data necessary for the injectivity of the computed function are disregared. We present examples of how programming languages based on reversible flowcharts can be designed, along with two code examples.

2   Reversible Flowcharts

Flowcharts have been used extensively in the study of programming languages. Most programming languages used today have a control flow, which can be easily modeled by flowcharts, making the latter important analytical tools in programming language theory (*e.g.*, [4,5,10,12,15]).

Reversible flowcharts. A *reversible flowchart F* is a finite directed graph with three kinds of nodes, each representing an *atomic operation* (Fig. 1): a *step* performs an elementary operation on the store specified by a transition function a; a *test* dispatches the control flow depending on the value of predicate e; and an *assertion* is a join point that passes incoming control flow through, depending on the value of predicate e. Computation in a flowchart proceeds sequentially along the directed graph of F. A *well-formed* flowchart has exactly one entry and one exit. An interpretation of a flowchart F consists of a domain X (*e.g.*, a store)

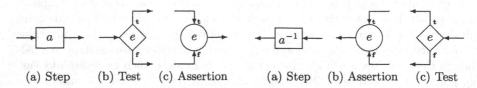

(a) Step   (b) Test   (c) Assertion   (a) Step   (b) Assertion   (c) Test

Fig. 1. Atomic operations of reversible flowcharts

Fig. 2. Inverted atomic operations of reversible flowcharts

and an appropriate association with the partial transition functions $(a : X \rightharpoonup X)$ and the predicates $(e : X \to Bool)$.

The transition function a of each step must be *locally invertible*, defined as having an inverse transition function a^{-1} that can be determined without referring to a's context or location in a flowchart.

The assertion operator is also new (Fig. 1(c)): Predicate e must be **true** when the control flow reaches the join point along the **true**-edge (labeled **t**) and **false** when the control flow reaches the join point along the **false**-edge (labeled **f**); otherwise, the operation is undefined. The operator is represented by a circle.

In classical flowcharts, the join points are not associated with a predicate and there is no information about the incoming control flow. Classical join points are sources of *backward non–determinism*, which break reversibility, as do non-invertible transition functions. Reversible flowcharts remove these sources.

Structured reversible flowcharts. Similar to classical flowcharts, reversible flowcharts allow unstructured control flow. Needless to say, it is easy to construct incomprehensible "spaghetti code" with unstructured reversible flowcharts.

A *structured control flow operator* (structured CFO) has exactly one entry and one exit. We define three structured reversible CFOs (Fig. 3): *sequence*, *selection*, and *loop*. A block B_i is either a locally invertible step (as above) or one of the three structured reversible CFOs. The latter can be nested any number of times. The constructs are all symmetric. A *structured reversible flowchart* is one constructed from locally invertible steps and structured reversible CFOs. Structured control flow makes a program modular and easier to verify.

The selection corresponds to an irreversible **if**-statement but has an exit assertion e_2. The loop is repeated as long as test e_1 and assertion e_2 are false. The loop corresponds to an irreversible **while** loop if B_1 is empty and to an irreversible **do-while** loop if B_2 is empty. In either case, the assertion at the loop entry and the test at the loop exit make the loop reversible.

Inverse flowcharts. Starting with a reversible flowchart, structured or unstructured, the following method can be used to generate an *inverse flowchart*: (1) change the direction of each arrow, (2) replace each transition function a with its inverse a^{-1}, and (3) replace each test by an assertion and each assertion by a test (the predicate e remains unchanged).

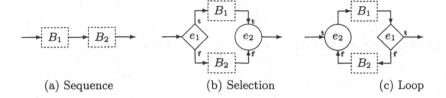

(a) Sequence   (b) Selection   (c) Loop

Fig. 3. Structured reversible CFOs

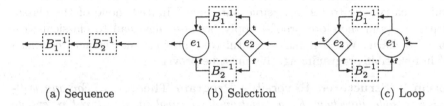

(a) Sequence   (b) Selection   (c) Loop

Fig. 4. Inverted structured reversible CFOs

Fig. 2 shows the inverse of each atomic operator in Fig. 1, where a^{-1} is the inverse transition function. Similarly, Fig. 4 shows the inverse of each CFO in Fig. 3, where B^{-1} is the inverse flowchart of B.

The flowchart resulting from inversion is also reversible, whether structured or unstructured. Inversion does not add or delete atomic operations or CFOs. Repeating the inversion once more restores the original reversible flowchart. The transformation is purely local and does not require global analyses or changes in control flow beyond changing the direction of the arrows. The ease with which reversible flowcharts are inverted is a unique property of this computation model and makes it an attractive analytical tool for program complexity [13]. In general, it is difficult to construct an inverse flowchart mechanically from a classical flowchart. Clearly, programming reversible flowcharts is quite different from programming classical flowcharts.

3   The Structured Reversible Program Theorem

Nowadays, it is easy to forget that the uses and benefits of structure in high-level programming languages were controversial. From a computational viewpoint, this debate was effectively closed by the *Structured Program Theorem* [4], which showed that structured and unstructured flowcharts have the same expressive power. Thus, the useful ancillary benefits of structured high-level languages, including their increased readability and being much easier to reason about, had no computational weaknesses.

The same question is relevant to the reversible computation paradigm. Reversible computing is sufficiently different from standard computational models

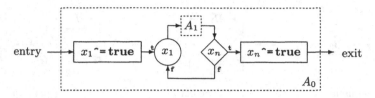

Fig. 5. Flowchart A_0 with main loop

that it is unclear whether results from classical (backward non-deterministic) computing carry over to the reversible paradigm.[1] Indeed, none of the classic constructions (and therefore proofs) apply because they lead to classical irreversible flowcharts. While it may be intuitively obvious that structure is also "free" in reversible programming, this must be proven.

Theorem 1 (Structured Reversible Program Theorem). *For any well-formed reversible flowchart F, a functionally equivalent structured reversible flowchart A_0, with at most a single reversible loop, can be constructed.*

Proof. Let F be a well-formed reversible flowchart and n be the number of edges in F. Let the domain of the transition functions and predicates of F be X. Below we construct a functionally equivalent *structured reversible flowchart A_0* over a trivial extension of X. For this, we label every edge in F uniquely by l_i $(1 \leq i \leq n)$. Without loss of generality, label the entry edge l_1, the exit edge l_n, and the two incoming edges of any assertion l_i and l_{i+1}.

The main idea of the proof is as follows. Each node and its incoming edges is translated to a structured equivalent. A main loop simulates the control flow of F one node at a time, by keeping track of which edge the execution follows in F. This *edge state* is modeled by adding a fresh Boolean variable x_i for each edge l_i to the domain X. The initial and final value of each x_i is **false**. The edge state is called i if x_i is **true** and all other x_j's are **false**. Thus, if in F the control flow is at edge l_i, then the edge state in A_0 should be i. The edge state is changed from i to j by using an injective transition function $P_{i,j}$ that executes $x_i\,\hat{}\texttt{=}\textbf{true}; x_j\,\hat{}\texttt{=}\textbf{true}$.[2] In F this corresponds to moving from edge l_i to edge l_j.

First, generate the main loop in Fig. 5 for entry edge l_1 and exit edge l_n. Flowchart A_0 is a reversible loop between an initial and a final step. The initial step sets x_1 to **true**. If test x_n is **true** the loop ends; otherwise, the loop continues. The path back to assertion x_1 is a skip operation.

Then build the structured reversible flowcharts A_i $(0 < i < n)$ by the rules in Fig. 6, where the atomic operation with incoming edge l_i (or l_i and l_{i+1}) in

---

[1] As an example, given a bounded store (*i.e.* a finite number of possible configurations), computations cannot be guaranteed to terminate for classical flowcharts. This is *not* true in reversible computing, where a bounded store *is* sufficient to obtain termination for well-formed reversible flowcharts.

[2] $x_i\,\hat{}\texttt{=}\textbf{true}$ is shorthand for $x_i := x_i \oplus \textbf{true}$, where $\oplus$ is logical *exclusive-or*. This is an injective (reversible) step.

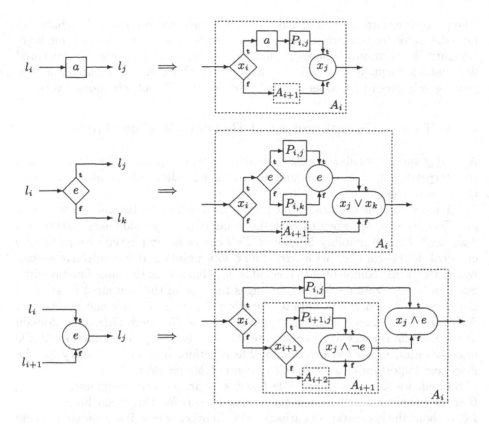

Fig. 6. Unstructured operations transformed into structured reversible flowcharts A_i

the left column is translated into the flowchart A_i in the right column. A dashed box A_j inside A_i stands for a well-formed flowchart simulating the execution of control flow along an edge l_j over exactly one node.

(1) A *step* with transition function a is executed in the translated flowchart only if the state is i. The state is then changed to j and x_j becomes **true**, simulating the control flow over the step. If the state is not i, then A_{i+1} is entered. By the unique numbering of edges, after executing A_{i+1} the state cannot be j, so an assertion of x_j is sufficient to distinguish between the two possibilities.

(2) A *test* is similar to a step. $P_{i,j}$ and $P_{i,k}$ change variables x_i, x_j and x_k. Thus, the value of predicate e in the test and the assertion must be the same, and depending on e the edge state is set to either j or k. By an argument analogous to the *step* case, the assertion $x_j \vee x_k$ is sufficient to distinguish this from A_{i+1}.

(3) Edges l_i and l_{i+1} of an *assertion* are translated simultaneously, so A_i contains A_{i+1} and A_{i+2}. Predicate e differentiates between the two possibilities.

Finally, for well-formedness, we insert a dummy step A_n (*e.g.*, an identity step) although it is never reached in a computation. The structured reversible flowchart A_0 generated by the rules in Fig. 5 and 6 thus simulates the execution of every step, test and assertion in the flowchart F. □

The proof is constructive, and shows that a structured reversible flowchart can be constructed from an arbitrary reversible flowchart with exactly the same functionality. Thus, from a computational viewpoint, structured and unstructured flowcharts are equally powerful, even in the reversible computing paradigm. The proof was inspired by Cooper's global proof sketch [5], but was more involved.

4 r-Turing Completeness of Reversible Flowcharts

At first glance, reversible flowcharts may not seem as powerful as their classical counterparts, which do not require assertions and allow any transformation on the store in steps.

First, we shall demonstrate that reversible flowcharts with unbounded space are *Turing-complete*, provided that the generation of *garbage data*, extraneous data needed for reversibility, is ignored. This can be accomplished by *injectivizing* classical flowcharts (*i.e.*, with irreversible join points and non-injective steps), which are Turing-complete, into reversible flowcharts with the same functionality. Such an injectivization effectively changes the *type* of the computed function: if the classical flowchart F computes function $f : X \rightarrow Y$, then the injectivized flowchart will compute a function $f_g : X \rightarrow Y \times G$, where G is some domain of garbage data, necessary to guarantee that f_g is an injective function. While injectivization works for any computable function, it is *not* necessary for the large and important class of *injective, computable functions*.[3]

Second, we show that reversible flowcharts are *r-Turing-complete*, meaning that they can compute the same functions as *reversible* Turing machines *cleanly*, *i.e.* without the generation of garbage data. In other words, if a reversible Turing machine (RTM) computes the injective function $f : X \rightarrow Y$, then f is computable without garbage data in reversible flowcharts (and by Thm. 1, in structured flowcharts). Since RTMs can cleanly compute *any* injective, computable function [3,13], so can reversible structured and unstructured flowcharts.

Theorem 2 (Injectivization of Classical Flowcharts). *For any well-formed classical flowchart F, a reversible flowchart F_h with the same functionality modulo the accumulation of garbage data can be constructed.*

Proof. As shown above, the irreversibility of classical flowchart is due to irreversible steps (non-injective transformations) and join points without assertions. To translate a classical flowchart into a reversible flowchart, the store will be extended with a *history stack* **h** to record the information required to reconstruct the previous computation state. The stack is associated with the two standard operations **push** and **pop**, which are inverse to each other. The operation **top** is used to check the top element of the stack. A join point is injectivized as follows.

---

[3] A classical computation example is *lossless* audio codecs. Every operation on a quantum state in a quantum computer must be *unitary*, and therefore injective.

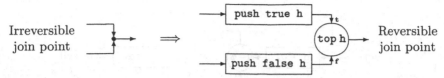

The injectivization of steps is similar: assume that every step is an assignment $x := e$, which overwrites x with the value of expression e. Replace every step with one computing $\mathtt{push}\ x\ \mathtt{h};\ x\ \verb|^|= e$, which saves the original value of x on $\mathtt{h}$. The resulting reversible flowchart F_h is an injectivized version of F. □

Corollary 1 (Input Embedding). *The* input embedding $f_i : x \mapsto (f(x), x)$ *of the function* f *computed by a classical flowchart* F *can be computed by a reversible flowchart* F_i.

Proof. Given a classical flowchart F computing f, (1) obtain a injectivized reversible flowchart F_h computing $f_h : x \mapsto (f(x), h)$, where h is the garbage (history) induced by Thm. 2. (2) Invert F_h to obtain F_h^{-1} which computes $f_h^{-1} : (f(x), h) \mapsto x$. (3) Construct F_i, which executes F_h, copies the values of all output variables into fresh variables, and executes F_h^{-1}. This rolls back the execution of F_h, clearing the history stack and restoring the initial values of all variables used by F_h. Flowchart F_i returns both the original input and the output of executing flowchart F, and therefore computes f_i. □

RTMs are usually defined using quadruple rules [2,16], instead of the more common quintuple rules. A quadruple TM is defined by a finite set of states Q, a finite set of symbols S, and a finite set of symbol rules $\langle q_1, s_1, s_2, q_2 \rangle$ and shift rules $\langle q_1, /, d, q_2 \rangle$. A symbol rules says that in state q_1 with the tape head reading symbol s_1, write s_2 and change into state q_2. A shift rule says that in state q_1, move the tape head in the direction $d \in \{-, 0, +\}$ (left, stay, right) and change into state q_2. For a TM to be *reversible* there must be both forward determinism (in the usual sense) and backward determinism. A quadruple TM is *backward deterministic* iff for any pair of distinct quadruples $\langle q_1, t_1, t_2, q_2 \rangle$ and $\langle q_1', t_1', t_2', q_2' \rangle$, if $q_2 = q_2'$ then $t_1 \neq /$, $t_1' \neq /$ and $t_2 \neq t_2'$.

Theorem 3 (r-Turing completeness). *Any reversible Turing machine can be simulated cleanly (without added garbage) by reversible flowcharts.*

Proof. The configuration of a Turing machine can be simulated as follows. $\mathtt{q}$ is a variable whose value is the current state, $\mathtt{s}$ holds the symbol under the tape head and $\mathtt{l}$ and $\mathtt{r}$ are stacks holding the left and right portions of the tape relative to the tape head, respectively.[4]

For a given RTM, assume that q_s and q_f are the start and finish states, respectively, and that the transition rules are numbered R_1 to R_n. Each transition rule R_i is translated into a functionally equivalent reversible flowchart C_i according to the rules shown in Fig. 8. The helper function Q_{q_1, q_2} consist of the

---

[4] For convenience we assume the stacks are infinitely deep. Finite stacks will work as well, although care must be taken to maintain reversibility.

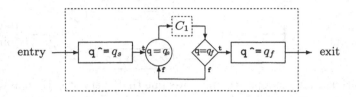

Fig. 7. Flowchart C_0 with main loop for RTM simulation

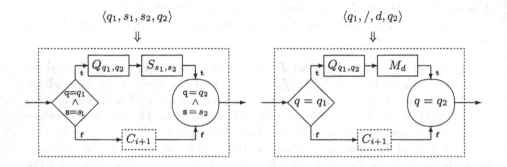

Fig. 8. Translation of RTM transition rules R_i into reversible flowcharts C_i

step $\mathsf{q\ \hat{}=}q_1$; $\mathsf{q\ \hat{}=}q_2$. This changes q's value from q_1 to q_2 reversibly, simulating changing the state of the RTM from q_1 to q_2. S_{s_1,s_2} is entirely analogous for the symbol variable s. The step M_d simulates moving the head in direction d. For example, M_+ is defined (reversibly) as $\mathsf{push\ s\ 1; pop\ s\ r}$. In the translation of both rule types, if rule R_i did not apply (enforced by the test predicate), control flows into C_{i+1}, the translation of rule R_{i+1}. Upon return from C_{i+1}, backward determinism of the RTM ensures that the given assertions are sufficient to differentiate between the two cases. In the translation of the final rule R_n, a dummy step is inserted in place of C_{n+1}. Execution of C_1 thus simulates the application of exactly the rule implied by the (simulated) configuration of the RTM.

C_1 can be embedded in a reversible loop C_0 that executes C_1 repeatedly, starting in state q_s until the final state q_f is reached (Fig. 7). C_0 thus computes the same function as the RTM, without the generation of garbage data.   □

5   Reversible Flowchart Programming Languages

The reversible flowchart computation model provides a theoretical justification for low-level unstructured *machine code* (*e.g.*, for a reversible microprocessor) as well as for high-level *block-structured reversible languages*. We give two programming languages as examples for both types and illustrate them with a garbage-free implementation of a lossless encoder for permutations given by Dijkstra [8].

Grammar of reversible language RL

$$
\begin{array}{lll}
p ::= b^+ & k ::= \text{from } l; & j ::= \text{goto } l; \\
b ::= l: k \; a^* \; j & \quad\mid\; \text{if } e \text{ from } l \text{ else } l; & \quad\mid\; \text{if } e \text{ goto } l \text{ else } l; \\
a ::= x \; \hat{} = e; & \quad\mid\; \text{entry}; & \quad\mid\; \text{exit};
\end{array}
$$

Grammar of structured reversible language SRL **Expressions**

$$
\begin{array}{ll}
p ::= b & b ::= a \\
a ::= x \; \hat{} = e; & \quad\mid\; b \; b \\
& \quad\mid\; \text{if } e \text{ then } b \text{ else } b \text{ fi } e \\
& \quad\mid\; \text{from } e \text{ do } b \text{ loop } b \text{ until } e
\end{array}
\qquad
\begin{array}{l}
e ::= c \mid x \mid o \; e \cdots e \\
o ::= + \mid * \mid \cdots
\end{array}
$$

Syntax Domains

$$
\begin{array}{llll}
p \in \text{Prog} & a \in \text{Assign} & j \in \text{Jump} & k \in \text{From} & l \in \text{Label} \\
b \in \text{BasicBlock} & e \in \text{Expr} & c \in \text{Const} & x \in \text{Var} & o \in \text{Op}
\end{array}
$$

Fig. 9. A family of reversible flowchart languages

The encoder implements an injective function. The decoder can be obtained from the encoder using the straightforward inversion of Sec. 2, and vice versa.

Unstructured Reversible Language. A program written in the unstructured reversible language RL is a sequence of basic blocks. A block consists of a label, an unconventional *from* construct, a sequence of assignments, and a jump. A jump may be unconditional (goto l), conditional (if e goto l_1 else l_2), or the exit from the program (exit). The values of all variables are initially zero. The syntax is shown in Fig. 9.

An assignment is a C-like exclusive-or assignment ($x \; \hat{} = e$), where variable x must not occur in expression e. This syntactic constraint makes the assignment self-inverse. In general, any reversible update can be used as assignment operator to the language (*e.g.*, the C-like assignment operators += and -=; see Sec. 2).

A from construct is an *unconditional assertion* (from l) that the control flow always comes from block l, a *conditional assertion* (if e from l_1 else l_2) that the control flow comes from block l_1 when predicate e is true and from block l_2 otherwise, or the entry of the program (entry). This construct makes the control flow of programs backward deterministic. Well-formed programs contain exactly one entry and one exit.

Structured Language. A program written in the structured reversible language SRL consists of one, possibly nested block. A block is an assignment, a sequence of blocks, a conditional (if e_1 then b_1 else b_2 fi e_2), or a loop (from e_1 do b_1 loop b_2 until e_2). They textually represent the reversible structured CFOs of Fig. 3. The syntax is shown in Fig. 9, with operational semantics rules omitted for space reasons.

Example: permutation-to-code. Consider the problem of translating an array x[] of length n, containing a permutation of the numbers $0, \ldots, n-1$, into an array where each index entry i counts the number of elements in x[] *smaller* than x[i] *preceding* the occurrence x[i] in x[]. For example, given the input permutation x[]={2,0,3,1,5,4}, we obtain the encoded array x[]={0,0,2,1,4,4}.

This is a fine program inversion example described by Dijkstra, who used an irreversible guarded commands language to write this program [8,11].

The structured reversible program in SRL is shown below in the left column. The right column shows the inverted program: a decoder that reconstructs the original permutation. Note that the encoder and decoder take the same number of steps on the corresponding input/output and that their space consumption is identical (due to the assumption that atomic step += and its inverse -= consume equal execution time and space). For simplicity, we use the reversible update operators += and -=, which can be simulated by ^= and auxiliary variables.

```
from  k=n                                      from  k=0
loop  k-=1                                      loop  j+=k
      from  j=0                                       from  j=k
      loop  if x[j]>x[k]                               loop  j-=1
            then x[j]-=1          ⟺                          if x[j]>=x[k]
            fi x[j]>=x[k]      program                       then x[j]+=1
            j+=1              inversion                      fi x[j]>x[k]
      until j=k                                       until j=0
      j-=k                                            k+=1
until k=0                                       until k=n
```

The same program can be expressed in the unstructured reversible language RL. The program can easily be inverted (omitted due to lack of space.)

```
10: entry;                          15: if x[j]>=x[k] from 14 else 13;
    goto 11;                            goto 16;
11: if k=n from 10 else 17;         16: from 15;
    k-=1;                               j+=1;
    goto 12;                            if j=k goto 17 else 12;
12: if j=0 from 11 else 16;         17: from 16;
    goto 13;                            j-=k;
13: from 12;                            if k=0 goto 18 else 11;
    if x[j]>x[k] goto 14 else 15;   18: from 17;
14: from 13;                            exit;
    x[j]-=1;
    goto 15;
```

Admittedly, both RL and SRL are small reversible programming languages. Their purpose is theoretical: to model unstructured control flow of low-level reversible machine code with jumps and register updates, high-level reversible languages with structured control flow and assignments and the clean translation and interpretation of these languages within the reversible computing paradigm (*e.g.*, garbage-free reversible self-interpretation [20]).

6   Conclusion

We introduced the concept of reversible flowcharts, and showed that structured and unstructured reversible flowcharts are equally expressive. We demonstrated an injection of classical flowcharts, and proved the r-Turing completeness of reversible flowcharts. The work presented here is part of a larger effort on the development of reversible programming systems, *e.g.* [1,7,9,19,20]. The results of this paper can be guidelines in designing new structured and unstructured reversible programming languages, independent of actual implementation.

Acknowledgments. An abstract of this paper was presented at the informal, unrefereed 18th Nordic Workshop on Programming Theory, 2006. Part of this work was supported by CREST, JST; and the FIRST research school.

References

1. Axelsen, H.B., Glück, R., Yokoyama, T.: Reversible machine code and its abstract processor architecture. In: Computer Science - Theory and Applications. Proceedings. LNCS, vol. 4649, pp. 56–69. Springer, Heidelberg (2007)
2. Bennett, C.H.: Logical reversibility of computation. IBM J. Res. Dev. 17(6), 525–532 (1973)
3. Bennett, C.H.: Time/space trade-offs for reversible computation. SIAM J. Comput. 18(4), 766–776 (1989)
4. Böhm, C., Jacopini, G.: Flow diagrams, Turing machines and languages with only two formation rules. Communications of the ACM 9(5), 366–371 (1966)
5. Cooper, D.C.: Böhm and Jacopini's reduction of flow charts. Communications of the ACM 10(8), 463–473 (1967)
6. Dahl, O.-J., Dijkstra, E.W., Hoare, C.A.R. (eds.): Structured Programming. Academic Press, London (1972)
7. De Vos, A., Van Rentergem, Y.: Reversible computing: from mathematical group theory to electronical circuit experiment. In: 2nd Conf. on Computing Frontiers, pp. 35–44. ACM Press, New York (2005)
8. Dijkstra, E.W.: Program inversion. In: Bauer, F.L., Broy, M. (eds.) Program Construction: Intl. Summer School. LNCS, vol. 69, pp. 54–57. Springer, Heidelberg (1978)
9. Frank, M.P.: Reversibility for Efficient Computing. PhD thesis. MIT, Cambridge (1999)
10. Gomard, C.K., Jones, N.D.: Compiler generation by partial evaluation: a case study. Structured Programming 12, 123–144 (1991)
11. Gries, D.: The Science of Programming, ch.21: Inverting Programs, Texts and Monographs in Computer Science. Springer, Heidelberg (1981)
12. Hatcliff, J.: An introduction to online and offline partial evaluation using a simple flowchart language. In: Hatcliff, J., Mogensen, T., Thiemann, P. (eds.) Partial Evaluation. Practice and Theory. LNCS, vol. 1706, pp. 20–82. Springer, Heidelberg (1999)
13. Jacopini, G., Mentrasti, P., Sontacchi, G.: Reversible Turing machines and polynomial time reversibly computable functions. SIAM Journal on Discrete Mathematics 3(2), 241–254 (1990)
14. Lutz, C.: Janus: a time-reversible language. Letter written to Landauer, R. (1986), http://www.cise.ufl.edu/~mpf/rc/janus.html
15. Manna, Z.: Mathematical Theory of Computation. McGraw-Hill, New York (1974)
16. Morita, K., Yamaguchi, Y.: A universal reversible Turing machine. In: Durand-Lose, J., Margenstern, M. (eds.) Machines, Computations, and Universality. Proceedings. LNCS, vol. 4664, pp. 90–98. Springer, Heidelberg (2007)

17. Munakata, T.: Beyond silicon: New computing paradigms. Special issue. Communications of the ACM 50(9), 30–72 (2007)
18. Toffoli, T.: Computation and construction universality of reversible cellular automata. J. Comput. Sys. Sci. 15, 213–231 (1977)
19. Yokoyama, T., Axelsen, H.B., Glück, R.: Principles of a reversible programming language. In: 5th Conf. on Computing Frontiers, pp. 43–54. ACM Press, New York (2008)
20. Yokoyama, T., Glück, R.: A reversible programming language and its invertible self-interpreter. In: Partial Evaluation and Program Manipulation. Proceedings, pp. 144–153. ACM Press, New York (2007)

Attribute Grammars and Categorical Semantics

Shin-ya Katsumata

Research Institute for Mathematical Sciences,
Kyoto University Kyoto, 606-8502, Japan
sinya@kurims.kyoto-u.ac.jp

Abstract. We give a new formulation of attribute grammars (AG for short) called *monoidal AGs* in traced symmetric monoidal categories. Monoidal AGs subsume existing domain-theoretic, graph-theoretic and relational formulations of AGs. Using a 2-categorical aspect of monoidal AGs, we also show that every monoidal AG is equivalent to a synthesised one when the underlying category is closed, and that there is a sound and complete translation from local dependency graphs to relational AGs.

1 Introduction

Attribute grammars are a mechanism to assign computation with bidirectional information flow to derivation trees of context free grammars [18]. Our intention is to give a categorical formulation of AGs. We employ *traced symmetric monoidal categories* (TSMC for short) as the underlying categories for the formulation.

The key notion that links AGs and TSMCs is the circular (or recursive) computation.

Circular computation is tightly related to the characteristic feature of AGs, namely computation with bidirectional information flow. To illustrate this, we consider the situation that an AG assigns to a derivation tree (top left of Figure 1) a computation with bidirectional information flow (top right of Figure 1). Boxes P, Q, R are computation units assigned by the AG to nodes p, q, r in the tree. Depending on the configuration of P, Q, R, the entire computation may involve circular computation. For instance, on the bottom of Figure 1 the box P feed-backs the input from R to Q so that the entire computation has a cycle.

It is therefore natural to formulate AGs in a mathematical theory that admits circular computation. In [16], Joyal et al introduced the concept of *traced monoidal categories*. From the viewpoint of computer science, they provide an

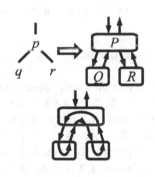

Fig. 1. Simple Description of Attribute Grammars

abstract account of feedback-loops, iteration and recursion in the models of computation, such as domain theory, iteration theory [5], Conway theory [6], relational models of flowcharts and networks [4], and so on.

The main observation of this paper is that by employing TSMCs as the underlying mathematical theory, we can achieve higher degree of abstraction in AGs. Following this observation, we propose a categorical formulation of AGs called *monoidal AGs*.

L. Aceto et al. (Eds.): ICALP 2008, Part II, LNCS 5126, pp. 271–282, 2008.
© Springer-Verlag Berlin Heidelberg 2008

The merit of this formulation is that we are free from concrete representation of data structures and computation. We show that three existing formulations of AGs: 1) Chirica and Martin's K-systems [8], 2) Dependency graphs in classical AGs and 3) Courcelle and Deransart's relational AGs [9] are formally related to the instances of monoidal AGs. Subsequently, by exploiting a 2-categorical aspect of monoidal AGs we show that in closed TSMCs every monoidal AG is equivalent to the one which does not use inherited attributes, and that there exists a sound and complete translation from local dependency graphs into relational AGs. The latter result, which technically hinges on Selinger's work [21], appears to be new.

Preliminaries. We adopt the standard algebraic treatment of CFGs. We regard a CFG $G = (T, N, S, P)$ as a many-sorted signature $\Sigma_G = (N, P)$ by identifying each production rule $p : X_0 \rightarrow v_0 X_1 v_1 \cdots X_n v_n \in P$ ($v_i \in T^*, X_i \in N, 0 \le i \le n$) and an operator $p : X_1 \cdots X_n \rightarrow X_0$. We also identify the set of derivation trees of G beginning with a non-terminal symbol $X \in N$ and the set $T_{\Sigma_G} X$ of closed Σ_G-terms of X. In this paper terminal symbols and the starting symbol do not play any role, so when declaring a CFG we just mention the set of nonterminal symbols and production rules.

The following concepts will be used in classical AGs. We fix a countably infinite set *Attr* of *attribute names*, and assume that it is closed under prefixing "*i.*" ($i \in \mathbf{N}$). A *named set* is a finite sequence of pairs of an attribute name and a set such that each attribute name in the sequence is different. For a named set $R = a_1 : V_1, \cdots, a_n : V_n$, by $|R|$ we mean $V_1 \times \cdots \times V_n$; for $x \in |R|$, by x_{a_i} we mean the i-th component of x; by $a(R)$ we mean $a_1, \cdots, a_n \in Attr^*$; by $n(R)$ we mean $\{a_1, \cdots, a_n\} \subseteq Attr$; by $i.R$ ($i \in \mathbf{N}$) we mean the named set $i.a_1 : V_1, \cdots, i.a_n : V_n$. For $l = a_1, \cdots, a_n \in Attr^*$ with distinct attribute names, by X^l we mean the named set $a_1 : X, \cdots, a_n : X$. For a pair of tuples a, b, by $a; b$ we mean the concatenation of them; for example, $(a, b); (c, d) = (a, b, c, d)$.

2   Classical AGs

We first informally describe the central idea of AGs. Let $G = (N, P)$ be a CFG. An AG $\mathcal{A}$ assigns a "computation unit" f_p (top of Figure 2) to each production rule $p : X_1 \cdots X_n \rightarrow X_0 \in P$. The computation unit has an I/O-port for X_0 at the top and n I/O-ports for $X_1 \cdots X_n$ at the bottom ($\mathcal{A}$ also specifies types of I/O ports, but we ignore them now). The unit processes all inputs and outputs simultaneously, regardless of direction. The information flowing downward is called *inherited attributes*, while the one flowing upward *synthesised attributes*. Given a derivation tree t of G, we construct a complex circuit by connecting computation units provided by $\mathcal{A}$ with each other according to the shape of t. The resulting circuit, which has an I/O port only at the top, is the computation assigned to t by the AG (bottom of Figure 2).

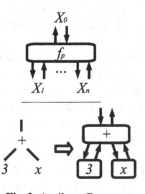

Fig. 2. Attribute Grammars

The above idea was proposed and formulated by in [18], where computation units were represented by set-theoretic functions. Fix a CFG $G = (N, P)$. A *classical AG* for G is the tuple $\mathcal{A} = (I, S, f)$ where

1. I and S are N-indexed family of named sets such that $n(I X) \cap n(S X) = \emptyset$ for each $X \in N$. Below we write U_p for the named set $1. S X_1, \cdots, n. S X_n, I X_0$ and D_p for the named set $S X_0, 1. I X_1, \cdots, n. I X_n$.
2. f is a P-indexed family of functions such that $f_p : |U_p| \rightarrow |D_p|$ for each $p : X_1 \cdots X_n \rightarrow X_0 \in P$. They are called *attribute calculation rule*. By expanding the definition of U_p and D_p, we may also see f_p as the following function:

$$f_p : |S X_1| \times \cdots \times |S X_n| \times |I X_0| \rightarrow |S X_0| \times |I X_1| \times \cdots \times |I X_n|. \tag{1}$$

The assignment of computation to derivation trees is done by *meaning functions*. Let $\mathcal{A}$ be an AG for G. A N-indexed family of functions

$$\mathcal{A}[\![-]\!]_X : T_{\Sigma_G} X \rightarrow (|I X| \rightarrow |S X|) \tag{2}$$

(subscript X is often dropped) is called the *meaning function* of $\mathcal{A}$ if it satisfies the following condition: for any $p : X_1 \cdots X_n \rightarrow X_0 \in P$, $t_i \in T_{\Sigma_G} X_i$ $(1 \leq i \leq n)$ and $x \in |I X_0|$, there exists $x_i \in |I X_i|$ $(1 \leq i \leq n)$ such that

$$\mathcal{A}[\![p(t_1, \cdots, t_n)]\!](x); x_1; \cdots ; x_n = f_p(\mathcal{A}[\![t_1]\!](x_1); \cdots ; \mathcal{A}[\![t_n]\!](x_n); x). \tag{3}$$

Example 1. Consider a CFG G_p for expressions over integers:

$$G_p = (\{V, n, +\}, \{E\}, E, \{c_n : E \rightarrow n, \mathsf{var} : E \rightarrow V, \mathsf{plus} : E \rightarrow E + E\}),$$

where n ranges over $\mathbf{Z}$. The following data gives an AG $\mathcal{A}_p = (I, S, f)$ for G_p:

$$I E = i : \mathbf{R}, \qquad S E = s : \mathbf{R}$$

$$f_{c_n}(i) = n, \quad f_{\mathsf{var}}(i) = i, \quad f_{\mathsf{plus}}(1.s, 2.s, i) = (1.s + 2.s, i, i).$$

If a meaning function $\mathcal{A}_p[\![-]\!] : T_{\Sigma_{G_p}} E \rightarrow \mathbf{R} \rightarrow \mathbf{R}$ exists, then from (3) it satisfies

$$\mathcal{A}_p[\![c_n]\!](i) = n, \quad \mathcal{A}_p[\![\mathsf{var}]\!](i) = i, \quad \mathcal{A}_p[\![\mathsf{plus}(t, t')]\!](i) = \mathcal{A}_p[\![t]\!](i) + \mathcal{A}_p[\![t']\!](i).$$

Thus the meaning function of $\mathcal{A}_p$ evaluates expressions over integers with real numbers.

The problem of classical AGs is that the existence of meaning functions is not always guaranteed. This is technically because the witnesses $x_1, \cdots, x_n$ ensuring (3) may not exist under some situation. Another way to look at the problem is that the computation of value $\mathcal{A}[\![p(t_1, \cdots, t_n)]\!]$ requires feed-backs of the output $x_1, \cdots, x_n$ of f_p to itself, but such circular computation can not be modelled in a naive way using set-theoretic functions.

To resolve this problem, we shall either i) seek for AGs that do not induce circular computation (such AGs are called *non-circular* or *well-formed* [18,10]) or ii) reformulate AGs within a mathematical theory that admits circular computation, such as domain theory. In this paper we take the latter option. For the mathematical foundation of the formulations of AGs, we employ *traced symmetric monoidal categories* [16,14], which are recently recognised as providing an abstract representation of circular computation.

3   Traced Symmetric Monoidal Categories and Int Construction

We assume that readers are familiar with *symmetric monoidal categories* (*SMC* for short), *symmetric monoidal functors* and *monoidal natural transformations*; see e.g. [20]. We fix a common method for taking tensors of multiple objects in SMCs. Every SMC is equivalent to a strict one (coherence theorem [20]), so we mainly talk about strict SMCs for legibility. We reserve notations $\mathbf{I}, \otimes$ and c for the unit, tensor product and symmetry for SMCs, respectively.

In a SMC $\mathbb{C}$, one can represent a computation with n inputs and m outputs as a $\mathbb{C}$-morphism $f : A_1 \otimes \cdots \otimes A_n \to B_1 \otimes \cdots \otimes B_m$. In order to express feedback loops / circular computation under this representation, we adopt the concept of *trace operators*. They were originally introduced to balanced monoidal categories (which subsume SMCs) by Joyal et al in [16]. The following formulation of trace operators on SMCs is due to Hasegawa [14].

$$\mathbf{Tr}_{A,B}^{\mathbf{I}}(f) = f$$
$$\mathbf{Tr}_{A,B}^{X \otimes Y}(f) = \mathbf{Tr}_{A,B}^{X}(\mathbf{Tr}_{A \otimes X, B \otimes X}^{Y}(f))$$
$$\mathbf{Tr}_{C \otimes A, C \otimes B}^{X}(C \otimes f) = C \otimes \mathbf{Tr}_{A,B}^{X}(f)$$
$$\mathbf{Tr}_{X,X}^{X}(c_{X,X}) = \mathrm{id}_X$$
$$\mathbf{Tr}_{A,B}^{X}(f \circ (g \otimes X)) = \mathbf{Tr}_{A',B}^{X}(f) \circ g$$
$$\mathbf{Tr}_{A,B'}^{X}((g \otimes X) \circ f) = g \circ \mathbf{Tr}_{A,B}^{X}(f)$$
$$\mathbf{Tr}_{A,B}^{X}((B \otimes g) \circ f) = \mathbf{Tr}_{A,B}^{Y}(f \circ (B \otimes g))$$

Fig. 3. Axioms for Trace Operators

Definition 1 ([16,14]). *A trace operator on a SMC $\mathbb{C}$ is a family of mappings* $\mathbf{Tr}_{A,B}^{X} : \mathbb{C}(A \otimes X, B \otimes X) \to \mathbb{C}(A, B)$ *that satisfies the axioms summarised in Figure 3 (see [16,14,2] for graphical presentations of the axioms). A traced symmetric monoidal category (TSMC) is a pair of a SMC and a trace operator on it.*

Let $\mathbb{C}, \mathbb{D}$ be TSMCs. A traced symmetric monoidal functor is a strong symmetric monoidal functor $(F : \mathbb{C} \to \mathbb{D}, m_I : \mathbf{I}_{\mathbb{D}} \xrightarrow{\cong} F\mathbf{I}_{\mathbb{C}}, m_{A,B} : FA \otimes_{\mathbb{D}} FB \xrightarrow{\cong} F(A \otimes_{\mathbb{C}} B))$ *that preserves the trace operator in the following sense:*

$$(\mathbf{Tr}_{\mathbb{D}})_{FA,FB}^{FX}(m_{A,B}^{-1} \circ Ff \circ m_{A,B}) = F((\mathbf{Tr}_{\mathbb{C}})_{A,B}^{X}(f)).$$

Besides trace operators, in [16] Joyal et al gave a construction of categories called **Int**. It was originally considered for the structure theorem for traced balanced monoidal categories. In this paper **Int** construction will be used for obtaining the categories where computation with bidirectional information flow can be naturally modeled.

Definition 2 ([16]). *Let $\mathbb{C}$ be a TSMC. We define a category* **Int**$(\mathbb{C})$ *by the following data: an object is a pair (A^-, A^+) of $\mathbb{C}$-objects[1], and a morphism $f : (A^-, A^+) \to (B^-, B^+)$ is a $\mathbb{C}$-morphism $f : A^+ \otimes B^- \to B^+ \otimes A^-$. The composition of f with $g : (B^-, B^+) \to (C^-, C^+)$ is defined to be the following morphism:*

$$\mathbf{Tr}_{A^+ \otimes C^-, C^+ \otimes A^-}^{B^-}((C^+ \otimes c) \circ (g \otimes A^-) \circ (B^+ \otimes c) \circ (f \otimes C^-) \circ (A^+ \otimes c)).$$

Consider a computation unit that has an input port A^+ and an output port A^- at the bottom, and an input port B^- and an output port B^+ at the top. This unit receives information from the bottom via A^+ and from the top via B^-, then outputs processed information to the bottom via A^- and to the top via B^+. In **Int**$(\mathbb{C})$ such a unit is expressed

---

[1] Compared to the original **Int** construction in [16], here the order of objects is swapped.

as a morphism $f : (A^-, A^+) \to (B^-, B^+)$, and its input-output relation is captured by a $\mathbb{C}$-morphism $f : A^+ \otimes B^- \to A^- \otimes B^+$. The definition of the composition in $\mathbf{Int}(\mathbb{C})$ is designed so that it correctly captures the input-output relation of the composition of two computation units ($\mathbf{Int}(\mathbb{C})$-morphisms); see [16,1,2] for graphical presentations of the composition.

Category $\mathbf{Int}(\mathbb{C})$ is a *compact closed category*, that is, a SMC such that every object has a left dual [17]. In this paper we only use the SMC structure of $\mathbf{Int}(\mathbb{C})$ given by

$$\mathbf{I}_{\mathbf{Int}(\mathbb{C})} = (\mathbf{I}, \mathbf{I}) \qquad (A^-, A^+) \otimes_{\mathbf{Int}(\mathbb{C})} (B^-, B^+) = (A^- \otimes B^-, A^+ \otimes B^+).$$

This tensor products correspond to combining I/O ports (and computation units) in parallel. For instance, the computation unit drawn on the top of Figure 2 can be expressed as an $\mathbf{Int}(\mathbb{C})$-morphism $f_p : (X_1^-, X_1^+) \otimes \cdots \otimes (X_n^-, X_n^+) \to (X_0^-, X_0^+)$.

Below we state the structure theorem for TSMCs. This is a specialisation of the one for traced balanced monoidal categories in [16].

Theorem 1. *The mapping $\mathbb{C} \mapsto \mathbf{Int}(\mathbb{C})$ can be extended to a left biadjoint to the forgetful functor from the 2-category of compact closed categories to that of TSMCs. The unit $N_{\mathbb{C}} : \mathbb{C} \to \mathbf{Int}(\mathbb{C})$ of this biadjunction, which maps a $\mathbb{C}$-object A to an $\mathbf{Int}(\mathbb{C})$-object $(\mathbf{I}, A)$, is full and faithful.*

4   Monoidal AGs

In this section we give a categorical formulation of AGs, called *monoidal AGs*. We first introduce the concept of Σ-algebras for SMCs, which are a monoidal version of set-theoretic many-sorted algebras. We note that the concept of algebras in SMCs are also related to *operads* [19].

Definition 3. *Let $\Sigma = (S, O)$ be a signature and $\mathbb{C}$ be a SMC. A Σ-algebra in $\mathbb{C}$ is a pair (A, α) such that A is a S-indexed family of $\mathbb{C}$-objects and α is a O-indexed family of $\mathbb{C}$-morphisms such that $\alpha_o : As_1 \otimes \ldots \otimes As_n \to As$ for each $o : s_1 \ldots s_n \to s \in O$.*

Let $\mathcal{A} = (A, \alpha)$ be a Σ-algebra in $\mathbb{C}$. The meaning function of $\mathcal{A}$ is a S-indexed family of mappings $\{\mathcal{A}[\![-]\!]_s : T_\Sigma s \to \mathbb{C}(\mathbf{I}, As)\}_{s \in S}$ such that the following holds for each $o : s_1 \cdots s_n \to s \in O$ (below we omit subscripts of meaning functions):

$$\mathcal{A}[\![o(t_1, \cdots, t_n)]\!] = \alpha_o \circ (\mathcal{A}[\![t_1]\!] \otimes \cdots \otimes \mathcal{A}[\![t_n]\!]).$$

Definition 4. *A* monoidal AG *for a CFG $G = (N, P)$ in a TSMC $\mathbb{C}$ is a Σ_G-algebra $\mathcal{A} = (A, \alpha)$ in $\mathbf{Int}(\mathbb{C})$.*

This short and simple formulation captures essential information of AGs. We compare monoidal AGs and classical AGs below.

1. The set of sorts of Σ_G is N; so A assigns to each nonterminal symbol $X \in N$ an $\mathbf{Int}(\mathbb{C})$-object, say (A^-X, A^+X). We regard them as the domains of inherited and synthesised attributes respectively; so A plays the role of both I and S.

2. To each production rule $p : X_1 \cdots X_n \to X_0 \in P$, α assigns an $\mathbf{Int}(\mathbb{C})$-morphism $\alpha_p : AX_1 \otimes \cdots \otimes AX_n \to AX_0$, which is the following $\mathbb{C}$-morphism by definition:

$$\alpha_p : A^+X_1 \otimes \cdots \otimes A^+X_n \otimes A^-X_0 \to A^+X_0 \otimes A^-X_1 \otimes \cdots \otimes A^-X_n.$$

One can see the similarity between the domain and codomain of α_p and those of attribute calculation rule (1); here tensor products are used instead of direct products (this is the reason of the name "monoidal" AG).

3. The meaning function of a monoidal AG $\mathcal{A}$ for G is a mapping $(X \in N)$

$$\mathcal{A}[\![-]\!] : T_{\Sigma_G}X \to \mathbf{Int}(\mathbb{C})(\mathbf{I}, (A^-X, A^+X)) \cong \mathbb{C}(A^-X, A^+X),$$

so it assigns to a derivation tree $t \in T_{\Sigma_G}X$ a computation from A^-X to A^+X expressed as a morphism in $\mathbb{C}$; compare this with (2).

To see the suitability of our categorical formulation of AGs, in the subsequent sections we compare instances of monoidal AGs and three existing formulations of AGs: i) Chirica and Martin's K-systems, ii) local dependency graphs in classical AGs and iii) Courcelle and Deransart's relational AGs.

4.1 Monoidal AGs in ωCPPO

The category $\omega\mathbf{CPPO}$ of ω-complete pointed partial orders and continuous functions is Cartesian closed and has the least fixpoint operator $\mathbf{fix}_D : [[D \to D] \to D]$, which determines a trace operator:

$$\mathbf{Tr}_{AB}^U(f)(a) = \pi(\mathbf{fix}_{B \times U}(\lambda(b, u) . f(a, u)));$$

so $\omega\mathbf{CPPO}$ is a traced CCC (for the above construction see [14]).

Monoidal AGs in $\omega\mathbf{CPPO}$ are related to domain-theoretic formulations of AGs. Among various such formulations, here we establish a formal connection between Chirica and Martin's *K-systems* [8] and monoidal AGs. Fix a CFG $G = (N, P)$.

Definition 5 ([8]). *A K-system for G is a tuple $\mathcal{D} = (D^-, D^+, f)$ such that*

- D^- *and* D^+ *are N-indexed family of ω-CPPOs called* inherited *and* synthesised *attribute domains, respectively. For each $X \in N$, we write DX for $D^-X \times D^+X$.*
- f *is a P-indexed family of continuous functions such that for each $p : X_1 \cdots X_n \to X_0 \in P$, $f_p : [DX_0 \times DX_1 \times \cdots \times DX_n \to D^+X_0 \times D^-X_1 \times \cdots \times D^-X_n]$.*

A K-system assigns a continuous function $D^t : [D^-X \to D^+X]$ to a derivation tree $t \in T_{\Sigma_G}X$ $(X \in N)$ as follows. We first recursively define a ω-CPPO D^t by

$$D^{p(t_1,\ldots,t_n)} = D^+X_0 \times D^-X_1 \times \cdots \times D^-X_n \times D^{t_1} \times \cdots \times D^{t_n} \quad (p : X_1 \cdots X_n \to X_0 \in P)$$

For $d \in D^t$, by $\pi(d)$ we mean the first projection of d. Next, we construct a continuous function $H^t : [D^-X \times D^t \to D^t]$ by induction on the structure of t:

$$H^{p(t_1,\ldots,t_n)}(i, (s, i_1, \ldots, i_n, w_1, \ldots, w_n))$$
$$= f_p((i, s), (i_1, \pi(w_1)), \ldots, (i_n, \pi(w_n))); (H^{t_1}(i_1, w_1), \ldots, H^{t_n}(i_n, w_n)).$$

This function congregates one-step computation of inherited and synthesised attributes at every node of t. We then define the continuous function $\mathcal{D}^t : [D^-X \rightarrow D^+X]$ that denotes the meaning of t by $\mathcal{D}^t(i) = \pi(\mathbf{fix}(\lambda x \in D^t . H^t(i, x)))$.

Let $\mathcal{D} = (D^-, D^+, f)$ be a K-system for G. We construct a monoidal AG $M(\mathcal{D}) = (D, \delta)$ for G in $\omega\mathbf{CPPO}$ as follows:

$$DX = (D^-X, D^+X)$$

$$\delta_p(s_1, \ldots, s_n, i) = \mathbf{fix}(\lambda(s, i_1, \ldots, i_n) . f_p((i, s), (i_1, s_1), \ldots, (i_n, s_n)))$$

where $X \in N$ and $p : X_1 \cdots X_n \rightarrow X_0 \in P$. On the other hand, every monoidal AG in $\omega\mathbf{CPPO}$ can be casted to a K-system in an obvious way. These constructions preserve the meanings of Σ_G-terms.

Theorem 2. *Let $\mathcal{D}$ be a K-system for a CFG $G = (N, P)$ and $\mathcal{A}$ be a monoidal AG for G in $\omega\mathbf{CPPO}$. Then for any $t \in T_{\Sigma_G}X$ ($X \in N$), we have $M(\mathcal{D})[\![t]\!] = \mathcal{D}^t$ and $(K(\mathcal{A}))^t = \mathcal{A}[\![t]\!]$.*

4.2 Monoidal AGs in Rel⁺

The category **Rel** of sets and relations has Cartesian (bi)products, which, at object level, takes the disjoint sum of given sets. In [16] it was shown that the following is a trace operator with respect to the Cartesian products:

$$\mathbf{Tr}^U_{AB}(R) = R_{AB} \cup R_{UB} \circ (R_{UU})^* \circ R_{AU},$$

where R_{XY} ($X \in \{A, U\}, Y \in \{B, U\}$) is the restriction R to the relation between X and Y, and $(R_{UU})^*$ is the transitive reflexive closure of R_{UU}. The same operation was also considered in [4]. We call this TSMC **Rel⁺**.

Monoidal AGs in **Rel⁺** are related to the concept of *local dependency graphs* (*LDG* for short) in classical AGs [18,10]. Let $\mathcal{A} = (I, S, f)$ be a classical AG for a CFG $G = (N, P)$. We look at the syntactic definition of f, and assign to each production rule $p : X_1 \cdots X_n \rightarrow X_0 \in P$ the following digraph α_p: the set of vertices is $n(U_p) \cup n(D_p)$, and there is an edge in α_p from $a \in n(U_p)$ to $a' \in n(D_p)$ if and only if the a'-component of the result of f_p depends on the a-component of f_p's input. We usually draw α_p so that $n(I X_0, S X_0)$ are placed at the top and $n(k. I X_k, k. S X_k)$ ($1 \le k \le n$) are placed at the bottom. The family α of digraphs constructed from $\mathcal{A}$ is called the *LDG* of $\mathcal{A}$. For instance, the LDG β of the classical AG $\mathcal{A}_p$ in Example 1 is at the top of Figure 4.

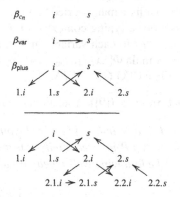

Fig. 4. LDG β of $\mathcal{A}_p$ (top) and an example of CDG (bottom)

LDGs are used to construct *compound dependency graphs* (*CDG* for short) of derivation trees. Let α be a LDG. For $t \in T_{\Sigma_G}X$ ($X \in N$), we recursively construct a graph $\mathrm{CDG}_\alpha(t)$ as follows: $\mathrm{CDG}_\alpha(p(t_1, \cdots, t_n))$ is the union of α_p and the graphs obtained by adding a prefix "k." ($1 \le k \le n$) to every node in $\mathrm{CDG}_\alpha(t_k)$. In the bottom of Figure 4 $\mathrm{CDG}_\beta(\mathsf{plus}(c_3, \mathsf{plus}(\mathsf{var}, c_2)))$ is drawn. CDGs are a primary tool for detecting circular computation in classical AGs; see [18,10].

By letting $AX = (n(\mathrm{I}\,X), n(\mathrm{S}\,X))$, each digraph α_p ($p : X_1 \cdots X_n \to X_0 \in P$) of a LDG α can be identified with a morphism in **Int(Rel$^+$)**:

$$\alpha_p \in \mathcal{P}(n(\mathrm{U}_p) \times n(\mathrm{D}_p)) \cong \mathbf{Int(Rel^+)}(AX_1 \times \cdots \times AX_n, AX_0).$$

Thus a local dependency graph α of a classical AG for a CFG G specifies a monoidal AG $\overline{\alpha} = (A, \alpha)$ for G in **Rel$^+$**.

Theorem 3. *Let α be a LDG of a classical AG $\mathcal{A}$ for a CFG $G = (N, P)$, and $t \in T_{\Sigma_G}X$ ($X \in N$). Then there exists a path from $i \in n(\mathrm{I}\,X)$ to $s \in n(\mathrm{S}\,X)$ in $\mathrm{CDG}_\alpha(t)$ if and only if $(i, s) \in \overline{\alpha}[\![t]\!]$.*

For example, $(i, s) \in \overline{\beta}[\![\mathsf{plus}(c_3, \mathsf{plus}(\mathsf{var}, c_2))]\!]$ as there is a path $i \to 2.i \to 2.1.i \to 2.1.s \to 2.s \to s$ in $\mathrm{CDG}_\beta(\mathsf{plus}(c_3, \mathsf{plus}(\mathsf{var}, c_2)))$, the graph on the bottom of Figure 4.

4.3 Monoidal AGs in Rel$^\times$

The category **Rel** has another symmetric monoidal structure given by $A \otimes B = A \times B$ (Cartesian products of sets). This is a part of the compact closed structure over **Rel**, so **Rel** is canonically traced [16]; we call this TSMC **Rel$^\times$**. The trace operator derived from the compact closed structure is the following:

$$\mathbf{Tr}^U_{AB}(R) = \{(a, b) \in A \times B \mid \exists u \in U \,.\, ((a, u), (b, u)) \in R\}.$$

Monoidal AGs in **Rel$^\times$** are related to Courcelle and Deransart's *relational AGs* [9]. We first fix a many-sorted first-order logic $\mathcal{L}$ with a standard set-theoretic interpretation $[\![-]\!]$. For a typing context Γ of $\mathcal{L}$, by $i.\Gamma$ we mean the context obtained by adding a prefix "$i.$" to each variable in Γ. For a well-typed formula $\Gamma \vdash \Phi$, we define $i.\Phi$ to be the formula $\Phi[i.x/x]_{x \in \mathrm{var}(\Gamma)}$ (where $\mathrm{var}(\Gamma)$ is the set of variables in Γ). Clearly $i.\Gamma \vdash i.\Phi$.

Fix a CFG $G = (N, P)$.

Definition 6 ([9]). *A relational AG for G in $\mathcal{L}$ is a tuple $\mathcal{R} = (\Gamma, \Phi)$ such that*

- *Γ is a N-indexed family of typing contexts and*
- *Φ is a P-indexed family of formuli such that for each $p : X_1 \cdots X_n \to X_0 \in P$, Φ_p is the following well-typed formula:*

$$1.\Gamma X_1, \ldots, n.\Gamma X_n, \Gamma X_0 \vdash \Phi_p.$$

Let $\mathcal{R} = (\Gamma, \Phi)$ be a relational AG for G. For any $t \in T_{\Sigma_G}X$ ($X \in N$), we recursively define a formula $\Gamma X \vdash \Phi_t$ by

$$\Phi_{p(t_1, \ldots, t_n)} = \exists 1.\Gamma X_1, \ldots, n.\Gamma X_n.\Phi_p \wedge 1.\Phi_{t_1} \wedge \ldots \wedge n.\Phi_{t_n} \quad (p : X_1 \cdots X_n \to X_0 \in P).$$

We also define the relation $\mathcal{R}_t$ to be $[\![\Phi_t]\!]$.

From a relational AG $\mathcal{R} = (\Gamma, \Phi)$ for G in $\mathcal{L}$, we construct a monoidal AG in $\mathbf{Rel}^{\times}$ as follows. By letting $AX = (1, [\![\Gamma X]\!])$ for $X \in N$, we notice that the relation $[\![\Phi_p]\!]$ for each $p : X_1 \cdots X_n \to X_0 \in P$ can be identified with a morphism in $\mathbf{Int}(\mathbf{Rel}^{\times})$:

$$[\![\Phi_p]\!] \in \mathcal{P}([\![1.\Gamma X_1, \ldots, n.\Gamma X_n, \Gamma X_0]\!]) \cong \mathbf{Int}(\mathbf{Rel}^{\times})(AX_1 \otimes \ldots \otimes AX_n, AX_0).$$

Therefore $\mathcal{R}$ specifies a monoidal AG $\overline{\mathcal{R}} = (A, \alpha)$ where $\alpha_p = [\![\Phi_p]\!]$.

Theorem 4. *Let $\mathcal{R}$ be a relational AG for a CFG $G = (N, P)$ in $\mathcal{L}$. Then for any $X \in N$ and $t \in T_{\Sigma_G} X$, we have $\mathcal{R}_t = \overline{\mathcal{R}} [\![t]\!]$.*

5 Relating Monoidal AGs

We next see that functors and natural transformations between TSMCs give translations of monoidal AGs and relations between such translations. We begin with some categorical aspects of algebras in SMCs. Fix a signature $\Sigma = (S, O)$.

Definition 7. *Let $\mathcal{A} = (A, \alpha)$ and $\mathcal{B} = (B, \beta)$ be Σ-algebras in a SMC $\mathbb{C}$. A Σ-algebra homomorphism from $\mathcal{A}$ to $\mathcal{B}$ is a S-indexed family $\{h_s : As \to Bs\}_{s \in S}$ of $\mathbb{C}$-morphisms satisfying $\beta_o \circ (h_{s_1} \otimes \ldots \otimes h_{s_n}) = h_s \circ \alpha_o$ for each $o : s_1 \cdots s_n \to s \in O$. We write $\mathbf{Alg}_\Sigma(\mathbb{C})$ for the category of Σ-algebras and Σ-algebra homomorphisms in $\mathbb{C}$.*

We write $\mathbf{SMC}$ for the 2-category of small SMCs, symmetric monoidal functors and monoidal natural transformations. One can easily check that the mapping $\mathbb{C} \mapsto \mathbf{Alg}_\Sigma(\mathbb{C})$ extends to a 2-functor $\mathbf{Alg}_\Sigma : \mathbf{SMC} \to \mathbf{Cat}$. We note that set-theoretic Σ-algebras are precisely captured by $\mathbf{Alg}_\Sigma(\mathbf{Set})$, where $\mathbf{Set}$ is the category of sets and functions with tensors given by Cartesian products.

The meaning function in Definition 3 can be seen as *initial algebra semantics*. We write $\mathcal{T}_\Sigma = (T_\Sigma, \iota)$ for the initial object in $\mathbf{Alg}_\Sigma(\mathbf{Set})$.

Definition 8. *1. Let $\mathbb{C}$ be a SMC. We define a symmetric monoidal functor $G_\mathbb{C} : \mathbb{C} \to \mathbf{Set}$ by $G_\mathbb{C} = \mathbb{C}(\mathbf{I}, -)$.*
2. Let $\mathbb{C}, \mathbb{D}$ be SMCs and $(F : \mathbb{C} \to \mathbb{D}, m_I : \mathbf{I}_\mathbb{D} \to F\mathbf{I}_\mathbb{C}, m_{A,B} : FA \otimes_\mathbb{D} FB \to F(A \otimes_\mathbb{C} B))$ be a symmetric monoidal functor. We define a monoidal natural transformation $G_F : G_\mathbb{C} \to G_\mathbb{D} \circ F$ by $(G_F)_\mathbb{C}(f) = Ff \circ m_I$.
3. For a Σ-algebra $\mathcal{A}$ in a SMC $\mathbb{C}$, by $|\mathcal{A}|$ we mean the underlying set-theoretic Σ-algebra $\mathbf{Alg}_\Sigma(G_\mathbb{C})(\mathcal{A})$. We say that two Σ-algebras in $\mathbb{C}$ are equivalent if their underlying algebras are isomorphic in $\mathbf{Alg}_\Sigma(\mathbf{Set})$.

The meaning function $\mathcal{A} [\![-]\!]$ of a Σ-algebra $\mathcal{A}$ in a SMC $\mathbb{C}$ is equal to the unique morphism $! : \mathcal{T}_\Sigma \to |\mathcal{A}|$ in $\mathbf{Alg}_\Sigma(\mathbf{Set})$. We regard two equivalent algebras as giving the same meaning to Σ-terms, because their meaning functions are equal modulo an isomorphism.

The general algebraic concepts above will be used as follows. Let G be a CFG. The mapping $\mathbf{AG}_G : \mathbb{C} \mapsto \mathbf{Alg}_{\Sigma_G}(\mathbf{Int}(\mathbb{C}))$ is the construction of the *category of monoidal AGs* for G in a TSMC $\mathbb{C}$. It extends to a 2-functor $\mathbf{AG}_G$ from the 2-category of TSMCs to $\mathbf{Cat}$, so relationships between TSMCs will immediately be reflected to those between different formulations of AGs. Below we apply this fact to show that i) in closed TSMCs every monoidal AGs are equivalent to those which do not use inherited attributes, and ii) there is a sound and complete translation from LDGs to relational AGs.

5.1 Equivalence between Monoidal AGs and Synthesised Ones

AGs that do not use inherited attributes are called *synthesised AG* (*S-AG* for short). In [8] Chirica and Martin showed that by using function spaces as attribute domains, every K-system can be reduced to the one which does not use inherited attributes. This technique was also applied to the encoding of AGs by higher-order catamorphisms [11].

In this section we further generalise these results to monoidal AGs. We introduce a monoidal version of synthesised AGs (*monoidal S-AGs*), and show that in *closed* TSMCs every monoidal AG is equivalent to a monoidal S-AG. Fix a CFG $G = (N, P)$.

Definition 9. *A monoidal S-AG for G in a TSMC $\mathbb{C}$ is a monoidal AG (A, α) for G such that $A^- X = \mathbf{I}$ for each $X \in N$.*

It is easy to see that a monoidal AG $\mathcal{A}$ for G in a TSMC $\mathbb{C}$ is S-AG if and only if there exists a Σ_G-algebra $\mathcal{A}'$ in $\mathbb{C}$ such that $\mathcal{A} = \mathbf{Alg}_{\Sigma_G}(\mathcal{N}_{\mathbb{C}})(\mathcal{A}')$.

To show that *every* monoidal AG is equivalent to a monoidal S-AG, we need an extra structure on $\mathbb{C}$. Recall that a SMC $\mathbb{C}$ is *closed* if $- \otimes B$ has a right adjoint $B \multimap -$ for every $\mathbb{C}$-object B. The key of the equivalence is the following theorem due to Hasegawa.

Theorem 5 ([13]). *Let $\mathbb{C}$ be a TSMC. Then $\mathcal{N}_{\mathbb{C}} : \mathbb{C} \to \mathbf{Int}(\mathbb{C})$ has a symmetric monoidal right adjoint $\mathcal{R}_{\mathbb{C}} : \mathbf{Int}(\mathbb{C}) \to \mathbb{C}$ if and only if $\mathbb{C}$ is closed.*

When $\mathbb{C}$ is closed, $\mathcal{R}_{\mathbb{C}}$ can be defined by $\mathcal{R}_{\mathbb{C}}(A^-, A^+) = A^- \multimap A^+$.

Theorem 6. *Let $\mathcal{A}$ be a monoidal AG for G in a closed TSMC $\mathbb{C}$. Then $\mathcal{A}$ is equivalent to the monoidal S-AG $\mathbf{Alg}_{\Sigma_G}(\mathcal{N}_{\mathbb{C}})(\mathbf{Alg}_{\Sigma_G}(\mathcal{R}_{\mathbb{C}})(\mathcal{A}))$.*

5.2 A Translation from Local Dependency Graphs to Relational AGs

Two TSMCs $\mathbf{Rel}^+$ and $\mathbf{Rel}^\times$, which provide the underlying category for local dependency graphs and relational AGs, are linked by the finite multiset endofunctor $\mathcal{M} : \mathbf{Rel} \to \mathbf{Rel}$ defined as follows. First, we define the set MA to be the set of finite multisets of A. We identify an element of MA and a function $f \in A \to \mathbf{N}$ that returns non-zero at finitely many elements in A (so $MA = A \to \mathbf{N}$ if A is finite). We write $\{a\}$ ($a \in A$) for the function that returns 1 only at a. The endofunctor $\mathcal{M}$ is then defined by

$$\mathcal{M}A = MA, \qquad \mathcal{M}R = \{(h_1, h_2) \mid h \in MR\}$$

where $h_1(a) = \sum_{b \in B, (a,b) \in R} h(a, b)$ and $h_2(b) = \sum_{a \in A, (a,b) \in R} h(a, b)$. For any relation $R \subseteq A \times B$, we have $(a, b) \in R$ if and only if $(\{a\}, \{b\}) \in \mathcal{M}R$; so functor $\mathcal{M}$ is faithful.

Theorem 7 ([21]). *The above data gives a TSM functor $\mathcal{M} : \mathbf{Rel}^+ \to \mathbf{Rel}^\times$.*

Fix a CFG $G = (N, P)$. We consider the functor $\mathbf{AG}_G(\mathcal{M}) : \mathbf{AG}_G(\mathbf{Rel}^+) \to \mathbf{AG}_G(\mathbf{Rel}^\times)$ that gives a translation between monoidal AGs for G.

Proposition 1. *Let $\mathcal{A}$ be a monoidal AG for G in* **Rel**$^+$. *Then $(i, s) \in \mathcal{A}[\![t]\!]$ if and only if $(\{i\}, \{s\}) \in (\mathbf{AG}_G(\mathcal{M})(\mathcal{A}))[\![t]\!]$.*

Functor $\mathbf{AG}_G(\mathcal{M})$ is the key of a sound and complete translation from LDGs to relational AGs. Let α be a LDG of a classical AG $\mathcal{A} = (\mathrm{I}, \mathrm{S}, f)$ for G. For each p : $X_1 \ldots X_n \to X_0 \in P$, we define A_p by $A_p = 1.\,\mathrm{I}\,X_1, 1.\,\mathrm{S}\,X_1, \cdots, n.\,\mathrm{I}\,X_n, n.\,\mathrm{S}\,X_n, \mathrm{I}\,X_0, \mathrm{S}\,X_0$. Then we can encode $\mathcal{M}\alpha_p$ as a set of vectors of natural numbers:

$$\mathcal{M}\alpha_p \cong \{f \in |\mathbf{N}^{a(A_p)}| \mid \exists h \in \alpha_p \to \mathbf{N}. \bigwedge_{a \in n(\mathrm{D}_p)} f_a = h_1(a) \wedge \bigwedge_{a \in n(\mathrm{U}_p)} f_b = h_2(a)\}.$$

The relation on the right hand side can be expressed in a first-order logic with natural numbers and the standard interpretation of them (we only need the sort nat for natural numbers, logical connectives $\exists$, $\wedge$, $\top$, $=$ and constants 0, $+$). Therefore from the LDG α we can construct a relational AG $\mathcal{R}_\alpha = (\Gamma, \Phi)$ as follows: for each $X \in N$, we define the context ΓX to be i_1 : nat, $\cdots$, i_n : nat, s_1 : nat, $\cdots$, s_m : nat where $i_1, \cdots, i_n = a(\mathrm{I}\,X)$ and $s_1, \cdots, s_m = a(\mathrm{S}\,X)$, and we define the formula Φ_p ($p : X_1 \ldots X_n \to X_0 \in P$) to be

$$\exists \{h_e\}_{e \in \alpha_p}. \left(\bigwedge_{a \in n(\mathrm{D}_p)} a = \sum_{b \in n(\mathrm{U}_p), (a,b) \in \alpha_p} h_{(a,b)} \right) \wedge \left(\bigwedge_{b \in n(\mathrm{U}_p)} b = \sum_{a \in n(\mathrm{D}_p), (a,b) \in \alpha_p} h_{(a,b)} \right).$$

For a named set R and $a \in n(R)$, by δ_a we mean the tuple $(0, \cdots, 0, \overset{a}{1}, 0, \cdots, 0) \in |\mathbf{N}^{a(R)}|$.

Theorem 8. *Let α be a LDG for a classical AG $\mathcal{A} = (\mathrm{I}, \mathrm{S}, f)$ of a CFG G $= (N, P)$. Then for any $t \in T_{\Sigma_G} X$ ($X \in N$), $i \in \mathrm{I}\,X$ and $s \in \mathrm{S}\,X$, $(\delta_i, \delta_s) \in (\mathcal{R}_\alpha)_t$ if and only if there exists a path from i to s in $\mathrm{CDG}_\alpha(t)$.*

6 Related Work

We have seen that our categorical formulation of AGs, namely monoidal AGs, are related to three existing formulations of AGs; K-systems [8], local dependency graphs [18] and Courcelle and Deransart's relational AGs [9]. However, there are many other formulations of AGs [7,22,15] that are not covered in this paper. The study of relationships between such AGs and monoidal AGs is left to the future work.

In [12] Girard proposed a novel interpretation of cut elimination called geometry of interaction (GoI), which was later analysed by researchers including Abramsky, Haghverdi, Jagadeesan and Scott [1,2,3]. It was revealed that TSMCs and **Int** construction were the key for an axiomatic account of GoI, and as a by-product many concrete TSMCs were investigated; see e.g. [1,2]. It is interesting to examine if monoidal AGs in the TSMCs discovered in the study of GoI are useful in the applications of AGs, such as compiler constructions, program transformations and XML processing.

Acknowledgment. I am indebted to Masahito Hasegawa for technical advises and stimulating discussions. I am also grateful to Susumu Nishimura, Keisuke Nakano, Kazuyuki Asada, Naohiko Hoshino and Ichiro Hasuo for valuable discussions.

References

1. Abramsky, S.: Retracting some paths in process algebra. In: CONCUR 1996. LNCS, vol. 1119, pp. 1–17. Springer, Heidelberg (1996)
2. Abramsky, S., Haghverdi, E., Scott, P.J.: Geometry of interaction and linear combinatory algebras. Math. Struct. in Comput. Sci. 12(5), 625–665 (2002)
3. Abramsky, S., Jagadeesan, R.: New foundations for the geometry of interaction. Inf. Comput. 111(1), 53–119 (1994)
4. Bainbridge, E.S.: Feedbacks and generalized logic. Inf. Control 31(1), 75–96 (1976)
5. Bloom, S.L., Ésik, Z.: Iteration theories; the equational logic of iterative processes. Springer, Heidelberg (1993)
6. Bloom, S.L., Ésik, Z.: Fixed-point operations on ccc's. part I. Theor. Comput. Sci. 155(1), 1–38 (1996)
7. Boyland, J.: Conditional attribute grammars. ACM Trans. Program. Lang. Syst. 18(1), 73–108 (1996)
8. Chirica, L.M., Martin, D.F.: An order-algebraic definition of Knuthian semantics. Math. Sys. Theory 13, 1–27 (1979)
9. Courcelle, B., Deransart, P.: Proofs of partial correctness for attribute grammars with applications to recursive procedures and logic programming. Inf. Comput. 78(1), 1–55 (1988)
10. Deransart, P., Jourdan, M., Lorho, B.: Attribute Grammars. LNCS. vol. 323. Springer, Heidelberg (1988)
11. Fokkinga, M., Jeuring, J., Meertens, L., Meijer, E.: A translation from attribute grammars to catamorphisms. The Squiggolist 2(1), 20–26 (1991)
12. Girard, J.-Y.: Geometry of Interaction I: Interpretation of System F. In: Ferro, R., et al. (eds.) Logic Colloquium 1988. North-Holland, Amsterdam (1989)
13. Hasegawa, M.: On traced monoidal closed categories. Invited talk at Traced Monoidal Categories, Network Algebras, and Applications (2007)
14. Hasegawa, M.: Models of Sharing Graphs: A Categorical Semantics of let and letrec. Springer, Heidelberg (1999)
15. Jacobs, B., Uustalu, T.: Semantics of grammars and attributes via initiality. In: Reflections on Type Theory, Lambda Calculus, and the Mind. Essays Dedicated to Henk Barendregt on the Occasion of his 60th Birthday, pp. 181–196. Radboud University (2007)
16. Joyal, A., Street, R., Verity, D.: Traced monoidal categories. Mathematical Proceedings of the Cambridge Philosophical Society 119(3), 447–468 (1996)
17. Kelly, G.M., Laplaza, M.L.: Coherence for compact closed categories. Journal of Pure and Applied Algebra 19, 193–213 (1980)
18. Knuth, D.E.: Semantics of context-free languages. Math. Sys. Theory 2(2), 127–145 (1968); See Math. Sys. Theory, 5(1) 95–96, 1971 for a correction
19. Leinster, T.: Higher Operads, Higher Categories. London Math. Soc. Lecture Note Series, vol. 298. Cambridge University Press, Cambridge (2004)
20. MacLane, S.: Categories for the Working Mathematician, 2nd edn. Graduate Texts in Mathematics, vol. 5. Springer, Heidelberg (1998)
21. Selinger, P.: A note on Bainbridge's power set construction (manuscript, 1998)
22. Swierstra, S.D., Vogt, H.: Higher order attribute grammars. In: Alblas, H., Melichar, B. (eds.) SAGA School 1991. LNCS, vol. 545, pp. 256–296. Springer, Heidelberg (1991)

A Domain Theoretic Model of Qubit Channels

Keye Martin

Naval Research Laboratory
Center for High Assurance Computer Systems
Washington, DC 20375
keye.martin@nrl.navy.mil

Abstract. We prove that the spectral order provides a domain theoretic model of qubit channels. Specifically, we show that the spectral order is the unique partial order on quantum states whose least element is the completely mixed state, which satisfies the mixing law and has the property that a qubit channel is unital iff it is Scott continuous and has a Scott closed set of fixed points. This result is used to show that the Holevo capacity of a unital qubit channel is determined by the largest value of its informatic derivative. In particular, these channels always have an informatic derivative that is necessarily not a classical derivative.

1 Introduction

The study of measurement was initiated within the context of computation [3]. In [5], it is shown that measurement can be used to prove fixed point theorems for mappings that are not monotone and unique fixed point theorems for mappings that are monotone. Results like these can be used to provide a unified view of numerical algorithms, for instance. In such applications, we are primarily concerned with operators f whose iterates $f^n(x)$ converge to a fixed point p. The informatic derivative $df_\mu(p)$ then measures the *rate* at which f converges to p.

The view of computation taken in the study of measurement, that a computation is a 'process' that evolves on a space of informatic objects, and that as it evolves we can measure the amount of information lost or gained, in retrospect lends itself very naturally to considerations in other areas, such as physics or the study of communication. In [2], it was discovered that natural domain theoretic structure existed in quantum mechanics. And developments such as [7] and [6] establish the importance of domains and measurements in classical information theory.

In this paper, we establish the significance of domain theory and measurement in *quantum* information theory. We first show that a classical binary channel is Scott continuous and has a Scott closed set of fixed points iff it is a binary symmetric channel, while a qubit channel is Scott continuous and has a Scott closed set of fixed points iff it is *unital*. The binary symmetric channels are exactly the entropy increasing binary channels; the unital qubit channels are exactly the entropy increasing qubit channels. One reason such channels are important is that

L. Aceto et al. (Eds.): ICALP 2008, Part II, LNCS 5126, pp. 283–297, 2008.
© Springer-Verlag Berlin Heidelberg 2008

they provide effective ways of interrupting communication. For instance, assuming all inputs are equally likely, the best way to interrupt communication for a fixed probability of error is to use a binary symmetric channel. The class of unital qubit channels includes most of the models used to describe noise: bit flipping, phase flipping, bit-phase flipping, phase damping ("decoherence"), depolarization, unitary channels and projective measurements.

In fact, the connection between entropy increasing channels and Scott continuous channels with Scott closed sets of fixed points also turns out to *uniquely* determine the spectral order on quantum states. We have known since [2] that the unique partial order on $\Delta^2 = \{(x,y) \in [0,1] : x + y = 1\}$ that satisfies the mixing law and has a least element of $\perp = (1/2, 1/2)$ is the Bayesian order. What we prove in this paper is the quantum analogue of this result: the spectral order is the unique partial order on two dimensional mixed quantum states Ω^2 that satisfies the mixing law, has least element $\perp = I/2$ and the *additional property* that every unital channel is Scott continuous and has a Scott closed set of fixed points. This additional property is trivially satisfied in the classical case Δ^2.

Finally, we use these results to give a method for calculating the *Holevo capacity* of a unital qubit channel. Surprisingly, each unital qubit channel has an informatic derivative defined everywhere except $\perp$. The largest value of a channel's informatic derivative determines its Holevo capacity. This informatic derivative is not a classical derivative. This demonstrates a completely new use for informatic rates of change.

2   The Domains of Classical and Quantum States

We review the basic ideas in the study of domains and measurements, and then the two examples of domains that are of interest in this paper.

2.1   Domain Theory and Measurement

A *domain* is a partially ordered set with intrinsic notions of completeness and approximation defined by the order. A *measurement* is a function μ that to each informative object x assigns a number μx which measures the information content of the object x. We now define each of these terms precisely before discussing them further.

The intrinsic notion of completeness that a domain has is that it forms a dcpo:

Definition 1. Let $(P, \sqsubseteq)$ be a partially ordered set or *poset*. A nonempty subset $S \subseteq P$ is *directed* if $(\forall x, y \in S)(\exists z \in S)\, x, y \sqsubseteq z$. The *supremum* $\bigsqcup S$ of $S \subseteq P$ is the least of its upper bounds when it exists. A *dcpo* is a poset in which every directed set has a supremum.

The intrinsic notion of approximation possessed by a domain is formalized by continuity:

Definition 2. Let $(D, \sqsubseteq)$ be a dcpo. For elements $x, y \in D$, we write $x \ll y$ iff for every directed subset S with $y \sqsubseteq \bigsqcup S$, we have $x \sqsubseteq s$, for some $s \in S$. We set

- $\downarrow x := \{y \in D : y \ll x\}$ and $\uparrow x := \{y \in D : x \ll y\}$
- $\downarrow x := \{y \in D : y \sqsubseteq x\}$ and $\uparrow x := \{y \in D : x \sqsubseteq y\}$

and say D is *continuous* if $\downarrow x$ is directed with supremum x for each $x \in D$.

Definition 3. A *domain* is a continuous dcpo. A *Scott domain* is a continuous dcpo in which any pair of elements with an upper bound has a supremum.

Definition 4. The *Scott topology* on a continuous dcpo D has as a basis all sets of the form $\uparrow x$ for $x \in D$. A set $S \subseteq D$ is *Scott closed* if it is a lower set that is closed under directed suprema.

A function $f : D \to E$ between domains is *Scott continuous* if the inverse image of a Scott open set in E is Scott open in D. This is equivalent [1] to saying that f is *monotone*,

$$(\forall x, y \in D)\, x \sqsubseteq y \Rightarrow f(x) \sqsubseteq f(y),$$

and that it *preserves directed suprema*:

$$f\left(\bigsqcup S\right) = \bigsqcup f(S),$$

for all directed $S \subseteq D$. In particular, for the domain $[0, \infty)^*$ of nonnegative reals in their opposite order, a Scott continuous function $\mu : D \to [0, \infty)^*$ will satisfy

1. For all $x, y \in D$, $x \sqsubseteq y \Rightarrow \mu x \geq \mu y$, and
2. If (x_n) is an increasing sequence in D, then

$$\mu\left(\bigsqcup_{n \geq 1} x_n\right) = \lim_{n \to \infty} \mu x_n.$$

This is the case of Scott continuity that pertains to measurements:

Definition 5. A Scott continuous $\mu : D \to [0, \infty)^*$ is said to *measure the content* of $x \in D$ if for all Scott open sets $U \subseteq D$,

$$x \in U \Rightarrow (\exists \varepsilon > 0)\, x \in \mu_\varepsilon(x) \subseteq U$$

where

$$\mu_\varepsilon(x) := \{y \in D : y \sqsubseteq x \ \& \ |\mu x - \mu y| < \varepsilon\}$$

are called the ε-*approximations* of x.

We often refer to μ as 'measuring' $x \in D$ or as measuring $X \subseteq D$ when it measures each element of X.

Definition 6. A *measurement* $\mu : D \to [0, \infty)^*$ is a Scott continuous map that measures the content of $\ker(\mu) := \{x \in D : \mu x = 0\}$.

In this paper, all measurements μ we work with measure all of D. This implies [5] that they are strictly monotone:

$$x \sqsubseteq y \ \& \ \mu x = \mu y \Rightarrow x = y$$

This property enables definition of the informatic derivative:

Definition 7. Let (D, μ) be a domain with a measurement μ that measures all of D. If $f : D \to D$ is a function and $p \in D$ is not compact, then

$$df_\mu(p) = \lim_{x \to p} \frac{\mu f(x) - \mu f(p)}{\mu x - \mu p}$$

is called the *informatic derivative* of f, provided that this limit exists.

2.2   The Bayesian Order on Classical States

The set of classical states

$$\Delta^2 := \{(x, y) \in [0, 1]^2 : x + y = 1\}$$

has a natural domain theoretic structure introduced in [2]:

Definition 8. For $x, y \in \Delta^2$,

$$x \sqsubseteq y \equiv (y_1 \leq x_1 \leq 1/2) \text{ or } (1/2 \leq x_1 \leq y_1).$$

The relation $\sqsubseteq$ on Δ^2 is called the *Bayesian order*.

This order is derived from the graph of entropy $H(x) = -x \log_2(x) - (1 - x) \log_2(1 - x)$ as follows:

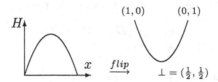

Theorem 1 ([2]). $(\Delta^2, \sqsubseteq)$ *is a Scott domain with maximal elements*

$$\max(\Delta^2) = \{(0, 1), (1, 0)\}$$

and least element $\bot = (1/2, 1/2)$. *The Shannon entropy* $H : \Delta^2 \to [0, \infty)^*$, *given by*

$$H(x) = -x_1 \log_2(x_1) - x_2 \log_2(x_2)$$

is a measurement.

2.3   The Spectral Order on Quantum States

Let $\mathcal{H}^2$ denote an two dimensional complex Hilbert space with specified inner product $\langle \cdot | \cdot \rangle$.

Definition 9. A *quantum state* is a density operator $\rho : \mathcal{H}^2 \to \mathcal{H}^2$, i.e., a self-adjoint, positive, linear operator with $\mathrm{tr}(\rho) = 1$. The quantum states on $\mathcal{H}^2$ are denoted Ω^2.

Quantum states are also sometimes call density operators or *mixed states*. The set of eigenvalues of an operator ρ, called the *spectrum* of ρ, is denoted $\mathrm{spec}(\rho)$.

Definition 10. A quantum state ρ on $\mathcal{H}^2$ is *pure* if

$$\mathrm{spec}(\rho) \subseteq \{0, 1\}.$$

The set of pure states is denoted Σ^2. They are in bijective correspondence with the one dimensional subspaces of $\mathcal{H}^2$.

Classical states are distributions on the set of pure states $\max(\Delta^2)$. An analogous result holds for quantum states: density operators encode distributions on the set of pure states Σ^2.

Definition 11. A *quantum observable* is a self-adjoint linear operator $e : \mathcal{H}^2 \to \mathcal{H}^2$.

Now, if we have the operator e representing the energy observable of a system (for instance), then its spectrum $\mathrm{spec}(e)$ consists of the actual energy values a system may assume. If our knowledge about the state of the system is represented by density operator ρ, then quantum mechanics predicts the probability that a measurement of observable e yields the value $\lambda \in \mathrm{spec}(e)$. It is

$$\mathrm{pr}(\rho \to e_\lambda) := \mathrm{tr}(p_e^\lambda \cdot \rho),$$

where p_e^λ is the projection corresponding to eigenvalue λ and e_λ is its associated eigenspace in the *spectral representation* of e.

Definition 12. Let e be an observable on $\mathcal{H}^2$ with $\mathrm{spec}(e) = \{1, 2\}$. For a quantum state ρ on Ω^2,

$$\mathrm{spec}(\rho|e) := (\mathrm{pr}(\rho \to e_1), \mathrm{pr}(\rho \to e_2)) \in \Delta^2.$$

We assume that all observables e have $|\mathrm{spec}(e)| = 2$. Intuitively, then, e is an experiment on a system which yields one of 2 different outcomes; if our a priori knowledge about the state of the system is ρ, then our knowledge about what the result of experiment e *will be* is $\mathrm{spec}(\rho|e)$. Thus, $\mathrm{spec}(\rho|e)$ determines our ability to *predict* the result of the experiment e.

Let $[a, b] = ab - ba$ denote the commutator of operators.

Definition 13. For quantum states $\rho, \sigma \in \Omega^2$, we have $\rho \sqsubseteq \sigma$ iff there is an observable $e : \mathcal{H}^2 \to \mathcal{H}^2$ such that $[\rho, e] = [\sigma, e] = 0$ and $\mathrm{spec}(\rho|e) \sqsubseteq \mathrm{spec}(\sigma|e)$ in Δ^2.

This is called the *spectral order* on quantum states.

Theorem 2 ([2]). $(\Omega^2, \sqsubseteq)$ *is a Scott domain with maximal elements*

$$\max(\Omega^2) = \Sigma^2$$

and least element $\perp = I/2$, *where* I *is the identity matrix. The von Neumann entropy* $S : \Omega^2 \to [0, \infty)^*$ *given by* $S(\rho) = -\mathrm{tr}(\rho \log_2(\rho))$ *is a measurement.*

The Hilbert space formalism makes things seem much more complicated than they really are in this case: the spectral order on Ω^2 has a much simpler description which we now consider.

There is a 1-1 correspondence between density operators on a two dimensional state space and points on the unit ball $\mathbb{B}^3 = \{x \in \mathbb{R}^3 : |x| \leq 1\}$: each density operator $\rho : \mathcal{H}^2 \to \mathcal{H}^2$ can be written uniquely as

$$\rho = \frac{1}{2} \begin{pmatrix} 1 + r_z & r_x - i r_y \\ r_x + i r_y & 1 - r_z \end{pmatrix}$$

where $r = (r_x, r_y, r_z) \in \mathbb{R}^3$ satisfies $|r| = \sqrt{r_x^2 + r_y^2 + r_z^2} \leq 1$. The vector $r \in \mathbb{B}^3$ is called the *Bloch vector* associated to ρ. Bloch vectors have a number of aesthetically pleasing properties.

If ρ and σ are density operators with respective Bloch vectors r and s, then (i) the eigenvalues of ρ are $(1 \pm |r|)/2$, (ii) the von Neumann entropy of ρ is $S\rho = H((1 + |r|)/2) = H((1 - |r|)/2)$, where $H : [0, 1] \to [0, 1]$ is the base two Shannon entropy, (iii) if ρ and σ are pure states and $r + s = 0$, then ρ and σ are orthogonal, and thus form a basis for the state space; conversely, the Bloch vectors associated to a pair of orthogonal pure states form antipodal points on the sphere, (iv) the Bloch vector for a convex sum of mixed states is the convex sum of the Bloch vectors, (v) the Bloch vector for the completely mixed state $I/2$ is $0 = (0, 0, 0)$.

Because of the correspondence between Ω^2 and $\mathbb{B}^3$, we regard the two as equal for the rest of the paper.

Example 1. From [2], using the Bloch representation of density operators, the spectral order on Ω^2 is given by $x \sqsubseteq y$ iff the line from the origin $\perp$ to y passes through x. That is,

$$x \sqsubseteq y \equiv (\exists p \in [0, 1]) \, x = py$$

for all $x, y \in \Omega^2$.

3   Classical and Quantum Channels

We review classical binary channels, qubit channels and then a special subclass of each of them: the entropy increasing channels.

3.1   Classical Channels

A binary channel has two inputs ("0" and "1") and two outputs ("0" and "1"). An input is sent through the channel to a receiver. Because of noise in the channel, what arrives may not necessarily be what the sender intended. The effect of noise on input data is modelled by a noise matrix u. If data is sent through the channel according to the distribution x, then the output is distributed as $y = x \cdot u$. The noise matrix u is given by

$$u = \begin{pmatrix} a & \bar{a} \\ b & \bar{b} \end{pmatrix}$$

where $a = P(0|0)$ is the probability of receiving 0 when 0 is sent and $b = P(0|1)$ is the probability of receiving 0 when 1 is sent and $\bar{x} := 1 - x$ for $x \in [0, 1]$. Thus, the noise matrix of a binary channel can be represented by a point (a, b) in the unit square $[0, 1]^2$ and all points in the unit square represent the noise matrix of some binary channel.

The noise matrix u of a binary channel defines a function $f : \Delta^2 \to \Delta^2$, given by $f(x) = x \cdot u$, which maps an input distribution $x \in \Delta^2$ to an output distribution $f(x) \in \Delta^2$.

3.2   Quantum Channels

A classical binary channel $f : \Delta^2 \to \Delta^2$ takes an input distribution to an output distribution. In a similar way, a qubit channel is a function of the form $\varepsilon : \Omega^2 \to \Omega^2$. Specifically,

Definition 14. A *qubit channel* is a function $\varepsilon : \Omega^2 \to \Omega^2$ that is convex linear and completely positive.

To say that ε is convex linear means that ε preserves convex sums i.e. sums of the form $x \cdot \rho + (1 - x) \cdot \sigma$. Complete positivity, defined in [8], is a condition which ensures that the definition of a qubit channel is compatible with natural intuitions about joint systems. For our purposes, there is no need to get lost in too many details of the Hilbert space formulation: thankfully, qubit channels also have a Bloch representation.

Definition 15. For a qubit channel $\varepsilon : \Omega^2 \to \Omega^2$, the mapping it induces on the Bloch sphere $f_\varepsilon : \mathbb{B}^3 \to \mathbb{B}^3$ is called the *Bloch representation* of ε.

The set of qubit channels is closed under convex sum and composition. If ε is a qubit channel and f_ε is its Bloch representation, then (i) the function f_ε is convex linear, (ii) composition of quantum channels corresponds to composition of Bloch representations: for channels $\varepsilon_1, \varepsilon_2$, we have $f_{\varepsilon_1 \circ \varepsilon_2} = f_{\varepsilon_1} \circ f_{\varepsilon_2}$, (iii) convex sum of quantum channels corresponds to convex sum of Bloch representations: for channels $\varepsilon_1, \varepsilon_2$ and $x \in [0, 1]$, we have $f_{x\varepsilon_1 + \bar{x}\varepsilon_2} = x f_{\varepsilon_1} + \bar{x} f_{\varepsilon_2}$.

To illustrate how these properties make it simple to calculate the Bloch representation of a qubit channel, consider the "bit flipping" channel,

$$\varepsilon(\rho) = (1 - p)\varepsilon_I(\rho) + p \cdot \varepsilon_x(\rho)$$

where ε_I is the identity channel and $\varepsilon_x(\rho) = \sigma_x \rho \sigma_x$, with σ_x being the spin operator $\sigma_x = \begin{pmatrix} 0 & 1 \\ 1 & 0 \end{pmatrix}$.

The Bloch representation of ε_I is $f_{\varepsilon_I}(r) = r$. Using the correspondence between density operators and Bloch vectors, we calculate directly that the Bloch representation of ε_x is $f_{\varepsilon_x}(r_x, r_y, r_z) = (r_x, -r_y, -r_z)$. Thus, by property (iii) of Bloch representations,

$$f_\varepsilon(r_x, r_y, r_z) = (1-p)(r_x, r_y, r_z) + p(r_x, -r_y, -r_z) = (r_x, (1-2p)r_y, (1-2p)r_z)$$

Notice that states of the form $(r_x, 0, 0)$ are unchanged by this form of noise, they are all *fixed points* of f_ε.

3.3   Entropy Increasing Channels

The classical channels $f : \Delta^2 \to \Delta^2$ which increase entropy $(H(f(x)) \geq H(x))$ are exactly those f with $f(\bot) = \bot$. They are the *strict* mappings of domain theory, which are also known as *binary symmetric channels* in information theory.

Similarly, the entropy increasing qubit channels are exactly those ε for which $\varepsilon(\bot) = \bot$. These are called *unital* in quantum information theory.

Definition 16. A qubit channel $\varepsilon : \Omega^2 \to \Omega^2$ is *unital* if $\varepsilon(\bot) = \bot$.

A qubit channel ε is unital iff its Bloch representation f_ε satisfies $f_\varepsilon(0) = 0$. Let us consider a few important examples of unital channels.

Example 2. Unitary channels. If U is a unitary operator on $\mathcal{H}^2$, then $\varepsilon(\rho) = U\rho U^\dagger$ is unital since $UU^\dagger = I$. The Bloch representation f_ε is given by $f(r) = Mr$ where M is a 3×3 orthogonal matrix with positive determinant, a *rotation*.

Example 3. Projective measurements. If $\{P_0, P_1\}$ are projections with $P_0 + P_1 = I$, then

$$\varepsilon(\rho) = P_0 \rho P_0 + P_1 \rho P_1$$

is a unital channel since $P_0^2 = P_0$ and $P_1^2 = P_1$. In this case, the Bloch representation f_ε satisfies $f_\varepsilon^2 = f_\varepsilon$.

Just as with qubit channels, unital channels are also closed under convex sum and composition: if ε_1 and ε_2 are unital channels, then $\varepsilon_1 \circ \varepsilon_2$ and $p \cdot \varepsilon_1 + (1-p) \cdot \varepsilon_2$ are unital for $p \in [0, 1]$.

Example 4. Let σ_x, σ_y and σ_z denote the spin operators

$$\sigma_x = \begin{pmatrix} 0 & 1 \\ 1 & 0 \end{pmatrix} \qquad \sigma_y = \begin{pmatrix} 0 & -i \\ i & 0 \end{pmatrix} \qquad \sigma_z = \begin{pmatrix} 1 & 0 \\ 0 & -1 \end{pmatrix}$$

Each is unitary and self-adjoint.

(i) Each spin operator σ_i defines a unital channel $\varepsilon_i(\rho) = \sigma_i \rho \sigma_i$. For a Bloch vector $r = (r_x, r_y, r_z)$, the respective Bloch representations s_x, s_y, s_z are $s_x(r) = (r_x, -r_y, -r_z)$, $s_y(r) = (-r_x, r_y, -r_z)$ and $s_z(r) = (-r_x, -r_y, r_z)$.

(ii) The Bit flipping channel $\varepsilon = (1 - p)\varepsilon_I + p\,\varepsilon_x$ is unital.
(iii) The phase flipping channel $\varepsilon = (1 - p)\varepsilon_I + p\,\varepsilon_z$ is unital.
(iv) The bit-phase flip channel $\varepsilon = (1 - p)\varepsilon_I + p\,\varepsilon_y$ is unital.
(v) The depolarization channel

$$d(x) = p \cdot \bot + (1 - p)x$$

is unital, for a fixed $p \in [0, 1]$.

Not all qubit channels are unital of course. Amplitude damping provides a well-known example of a qubit channel that is not.

4 Scott Continuity of Unital Channels

Our first result establishes that from the domain theoretic perspective, unital qubit channels are the quantum analogue of binary symmetric channels in the classical case.

Theorem 3

- A classical channel $f : \Delta^2 \to \Delta^2$ is binary symmetric iff it is Scott continuous and its set of fixed points is Scott closed.
- A quantum channel $f : \Omega^2 \to \Omega^2$ is unital if and only if it is Scott continuous and its set of fixed points is Scott closed.

Proof. First consider the classical case. If a classical channel f is Scott continuous, then it has a least fixed point, and since the set of fixed points is Scott closed, $\bot = (1/2, 1/2)$ must be a fixed point. This implies that f is binary symmetric since

$$(1/2, 1/2) \begin{pmatrix} a & \bar{a} \\ b & \bar{b} \end{pmatrix} = ((a + b)/2, (\bar{a} + \bar{b})/2) = (1/2, 1/2)$$

Conversely, suppose that f is binary symmetric. Then it can be written as

$$f(a, b) = (1 - p) \cdot (a, b) + p \cdot (b, a)$$

for some $p \in [0, 1]$. First we show that f is Scott continuous. For the monotonicity of f, let $x, y \in \Delta^2$ with $x \sqsubseteq y$. Then we want to show $f(x) \sqsubseteq f(y)$. Writing $x = (x_1, x_2)$ and $y = (y_1, y_2)$, we have

$$(y_1 \leq x_1 \leq 1/2) \quad \text{or} \quad (1/2 \leq x_1 \leq y_1)$$

by the definition of $\sqsubseteq$ on Δ^2; we seek to establish

$$(f_1(y) \leq f_1(x) \leq 1/2) \quad \text{or} \quad (1/2 \leq f_1(x) \leq f_1(y))$$

where we have written $f(x) = (f_1(x), f_2(x))$ and $f(y) = (f_1(y), f_2(y))$. Notice that

$$f_1(x) = (1 - 2p)x_1 + p \quad \text{and} \quad f_1(y) = (1 - 2p)y_1 + p.$$

We consider the cases $y_1 \leq x_1 \leq 1/2$ and $1/2 \leq x_1 \leq y_1$ separately.

In the first case, $f_1(y) \leq f_1(x) \leq 1/2$ holds when $p \leq 1/2$, and $1/2 \leq f_1(x) \leq f_1(y)$ holds when $p \geq 1/2$. In the second case, $1/2 \leq f_1(x) \leq f_1(y)$ holds when $p \leq 1/2$, and $f_1(y) \leq f_1(x) \leq 1/2$ holds for $p \geq 1/2$. Thus, $f(x) \sqsubseteq f(y)$, which proves f is monotone. Its Scott continuity now follows from its Euclidean continuity and the fact that suprema in the Bayesian order coincide with limits in the Euclidean topology.

Now we show that the fixed points of f form a Scott closed set. If $p = 0$, then f is the identity mapping, in which case its set of fixed points is Scott closed. Otherwise, its only fixed point is $\bot$, since for $p > 0$,

$$(a,b) = f(a,b) \implies (a,b) = (b,a) \implies (a,b) = (1/2, 1/2) = \bot.$$

Either way, the fixed points of f form a Scott closed subset of Δ^2.

In the quantum case, any channel f that is Scott continuous and has a Scott closed set of fixed points must have $\bot$ as a fixed point, and so must be unital. For the converse, we first show that any unital f is Scott continuous. Recall that f can be written in Bloch form as $f(r) = M \cdot r$ for some 3×3 real matrix M. Then f is Euclidean continuous, and since suprema in the spectral order are limits in the Euclidean topology, f is Scott continuous in the spectral order provided it is monotone.

For the monotonicity of f, let $r \sqsubseteq s$ in the spectral order on Ω^2. Then the straight line segment $\pi_{\bot s} : [0,1] \to \Omega^2$ from $\bot$ to s, given by $\pi_{\bot s}(t) = t \cdot s$ for $t \in [0,1]$, must pass through r. To show that $f(r) \sqsubseteq f(s)$, we must show that the line from $\bot$ to $f(s)$ passes through $f(r)$. But this much is clear since $f(\pi_{\bot s}(t)) = M(t \cdot s) = t \cdot f(s) = \pi_{\bot f(s)}(t)$. Thus, all unital channels are Scott continuous.

To see that the set of fixed points $\text{fix}(f)$ is Scott closed, we first show that it is a lower set. If $s \in \text{fix}(f)$ and $r \sqsubseteq s$, then r lies on the line segment that joins $\bot$ to s. But any point on this line is a fixed point of f since $f(\pi_{\bot s}(t)) = \pi_{\bot f(s)}(t) = \pi_{\bot s}(t)$. In particular, $r \in \text{fix}(f)$. The set $\text{fix}(f)$ is closed under directed suprema by the Scott continuity of f. Thus, $\text{fix}(f)$ is Scott closed. $\square$

Selfmaps on Hausdorff spaces have closed sets of fixed points. But the Scott topology is not Hausdorff, so the result above is meaningful. The fact that the set of fixed points is Scott closed also has experimental significance: in attempting to prepare $|0\rangle$ during QKD, Alice actually prepares $(1 - \varepsilon)|0\rangle\langle 0| + \varepsilon|1\rangle\langle 1|$ for some small $\varepsilon > 0$. Then this too is a fixed point of the noise operator, provided $|0\rangle$ is, so the only reduction in capacity is due solely to error in preparation – Alice does not suffer 'more noise' simply because she cannot prepare a qubit exactly.

5   Uniqueness of the Spectral Order

The order on Δ^2 is canonical as follows:

Theorem 4 ([2]). *There is a unique partial order on Δ^2 that satisfies the mixing law*

$$x \sqsubseteq y \text{ and } p \in [0,1] \;\Rightarrow\; x \sqsubseteq (1-p)x + py \sqsubseteq y$$

and has $\perp := (1/2, 1/2)$ as a least element. It is the Bayesian order on classical two states.

Because of the simplicity of Δ^2, it then follows that the binary symmetric channels are exactly the classical channels that are Scott continuous and have a Scott closed set of fixed points. In this section, we prove the analogous result for the spectral order.

The special orthogonal group SO(3) is the set of 3×3 orthogonal real matrices M with a positive determinant i.e. those matrices M such that $M^{-1} = M^t$ and $\det(M) = +1$. Such matrices are called *rotations*.

Lemma 1

(i) *Every rotation $f \in$ SO(3) is the Bloch representation of a unital channel,*
(ii) *For any $x \in \max(\Omega^2)$, there is a rotation $f \in$ SO(3) such that $f(x) = (0, 0, 1)$.*

Proof. (i) This is a folklore result, see [8] for instance.

(ii) This follows from the fact that SO(3) is a transitive group action on S^2. However, we want to write a self-contained paper, so let us give a simpler proof. Every unit vector x appears as the third column $M(0,0,1)$ of some orthogonal matrix M since by the Gram–Schmidt process we can always find an orthonormal basis $\{v_1, v_2, v_3\}$ whose first vector is $v_1 = x$. Given such an orthonormal basis, we construct an orthogonal matrix f whose column vectors are the vectors in the orthonormal basis with the third column being x.

So let us take an orthogonal matrix M such that $M(0,0,1) = x$. Then M^{-1} is an orthgonal matrix with $M^{-1}(x) = (0,0,1)$. If $\det(M^{-1}) = -1$, we set

$$f = \begin{pmatrix} -1 & 0 & 0 \\ 0 & 1 & 0 \\ 0 & 0 & 1 \end{pmatrix} \cdot M^{-1}$$

then f is a rotation with $f(x) = (0,0,1)$. Otherwise, $\det(M^{-1}) = +1$, in which case we set $f = M^{-1}$. $\square$

Theorem 5. *There is a unique partial order on Ω^2 with the following three properties:*

(i) *It has least element $\perp = I/2$,*
(ii) *It satisfies the mixing law: if $r \sqsubseteq s$, then $r \sqsubseteq tr + (1-t)s \sqsubseteq s$, for all $t \in [0,1]$,*
(iii) *Every unital channel $f : \Omega^2 \to \Omega^2$ is monotone and has a lower set of fixed points.*

It is the spectral order, and gives Ω^2 the structure of a Scott domain on which all unital channels are Scott continuous and have a Scott closed set of fixed points.

Proof. In this proof, we work with Bloch representations. By the mixing law, the depolarization channel $d_t(x) = tx + (1 - t)\bot = tx$ is deflationary, so $tx \sqsubseteq x$ for each $t \in [0, 1]$. Thus, $\sqsubseteq$ contains the spectral order.

Now suppose $r \sqsubseteq s$. We want to show that r precedes s on the line that travels from $\bot$ to s and on to a pure state. Draw the line $\pi_{\bot a}$ from $\bot$ to r until it hits the boundary of the Bloch sphere at a point a. Similarly, let $\pi_{\bot b}$ denote the line from $\bot$ to s and on to a pure state b. Since $r \sqsubseteq s$ and $s \sqsubseteq b$, we have $r \sqsubseteq b$ by transitivity and thus $r \sqsubseteq a, b$.

Let f be a rotation such that $f(b) = (0, 0, 1)$. Then $f(-b) = (0, 0, -1)$. Let p be the Bloch representation of a projective measurement in the basis whose Bloch vectors are $\{f(b), f(-b)\}$. Then

$$\text{Im}(p) = \{(0, 0, t) : t \in [-1, 1]\}$$

Since $r \sqsubseteq b$, $f(r) \sqsubseteq f(b)$ and thus $p(f(r)) \sqsubseteq p(f(b))$. But, $p(f(b)) = f(b)$, so $f(r)$ is also a fixed point, since the fixed points of p are Scott closed. Then $f(r), f(b) \in \text{Im}(p)$. This means $f(r)$ and $f(b)$ lie on a line that joins a pure state to its antipode. Because f is a rotation, the same is true of r and b. However, by the mixing law, the line from r to b, which increases with respect to $\sqsubseteq$, does not pass through $\bot$ since $\bot$ is the least element (otherwise, $r = \bot$ and the proof is finished). Then $a = b$, which means r and s lie on a line that joins $\bot$ to a pure state a.

So let us write $r = xa$ and $s = ya$ for $x, y \in [0, 1]$. If $x \le y$, the proof is done. If $x > y$, then $s = (y/x)r \sqsubseteq r$ using the depolarization operator $d_{y/x}$. But since $r \sqsubseteq s$, we have $r = s$ by antisymmetry of $\sqsubseteq$. $\square$

Notice that the discrete order on $\Omega^2 \setminus \{0\}$ with 0 adjoined as the least element gives a domain that makes all unital channels Scott continuous with a Scott closed set of fixed points, so requiring the mixing law is essential in uniquely characterizing the spectral order.

6   Holevo Capacity from the Informatic Derivative

A standard way of measuring the capacity of a quantum channel in quantum information is the Holevo capacity; it is sometimes called the product state capacity since input states are not allowed to be entangled across two or more uses of the channel.

Definition 17. For a trace preserving quantum operation f, the *Holevo capacity* is given by

$$C(f) = \sup_{\{x_i, \rho_i\}} \left[S\left(f\left(\sum_i x_i \rho_i \right) \right) - \sum_i x_i \cdot S(f(\rho_i)) \right]$$

where the supremum is taken over all ensembles $\{x_i, \rho_i\}$ of possible input states ρ_i to the channel.

The possible input states ρ_i to the channel are in general mixed and the x_i are probabilities with $\sum_i x_i = 1$. If f is the Bloch representation of a qubit channel, the Holevo capacity of f is given by

$$C(f) = \sup_{\{x_i, r_i\}} \left[H\left(\frac{1 + |f(\sum_i x_i r_i)|}{2} \right) - \sum_i x_i \cdot H\left(\frac{1 + |f(r_i)|}{2} \right) \right]$$

where r_i are Bloch vectors for density operators in an ensemble, and we recall that eigenvalues of a density operator with Bloch vector r are $(1 \pm |r|)/2$.

Theorem 6. *Let $\mu(x) = 1 - |x|$ denote the standard measurement on Ω^2. For any unital channel f and any $p \in \Omega^2$ different from $\bot$,*

$$df_\mu(p) = \frac{|f(p)|}{|p|}$$

Thus, the Holevo capacity of f is determined by the largest value of its informatic derivative. Explicitly,

$$C(f) = 1 - H\left(\frac{1}{2} + \frac{1}{2} \sup_{x \in \ker(\mu)} df_\mu(x) \right)$$

Proof. Since $x \sqsubseteq p$ iff $x = tp$ for some $t \in [0, 1]$, $x \to p$ in the μ topology iff $t \to 1^-$, so

$$df_\mu(p) = \lim_{x \to p} \frac{\mu f(x) - \mu f(p)}{\mu x - \mu p} = \lim_{t \to 1^-} \frac{\mu f(tp) - \mu f(p)}{\mu(tp) - \mu p}$$

$$= \lim_{t \to 1^-} \frac{|f(p)| - |f(tp)|}{|p| - |tp|}$$

$$= \lim_{t \to 1^-} \frac{|f(p)|(1 - |t|)}{|p|(1 - |t|)} \qquad \text{(Linearity of } f\text{)}$$

$$= \frac{|f(p)|}{|p|}$$

Now we show that the Holevo capacity is determined by the largest value of its informatic derivative. By the Euclidean continuity of $|f|$, there is a pure state $r \in \Omega^2$ for which

$$|f(r)| = \max_{|x|=1} |f(x)| = m^+$$

Setting $r_1 = r$, $r_2 = -r$ and $x_1 = x_2 = 1/2$ defines an ensemble for which the expression maximized in the definition of $C(f)$ reduces to $1 - H((1 + m^+)/2)$. Notice that in this step we explicitly make use of the fact that f is unital: $f(0) = 0$. This proves $1 - H((1 + m^+)/2) \leq C(f)$.

For the other inequality, any term in the supremum is bounded from above by

$$1 - \sum_i x_i \cdot H\left(\frac{1 + |f(r_i)|}{2} \right)$$

since $H(x) \leq 1$. For each r_i, there is a pure state $p_i \in \max(\Omega^2)$ with $r_i \sqsubseteq p_i$. By the Scott continuity of f,

$$|f(r_i)| \leq |f(p_i)| \leq \sup_{|x|=1} |f(x)| = m^+,$$

so we have

$$H\left(\frac{1 + |f(r_i)|}{2}\right) \geq H\left(\frac{1 + m^+}{2}\right)$$

which then gives $C(f) \leq 1 - H((1 + m^+)/2)$.   □

Thus, we see that $C(f) = 1$ for *any* rotation f since $df_\mu = 1$. Notice that $df_\mu \equiv 1$ iff f is a rotation. For each $p \in [0, 1]$, the unique channel $f \sqsubseteq 1$ with $df_\mu = p$ is the depolarization channel $f = d_p$, so that $C(d_p) = 1 - H((1+p)/2)$. In fact, the map $(p, 1 - p) \mapsto d_{1-2p}$ defines an isomorphism from the nonnegative classical binary symmetric channels onto the depolarization channels that *preserves capacity*. The only unital qubit channel with capacity zero is 0 itself.

Example 5. The two Pauli channel in Bloch form is

$$\varepsilon(r) = p\,r + \left(\frac{1-p}{2}\right) s_x(r) + \left(\frac{1-p}{2}\right) s_y(r)$$

where s_x and s_y are the Bloch representations of the unitary channels ε_x and ε_y. This simplifies to

$$\varepsilon(r_x, r_y, r_z) = (pr_x, pr_y, -(1-p)r_z)$$

The matrix associated to ε is diagonal, so the diagonal element (eigenvalue) that has largest magnitude also yields the largest value of its informatic derivative. The capacity of the two Pauli channel is then

$$1 - H\left(\frac{1 + \max\{p, 1-p\}}{2}\right)$$

where $p \in [0, 1]$.

The set of unital channels $\mathcal{U}$ is compact hence closed and thus forms a dcpo as a subset of the domain $[\Omega^2 \to \Omega^2]$.

Corollary 1. *The Holevo capacity* $C : \mathcal{U} \to [0, 1]$ *is Scott continuous.*

7   Closing

The set of unital qubit channels $\mathcal{U}$ is a convex monoid and a dcpo with respect to which the Holevo capacity is monotone. In a similar way, the interval domain $\mathbf{I}[0, 1]$, which models classical binary channels, is a convex monoid and a dcpo with respect to which the Shannon capacity is monotone [7].

References

1. Abramsky, S., Jung, A.: Domain theory. In: Abramsky, S., Gabbay, D.M., Maibaum, T.S.E. (eds.) Handbook of Logic in Computer Science, vol. III, Oxford University Press, Oxford (1994)
2. Coecke, B., Martin, K.: A partial order on classical and quantum states. Oxford University Computing Laboratory, Research Report PRG-RR-02-07 (August 2002), http://web.comlab.ox.ac.uk/oucl/publications/tr/rr-02-07.html
3. Martin, K.: A foundation for computation. Ph.D. Thesis, Tulane University, Department of Mathematics (2000)
4. Martin, K.: Entropy as a fixed point. In: Díaz, J., Karhumäki, J., Lepistö, A., Sannella, D. (eds.) ICALP 2004. LNCS, vol. 3142. Springer, Heidelberg (2004)
5. Martin, K.: The measurement process in domain theory. In: Welzl, E., Montanari, U., Rolim, J.D.P. (eds.) ICALP 2000. LNCS, vol. 1853. Springer, Heidelberg (2000)
6. Martin, K.: Topology in information theory in topology. Theoretical Computer Science (to appear)
7. Martin, K., Moskowitz, I.S., Allwein, G.: Algebraic information theory for binary channels. In: Proceedings of MFPS 2006. Electronic Notes in Theoretical Computer Science, vol. 158, pp. 289–306 (2006)
8. Nielsen, M., Chuang, I.: Quantum computation and quantum information. Cambridge University Press, Cambridge (2000)
9. Shannon, C.E.: A mathematical theory of communication. Bell Systems Technical Journal 27, 379–423, 623–656 (1948)

Interacting Quantum Observables

Bob Coecke and Ross Duncan

Oxford University Computing Laboratory

Abstract. We formalise the constructive content of an essential feature of quantum mechanics: the interaction of complementary quantum observables, and information flow mediated by them. Using a general categorical formulation, we show that pairs of mutually unbiased quantum observables form bialgebra-like structures. We also provide an abstract account on the quantum data encoded in complex phases, and prove a normal form theorem for it. Together these enable us to describe all observables of finite dimensional Hilbert space quantum mechanics. The resulting equations suffice to perform computations with elementary quantum gates, translate between distinct quantum computational models, establish the equivalence of entangled quantum states, and simulate quantum algorithms such as the quantum Fourier transform. All these computations moreover happen within an intuitive diagrammatic calculus.

1 Introduction

Complementary quantum observables such as position and momentum cannot be assigned sharp values at the same time. This fact constitutes the heart of quantum physics. That the self-adjoint operators which characterise these don't commute, motivated the study of non-commutative C^*-algebras, and that their propositional lattices are not distributive resulted in Birkhoff-von Neumann quantum logic. Neither of these axiomatic approaches unveils the true *capabilities* which these complementary observables provide. They merely involve weakening the commutativity/distributivity equation, rendering them essentially useless for any quantum informatic purpose. In this paper we provide an axiomatic account of complementary quantum observables which enables us to tackle problems of actual interest to quantum informatics: algorithm design, identifying the capabilities of multi-partite entanglement, translation between distinct quantum computational models etc. Our starting point is the axiomatisation of quantum observables proposed by Pavlovic and one of the authors in [5] which substantially relied on Carboni and Walters' cartesian bicategories [2]. This notion of quantum observable strongly improves on the one due to Abramsky and one of the authors in [1], the paper which initiated categorical quantum axiomatics, in that it axiomatises quantum observables in terms of dagger symmetric monoidal structure only, allowing for an operational interpretation, a diagrammatic calculus, as well as the 'necessary' higher level of abstraction.[1]

---

[1] For a detailed discussion of this necessity see [3,12].

L. Aceto et al. (Eds.): ICALP 2008, Part II, LNCS 5126, pp. 298–310, 2008.

Somewhat ironically, while classical structures were crafted to reason about classical control, this paper shows that considering a pair of interacting classical structures—corresponding to complementary quantum observables—are a powerful vehicle to specify and reason about pure quantum states and operations, with many applications. We formalise this notion of complementarity through a set of equations which axiomatise copyability of classical states and the information flow through incompatible classical structures. Surprisingly the relevant equations are almost exactly those of a bialgebra [13], differing only by scalar factors. We show that the axioms of this structure, a *scaled bialgebra*, express the essential features of quantum mechanics in very direct yet usable fashion.

2   Categories of Quantum States and Processes

A †-*symmetric monoidal category* (†-SMC) [12] is a symmetric monoidal category $(\mathbf{C}, \otimes, I)$ together with an identity-on-objects contravariant endofunctor $^\dagger : \mathbf{C} \to \mathbf{C}$ which preserves the monoidal structure. An elementary account of †-SMCs and their graphical representations is in [12]. Well known examples include **Rel**, the category of sets and relations, and **FdHilb**, the category of finite dimensional Hilbert spaces and linear maps.

However quantum states are not vectors in a Hilbert space: they are one-dimensional subspaces. To articulate this fact we will use the word *state* exclusively to refer to such one-dimensional subspaces. Similarly, a non-degenerate observable does not correspond to a basis, but rather a maximal family of mutually orthogonal states. The move from a linear to a projective setting is formalised using a pair of categories, **FdHilb**$_p$ and **FdHilb**$_{wp}$. The category **FdHilb**$_p$ has the same objects as **FdHilb** but its morphisms are equivalence classes of **FdHilb**-morphisms for the congruence

$$f \sim g \;\Leftrightarrow\; \exists c \in \mathbb{C} \setminus \{0\} \; s.t. \; f = c \cdot g.$$

This quotient reduces the scalar monoid to a two-element set, hence the capacity for probabilistic reasoning is lost. The solution consists of enriching **FdHilb**$_p$ with *probabilistic weights* i.e. to consider morphisms of the form $r \cdot f$ where $r \in \mathbb{R}^+$ and f a morphism in **FdHilb**$_p$. Therefore, let **FdHilb**$_{wp}$ be the category whose objects are those of **FdHilb** and whose morphisms are equivalence classes of **FdHilb**-morphisms for the congruence

$$f \sim g \;\Leftrightarrow\; \exists \alpha \in [0, 2\pi) \text{ s.t. } f = e^{i\alpha} \cdot g.$$

We regain the absolute values of the inner-product, and thus the probabilistic distance between states.[2] These three categories are related via inclusions:

$$\mathbf{FdHilb}_p \xleftarrow{\hspace{1cm}} \underset{r \in \mathbb{R}^+}{\hookrightarrow} \mathbf{FdHilb}_{wp} \xleftarrow{\hspace{1cm}} \underset{\alpha \in [0, 2\pi)}{\hookrightarrow} \mathbf{FdHilb}$$

---

[2] A detailed categorical account on **FdHilb**$_{wp}$ is in [3]; in particular, neither **FdHilb**$_p$ nor **FdHilb**$_{wp}$ has biproducts, so the approach to measurements taken in [1] will not work here.

We write $|\psi\rangle$ (or rarely ψ) to denote vectors; $\|\psi\rangle\rangle$ denotes a state spanned by this vector. Similarly, $|\sum_i c_i|i\rangle\rangle$ is the state spanned by vector $\sum_i c_i|i\rangle$. We take as given a canonical basis for the Hilbert space $\mathbb{C}^n$, which we write $\{|i\rangle\}_i$. This basis then fixes a canonical observable given by the states $\{\|i\rangle\rangle\}_i$. The hom-set $\mathbf{FdHilb}_p(\mathbb{C},\mathbb{C}^2)$—that is, the space of linear maps $\mathbb{C} \to \mathbb{C}^2$— corresponds to the the points of the *Bloch sphere*.[3] The unitaries in $\mathbf{FdHilb}_p(\mathbb{C},\mathbb{C}^2)$, i.e. those maps U satisfying $UU^\dagger = U^\dagger U = 1$ correspond to rotations of the Bloch sphere.

3   Classical Structures and the Spider Theorem

A *classical structure* [5] in a †-SMC is an internal cocommutative comonoid $(A, \delta : A \to A \otimes A, \epsilon : A \to \mathrm{I})$ with δ both *isometric* and *Frobenius*, that is,

$$\delta^\dagger \circ \delta = 1_A \qquad \text{and} \qquad \delta \circ \delta^\dagger = (\delta^\dagger \otimes 1_A) \circ (1_A \otimes \delta),$$

respectively. The unit object I canonically comes with classical structure $\lambda_{\mathrm{I}} : \mathrm{I} \simeq \mathrm{I} \otimes \mathrm{I}$ and 1_{I}. An orthonormal base $\{\psi_i\}_i$ for Hilbert space $\mathcal{H}$ induces

$$\delta : \mathcal{H} \to \mathcal{H} \otimes \mathcal{H} :: \psi_i \mapsto \psi_i \otimes \psi_i \quad \text{and} \quad \epsilon : \mathcal{H} \to \mathbb{C} :: \psi_i \mapsto 1 \qquad (1)$$

as a classical structure. Conversely, each classical structure in $\mathbf{FdHilb}$ arises in this way [6]. Hence, classical structure axiomatises the concept of (orthonormal) base in a Hilbert space. Obviously, these classical structures are inherited by $\mathbf{FdHilb}_{wp}$, and passing to $\mathbf{FdHilb}_{wp}$ clarifies what the data which specify a classical structure represent. A state $\|\psi\rangle\rangle$ is *unbiased* for some observable $\{\|\phi_i\rangle\rangle\}_i$ if for all i we have that $|\langle\psi \mid \phi_i\rangle|^2 = 1/\dim(\mathcal{H})$ whenever ψ and ϕ_i are unit vectors. Two observables $\{\|\psi_i\rangle\rangle\}_i$ and $\{\|\phi_i\rangle\rangle\}_i$ are *complementary* whenever $\|\psi_i\rangle\rangle$ is unbiased for $\{\|\phi_j\rangle\rangle\}_j$ for all i.

Proposition 1. *In $\mathbf{FdHilb}_{wp}$ each pair consisting of an observable $\{\|\psi_i\rangle\rangle\}_i$ on a Hilbert space $\mathcal{H}$ and another state $\|\varepsilon\rangle\rangle$ of $\mathcal{H}$ which is unbiased for $\{\|\psi_i\rangle\rangle\}_i$ defines a unique classical structure by setting, for all i,*

$$\delta(\|\psi_i\rangle\rangle) = \|\psi_i \otimes \psi_i\rangle\rangle \qquad \text{and} \qquad |\epsilon(-)| = |\langle\varepsilon, -\rangle|$$

Conversely, all classical structures in $\mathbf{FdHilb}_{wp}$ arise in this way.

The crux to this result is the fact that a set of n base 'vectors' $\{|\psi_i\rangle\}_i$ of a Hilbert space, up to a *common* global phase, is faithfully represented by the $n+1$ 'states' $\{\|\psi_i\rangle\rangle\}_i \cup \{\|\sum_i \psi_i\rangle\rangle\}$. On the Bloch sphere an observable $\{\|\psi_0\rangle\rangle, \|\psi_1\rangle\rangle\}$, e.g. $\{\|0\rangle\rangle, \|1\rangle\rangle\}$, comprises two antipodal points, while $\|\varepsilon\rangle\rangle$, e.g. $\|+\rangle\rangle$, lies on the corresponding equator, together making up a T-shape:

---

[3] In any monoidal category maps of the type $I \to A$ are called points.

It is standard to interpret the *eigenstates* $\|\psi_i\rangle\rangle$ for an observable $\{\|\psi_i\rangle\rangle\}_i$ as classical data. Hence, in **FdHilb**$_{wp}$, the operation δ of a classical structure copies the eigenstates $\|\psi_i\rangle\rangle$ of the observable it is associated with. We can interpret $\|\varepsilon\rangle\rangle$ as the state which uniformly deletes these eigenstates: by unbiasedness the probabilistic distance of each eigenstate $\|\psi_i\rangle\rangle$ to $\|\varepsilon\rangle\rangle$ is equal. Therefore we will refer to δ as (classical) copying and to ϵ as (classical) erasing. A crucial point here is that given an observable there is a choice involved in picking ϵ.

Within graphical calculus for †-SMCs (see [12]) we depict the morphisms δ and ϵ by ⬤ and ⬤, and their adjoints, $\delta^\dagger$ and $\epsilon^\dagger$ by ⬤ and ⬤, When taking the monoidal structure to be strict—which we will do throughout this paper— classical structures obey the following remarkable theorem [4].[4]

Theorem 1. *Let* $f, g : A^{\otimes n} \to A^{\otimes m}$ *be two morphisms generated from classical structure* (A, δ, ϵ) *and the dagger symmetric monoidal structure. If the graphical representation both of* f *and* g *is connected then* $f = g$.

Hence, such a morphism only depends on the object A and the number of inputs and outputs. We represent this morphism as an $n + m$-*legged spider*

Theorem 1 allows the dots representing δ, ϵ, $\delta^\dagger$ and $\epsilon^\dagger$ to 'fuse' into a single dot, provided all the dots are connected. Note that, conversely, the axioms of classical structure are consequences of this fusing principle.

Classical structure refines the †-compact structure which was used in [1,12], provided the latter is self-dual. Graphical reasoning in compact structure by 'yanking' is subsumed by reasoning in terms of the above 'spider theorem'. This will become clear in the first example of §6.3. We can define the *conjugate* $f_* :$ $A \to B$ of a morphism $f : A \to B$ relative to classical structures $(A, \delta_A, \epsilon_A)$ and $(B, \delta_B, \epsilon_B)$ to be $f_* := (1_B \otimes \eta_A^\dagger) \circ (1_B \otimes f^\dagger \otimes 1_A) \circ (\eta_B \otimes 1_A)$ where $\eta_X := \delta_X \circ \epsilon_X^\dagger$. In **FdHilb** the linear function f_* is obtained by conjugating the entries of the matrix of f when expressed in the classical structure bases. The *dimension* of A is $\dim(A) := \eta_A^\dagger \circ \eta_A$ represented graphically by a circle.

4   A Generalised Spider Theorem and Abstract Phase Data

Let (A, δ, ϵ) be a classical structure in a †-SMC. On *points* $\psi, \phi : I \to A$ we define

$$\psi \odot \phi = \delta^\dagger \circ (\psi \otimes \phi) \qquad \text{i.e.}$$

---

[4] Similar results are known for concrete dagger Frobenius algebras, e.g. 2D topological quantum field theories, as well as in more abstract categorical settings [11].

Since $(A, \delta^\dagger, \epsilon^\dagger)$ forms a commutative monoid, this operation is immediately associative and commutative, with unit $\epsilon^\dagger$. Now define

$$\Lambda : \mathbf{C}(I, A) \to \mathbf{C}(A, A) :: \psi \mapsto \delta^\dagger \circ (\psi \otimes 1_A) \qquad \text{i.e.}$$

From the properties of $\delta^\dagger$ it immediately follows that Λ is a homomorphism of monoids, and that for every ψ

$$\Lambda(\psi) \circ \delta^\dagger = \delta^\dagger \circ (1_A \otimes \Lambda(\psi)) = \delta^\dagger \circ (\Lambda(\psi) \otimes 1_A) \quad \text{i.e.}$$

Since $\odot$ is commutative, we also have

$$\Lambda(\psi) \circ \Lambda(\phi) = \Lambda(\phi) \circ \Lambda(\psi) \qquad \text{i.e.}$$

and since $\Lambda(\psi)^\dagger = (\psi^\dagger \otimes 1_A) \circ \delta$, the spider theorem yields $\Lambda(\psi)^\dagger = \Lambda(\psi_*)$. Now let $\delta_n : A \to A^{\otimes n}$ be defined by the recursion $\delta_0 = \epsilon$, $\delta_1 = 1_A$ and $\delta_n = \delta \circ (\delta_{n-1} \otimes 1_A)$.

Theorem 2. *Let $f : A^{\otimes n} \to A^{\otimes m}$ be a morphism generated from classical structure (A, δ, ϵ), points $\psi_i : I \to A$ (not necessarily all distinct), and dagger symmetric monoidal structure. If the graphical representation of f is connected then*

$$f = \delta_m \circ \Lambda\left(\bigodot_i \psi_i\right) \circ \delta_n^\dagger. \tag{2}$$

This is a strict generalisation of Theorem 1: besides the number of inputs and outputs there is now also the *product of all points* which distinguishes classes of equal diagrams. We obtain a *decorated spider*:

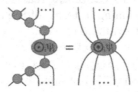

In graphical terms, Theorem 2 allows arbitrary decorated dots of the same colour to 'fuse' together provided we 'multiply their decorations'.

In $\mathbb{C}^n$, consider $|\psi\rangle = \sum_i c_i |i\rangle$; when written in the basis fixed by (δ, ϵ), $\Lambda(\psi)$ consists of the diagonal $n \times n$ matrix with $c_1, \ldots, c_n$ on the diagonal. Hence, $\Lambda(\psi)$ is unitary, upto a normalisation factor, if and only if $\|\psi\rangle\!\rangle$ is unbiased for $\{\||1\rangle\!\rangle, \ldots, \||n\rangle\!\rangle\}$. This fact admits generalisation to arbitrary †-SMCs.

Definition 1. We call $\psi : I \to A$ *unbiased relative to* δ if $\Lambda(\psi)$ is unitary.

Proposition 2. *The set of points which are unbiased relative to a classical structure forms a group under $\odot$ with $(-)_*$ as the inverse.*

For ψ unbiased relative to δ, by Theorem 2 and Proposition 2, we have

$$\quad = (\Lambda(\psi) \otimes \Lambda(\psi)) \circ \delta \circ \Lambda(\psi)^\dagger \qquad \text{and} \qquad \quad = \epsilon \circ \Lambda(\psi)^\dagger$$

and again by the generalised spider theorem it then follows that these morphisms define a classical structure. We call it a *phase shift* of (A, δ, ϵ). In $\mathbb{C}^2$, these phased variants to the classical structure $\{\|0\rangle\!\rangle, \|1\rangle\!\rangle, \|+\rangle\!\rangle\}$ (cf. Proposition 1) are those obtained by varying the choice of $\|\varepsilon\rangle\!\rangle$ on the equator of the Bloch sphere:

The states which are unbiased relative to $\{\|0\rangle\!\rangle, \|1\rangle\!\rangle\}$ are of the form $\|+_\theta\rangle\!\rangle :=$ $\|0\rangle + e^{i\theta}|1\rangle\!\rangle$ so form a family parameterised by a phase θ. In particular, we have $\|+_{\theta_1}\rangle\!\rangle \odot \|+_{\theta_2}\rangle\!\rangle = \|+_{\theta_1+\theta_2}\rangle\!\rangle$, that is, the operation $\odot$ boils down to *adding up* phases modulo 2π, which is an abelian group with minus as inverse.

5 Complementary Observables as Scaled Bialgebras

The goal of this section is to show that each pair of complementary observables in $\mathbf{FdHilb}_{wp}$ defines a *scaled bialgebra*. In the next section we will then use this scaled algebra structure together with the generalised spider theorem for phase data to reason about quantum informatics. First we define and study an abstract notion of complementary observables. We then derive a general scaled bialgebra law in categories with 'enough points' such as $\mathbf{FdHilb}$. This result then carries over to $\mathbf{FdHilb}_{wp}$ where it takes a much simpler form.

5.1 Complementary Classical Structures (CCSs)

In eq.(1) we described classical structures in $\mathbf{FdHilb}$ as maps which copy basic vectors, and hence also the corresponding states in $\mathbf{FdHilb}_{wp}$. We introduce an abstract counterpart to these 'copy-able' points. We assume as given a classical structure (A, δ, ϵ) in a †-SMC.

Recall that if a †-SMC has a classical structure on an object A then the monoidal subcategory generated by A is †-compact, and hence we can define the *dimension* of A by $\dim(A) = \epsilon \circ \delta^\dagger \circ \delta \circ \epsilon^\dagger$.[5] For brevity, we define $D = \dim(A)$. We will, in addition, assume the existence of a self-adjoint scalar $\sqrt{D}$, which we we denote graphically as ◆. As the notation suggests, $\sqrt{D}$ satisfies

$$\sqrt{D} \otimes \sqrt{D} = D = \dim(A) \qquad \text{or, graphically:} \qquad \text{◆ ◆} = \bigcirc .$$

Notation. We represent all the points $a : I \to A$ which are unbiased with respect to (A, δ, ϵ) by dots of the same green (light grey) colour used before. Those points which are 'copied' by δ in the sense of the definition below we mark by a different colour, here red, or darker grey. Any other points are marked in black. In light of the special role played by unbiased points, we will use the spider notation only for these.

---

[5] One can show that $\dim(A)$ does not depend on the choice of classical structure.

Definition 2. *We call a point* $a_i : I \to A$ *classical relative to* (A, δ, ϵ) *if both* $\sqrt{D} = \epsilon \circ a_i$ *and* $\sqrt{D} \cdot (\delta \circ a_i) = a_i \otimes a_i$ *hold, that is, graphically,*

and .

The classical points for a classical structure in $\mathbf{FdHilb}_{wp}$ are of course the states $\{\|\psi_i\rangle\rangle\}_i$ of Proposition 1.

The abstract conception of a classical point allows the concrete notion of unbiasedness to be derived from the abstract formulation of Definition 1:

Lemma 1. *If* $a_i : I \to A$ *is classical and* $\alpha : I \to A$ *unbiased for* (A, δ, ϵ) *then*

$$(\alpha^\dagger \circ a_i) \cdot (\alpha^\dagger \circ a_i)^\dagger = D \qquad \text{i.e.} \qquad \text{}.$$

The classical points are "eigenvectors" in a suitable sense:

Lemma 2. *If* $a_i : I \to A$ *is classical for* (A, δ, ϵ) *and* $\psi : I \to A$ *arbitrary then*

$$\sqrt{D} \cdot (\Lambda(\psi) \circ a_i) = (\psi_*^\dagger \circ a_i) \cdot a_i \qquad \text{i.e.} \qquad \text{}.$$

The monoid multiplication on points carries over to scalars:

Lemma 3. *Let* $a_i : I \to A$ *be classical and* $\psi, \phi : I \to A$ *arbitrary then*

$$\sqrt{D} \cdot (a_i^\dagger \circ (\psi \odot \phi)) = (a_i^\dagger \circ \psi) \cdot (a^\dagger \circ \phi) \qquad \text{i.e.} \qquad \text{}.$$

Remark 1. The reader may find the scalar factors in the above equations mysterious, not to say vexing. But recall that in $\mathbf{FdHilb}$ the equation $|\epsilon^\dagger| = \sqrt{D}$ is required to satisfy the comonoid laws; this scalar factor reappears here.[6]

Definition 3. Two classical structures $(A, \delta_X, \epsilon_X)$ and $(A, \delta_Z, \epsilon_Z)$ in a †-SMC are called *complementary* if they obey the following rules:

- whenever $z_i : I \to A$ is classical for (δ_X, ϵ_X) it is unbiased for (δ_Z, ϵ_Z);
- whenever $x_j : I \to A$ is classical for (δ_Z, ϵ_Z) it is unbiased for (δ_X, ϵ_X);
- $\epsilon_X^\dagger$ is classical for (δ_Z, ϵ_Z) and $\epsilon_Z^\dagger$ is classical for (δ_X, ϵ_X).

We abbreviate *complementary classical structure* as CCS.

Notation. The reason that we refer to the classical points of $(A, \delta_X, \epsilon_X)$ by z_i, and vice versa, is because z_i is unbiased to (δ_Z, ϵ_Z) and hence can participate in the generalised spider theorem for the classical structure (δ_Z, ϵ_Z).

For any non-degenerate quantum observable we can find a pair of complementary classical structures in $\mathbf{FdHilb}_{wp}$ merely by picking $\|\epsilon\rangle\rangle$ for one observable from among the eigenstates of the other observable.

---

[6] An alternative would be to replace $(\epsilon \otimes 1_{\mathcal{H}}) \circ \delta = 1_{\mathcal{H}}$ with $(\epsilon \otimes 1_{\mathcal{H}}) \circ \delta = \frac{1}{\sqrt{D}} 1_{\mathcal{H}}$.

5.2   Derivation of the Scaled Bialgebra Law from Abstract Bases

Definition 4. A set of points $\{a_i\}_i$ is a *basis* for an object A if for all $f, g :$ $A \rightarrow B$, if $f \circ a_i = g \circ a_i$ for all a_i, then $f = g$. A basis is *classical*, or *unbiased*, with respect to some classical structure (A, δ, ϵ) if its elements are respectively classical, or unbiased, with respect to this structure. An unbiased basis is called *closed* if for all a_i, a_j there exists a_k such that $a_i \odot a_j = a_k$ and $a_0 = \epsilon^\dagger$. We say that a $\dagger$-SMC has *monoidal bases* when, for each basis $\{a_i\}_i$ for A, and each basis $\{b_j\}_j$ for B, the set $\{a_i \otimes b_j\}_{ij}$ is a basis for $A \otimes B$.

An immediate consequence of this definition is that whenever b is an element of a closed unbiased basis $\{b_i\}_i$, then $\Lambda(b)$ is a permutation on the set $\{b_i\}_i$. Further, by Lemma 2 every classical point is an eigenvector of $\Lambda(b)$.

Lemma 4. *Let $\{a_i\}_i$ be a classical basis for A suppose $p : A \rightarrow A$ acts as a permutation on this set; then*

$$(p \otimes p) \circ \delta = \delta \circ p \qquad \text{i.e.} \qquad \vcenter{\hbox{figure}}.$$

Lemma 5. *Let (δ_X, ϵ_X), (δ_Z, ϵ_Z) be CCSs, let x be in a closed classical basis of (δ_Z, ϵ_Z) and let z be unbiased for (δ_Z, ϵ_Z), then*

$$\sqrt{D} \cdot (\Lambda^X(x) \circ \Lambda^Z(z)) = (x^\dagger \circ z) \cdot (\Lambda^Z(z) \circ \Lambda^X(x)) \qquad \text{i.e.} \qquad \vcenter{\hbox{figure}}.$$

The Pauli matrices provide an example of these commutation relations.

Lemma 6. *Let (δ_Z, ϵ_Z) and (δ_X, ϵ_X) be CCSs and let U_Z denote all the unbiased points and C_Z a basis of classical points for (δ_Z, ϵ_Z). Suppose $x \in C_Z$ and let $\mathsf{X} = \Lambda^X(x)$. If C_Z is closed under $\odot_X$ then:*

- X *is a permutation on C_Z;*
- X *is an automorphism on U_Z such that*

$$\mathsf{X} \circ (\alpha \odot_Z \beta) = (\mathsf{X} \circ \alpha) \odot_Z (\mathsf{X} \circ \beta); \quad , \quad \mathsf{X} \circ \epsilon_Z^\dagger = \epsilon_Z^\dagger \quad \text{and} \quad (\mathsf{X} \circ \alpha)^{-1} = \mathsf{X}^\dagger \circ \alpha^{-1}.$$

Corollary 1. *$(C_Z, \odot_X)$ is an abelian group with a group action on U_Z defined by $(x, z) \mapsto \Lambda^X(x) \circ z$.*

Lemma 7. *Consider a $\dagger$-SMC with monoidal bases and let σ be the monoidal symmetry. Let $(A, \delta_X, \epsilon_X)$ and $(A, \delta_Z, \epsilon_Z)$ be CCSs with classical bases $\{z_j\}_j$ and $\{x_i\}_i$; $\{x_i\}_i$ is closed if and only if*

$$D \cdot (\delta_X^\dagger \otimes \delta_X^\dagger) \circ (1_A \otimes \sigma \otimes 1_A) \circ (\delta_Z \otimes \delta_Z) = \sqrt{D} \cdot \delta_Z \circ \delta_X^\dagger \qquad \text{i.e.} \qquad \vcenter{\hbox{figure}}.$$

Corollary 2. *In the above situation $\{x_i\}_i$ is closed if and only if $\{z_i\}$ is.*

Theorem 3. *Let (δ_X, ϵ_X) and (δ_Z, ϵ_Z) be CCSs with closed bases including the points z and x respectively. Then, graphically,*

$$\text{(diagram)}$$

$$\tag{3}$$

We call the morphisms obeying eq.(3) a 'scaled bialgebra'.

Proposition 3. If (δ_X, ϵ_X) and (δ_Z, ϵ_Z) form a scaled bialgebra then

$$\text{(diagram)}$$

i.e. it is a 'scaled Hopf algebra' with $\dim(A) \cdot 1_A$ as its 'antipode'.

5.3 Complementary Classical Observables in FdHilb$_{wp}$

Classical structures in **FdHilb** 'are' bases [6] so complementary pairs of bases which satisfy the closedness condition of Definition 4 induce scaled bialgebras in the sense of Theorem 3. These scaled bialgebra laws carry over to the CCSs in **FdHilb**$_{wp}$ consisting of the states spanned by the basis vectors. Moreover, in **FdHilb**$_{wp}$, since all scalars are positive reals, all scalars in eqs.(3) coincide, so cancellation simplifies eqs.(3) to

$$\text{(diagram)} \tag{4}$$

Conversely, any pair of complementary observables yields a family of CCSs in **FdHilb**$_{wp}$ mediated by a group of permutations on the respective sets of classical states, and one can always construct a corresponding underlying family of CCSs in **FdHilb**. What we so far failed to prove is that in general we can always construct a corresponding underlying family of 'closed' CCSs. However: **(i)** CCSs in **FdHilb** on $\mathbb{C}^2$ and $\mathbb{C}^3$ 'are' closed; **(ii)** CCSs can be chosen to be closed for all (to us) known constructions of mutually unbiased bases (e.g. [10]); **(iii)** we constructed closed CCSs on $\mathbb{C}^n$ in **FdHilb**$_{wp}$ for all n. Hence for all practical situations involving complementary observables eq.(4) hold. We conjecture that closed CCSs can be derived from any pair of mutually unbiased observables.[7]

6   Applications and Examples in Quantum Informatics

Inevitably, the examples from this field are constructed from the ubiquitous qubit i.e. $\mathbb{C}^2$. Take the 'green' classical structure (δ_Z, ϵ_Z) as in eq.(1) for $\{|0\rangle, |1\rangle\}$. The unbiased points for (δ_Z, ϵ_Z) are of the form $|\alpha_Z\rangle = |0\rangle + e^{i\alpha}|1\rangle$, and $|\alpha_Z\rangle \odot_Z$

---

[7] The study of mutually unbiased bases is an active area of research; characterisation of the maximal number of mutually unbiased bases is one of the important open problems in quantum informatics.

$|\beta_Z\rangle = |\alpha + \beta_Z\rangle$. Further, $\Lambda^Z(\alpha) = \begin{pmatrix} 1 & 0 \\ 0 & e^{i\alpha} \end{pmatrix}$, in particular, $\Lambda^Z(\pi) = \mathsf{Z}$. Notice that $\epsilon_Z^\dagger = |0\rangle + |1\rangle$ and $|\pi_Z\rangle = |0\rangle - |1\rangle$ form a basis, which is closed and unbiased with respect to (δ_Z, ϵ_Z) and define a complementary 'red' classical structure (δ_X, ϵ_X). The unbiased points for (δ_X, ϵ_X) have the form $|\theta_X\rangle = \sqrt{2}(\cos\frac{\theta}{2}|0\rangle + \sin\frac{\theta}{2}|1\rangle)$ and $\Lambda^X(\theta) = \begin{pmatrix} \cos\frac{\theta}{2} & \sin\frac{\theta}{2} \\ \sin\frac{\theta}{2} & \cos\frac{\theta}{2} \end{pmatrix}$, in particular, $\Lambda^Z(\pi) = \mathsf{X}$. We have $\mathsf{Z} \circ |\theta_X\rangle = |-\theta_X\rangle$, and, upto a global phase, $\mathsf{X} \circ |\alpha_Z\rangle = |-\alpha_Z\rangle$. In the language of Lemma 6 we have: if $|C_Z| = 2$, then $(C_Z, \odot_X)$ is the symmetric group S_2, its unique non-identity element X is self-adjoint, and for $\alpha \in U_Z$ we have $\mathsf{X} \circ \alpha = \alpha_*$ i.e. X assigns the inverses for the group U_Z.

For some of the examples below it will also be convenient to explicitly have the unitary operation which changes the green dots into red dots, that is, concretely, the unitary operation which establishes the corresponding change of basis. In the case of (δ_Z, ϵ_Z) and (δ_X, ϵ_X) given above, the two structures are connected via the familiar Hadamard map H. As well as being unitary, $H = H^\dagger$ so this map is particularly well behaved. We will introduce H into the graphical language with the following equations and .

Below we disregard scalar factors which only distract from the essential point.

6.1   Quantum Gates, Circuits, and Algorithms

Above we introduced 1-qubit unitaries $\Lambda^Z(\alpha)$ and $\Lambda^X(\beta)$ corresponding to rotations in the X-Y and the Y-Z planes respectively; these suffice to represent all 1-qubit unitaries, and their basic equational properties follow from the various lemmas introduced in the preceding sections. We demonstrate how to define the $\wedge\mathsf{X}$ and $\wedge\mathsf{Z}$ gates, and prove two elementary equations involving them. The addition of these gates will provide a computationally universal set of gates.

Example 1 ($\wedge\mathsf{X}$ gate). Setting one verifies by concrete calculation that $\wedge\mathsf{X} = $. We can also give an abstract proof. Let $|i\rangle$ be a classical point for the green classical structure; by evaluating it with an input to its control qubit (the green end) we have , which for $|i\rangle = |0_X\rangle$ is the identity, and in the binary case, for $|i\rangle = |1_X\rangle$, is the unique operation X. By applying it three times, alternating the target and control input, we obtain, , i.e. σ. while this is a well-known property of $\wedge\mathsf{X}$, our proof uses only the bialgebra structure hence it will hold in much greater generality than just for qubits.

Example 2 ($\wedge Z$ gate). One can derive the $\wedge Z$ from that of $\wedge X$ by augmenting the target qubit of the $\wedge X$ with H gates i.e. $=$ ⊦●⊦●⊦ $=$ ●⊞●. We have that

$\wedge Z \circ \wedge Z = 1$ since ●⊞● $=$ ●⊞● $=$ ●⊞● $=$ ● ● $=$ │ │ $=$ │ │.

Example 3 (An algorithm : the quantum Fourier transform). The quantum Fourier transform is one of the most important quantum algorithms, lying at the centre of Shor's famous factoring algorithm. The equations we have enable this algorithm to be simulated in the diagrammatic language. Unlike the preceding examples, here we require the interaction between the two phase groups.

In our language the $\wedge Z_\alpha$ gate is [diagram with $-\alpha/2$ and $\alpha/2$] and the circuit [diagram with $-\pi/4$ and $\pi/4$] involving it realises the quantum Fourier transform for 2 qubits. The algorithm can be simulated graphically, as shown below:

This example makes use of classical values coded as quantum states to control the interference of phases: this is the archetypal behaviour of quantum algorithms.

6.2   Multi-partite Entanglement

In our graphical language, a quantum state is nothing more than a circuit with no inputs; output edges correspond to the individual qubits making up the state. The interior of the diagram, i.e. its graph structure, describes how these qubits are related. Hence this notation is ideal for representing large entangled states.

Example 4. The cluster states used in measurement-based quantum computing, can be prepared in several ways; the graphical calculus provides short proofs of their equivalence. For example, the original scheme describes a $\wedge Z$ interaction between qubits initially prepared in the state $|+\rangle$; in our notation this is $|0_Z\rangle$, or ●. So 1D cluster arises as ····· [boxed diagrams] ····· where the boxes delineate the individual $|+\rangle$ preparations and $\wedge Z$ operations. Alternatively, the cluster state can be prepared by fusion of states of the form $|0+\rangle + |1-\rangle$. Our $\delta_Z^\dagger$ is in fact this fusion operation, so a 1D cluster arises as

..... ⟨diagram⟩ ...... Using the spider theorem, these are equivalent ⟨diagram⟩ = ⟨diagram⟩ = ⟨diagram⟩. Ongoing work seeks to classify multipartite entangled states in terms of their graphical representatives, and to formalises general matrix product states.

6.3 Properties of Quantum Computational Models

Our formalism axiomatises two key features of quantum mechanics: the underlying monoidal structure and the interaction of complementary observables. Furthermore it is a semantic, which is to say *extensional*, framework which makes it ideal for unifying various approaches to quantum computation. E.g. we can demonstrate equivalence between different quantum computational models.

Example 5 (Verifying one-way quantum computations). We show how to verify some example programs for the one-way model, taken from [7], by translation to equivalent quantum circuits. Post-selected qubit measurements[8] can be represented by copoints such a ⟨symbol⟩. The spider theorem allows the post-selected one-way program ⟨diagram⟩ = ⟨diagram⟩ = ⟨diagram⟩ implementing a $\wedge X$ operation upon its inputs to be rewritten to a $\wedge X$ gate in no more than two steps . Now recall that any single qubit unitary map U has an Euler decomposition as such that $U = Z_\alpha X_\beta Z_\gamma$. In our notation this is $Z_\alpha = \Lambda^Z(\alpha)$ and $X_\alpha = \Lambda^X(\alpha)$. Again a sequence of simple rewrites shows that the one-way program ⟨diagram⟩ = ⟨diagram⟩ = ⟨diagram⟩ to compute such a unitary indeed computes the desired map.

References

1. Abramsky, S., Coecke, B.: A categorical semantics of quantum protocols. In: Abramsky, S., Coecke, B. (eds.) Proceedings of the 19th Annual IEEE Symposium on Logic in Computer Science (LiCS). IEEE Computer Science Press, Los Alamitos (2004); Abstract physical traces. Theory and Applications of Categories 14, 111–124 (2005)
2. Carboni, A., Walters, R.F.C.: Cartesian bicategories I. Journal of Pure and Applied Algebra 49, 11–32 (1987)
3. Coecke, B.: De-linearizing linearity: projective quantum axiomatics from strong compact closure. ENTCS 170, 49–72 (2007)

---

[8] Classical structures were initially introduced in [5] to represent classical control structure so this example can easily be extended with the required unitary corrections.

4. Coecke, B., Paquette, E.O.: POVMs and Naimark's theorem without sums (2006) (to appear in ENTCS), arXiv:quant-ph/0608072
5. Coecke, B., Pavlovic, D.: Quantum measurements without sums. In: Chen, G., Kauffman, L., Lamonaco, S. (eds.) Mathematics of Quantum Computing and Technology, pp. 567–604. Taylor and Francis, Abington (2007)
6. Coecke, B., Pavlovic, D., Vicary, J.: Dagger Frobenius algebras in FdHilb are bases. Oxford University Computing Laboratory Research Report RR-08-03 (2008)
7. Danos, V., Kashefi, E., Panangaden, P.: The measurement calculus. Journal of the ACM 54(2) (2007), arXiv:quant-ph/0412135
8. Joyal, A., Street, R.: The Geometry of tensor calculus I. Advances in Mathematics 88, 55–112 (1991)
9. Kelly, G.M., Laplaza, M.L.: Coherence for compact closed categories. Journal of Pure and Applied Algebra 19, 193–213 (1980)
10. Klappenecker, A., Rötteler, M.: Constructions of mutually unbiased bases. LNCS, vol. 2948, pp. 137–144. Springer, Heidelberg (2004)
11. Kock, J.: Frobenius Algebras and 2D Topological Quantum Field Theories. In: Composing PROPs. Theory and Applications of Categories, vol. 13, pp. 147–163. Cambridge University Press, Cambridge (2003)
12. Selinger, P.: Dagger compact closed categories and completely positive maps. ENTCS, 170, 139–163 (2005),
www.mathstat.dal.ca/~selinger/papers.htmldagger
13. Street, R.: Quantum Groups: A Path to Current Algebra, Cambridge UP (2007)

Perpetuality for Full and Safe Composition
(in a Constructive Setting)

Delia Kesner

PPS, Université Paris Diderot

Abstract. We study perpetuality in calculi with explicit substitutions having *full composition*. A simple perpetual strategy is used to define strongly normalising terms inductively. This gives a simple argument to show preservation of β-strong normalisation as well as strong normalisation for typed terms. Particularly, the strong normalisation proof is based on implicit substitution rather than explicit substitution, so that it turns out to be modular w.r.t. the well-known proofs for typed lambda-calculus. All the proofs we develop are constructive.

1   Introduction

In calculi with *explicit* substitutions (ES) without composition rules, such as λx [20,23], outermost substitutions must be delayed until the total execution of all the innermost substitutions appearing in the same environment. Thus for example, the outermost substitution $[x/v]$ in the term $(zyx)[y/xx][x/v]$ must be delayed until $[y/xx]$ is first executed on zyx. This can be recovered by the use of composition rules which allow to propagate substitutions through (non pure) terms. Thus, $(zyx)[y/xx][x/v]$ can be reduced to $(zyx)[x/v][y/(xx)[x/v]]$, which can be further reduced to $(zyv)[y/vv]$, a term equal to $(zyx)[y/xx]\{x/v\}$, where $\{x/v\}$ denotes the standard *meta/implicit* substitution (on non pure terms) that the *explicit* substitution $[x/v]$ is supposed to implement.

Composition rules for ES first appeared in $\lambda\sigma$ [1]. They are used to get confluence on open terms [10,11] when implementing higher-order unification [7] or functional abstract machines [19]. They guarantee a property, called *full composition*, that calculi without composition do not enjoy: any term of the form $t[x/u]$ can be reduced to $t\{x/u\}$, *i.e.* explicit substitution implements the implicit one.

Many calculi with ES such as $\lambda\sigma$ [1], $\lambda\sigma_{\Uparrow}$ [10], λ_{sub} [22], $\lambda\mathtt{lxr}$ [14] and $\lambda\mathtt{es}$ [11] enjoy full composition. However, $\lambda\sigma$ and $\lambda\sigma_{\Uparrow}$ do not enjoy neither strong normalisation (SN) for *typed* terms, nor preservation of β-strong normalisation (PSN) for *untyped* terms, a result which is a consequence of Melliès' counter-example [21]. But full composition and normalisation can live together, leading to a notion of *safe* composition; this is for example the case of λ_{sub}, $\lambda\mathtt{es}$ and $\lambda\mathtt{lxr}$. The available SN proofs for calculi with composition are indirect: either one simulates reduction by means of another well-founded relation, or SN is deduced from a sufficient property, as for example PSN. Proofs using the first technique are for example those for λ_{ws} [6] and $\lambda\mathtt{lxr}$ [14], based on the well-foundedness of

L. Aceto et al. (Eds.): ICALP 2008, Part II, LNCS 5126, pp. 311–322, 2008.

the reduction relation for multiplicative exponential linear logic (MELL) proof-nets [9]. An example of SN proof using the second technique is that for λes, where PSN is obtained by two consecutive translations, one from λes into a calculus with ES and weakenings, the second one from this intermediate calculus into the Church-Klop's Λ_I-calculus [16]. In both cases the proofs are long, but especially not self-contained.

Since nothing indicates that calculi with safe composition could be only understood in terms of MELL proof-nets or the Λ_I-calculus, it will be then significant to provide independent arguments to prove normalisation properties for them. This would be useful, particularly, when integrating them inside other richer frameworks such as type theory.

The aim of this paper is to understand safe composition. For that, we choose to work with a simple calculus, that we call λex, obtained by extending λx with one rewriting rule for composition of *dependent* substitutions and one equation for commutation of *independent* substitutions. A similar calculus is studied in [24], where our equation is treated as a non-terminating reduction rule. The λex-calculus uses *unary* constructors for substitutions but has the same expressive power than calculi with *n-ary* substitutions: thus for example $(xy)[y/x, x/y]$ can be implemented by the α-equivalent term $(wy)[y/x][w/y]$. Indeed, while simultaneous substitutions are specified by *lists* (given by n-ary substitutions) in calculi like $\lambda\sigma$, they are modelled by *sets* (given by commutation of independent unary substitutions) in λex. The λex-calculus is conceptually simple, it enjoys full composition and confluence on open-terms.

The technical tools used in this paper are the following. We first define a *perpetual* reduction strategy for λex: if $t \notin \mathcal{SN}_{\lambda\text{ex}}$ and t reduces to t' by the strategy, then $t' \notin \mathcal{SN}_{\lambda\text{ex}}$. In particular, since the perpetual strategy reduces $t[x/u]$ to $t\{x/u\}$, one has to show that normalisation of **I**mplicit substitution implies normalisation of **E**xplicit substitution:

$$(\textbf{IE}) \quad u \in \mathcal{SN}_{\lambda\text{ex}} \ \& \ t\{x/u\} \in \mathcal{SN}_{\lambda\text{ex}} \text{ imply } t[x/u] \in \mathcal{SN}_{\lambda\text{ex}}$$

In other words, ES implements implicit substitutions but nothing more than that, otherwise one may get calculi such as $\lambda\sigma$ where $t[x/u]$ does much more than $t\{x/u\}$ since it is able to behave like $t\{x/u\}$ but also to behave differently (for example by looping) before reducing to $t\{x/u\}$. A consequence of **(IE)** is that standard techniques to show SN based on *meta*-substitution can also be applied to calculi with ES, thus considerably simplifying the reasoning. Indeed, the perpetual strategy is used to give an inductive characterisation of the set $\mathcal{SN}_{\lambda\text{ex}}$ by means of just four inference rules. This characterisation is then used to show that untyped terms enjoy PSN and typed terms enjoy SN. In particular, SN is shown by using arithmetical arguments: the proof is the one for simply typed λ-calculus but just adds the new case $t[x/u]$. In that sense we can say that our SN proof is modular w.r.t. the SN proof for typed lambda-terms. All our proofs are constructive in the sense that neither excluded middle nor double negation elimination are used. At the end of the paper we also show how SN of other calculi (with or without) full composition can be obtained from SN of λex.

Perpetual strategies are studied for the non equational systems $\lambda\mathbf{x}$ in [3,15,18], and λ_{ws} in [2]. No abstract use of full composition can be done there. Current investigations carried out in [29] show PSN for different calculi with (full or not) composition. The approach is based on proofs by contradiction which analyse some *minimal* not terminating reduction sequence of the underlying calculus.

The paper is organised as follows. Section 2 introduces syntax and reduction rules. Perpetuality is studied in Section 3 and normalisation proofs are given in Section 4. Section 5 presents the labelling technique to show the **(IE)** property. In Section 6 we explain how to infer SN for other calculi with ES from our result in Section 4. We conclude and give directions for further work in Section 7.

Full details of the proofs in the paper are available on [12].

2 Syntax

The $\lambda\mathbf{ex}$-calculus can be viewed as the $\lambda\mathbf{x}$-calculus together with a safe composition rule for dependent substitutions and a commutativity equation for independent substitutions. The set of x-terms is defined by:

$$\mathcal{T}_\mathbf{x} ::= x \mid \mathcal{T}_\mathbf{x} \, \mathcal{T}_\mathbf{x} \mid \lambda x.\mathcal{T}_\mathbf{x} \mid \mathcal{T}_\mathbf{x}[x/\mathcal{T}_\mathbf{x}]$$

Free and bound variables are defined as usual by assuming the terms $\lambda x.t$ and $t[x/u]$ bind x in t. The congruence generated by renaming of bound variables is called α-conversion. Thus for example $(\lambda y.x)[x/y] =_\alpha (\lambda z.w)[w/y]$. We use the notation $\overline{t_n}$ for a list of terms $t_1,\ldots,t_n$ and $\overline{ut_n}$ for $ut_1\ldots t_n$ which is an abbreviation of $(\ldots((ut_1)t_2)\ldots\ldots t_n)$.

Meta-substitution on x-terms is defined modulo α-conversion in such a way that capture of variables is avoided:

$$x\{x/v\} := v \qquad\qquad (tu)\{x/v\} := t\{x/v\}u\{x/v\}$$
$$y\{x/v\} := y \text{ if } y \neq x \qquad (\lambda y.t)\{x/v\} := \lambda y.t\{x/v\}$$
$$t[y/u]\{x/v\} := t\{x/v\}[y/u\{x/v\}]$$

Thus for example $(\lambda y.x)\{x/y\} = \lambda z.y$. Note that $t\{x/u\} = t$ if $x \notin \mathbf{fv}(t)$.

Besides α-conversion, we consider the following equations and rules.

Equations :			
$t[x/u][y/v]$	$=_\mathbf{C}$	$t[y/v][x/u]$	if $y \notin \mathbf{fv}(u)$ & $x \notin \mathbf{fv}(v)$
Rules :			
$(\lambda x.t)\, u$	$\to_\mathbf{B}$	$t[x/u]$	
$x[x/u]$	$\to_\mathbf{Var}$	u	
$t[x/u]$	$\to_\mathbf{Gc}$	t	if $x \notin \mathbf{fv}(t)$
$(tu)[x/v]$	$\to_\mathbf{App}$	$t[x/v]\, u[x/v]$	
$(\lambda y.t)[x/v]$	$\to_\mathbf{Lamb}$	$\lambda y.t[x/v]$	
$t[x/u][y/v]$	$\to_\mathbf{Comp}$	$t[y/v][x/u[y/v]]$	if $y \in \mathbf{fv}(u)$

The *rewritingrelation* generated by all the previous rules except B is denoted by x. We write Bx for B$\cup$x. The *equivalence relation* generated by the conversions

α and C is written e. The *reduction relation* generated by the *rewriting relations modulo* e specify rewriting of e-equivalence classes:

$$t \to_{ex} t' \quad \text{iff} \quad \exists\, s, s' \text{ s.t. } t =_e s \to_x s' =_e t'$$
$$t \to_{\lambda ex} t' \quad \text{iff} \quad \exists\, s, s' \text{ s.t. } t =_e s \to_{Bx} s' =_e t'$$

Note that all the equations and rules are assumed to avoid capture of variables by α-conversion. Thus for example we have $y \neq x$ and $y \notin \mathtt{fv}(v)$ in rule Lamb. Same kind of assumptions are done for Comp and C.

The notation $\to_{\lambda ex}^{*}$ (resp. $\to_{\lambda ex}^{+}$) is used for the reflexive and transitive (resp. transitive) closure of $\to_{\lambda ex}$. Thus, if $t \to_{\lambda ex}^{*} t'$ in 0 reduction steps, then $t =_e t'$.

A term t is said to be in λex-normal form, written $t \in \mathcal{NF}_{\lambda ex}$, if there is no s such that $t \to_{\lambda ex} s$. A term t is said to be λex-strongly normalising, written $t \in \mathcal{SN}_{\lambda ex}$, if there is no infinite λex-reduction sequence starting at t, in which case $\eta_{\lambda ex}(t)$ denotes the maximal length of a λex-reduction sequence starting at t. A standard inductive definition of $\mathcal{SN}_{\lambda ex}$ can be given by:

$$t \in \mathcal{SN}_{\lambda ex} \text{ iff } \forall s\ (t \to_{\lambda ex} s \text{ implies } s \in \mathcal{SN}_{\lambda ex})$$

The following basic properties of λex-reduction can be shown by a straightforward induction on the λex-reduction relation.

Lemma 1 (Basic Properties).

- *If $t \to_{\lambda ex} t'$, then $\mathtt{fv}(t') \subseteq \mathtt{fv}(t)$.*
- *For $\mathcal{R} \in \{ex, \lambda ex\}$, if $t \to_{\mathcal{R}} t'$, then $u\{x/t\} \to_{\mathcal{R}}^{*} u\{x/t'\}$ and $t\{x/u\} \to_{\mathcal{R}} t'\{x/u\}$. Thus in particular $t\{x/u\} \in \mathcal{SN}_{\mathcal{R}}$ implies $t \in \mathcal{SN}_{\mathcal{R}}$.*

The rule Comp and the equation C guarantee the following property:

Lemma 2 (Full composition). $t[x/u] \to_{ex}^{+} t\{x/u\}$.

Proof. By induction on t. The interesting case is $t = s[y/v]$. If $x \in \mathtt{fv}(v)$, then $s[y/v][x/u] \to_{Comp} s[x/u][y/v[x/u]] \to_{ex\ (i.h.)}^{*} s\{x/u\}[y/v\{x/u\}] = t\{x/u\}$. If $x \notin \mathtt{fv}(v)$, then $s[y/v][x/u] =_C s[x/u][y/v] \to_{ex\ (i.h.)}^{*} s\{x/u\}[y/v] = t\{x/u\}$.

Lemma 3 (Confluence). *The reduction relation is confluent on open terms.*

Proof. This can be proved by the Tait and Martin Löf technique. The proof proceeds similarly to that of the λes-calculus given in [11].

3   Perpetuality

A *perpetual* strategy gives an *infinite* reduction sequence for a term, if one exists, otherwise, it gives a finite reduction sequence leading to some normal form. Perpetual strategies can be seen as antonyms of normalising strategies, they are particularly used to obtain normalisation results. For a survey about perpetual strategies we refer the reader to [30].

In contrast to *one-step* strategies for ES given for example in [18,3], we now define a *many-step* strategy for x-terms which preserves λex-normal forms and gives a $\rightarrow^+_{\lambda\text{ex}}$-reduct for any $t \notin \mathcal{NF}_{\lambda\text{ex}}$.

This is done according to the following cases. If $t = xt_1 \ldots t_n$, rewrite the *left-most* t_i which is reducible. If $t = \lambda x.u$, rewrite u. If $t = (\lambda x.s)u\overline{v_n}$, rewrite the head redex. If $t = s[x/u]\overline{v_n}$ and $u \notin \mathcal{SN}_{\lambda\text{ex}}$, rewrite u. If $t = s[x/u]\overline{v_n}$ and $u \in \mathcal{SN}_{\lambda\text{ex}}$, rewrite the head redex using full composition. Formally,

Definition 1. *The* strategy $\leadsto$ *on* x-terms *is given by an inductive definition.*

$$
\frac{\overline{u_n} \in \mathcal{NF}_{\lambda\text{ex}} \quad t \leadsto t'}{x\overline{u_n}t\overline{v_m} \leadsto x\overline{u_n}t'\overline{v_m}} \text{ (p-var)} \quad \frac{t \leadsto t'}{\lambda x.t \leadsto \lambda x.t'} \text{ (p-abs)} \quad \frac{}{(\lambda x.t)u\overline{u_n} \leadsto t[x/u]\overline{u_n}} \text{ (p-B)}
$$

$$
\frac{u \in \mathcal{SN}_{\lambda\text{ex}}}{t[x/u]\overline{u_n} \leadsto t\{x/u\}\overline{u_n}} \text{ (p-subs1)} \quad \frac{u \notin \mathcal{SN}_{\lambda\text{ex}} \quad u \leadsto u'}{t[x/u]\overline{u_n} \leadsto t[x/u']\overline{u_n}} \text{ (p-subs2)}
$$

The strategy is deterministic so that $t \leadsto u$ and $t \leadsto v$ implies $u = v$. Moreover, the strategy is not necessarily leftmost-outermost or left-to-right because of the (p-subs1) rule: substitution propagation can be performed in any order. Note also that the strategy is not effective since it is based on an undecidable predicate. The strategy is perpetual: if $t \notin \mathcal{SN}_{\lambda\text{ex}}$ and $t \leadsto t'$, then $t' \notin \mathcal{SN}_{\lambda\text{ex}}$. This will be used later to give an inductive characterisation of the set $\mathcal{SN}_{\lambda\text{ex}}$.

Lemma 4. *If* $t \leadsto t'$, *then* $t \rightarrow^+_{\lambda\text{ex}} t'$.

Proof. By induction on the strategy $\leadsto$ using Lemma 2.

Theorem 1 (Perpetuality). *If* $t \leadsto t'$ *and* $t' \in \mathcal{SN}_{\lambda\text{ex}}$, *then* $t \in \mathcal{SN}_{\lambda\text{ex}}$.

Proof. By induction on the strategy $\leadsto$. We only treat the non trivial cases.

(p-B) $t = (\lambda x.s)u\overline{u_n} \leadsto s[x/u]\overline{u_n} = t'$. If $s[x/u]\overline{u_n} \in \mathcal{SN}_{\lambda\text{ex}}$, then $s, u, \overline{u_n} \in \mathcal{SN}_{\lambda\text{ex}}$. We thus show by induction on $\eta_{\lambda\text{ex}}(s) + \eta_{\lambda\text{ex}}(u) + \Sigma_i \eta_{\lambda\text{ex}}(u_i)$ that every λex-reduct of $(\lambda x.s)u\overline{u_n}$ is in $\mathcal{SN}_{\lambda\text{ex}}$. Conclude $(\lambda x.s)u\overline{u_n} \in \mathcal{SN}_{\lambda\text{ex}}$.

(p-subs2) $t = s[x/u]\overline{u_n} \leadsto s[x/u']\overline{u_n} = t'$, $u \notin \mathcal{SN}_{\lambda\text{ex}}$ and $u \leadsto u'$. If $s[x/u']\overline{u_n} \in \mathcal{SN}_{\lambda\text{ex}}$ then in particular $u' \in \mathcal{SN}_{\lambda\text{ex}}$, thus $u \in \mathcal{SN}_{\lambda\text{ex}}$ by the i.h. From $u \notin \mathcal{SN}_{\text{ex}}$ and $u \in \mathcal{SN}_{\lambda\text{ex}}$ we get (constructively) any proposition, so in particular $t \in \mathcal{SN}_{\lambda\text{ex}}$.

(p-subs1) $t = s[x/u]\overline{u_n} \leadsto s\{x/u\}\overline{u_n} = t'$ and $u \in \mathcal{SN}_{\lambda\text{ex}}$. Then the **(IE)** property (Lemma 8) allows to conclude.

4 Normalisation Properties

To show that untyped x-terms enjoy PSN and typed x-terms are λex-strongly normalising we proceed in two different steps. We first define an inductive set $\mathcal{ISN}$ which turns out to be equal to $\mathcal{SN}_{\lambda\text{ex}}$. PSN can then be easily proved by using the inductive definition of $\mathcal{ISN}$. To show SN, we can then choose at least two different ways to proceed. We include in Section 4.1 the shortest one which is based on simple arithmetical arguments [27], and we refer the reader to [12] for the second one which uses standard reducibility technology [8,26].

Inductive characterisations of SN terms are useful, for instance, in constructive SN proofs. An inductive definition of SN terms for the λ-calculus is given for example in [28]. It was then extended in [3,18] for calculi with ES, but using many different inference rules to characterise SN terms of the form $t[x/u]$. We just give here one inference rule for each possible x-term.

Definition 2. *The* inductive set $\mathcal{ISN}$ *is defined as follows:*

$$\frac{t_1,\ldots,t_n \in \mathcal{ISN} \quad n \geq 0}{xt_1\ldots t_n \in \mathcal{ISN}}\,(\text{var}) \qquad \frac{u[x/v]t_1\ldots t_n \in \mathcal{ISN} \quad n \geq 0}{(\lambda x.u)vt_1\ldots t_n \in \mathcal{ISN}}\,(\text{app})$$

$$\frac{u\{x/v\}t_1\ldots t_n \in \mathcal{ISN} \quad v \in \mathcal{ISN} \quad n \geq 0}{u[x/v]t_1\ldots t_n \in \mathcal{ISN}}\,(\text{subs}) \qquad \frac{u \in \mathcal{ISN}}{\lambda x.u \in \mathcal{ISN}}\,(\text{abs})$$

Proposition 1. $\mathcal{SN}_{\lambda\text{ex}} = \mathcal{ISN}$.

Proof. If $t \in \mathcal{SN}_{\lambda\text{ex}}$, $t \in \mathcal{ISN}$ is proved by induction on the pair $\langle \eta_{\lambda\text{ex}}(t),$ $\text{size}(t)\rangle$. If $t \in \mathcal{ISN}$, $t \in \mathcal{SN}_{\lambda\text{ex}}$ is proved by induction on $t \in \mathcal{ISN}$ using Theorem 1.

Theorem 2 (PSN for λ-terms). *If* $t \in \mathcal{SN}_\beta$, *then* $t \in \mathcal{SN}_{\lambda\text{ex}}$.

Proof. By induction on the definition of $\mathcal{SN}_\beta$ [28] using Prop. 1. If $t = x\overline{t_n}$ with $t_i \in \mathcal{SN}_\beta$, then $t_i \in \mathcal{SN}_{\lambda\text{ex}}$ by the i.h. so that the (var) rule allows to conclude. The case $t = \lambda x.u$ is similar. If $t = (\lambda x.u)v\overline{t_n}$, with $u\{x/v\}\overline{t_n} \in \mathcal{SN}_\beta$ and $v \in \mathcal{SN}_\beta$, then both terms are in $\mathcal{SN}_{\lambda\text{ex}}$ by the i.h. so that the (subs) rule gives $u[x/v]\overline{t_n} \in \mathcal{SN}_{\lambda\text{ex}}$ and the (app) rule gives $(\lambda x.u)v\overline{t_n} \in \mathcal{SN}_{\lambda\text{ex}}$.

We now give a type system for x-terms. Richer type systems with intersection types could also be given to characterise the set $\mathcal{SN}_{\lambda\text{ex}}$ in terms of typed terms (see [13,18] for details).

Types are built over a set of atomic types and the $\rightarrow$ constructor. An *environment* is a finite set of pairs $x : A$. A *sequent* $\Gamma \vdash t : A$ is formed by an environment Γ, a term t and a type A. *Derivations* of sequents are obtained by application of the following typing rules.

$$\frac{}{\Gamma, x : A \vdash x : A} \qquad \frac{\Gamma \vdash t : A \rightarrow B \quad \Gamma \vdash u : A}{\Gamma \vdash tu : B}$$

$$\frac{\Gamma, x : A \vdash t : B}{\Gamma \vdash \lambda x.t : A \rightarrow B} \qquad \frac{\Gamma \vdash u : B \quad \Gamma, x : B \vdash t : A}{\Gamma \vdash t[x/u] : A}$$

A term t *of type* A, written t^A, is a term s.t. $\Gamma \vdash t : A$ is derivable for some Γ. A *typed* term t is a term of type A for some type A.

Induction on type derivations together with weakening/strengthening allow us to show the following stability properties.

Lemma 5 (Stability of Typed Terms)

(by substitution) *If $\Gamma \vdash u : B$ & $\Gamma, x : B \vdash t : A$, then $\Gamma \vdash t\{x/u\} : A$.*

(by reduction) *If $\Gamma \vdash t : A$ & $t \to_{\lambda \text{ex}} t'$, then $\Gamma \vdash t' : A$.*

4.1   The Arithmetical Technique

This technique is based on van Daalen's strong normalisation proof for the typed lambda-calculus [27], and is extremely short.

Lemma 6. *If $t^A, u^B \in \mathcal{SN}_{\lambda \text{ex}}$, then $t\{x^B/u^B\} \in \mathcal{SN}_{\lambda \text{ex}}$.*

Proof. By induction on $\langle B, \eta_{\lambda \text{ex}}(t), \texttt{size}(t) \rangle$.

- The cases $t = x$, $t = \lambda y.v$ and $y\overline{v_n}$ are straightforward.
- $t = xv\overline{v_n}$. The i.h. gives $V = v\{x/u\}$ and $V_i = v_i\{x/u\}$ in $\mathcal{SN}_{\lambda \text{ex}}$. To show $t\{x/u\} = uV\overline{V_n} \in \mathcal{SN}_{\lambda \text{ex}}$ we show that all its reducts are in $\mathcal{SN}_{\lambda \text{ex}}$. We reason by induction on $\eta_{\lambda \text{ex}}(u) + \eta_{\lambda \text{ex}}(V) + \Sigma_{i \in 1 \ldots n} \, \eta_{\lambda \text{ex}}(V_i)$.
  If reduction takes place in a subterm of $uV\overline{V_n}$, we conclude by the i.h. Suppose $u = \lambda y.U$ and $(\lambda y.U)V\overline{V_n} \to U[y/V]\overline{V_n}$. Then $\texttt{type}(V) = \texttt{type}(v) < \texttt{type}(u) = \texttt{type}(x)$ so that $U\{y/V\} \in \mathcal{SN}_{\lambda \text{ex}}$ by the i.h. Write $U\{y/V\}\overline{V_n} = (z\overline{V_n})\{z/U\{y/V\}\}$. We have $\texttt{type}(U\{y/V\}) = \texttt{type}(U) < \texttt{type}(u)$ so that $U\{y/V\}\overline{V_n} \in \mathcal{SN}_{\lambda \text{ex}}$ by the i.h. We conclude $U[y/V]\overline{V_n} \in \mathcal{SN}_{\lambda \text{ex}}$ by Prop. 1.
- $t = (\lambda y.s)v\overline{v_n}$. The i.h. gives $S = s\{x/u\}$, $V = v\{x/u\}$ and $V_i = v_i\{x/u\}$ in $\mathcal{SN}_{\lambda \text{ex}}$. To show $t\{x/u\} = (\lambda y.S)V\overline{V_n} \in \mathcal{SN}_{\lambda \text{ex}}$ we show that all its reducts are in $\mathcal{SN}_{\lambda \text{ex}}$. We reason by induction on $\eta_{\lambda \text{ex}}(S) + \eta_{\lambda \text{ex}}(V) + \Sigma_{i \in 1 \ldots n} \, \eta_{\lambda \text{ex}}(V_i)$.
  If reduction takes place in a subterm of $(\lambda y.S)V\overline{V_n}$, we conclude by the i.h. Otherwise suppose $(\lambda y.S)V\overline{V_n} \to S[y/V]\overline{V_n}$. Now, take $T = s\{y/v\} \, \overline{v_n}$. Since $\eta_{\lambda \text{ex}}(T) < \eta_{\lambda \text{ex}}(t)$, then the i.h. gives $T\{x/u\} \in \mathcal{SN}_{\lambda \text{ex}}$. We write $S\{y/V\}\overline{V_n} = T\{x/u\}$ so that Prop. 1 gives $S[y/V]\overline{V_n} \in \mathcal{SN}_{\lambda \text{ex}}$. Thus all the reducts of $t\{x/u\}$ are $\mathcal{SN}_{\lambda \text{ex}}$ and we can conclude $t\{x/u\} \in \mathcal{SN}_{\lambda \text{ex}}$.
- $t = s[y/v]\overline{v_n}$. The proof proceeds as in the previous case.

Theorem 3 (SN for λex). *If t is a typed term, then $t \in \mathcal{SN}_{\lambda \text{ex}}$.*

Proof. By induction on t. The cases $t = x$ and $t = \lambda x.u$ are straightforward. If $t = uv$, then u, v are typed and by the i.h. $u, v \in \mathcal{SN}_{\lambda \text{ex}}$. We write $t = (z \, v)\{z/u\}$, where $z \, v$ is $\mathcal{SN}_{\lambda \text{ex}}$ by Definition 2 and appropriately typed. Lemma 6 then gives $t \in \mathcal{SN}_{\lambda \text{ex}}$. If $t = u[x/v]$, then u, v are typed and by the i.h. $u, v \in \mathcal{SN}_{\lambda \text{ex}}$ so that Lemma 6 gives $u\{x/v\} \in \mathcal{SN}_{\lambda \text{ex}}$. Prop. 1 allows us to conclude $u[x/v] \in \mathcal{SN}_{\lambda \text{ex}}$.

5 The (IE) Property

The aim of this section is to show the key argument used to guarantee that our strategy (Definition 1) is perpetual. More precisely, we show that normalisation of **I**mplicit substitution implies normalisation of **E**xplicit substitution:

$$\textbf{(IE)}\ u \in \mathcal{SN}_{\lambda\text{ex}}\ \&\ t\{x/u\} \in \mathcal{SN}_{\lambda\text{ex}}\ \text{imply}\ t[x/u] \in \mathcal{SN}_{\lambda\text{ex}}$$

To show the **(IE)** property we adapt the labelling technique [4,2] to the equational case. Given a set of variables $\mathbb{S}$, the $\mathbb{S}$-*labelled terms* (or simply *labelled terms* if $\mathbb{S}$ is clear from the context), are given by:

$$\mathcal{T}_{\mathbb{S}} ::= x \mid \mathcal{T}_{\mathbb{S}}\,\mathcal{T}_{\mathbb{S}} \mid \lambda x.\mathcal{T}_{\mathbb{S}} \mid \mathcal{T}_{\mathbb{S}}[x/\mathcal{T}_{\mathbb{S}}] \mid \mathcal{T}_{\mathbb{S}}[\![x/v]\!]\ (v \in \mathcal{SN}_{\lambda\text{ex}}\ \&\ \texttt{fv}(v) \subseteq \mathbb{S})$$

Thus, labelled substitutions can only contain x-terms so in particular they cannot contain other labelled substitutions inside them.

Note that we can always assume that subterms $u[x/v]$ and $u[\![x/v]\!]$ inside $t \in \mathcal{T}_{\mathbb{S}}$ are s.t. $x \notin \mathbb{S}$. Indeed, α-conversion allows to choose names outside $\mathbb{S}$ for the bound variables of $\mathbb{S}$-terms. The idea behind the operational semantics of $\mathbb{S}$-terms, specified by the following set of equations and reduction rules, is that labelled substitutions may commute/traverse ordinary substitutions but these last ones cannot traverse the labelled ones. This behaviour of labelled substitutions is later used to simulate application of implicit substitution.

Equations :			
$t[y/u][\![x/v]\!]$	$=_{\underline{c}}$	$t[\![x/v]\!][y/u]$	if $x \notin \texttt{fv}(u)$ & $y \notin \texttt{fv}(v)$
$t[\![y/u]\!][\![x/v]\!]$	$=_{\underline{c}}$	$t[\![x/v]\!][\![y/u]\!]$	if $x \notin \texttt{fv}(u)$ & $y \notin \texttt{fv}(v)$
Rules :			
$x[\![x/v]\!]$	$\rightarrow_{\textsf{Var}}$	v	
$t[\![x/v]\!]$	$\rightarrow_{\textsf{Gc}}$	t	if $x \notin \texttt{fv}(t)$
$(tu)[\![x/v]\!]$	$\rightarrow_{\textsf{App}}$	$t[\![x/v]\!]\,u[\![x/v]\!]$	
$(\lambda y.t)[\![x/v]\!]$	$\rightarrow_{\textsf{Lamb}}$	$\lambda y.t[\![x/v]\!]$	
$t[y/u][\![x/v]\!]$	$\rightarrow_{\textsf{Comp}}$	$t[\![x/v]\!][y/u[\![x/v]\!]]$	if $x \in \texttt{fv}(u)$

The $\underline{x}$ (resp. $\mathbb{EX}$) reduction relation is generated by the previous rules modulo α (resp. $\alpha \cup \underline{c}$) conversion. In particular, they enjoy termination.

As expected, reduction on labelled terms can be simulated by reduction on their underlying x-terms.

Definition 3. *Unlabelled of* $\mathbb{S}$-*terms are* x-*terms defined by induction.*

$$\texttt{U}(x) := x \qquad \texttt{U}(\lambda x.t) := \lambda x.\texttt{U}(t) \qquad \texttt{U}(t[x/u]) := \texttt{U}(t)[x/\texttt{U}(u)]$$
$$\texttt{U}(tu) := \texttt{U}(t)\texttt{U}(u) \qquad \texttt{U}(t[\![x/u]\!]) := \texttt{U}(t)[x/u]$$

Consider the relation $\lambda\underline{\text{ex}} = \lambda\text{ex} \cup \mathbb{EX}$ on labelled terms.

Lemma 7. *Let* $t \in \mathcal{T}_{\mathbb{S}}$. *If* $t \in \mathcal{SN}_{\lambda\underline{\text{ex}}}$, *then* $\texttt{U}(t) \in \mathcal{SN}_{\lambda\text{ex}}$.

Proof. We first prove by induction on $\to_{\lambda\mathrm{ex}}$ the following: for $t \in \mathcal{T}_{\mathbb{S}}$, if $\mathtt{U}(t) \to_{\lambda\mathrm{ex}}$ u', then $\exists\, u \in \mathcal{T}_{\mathbb{S}}$ s.t. $t \to_{\lambda\underline{\mathrm{ex}}} u$ and $\mathtt{U}(u) = u'$. To conclude, we prove that every $\lambda\mathrm{ex}$-reduct of $\mathtt{U}(t)$ is in $\mathcal{SN}_{\lambda\mathrm{ex}}$ by induction on $\eta_{\lambda\underline{\mathrm{ex}}}(t)$ using the first property.

Taking $\mathbb{S} = \mathtt{fv}(u)$ and transforming the x-term $s[x/u]\overline{u_n}$ into the $\lambda\underline{\mathrm{ex}}$-term $s[\![x/u]\!]\overline{u_n}$ we have the following special case.

Corollary 1. *If* $s[\![x/u]\!]\overline{u_n} \in \mathcal{SN}_{\lambda\underline{\mathrm{ex}}}$, *then* $s[x/u]\overline{u_n} \in \mathcal{SN}_{\lambda\mathrm{ex}}$.

We now split $\lambda\underline{\mathrm{ex}}$ in two disjoint relations $\lambda\underline{\mathrm{ex}}^i$ and $\lambda\underline{\mathrm{ex}}^e$ which will be projected into $\lambda\mathrm{ex}$-reduction sequences differently.

Definition 4. *The* internal *reduction relation* $\lambda\underline{\mathrm{ex}}^i$ *is given by* EX-*reduction together with* $\lambda\mathrm{ex}$-*reduction in the bodies of labelled substitutions. The* external *reduction relation* $\lambda\underline{\mathrm{ex}}^e$ *is given by* $\lambda\mathrm{ex}$-*reduction everywhere except inside bodies of labelled substitutions.*

We will also use the following function $\mathtt{xc}$ from labelled terms to x-terms.

$\mathtt{xc}(x) := x$		$\mathtt{xc}(tu) := \mathtt{xc}(t)\mathtt{xc}(u)$
$\mathtt{xc}(\lambda y.t) := \lambda y.\mathtt{xc}(t)$		$\mathtt{xc}(t[x/u]) := \mathtt{xc}(t)[x/\mathtt{xc}(u)]$
		$\mathtt{xc}(t[\![x/v]\!]) := \mathtt{xc}(t)\{x/v\}$

Corollary 2. *Let* t *be a labelled term. If* $\mathtt{xc}(t) \in \mathcal{SN}_{\lambda\mathrm{ex}}$, *then* $t \in \mathcal{SN}_{\lambda\underline{\mathrm{ex}}}$.

Proof. Prove (using Lemma 1) that $\lambda\underline{\mathrm{ex}}$ can be projected into $\lambda\mathrm{ex}$ as follows:

1. $t \to_{\lambda\underline{\mathrm{ex}}^i} t'$ implies $\mathtt{xc}(t) \to_{\lambda\mathrm{ex}}^* \mathtt{xc}(t')$.
2. $t \to_{\lambda\underline{\mathrm{ex}}^e} t'$ implies $\mathtt{xc}(t) \to_{\lambda\mathrm{ex}}^+ \mathtt{xc}(t')$.

Then show that $\lambda\underline{\mathrm{ex}}^i$ is terminating (see [12] for details). Last, apply the abstract Theorem 4 given at the end of this section by taking $\mathtt{a}_1 = \lambda\underline{\mathrm{ex}}^i$, $\mathtt{a}_2 = \lambda\underline{\mathrm{ex}}^e$, $\mathtt{A} = \lambda\mathrm{ex}$ and $u\,\mathcal{R}\,U$ iff $\mathtt{xc}(u) = U$. We get that $\mathtt{xc}(t) \in \mathcal{SN}_{\lambda\mathrm{ex}}$ implies (constructively) $t \in \mathcal{SN}_{\lambda\underline{\mathrm{ex}}^i \cup \lambda\underline{\mathrm{ex}}^e} = \mathcal{SN}_{\lambda\underline{\mathrm{ex}}}$ so we thus conclude.

The previous corollary allows us to conclude with the main property required in the proof of the Perpetuality Theorem:

Lemma 8 (IE Property). *If* $u, s\{x/u\}\overline{u_n} \in \mathcal{SN}_{\lambda\mathrm{ex}}$, *then* $s[x/u]\overline{u_n} \in \mathcal{SN}_{\lambda\mathrm{ex}}$.

Proof. Let $s\{x/u\}\overline{u_n} \in \mathcal{SN}_{\lambda\mathrm{ex}}$, define $\mathbb{S} = \mathtt{fv}(u)$ and consider the $\mathbb{S}$-labelled term $s[\![x/u]\!]\overline{u_n}$. Then $\mathtt{xc}(s[\![x/u]\!]\overline{u_n}) = \mathtt{xc}(s)\{x/u\}\overline{\mathtt{xc}(u_n)} = s\{x/u\}\overline{u_n}$ so that $\mathtt{xc}(s[\![x/u]\!]\overline{u_n}) \in \mathcal{SN}_{\lambda\mathrm{ex}}$. We get $s[\![x/u]\!]\overline{u_n} \in \mathcal{SN}_{\lambda\underline{\mathrm{ex}}}$ by Corollary 2 and $s[x/u]\overline{u_n} \in \mathcal{SN}_{\lambda\mathrm{ex}}$ by Corollary 1.

Theorem 4. *Let* $\mathtt{a}_1$ *and* $\mathtt{a}_2$ *be two reduction relations on* $\mathtt{s}$ *and let* $\mathtt{A}$ *be a reduction relation on* $\mathtt{S}$. *Let* $\mathcal{R} \subseteq \mathtt{s} \times \mathtt{S}$. *Suppose* $\mathtt{a}_1$ *is well-founded and also*

- *For every* u, v, U *(u* $\mathcal{R}$ *U & u* $\mathtt{a}_1$ *v imply* $\exists V$ *s.t. vRV and U* $\mathtt{A}^*$ *V).*
- *For every* u, v, U *(u* $\mathcal{R}$ *U & u* $\mathtt{a}_2$ *v imply* $\exists V$ *s.t. v* $\mathcal{R}$ *V and U* $\mathtt{A}^+$ *V).*

Then, $t\,\mathcal{R}\,T$ *&* $T \in \mathcal{SN}_{\mathtt{A}}$ *imply* $t \in \mathcal{SN}_{\mathtt{a}_1 \cup \mathtt{a}_2}$.

Proof. A constructive proof of this theorem can be found in Corollary 26 of [17].

6   Deriving SN for Other Calculi

We now informally derive SN for other calculi with ES (having or not safe composition) from SN of λex, thus suggesting the existence of self-contained SN proofs also for them. However, while the correspondence/translation between yet another calculi without composition and λex seems to be unproblematic, the relation with *all* possible forms of safe composition is not claimed.

- The λx-calculus [20,23] is a sub-calculus of λex. The fact that $t \rightarrow_{\lambda x} t'$ implies $t \rightarrow^+_{\lambda ex} t'$ is then straightforward. Since typed terms in both calculi are the same, we thus deduce that typed x-terms are λx-strongly normalising.

- The λes-calculus [11] can be seen as a refinement of λex, where propagation of substitution with respect to application and substitution is done in a controlled way. We refer the reader to [11] for details on the rules. The fact that $t \rightarrow_{\lambda es} t'$ implies $t \rightarrow^+_{\lambda ex} t'$ is straightforward. Typed terms in both calculi are the same, we thus deduce that typed x-terms are λes-strongly normalising.

- Milner's calculus with partial substitution [22], called λ_{sub}, is able to encode λ-calculus in terms of a bigraphical reactive system. Syntax of λ_{sub} is given by x-terms and reduction rules *completely* propagate a substitution $[x/u]$ only on one occurrence of x at a time (see for example [22] for details). In [13] it is shown that there exist a translation T from x-terms to x-terms such that $t \rightarrow_{\lambda_{sub}} t'$ implies $T(t) \rightarrow^+_{\lambda es} T(t')$. Since translation T preserves typability, we conclude that typed x-terms are λ_{sub}-strongly normalising from the previous point.

- A λ-calculus with *partial* β-steps appears in [5]. Syntax is given by pure λ-terms and semantics is very similar to that of λ_{sub}. Similarly to [13], a translation T from λ-terms to x-terms can be defined to project one-step reduction in λ_{β_p} into at least one-step reduction in λ_{sub}. Since typed λ-terms translate to typed x-terms, then typed λ-terms are λ_{β_p}-strongly normalising from the previous point.

- David and Guillaume [4] defined a calculus with *labels*, called λ_{ws}, which allows *controlled* composition of ES without losing PSN. The calculus λ_{ws} has a strong form of composition which is safe but not full. Its (typed) named notation can be translated into (typed) x-terms in such a way that SN for typed terms in λ_{ws} is a consequence of SN for typed λex.

- A calculus with a safe notion of composition in director string notation is defined in [25]. Its named version can be understood as λx together with a composition rule $t[x/u][y/v] \rightarrow t[x/u[y/v]]$ where $y \in \mathtt{fv}(u)$ & $y \notin \mathtt{fv}(t)$. The calculus can be easily simulated in λex by rules Comp and Gc. Thus, again, typed x-terms are strongly normalising.

- The λesw-calculus [11] was used as a technical tool to show PSN for λes. The syntax extends x-terms with weakening constructors. It is then straightforward to define a translation T from λesw-terms to x-terms which forgets these weakening operators. The reduction relation λesw can be split into an equational system $\mathcal{E}$ and two rewriting relations $\mathcal{L}_1$ and $\mathcal{L}_2$ s.t. $t =_{\mathcal{E}} t'$ or $t \rightarrow_{\mathcal{L}_1} t'$ implies $T(t) =_\mathtt{c} T(t')$ and $t \rightarrow_{\mathcal{L}_2} t'$ implies $T(t) \rightarrow^+_{\lambda ex} T(t')$.

The reduction relation generated by the rules $\mathcal{L}_1$ modulo the equations $\mathcal{E}$ can be easily shown to be terminating. Therefore, every infinite λesw-reduction sequence must contain infinitely many reduction steps generated by the rules

$\mathcal{L}_2$ modulo the equations $\mathcal{E}$, so that we obtain, via the translation T, an infinite λex-reduction sequence. Also, typed λesw-terms trivially translate via T to typed x-terms. A consequence is that typed xw-terms are λesw-strongly normalising.

7   Conclusion

We define a simple perpetual strategy for a calculus with ES enjoying full composition. We use this strategy to provide an inductive definition of SN terms which is then used to prove that untyped terms enjoy PSN and typed terms are SN. The proofs are simple, but especially self-contained, no simulation of the source calculus into another SN calculus is used. The inductive characterisation of SN terms and the SN theorem are extremely simple w.r.t. other proofs in the literature [3,18] for ES. Last but not least, our development is constructive as we make no use of classical logic reasoning.

Some remarks about the application of this method to other calculi might be interesting. First of all, it is worth noticing that full composition alone is not sufficient to achieve the SN proof, otherwise the $\lambda\sigma$-calculus [1], which is known to *not* being strongly normalising [21], could be treated. Indeed, our strategy $\rightsquigarrow$ for λex is not perpetual for $\lambda\sigma$: Melliès' counter-example is based on an infinite $\lambda\sigma$-reduction sequence starting from a typed term which is not reached by our perpetual strategy. In other words, $\rightsquigarrow$ is incomplete for $\lambda\sigma$. The definition of a perpetual strategy for $\lambda\sigma$ remains open. The definition of a one-step perpetual strategy (eventually effective) for λex also deserves future attention.

We believe that a de Bruijn version of λex could be useful in real implementations. This could be achieved by using for example $\lambda\sigma$ technology (so that equation C can be eliminated) together with the control of composition needed to guarantee strong normalisation.

Acknowledgements. I am grateful to M. Fernández, S. Lengrand, F. Renaud, F. R. Sinot, V. van Oostrom and the anonymous referees for useful comments.

References

1. Abadi, M., Cardelli, L., Curien, P.L., Lévy, J.-J.: Explicit substitutions. Journal of Functional Programming 4(1), 375–416 (1991)
2. Arbiser, A., Bonelli, E., Ríos, A.: Perpetuality in a lambda calculus with explicit substitutions and composition. WAIT, JAIIO (2000)
3. Bonelli, E.: Perpetuality in a named lambda calculus with explicit substitutions. Mathematical Structures in Computer Science 11(1), 47–90 (2001)
4. David, R., Guillaume, B.: A λ-calculus with explicit weakening and explicit substitution. Mathematical Structures in Computer Science 11, 169–206 (2001)
5. de Bruijn, N.G.: Generalizing Automath by Means of a Lambda-Typed Lambda Calculus. In: Mathematical Logic and Theoretical Computer Science. Lecture Notes in Pure and Applied Mathematics, vol. 106 (1987)
6. Di Cosmo, R., Kesner, D., Polonovski, E.: Proof nets and explicit substitutions. Mathematical Structures in Computer Science 13(3), 409–450 (2003)
7. Dowek, G., Hardin, T., Kirchner, C.: Higher-order unification via explicit substitutions. Information and Computation 157, 183–235 (2000)

8. Girard, J.-Y.: Interprétation fonctionelle et élimination des coupures dans l'arithmétique d'ordre supérieure. Thèse de doctorat d'état, Univ. Paris VII (1972)
9. Girard, J.-Y.: Linear logic. Theoretical Computer Science 50 (1987)
10. Hardin, T., Lévy, J.-J.: A confluent calculus of substitutions. In: France-Japan Artificial Intelligence and Computer Science Symposium, Izu, Japan (1989)
11. Kesner, D.: The theory of calculi with explicit substitutions revisited. In: Duparc, J., Henzinger, T.A. (eds.) CSL 2007. LNCS, vol. 4646, pp. 238–252. Springer, Heidelberg (2007)
12. Kesner, D.: Perpetuality for full and safe composition (in a constructive setting) (2008), http://www.pps.jussieu.fr/~kesner/papers/
13. Kesner, D., Ó Conchúir, S.: Fundamental properties of Milner's non-local explicit substitution calculus, http://www.pps.jussieu.fr/~kesner/papers/
14. Kesner, D., Lengrand, S.: Resource operators for lambda-calculus. Information and Computation 205(4), 419–473 (2007)
15. Khasidashvili, Z., Ogawa, M., van Oostrom, V.: Uniform Normalisation beyond Orthogonality. In: Middeldorp, A. (ed.) RTA 2001. LNCS, vol. 2051, pp. 122–136. Springer, Heidelberg (2001)
16. Klop, J.-W.: Combinatory Reduction Systems. PhD thesis, Mathematical Centre Tracts 127. CWI, Amsterdam (1980)
17. Lengrand, S.: Normalisation and Equivalence in Proof Theory and Type Theory. PhD thesis, University Paris 7 and University of St Andrews (November 2006)
18. Lengrand, S., Lescanne, P., Dougherty, D., Dezani-Ciancaglini, M., van Bakel, S.: Intersection types for explicit substitutions. Information and Computation 189(1), 17–42 (2004)
19. Lévy, J.-J., Maranget, L.: Explicit substitutions and programming languages. In: Pandu Rangan, C., Raman, V., Ramanujam, R. (eds.) FST TCS 1999. LNCS, vol. 1738, pp. 181–200. Springer, Heidelberg (1999)
20. Lins, R.: A new formula for the execution of categorical combinators. In: Siekmann, J.H. (ed.) CADE 1986. LNCS, vol. 230, pp. 89–98. Springer, Heidelberg (1986)
21. Melliès, P.-A.: Typed λ-calculi with explicit substitutions may not terminate. In: TLCA. LNCS, vol. 902, pp. 328–334. Springer, Heidelberg (1995)
22. Milner, R.: Local bigraphs and confluence: two conjectures. In: EXPRESS. ENTCS vol. 175 (2006)
23. Rose, K.: Explicit cyclic substitutions. In: CTRS. LNCS, vol. 656. Springer, Heidelberg (1992)
24. Sakurai, T.: Strong normalizability of calculus of explicit substitutions with composition, http://www.math.s.chiba-u.ac.jp/~sakurai/papers.html
25. Sinot, F.-R., Fernández, M., Mackie, I.: Efficient reductions with director strings. In: Nieuwenhuis, R. (ed.) RTA 2003. LNCS, vol. 2706, pp. 46–60. Springer, Heidelberg (2003)
26. Tait, W.: Intensional interpretation of functionals of finite type I. Journal of Symbolic Logic 32 (1967)
27. van Daalen, D.T.: The language theory of automath. PhD thesis, Technische Hogeschool Eindhoven (1977)
28. van Raamsdonk, F.: Confluence and Normalization for Higher-Order Rewriting. PhD thesis, Amsterdam University, Netherlands (1996)
29. Sinot, F.-R., van Oostrom, V.: Preserving termination of the λ-calculus or not (unpublished note) (2007)
30. van Raamsdonk, F., Severi, P., Sorensen, M.H., Xi, H.: Perpetual reductions in λ-calculus. Information and Computation 149(2) (1999)

A System F with Call-by-Name Exceptions

Sylvain Lebresne

[1] Preuves, Programmes et Systèmes (PPS), CNRS, Université Paris 7, Paris, France
[2] Projet Logical, LIX, École Polytechnique, Palaiseau, France

Abstract. We present an extension of System F with call-by-name exceptions. The type system is enriched with two syntactic constructs: a union type $A \uplus \{\varepsilon\}$ for programs of type A whose execution may raise the exception ε at top level, and a *corruption type* $A^{\{\varepsilon\}}$ for programs that may raise the exception ε in any evaluation context (not necessarily at top level). We present the syntax and reduction rules of the system, as well as its typing and subtyping rules. We then study its properties, such as confluence. Finally, we construct a realizability model using orthogonality techniques, from which we deduce that well-typed programs are weakly normalizing and that the ones who have the type of natural numbers really compute a natural number, without raising exceptions.

1 Introduction

Exceptions are a convenient mechanism for handling errors in programming languages. Most modern languages use them: Java, Objective Caml, C++, ... The main computational features of exceptions are:

1. You can raise an exception instead of any other expression (or instruction);
2. It propagates automatically by default;
3. Programmers can catch it only when (s)he needs to.

Exceptions have long been confined to call-by-value languages and are usually presented as a mechanism which "cuts through" the normal control flow of a program when raised. Adding exceptions to lazy languages is more difficult since an expression is evaluated only when needed and thus, programs do not have a readily-predictable control flow. As noted by S. Peyton Jones *et al.* [8], "(...) the only productive way to think about an expression is to consider the *value it computes*, not the *way in which the value is computed*". That is why they proposed the idea of exceptions-as-values: a value (of any type) is either a "normal" value, or it is an "exceptional" one.

Exceptions have been less studied in type theoretical settings (in a broad sense). Indeed, exceptions are more a practical facility than a theoretically useful tool. However, with the development of proof assistants, there is a higher demand of users for an exceptions mechanism in these tools.

But adding exceptions to type theoretical frameworks raises new difficulties, at least for two reasons: first, because these languages are independent from any reduction strategy (thus a notion of control-flow has no sense), and second, it is undesirable in such frameworks to give all the possible types to an exception.

L. Aceto et al. (Eds.): ICALP 2008, Part II, LNCS 5126, pp. 323–335, 2008.

The independence towards reduction strategy (see Coq [2] for example) is a consequence of the fact that type theoretical languages are usually "pure" (they do not allow side effects). Hence, the idea of exceptions-as-values seems also well fitted to these languages.

Moreover, in type theoretical languages, exceptions cannot be in all types for consistency reasons. More generally, these languages are expected to capture more properties than usual functional languages. In particular, the exceptions raised by an expression should be reflected in the type of this expression.

In this paper, we propose an extension of System F with exceptions-as-values and a type system that allows static detection of uncaught exceptions using a notion of *corruption*. This notion, by using subtyping, avoids any extra clutter for the programmer and allows for modularity. Here, we use System F as a first step towards more elaborate type theoretical frameworks.

The remaining of the paper is organized as follows. We explain our design in Section 2: we justify the kind of exception-as-values we use and describe the three levels of corruption our type system distinguishes. We formally present our calculus in Section 3 and state the properties it enjoys. Then, the Section 4 provides some examples. We design in Section 5 a realizability model of our calculus that gives some insight on the meaning of corruption. Finally, we present in Section 6 some related works before concluding in Section 7 with future works. Due to space constraints, proofs of the results of this paper are not given[1].

2   Design of the System

2.1   Which Exceptions-as-Values?

There are essentially two designs for exceptions as values: either we encode explicitly exceptions in the language, or we make them primitives.

Encoding explicitly exceptions is an old idea [12, 9]: to each type A is associated a type `Maybe` A which is either values of A tagged as correct values or exceptional values (this idea is nicely explained, for the Haskell programming language, in [8]). It has later been realized that the `Maybe` type constructor forms a *monad* [6, 13]. And P. Wadler and P. Thiemann proposed in [14] to add effects to monads, allowing for the detection of uncaught exceptions in such monadic encoding. However, this approach has some drawbacks:

- Terms using exceptions are crippled by extra clutter. For example, in Haskell, to apply a function `f :: Int -> Int` to a value `x :: Maybe Int` we are forced to write:

      do a <- x
         return (f a)

  Using exceptions is not as transparent for the programmer as it is in call-by-value languages;

---

[1] An extended version of this paper with sketches of proofs is available at http://www.pps.jussieu.fr/~lebresne/SystemFWithExceptions.pdf.

- As remarked in [8], modularity and code re-use are compromised, especially for higher order functions. Consider a sorting function taking a comparison function (returning a boolean) as argument. Then, the sorting function cannot be applied to a comparison function which raises exceptions;
- Monads force the evaluation of arguments (in the example above, the evaluation of x is forced before the application to f). One could not see that as an inconvenience, and this is indeed desirable for most uses of monads, but nonetheless, we think that it can be avoided for exceptions.

This leads us to the second design choice: making exceptions primitives. This has been first proposed by S. Peyton Jones *et al.* [8] with *imprecise exceptions*. The idea is that a value of *any* type is either a "normal" value, or an "exceptional" one. The resulting mechanism allows exceptions to be used in place of any other term (as for more traditional "call-by-value" exceptions). Note that since values may be exceptional, we can have for instance, a list, which is fully defined but for which some elements are exceptional values (see Section 4). These exceptions are raised only when (and if) the list is evaluated.

Our system, named *Fx*, adapts this idea to System F, adding it two new term constructions: raise and try. But while the exceptions of [8] are not precisely typed (the raising operation is in all types), we propose a type system where the type of an expression indicates which exceptions the expression may raise.

2.2   Expected Properties

The type system we will present enjoys the following properties:

- If a term can raise an exception, its type indicates it. In particular, programs of type ℕ are not able to raise exceptions;
- Programmers can use a term raise ε in place of any other term. In particular, raise ε type as a function;
- Exceptions and their typing discipline do not jeopardize modularity and code re-use. A function defined without exceptions in mind still accepts exceptional arguments and behave in a sensible way. Moreover, this is done without knowing the actual code of the function.

2.3   Three Levels of Corruption

We call *corrupted*, a term that may mention exceptions. Given a type A (say the type ℕ of natural numbers), we distinguish three levels of *corruptions* for the terms related with this type:

- Terms of A. They are not corrupted, either they do not mention exceptions or catch them all;
- Terms of $A \uplus \{\varepsilon\}$. They are terms of A or terms that reduce to the exception ε (they raise it).

– Terms of $A^{\{\varepsilon\}}$. They are terms of A that may mention the exception ε but do not necessarily reduce to it (for instance, if S is the successor function, S (**raise** ε) has type $\mathbb{N}^{\{\varepsilon\}}$, but not type $\mathbb{N} \uplus \{\varepsilon\}$ since it has not type $\mathbb{N}$ nor does it reduce to **raise** ε).

Moreover, to handle the properties of corruption, we use a subtyping relation, and have in particular the subtyping: $A \le A \uplus \{\varepsilon\} \le A^{\{\varepsilon\}}$.

2.4   Why We Need to Distinguish These Three Levels

The construction $A \uplus \{\varepsilon\}$ is really needed because of the typing of the **try** operation, since for a **try** to catch an exception in its body, this body has to reduce to the exception.

But because we do not want to change the typing rule of application, the construction $A \uplus \{\varepsilon\}$ clearly does not fulfill all our needs. Firstly, we cannot use it to type S (**raise** ε). And secondly, there remains terms that we cannot substitute by an exception: we can write $(\lambda x.\, S\; x)\, 0$ but not $(\lambda x.\, S\; x)$ (**raise** ε).

To solve these problems, we use a second type construction, the *corruption* of a type A by an exception of name ε, denoted $A^{\{\varepsilon\}}$. The main property that the corruption enjoys is a good behavior with respect to arrow types:

$$(A \to B)^{\{\varepsilon\}} \quad = \quad A^{\{\varepsilon\}} \to B^{\{\varepsilon\}}$$

This subtyping equality may seem paradoxical with the usual subtyping rule of arrow (contra-variance to the left, co-variance to the right). This is however justified by the realizability model of Section 5.

Intuitively, terms of type $A^{\{\varepsilon\}}$ should be seen as terms of type A where some sub-terms may have been replaced by **raise** ε (hence, programmers can use **raise** ε wherever they want, which, in turns, corrupts the resulting type). Equivalently, while terms of $A \uplus \{\varepsilon\}$ are terms that may reduce to **raise** ε at top-level, terms of $A^{\{\varepsilon\}}$ are the ones that may reduce to **raise** ε in *any* evaluation context.

Now, with corruption, we can apply a function $f : A \to B$ to a potentially exceptional term. Indeed, we have that

$$A \to B \;\le\; (A \to B) \uplus \{\varepsilon\} \;\le\; (A \to B)^{\{\varepsilon\}} \;=\; A^{\{\varepsilon\}} \to B^{\{\varepsilon\}}.$$

Remark that since we use subtyping, there is no need to actually know the term f. This allows for modularity and, in particular, this is convenient for primitive functions like S, allowing to type-check S (**raise** ε) with the type $\mathbb{N}^{\{\varepsilon\}}$.

2.5   Typing the Recursion Operator

So, F_x uses primitive natural numbers and hence, provides the usual recursion operator (denoted **rec**). Moreover, a corrupted natural number is an integer where some sub-terms may have been replaced by an exception. Hence, computationally, it is not difficult to write, using **rec**, a function that, given a corrupted integer, return either its argument if it is a well formed natural number, or the corrupting exception otherwise (we will give such a function in Section 4). But to give it the type we expect, *i.e.* $N^\Delta \to N \uplus \Delta$, we will have to give to the recursion operator a type more precise than the one it is usually given.

3 Formal Presentation

3.1 Syntax, Reductions and Associated Properties

Syntax of Terms. We consider a countable set $\mathcal{E}$ of names of exceptions (we can use more than one exception in Fx) and a distinguished set of variables $\mathcal{V}$. Terms of Fx are defined by:

$$M, N ::= x \mid \lambda x. M \mid M\ N \mid \mathbf{raise}\, \varepsilon \mid \mathbf{try}\, M \,\mathbf{with}\, \varepsilon \mapsto N \mid 0 \mid S \mid \mathbf{rec}$$

In this definition, variables are ranged over by $x, y, \ldots$ while exception names are ranged over by $\varepsilon, \varepsilon', \ldots$. Notions of free and bound variables are defined as usual, as well as the external operation of substitution (written $M\{x := N\}$). The set of all closed terms is denoted $\mathcal{T}$ and terms are considered up to α-equivalence. Note that the construction $\mathbf{try}\, M \,\mathbf{with}\, \varepsilon \mapsto N$ does not bind the occurrences of ε. The term $\mathbf{raise}\, \varepsilon$ is called an *exception*, ε being its name, but, as an abuse of terminology, we also call ε an exception. In the term $\mathbf{try}\, M \,\mathbf{with}\, \varepsilon \mapsto N$ we will sometimes call M the body and N the handler of the $\mathbf{try}$ construction.

Computation in Fx. Values are the terms of Fx defined by

$$V ::= \lambda x. M \mid 0 \mid S \mid S\ N \mid \mathbf{rec} \mid \mathbf{rec}\ M \mid \mathbf{rec}\ M\ N$$

The notion of reduction for the calculus is given by the rules of Figure 1.

$(\lambda x. M)\ N \ >\ M\{x := N\}$	$(\mathbf{raise}\, \varepsilon)\ M \ >\ \mathbf{raise}\, \varepsilon$
$\mathbf{try}\,(\mathbf{raise}\, \varepsilon)\,\mathbf{with}\, \varepsilon \mapsto N \ >\ N$	$\mathbf{rec}\ X\ Y\ 0 \ >\ X$
$\mathbf{try}\,(\mathbf{raise}\, \varepsilon')\,\mathbf{with}\, \varepsilon \mapsto N \ >\ \mathbf{raise}\ \varepsilon'$	$\mathbf{rec}\ X\ Y\ (S\ N) \ >\ Y\ N\ (\mathbf{rec}\ X\ Y\ N)$
$\mathbf{try}\, V \,\mathbf{with}\, \varepsilon \mapsto N \ >\ V$	$\mathbf{rec}\ X\ Y\ (\mathbf{raise}\, \varepsilon) \ >\ \mathbf{raise}\, \varepsilon$

Fig. 1. Notion of reduction for Fx

Computation in Fx is defined by the relation of reduction $\succ$ as the least congruence containing $>$.

Note that, as usual, the scope of capture of the $\mathbf{try}$ construction is dynamic: in the term $(\lambda x. \mathbf{try}\, x \,\mathbf{with}\, \varepsilon \mapsto 0)\ (\mathbf{raise}\, \varepsilon)$, the exception is caught during reduction and the whole term reduces to 0.

We say that a term M *raises the exception* ε if $M \succ^* \mathbf{raise}\, \varepsilon$ (that is, if M reduces to the exception named ε).

Adding $\mathbf{raise}$ and $\mathbf{try}$ does not break the confluence of the calculus:

Theorem 1 (Confluence). *If M, N and N' are terms such that $M \succ^* N$ and $M \succ^* N'$, then there exists a term P such that $N \succ^* P$ and $N' \succ^* P$.*

3.2   The Type System

As stressed in Section 2.3, Fx uses a subtyping relation $\leq$. Thus, Fx is in fact an extension of System $F\eta$ (System F with subtyping [5, 15]). Besides, we can use more than one exception so that type constructions handle sets of exceptions names.

The syntax of types for Fx is built upon the one of System F:

$$A, B ::= \alpha \mid \mathbb{N} \mid A \to B \mid \forall \alpha.\, A \mid A \uplus \Delta \mid A^{\Delta}$$

In $A \uplus \Delta$ and A^{Δ}, Δ is a finite set of exceptions names ($\Delta \subseteq \mathcal{E}$). Moreover, α stands for a type variable taken from the set of type variables $\mathcal{A}$. Notions of free and bound type variable are defined as usual, as well as the external operation of substitution (written $A\{\alpha := B\}$). We denote by $FV(A)$ the set of all the free type variables of the type A. Types are considered up to α-equivalence. Precedences for the arrow construction and the universal quantifier are the usual ones; the precedences of $A \uplus \Delta$ and A^{Δ} being higher.

Typing. A *typing context* Γ is a finite set of declarations having the form $\Gamma \equiv x_1 : A_1, \ldots, x_n : A_n$ where $x_1, \ldots, x_n$ are pairwise distinct term variables and where $A_1, \ldots, A_n$ are arbitrary types. The set $FV(\Gamma)$ denotes the union of the sets of free type variables for the types used in Γ. The type system of Fx is defined from the *typing judgment*

$$\Gamma \vdash M : A$$

that reads 'in the typing context Γ, the term M has type A'. This judgment is inductively defined by the rules of Figure 2. Remark that the typing rules for System $F\eta$ are unchanged, we simply add rules. Also note that the usual typing rule for the recursion operator can be retrieved from *(rec)* by taking $\Delta = \emptyset$ (and the *(rec)* rule is in fact a typing scheme).

System Fη typing rules:

$$\frac{(x : A) \in \Gamma}{\Gamma \vdash x : A} \; (ax) \qquad \frac{\Gamma, x : A \vdash M : B}{\Gamma \vdash \lambda x.\, M : A \to B} \; (abs) \qquad \frac{\Gamma \vdash M : A \to B \quad \Gamma \vdash N : A}{\Gamma \vdash M\, N : B} \; (app)$$

$$\frac{\Gamma \vdash M : A \quad \alpha \notin FV(\Gamma)}{\Gamma \vdash M : \forall \alpha.\, A} \; (gen) \qquad \frac{\Gamma \vdash M : A \quad A \leq B}{\Gamma \vdash M : B} \; (subs)$$

Natural numbers typing rules:

$$\frac{}{\Gamma \vdash 0 : \mathbb{N}} \; (zero) \qquad \qquad \frac{}{\Gamma \vdash S : \mathbb{N} \to \mathbb{N}} \; (succ)$$

$$\frac{}{\Gamma \vdash \mathbf{rec} : \forall \alpha.\, \alpha \uplus \Delta \to (\mathbb{N}^{\Delta} \to \alpha \uplus \Delta \to \alpha \uplus \Delta) \to \mathbb{N}^{\Delta} \to \alpha \uplus \Delta} \; (rec)$$

Exceptions handling typing rules:

$$\frac{}{\Gamma \vdash \mathbf{raise}\, \varepsilon : \forall \alpha.\, \alpha \uplus \{\varepsilon\}} \; (raise) \qquad \frac{\Gamma \vdash M : A \uplus \{\varepsilon\} \quad \Gamma \vdash N : A}{\Gamma \vdash \mathbf{try}\, M \,\mathbf{with}\, \varepsilon \mapsto N : A} \; (try)$$

Fig. 2. Typing judgments

Subtyping. The *subtyping* relation between two types A and B, written $A \leq B$, is inductively defined by the rules of Figure 3. The equality $A = B$ is defined as short for "$A \leq B$ and $A \geq B$", and the inference rules with an equality on conclusion is a notation for the two expected inference rules.

System Fη rules :

$$\frac{}{A \leq A} \; (st\text{-}id) \qquad \frac{A \leq B \quad B \leq C}{A \leq C} \; (st\text{-}trans) \qquad \frac{A' \leq A \quad B \leq B'}{A \to B \leq A' \to B'} \; (st\text{-}arrow)$$

$$\frac{A \leq B \quad \alpha \notin FV(A)}{A \leq \forall \alpha.\, B} \; (f\text{-}gen) \qquad \frac{}{\forall \alpha.\, A \leq A\{\alpha := B\}} \; (f\text{-}inst)$$

$$\frac{\alpha \notin FV(A)}{\forall \alpha.\, (A \to B) \leq A \to \forall \alpha.\, B} \; (f\text{-}arr)$$

Exception related rules :

$$\frac{}{A \uplus \emptyset \leq A} \; (ex\text{-}noexu) \qquad \frac{}{A^{\emptyset} \leq A} \; (ex\text{-}noexc) \qquad \frac{}{(A \to B) \uplus \Delta \leq A \to B \uplus \Delta} \; (ex\text{-}arru)$$

$$\frac{A \leq B}{A \uplus \Delta \leq B \uplus \Delta} \; (ex\text{-}ctx) \qquad \frac{}{A \leq A \uplus \Delta} \; (ex\text{-}uni) \qquad \frac{}{A \uplus \Delta \leq A^{\Delta}} \; (ex\text{-}corrupt)$$

$$\frac{}{\forall \alpha.\, A^{\Delta} \leq (\forall \alpha.\, A)^{\Delta}} \; (ex\text{-}fallc) \qquad \frac{}{\forall \alpha.\, (A \uplus \Delta) \leq (\forall \alpha.\, A) \uplus \Delta} \; (ex\text{-}fallu)$$

Exception related equality rules :

$$\frac{}{(A \uplus \Delta) \uplus \Delta' = A \uplus (\Delta \cup \Delta')} \; (eq\text{-}uu) \qquad \frac{}{(A^{\Delta})^{\Delta'} = A^{(\Delta \cup \Delta')}} \; (eq\text{-}cc)$$

$$\frac{}{(A \uplus \Delta)^{\Delta'} = A^{\Delta'} \uplus (\Delta - \Delta')} \; (eq\text{-}uc) \qquad \frac{}{(A \to B)^{\Delta} = A^{\Delta} \to B^{\Delta}} \; (eq\text{-}arrc)$$

Fig. 3. The subtyping relation

The subtyping rules of Fη are unchanged. The rules *(ex-noexu)*, *(ex-noexc)*, *(eq-uu)* and *(eq-cc)* dealt with sets of exceptions. The hierarchy of corruption (see 2.3) is implemented by *(ex-uni)* and *(ex-corrupt)*. The rules *(ex-fallc)* and *(ex-fallu)* are justified by the absence of computational content of the universal quantification. Moreover, corruption and union commutes *(eq-uc)*.

The subtyping is stable by union *(ex-ctx)*, but also by corruption (this can be proved by simultaneous structural inductions on A and B). Rule *(ex-arru)* simply says that, since a term M of type $(A \to B) \uplus \Delta$ is either a term of type $A \to B$ or an exception of Δ, it can always be applied to a term of type A, resulting in a term of type B (if M is a true function) or an exception of Δ (if so is M).

Finally, as discussed in Section 2.4, the rule *(eq-arrc)* is the main rule of corruption and allows exceptions to be used anywhere. Note that we really need an equality here on pain of losing the subject-reduction property.

3.3 Properties of Typing

We define the relation $\sqsubseteq_\Delta$ between terms by : $M \sqsubseteq_\Delta N$ if and only if N is obtained from M by replacing some sub-terms in any position by $\mathbf{raise}\,\varepsilon$, ε belonging to Δ. Then, Theorem 2 formally states that, in term of programming, exceptions can be used in any place, but with the added cost of corrupting the type.

Theorem 2 (corruption). *If M and N are two terms, A a type and Δ a set of exceptions such that $\Gamma \vdash M : A$ and $M \sqsubseteq_\Delta N$, then $\Gamma \vdash N : A^\Delta$.*

4 Examples

A simple yet classical function on natural numbers which can raise an exception is the predecessor function. In Fx, we can define:

$$\mathbf{pred} \equiv \mathbf{rec}\,(\mathbf{raise}\,\varepsilon)\,(\lambda x.\,\lambda y.\,x) \quad : \quad \mathbb{N} \to \mathbb{N} \uplus \{\varepsilon\}$$

It has the expected reductions, i.e. $\mathbf{pred}\ 0 \succ^* \mathbf{raise}\,\varepsilon$ and $\mathbf{pred}\ (S\ N) \succ^* N$. We can then define a "safe" predecessor $\mathbf{pred}'$ from $\mathbf{pred}$ which returns 0 when applied to 0:

$$\mathbf{pred}' \equiv \lambda n.\,\mathbf{try}\,(\mathbf{pred}\ n)\,\mathbf{with}\,\varepsilon \mapsto 0 \quad : \quad \mathbb{N} \to \mathbb{N}$$

As Fx is an extension of System F, we can define lists using second-order encodings. Let us recall such encodings of $\mathtt{list}$:

$$\mathtt{list} \equiv \forall\beta.\,\forall\alpha.\,(\alpha \to (\beta \to \alpha \to \alpha) \to \alpha)$$
$$\mathtt{nil} \equiv \lambda n.\,\lambda c.\,n \ : \ \mathtt{list}$$
$$\mathtt{cons} \equiv \lambda i.\,\lambda l.\,\lambda n.\,\lambda c.\,c\,i\,(l\,z\,c) \ : \ \forall\beta.\,\beta \to \mathtt{list}(\beta) \to \mathtt{list}(\beta)$$

where we use the shortcut notation $\mathtt{list}(A) \equiv \forall\alpha.\,(\alpha \to (A \to \alpha \to \alpha) \to \alpha)$.

We can now define $\mathbf{head}$ and $\mathbf{tail}$ functions that raise an exception when applied to the empty list. Notice that the code of the $\mathbf{tail}$ function relies on the same "trick" than the one of the predecessor for natural numbers in their second-order encoding version:

$$\mathbf{head} \equiv \lambda l.\,l\,(\mathbf{raise}\,\varepsilon)\,(\lambda i.\,\lambda r.\,i) \ : \ \mathtt{list} \to \forall\beta.\,\beta \uplus \{\varepsilon\}$$
$$\mathbf{tail}' \equiv \lambda l.\,\lambda n.\,\lambda c.\,(l\,(\lambda x.\,n)\,(\lambda e.\,\lambda x.\,\lambda y.\,y\,i\,(x\,c)))\,(\lambda x.\,\lambda y.\,y)$$
$$\mathbf{tail} \equiv \lambda l.\,l\,(\mathbf{raise}\,\varepsilon)\,(\lambda n.\,\lambda c.\,\mathbf{tail}'\,l) \ : \ \mathtt{list} \to \mathtt{list} \uplus \{\varepsilon\}$$

We now define the mapping of a function to a list of integers:

$$\mathbf{map} \equiv \lambda f.\,\lambda l.\,\lambda n.\,\lambda c.\,l\,n\,(\lambda i.\,\lambda r.\,c\,(f\,i)\,r) \ : \ (\mathbb{N} \to \mathbb{N}) \to \mathtt{list}(\mathbb{N}) \to \mathtt{list}(\mathbb{N})$$

Then we can define a (not very useful) function mapping to a list a function that take the successor of the predecessor of the elements:

$$\mathbf{foo} \equiv \mathbf{map}\,(\lambda e.\,S\,(\mathbf{pred}\ e)) \ : \ \mathtt{list}(\mathbb{N}) \to \mathtt{list}(\mathbb{N}^{\{\varepsilon\}})$$

Now, if given a list l, $\mathbf{foo}\ l$ has type $\mathtt{list}(\mathbb{N}^{\{\varepsilon\}})$ and we can still get the head of the list with $\mathbf{head}\,(\mathbf{foo}\ l)$. However, this natural number can be corrupted (if the first element of l is 0) and if we want to check for the corruption, we can use the following function, that 'uncorrupts' integers :

$$eval \equiv \lambda n.\,(\texttt{rec}\ (\lambda a.\,a)\ (\lambda m.\,\lambda r.\,\lambda a.\,r\ (S\ a))\ n)\ 0 \quad : \quad \mathbb{N}^\Delta \to \mathbb{N} \uplus \Delta$$

To type this function, we instantiate the type of the recursion operator by the type $(\mathbb{N} \to \mathbb{N} \uplus \Delta) \uplus \Delta$.

5   Realizability Model

5.1   Daimon, Weak Head Reduction and Contexts

We add a *daimon*, ✠, similar to the one of [3]. It has no typing rules and computationally behaves like an uncatchable exception.

A (closed) term is in *weak head normal form*, if it is in one of the following forms (where V is a value):

$$whnf ::= V \mid \texttt{raise}\,\varepsilon \mid ✠$$

Rules for *weak head reduction* ($\succ_h$) are given in Figure 4. The transitive and reflexive closure of $\succ_h$ is noted $\succ_h^*$.

$$\frac{M > M'}{M \succ_h M'} \qquad \frac{M \succ_h M'}{M\ N \succ_h M'\ N}$$

$$\frac{M \succ_h M' \quad N \succ_h N'}{\texttt{try}\,M\,\texttt{with}\,\varepsilon \mapsto N \succ_h \texttt{try}\,M'\,\texttt{with}\,\varepsilon \mapsto N'} \qquad \frac{M \succ_h M'}{\texttt{rec}\ X\ Y\ M \succ_h \texttt{rec}\ X\ Y\ M'}$$

Fig. 4. Weak head reduction

A *context* is a term with a hole (denoted by [ ]) and is defined by:

$$C ::= [\,] \mid C\ N \mid \texttt{try}\,C\,\texttt{with}\,\varepsilon \mapsto ✠ \mid \texttt{rec}\ M\ N\ C$$

The set of all contexts is noted $\mathcal{C}$ and the term obtained by filling the hole of a context C with the term M is noted $C[M]$. Note the restriction in the handler of **try** to ✠. In fact, we consider contexts up to the following equivalence relation:

$$\texttt{try}\,(\texttt{try}\,[\,]\,\texttt{with}\,\varepsilon_1 \mapsto ✠)\,\texttt{with}\,\varepsilon_2 \mapsto ✠ \equiv \texttt{try}\,(\texttt{try}\,[\,]\,\texttt{with}\,\varepsilon_2 \mapsto ✠)\,\texttt{with}\,\varepsilon_1 \mapsto ✠$$

Then, Δ being the set of exception $\{\varepsilon_0, \ldots, \varepsilon_n\}$, we denote by $\texttt{try}\,[\,]\,\texttt{with}\Delta \mapsto ✠$ the context $\texttt{try}\ldots\texttt{try}\,[\,]\,\texttt{with}\,\varepsilon_0 \mapsto ✠\ldots\texttt{with}\,\varepsilon_n \mapsto ✠$.

5.2   Operations on Sets

We define some operations on sets of contexts :

$$S^\perp = \{\,M \mid \forall C \in S,\ C[M] \succ^* ✠\,\}$$

$$A \cdot S = \{\,C[[\,]\ N] \mid C \in S,\ N \in A\,\} \qquad \downarrow_\Delta S = S \circ \{\,\texttt{try}\,[\,]\,\texttt{with}\,\Delta \mapsto ✠\,\}$$

$$S \circ T = \{\,C[D[\,]] \mid C \in S,\ D \in T\,\} \qquad \uparrow_\Delta S = \{\,\texttt{try}\,[\,]\,\texttt{with}\,\Delta \mapsto ✠\,\} \circ S$$

5.3 A Model for Fx

We define a realizability model for Fx using techniques of orthogonality (see [11] for examples of use of such techniques). We call valuation function any function ρ from type variables to the powerset of $\mathcal{C}$ minus the empty set ($\rho : \mathcal{A} \to (\mathcal{P}(\mathcal{C}))^+$). To each type A we associate two sets:

A set of contexts $|A|_\rho \subseteq \mathcal{C}$
A set of terms $[\![A]\!]_\rho \subseteq \mathcal{T}$

The set $[\![A]\!]_\rho$ is uniformly defined from $|A|_\rho$ by

$$[\![A]\!]_\rho = |A|_\rho^\perp = \{\, M \mid \forall C \in |A|_\rho, \; C[M] \succ^* \maltese \,\}.$$

The set $|A|_\rho$ is defined by induction on A by:

$$
\begin{aligned}
|\alpha|_\rho &= \rho(\alpha) \\
|\mathbb{N}|_\rho &= \{\, \mathtt{rec}\ \maltese\ (\lambda y.\lambda x.\, x)\ [\,]\,\} \\
|A \uplus \Delta|_\rho &= \downarrow_\Delta |A|_\rho \\
|A^\Delta|_\rho &= \uparrow_\Delta |A|_\rho
\end{aligned}
\qquad
\begin{aligned}
|A \to B|_\rho &= \bigcup_{\Delta \subseteq \mathcal{E}} (|A^\Delta|_\rho)^\perp \cdot |B^\Delta|_\rho \\
|\forall \alpha.\, A|_\rho &= \bigcup_{S \subseteq \mathcal{C}^+} |A|_{\rho;\,\alpha \leftarrow S}
\end{aligned}
$$

Note that the interpretation in the model of the construction $A \uplus \Delta$ and A^Δ follows, to some extends, the idea that terms of type $A \uplus \Delta$ are terms that may raise an exception only at top level, where terms of A^Δ are those that may raise an exception in any evaluation context. This is emphasized by the "opposition" of the operations $\downarrow_\Delta$ and $\uparrow_\Delta$.

The other interesting point of the model is the interpretation of arrow types. In Fx, a function f who has type $A \to B$ has also all the types $A^\Delta \to B^\Delta$ for any Δ. Our arrow type is thus smaller than the usual realizability one and so, functions of Fx are in particular realizability functions.

We define the interpretation of a typing context Γ by:

$$[\![\Gamma]\!]_\rho \quad = \quad \{\, \sigma \mid \forall (x : A) \in \Gamma, \; \sigma(x) \in [\![A]\!]_\rho \,\}$$

Moreover, if σ is a substitution of term variables and M is a term, we use the notation $M[\sigma]$ for the *parallel substitution* of M by σ, which consists in applying σ to all free variables of M in parallel. We can now show that our interpretation is sound with respect to typing:

Theorem 3 (Model soundness). *If M is a term, A a type and Γ a typing context such that $\Gamma \vdash M : A$, then for all valuation function ρ and for all substitution $\sigma \in [\![\Gamma]\!]_\rho$ we have $M[\sigma] \in [\![A]\!]_\rho$.*

Note that in this model, we only consider closed terms by construction. For this very reason, we cannot establish a strong normalization theorem using this model. But, from the model, we obtain a weak head normalization theorem:

Theorem 4 (Weak head normalization). *If M is a closed term, A a type and Γ a typing context such that $\Gamma \vdash M : A$, then M has a weak head normal form.*

The model allows us to prove that our typing of exceptions is safe for the primitive data types, the natural numbers:

Lemma 1 (type safety for natural numbers). *If M is a term such that $\vdash M : \mathbb{N}$, then $M \succ^* S^n\ 0$ for some $n \geq 0$.*

Hence, if a program is of the type of the natural numbers, we assure that it will compute a true natural number without producing errors.

6   Related Works

Many works about the static detection of uncaught exceptions have been done, based on typing or not. For instance, for the OCaml languages, J.C. Guzmn and A. Surez [4] have proposed an extension of the type system where arrows are annotated by which exceptions a function can raise. Later, X. Leroy and F. Pessaux [7] proposed a similar system but add polymorphism over these annotations. Their solution is efficient and covers all the Ocaml language, including modularity. However, all these works consider exceptions in call-by-value languages and rely heavily on the exceptions-as-control-flow paradigm.

In the literature, exceptions are often considered as control operators [10]. However and contrarily to most control operators, the typing of exceptions does not necessarily lift the logic to a classical one. Besides, in this paper, we address the problem of the static detection of uncaught exceptions. We do not know of previous works on control operators dealing with this particular problem.

Exceptions in type theoretical settings have been less studied. However, R. David and G. Mounier [1] have designed a typed mechanism of exceptions for the language AF2. But their exceptions are restricted in the sense that only data types can carry exceptions.

7   Conclusion and Future Works

We have presented the Fx calculus, an extension of System F with typed exceptions. We have presented a mechanism of exceptions that does not force a particular β-reduction strategy for the calculus. We have also provided a type system for this mechanism that performs static detection of uncaught exceptions. This type system is modular and allows the use and propagation of exceptions to be transparent for the programmer. Finally, we have justified the semantic of our calculus by exhibiting a realizability model.

We believe that our calculus can be extended in the following ways:

- We conjecture, but have not proved yet, the subject-reduction property for Fx. The difficulty lies in the interaction between subtyping and implicit polymorphism. However, we do have proved that the restriction of Fx to first-order types have the subject-reduction property, and we believe that in adapting our exceptions to languages with explicit polymorphism (like the dependent product), we will not encounter this problem. Besides, for Fx,

we have shown that our model allows us to state a type safety lemma that makes the proof of the subject-reduction property less urgent.
- Our realizability model only allows to state a weak normalization theorem. To turn it into a strong normalization one, we need to find a suitable notion of saturated sets (that can handle open terms).
- We can extend the calculus to allow exceptions to carry informations.
- A natural extension would be to add dependent product to our calculus, and we have good hopes that such an extension can be done. At least, we already know how to extend our realizability model to handle the dependent product: if T is a type and U_x a type family indexed by x, we will take

$$| \Pi x : T.U |_\rho = \bigcup_{\Delta \subseteq \mathcal{E}} \{ M \cdot C \mid M \in [\![T^\Delta]\!]_\rho \wedge C \in | U_M^\Delta |_\rho \}$$

- Type inference in $F x$ is obviously undecidable [16]. However, type inference for the restriction of $F x$ to first-order types remains to be studied, and we have good hopes that it is decidable, since we know that in such a restriction, the subtyping relation is decidable.

Acknowledgments. This work benefited from several discussions with and suggestions from Hugo Herbelin and Alexandre Miquel.

References

1. David, R., Mounier, G.: An intuitionistic λ-calculus with exceptions. Journal of Functional Programming 15(01), 33–52 (2004)
2. The Coq development team. The Coq Proof Assistant Reference Manual v8.1 (2006)
3. Girard, J.Y.: Locus Solum: From the rules of logic to the logic of rules. Mathematical Structures in Computer Science 11(03), 301–506 (2001)
4. Guzman, J., Suarez, A.: An extended type system for exceptions. In: Proceedings of the ACM SIGPLAN Workshop on ML and its Applications, pp. 127–135 (1994)
5. Mitchell, J.C.: Polymorphic type inference and containment. Information and Computation 76(2-3), 211–249 (1988)
6. Moggi, E.: Notions of computation and monads. INF. COMPUT. 93(1), 55–92 (1991)
7. Pessaux, F., Leroy, X.: Type-based analysis of uncaught exceptions. In: Proceedings of the 26th ACM SIGPLAN-SIGACT symposium on Principles of programming languages, pp. 276–290 (1999)
8. Peyton Jones, S., Reid, A., Henderson, F., Hoare, T., Marlow, S.: A semantics for imprecise exceptions. ACM SIGPLAN Notices 34(5), 25–36 (1999)
9. Spivey, M.: A functional theory of exceptions. Science of Computer Programming 14(1), 25–42 (1990)
10. Thielecke, H.: Comparing Control Constructs by Double-Barrelled CPS. Higher-Order and Symbolic Computation 15(2), 141–160 (2002)
11. Vouillon, J., Melliès, P.A.: Semantic types: a fresh look at the ideal model for types. In: Proceedings of the 31st ACM SIGPLAN-SIGACT symposium on Principles of programming languages, pp. 52–63 (2004)

12. Wadler, P.: How to Replace Failure by a List of Successes A method for exception handling, backtracking, and pattern matching. Functional Programming Languages and Computer Architecture (1985)
13. Wadler, P.: Comprehending monads. In: Proceedings of the 1990 ACM conference on LISP and functional programming, pp. 61–78 (1990)
14. Wadler, P., Thiemann, P.: The marriage of effects and monads. ACM Transactions on Computational Logic (TOCL) 4(1), 1–32 (2003)
15. Wells, J.B.: The undecidability of Mitchells subtyping relation. Technical Report 95-019, Boston University, Boston, Massachusetts (1995)
16. Wells, J.B.: Typability and type checking in System F are equivalent and undecidable. Annals of Pure and Applied Logic 98(1-3), 111–156 (1999)

Linear Logical Algorithms

Robert J. Simmons and Frank Pfenning

Carnegie Mellon University
{rjsimmon,fp}@cs.cmu.edu

Abstract. Bottom-up logic programming can be used to declaratively specify many algorithms in a succinct and natural way, and McAllester and Ganzinger have shown that it is possible to define a cost semantics that enables reasoning about the running time of algorithms written as inference rules. Previous work with the programming language Lollimon demonstrates the expressive power of logic programming with linear logic in describing algorithms that have imperative elements or that must repeatedly make mutually exclusive choices. In this paper, we identify a bottom-up logic programming language based on linear logic that is amenable to efficient execution and describe a novel cost semantics that can be used for complexity analysis of algorithms expressed in linear logic.

Keywords: Bottom-up logic programming, forward reasoning, linear logic, deductive databases, cost semantics, abstract running time.

1 Introduction

Logical inference rules are a concise and powerful tool for expressing many algorithms in a declarative way. In the last decade, several lines of work have advanced the argument that it is not only possible but convenient to formally reason about the *running time* of algorithms expressed as inference rules.

Work on this topic can be broadly categorized into two groups: work that takes a language similar to the pure bottom up logic programming language presented by McAllester [1] and automates reasoning about the complexity of algorithms expressed in that language [2,3], and work aimed at allowing analysis for logic programming languages with richer features [4,5,6].

This paper falls into the second category; we present a bottom-up logic programming language based on intuitionistic linear logic [7] that cleanly integrates a notion of state transition with the saturating forward reasoning present in bottom-up logic programming. We follow the two-part approach taken by McAllester and Ganzinger in [1,4,5]. First, we give the language a dynamic cost semantics called the *abstract running time* that looks at a chain of logical inferences as a computation and defines the cost of that computation, and then we describe an interpreter that can be shown to execute those computations in time proportional to the abstract running time. Both of these concepts are critical – without the interpreter, there is no reason to believe that the notion of abstract

L. Aceto et al. (Eds.): ICALP 2008, Part II, LNCS 5126, pp. 336–347, 2008.
© Springer-Verlag Berlin Heidelberg 2008

$$\frac{\mathsf{edge}(x,y)}{\mathsf{edge}(y,x)} \ r1 \qquad \frac{\mathsf{edge}(x,y)}{\mathsf{path}(x,y)} \ r2 \qquad \frac{\begin{array}{c}\mathsf{edge}(x,y)\\\mathsf{path}(y,z)\end{array}}{\mathsf{path}(x,z)} \ r3$$

Fig. 1. A simple pure, bottom-up program for computing graph connectivity

running time is based in reality, and without the definition of abstract running time, reasoning about the complexity of algorithms requires understanding the intricacies of the interpreter's implementation.

We start by briefly describing a pure logic programming language [1] in which various graph algorithms and program analyses can be expressed concisely and executed efficiently. One example is the program in Fig. 1 that computes connectivity over an undirected graph.

Given a graph $G = (E, V)$, this algorithm starts with a database that has a fact $\mathsf{edge}(\mathsf{a}, \mathsf{b})$ for every edge $(\mathsf{a}, \mathsf{b}) \in E$. The intended meaning of this program is that $\mathsf{path}(\mathsf{a}, \mathsf{b})$ should hold if and only if there is a path between vertex a and b in graph G. Here, and throughout the paper, we will represent constants as $\mathsf{a}, \mathsf{b}, \mathsf{c}, \ldots$ and variables as $x, y, z, \ldots$, and we will insist that all the terms in our database be *ground*, meaning that they contain no free variables, and that all rules be *range-restricted*, meaning that the variables in the conclusions (below the line) are a subset of the variables in the premises (above the line). This last restriction ensures that the database continues to contain only ground facts as new facts are derived.

In order to calculate the path relation, rules are applied exhaustively in the forward direction until *saturation* is reached; that is, until no possible forward inference can cause us to learn anything new. The *closure* of an initial database Γ under the rules in a program P (written $\mathsf{Clo}_P(\Gamma)$ or just $\mathsf{Clo}(\Gamma)$) is the smallest set containing Γ closed under the rules in P. Unlike Datalog, the language contains function symbols, so the closure may be infinite; however, we are interested only in programs with a finite closure.

The pure bottom-up logic programming language sketched above and described fully in [1] and elsewhere has great expressive power but also some obvious limitations. We will briefly mention related work on efforts related to our own.

Consider the way we encoded the graph G in Fig. 1. The collection of edges was represented not as a matrix or an adjacency list, but merely as a collection of facts – the data structure that we were working over was implicit in the database. This idiom of *database-as-data-structure* is a strength of this declarative style of programming, as details of underlying data structures can be omitted. However, because the notion of database we use is one that incrementally "learns" all derivable facts in an unspecified manner, it is difficult to describe algorithms that have distinct states or phases. Several attempts at addressing this problem amount to the identification of reasonable forms of locally stratified negation, such as temporal [8] or XY [9] stratification. However, stratified negation cannot easily describe algorithms that must repeatedly take only one of a number of possible steps, and this can make specifying greedy algorithms difficult [6].

Several disconnected lines of research have approached this problem. Greco and Zaniolo describe a variant of Datalog with an intrinsic notion of *choice* that has a semantics based on stable models and can naturally express a number of greedy algorithms [6]. They define an execution model for their system and show a number of complexity results, but they do not give a cost semantics, so all complexity results are based on directly reasoning about the interpreter.

Ganzinger and McAllester [5] do not explicitly consider the applicability of their system to greedy algorithms, but they demonstrate that their system, based on *deletion* of facts and *priorities* on rules, can express many of the same algorithms that motivated Greco and Zaniolo, such as algorithms computing minimum spanning trees and shortest paths. Unfortunately, the expressiveness of their system is hard to determine because they define an unusual notion of deletion that does not have any clear logical justification.

Pfenning and López et al. [10,11] propose linear logic as a more principled foundation of Ganzinger and McAllester's work. They show that their implementation of a linear logic programming language, Lollimon, is powerful enough to express many of the algorithms shown in Ganzinger and McAllester's previous work. However, they cannot reason about the running time of such algorithms, only their correctness, and complexity results would seem to be very difficult to obtain in a language such as Lollimon that allows for almost arbitrary integration of forward and backward chaining.

The primary contribution of this paper is the presentation of a programming language based on bottom-up reasoning in linear logic – essentially a first-order, Horn-like fragment of Lollimon – that is both useful for the specification of algorithms and in the analysis of their running time. To our knowledge, this is the first such result for a programming language based on linear logic. Section 2 describes the use of first-order linear logic in specifying a number of simple algorithms. Section 3 defines the operational semantics and cost semantics of the language, demonstrates the use of cost semantics in reasoning about complexity, and briefly describes the interpreter that demonstrates that the cost semantics are reasonable. Section 4 concludes and mentions a number of possible extensions to the basic, pure language considered here.

2 Bottom-Up Programming in Linear Logic

The pure bottom-up logic programming language introduced in the previous section is built from *atomic propositions* like $\mathsf{nat}(n)$, $\mathsf{edge}(a, b)$, and $\mathsf{path}(v, u)$. These facts represent truth in the usual, mathematical sense - the rule $r2$ in Fig. 1 says that if we know that there is an edge between some vertices a and b, we can also know that there is a path between them. However, after we learn $\mathsf{path}(\mathsf{a}, \mathsf{b})$, we still know $\mathsf{edge}(\mathsf{a}, \mathsf{b})$, because we treat truth as *persistent*.

Linear logic has a notion of persistent truth, but also has a notion of truth that describes the current (and possibly changing) state of the world. We refer to this notion of "truth in the current state of the system" as *ephemeral truth*, and in addition to the persistent atomic propositions that we have previously

$$\frac{\underline{\mathsf{wins}}(x, n) \quad \underline{\mathsf{wins}}(y, n)}{\underline{\mathsf{wins}}(x, \mathsf{s}(n)) \quad \mathsf{won}(x, y, n)}$$

Fig. 2. A simple linear logic program describing arbitrary single-elimination tournaments

seen, we introduce ephemeral (or *linear*) atomic propositions that we distinguish from persistent propositions by using an underline: $\underline{\mathsf{linear}}(x)$.

Rules with ephemeral propositions as premises introduce the possibility of changing the state of the world. The rule given in Fig. 2 describes a single-elimination tournament in which any team can play another team that has the same number of wins. If we have two teams a and c that have both won zero games, we can represent this as the two linear atomic propositions $\underline{\mathsf{wins}}(\mathsf{a}, \mathsf{z})$ and $\underline{\mathsf{wins}}(\mathsf{c}, \mathsf{z})$. These atomic propositions satisfy the two premises of the rule in Fig. 2. If we arbitrarily let $x = \mathsf{c}$ and $y = \mathsf{a}$, treating c as the "winning team," the rule represents the possibility of transitioning from a state where both teams a and c have won zero games and are still in the running to a state where team c has won one game and where team a is out of the running. There is no $\underline{\mathsf{wins}}(y, n)$ in the conclusion because the tournament is single-elimination – after losing, a team cannot play any other teams. Applying this rule requires *consuming* the two linear propositions we had before and replacing them with a single new linear atomic proposition $\underline{\mathsf{wins}}(\mathsf{c}, \mathsf{s}(\mathsf{z}))$. Applying the rule also adds the persistent atomic proposition $\mathsf{won}(\mathsf{c}, \mathsf{a}, \mathsf{z})$ to the database, which represents a persistent record of the fact that c defeated a in round z.

Changes to the state of a system are not necessarily reversible. While we could imagine a backtracking semantics that would eventually consider team a beating team c, or consider them playing other teams in the first round, we instead read rules with linear premises as describing a *committed choice* – once we apply a rule that consumes an ephemeral proposition, we will never consider any other way that proposition could have been consumed. Put another way, while our rules may describe a system that can evolve in many ways from an initial state, when reading our rules as an algorithm, the algorithm will follow *one* particular evolution of that system in a *don't-care* nondeterministic manner.

We can use these ephemeral atomic propositions to support algorithms that require certain actions to happen a fixed number of times, as well as algorithms that require some actions to be mutually exclusive. The example in Fig. 3 is a linear algorithm to compute a spanning tree of a connected, undirected graph $G = (E, V)$ that has some distinguished vertex $\mathsf{root} \in V$. The input to the algorithm is a persistent atomic proposition $\mathsf{edge}(\mathsf{a}, \mathsf{b})$ for every edge $(\mathsf{a}, \mathsf{b}) \in E$ and a single ephemeral atomic proposition $\underline{\mathsf{vert}}(\mathsf{a})$ for every vertex $\mathsf{a} \in V$. We view the relation tree as a directed subgraph of G where $\mathsf{tree}(\mathsf{a}, \mathsf{b})$ is true iff there is an edge from a to b in the tree.

Correctness of this spanning tree algorithm follows from invariants maintained by the rules. Take V' to be the set of all x such that $\mathsf{intree}(x)$ holds, and take E'

$$\frac{\mathsf{edge}(x,y)}{\mathsf{edge}(y,x)}\ r1 \qquad \frac{\mathsf{vert}(\mathsf{root})}{\mathsf{intree}(\mathsf{root})}\ r2 \qquad \frac{\begin{array}{c}\mathsf{edge}(x,y)\\ \mathsf{intree}(x)\\ \mathsf{vert}(y)\end{array}}{\begin{array}{c}\mathsf{tree}(x,y)\\ \mathsf{intree}(y)\end{array}}\ r3$$

Fig. 3. Finding a rooted spanning tree of an undirected graph

to be the the set of all ordered pairs (x, y) such that $\mathsf{tree}(x, y)$ holds. We have two state invariants, maintained by rule application:

1. E' is a subgraph of E and a spanning tree over V'.
2. The set V/V' is always the set of variables x' such that $\underline{\mathsf{vert}}(x')$ holds.

The two examples in Fig. 4 are more imperative in nature; both take as inputs some multiset of items represented by linear atomic propositions of the form $\underline{\mathsf{item}}(x)$ and place them into a data structure. The program on the left requires an additional input of the form $\underline{\mathsf{list}}(\mathsf{nil})$ and collects items into a list represented by a structured term, using $x :: l$ as a shorthand for $\mathsf{cons}(x, l)$. The program on the right requires no additional inputs, and collects items into a forest of binary-heap-like trees. Trees are represented as linear atomic propositions of the form $\underline{\mathsf{tree}}(n, t)$, where n is a natural number expressing the depth of the tree and t is a structured term representing the actual tree, a term consisting of an item and a list of subtrees.

These examples bring up another important property of linear/ephemeral propositions. For the purposes of bottom-up logic programming, deriving a persistent proposition twice is not any different than deriving it once; however, with linear propositions we are concerned with the *multiplicity* of those propositions: having two copies of $\underline{\mathsf{item}}(\mathsf{a})$ is different than having one. We will ensure that we can unambiguously refer to linear propositions by labelling them uniquely. For instance, the list-collection example on the left side of Fig. 4 could, from the multiset of atomic propositions $\{l_0 : \underline{\mathsf{list}}(\mathsf{nil}), l_1 : \underline{\mathsf{item}}(\mathsf{a}), l_2 : \underline{\mathsf{item}}(\mathsf{a}), l_3 : \underline{\mathsf{item}}(\mathsf{b})\}$, derive $\underline{\mathsf{list}}(\mathsf{a} :: \mathsf{b} :: \mathsf{a} :: \mathsf{nil})$ and $\underline{\mathsf{list}}(\mathsf{a} :: \mathsf{a} :: \mathsf{b} :: \mathsf{nil})$, but not $\underline{\mathsf{list}}(\mathsf{a} :: \mathsf{a} :: \mathsf{a} :: \mathsf{nil})$, because there are only two linear resources $\underline{\mathsf{item}}(\mathsf{a})$ and the derivation of that proposition requires three such resources. Committed choice ensures that we will only compute *one* of the three possible lists, or more generally one of the $!n$ possible lists given n distinct items.

These examples demonstrate the power of linear logic to express algorithms that would be difficult or inelegant to code in a system without linear resources.

$$\frac{\begin{array}{c}\underline{\mathsf{item}}(x)\\ \underline{\mathsf{list}}(l)\end{array}}{\underline{\mathsf{list}}(x :: l)} \qquad\Bigg| \qquad \frac{\underline{\mathsf{item}}(x)}{\underline{\mathsf{tree}}(\mathsf{z}, \mathsf{node}(x, \mathsf{nil}))} \qquad \frac{\begin{array}{c}\underline{\mathsf{tree}}(n, \mathsf{node}(x, ts))\\ \underline{\mathsf{tree}}(n, t)\end{array}}{\underline{\mathsf{tree}}(\mathsf{s}(n), \mathsf{node}(x, t :: ts))}$$

Fig. 4. Arbitrarily collecting items in a list (left) or in a forest of trees (right)

In the next section, we will make this more formal by defining an operational semantics based on linear logic and a cost semantics that allows us to reason about running time and complexity without knowing the details of an interpreter for the language.

3 Language Semantics

In this section, we will develop the tools for reasoning about the algorithms we began to specify in the previous section. Two of the fundamental properties of an algorithm are its run time behavior, specified by an operational semantics, and its running time, specified by a cost semantics. We will describe both.

We have already presented a number of programs, but we will formally define a program P as a series of rules. Each rule has one or more atomic propositions $A_0, \ldots, A_{n-1}$ as premises and zero or more atomic propositions $C_0, \ldots, C_{m-1}$ as conclusions; for example, in clause $r2$ of Fig. 3, $A_0 = \underline{\text{vert}}(\text{root})$ and $C_0 = \text{intree}(\text{root})$. There are two additional restrictions on the form of rules:

- *Range restriction.* The free variables in the conclusion must be a subset of the free variables in the premises. This ensures that a ground database will remain ground when inference rules are applied.
- *Separation.* The program must consistently identify some propositions as linear and some as persistent; this was indicated before by writing linear propositions as **prop** and persistent propositions as prop. Separation also requires that in any rule with linear atomic propositions among the conclusions $C_0, \ldots, C_{m-1}$, at least one of the premises $A_0, \ldots, A_{n-1}$ must be a linear atomic proposition. This requirement helps ensure that we will not "flood" the database with unlimited copies of ephemeral propositions, and also allows us to implement the saturation function Clo effectively.

3.1 Operational Semantics

In this section, we describe an operational semantics for the language we have defined, noting that the operational semantics does not make much sense as an implementation, as it makes transparently bad choices like running saturating forward chaining redundantly. The input is a finite *initial state* $\langle \Gamma_0, \Delta_0 \rangle$ where Γ_0 is a set of persistent propositions and Δ_0 is a set of labeled linear propositions; a *program trace* is a finite list of states $\langle \Gamma_0, \Delta_0 \rangle \ldots \langle \Gamma_m, \Delta_m \rangle$.

For each state $\langle \Gamma_i, \Delta_i \rangle$, the operational semantics calculates the saturated database $\text{Clo}(\Gamma_i)$ of all the persistent atomic facts that are implied by Γ_i in P by exhaustive forward reasoning, not involving linear propositions in any way. Assuming the process of saturated forward inference terminates, the operational semantics picks an arbitrary rule $r \in P$ and a grounding substitution σ (that is, a substitution that maps every free variable x in the rule to a variable-free term t) such that, for each premise A_i of the rule r, $A_i \sigma$ is in $\text{Clo}(\Gamma_i)$ (if it is persistent) or Δ_i (if it is linear). Applying that rule removes one or more linear resources from Δ_i and adds each conclusion $C_i\sigma$ to either Γ_i or Δ_i depending on whether C_i is linear or persistent. This results in a new state $\langle \Gamma_{i+1}, \Delta_{i+1} \rangle$, and the trace is extended. If there is no rule r and substitution σ satisfying the conditions described above, then the trace cannot be extended and is called a *complete* trace.

The operational semantics separates treatment of monotonic deduction that involves only persistent propositions and the committed choice reasoning that involves consuming ephemeral propositions. This distinction will be reflected in the definition of abstract running time, but we can already see that it is reflected in the arguments about the *termination* of algorithms. It was mentioned in Section 1 that we have to give an argument that the closure will be finite, as this is not true in general; we also have to give a termination argument bounding the length of the program trace by bounding the number of possible applications of rules with linear premises.

3.2   Linear Logic

While many details are beyond the scope of this paper, we will sketch the description of the language and operational semantics in terms of intuitionistic linear logic; our system is a fragment of the judgmental reconstruction of first order intuitionistic linear logic described in [7,12]. The necessary fragment of linear logic is roughly analogous to the Horn fragment of standard intuitionistic logic.

$$\begin{array}{lll}
\text{Atomic propositions} & A & \\
\text{Basic propositions} & Q & ::= A \mid !A \\
\text{State propositions} & S & ::= Q \mid \mathbf{1} \mid S \otimes S \\
\text{State transitions} & R & ::= S \mid S \multimap R \mid \forall x.R \\
\text{Persistent hypotheses} & \Gamma & ::= \cdot \mid \Gamma, R \; pers \\
\text{Ephemeral hypotheses} & \Delta & ::= \cdot \mid \Delta, R \; eph
\end{array}$$

The translation of a rule r with premises $A_0, \ldots, A_{n-1}$ and conclusions $C_0, \ldots, C_{m-1}$ is the persistent proposition

$$r : \forall \mathbf{x_0}.Q_0 \multimap \ldots \multimap \forall \mathbf{x_{n-1}}.Q_{n-1} \multimap (Q_0' \otimes \ldots \otimes Q_{m-1}')$$

with $Q_i = A_i$ if A_i is an ephemeral atomic proposition and $Q_i = !A_i$ if A_i is a persistent atomic proposition, and similarly for Q_i' and C_i. The "curried" form is intended to clarify that the variables $\mathbf{x_i}$ first occur in the premise Q_i.

Judgments in the sequent calculus presentation of intuitionistic linear logic have the form $\Gamma; \Delta \vdash R \; eph$; we write $R \; pers$ to indicate that R is persistent, and we write $R \; eph$ to indicate that R is ephemeral. We omit writing the translated rules from the program P that are tacitly included in Γ.

The concepts of *polarity* and *focusing* as described in [13] are useful in describing logic programming from a proof theoretic perspective. In particular, focusing allows us to define *derived rules* in linear logic for any formula in the fragment described above. If we have a rule with premises $\underline{a}$ and b, and with conclusions c and d, we express that rule in linear logic as a $\multimap$!b $\multimap$ (c $\otimes$!d). If we treat every atomic proposition as having positive polarity, focusing on the persistent proposition a $\multimap$!b $\multimap$ (c $\otimes$!d) gives this derived inference rule:

$$\frac{\Gamma; \cdot \vdash \mathsf{b} \; eph \qquad \Gamma, \mathsf{d} \; pers; \Delta, \mathsf{c} \; eph \vdash \gamma}{\Gamma; \Delta, \mathsf{a} \; eph \vdash \gamma}$$

where γ is an arbitrary conclusion.

What we see from these rules is that our "next state" actually appears in the *premise* of the derived rule; this may seem a bit unnatural, but it is consistent with Lollimon and other linear logic programming languages [11].

We have left out the details that allow us to actually prove the following theorems, but we can still state the soundness and (non-deterministic) completeness of our language with respect to linear logic.

Theorem 1 (Soundness of operational semantics)
For any (separated and range-restricted) program P and for any program trace $\langle \Gamma_0, \Delta_0 \rangle \ldots \langle \Gamma_m, \Delta_m \rangle$, given a sequent of the form $\Gamma_m; \Delta_m \vdash \gamma$ for an arbitrary γ, there exists a derivation of $\Gamma_0; \Delta_0 \vdash \gamma$.

Proof By induction on the length of the abstract trace. We need a lemma that if $A \in \mathsf{Clo}(\Gamma)$, then $\Gamma; \cdot \vdash A \; eph$.

Theorem 2 (Nondeterministic completeness of operational semantics)
For any (separated and range-restricted) program P, if the sequent $\Gamma_0; \Delta_0 \vdash \gamma$ is derivable using the sequent $\Gamma_m; \Delta_m \vdash \gamma$, where Γ_0, Δ_0, Γ_m, and Δ_m contain only ground, atomic propositions and γ is an arbitrary conclusion, then there exists some program trace $\langle \Gamma_0, \Delta_0 \rangle \ldots \langle \Gamma'_m, \Delta_m \rangle$ where $\mathsf{Clo}(\Gamma_m) = \mathsf{Clo}(\Gamma'_m)$.

Proof. By induction on focused derivations. We need a lemma that if $\Gamma; \cdot \vdash A \; eph$, then $A \in \mathsf{Clo}(\Gamma)$.

Theorem 2 says that if we can "work on the left" in linear logic from a sequent $\Gamma_0; \Delta_0 \vdash \gamma$ to a sequent $\Gamma_m; \Delta_m \vdash \gamma$, then some trace obeying the operational semantics follows an equivalent path; however, because the operational semantics allows an arbitrary choice of which applicable rule with linear premises to apply, a correct implementation of the operational semantics might never take such a path.

3.3   Cost Semantics

We define, following Ganzinger and McAllester [4,5], a cost semantics called the *abstract running time*. This cost semantics will allow us to reason about algorithms written in this language, such as the ones in in Section 2, without considering the details of the implementation. The abstract running time of a trace $\langle \Gamma_0, \Delta_0 \rangle \ldots \langle \Gamma_m, \Delta_m \rangle$ is the sum of four components: $|\Gamma_0| + |\Delta_0| + m + \Phi$. The first two components, $|\Gamma_0|$ and $|\Delta_0|$, are just the number of persistent and linear resources (respectively) given as input. The other two components are m, the number of transitions involving rules with linear premises, and Φ, the number of *unique prefix firings* – a quantity we will now define.

Definition 1 (Prefix firing). *Let* $\langle \Gamma_0, \Delta_0 \rangle \ldots \langle \Gamma_m, \Delta_m \rangle$ *be a program trace of a program* P. *A* prefix firing *is a triple* $\langle r, \sigma, [l_0, \ldots, l_{k-1}] \rangle$ *such that*

- *There is a rule* r *in* P *with premises* $A_0, \ldots, A_{n-1}$.
- *The substitution* σ *assigns a ground term for every free variable in the premises* $A_0, \ldots, A_{k-1}$.
- *There is some state* $\langle \Gamma_i, \Delta_i \rangle$ *where for all* $0 \leq j < k$, *either* $A_j \sigma \in \mathsf{Clo}(\Gamma_i)$ *or* $l_j : A_j \sigma \in \Delta_i$, *and either*
  - *All of* $A_0, \ldots, A_{k-1}$ *are persistent atomic propositions, or else*
  - $k < n$ *and there is* **no** *substitution* σ' *that assigns the same terms as* σ *to the free variables in* $A_0, \ldots, A_{k-1}$ *and additionally assigns ground terms to all the free variables in* A_k *such that* $A_k \sigma' \in \mathsf{Clo}(\Gamma_i)$ *or such that* $l : A_k \sigma' \in \Delta_i'$, *where* Δ_i' *is* Δ_i *with all the linear propositions labeled* $l_0, \ldots, l_{k-1}$ *removed.*

As mentioned previously, if multiple instances of the ground linear proposition appear in Δ_i, they have distinct labels and can be used to form distinct prefix firings. Because we don't care about labels of persistent atomic propositions, and the definition doesn't use them, we write them as an underscore "_".

The majority of the definition just expresses the fact that the order of premises matters; the last bullet point is the complicated one. It describes the conditions where we can ignore would-be prefix firings that include linear propositions; we can do so if we know that, in every state, we will always be able to expand the prefix firing to a larger one.

3.4   Using the Abstract Running Time

Because the operational semantics is quite nondeterministic, and because our cost semantics depends on the number of steps taken using of rules with linear premises, we can expect reasoning in general about the running time of programs to be undecidable. However, for well-designed programs it is usually still possible to effectively reason about both the number of prefix firings and the length of the program trace in order to get an informative abstract running time. We give an example in this section, and the extended technical report [14] shows a similar analysis that gives the list collection and heap collection example in Fig. 4 running times in $O(n)$ and $O(n \log n)$, respectively, where n is the number of input items.

We will show that the spanning tree algorithm in Fig. 3 has an abstract running time in $O(|E| + |V|)$, that is, proportional to the number of edges plus the number of vertices. The abstract running time is $|\Gamma_0| + |\Delta_0| + m + \Phi$. It is obvious that $|\Gamma_0| = |E|$ and $|\Delta_0| = |V|$ based on how the problem is set up; also, because every linear transition consumes a linear resource corresponding to some $v \in V$, an abstract trace can have at most $|V|$ transitions, which is to say that m is bounded by $|V|$.

We consider the prefix firings for each of the three rules in turn. Rule $r1$ can have at most $2|E|$ prefix firings, as every edge $(\mathsf{a}, \mathsf{b}) \in E$ leads to two facts:

edge(a, b) and edge(b, a). Rule $r2$ has no prefix firings, as it has one linear proposition that is either there or not. Rule $r3$ can have at most $4|E|$ prefix firings. The first premise edge(x, y) effectively "grounds" the rest of the premises, leading to $2|E|$ prefix firings of the form $\langle r3, \sigma, [\_]\rangle$, and the final state will include intree(a) for every vertex a, resulting in at most $2|E|$ prefix firings of the form $\langle r3, \sigma, [\_, \_]\rangle$. However, there are no prefix firings of the form $\langle r3, \sigma, [\_, \_, l]\rangle$, because a prefix firing that covers all the premises does not meet the condition that $k < n$. This gives us an abstract running time bounded by $2|V| + 7|E|$, so the abstract running time is in $O(|E| + |V|)$.

3.5   Implementing the Operational Semantics

This theorem describes the relationship between the operational semantics, the cost semantics, and the interpreter; it is a is a close analogue to the comparable theorem in [5].

Theorem 3. *For any terminating program P, there exists an interpreter running on a RAM machine extended with constant time hash table operations such that for any initial state $\langle \Gamma_0, \Delta_0\rangle$ the interpreter executes a complete trace $\langle \Gamma_0, \Delta_0\rangle, \ldots, \langle \Gamma_m, \Delta_m\rangle$ and returns $\mathsf{Clo}(\Gamma_m)$ and Δ_m in time proportional to the abstract running time of the trace.*

Theorem 3 establishes that reasoning about the behavior of algorithms described in the language we have presented is a three-part process. First, we must demonstrate that, for a given program, $\mathsf{Clo}(\Gamma)$ is always finite that no trace of the operational semantics can have unbounded length. Second, we must give a bound to the abstract running time of *all* possible complete traces in terms of the initial state. Having done so, Theorem 3 ties the knot by ensuring that the implementation will execute one of the possible complete traces and will do so in time proportional to the abstract running time of that trace. Because we have bounded the abstract running time of any arbitrary trace, we know that the trace actually executed by the interpreter has an abstract running time within that bound and is therefore executed in time proportional to that bound.

The interpreter that establishes Theorem 3 is sketched here and described fully in the extended version of this paper [14]. Given a program, we create a derived program where for each rule r with n premises in the original program, the derived program has $2n$ rules and introduces $2n$ new atomic propositions (referred to as derived propositions), one for each premise A_i and one for each prefix $A_0, \ldots, A_i$. The derived propositions expose variables that are shared between premises of a rule, allowing an index to efficiently discover premises with matching instantiations of those variables. Two work lists (queues) – one dealing with persistent propositions and one dealing with ephemeral propositions – together contain all the immediate consequences of the facts in the index.

The portion of the interpreter dealing with purely persistent propositions is similar to the interpreter in [1]. When a fact is removed from the persistent work queue and added to the index, the index is used to find all immediate consequences of that fact and those already in the index; these consequences are added to the queue. The treatment of linear atomic propositions is novel. The index and linear work queues are allowed to temporarily contain multiple derived propositions that are all consequences of Δ_i, the current multiset of non-derived atomic propositions, even if some cannot simultaneously be consequences of Δ_i because they require consuming the same ephemeral propositions. These ephemeral propositions are only consumed when a rule from the original program is applied, in the process removing all the derived propositions that depended on the consumed propositions.

In order to avoid unnecessarily declaring and then deleting atomic propositions, upon removing an item from the linear work queue the index is recursively used to find the *first* atomic proposition implied by the program and the dequeued proposition, find the first atomic proposition implied by that proposition, and so on. Either this will succeed until we have shown that $A_0, \ldots, A_{n-1}$ are all derivable, in which case we apply that rule, or it will fail, in which case backtracking, depth-first search looks for a different way to fully apply the rule. Each failure corresponds to a prefix that cannot be extended; therefore, each successful search can be charged against the number of linear transitions, and each unsuccessful search can be charged against the number of prefix firings resulting from non-extendable prefixes that include linear propositions.

4   Conclusion and Future Work

We have described a bottom-up logic programming language that has a notion of ephemeral truth as well as the more familiar notion of persistent truth, and we have defined a cost semantics that allows for reasoning about the running time of programs written in this language. The language can be used to express and analyze a number of algorithms that have a notion of stateful change or nondeterministic update, and other algorithms are described in the extended version [14]. Our system is unique among similar work in having a proof-theoretic semantics based on focusing and linear logic. In the future, we are interested in pursuing a number of extensions to the language described here, including priorities similar to those in [5], temporal stratification and stratified negation similar to [8], and a notion of equality to describe algorithms that use union-find.

Acknowledgments. We would like to thank Michael Ashley-Rollman, Dan Licata, and the three anonymous reviewers for their comments on earlier drafts of this paper. This material is based upon work supported under a National Science Foundation Graduate Research Fellowship by the first author.

We wish to dedicate this paper to Harald Ganzinger, with whom the second author discussed some of the core ideas presented here, and whose untimely passing prevented him from participating further in this research.

References

1. McAllester, D.A.: On the complexity analysis of static analyses. J. ACM 49(4), 512–537 (2002)
2. Nielson, F., Nielson, H.R., Seidl, H.: Automatic complexity analysis. In: Le Métayer, D. (ed.) ESOP 2002 and ETAPS 2002. LNCS, vol. 2305, pp. 243–261. Springer, Heidelberg (2002)
3. Liu, Y.A., Stoller, S.D.: From Datalog rules to efficient programs with time and space guarantees. In: PPDP 2003: Proceedings of the 5th ACM SIGPLAN international conference on Principles and practice of declaritive programming, pp. 172–183. ACM, New York (2003)
4. Ganzinger, H., McAllester, D.A.: A new meta-complexity theorem for bottom-up logic programs. In: Goré, R.P., Leitsch, A., Nipkow, T. (eds.) IJCAR 2001. LNCS (LNAI), vol. 2083, pp. 514–528. Springer, Heidelberg (2001)
5. Ganzinger, H., McAllester, D.A.: Logical algorithms. In: Stuckey, P.J. (ed.) ICLP 2002. LNCS, vol. 2401, pp. 209–223. Springer, Heidelberg (2002)
6. Greco, S., Zaniolo, C.: Greedy algorithms in Datalog. Theory Pract. Log. Program. 1(4), 381–407 (2001)
7. Chang, B.Y.E., Chaudhuri, K., Pfenning, F.: A judgmental analysis of linear logic. Technical Report CMU-CS-03-131, Carnegie Mellon University (April 2003)
8. Nomikos, C., Rondogiannis, P., Gergatsoulis, M.: Temporal stratification tests for linear and branching-time deductive databases. Theor. Comput. Sci. 342(2-3), 382–415 (2005)
9. Arni, F., Ong, K., Tsur, S., Wang, H., Zaniolo, C.: The Deductive Database System LDL++. Theory Pract. Log. Program. 3(1), 61–94 (2003)
10. Pfenning, F.: Linear logical algorithms. In: Workshop on Programming Logics in memory of Harald Ganzinger, Saarbrücken (June 2005) (invited talk)
11. López, P., Pfenning, F., Polakow, J., Watkins, K.: Monadic concurrent linear logic programming. In: PPDP 2005: Proceedings of the 7th ACM SIGPLAN International Conference on Principles and Practice of Declarative Programming, pp. 35–46. ACM, New York (2005)
12. Chaudhuri, K.: The Focused Inverse Method for Linear Logic. PhD thesis, Carnegie Mellon University (December 2006)
13. Chaudhuri, K., Pfenning, F., Price, G.: A Logical Characterization of Forward and Backward Chaining in the Inverse Method. In: Automated Reasoning, vol. 4130, pp. 97–111. Springer, Heidelberg (2006)
14. Simmons, R.J., Pfenning, F.: Linear Logical Algorithms. Technical Report CMU-CS-08-104, Carnegie Mellon University (May 2008)

A Simple Model of Separation Logic for Higher-Order Store

Lars Birkedal[1], Bernhard Reus[2], Jan Schwinghammer[3], and Hongseok Yang[4]

[1] IT University of Copenhagen
[2] University of Sussex, Brighton
[3] Saarland University, Saarbrücken
[4] Queen Mary, University of London

Abstract. Separation logic is a Hoare-style logic for reasoning about pointer-manipulating programs. Its core ideas have recently been extended from low-level to richer, high-level languages. In this paper we develop a new semantics of the logic for a programming language where code can be stored (i.e., with higher-order store). The main improvement on previous work is the simplicity of the model. As a consequence, several restrictions imposed by the semantics are removed, leading to a considerably more natural assertion language with a powerful specification logic.

1   Introduction

Higher-order store is included in modern programming languages in the form of code pointers and storable objects. "Higher-order" here refers to the fact that one can keep not only data in the store but also procedures or commands that manipulate the store themselves. It is widely used in systems code, such as operating system kernels, device drivers and web servers. For instance, the Linux kernel keeps multiple linked lists whose nodes store code fragments, and calls those fragments in response to external events, such as a signal from a printer.

However, formal reasoning about higher-order store is still an open problem. Although several sound program logics for higher-order store have been proposed, they either are intended for machine code [4] or they fail to combine local reasoning with intuitive rules for stored code while maintaining the simplicity of Hoare logic for first-order store [6,15]. The difficulty is that a logic for higher-order store should accommodate reasoning about "recursion through the store", a tricky implicit recursion implemented by stored procedures.

The goal of our research is to solve the problem of reasoning about higher-order store using separation logic. Separation logic is a program logic for reasoning modularly about programs with pointers. It has been demonstrated that the logic substantially simplifies formal program verification in low-level C-like programming languages as well as richer, higher-level languages [1,2,3,7,8,9,12,13,17]. Our aim is to design program logics for higher-order store that keep all the benefits of

L. Aceto et al. (Eds.): ICALP 2008, Part II, LNCS 5126, pp. 348–360, 2008.

separation logic, such as (higher-order) frame rules, while providing efficient, sound proof rules for recursion through the store.

In this paper, we investigate the semantic foundations for developing separation logic for higher-order store. We build on previous work of Reus and Schwinghammer [15], which identified key semantic challenges for such a logic, and provided fairly sophisticated solutions based on functor categories. In this paper, we take different approaches to the various problems, and as a result obtain a more powerful logic and a substantially simpler semantic model.

We now give an overview of two key semantic challenges that are involved in developing separation logic for higher-order store. We outline how those challenges were addressed in earlier work [15], and compare with our new model.

The first challenge is to find a model that validates the frame rule known from separation logic [17]. In traditional models of separation logic [10], the soundness of the frame rule relies on programs satisfying a frame property, which says that the meaning of each program phrase only relies on its "footprint". To ensure that all program phrases – in particular, memory allocation – satisfy the frame property, the models interpret commands as relations (i.e., functions from input states to *sets* of output states), and memory allocation denotes a function that nondeterministically picks new memory. Now, in a language with higher-order store, the semantics involves solving recursive domain equations. With nondeterministic memory allocation, one is naturally led to recursive domain equations using powerdomains. These are problematic not only because it is unclear whether they can be used to show the existence of recursive properties of the heap but also because programs would no longer denote ω-continuous functions, due to the *countable* nondeterminism arising from memory allocation. Instead, Reus and Schwinghammer considered a functor category, indexed over finite sets of locations, which made it possible to prove that programs obeyed a frame property without relying on a nondeterministic allocator. However, this involved two non-trivial aspects. First, recursive domain equations now had to be solved not in an ordinary category of domains, but in the functor category. Second, the frame property became a recursively defined property, whose existence required a separate non-trivial proof. While Reus and Schwinghammer succeeded in defining a model that validates the frame rule, the technical complications involved make it difficult to scale the ideas to richer languages and richer logics, e.g., with higher-order frame rules [2,3,11].

In this paper we validate the frame rule without relying on the frame property of programs. Instead, we "bake-in" the frame rule into the interpretation of Hoare triples, using an idea from [3]. (This is described in detail in Section 4.) In particular, this approach allows us to model memory allocation by a simple deterministic allocator, so that we can model the programming language using ordinary recursively defined domains, avoiding the complications in [15]. Furthermore, the approach also allows us to validate a whole range of higher-order frame rules and to include pointer arithmetic.

The second challenge is to validate proof rules for recursion through the store [16]. Such rules essentially amount to having recursively defined

$e \in \text{EXP} ::= \ldots \mid \text{`}C\text{'}$ quote (command as expression)

$C \in \text{COM} ::= \text{skip} \mid C_1;C_2 \mid \text{if } (e_1=e_2) \text{ then } C_1 \text{ else } C_2$ no op, sequencing, conditional
$\qquad\quad \mid \text{ let } x=\text{new } (e_1,\ldots,e_n) \text{ in } C \mid \text{free } e$ allocation, disposal
$\qquad\quad \mid [e_1]:=e_2 \mid \text{let } y=[e] \text{ in } C \mid \text{eval } [e]$ assignment, lookup, unquote

Fig. 1. Syntax of expressions and commands

specifications, which denote recursive properties of the domain for commands. It is well-known that to establish the existence of such recursive properties of domains one needs additional conditions involving, in particular, admissibility and certain forms of downward closure [14]. In [15], these conditions were ensured by restricting the assertion language of the logic. In the present paper, we avoid such restrictions by changing the interpretation of triples and slightly modifying the recursion rules. In particular, we use an admissible and downwards closure of the post-condition, similar to the use of ⊥⊥-closure in [3] (see Section 4).

2   Programs, Assertions and Specifications

Programs. The abstract syntax of the programming language is presented in Fig. 1. It is essentially as in [15], with dynamic allocation (but here we assume a more realistic, deterministic memory allocator) and storable, parameterless procedures. The language is deliberately kept simple so that we can study higher-order store without distraction. We point out two features of the language which proved problematic for the semantics given in *loc. cit.* First, the language assumes that addresses are natural numbers, so that it is possible to apply arithmetic operations on addresses. Next, the language includes an allocator that deterministically picks n-consecutive cells.

Assertions. The assertions $P, Q, \ldots$ used in Hoare triples are built from the formulas of classical predicate logic and the additional separation logic assertions that describe the heap ($e \mapsto e'$, **emp**, $P * Q$ and $P \mathbin{-\!*} Q$; cf. [17]). Note that expressions in formulas can point to quoted code, as in $x \mapsto \text{`}C\text{'}$, so that they can be used to specify properties of stored procedures. We use two abbreviations:

$$e \mapsto \_ \;\overset{def}{=}\; \exists x'. e \mapsto x', \qquad e \mapsto e_1,..,e_n \;\overset{def}{=}\; e \mapsto e_1 * e+1 \mapsto e_2 * .. * e+n-1 \mapsto e_n.$$

where $x' \notin fv(e)$. We write $\Gamma \vdash P (: \text{Assert})$ for some finite set of variables Γ, when the assertion P contains only free variables in Γ.

Specifications. Specifications are formulas of first-order intuitionistic logic with equality. In addition, it includes Hoare triples as atomic formulas and invariant extensions $\varphi \otimes P$ (from [3]):

$$\varphi, \psi ::= e_1=e_2 \mid \{P\}C\{Q\} \mid \varphi \otimes P \mid \mathbf{T} \mid \mathbf{F} \mid \varphi \wedge \psi \mid \varphi \vee \psi \mid \varphi \Rightarrow \psi \mid \exists x.\varphi \mid \forall x.\varphi$$

PROOF RULES FOR STORED CODE

$$((\forall \vec{y}.\{P\}\text{eval }[e]\{Q\}) \;\Rightarrow\; \forall \vec{y}.\{P\}C\{Q\}) \;\Rightarrow\; \forall \vec{y}.\{P * e \mapsto {`C'}\}\text{eval }[e]\{Q * e \mapsto {`C'}\}$$
$$\text{(where } \vec{y} \notin fv(e, C))$$

$$(\forall x.\,(\forall \vec{y}.\{P * e \mapsto x\}\text{eval }[e]\{Q * e \mapsto x\}) \;\Rightarrow\; \forall \vec{y}.\{P * e \mapsto x\}C\{Q * e \mapsto x\})$$
$$\Rightarrow \; \forall \vec{y}.\{P * e \mapsto {`C'}\}\text{eval }[e]\{Q * e \mapsto {`C'}\} \quad \text{(where } x \notin fv(P, Q, \vec{y}, e, C), \vec{y} \notin fv(e, C))$$

$$(\forall x.\,(\forall \vec{y}.\{P * e \mapsto x\}\text{eval }[e]\{Q\}) \;\Rightarrow\; \forall \vec{y}.\{P * e \mapsto x\}C\{Q\})$$
$$\Rightarrow \; \forall \vec{y}.\{P * e \mapsto {`C'}\}\text{eval }[e]\{Q\} \qquad \text{(where } x \notin fv(P, Q, \vec{y}, e, C), \vec{y} \notin fv(e, C))$$

PROOF RULES FOR HOARE TRIPLES

$$(\forall x.\{P * x \mapsto e\}C\{Q\}) \;\Rightarrow\; \{P\}\text{let } x = \text{new } e \text{ in } C\{Q\} \qquad \text{(where } x \notin fv(P, Q, e))$$
$$(\forall x.\{P * e \mapsto x\}C\{Q\}) \;\Rightarrow\; \{\exists x.\, P * e \mapsto x\}\text{let } x = [e] \text{ in } C\{Q\} \quad \text{(where } x \notin fv(Q, e))$$

$$\{e \mapsto \_\}\text{free}(e)\{\mathbf{emp}\} \qquad \{e \mapsto \_\}[e] := e'\{e \mapsto e'\}$$

$$\frac{[\![P]\!]_\eta^{\mathcal{A}} \subseteq [\![P']\!]_\eta^{\mathcal{A}} \text{ and } [\![Q']\!]_\eta^{\mathcal{A}} \subseteq [\![Q]\!]_\eta^{\mathcal{A}} \text{ for all } \eta \in [\![\Gamma]\!]}{\Gamma \;\vdash\; \{P'\}C\{Q'\} \Rightarrow \{P\}C\{Q\}}$$

PROOF RULES FOR INVARIANT EXTENSION $- \otimes P$

$$\varphi \;\Rightarrow\; \varphi \otimes P \qquad\qquad \{P\}C\{P'\} \otimes Q \;\Leftrightarrow\; \{P * Q\}C\{P' * Q\}$$
$$(e_0 = e_1) \otimes Q \;\Leftrightarrow\; e_0 = e_1 \qquad\qquad (\varphi \otimes P) \otimes Q \;\Leftrightarrow\; \varphi \otimes (P * Q)$$
$$(\varphi \oplus \psi) \otimes P \;\Leftrightarrow\; (\varphi \otimes P) \oplus (\psi \otimes P) \qquad (\kappa x.\,\varphi) \otimes P \;\Leftrightarrow\; \kappa x.\,\varphi \otimes P$$
$$\text{(where } \oplus \in \{\Rightarrow, \wedge, \vee\}) \qquad\qquad \text{(where } \kappa \in \{\forall, \exists\}, x \notin fv(P))$$

Fig. 2. Some proof rules

While assertions express properties of states, specifications describe properties of programs (sometimes using assertions inside Hoare triples). For a finite set Γ of variables, $\Gamma \vdash \varphi(:\mathsf{Spec})$ means that Γ includes all free variables of φ.

Proof rules. Our specification logic includes all the usual proof rules of intuitionistic first-order logic with equality, and special rules for Hoare triples and invariant extension $\varphi \otimes P$. Fig. 2 lists some of those, where the context Γ for each specification is omitted. Note that the consequence rule uses semantically valid implications for assertions, some of which can be proved using the proof rules from classical logic and the logic of Bunched Implications. In this way, the consequence rule embeds reasoning about assertions into the specification logic without the need to commit to a specific proof system for assertions.

Most of the rules in the figure are standard and known from separation logic. The only exceptions are the three proof rules for stored procedures.[1] These rules

---

[1] For simplicity, we do not consider mutually recursive stored procedures here, but it is straightforward to generalize our rules to handle them. Also, the first rule for stored procedures can be derived from the second and the higher-order frame rules. We include it in order to point out the subtleties of reasoning about stored procedures.

are similar to the rule for calling a parameterless, recursive procedure p declared as $p \Leftarrow C$, where C is the body of p that may contain a recursive call to p:

$$\frac{(\forall \vec{y}.\,\{P\}\mathsf{call}\ p\{Q\}) \vdash \forall \vec{y}.\,\{P\}C\{Q\}}{\forall \vec{y}.\,\{P\}\mathsf{call}\ p\{Q\}} \tag{1}$$

This rule is usually proved sound via fixpoint induction (note that p in the premiss semantically refers to any procedure with the required properties, whereas in the conclusion p refers to the declared procedure). For the language of Fig. 1, the fact that stored procedures are in use means that the declaration of a procedure is now expressed by an assertion $e \mapsto {}'C'$, stating that e is a reference to the procedure with body C.

In Fig. 2 there are three rules for stored code that might call itself recursively, establishing partial correctness (and hence do not feature in the logic for *total* correctness of [6]). The first rule prohibits any access to the storing location e except through eval $[e]$, whereas the second and third are more permissive. Note also that only the first two rules establish that the stored procedure called has not been altered. This is important in cases where the procedure gets updated. Updating code after its first call is a general pattern of usage of stored code, found e.g. in device drivers [5]: the first call is used for initialisation that further calls rely on.

It would be preferable to have just one rule for recursive procedures, e.g.

$$(\forall x.\ (\forall \vec{y}.\,\{P * e \mapsto x\}\mathsf{eval}\ [e]\{Q\}) \ \Rightarrow \ \forall \vec{y}.\,\{P * e \mapsto x\}C\{Q\})$$
$$\Rightarrow \ \forall \vec{y}.\,\{P * e \mapsto {}'C'\}\mathsf{eval}\ [e]\{Q['C'/x]\}\,(\text{where } x \notin \mathit{fv}(P, \vec{y}, e, C), \vec{y} \notin \mathit{fv}(e, C)) \,.$$

Alas we cannot easily prove such a rule sound. Our soundness proof for the recursion rules relies on pre- and post-condition satisfying properties (a) and (b), resp., as stated in the proof of Lemma 5 (see Section 4), and an arbitrary post-condition might violate property (b). We achieve soundness by restricting the shape of the post-condition to $Q * e \mapsto x$ and stipulating the side condition $x \notin \mathit{fv}(Q)$. As a consequence, we cannot instantiate the recursion rules with post-condition $e \mapsto x * e' \mapsto x$, as needed for a self-copying command let $x=[e]$ in $[e'] := x$ stored at e. Yet, also $Q * e \mapsto x * e' \mapsto x$ satisfies property 2 mentioned above, and soundness of a corresponding recursion rule could be established analogously to Lemma 5. This means the logic is incomplete; our objective has been to find rules that are easy to apply on programs with "common" use of stored procedures.

Alternatively, we could have given a single rule for stored procedures, but with a *semantic* side-condition to rule out unsuitable post-conditions.

Example 1 (Factorial). Consider the following specification and implementation:

$$F_o \stackrel{def}{=} \mathsf{let}\ x=[o]\ \mathsf{in}\ \mathsf{let}\ r=[o+1]\ \mathsf{in}$$
$$\qquad \mathsf{if}\ (x{=}0)\ \mathsf{then}\ \mathsf{skip}\ \mathsf{else}\ ([o+1]:=r \cdot x;\ [o]:=x-1; \mathsf{eval}\ [o+2])$$
$$C \stackrel{def}{=} [o+2]:={}'F_o';\mathsf{eval}\ [o+2] \qquad\qquad o \vdash \{o \mapsto 5, 1, \_\}C\{o \mapsto 0, 5!, {}'F_o'\}$$

The command C implements the factorial function in an object-oriented style, using three consecutive cells $(o, o+1, o+2)$. The first two cells represent fields arg and res, and the third cell denotes a method that computes the factorial of arg (decrementing it as a side effect) and multiplies this onto res. Note that the procedure F_o stored in $o+2$ calls itself by recursion through the store; see the last instruction eval $[o+2]$ of F_o.

The specification expresses that C computes 5! and stores it in cell $o+1$. The key step of the proof is the derivation

$$\frac{o \vdash (\forall ij.\{o \mapsto i, j\}\text{eval}\,[o+2]\{o \mapsto 0, j \cdot i!\}) \Rightarrow (\forall ij.\{o \mapsto i, j\}F_o\{o \mapsto 0, j \cdot i!\})}{o \vdash \forall ij.\{o \mapsto i, j, {}^{\prime}F_o{}^{\prime}\}\text{eval}\,[o+2]\{o \mapsto 0, j \cdot i!, {}^{\prime}F_o{}^{\prime}\}}$$

which shows the correctness of the stored procedure F_o. This step applies the first rule for stored procedures, and it illustrates the benefit of the rule. Here, the rule lets us hide the cell $o+2$ for code F_o in the premise, thereby giving a simple specification to discharge. The derivation of this specification itself is omitted; it involves only routine applications of standard separation logic proof rules.   □

Example 2. Next, we illustrate the typical use of the three rules for stored procedures with program C_n's below:

$F_1 \stackrel{def}{=}$ let $j=[i]$ in $\left(\text{if } j=0 \text{ then skip else } ([i]:=j-1; \text{eval}\,[i+1])\right)$

$F_2 \stackrel{def}{=}$ let $j=[i]$ in let $f=[i+1]$ in $\left([i]:=f; \text{if } j=0 \text{ then } [i]:=0 \text{ else } ([i]:=j-1; \text{eval}\,[i+1])\right)$

$F_3 \stackrel{def}{=}$ let $j=[i]$ in $\left(\text{if } j=0 \text{ then } ([i+1]:={}^{\prime}\text{skip}{}^{\prime}) \text{ else } ([i]:=j-1; \text{eval}\,[i+1])\right)$

$C_n \stackrel{def}{=} [i+1]:=F_n; \text{eval}\,[i+1]$

All of the C_n's decrease the value of i to zero (rather inefficiently), using recursion through the store. Additionally, C_2 dereferences cell $i+1$ to get the stored procedure F_2 and copy it to cell i temporarily. C_3 replaces the stored procedure in $i+1$ by skip at the end of the execution. For these programs, we want to prove:

$$i \vdash \{i \mapsto \_, \_\}C_1\{i \mapsto 0, {}^{\prime}F_1{}^{\prime}\} \quad i \vdash \{i \mapsto \_, \_\}C_2\{i \mapsto 0, {}^{\prime}F_2{}^{\prime}\} \quad i \vdash \{i \mapsto \_, \_\}C_3\{i \mapsto 0, {}^{\prime}\text{skip}{}^{\prime}\}.$$

The major step of the proof of C_1 is the use of the first rule for stored procedures:

$$\frac{i \vdash \{i \mapsto \_\}\text{eval}\,[i+1]\{i \mapsto 0\} \Rightarrow \{i \mapsto \_\}F_1\{i \mapsto 0\}}{i \vdash \{i \mapsto \_, {}^{\prime}F_1{}^{\prime}\}\text{eval}\,[i+1]\{i \mapsto 0, {}^{\prime}F_1{}^{\prime}\}}$$

which shows a property of the stored procedure F_1. Note that the first rule successfully hides cell $i+1$ in the premise, giving us a simple subgoal to discharge. Similarly, the application of rules for stored procedures form the major steps of the proofs of the remaining triples for C_2 and C_3:

$$\frac{i \vdash \forall x.\{i \mapsto \_, x\}\text{eval}\,[i+1]\{i \mapsto 0, x\} \Rightarrow \{i \mapsto \_, x\}F_2\{i \mapsto 0, x\}}{i \vdash \{i \mapsto \_, {}^{\prime}F_2{}^{\prime}\}\text{eval}\,[i+1]\{i \mapsto 0, {}^{\prime}F_2{}^{\prime}\}}$$

$$\frac{i \vdash \forall x.\{i \mapsto \_, x\}\text{eval}\,[i+1]\{i \mapsto 0, {}^{\prime}\text{skip}{}^{\prime}\} \Rightarrow \{i \mapsto \_, x\}C_3\{i \mapsto 0, {}^{\prime}\text{skip}{}^{\prime}\}}{i \vdash \{i \mapsto \_, {}^{\prime}F_3{}^{\prime}\}C_3\{i \mapsto 0, {}^{\prime}\text{skip}{}^{\prime}\}}$$

Since F_2 directly accesses cell $i+1$, which stores the procedure, and F_3 updates the storing cell, we have used the second rule for C_2 and the third for C_3.   □

3   Semantics of Programs and Assertions

Our interpretation of the programming language is based on a solution of a recursive domain equation, which is defined in the category **Cppo** of directed complete pointed partial orders (in short, cppos) and strict continuous functions.

Let $Nats^+$ be the set of positive natural numbers, ranged over by ℓ and n, and for $n \in Nats^+$, write $[n]$ for the set $\{1, \ldots, n\}$. For a cppo A, we consider a cppo of $Nats^+$-*labelled records* with entries from A (i.e. a labelled smash product of arbitrary finite arity), which will be used to model *heaps*. Its underlying set is $Rec(A) = \left(\sum_{N \subseteq_{fin} Nats^+} (N \to A_\downarrow) \right)_\perp$, where $(N \to A_\downarrow)$ denotes the cpo of maps from the finite address set N to the cpo $A_\downarrow = A - \{\perp\}$ of non-bottom elements of A. For $\perp \neq \iota_N(r) \in Rec(A)$ we write $\mathsf{dom}(r) = N$ and use record notation $\{\!\!| \ell_1 = a_1, \ldots, \ell_n = a_n |\!\!\}$ if $N = \{\ell_1, \ldots, \ell_n\}$ and $r(\ell_i) = a_i$ for all $i \in [n]$. Note that field selection is actually application if the label is in the domain of the record (for our semantic definitions this restricted form of field selection will be sufficient). We shall also write $r[\ell \mapsto a]$ for the record that maps ℓ to a and all other $\ell' \in \mathsf{dom}(r)$ to $r(\ell')$ (assuming $\ell' \in \mathsf{dom}(r)$). In case that r is $\perp$, we define $r[\ell \mapsto a]$ to be $\perp$. The ordering on $Rec(A)$ is given by

$$r \sqsubseteq r' \overset{\text{def}}{\Leftrightarrow} r \neq \perp \Rightarrow (\mathsf{dom}(r) = \mathsf{dom}(r') \wedge \forall \ell \in \mathsf{dom}(r). r(\ell) \sqsubseteq r'(\ell)).$$

The *disjointness predicate* $r \# r'$ on records holds if $r, r' \neq \perp$ and $\mathsf{dom}(r) \cap \mathsf{dom}(r') = \emptyset$, and a continuous *(partial) combining operation* $r \bullet r'$ is defined by $r \bullet r' \overset{\text{def}}{=}$ if $(r \# r')$ then $(r \cup r')$ else (if $(r = \perp \vee r' = \perp)$ then $\perp$ else undefined).

The semantics of the programming language is given by a solution for the following domain equation:

$$Val = Integers_\perp \oplus Com_\perp \qquad Heap = Rec(Val) \qquad Com = Heap \multimap T_{err}(Heap)$$

where $T_{err}(D) = D \oplus \{error\}_\perp$ is the error monad. We usually omit the tags and (for $h \in Heap$) will simply write $h \in T_{err}(Heap)$ and $error \in T_{err}(Heap)$, resp. Recall that a solution $i : F_{Com}(Com, Com) \cong Com$ can be obtained by the usual inverse limit construction, where F_{Com} is the evident locally continuous functor obtained by separating negative and positive occurrences of Com in the right-hand sides of the three equations above.[2] Moreover, such a solution is a *minimal invariant*, in the sense that $id_{Com} = lfp(\lambda e : Com \multimap Com. i \circ F_{Com}(e, e) \circ i^{-1})$ [14]. The soundness proof of the rules for stored procedures exploits this fact.

Interpretation of the programming language Fig. 3 gives the interpretation $[\![C]\!]_\eta$ of commands in $Heap \multimap T_{err}(Heap)$ (which is isomorphic to Com), where $\eta \in Env \overset{\text{def}}{=} (Var \to Val_\downarrow)$ is an environment mapping identifiers to (non-bottom) values in Val. An interpretation function for expressions $[\![e]\!]_\eta^{\mathcal{E}} \in Val_\downarrow$ is assumed, where the only non-standard cases are quoted commands. $[\![`C']\!]_\eta^{\mathcal{E}}$ is defined to be $i([\![C]\!]_\eta)$ (which implicitly makes use of the embedding of Com into Val). In the defining equations in Fig. 3 we assume that $h \neq \perp$, and set $[\![C]\!]_\eta \perp = \perp$

---

[2] Formally, $F_{Com}(X, Y)$ is $Rec(Integers_\perp \oplus X_\perp) \multimap T_{err}(Rec(Integers_\perp \oplus Y_\perp))$.

$$[\![\mathsf{skip}]\!]_\eta\, h \overset{def}{=} h$$

$$[\![C_1;C_2]\!]_\eta\, h \overset{def}{=} \text{if } [\![C_1]\!]_\eta\, h \in \{\bot,\, error\} \text{ then } [\![C_1]\!]_\eta \text{ else } [\![C_2]\!]_\eta\, ([\![C_1]\!]_\eta\, h)$$

$$[\![\text{if } e{=}e' \text{ then } C_1 \text{ else } C_2]\!]_\eta\, h \overset{def}{=} \text{if } \{[\![e_1]\!]_\eta^{\mathcal{E}},\, [\![e_2]\!]_\eta^{\mathcal{E}}\} \subseteq Com \text{ then } \bot$$
$$\text{else if } ([\![e]\!]_\eta^{\mathcal{E}} = [\![e']\!]_\eta^{\mathcal{E}}) \text{ then } [\![C_1]\!]_\eta\, h \text{ else } [\![C_2]\!]_\eta\, h$$

$$[\![\text{let } x{=}\mathsf{new}\, e_1, ..., e_n \text{ in } C]\!]_\eta\, h \overset{def}{=} \text{let } \ell = \min\{\ell \mid \forall \ell'.\, (\ell{\le}\ell'{<}\ell{+}n) \Rightarrow \ell' \notin dom(h)\}$$
$$\text{in } [\![C]\!]_{\eta[x \mapsto \ell]}\, (h \bullet \{\!|\ell{=}[\![e_1]\!]_\eta^{\mathcal{E}},\, \ldots,\, \ell{+}n{-}1{=}[\![e_n]\!]_\eta^{\mathcal{E}}|\!\})$$

$$[\![\mathsf{free}\, e]\!]_\eta\, h \overset{def}{=} \text{if } [\![e]\!]_\eta^{\mathcal{E}} \notin dom(h) \text{ then } error$$
$$\text{else (let } h' \text{ s.t. } h = h' \bullet \{\!|[\![e]\!]_\eta^{\mathcal{E}}{=}h([\![e]\!]_\eta^{\mathcal{E}})|\!\} \text{ in } h')$$

$$[\![[e_1]{:=}e_2]\!]_\eta\, h \overset{def}{=} \text{if } [\![e_1]\!]_\eta^{\mathcal{E}} \notin dom(h) \text{ then } error \text{ else } (h[[\![e_1]\!]_\eta^{\mathcal{E}} \mapsto [\![e_2]\!]_\eta^{\mathcal{E}}])$$

$$[\![\text{let } x{=}[e] \text{ in } C]\!]_\eta\, h \overset{def}{=} \text{if } [\![e]\!]_\eta^{\mathcal{E}} \notin dom(h) \text{ then } error \text{ else } [\![C]\!]_{\eta[x \mapsto h([\![e]\!]_\eta^{\mathcal{E}})]}\, h$$

$$[\![\mathsf{eval}\, [e]]\!]_\eta\, h \overset{def}{=} \text{if } ([\![e]\!]_\eta^{\mathcal{E}} \notin dom(h) \vee h([\![e]\!]_\eta^{\mathcal{E}}) \notin Com) \text{ then } error$$
$$\text{else } i^{-1}(h([\![e]\!]_\eta^{\mathcal{E}}))(h)$$

Fig. 3. Interpretation of commands $[\![C]\!]_\eta \in Heap \multimap T_{err}(Heap)$

$$[\![e_1 \le e_2]\!]_\eta^{\mathcal{A}} \overset{def}{=} \{h \in Heap \mid h \neq \bot \Rightarrow [\![e_i]\!]_\eta^{\mathcal{E}} \in Integers \wedge [\![e_1]\!]_\eta^{\mathcal{E}} \le [\![e_2]\!]_\eta^{\mathcal{E}}\}$$

$$[\![e_1 = e_2]\!]_\eta^{\mathcal{A}} \overset{def}{=} \{h \in Heap \mid h \neq \bot \Rightarrow [\![e_1]\!]_\eta^{\mathcal{E}} = [\![e_2]\!]_\eta^{\mathcal{E}}\}$$

$$[\![\forall x.\, P]\!]_\eta^{\mathcal{A}} \overset{def}{=} \bigcap \{[\![P]\!]_{\eta[x \mapsto v]}^{\mathcal{A}} \mid v \in Val\} \qquad [\![\mathsf{emp}]\!]_\eta^{\mathcal{A}} \overset{def}{=} \{\{\!|\,|\!\},\, \bot\} \qquad [\![P * Q]\!]_\eta^{\mathcal{A}} \overset{def}{=} [\![P]\!]_\eta^{\mathcal{A}} * [\![Q]\!]_\eta^{\mathcal{A}}$$

$$[\![e \mapsto e']\!]_\eta^{\mathcal{A}} \overset{def}{=} \{h \in Heap \mid h \neq \bot \Rightarrow dom(h) = \{[\![e]\!]_\eta^{\mathcal{E}}\} \wedge h([\![e]\!]_\eta^{\mathcal{E}}) = [\![e']\!]_\eta^{\mathcal{E}}\}$$

Fig. 4. Interpretation $[\![P]\!]^{\mathcal{A}} : Env \to \mathcal{P}$ of assertions

for all C and η. Note that the conditional only permits restricted comparison of expressions, so that commands denote continuous functions.

Interpretation of assertions Let $\mathcal{P}$ be the set of predicates $p \subseteq Heap$ that contain $\bot$. The separating conjunction for these predicates, known from separation logic [17], is defined by: $h \in p_1 * p_2 \overset{def}{\Leftrightarrow} \exists h_1, h_2.\ h = h_1 \bullet h_2 \wedge h_1 \in p_1 \wedge h_2 \in p_2$. Note that $p * q \in \mathcal{P}$ whenever $p \in \mathcal{P}$ and $q \in \mathcal{P}$. Clearly '$*$' is associative and commutative, since '$\bullet$' is, and if $p \subseteq p'$ and $q \subseteq q'$ then $p * q \subseteq p' * q'$.

The poset $(\mathcal{P}, \subseteq)$ forms a complete boolean BI algebra.[3] Thus, we get a canonical BI hyperdoctrine $\mathbf{Set}(-, \mathcal{P})$, which soundly models classical (higher-order) predicate BI [1]. In particular this yields an interpretation for the quantifiers. Some cases of this interpretation of assertions are spelled out in Fig. 4.

4   Semantics of Specifications

We now define the interpretation of specifications, and show how it addresses the two key challenges described in the introduction. The most interesting

---

[3] The negation and false of this boolean algebra are slightly unusual, and are defined by $\neg p \overset{def}{=} (Heap - p) \cup \{\bot\}$ and $false \overset{def}{=} \{\bot\}$. Conjunction, disjunction and true are defined as in the usual powerset boolean algebra.

components of our interpretation are semantic Hoare triples, which we will use to interpret (syntactic) Hoare triples. For each predicate $p \in \mathcal{P}$, let $\mathrm{Ad}(p)$ be the admissible, downward closure of p in $T_{err}(Heap)$ (i.e., the smallest admissible, downward-closed subset of $T_{err}(Heap)$ that includes p; it may be obtained as the intersection of all admissible, downward-closed subsets of $Heap$ that include p).

Definition 1 (Semantic triple). *A semantic Hoare triple is a triple of predicates $p, q \in \mathcal{P}$ and function $c \in F_{Com}(Com, Com)$, written $\{p\}c\{q\}$. A semantic triple $\{p\}c\{q\}$ is valid, denoted $\models \{p\}c\{q\}$, if and only if, for all $r \in \mathcal{P}$ and all $h \in Heap$, we have that $h \in p * r \Rightarrow c(h) \in \mathrm{Ad}(q * r)$.*

Intuitively, a semantic triple $\{p\}c\{q\}$ specifies that c should transform an input state in p to an output state in q. Furthermore, the triple says that this transformation should modify only the portion of memory for p (because, otherwise, it would not preserve some invariant r when r was $*$-attached to the precondition p). Note that $\models \{p\}c\{q\}$ ensures the absence of memory errors for inputs in $p * r$ for all r, because $\mathrm{Ad}(q * r)$ cannot contain *error*.

We point out two important aspects of valid semantic Hoare triples and their relationships to the points raised in the introduction. First, the definition of validity includes a universal quantification over $*$-added invariants r. Since we will interpret (syntactic) Hoare triples using the validity of semantic triples, this universal quantification means that Hoare triples in our logic impose a stronger requirement on commands than the ones in standard separation logic. In particular, the requirement is strong enough to imply the frame rule:

Lemma 1 (Frame rule). *If $\models \{p\}c\{q\}$, then $\models \{p * r\}c\{q * r\}$ for all $r \in \mathcal{P}$.*

In this way, our model addresses the first challenge in the introduction regarding the soundness of the frame rule. Second, the definition of the validity takes the admissible, downward closure $\mathrm{Ad}(q * r)$ of post-conditions. As a result, whenever we define a subset of $F_{Com}(Com, Com)$ using a semantic Hoare triple, it is guaranteed that the resulting set is admissible and downward-closed:

Lemma 2. *For all $p, q \in \mathcal{P}$, the subset $\{c \mid \{p\}c\{q\}$ is valid$\}$ is an admissible, downward-closed subset of $F_{Com}(Com, Com)$.*

It is this property that lets us prove the soundness of the proof rules for stored procedures, without requiring any additional conditions, such as a syntactic restriction on assertions [15].

We interpret specifications following the usual Kripke semantics of intuitionistic logic. Our interpretation uses a particular Kripke structure that lets us validate all the higher-order frame rules, i.e., rules for invariant extension $\varphi \otimes P$. Concretely, the Kripke structure is the preorder $(\mathcal{P}, \sqsubseteq)$ where the relation $\sqsubseteq$ is defined by: $p \sqsubseteq q \overset{def}{\Leftrightarrow} \exists r \in \mathcal{P}. \ p * r = q$. Each world p in this Kripke structure should be thought of as an invariant to be added by (higher-order) frame rules, and the preorder $p \sqsubseteq q$ denotes that q is obtained by extending p with some disjoint invariant r. This Kripke structure has been studied in [3], and we will use the results from that paper.

$$\eta, p \models \varphi \wedge \psi \stackrel{def}{\Leftrightarrow} \eta, p \models \varphi \text{ and } \eta, p \models \psi$$

$$\eta, p \models \varphi \Rightarrow \psi \stackrel{def}{\Leftrightarrow} \text{ for all } r \in \mathcal{P}, \text{ if } p \sqsubseteq r \text{ and } \eta, r \models \varphi, \text{ then } \eta, r \models \psi$$

$$\eta, p \models \varphi \otimes P \stackrel{def}{\Leftrightarrow} \eta, p * \llbracket P \rrbracket_\eta^{\mathcal{A}} \models \varphi$$

$$\eta, p \models \{P\}C\{Q\} \stackrel{def}{\Leftrightarrow} \models \{\llbracket P \rrbracket_\eta^{\mathcal{A}} * p\} \llbracket C \rrbracket_\eta \{\llbracket Q \rrbracket_\eta^{\mathcal{A}} * p\}$$

Fig. 5. Interpretation $\eta, p \models \varphi$ of specifications

Some cases of the definition of the satisfaction relation $\models$ are shown in Fig. 5. Note that Hoare triples are interpreted using the validity of semantic triples.

Soundness We recall one consequence of our semantics, which is discussed in more detail in [3]. It is the soundness of the generalized frame rule: $\varphi \Rightarrow \varphi \otimes P$. Since the interpretation follows the standard Kripke semantics, every formula φ satisfies the usual Kripke monotonicity: $\forall \eta, r, q. (\eta, r \models \varphi) \wedge (r \sqsubseteq q) \Rightarrow (\eta, q \models \varphi)$. Since $r \sqsubseteq q$ just means that $q = r*p$ for some p, the above monotonicity condition is equivalent to $\forall \eta, r, p. (\eta, r \models \varphi) \Rightarrow (\eta, r * p \models \varphi)$. This just means that adding an invariant p for each specification maintains the truth of a specification, and explains why the generalized frame rule is sound in our semantics.

Lemma 3 (Invariants, [3]). *All the axioms for invariant extensions are sound.*

Our semantics validates all the proof rules for specifications. In the following, we focus on the second rule for stored procedures.

Lemma 4 (Recursion). *The second rule for stored procedures is sound.*

Proof. For each $\eta \in \llbracket \Gamma \rrbracket$ and $r \in \mathcal{P}$, define a predicate $A_{\eta,r}$ on $Com \times Com$ by

$$A_{\eta,r}(c,d) \stackrel{def}{\Leftrightarrow} \forall \vec{v} \in Val^n. \models \{\llbracket P * e \mapsto x \rrbracket_{\eta_1}^{\mathcal{A}} * r\} i^{-1}(d) \{\llbracket Q * e \mapsto x \rrbracket_{\eta_1}^{\mathcal{A}} * r\}$$

where $\eta_1 = \eta[\vec{y} \mapsto \vec{v}, x \mapsto c]$. Pick any $\eta \in \llbracket \Gamma \rrbracket$ and $r \in \mathcal{P}$. By the definition of $\llbracket eval [e] \rrbracket$ and the usual substitution lemma (which holds for our interpretation), the soundness of the rule boils down to proving the following implication.

$$\left(\forall c \in Com. \forall r' \sqsupseteq r. \ A_{\eta,r'}(c,c) \Rightarrow A_{\eta,r'}(c, \llbracket 'C' \rrbracket_\eta) \right) \Rightarrow A_{\eta,r}(\llbracket 'C' \rrbracket_\eta, \llbracket 'C' \rrbracket_\eta).$$

Suppose that there is a predicate $S_{\eta',r'}$ on Com parameterized by (η', r'), such that (1) $S_{\eta',r'}(c) \Leftrightarrow (\forall d \in Com. \ S_{\eta',r'}(d) \Rightarrow A_{\eta',r'}(d,c))$. Then, we have that (2) $\forall c. S_{\eta,r}(c) \Rightarrow A_{\eta,r}(c,c)$. Hence, assuming the precondition $\forall c. \forall r' \sqsupseteq r. A_{\eta,r'}(c,c) \Rightarrow A_{\eta,r'}(c, \llbracket 'C' \rrbracket_\eta)$, we obtain $\forall c. \ S_{\eta,r}(c) \Rightarrow A_{\eta,r}(c, \llbracket 'C' \rrbracket_\eta)$ and therefore $S_{\eta,r}(\llbracket 'C' \rrbracket_\eta)$ by (1). But then (2) shows $A_{\eta,r}(\llbracket 'C' \rrbracket_\eta, \llbracket 'C' \rrbracket_\eta)$, as required. It remains to establish the existence of a predicate $S_{\eta',r'}$ satisfying (1). This is done in the following Lemma 5. $\square$

Lemma 5 (Existence). *For all η, r, there exists $S_{\eta,r} \subseteq Com$ such that $S_{\eta,r}(c)$ holds iff $\forall d. \ S_{\eta,r}(d) \Rightarrow A_{\eta,r}(d,c)$, where $A_{\eta,r}$ is as in the proof of Lemma 4.*

Proof. The proof builds on the same technique as used in [16], but many details have changed. Let $\mathcal{C}$ denote the set of admissible subsets of Com, which forms a complete lattice when ordered by $\subseteq$. Pick η and $r \in \mathcal{P}$. We define an operation $\Phi \colon \mathcal{C}^{op} \to \mathcal{C}$, by $S \;\mapsto\; \{c \in Com \mid \forall d.\, d \in S \Rightarrow A_{\eta,r}(d,c)\}$. That $\Phi(S)$ is admissible follows from the admissibility of $A_{\eta,r}(d,-)$, which itself comes from Lemma 2. The symmetrisation $\Phi^{\S}(S,T) \overset{def}{=} \langle \Phi(T), \Phi(S)\rangle$ of Φ is a monotonic map on the complete lattice $\mathcal{C}^{op} \times \mathcal{C}$ and thus has a least (pre-) fixed point (S^-, S^+), by Tarski's fixed point theorem. Then (S^+, S^-) is also a fixed point of $\Phi^{\S}$, so one obtains $S^+ \subseteq S^-$. A predicate $S_{\eta,r} \in \mathcal{C}$ with the required property $S_{\eta,r} = \Phi(S_{\eta,r})$ is obtained by proving the opposite inclusion.

To this end, for $l \sqsubseteq id_{Com}$ and $S_1, S_2 \in \mathcal{C}$, define $l : S_1 \subset S_2$ to mean that $\forall c \in S_1.\, l(c) \in S_2$. Note that from

$$(1) \qquad l : S_1 \subset S_2 \;\Rightarrow\; (i \circ F_{Com}(l,l) \circ i^{-1}) : \Phi(S_2) \subset \Phi(S_1)$$

for all $l \sqsubseteq id_{Com}$, it follows by fixed point induction that $lfp(\lambda l.\, i \circ F_{Com}(l,l) \circ i^{-1}) : S^- \subset S^+$. This is equivalent to $id_{Com} : S^- \subset S^+$, i.e., $S^- \subseteq S^+$, because $lfp(...)$ is id_{Com} by the minimal invariant property of Com.

It remains to prove (1). For this, one needs only prove the following two properties. Let $\text{Cl}^{\downarrow}(p)$ be the downward closure of a predicate p. For all environments η', heaps h and functions l with $l \sqsubseteq id_{Com}$, if $j \overset{def}{=} Rec(\hat{l})$,

(a) $h \in [\![P * e \mapsto x]\!]^{\mathcal{A}}_{\eta'}$ implies $j(h) \in \text{Cl}^{\downarrow} [\![P * e \mapsto x]\!]^{\mathcal{A}}_{\eta'[x \mapsto l(\eta'(x))]}$,
(b) $h \in [\![Q * e \mapsto x]\!]^{\mathcal{A}}_{\eta'[x \mapsto l(\eta'(x))]}$ implies $j(h) \in \text{Cl}^{\downarrow} [\![Q * e \mapsto x]\!]^{\mathcal{A}}_{\eta'}$.

To see why it is suffices to prove (a) and (b), suppose $l \sqsubseteq id_{Com}$ satisfies $l : S_1 \subset S_2$. Pick $c \in \Phi(S_2)$. We have to show $(i \circ F_{Com}(l,l) \circ i^{-1})(c) \in \Phi(S_1)$. Thus, for all $d \in S_1$, we must show that $A_{\eta,r}(d, (i \circ F_{Com}(l,l) \circ i^{-1})(c))$ holds, i.e., for all $\vec{v} \in Val^n$

$$(2) \quad \models \{ [\![P * e \mapsto x]\!]^{\mathcal{A}}_{\eta[\vec{y} \mapsto \vec{v}, x \mapsto d]} * r \} F_{Com}(l,l)(i^{-1}(c)) \{ [\![Q * e \mapsto x]\!]^{\mathcal{A}}_{\eta[\vec{y} \mapsto \vec{v}, x \mapsto d]} * r \}.$$

For this, pick $d \in S_1$ and $\vec{v} \in Val^n$. Since $l : S_1 \subset S_2$, we have that $l(d) \in S_2$, and since $c \in \Phi(S_2)$, it must be the case that

$$(3) \qquad \models \{ [\![P * e \mapsto x]\!]^{\mathcal{A}}_{\eta[\vec{y} \mapsto \vec{v}, x \mapsto l(d)]} * r \} i^{-1}(c) \{ [\![Q * e \mapsto x]\!]^{\mathcal{A}}_{\eta[\vec{y} \mapsto \vec{v}, x \mapsto l(d)]} * r \}.$$

We will now prove that (3) implies (2).

To simplify notation, we assume without loss of generality that η is such that $\eta(\vec{y}) = \vec{v}$. Pick $r' \in \mathcal{P}$ and $h \in [\![P * e \mapsto x]\!]^{\mathcal{A}}_{\eta[x \mapsto d]} * r * r'$. Let j be $Rec(\hat{l})$. Then, we have to show the set membership below:

$$F_{Com}(l,l)(i^{-1}(c))(h) = T_{err}(j)(i^{-1}(c)(j(h))) \in \text{Ad}([\![Q * e \mapsto x]\!]^{\mathcal{A}}_{\eta[x \mapsto d]} * r * r').$$

By property (a) and definition of j, $j(h)$ is in $\text{Cl}^{\downarrow}([\![P * e \mapsto x]\!]^{\mathcal{A}}_{\eta[x \mapsto l(d)]} * r * r')$. So, we have (4) $i^{-1}(c)(j(h)) \in \text{Ad}([\![Q * e \mapsto x]\!]^{\mathcal{A}}_{\eta[x \mapsto l(d)]} * r * r')$, because of (3) and the monotonicity of $i^{-1}(c)$. Note that by the property (b) and the definition of j, $T_{err}(j)$ should map heaps in $([\![Q * e \mapsto x]\!]^{\mathcal{A}}_{\eta[x \mapsto l(d)]} * r * r')$ to those in $\mathrm{Ad}([\![Q * e \mapsto x]\!]^{\mathcal{A}}_{\eta[x \mapsto d]} * r * r')$. Furthermore, for all continuous functions f on $T_{err}(Heap)$, if f maps every heap in a predicate p into $\mathrm{Ad}(q)$, it also maps all heaps in $\mathrm{Ad}(p)$ into $\mathrm{Ad}(q)$. Thus, since $T_{err}(j)$ is continuous, it maps heaps in $\mathrm{Ad}([\![Q * e \mapsto x]\!]^{\mathcal{A}}_{\eta[x \mapsto l(d)]} * r * r')$ into $\mathrm{Ad}([\![Q * e \mapsto x]\!]^{\mathcal{A}}_{\eta[x \mapsto d]} * r * r')$. By (4), this means that $T_{err}(j)(i^{-1}(c)(j(h)))$ belongs to $\mathrm{Ad}([\![Q * e \mapsto x]\!]^{\mathcal{A}}_{\eta[x \mapsto d]} * r * r')$.   □

5   Conclusion and Future Work

We have developed a simple model of separation logic for a language with higher-order store. The model validates proof rules for recursion through the store and a wide range of higher-order frame rules. Future work includes extending the model to richer programming languages, in particular to languages with higher-order functions. In order to obtain modularity it is also necessary to develop a version of the logic where assertions do not contain code explicitly but rather abstract specifications of its behaviour. We are confident that the simplicity of the present model will make that possible. In future work we also plan to extend the relationally parametric model of separation logic in [3] to higher-order store.

References

1. Biering, B., Birkedal, L., Torp-Smith, N.: BI-hyperdoctrines, higher-order separation logic, and abstraction. ACM TOPLAS 29(5) (2007)
2. Birkedal, L., Torp-Smith, N., Yang, H.: Semantics of separation-logic typing and higher-order frame rules for algol-like languages. LMCS 2(5-1) (2006)
3. Birkedal, L., Yang, H.: Relational parametricity and separation logic. In: Seidl, H. (ed.) FOSSACS 2007. LNCS, vol. 4423. Springer, Heidelberg (2007)
4. Cai, H., Shao, Z., Vaynberg, A.: Certified self-modifying code. In: Proc. PLDI 2007, pp. 66–77 (2007)
5. Corbet, J., Rubini, A., Kroah-Hartman, G.: Linux Device Drivers, 3rd edn. O'Reilly, Sebastopol (2005)
6. Honda, K., Yoshida, N., Berger, M.: An observationally complete program logic for imperative higher-order functions. In: Proc. LICS 2005, pp. 270–279 (2005)
7. Krishnaswami, N., Aldrich, J., Birkedal, L.: Modular verification of the subject-observer pattern via higher-order separation logic. In: FTfJP 2007 (2007)
8. Nanevski, A., Ahmed, A., Morrisett, G., Birkedal, L.: Abstract predicates and mutable ADTs in Hoare type theory. In: De Nicola, R. (ed.) ESOP 2007. LNCS, vol. 4421, pp. 189–204. Springer, Heidelberg (2007)
9. Nanevski, A., Morrisett, G., Birkedal, L.: Polymorphism and separation in Hoare type theory. In: Proc. ICFP 2006, pp. 62–73 (2006)
10. O'Hearn, P.W., Reynolds, J.C., Yang, H.: Local reasoning about programs that alter data structures. In: Fribourg, L. (ed.) CSL 2001 and EACSL 2001. LNCS, vol. 2142, pp. 1–18. Springer, Heidelberg (2001)
11. O'Hearn, P.W., Yang, H., Reynolds, J.C.: Separation and information hiding. In: Proc. of 31st POPL, pp. 268–280 (2004)

12. Parkinson, M.: When separation logic met Java. In: FTfJP 2006 (2006)
13. Parkinson, M., Bierman, G.: Separation logic, abstraction and inheritance. In: Proc. 35th POPL (2008)
14. Pitts, A.M.: Relational properties of domains. Information and Computation 127, 66–90 (1996)
15. Reus, B., Schwinghammer, J.: Separation logic for higher-order store. In: Ésik, Z. (ed.) CSL 2006. LNCS, vol. 4207, pp. 575–590. Springer, Heidelberg (2006)
16. Reus, B., Streicher, T.: About Hoare logics for higher-order store. In: Caires, L., Italiano, G.F., Monteiro, L., Palamidessi, C., Yung, M. (eds.) ICALP 2005. LNCS, vol. 3580, pp. 1337–1348. Springer, Heidelberg (2005)
17. Reynolds, J.C.: Separation logic: A logic for shared mutable data structures. In: Proc. LICS 2002, pp. 55–74 (2002)

Open Implication*

Karin Greimel[1], Roderick Bloem[1], Barbara Jobstmann[2],
and Moshe Vardi[3]

[1]Graz University of Technology
[2]EPFL
[3]Rice University

Abstract. We argue that the usual trace-based notions of implication
and equivalence for linear temporal logics are too strong and should be
complemented by the weaker notions of open implication and open equivalence. Although open implication is harder to compute, it can be used to
advantage both in model checking and in synthesis. We study the difference between trace-based equivalence and open equivalence and describe
an algorithm to compute open implication of Linear Temporal Logic formulas with an asymptotically optimal complexity. We also show how to
compute open implication while avoiding Safra's construction. We have
implemented an open-implication solver for Generalized Reactivity(1)
specifications. In a case study, we show that open equivalence can be
used to justify the use of an alternative specification that allows us to
synthesize much smaller systems in far less time.

1 Introduction

A recent verification project at STMicroelectronics [17] considered an arbiter
that receives requests and provides acknowledgments. Two of the requirements
for the design read: (R_1) From some time on, the difference between the total
number of requests and the total number of acknowledgments is zero, and (R_2)
the total number of acknowledgments never exceeds the total number of requests.
Requirement R_1 does not imply R_2: a trace that contains an acknowledgment
followed by a request with no further acknowledgments or requests thereafter
fulfills R_1 but not R_2. Nevertheless, because one can not predict the number of
requests that will come, the only way to implement R_1 is to always wait for a
request before sending an acknowledge. Thus, any implementation that fulfills
R_1 also fulfills R_2. We say that R_1 *open-implies* R_2. Thus, it suffices to make
sure that R_1 holds; R_2 follows. Likewise, we say that two specifications are *open
equivalent* if they are fulfilled by the same implementations.

Traditionally, for linear specification formalisms such as Linear Temporal
Logic (LTL) [20] or Büchi automata [5], only trace implication and trace equivalence have been studied. Intuitively, trace implication and trace equivalence

---

* This work was supported by EU grant 217069 (COCONUT), the Swiss National
 Science Foundation (Indo-Swiss Research Program and NCCR MICS), NSF grants
 CCF-0613889, ANI-0216467, and CCF-0728882, BSF grant 9800096, and a gift from
 Intel. This paper is based on the MS thesis of the first author [8].

L. Aceto et al. (Eds.): ICALP 2008, Part II, LNCS 5126, pp. 361–372, 2008.

are defined with respect to all systems. In contrast, open implication and open equivalence are defined with respect to open systems only. In open systems we distinguish between inputs and outputs and we require that the system be *receptive* to all inputs [9], the intuition being that the system cannot block the actions of the environments.

The notions of open implication and open equivalence have not been studied in the literature. We argue here that these are important notions. When model checking open systems, a specification can always be substituted by an open-equivalent one: it is fulfilled by the same open systems. Likewise, for automatic synthesis of open systems from specifications [21], one may replace the specification by any realizable specification that open-implies it. The stronger specification may be easier to synthesize. Consider for instance, a simplified specification of an arbiter with input r for request and output a for acknowledgement. The specification reads $\varphi = (\mathsf{G}\,\mathsf{F}\,r) \rightarrow \mathsf{G}(a \rightarrow \mathsf{X}(\neg a\,\mathsf{U}\,r))$. Now consider $\varphi' = \mathsf{G}(a \rightarrow \mathsf{X}(\neg a\,\mathsf{W}\,r))$. We have that φ and φ' are open equivalent but not trace equivalent. Moreover, the language of φ' can be represented by a weak automaton and is thus both easier to model check [4,15] and (much) easier to synthesize [10,16].

In this paper, we show that the inability to predict the future is the underlying cause for the difference between open implication and trace implication. Then, we consider the problem of deciding whether φ open-implies ψ for LTL formulas φ and ψ. We provide an algorithm that runs in 2EXPTIME in $|\varphi|$ and PSPACE in $|\psi|$, matching the lower bounds. This algorithm uses Safra's intricate determinization construction. We complement this with an algorithm that avoids Safra's construction, is much easier to implement, and may be far more efficient when the specifications are not equivalent. Additionally, we consider *Generalized Reactivity(1)* formulas [19]. Although less expressive than LTL, such formulas suffice to conveniently describe most properties that occur in practice. Efficient synthesis tools for this subset have been used on realistic examples [2,3]. We present an implementation of open implication based on this approach and show that it can be used to significantly simplify the synthesis of an arbiter for an industrial bus.

2   Preliminaries

We consider systems with input signals I and output signals O. We define $AP = I \cup O$, which is the set of atomic propositions in the logic specifications defined below. Our input alphabet is thus $D = 2^I$, the output alphabet is $\Sigma = 2^O$, and we define $\mathcal{A} = 2^{AP}$.

Transducers and Trees. We use transducers to represent open systems. A *(possibly infinite) transducer* with inputs D and outputs Σ is a tuple $T = (Q, q_0, \delta, \lambda)$, where Q is the (possibly infinite) state space, $q_0 \in Q$ is the initial state, $\delta : Q \times D \rightarrow Q$ is the transition function, and $\lambda : Q \rightarrow \Sigma$ is the output function. In each state, the transducer outputs a letter in Σ, then reads a letters in D, and moves to the next state. Transducers correspond to Moore machines. A transducer is *finite* if Q is finite. The *run* of T on a sequence $d = d_0 d_1 \cdots \in D^\omega$ is a sequence $\rho_0 \rho_1 \cdots \in Q^\omega$, where $\rho_0 = q_0$ and $\rho_{i+1} = \delta(\rho_i, d_i)$. The corresponding *word* is $\lambda(\rho) = w_0 w_1 \cdots \in \mathcal{A}^\omega$ such that $w_i = \lambda(\rho_i) \cup d_i$. The set $L(T)$ denotes the words corresponding to some run of T and is called the *language of* T.

We use *trees* to represent transducers and runs of alternating automata (below). A Σ-labeled D-tree is a tuple (τ, λ), where τ is the set of nodes, a prefix closed subset of D^*, and $\lambda : \tau \to \Sigma$ is the labeling function. If $\tau = D^*$, τ is *complete*. The node ε is the *root* of the tree and a node $t \cdot d$ is a *successor* of t. A *path* π in τ is a maximal sequence of nodes $t_0 t_1 \ldots$ such that $t_0 = \varepsilon$ and there are $d_0 d_1 \ldots$ such that $t_{i+1} = t_i \cdot d_i$. Paths can be finite or infinite. We assign to each path $\pi = t_0 t_1 \ldots$ a word $\lambda(\pi) = w_0 w_1 \ldots$ such that $w_i = \lambda(t_i) \cup d_i$ for all $i \geq 0$.

The *unrolling* of a transducer $T = (Q, q_0, \delta, \lambda)$ is a complete Σ-labeled D-tree (τ, λ), such that each run ρ of T is mapped to an infinite path π in (τ, λ) with $\lambda(\rho) = \lambda(\pi)$. A tree is *regular* if it is the unrolling of some finite transducer. We denote the set of all regular Σ-labeled trees with directions D by $\mathcal{T}$.

Temporal Logics. We write specifications in *Linear Temporal Logic (LTL)* [20]. The syntax of LTL is defined in negation normal form as $\varphi ::= \text{true} \mid \text{false} \mid p \mid \neg p \mid \varphi \vee \varphi \mid \varphi \wedge \varphi \mid \mathsf{X}\varphi \mid \varphi \mathsf{U} \varphi \mid \varphi \mathsf{R} \varphi$ with $p \in AP$. We use the usual semantics of LTL for words in $\mathcal{A}^\omega$. The set of words that satisfies φ is denoted by $L(\varphi) \subseteq \mathcal{A}^\omega$. A Σ-labeled D-tree t satisfies φ if for all paths π of t, $\pi \models \varphi$. A transducer T satisfies φ ($T \models \varphi$) if its unrolling does. A formula φ is *satisfiable* if $L(\varphi) \neq \emptyset$, it is *tautologous* if $L(\varphi) = \mathcal{A}^\omega$ and it is *realizable* if there is a tree t such that $t \models \varphi$.

Automata. Let $\mathbb{B}^+(X)$ denote the set of Boolean formulas without negations over X. We say that a set $C \subseteq 2^X$ *satisfies* $\varphi \in \mathbb{B}^+(X)$ (written $C \models \varphi$) if φ evaluates to true after replacing all occurrences of $c \in C$ ($c \notin C$) in φ by true (false, resp.). Set C is *minimal* if forall $c \in C$, $(C \setminus \{c\}) \not\models \varphi$.

An *alternating parity tree automaton* for Σ-labeled D-trees is a tuple $A = (Q, q_0, \delta, F)$, where Q is a finite set of states, $q_0 \in Q$ is the initial state, $\delta : Q \times \Sigma \to \mathbb{B}^+(Q \times D)$ is the transition relation and the acceptance condition $F = (F_1, \ldots, F_k)$ is a partition of Q, where k is the *index* of A. We use A^q to denote the automaton A with initial state q.

We say that an alternating tree automaton is *nondeterministic* if it does not force multiple copies to one child. That is, for all $q \in Q$ and $\sigma \in \Sigma$, if $C \models \delta(q, \sigma)$ and C is minimal then for all $(q, d) \in C$ and $(q', d') \in C$, if $d = d'$ then $q = q'$. The automaton is *universal* if all formulas are conjunctions and it is *deterministic* if it is both nondeterministic and universal. For deterministic automata we can assume, without loss of generality, that the transition relation is of the form $\delta : Q \times \Sigma \times D \to Q$. An automaton is a *co-Büchi* automaton if $k = 2$ and a *Büchi* automaton if $k = 3$ and $F_1 = \emptyset$. Tree automata run on trees with directions D. If $|D| = 1$, we say the automaton runs on *words* (over Σ) and omit D.

A *run* of an alternating tree automaton A on a tree (τ_I, λ_I) is a tree $T_\rho = (\tau_\rho, \lambda_\rho)$ with $\tau_\rho \subseteq \mathbb{N}^*$ and $\lambda_\rho : \tau_\rho \to (Q \times \tau_I)$ for which (1) $\lambda_\rho(\varepsilon) = (q_0, \varepsilon)$

and (2) If t_ρ is a node of T_ρ with label (q, t_I) and the children of t_ρ are labeled $(q_1, t_1), \ldots, (q_n, t_n)$, then for all $i \in \{1, \ldots, n\}$ there is a $d_i \in D$ such that $t_i = t_I \cdot d_i$ and $\{(q_1, d_1), \ldots, (q_n, d_n)\} \models \delta(q, \lambda_I(t_I))$. (Not all directions must appear in $\{(q_1, d_1), \ldots, (q_n, d_n)\}$.) Let $\pi = t_0 t_1 \ldots$ be an infinite path in $(\tau_\rho, \lambda_\rho)$, then $\inf(\pi) = \{q \in Q \mid$ there exist infinitely many nodes $t \in \pi$ with $\lambda_\rho(t) = (q, t_I)$ for some $t_I\}$. A path is *accepting* if the minimal $i \in \{1, \ldots, k\}$ for which $\inf(\pi) \cap F_i \neq \emptyset$ is even. A run is accepting if all infinite paths are accepting. An automaton accepts an input tree (τ_I, λ_I), if there exists an accepting run on (τ_I, λ_I). We call the set of trees accepted by A the *language of A* and denote it by $L(A)$.

We use three letter acronyms for automata, where the first denotes the branching mode of the automaton (nondeterministic, universal, deterministic, or alternating), the second describes the acceptance condition (parity, Büchi or co-Büchi), and the third letter indicates the input elements (words or trees). For instance, a UPT is a universal parity tree automaton.

3 Open Implication

3.1 Definitions, Characteristics, and Lower Bounds

Definitions. Let us first recall the standard notions of implication and equivalence between two LTL formulas and then define open implication and open equivalence.

Definition 1. *Given two LTL formulas φ and ψ, φ trace-implies ψ if $L(\varphi) \subseteq L(\psi)$. Formula φ is trace equivalent to ψ if $L(\varphi) = L(\psi)$.*

Definition 2. *Given two LTL formulas φ and ψ, φ open-implies ψ, denoted by $\varphi \multimap \psi$, if for all (infinite) transducers T we have that $T \models \varphi$ implies $T \models \psi$. Likewise, $\varphi \multimap\!\!\!\multimap \psi$ (φ is open equivalent to ψ) if $\varphi \multimap \psi$ and $\psi \multimap \varphi$.*

Theorem 1. *If for all finite transducers T, $T \models \varphi$ implies $T \models \psi$, then $\varphi \multimap \psi$.*

Proof. We prove the converse. If $\varphi \not\multimap \psi$, then there is a (possibly infinite) transducer T such that $T \models \varphi$, but $T \not\models \psi$. The unrolling of T is accepted by the deterministic Streett automaton A that accepts all trees (of the proper arity) satisfying the CTL* formula $\chi = \mathsf{A}\varphi \wedge \neg\mathsf{A}\psi$ [7]. Since the language of A is not empty, there exists a finite transducer generating a tree accepted by A [22]. Thus, there exists a finite transducer T' such that $T' \models \varphi$, but $T' \not\models \psi$.

Without loss of generality, we refer to finite transducers in the remainder of the paper.

Open versus Trace Equivalence. If two specifications are open equivalent but not trace equivalent, then the traces in which the specifications differ cannot be produced by a transducer because they require knowledge of the future.

Rosner [23] distinguishes two reasons for unrealizability. First, a specification may be unrealizable because there is an infinite input word that cannot be paired with an output word. The second reason is that some specifications require clairvoyance. For instance, for the specification $a \leftrightarrow \mathsf{X}r$, where a is an output and r is an input, there exists a valid output word for every input word. Lack of knowledge of the future input prevents an implementation. (See also [29].)

Formally, given a specification φ, we call $w \in \mathcal{A}^\omega$ φ-clairvoyant if $w \models \varphi$ but for some prefix $w' \cdot (i \cup o)$ there is no transducer T that outputs o in the initial state, such that for all words v of T, $w' \cdot v \models \varphi$. That is, the word cannot be used in a transducer because after some point, there is no correct reaction to all future inputs. Note that only clairvoyant words satisfy $a \leftrightarrow \mathsf{X}r$. A word that is not φ-clairvoyant is called φ-secure.

If two realizable specifications φ and ψ are open equivalent, then the set of φ-secure words and the set of ψ-secure words are equal.

Theorem 2. *We have $\varphi \multimap \psi$ iff $L(\varphi) \setminus L(\psi)$ consists of φ-clairvoyant words.*

Proof. The key insight is that w is φ-secure iff there is a transducer T such that $T \models \varphi$ and $w \in L(T)$.

Let $w \in L(\varphi) \setminus L(\psi)$. If w is φ-secure then there is a transducer that satisfies φ, contains w, and thus does not satisfy ψ, so $\varphi \not\multimap \psi$. Vice-versa, suppose that $\varphi \not\multimap \psi$. Then there is a transducer T that satisfies φ and not ψ. This transducer contains a word w that satisfies φ and not ψ and this word is φ-secure.

Extending our notation to ω-regular languages, we have that for every ω-regular language L there is an ω-regular language L' that consists of the L-secure words in L. Language L' is the unique minimal representative of the open-equivalence class of L and precisely characterizes all transducers that satisfy L. The language can be constructed from a DPW A with language L by removing all edges $(q, o \cup i, q')$ such that there is an i' with $(q, o \cup i', q'') \in \delta$ and $L(A^{q''})$ is not realizable.

Lower Bounds. An obvious solution to deciding open implication is to apply the approach suggested in the proof of Theorem 1 by checking nonemptiness of the tree automaton for the formula χ. The problem with this algorithm is that it is doubly exponential in both $|\varphi|$ and $|\psi|$. After discussing the lower-bound complexity of deciding open equivalence and open implication between two LTL formulas, we describe an asymptotically-optimal algorithm.

Theorem 3. *Let φ and ψ be two LTL formulas. (1) Deciding whether $\varphi \multimap \psi$ is 2EXPTIME-hard, so is deciding whether $\varphi \leftrightarrow\!\!\!\multimap \psi$ and (2) Deciding whether $\varphi \multimap \psi$ is 2EXPTIME-hard for a fixed ψ and PSPACE-hard for a fixed φ.*

Proof. We have that φ is unrealizable iff $\varphi \multimap$ false iff $\varphi \leftrightarrow\!\!\!\multimap$ false and LTL-realizability is 2EXPTIME-complete [21]. This proves 2EXPTIME-hardness.

We prove that ψ is tautologous iff true $\multimap \psi$. Because deciding validity of LTL formulas is PSPACE-complete [25], this proves that open implication is PSPACE-hard in ψ. The forward direction is trivial. For the other direction, assume that ψ is not tautologous, then $\exists w \in \mathcal{A}^\omega : w \not\models \psi$. Since we can choose w as a finite prefix followed by a finite cycle [27], we can construct a transducer T such that w is a word of T. We have that $T \models$ true but $T \not\models \psi$, so true $\not\multimap \psi$.

3.2   Algorithm and Upper Bounds

We show an algorithm to decide whether $\varphi \looparrowright \psi$ that runs in time doubly exponential in φ and in space polynomial in ψ. We first describe an algorithm that is exponential in ψ, and then show how to obtain optimal space complexity. In the following, we fix $n = |\varphi|$ and $m = |\psi|$.

The algorithm proceeds as follows:

1. Construct a DPT $A_{\mathrm{DPT}} = (Q_{\mathrm{DPT}}, q_{0\mathrm{DPT}}, \delta_{\mathrm{DPT}}, F_{\mathrm{DPT}})$ such that $L(A_{\mathrm{DPT}}) = \{t \in \mathcal{T} \mid t \models \varphi\}$ with at most $2^{n2^{2n+2}+4n}$ states and index $i_{\mathrm{DPT}} = 2^{2n+1}$ [18,27].

2. Compute the set $W_\varphi = \{q \in Q_{\mathrm{DPT}} \mid L(A_{\mathrm{DPT}}{}^q) \neq \emptyset\}$ in doubly exponential time in n [7].

3. Construct a DPW $A_{\mathrm{DPW}} = (Q_{\mathrm{DPW}}, q_{0\mathrm{DPW}}, \delta_{\mathrm{DPW}}, F_{\mathrm{DPW}})$ over AP with $|Q_{\mathrm{DPT}}|$ states and index i_{DPT} such that $\sigma \in L(A_{\mathrm{DPW}})$ iff $\sigma = \lambda(\pi)$ for some path π of a tree $t \in L(A_{\mathrm{DPT}})$ (see below).

4. Construct a NBW $A_{\mathrm{NBW}} = (Q_{\mathrm{NBW}}, q_{0\mathrm{NBW}}, \delta_{\mathrm{NBW}}, F_{\mathrm{NBW}})$ with at most $2|Q_{\mathrm{DPT}}|i_{\mathrm{DPT}}$ states, such that $L(A_{\mathrm{NBW}}) = L(A_{\mathrm{DPW}})$ [12].

5. Construct an NBW B_{NBW} with at most 2^{2m} states that accepts all words in $L(\neg\psi)$ [27].

6. Check if $L(A_{\mathrm{NBW}}) \cap L(B_{\mathrm{NBW}}) = \emptyset$ in time linear in the size of A_{NBW} and B_{NBW} [27].

The DPW $A_{\mathrm{DPW}} = (Q_{\mathrm{DPW}}, q_{0\mathrm{DPT}}, \delta_{\mathrm{DPW}}, F_{\mathrm{DPW}})$ is constructed as follows. We have $Q_{\mathrm{DPW}} = W_\varphi$, $\delta_{\mathrm{DPW}}(q, o \cup i) = \delta_{\mathrm{DPT}}(q, o, i)$ if $\forall j \in I : \delta_{\mathrm{DPT}}(q, o, j) \in W_\varphi$, and F_{DPW} equals F_{DPT} restricted to states in Q_{DPW}.

Lemma 1. $\sigma \in L(A_{\mathrm{DPW}})$ iff $\exists t \in L(A_{\mathrm{DPT}})$ with a path π such that $\lambda(\pi) = \sigma$.

Proof. Let $\sigma = \sigma_0\sigma_1 \cdots \in L(A_{\mathrm{DPW}})$ and suppose that $\sigma_j = i_j \cup o_j$ with $i_j \in D$ and $o_j \in \Sigma$. Then $\forall i \in D$, the run of the DPT for $\sigma_0\sigma_1 \ldots \sigma_{j-1}(o_j \cup i)$ ends in a state in W_φ, whence we can extend the path to an accepted tree that includes the word σ. Vice versa, if there exists a tree $t \in L(A_{\mathrm{DPT}})$ with a path π such that $\lambda(\pi) = \sigma$ then, by construction, σ is accepted by A_{DPW}.

Theorem 4. $\varphi \not\looparrowright \psi$ iff $L(A_{\mathrm{NBW}}) \cap L(B_{\mathrm{NBW}}) \neq \emptyset$.

Proof. If $\varphi \not\looparrowright \psi$ then there is a transducer T such that $T \models \varphi$ and $T \not\models \psi$, so for some path $\pi \in t$ where t is the unrolling of T, we have $\lambda(\pi) \not\models \psi$ and $\lambda(\pi) \models \varphi$. Thus, $\lambda(\pi)$ is in $L(A_{\mathrm{NBW}}) \cap L(B_{\mathrm{NBW}})$. Similarly, if there is a word σ in $L(A_{\mathrm{NBW}}) \cap L(B_{\mathrm{NBW}})$ then there is a regular tree $t \in \mathcal{T}$ satisfying φ with a path $\pi \in t$ such that $\lambda(\pi) = \sigma$. The transducer T generating t models φ and violates ψ (because $\sigma \not\models \psi$), so $\varphi \not\looparrowright \psi$ holds.

Theorem 5. *Deciding* $\varphi \looparrowright \psi$ *is 2EXPTIME-complete and PSPACE-complete when φ is fixed. Deciding open equivalence is 2EXPTIME-complete.*

Proof. Hardness was shown in Theorem 3. The algorithm runs in time $2^{2^{O(n)}}2^{O(m)}$. The first four steps of the algorithm use time and space $2^{2^{O(n)}}$.

Deciding whether $L(A_{NBW}) \cap L(B_{NBW}) = \emptyset$ can be done within the resources required. The key is avoiding an explicit construction of B_{NBW}, rather, constructing its state while performing an on-the-fly search. We check whether there is a word that is accepted by the NBW A_{NBW} and the NBW B_{NBW} by nondeterministically guessing a word $\sigma \in A^\omega$ and simultaneously keeping track of the corresponding runs in both automata. We only have to store two states of the NBW A_{NBW} and two states of the NBW B_{NBW} at each step of the algorithm. Since each state of A_{NBW} has size $2^{O(n)}$ and each state of B_{NBW} has size $O(m)$, nonemptiness for $L(A_{NBW}) \cap L(B_{NBW})$ can be checked using $2^{O(n)} + O(m)$ nondeterministic space. By [24], this can be done using $2^{O(n)} + O(m^2)$ deterministic space. The time requirement is exponential in the space requirement, so it is $2^{2^{O(n)}} 2^{O(m^2)}$.

Altogether, the algorithm uses doubly exponential time in n and polynomial space in m.

Open implication can be viewed as a simultaneous realizability testing for the implicate (left-hand-side of implication) and validity testing for the implicant (right-hand-side of the implication). For a fixed implicant, open implication is 2EXPTIME-complete, just like realizability[1], and for a fixed implicate open implication is PSPACE-complete, just like validity.

For the next section, we need a bound on the size of the witness for $\varphi \not\Rightarrow \psi$.

Lemma 2. *If* $L(A_{NBW}) \cap L(B_{NBW}) \neq \emptyset$ *then there exists a word* $uv \in A^*$ *of length at most* $2^{2m+1}|Q_{NBW}|$ *such that* $uv^\omega \in L(A_{NBW}) \cap L(B_{NBW})$.

Proof. The product automaton C_{NBW} of A_{NBW} and B_{NBW} has at most $2 \cdot 2^{2m}|Q_{NRW}|$ states. If $L(C_{NBW}) \neq \emptyset$ then there exists a word $uv \in A^*$ whose length is at most the number of states in C_{NBW} such that $uv^\omega \in L(C_{NBW})$.

Theorem 6. *If* $\varphi \not\Rightarrow \psi$, *there is a transducer* T *with at most* $2^{2m+1}|Q_{NBW}||Q_{DPT}|$ *states such that* $T \models \varphi$ *but* $T \not\models \psi$.

Proof. Let $\pi = (i_0 \cup o_0) \ldots (i_{k-1} \cup o_{k-1})\big((i_k \cup o_k) \ldots (i_{l-1} \cup o_{l-1})\big)^\omega$ in $L(A_{NBW}) \cap L(B_{NBW})$ with $l \leq 2^{2m+1}|Q_{NBW}|$. The transducer is $T = (Q, q_0, \delta, \lambda)$ with $Q = W_\varphi \times \{0, \ldots, l-1, \bot\}$, where the second element keeps track of whether and where we are in π. From A_{DPT} we can derive a transducer T' with state space W_φ that satisfies φ. Our transducer T behaves like T' for all states in $Q_{DPT} \times \{\bot\}$. For $j \in \{0, \ldots, l-1\}$, we have $\lambda((q, j)) = o_j$ and $\delta((q, j), i) = (\delta_{DPT}(q, o_j, i), j')$, where $j' = \bot$ if $i \neq i_j$ and if $i = i_j$ then $j' = j+1$ if $j < l-1$ and k otherwise. Note that $\delta_{DPT}(q, o_j, i) \in W_\varphi$ because π is accepted by A_{NBW} and thus by A_{DPW}. The transducer violates ψ when the input sequence is as in π and satisfies φ. The number of states of T is at most $2^{2m+1}|Q_{NBW}||Q_{DPT}| = 2^{n2^{2n+3} + 10n + 3 + 2m}$

---

[1] In spite of the doubly exponential lower bound, there have been recently encouraging developments regarding the practicality of realizability checking [10,16,19].

Notation: Let $witn(n, m) = 2^{n2^{2n+3}+10n+3+2m}$.

Note that it is possible to avoid constructing A_{NBW} in our algorithm, if we check language emptiness of $L(A_{DPW}) \cap L(B_{NBW})$ directly. This leads to a slightly better upper bound in Theorem 6.

Our proof techniques can be extended to other linear specification formalisms that allow a translation of the specification into an NBW. Two popular formalisms falling into that class are QPTL [26] and the industrial PSL [6]. The algorithm follows the approach described above, adapting Step 1 and 5 to the formalism used. The complexity of the algorithm depends on the cost of translating the specification into an NBW. For QPTL and PSL it is possible to find an algorithm whose complexity matches the lower bounds for realizability and validity of the respective logics.

Note that the use of quantifiers allows us to check open equivalence between specifications at different levels of abstraction, e.g., a specification can be checked against a refined version that includes variables encoding implementation details. This is particular useful for synthesis of Generalized Reactivity(1) (cf. Section 4), which introduces additional variables to encode LTL specifications.

3.3   Avoiding Safra's Construction

In this section, we present another algorithm to decide if $\varphi \twoheadrightarrow \psi$, based on [16], which avoids Safra's intricate determinization construction and parity games and lends itself to implementation [10].

In [16], Kupferman and Vardi provide an approach to decide the realizability problem for LTL. Given an LTL formula φ, they construct a UCT U that accepts exactly all trees that are solutions to the realizability problem of φ.

Theorem 7. [16] *The realizability problem for an LTL formula φ can be reduced to the nonemptiness problem for a UCT with at most $4^{|\varphi|}$ states.*

In order to check if the language of U is empty, U is translated into a corresponding NBT N.

Theorem 8. [16] *Let U be a UCT with p states. For each $k > 0$ we can construct an NBT N_k with $2^{p(\log(k)+2)}$ states such that a tree generated by a transducer with at most k states is accepted by U iff it is accepted by N_k.*

Intuitively, the size of N_k is bounded by the size of the transducers generating the trees N_k has to accept. (Note that in general one cannot translate a UCT to an equivalent NBT.)

Since we are looking for a transducer that fulfills φ and violates ψ, Theorem 6 provides a bound on the size of the transducers of interest, which is $witn(n, m)$. We can replace the algorithm of Section 3.2 by the following algorithm:

1. Construct a UCT A_{UCT} of size 4^n such that $L(A_{UCT}) = \{t \in \mathcal{T} \mid t \models \varphi\}$. From A_{UCT} construct the NBT $N_{witn(n,m)}$ (Theorem 8). The number of states of this NBT is $2^{O(m)2^{O(n)}}$.

2. Compute the set W_φ of states q of $N_{witn(n,m)}$, such that $N^q_{witn(n,m)}$ accepts some tree, in quadratic time [28].
3. From $N_{witn(n,m)}$ construct an NBW A_{NBW} such that $\sigma \in L(A_{NBW})$ if $\sigma = \lambda(\pi)$ for some path π of a tree $t \in L(N_{witn(n,m)})$.
4. Construct an NBW B_{NBW} with at most 4^m states that accepts all words in $L(\neg\psi)$ [27].
5. Check if $L(A_{NBW}) \cap L(B_{NBW}) = \emptyset$ in time linear in the size of A_{NBW} and B_{NBW} [27].

Theorem 9. *Deciding if $\varphi \multimap \psi$ can be reduced to the language emptiness check of the product between A_{NBW} and B_{NBW}.*

The revised algorithm is doubly exponential in φ and exponential in ψ. We do not attempt to be space efficient here, as the automaton $N_{witn(n,m)}$ is already exponential in ψ. Nevertheless, this approach is useful as it avoids Safra's construction and parity games. It is particularly suitable for finding counterexamples to open implication, since it can be implemented incrementally by increasing the size of the transducers we are looking for. This may allow us to find counterexamples using much smaller automata than the full deterministic parity automaton [16,14].

4   Generalized Reactivity

Generalized Reactivity(1), or GR(1) for short, is a specification formalism that has been proposed in [19] for synthesis. GR(1) specifications consist of two sets of symbolically represented DBWs, one for the environment and one for the system. This formalism avoids the determinization step normally required for synthesis; it has a symbolic synthesis algorithm consisting of a triply nested fixpoint computation [19]. Experience shows that the formalism can be used to synthesize modest sized industrial circuits from their specifications, and that the restriction to GR(1) specifications is not overly restrictive [2,3].

We briefly recapitulate the construction of [19]. A DBW over AP with n states can be symbolically represented by an LTL formula φ by using a set V of $\lceil \lg(n) \rceil$ new atomic propositions. The formula is a conjunction of three parts: (1) φ^i is a propositional formula over V denoting the initial state, (2) φ^t is a formula of the form $G \bigwedge_i (\chi_i \to X \xi_i)$ representing the complete, deterministic transition relation, where χ_i and ξ_i are propositional formulas over $AP \cup V$ and V, respectively, and (3) φ^f is a formula of the form $G F \chi$, where χ is a propositional formula over V, representing the fairness condition. For instance, we represent $G(r \to F a)$ with $V = \{s\}$ by $\psi = \neg s \wedge G((\neg s \wedge r \wedge \neg a \to X s) \wedge (\neg s \wedge (\neg r \vee a) \to X \neg s) \wedge (s \wedge \neg a \to X s) \wedge (s \wedge a \to X \neg s)) \wedge G F \neg s$. The DBW is shown in Figure 1.

A GR(1) specification has the form $\varphi = (\bigwedge_j \varphi_{e,j}) \to (\bigwedge_j \varphi_{s,j})$, where environment assumptions $\varphi_{e,j}$ and the system guarantees $\varphi_{s,j}$ represent DBWs. In the sequel, let $\varphi_b^a = \bigwedge_j \varphi_{b,j}^a$ for $a \in \{i, t, f\}$ and $b \in \{s, e\}$. GR(1) formulas are intended to describe Mealy machines, not Moore machines, which leads to small technical differences with the previous presentation. Also, in keeping with [19],

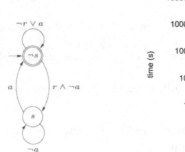

Fig. 1. DBW for $\mathsf{G}(r \to \mathsf{F}\, a)$ **Fig. 2.** Time needed for calculations

we use game-based terminology here. A *game* corresponds to a tree automaton and a *winning state* corresponds to a state with a nonempty language.

In order to decide realizability of a GR(1) formula, a game graph G_φ is built. The transition structure of the game graph is given by the combination of φ_e^t and φ_s^t, the initial state is $\varphi_e^i \wedge \varphi_s^i$, and the winning condition is $\varphi_e^f \to \varphi_s^f$. The winning region W_φ of the game is computed symbolically by a triply nested fixpoint formula, and the formula is realizable if the initial state is winning [19].

We now describe how to decide open implication. Suppose we have two GR(1) specifications, $\varphi = \bigwedge_i \varphi_{e,i} \to \bigwedge_j \varphi_{s,j}$ and $\psi = \bigwedge_k \psi_{e,k} \to \bigwedge_l \psi_{s,l}$. We check whether $\varphi \multimap \psi$ as follows.

1. Construct the game graph G_φ and compute the winning region W_φ.
2. Construct the game graph G_ψ and the product G of G_φ and G_ψ.
3. Check if there is a path in G that (i) stays within W_φ, (ii) satisfies $\varphi_e^f \to \varphi_s^f$, and (iii) violates $\psi_e^f \to \psi_s^f$.

Note that this algorithm is similar to the one described above, although removal of the losing states (the states with an empty language) has been replaced by the requirement that the path remain in the set of winning states. Thus, we are looking for a path within W_φ that satisfies all of the $\psi_{e,k}^f$, violates one of the $\psi_{s,l}^f$, and either violates one if the $\varphi_{e,i}^f$ or satisfies all of the $\varphi_{s,j}^f$. This is expressed by the μ-calculus [13] formula $\gamma = \mu Y . W_\varphi \wedge (\gamma' \vee pre(Y))$, where
$$\gamma' = \bigvee_{i,l}(\nu Y .(W_\varphi \wedge \neg\varphi_{e,i}^f \wedge E_\psi \wedge \neg\psi_{s,l}^f)) \vee \bigvee_l(\nu Y .(W_\varphi \wedge S_\varphi \wedge E_\psi \wedge \neg\psi_{s,l}^f)),$$
$E_\psi = \bigwedge_k pre(\mu Z . Y \wedge (\psi_{e,k}^f \vee pre(Z)))$, and $S_\varphi = \bigwedge_j pre(\mu Z . Y \wedge (\varphi_{s,j}^f \vee pre(Z)))$.

The complexity of a symbolic algorithm can be given in terms of the number of symbolic steps [1], where steps in this case are preimage computations and computations of the force operator used for games [19]. Computing the winning region of G_φ requires a cubic number of steps in terms of the number of states in G_φ. Computing γ, a doubly-nested fixpoint, and thus requires only a quadratic number of steps in terms of the size of G.

Theorem 10. *We have that $\varphi \multimap \psi$ iff the initial state of G is in the set γ. This computation uses a number of symbolic steps cubic in G_φ and quadratic in G_ψ.*

4.1   Experimental Results

We have implemented the algorithm for open implication of GR(1) formulas in ANZU [11], a synthesis tool for GR(1) specifications. We have tested our implementation on specifications of an arbiter for ARM's AMBA AHB bus used in [2,3]. In Figure 2, we show the time ANZU takes to synthesize the specifications and the time needed to calculate open implication. The *old* specification, which was used in [2] can only be synthesized for up to 7 masters. ANZU runs out of memory for larger instances. In [3] an improved version of the specification was presented, but it was not proven that the *old* and *new* specifications are equivalent. The *new* specification can be synthesized for up to 15 masters[2]. (2GB of memory were available.) Using the algorithm presented above, we can show that the *new* specification open-implies the *old* one and can thus be used in its stead. Figure 2 also shows that the combined time needed to calculate open implication and to synthesize the *new* specification is less than the time needed to synthesize the *old* specification, when that is possible. It should be noted that the circuits that result from the *new* specification are much smaller than those resulting from the *old* specification.

5   Conclusions

We have argued that open implication is an important concept both in model checking and in synthesis. We have given algorithms to compute open implication and open equivalence for the specification formalisms LTL and GR(1). For LTL, we have shown an algorithm that runs in time that is doubly exponential in the size of the implicate and space that is polynomial in the size of the implicant, matching the lower bounds. We have also shown how to implement the algorithm while avoiding Safra's construction. Finally, we implemented the approach for GR(1) specifications and showed that it can be used to show the correctness of simple specifications and, thus, to synthesize circuits that would otherwise be out of reach.

References

1. Bloem, R., Gabow, H., Somenzi, F.: An algorithm for strongly connected component analysis in $n \log n$ symbolic steps. Formal Methods in System Design (2006)
2. Bloem, R., Galler, S., Jobstmann, B., Piterman, N., Pnueli, A., Weiglhofer, M.: Automatic hardware synthesis from specifications: A case study. In: DATE (2007)
3. Bloem, R., Galler, S., Jobstmann, B., Piterman, N., Pnueli, A., Weiglhofer, M.: Specify, compile, run: Hardware from PSL. In: 6th International Workshop on Compiler Optimization Meets Compiler Verification, pp. 3–16 (2007)
4. Bloem, R., Ravi, K., Somenzi, F.: Efficient decision procedures for model checking of linear time logic properties. In: Halbwachs, N., Peled, D.A. (eds.) CAV 1999. LNCS, vol. 1633. Springer, Heidelberg (1999)

---

[2] Our specifications are slightly different from those used in [2,3]

372   K. Greimel et al.

5. Büchi, J.R.: On a decision method in restricted second order arithmetic. In: International Congress on Logic, Methodology, and Philosophy of Science (1962)
6. Eisner, C., Fisman, D.: A Practical Introduction to PSL. Springer, Heidelberg (2006)
7. Emerson, E.A., Jutla, C.S.: The complexity of tree automata and logics of programs (extended abstract). In: Proc. Foundations of Computer Science (1988)
8. Greimel, K.: Open implication. Master's thesis, Graz University of Technology (2007)
9. Harel, D., Pnueli, A.: On the development of reactive systems. In: Logics and Models of Concurrent Systems, pp. 477–498 (1985)
10. Jobstmann, B., Bloem, R.: Optimizations for LTL synthesis. In: 6th Conference on Formal Methods in Computer Aided Design (FMCAD 2006), pp. 117–124 (2006)
11. Jobstmann, B., Galler, S., Weiglhofer, M., Bloem, R.: Anzu: A tool for property synthesis. In: Computer Aided Verification, pp. 258–262 (2007)
12. King, V., Kupferman, O., Vardi, M.Y.: On the complexity of parity word automata. In: Foundations of Software Science and Computation Structures (2001)
13. Kozen, D.: Results on the propositional μ-calculus. Theoretical Computer Science 27, 333–354 (1983)
14. Kupferman, O., Piterman, N., Vardi, M.Y.: Safraless compositional synthesis. In: Ball, T., Jones, R.B. (eds.) CAV 2006. LNCS, vol. 4144, pp. 31–44. Springer, Heidelberg (2006)
15. Kupferman, O., Vardi, M.Y.: Relating linear and branching model checking. In: IFIP Working Conference on Programming Concepts and Methods (1998)
16. Kupferman, O., Vardi, M.Y.: Safraless decision procedures. In: Symposium on Foundations of Computer Science (FOCS 2005), pp. 531–542 (2005)
17. McIsaac, A.: Personal Communication (November 2006)
18. Piterman, N.: From nondeterministic Büchi and Streett automata to deterministic parity automata. In: Logic in Computer Science (LICS 2006), pp. 255–264 (2006)
19. Piterman, N., Pnueli, A., Sa'ar, Y.: Synthesis of reactive(1) designs. In: Proc. Verification, Model Checking and Abstract Interpretation, pp. 364–380 (2006)
20. Pnueli, A.: The temporal logic of programs. In: IEEE Symposium on Foundations of Computer Science, Providence, RI, pp. 46–57 (1977)
21. Pnueli, A., Rosner, R.: On the synthesis of a reactive module. In: Proc. Symposium on Principles of Programming Languages (POPL 1989), pp. 179–190 (1989)
22. Rabin, M.O.: Automata on Infinite Objects and Church's Problem. In: Regional Conference Series in Mathematics. American Mathematical Society, Providence (1972)
23. Rosner, R.: Modular Synthesis of Reactive Systems. PhD thesis, Weizmann Institute of Science (1992)
24. Savitch, W.J.: Relationships between nondeterministic and deterministic tape complexities. Journal of Computer and System Sciences 4(2), 177–192 (1970)
25. Sistla, A.P., Clarke, E.M.: The complexity of propositional linear temporal logic. Journal of the ACM 3, 733–749 (1985)
26. Sistla, A.P., Vardi, M.Y., Wolper, P.: The complementation problem for Büchi automata with applications to temporal logic. Theoretical Computer Science (1987)
27. Vardi, M., Wolper, P.: Reasoning about infinite computations. Information and Computation 115, 1–37 (1994)
28. Vardi, M.Y., Wolper, P.: Automata-theoretic techniques for modal logics of programs. Journal of Computer and System Sciences 32(2), 182–221 (1986)
29. Yoshiura, N.: Finding the causes of unrealizability of reactive system formal specifications. In: Proc. Software Engineering and Formal Methods (SEFM 2004) (2004)

ATL* Satisfiability Is 2EXPTIME-Complete*

Sven Schewe

Universität des Saarlandes, 66123 Saarbrücken, Germany

Abstract. The two central decision problems that arise during the design of safety critical systems are the satisfiability and the model checking problem. While model checking can only be applied after implementing the system, satisfiability checking answers the question whether a system that satisfies the specification exists. Model checking is traditionally considered to be the simpler problem – for branching-time and fixed point logics such as CTL, CTL*, ATL, and the classical and alternating time μ-calculus, the complexity of satisfiability checking is considerably higher than the model checking complexity. We show that ATL* is a notable exception of this rule: Both ATL* model checking and ATL* satisfiability checking are 2EXPTIME-complete.

1 Introduction

One of the main challenges in system design is the construction of correct implementations from their temporal specifications. Traditionally, system design consists of three separated phases, the specification phase, the implementation phase, and the validation phase. From a scientific point of view it seems inviting to overcome the separation between the implementation and validation phase, and replace the manual implementation of a system and its subsequent validation by a push-button approach, which automatically synthesizes an implementation that is correct by construction. Automating the system construction also provides valuable additional information: we can distinguish *unsatisfiable* system specifications, which otherwise would go unnoticed, leading to a waste of effort in the fruitless attempt of finding a correct implementation.

One important choice on the way towards the ideal of fully automated system construction is the choice of the specification language. For temporal specifications, three different types of logics have been considered: Linear-time logics [1], branching-time logics [2], and alternating-time logics [3]. The different paradigms are suitable for different types of systems and different design phases. Linear-time logic can only reason about properties of all possible runs of the system. Consequently, it cannot express the existence of different runs. A constructive non-emptiness test of an LTL specification is therefore bound to create a system that has exactly one possible run. Branching-time logics [2], on the other hand,

---

* This work was partly supported by the German Research Foundation (DFG) as part of the Transregional Collaborative Research Center "Automatic Verification and Analysis of Complex Systems" (SFB/TR 14 AVACS).

L. Aceto et al. (Eds.): ICALP 2008, Part II, LNCS 5126, pp. 373–385, 2008.

Logic	Model Checking (Structure)	Model Checking	Satisfiability Checking
LTL	NLOGSPACE [4]	PSPACE [5]	PSPACE [4]
CTL	NLOGSPACE [6]	PTIME [5]	EXPTIME [2]
CTL*	NLOGSPACE [6]	PSPACE [5]	2EXPTIME [4]
ATL	PTIME [3]	PTIME [3]	EXPTIME [7]
ATL*	PTIME [3]	2EXPTIME [3]	**2EXPTIME**

Fig. 1. For all previously considered branching-time temporal specifications, satisfiability checking is at least exponentially harder than model checking (in the specification). We show that ATL* is an interesting exception to this rule.

can reason about *possible* futures, but they refer to *closed systems* [3] that do not distinguish between different participating agents. Finally, alternating-time logics [3] can reason about the strategic capabilities of groups of agents to cooperate to obtain a temporal goal. The following example illustrates the differences.

Consider a vending machine that offers coffee and tea. Ultimately, we want the machine to react on the requests of a customer, who shall be provided with coffee or tea upon her wish. In alternating-time logic, we have a natural correspondence to this requirement: we simply specify that the customer has the strategic capability to get coffee or tea from the machine, without the need of cooperation, written $\langle\langle customer \rangle\rangle \bigcirc get_{coffee}$ or $\langle\langle customer \rangle\rangle \bigcirc get_{tea}$, respectively.

In branching-time logic, there is no natural correspondence to the property. The typical approximation is to specify the *possibility* that coffee or tea is provided, written $E \bigcirc get_{coffee}$ or $E \bigcirc get_{tea}$, respectively. However, this does no longer guarantee that the customer can choose; the specification is also fulfilled if the health insurance company can override the decision for coffee. A workaround may be to introduce a dedicated communication interface between the vending machine and the customer, and represent the desire for coffee by a $desire_{coffee}$ bit controlled by the customer. The property may then be approximated by $E \bigcirc desire_{coffee}$ and $desire_{coffee} \rightarrow A \bigcirc get_{coffee}$. In LTL, the possibility of different system behaviors cannot be expressed within the logic. Here, in addition to specifying an interface, we would have to distinguish between parts of the system under our control (the vending machine) and parts outside of our control (the customer). The most likely approximation would be $desire_{coffee} \rightarrow \bigcirc get_{coffee}$, with the addition that there is not only the need to design an interface to the customer beforehand, but also to make assumptions about her behavior.

Using the workarounds for branching-time or linear-time logic requires solving the central design problem of designing interfaces in an early specification phase, instead of starting with an abstract view on the system. Especially in the case that synthesis fails, we could no longer distinguish if we made an error in designing the interfaces, or if the specification is unrealizable.

As an example for this effect, we could consider to use alternating-time logic for the specifications of protocol fairness. ATL* has, for example, been used to express the fairness requirement "Bob cannot obtain Alice's signature unless Alice can obtain Bob's signature as well" [8] in contract signing protocols.

Using satisfiability checking techniques for alternating-time logics, we can automate [9] the proof that fair contract signing is not possible without a trusted third party [10] (under the presence of other standard requirements).

The alternating-time temporal logic ATL* [3] extends the classic branching-time temporal logic CTL* [4] with path quantifiers that refer to the strategic capabilities of groups of agents. An ATL* specification $\langle\!\langle A'\rangle\!\rangle\varphi$ requires that the group A' of agents can cooperate to enforce the path formula φ. ATL* formulas are interpreted over concurrent game structures, a special type of labeled transition systems, where each transition results from a set of decisions, one for each agent. When interpreted over a concurrent game structure C, $\langle\!\langle A'\rangle\!\rangle\varphi$ holds true in a state s of C if the agents in A' can win a two player game against the agents not in A'. In this game, the two groups of agents take turns in making their decisions (starting with the agents in A'), resulting in an infinite sequence $ss_1s_2\ldots$ of states of the concurrent game structure C. The agents in A' win this game, if the infinite sequence $ss_1s_2\ldots$ satisfies the path formula φ.

Since ATL* specifications can canonically be transformed into alternating-time μ-calculus (ATM) formulas [11,3], ATL* inherits the decidability and finite model property from ATM [9]. This translation from ATL* to ATM, comprises a doubly exponential blow-up, which is in line with the doubly exponential model checking complexity of ATL* [11,3]. The *complexity* of the ATL* satisfiability and synthesis problem, on the other hand, has been an interesting open challenge since its introduction [7]: While the complexity of the satisfiability problem is known to be EXPTIME-complete for the least expressive alternating-time logic ATL [12,7] as well as for the most expressive alternating-time logic ATM [9], the complexity of the succinct and intuitive temporal logic ATL* has only been known to be in 3EXPTIME [9,11], and to inherit the 2EXPTIME hardness from CTL*, leaving an exponential gap between both bounds.

Outline. In this paper, we introduce an automata-theoretic decision procedure to demonstrate that deciding the satisfiability of an ATL* specification and, for satisfiable specifications, constructing a correct model of the specifications is no more expensive than model checking: both problems are 2EXPTIME-complete in the size of the specification. To the contrary, the cost of model checking a concurrent game structure against an ATL* specification is also polynomial in the size of the concurrent game structure. While polynomial conveys the impression of feasibility, the degree of this polynomial is, for known algorithms, exponential in the size of the specification [3,11].

On first glance, an automata-theoretic construction based on automata over concurrent game structures (ACGs) [9] – the alternating-time extension of symmetric alternating-automata [13] – does not seem to be a promising starting point for the construction of a 2EXPTIME algorithm, because synthesis procedures based on alternating automata usually shift all combinatorial difficulties to testing their non-emptiness [14]. Using a doubly exponential translation from ATL* through ATM to an equivalent ACG suffices to proof the finite model property of ATL* [9], but indeed leads to a *triply* exponential construction.

In order to show that a constructive non-emptiness test for ATL* specifications can be performed in doubly exponential time, we combine two concepts: We first show that every model can be transformed into an *explicit* model that includes a certificate of its correctness. For this special kind of model, it suffices to build an ACG that only checks the correctness of the certificate. Finally, we show that we can construct such an automaton, which is only singly exponential in the size of the specification. Together with the exponential cost of a constructive non-emptiness test of ACGs [9], we can provide a 2EXPTIME synthesis algorithm for ATL* specifications that returns a model together with a correctness certificate. 2EXPTIME-completeness then follows with the respective hardness result for the syntactic sublogic CTL* [4] of ATL*.

2   Logic, Models and Automata

In this section we recapture the logic ATL* [3], concurrent game structures, over which ATL* specifications are interpreted, and automata over concurrent game structures [9], which are used to represent alternating-time specifications.

2.1   Concurrent Game Structures

Concurrent game structures [3] generalize labeled transition systems (or pointed Kripke structures) to a setting with multiple agents. A *concurrent game structure* (CGS) is a tuple $\mathcal{C} = (P, A, S, s_0, l, \Delta, \tau)$, where

- P is a finite nonempty set of atomic propositions,
- A is a finite nonempty set of agents,
- S is a nonempty set of states, with a designated initial state $s_0 \in S$,
- $l : S \to 2^P$ is a labeling function that decorates each state with a subset of the atomic propositions,
- Δ is a nonempty set of possible decisions for every agent, and
- $\tau : S \times \Delta^A \to S$ is a transition function that maps a state and the decisions of the agents to a new state.

For a CGS $\mathcal{C}$, a *strategy* for a set $A' \subseteq A$ of agents is a mapping $f_{A'} : S^* \to \Delta^{A'}$ from finite traces to decisions of the agents in A', and a *counter strategy* is a mapping $f^c_{A \smallsetminus A'} : S^* \times \Delta^{A'} \to \Delta^{A \smallsetminus A'}$ from finite traces and decisions of the agents in A' to decisions of the agents in $A \smallsetminus A'$. For a given strategy $f_{A'}$ and counter strategy $f^c_{A \smallsetminus A'}$, the set of *plays* starting at a position s_1 is defined as

$$plays(s_1, f_{A'}) = \{s_1 s_2 s_3 \ldots \mid \forall i {\geq} 1\, \exists d' {\in} \Delta^{A \smallsetminus A'}.\, s_{i+1} = \tau(s_i, (f_{A'}(s_1 \ldots s_i), d'))\},$$
$$plays(s_1, f^c_{A \smallsetminus A'}) = \{s_1 s_2 s_3 \ldots \mid \forall i {\geq} 1\, \exists d {\in} \Delta^A.\, s_{i+1} {=} \tau(s_i, (f^c_{A \smallsetminus A'}(s_1 \ldots s_i, d), d))\}.$$

2.2   ATL*

ATL* extends the classical branching-time logic CTL* by path quantifiers that allow for reasoning about the strategic capability of groups of agents.

ATL* Syntax. ATL* contains formulas $\langle\!\langle A' \rangle\!\rangle \psi$, expressing that the group $A' \subseteq A$ of agents can enforce that the path formula ψ holds true. Formally, the state formulas (Φ) and path formulas (Π) of ATL* are given by the following grammar (where $p \in P$ is an atomic proposition, and $A' \subseteq A^1$ is a set of agents).

$$\Phi := true \mid p \mid \Phi \wedge \Phi \mid \Phi \vee \Phi \mid \neg\Phi \mid \langle\!\langle A' \rangle\!\rangle \Pi, \text{ and}$$
$$\Pi := \Phi \mid \Pi \wedge \Pi \mid \Pi \vee \Pi \mid \neg\Pi \mid \bigcirc\Pi \mid \Pi \, U \, \Pi.$$

Every state formula is an ATL* formula. We call an ATL* formula *basic* iff it starts with a path quantifier $\langle\!\langle A' \rangle\!\rangle$.

Semantics. An ATL* specification with atomic propositions $\subseteq P$ is interpreted over a CGS $\mathcal{C} = (P, A, S, s_0, l, \Delta, \tau)$. $\|\varphi\|_{\mathcal{C}} \subseteq S$ denotes the set of states where φ holds. A CGS $\mathcal{C} = (P, A, S, s_0, l, \Delta, \tau)$ is a *model* of a specification φ ($\mathcal{C} \models \varphi$) with atomic propositions P iff φ holds in the initial state ($s_0 \in \|\varphi\|_{\mathcal{C}}$).

For each state s of $\mathcal{C}$, $path(s)$ denotes all paths in $\mathcal{C}$ that originate from s, and $path(\mathcal{C}) = \bigcup\{path(s) \mid s \in S\}$ denotes the set of all paths in $\mathcal{C}$.

An ATL* formula is evaluated along the structure of the formula.

- Atomic propositions and Boolean connectives are interpreted as usual:
  $\|true\|_{\mathcal{C}} = S$, $\|p\|_{\mathcal{C}} = \{s \in S \mid p \in l(s)\}$, and
  $\|\varphi \wedge \psi\|_{\mathcal{C}} = \|\varphi\|_{\mathcal{C}} \cap \|\psi\|_{\mathcal{C}}$, $\|\varphi \vee \psi\|_{\mathcal{C}} = \|\varphi\|_{\mathcal{C}} \cup \|\psi\|_{\mathcal{C}}$, and $\|\neg\varphi\|_{\mathcal{C}} = S \smallsetminus \|\varphi\|_{\mathcal{C}}$.
- Basic formulas $\varphi = \langle\!\langle A' \rangle\!\rangle \psi$ hold true in a state s if the agents in A' have a strategy which ensures that all plays starting in s satisfy the path formula ψ:
  $s \in \|\varphi\|_{\mathcal{C}} \Leftrightarrow \exists f_{A'} : S^* \to \Delta^{A'} . plays(s, f_{A'}) \subseteq \|\psi\|_{\mathcal{C}}^{path}$.

For a path formula φ and a CGS $\mathcal{C}$, $\|\varphi\|_{\mathcal{C}}^{path} \subseteq path(\mathcal{C})$ denotes the set of paths of $\mathcal{C}$ where φ holds. Path formulas are interpreted as follows:

- For state formulas φ, $\|\varphi\|_{\mathcal{C}}^{path} = \bigcup\{path(s) \mid s \in \|\varphi\|_{\mathcal{C}}\}$.
- Boolean connectives are interpreted as usual: $\|\varphi \wedge \psi\|_{\mathcal{C}}^{path} = \|\varphi\|_{\mathcal{C}}^{path} \cap \|\psi\|_{\mathcal{C}}^{path}$,
  $\|\varphi \vee \psi\|_{\mathcal{C}}^{path} = \|\psi\|_{\mathcal{C}}^{path} \cup \|\varphi\|_{\mathcal{C}}^{path}$, and $\|\neg\varphi\|_{\mathcal{C}}^{path} = path(\mathcal{C}) \smallsetminus \|\varphi\|_{\mathcal{C}}^{path}$.
- A path $\pi = s_1, s_2, s_3, s_4 \ldots$ satisfies $\bigcirc\varphi$ (read: *next* φ), if the path $s_2, s_3, s_4 \ldots$ obtained by deleting the first letter of π satisfies φ:
  $\|\bigcirc\varphi\|_{\mathcal{C}}^{path} = \{s_1, s_2, s_3, s_4 \ldots \in path(\mathcal{C}) \mid s_2, s_3, s_4 \ldots \in \|\varphi\|_{\mathcal{C}}^{path}\}$.
- A path $\pi = s_1, s_2, s_3, s_4 \ldots$ satisfies $\varphi \, U \, \psi$ (read: φ *until* ψ), if there is a natural number $n \in \mathbb{N}$ such that
  (1) the path $s_n, s_{n+1}, s_{n+2} \ldots$ obtained by deleting the initial sequence $s_1, s_2, s_3 \ldots s_{n-1}$ of π satisfies the path formula ψ, and
  (2) for all $i < n$, the path $s_i, s_{i+1}, s_{i+2} \ldots$ obtained by deleting the initial sequence $s_1, s_2, s_3 \ldots s_{i-1}$ of π satisfies the path formula φ:
  $\|\varphi \, U \, \psi\|_{\mathcal{C}}^{path} = \{s_1, s_2, s_3, s_4 \ldots \in path(\mathcal{C}) \mid$
  $\exists n \in \mathbb{N} . (s_n, s_{n+1}, s_{n+2} \ldots \in \|\psi\|_{\mathcal{C}}^{path} \wedge \forall i < n. s_i, s_{i+1}, s_{i+2} \ldots \in \|\varphi\|_{\mathcal{C}}^{path})\}$.

Note that the validity of basic formulas $\langle\!\langle A' \rangle\!\rangle \psi$ is implicitly defined by the outcome of a two player game with an ω-regular (LTL) objective. Such games are determined [11]. Consequently, there is a counter strategy $f_{A \smallsetminus A'}^c : S^* \times \Delta^{A'} \to \Delta^{A \smallsetminus A'}$ such that $plays(s, f_{A \smallsetminus A'}^c) \subseteq \|\neg\psi\|_{\mathcal{C}}^{path}$ if and only if $s \notin \|\langle\!\langle A' \rangle\!\rangle \psi\|_{\mathcal{C}}$.

---

[1] We assume that the set A of agents is known and fixed. For satisfiability checking, one could argue that this is not necessarily the case. However, we can assume without loss of generality that there is at most one agent that does not occur in the formula [7].

2.3   Automata over Concurrent Game Structures

Automata over concurrent game structures (ACGs) [9] provide an automata-theoretic framework for alternating-time logics. Generalizing symmetric automata [13], ACGs contain *universal atoms* $(\Box, A')$, which refer to *all* successor states for *some* decision of the agents in A', and *existential atoms* $(\Diamond, A')$, which refer to *some* successor state for *each* decision of the agents *not* in A'. ACGs can run on CGSs with arbitrary sets Δ of decisions. For the purpose of this paper, it suffices to consider ACGs with a Co-Büchi acceptance condition.

An ACG is a tuple $\mathcal{A} = (\Sigma, Q, q_0, \delta, F)$, where Σ is a finite alphabet, Q is a finite set of states, $q_0 \in Q$ is a designated initial state, δ is a transition function, and $F \subseteq Q$ is a set of final states. The transition function $\delta : Q \times \Sigma \to \mathbb{B}^+(Q \times (\{\Box, \Diamond\} \times 2^A))$ maps a state and an input letter to a positive Boolean combination of universal atoms – $(q, \Box, A')$ – and existential atoms – $(q, \Diamond, A')$.

A run tree $\langle R, r : R \to Q \times S \rangle$ on a given CGS $\mathcal{C} = (P, A, S, s_0, l, \Delta, \tau)$ is a $Q \times S$-labeled tree where the root is labeled with (q_0, s_0) and where, for a node n with a label (q, s) and a set $L = \{r(n \cdot \rho) \mid n \cdot \rho \in R\}$ of labels of its successors, there is a set $\mathfrak{A} \subseteq Q \times (\{\Box, \Diamond\} \times 2^A)$ of atoms satisfying $\delta(q, l(s))$ such that

- for all universal atoms $(q', \Box, A')$ in $\mathfrak{A}$, there exists a decision $d \in \Delta^{A'}$ of the agents in A' such that, for all counter decisions $d' \in \Delta^{A \smallsetminus A'}$, $(q', \tau(s, (d, d'))) \in L$, and
- for all existential atoms $(q', \Diamond, A')$ in $\mathfrak{A}$ and all decisions $d' \in \Delta^{A \smallsetminus A'}$ of the agents not in A', there exists a counter decision $d \in \Delta^{A'}$ such that $(q', \tau(s, (d, d'))) \in L$.

A run tree is accepting iff all paths satisfy the Co-Büchi condition that only finitely many positions on the path are labeled with a final state (or rather: with an element of $F \times S$), and a CGS is accepted iff it has an accepting run tree.

The *atoms* of an ACG $\mathcal{A}$ are the elements of the set $atom(\mathcal{A}) \subseteq Q \times (\{\Box, \Diamond\} \times 2^A)$ of atoms that actually occur in some Boolean function $\delta(q, \sigma)$, and the *size* $|Q| + |atom(\mathcal{A})|$ of $\mathcal{A}$ is the sum of the number of its states and atoms.

Theorem 1. *[9] A constructive non-emptiness test of an ACG can be performed in time exponential in the size of the ACG.* $\Box$

An automaton is called *universal* if all occurring Boolean functions $\delta(q, \sigma)$ are conjunctions of atoms in $Q \times \{(\Box, \emptyset)\}$.

3   From General to Explicit Models

In this section we show that every model of a specification can be transformed into an *explicit model*, which makes both the truth of each basic subformula in the respective state and a (counter) strategy that witnesses the validity or invalidity of this basic subformulas explicit. This result is exploited in the following section by constructing a small ACG $\mathcal{A}_\varphi$ that accepts the explicit models of φ. Constructing an explicit model from a general model consists of three steps:

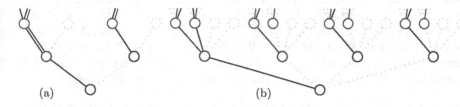

Fig. 2. In the central third step of the transformation of an arbitrary model into an explicit CGT, a CGT is *widened* in order to enable a finite encoding of witness strategies for the (in)validity of basic subformulas in the labels. Figure 2a shows a CGT for a single agent a and a binary set $\Delta = \{left, right\}$, where $\langle\!\langle a \rangle\!\rangle\varphi$ holds in every position. The color coding maps a witness strategy for $\langle\!\langle a \rangle\!\rangle\varphi$ to every position p – in the single agent case an infinite path rooted in p that satisfies φ. In Figure 2a, the path that always turns left is a witness strategy for the validity of $\langle\!\langle a \rangle\!\rangle\varphi$ in the root, indicated by coloring this path and the root of the tree in the same color (red). In general, witness strategies cannot be finitely encoded in the labels of a CGT, because there is no bound on the number of paths a position belongs to. The tree is therefore *widened* by extending Δ to $\Delta' = \{(left, new), (left, cont), (right, new), (right, cont)\}$ (Figure 2b). Witness strategies for the resulting CGT are constructed from witness strategies for the original CGT by turning first to a *new*, and henceforth to a *cont* direction, avoiding the unbounded overlap of witness strategies – for $\langle\!\langle a \rangle\!\rangle\varphi$, every position p occurs in at most one witness strategy that does not start in p – allowing for their finite representation.

1. In a first step, we add a fresh atomic proposition b for each basic subformula b of φ, and extend the labeling function such that $b \in l(s) \Leftrightarrow s \in \|b\|_\mathcal{C}$.
2. In a second step, we unravel the model obtained in the first step to a tree. Using trees guarantees that no position can be part of infinitely many witnesses. However, the number of witness strategies a position might belong to remains unbounded. (Or: May be linear in the number of its predecessors.)
3. In a final step, we *widen* the tree by adding a single Boolean decision to the set Δ of decisions available to every agent (cf. Figure 2).
   This widening allows us to map arbitrary but fixed witness (counter) strategies from the original tree to witness (counter) strategies in the widened tree such that witnesses for the validity of the same basic subformula b (or its negation $\neg b$) in different states do not overlap. (With the exception of the trivial case that the witness strategy must cover all successors.) This allows us to explicitly encode the witnesses in the widened strategy trees.

From Models to Basic Models. For a given ATL* specification φ, we denote with B_φ the set of its basic subformulas. We call a model $\mathcal{C} = (P \uplus B_\varphi, A, S, s_0, l, \Delta, \tau) \models \varphi$ of an ATL* formula φ *basic* if, for all basic subformulas $b \in B_\varphi$ of φ and all states $s \in S$ of $\mathcal{C}$, $b \in l(s) \Leftrightarrow s \in \|b\|_\mathcal{C}$. Since the additional propositions B_φ do not occur in the specification, the following lemma holds trivially:

Lemma 1. *An ATL* formula is satisfiable iff it has a basic model.* □

From Models to Tree Models. We call a CGS $\mathcal{C} = (P, A, S, s_0, l, \Delta, \tau)$ a *concurrent game tree* (CGT) if $S = (\Delta^A)^*$, $s_0 = \varepsilon$, and $\tau(s, d) = s \cdot d$. For a CGS $\mathcal{C} = (P, A, S, s_0, l, \Delta, \tau)$, we call $\mathcal{T_C} = (P, A, (\Delta^A)^*, \varepsilon, l \circ u, \Delta, \tau')$ where $\tau'(s, d) = s \cdot d$, and where the *unraveling function* $u : (\Delta^A)^* \to S$ is defined recursively by $u(\varepsilon) = s_0$, and $u(s) = s' \Rightarrow u(s \cdot d) = \tau(s', d)$, the *unraveling* of $\mathcal{C}$. We extend u to finite and infinite paths $(u(s_0 s_1 s_2 \ldots) = u(s_0) u(s_1) u(s_2) \ldots)$.

Lemma 2. *A CGS $\mathcal{C}$ is a (basic) model of a specification φ if and only if its unraveling $\mathcal{T_C}$ is a (basic) model of φ.*

Proof. By induction over the structure of φ, it is easy to prove that $s \in \|\varphi\|_{\mathcal{T_C}} \Leftrightarrow u(s) \in \|\varphi\|_{\mathcal{C}}$, and $\pi \in \|\varphi\|_{\mathcal{T_C}}^{path} \Leftrightarrow u(\pi) \in \|\varphi\|_{\mathcal{C}}^{path}$. The only non-trivial part in the induction is the transformation of the witness strategies for basic formulas ($\varphi = \langle\!\langle A' \rangle\!\rangle \psi$). However, we can simply use the unraveling function u to transform a witness (counter) strategy $f_{A'}$ or $f_{A \smallsetminus A'}^c$ for $\mathcal{C}$ into a witness (counter) strategy $f'_{A'}$ or $f_{A \smallsetminus A'}^c{}'$, respectively, for $\mathcal{T_C}$. For this, we fix $f'_{A'}(\pi) = f_{A'}(u(\pi))$ or $f_{A \smallsetminus A'}^c{}'(\pi, d) = f_{A \smallsetminus A'}^c(u(\pi), d)$, respectively. This ensures that $plays(u(s), f_{A'}) = u(plays(s, f'_{A'})) := \{u(\pi) \mid \pi \in plays(s, f'_{A'})\}$, or $plays(u(s), f_{A \smallsetminus A'}^c) = u(plays(s, f_{A \smallsetminus A'}^c{}'))$. Using the induction hypothesis, we get $plays(s, f'_{A'}) \subseteq \|\psi\|_{\mathcal{C}}^{path}$ or $plays(s, f_{A \smallsetminus A'}^c{}') \subseteq \|\neg\psi\|_{\mathcal{C}}^{path}$, respectively. $\qquad\square$

From Tree Models to Explicit Tree Models. For a CGT $\mathcal{T} = (P, A, (\Delta^A)^*, \varepsilon, l, \Delta, \tau)$, we call the CGT $\mathcal{T}_w = (P, A, (\Delta'^A)^*, \varepsilon, l \circ h, \Delta', \tau')$, where $\Delta' = \Delta \times \{new, cont\}$, $h : (\Delta'^A)^* \to (\Delta^A)^*$ is a hiding function that hides the $\{new, cont\}$ part of a trace position-wise, and $\tau'(s, d) = s \cdot d$ is the usual transition function of trees, the (Boolean) *widening* of $\mathcal{T}$.

Lemma 3. *A CGT $\mathcal{T}$ is a (basic) model of a specification φ if and only if its (Boolean) widening $\mathcal{T}_w$ is a (basic) model of φ.*

Proof. By induction over the structure of φ. Again, the only non-trivial part is the transformation of the witness strategies for basic formulas ($\varphi = \langle\!\langle A' \rangle\!\rangle \psi$). For this part, we can use the hiding function h to transform a witness strategy $f_{A'}$ in $\mathcal{T}$ into a witness strategy $f'_{A'}$ in its widening $\mathcal{T}_w$ by choosing $f'_{A'}(\pi) = (f_{A'}(h(\pi)), *)$, where $* \in \{new, cont\}$ can be chosen arbitrarily. This ensures $plays(h(s), f_{A'}) = h(plays(s, f'_{A'})) := \{h(\pi) \mid \pi \in plays(s, f'_{A'})\}$. Using the induction hypothesis, we get $plays(s, f'_{A'}) \subseteq \|\psi\|_{\mathcal{C}}^{path}$. As in the previous lemma, we get the analogous result for the transformation of a witness counter strategy. $\qquad\square$

Let, for a basic subformula $B_\varphi \ni b = \langle\!\langle A' \rangle\!\rangle \varphi_b$ of a specification φ, $a(b) = A'$ and $a(\neg b) = A \smallsetminus A'$ denote the set of agents that cooperate to ensure φ_b and the set of their opponents, respectively, and let $E_\varphi = \{(b, new), (b, cont), (\neg b, new), (\neg b, cont) \mid b \in B_\varphi\}$ denote an extended set of subformulas. We call a concurrent game structure $\mathcal{C} = (P \uplus B_\varphi \uplus E_\varphi, A, S, s_0, l, \Delta, \tau)$ *well-formed* if it satisfies the following requirements:

- $\forall s \in S.\ b \notin l(s) \Rightarrow \forall d \in \Delta^{a(b)} \exists d' \in \Delta^{a(\neg b)}.\ (\neg b, new) \in l(\tau(s, (d, d')))$,
- $\forall s \in S.\ (\neg b, new) \in l(s) \Rightarrow \forall d \in \Delta^{a(b)} \exists d' \in \Delta^{a(\neg b)}.\ (\neg b, cont) \in l(\tau(s, (d, d')))$,
- $\forall s \in S.\ (\neg b, cont) \in l(s) \Rightarrow \forall d \in \Delta^{a(b)} \exists d' \in \Delta^{a(\neg b)}.\ (\neg b, cont) \in l(\tau(s, (d, d')))$,
- $\forall s \in S.\ b \in l(s) \Rightarrow \exists d \in \Delta^{a(b)} \forall d' \in \Delta^{a(\neg b)}.\ (b, new) \in l(\tau(s, (d, d')))$,
- $\forall s \in S.\ (b, new) \in l(s) \Rightarrow \exists d \in \Delta^{a(b)} \forall d' \in \Delta^{a(\neg b)}.\ (b, cont) \in l(\tau(s, (d, d')))$, and
- $\forall s \in S.\ (b, cont) \in l(s) \Rightarrow \exists d \in \Delta^{a(b)} \forall d' \in \Delta^{a(\neg b)}.\ (b, cont) \in l(\tau(s, (d, d')))$.

For a basic subformula $B_\varphi \ni b = \langle\!\langle a(b) \rangle\!\rangle \varphi_b$ of φ and its negation $\neg b$, we call the set of traces $witness(s, b) = \{ss_1s_2s_3 \ldots \in path(s) \mid b \in l(s),\ (b, new) \in l(s_1)$ and $\forall i \geq 2.\ (b, cont) \in l(s_i)\}$ and $witness(s, \neg b) = \{ss_1s_2s_3 \ldots \in path(s) \mid b \notin l(s),\ (\neg b, new) \in l(s_1)$ and $\forall i \geq 2.\ (\neg b, cont) \in l(s_i)\}$ the explicit witnesses for b and $\neg b$ in s. $\mathcal{C}$ is called an *explicit model* of φ if the explicit witnesses are contained in the set of paths that satisfy φ_b and $\neg \varphi_b$, respectively. ($witness(s, b) \subseteq \|\varphi_b\|_{\mathcal{C}}^{path}$ and $witness(s, \neg b) \subseteq \|\neg \varphi_b\|_{\mathcal{C}}^{path}$ for all $s \in S$ and $b \in B_\varphi$.) Note that explicit models of φ are in particular basic models of φ.

Lemma 4. *Given a CGT $\mathcal{T}$ that is a basic model of an ATL* formula φ and a set of witness strategies for $\mathcal{T}$, we can construct an explicit model of φ.*

Proof. In the proof of the previous lemma, we showed that the widening $\mathcal{T}_w$ of a basic tree model $\mathcal{T}$ of φ is a basic model of φ. Moreover, we showed that, for the translation of witness (counter) strategies that demonstrate the (in)validity of a subformula $b \in B_\varphi$ of φ in a state s of $\mathcal{T}_w$, *any* extension $* \in \{new, cont\}$ can be chosen. In particular, the agents in $a(b)$ or $a(\neg b)$, respectively, can choose to first pick the new extension, and henceforth to pick the extension $cont$. For non-universal specifications, that is, for the case $a(b) \neq \emptyset$ or $a(\neg b) \neq \emptyset$, respectively, this particular choice provides the guarantee that states reachable under the new strategy $f'_{a(b)}$ or counter strategy $f^c_{a(\neg b)}{}'$, respectively, from different states in $\mathcal{T}_w$ are disjoint. ($\forall s_1, t_1 \in (\Delta'^A)^* \forall i, j > 1.\ s_1s_2s_3 \ldots \in plays(s_1, f'_{a(b)}) \wedge t_1t_2t_3 \ldots \in plays(t_1, f'_{a(b)}) \wedge s_i = t_j \Rightarrow s_1 = t_1$, and the analogous result for $f^c_{a(\neg b)}{}'$.)

For universal specifications, that is, for the case $a(b) = \emptyset$ or $a(\neg b) = \emptyset$, respectively, the respective player intuitively has no choice, and the (counter) strategy $f'_{a(b)}$ or $f^c_{a(\neg b)}{}'$ is well defined.

In both cases, we mark the positions reachable under $f'_{a(b)}$ in one step from a position s_1 with $b \in l(s_1)$ by (b, new) and positions reachable under $f^c_{a(\neg b)}{}'$ in one step from a position s_1 with $b \notin l(s_1)$ by $(\neg b, new)$, and we mark positions reachable in *more* than one step by $(b, cont)$ and $(\neg b, cont)$, respectively.

By construction, the resulting CGT $\mathcal{T}_w$ is well-formed, and $b \in l(s) \Rightarrow witness(s, b) = plays(s, f'_{a(b)})$ and $b \notin l(s) \Rightarrow witness(s, \neg b) = plays(s, f^c_{a(\neg b)}{}')$ hold. By Lemma 3, we also get $b \in l(s) \Rightarrow plays(s, f'_{a(b)}) \subseteq \|\varphi_b\|_{\mathcal{C}}^{path}$ and $b \notin l(s) \Rightarrow plays(s, f^c_{a(\neg b)}{}') \subseteq \|\neg \varphi_b\|_{\mathcal{C}}^{path}$. $\qquad\square$

Theorem 2. *A specification has an explicit model if and only if it has a model.*

Proof. The 'if' direction is implied by the Lemmata 1–4. For the 'only if' direction, it is obvious that, for a given explicit model $(P \uplus B_\varphi \uplus E_\varphi, A, S,$

$s_0, l, \Delta, \tau)$ of an ATL* formula φ, and for the projection of the labeling function to the atomic propositions $(l'(s)=l(s)\cap P)$, $(P, A, S, s_0, l', \Delta, \tau)$ is a model of φ.   □

4   ATL* Satisfiability Is 2EXPTIME-Complete

We exploit the explicit model theorem by constructing an ACG $\mathcal{A}_\varphi$ from an ATL* specification φ that accepts only the explicit models of φ. Testing if a CGS is a model of φ is considerably harder than testing if it is an explicit model. The latter only comprises two simple tests: Checking the well-formedness criterion can be performed by a (safety) ACG with $O(|B_\varphi|)$ states, while, for all basic subformulas $b \in B_\varphi$ of φ, testing if all paths in $witness(s, b)$ satisfy the path formula φ_b and if all paths in $witness(s, \neg b)$ satisfy the path formula $\neg \varphi_b$ can be performed by a universal ACG that is exponential in φ_b.

Automata that check the (much weaker) model property, on the other hand, need to guarantee consistency of the automaton decisions, which is usually solved by using *deterministic* word automata to represent the single φ_b, leading to an exponentially larger ACG (with parity acceptance condition and a number of colors exponential in the length of φ).

We call a CGS $\mathcal{C}$ *plain* if all states in $\mathcal{C}$ are reachable from the initial state. We can restrict our focus without loss of generality to plain concurrent game structures, because unreachable states have no influence on the model property (nor are they traversed by an automaton).

Lemma 5. *For a specification φ, we can build an ACG $\mathcal{A}_w$ with $O(|E_\varphi|)$ states that accepts a plain CGS $\mathcal{C} = (P \uplus B_\varphi \uplus E_\varphi, A, S, s_0, l, \Delta, \tau)$ iff it is well-formed.*

Proof. We can simply set $\mathcal{A}_w = (\Sigma_w, Q_w, q_0^w, \delta, \emptyset)$ with $\Sigma_w = 2^{B_\varphi \uplus E_\varphi}$ (the atomic propositions P are not interpreted), $Q_w = \{q_0^w\} \uplus E_\varphi$, and

- $\delta(q_0^w, \sigma) = (q_0^w, \Box, \emptyset) \wedge \bigwedge_{b \in \sigma \cap B_\varphi}((b, new), \Box, a(b)) \wedge \bigwedge_{b \in B_\varphi \smallsetminus \sigma}((\neg b, new), \Diamond, a(b))$
  $\wedge \bigwedge_{(b,*) \in \sigma \cap E_\varphi}((b, cont), \Box, a(b)) \wedge \bigwedge_{(\neg b,*) \in \sigma \cap E_\varphi}((\neg b, cont), \Diamond, a(b))$, and
- for all $e \in E_\varphi$, $\delta(e, \sigma) = true$ if $e \in \sigma$, and $\delta(e, \sigma) = false$ otherwise.

The $(q_0^w, \Box, \emptyset)$ part of the transition function guarantees that every reachable position in the input CGS is traversed, and the remainder of the transition function simply reflects the well-formedness constraints.   □

Theorem 3. *[4] Given an LTL formula φ, we can build an equivalent universal Co-Büchi word automaton whose size is exponential in the length of φ.*   □

In the context of this paper, the equivalent universal word automaton is read as a universal ACG $\mathcal{U}$ that accepts exactly those words that satisfy the LTL formula. (Words can be viewed as special concurrent game structures with a singleton set of decisions ($|\Delta| = 1$) or an empty set of agents ($A = \emptyset$).)

Let, for a path formula ψ, $\widehat{\psi}$ denote the formula obtained by replacing all occurrences of direct basic subformulas $b \in B_\psi$ by b (read as atomic proposition).

Lemma 6. *For a specification φ and every $B_\varphi \ni b = \langle\langle a(b)\rangle\rangle\varphi_b$ we can build two universal ACGs $\mathcal{A}_b$ and $\mathcal{A}_{\neg b}$ whose size is exponential in the size of $\widehat{\varphi_b}$ and that accept a plain CGS $\mathcal{C} = (P \uplus B_\varphi \uplus E_\varphi, A, S, s_0, l, \Delta, \tau)$ iff $witness(s, b) \subseteq \|\widehat{\varphi_b}\|_{\mathcal{C}}^{path}$ and $witness(s, \neg b) \subseteq \|\neg\widehat{\varphi_b}\|_{\mathcal{C}}^{path}$, respectively, hold true.*

Proof. By Theorem 3 we can translate the LTL formula $\widehat{\varphi_b}$ into an equivalent universal ACG $\mathcal{U}_b = (P \uplus B_\varphi, Q_b, q_0^b, \delta_b, F_b)$ whose size is exponential in the length of $\widehat{\varphi_b}$. From $\mathcal{U}_b$, we infer the universal ACG $\mathcal{A}_b = (P \uplus B_\varphi \uplus E_\varphi, Q_b \times \{new, cont\} \uplus \{q_b\}, q_b, \delta, F_b \times \{cont\})$ with the following transition function:

- $\delta(q_b, \sigma) = (q_b, \Box, \emptyset)$ if $b \notin \sigma$ and
- $\delta(q_b, \sigma) = (q_b, \Box, \emptyset) \wedge \bigwedge_{q \in \delta_b(q_0^b, \sigma)}((q, new), \Box, \emptyset)$ otherwise,
- $\delta((q, new), \sigma) = true$ if $(b, new) \notin \sigma$ and
- $\delta((q, new), \sigma) = \bigwedge_{q' \in \delta_b(q, \sigma)}((q', cont), \Box, \emptyset)$ otherwise, and
- $\delta((q, cont), \sigma) = true$ if $(b, cont) \notin \sigma$ and
- $\delta((q, cont), \sigma) = \bigwedge_{q' \in \delta_b(q, \sigma)}((q', cont), \Box, \emptyset)$ otherwise.

δ again uses the $(q_b, \Box, \emptyset)$ part of the transition function to traverse every reachable position in the input CGS. The assignments $\delta((q, *), \sigma) = true$ ensure that, starting in any reachable state s, only the infinite paths in $witness(s, b)$ are traversed. The remaining transitions reflect the requirement that, for all reachable positions s, all paths in $witness(s, b)$ must satisfy the path formula $\widehat{\varphi_b}$.

$\mathcal{A}_{\neg b}$ can be constructed analogously. □

Theorem 4. *For a given ATL* specification φ, we can construct an ACG $\mathcal{A}_\varphi$ that is exponential in the size of φ and that accepts a plain CGS if and only if it is an explicit model of φ.*

Proof. We build the automaton $\mathcal{A}_\varphi = (2^{P \uplus B_\varphi \uplus E_\varphi}, \{q_0\} \uplus Q_w \uplus \biguplus_{b \in B_\varphi}\{q_b, q_{\neg b}\} \uplus (Q_b \uplus Q_{\neg b}) \times \{new, cont\}, q_0, \delta, \biguplus_{b \in B_\varphi}(F_b \uplus F_{\neg b}) \times \{cont\})$ that consists of the states of the ACG $\mathcal{A}_w$ and, for every basic subformula $b \in B_\varphi$ of φ, of the ACGs $\mathcal{A}_b$ and $\mathcal{A}_{\neg b}$, and a fresh initial state q_0. The transition function for the non-initial states is simply inherited from the respective ACG, and for the initial state we set $\delta(q_0, \sigma) = false$ if σ does not satisfy φ (when read as a Boolean formula over atomic propositions and basic subformulas), and $\delta(q_0, \sigma) = \delta(q_0^w, \sigma) \wedge \bigwedge_{b \in B_\varphi} \delta(q_b, \sigma) \wedge \delta(q_{\neg b}, \sigma)$ otherwise.

The lemmata of this section imply that $\mathcal{A}_\varphi$ is exponential in the size of φ, and accepts a plain CGS if and only if it is an explicit model of φ. □

It is only a small step from the non-emptiness preserving reduction of ATL* to a 2EXPTIME algorithm for ATL* satisfiability checking and synthesis.

Together, Theorems 1, 2 and 4 provide a 2EXPTIME algorithm for a constructive satisfiability test for an ATL* specification. The corresponding hardness result can be inferred from the 2EXPTIME completeness [4] of the satisfiability problem for the syntactic sublogic CTL* (and even for CTL+ [15]) of ATL*.

Corollary 1. *The ATL* satisfiability and synthesis problems are 2EXPTIME-complete.* □

5 Conclusions

We showed that the satisfiability and synthesis problem of ATL* specifications is 2EXPTIME-complete. This result is surprising: For the remaining branching-time temporal logics, the satisfiability problem is at least exponentially harder than the model checking problem [4,14] (in the size of the specification).

What is more, the suggested reduction indicates that ATL* synthesis may be feasible. The exponential blow-up in the construction of the ACG is the same blow-up that occurs when translating an LTL specification to a nonde-terministic word automaton. While this blow-up is unavoidable in principle, it is also known that no blow-up occurs in most practical examples. This gives rise to the assumption that, for most practical ATL* specifications φ, the size of the emptiness equivalent Co-Büchi ACG $\mathcal{A}_\varphi$ will be small. Moreover, $\mathcal{A}_\varphi$ is essentially universal (plus a few simple local constraints), and synthesis proce-dures for universal Co-Büchi automata have recently seen a rapid development (cf. [16,17,18]).

ATL* specifications thus seem to be particularly well suited for synthesis: They form one of the rare exceptions of the rule that testing (model-checking) is simpler than constructing a solution.

References

1. Wolper, P.: Synthesis of Communicating Processes from Temporal-Logic Specifi-cations. PhD thesis, Stanford University (1982)
2. Clarke, E.M., Emerson, E.A.: Design and synthesis of synchronization skeletons us-ing branching time temporal logic. In: Proc. IBM Workshop on Logics of Programs, pp. 52–71. Springer, Heidelberg (1981)
3. Alur, R., Henzinger, T.A., Kupferman, O.: Alternating-time temporal logic. Jour-nal of the ACM 49, 672–713 (2002)
4. Emerson, E.A.: Temporal and modal logic, pp. 995–1072. MIT Press, Cambridge (1990)
5. Clarke, E.M., Emerson, E.A., Sistla, A.P.: Automatic verification of finite-state concurrent systems using temporal logic specifications. Transactions On Program-ming Languages and Systems 8, 244–263 (1986)
6. Kupferman, O., Vardi, M.Y., Wolper, P.: An automata-theoretic approach to branching-time model checking. Journal of the ACM 47, 312–360 (2000)
7. Walther, D., Lutz, C., Wolter, F., Wooldridge, M.: Atl satisfiability is indeed exptime-complete. Journal of Logic and Computation 16, 765–787 (2006)
8. Kremer, S., Raskin, J.F.: A game-based verification of non-repudiation and fair exchange protocols. Journal of Computer Security 11, 399–430 (2003)
9. Schewe, S., Finkbeiner, B.: Satisfiability and finite model property for the alternating-time μ-calculus. In: Proc. CSL, pp. 591–605. Springer, Heidelberg (2006)
10. Even, S., Yacobi, Y.: Relations among public key signature systems. Technical Report 175, Technion, Haifa, Israel (1980)
11. de Alfaro, L., Henzinger, T.A., Majumdar, R.: From verification to control: Dy-namic programs for omega-regular objectives. In: Proc. LICS, pp. 279–290. IEEE Computer Society Press, Los Alamitos (2001)

12. van Drimmelen, G.: Satisfiability in alternating-time temporal logic. In: Proc. LICS, pp. 208–217. IEEE Computer Society Press, Los Alamitos (2003)

13. Wilke, T.: Alternating tree automata, parity games, and modal μ-calculus. Bull. Soc. Math. Belg. 8 (2001)

14. Kupferman, O., Vardi, M.Y.: Church's problem revisited. The bulletin of Symbolic Logic 5, 245–263 (1999)

15. Wilke, T.: CTL$^+$ is exponentially more succinct than CTL. In: Pandu Rangan, C., Raman, V., Ramanujam, R. (eds.) FST TCS 1999. LNCS, vol. 1738, pp. 110–121. Springer, Heidelberg (1999)

16. Kupferman, O., Vardi, M.: Safraless decision procedures. In: Proc. 46th IEEE Symp. on Foundations of Computer Science, Pittsburgh, pp. 531–540 (2005)

17. Kupferman, O., Piterman, N., Vardi, M.Y.: Safraless compositional synthesis. In: Ball, T., Jones, R.B. (eds.) CAV 2006. LNCS, vol. 4144, pp. 31–44. Springer, Heidelberg (2006)

18. Schewe, S., Finkbeiner, B.: Bounded synthesis. In: Namjoshi, K.S., Yoneda, T., Higashino, T., Okamura, Y. (eds.) ATVA 2007. LNCS, vol. 4762, pp. 474–488. Springer, Heidelberg (2007)

Visibly Pushdown Transducers*

Jean-François Raskin[1] and Frédéric Servais[2]

[1] Computer Science Department,
[2] Department of Computer & Decision Engineering (CoDE),
Université Libre de Bruxelles (U.L.B.)

Abstract. Visibly pushdown automata have been recently introduced by Alur and Madhusudan as a subclass of pushdown automata. This class enjoys nice properties such as closure under all Boolean operations and the decidability of language inclusion. Along the same line, we introduce here visibly pushdown transducers as a subclass of pushdown transducers. We study properties of those transducers and identify subclasses with useful properties like decidability of type checking as well as preservation of regularity of visibly pushdown languages.

1 Introduction

Visibly pushdown languages (VPL) have recently been proposed by Alur and Madhusudan in [3] as a subclass of *context-free languages* (CFL) with interesting closure and decidability properties. While CFL are not closed under intersection nor under complementation, VPL are closed under all Boolean operations, and the language inclusion problem is decidable. VPL are expressive enough to model a large number of relevant problems, for example those related to the analysis of programs with procedure calls or to the formalization of structured documents (like XML documents). As a consequence, VPL offer an appropriate theoretical framework to unify many known decidability results in those fields as well as opportunities to solve new problems. In [3], visibly pushdown automata (VPA) are defined as a subclass of the pushdown automata where stack operations are restricted by the input word. VPA operate on words over a *tagged alphabet* $\hat{\Sigma} = \Sigma^c \uplus \Sigma^r \uplus \Sigma^i$ where Σ^c are *call* symbols, Σ^r are *return* symbols, and Σ^i are *internal* symbols. Each time a call symbol is read, the automaton has to push a symbol on the stack; each time a return symbol is read, the automaton has to pop a symbol from the stack; and each time an internal symbol is read, the automaton must leave the stack unchanged. VPA exactly recognize VPL.

Transducers are machines that model relations between words, i.e. they recognize sets of pairs of words. Transducers transform languages into languages: let L be a set of words, T a transducer then $T(L) = \{w \mid \exists v \in L : T \text{ accepts the pair } (v, w)\}$. There are many important applications of transducers. For example, while languages are useful to formalize sets of XML documents (i.e. XML document types), transducers are useful to formalize XML document transformations (e.g., XSLT) [9]. Motivated by

---

* This research was supported by the Belgian FNRS grant 2.4530.02 of the FRFC project "Centre Fédéré en Vérification" and by the project "MoVES", an Interuniversity Attraction Poles Programme of the Belgian Federal Government.

L. Aceto et al. (Eds.): ICALP 2008, Part II, LNCS 5126, pp. 386–397, 2008.

this application, the *type checking problem* asks if all the words of L_1 are translated into words of L_2 under a transducer T, i.e. whether $T(L_1) \subseteq L_2$. Transducers have also been intensively used in the so-called regular model-checking [1,5]. In that setting, the states of a system are modeled by words, state sets by languages and state transitions by transducers. So far, the concept of regular model-checking has only been applied to regular languages (with the notable exception of [6]). Unfortunately some parametric systems cannot be modeled in this setting and more powerful classes of transducers with good decidability and closure properties are needed.

In this paper, we study several subclasses of *pushdown transducers*. In the spirit of [3], we define subclasses of pushdown transducers by imposing restrictions on the use of the stack and the transition relation. We study three main classes of pushdown transducers. First, *visibly pushdown transducers* are pushdown transducers that operate over pairs of words defined on a tagged alphabet $\hat{\Sigma}$. Those transducers respect two restrictions: (i) along the reading of a pair of words, either the head is moved only in one of the two words (allowing deletion and insertion), or it is moved over a pair of symbols of the same type (two calls, two returns, or two internals), (ii) when reading internals the transducer leaves the stack unchanged, when reading calls it pushes a symbol on the stack, when reading returns it pops a symbol from the stack. We show here that unfortunately the type checking is undecidable for this class even if L_1 and L_2 are VPL. They are not closed under composition and they do not preserve VPL, i.e. the transduction of a VPL is not necessarily a VPL. Second, *synchronized visibly pushdown transducers* are obtained from visibly pushdown transducers by imposing the following additional restrictions: (i) when a call is deleted then the *matching* return is deleted, (ii) when a call is inserted then a matching return is inserted, and (iii) when a call is copied then the matching return is also copied. We show that this class of pushdown transducers has a decidable type checking problem for VPL. This result is not trivial as we also show that the transduction of a VPL with a synchronized visibly pushdown transducer is not necessarily a VPL. This class of transducers is well suited to formally validate XML document transformations. Indeed, opening and closing tags are modeled by calls and returns respectively, and a transformation that inserts (respectively deletes) a new opening tag will usually also insert (respectively delete) the corresponding closing tag. The synchronized restriction to our transducer is therefore very natural in that context. Finally, we define the class of *fully synchronized visibly pushdown transducers* as a subclass of synchronized visibly pushdown transducers that, in addition to having a decidable type checking problem, preserve VPL, and are closed under composition and inverse. This class of transducers has all the properties required to extend the techniques used in regular model-checking from regular languages to VPL.

2   Preliminaries

Basics. An *alphabet* Σ is a finite set of symbols[1], we note Σ_ϵ for $\Sigma \cup \{\epsilon\}$ (the alphabet Σ together with the empty word symbol ϵ). The *tagged alphabet* over Σ is an alphabet, noted $\hat{\Sigma}$, which is equal to $\Sigma^c \uplus \Sigma^r \uplus \Sigma^i$, where $\Sigma^c = \{\bar{a} \mid a \in \Sigma\}$, $\Sigma^r = \{\underline{a} \mid$

---

[1] For technical reasons, we assume that all alphabets Σ in this paper are such that $|\Sigma| \geq 2$.

$a \in \Sigma\}$ and $\Sigma^i = \{a \mid a \in \Sigma\}.$[2] A *word* over Σ is a finite sequence of symbols in Σ. A *language* over Σ is a set of words over Σ. In the rest of the paper, given any alphabet Σ, we note RL(Σ), respectively CFL(Σ), the set of *regular*, respectively *context-free*, languages over Σ. Let π be the function from $\hat{\Sigma}$ into Σ defined as follows: $\pi(a) = \pi(\bar{a}) = \pi(\underline{a}) = a$. We extend π to words: for $w = a_1 a_2 \ldots a_n$, $\pi(w) = \pi(a_1)\pi(a_2)\ldots\pi(a_n)$, and to languages: $\pi(L) = \{\pi(w) \mid w \in L\}$. Let $\Sigma_1 \subseteq \Sigma_2$, for $w \in \Sigma_2^*$, $\downarrow^{\Sigma_1}(w) \in \Sigma_1^*$ returns the word w where the occurences of symbols in $\Sigma_2 \setminus \Sigma_1$ have been erased. Finally, a *stack alphabet* Γ is a finite set of symbols that contains a special symbol, noted $\perp$, called the *bottom-of-stack symbol*.

Visibly pushdown languages. A *visibly pushdown automaton* (VPA) [3] on finite words over the tagged alphabet $\hat{\Sigma} = \Sigma^c \uplus \Sigma^r \uplus \Sigma^i$ is a tuple $A = (Q, Q_0, Q_f, \Gamma, \delta)$ where Q is a finite set of states, $Q_0 \subseteq Q$, respectively $Q_f \subseteq Q$, the set of initial states, respectively final states, Γ the stack alphabet, and $\delta = \delta_c \uplus \delta_r \uplus \delta_i$ where $\delta_c \subseteq Q \times \Sigma^c \times Q \times (\Gamma \setminus \{\perp\})$ are the *call transitions*, $\delta_r \subseteq Q \times \Gamma \times \Sigma^r \times Q$ are the *return transitions*, and $\delta_i \subseteq Q \times \Sigma^i \times Q$ are the *internal transitions*. On a call transition $(q, a, q', \gamma) \in \delta_c$, γ is pushed onto the stack and the control goes from q to q'. On a return transition $(q, \gamma, a, q') \in \delta_r$, γ is popped from the stack (note that if $\perp$ is the top of the stack then it is read but not popped). Finally, on an internal transition $(q, a, q') \in \delta_i$, there is no stack operation. Accordingly, a *run* of a visibly pushdown automaton A over the word $w = a_1 \ldots a_l$ is a sequence $\{(q_k, \sigma_k)\}_{0 \le k \le l}$, where q_k is the state and $\sigma_k \in \Gamma^*$ is the stack at step k, such that $q_0 \in Q_0$, $\sigma_0 = \perp$, and for each $k < l$, we have either: (i) $(q_k, a_{k+1}, q_{k+1}, \gamma) \in \delta_c$ and $\sigma_{k+1} = \gamma\sigma_k$; (ii) $(q_k, \gamma, a_{k+1}, q_{k+1}) \in \delta_r$ and if $\gamma \ne \perp$ then $\sigma_k = \gamma\sigma_{k+1}$ else $\sigma_k = \sigma_{k+1} = \perp$; or (iii) $(q_k, a_{k+1}, q_{k+1}) \in \delta_i$ and $\sigma_k = \sigma_{k+1}$. A run is *accepting* if $q_l \in Q_f$. A word w is *accepted* by A if there exists an accepting run of A over w. $L(A)$, the *language* of A, is the set of words accepted by A. A language L over a tagged alphabet $\hat{\Sigma}$ is a *visibly pushdown language* if there is a VPA A over $\hat{\Sigma}$ such that $L(A) = L$. We note VPL($\hat{\Sigma}$) for the set of VPL over the tagged alphabet $\hat{\Sigma}$.

Example 1. $V_{2n} = \{\bar{a}^n \underline{b}^n \mid n \ge 0\}$ is a VPL($\hat{\Sigma}$), while $C_{2n} = \{a^n b^n \mid n \ge 0\}$ is not.

Proposition 1 ([3]). *Here are the main properties of* VPL *and* VPA.

1. *The class of* VPL *is closed under all Boolean operations.*[3] *In particular, given* $A, A_1, A_2 \in$ VPA *we can compute in polynomial time a* VPA B *such that* $L(B) = L(A_1) \cap L(A_2)$, *and in exponential time a* VPA C *such that* $L(C) = \overline{L(A)}$.
2. *Given* $A_1, A_2 \in$ VPA, *the problem of deciding whether* $L(A_1) \subseteq L(A_2)$ *is* EXPTIME-COMPLETE, *when* A_2 *is deterministic the problem is* PTIME-COMPLETE.
3. *Given* $A \in$ VPA, *we can decide, in polynomial time, whether* $L(A) = \emptyset$.
4. *Let* $C \in$ CFL(Σ), *then there exists* $V \in$ VPL($\hat{\Sigma}$) *such that* $\pi(V) = C$.

---

[2] We sometimes write a^c for $\bar{a}$, a^r for $\underline{a}$ and a^i for a. We may also write a when the type of a is clear from the context.

[3] This is in sharp contrast with CFL that are not closed under intersection nor complement.

The following result states the undecidability of checking inclusion between a CFL and VPL. To the best of our knowledge, the direction CFL into VPL is not established in the literature. We give a proof of the theorem in [10].

Theorem 1. *Let $C \in$ CFL and $V \in$ VPL then checking whether $C \subseteq V$, and checking whether $V \subseteq C$ are undecidable problems.*

Transduction relations and the type-checking problem. A relation $R \subseteq \Sigma^* \times \Sigma^*$ is a *transduction relation*, or simply a *transduction*, over Σ, i.e. a set of pairs of words over Σ. When $R(v, w)$ holds, we sometimes call v the *input* and w the *output* of the transduction. The *transduction of a word v* over Σ by a transduction relation $R \subseteq \Sigma^* \times \Sigma^*$ is the language $\{w \mid R(v, w)\}$, noted $R(v)$. The *transduction of a language L* over Σ by a transduction relation $R \subseteq \Sigma^* \times \Sigma^*$ is the language $\{w \mid \exists v \in L : R(v, w)\}$, noted $R(L)$. The *type checking problem* asks, given an effective representation of two languages L_1 and L_2, and an effective representation of a transduction relation R, to establish if $R(L_1) \subseteq L_2$.

3   Visibly Pushdown Transducers

VPA are pushdown automata such that the input restrict the stack operations. Similarly we define *visibly pushdown transducers* as pushdown transducers such that input and output restrict the stack operations. Such a transducer will push, respectively pop, onto the stack when it reads and/or write a call, respectively a return.

Definition 1 (VPT). A *visibly pushdown transducer* on finite words over $\hat{\Sigma}$ is a tuple $T = (Q, Q_0, Q_f, \Gamma, \delta)$ where Q is a finite set of states, $Q_0 \subset Q$, respectively $Q_f \subseteq Q$, the set of initial states, respectively final states, Γ the stack alphabet, and $\delta = \delta_c \uplus \delta_r \uplus \delta_i$, with $\delta_c \subseteq Q \times \Sigma_\epsilon^c \times \Sigma_\epsilon^c \times Q \times (\Gamma \setminus \{\bot\})$, $\delta_r \subseteq Q \times \Gamma \times \Sigma_\epsilon^r \times \Sigma_\epsilon^r \times Q$, $\delta_i \subseteq Q \times \Sigma_\epsilon^i \times \Sigma_\epsilon^i \times Q$. Moreover if $(q, \alpha, \beta, q', \gamma) \in \delta_c$, $(q, \gamma, \alpha, \beta, q') \in \delta_r$ or $(q, \alpha, \beta, q') \in \delta_i$ then $\alpha \neq \epsilon$ or $\beta \neq \epsilon$. The class of visibly pushdown transducer is noted VPT.[4]

Definition 2 (Run of a VPT). A *run* of a VPT T over (v, w), where $v = a_1 \ldots a_l$ and $w = b_1 \ldots b_m$ are words on $\hat{\Sigma}$, is a sequence $\{(q_k, i_k, j_k, \sigma_k)\}_{0 \leq k \leq n}$, where q_k is the state at step k, i_k, respectively j_k, are the index of the last letter of v, respectively w, the transducer has reached, and $\sigma_k \in \Gamma^*$ is the stack, such that $q_0 \in Q_0, i_0 = 1, j_0 = 1$, $\sigma_0 = \bot$, and for all $k < n$, let $\alpha = \epsilon$ or $\alpha = a_{i_k}$ and $\beta = \epsilon$ or $\beta = b_{j_k}, i_{k+1} = i_k + |\alpha|$, $j_{k+1} = j_k + |\beta|$, and we have either: (i) $(q_k, \alpha, \beta, q_{k+1}, \gamma) \in \delta_c$ and $\sigma_{k+1} = \gamma \sigma_k$, (ii) $(q_k, \gamma, \alpha, \beta, q_{k+1}) \in \delta_r$ and if $\gamma \neq \bot$ then $\sigma_k = \gamma \sigma_{k+1}$, else $\sigma_k = \sigma_{k+1} = \bot$, (iii) $(q_k, \alpha, \beta, q_{k+1}) \in \delta_i$ and $\sigma_k = \sigma_{k+1}$. A run is *accepting* if $q_n \in Q_f, i_n = |v| + 1$, and $j_n = |w| + 1$.

---

[4] Note that we define transducers that operate over pairs of words defined on the same alphabet. This is not restrictive: a transducer from words on an alphabet Σ_1 to words on an alphabet Σ_2 can be seen as a transducer from $\Sigma_1 \cup \Sigma_2$ to $\Sigma_1 \cup \Sigma_2$. In the following, we will abuse notations and sometimes we will define transducers where the input and output alphabets differ.

We note $[\![T]\!]$ the transduction induced by T, it is the set of pairs $(v, w) \in \hat{\Sigma}^* \times \hat{\Sigma}^*$ such that there exists an accepting run of T on $(v, w)^5$. A transduction relation $R \subseteq \hat{\Sigma}^* \times \hat{\Sigma}^*$ is a *visibly pushdown transduction* if there exists $T \in$ VPT such that $R = [\![T]\!]$.

Example 2. The transducer T_{del} of Fig. 1(a) deletes the calls $\bar{a}$, respectively the returns $\underline{b}$, and replaces them with the internals a, respectively b, it further verifies that the number of deleted calls is equal to the number of deleted returns. Clearly, T_{del} is a VPT that transduces V_{2n} into C_{2n} (defined in Example 1), which is also obtained when T_{del} is applied on $\hat{\Sigma}^*$. The transducer T_{ins} of Fig. 1(b) copies the calls $\bar{a}$ it encounters and then inserts the same number of returns $\underline{b}$, finally it renames the remaining returns $\underline{b}$ into $\underline{c}$. Then T_{ins} is a VPT that transduces V_{2n} into the language $S_{3n} = \{\bar{a}^n \underline{b}^n \underline{c}^n \mid n \geq 0\}$.

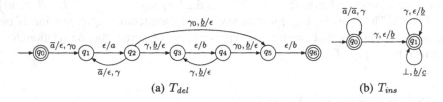

(a) T_{del} (b) T_{ins}

Fig. 1. Examples of VPT

Definition 3 (Inverse transducer). Given a VPT $T = (Q, Q_0, Q_f, \Gamma, \delta)$, we define its *inverse* $T^{-1} = (Q, Q_0, Q_f, \Gamma, \delta')$ with (i) $(q_1, \beta, \alpha, q_2, \gamma) \in \delta'_c \Leftrightarrow (q_1, \alpha, \beta, q_2, \gamma) \in \delta_c$, (ii) $(q_1, \gamma, \beta, \alpha, q_2) \in \delta'_r \Leftrightarrow (q_1, \gamma, \alpha, \beta, q_2) \in \delta_r$, and (iii) $(q_1, \beta, \alpha, q_2) \in \delta'_i \Leftrightarrow (q_1, \alpha, \beta, q_2) \in \delta_i$.

Proposition 2 (Inverse transduction). *Let $T \in$ VPT, then $[\![T^{-1}]\!] = [\![T]\!]^{-1}$.*

Proof. Any run of T on (v, w) can easily be transformed in a run of T^{-1} on (w, v) by interchanging α with β and i_k with j_k. □

Lemma 1. *For all $C \in$ CFL$(\hat{\Sigma})$, there exist $T \in$ VPT$(\hat{\Sigma})$ and $V \in$ VPL$(\hat{\Sigma})$ such that $T(V) = C$.*

Proof. In this proof we use the alphabet $\hat{\Sigma}$ which is the set $\{(a^x)^y \mid a \in \Sigma \land x, y \in \{c, r, i\}\}$ and we make the hypothesis that Σ contains the letters c, r, and i. This is without lost of generality as we make the hypothesis that our alphabets always contain at least two letters.

First, by Proposition 1, there exists $V' \in$ VPL$(\hat{\Sigma})$ such that $\pi(V') = C$. With the notations above, π is defined as follows: $\pi((a^x)^y) = a^x$. Second, let us consider the function $\tau_1 : \hat{\Sigma} \to \hat{\Sigma} \times \hat{\Sigma}$ defined as $\tau_1((a^x)^y) = x^i a^y$. This function codes any character of $\hat{\Sigma}$ into a sequence of two characters of $\hat{\Sigma}$. We extend the function τ_1 to words as follows: let $w = a_1 \ldots a_n \in \hat{\Sigma}^*$, $\tau_1(w) = \tau_1(a_1) \ldots \tau_1(a_n)$. Given A' a VPA on $\hat{\Sigma}$ for V', it is easy to construct A a VPA on $\hat{\Sigma}$ such that $L(A) = \tau_1(L(A'))$, since

---

^{5} In the sequel, we sometimes say that the transducer read the input v and write the output w.

τ_1 maps a call on an internal followed by a call, maps a return on an internal followed by a return, and maps an internal on two internals.

Third, let us consider the function $\tau_2 : \{c^i, r^i, i^i\} \times \hat{\Sigma} \to \hat{\Sigma}$ defined by: $\tau_2(x^i a^y) = a^x$. Clearly, for any word $w \in \hat{\Sigma}^*$, $\pi(w) = \tau_2(\tau_1(w))$. We are left to show that τ_2 can be defined as a VPT T. Here is the construction. First, T, when in state q, reads an internal x^i which determines the type of the ouput: a call if $x = c$, a return if $x = r$, and an internal if $x = i$. Accordingly, it goes into q^c, q^r or q^i respectively using the transitions $(q, c^i, \epsilon, q^c) \in \delta_i$, $(q, r^i, \epsilon, q^r) \in \delta_i$ or $(q, i^i, \epsilon, q^i) \in \delta_i$. Note that those transitions do not move the head on the output (so erasing the internal x^i). Then, T reads the next letter a^y and rewrites it into the output type defined by its current state, that is if the state is q^c then it writes (imposes to read) a^c on the output, etc. There are nine cases to consider, (i) read a call write a call, (ii) read a call write a return, (iii) read a call write an internal, (iv) read a return write a call, and so on. For translation of one type of character to another, we need to use two transitions that use first epsilon on output and then epsilon on input. Here are two representative cases over the nine cases: (i) for a^c into a^c: $(q^c, a^c, a^c, q, \gamma) \in \delta_c$, (ii) for a^c into a^r: $(q^r, a^c, \epsilon, q_a^r, \gamma) \in \delta_c$, $(q_a^r, \bot, \epsilon, a^r, q) \in \delta_r$ and $(q_a^r, \gamma, \epsilon, a^r, q) \in \delta_r$. Clearly, T is a VPT. To complete the proof, a simple induction shows that $T(V) = \tau_2(V) = \pi(V') = C$. Note that the VPT is not using the stack: only one character is pushed on the stack and return transitions can always use this character or the bottom character. As a matter of fact, the transduction τ_2 is definable by a finite state transducer on $\hat{\Sigma}$. $\square$

In the next proposition, we establish that the transduction and inverse transduction of a VPL by a VPT is not necessarily a VPL nor even a CFL, and that the transduction and inverse transduction of a RL by a VPT is not necessarily a VPL but it is always a CFL. We note VPT(RL) $= \{T(R) \mid T \in$ VPT$, R \in$ RL$\}$ and VPT(VPL) $= \{T(V) \mid T \in$ VPT$, V \in$ VPL$\}$.

Proposition 3. VPL $\subsetneq$ VPT(RL) $\subseteq$ CFL $\subsetneq$ VPT(VPL).

Proof. First, we know that VPT(RL) $\subseteq$ CFL since it is true for the class of pushdown transducers (which contains VPT). Second, to show that VPL $\subseteq$ VPT(RL), for any $V \in$ VPL we construct a VPT that first ignores the input (taking only transitions that are labelled by ϵ for the input), checks that the output is in V by simulating the VPA that accepts V, and when it reaches the end of the output, it reads the input without constraining the output using ϵ transitions on the output. When executing this transducer on $\hat{\Sigma}^*$, we get V. Third, to show that VPL $\neq$ VPT(RL), we consider T_{del} of Example 2: when executed on $\hat{\Sigma}^*$ it returns C_{2n}, a CFL which is not a VPL. Fourth, to prove CFL $\subsetneq$ VPT(VPL), first by Lemma 1 we get CFL $\subseteq$ VPT(VPL), second we consider the transducer T_{ins} of Example 2, it transduces $V_{2n} \in$ VPL into $S_{3n} \notin$ CFL. $\square$

In the next result states that the class of VPT is not closed under composition.

Corollary 1 (Composition). *There exists* $T, T' \in$ VPT *such that* $[\![T]\!] \circ [\![T']\!]$ *is not a visibly pushdown transduction.*

Proof. From Proposition 3, there are $V \in$ VPL and $T \in$ VPT such that $T(V) \notin$ CFL. Also there exist $R \in$ RL and $T' \in$ VPT such that $T'(R) = V$ since VPL $\subseteq$ VPT(RL). So $[\![T]\!] \circ [\![T']\!](R) \notin$ CFL but then it cannot be a VPT as VPT(RL) $\subseteq$ CFL. $\square$

The next theorem shows that the type checking problem of VPT against VPL is undecidable.

Theorem 2. *For $A_1, A_2 \in$ VPA and $T \in$ VPT, it is undecidable whether $T(L(A_1)) \subseteq L(A_2)$.*

Proof. Let $C \in$ CFL($\hat{\Sigma}$), by Lemma 1 there exist $V \in$ VPL($\hat{\Sigma}$) and $T \in$ VPT such that $T(V) = C$. Therefore we have that $T(V) \subseteq V'$ iff $C \subseteq V'$ which is undecidable as established in Theorem 1.   □

4   Synchronized Visibly Pushdown Transducers

We define here a restricted class of transducers that allow typechecking. The idea is to
synchronize the insertion, respectively the deletion, of a call with the insertion, respectively the deletion, of the matching return.

Definition 4 (SVPT). A *synchronized visibly pushdown transducer* is a VPT such that
$\Gamma = \Gamma_{copy} \uplus \Gamma_{del} \uplus \Gamma_{ins} \uplus \{\bot\}$ and such that if $(q, \alpha, \beta, q', \gamma) \in \delta_c$ or $(q, \gamma, \alpha, \beta, q') \in \delta_r$ then either: (i) $\alpha = \epsilon, \beta \neq \epsilon$ and $\gamma \in \Gamma_{ins} \cup \{\bot\}$, (ii) $\alpha \neq \epsilon, \beta = \epsilon$ and $\gamma \in \Gamma_{del} \cup \{\bot\}$, or (iii) $\alpha \neq \epsilon, \beta \neq \epsilon$ and $\gamma \in \Gamma_{copy} \cup \{\bot\}$.[6] The set of synchronized visibly pushdown transducer is noted SVPT.

Example 3. T_{del} of Example 2 is a SVPT with $\Gamma_{del} = \Gamma, \Gamma_{ins} = \emptyset$ and $\Gamma_{copy} = \emptyset$.
On the other hand, T_{ins} is not a SVPT since γ is used for inserting, see transition
$(q_0, \gamma, \epsilon, \underline{b}, q_1)$, and for copying, see transition $(q_0, \bar{a}, \bar{a}, q_0, \gamma)$.

The next proposition states the class SVPT is closed by inverse.

Proposition 4. *Let $T \in$ SVPT then $T^{-1} \in$ SVPT.*

Proof. T^{-1} is a VPT (Proposition 2). Moreover, with $\Gamma = \Gamma'_{copy} \uplus \Gamma'_{del} \uplus \Gamma'_{ins} \uplus \{\bot\}$ where $\Gamma'_{copy} = \Gamma_{copy}, \Gamma'_{del} = \Gamma_{ins}$ and $\Gamma'_{ins} = \Gamma_{del}$, this transducer is synchronized.   □

In the next proposition, we establish that the transduction or inverse transduction of a
VPL by a SVPT is not always a VPL. We note SVPT(RL) = $\{S(R) \mid S \in$ SVPT, $R \in$ RL$\}$ and SVPT(VPL) = $\{S(V) \mid S \in$ SVPT, $V \in$ VPL$\}$.

Proposition 5. VPL $\subsetneq$ SVPT(RL) = SVPT(VPL) $\subsetneq$ CFL.

Proof. First, for VPL $\subsetneq$ SVPT(RL), consider T_{del} of Example 2, it transduces a RL into
a CFL that is not a VPL (see Proposition 3), T_{del} is a SVPT. Second, for SVPT(RL) =
SVPT(VPL), consider any $S \in$ SVPT and $A \in$ VPA. Let $V = L(A)$. We construct
$S' \in$ SVPT such that $S'(\hat{\Sigma}^*) = S(V)$. More concretely, we impose that, for all $w \in \hat{\Sigma}^*$ we have that $S'(w) = S(w)$ when $w \in V$ and $S'(w) = \emptyset$ otherwise. To achieve
that, S' simulates S and A: it translates w as S does and, in parallel, it simulates A on w.
A run of S' is accepting if the corresponding runs in S and A are accepting. It is crucial
to note that the parallel simulation of the stacks of S and A is only possible because S
is a SVPT: each time that it copies, respectively deletes or inserts, a call, it will copy,
respectively delete or insert the *matching* return. As a consequence the content of the
stack of S and A can be represented as pairs of symbols as follows:

---

[6] As SVPT are VPT, call transitions are not allowed to push $\bot$.

- *call-return copy*: when A and S are moving and pushing a symbol γ and $\gamma' \in \Gamma_{copy}$ on their respective stack, S' pushes the symbol (γ, γ'). As S is a SVPT and $\gamma' \in \Gamma_{copy}$, this ensures that when we reach the *matching* return, S copies the return, and the pair (γ, γ') will be popped from the stack. This simulates the behavior of the stacks of A and S. From there, S' continues the parallel simulation of A and S.
- *call-return delete*: when A and S are moving and pushing a symbol γ and $\gamma' \in \Gamma_{del}$ on their respective stack, S' pushes the symbol (γ, γ'). As S is a SVPT and $\gamma' \in \Gamma_{del}$, this ensures that when we reach the *matching* return, S will delete the return, and the pair (γ, γ') will be popped from the stack. This simulates the behavior of the stacks of A and S. From there, S' continues the parallel simulation of A and S.
- *call-return insert*: on a call-return insert, only S is moving. It pushes a symbol $\gamma' \in \Gamma_{ins}$ on its stack. To simulate this, S' pushes the pair $(\gamma_\epsilon, \gamma')$ on its stack, γ_ϵ being a new stack symbol that does not belong to the stack symbols of A. As γ' belongs to Γ_{ins}, we know that the *matching* return will be inserted (so no input will be read and A will not move), at that time S' will pop the pair $(\gamma_\epsilon, \gamma')$, not moving on the input. This simulates the behavior of the stacks of A and S. From there, S' continues the parallel simulation of A and S.
- Other cases are treated similarly.

Third, SVPT(VPL) $\subseteq$ CFL is a consequence of the facts that SVPT(RL) $\subseteq$ VPT(RL) $\subseteq$ CFL and SVPT(RL) $=$ SVPT(VPL). Finally, SVPT(VPL) $\neq$ CFL is a consequence of the fact that typechecking SVPT against VPL is decidable (Theorem 3, see below) and the undecidability of checking the inclusion of a CFL into a VPL (Theorem 1). □

Non-deleting and non-inserting transducers. Two important subclasses of SVPT are the class of transducers that do not insert and the ones that do not delete.

Definition 5. A *non-inserting* SVPT $T = (Q, Q_0, Q_f, \Gamma, \delta)$ is a SVPT such that (i) $\delta_c \subseteq Q \times \Sigma^c \times \Sigma_\epsilon^c \times Q \times \Gamma$, (ii) $\delta_r \subseteq Q \times \Gamma \times \Sigma^r \times \Sigma_\epsilon^r \times Q$ and (iii) $\delta_i \subseteq Q \times \Sigma^i \times \Sigma_\epsilon^i \times Q$ (and thus $\Gamma_{ins} = \emptyset$). This class is noted SVPT$_{ni}$. A *non-deleting* SVPT $T = (Q, Q_0, Q_f, \Gamma, \delta)$ is a SVPT such that (i) $\delta_c \subseteq Q \times \Sigma_\epsilon^c \times \Sigma^c \times Q \times \Gamma$, (ii) $\delta_r \subseteq Q \times \Gamma \times \Sigma_\epsilon^r \times \Sigma^r \times Q$ and (iii) $\delta_i \subseteq Q \times \Sigma_\epsilon^i \times \Sigma^i \times Q$ (and thus $\Gamma_{del} = \emptyset$). This class is noted SVPT$_{nd}$.

Proposition 6. *Let* $T \in$ SVPT,

1. $T \in$ SVPT$_{nd}$ *iff* $T^{-1} \in$ SVPT$_{ni}$,
2. *if* $T \in$ SVPT$_{nd}$ *and* $V \in$ VPL *then* $T(V) \in$ VPL,
3. *if* $T \in$ SVPT$_{ni}$ *and* $V \in$ VPL *then* $T^{-1}(V) \in$ VPL.

Proof. The first assertion is a direct consequence of Proposition 2 stating that T^{-1} is also a VPT and the fact that the inverse transducer of a non-inserting, respectively non-deleting, transducer is obviously a non-deleting, respectively non-inserting, transducer.

Our proof for the second assertion is constructive. Given the SVPT$_{nd}$ T, and the VPA A^{in} for V, we construct a VPA A^{out} that accepts $T(V)$. We sketch here the main arguments of the proof, the full detailed proof is given in [10]. On a word w, A^{out} guesses a word v and checks that the pair $(v, w) \in [\![T]\!]$ and $v \in V$. For that, the VPA

A^{out} simulates in parallel the execution of A^{in} on v and the execution of T on the pair (v, w), its run is accepting if the simulated runs in A^{in} and T are accepting. The main delicate part of the proof is to show that A^{out} can simulate the two stacks while respecting the restrictions imposed to a VPA. The parallel simulation of the stack of A^{in} and T is possible because T is a SVPT$_{nd}$: each time that it copies, respectively inserts, a call, it will copy, respectively insert, the *matching* return. As a consequence, the content of the stacks of A^{in} and T can be represented as pairs of symbols as follows:

- *call-return copy*: when A^{in} and T are moving and pushing a symbol γ^{in} and $\gamma^T \in \Gamma^T_{copy}$ on their respective stack, A^{out} pushes the symbol (γ^{in}, γ^T) and reads in w the same symbol as written by T. As T is a SVPT$_{nd}$ and $\gamma^T \in \Gamma^T_{copy}$, this ensures that when we reach the *matching* return in v (and A^{in} pop γ^{in}), T copies the return in w and pop γ^T. At that time, A^{out} pops the pair (γ^{in}, γ^T) from its stack and reads in w the same symbol as written by T. This simulates the behavior of the stacks of A^{in} and T. From there, A^{out} continues the parallel simulation of A^{in} and T.

- *call-return insert*: on a call-return insert, only T is moving. It pushes a symbol $\gamma^T \in \Gamma^T_{ins}$ on its stack and write β. To simulate this, A^{out} reads β and pushes the pair $(\gamma_\epsilon, \gamma^T)$ on its stack, γ_ϵ being a new stack symbol that does not belong to the stack symbols of A^{in}. As γ^T belongs to Γ^T_{ins}, we know that the *matching* return in w, say β', will be inserted (no letter will be read by T and so A^{in} will not move), at that time A^{out} will pop the pair $(\gamma_\epsilon, \gamma^T)$ when reading β'. This simulates the behavior of the stacks of A^{in} and T. From there, A^{out} continues the parallel simulation of A^{in} and T.

Note that A^{out} could not simulate a transducer that deletes matching calls and returns as it would have to modify its stack while not reading any letter, which is not allowed in a VPA. The last assertion is a direct consequence of the first and the second.   □

We can now prove that type checking is decidable for SVPT.

Theorem 3. *Let* $A_1, A_2 \in$ VPA *and* $T \in$ SVPT, *the problem of checking if* $T(L(A_1)) \subseteq L(A_2)$ *is* EXPTIME-COMPLETE, *the problem is* PTIME-COMPLETE *when* A_2 *is deterministic.*

Proof. We know that checking inclusion between two VPL is EXPTIME-HARD (Proposition 1), if we choose T to be the identity transducer (which is a SVPT), we obtain the hardness part. For the easiness part, we first show that T is equivalent to the composition of two transducers: $[\![T]\!] = [\![T_{ni}]\!] \circ [\![T_{nd}]\!]$, which are respectively *non-inserting* and *non-deleting*.

T_{nd} will behave as T with the essential difference that whenever T deletes a *call*, a *return*, respectively an *internal*, T_{nd} replaces it with ϵ^c, ϵ^r, respectively ϵ^i which are new call, return, respectively internal symbols that do not belong to the alphabet $\hat{\Sigma}$. More formally, let $T = (Q, Q_0, Q_f, \Gamma, \delta)$ over $\hat{\Sigma}$, T_{nd} is a transducer from $\hat{\Sigma}$ into $\hat{\Sigma}_{nd} = \Sigma^c_{nd} \uplus \Sigma^r_{nd} \uplus \Sigma^i_{nd}$ where $\Sigma^c_{nd} = \Sigma^c \uplus \{\epsilon^c\}$, $\Sigma^r_{nd} = \Sigma^r \uplus \{\epsilon^r\}$, $\Sigma^i_{nd} = \Sigma^i \uplus \{\epsilon^i\}$. We define $T_{nd} = (Q, Q_0, Q_f, \Gamma^{nd}, \delta_{nd})$, such that $\Gamma^{nd} = \Gamma^{nd}_{copy} \uplus \Gamma^{nd}_{ins} \uplus \{\perp\}$, where $\Gamma^{nd}_{copy} = \Gamma_{copy} \uplus \Gamma_{del}$, and $\Gamma^{nd}_{ins} = \Gamma_{ins}$, $\delta_{nd} = \delta'_c \cup \delta'_r \cup \delta'_i$, $\delta'_c =$

$\{(q, \alpha, b, q', \gamma) \in \delta_c \mid \alpha \in \hat{\Sigma}_{\epsilon}^c, b \in \hat{\Sigma}^c\} \cup \{(q, a, \epsilon^c, q', \gamma) \mid (q, a, \epsilon, q', \gamma) \in \delta_c, a \in \hat{\Sigma}^c\}, \delta'_r = \{(q, \gamma, \alpha, b, q') \in \delta_r \mid \alpha \in \hat{\Sigma}_{\epsilon}^r, b \in \hat{\Sigma}^r\} \cup \{(q, \gamma, a, \epsilon^r, q') \mid (q, \gamma, a, \epsilon, q') \in \delta_r, a \in \hat{\Sigma}^r\}, \delta'_i = \{(q, \alpha, b, q') \in \delta_i \mid \alpha \in \hat{\Sigma}_{\epsilon}^i, b \in \hat{\Sigma}^i\} \cup \{(q, a, \epsilon^i, q') \mid (q, a, \epsilon, q') \in \delta_i, a \in \hat{\Sigma}^i\}$. Note that T_{nd} does not erase and so the partition of the stack symbols is different from the one for T, and clearly we have $T_{nd} \in$ SVPT$_{nd}$. Furthermore, by construction, we have that for all $(w_1, w_2) \in [\![T]\!]$, there exists w_3 such that $(w_1, w_3) \in [\![T_{nd}]\!]$ and $w_2 = \downarrow^{\hat{\Sigma}}(w_3)$, and conversely, for all $(w_1, w_3) \in [\![T_{nd}]\!]$, $(w_1, \downarrow^{\hat{\Sigma}}(w_3)) \in [\![T]\!]$. As T is a SVPT, for all $w \in T_{nd}(\hat{\Sigma}^*)$, w is such that the matching return of every ϵ^c is an ϵ^r, and conversely. This property is easily proved by induction on the length of runs of T_{nd}. We say that those words are *synchronized* on the pair (ϵ^c, ϵ^r).

We define the transducer T_{ni}. For all w_1 that are synchronized on the pair (ϵ^c, ϵ^r), the transducer accepts the pairs $(w_1, \downarrow^{\hat{\Sigma}}(w_1)) \in \hat{\Sigma}_{nd} \times \hat{\Sigma}$. Clearly, this transduction relation is realized by the following transducer $T_{ni} = (\{q\}, \{q\}, \{q\}, \{\gamma_{copy}, \gamma_{del}, \perp\}, \delta_{ni})$, on $(\hat{\Sigma} \cup \{\epsilon^c, \epsilon^r, \epsilon^i\}) \times \hat{\Sigma}$, such that $\delta_{ni} = \delta''_c \cup \delta''_r \cup \delta''_i$, $\delta''_c = \{(q, a, a, q, \gamma_{copy}) \mid a \in \Sigma^c\} \cup \{(q, \epsilon^c, \epsilon, q, \gamma_{del})\}$, $\delta''_r = \{(q, \gamma_{copy}, a, a, q) \mid a \in \Sigma^r\} \cup \{(q, \gamma_{del}, \epsilon^r, \epsilon, q)\} \cup \delta''_r = \{(q, \perp, a, a, q) \mid a \in \Sigma^r\} \cup \{(q, \perp, \epsilon^r, \epsilon, q)\}$, and $\delta''_i = \{(q, a, a, q) \mid a \in \Sigma^i\} \cup \{(q, \epsilon^i, \epsilon, q)\}$ which is in the class SVPT$_{ni}$.[7] Clearly, $[\![T]\!] = [\![T_{ni}]\!] \circ [\![T_{nd}]\!]$.

To finish the proof, we consider the following equivalence: $T(L(A_1)) \subseteq L(A_2) \Leftrightarrow T_{nd}(L(A_1)) \cap T_{ni}^{-1}(\overline{L(A_2)}) = \emptyset$. The proof of Proposition 6 tells us that we can construct, in deterministic polynomial time in the size of T_{nd} and of A_1, a VPA B_1 that accepts the language $T_{nd}(L(A_1))$. Also, Proposition 1 tells us that we can compute, in deterministic exponential time in the size of A_2, an automaton B_2 that accepts $\overline{L(A_2)}$ (in polynomial time if A_2 is deterministic), and we can construct, in deterministic polynomial time in the size of B_2 and T_{ni}^{-1}, a VPA B_3 that accepts $T_{ni}^{-1}(\overline{L(A_2)})$. Finally, checking emptiness of intersection between two VPA can be done in deterministic polynomial time (Proposition 1). This concludes our proof of EXPTIME-EASINESS (PTIME-EASINESS if A_2 is deterministic). □

The following proposition states that any CFL can be obtained by applying two SVPT on a VPL.

Proposition 7. *For all* $C \in$ CFL$(\hat{\Sigma})$, *there exist* $V \in$ VPL$(\hat{\Sigma})$, $T_1, T_2 \in$ SVPT *such that* $T_2(T_1(V)) = C$.

Proof. First, the proof of Lemma 1 tells us that there exists $V \in$ VPL$(\hat{\Sigma})$ such that $\tau_2(V) = C$, where $\tau_2 : \{c^i, r^i, i^i\} \times \hat{\Sigma} \to \hat{\Sigma}$ is defined as: $\tau_2(x^i a^y) = a^x$. We now show that τ_2 can be expressed as the composition of two SVPT. We decompose τ_2 into the following two functions. First, $\tau_3 : \{c^i, r^i, i^i\} \times \hat{\Sigma} \to \{c^i, r^i, i^i\} \times \hat{\Sigma}^i$ defined as: $\tau_3(x^i a^y) = x^i a^i$. Second, $\tau_4 : \{c^i, r^i, i^i\} \times \hat{\Sigma}^i \to \hat{\Sigma}$ defined as: $\tau_4(x^i a^i) = a^x$. Clearly, those two functions can be expressed as SVPT and $\tau_2 = \tau_4 \circ \tau_3$. □

As a consequence of Proposition 7 and Theorem 1, we cannot type check the composition of two SVPT against VPL.

---

[7] Note that without the hypothesis of *synchronized* on the pair (ϵ^c, ϵ^r), there is no SVPT$_{ni}$ that realizes $\downarrow^{\hat{\Sigma}}$, that is the reason why this construction can not be generalized when T is a VPT.

Theorem 4. *Let* $A_1, A_2 \in$ VPA *and* $T_1, T_2 \in$ SVPT, *it is undecidable whether* $T_1(T_2(L(A_1))) \subseteq L(A_2)$.

Fully synchronized visibly pushdown transducers. We finish this section by introducing a class of VPT that maintain regularity, are closed under inverse and under composition and for which type checking is decidable.

Definition 6 (FSVPT). A *fully synchronized visibly pushdown transducer* is a synchronized visibly pushdown transducer which is both non-inserting and non-deleting. This class is noted FSVPT.

Theorem 5. *Let* $T \in$ FSVPT, *then:*

1. VPL *preservation: for any* $V \in$ VPL, $T(V) \in$ VPL;
2. *Inverse:* $T^{-1} \in$ FSVPT;
3. *Composition: for any* $T_1 \in$ FSVPT *there exists* $T_2 \in$ FSVPT *such that* $[\![T_2]\!] = [\![T_1]\!] \circ [\![T]\!]$;
4. *Decidable type-checking: given two* VPA A_1, A_2, *deciding* $T(L(A_1)) \subseteq L(A_2)$ *is* EXPTIME-COMPLETE.

Note that FSVPT $=$ SVPT$_{nd}$∩SVPT$_{ni}$, this class is exactly the class of VPT that do not delete nor insert. Moreover, we could define *finite state transducers* to transduce words on $\hat{\Sigma}$, such that calls are mapped on calls, returns on returns, and internals on internals. This class would be a strict subclass of FSVPT as such automata would translate languages from RL($\hat{\Sigma}$) into RL($\hat{\Sigma}$) while FSVPT can transduce languages from RL($\hat{\Sigma}$) into languages that are not in RL($\hat{\Sigma}$). Finally, if T is a FSVPT then it can be seen as a VPA that works on pairs of symbols (of the same type), and so, equivalence between FSVPT is EXPTIME-COMPLETE.

5   Conclusion

In this paper, we have identified two interesting sub-classes of pushdown transducers. SVPT (synchronized visibly pushdown transducer) is a powerful subclass with a decidable (EXPTIME-COMPLETE) type checking problem against VPL. This positive result is surprising as we have shown that SVPT do not preserve VPL. Also, the class of SVPT is not closed under composition. This has triggered the definition of FSVPT (fully synchronized visibly pushdown transducers), this class of transducers enjoys nice properties like preservation of VPL, closure to composition and decidable (EXPTIME-COMPLETE) type checking problem against VPL.[8]

Alur and Madushudan have shown in [4] that VPL are equivalent to regular languages of *nested words*. Our results can be rephrased in this setting as well. In [2], Alur has studied the relation between VPL and tree languages. In future work, we will study in details the relation between the transducers defined on regular tree languages, as

---

[8] A. Thomo et al. defined in [11] a class of visibly pushdown transducers equivalent to ours. However, their article does not study this class of transducers per se and they incorrectly states that VPT maintains VPL in contradiction with our Proposition 3.

defined in [7,8], and our transducers. It seems pretty clear that their expressive power are incomparable but a fine comparison requires a large effort of formalization and is beyond the subject of this paper. As already said, those works on transducers were often motivated by the application in XML, we will study the practical advantages and drawbacks of our transducers for that application in future work.

In [6], Fisman and Pnueli use context-free languages for extending regular model-checking. CFL are used to model the set of initial states of the system, the transition relation as well as the specification (the set of *good* states) are given by finite state transducers and automata, respectively. We conjecture that FSVPT can be used to rephrase and extend those results by offering an unified framework for regular model-checking in the context of VPL, as FSVPT are preserving VPL. We will investigate this important application in future work.

Acknowledgement. We want to thank Laurent Van Begin for suggesting the main idea of the proof for Theorem 1 and Ahmed Bouajjani for pointing us the paper of Fisman and Pnueli.

References

1. Abdulla, P., Legay, A., d'Orso, J., Rezine, A.: Tree regular model checking: A simulation-based approach. J. Log. Algebr. Program. 69(1-2), 93–121 (2006)
2. Alur, R.: Marrying words and trees. In: PODS, pp. 233–242 (2007)
3. Alur, R., Madhusudan, P.: Visibly pushdown languages. In: STOC, pp. 202–211 (2004)
4. Alur, R., Madhusudan, P.: Adding nesting structure to words. In: H. Ibarra, O., Dang, Z. (eds.) DLT 2006. LNCS, vol. 4036, pp. 1–13. Springer, Heidelberg (2006)
5. Bouajjani, A., Jonsson, B., Nilsson, M., Touili, T.: Regular model checking. In: CAV, pp. 403–418 (2000)
6. Fisman, D., Pnueli, A.: Beyond regular model checking. In: Hariharan, R., Mukund, M., Vinay, V. (eds.) FSTTCS 2001. LNCS, vol. 2245, pp. 156–170. Springer, Heidelberg (2001)
7. Maneth, S., Berlea, A., Perst, T., Seidl, H.: XML type checking with macro tree transducers. In: PODS, pp. 283–294 (2005)
8. Martens, W., Neven, F.: Typechecking top-down uniform unranked tree transducers. In: Calvanese, D., Lenzerini, M., Motwani, R. (eds.) ICDT 2003. LNCS, vol. 2572, pp. 64–78. Springer, Heidelberg (2002)
9. Milo, T., Suciu, D., Vianu, V.: Typechecking for XML transformers. In: PODS, pp. 11–22 (2000)
10. Raskin, J.-F., Servais, F.: Visibly pushdown transducers. Technical Report 2008.100, Federated Center for Verification - Université Libre de Bruxelles (2008),
    http://www.ulb.ac.be/di/ssd/cfv/publications.html
11. Thomo, A., Venkatesh, S., Ying Ye, Y.: Visibly pushdown transducers for approximate validation of streaming XML. In: FoIKS, pp. 219–238 (2008)

The Non-deterministic Mostowski Hierarchy and Distance-Parity Automata

Thomas Colcombet[1],[*] and Christof Löding[2]

[1] LIAFA/CNRS, France
[2] RWTH Aachen, Germany

Abstract. Given a Rabin tree-language and natural numbers i, j, the language is said to be i, j-feasible if it is accepted by a parity automaton using priorities $\{i, i+1, ..., j\}$. The i, j-feasibility induces a hierarchy over the Rabin-tree languages called the Mostowski hierarchy.

In this paper we prove that the problem of deciding if a language is i, j-feasible is reducible to the uniform universality problem for distance-parity automata. Distance-parity automata form a new model of automata extending both the nested distance desert automata introduced by Kirsten in his proof of decidability of the star-height problem, and parity automata over infinite trees. Distance-parity automata, instead of accepting a language, attach to each tree a cost in $\omega + 1$. The uniform universality problem consists in determining if this cost function is bounded by a finite value.

1 Introduction

Finite automata running on infinite trees, originally introduced by Rabin in his seminal work [15] are now widely considered as one of the key paradigms for understanding many logics relevant to verification. Those automata are known to be effectively equivalent to monadic second-order logic, μ-calculus, and to subsume all the standard temporal logics.

An important parameter in the definition of the automaton model is the acceptance condition. This acceptance condition determines, given a run of the automaton, whether it is accepting or not. Different (often equivalent) choices of acceptance conditions are known from the literature such as Büchi, Rabin, Muller, or Streett conditions (cf. [16]). Though all possess their own interest, the *parity condition* has emerged for many reasons as the central condition in the theories of automata, logic and games.

When using a parity condition, each state of the automaton is labelled by a natural number – called a priority – belonging to a fixed finite interval $[i, j]$. A run is accepting if on every branch the highest priority seen infinitely often is even. A language is said to be i, j-feasible if there exists a finite automaton using the interval of priorities $[i, j]$ accepting this language. Of course, the language does not change if we shift all priorities by steps of 2 or -2. This is why we

[*] Supported by the AutoMathA program of the ESF.

L. Aceto et al. (Eds.): ICALP 2008, Part II, LNCS 5126, pp. 398–409, 2008.
© Springer-Verlag Berlin Heidelberg 2008

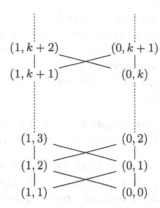

Fig. 1. Hierarchy of Mostowski indices

can restrict ourselves to $i = 0$ or $i = 1$. It is also clear that the bigger the interval $[i, j]$ is, the more tree languages are i, j-feasible. Mostowski first studied this parameter [10], and the corresponding ladder-shaped hierarchy – depicted Figure 1 – is named after him.

This Mostowski hierarchy exists in different variants according to the nature of the transition relation used by the automaton. Over trees the hierarchy is strict for all standard models of automata: deterministic [18] (even over words), non-deterministic [11], and alternating [1] (in combination with [4,9]), see also [2]. The hierarchy collapses over words for non-deterministic automata to the Büchi level $(1, 2)$, and in the alternating case to the intersection of levels $(0, 1)$ and $(1, 2)$.

The next step in the study of this hierarchy is the question of decidability. The problem is the following: given a regular language of infinite trees L (i.e., accepted by a non-deterministic or alternating automaton), and natural numbers $i \leq j$, is the language i, j-feasible? This question is parameterised both by the nature of the language L, that we call the *input* language, and the nature of the class of automata for which we test the i, j-feasibility, the *output* class.

In the case of any non-deterministic automaton as input, the $1, 1$-feasibility and the $0, 0$-feasibility in the non-deterministic Mostowski hierarchy is decidable for simple reasons: a language is at level $(1, 1)$ iff it consists solely of finite trees, and a language is at level $(0, 0)$ iff it is closed for the standard topology over infinite trees. Those two properties are easily shown to be decidable. Note also that the non-deterministic and the alternating hierarchy coincide over those two levels.

More interestingly, the problem is known to be decidable in the Mostowski hierarchy of languages accepted by deterministic automata. The problem is decidable if both the input language and the output class are deterministic [13] (see also [14] for more details). The problem is also known to be decidable if the input language is deterministic, and the output class is non-deterministic [14] (which is a refinement of the case of a path language as input [13]). The special case of deciding if a deterministic language is accepted by a non-deterministic Büchi automaton (i.e., $1, 2$-feasible in the non-deterministic Mostowski hierarchy) was formerly settled in [17]. Let us finally remark that every deterministic language is alternating co-Büchi, i.e., that every deterministic language falls in the level $(0, 1)$ of the alternating Mostowski hierarchy. And hence the case of a deterministic language as input and an alternating one as output is also settled.

This paper is the first part in an attempt to show the decidability of the following problem.

Problem 1. Given a regular tree language L and natural numbers $i \leq j$, answer whether L is i, j-feasible in the non-deterministic Mostowski hierarchy or not.

The scheme of the proof is inspired from the proof of decidability of the (restricted) star-height problem due to Kirsten [7] (the problem was originally solved by Hashiguchi [6]). The star-height problem is the following: given a regular language of finite words and a natural number k, is it possible to describe the language by a regular expression using at most k nesting of Kleene stars?

For showing the decidability of the star-height problem, Kirsten introduces a new class of automata: nested distance desert automata. Those are non-deterministic finite automata running on finite words and equipped with some counting features. The semantics of such automata is either to reject a word, or to compute a natural number for it, that we can see as the price to pay for accepting it. Hence a nested distance desert automaton defines a partial mapping from words to natural numbers. The proof then goes in two steps.

- Reduce the star-height problem to the limitedness problem for nested distance desert automata (the limitedness is the problem of determining if the partial mapping defined by the automaton is bounded).
- Solve the limitedness problem for nested distance desert automata.

We want to follow exactly the same scheme to solve Problem 1. We introduce the family of distance-parity automata running over infinite trees. Those automata combine the features of nested distance desert automata and of parity automata. If we restrict a distance-parity automaton to run on finite words, we fall back to the class of automata defined by Kirsten. If we restrict those automata to infinite words, we get a family of automata equivalent to the hierarchical ωB-automata in [3]. The limitedness problem still makes sense for the distance-parity automata, but we prefer to work with the uniform universality problem: a distance-parity automaton is uniformly universal if the function it defines is both total and bounded. This problem is decidable for finite words from [7], and can be shown to be decidable over infinite words using [3].

Our proof scheme then goes as follows in two steps:

- Reduce the i, j-feasibility problem to the uniform universality problem for distance-parity automata (Theorem 2).
- Show the decidability of the uniform universality problem for distance-parity automata (open).

This paper is concerned with the first item. This reduction part is very different from the one in the proof of Kirsten. Indeed, there is an intrinsic difficulty in the study of regular languages of infinite trees: there is no known notion of a canonical object describing a language. When dealing with finite words, one can use the minimal deterministic automaton or the syntactic monoid. When dealing with infinite words, one can use the syntactic ω-semigroup. When dealing with finite trees, there exist a minimal bottom-up deterministic automaton, and a corresponding algebraic presentation. Most methods for characterising classes of languages begin by taking such a canonical presentation. But for infinite trees no canonical type of acceptor is known. This problem is very deep as it can be witnessed by the following fact: some languages are inherently ambiguous, i.e., it is impossible to provide an automaton that would possess a single accepting run over every accepted input (the proof of this result is given in [12] but has not been published; see also [5]). One contribution of this paper is the notion of a *guidable automaton*, i.e., an automaton that is able to 'mimic' the behaviour of every automaton accepting the same language. This guidable automaton plays the role of the canonical presentation in our reduction. We show that every regular tree language is accepted by a guidable automaton (Theorem 1).

The second part of our reduction is an 'on the fly' optimisation of the guidable automaton. This optimisation process makes use of the distance features of distance-parity automata. It is shown to be both correct, and optimal.

The remainder of the paper is organised as follows. Section 2 is devoted to definitions, in particular automata and the acceptance conditions we use. Section 3 presents the notion of guidable automaton and we prove that such an automaton exists for each regular language of trees. In Section 4 we establish Theorem 2 reducing the i, j-feasibility problem to the uniform universality problem.

2   Definitions

Words are finite sequences of letters. The set of words over an alphabet A is denoted by A^*. The empty word is ε, and uv is the concatenation of u and v. We denote by $\sqsubseteq$ the prefix relation over words and by $\sqsubset$ its strict variant. The length of a word u is $|u|$, and $|u|_a$ for $a \in A$ is the number of occurrences of a in u. An ω-word is an infinite sequence of letters and by A^ω we denote the set of infinite words over A. The ordered set of natural numbers is written as ω, and $\omega + 1$ is ω augmented with the maximal value ω.

For the sake of simplicity, we assume that trees are binary and complete (i.e., with no leaves)[1]. A *tree* labelled by a finite *alphabet* A (we also say an A-tree) is a mapping from $\{0,1\}^*$ to A. The elements of $\{0,1\}^*$ are called *nodes*. A *branch* is a maximal totally ordered set of nodes. A branch naturally induces an ω-word in A^ω. It is sometimes convenient to identify a branch with this ω-word.

An *automaton* is a tuple $\mathcal{A} = (Q, A, I, \Delta, col)$ in which:

---

[1] We can code leaves in an infinite binary tree by marking all nodes below by a special dummy symbol. It is easy to show that if the interval $[i, j]$ contains an even priority, then the original language is i, j-feasible iff the one after coding is i, j-feasible. This means that we cannot treat the case of 1, 1-feasibility. However we have seen that 1, 1-feasibility is easy by other arguments.

– Q is a finite set of *states*, $I \subseteq Q$ is the set of *initial states*,
– A is the alphabet,
– $\Delta \subseteq Q \times A \times Q \times Q$ is the *transition relation*,
– $col : Q \to Cols$ is a *colour mapping* to some finite set $Cols$ of *colours*.

A *run* ρ of an automaton over an A-tree t is a Q-tree such that $\rho(\varepsilon) \in I$ and for all $v \in \{0,1\}^*$, $(\rho(v), t(v), \rho(v0), \rho(v1)) \in \Delta$. The $Cols$-tree $col(\rho)$ denotes $(col \circ \rho)$. Each automaton – depending on its nature – comes with a mapping val from $Cols^\omega$ to $\omega + 1$. The *value* $val(\rho)$ of a run ρ is the supremum of $val(\beta)$ over all branches β of $col(\rho)$. The *value* $\mathcal{A}(t)$ of an A-tree t is the minimum of $val(\rho)$ for ρ ranging over all runs over t (by default ω if there is no such run).

We are now ready to introduce the different value mappings used throughout the paper. The *parity* condition corresponds to an interval $[i,j]$ of natural numbers – called *priorities* – as set of colours. Given an infinite sequence of priorities $u \in [i,j]^\omega$, $val(u)$ is 0 if the maximal priority appearing infinitely often is even, else it is ω. A *parity automaton* is an automaton using a parity condition. A *tree* t is accepted by such an automaton iff $\mathcal{A}(t) = 0$. The *parity index* (or Mostowski index) of a parity automaton is the pair (i,j) of the minimal and maximal priorities used in the automaton. We designate by $L(\mathcal{A})$ the set of trees that are accepted by the parity automaton $\mathcal{A}$. A language is *regular* if it is equal to $L(\mathcal{A})$ for some parity automaton $\mathcal{A}$, and is *i,j-feasible* if furthermore $\mathcal{A}$ has parity index (i,j).

A *distance condition* (corresponding to nested distance desert automata in [7], and hierarchical B-automata in [3]) is defined for a totally ordered set of colours $D = \{d_1, r_1, \ldots, d_k, r_k\}$ with the order $d_1 < r_1 < \cdots < d_k < r_k$. The colours $d_1, \ldots, d_k$ are called *distance colours*, while the colours $r_1, \ldots, r_k$ are *reset colours*. Given an infinite sequence $u \in D^\omega$, its value $val(u)$ is the supremum of $|v|_{d_i}$ where v ranges over all finite factors of u such that the maximal colour of v is d_i. One can see this as having k counters numbered from 1 to k. When seeing d_i the corresponding counter is incremented and all the counters below are reset. When seeing r_i, all the counters up to i are reset. The value of a sequence is the supremum of all values of counters seen during this process (starting from 0).

We can derive from the two previous mappings a last one, the *distance-parity* mapping, which can be seen as a conjunction of a distance and a parity condition. It is described for a set of colours of the form $Cols = D \times [i,j]$ where D is the ordered set of colours of a distance condition and $[i,j]$ is a finite interval of natural numbers. For a sequence $u \in Cols^\omega$ one derives the two corresponding sequences $u_1 \in D^\omega$ and $u_2 \in [i,j]^\omega$ obtained by projection to the first and second components of the elements in u, respectively. The value $val(u)$ is the maximum of $val(u_1)$ (as a distance condition) and $val(u_2)$ (as a parity condition).

Given a distance-parity automaton $\mathcal{A}$, we define for each $N < \omega$ the language $L^{(N)}(\mathcal{A}) := \{t \; : \; \mathcal{A}(t) \leq N\}$. In this way $\mathcal{A}$ defines a non-decreasing ω-sequence of languages, i.e., an ω-chain of languages. It is easy to observe that for each N, $L^{(N)}(\mathcal{A})$ is a regular language: For a fixed N the counters for the distance colours can be coded in the states of the automaton. This construction gives a parity automaton of index (i,j) for $L^{(N)}(\mathcal{A})$, hence we define the *parity index* of a distance-parity automaton to be the parity index of the underlying parity condition. From the above explanation we easily conclude the following.

Fact 1. *Given a distance-parity automaton A of parity index (i, j) and a natural number N, the language $L^{(N)}(A)$ is i, j-feasible.*

The *limitedness problem* is the following: given a distance-parity automaton A, determine if $L^{(N)}(A)$ is ultimately constant (in this case, the automaton is said to be *limited*). In this paper we prefer the *uniform universality problem*: given an automaton A, determine if $L^{(N)}(A)$ is equal to the set of all trees for some natural number N (in this case the automaton is said to be *uniformly universal*).

3   Guidable Automata

As mentioned in the introduction, the first step of the proof is to construct a so-called guidable automaton accepting the language L. The underlying idea is that we want to be able to relate accepting runs of an automaton for L that has a minimal number of priorities to accepting runs of our guidable automaton.

Definition 1. *A parity automaton $A = (Q_A, A, \{q_0\}, \Delta_A, col_A)$ is guidable if for every parity automaton $B = (Q_B, A, I_B, \Delta_B, col_B)$ such that $L(B) \subseteq L(A)$ there exist a mapping $g : Q_A \times \Delta_B \to \Delta_A$ with the following properties:*

- $g(p, (q, a, q', q'')) = (p, a, p', p'')$ *for some* $p', p'' \in Q_A$.
- *For every accepting run ρ of B over a tree t, $g(\rho)$ is an accepting run of A over t, where $g(\rho) = \rho'$ is the unique run such that $\rho'(\varepsilon) = q_0$, and for all $u \in \{0, 1\}^*$:*

$$(\rho'(u), t(u), \rho'(u0), \rho'(u1)) = g(\rho'(u), (\rho(u), t(u), \rho(u0), \rho(u1))) .$$

In this case we say that (B, g) guides A.

One way to see this definition is that g is a deterministic transducer with state set Q_A that takes as input a run ρ of B and outputs a run ρ' of A such that if ρ is accepting (for B), then ρ' is accepting (for A).

An example of an automaton over finite words that is not guidable (the definition of guidable automaton can easily be translated by the reader to the case of finite words) is the automaton that accepts all $\{a, b\}$-words by guessing in the first step of the run if the last letter is a or b, and then proceeds to two subautomata, one accepting words ending with a, the other accepting words ending with b. It is quite clear that it is not possible to guide such an automaton. This example carries the important intuition that an automaton is guidable if it is never forced to make an unnecessary guess concerning the remaining input. Also consistent with this intuition, note that every deterministic automaton is obviously guidable.

In our context of infinite trees, the only way we use the property of being guidable is by the following *simultaneous pumping* argument.

Lemma 2. *Suppose $(\mathcal{B}, g)$ guides $\mathcal{A}$ and consider accepting runs ρ, ρ' as in Definition 1. Let $u \sqsubset v$ be nodes such that $\rho(u) = \rho(v)$ and $\rho'(u) = \rho'(v)$. If the maximal priority in ρ between u and v is even, then the maximal priority in ρ' between u and v is also even.*

Proof. Consider the run τ obtained from ρ by repeating infinitely often the part between u and v (i.e., positions x such that $u \sqsubseteq x$ and $v \not\sqsubseteq x$), and τ' be obtained from ρ' in the same way. If the maximal priority between u and v in ρ is even, then the run τ is accepting. But it is not difficult to see that $g(\tau) = \tau'$ and hence by definition of guidable automata, τ' is also accepting. Thus, the maximal priority n appearing infinitely often on the infinite branch obtained by pumping is even. Since this branch is obtained by pumping ρ' between u and v, the priority n appears as the maximal one between u and v in ρ'. $\square$

The main result of this section is that for each regular tree language we can construct a guidable automaton.

Theorem 1. *For each regular tree language L there exists effectively a guidable parity automaton $\mathcal{A}$ accepting L.*

Proof. We start from a parity automaton $\mathcal{C} = (Q_\mathcal{C}, A, I_\mathcal{C}, \Delta_\mathcal{C}, col_\mathcal{C})$ accepting the complement of L. What we show is that applying a standard complementation procedure to $\mathcal{C}$ – as it can be found, e.g., in [16] – yields a guidable automaton $\mathcal{A}$ (which obviously accepts L). The rough idea why the automaton is guidable is that a positional winning strategy in a game witnessing that an automaton $\mathcal{B}$ has empty intersection with $\mathcal{C}$ can be rewritten into a mapping g which guides $\mathcal{A}$. This is a reasonable approach because both the mapping g and the winning strategy in the above game witness that $L(\mathcal{B}) \subseteq L(\mathcal{A})$.

We now formally describe the automaton $\mathcal{A}$ and then show that it is guidable. By G we denote the set of all mappings $f : \Delta_\mathcal{C} \to \{0, 1\}$. Each path in an $A \times G$ labelled tree corresponds to an infinite sequence over $(A \times G \times \{0, 1\})$, where the $\{0, 1\}$-component indicates the direction the path takes in each step. We say that such a sequence

$$(a_0, f_0, i_0)(a_1, f_1, i_1) \cdots \in (A \times G \times \{0, 1\})^\omega$$

is $\mathcal{C}$-accepting if there is a transition sequence $\tau_0 \tau_1 \cdots \in \Delta_\mathcal{C}^\omega$ such that

- for each j the transition τ_j is of the form $(q_j, a_j, q_j^{(0)}, q_j^{(1)})$ with $f_j(\tau_j) = i_j$ and $q_j^{(i_j)} = q_{j+1}$,
- q_0 is an initial state of $\mathcal{C}$, and
- the acceptance condition of $\mathcal{C}$ is satisfied by $q_0 q_1 \cdots$.

It is easy to see that the set of all $\mathcal{C}$-accepting sequences over $(A \times G \times \{0, 1\})$ is a regular language of infinite words (a non-deterministic automaton can guess the transition sequence of $\mathcal{C}$ and verify the local properties). Hence, the set of all sequences that are not $\mathcal{C}$-accepting is also a regular language, and from a deterministic parity word automaton for this language one constructs a deterministic parity tree automaton (with a single initial state) $\mathcal{A}' = (Q_\mathcal{A}, A \times$

$G, \{q_I^A\}, \Delta_{A'}, col_A)$ accepting all $(A \times G)$-trees in which all paths correspond to non-C-accepting sequences. By projecting away the G-component one obtains the automaton $A = (Q_A, A, \{q_I^A\}, \Delta_A, col_A)$ for the language L. Note that A' and A only differ on the inputs in the transitions.

We show that A is indeed a guidable automaton. Let $B = (Q_B, A, I_B, \Delta_B, col_B)$ be a tree automaton with $L(B) \subseteq L(A)$. The mapping g is constructed from a strategy in the emptiness game for the language $L(B) \cap L(C)$. In this game, Adam wants to verify that $L(B) \cap L(C) \neq \emptyset$ and Eve wants to show the contrary. In other words, Adam plays for constructing both a run of B and a run of C corresponding to the same tree, while Eve wants to show the failure of this construction by witnessing an invalid branch (rejectin for B or for C). The rules of the game are as follows:

1. Adam starts by choosing a starting position $(p_I, q_I) \in I_B \times I_C$.
2. From a position $(p, q) \in Q_B \times Q_C$, Adam picks two transitions $(p, a, p^{(0)}, p^{(1)})$ $\in \Delta_B$ and $(q, a, q^{(0)}, q^{(1)}) \in \Delta_C$ from these states (both using the same input letter a). The game is now in position $((p, a, p^{(0)}, p^{(1)}), (q, a, q^{(0)}, q^{(1)}))$.
3. Eve chooses a direction $i \in \{0, 1\}$ and the game moves to $(p^{(i)}, q^{(i)})$.

The result of the game (the part that is interesting for the winning condition) is an infinite sequence $(p_0, q_0)(p_1, q_1) \cdots \in (Q_B \times Q_C)^\omega$. Eve wins if either $p_0 p_1 \cdots$ does not satisfy the parity condition of B or $q_0 q_1 \cdots$ does not satisfy the parity condition of C.

This is the standard game for verifying emptiness of tree automata (see [16]) and Eve has a winning strategy iff $L(B) \cap L(C) = \emptyset$. Because $L(B) \subseteq L$, we know that Eve has indeed a winning strategy (C accepts the complement of L).

The winning condition for Eve is the disjunction of two parity conditions and hence can be written as a Rabin condition. Therefore Eve has a positional winning strategy (see [8,19]). Such a positional winning strategy is a mapping $\sigma_E : \Delta_B \times \Delta_C \to \{0, 1\}$ assigning to each pair of transitions a direction (for the transition pairs that do not correspond to valid game positions, an arbitrary value can be chosen). It can be equivalently written as a mapping $\sigma_E : \Delta_B \to (\Delta_C \to \{0, 1\})$ assigning to each B-transition a mapping from the set of C-transitions to $\{0, 1\}$. From this we first define $g' : Q_A \times \Delta_B \to \Delta_{A'}$ by $g'(p, (q, a, q', q'')) = (p, (a, f), p', p'')$ for the unique $p', p'' \in Q_A$ such that $(p, (a, f), p', p'') \in \Delta_{A'}$ with $f = \sigma_E(q, a, q', q'')$. The mapping g is then obtained from g' by projecting away the G-component.

We need now to show that g translates accepting B-runs into accepting A-runs. Let ρ be an accepting run of B on some tree t. Assume that $g(\rho)$ is rejecting. Then also the run $g'(\rho)$ is rejecting, where $g'(\rho)$ is a run over the input tree t' that is obtained from t by adding the G-components of the transitions used in $g'(\rho)$ to the node labels of t. This means that there is an infinite branch $i_0 i_1 \cdots$ with labels $(a_0, f_0)(a_1, f_1) \cdots$ in t' such that the sequence $(a_0, f_0, i_0)(a_1, f_1, i_1) \cdots$ is C-accepting. Let $\tau_0 \tau_1 \cdots \in \Delta_C^\omega$ be the transition sequence from the definition of C-acceptance.

Let $\tau_0' \tau_1' \cdots \in \Delta_B^\omega$ be the transition sequence of B along the path $i_0 i_1 \cdots$ in the run ρ. Now assume that in the emptiness game described above Eve plays according to σ_E. One can verify that Adam can play the transition pairs $(\tau'_0, \tau_0), (\tau'_1, \tau_1), \ldots$ against σ_E (because $\sigma_E(\tau'_j, \tau_j) = f_j(\tau_j) = i_j$) and thus wins against σ_E. This contradicts the choice of σ_E as a winning strategy for Eve.   □

4   Reduction from Parity Rank to Uniform Universality

In this section we describe how to reduce the problem of deciding whether a regular tree language is i, j-feasible to the problem of deciding the uniform universality of a distance-parity automaton. We fix a regular language L and an interval $[i, j]$ of natural numbers. We also fix a guidable parity automaton $\mathcal{A} = (Q, A, q_0, \delta, col)$ with $L(\mathcal{A}) = L$ using priorities in some interval P. In the following, we often identify runs ρ of $\mathcal{A}$ with their colouring $col(\rho)$, i.e., with P-trees.

The idea is to construct a distance-parity automaton $\mathcal{T}_{i,j}$ of parity index (i, j) reading P-trees such that:

(Correctness). For all P-trees t, if $\mathcal{T}_{i,j}(t) < \omega$ then t is accepting (Lemma 4).
(Completeness). For every i, j-feasible language $K \subseteq L$, there exists $M \in \omega$
   such that for all trees $t \in K$, there exists an accepting run ρ of $\mathcal{A}$ over t
   with $\mathcal{T}_{i,j}(col(\rho)) \leq M$ (Lemma 5).

Then we can cascade the automata $\mathcal{T}_{i,j}$ and $\mathcal{A}$ into a single one denoted $\mathcal{A}_{i,j}$ using the same distance-parity condition as $\mathcal{T}_{i,j}$: this automaton guesses a run of $\mathcal{A}$ and applies the automaton $\mathcal{T}_{i,j}$ on the colouring of this run. This resulting distance-parity automaton is such that $L^{(N)}(\mathcal{A}_{i,j}) \subseteq L$ for all N (correctness), and such that for all i, j-feasible $K \subseteq L$, $K \subseteq L^{(M)}(\mathcal{A}_{i,j})$ for some M (completeness). Let us now construct a distance-parity automaton $\mathcal{U}_{i,j}$ which at the beginning non-deterministically decides either to execute the automaton $\mathcal{A}_{i,j}$ or an automaton $\mathcal{C}$ accepting the complement of L.

Lemma 3. *The automaton $\mathcal{U}_{i,j}$ is uniformly universal iff L is i, j-feasible.*

Proof. Assume that $L^{(N)}(\mathcal{U}_{i,j})$ contains all trees for some N. This means that $L^{(N)}(\mathcal{A}_{i,j}) \cup L(\mathcal{C})$ contains all trees, and thus $L \subseteq L^{(N)}(\mathcal{A}_{i,j})$. Since furthermore $L^{(N)}(\mathcal{A}_{i,j}) \subseteq L$ (correctness), we have $L = L^{(N)}(\mathcal{A}_{i,j})$. And by Fact 1, L is i, j-feasible. Conversely, assume that L is i, j-feasible. Then by completeness, there exists M such that $L \subseteq L^{(M)}(\mathcal{A}_{i,j})$. Hence $L \cup L(\mathcal{C}) \subseteq L^{(M)}(\mathcal{U}_{i,j})$ contains all trees. The automaton $\mathcal{U}_{i,j}$ is uniformly universal.   □

From this we directly get our main theorem:

Theorem 2. *The problem of deciding the i, j-feasibility of a regular tree language is reducible to the uniform universality of distance-parity automata.*

We now describe the automaton $\mathcal{T}_{i,j}$ in detail. Intuitively, $\mathcal{T}_{i,j}$ maps each priority in P to a priority in $[i, j]$. When reading a priority $k \in P$ it "outputs" the priority associated to k. To implement this idea, the main objects used by $\mathcal{T}_{i,j}$ are partial mappings from P to $[i, j]$ (for technical reasons we allow some priorities to be undefined). To ensure correctness of $\mathcal{T}_{i,j}$ (in the sense mentioned above), these mappings have to respect some conditions. First of all, odd priorities of P should be mapped to odd priorities of $[i, j]$. Second, the ordering of the priorities should be respected in the following sense: the image of every odd priority is required to dominate the image of all even priorities below it.

Formally, let S be the set of partial mappings $s : P \to [i, j]$ such that for all k for which $s(k)$ is defined:

1. If k is odd, then $s(k)$ is odd.
2. If k is even, then for all odd $l > k$ such that $s(l)$ is defined, $s(l) \geq s(k)$.

The transitions of $\mathcal{T}_{i,j}$ allow to change the mapping s in a safe way. It is not very difficult to observe that on reading priority $k \in P$, $\mathcal{T}_{i,j}$ can safely change the values for P-priorities strictly below k. Additionally, we also allow a bounded number of arbitrary changes of the values. This bound is not fixed a priori and is controlled by the distance condition.

We now proceed to the formal definition of the distance-parity automaton $\mathcal{T}_{i,j} = (S \times P \times S, P, S_I, \delta, col)$.

- The set of initial states is $S_I = \{\langle s_\perp, k, s \rangle \mid s \in S, \ k \in P\}$, where $s_\perp$ is the mapping that is undefined everywhere.
- For all $s', s, s_0, s_1 \in S$, $k, k_0, k_1 \in P$ we allow the transition

$$\big(\langle s', k, s \rangle, \ k, \ (0, \langle s, k_0, s_0 \rangle), \ (1, \langle s, k_1, s_1 \rangle) \big) .$$

In fact, every transition is allowed, provided the first component is equal to the last one used at the parent node, and the second component remembers the label of the P-tree at the current node. As a consequence, a run of $\mathcal{T}_{i,j}$ on a given input tree is completely determined by the tree and the last components of its states.
- Distance colours are $D = \{d_k, r_{k-1} \ : \ k \in P\}$, and the priorities lie in $[i, j]$.
- For all $q \in S \times P \times S$ we set $col(q) = (dst(q), pri(q))$ with:

$$dst(\langle s', k, s \rangle) = \begin{cases} r_{k-1} & \text{if } s(k) \text{ is defined and } s'(l) = s(l) \text{ for all } l \geq k \\ d_k & \text{if } s(k) \text{ is undefined and } s'(l) = s(l) \text{ for all } l \geq k \\ d_l & \text{for the maximal } l \text{ with } s'(l) \neq s(l) \text{ otherwise.} \end{cases}$$

$$pri(\langle s', k, s \rangle) = \begin{cases} s(k) & \text{if } s(k) \text{ is defined,} \\ i & \text{if } s(k) \text{ is undefined.} \end{cases}$$

We now establish that $\mathcal{T}_{i,j}$ satisfies the correctness condition mentioned at the beginning of this section.

Lemma 4 (correctness). *For all P-trees t, if $\mathcal{T}_{i,j}(t) < \omega$ then t satisfies the parity condition of P on each branch.*

Proof. Let t be a P-tree and let ρ be a run of $\mathcal{T}_{i,j}$ on t. Consider a branch B and let k be the maximal P-priority that appears infinitely often on B in t. We show that if k is odd, then B is rejecting in ρ.

In the following, the s-value of a priority in P refers to the value it is mapped to by the sates of $T_{i,j}$. First note that r_{k-1} is the maximal release that can occur infinitely often on B in ρ. Hence, if the s-value of some $l \geq k$ changes infinitely often along B, then the distance condition is not satisfied and B is rejecting. Otherwise, the s-value of k becomes stable eventually on B. This value is odd because k is odd (property 1 of S), and furthermore it is bigger than the s-values of the smaller even priorities (property 2 of S). Hence, the maximal priority assumed infinitely often in ρ along B is odd and thus B is rejecting in ρ.   □

Lemma 5 (completeness). *For every i,j-feasible language $K \subseteq L$, there exists $M \in \omega$ such that for all trees $t \in K$, there exists an accepting run ρ of $\mathcal{A}$ over t with $T_{i,j}(col(\rho)) \leq M$.*

This is the difficult part of the proof. The principle is the following. Consider an automaton $\mathcal{B}$ of parity index (i,j) that accepts K and a mapping g such that $(\mathcal{B}, g)$ guides $\mathcal{A}$ (this is possible since $\mathcal{A}$ is guidable). Let us consider an accepting run τ of $\mathcal{B}$ over a tree $t \in K$. We set ρ to be $g(\tau)$, and we aim at constructing a run of $T_{i,j}$ witnessing that $T_{i,j}(col(\rho)) \leq M$ for a bound M which does not depend on t (but depends on $\mathcal{B}$).

The very informal idea for constructing the run of $T_{i,j}$ is to try to mimic the priorities used by the run τ. For defining the states of $T_{i,j}$ used at each position of the run, we heavily rely on Lemma 2 which relates the use of priorities in τ to the use of priorities in ρ. This lemma is only usable in presence of loops of $\mathcal{B}$ in τ. Therefore, in parts of τ where $\mathcal{B}$ has not yet entered a loop or enters a new strongly connected component of its transition graph, the states of $T_{i,j}$ map some priorities to an undefined value. The distance part of the condition is exactly used as a counter of such kind of "errors". But as $\mathcal{B}$ can only finitely often change the strongly connected component, one can imagine that the number of those errors is bounded by some M depending solely on the size of $\mathcal{B}$.

5   Conclusion

We have shown in this paper that the problem of deciding the levels of the non-deterministic Mostowski hierarchy can be reduced to the problem of uniform universality for distance-parity automata. The next step is of course to show the decidability of this latter problem. We have already obtained partial results showing that uniform universality is decidable for special classes of distance-parity automata (over trees). We expect the decidability of the general problem.

A key tool in our reduction is the notion of guidable automaton. We have shown that each regular language of infinite trees can be accepted by such an automaton. This model is interesting in its own right because it somehow shows that there is a canonical way of using non-determinism for accepting a language of infinite trees. We plan to investigate this model further and to see if it can be applied in other contexts.

References

1. Arnold, A.: The μ-calculus alternation-depth hierarchy is strict on binary trees. Informatique Théorique et Applications 33(4/5), 329–340 (1999)
2. Arnold, A., Duparc, J., Murlak, F., Niwiński, D.: On the topological complexity of tree languages. In: Flum, J., Grädel, E., Wilke, T. (eds.) Logic and automata: History and Perspectives, pp. 9–28. Amsterdam University Press (2007)
3. Bojańczyk, M., Colcombet, T.: Bounds in ω-regularity. In: Proceedings of LICS 2006, pp. 285–296. IEEE Computer Society Press, Los Alamitos (2006)
4. Bradfield, J.C.: The modal μ-calculus alternation hierarchy is strict. Theor. Comput. Sci. 195(2), 133–153 (1998)
5. Carayol, A., Löding, C.: MSO on the infinite binary tree: Choice and order. In: Duparc, J., Henzinger, T.A. (eds.) CSL 2007. LNCS, vol. 4646, pp. 161–176. Springer, Heidelberg (2007)
6. Hashiguchi, K.: Algorithms for determining relative star height and star height. Inf. Comput. 78(2), 124–169 (1988)
7. Kirsten, D.: Distance desert automata and the star height problem. RAIRO – Theoretical Informatics and Applications 3(39), 455–509 (2005)
8. Klarlund, N.: Progress measures, immediate determinacy, and a subset construction for tree automata. Annals of Pure and Applied Logic 69(2–3), 243–268 (1994)
9. Lenzi, G.: A hierarchy theorem for the μ-calculus. In: Meyer auf der Heide, F., Monien, B. (eds.) ICALP 1996. LNCS, vol. 1099, pp. 87–97. Springer, Heidelberg (1996)
10. Mostowski, A.W.: Regular expressions for infinite trees and a standard form of automata. In: Skowron, A. (ed.) SCT 1984. LNCS, vol. 208, pp. 157–168. Springer, Heidelberg (1985)
11. Niwiński, D.: On fixed-point clones. In: Kott, L. (ed.) ICALP 1986. LNCS, vol. 226, pp. 464–473. Springer, Heidelberg (1986)
12. Niwiński, D., Walukiewicz, I.: Ambiguity problem for automata on infinite trees (unpublished note)
13. Niwiński, D., Walukiewicz, I.: Relating hierarchies of word and tree automata. In: Morvan, M., Meinel, C., Krob, D. (eds.) STACS 1998. LNCS, vol. 1373, pp. 320–331. Springer, Heidelberg (1998)
14. Niwiński, D., Walukiewicz, I.: Deciding nondeterministic hierarchy of deterministic tree automata. Electr. Notes Theor. Comput. Sci. 123, 195–208 (2005)
15. Rabin, M.O.: Decidability of second-order theories and automata on infinite trees. Transactions of the American Mathematical Society 141, 1–35 (1969)
16. Thomas, W.: Languages, automata, and logic. In: Handbook of Formal Language Theory, vol. III, pp. 389–455. Springer, Heidelberg (1997)
17. Urbański, T.F.: On deciding if deterministic Rabin language is in Büchi class. In: Welzl, E., Montanari, U., Rolim, J.D.P. (eds.) ICALP 2000. LNCS, vol. 1853, pp. 663–674. Springer, Heidelberg (2000)
18. Wagner, K.W.: Eine topologische Charakterisierung einiger Klassen regulärer Folgenmengen. J. Inf. Process. Cybern. EIK 13(9), 473–487 (1977)
19. Zielonka, W.: Infinite games on finitely coloured graphs with applications to automata on infinite trees. Theoretical Computer Science 200(1–2), 135–183 (1998)

Analyzing Context-Free Grammars Using an Incremental SAT Solver

Roland Axelsson[1], Keijo Heljanko[2,*], and Martin Lange[1]

[1] Institut für Informatik, Ludwig-Maximilians-Universität München, Germany
[2] Department of Information and Computer Science,
Helsinki University of Technology (TKK), Finland

Abstract. We consider bounded versions of undecidable problems about context-free languages which restrict the domain of words to some finite length: inclusion, intersection, universality, equivalence, and ambiguity. These are in (co)-NP and thus solvable by a reduction to the (un-)satisfiability problem for propositional logic. We present such encodings – fully utilizing the power of incrementat SAT solvers – prove correctness and validate this approach with benchmarks.

1 Introduction

Context-free grammars (CFG) and languages (CFL) have been used intensively by computer scientists and linguists since Chomsky formalized them in 1956. They have applications in compiler design, speech processing, bioinformatics, static program analysis, XML processing, etc.

The word problem for CFGs is decidable in cubic time and quadratic space and the Pumping Lemma for CFLs [1] provides a criterion by which the emptiness problem becomes decidable as well. However, it has since long been known that the following problems are undecidable: *universality* (given a CFG G over some alphabet Σ, is $L(G) = \Sigma^*$?); *inclusion, intersection,* and *equivalence* (given two CFG G_1 and G_2, is $L(G_1) \subseteq L(G_2)$, is $L(G_1) \cap L(G_2) = \emptyset$, and is $L(G_1) = L(G_2)$?)

Another very important undecidable problem is *ambiguity* – is there a word which has at least two different parse trees w.r.t. a given CFG? After seeing little progress for many years, this problem has recently attracted attention again [3,8] which is, e.g., due to its importance in compiler design and bioinformatics.

Due to decidability of the word problem these problems are all (co)-semi-decidable through an enumeration of Σ^*. Hence *bounded* versions of these problems become decidable. For example, the *bounded universality* problem is: given a CFG G and a $k \in \mathbb{N}$, does $L(G)$ contain all words of length $\leq k$? Since the word problem is even decidable in polynomial time, they are in (co-)NP and can therefore be solved by a polynomial reduction to (UN-)SAT, provided that k is given in unary coding.

* Work financially supported by Academy of Finland (project 112016) and Technology Industries of Finland Centennial Foundation.

L. Aceto et al. (Eds.): ICALP 2008, Part II, LNCS 5126, pp. 410–422, 2008.
© Springer-Verlag Berlin Heidelberg 2008

The use of modern SAT solvers such as zChaff [6] has proved to be extremely beneficial in areas like computer aided verification, AI planning, theorem proving, cryptanalysis, electronic design automation, etc. Here we show that, apparently, formal language theory is also among them.

The observation about decidability of the above bounded problems is not new although we are not aware of any work that exploits this idea thoroughly in order to tackle the unbounded (and undecidable) problems. Here we present optimized reductions of these bounded problems to SAT s.t. a SAT solver can find witnesses, resp. counterexamples for these problems. The basis for the reduction is the well-known CYK algorithm [9]. We generate propositional logic constraints encoding which nonterminals may/must occur at certain positions in a CYK parse table. A slightly different approach to encoding CYK parsing into SAT has been independently discovered in [7]. However, the textbook version of CYK is unsuitable since it requires the CFG to be in Chomsky Normal Form (CNF) which may incur an exponential blow-up in the grammar. That would clearly be counterproductive in the search for optimized reductions. We therefore develop and use an optimized version of CYK which may be known in the community but does not seem to have made it into the literature. When it comes to ambiguity, we even must. Note that the transformation into CNF does not preserve ambiguity. We therefore use a different normal form without these deficiencies.

The crucial difference between our symbolic encoding and an explicit execution of the CYK algorithm though, is the absence of an input word. This has to be "guessed" by the SAT solver and the constraints will ensure (non)-inclusion in the languages of some CFGs. Thus, we do not only get that the SAT formula is satisfiable iff the language of the CFG, say, contains an ambiguous word of length $\leq k$. The satisfying assignment also encodes this word as well as two different parse trees.

Note that CYK tables are in some sense closed under extension "to the right": the triangular table of size $(k + 1) \times (k + 1)$ can be obtained from the one of size $k \times k$ by adding a column of length $k + 1$. This is what makes *incremental* SAT solving predestined for solving bounded CFL problems. If a witness/counterexample of size $\leq k$ is not found, additional constraints for a greater bound plus a few changes in the current constraints yield the new formula. Incremental SAT solvers maximally utilize information gathered in solving a SAT instance to solve the next "bigger" but structurally very similar one. Such solvers are therefore of particular interest for our setting.

This is clearly just a semi-decision procedure for the unbounded versions of the considered problems. But it has distinct advantages over approximation approaches for ambiguity [3,8]. While the accuracy of answers given by those depends on the quality of the approximation (that may produce false-positives), our approach is only limited by time and available memory; the structure of the produced formula does not pose any difficulty to the SAT solver. A report on an empirical evaluation is included after some preliminary definitions, and the presentation as well as exemplary correctness proofs of encodings for the above problems. On the other hand, our approach is clearly not complete and not meant to replace approximation approaches to ambiguity. Instead, they could also be combined, e.g. to provide a search for smallest witnesses to the problems of horizontal and vertical ambiguity in [3] for instance.

2   Preliminaries

Let Σ be an alphabet. As usual, we write $|w|$ for the length of a word w, Σ^* for the set of finite words over Σ, $\Sigma^{\leq k}$ for $\{w \in \Sigma^* \mid |w| \leq k\}$ for any $k \in \mathbb{N}$, ϵ for the empty word and uv or LL' to denote concatenation of words, sets, languages, etc. If $w = a_1 \ldots a_n$ we write $w^{i \cdots j}$ for its subword $a_i \ldots a_j$ of length $j - i + 1$. A *context-free grammar* is a tuple $G = (N_G, \Sigma, P_G, S_G)$ where N_G is a finite set of non-terminals, Σ is an alphabet, $N_G \cap \Sigma = \emptyset$, $S_G \in N_G$ is the starting symbol and $P_G \subset N_G \times (N_G \cup \Sigma)^*$ is a finite set of production rules. We use infix notation $A \to \alpha$ to denote $(A, \alpha) \in P_G$. As the size of G we define $|G| := |N| + \sum_{A \to \alpha} |\alpha|$.

The *derivation* relation $\Rightarrow_G \subseteq (N \cup \Sigma)^+ \times (N \cup \Sigma)^*$ is defined as $\alpha A \beta \Rightarrow_G \alpha \gamma \beta$ iff $A \to \gamma \in P$ for $\alpha, \beta, \gamma \in (N \cup \Sigma)^*, A \in N$. We will drop the index G if it becomes clear from the context.

$L(G) := \{w \in \Sigma^* \mid S_G \Rightarrow^* w\}$ is the *language* of G. An alternative way to define the derivation of a word is via the existence of *parse trees*. We assume the reader familiar with these fundamental concepts and refer to [4] for details.

A word w is *ambiguous* w.r.t. a CFG G, if there are two different parse trees for w w.r.t. G. For a CFG G let $amb(G)$ denote the set of words that are ambiguous w.r.t. G. G itself is called *ambiguous* if $amb(G) \neq \emptyset$.

In the following we will always assume the context-free languages under consideration not to contain the empty word ϵ. This is not a restriction but simplifies the presentation.

Definition 1. A CFG is in *binary normal form* (2NF), if for all $(A \to \alpha) \in P_G$ we have $\alpha \in \{\epsilon\} \cup \Sigma \cup N \cup NN$. It is *acyclic* if for all $A \neq B \in N$ we have, if $A \Rightarrow^* B$ then $B \not\Rightarrow^* A$. It is *reduced* if all nonterminals are *reachable* and *productive*, i.e. for all $A \in N$ there are sentential forms α, β and a word $w \in \Sigma^*$ s.t. $S \Rightarrow^* \alpha A \beta$ and $A \Rightarrow^* w$.

Lemma 1. *For every CFG G there is a CFG G' in reduced acyclic 2NF, computable in time $\mathcal{O}(|G|^2)$ s.t. $L(G) = L(G')$ and $|G'| = \mathcal{O}(|G|)$.*

Lemma 2. *Let G be an acyclic grammar. There exists a well-founded strict partial order $> \subseteq N_G^2$, s.t. if $A \Rightarrow^* B$ and $A \neq B$ then $A > B$.*

3   The Encoding

The task is now to create propositional logic constraints from a CFG G and a $k \in \mathbb{N}$ that are satisfiable iff $L(G)$ is (k-bounded) universal, ambiguous, etc. Let $G = (N, \Sigma, P, S)$ in reduced, acyclic 2NF and k be fixed. We use two kinds of propositional variables: X_i^a for every $a \in \Sigma$ and every $1 \le i \le k$ stating that the i-th symbol of a witnessing word is a. An assignment to these variables corresponds to the choice of a witness w. The other kind is $X_{i,j}^A$ and states that nonterminal A derives the subword $w^{i..j}$. Let $\mathcal{P} = \{X_i^a, X_{i,j}^A \mid a \in \Sigma, A \in N, 1 \le i \le j \le k\}$. In the following, η will denote an assignment of these variables to $\{\mathtt{tt}, \mathtt{ff}\}$, and we write $\eta(\mathcal{C}) = \mathtt{tt}$ if η satisfies all the constraints $\mathcal{C}$ under the usual interpretation of the operators in propositional logic.

With incrementality in mind we define constraint sets w.r.t. some p, p' s.t. $1 \le p \le p' \le k$. Intuitively, these contain the constraints for the columns $p, \ldots, p'$ of a CYK table. Note that they may use variables with indices below p.

Let $\Phi = \{\varphi_1, \ldots, \varphi_n\}$ be an ordered set of propositional formulas. There is a standard trick to state that at most one of them holds by introducing auxiliary variables Y_i for $1 \le i \le n+1$.

$$One(\Phi) \;:=\; \{\, (\varphi_i \to \neg Y_i \wedge Y_{i+1}) \wedge (Y_i \to Y_{i+1}) \mid 1 \le i \le n \,\}$$

With this macro, we can easily state that each position in a witnessing word is occupied by a unique symbol.

$$\mathcal{W}(p, p') \;:=\; \{\, One(\{X_i^a \mid a \in \Sigma\}) \wedge \bigvee_{a \in \Sigma} X_i^a \mid p \le i \le p' \,\}$$

Lemma 3. For any η, $\eta(\mathcal{W}(p, p')) = \mathtt{tt}$ iff there exists a unique sequence $b_p, \ldots, b_{p'}$, s.t. $\eta(X_i^{a_i}) = \mathtt{tt}$ iff $b_i = a_i$ for all $p \le i \le p', a_i \in \Sigma$.

We will therefore simply write $w_\eta^{p,p'}$ for the unique $w^{p..p'}$ induced by η. We encode a derivation with the help of constraints $\mathcal{R}(p, p') :=$

$$\{\, X_{i,j}^A \leftrightarrow \underset{\substack{A \to a \\ if\, i=j}}{\bigvee} X_i^a \vee \underset{\substack{A \to B \\ A \neq B}}{\bigvee} X_{i,j}^B \vee \underset{\substack{A \to BC\, or \\ A \to CB \\ A \neq B, C \Rightarrow^* \epsilon}}{\bigvee} X_{i,j}^B \vee \underset{A \to BC\ h=i}{\bigvee} \overset{j-1}{\bigvee} (X_{i,h}^B \wedge X_{h+1,j}^C)$$

$$\mid A \in N, p \le i \le j \le p'\}$$

This encoding splits up the derivation of $w_\eta^{i,j}$ by non-terminal A into the following four cases (marked by the big disjunctions): derivation of a single terminal, two cases of single non-terminal derivations and the derivation of composites. Note that pre-computing the set of all nonterminals C s.t. $C \Rightarrow^* \epsilon$ can be done in time $\mathcal{O}(|G|)$. It is also a necessary preliminary step during the transformation into 2NF. So far, $\mathcal{R}(p, p')$ contains a bi-implication. However, for some problems, implications in one direction only will suffice. For example, when encoding bounded emptiness, the $\leftarrow$-parts are unnecessary. In general, the $\to$-parts express soundness of the encoding and are used to express that something is derivable; the $\leftarrow$-parts encode completeness and can be used to express that a word is *not* derivable. We write $\mathcal{R}^\to(p, p')$ and $\mathcal{R}^\leftarrow(p, p')$ for the soundness, resp. completeness parts only.

Lemma 4. Let $k > 0$, η be an assignment s.t. $\eta(\mathcal{R}^\to(1, k) \cup \mathcal{W}(1, k)) = \mathtt{tt}$ and $w = w_\eta^{1,k}$. Then for all $A \in N_G$, all $1 \le i \le j \le k$ we have: if $\eta(X_{i,j}^A) = \mathtt{tt}$ then $A \Rightarrow^* w^{i..j}$.

Proof. Suppose $\eta(X_{i,j}^A) = \text{tt}$. We prove the claim by induction on $j - i$ where we refer to the four different (big) disjunctions in $\mathcal{R}^{\rightarrow}(1,k)$ as "blocks 1–4".

Base case (i = j): Clearly, at least one variable from blocks 1-3 has to be evaluated to tt. Block 4 evaluates to ff for $i = j$.

1. $\eta(X_i^a) = \text{tt}$. There must be a rule $A \to w^i$ and therefore $A \Rightarrow^* w^i$.
2. $\eta(X_{i,i}^B) = \text{tt}$. We proceed by well-founded induction on $>$. Suppose for all $B < A$, we have $B \Rightarrow^* w^i$ if $\eta(X_{i,i}^B) = \text{tt}$. Because of the rule $A \to B$ we also have $A \Rightarrow^* w^i$.
3. $\eta(X_{i,i}^B) = \text{tt}$. Analogous to (2).

Inductive case (i < j): Block 1 evaluates to ff for $i < j$, so at least one disjunct from blocks 2–4 has to evaluate to tt.

1. $\eta(X_{i,j}^B) = \text{tt}$. Same as in the base case.
2. $\eta(X_{i,h}^B) = \eta(X_{h+1,j}^C) = \text{tt}$ (4). In particular, h, B, C exist and $i \leq h < j$. Clearly $h - i \leq j - i$ and $j - (h+1) \leq j - i$ and therefore by induction hypothesis $B \Rightarrow^* w^{i..h}$ and $C \Rightarrow^* w^{h+1..j}$. As $A \to BC$ it follows that $A \Rightarrow^* w^{i..j}$. □

Lemma 5. *Let $k > 0$, η be an assignment s.t. $\eta(\mathcal{R}^{\leftarrow}(1,k) \cup \mathcal{W}(1,k)) = \text{tt}$ and $w = w_\eta^{1,k}$. Then for all $A \in N_G$, all $1 \leq i \leq j \leq k$ we have: if $A \Rightarrow^* w^{i..j}$ then $\eta(X_{i,j}^A) = \text{tt}$.*

Proof. Again, we prove this by induction on $j - i$. Let $w = w_\eta$. In the base case suppose $A \Rightarrow^* w^{i..i} = a$. Thus, $\eta(X_i^a) = \text{tt}$. Furthermore, there is a derivation tree with root A and leaf front a. Clearly, whenever a node in this tree has two successors labeled B and C then $B \Rightarrow^* \epsilon$ or $C \Rightarrow^* \epsilon$. Because of 2NF, a must be generated by some rule $B \to a$, and because of block 1 we have $\eta(X_{i,i}^B) = \text{tt}$. A separate induction on the height of the tree – using blocks 2–4 – shows that we have $\eta(X_{i,i}^C) = \text{tt}$ for all predecessors of this B in this tree, including the root A. The crucial insight to the applicability of this induction is the fact that in this parse tree the node labels on the path from the root to the leaf a are strictly decreasing w.r.t. $>$ according to Lemma 2.

Now assume $j > i$ and $A \Rightarrow^* w^{i..j}$. Hence, we have a parse tree t with root A whose leaf front is $w^{i..j}$. For a node n in t we write $w(n)$ for the subword of $w^{i..j}$ that constitutes of the leaf labels in the subtree under n. Furthermore, for two words u, v we write $u \prec v$ if u is a genuine subword of v.

Note that $|w^{i..j}| \geq 2$, and – because of 2NF – leaves in this tree have a direct predecessor n that can only have a single successor. Therefore, for each such n we have $w(n) \prec w^{i..j}$. Note that $w(n_0) = w^{i..j}$ for n_0 the root of t. Hence, there must be a highest (closest to the root) node n in this tree, that is labeled with some $B \in N_G$ and has two successors n_1 and n_2 labeled with some C, resp. D, s.t. $w(n_1) \prec w(n) = w(n_0)$ and $w(n_2) \prec w(n) = w(n_0)$. Hence, $w(n_1) = w^{i..h}$ and $w(n_2) = w^{h+1..j}$ for some $i \leq h < j$. But then we have $C \Rightarrow^* w^{i..h}$, $D \Rightarrow^* w^{h+1..j}$, and, by hypothesis, $\eta(X_{i,h}^C) = \eta(X_{h+1,j}^D) = \text{tt}$. Since $B \to CD$ we have $\eta(X_{i,j}^B) = \text{tt}$ by block 4. Finally, the path from the node labeled B to

$\mathcal{C}$	$\mathcal{R}_C(p,p')$			
	$\mathcal{R}_G^{\rightarrow}(p,p')$	$\mathcal{R}_G^{\leftarrow}(p,p')$	$\mathcal{R}_{G'}^{\rightarrow}(p,p')$	$\mathcal{R}_{G'}^{\leftarrow}(p,p')$
bINCL$_{G,G'}$	X			X
bUNIV$_G$		X		
bISECT$_{G,G'}$	X		X	
bEQUIV$_{G,G'}$	X	X	X	X

Fig. 1. How to use the $\mathcal{R}$-constraints

the root A must be strictly increasing w.r.t. $<$ again, and an induction on its length eventually shows $\eta(X_{i,j}^A) = \text{tt}$ using blocks 1–3. □

3.1 Constraints for Particular Problems

We will now assemble the above constraints in order to obtain encodings of the following problems. Let G, G' be CFG and $k > 0$.

- *Bounded Inclusion* (bINCL): does $\forall w \in \Sigma^{\leq k} : w \in L(G) \Rightarrow w \in L(G')$ hold?
- *Bounded Universality* (bUNIV): is $\Sigma^{\leq k} \subseteq L(G)$?
- *Bounded Intersection* (bISECT): is there a $w \in \Sigma^k \cap L(G) \cap L(G')$?
- *Bounded Equivalence* (bEQUIV): does $\forall w \in \Sigma^{\leq k} : w \in L(G) \Leftrightarrow w \in L(G')$ hold?

For those that take two CFG G, G' as input we write $\mathcal{R}_G$ to clarify which CFG the constraints refer to.

The following is not hard to prove using the fact that the word problem for a CFG can be solved in polynomial time. Note that bounded ambiguity is missing. It will be treated separately below.

Proposition 1. *For unarily encoded* $k \in \mathbb{N}$, *the problems* bINCL, bUNIV, bEQUIV *are in co-NP, and* bISECT *is in NP.*

All of these encodings have a similar structure: they take some form of the $\mathcal{R}$-constraints plus a single problem specific one constraining the grammar's starting symbols. We therefore define

$$\mathcal{C}(p,p') = \mathcal{W}(p,p') \cup \mathcal{R}_C(p,p') \cup \mathcal{S}_C(p,p')$$

for $\mathcal{C} \in \{\text{bINCL, bUNIV, bISECT, bEQUIV}\}$. The $\mathcal{R}$-parts can be obtained from the table in Fig. 1. The $\mathcal{S}$-part is always a single constraint $\mathcal{S}_C := \{\bigvee_{j=p}^{p'} T_C(j)\}$ with

$$T_{\text{bINCL}}(j) := X_{1,j}^{S_G} \wedge \neg X_{1,j}^{S_{G'}} \qquad T_{\text{bUNIV}}(j) := \neg X_{1,j}^{S_G}$$
$$T_{\text{bISECT}}(j) := X_{1,j}^{S_G} \wedge X_{1,j}^{S_{G'}} \qquad T_{\text{bEQUIV}}(j) := X_{1,j}^{S_G} \leftrightarrow \neg X_{1,j}^{S_{G'}}$$

We write bINCL(p) for bINCL($1, p$), etc. The following theorem confirms the introductory statement about the reductions from these bounded problems to SAT being polynomial. Its proof is straight-forwardly based on standard techniques for obtaining conjunctive normal form and therefore not presented here.

Proposition 2. *Let G, G' be CFGs, $k > 0$. For any set of constraints $C \in \{$ bINCL(k), bUNIV(k), bISECT(k), bEQUIV(k) $\}$ there is an equivalent propositional formula Φ_C in conjunctive normal form over $\mathcal{O}(|N_G \cup N_{G'}| \cdot k^2)$ many variables s.t. $|\Phi_C| = \mathcal{O}((|G| + |G'|) \cdot k^3)$.*

We will prove correctness of one of these reductions, namely for bINCL. The others are proved in a similar way.

Theorem 1. *Let G, G' be CFGs in reduced acyclic 2NF, $k > 0$. Then bINCL$_{G,G'}(k)$ is satisfiable iff there is a $w \in \Sigma^{\leq k}$ s.t. $w \in L(G) \setminus L(G')$.*

Proof. ($\Rightarrow$) Suppose η is a satisfying evaluation of bINCL$_{G,G'}(1, k)$. Let $w = w_\eta^{1,k}$ according to Lemma 3. We will show that there is a $k' \leq k$ s.t. $S_G \Rightarrow^* w^{1..k'}$ and $S_{G'} \not\Rightarrow^* w^{1..k'}$. Let k' be the least j s.t. $\eta(X_{1,j}^{S_G}) = \text{tt}$ and $\eta(X_{1,j}^{S_{G'}}) = \text{ff}$. Its existence is guaranteed by the specific constraints for bINCL. The rest follows immediately from Lemmas 4 and 5.

($\Leftarrow$) W.l.o.g we assume that the counterexample w is of minimal length k, i.e. that bINCL$_{G,G'}(k')$ is unsatisfiable for any $k' < k$. We construct an evaluation η of bINCL$_{G,G'}(1, k)$ as follows.

$$\eta(X_i^a) = \text{tt} \text{ iff } w^i = a \qquad \eta(X_{i,j}^A) = \text{tt} \text{ iff } A \Rightarrow^* w^{i..j}$$

for all $A \in N_G \cup N_{G'}$, $1 \leq i \leq j \leq k$. A simple inspection of the constraints in bINCL(k) shows that they are all fulfilled by η. $\qquad\qquad\square$

Theorem 2. *Let G, G' be CFGs in 2NF, $k > 0$. Then we have*

- *bUNIV$_G(k)$ is satisfiable iff there is a $w \in \Sigma^{\leq k}$ s.t. $w \notin L(G)$.*
- *bISECT$_{G,G'}(k)$ is satisfiable iff there is a $w \in \Sigma^{\leq k}$ s.t. $w \in L(G) \cap L(G')$.*
- *bEQUIV$_{G,G'}(k)$ is satisfiable iff there is a $w \in \Sigma^{\leq k}$ s.t. $w \in L(G) \setminus L(G')$ or $w \in L(G') \setminus L(G)$.*

A counterexample for the universality problem could therefore be found by iteratively checking the constraint sets bUNIV(1), bUNIV(2), ... for satisfiability. Note that bUNIV($k + 1$) contains many constraints already present in bUNIV(k). In fact, for all of the above problems we have the following relation. Let $0 < k < k'$.

$$C(k') = (C(k) \setminus \bigcup_{1 \leq p \leq p' \leq k} S_C(p, p')) \cup C(k+1, k') \cup S_C(k+1, k')$$

Hence, these constraints support incrementality in the sense that the wider range $\Sigma^{\leq k'}$ can be checked by modifying the constraints for the smaller range $\Sigma^{\leq k}$. Furthermore, the increase need not take place in steps of size 1 only.

3.2   Ambiguity

We define the bounded ambiguity problem bAMB for a grammar G in reduced acyclic 2NF and a $k \geq 1$ in a non-obvious way: is there a nonterminal $A \in N_G$ and a word $v \in \Sigma^{\leq k}$ s.t. v has at least two different parse trees with roots labeled A that differ in a node on level 1?

Note that a word w is ambiguous in the original sense w.r.t. a grammar G iff it has two different parse trees that differ in a node (determined by the derived subword under that node and the node's label) which is not the root. Therefore, these trees must have a subtree each with equally labeled roots and equal derived subwords that differ on level 1. In other words, a derivation for w derives a subword v from a nonterminal A by using two different rules for A or using one rule in two different ways.

By not looking for ambiguous words, but ambiguous subwords, found witnesses explain the reason for ambiguity more clearly. For example, if the examined grammar was an ambiguous one for Java, then the witness may not be a whole Java program but just an ambiguous Java expression. Furthermore, this definition of bounded ambiguity allows for much more compact encodings. Finally, if a CFG is reduced, i.e. all terminals are reachable and productive then we have the following property: if (v, A) is an instance of bAMB for a CFG G as defined above, then there is an ambiguous $w \in L(G)$ s.t. $w = uvz$ for some $u, z \in \Sigma^*$. The converse direction holds trivially. Thus, bounded ambiguity in our sense is just a more detailed description of bounded ambiguity as one may expect it.

Proposition 3. *The problem* bAMB *is solvable in NP for unarily encoded* $k \in \mathbb{N}$.

Before we can present the encoding we need to reconsider the transformation of a CFG into reduced acyclic 2NF. Remember that acyclicity is necessary for the $\mathcal{R}$-constraints to be correct. However, it requires the removal of productions of the form $A \to A$ after replacing nonterminals with equivalence class representatives in the construction of Lemma 1. But then the transformation does not preserve ambiguity anymore, because such a cyclic rule can be its cause.

Definition 2. An extended CFG is a tuple $G = (N, \Sigma, P, S, M, E)$ like a CFG with $E \subseteq M \subseteq N$ called the *ambiguously nullable nonterminals* and the *ambiguous nonterminals*. The notions of language, derivability, 2NF, acyclicity, reducedness etc. are defined as for a CFG. However, we define $amb(G) = \{w \mid$ there are two different parse trees for w, or there is one parse tree containing a nonterminal $A \in M \}$.

Then we can reformulate Lemma 1 for the new purpose as follows.

Lemma 6. *For every CFG G there is an extended CFG G' in acyclic and reduced 2NF, computable in time $\mathcal{O}(|G|^2)$ s.t. $L(G') = L(G)$, and $|G'| = \mathcal{O}(|G|)$. Moreover, we have $amb(G') = amb(G)$, and $A \in E$ iff there are two different parse trees with root A and leaf front ϵ.*

Proof. Let $G = (N, \Sigma, P, S)$ be a CFG. It can be reduced and transformed into 2NF in time $\mathcal{O}(|G|)$. Define $G' := (\tilde{N}, \Sigma, \tilde{P}, \tilde{S}, M, E)$ as the canonical factorisation of G under the equivalence relation $A \sim B$ iff $A \Rightarrow^* B \Rightarrow^* A$. I.e. its non-terminals are equivalence classes $\tilde{A}$ under this relation, and the production rules of G' are canonically derived from those in G. It should be clear that G' is also reduced. Let E consist of all $\tilde{A}$ that can derive ϵ in at least two different ways. This can be computed in time $\mathcal{O}(|G'|)$. Define $M := E \cup \{\tilde{A} \mid A \Rightarrow^+ A\}$. Note that M can be computed in time $\mathcal{O}(|G|^2)$. In order to make G' acyclic, simply remove all productions of the form $\tilde{A} \to \tilde{A}$.

It is not hard to see that $amb(G') \subseteq amb(G)$ holds. For the converse direction, assume that $t_1 \neq t_2$ are two parse trees for some w w.r.t. G. Let $\tilde{t}_1$ and $\tilde{t}_2$ result from them by replacing every node label A with $\tilde{A}$ and collapsing edges of the form $\tilde{A} \to \tilde{A}$. Note that these are parse trees for w w.r.t. G'. If $\tilde{t}_1 \neq \tilde{t}_2$ then $w \in amb(G')$. Otherwise, if $\tilde{t}_1 = \tilde{t}_2$ then either they coincide because of a collapsed edge in some t_i. In this case, $\tilde{t}_1$ must contain some $\tilde{A} \in M$ and therefore $w \in amb(G')$. Or there are nodes with labels A and B in t_1 and t_2 that get mapped to the same node $\tilde{A}$ in $\tilde{t}_1$, i.e. $A \sim B$ and therefore $\tilde{A} \in M$. $\square$

We are now ready to describe the SAT encoding of bounded ambiguity. As above, we assume a macro $Two(\Phi)$ which, for an ordered set Φ of propositional formulas, is satisfiable iff there is an assignment satisfying at least two formulas out of Φ. It can easily be constructed by introducing at most $2 \cdot |\Phi| + 2$ new variables, c.f. the construction of One above.

Let G be an extended CFG in reduced, acyclic 2NF. The $\mathcal{W}$-constraints remain the same. Since "having two different parse trees" entails being derivable, we also add the $\mathcal{R}^{\to}$ constraints defined above. Finally, we simply have to state that there is a nonterminal which forms the root of the parse (sub)tree which is either an ambiguous nonterminal or to which two different productions apply.

$$\mathrm{bAMB}(k, k') := \mathcal{W}(k, k') \cup \mathcal{R}^{\to}(k, k') \cup$$

$$\{ \bigvee_{j=k}^{k'} \Big(\bigvee_{A \in M_G} X_{1,j}^A \vee \bigvee_{\substack{A \to BC \text{ or} \\ A \to CB \\ A \neq B, C \in E_G}} (X_{1,j}^A \wedge X_{1,j}^B) \vee \bigvee_{A \in N_G \setminus M_G} (X_{1,j}^A \wedge Two(\mathcal{P}_{A,j})) \Big) \}$$

where

$$\mathcal{P}_{A,j} := \{ X_{1,j}^B \mid A \to \alpha \in \{B, BC, CB\} \text{ with } C \Rightarrow^* \epsilon \}$$
$$\cup \{ X_1^a \mid A \to a \text{ and } j = 1 \}$$
$$\cup \{ X_{1,h}^b \wedge X_{h+1,j}^C \mid A \to BC, 1 \leq h < j \}$$

encodes all the different productions that can be made at the root labeled A of a parse tree for a word of length j. Again, let $\mathrm{bAMB}(k) := \mathrm{bAMB}(1, k)$. It is not difficult to see that the encoding of this problem supports incrementality as well. In each increment, the $\mathcal{W}$- and $\mathcal{R}^{\to}$-constraints remain, the other one has to be deleted, etc.

Lemma 6 together with an argument similar to that in the proof of Thm. 1 yields correctness of the encoding.

Theorem 3. *Let G be a CFG in 2NF, $k > 0$. Then* bAMB(k) *is satisfiable iff there are $u, v, z \in \Sigma^*$ s.t. $uvz \in L(G)$, $|v| \leq k$ and there are two different parse trees for w that differ on level 1 of the subtree for v.*

Proposition 4. *Let G be a CFG, $k > 0$. Then* bAMB(k) *can be equivalently translated into a propositional formula Φ in conjunctive normal form over $\mathcal{O}(|N_G| \cdot k^2)$ many variables s.t. $|\Phi| = \mathcal{O}(|G| \cdot k^3)$.*

4 Comparison

A prototype implementation of the reduction approach (`cfganalyzer`) has been implemented for all 5 bounded problems mentioned above. It is written in OCaml 3.09.3, uses zChaff version 2007.03.12 as a linked-in incremental SAT solver and is available online.[1]

Of the problems discussed here, ambiguity is the one to which most attention has been paid and for which a number of tools is available. These basically split up into three different approaches: (1) brute-force ambiguity detection, (2) LRR detection and (3) language approximation.

(1) Brute-force ambiguity detection systematically generates parse trees of a certain maximal size and looks for double appearances of the derived words. Ambiguous words which exceed the bound are not found – as in our approach. The crucial difference though is the use of a high-performance SAT solver as a back-end. While brute-force ambiguity detectors need to generate all parse trees for a certain bound one-by-one, our reduction covers all parse trees for that bound at once, and it is up to the SAT solver to find two in its solution space. In terms of complexity: we use a polynomial reduction to an NP-problem while (1) is an exponential reduction to a problem in P (finding equal strings in lists). The performance discrepancies between derivation generators and `cfganalyzer` can be seen by comparing Fig. 2 to the results of AMBER in [2]; `cfganalyzer` is more than 1000 times faster on subwords of the same size as words in e.g. the Pascal grammar and capable of pushing the bounds to $k = 25$ in reasonable time where AMBER is already at 100.000 sec for $k = 17$.

(2) LRR or LR-regularity is a generalisation of the well-known LR(k) grammar classes [5]. Instead of a k-symbol lookahead, an LRR parser considers regular equivalence classes on the remaining input and reports parsing conflicts. LRR detectors rely on the fact that every LRR-grammar is unambiguous and simply check a given grammar for this property. But since not every unambiguous grammar is LRR this method is of course also incomplete. Although being relatively fast, common LR(k) parsers such as `yacc` often reveal little about the causes of conflicts. Another positive effect of our approach is that it does always offer a detailed report on the cause of the ambiguity upon termination, i.e. provides two parsetrees for the ambiguous subword.

---

[1] http://www.tcs.ifi.lmu.de/~mlange/cfganalyzer,
We would also like to thank Harri Haanpää (TKK) and Anders Møller (Århus) for kindly providing us with benchmark CFGs.

(3) Methods of the third kind usually are complete but not sound by over-approximating the grammar. False-positives can occur because the language of the approximated grammar is a superset of the original one. Examples are the ACLA framework [3] or Schmitz's method [8]. Our approach does not easily compare to those since it is an under-approximation: it is sound, and complete only in the sense that it produces no spurious reports. It does however not terminate on unambiguous inputs. Hence, the situation is dual to that of the over-approximation approaches which can reliably report unambiguity. Because of this duality these two approaches combine well: a reported potential ambiguity of an over-approximation tool may be confirmed as a fact by `cfganalyzer` and a seemingly non-terminating run of it can be verified as unambiguous by such a tool.

To measure the performance of `cfganalyzer` it was run on 81 ambiguous grammars from bioinformatics, ambiguous variants of programming languages as well as on a larger number of toy examples from [2]. Note that unambiguous ones are meaningless benchmarks here. Crucial for the performance of the tool on ambiguous examples is of course the size of the grammar as it directly influences the bound k up to which witnesses are found before the SAT solver runs out of memory. Their number of rules varies between 3 and 862 (in 2NF). Most grammars have less than 200 rules, but among the 13 grammars with number of rules above 200, there are such prominent examples as C (413 rules), SML (304), Pascal (337), Elsa C++ (862) and SQL (202). Fig. 2 gives an overview of the performance on these in relation to the witness size k for ambiguous subwords. All ambiguities in the given grammars were confirmed by `cfganalyzer`.

We have also examined `cfganalyzer`'s performance on the bounded equivalence problem. It not only is the most difficult of the other problems but it also has obvious applications in CFG design whenever one grammar serves as a specification and another as an implementation, and one wants to ensure that they generate the same language. The following scenario provides a nice test suite. At Helsinki University of Technology, students are given veral descriptions of CFLs

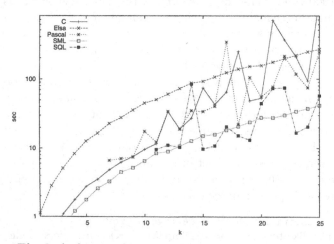

Fig. 2. Ambiguity detection of subwords with length k

and their task is to come up with CFGs which generate them. An automatized homework grading system has collected approx. 2000 student submissions for 40 different CFLs. Currently, unequivalence is only tested by sampling random words and checking that they are in both or neither of the two languages.

For running qualitatively better tests – cfganalyzer will not miss counterexamples up to the given bound unlike the testing approach – we have checked each of the 2000 grammars against all the sample solutions over the same alphabet (which of course makes the equivalence test fail for a large percentage). First, a coarse mapping of the submissions to the solution grammars was made, sorting out all submissions which were inequivalent to all solutions within a bound of $k \leq 15$ already (less than 0.1s in most tests). The remaining 251 grammars which potentially matched a solution were given a more thorough check by setting the maximum bound up to $k = 50$. Checking this range took on average 23.41s which is well below the time it would take to test all $|\Sigma|^{51} - 1$ words of length upto 50. This confirms cfganalyzer's feasibility and usefulness in set-ups that have to deal with CFGs in an automatic fashion.

5 Conclusion

The previous section shows that undecidable problems of CFLs can be (under-) approximated by bounding the search space of witnesses / counterexamples and using an incremental SAT solver for finding them. This approach is sound and "complete upto termination": it does not yield false-positives but, while unambiguity for example cannot be proved but only insinuated by the lack of found witnesses. This complements other work on ambiguity detection, in particular over-approximations which are complete – they can prove unambiguity – but not sound. The prototype implementation cfganalyzer shows feasibility of this approach: it has found ambiguity of large real-world grammars in short time. It also shows that this approach by far outperforms other existing and comparable approaches, e.g. under-approximations like the brute-force enumeration of parse trees of bounded length.

References

1. Bar-Hillel, Y., Perles, M., Shamir, E.: On formal properties of simple phrase structure grammars. Zeitschrift für Phonologie, Sprachwissenschaft und Kommunikationsforschung 14, 113–124 (1961)
2. Basten, H.J.S.: The usability of ambiguity detection methods for context-free grammars. In: Johnstone, A., Vinju, J. (eds.) Eighth Workshop on Language Descriptions, Tools, and Applications (LDTA 2008), Budapest, Hungary (April 2008)
3. Brabrand, C., Giegerich, R., Møller, A.: Analyzing ambiguity of context-free grammars. In: Holub, J., Žďárek, J. (eds.) CIAA 2007. LNCS, vol. 4783, pp. 214–225. Springer, Heidelberg (2007)
4. Hopcroft, J., Ullman, J.: Introduction to Automata Theory, Languages, and Computation. Addison-Wesley, Reading (1979)

5. Culik II, K., Cohen, R.S.: LR-regular grammars - an extension of LR(k) grammars. Journal of Computer and System Sciences 7(1), 66–96 (1973)
6. Moskewicz, M.W., Madigan, C.F., Zhao, Y., Zhang, L., Malik, S.: Chaff: Engineering an efficient sat solver. In: DAC, pp. 530–535. ACM, New York (2001)
7. Quimper, C.-G., Walsh, T.: Decomposing global grammar constraints. In: Bessière, C. (ed.) CP 2007. LNCS, vol. 4741, pp. 590–604. Springer, Heidelberg (2007)
8. Schmitz, S.: Conservative ambiguity detection in context-free grammars. In: Arge, L., Cachin, C., Jurdziński, T., Tarlecki, A. (eds.) ICALP 2007. LNCS, vol. 4596, pp. 692–703. Springer, Heidelberg (2007)
9. Younger, D.H.: Recognition and parsing of context-free languages in time n^3. Information and Control 10(2), 372–375 (1967)

Weak Pseudorandom Functions in Minicrypt

Krzysztof Pietrzak[1] and Johan Sjödin[2],[*]

[1] CWI Amsterdam
[2] ETH Zurich

Abstract. A family of functions is *weakly* pseudorandom if a random member of the family is indistinguishable from a uniform random function when queried on *random* inputs. We point out a subtle ambiguity in the definition of weak PRFs: there are natural weak PRFs whose security breaks down if the randomness used to sample the inputs is revealed. To capture this ambiguity we distinguish between *public-coin* and *secret-coin* weak PRFs.

We show that the existence of a secret-coin weak PRF which is *not* also a public-coin weak PRF implies the existence of two pass key-agreement (i.e. public-key encryption). So in Minicrypt, i.e. under the assumption that one-way functions exist but public-key cryptography does not, the notion of public- and secret-coin weak PRFs coincide.

Previous to this paper all positive cryptographic statements known to hold exclusively in Minicrypt concerned the adaptive security of constructions using non-adaptively secure components. Weak PRFs give rise to a new set of statements having this property. As another example we consider the problem of range extension for weak PRFs. We show that in Minicrypt one can beat the best possible range expansion factor (using a fixed number of distinct keys) for a very general class of constructions (in particular, this class contains all constructions that are known today).

1 Introduction

1.1 Weak Pseudorandom Functions

Informally, a pseudorandom function (PRF) is a function which cannot be distinguished from a uniform random function by any efficient distinguisher. PRFs have a wide range of applications in cryptography. Sometimes, however, the full power of a PRF is not needed and it is sufficient when the function cannot be distinguished when queried on random values. Such objects are referred to as *weak* PRFs.[1]

PRFs are black-box reducible to one-way functions. In particular, it was shown in [9] how to construct a PRF from any pseudorandom generator and in [11] a construction of a pseudorandom generator from any one-way function was introduced. Unfortunately those (black-box) reductions are not efficient enough

---

[*] This work was partially supported by the Zurich Information Security Center. It represents the views of the authors.

[1] Sometimes they are called PRFs secure under a *known-plaintext* attack (KPA).

L. Aceto et al. (Eds.): ICALP 2008, Part II, LNCS 5126, pp. 423–436, 2008.

to be practical[2]. In [23], Naor and Reingold gave a quite efficient construction of a PRF relying on a number-theoretic assumption. More precisely, assuming that the so called decisional Diffie-Hellman (DDH) assumption holds in some cyclic group $G = \langle g \rangle$ of order q, they proposed a PRF $(\mathbb{Z}_q)^{n+1} \times \{0,1\}^n \to G$ defined by

$$((k_0, \ldots, k_n), a) \mapsto g^{k_0 \cdot \Pi_{a_i=1} k_i},$$

where $(k_0, \ldots, k_n)$ is the secret key and a_i denotes the i'th bit of a. As observed in [22], if one just wants to construct a weak PRF, then even a much simpler construction exists: let G be a cyclic group of order q, then

$$\exp : \mathbb{Z}_q \times G \to G \quad \text{defined as} \quad \exp(k, a) = a^k$$

is a weak PRF if the DDH assumption holds in G (the exponent k is the secret key). Note that compared to the Naor-Reingold construction, this construction has a much shorter key ($\lceil \log(q) \rceil$ compared to $(n+1)\lceil \log(q) \rceil$) and is more efficient (each evaluation requires one exponentiation, compared to one exponentiation and up to n multiplications).

1.2   Public-Coin vs. Secret-Coin Weak PRFs

A standard choice for the group G used in $\exp$ would be a large subgroup of $\mathbb{Z}_p^*$ (where p is some large prime) of prime order q (which exists if and only if q divides the order $p-1$ of $\mathbb{Z}_p^*$). The DDH assumption is believed to hold in such groups if q is sufficiently large.[3] A natural way to sample an element from G is to choose an element $r \in \mathbb{Z}_q$ uniformly at random and set $a = g^r$ for some generator g of G.[4]

For $a_1, a_2, \ldots$ sampled this way, the tuples $(a_1, v_1), (a_2, v_2), \ldots$ computed by the weak PRF as $v_i = \exp(k, a_i) = a_i^k$ are indistinguishable from (a_1, u_1), $(a_2, u_2), \ldots$ where each u_i is a uniform random element in G. Now, assume that the distinguisher also gets to see the randomness used to sample the a_i's. Say r_1, r_2 are such that $a_1 = g^{r_1}$ and $a_2 = g^{r_2}$. Then one can easily distinguish $v_1 = a_1^k, v_2 = a_2^k$ from u_1, u_2 as $v_1^{r_2} = v_2^{r_1}$ ($= g^{r_1 r_2}$) but $u_1^{r_2} = u_2^{r_1}$ only holds with probability $1/q$.

Thus the security of the weak PRF $\exp$ completely breaks down if the randomness used to sample the random inputs is revealed. We will call such a weak PRF a *secret-coin* weak PRF, as opposed to a *public-coin* weak PRF which stays secure even if the random coins used to sample the inputs are revealed. Whether a weak PRF is a public-coin or just secret-coin weak PRF depends on the input-sampling algorithm, which hence must be part of the definition of the weak PRF.

---

[2] Even though progress has recently been made [10,14] for achieving more efficient reductions of PRGs from one-way functions.

[3] Say $p - 1 = 2q$ where $\log(p)$ is at least our security parameter.

[4] Alternatively one could sample random elements from $\mathbb{Z}_p^*$ until an element a is found where $a^q = a$, but this is less efficient (the expected number of tries is $q/(p-1)$), and only works if q^2 does not divide $p - 1$ as otherwise there is more than just one subgroup of order q.

1.3   Public-Coins=Secret-Coins in Minicrypt

As the function exp shows, the distinction between public- and secret-coin weak PRFs is meaningful and one can imagine many situations where the notion of a secret-coin weak PRF is not sufficient. In particular, this will always be the case when public randomness is used to sample the inputs.

It is not hard to see, that if a weak PRF is secret- but not public-coin, then the input-sampling algorithm must be a distributional one-way function [17].[5] We further show that any secret-coin weak PRF which is not also a public-coin weak PRF must be very artificial, in the sense that one can construct a public-key encryption scheme from it:

Theorem 1. *If there exists a secret-coin weak PRF which is not a public-coin weak PRF, then a (IND-CPA) secure public-key encryption scheme exists (see Section 1.4 for the disclaimer).*

Thus it is not surprising that the DDH assumption required for the security of exp implies public-key encryption [4,7]. So in Minicrypt (a name coined by Impagliazzo to denote the hypothetical world where one-way functions exist, but public-key cryptography does not [16]), the notion of public- and secret-coin weak PRFs coincide.

In order to prove Theorem 1, we show how to construct a public-key encryption scheme from any secret-coin weak PRF F and a distinguisher D which can distinguish F from a uniform random function when additionally to the random input/output pairs it is provided with the randomness used by the input sampling algorithm S (such a D exists as by assumption F is not a public-coin weak PRF).

1.4   Uniform vs. Non-uniform and Negligible vs. Noticeable

There is a gap between what is generally considered a successful distinguisher (or any other kind of an adversary) and what one expects from a protocol like an encryption scheme: a system is usually considered broken even if only a *non-uniform* adversary exists, whereas a protocol should be *uniform* and achieve its task with *overwhelming*[6] probability to be considered useful. The encryption scheme we construct in order the prove Theorem 1 uses the distinguisher D as a black-box, and only if D is *uniform* and has *noticeable* advantage in distinguishing F from a uniform random function, we will get a useful (as described above) key-agreement protocol. But if D is non-uniform, also the key-agreement protocol will be non-uniform. Furthermore, if D has only *non-negligible* (but not noticeable) advantage, then our encryption scheme will only be secure for infinitely many values of the security parameter (and not as usual for all sufficiently large ones).

---

[5] A distributional one-way function is a function where no efficient algorithm can find a *random* pre-image. Distributional one-way functions are equivalent to one-way functions.

[6] See Section 2 for a definition of negligible, noticeable and overwhelming.

1.5   Range Extension for Weak PRFs

The problem of range extension for weak PRFs is the following: given a weak PRF $f : \mathcal{K} \times \mathcal{X} \rightarrow \mathcal{X}$, construct a weak PRF $F : \mathcal{K}^t \times \mathcal{X} \rightarrow \mathcal{X}^s$ where F uses f as a black-box, t is the number of keys, and s is the so-called *expansion factor*. A trivial solution is to set

$$F((k_1, \ldots, k_t), x) = [f(k_1, x), \ldots, f(k_t, x)],$$

but this is not very satisfying as the expansion factor is only the number of keys (i.e. $s = t$). All efficient constructions for range extension of weak PRFs, that we are aware of [2,21,20], are of the form that the i'th output block y_i is computed as

$$y_i = f_{k_{i_q}} \circ f_{k_{i_{q-1}}} \circ \cdots \circ f_{k_{i_1}}(x)$$

for some $i_1, \ldots, i_q \in \{1, \ldots, t\}$. The construction from [20] achieves an expansion factor of $s = 2^t - 1$, and in [25] it is shown that this is tight: no construction (of the form as described above) with an expansion factor greater than $2^t - 1$ can be proven secure via a black-box reduction. We show that it is possible to beat this bound in Minicrypt. For this, consider the following construction which uses two keys and has an expansion factor of 4 (which is more than $2^t - 1 = 3$)

$$F((k_1, k_2), x) = [f(k_1, x), f(k_2, x), f(k_2, f(k_1, x)), f(k_1, f(k_2, x))]. \tag{1}$$

In [21], it has been claimed that this function is indeed a weak PRF, but in fact it is not as observed in [20]. The function $f = \exp$ is a simple counterexample, as in this case the last two values of the output in (1) are identical, namely

$$[\exp(k_2, \exp(k_1, x)), \exp(k_1, \exp(k_2, x))] = [x^{k_1 k_2}, x^{k_1 k_2}]. \tag{2}$$

In Section 4, we prove the following theorem:

Theorem 2. *If there exists a weak PRF f with superpolynomial domain size for which the construction given by equation (1) is not a weak PRF, then a secure public-key encryption scheme exists (disclaimer is below).*

The requirement that the domain size must be superpolynomial is necessary as otherwise the construction given in (1) is not a weak PRF even if f is a uniform random function. The reason is that the last two values of the output in (1), i.e. $[f(k_2, f(k_1, x)), f(k_1, f(k_2, x))]$, collide twice as often as for random elements. If now the domain size is not superpolynomial, collisions will occur with high probability after polynomially many input-output samples (allowing us to distinguish F from a uniform random function).

In order to prove Theorem 2, we show how to construct a two-pass key-agreement protocol from any weak PRF f and a distinguisher D which can (for random keys k_1, k_2) distinguish tuples computed as $[f(k_2, f(k_1, .)), f(k_1, f(k_2, .))]$

from random tuples. For this reduction, we have the same issue with uniform vs. non-uniform and negligible vs. noticeable, as discussed in Section 1.4.

1.6 Related Work

ADAPTIVE SECURITY IN MINICRYPT. The first positive cryptographic result proven to hold only in Minicrypt stated that the cascade of non-adaptively secure PRFs gives a construction with some weak form of adaptive security [24]. We have found more results since then, but all were about the adaptive security of some construction based on non-adaptively secure components.[7] The results of this paper show that weak PRFs give rise to a completely new class of statements that hold exclusively in Minicrypt.

OTHER WORLDS. Wee [26] shows that some cryptography (i.e. "non-trivial" argument systems) is even possible in Pessiland, which is another of Impagliazzo's possible worlds [16] where not even one-way functions exist. Dent [3] explores the limits of cryptography in universes whose existence is conjectured in popular Science Fiction literature.

WIN-WIN. Our results can be viewed as "win-win" statements, where one shows that (at least) one of two "positive" cryptographic statements is true. For example we show that either every secret-coin weak PRF is a public-coin weak PRF or public-key crypto exists. Results of similar flavour have been given before, in particular Dziembowski shows that either "forward-secure storage" is possible or (a weak form of) oblivious transfer exists [6]. Dubrov and Ishai show that either every efficiently samplable distribution can be sampled using few random bits, or one-way functions imply collision-resistant hashing [5].

PUBLIC VS. SECRET COINS. That differentiating between the public- and secret-coin variants of primitives is meaningful and important has been shown for at least two important primitives, namely collision resistant hash functions (CRHF) and trapdoor permutations (TDP).

In [15], Hsiao and Reyzin define public- and secret-coin families of CRHFs. The collision resistance of the latter requires the coins used to sample the function to be kept secret. They show that no black-box reduction from secret-coin to public-coin CRHFs exists.

The classical definition of a TDP states that it is hard to invert the permutation given a random element from the range of the permutation. In [8], Goldreich observes that for many applications this is not enough, and in fact *enhanced* TDPs are needed. Those have the property that it is hard to invert a random element even when given the random coins used to sample this element.

---

[7] For example, in [19] it was shown that the four round Feistel-network with non-adaptively secure round functions is not a pseudorandom permutation in general. To be more precise, it was shown that there exists a non-adaptively secure function f (whose security is based on the *inverse* DDH assumption [1]) for which the four-round Feistel-network with f as round function can be distinguished from a random permutation with two adaptive queries. Here, one can show that any such counterexample implies a three round key-agreement protocol (this is unpublished).

2 Basic Definitions

Throughout, let $n \in \mathbb{N}$ denote a security parameter. An entity (e.g. adversary) is efficient and uniform if it can be implemented by a probabilistic Turing machine whose running time is polynomial in the input length (which for us will always mean polynomial in n). It is efficient and non-uniform if it can be realized by a sequence of circuits (one for each n) of polynomial (in n) size.

For a set $\mathcal{X}$, let $x \xleftarrow{\$} \mathcal{X}$ denote that x is assigned a value from $\mathcal{X}$ uniformly at random. Let x^q denote the sequence $x_1, \ldots, x_q$. For a function $f : \mathcal{X} \to \mathcal{Y}$ and $x^q \in \mathcal{X}^q$, let $f(x^q)$ denote $f(x_1), \ldots, f(x_q)$.

UNIFORM RANDOM FUNCTIONS. $\mathbf{R}_{\mathcal{X}, \mathcal{Y}}$ denotes a uniform random function $\mathcal{X} \to \mathcal{Y}$.

NEGLIGIBLE. A function $\mu : \mathbb{N} \to [0, 1]$ is negligible if for any $c > 0$ there is an n_0 such that $\mu(n) \le 1/n^c$ for all $n \ge n_0$. To the contrary, μ is non-negligible if for some $c > 0$ we have $\mu(n) \ge 1/n^c$ for infinitely many n. Throughout, $\mathsf{negl}(n)$ denotes a negligible function in n.

OVERWHELMING. A function $\tau(\cdot) : \mathbb{N} \to [0, 1]$ is overwhelming if $1 - \tau(\cdot)$ is negligible.

NOTICEABLE. A function $\phi : \mathbb{N} \to [0, 1]$ is noticeable if for some $c > 0$ there is an n_0 such that $\phi(n) \ge 1/n^c$ for all $n \ge n_0$.

Note that non-negligible is not the same as noticeable. For example $\mu(n) \stackrel{\text{def}}{=} n \bmod 2$ is non-negligible but not noticeable.

BIT-AGREEMENT. Bit-agreement is a protocol between two efficient parties, which we refer to as Alice and Bob . They get the security parameter n in unary (denoted 1^n) as a common input and can communicate over an authentic channel. Finally, Alice and Bob output a bit b_A and b_B, respectively. The protocol has correlation ϵ if for all n

$$P[b_A = b_B] \ge (1 + \epsilon(n))/2.$$

Furthermore, the protocol is δ-secure if for any efficient adversary E, which can observe the whole communication C, and for all n

$$P[E(1^n, C) = b_B] \le 1 - \delta(n)/2.$$

KEY-AGREEMENT. If $\epsilon(\cdot)$ and $\delta(\cdot)$ are overwhelming then such a protocol achieves key-agreement. Any protocol which achieves bit-agreement with noticeable correlation $\epsilon(\cdot)$ and overwhelming security $\delta(\cdot)$ can be turned into a key-agreement protocol without increasing the number of rounds using parallel repetition and privacy amplification [12,13].

If $\epsilon(\cdot)$ is only non-negligible (i.e. for any constant $c > 0$, $\epsilon(n) \ge 1/n^c$ for infinitely many n), then also the key-agreement protocol will only achieve correctness for infinitely many (and not for all sufficiently large) choices of the security parameter.

3 Public-Coin vs. Secret-Coin Weak PRFs

Definition 1 (weak PRFs). *Consider a pair of efficient algorithms* F, KeyGen *where for any* $n \in \mathbb{N}$ *we have*

$$\text{KeyGen} : 1^n \to \mathcal{K}_n \qquad \text{F} : \mathcal{K}_n \times \mathcal{X}_n \to \mathcal{Y}_n.$$

KeyGen *is the randomized key-generation algorithm which on input a security parameter n (and some uniform random bits) outputs a key from the keyspace* $\mathcal{K}_n$. *Let the random variables* X_i, Y_i, *and* Z_i *for* $1 \leq i \leq \ell$ *be defined by first sampling a key* $k \leftarrow \text{KeyGen}(1^n)$ *and then setting (below we use the same random function* $\mathbf{R}_{\mathcal{X}_n, \mathcal{Y}_n}$ *for all* i)

$$X_i \stackrel{\$}{\leftarrow} \mathcal{X}_n \qquad Y_i \leftarrow \text{F}(k, X_i) \qquad Z_i \leftarrow \mathbf{R}_{\mathcal{X}_n, \mathcal{Y}_n}(X_i).$$

F *is a* weak pseudorandom function *secure if for every efficient distinguisher* D *and any polynomial* $\ell = \ell(n)$

$$|\text{P}[\text{D}(X^\ell, Y^\ell) = 1] - \text{P}[\text{D}(X^\ell, Z^\ell) = 1]| = \text{negl}(n).$$

Definition 2 (public-coin and secret-coin weak PRFs). *Let* F, KeyGen *be efficient algorithms as in the previous definition, and let* Sample : $\{0,1\}^{s(n)} \to \mathcal{X}_n$ *be an efficient input sampling algorithm.*

Let the random variables R_i, X_i, Y_i, *and* Z_i *be defined for* $1 \leq i \leq \ell$ *by first sampling a key* $k \leftarrow \text{KeyGen}(1^n)$ *and then setting*

$$R_i \stackrel{\$}{\leftarrow} \{0,1\}^{s(n)} \qquad X_i \leftarrow \text{Sample}(R_i) \qquad Y_i \leftarrow \text{F}(k, X_i) \qquad Z_i \leftarrow \mathbf{R}_{\mathcal{X}_n, \mathcal{Y}_n}(X_i).$$

The three algorithms F, KeyGen, Sample *are a* public-coin *weak PRF if for all efficient* D *and any polynomial* $\ell = \ell(n)$

$$|\text{P}[\text{D}(R^\ell, X^\ell, Y^\ell) = 1] - \text{P}[\text{D}(R^\ell, X^\ell, Z^\ell) = 1]| = \text{negl}(n)$$

(i.e. the weak PRF, by Definition 1, stays secure even if the randomness used to sample the inputs is revealed). Furthermore, F, KeyGen, Sample *are referred to as a* secret-coin *weak PRF if for all efficient* D *and any polynomial* $\ell = \ell(n)$

$$|\text{P}[\text{D}(X^\ell, Y^\ell) = 1] - \text{P}[\text{D}(X^\ell, Z^\ell) = 1]| = \text{negl}(n).$$

Clearly, every public-coin weak PRF is a secret-coin weak PRF. Also if F, KeyGen is a weak PRF (by Definition 1) and the output of Sample is close to uniform, then F, KeyGen, Sample is a secret-coin weak PRF. Note that in the definitions above, secure means secure against efficient uniform adversaries. To get a (stronger) notion which implies security against non-uniform adversaries, one must just consider a sequence of poly-size circuits instead of the poly-time bounded Turing-machine D (cf. Section 1.4).

PROTOCOL BITAGREEMENT(n)

Alice   Bob
$$b_B \xleftarrow{\$} \{0,1\}$$
$$k \leftarrow \mathsf{KeyGen}(1^n)$$

for $i = 1, \ldots, q = q(n)$ do

$r_i \xleftarrow{\$} \{0,1\}^n$ $x_i \leftarrow \mathsf{Sample}(r_i)$ od; $\xrightarrow{x^q}$ for $i = 1, \ldots, q$ do

if $b_B = 0$ then $z_i \leftarrow \mathsf{F}(k, x_i)$

elseif $b_B = 1$ then $z_i \leftarrow \mathbf{R}_{\mathcal{X}_n, \mathcal{Y}_n}(x_i)$ od;

$\xleftarrow{z^q}$

$b_A \leftarrow \mathsf{D}(r^q, x^q, z^q)$

Fig. 1. A bit-agreement protocol from a secret-coin weak PRF which is not a public-coin weak PRF

3.1 The Reduction

Let $(\mathsf{F}, \mathsf{KeyGen}, \mathsf{Sample})$ be a secret-coin weak PRF which is not a public-coin weak PRF. For $i = 1, 2, \ldots$ consider the random variables r_i, x_i, y_i, and u_i, defined by $k \leftarrow \mathsf{KeyGen}(1^n)$, $r_i \xleftarrow{\$} \{0,1\}^{s(n)}$, $x_i \leftarrow \mathsf{Sample}(r_i)$, $y_i \leftarrow \mathsf{F}(k, x_i)$, and $u_i \xleftarrow{\$} \mathcal{Y}_n$ (i.e. u_i is a random element from the range of F). Now, as F is not a public-coin weak PRF, there exist an efficient distinguisher D, a polynomial $q(.)$, and a non-negligible function $\phi(.)$ such that

$$\Pr[\mathsf{D}(r^q, x^q, y^q) = 1] - \Pr[\mathsf{D}(r^q, x^q, u^q) = 1] \geq \phi(n). \tag{3}$$

Further, as F is a secret-coin weak PRF we have for any efficient E that

$$|\Pr[\mathsf{E}(x^q, y^q) = 1] - \Pr[\mathsf{E}(x^q, u^q) = 1]| = \mathsf{negl}(n). \tag{4}$$

In order to prove Theorem 1, we must construct a two-pass public-key encryption scheme from the weak PRF. As discussed in Section 2, it is sufficient to construct a two-pass bit-agreement protocol with non-negligible (or noticeable, see the discussion in Section 1.4) correlation and overwhelming security. Such a protocol BITAGREEMENT is shown in Figure 1. The idea behind the protocol is quite simple: First, Alice samples some random strings, on which she invokes Sample to get random inputs to the secret-coin weak PRF F. Then she sends the inputs to Bob, who either return the outputs of F on these inputs or random values depending on his randomly chosen bit b_B. As Alice knows the randomness she used to sample the inputs, she can use the distinguisher D to get a guess b_A on b_B with non-negligible correlation, as shown in Claim 1 below. Furthermore, an adversary who does not know the randomness used to sample the inputs cannot distinguish the cases where Bob sends random values or values computed by F, as shown in Claim 2 below.

Claim 1. BITAGREEMENT(n) *has correlation* $\phi(n)$, *with* ϕ *as in (3).*

Proof.

$$\Pr[b_A = b_B] = \Pr[b_B = 1] \cdot \Pr[b_A = 1 | b_B = 1] + \Pr[b_B = 0] \cdot \Pr[b_A = 0 | b_B = 0]$$
$$= \frac{1}{2} + \frac{\Pr[b_A = 1 | b_B = 1] - \Pr[b_A = 1 | b_B = 0]}{2} \geq \frac{1}{2} + \frac{\phi(n)}{2}$$

$\square$

Claim 2. BITAGREEMENT(n) *is* $1 - \mathsf{negl}(n)$ *secure.*

Proof. For any efficient adversary E

$$\Pr[\mathsf{E}(x^q, z^k) = b_B] = \Pr[b_B = 1] \cdot \Pr[\mathsf{E}(x^q, z^k) = 1 | b_B = 1] + \Pr[b_B = 0] \cdot \Pr[\mathsf{E}(x^q, z^k) = 0 | b_B = 0]|$$
$$= \frac{1}{2} + \frac{\Pr[\mathsf{E}(x^q, z^k) = 1 | b_B = 1] - \Pr[\mathsf{E}(x^q, z^k) = 1 | b_B = 0]}{2} = \frac{1}{2} + \mathsf{negl}(n),$$

where the last step follows by (4). $\square$

Proof (of Theorem 1). The theorem follows from Claim 1 and 2 and the fact that one can construct a key-agreement protocol from any bit-agreement protocol which has noticeable correlation and overwhelming security without increasing the number of rounds (via parallel repetition and privacy amplification [12,13]).

4   Range Extension for Weak PRFs

Let $\mathsf{F} : \mathcal{K}_n \times \mathcal{X}_n \to \mathcal{X}_n, \mathsf{KeyGen} : 1^n \to \mathcal{K}_n$ denote a weak PRF as in Definition 1, with the additional property that the domain and range are identical, i.e. $\mathcal{X}_n = \mathcal{Y}_n$, and that the domain is of superpolynomial size, i.e. for all c there is an n_0 such that $|\mathcal{X}_n| \geq n^c$ for all $n > n_0$. In order to prove Theorem 2, we must show that if $\mathsf{G} : \mathcal{K}_n^2 \times \mathcal{X}_n \to \mathcal{X}_n^4$ defined as

$$\mathsf{G}((k, k'), x) = [\mathsf{F}(k, x), \mathsf{F}(k', x), \mathsf{F}(k', (\mathsf{F}(k, x))), \mathsf{F}(k, (\mathsf{F}(k', x)))] \tag{5}$$

is *not* a weak PRF, then a two-pass key-agreement protocol exists. If G is not a weak PRF, there exists an efficient distinguisher D, a polynomial $q(.)$, and a non-negligible function $\phi(.)$, such that for $x^q \xleftarrow{\$} \mathcal{X}_n^q, u^q \leftarrow \mathbf{R}_{\mathcal{X}_n, \mathcal{X}_n^4}(x^q), k \leftarrow \mathsf{KeyGen}(1^n)$, and $k' \leftarrow \mathsf{KeyGen}(1^n)$

$$\Pr[\mathsf{D}(x^q, \mathsf{G}((k, k'), x^q)) = 1] - \Pr[\mathsf{D}(x^q, u^q) = 1] \geq \phi(n). \tag{6}$$

We now define three other functions $\widetilde{\mathsf{G}}, \mathsf{H}, \widetilde{\mathsf{H}}$ which will be used in the proof. The function $\widetilde{\mathsf{G}}$ is defined almost as G, but with $\mathsf{F}(k', .)$ replaced by a URF. The systems H and $\widetilde{\mathsf{H}}$ are defined like G and $\widetilde{\mathsf{G}}$ respectively, but without the last term. For the rest of this section, we let $\mathbf{R}$ denote $\mathbf{R}_{\mathcal{X}_n, \mathcal{X}_n}$.

$$\widetilde{\mathsf{G}}((k, k'), x) = [\mathsf{F}(k, x), \mathbf{R}(x), \mathbf{R}(\mathsf{F}(k, x)), \mathsf{F}(k, \mathbf{R}(x))]$$
$$\mathsf{H}((k, k'), x) = [\mathsf{F}(k, x), \mathsf{F}(k', x), \mathsf{F}(k', \mathsf{F}(k, x))]$$
$$\widetilde{\mathsf{H}}((k, k'), x) = [\mathsf{F}(k, x), \mathbf{R}(x), \mathsf{F}(k', \mathbf{R}(x))]$$

Protocol BitAgreement2(n)

Alice   Bob

$$b_\mathsf{B} \xleftarrow{\$} \{0,1\}$$

$$k_\mathsf{A} \leftarrow \mathsf{KeyGen}(1^n) \qquad\qquad k_\mathsf{B} \leftarrow \mathsf{KeyGen}(1^n)$$

for $i = 1, \ldots, q = q(n)$ do

$$r_i \xleftarrow{\$} \mathcal{X}_n \ \ s_i \leftarrow \mathsf{F}(k_\mathsf{A}, r_i) \text{ od;} \xrightarrow{r^q, s^q} \text{ for } i = 1, \ldots, q \text{ do}$$

if $b_\mathsf{B} = 0$ then $t_i \leftarrow \mathsf{F}(k_\mathsf{B}, r_i)$

$y_i \leftarrow \mathsf{F}(k_\mathsf{B}, s_i)$

elseif $b_\mathsf{B} = 1$ then $t_i \xleftarrow{\$} \mathbf{R}_{\mathcal{X}_n, \mathcal{X}_n}(r_i)$

$y_i \xleftarrow{\$} \mathbf{R}_{\mathcal{X}_n, \mathcal{X}_n}(s_i)$ od;

$$\xleftarrow{t^q, y^q}$$

for $i = 1, \ldots, q$ do $z_i \leftarrow \mathsf{F}(k_\mathsf{A}, t_i)$ od;

$$b_\mathsf{A} \leftarrow \mathsf{D}(r^q, s^q, t^q, y^q, z^q)$$

Fig. 2. A bit-agreement protocol from a weak PRF $\mathsf{F} : \mathcal{K}_n \times \mathcal{X}_n \to \mathcal{X}_n$ where the construction given by (1) is not a weak PRF

Below we will show that if F is a weak PRF, then also $\widetilde{\mathsf{G}}$, H, and $\widetilde{\mathsf{H}}$ are weak PRFs. The idea behind the bit-agreement protocol given in Figure 2 is now quite simple: First, Alice and Bob each sample a random key for F. Then Bob flips a random coin b_B. If $b_\mathsf{B} = 0$, Alice and Bob together simulate an attack on G, and if $b_\mathsf{B} = 1$, they simulate an attack on $\widetilde{\mathsf{G}}$. As Alice can distinguish G from random (and thus from $\widetilde{\mathsf{G}}$), she can learn b_B with non-negligible advantage, as shown in Claim 5. However, an adversary Eve does not see the last term of G or $\widetilde{\mathsf{G}}$, as they are computed by Alice and not sent over to Bob. Hence, Eve only sees the outputs as they are given by H if $b_\mathsf{B} = 0$ and by $\widetilde{\mathsf{H}}$ if $b_\mathsf{B} = 1$. As H and $\widetilde{\mathsf{H}}$ are weak PRFs, Eve only has negligible advantage in distinguishing those two cases (and thus also in guessing b_B), as shown in Claim 6.

Claim 3. *If* F *is a weak PRF, then* $\widetilde{\mathsf{G}}$ *is a weak PRF.*

Proof. We have to show that for any polynomial $q(.)$ and $x_1, \ldots, x_q \xleftarrow{\$} \mathcal{X}_n$ ($q = q(n)$) the q four-tuples

$$[x_i, \mathsf{F}(k, x_i), \mathbf{R}(x_i), \mathbf{R}(\mathsf{F}(k, x_i)), \mathsf{F}(k, \mathbf{R}(x_i))] \tag{7}$$

are indistinguishable from random. Sample $x'_1, \ldots, x'_q \xleftarrow{\$} \mathcal{X}_n$, $x''_1, \ldots, x''_q \xleftarrow{\$} \mathcal{X}_n$, and consider the distribution

$$[x_i, \mathsf{F}(k, x_i), x'_i, x''_i, \mathsf{F}(k, x'_i)]. \tag{8}$$

As F is a weak PRF, the five tuples given by (8) are indistinguishable from random. We will now show that (8) is indistinguishable from (7). First note that $\mathbf{R}(x_1), \ldots, \mathbf{R}(x_q)$ has the same distribution as $x'_1, \ldots, x'_q$, unless $x_i = x_j$ for some $i \neq j$, as q is polynomial and $|\mathcal{X}|$ is superpolynomial, the probability of this event is negligible. So we can safely replace $\mathbf{R}(x_i)$ in (7) with x'_i in (8). Similarly, we can replace $\mathbf{R}(\mathsf{F}(k, x_i))$ with x''_i as this will make no difference unless $\mathsf{F}(k, x_i) = \mathsf{F}(k, x_j)$ or $\mathsf{F}(k, x_i) = x_j$ for some $i \neq j$, which only happens with negligible probability. □

Claim 4. *If* F *is a weak PRF, then* H *and* $\widetilde{\mathsf{H}}$ *are weak PRFs.*

Proof. That $\widetilde{\mathsf{H}}$ is weakly pseudorandom follows directly from the fact that $\widetilde{\mathsf{G}}$ is weakly pseudorandom (as shown by the previous claim). To show that H is weakly pseudorandom, we show that, for random $x_1, \ldots, x_q \in \mathcal{X}_n$, the tuples

$$[x_i, \mathsf{F}(k, x_i), \mathsf{F}(k', x_i), \mathsf{F}(k', \mathsf{F}(k, x_i))] \tag{9}$$

are indistinguishable from random. For this, it is sufficient by the triangle inequality, to show the following two facts. First, for random $x'_1, \ldots, x'_q \in \mathcal{X}_n$, the tuples

$$[x_i, x'_i, \mathsf{F}(k', x_i), \mathsf{F}(k', x'_i)] \tag{10}$$

are indistinguishable from (9), since F is a weak PRF and thus $\mathsf{F}(k, x_i)$ can be replaced by a random x'_i. That (10) is indistinguishable from random follows directly from the fact that F is a weak PRF. □

Claim 5. BitAgreement2(n) *has non-negligible correlation* $\phi(n) - \mathsf{negl}(n)$, *with* ϕ *as in (6).*

Proof. For $x^q \overset{\$}{\leftarrow} \mathcal{X}_n^q$ and $u^q \leftarrow \mathbf{R}_{\mathcal{X}_n, \mathcal{X}_n^q}(x^q)$

$$\Pr[b_\mathsf{A} = b_\mathsf{B}] = \Pr[b_\mathsf{B} = 1] \cdot \Pr[b_\mathsf{A} = 1 | b_\mathsf{B} = 1] + \Pr[b_\mathsf{B} = 0] \cdot \Pr[b_\mathsf{A} = 0 | b_\mathsf{B} = 0]$$

$$= \frac{1}{2} + \frac{\Pr[b_\mathsf{A} = 1 | b_\mathsf{B} = 1] - \Pr[b_\mathsf{A} = 1 | b_\mathsf{B} = 0]}{2}$$

$$= \frac{1}{2} + \frac{\Pr[\mathsf{D}(x^q, \mathsf{G}((k_\mathsf{A}, k_\mathsf{B}), x^q)) = 1] - \Pr[\mathsf{D}(x^q, \widetilde{\mathsf{G}}((k_\mathsf{A}, k_\mathsf{B}), x^q)) = 1]}{2}$$

$$= \frac{1}{2} + \frac{\Pr[\mathsf{D}(x^q, \mathsf{G}((k_\mathsf{A}, k_\mathsf{B}), x^q)) = 1] - \Pr[\mathsf{D}(x^q, u^q) = 1] \pm \mathsf{negl}(n)}{2}$$

$$\geq \frac{1}{2} + \frac{\phi(n) \pm \mathsf{negl}(n)}{2},$$

where the second last step follows as $\widetilde{\mathsf{G}}$ is a weak PRF (as shown in Claim 3). □

Claim 6. BitAgreement2(n) *is* $1 - \mathsf{negl}(n)$ *secure.*

Proof. Consider any efficient adversary E who can observe the communication $C = \{r^q, s^q, t^q, y^q\}$ between Alice and Bob . If $b_\mathsf{B} = 0$, then C has the same distribution as generated by H, and if $b_\mathsf{B} = 1$, then C has the same distribution as generated by $\widetilde{\mathsf{H}}$. The security now follows as an efficient E cannot distinguish H from $\widetilde{\mathsf{H}}$, since these are both weak PRFs (as shown in Claim 4). More formally, for $x^q \xleftarrow{\$} \mathcal{X}_n^q$

$$\Pr[\mathsf{E}(C) = b_\mathsf{B}] = \Pr[b_\mathsf{B} = 1] \cdot \Pr[\mathsf{E}(C) = 1 | b_\mathsf{B} = 1] + \Pr[b_\mathsf{B} = 0] \cdot \Pr[\mathsf{E}(C) = 0 | b_\mathsf{B} = 0]|$$
$$= \frac{1}{2} + \frac{\Pr[\mathsf{E}(C) = 1 | b_\mathsf{B} = 1] - \Pr[\mathsf{E}(C) = 1 | b_\mathsf{B} = 0]}{2}$$
$$= \frac{1}{2} + \frac{\Pr[\mathsf{E}(x^q, \tilde{\mathsf{H}}(k_\mathsf{A}, x^q)) = 1] - \Pr[\mathsf{E}(x^q, \mathsf{H}(k_\mathsf{A}, x^q)) = 1]}{2} = \frac{1}{2} \pm \mathsf{negl}(n).$$

□

Proof (of Theorem 2). The theorem follows from Claim 5 and 6, and the fact that one can construct a key-agreement protocol from any bit-agreement protocol which has noticeable correlation and overwhelming security without increasing the number of rounds (via parallel repetition and privacy amplification [12,13]).

5   Can We Efficiently Deconstruct "Useful" Properties?

BLACK BOX FALSIFICATION. What does it mean that some statement "holds in Minicrypt"? Trivially, it means that in order to falsify the statement, we must assume the existence of something at least as strong as key-agreement. As observed in [24], the fact that no black-box reduction one-way functions to key-agreement exists [18], implies that no "black-box falsification" for such statements can exist (in [24] this was called a "black-box break"). E.g. by Theorem 2, there is no black-box reduction from one-way functions to a weak PRF and a distinguisher, such that the distinguisher breaks the security of the construction given by (1) when instantiated with this weak PRF.[8]

DECONSTRUCTING THE HOMOMORPHIC PROPERTY. We showed that the statements of the theorems from this paper are non-trivial, by showing that they do no longer hold outside Minicrypt, or more precisely under the standard DDH assumption (which is false in Minicrypt as it implies key-agreement). These counterexamples use the homomorphic property of the group, i.e. that $(x^a)^b = (x^b)^a$. This is eminent in (2) which shows that the construction given by (1) does not give secure range extension outside of Minicrypt. Usually, a weak PRF which is homomorphic is a very useful thing to have, but clearly not if we want to use this PRF in the construction given in (1). In order to safely use a weak PRF in (1), all we have to make sure is, that the protocol from Figure 2 is NOT a secure bit-agreement protocol. Intuitively, that should not be too hard.

Open Problem 1. *Is there an efficient construction ϕ, such that for any weak PRF F, $\phi(\mathsf{F})$ is a weak PRF but the protocol from Figure 2 is NOT a secure bit-agreement protocol when instantiated with $\phi(\mathsf{F})$ (and thus (1) is a secure range extension for $\phi(\mathsf{F})$).*

We can solve the above problem by first constructing a PRG from the weak PRF and then using the GGM reduction [9] to get a regular PRF (note that (1) is trivially secure for PRFs), but this is not really efficient: each evaluation of the PRF

---

[8] Let us stress that black-box reductions for one-way functions to PRFs do exist [11,9].

would make a linear (in the input length) number of invocations to the weak PRF. The above problem is somewhat antipodal to questions usually asked in cryptography, where one tries to construct something useful, like asking "can we construct key-agreement from PRFs via black-box reductions" (the answer is no [18]). Whereas here we are looking for a way to make some particular construction insecure. Being more ambitious, we can ask if we can efficiently destroy any property of a weak PRF which could be used to get a key-agreement protocol.

Open Problem 2. *Is there an efficient construction ϕ, such that for any weak PRF F, $\phi(F)$ is a weak PRF and any construction which is secure in Minicrypt when instantiated with a weak PRF, is secure (in the real world) when instantiated with $\phi(F)$?*

The general problem can be summarized as follows. We know several constructions for extending the range of weak PRFs, getting public-coin weak PRFs from secret-coin weak PRFs[9], and achieving adaptive security from non-adaptive PRFs [24,19], that one can show to be secure in Minicrypt, but which are (under standard assumptions) not secure in the real world. As the constructions are secure in Minicrypt, each weak/non-adaptive PRF for which the construction actually is insecure can be used to construct a key-agreement protocol (via some particular black-box construction), and must hence have a lot of structure. The question is whether there is an efficient way to modify the weak/non-adaptive PRF, such that it still keeps its original security guarantee (of being a weak/non-adaptive PRF), but cannot be used anymore for the key-agreement protocols. Then this modified PRF can safely be used in the efficient constructions that have been proven to be secure in Minicrypt.

References

1. Bao, F., Deng, R.H., Zhu, H.: Variations of Diffie-Hellman problem. In: Qing, S., Gollmann, D., Zhou, J. (eds.) ICICS 2003. LNCS, vol. 2836. Springer, Heidelberg (2003)
2. Damgård, I., Nielsen, J.B.: Expanding pseudorandom functions; or: From known-plaintext security to chosen-plaintext security. In: Yung, M. (ed.) CRYPTO 2002. LNCS, vol. 2442, pp. 449–464. Springer, Heidelberg (2002)
3. Dent, A.: Cryptography in a hitchhiker's universe. Journal of Craptology 4 (2007)
4. Diffie, W., Hellman, M.: New directions in cryptography. IEEE Transactions on Information Theory IT-22(6), 644–654 (1976)
5. Dubrov, B., Ishai, Y.: On the randomness complexity of efficient sampling. In: Proc. 38th ACM Symposium on the Theory of Computing (STOC), pp. 711–720 (2006)
6. Dziembowski, S.: On forward-secure storage. In: Dwork, C. (ed.) CRYPTO 2006. LNCS, vol. 4117, pp. 251–270. Springer, Heidelberg (2006)
7. El-Gamal, T.: A public key cryptosystem and a signature scheme based on discrete logarithms. IEEE Transactions on Information Theory 31(4), 469–472 (1985)

---

[9] This construction simply uses the secret-coin weak PRF as the public-coin weak PRF.

8. Goldreich, O.: Foundations of Cryptography. Basic Applications, vol. II. Cambridge University Press, Cambridge (2004)
9. Goldreich, O., Goldwasser, S., Micali, S.: How to construct random functions. J. ACM 33(4), 792–807 (1986)
10. Haitner, I., Harnik, D., Reingold, O.: Efficient pseudorandom generators from exponentially hard one-way functions. In: Bugliesi, M., Preneel, B., Sassone, V., Wegener, I. (eds.) ICALP 2006. LNCS, vol. 4052, pp. 228–239. Springer, Heidelberg (2006)
11. Håstad, J., Impagliazzo, R., Levin, L.A., Luby, M.: A pseudorandom generator from any one-way function. SIAM J. Comput. 28(4), 1364–1396 (1999)
12. Holenstein, T.: Personal Communication (2005)
13. Holenstein, T.: Immunization of key-agreement schemes. PhD thesis, ETH Zürich (2006) ISBN 3-86628-088-2
14. Holenstein, T.: Pseudorandom generators from one-way functions: A simple construction for any hardness. In: Halevi, S., Rabin, T. (eds.) TCC 2006. LNCS, vol. 3876, pp. 443–461. Springer, Heidelberg (2006)
15. Hsiao, C.-Y., Reyzin, L.: Finding collisions on a public road, or do secure hash functions need secret coins? In: Franklin, M. (ed.) CRYPTO 2004. LNCS, vol. 3152, pp. 92–105. Springer, Heidelberg (2004)
16. Impagliazzo, R.: A personal view of average-case complexity. In: Structure in Complexity Theory Conference, pp. 134–147 (1995)
17. Impagliazzo, R., Luby, M.: One-way functions are essential for complexity based cryptography (extended abstract). In: IEEE Symposium on the Foundations of Computer Science (FOCS) 1989, pp. 230–235 (1989)
18. Impagliazzo, R., Rudich, S.: Limits on the provable consequences of one-way permutations. In: Proc. 21th ACM Symposium on the Theory of Computing (STOC), pp. 44–61 (1989)
19. Maurer, U., Oswald, Y.A., Pietrzak, K., Sjödin, J.: Luby-Rackoff ciphers from weak round functions? In: Vaudenay, S. (ed.) EUROCRYPT 2006. LNCS, vol. 4004, pp. 391–408. Springer, Heidelberg (2006)
20. Maurer, U.M., Sjödin, J.: A fast and key-efficient reduction of chosen-ciphertext to known-plaintext security. In: Naor, M. (ed.) EUROCRYPT 2007. LNCS, vol. 4515, pp. 498–516. Springer, Heidelberg (2007)
21. Minematsu, K., Tsunoo, Y.: Expanding weak PRF with small key size. In: Won, D.H., Kim, S. (eds.) ICISC 2005. LNCS, vol. 3935, pp. 284–298. Springer, Heidelberg (2006)
22. Naor, M., Pinkas, B., Reingold, O.: Distributed pseudo-random functions and KDCs. In: Stern, J. (ed.) EUROCRYPT 1999. LNCS, vol. 1592, pp. 327–346. Springer, Heidelberg (1999)
23. Naor, M., Reingold, O.: Number-theoretic constructions of efficient pseudo-random functions. J. of the ACM 51(2), 231–262 (2004)
24. Pietrzak, K.: Composition implies adaptive security in minicrypt. In: Vaudenay, S. (ed.) EUROCRYPT 2006. LNCS, vol. 4004, pp. 328–338. Springer, Heidelberg (2006)
25. Pietrzak, K., Sjödin, J.: Domain extension for weak PRFs; the good, the bad, and the ugly. In: Advances in Cryptology — EUROCRYPT 2007, vol. 4515, pp. 517–533. Springer, Heidelberg (2007)
26. Wee, H.: Finding pessiland. In: Halevi, S., Rabin, T. (eds.) TCC 2006. LNCS, vol. 3876, pp. 429–442. Springer, Heidelberg (2006)

On Black-Box Ring Extraction and Integer Factorization

Kristina Altmann, Tibor Jager, and Andy Rupp

Horst Görtz Institute for IT-Security
Ruhr-University Bochum
{kristina.altmann,tibor.jager}@nds.rub.de,
arupp@crypto.rub.de

Abstract. The black-box extraction problem over rings has (at least) two important interpretations in cryptography: An efficient algorithm for this problem implies (i) the equivalence of computing discrete logarithms and solving the Diffie-Hellman problem and (ii) the in-existence of secure ring-homomorphic encryption schemes.

In the special case of a finite field, Boneh/Lipton [1] and Maurer/ Raub [2] show that there exist algorithms solving the black-box extraction problem in subexponential time. It is unknown whether there exist more efficient algorithms.

In this work we consider the black-box extraction problem over finite rings of characteristic n, where n has at least two different prime factors. We provide a polynomial-time reduction from factoring n to the black-box extraction problem for a large class of finite commutative unitary rings. Under the factoring assumption, this implies the in-existence of certain efficient generic reductions from computing discrete logarithms to the Diffie-Hellman problem on the one side, and might be an indicator that secure ring-homomorphic encryption schemes exist on the other side.

1 Introduction

Informally speaking, the black-box extraction problem over an algebraic structure A (like a group, ring, or a field) can be described as follows: Given an explicit representation of A (e.g., the cyclic group $(\mathbb{Z}_n, +)$ with the canonical binary representation of elements) as well as access to a black-box resembling the structure of A and hiding an element $x \in A$, the challenge is to recover x in the given explicit representation. Algorithms that work on the black-box representation of an algebraic structure, and thus on *any* concrete representation, are called *generic* or *black-box* algorithms.

The black-box extraction problem has been studied in various variants and contexts, e.g., see [3,4,5,1,2]. The case where the algebraic structure is a *cyclic group* (with given representation $(\mathbb{Z}_n, +)$), and the extraction problem is better known as the discrete logarithm problem, was considered by Nechaev [3] and Shoup [4]. They showed that the expected running time of any generic algorithm for this problem is $\Omega(\sqrt{p})$, where p is the largest prime factor of the group order n. Here, the integer n as well as its factorization is assumed to be publicly known.

L. Aceto et al. (Eds.): ICALP 2008, Part II, LNCS 5126, pp. 437–448, 2008.
© Springer-Verlag Berlin Heidelberg 2008

Boneh and Lipton [1] considered the black-box extraction problem over *prime fields* $\mathbb{F}_p$. Based on a result due to Maurer [6] they developed an algorithm solving the problem in subexponential time (in $\log p$). Maurer and Raub [2] augmented this result to finite *extension fields* $\mathbb{F}_{p^k}$ by providing an efficient reduction from the black-box extraction problem over $\mathbb{F}_{p^k}$ to the black-box extraction problem over $\mathbb{F}_p$. Currently it is unknown whether there exist more efficient algorithms for black-box extraction over fields.

The black-box extraction problem over fields/rings has at least two important applications in cryptography. For $(\mathbb{Z}_n, +, \cdot)$ it can be interpreted as the problem of solving the discrete logarithm problem given access to an oracle for the Diffie-Hellman problem: $(\mathbb{Z}_n, +)$ forms a cyclic additive group. The black-box provides access to the common operations on this group as well as to the additional operation "$\cdot$". This extra operation can be interpreted as an oracle solving the Diffie-Hellman problem in the group $(\mathbb{Z}_n, +)$. Hence, an efficient algorithm for the black-box extraction problem over $(\mathbb{Z}_n, +, \cdot)$ would correspond to an efficient *generic* reduction from computing discrete logarithms to solving the Diffie-Hellman problem over cyclic groups of order n. Such reductions are known for groups where the group order is prime and meets certain properties [7], or if certain side information, depending on the respective group, is given [6]. It is also known that no efficient generic reduction exists for groups with orders containing a large multiple prime factor [8]. Bach [9] has presented a non-generic reduction from factoring n to computing discrete logarithms in the group $\mathbb{Z}_n^*$.

Furthermore, the analysis of the black-box extraction problem sheds light on the existence of secure ring/field-homomorphic encryption schemes. A homomorphic encryption scheme is a scheme where the encryption function is a homomorphism from the plaintext space to the ciphertext space. Several *group*-homomorphic encryption schemes are known, such as native RSA, native ElGamal or the Paillier encryption scheme. A natural question arising in this context is whether there exist secure *ring*-homomorphic encryption schemes, that is, schemes where the plaintext- and ciphertext space exhibit a ring structure, and the encryption function is a ring-homomorphism. The results by Boneh and Lipton [1] and Maurer and Raub [2] imply that for the special case of *field*-homomorphic encryption any such scheme can be broken in subexponential time. It is unknown whether there are more efficient algorithms. An efficient algorithm for the black-box ring problem would imply the in-existence of secure ring-homomorphic encryption.

1.1   Our Contribution

In this work we consider the black-box extraction problem over finite commutative rings with unity whose characteristic n is the product of at least two different primes. To the best of our knowledge, this case has not been treated in the literature yet. We present an efficient reduction from finding a non-trivial factor of n to the black-box extraction problem over virtually any ring R where computation is efficient. To this end, we extend a technique due to Leander and Rupp [10] which was originally used to prove the equivalence of breaking RSA and factoring regarding generic ring algorithms.

We first provide a reduction for the case $R = \mathbb{Z}_n$. This case is especially interesting since Boneh and Lipton pointed out that their subexponential time black-box extraction algorithm for finite fields can be extended to finite rings $\mathbb{Z}_n$ if n is squarefree, requiring that the factorization of n is known. Our result implies that there are no better algorithms than those that factorize n. Moreover, under the assumption that factoring n is hard, this implies the in-existence of efficient *generic* reductions from computing discrete logarithms to solving the Diffie-Hellman problem in cyclic groups of order n. Note that, in contrast to Bach [9] who presented a reduction from factoring n to computing discrete logarithms in $\mathbb{Z}_n^*$, i.e. where group elements are represented by integers modulo n and the group order is $\phi(n)$ with $\phi(\cdot)$ denoting the Euler totient function, we consider *generic* reductions in groups of order n, regardless of the representation of elements.

We extend our reduction to rings of the form $\mathbb{Z}_n[X_1, \ldots, X_t]/J$, where $t \geq 0$ and J is an ideal in $\mathbb{Z}_n[X_1, \ldots, X_t]$ for which a Gröbner basis is known. If computation (i.e., applying the ring operations including reduction, equality testing and random sampling) in R is efficient, then the same holds for our reduction from finding a factor of n to black-box extraction over R.

Boneh/Lipton [1] and Maurer/Raub [2] show that any field-homomorphic encryption scheme can be broken in subexponential time. It is an open question whether there exist more efficient generic algorithms. For a large class of rings we can negate this question, assuming that factoring the ring characteristic cannot be done better than in subexponential time. This might be seen as an indicator for the existence of secure ring-homomorphic encryption schemes.

It is possible to extend the results presented in this paper to rings given in basis representation and to direct products $R_1 \times \cdots \times R_l$ of rings whenever our proofs apply to at least one component R_i, $i \in \{1, \ldots, l\}$. However, we have to refer to the full version of this paper [11] for details.

2 The Black-Box Ring Extraction Problem

Informally, black-box ring algorithms are the class of algorithms that operate on the structure of an algebraic ring without exploiting specific properties of the respresentation of ring elements. We adopt Shoup's generic group model [4] to formalize the notion of black-box ring algorithms:

Let $(R, +, \cdot)$ be a finite commutative unitary ring and $S \subseteq \{0,1\}^{\lceil \log_2(|R|) \rceil}$ be a set of bit strings of cardinality $|R|$. Let $\sigma : R \to S$ be a bijective encoding function which assigns ring elements to bit strings, chosen at random among all possible bijections. A *black-box ring algorithm* is an algorithm that takes as input an *encoding list* $(\sigma(r_1), \ldots, \sigma(r_k))$, where $r_i \in R$. Note that depending on the particular problem the algorithm might take some additional data as input, such as the characteristic of R, for example. In order to be able to perform the ring operations on randomly encoded elements, the algorithm may query a *black-box ring oracle* $\mathcal{O}_{R,\sigma}$. The oracle takes two indices i, j into the encoding list and a symbol $\circ \in \{+, -, \cdot\}$ as input, computes $\sigma(r_i \circ r_j)$ and appends this bit string to the encoding list (to which the algorithm always has access). We capture the notion of a black-box ring representation by the following definition:

Definition 1 (Black-Box Ring Representation). *Let $(R, +, \cdot)$ be a finite ring. We call the tuple $(\sigma, \mathcal{O}_{R,\sigma})$ consisting of a randomly chosen encoding function $\sigma : R \to S$, and a corresponding black-box ring oracle $\mathcal{O}_{R,\sigma}$ a black-box ring representation for R and denote it by R^σ.*

For short, we sometimes call R^σ a black box ring (meaning that we consider a ring exhibiting the structure of R but whose elements are encoded by random bit strings). As an abuse of notation we occasionally write $\sigma(x) \in R^\sigma$ meaning that the unique encoding $\sigma(x)$ of an element $x \in R$ is given. Moreover, when we say in the following that an algorithm $\mathcal{A}$ performs operations on the black-box ring R^σ, we mean that $\mathcal{A}$ interacts with the black-box ring oracle as described above. Having formalized the notion of a black-box ring, we can define the black-box ring extraction problem:

Definition 2 (BBRE Problem). *Let R be an explicitly given finite commutative ring with unity 1 and known characteristic n. Furthermore, let $B :=$ $\{r_1, \ldots, r_t\}$ be an (explicitly given) generating set of R. The black-box ring extraction (BBRE) problem for R is the task of computing $x \in R$, where x is chosen uniformly random from R, given $\sigma(x), \sigma(1), \sigma(r_1), \ldots, \sigma(r_t) \in R^\sigma$.*

3   BBRE for $\mathbb{Z}_n$ and Integer Factorization

In this section we consider the BBRE problem for rings which are isomorphic to $\mathbb{Z}_n$, where n has at least two different prime factors. We provide a reduction from factoring n to the BBRE problem in the following sense: If there exists an efficient algorithm solving the BBRE problem for $\mathbb{Z}_n$ with non-negligible success probability, then there exists an efficient algorithm finding a factor of n with non-negligible probability.

Theorem 1. *Let $R := \mathbb{Z}_n$ for some integer n having at least two different prime factors. Let $\mathcal{A}$ be an algorithm for the BBRE problem that performs at most $m \leq n$ operations on R^σ. Assume that $\mathcal{A}$ solves the BBRE problem with probability ϵ. Then there is an algorithm $\mathcal{B}$ having white-box access to $\mathcal{A}$ that finds a factor of n with probability at least*

$$\frac{|\epsilon - \frac{1}{n}|}{m^2 + 3m + 2}$$

by running $\mathcal{A}$ once and performing an additional amount of $O\left(m^2\right)$ random choices and $O\left(m^3\right)$ operations on R as well as $O\left(m^2\right)$ gcd computations on $\log_2(n)$-bit numbers.

Proof. We replace the original black-box ring oracle $\mathcal{O}_{R,\sigma}$ with an oracle $\mathcal{O}_{\mathsf{sim}}$ that simulates $\mathcal{O}_{R,\sigma}$ without using the knowledge of the secret x. In order to make this step more comprehensible, let us first define a slightly modified but *equivalent* version of the original black-box ring oracle: Instead of using the ring

$R = \mathbb{Z}_n$ for the internal representation of ring elements, these elements are represented by polynomials in the variable X over R which are evaluated with x each time the encoding of a newly computed element must be determined.

Definition 3 (An Equivalent Oracle). *The oracle $\mathcal{O}$ has an input and an output port as well as a random tape and performs computations as follows.*
Input. *As input $\mathcal{O}$ receives the modulus n and an element $x \in_U R$.*
Internal State. *As internal state $\mathcal{O}$ maintains two lists $L \subset R[X]$ and $E \subset S_n$. For an index i let L_i and E_i denote the i-th element of L and E, respectively.*
Encoding of Elements. *Each time a polynomial P should be appended to the list L the following computation is triggered to determine the encoding of $P(x)$: $\mathcal{O}$ checks if there exists any index $1 \le i \le |L|$ such that*

$$(P - L_i)(x) \equiv 0 \bmod n .$$

If this equation holds for some i, then the respective encoding E_i is appended to E again. Otherwise the oracle chooses a new encoding $s \in_U S \backslash E$ and appends it to E.
The computation of $\mathcal{O}$ starts with an initialization phase, which is run once, followed by the execution of the query-handling phase:
Initialization. *The list L is initialized with the polynomials $1, X$ and the list E is initialized with corresponding encodings.*
Query-handling. *Upon receiving a query $(\circ, i_1, i_2)$ on its input tape, where $\circ \in \{+, -, \cdot\}$ identifies an operation and i_1, i_2 are indices identifying the list elements the operation should be applied to, $\mathcal{O}$ appends the polynomial $P := L_{i_1} \circ L_{i_2}$ to L and the corresponding encoding to E.*

We say that an algorithm is successful in this game iff it outputs x, and denote this event with **S**. Note that $\epsilon = \Pr[\mathbf{S}]$.

A Simulation Game. Now we replace $\mathcal{O}$ by a simulation oracle $\mathcal{O}_{\mathsf{sim}}$. The simulation oracle is defined exactly like $\mathcal{O}$ except that it determines the encodings of elements in a different way in order to be independent of the secret x.

Each time a polynomial P is appended to the end of list L (during initialization or query-handling), $\mathcal{O}_{\mathsf{sim}}$ does the following: Let $L_j = P$ denote the last entry of the updated list. Then for each $i < j$ the simulation oracle chooses a *new* element $x_{i,j} \in R$ uniformly at random and checks whether

$$(L_i - L_j)(x_{i,j}) \equiv 0 \bmod n .$$

If the above equation is not satisfied for any i, the oracle chooses a new encoding $s \in_U S \backslash E$ and appends it to E. Otherwise, for the first i the equation is satisfied, the corresponding encoding E_i is appended to E again (i.e., $E_j := E_i$). The algorithm is successful in the simulation game if it outputs the element x (given as input to $\mathcal{O}_{\mathsf{sim}}$). We denote this event by $\mathbf{S}_{\mathsf{sim}}$.

Note that due to the modification of the element encoding procedure, it is now possible that both an element $L_i(x)$ is assigned to two or more different encodings and that different elements are assigned to the same encoding. In these cases the behavior of $\mathcal{O}_{\mathsf{sim}}$ differs from that of $\mathcal{O}$, what may allow to distinguish between the oracles. In the case of a differing behaviour the following failure event **F** occurred: There exist $i, j \in \{1, \ldots, |L|\}$ satisfying the equation

$$(L_i - L_j)(x) \equiv 0 \bmod n \text{ and } (L_i - L_j)(x_{i,j}) \not\equiv 0 \bmod n, \text{ or} \qquad (1)$$

$$(L_i - L_j)(x) \not\equiv 0 \bmod n \text{ and } (L_i - L_j)(x_{i,j}) \equiv 0 \bmod n. \qquad (2)$$

It is important to observe that the original game and the simulation game proceed identically unless **F** occurs. Thus the Difference Lemma [12] yields the inequality

$$|\Pr[\mathbf{S}] - \Pr[\mathbf{S}_{\mathsf{sim}}]| \leq \Pr[\mathbf{F}].$$

Bounding the Probability of Success in the Simulation Game. Since all computations are independent of the uniformly random element $x \in R$ the algorithm $\mathcal{A}$ can not do better than guessing x, i.e. $\Pr[\mathbf{S}_{\mathsf{sim}}] \leq \frac{1}{|R|} = \frac{1}{n}$.

Bounding the Probability of a Simulation Failure. Let $\mathfrak{D} := \{L_i - L_j | 1 \leq i < j \leq |L|\}$ denote the set of all non-trivial differences of polynomials in L after a run of $\mathcal{A}$. In the following we show how the probability that a polynomial $\Delta \in \mathfrak{D}$ causes a simulation failure is related to the probability of revealing a factor of n by simply evaluating Δ with a uniformly random element from R.

For fixed $\Delta \in \mathfrak{D}$ let $\mathbf{F}_\Delta$ denote the event that Δ causes a simulation failure as defined by Equations (1) and (2). Furthermore, let $\mathbf{D}_\Delta$ denote the event that $\gcd(n, \Delta(a)) \notin \{1, n\}$ when choosing an element a uniformly at random from R.

We are going to express the probabilities of both events using the same terms. Let $n = \prod_{i=1}^{k} p_i^{e_i}$ be the prime factor decomposition of n. Hence, R is isomorphic to $\mathbb{Z}_{p_1^{e_1}} \times \cdots \times \mathbb{Z}_{p_k^{e_k}}$ by the Chinese Remainder Theorem. Using the notation

$$\nu_i := \frac{|\{a \in R \mid \Delta(a) \equiv 0 \bmod p_i^{e_i}\}|}{|R|},$$

we can express the probability of $\mathbf{F}_\Delta$ by

$$\Pr[\mathbf{F}_\Delta] = 2 \Pr_{a \in_U R}[\Delta(a) \equiv 0 \bmod n] \left(1 - \Pr_{a \in_U R}[\Delta(a) \equiv 0 \bmod n]\right)$$
$$= 2 \left(\prod_{i=1}^{k} \nu_i\right) \left(1 - \prod_{i=1}^{k} \nu_i\right). \qquad (3)$$

Similarly, we can write the probability of $\mathbf{D}_\Delta$ as

$$\Pr[\mathbf{D}_\Delta] = 1 - \Pr_{a \in_U R}[\Delta(a) \equiv 0 \bmod n] - \prod_{i=1}^{k} \Pr_{a \in_U R}[\Delta(a) \not\equiv 0 \bmod p_i^{e_i}]$$
$$= 1 - \prod_{i=1}^{k} \nu_i - \prod_{i=1}^{k}(1 - \nu_i) \qquad (4)$$

Now, the key observation is that we have the following relation between the probabilities of the events $\mathbf{F}_\Delta$ and $\mathbf{D}_\Delta$:

Lemma 1. $\forall \Delta \in \mathfrak{D}: \ 2 \Pr[\mathbf{D}_\Delta] \geq \Pr[\mathbf{F}_\Delta]$

Proof (Sketch). The inequality

$$2 \left(1 - \prod_{i=1}^{k} \nu_i - \prod_{i=1}^{k}(1 - \nu_i)\right) \geq 2 \left(\prod_{i=1}^{k} \nu_i\right)\left(1 - \prod_{i=1}^{k} \nu_i\right)$$

is equivalent to $\left(1 - \prod_{i=1}^{k} \nu_i\right)^2 \geq \prod_{i=1}^{k}(1 - \nu_i)$. The latter is easily proven by complete induction on $k \geq 2$. □

The Factoring Algorithm. Consider an algorithm $\mathcal{B}$ that runs the BBRE algorithm $\mathcal{A}$ on an arbitrary instance of the BBRE problem over $\mathbb{Z}_n$. During this run it records the sequence of queries that $\mathcal{A}$ issues, i.e., it records the same list L of polynomials as the black-box ring oracle. Then for each $\Delta \in \mathfrak{D}$ the algorithm $\mathcal{B}$ chooses a new random element $a \in \mathbb{Z}_n$ and computes $\gcd(n, \Delta(a))$. There are at most $(m + 2)(m + 1)/2$ such polynomials, each can be evaluated using at most $m + 1$ ring operations (since it is given as a straight-line program of length at most m). Thus, $\mathcal{B}$ chooses $O\left(m^2\right)$ random elements and performs $O\left(m^3\right)$ operations on R as well as $O\left(m^2\right)$ gcd computations on $\log_2(n)$-bit numbers.

Let $\mathbf{D}$ denote the event that $\mathcal{B}$ finds a factor of n. It holds that $\Pr[\mathbf{D}] \geq \max\{\Pr[\mathbf{D}_\Delta] | \Delta \in \mathfrak{D}\}$. Thus the total probability of simulation failure $\Pr[\mathbf{F}]$ is upper bounded by

$$\Pr[\mathbf{F}] = \sum_{\Delta \in \mathfrak{D}} \Pr[\mathbf{F}_\Delta] \leq 2 \sum_{\Delta \in \mathfrak{D}} \Pr[\mathbf{D}_\Delta] \leq (m + 2)(m + 1) \Pr[\mathbf{D}].$$

Finally by applying the Difference Lemma [12] it holds that

$$\Pr[\mathbf{F}] \geq |\Pr[\mathbf{S}] - \Pr[\mathbf{S}_{\mathsf{sim}}]| = |\epsilon - \frac{1}{n}|. \qquad \square$$

4   An Extension to Multivariate Polynomial Rings

In this section we are going to lift our reduction from the special case $R = \mathbb{Z}_n$ to the case $R = \mathbb{Z}_n[X_1, \ldots, X_t]/J$, where $\mathbb{Z}_n[X_1, \ldots, X_t]$ denotes the ring of polynomials over $\mathbb{Z}_n$ in indeterminates $X_1, \ldots, X_t$ $(t \geq 0)$ and J is an ideal in this polynomial ring such that R is finite. Note that any finite commutative unitary ring has a representation of the above form:

Lemma 2. *Let R be a finite commutative unitary ring of characteristic n. Then there is an integer $t \leq \log_2 |R|$ and a finitely generated ideal J of $\mathbb{Z}_n[X_1, \ldots, X_t]$ such that $R \cong \mathbb{Z}_n[X_1, \ldots, X_t]/J$.*

Let $n = \prod_{i=1}^{k} p_i^{e_i}$ be the prime factor decomposition of n. This decomposition induces a decomposition of the ring $\mathbb{Z}_n[X_1, \ldots, X_t]/J$ into a direct product of rings with prime-power characteristic:

Lemma 3. *Let $F := \{f_1, \ldots, f_s\}$ be a set of polynomials generating an ideal J in $\mathbb{Z}_n[X_1, \ldots, X_t]$, denoted by $J = \langle F \rangle$. Let $R = \mathbb{Z}_n[X_1, \ldots, X_t]/\langle F \rangle$ and $n = \prod_{i=1}^{k} p_i^{e_i}$ be the prime factor decomposition of n. Then R is decomposable into a direct product of rings $R \cong R_1 \times \cdots \times R_k$, where $R_i := \mathbb{Z}_{p_i^{e_i}}[X_1, \ldots, X_t]/\langle F \rangle$.*

We call this way of decomposing R the *prime-power decomposition* of R.

4.1   Gröbner Bases for Polynomial Ideals over Rings

Roughly speaking, a Gröbner basis G is a generating set of an ideal J in a multivariate polynomial ring exhibiting the special property that reduction of polynomials from J modulo the set G always yields the residue zero. This property is not satisfied for arbitrary ideal bases. Gröbner bases were originally introduced for polynomial rings over finite fields, and extended to the case where the coefficients are elements of a Noetherian ring, such as $\mathbb{Z}_n$. In the following we adopt the notation and definitions from [13].

A *power product* in indeterminates $X_1, \ldots, X_t$ is a product of the form $\mathcal{X} = X_1^{a_1} \cdot \ldots \cdot X_t^{a_t}$ for some $(a_1, \ldots, a_t) \in \mathbb{N}_0^t$. Let $>$ be some arbitrary admissible ordering of power products (e.g. lexicographic). Let $f \in \mathbb{Z}_n[X_1, \ldots, X_t]$ with $f \neq 0$. Then we can write f as $f = c_1 \mathcal{X}_1 + \ldots + c_s \mathcal{X}_s$, where $c_1, \ldots, c_s \in \mathbb{Z}_n \backslash \{0\}$ and $\mathcal{X}_1 > \ldots > \mathcal{X}_s$. The *leading coefficient* $\mathrm{lc}(f)$ and the *leading power product* $\mathrm{lp}(f)$ of f with respect to $>$ are defined as $\mathrm{lc}(f) := a_1$ and $\mathrm{lp}(f) := \mathcal{X}_1$, respectively.

Definition 4 (Polynomial Reduction). *Let two polynomials f and h and a set of non-zero polynomials $F = \{f_1, \ldots, f_s\}$ in $\mathbb{Z}_n[X_1, \ldots, X_t]$ be given.*

(a) *We say that f can be reduced to h modulo F in one step, denoted by $f \xrightarrow{F} h$, if and only if $h = f - (c_1 \mathcal{X}_1 f_1 + \ldots + c_s \mathcal{X}_s f_s)$ for $c_1, \ldots, c_s \in R$ and power products $\mathcal{X}_1, \ldots, \mathcal{X}_s$, where $\mathrm{lp}(f) = \mathcal{X}_i \mathrm{lp}(f_i)$ for all i such that $c_i \neq 0$ and $\mathrm{lt}(f) = c_1 \mathcal{X}_1 \mathrm{lt}(f_1) + \ldots + c_s \mathcal{X}_s \mathrm{lt}(f_s)$.*

(b) *A polynomial h is called minimal with respect to F if h cannot be reduced. We denote the reduction of f to its minimal residue by $f \bmod F$.*

Using the above definition a Gröbner basis is characterized as follows.

Definition 5 (Gröbner Basis). *Let J be an ideal in $\mathbb{Z}_n[X_1, \ldots, X_t]$ and $G = \{g_1, \ldots, g_s\}$ be a set of non-zero polynomials such that $\langle G \rangle = J$. Then G is called a Gröbner basis for J if for any polynomial $f \in \mathbb{Z}_n[X_1, \ldots, X_t]$ we have*

$$f \in J \iff f \bmod G = 0.$$

The following lemma requires that a Gröbner basis for the given ring ideal is known. The lemma is crucial for proving that (similar to the $\mathbb{Z}_n$-case) an element $f \in R \cong R_1 \times \cdots \times R_k$ that is congruent to zero over a component R_i but not congruent to zero over another component R_j (cf. Lemma 3) helps in factoring n. Observe that Lemma 4 requires that the leading coefficients of all given Gröbner basis elements are units. For our purposes this is not a restriction but a reasonable assumption, since otherwise the given representation of R would immediately reveal a factor of n. A proof for this lemma based on the notion of syzygies can be found in the full version of this paper [11].

Lemma 4. *Let* $A = \mathbb{Z}_n[X_1, \ldots, X_t]$ *and* $n = \prod_{i=1}^{k} p_i^{e_i}$. *Furthermore, let* $G = \{g_1, \ldots, g_s\}$ *be a Gröbner basis for the ideal* $J = \langle g_1, \ldots, g_s \rangle$ *in* A *such that* $\mathrm{lc}(g_l) \in \mathbb{Z}_n^*$ *for all* $l \in \{1, \ldots, s\}$. *Then for each* $i \in \{1, \ldots, k\}$ *the set* $G_i = \{p_i^{e_i}, g_1, \ldots, g_s\}$ *is a Gröbner basis for the ideal* $\langle p_i^{e_i}, g_1, \ldots, g_s \rangle$ *in* A.

4.2 BBRE for $\mathbb{Z}_n[X_1, \ldots, X_t]/J$ and Integer Factorization

In the case $R = \mathbb{Z}_n$ our factoring algorithm is successful when it is able to find an element $a \in R$ such that $a \in \langle p_i^{e_i} \rangle$ and $a \notin \langle p_j^{e_j} \rangle$ for some $i, j \in \{1, \ldots, k\}$. The following lemma shows that a generalization of this fact holds for rings of the form $\mathbb{Z}_n[X_1, \ldots, X_t]/J$ where J is given as a Gröbner basis.

Lemma 5. *Let* $A = \mathbb{Z}_n[X_1, \ldots, X_t]$ *where* $n = \prod_{i=1}^{k} p_i^{e_i}$ *and* $k \geq 2$. *Furthermore, let* $G = \{g_1, \ldots, g_s\}$ *be a Gröbner basis for the ideal* $J = \langle g_1, \ldots, g_s \rangle$ *in* A *such that* $\mathrm{lc}(g_l) \in \mathbb{Z}_n^*$ *for all* $l \in \{1, \ldots, s\}$. *Assume an element* $f \in A$ *is given, such that* $f \in J_i = \langle p_i^{e_i}, g_1, \ldots, g_s \rangle$ *and* $f \notin J_j = \langle p_j^{c_j}, g_1, \ldots, g_s \rangle$ *for some* $i, j \in \{1, \ldots, k\}$. *Then computing* $\gcd(\mathrm{lc}(r), n)$, *where* $r = f \bmod G$, *yields a non-trivial factor of* n.

Proof (Sketch). r cannot be zero, since $f \notin J$. However, since $r \in \langle p_i^{e_i}, g_1, \ldots, g_s \rangle$ and $G_i := \{p_i^{e_i}, g_1, \ldots, g_s\}$ is a Gröbner basis by Lemma 4, r must be reducible w.r.t. G_i. Since r is minimal w.r.t. G and the coefficients of elements of G are units in $\mathbb{Z}_n$, it follows that there is no $g_i \in G$ such that $\mathrm{lp}(g_i)|\mathrm{lp}(r)$. Thus the leading coefficient $\mathrm{lc}(r)$ must be divisible by $p_i^{e_i}$. Computing $\gcd(\mathrm{lc}(r), n)$ yields a non-trivial factor of n. □

The above fact allows us to formulate and prove a theorem similar to Theorem 1.

Theorem 2. *Let* $R := \mathbb{Z}_n[X_1, \ldots, X_t]/J$ *for some integer* n *having at least two different prime factors. Assume a Gröbner basis* $G = \{g_1, \ldots, g_s\}$ *for* J *is given such that* $\mathrm{lc}(g_l) \in \mathbb{Z}_n^*$ *for* $l \in \{1, \ldots, s\}$. *Let* $\mathcal{A}$ *be an algorithm for the BBRE problem that performs at most* $m \leq |R|$ *operations on* R^σ *and solves the BBRE problem with probability* ϵ. *Then there is an algorithm* $\mathcal{B}$ *having white-box access to* $\mathcal{A}$ *that finds a factor of* n *with probability at least*

$$\frac{|\epsilon - \frac{1}{n}|}{(m + t + 2)(m + t + 1)}$$

by running $\mathcal{A}$ *once and performing an additional amount of* $O\left((m + t)^2\right)$ *random choices and* $O\left(m(m + t)^2\right)$ *operations on* R *as well as* $O\left((m + t)^2\right)$ *gcd computations on* $\log_2(n)$*-bit integers.*

Proof (Sketch). We adopt the proof of Theorem 1. The description of the original and the simulation game almost carries over by setting $R := \mathbb{Z}_n[X_1, \ldots, X_t]/J$. There are only two technical differences concerning the oracles $\mathcal{O}$ and $\mathcal{O}_{\mathrm{sim}}$ considered in the original game and the simulation game: (i) the list L maintained by both oracles is initialized with the $t + 1$ generating elements $1, X_1, \ldots, X_t$ of

R, and with the variable X. As before, ring elements are representend by polynomials in $R[X] = (\mathbb{Z}_n[X_1, \ldots, X_t]/J)[X]$. (ii) whenever an element $P \in R[X]$ is appended to the list L, say as element $L_j = P$, $\mathcal{O}$ checks whether there exists an element $L_i \in L$ such that $(L_i - L_j)(x) \in J$ which is equivalent to checking whether the minimal residue $r = (L_i - L_j)(x) \bmod G$ is the zero polynomial. Instead of using the given secret x in the above evaluation, the simulation oracle $\mathcal{O}_{\mathsf{sim}}$ performs this check using a *new* random element $x_{i,j} \in R$ for each difference polynomial $L_i - L_j$ $(i < j \in \{1, \ldots, |L|\})$.

The rest of the description of the games applies unchanged. Let the events $\mathbf{S}$, $\mathbf{S}_{\mathsf{sim}}$ and $\mathbf{F}$ be defined analogously to the case $R = \mathbb{Z}_n$. As before we obtain the bound $\Pr[\mathbf{S}_{\mathsf{sim}}] \leq \frac{1}{|R|} \leq \frac{1}{n}$ for success in the simulation game.

To derive a bound on $\Pr[\mathbf{F}]$ we proceed as follows. Let $\mathfrak{D} := \{L_i - L_j | 1 \leq i < j \leq |L|\}$ denote the set of all non-trivial differences of polynomials in L after a run of $\mathcal{A}$, and let $\Delta \in \mathfrak{D}$. Let $n = \prod_{i=1}^{k} p_i^{e_i}$ be the prime factor decomposition of n, then R has a prime power decomposition into

$$R \cong \mathbb{Z}[X_1, \ldots, X_t]/\langle p_1^{e_1}, G \rangle \times \cdots \times \mathbb{Z}[X_1, \ldots, X_t]/\langle p_k^{e_k}, G \rangle$$

by Lemma 3. Let $\nu_i := \frac{|\{a \in R \mid \Delta(a) \in \langle p_i^{e_i}, G \rangle\}|}{|R|}$ be the probability that $\Delta(a) \in \langle p_i^{e_i}, I \rangle$ for a uniformly random element $a \in R$. Using this redefinition of the probabilities ν_i, the probability $\Pr[\mathbf{F}_\Delta]$ that Δ causes a simulation failure is given by Equation 3.

By Lemma 5, Δ reveals a factor of n if we can find an element $a \in R$ such that $\Delta(a) \in \langle p_i^{e_i}, G \rangle$ and $\Delta(a) \notin \langle p_j^{e_j}, G \rangle$ for some $i, j \in \{1, \ldots, k\}$. In this case computing $\gcd(\mathrm{lc}(\Delta(a) \bmod G), n)$ yields a non-trivial factor of n. The probability $\Pr[\mathbf{D}_\Delta]$ of finding such an element by evaluating Δ with some random $a \in_U R$ is given by Equation 4.

The Factoring Algorithm. For each $\Delta \in \mathfrak{D}$ the algorithm $\mathcal{B}$ chooses a random $a \in_U R$ and computes $\gcd(\mathrm{lc}(\Delta(a) \bmod G), n)$. There are at most $(m + t + 2)(m + t + 1)/2$ polynomials in $\mathfrak{D}$, each can be evaluated using at most $m + 1$ ring operations.

Let $\mathbf{D}$ denote the event that $\mathcal{B}$ finds a factor of n. Since the equations describing $\Pr[\mathbf{D}_\Delta]$ and $\Pr[\mathbf{F}_\Delta]$ are the same as in the $\mathbb{Z}_n$ case, the relationship $2\Pr[\mathbf{D}_\Delta] \geq \Pr[\mathbf{F}_\Delta]$ still holds. Using the fact that $\Pr[\mathbf{D}] \geq \max\{\Pr[\mathbf{D}_\Delta] \mid \Delta \in \mathfrak{D}\}$ the total probability of simulation failure $\Pr[\mathbf{F}]$ is upper bounded by

$$\Pr[\mathbf{F}] = \sum_{\Delta \in \mathfrak{D}} \Pr[\mathbf{F}_\Delta] \leq 2 \sum_{\Delta \in \mathfrak{D}} \Pr[\mathbf{D}_\Delta] \leq (m + t + 2)(m + t + 1)\Pr[\mathbf{D}].$$

Therefore the probability of finding a factor of n with this algorithm is at least

$$\Pr[\mathbf{D}] \geq \frac{|\Pr[\mathbf{S}] - \frac{1}{n}|}{(m + t + 2)(m + t + 1)}. \qquad \square$$

5   Implications for General Rings

Unfortunately, despite that fact that for any finite commutative unitary ring R there exists a polynomial representation (cf. Lemma 2), our result does not immediately carry over to any such ring. This is because to make our reduction work the explicit polynomial representation of R must be known, and we require that a Gröbner basis for the ideal J is known. If we are given a basis other than a Gröbner basis as a description for J, we could compute a Gröbner basis from this input, for example using a variant of Buchberger's algorithm for Noetherian rings [13]. However, for Gröbner basis algorithms still no upper bound on the running time is known. It is known that there are instances where constructing a Gröbner basis takes time in the order of $2^{2^{O(t)}}$ [14]. Thus, we cannot give the factoring algorithm described in our reduction an arbitrary basis of the ideal J as input and let it first compute a Gröbner basis, since there are cases where this computation may easily exceed the time needed to factorize n directly. Hence, our result only holds for families of rings R where a Gröbner basis for J is given or known to be efficiently computable.

However, the "good" news is that our results seem to cover virtually any representation of a finite commutative unitary ring R where computation is efficient: Known representations for finite commutative rings with identity are *table representation*, *basis representation*, and *polynomial representation* (cf. [15]). A table representation of a ring R requires $O\left(|R|^2\right)$ space, which is clearly too much for ring sizes of cryptographic interest. A basis representation requires $O((\log|R|)^3)$ space, thus might be interesting for cryptographic applications. Our result can be extended to rings given in basis representation, however, we have to refer to the full version of this paper [11] for the proof. For rings given in polynomial representation, the ideal J is specified by some set of generating polynomials. In order to be able to perform equality checks between ring elements efficiently — which corresponds to solving instances of the ideal membership problem — there is currently no other way than providing a Gröbner basis for the ideal J. Thus we may conclude that our work seems to cover virtually any ring representation of cryptographic interest.

Acknowledgements. We would like to thank Roberto Avanzi, Lothar Gerritzen, and Gregor Leander for helpful discussions, and Dan Brown for pointing out an error in a previous version.

References

1. Boneh, D., Lipton, R.J.: Algorithms for black-box fields and their application to cryptography (extended abstract). In: Koblitz, N. (ed.) CRYPTO 1996. LNCS, vol. 1109, pp. 283–297. Springer, Heidelberg (1996)
2. Maurer, U., Raub, D.: Black-box extension fields and the inexistence of field-homomorphic one-way permutations. In: Kurosawa, K. (ed.) ASIACRYPT 2007. LNCS, vol. 4833, pp. 427–443. Springer, Heidelberg (2007)

3. Nechaev, V.I.: Complexity of a determinate algorithm for the discrete logarithm. Mathematical Notes 55(2), 165–172 (1994)
4. Shoup, V.: Lower bounds for discrete logarithms and related problems. In: Fumy, W. (ed.) EUROCRYPT 1997. LNCS, vol. 1233, pp. 256–266. Springer, Heidelberg (1997)
5. Maurer, U.: Abstract models of computation in cryptography. In: Smart, N.P. (ed.) IMA Int. Conf. LNCS, vol. 3796, pp. 1–12. Springer, Heidelberg (2005)
6. Maurer, U.: Towards the equivalence of breaking the Diffie-Hellman protocol and computing discrete algorithms. In: Desmedt, Y. (ed.) CRYPTO 1994. LNCS, vol. 839, pp. 271–281. Springer, Heidelberg (1994)
7. den Boer, B.: Diffie-Hellman is as strong as discrete log for certain primes. In: Goldwasser, S. (ed.) CRYPTO 1988. LNCS, vol. 403, pp. 530–539. Springer, Heidelberg (1990)
8. Maurer, U.M., Wolf, S.: Lower bounds on generic algorithms in groups. In: Nyberg, K. (ed.) EUROCRYPT 1998. LNCS, vol. 1403, pp. 72–84. Springer, Heidelberg (1998)
9. Bach, E.: Discrete logarithms and factoring. Technical Report UCB/CSD-84-186, EECS Department, University of California, Berkeley (June 1984)
10. Leander, G., Rupp, A.: On the equivalence of RSA and factoring regarding generic ring algorithms. In: Lai, X., Chen, K. (eds.) Advances in Cryptology — ASIACRYPT 2007. LNCS, vol. 4284, pp. 241–251. Springer, Heidelberg (2006)
11. Altmann, K., Jager, T., Rupp, A.: On black-box ring extraction and integer factorization. Cryptology ePrint Archive, Report 2008/156 (2008), http://eprint.iacr.org/
12. Shoup, V.: Sequences of games: a tool for taming complexity in security proofs. Cryptology ePrint Archive, Report 2004/332 (2004), http://eprint.iacr.org/
13. Adams, W., Loustaunau, P.: An introduction to Gröbner bases. Graduate Studies in Math, vol. 3. Oxford University Press, Oxford (1994)
14. Mayr, E.W., Meyer, A.: The complexity of the word problems for commutative semigroups and polynomial ideals. Advances in Mathematics 46, 305–329 (1982)
15. Agrawal, M., Saxena, N.: Automorphisms of finite rings and applications to complexity of problems. In: Diekert, V., Durand, B. (eds.) STACS 2005. LNCS, vol. 3404, pp. 1–17. Springer, Heidelberg (2005)

Extractable Perfectly One-Way Functions

Ran Canetti[1],[*] and Ronny Ramzi Dakdouk[2],[**]

[1] IBM T. J. Watson Research Center, Hawthorne, NY
canetti@watson.ibm.com
[2] Yale University, New Haven, CT
dakdouk@cs.yale.edu

Abstract. We propose a new cryptographic primitive, called **extractable perfectly one-way (EPOW) functions**. Like perfectly one-way (POW) functions, EPOW functions are probabilistic functions that reveal no information about their input, other than the ability to verify guesses. In addition, an EPOW function, f, guarantees that any party that manages to compute a value in the range of f "knows" a corresponding preimage.

We capture "knowledge of preimage" by way of algorithmic extraction. We formulate two main variants of extractability, namely non-interactive and interactive. The noninteractive variant (i.e., the variant that requires non-interactive extraction) can be regarded as a generalization from specific knowledge assumptions to a notion that is formulated in general computational terms. Indeed, we show how to realize it under several different assumptions. The interactive-extraction variant can be realized from certain POW functions.

We demonstrate the usefulness of the new primitive in two quite different settings. First, we show how EPOW functions can be used to capture, in the standard model, the "knowledge of queries" property that is so useful in the Random Oracle (RO) model. Specifically, we show how to convert a class of CCA2-secure encryption schemes in the RO model to concrete ones by simply replacing the Random Oracle with an EPOW function, without much change in the logic of the original proof. Second, we show how EPOW functions can be used to construct 3-round ZK arguments of knowledge and membership, using weaker knowledge assumptions than the corresponding results due to Hada and Tanaka (Crypto 1998) and Lepinski (M.S. Thesis, 2004). This also opens the door for constructing 3-round ZK arguments based on other assumptions.

1 Introduction

The Random Oracle methodology [15,4] consists of two steps. The first step involves designing a protocol and proving security in an idealized model called the Random Oracle (RO) model. In the RO model, all parties have oracle access to a public random function, O. The oracle answers are uniform and independent with only one constraint, specifically, that all answers to the same query are identical. The second step involves "moving" the protocol from this idealized model to the real world. This is done by "replacing" the RO with a cryptographic hash function such as SHA1 [16] or MD5 [26]. In other words, every oracle call is replaced by a function call to some publicly

* Supported by NSF grant CFF-0635297 and US-Israel Binational Science Foundation Grant 2006317.
** Work supported by NSF grant #0331548.

L. Aceto et al. (Eds.): ICALP 2008, Part II, LNCS 5126, pp. 449–460, 2008.
© Springer-Verlag Berlin Heidelberg 2008

known cryptographic hash function. This transformation is known as an instantiation of Random Oracles.

Although the first step of the RO methodology is rigorous, the second step remains a heuristic for the most part. While most results in this area provide proofs in the RO model, they lack even informal justification as to why the instantiated protocols may be secure. Such justification is of dire need given the fact that the RO methodology is not sound in general. Specifically, it was shown that there are schemes secure in the RO model without any secure instantiations [9,24,17]. Furthermore, there exist natural primitives that are realizable in the RO model but can not be realized at all in the standard model, regardless of the computational assumptions used [25].

Given the general impossibility results mentioned above, one may resort to considering a proof in the RO model as a "stepping stone" towards a proof in the standard model. However, there is a severe flaw with this point of view: When it comes to security properties, proofs in the RO model use the Random Oracle somewhat like a Swiss Army knife. Random Oracles satisfy many cryptographic properties including collision resistance (it is hard to find two queries with the same RO answer), uniformity (the answer to any query is uniformly distributed), unpredictability or correlation intractability [9], programmability [25] and knowledge of queries (any machine that computes $O(q)$ "knows" q). Furthermore, works that use the RO methodology do not often highlight the specific properties of Random Oracles that are used or needed for the current proof. This makes translating a proof from the RO model to the standard model a harder task. And indeed, proofs in the RO model usually follow different lines from the corresponding ones in the standard model. This is contrary to the intuition behind the RO methodology, which is to use the randomness in the RO model to come up with simple proofs and then replace the Random Oracle by an appropriate function while maintaining the overall proof structure.

In light of the above discussion, it is interesting to identify specific properties of Random Oracles that are essential for the security of specific protocols. Once these properties are identified, it may then be possible to capture them with concrete functions that can be used to replace Random Oracles. Such an approach motivated the introduction of perfectly one-way (POW) functions in [7] as functions that capture the hiding property of Random Oracles and are then used to instantiate Random Oracles in a semantically-secure encryption scheme.[1] In another attempt, Boldyreva and Fischlin [6] introduce a strong variant of pseudorandom generators geared towards instantiating OAEP.

However, attempts at direct instantiation of encryption schemes secure against chosen ciphertext attacks (IND-CCA2) in the RO model have failed. It seems that one main problem is to translate a central property of Random Oracles, namely knowledge of queries, to the standard model. This property proves essential for the security proof in the RO model but it has not been previously formalized and captured by concrete functions.

1.1 Our Work

We formalize the "knowledge of queries" property mentioned above and cast it on a concrete object in the standard model. We call the new object an extractable perfectly

---

[1] Informally, POW functions are probabilistic functions that hide all partial information about the input.

one-way (EPOW) function. Then, we use EPOW functions not only to instantiate such schemes but also use a proof of security that follows similar logic as the original proof. The intended goal in this instantiation is not to try to achieve a more efficient construction than the existing ones in the literature but rather identify and realize the needed properties of the random oracle so that the proof of security remains the same in the standard model in both its logic and simplicity. In addition, we show that EPOW functions are useful in other contexts. We go into more detail shortly.

Extractable perfectly one-way functions. In the RO model, the knowledge of queries property means that any machine that computes an RO answer, $O(q)$, "knows" q. Even though such a property is easy to formalize and satisfy in the RO model if the range of O is sparse, defining it in the standard model while maintaining hiding properties is tricky. Towards this end, we build on the notion of perfect one-wayness presented in [7] to introduce a new class of functions called **extractable perfectly one-way** (EPOW) functions. These are functions that hide all information about the input but any machine that computes a valid image, "knows" a corresponding preimage. We also require a similar property to hold with respect to auxiliary information which may include other images. The corresponding statement is any machine that computes a *new* valid image, even in the presence of other images, knows a corresponding preimage. Although using extractability with a weaker hiding property may be sufficient for certain applications, it is of particular interest when combined with POW functions since it gives a better approximation of the properties expected from a Random Oracle.

From one angle, extractability can be interpreted as saying that the only way to produce a point in the range of this function is by taking a point in the input domain and then applying the algorithm that computes this function to the input. From another perspective, an EPOW function is an obfuscation of a point function [1,30] with the additional property that the original source program, that computes the point function in the clear, can be extracted from the view of any potentially adversarial obfuscator. This property can in fact be defined with respect to any function family.

We define two variants of EPOW functions, namely noninteractive and interactive. Noninteractive extraction is captured by the existence of a (nonblackbox) preimage extractor. In more detail, every adversary, that tries to output a point in the range, has a corresponding extractor that gets the view of the adversary and outputs a preimage. We emphasize that the extractor gets the view of the adversary *including any private random coins.* The interactive variant is described later on.

On the relation between noninteractive EPOW functions and NIZK. Superficially, EPOW functions resemble noninteractive zero-knowledge (NIZK) arguments of knowledge [29,28] in that an image can be viewed as a proof of preimage knowledge. However, EPOW functions and NIZK arguments of knowledge differ in several ways. First, NIZK secrecy, i.e., zero knowledge, holds over the choices of the Common Reference String (CRS) while EPOW functions require secrecy to hold without a CRS. Second, EPOW functions are not required to have efficient verification, that is deciding whether a given point belongs to the range of the function. (Not to be confused with the verification requirement on POW functions, where it is easy to check that a given output is an image of a given input.) We mention that our noninteractive EPOW constructions satisfy a weaker form of verification, which seems to be needed for the ZK application but not for our Random Oracle instantiation. On the other hand, our interactive EPOW constructions are not known to satisfy this form of verification. Third, NIZK arguments of knowledge require a *universal blackbox* extractor to recover a witness with the help of auxiliary information about the CRS. On the other hand, EPOW functions only require a nonblackbox extractor for every adversary. However, this extractor has to recover a preimage from the view of the adversary *without any extra information that is not given to the adversary*. The latter formulation may better capture our intuition about knowledge because it clearly demonstrates that an adversary "knows" a preimage by recovering it from its view alone.

On the relation between noninteractive EPOW functions and other knowledge assumptions. From another angle, extractable functions look similar to other knowledge assumptions such as the knowledge of exponent (KE) assumption [12,20] and the proof of knowledge (POK) assumption [23]. In fact, we view *extractable functions as an abstraction away from specific knowledge assumptions, much like a one-way function is an abstraction of specific one-way assumptions, such as the discrete logarithm (DL) assumption*. In other words, the DL assumption gives us a one-way function but it may even give us more, e.g., a one-way permutation in certain group or certain algebraic properties. However, we abstract away from these particularities and identify the essential property needed. Likewise, we use extractable functions as a step towards capturing the abstract knowledge assumption - it provides a relatively simple primitive that is defined only in terms of its general computational properties, that seems to be useful in a number of places, and that can be realized by a number of different assumptions. (We show later that either the KE or the POK assumption, when combined with a hardness assumption such as the DDH assumption, is sufficient for constructing EPOW functions).

On the constructions. We give three simple constructions of EPOW functions. The first one uses a POW function and a "strong" notion of NIZK proof of preimage knowledge. In addition, we provide another construction from the POW construction in [7] and the KE assumption. At a high level, the KE assumption guarantees preimage extraction, while hiding can be based on a strong variant of the DDH assumption. The third construction is similar to the second one but it uses the POK assumption (with the same DDH assumption mentioned above). However, none of these constructions satisfies all of our requirements (see [8] for more details). Thus, we turn our attention to EPOW functions with interactive extraction.

Interactive EPOW Functions. These are POW functions with interactive extraction. Informally, interactive extraction means that if a party interacts *consistently* with a challenger, then it "knows" a preimage. Interaction between the prover and the challenger is restricted to Arthur-Merlin games. Furthermore, the messages sent by the prover are restricted to images of the interactive EPOW function. For instance, in a 3-round game of this type, the prover computes hashes of the preimage using different random coins for the EPOW function, $\mathbb{H}$, chosen by the challenger. In more detail, the prover sends $y = H_k(x, r_0)$ in the first round, the challenger then responds with a uniform string, r_1, and the prover sends the corresponding image, $H_k(x, r_1)$, in the last round. Here, extractability means that if the images in the first and third round share a common preimage, then the prover knows it. Similar to the noninteractive setting, knowledge of preimage is captured by the existence of a preimage extractor.

We show how to transform POW functions to interactive EPOW functions. Informally, our transformation imposes a structure on the new function so that a preimage can be recovered from any two "related" images. For clarity, consider a toy construction to recover the first bit only. Specifically, if $\mathbb{H}$ is the old POW function and x is the input, then $\mathbb{H}'$ is defined as $H'_k(x, (r_1, r_2)) = H_k((x, 1), r_1), H_k((x, x_1), r_2)$, where x_1 is the first bit of x. To recover x_1, the extractor asks the prover to compute $H'_k(x, (r_1, r_2))$, then it rewinds the protocol, and forces the prover to compute $H'_k(x, (r'_1, r_1))$ using the same r_1 as before. Note that x_1 can be recovered (by simple comparison) from $H_k((x, 1), r_1)$ and $H_k((x, x_1), r_1)$ computed in the first and second game respectively.

We remark that a slightly weaker notion of interactive EPOW functions can be constructed from any POW function and a corresponding Σ-protocol [5,11] for proving preimage knowledge.

1.2 Applications

Using EPOW functions to instantiate Random Oracles in Encryption Schemes. As mentioned before, POW functions are used in [7] to capture and realize CPA-security of the encryption scheme in [4]. However, this is not sufficient for CCA2-security as POW functions may not guarantee extractability. So, an EPOW function provides the missing link, namely extractability, for replacing a Random Oracle by a POW function. Here, we use EPOW functions to instantiate the second encryption scheme in [4] (recalled shortly), and translate the proof to the standard model in a straightforward way. This scheme uses a trapdoor permutation, M, and two Random Oracles, O_1, O_2, to encrypt a message, m, as $c = (M(r), O_1(r) \oplus m, O_2(r, m))$, where r is uniform. At a high level, it is CCA2-secure because the hiding property of Random Oracles gives us semantic security while knowledge of queries gives us knowledge of plaintext (the latter property is what enables proving CCA2-security). Thus, if we replace the Random Oracle by an EPOW function in the previous scheme we get a CCA2-secure encryption scheme in the standard model. This scheme can be either noninteractive or 3-round depending on whether the EPOW function is noninteractively or interactively extractable.

This approach can be utilized to realize other encryption schemes in the RO model. In particular, we show how to instantiate some schemes that provably cannot be instantiated using the standard instantiation prescribed in the RO methodology [9,24], where each RO query is replaced with a call to a specific function. Thus, the aforementioned instantiation is different from the standard one and does not contradict the impossibility results mentioned above. A detailed presentation of this result appears in [8].

On the connection to other approaches and CCA2 schemes. We remark that generic transformations from any semantically-secure scheme to a CCA2-secure one have been studied before [14,27]. Also, the KE assumption has been used to prove that certain encryption schemes are plaintext-aware, which when coupled with semantic security gives CCA-secure schemes [3,13]. Moreover, Katz [22] used the notion of proofs of plaintext knowledge to construct efficient 3-round CCA2-secure schemes. We emphasize that the contributions of this work are not in giving better or more efficient constructions than existing ones in the literature, but rather in the methodology of replacing Random Oracles as described above.

Using EPOW functions to construct 3-round ZK protocols. We give one more application of EPOW functions in the context of Zero-Knowledge (ZK) systems. Current 3-round ZK arguments and proofs use strong and very specific number theoretic assumptions [20,21,23,3]. On the other hand, we construct 3-round ZK arguments of knowledge and membership assuming only the existence of a variant of EPOW functions and noninteractive witness-indistinguishable (WI) arguments [2,19]. This allows for abstracting from specific number theoretic assumptions and opens the door for basing 3-round ZK arguments on other assumptions sufficient for constructing this variant of EPOW functions. On the one hand, the existence of EPOW functions is an assumption that is stated in general computational terms without resorting to specific algebraic constructs. On the other hand, the assumption seems rather basic and in particular less specific than current knowledge assumptions.

As a concrete example, we use our second EPOW construction to build such ZK arguments. We remark that the KE assumption used here is weaker than the corresponding knowledge assumptions used for constructing 3-round ZK arguments in [20,21,3]. Specifically, we eliminate the need for the second KE assumption in [21] and later updated in [3]. We note that both simulation and extraction are nonblackbox.

Organization. We introduce and define extractable functions in Section 3. We then highlight one noninteractive and one interactive constructions in Sections 4 and 5. The last two sections discuss applications to Random Oracle instantiation and 3-round ZK protocols, respectively. A more detailed presentation, common definitions, and proofs appear in [8].

2   Preliminaries

A function, μ, is called negligible if it decreases faster than any inverse polynomial. Formally, for any polynomial p, there exists an N_p such that, for all $n \geq N_p$: $\mu(n) < \frac{1}{p(n)}$. We reserve μ to denote negligible functions. A distribution is called **well-spread** if it has superlogarithmic min-entropy, i.e., $max_k Pr[X_n = k]$ is a negligible function in n. A probabilistic function family is a set of efficient probabilistic functions having common input and output domains. Formally, $\mathbf{H}^n = \{H_k\}_{k \in K_n}$ is a function family with key space K_n and randomness domain R_n if, for all $k \in K_n, H_k : I_n \times R_n \rightarrow O_n$. A probabilistic function family has **public randomness** if for all k, $H_k(x, r) = r, H'_k(x, r)$ for some deterministic function H'_k. A family ensemble is a collection of function families, i.e., $\mathbb{H} = \{\mathbf{H}^n\}_{n \in \mathbb{N}}$. An uninvertible function, f, with respect to a well-spread distribution, $\mathbb{X}$, is an efficiently computable function that is hard to invert on $\mathbb{X}$. Formally, for any PPT, A, $Pr[x \leftarrow X_n, A(f(x)) = x] < \mu(n)$. If f is uninvertible with respect to *any* well-spread distribution, then it is called uninvertible.

Perfectly One-way Probabilistic Functions. A perfectly one-way (POW) function is a probabilistic function that satisfies collision resistance and hides all information about its input. Due to its probabilistic nature, such a function is coupled with an efficient **verification** scheme that determines whether a given string is a valid hash of some given input [10].

One formulation of information hiding requires hardness of indistinguishability between hashes of the same input and hashes of different inputs [10], where the former is taken from a well-spread distribution and the latter inputs are uniform and independent. We also consider the presence of auxiliary information, which is represented as an uninvertible function of the input. A notable special case of indistinguishability is pseudorandomness, i.e., hashes of the same input are indistinguishable from uniform. Moreover, the **statistical** version of both definitions can be obtained by dropping the requirements of auxiliary information and efficiency of the adversary. The formal definitions appear in [8].

3 Extractable Functions

An extractable function is one for which any machine that "computes" a point in the range, "knows" a corresponding preimage. As a starting point, we can formulate this notion by requiring any efficient machine that computes an image *without auxiliary input* to "know" a preimage. Although, this requirement seems reasonable, it is not sufficient for applications where auxiliary information is present. On the other hand, formulating this notion in the presence of auxiliary information is tricky. As a toy example, A can be a machine that receives an image as an input and copies it to its output. Moreover, A may still receive an image hidden in its auxiliary input in a subtle way but can be efficiently extracted from it. Yet, we do not think that this captures our intuition because A does not really compute the function, rather it decodes the image syntactically from its input. Thus, we need a meaningful way of telling apart "copying" an image from "computing" an image.

Following [18], we consider two types of auxiliary information. The first one, called **independent auxiliary information**, consists of auxiliary information computed before a function is sampled from a family ensemble, $\mathbb{H}$. We stress that this input is independent of the particular function currently used. This prevents hiding images in this type of input. The second type, called **dependent auxiliary information**, is restricted to images under $\mathbb{H}$. This is a restricted form of dependent auxiliary information but it is sufficient for our applications. Given these two types of inputs, we require that no adversary can come up with a *new* image without knowing a corresponding preimage. We capture knowledge of a preimage by requiring for every A, that computes a new image, a corresponding extractor, $\mathcal{K}_A$, that has access to the private input of A and computes a preimage. We emphasize that $\mathcal{K}_A$ has to compute the preimage from the view of A without any additional information.

For clarity, we first formalize this notion in the presence of independent auxiliary information alone before addressing the general case.

Definition 1. *Let $\mathbb{H} = \{\mathbf{H}^n\}_{n \in \mathbb{N}}$ be any verifiable family ensemble (with verifier $V_{\mathbb{H}}$). Then, $\mathbb{H}$ is called **noninteractively extractable** if for any PPT, A (with private random coins denoted by r_A), and polynomial, p, there exists a PPT, $\mathcal{K}_A$, such that for any auxiliary information, z:*

$$Pr[k \leftarrow K_n, \ y = A(k, z, r_A), \ x \leftarrow \mathcal{K}_A(k, z, r_A) : V_{\mathbb{H}}(x, y) = 1 \ or \ (\forall x', V_{\mathbb{H}}(x', y) \neq 1)]$$

$$> 1 - \frac{1}{p(n)} - \mu(n).$$

Note that we allow a noticeable extraction error. The constructions from the KE or POK assumption have a negligible error. However, the error in our interactive constructions is not known to be negligible. So, for uniformity, we adopt the weaker notion.

There are two possible ways to introduce dependent auxiliary information into Definition 1. One can allow this auxiliary information to be images of any input while the more restrictive way forces the images to correspond to inputs chosen from well-spread distributions. Even though the former is more general, the latter is sufficient for our applications. Thus, we use the latter notion in this work. The formal definitions are not presented here due to space constraints.

Interactive extraction. In the interactive setting, we force an adversary, A, to compute not only one image but a large fraction of the images of x (recall, the function is probabilistic). We say that if A can do so, then x is extractable. We achieve the first property by forcing A to use random coins for the probabilistic function that are chosen by an external challenger. In more detail, we define a 3-round game between A and a challenger (or knowledge extractor). [2] At the end of the game, if the interaction is consistent (we say shortly what this means) then extraction is possible.

The game starts with A sending an image, y_0. The challenger sends uniform strings, $r_1, ..., r_n$, and A has to answer with images, $y_1, ..., y_n$, using $r_1, ..., r_n$ as random coins for $\mathbb{H}$. We call an interaction consistent if there is a common preimage, x, of $y_0, ..., y_n$ with $r_1, ..., r_n$ as random coins for the last n images. We then say $\mathbb{H}$ is interactively extractable if for any adversary that plays this game consistently, there is a corresponding extractor that recovers a common preimage of $y_0, ..., y_n$. We also allow A to receive an auxiliary input that can depend, in an arbitrary way, on the choice of the function from the ensemble, $\mathbb{H}$.

4   A Noninteractive EPOW Construction

Before we present the EPOW construction, we show a simpler construction that achieves extractability but satisfies a weaker notion of computational hardness, namely one-wayness. Both constructions use the KE assumption to satisfy extractability. Informally, the KE assumption says that it is hard to compute, on input p, q, g, g^a, a pair of elements (g^r, g^{ra}) without knowing r, where p and q are primes, $p = 2q + 1$, and g is a generator for the quadratic residue group modulo p. This assumption can be formulated with or without independent auxiliary information (it can be shown that it does not hold with respect to auxiliary information that depends on (p, q, g, g^a)).

Note that the KE and discrete-log (DL) assumptions imply that the family ensemble, $\mathbb{F} = \{\{f_{p,q,g,g^a}\}_{(p,q,g,g^a) \in PQGA_n}\}_{n \in \mathbb{N}}$, where $f_{p,q,g,g^a}(x) = g^x, (g^a)^x$, is an extractable one-way (EOW) family ensemble. We strengthen the previous construction into a POW function by masking x with a uniform element r as in [7]. Formally, $H_{p,q,g,g^a}(x, r) = g^r, g^{ar}, g^{rx}, g^{arx}$.

Preimage extraction. If the KE assumption holds without auxiliary information then for any PPT, A, that outputs a valid image $(g^r, g^{ar}, g^{rx}, g^{arx})$, there are two PPT, $\mathcal{K}_1$ and $\mathcal{K}_2$, such that $\mathcal{K}_1$ extracts r and $\mathcal{K}_1$ extracts rx. Consequently, $\mathbb{H}$ is extractable. Moreover, if the KE assumption holds with respect to auxiliary information, then $\mathbb{H}$ is extractable with respect to *independent* auxiliary information. However, $\mathbb{H}$ is not extractable in the presence of dependent auxiliary information. Note that extraction occurs here with negligible error.

Information hiding. The secrecy of this construction is similar to that of the corresponding one in [7], specifically $H_{p,q,g}(x, r) = g^r, g^{rx}$. In particular, secrecy of both constructions is based on a stronger version of the DDH assumption. Informally, g^a, g^b, g^{ab} is indistinguishable from g^a, g^b, g^z where a is drawn from a *well-spread* distribution instead of uniform. However, these secrecy notions differ in two ways. First, the [7]

---

[2] In the full version of the paper, we define a 2-round version. However, realizing this notion seems to require stronger assumptions.

construction is pseudorandom while this one is an indistinguishable POW function. Second, secrecy in [7] holds for a randomly chosen function while we use secrecy that holds for any function. While the former is sufficient in some applications, such as Random Oracle instantiation in encryption schemes, the latter is needed in the ZK protocol (Section 7). Consequently, following [20], the DDH assumption used here is assumed to hold for any (p, q, g) instead of a randomly chosen one.

5 Construction of Interactive EPOW Functions

The construction presented here is based on hardness assumptions and achieves both interactive extraction and perfect one-wayness. However, it does not achieve perfect one-wayness with auxiliary information. A second construction that satisfies the latter property appears in the full version of the paper.

The idea behind both constructions is to have pairs of related images satisfy the property that it is easy to compute a preimage if both of them are available. In more detail, we identify for every r, a related $\hat{r}$, such that $O(x, r), O(x, \hat{r})$ reveals x. However, $O(x, r), O(x, \hat{r})$ is unlikely to appear in a single execution of the extraction game. So, the extractor can recover a preimage by sending r in the second round of the game to get $O(x, r)$, rewinding A, and then sending $\hat{r}$ to get $O(x, \hat{r})$. More details appear after the construction.

Construction 1. *Let* $\mathbb{H} = \{H^n\}_{n \in \mathbb{N}}$ *and* $\mathbb{G} = \{G^n\}_{n \in \mathbb{N}}$ *be two family ensembles. Denote by* $\mathbb{O} = \{O^n\}_{n \in \mathbb{N}}$ *the family ensemble defined as:*

$$O_{k=(k_1, k_2, k_3)}(x, (r_0^1, r_0^2, r_0^3, r_1 ..., r_n, r_G)) =$$

$$r_0^2, r_0^3, H_{k_1}(x, r_0^1), H_{k_2}(t_1, r_1), ..., H_{k_2}(t_n, r_n), G_{k_3}(x, r_G),$$

where for all i, $t_i = H_k(x, r_0^2)$ *if* $x_i = 1$, *and* $t_i = H(x, r_0^3)$ *otherwise.*

Primage extraction. For simplicity, and to see why Construction 1 is extractable assume that A receives only a single challenge, r^O, in the second round of the extraction game. Informally, $\mathcal{K}$ tries to make A output two "related" hashes that allows it to recover x. In more detail, $\mathcal{K}$ sends $r_0^1, r_0^2, r_0^3, r_1, ..., r_n$ to A in the first execution of the game, where all strings are uniform. $\mathcal{K}$ then rewinds A and starts a new game. In the second game, $\mathcal{K}$ sends $u_0^1, \boldsymbol{r_0^2}, u_0^3, u_1, ..., u_n$, where $u_0^1, u_0^3, u_1, ..., u_n$ are chosen uniformly but r_0^2 (the string in bold font) is the same as the one used in the first interaction. If A answers both challenges consistently, then $\mathcal{K}$ can recover x. This is so because the message in the last round of the first game contains $t = H_{k_1}(x, r_0^1)$ in the clear, while the message in the last round in the second game contains $H_{k_2}(t, u_i)$ if and only if the ith bit of x is 1. We remark that the technical proof requires that $\mathbb{H}$ satisfies a strong form of collision resistance. The formal definition and proof of extraction appears in the full version of the paper.

Information hiding. This construction uses two functions, $\mathbb{H}$ and $\mathbb{G}$, instead of one due to the properties needed to prove perfect one-wayness and extractability. Specifically, our proof of perfect one-wayness uses the assumption that $\mathbb{H}$ is *statistically* perfectly one-way. On the other hand, extractability assumes that $\mathbb{H}$ satisfies strong collision resistance. Currently, we do not know of any class of functions that satisfies this requirement except statistically binding functions. However, no single function can be both statistically pseudorandom (hiding) and statistically binding. Therefore, we use two functions. We assume that $\mathbb{G}$ is strongly collision resistant, e.g., statistically binding, so that $\mathbb{O}$ is strongly collision resistance and consequently extractable. On the other hand, $\mathbb{H}$ is assumed to be a statistically POW function. Therefore, if $\mathbb{G}$ is computationally perfectly one-way with auxiliary information (it is sufficient that the auxiliary information be only a statistically hiding function), then $\mathbb{O}$ is a *computationally* POW function. We emphasize that $\mathbb{O}$ is a POW function but not necessarily with respect to auxiliary information. In the full version of the paper, we modify the construction to meet this requirement based on a strong POW assumption.

6   Instantiating the Second Encryption Scheme of [4]

We use EPOW functions to instantiate Random Oracles in the second encryption scheme of [4] while maintaining a similar proof of security. Extractable POW functions allow us to do so because they capture two properties of Random Oracles essential for the original proof, namely, pseudorandomness and knowledge of queries.

The original scheme uses a family ensemble of trapdoor permutations, $\mathbb{M}$, with key space PK_n and trapdoor SK_n, and two random oracles O_1 and O_2. The encryption of a message, m, is $c = M_{pk}(q), O_1(q) \oplus m, O_2(m, q)$, where q is uniform.

Informally, this scheme is IND-CCA2 because it is IND-CPA and the decryption oracle does not help the adversary, A. In more detail, without access to the decryption oracle, A has a negligible advantage because M is one-way. On the other hand, any *valid* decryption query, c_1, c_2, c_3, that A makes must be preceded by two Random Oracle queries, $M_{sk}(c_1)$ and $M_{sk}(c_1), O_1(M_{sk}(c_1)) \oplus c_2$. However, if A makes any of these two queries it can compute the plaintext on its own *without the decryption oracle*.

Interactive instantiation. In the interactive setting, each oracle query is replaced by a call to a function, $\mathbb{H}$. Moreover, to encrypt a message, m, E sends a hash of a uniform string, q, in the first round. D responds by sending random strings $r_1, ..., r_n$. In the last round, E sends n hashes of q using $r_1, ..., r_n$ as random coins for $\mathbb{H}$. E also sends the ciphertext of m using the original scheme (with $\mathbb{H}$ in place of the Random Oracle) with the same q as the one used in the first round. We note that the first two messages are independent of the plaintext and thus can be sent ahead of time.

The idea behind this instantiation is to make use of interaction to verify that the sender actually knows q. This utilizes the fact that $\mathbb{H}$ satisfies interactive preimage extraction. So that any adversary communicating with the decryption oracle knows what the plaintext is. Hence, the decryption oracle does not really help the adversary. Therefore, IND-CCA2 can be reduced to IND-CPA. Since this scheme can be shown to be IND-CPA, it is IND-CCA2 in the interactive setting.

Noninteractive instantiation. A similar relation can be drawn between the existence of noninteractive EPOW functions and noninteractive instantiation of this scheme. Specifically, if $\mathbb{M}$ is a trapdoor permutation and $\mathbb{H}$ is an extractable (with dependent auxiliary information) and pseudorandom POW function with public randomness, then the scheme, $E(m, pk' = (pk, k_1)) = r_1, M_{pk}(q), y \oplus m, H_{k_1}(q, m, r_2)$, where $H_{k_1}(q, r_1) = r_1, y$, is IND-CCA2.[3]

---

[3] The construction in Section 4 is an indistinguishable POW function but is not known to be pseudorandom. Realizing the latter requirement with noninteractive extraction remains open.

7 Overview of the 3-Round Zero-Knowledge Protocol

EPOW functions can also be used to construct 3-round ZK argument systems. Such functions allow us to do so because of their knowledge and secrecy properties. Informally, the protocol starts with the prover sending an EPOW function. The verifier responds with a corresponding image of a uniform string. The protocol ends with the prover sending a noninteractive witness-indistinguishable (WI) proof that either the theorem is true or the prover "knows" a preimage of the verifier's message. Intuitively, this protocol is sound because the verifier's message completely hides its preimage. Thus, the (polynomially-bounded) prover does not "know" a preimage. Consequently, if the verifier accepts the conversation then by the soundness property of the WI proof, the theorem has to be true. On the other hand, this protocol is zero-knowledge because the verifier "knows" a preimage of its message. In other words, a simulator can use the extractor for the EPOW function to recover a preimage and produce a WI proof using this preimage as a witness. In more detail, the simulator sends a random EPOW function in the first round. The verifier responds with an image under this function, and the simulator uses the extractor to recover a corresponding preimage, and then uses it as a witness in computing the noninteractive WI proof.

We emphasize that when using the construction of Section 4 in the above ZK protocol, we do not use any algebraic property of the discrete log in a direct way. This opens the door for basing 3-round ZK arguments on assumptions other than the KE assumption as long as such assumptions prove sufficient for constructing such EPOW functions.

We remark that using EPOW functions with arbitrary small but noticeable extraction failure probability gives weak simulation, i.e., simulation fails with arbitrary small but noticeable probability. On the other hand, if an EPOW function, such as construction of Section 4, has negligible extraction error then simulation succeeds overwhelmingly.

Acknowledgements. We are grateful to Joan Feigenbaum for many enlightening discussions and suggestions. We also thank the anonymous referees for their helpful comments and remarks.

References

1. Barak, B., Goldreich, O., Impagliazzo, R., Rudich, S., Sahai, A., Vadhan, S., Yang, K.: On the (im)possibility of obfuscating programs. In: Kilian, J. (ed.) CRYPTO 2001. LNCS, vol. 2139. Springer, Heidelberg (2001)
2. Barak, B., Ong, S., Vadhan, S.: Derandomization in cryptography. In: Galbraith, S.D. (ed.) Cryptography and Coding 2007. LNCS, vol. 4887. Springer, Heidelberg (2007)
3. Bellare, M., Palacio, A.: The knowledge-of-exponent assumptions and 3-round zero-knowledge protocols. In: Franklin, M. (ed.) CRYPTO 2004. LNCS, vol. 3152. Springer, Heidelberg (2004)
4. Bellare, M., Rogaway, P.: Random oracles are practical:a paradigm for designing efficient protocols. In: CCS 1993 (1993)
5. Blum, M.: How to prove a theorem so no one else can claim it. In: Proceedings of the International Congress of Mathematicians (1986)
6. Boldyreva, A., Fischlin, M.: On the security of OAEP. In: Lai, X., Chen, K. (eds.) ASIACRYPT 2006. LNCS, vol. 4284. Springer, Heidelberg (2006)

7. Canetti, R.: Towards realizing random oracles:hash functions that hide all partial information. In: Kaliski Jr., B.S. (ed.) CRYPTO 1997. LNCS, vol. 1294. Springer, Heidelberg (1997)
8. Canetti, R., Dakdouk, R.R.: Extractable perfectly one-way functions. eprint (2008)
9. Canetti, R., Goldreich, O., Halevi, S.: The random oracle methodology, revisited. In: STOIC 1998 (1998)
10. Canetti, R., Micciancio, D., Reingold, O.: Perfectly one-way probabilistic hash functions. In: STOIC 1998 (1998)
11. Cramer, R., Damgard, I., Nielsen, J.B.: Multiparty computation from threshold homomorphic encryption. In: Pfitzmann, B. (ed.) EUROCRYPT 2001. LNCS, vol. 2045. Springer, Heidelberg (2001)
12. Damgard, I.: Towards practical public key systems secure against chosen ciphertext attacks. In: Crypto 1992 (1992)
13. Dent, A.: The cramer-shoup encryption scheme is plaintext aware in the standard model. In: Vaudenay, S. (ed.) EUROCRYPT 2006. LNCS, vol. 4004. Springer, Heidelberg (2006)
14. Dolev, D., Dwork, C., Naor, M.: Nonmalleable cryptography. SIAM Journal on Computing 30 (2000)
15. Fiat, A., Shamir, A.: How to prove yourself:practical solutions to identification and signature problems. In: Crypto 1986 (1986)
16. Federal Information Processing Standard (FIPS). Secure hash standard. NIST, FIPS publication 180 (1993)
17. Goldwasser, S., Kalai, Y.T.: On the (in)security of the fiat-shamir paradigm. In: FOCS 2003 (2003)
18. Goldwasser, S., Kalai, Y.T.: On the impossibility of obfuscation with auxiliary input. In: FOCS 2005 (2005)
19. Groth, J., Ostrovsky, R., Sahai, A.: Non-interactive zaps and new techniques for NIZK. In: Dwork, C. (ed.) CRYPTO 2006. LNCS, vol. 4117. Springer, Heidelberg (2006)
20. Hada, S., Tanaka, T.: On the existence of 3-round zero-knowledge protocols. In: Krawczyk, H. (ed.) CRYPTO 1998. LNCS, vol. 1462. Springer, Heidelberg (1998)
21. Hada, S., Tanaka, T.: On the existence of 3-round zero-knowledge protocols (eprint) (1999)
22. Katz, J.: Efficient and non-malleable proofs of plaintext knowledge and applications. In: Eurocrypt 2003 (2003)
23. Lepinski, M.: On the existence of 3-round zero-knowledge proofs. M.S. Thesis (2002)
24. Maurer, U., Renner, R., Holenstein, C.: Indifferentiability, impossibility results on reductions, and applications to the random oracle methodology. In: Naor, M. (ed.) TCC 2004. LNCS, vol. 2951. Springer, Heidelberg (2004)
25. Nielsen, J.: Separating random oracle proofs from complexity theoretic proofs:the non-committing encryption case. In: Yung, M. (ed.) CRYPTO 2002. LNCS, vol. 2442. Springer, Heidelberg (2002)
26. Rivest, R.: The MD5 message-digest algorithm. IETF Network Working Group, RFC 1321 (1992)
27. Sahai, A.: Non-malleable non-interactive zero knowledge and adaptive chosen-ciphertext security. In: FOCS 1999 (1999)
28. De Santis, A., Di Crescenzo, G., Ostrovsky, R., Persiano, G., Sahai, A.: Robust non-interactive zero knowledge. In: Kilian, J. (ed.) CRYPTO 2001. LNCS, vol. 2139. Springer, Heidelberg (2001)
29. De Santis, A., Persiano, G.: Zero knowledge proofs of knowledge without interaction. In: FOCS 1992 (1992)
30. Wee, H.: On obfuscating point functions. In: STOIC 2005 (2005)

Error-Tolerant Combiners
for Oblivious Primitives*

Bartosz Przydatek[1] and Jürg Wullschleger[2]

[1] Google Switzerland, Zurich, Switzerland
przydatek@google.com
[2] University of Bristol Bristol, United Kingdom
j.wullschleger@bristol.ac.uk

Abstract. A *robust combiner* is a construction that combines several implementations of a primitive based on different assumptions, and yields an implementation guaranteed to be secure if at least *some* assumptions (i.e. sufficiently many but not necessarily all) are valid.

In this paper we generalize this concept by introducing *error-tolerant* combiners, which in addition to protection against insecure implementations provide tolerance to functionality failures: an error-tolerant combiner guarantees a secure and correct implementation of the output primitive even if some of the candidates are insecure or faulty. We present simple constructions of error-tolerant robust combiners for oblivious linear function evaluation. The proposed combiners are also interesting in the regular (not error-tolerant) case, as the construction is much more efficient than the combiners known for oblivious transfer.

1 Introduction

The security of many cryptographic schemes is based on unproven assumptions about difficulty of some computational problems, like factoring integer numbers or computing discrete logarithms. Even though some standard assumptions are supported by years of extensive study and attacks, there is still no guarantee of their validity. Moreover, for many newer assumptions even such supporting evidence is missing, and so in general it is unclear how to decide which assumptions are trustworthy. Therefore, when given several implementations of some cryptographic primitive, each based on a different assumption, it is difficult to decide which implementation is the most secure one.

Robust combiners offer a method of coping with such difficulties: they take as input several candidate schemes based on different assumptions, and construct a scheme whose security is guaranteed if at least *some* candidates are secure. Such an approach provides tolerance against wrong assumptions since even a breakthrough algorithm for breaking one (or some) of the assumptions does not necessarily make the combined scheme insecure. The concept of robust combiners is actually not so new, but a more formal treatment of robust combiners was

---

* Work done in part at ETH Zurich. JW is supported by EPSRC.

L. Aceto et al. (Eds.): ICALP 2008, Part II, LNCS 5126, pp. 461–472, 2008.

initiated quite recently [14,13]. Combiners for some primitives, like one-way functions or pseudorandom generators, are rather simple, while for others, e.g., for oblivious transfer, the construction of combiners seems considerably harder [13].

Most constructions of robust combiners proposed so far assume the correct functionality of the candidate implementations and focus on protecting security against wrong assumptions only.[1] Such an approach is justified by the fact that often it is possible to test or to verify the correctness of the candidates prior to their use in the combiner [13]. However, even though in principle the correctness tests and verification can be performed efficiently, in practice they require considerable effort and expert knowledge. Moreover, such a testing-based approach provides no protection when the errors of the candidates are controlled by an adversary or are caused by sporadic failures after the initial tests, e.g. due to aging hardware or external noise.

Contributions. We propose a systematic approach to coping with erroneous candidates input to a combiner. That is, we introduce *error-tolerant* combiners, which in addition to protection against insecure assumptions and implementations provide tolerance to functionality errors of the candidates. In other words, an error-tolerant robust combiner guarantees a secure and correct implementation of the output primitive even if some of the candidates are insecure or faulty. To exemplify the proposed notion of error-tolerant robust combiners we focus on two-party oblivious primitives. We present constructions of error-tolerant combiners for oblivious linear function evaluation (OLFE). The primitive of OLFE can be viewed as a generalization of oblivious transfer (OT), and so is complete for arbitrary secure distributed computations [16,11,28].

The proposed combiners are optimal in terms of candidates' use (they use every candidate implementation only once), and in the honest-but-curious case work as soon as the input is non-trivial. For OT, no such reduction is known [7]. Additionally, the presented OLFE-combiner is also interesting in the regular (not error-tolerant) case. In particular, the construction is much more efficient than the combiners known for oblivious transfer.

Related work. As mentioned above, there are numerous implicit uses and constructions of combiners in the literature (e.g., [1,9,17,8,15]), but a more rigorous study of robust combiners was initiated only recently, by Herzberg [14] and by Harnik *et al.* [13], who have formalized the notion of combiners, and have shown constructions of combiners for various primitives. In particular Harnik *et al.* [13] have proposed a combiner for key agreement, which tolerates some failures of the candidates (cf. Sect. 2), and also have shown that not all primitives are easy to combine. In [18] robust combiners for private information retrieval (PIR) were proposed, and also *cross-primitive* combiners have been studied, which can be viewed as generalized reductions between primitives. This generalization lead to a partial separation of PIR from OT. In [19] generalized definitions of combiners for two-party primitives have been proposed, leading to constructions strictly stronger than the ones known before, and allowing for easier impossibility proofs.

---

[1] A notable exception is a combiner for key agreement due to Harnik *et al.* [13], which can handle some failures of the candidates (cf. discussion in Sect. 2).

OLFE has been studied in [26], where its symmetry has been shown. Rivest [24] has shown that with one run of OLFE a very simple commitment scheme can be implemented. In a recent work [12], Harnik *et al.* improve the efficiency of previous OT-combiners, and show that it is possible to construct them with a *constant rate*. While this is a great improvement for OT-combiners, it is only *asymptotically* efficient, and even then by a (probably big) constant factor less efficient than our OLFE-combiner, which is not only much simpler, but also uses every candidate exactly *once*, which is certainly optimal.

2   Preliminaries and Definitions

Primitives. We review shortly the primitives relevant in this work. For more formal definitions we refer to the literature. The parties participating in the protocols and the adversary are assumed to be probabilistic polynomial time Turing machines, (PPTMs).

Oblivious transfer[2] (OT) is a protocol between a sender holding two bits b_0 and b_1, and a receiver holding a choice-bit c. The protocol allows the receiver to get bit b_c so that the sender does not learn any information about receiver's choice c, and the receiver does not learn any information about bit b_{1-c}.

Oblivious Linear Function Evaluation (OLFE) over a finite field $\mathbb{F}$ is a natural generalization of oblivious transfer for domains larger than one bit [26], and is a special case of *oblivious polynomial evaluation* [20]. In OLFE over $\mathbb{F}$ the sender's input is a linear function $f(x) = a_1 x + a_0$, where $a_0, a_1, x \in \mathbb{F}$, and the receiver's input is an argument $c \in \mathbb{F}$. The goal of OLFE is that the receiver learns the value of sender's function at the argument of his choice, i.e. he learns $y = f(c)$ (and nothing else), and the sender learns nothing. Oblivious transfer is indeed a special case of OLFE: the output bit b_c of OT with inputs (b_0, b_1) and c respectively, can be interpreted as an evaluation of the linear function $f(c) = a_1 \cdot c + a_0$ over $\mathbb{F}_2$, since

$$b_c = \underbrace{(b_0 + b_1)}_{\equiv a_1} \cdot c + \underbrace{b_0}_{\equiv a_0} \ .$$

Many protocols based on OT can be generalized to OLFE, and thereby increasing their efficiency. For more applications, see [20].

Secret sharing [3,25] allows a party to distribute a secret among a group of parties, by providing each party with a *share*, such that only authorized subsets of parties can collectively reconstruct the secret from their shares. We say that a sharing among n parties is a k-out-of-n secret sharing, if any k correct shares are sufficient to reconstruct the secret, but any subset of less than k shares gives

---

[2] The version of oblivious transfer described here and used in this paper is more precisely denoted as *1-out-of-2 bit-OT* [10]. There are several other versions of OT, e.g., *Rabin's OT*, *1-out-of-n bit-OT*, or *1-out-of-n string-OT*, but all are known to be equivalent [23,4,5].

no information about the secret. A simple method for k-out-of-n secret sharing was proposed by Shamir [25]: a party P having a secret $s \in \mathbb{F}_q$, where $q > n$, picks a random polynomial $f(x)$ over $\mathbb{F}_q$, such that $f(0) = s$ and the degree of $f(x)$ is (at most) $k - 1$. A share for party P_i is computed as $s_i := f(z_i)$, where $z_1, \ldots, z_n$ are fixed, publicly known, distinct non-zero values from $\mathbb{F}_q$. Since the degree of $f(x)$ is at most $k - 1$, any k shares are sufficient to reconstruct $f(x)$ and compute $s = f(0)$ (via Lagrange interpolation). On the other hand, any $k - 1$ or fewer shares give no information about s, since they can be completed consistently to yield a sharing of any arbitrary $\bar{s} \in F[q]$, and the number of possible completions is the same for every $\bar{s}$.

Robust Combiners. In this section we recall definitions of robust combiners, and present generalizations for combiners of two-party primitives, which capture constructions that can tolerate even incorrect candidates. We say that a candidate implements a primitive *correctly*, if it produces the correct output when both players are honest, and we say that a candidate is *secure* for Alice (Bob), if the action of a dishonest Bob (Alice) can be simulated in an ideal setting (for more details cf. [22]).

Definition 1 (($(k; n)$-robust $\mathcal{F}$-combiner [13]). *Let $\mathcal{F}$ be a cryptographic primitive. A $(k; n)$-robust $\mathcal{F}$-combiner is a PPTM which gets n candidates implementing $\mathcal{F}$ as inputs and implements $\mathcal{F}$ while satisfying the following properties:*

1. *If at least k candidates securely implement $\mathcal{F}$, then the combiner securely implements $\mathcal{F}$.*
2. *The running time of the combiner is polynomial in the security parameter κ, in n, and in the lengths of the inputs to $\mathcal{F}$.*

To capture the error-tolerance of a combiner we introduce a parameter γ, which denotes the number of candidates that are assumed to provide correct functionality. The following definition of error-tolerant combiners is based on a generalization of Definition 1, proposed recently in [19].

Definition 2 (($(\alpha, \beta, \gamma; n)$-robust $\mathcal{F}$-combiner). *Let $\mathcal{F}$ be a cryptographic primitive for two parties Alice and Bob. A $(\alpha, \beta, \gamma; n)$-robust $\mathcal{F}$-combiner is a PPTM which gets n candidates implementing $\mathcal{F}$ as inputs and implements $\mathcal{F}$ while satisfying the following properties:*

1. *If at least γ candidates implement $\mathcal{F}$ correctly, at least α candidates implement $\mathcal{F}$ securely for Alice, and at least β candidates implement $\mathcal{F}$ securely for Bob, then the combiner securely implements $\mathcal{F}$.*
2. *The running time of the combiner is polynomial in the security parameter κ, in n, and in the lengths of the inputs to $\mathcal{F}$.*

Combiners which are $(\alpha, \beta, n; n)$-robust, that is constructions working only when all candidates provide correct functionality, are called in short just $(\alpha, \beta; n)$-*robust*. Note that any $(k, k; n)$-robust combiner is always also a $(k; n)$-robust combiner, but a $(k; n)$-robust combiner may *not* be a $(k, k; n)$-robust combiner.

Motivated by an observation that the constructions of $(\alpha, \beta; n)$-robust combiners can be "non-uniform", with explicit dependence on α, β, a stronger notion of *uniform* combiners was introduced in [19]. Intuitively, an $\{\delta; n\}$-robust *uniform* combiner is a single construction that is *simultaneously* a $(\alpha, \beta; n)$-robust combiner for all $\alpha, \beta \in \{0, \ldots, n\}$ satisfying $\alpha + \beta \geq \delta$. In particular, a uniform combiner doesn't obtain the values of α, β as parameters. The next definition generalizes the notion of uniform combiners to the error-tolerant setting.

Definition 3 ($\{\delta, \gamma; n\}$-robust uniform $\mathcal{F}$-combiner). *Let $\mathcal{F}$ be a two-party primitive. We say that an $\mathcal{F}$-combiner is a $\{\delta, \gamma; n\}$-robust uniform $\mathcal{F}$-combiner if it is a $(\alpha, \beta, \gamma; n)$-robust $\mathcal{F}$-combiner, simultaneously for all $\alpha, \beta \in \{0, \ldots, n\}$ satisfying $\alpha + \beta \geq \delta$.*

Uniform combiners which are $\{\delta, n; n\}$-robust, that is constructions working only when all candidates provide functionality, correspond to the original notion $\{\delta; n\}$-robust uniform combiners. Note that the parameter δ is a bound on *the sum* of the number of candidates secure for Alice and the number of candidates secure for Bob, hence given n candidates δ is from the range $0 \ldots 2n$. As an example consider a $\{4; 3\}$-robust *uniform* combiner: it is a (regular) $(2; 3)$-robust combiner, but at the same time it is also a $(3, 1; 3)$-robust combiner and a $(1, 3; 3)$-robust combiner.

Error Tolerance of Robust Combiners. So far research on robust combiners focused on protection against wrong computational assumptions, and the proposed constructions usually required that the candidates input to a combiner provide the desired functionality. In general, this approach is justified by the fact that in cryptographic schemes usually the security is based on some assumptions, while the functionality properties are straightforward and hold unconditionally. Moreover, Harnik *et al.* [13] suggested that sometimes a possible way of dealing with unknown implementations of primitives is to test them for the desired functionality before combining them.

However, in many realistic scenarios this functionality assumption is not always justified. For example, if the candidate implementations are given as black-boxes, their failures might be time-dependent, e.g. caused by some transient external factors, or even controlled by an adversary, due to some malicious "features" (malware). In such a case, the testing could be successful, but the actual run within a combiner would produce wrong results. Moreover, even if the candidate implementations are given as programs or software packages, it is sometimes unreasonable to assume that a user will be able to check their functionality, as such a check requires substantial effort and expert knowledge.

A notable exception of the above functionality assumption is a robust combiner for key agreement (KA) proposed by Harnik *et al.* [13]. This combiners works by constructing a relaxed KA (where agreement is achieved with relatively high probability), and then by using it together with an error correcting code to increase the probability of agreement. However, this construction is also partially based on testing (to construct a relaxed KA), and is tolerant only against stochastic errors. In particular, the construction fails when the errors are caused by a malicious adversary which has ability of introducing arbitrary errors in a single candidate. Such an adversary can easily force incorrect final results, by causing errors at selected locations in the transmitted codeword, which lead to wrong error correction.

We suggest a more systematic approach to error tolerance. In particular, the proposed definitions of error-tolerant combiners require that constructions guarantee correctness and security of the resulting implementation even when some of the input candidates are under full control of an adversary.

3   Robust Combiners for OLFE

In this section, we propose a number of robust combiners for *oblivious linear function evaluation* (OLFE). In particular, we present a $(\alpha, \beta; n)$-robust OLFE-combiner works for any $\alpha + \beta > n$. The proposed construction achieves optimal robustness and is much more efficient than the most efficient OT-combiners with the same robustness [19]: it uses any candidate instance only *once*, which is optimal, and therefore might be preferable in some scenarios. Moreover, we exploit the symmetry of OLFE to construct efficient *uniform* OLFE-combiners. Subsequently we present error-tolerant combiners for OLFE, which work even if some of the input candidates are incorrect and do not guarantee the OLFE-functionality. These combiners are optimal both terms of candidates' use, and in the achieved bound on α and β. For OT, no such reduction is known [7,27].

3.1   OLFE-Combiner

We start by presenting a robust OLFE-combiner, which does not provide error-tolerance, but is of interest for at least two reasons. First, it constitutes a basis for error-tolerant constructions presented later. Second, it is very efficient, and can be advantageous in scenarios where otherwise an OT-combiner would be used. For example, if we have a protocol for secure function evaluation based on OLFE, a protocol for OLFE based on OT, and three implementations of OT from which we assume two to be correct, it will be more efficient to use the OLFE-combiner we present in this section than the OT-combiner: the construction would require only half as many calls to each OT-instance. Furthermore, the construction is perfect (its error probability is equal zero).

Recall that OLFE is a primitive defined over some finite field $\mathbb{F}_q$, where the sender's input is a linear function $f(x) = a_1 x + a_0$, with $a_0, a_1, x \in \mathbb{F}_q$, and the receiver's input is an argument $c \in \mathbb{F}_q$. The receiver learns the value of the function on his input, $y = f(c)$, and the sender learns nothing.

In our OLFE-combiner we use Shamir secret sharing scheme [25] for both players at the same time to protect the privacy of the inputs. Given inputs $f(x) := a_1 x + a_0$ resp. $c \in \mathbb{F}_q$, the sender and the receiver proceed as follows. The sender picks two random polynomials $A_0(z)$ of degree $n - 1$ and $A_1(z)$ of degree $n - \alpha$, such that $A_0(0) = a_0$ and $A_1(0) = a_1$. The receiver picks a random polynomial $C(z)$ of degree $n - \beta$, such that $C(0) = c$. Then the parties evaluate

---

Protocol OLFE-rc

SENDER'S INPUT: linear function $f(x) := a_1 x + a_0$, with $a_0, a_1 \in \mathbb{F}_q$
RECEIVER'S INPUT: evaluation point $c \in \mathbb{F}_q$
INPUT OLFE PROTOCOLS: $\mathsf{OLFE}_1, \ldots, \mathsf{OLFE}_n$
parameters: $n < q; \alpha, \beta$; distinct non-zero constants $z_1, \ldots, z_n \in \mathbb{F}_q$

1. Sender picks two random polynomials:
   $A_0(z)$ of degree $n - 1$ such that $A_0(0) = a_0$, and
   $A_1(z)$ of degree $n - \alpha$ such that $A_1(0) = a_1$.
2. Receiver picks a random polynomial $C(z)$ of deg. $n-\beta$, s.t. $C(0) = c$.
3. $\forall i \in [n]$ the parties run OLFE_i, with sender holding input
   $f_i(x) = A_1(z_i)x + A_0(z_i)$, and receiver holding input $c_i = C(z_i)$.
4. Receiver uses the values $\{(z_1, f_1(c_1)), \ldots, (z_n, f_n(c_n))\}$ to interpolate a polynomial $\tilde{h}(z)$ of degree $n - 1$, and outputs $y = \tilde{h}(0)$.

Fig. 1. OLFE-rc: A $(\alpha, \beta; n)$-robust OLFE-combiner for $\alpha + \beta > n$

locally these polynomials for n distinct non-zero values $z_1, \ldots, z_n \in \mathbb{F}_q$, and use the resulting values as input to the instances of OLFE.

More precisely, we can view the two polynomials of the sender, $A_0(z)$ and $A_1(z)$ as parts of a two-dimensional polynomial $F(x, z)$ of degree 1 in x and degree $n - 1$ in z, $F(x, z) := A_1(z) \cdot x + A_0(z)$, which satisfies $F(x, 0) = f(x)$. Note that for any constant $z_i \in \mathbb{F}_q$, the polynomial $f_i(x) := F(x, z_i) = A_1(z_i)x + A_0(z_i)$ is just a linear function. Furthermore we define a polynomial $h(z) := F(C(z), z) = A_1(z) \cdot C(z) + A_0(z)$, which clearly satisfies $h(0) = f(c)$, i.e., $h(0)$ is the value the receiver should obtain. Hence the goal is now to allow the receiver to learn $h(0)$. To achieve this, the parties run OLFE candidates, through which the receiver obtains sufficiently many points on $h(z)$ to enable its interpolation and the computation of $h(0)$. To evaluate through OLFE_i the polynomial $h(z)$ at any particular value $z_i \in \mathbb{F}_q$, the sender's input is $f_i(x) := F(x, z_i)$, and the receiver's input is $C(z_i)$.

Intuitively, the privacy of the users in this construction is protected by the degrees of the polynomials we use. Since we have to be prepared that in worst case only α of the n input OLFE-protocols are secure for the sender, and only β are secure for the receiver, the polynomials $A_0(z)$ and $A_1(z)$ must have degree at least $n-\alpha$, and the degree of $C(z)$ must be at least $n-\beta$. On the other hand, note that $h(z)$ is of degree $\max\{\deg(A_0), \deg(A_1) + \deg(C)\} = \max\{n-1, 2n-\alpha-\beta\}$. Since we're using n evaluation points, this degree must be at most $n-1$, otherwise interpolation is not possible. This implies, that $\alpha + \beta > n$ must be satisfied. Indeed, this construction works for any α, β with $\alpha + \beta > n$, which is optimal. Figure 1 presents the combiner in full detail, and its analysis is given below.

Lemma 1. *Protocol OLFE-rc is correct if $\alpha + \beta > n$.*

Proof. By construction, we have $f(x) = F(x,0)$ and $c = C(0)$. Hence $f(c) = F(C(0),0) = h(0)$. Further, we have $y_i := f_i(c_i) = f(g(z_i), z_i) = h(z_i)$. Since h has degree $\max\{n-1, 2n-\alpha-\beta\} < n$, the interpolated of y_i's will result in a polynomial $\tilde{h}(z)$ identical to $h(z)$. Hence, we have $y = \tilde{h}(0) = h(0) = f(c)$.   □

Lemma 2. *Protocol* OLFE-rc *is secure against a malicious sender if at least β input instances of OLFE are secure against a malicious sender.*

Proof. *(sketch)* Note that the values c_i form a $(n-\beta+1)$-out-of-n secret sharing of c, hence a malicious sender that sees at most $(n-\beta)$ values of c_i (of his choice) does not get any information about c.   □

Lemma 3. *Protocol* OLFE-rc *is secure against a malicious receiver if at least α input instances of OLFE are secure against a malicious receiver.*

A proof of this lemma is given in [22]. From the above lemmas, and from lower bounds on robustness of OT-combiners [19] we obtain the following theorem:

Theorem 1. *For every $n > 1$ and α, β satisfying $\alpha + \beta > n$ there exists an efficient third-party black-box $(\alpha, \beta; n)$-robust combiner for OLFE over $\mathbb{F}_q$ with $q > n$. The combiner is perfect, achieves optimal robustness, and uses only one run of each candidate instance of OLFE, which is also optimal.*

3.2   Uniform OLFE-Combiner Based on Symmetry of OLFE

A two-party primitive is *symmetric* if it can be *logically* reversed, i.e. if any implementation of a primitive $\mathcal{F}$ between Alice and Bob it can be logically transformed into an implementation of $\mathcal{F}$ with reversed roles of Alice and Bob. Such a logical reversal differs from *physical* reversal, in which the parties just swap their roles when executing the corresponding protocol. In a recent work [19] it was shown, that the symmetry of oblivious transfer [6,21,26] can be used to construct efficient *universal* OT-combiners, which are based on an observation that a physical reversal of a symmetric primitive followed by a logical reversal (so-called swap-operation) yields an implementation with swapped security properties. Since OLFE is also symmetric [26], we can use the same trick as in [19] to obtain the following theorem.

Theorem 2. *For any $n > 1$ and any $\delta > n$ there exists a third-party black-box $\{\delta; n\}$-robust uniform combiner for OLFE over $\mathbb{F}_q$ with $q > 2n$, using the swap-operation. The combiner is perfect and uses only two runs of each candidate instance of OLFE.*

3.3   Error-Tolerant OLFE-Combiners

The OLFE-combiners described so far are very efficient, but break if any of the candidates is incorrect, i.e. if a candidate provides incorrect output to the receiver. In this section we present two very efficient constructions of error-tolerant OLFE combiners, which are both robust against *insecure* candidates and tolerate *erroneous* candidates. However, the proposed constructions differ in the error-tolerance and robustness bounds, and also in the guaranteed level of security. The first construction achieves better error-tolerance and robustness, but guarantees security against an honest-but-curious receiver only. The second achieves security against malicious parties, at a cost of lower error-tolerance and robustness. Both combiners are based on the combiner OLFE-rc from Sect. 3.1.

OLFE-combiner with Honest-but-curious Receiver. We modify the construction OLFE-rc to enable error correction while still preserving privacy of the participants. To allow for error correction (using for example Berlekamp-Welch algorithm [2]), we introduce additional redundancy in the information obtained by the receiver, by decreasing the degree of the polynomial $h(z)$. In particular, the degree of the polynomial $A_0(z)$ (which shares the coefficient a_0) is decreased by 2ε, where ε denotes the number of erroneous candidates tolerated by the combiner. This degree reduction of polynomial $A_0(z)$ in OLFE-rc, together with error-correction added in Step 4 (before interpolation of a polynomial $\tilde{h}(z)$ of degree $n - 1 - 2\varepsilon$), are already all modifications we need to obtain an error-tolerant combiner. Let OLFE-rc-hr denote the resulting construction. Since the degree of $h(z)$ is given by $\max\{\deg(A_0), \deg(A_1) + \deg(C)\} = \max\{n-1-2\varepsilon, 2n-\alpha-\beta\}$, to ensure correction of up to ε errors it must hold that $n - 1 - 2\varepsilon \geq 2n - \alpha - \beta$. Using $\varepsilon = n - \gamma$, this implies that successful error correction is possible if $\alpha + \beta + 2\gamma > 3n$, i.e. we obtain the following lemma.

Lemma 4. *Protocol* OLFE-rc-hr *is correct if* $\alpha + \beta + 2\gamma > 3n$.

Since the modifications introduced to construct OLFE-rc-hr do not affect the information that can possibly leak (due to insecure candidates) from the receiver to the sender, the security of the receiver remains unchanged.

Lemma 5. *Protocol* OLFE-rc-hr *is secure against a malicious sender if at least* β *input instances of OLFE are secure against a malicious sender.*

To complete the security analysis of OLFE-rc-hr it remains to argue the security of the sender.

Lemma 6. *Protocol* OLFE-rc-hr *is secure against an honest-but-curious receiver if at least* α *input instances of OLFE are secure against a malicious receiver, and if* $\alpha + \beta + 2\gamma > 3n$ *holds.*

A proof of this lemma is given in [22]. From the above analysis we obtain the following theorem:

Theorem 3. *Let* $n > 1$ *and* α, β, γ *be such that* $\alpha + \beta + 2\gamma > 3n$. *There exists an efficient perfect* $(\alpha, \beta, \gamma; n)$-*combiner for OLFE over* $\mathbb{F}_q$ *with* $q > n$, *secure against an honest-but-curious receiver. The combiner uses only one run of each candidate OLFE protocol.*

OLFE-combiner Secure Against Malicious Parties. It is not hard to see that combiner OLFE-rc-hr does not guarantee privacy of the sender when the receiver is malicious. For example,[3] when $n = 4, \alpha = 3, \beta = 4$, and $\gamma = 3$, the degree of the corresponding $h(z)$ is equal 1, i.e. only two points are needed for interpolation of $h(z)$. A malicious receiver can choose the evaluation points c_i arbitrarily (instead of a constant representing his polynomial $C(z)$ of degree $n - \beta = 0$), and so interpolate two various polynomials $h'(z)$ and $h''(z)$ of degree 1, corresponding to two values on the senders input function $f(x) = a_1 x + a_0$.

An additional simple modification is sufficient to avoid the above problem: to ensure the privacy of the sender we increase the degree of the sharing of the coefficient a_1 by 2ε. That is, the new error-tolerant OLFE-combiner secure also against malicious receivers, called OLFE-rc-et in the following, works as the combiner OLFE-rc-hr, but the degree of the polynomial $A_1(z)$ is $n - \alpha + 2\varepsilon$ rather than only $n - \alpha$. Since now the degree degree of $h(z)$ is given by $\max\{n - 1 - 2\varepsilon, 2n - \alpha - \beta + 2\varepsilon\}$, to ensure correction of up to ε errors the following condition must then hold

$$n - 1 - 2\varepsilon \geq 2n - \alpha - \beta + 2\varepsilon \,,$$

and we obtain the following lemma.

Lemma 7. *Protocol OLFE-rc-et is correct if $\alpha + \beta + 4\gamma > 5n$.*

As previously, the modifications introduced to construct OLFE-rc-et do not affect the information that can possibly leak to the sender the security of the receiver remains unchanged.

Lemma 8. *Protocol OLFE-rc-et is secure against a malicious sender if at least β input instances of OLFE are secure against a malicious sender.*

Finally, we have to argue the security of the sender.

Lemma 9. *Protocol OLFE-rc-et is secure against a malicious receiver if at least α input instances of OLFE are secure against a malicious receiver, and if $\alpha + \beta + 4\gamma > 5n$ holds.*

A proof of this lemma is given in [22], and we obtain the following theorem:

Theorem 4. *Let $n > 1$ and α, β, γ be such that $\alpha + \beta + 4\gamma > 5n$. There exists an efficient perfect $(\alpha, \beta, \gamma; n)$-combiner for OLFE over $\mathbb{F}_q$ with $q > n$, secure against malicious parties. The combiner uses only one run of each candidate OLFE protocol.*

Uniform Error-tolerant OLFE-combiners. As with the combiner OLFE-rc, we can exploit the symmetry of OLFE and use the swap-operation (cf. Sect. 3.2) to obtain *uniform* error-tolerant OLFE-combiners based on OLFE-rc-hr and OLFE-rc-et. For details see [22].

---

[3] Note, that these values satisfy the condition $\alpha + \beta + 2\gamma > 3n$.

4   Conclusions

Robust combiners are a useful tool for dealing with implementations of cryptographic primitives based on various computational assumptions. In this paper we have studied robust combiners for oblivious primitives. In particular, we have introduced error-tolerant combiners, which offer protection not only against insecure but also against erroneous candidate implementations. We have presented a number of robust combiners for OLFE, both regular and error-tolerant. The proposed constructions differ in the achieved efficiency and robustness/error-tolerance, and offer a trade-off between these two measures.

While the presented constructions of error-tolerant combiners are optimally efficient, they are not optimal in every respect. For example, there is a gap in the robustness achieved by the proposed error-tolerant OLFE-combiners secure against honest-but-curious and malicious parties. Of particular interest are also regular OLFE-combiners, as they are both optimally efficient and optimally robust. Moreover, we believe that OLFE is also interesting as a stand-alone primitive, and not only as a special case of oblivious polynomial evaluation. In particular, OLFE is symmetric, which allows for construction of uniform OLFE-combiners.

References

1. Asmuth, C., Blakely, G.: An effcient algorithm for constructing a cryptosystem which is harder to break than two other cryptosystems. Computers and Mathematics with Applications 7, 447–450 (1981)
2. Berlekamp, E.R., Welch, L.R.: Error correction for algebraic block codes, U.S. Patent 4 633 470 (1986)
3. Blakley, G.R.: Safeguarding cryptographic keys. In: Proceedings of the National Computer Conference. American Federation of Information Processing Societies, pp. 313–317 (1979)
4. Crépeau, C.: Equivalence between two flavours of oblivious transfers. In: Pomerance, C. (ed.) CRYPTO 1987. LNCS, vol. 293, pp. 350–354. Springer, Heidelberg (1988)
5. Crépeau, C., Kilian, J.: Achieving oblivious transfer using weakened security assumptions (extended abstract). In: Proc. IEEE FOCS 1988, pp. 42–52 (1988)
6. Crépeau, C., Sántha, M.: On the reversibility of oblivious transfer. In: Davies, D.W. (ed.) EUROCRYPT 1991. LNCS, vol. 547, pp. 106–113. Springer, Heidelberg (1991)
7. Damgård, I., Kilian, J., Salvail, L.: On the (im)possibility of basing oblivious transfer and bit commitment on weakened security assumptions. In: Stern, J. (ed.) EUROCRYPT 1999. LNCS, vol. 1592, pp. 56–73. Springer, Heidelberg (1999)
8. Dodis, Y., Katz, J.: Chosen-ciphertext security of multiple encryption. In: Kilian, J. (ed.) TCC 2005. LNCS, vol. 3378, pp. 188–209. Springer, Heidelberg (2005)
9. Even, S., Goldreich, O.: On the power of cascade ciphers. ACM Trans. Comput. Syst. 3(2), 108–116 (1985)
10. Even, S., Goldreich, O., Lempel, A.: A randomized protocol for signing contracts. Communications of the ACM 28(6), 637–647 (1985)

11. Goldreich, O., Micali, S., Wigderson, A.: How to play any mental game — a completeness theorem for protocols with honest majority. In: Proc. 19th ACM STOC, pp. 218–229 (1987)
12. Harnik, D., Ishai, Y., Kushilevitz, E., Nielsen, J.B.: OT-combiners via secure computation. In: Proc. TCC 2008. LNCS, Springer, Heidelberg (2008)
13. Harnik, D., Kilian, J., Naor, M., Reingold, O., Rosen, A.: On robust combiners for oblivious transfer and other primitives. In: Cramer, R.J.F. (ed.) EUROCRYPT 2005. LNCS, vol. 3494, pp. 96–113. Springer, Heidelberg (2005)
14. Herzberg, A.: On tolerant cryptographic constructions. In: Menezes, A. (ed.) CT-RSA 2005. LNCS, vol. 3376, pp. 172–190. Springer, Heidelberg (2005)
15. Hohenberger, S., Lysyanskaya, A.: How to securely outsource cryptographic computations. In: Kilian, J. (ed.) TCC 2005. LNCS, vol. 3378, pp. 264–282. Springer, Heidelberg (2005)
16. Kilian, J.: Founding cryptography on oblivious transfer. In: Proc. 20th ACM STOC, pp. 20–31 (1988)
17. Maurer, U., Massey, J.L.: Cascade ciphers: The importance of being first. Journal of Cryptology 6(1), 55–61 (1993). Preliminary version. In: Proc. IEEE Symposium on Information Theory (1990)
18. Meier, R., Przydatek, B.: On robust combiners for private information retrieval and other primitives. In: Dwork, C. (ed.) CRYPTO 2006. LNCS, vol. 4117, pp. 555–569. Springer, Heidelberg (2006)
19. Meier, R., Przydatek, B., Wullschleger, J.: Robuster combiners for oblivious transfer. In: Vadhan, S.P. (ed.) TCC 2007. LNCS, vol. 4392, pp. 404–418. Springer, Heidelberg (2007)
20. Naor, M., Pinkas, B.: Oblivious polynomial evaluation. SIAM J. Comput. 35(5), 1254–1281 (2006)
21. Ostrovsky, R., Venkatesan, R., Yung, M.: Fair games against an all-powerful adversary. In: Advances in Computational Complexity Theory. AMS DIMACS Series in Discrete Mathematics and Theoretical Computer Science, vol. 13, pp. 155–169. AMS (1993)
22. Przydatek, B., Wullschleger, J.: Error-tolerant combiners for oblivious primitives, full version of this paper, Cryptology ePrint Archive, eprint.iacr.org (2008)
23. Rabin, M.O.: How to exchange secrets by oblivious transfer, Tech. Memo TR-81, Aiken Computation Laboratory (1981), eprint.iacr.org/2005/187
24. Rivest, R.L.: Unconditionally secure commitment and oblivious transfer schemes using private channels and a trusted initializer (unpublished manuscript) (1999)
25. Shamir, A.: How to share a secret. Commun. ACM 22(11), 612–613 (1979)
26. Wolf, S., Wullschleger, J.: Oblivious transfer is symmetric. In: Vaudenay, S. (ed.) EUROCRYPT 2006. LNCS, vol. 4004, pp. 222–232. Springer, Heidelberg (2006)
27. Wullschleger, J.: Oblivious-transfer amplification. In: Naor, M. (ed.) EUROCRYPT 2007. LNCS, vol. 4515. Springer, Heidelberg (2007), arxiv.org/abs/cs.CR/0608076
28. Yao, A.C.-C.: How to generate and exchange secrets (extended abstract). In: Proc. 27th IEEE FOCS, pp. 162–167 (1986)

Asynchronous Multi-Party Computation
with Quadratic Communication

Martin Hirt[1], Jesper Buus Nielsen[2], and Bartosz Przydatek[3],[*]

[1] Dept. of Computer Science, ETH Zurich, Switzerland
[2] Dept. of Computer Science, University of Aarhus, Denmark
[3] Google Switzerland, Zurich, Switzerland

Abstract. We present an efficient protocol for secure multi-party computation in the asynchronous model with optimal resilience. For n parties, up to $t < n/3$ of them being corrupted, and security parameter κ, a circuit with c gates can be securely computed with communication complexity $\mathcal{O}(cn^2\kappa)$ bits, which improves on the previously known solutions by a factor of $\Omega(n)$. The construction of the protocol follows the approach introduced by Franklin and Haber (Crypto'93), based on a public-key encryption scheme with threshold decryption. To achieve the quadratic complexity, we employ several techniques, including circuit randomization due to Beaver (Crypto'91), and an abstraction of *certificates*, which can be of independent interest.

1 Introduction

Secure multi-party computation. Secure multi-party computation (MPC) allows a set of n parties (players) to evaluate an agreed function of their inputs in a secure way, where security means that an adversary corrupting some of the parties, cannot achieve more than controlling their inputs and outputs. In particular, the adversary does not learn the inputs of the uncorrupted parties, and she cannot influence the outputs of the uncorrupted parties, except by selecting the inputs of the corrupted players. We focus on *asynchronous communication*, i.e., the messages in the network can be delayed for an arbitrary amount of time (but eventually, all messages are delivered). As a worst-case assumption, we give the ability of controlling the delay of messages to the adversary. Asynchronous communication models real-world networks, like the Internet, much better than synchronous communication. However, it turns out that MPC protocols for asynchronous networks are significantly more involved than their synchronous counterparts. One reason for this is that a player in an asynchronous network waiting for a message cannot distinguish whether the sender is corrupted and did not send the message, or the message was sent but delayed in the network. This implies also that in a fully asynchronous setting it is impossible to consider the inputs of *all* uncorrupted players when evaluating the function — inputs of up to t (potentially honest) players have to be ignored.

History and related work. The MPC problem was first proposed by Yao [26] and solved by Goldreich, Micali, and Wigderson [18] for computationally bounded adversaries and

---

[*] Work done in part at ETH Zurich.

L. Aceto et al. (Eds.): ICALP 2008, Part II, LNCS 5126, pp. 473–485, 2008.
© Springer-Verlag Berlin Heidelberg 2008

by Ben-Or, Goldwasser, and Wigderson [5] and independently by Chaum, Crépeau, and Damgård [12] for computationally unbounded adversaries. All these protocols considered a synchronous network with a global clock. The first MPC protocol for the asynchronous model (with unconditional security) was proposed by Ben-Or, Canetti, and Goldreich [4]. Extensions and improvements, still in the unconditional model, were proposed in [6,24]. A great overview of asynchronous MPC with unconditional security is given in [8]. The most efficient asynchronous protocol up to date [19] communicates $\mathcal{O}(n^3\kappa)$ bits per multiplication gate, where κ is a security parameter.

Contributions. We present an asynchronous MPC protocol, cryptographically secure with respect to an active adversary corrupting up to $t < n/3$ players (this is optimal in an asynchronous network). Once the inputs are distributed, the protocol requires $\mathcal{O}(c_M n^2 \kappa)$ bits of communication to evaluate a circuit with c_M multiplication gates and with security parameter κ. This improves on the communication complexity of the most efficient optimally-secure asynchronous MPC protocol by a factor of $\Omega(n)$. The new protocol, similarly as [19], uses the approach based on threshold encryption [17, 13], but introduces several modifications, which result in both conceptual simplification and improved efficiency. In particular, we use a notion of *certificates*, which greatly simplify the description of the protocol on an abstract level.

2   Formal Model and Preliminaries

Notation. We use n to denote the number of players (i.e., parties) participating in the MPC protocol, we use $P_1, \ldots, P_n$ to denote the players, and we use $\mathcal{P}$ to denote the set of all players. For an integer $m > 0$ we write $[m]$ to denote the set $\{1, \ldots, m\}$. Our constructions are parametrized by a security parameter κ.

Communication Model. We consider an *asynchronous* communication network, with point-to-point secure channels, but without guaranteed delivery of messages. An n-player protocol is a tuple $\pi = (P_1, \ldots, P_n, \text{init})$, where each P_i is a probabilistic interactive Turing machine, and init is an *initialization function*, used for the usual set-up tasks, like initialization, setting up cryptographic keys, etc. The players communicate over a network in which the delay between sending and delivery of a message is unbounded. We measure the communication complexity by the worst case number of bits sent by the honest parties.

Security Model. We use the model of asynchronous protocols proposed by Canetti [9]. Formally our model for running a protocol is a hybrid model with a functionality init for distributing initial cryptographic keys among the parties. We consider a poly-time adversary, which can corrupt up to $t < n/3$ parties *before* the execution of the protocol, i.e., we consider a static adversary, and corrupted parties are under full control of the adversary. The adversary schedules the delivery of the messages arbitrarily, except that it must eventually deliver all message sent be honest parties.

The security of a protocol is defined relative to an ideal evaluation of the circuit: for any adversary attacking the protocol must exist a simulator which simulates the attack of the adversary to any environment, given only an ideal process for evaluating the circuit. The simulator has very restricted capabilities: It sees the inputs of the corrupted parties.

Then it picks a subset $\mathcal{W} \subseteq [n]$ of the parties to be the input providers, s.t. $|\mathcal{W}| \geq n - t$. The adversary determines the inputs of the corrupted parties. The input gates of Circ belonging to the parties from $\mathcal{W}$ are assigned the inputs of the corresponding parties, and the remaining input gates are assigned default values. Then Circ is evaluated and the outputs of the corrupted parties are shown to the simulator, which must then simulate the entire view of an execution of the protocol.

2.1 Cryptographic Primitives and Protocols

In the proposed MPC protocols we employ a number of standard primitives and sub-protocols. We introduce the required notation and tools with their essential properties, and then we point to the literature to example implementations.

Homomorphic Encryption with Threshold Decryption. We assume the existence of a semantically secure public-key encryption scheme, which additionally is homomorphic and enables threshold decryption, as specified below.

Encryption and decryption. For an encryption key e and a decryption key d, let $\mathscr{E}_e : \mathbb{M} \times \mathbb{R} \rightarrow \mathbb{C}$ denote the encryption function mapping a plaintext $x \in \mathbb{M}$ and a randomness $r \in \mathbb{R}$ to a ciphertext $X \in \mathbb{C}$, and let $\mathscr{D}_d : \mathbb{C} \rightarrow \mathbb{M}$ denote the corresponding decryption function, where $\mathbb{M}, \mathbb{R}, \mathbb{C}$ are algebraic structures, as specified below. We require that $\mathbb{M}$ is a ring $\mathbb{Z}_M$ for some $M > 1$, and we use "$\cdot$" to denote multiplication in $\mathbb{M}$. We often use capital letters to denote encryptions of the plaintexts denoted by the corresponding lower-case letters. When keys are understood, we write $\mathscr{E}$, $\mathscr{D}$ instead of $\mathscr{E}_e$, $\mathscr{D}_d$, and we often omit the explicit mentioning of the randomness in the encryption function $\mathscr{E}$.

Homomorphic property. We require that there exist (efficiently computable) binary operations $+$, $*$, $\oplus$, such that $(\mathbb{M}, +)$, $(\mathbb{R}, *)$, $(\mathbb{C}, \oplus)$ are algebraic groups, and that $\mathscr{E}_e$ is a group homomorphism, i.e. $\mathscr{E}(a, r_a) \oplus \mathscr{E}(b, r_b) = \mathscr{E}(a + b, r_a * r_b)$. We use $A \ominus B$ to denote $A \oplus (-B)$, where $-B$ denotes the inverse of B in the group $\mathbb{C}$. For an integer a and $B \in \mathbb{C}$ we use $a \star B$ to denote the sum of B with itself a times in $\mathbb{C}$.

Ciphertext re-randomization. For $X \in \mathbb{C}$ and $r \in \mathbb{R}$ we let $\mathscr{R}_e(X, r) = X \oplus \mathscr{E}_e(0, r)$. We use $X' = \mathscr{R}_e(X)$ to denote $X' = \mathscr{R}_e(X, r)$ for a uniformly random $r \in \mathbb{R}$. We call $X' = \mathscr{R}_e(X)$ a re-randomization of X. Note that X' is a uniformly random encryption of $\mathscr{D}_d(X)$.

Threshold decryption. We require a threshold function sharing of decryption $\mathscr{D}_d$ among n parties, i.e. that for a construction threshold $t_{\mathsf{D}} = t + 1$, there is a sharing $(d_1, \ldots, d_n)$ of the decryption key d (where d_i is intended for party P_i), satisfying the following conditions. Given the decryption shares $x_i = \mathscr{D}_{i, d_i}(X)$ for t_{D} distinct decryption-key shares d_i, it is possible to efficiently compute x such that $x = \mathscr{D}_d(X)$. When keys are understood, we write $\mathscr{D}_i(X)$ to denote the function computing decryption share of party P_i for ciphertext X, and $x = \mathscr{D}(X, \{x_i\}_{i \in I})$ to denote the process of combining the decryption shares $\{x_i\}_{i \in I}$ to a plaintext x.

Security. We require the usual security of the threshold cryptosystem, cf. [13], and in particular require that there exists an efficient two-party zero-knowledge protocol for proving the correctness of decryption shares.

Digital Signatures. We assume the existence of a digital signature scheme unforgeable against an adaptive chosen message attack. For a signing key s and a verification key v, let $\mathsf{Sign}_s : \{0,1\}^* \to \{0,1\}^\kappa$ denote the signing function, and let $\mathsf{Ver}_v : \{0,1\}^* \times \{0,1\}^\kappa \to \{0,1\}$ denote the verification function, where $\mathsf{Ver}_v(x,\sigma) = 1$ indicates that σ is a valid signature on the message x. We write $\mathsf{Sign}_i/\mathsf{Ver}_i$ to denote the signing/verification operation of party P_i.

Threshold Signatures. We assume the existence of a *threshold* signature scheme, which is unforgeable against an adaptive chosen message attack. For a signing key s and a verification key v, let $\mathscr{S}_s : \{0,1\}^* \to \{0,1\}^\kappa$ denote the signing function, and let $\mathscr{V}_v : \{0,1\}^* \times \{0,1\}^\kappa \to \{0,1\}$ denote the verification function, where $\mathscr{V}_v(m,\sigma) = 1$ indicates that σ is a valid signature on m.

Threshold signing. We require that there exists a threshold sharing of $\mathscr{S}_s$ among n parties, i.e. that for a given signing threshold t_S, $1 < t_S \le n$, there exists a sharing $(s_1, \ldots, s_n)$ of the signing key s (where s_i is intended for P_i), such that given signature shares $\sigma_i = \mathscr{S}_{i,s_i}(x)$ for t_S distinct signing-key shares s_i, it is possible to efficiently compute a signature σ satisfying $\mathscr{V}_v(x,\sigma) = 1$. We will always have $t_S = n - t$. When keys are understood, we use $\mathscr{S}_i(x)$ to denote the function computing P_i's signature share for the message x, and $\sigma = \mathscr{S}(x, \{\sigma_i\}_{i \in I})$ to denote the process of combining the signature shares $\{\sigma_i\}_{i \in I}$ to a signature σ.

Security. The scheme should be unforgeable against adaptive chosen message attack when the adversary is given $(t_S - 1)$ signing-key shares, and we require that there exists an efficient two-party zero-knowledge protocol for proving the correctness of signature shares.

Byzantine Agreement. We require a Byzantine Agreement (BA) protocol: Each P_i has input $v_i \in \{0,1\}$ and output $w_i \in \{0,1\}$, where: *Termination*: If all honest parties enter the BA, then the BA eventually terminates. *Consistency*: Upon termination the outputs of all honest players are equal, i.e. $w_i = w$ for some $w \in \{0,1\}$. *Validity*: If all honest parties have input $v_i = w$, then the output is w.

Cryptographic Assumptions & Instantiations of Tools. All the above tools can be instantiated in the standard (random oracle devoid) model using known results from [23, 16, 14, 13, 3, 25, 22, 7]. For details see [20].

3   Certificates

In order to achieve robustness we require every party to prove (in zero-knowledge) the correctness of essentially every value she provides during the protocol execution. To implement this process efficiently we introduce *certificates*, which are used for certifying the truth of claims. Any party can verify the correctness of a certificate locally, without any interaction. Moreover, a certificate should provide no other information than the truth of the claim. Finally, a party can convince any other party about the truth of the corresponding claim by sending the certificate. More formally, we say that a bit-string α *is a certificate for claim* m if there exits a publicly known, efficiently computable verification procedure V, such that the following conditions are satisfied, except with negligible probability: if $V(\alpha, m) = 1$ then claim m is true (soundness), and α gives no other information than the truth of the claim m (zero-knowledge). Moreover we require completeness, i.e. the ability of generating certificates for true claims needed in our protocols, like for example:

(i) «P_i knows the plaintext of X_i»   (iii) «the plaintext of X_i is in the set $\{0, 1, 2\}$»
(ii) «X_i is the unique input of P_i»   (iv) «at least $n - t$ parties have received X_i»

If X is some value, and α is a certificate for some claim m about X (e.g., claim (iii) above), then we say that X *is a value certified (by α) for claim m.*

We often require also that the certificates for correctness/validity of some data X imply also *uniqueness* of the data, i.e. that it is not possible to obtain two valid certificates for two different values for the same claim. This can be achieved by assigning unique identifiers to every gate, every wire and every step in the protocols, and requiring that the identifiers are parts of the claims, e.g. «X_i is input of P_i for wire id», and that parties participate in construction of at most one certificate for a particular claim. Occasionally, to clarify the issues, we explicitly specify the identifiers, but for simplicity the use of identifiers is usually implict.

Constructing certificates. Certificates can be implemented in a simple way using any signature scheme (Sign, Ver): a certificate α for claim m is just a set of at least $n - t$ correct signatures: $\alpha := \{\sigma_i\}_{i \in I}$, where $|I| \geq n - t$ and each σ_i is a signature of party P_i on message m. To create *short* certificates we employ a *threshold* signature scheme $(\mathscr{S}, \mathscr{V})$ with a threshold $t_S = n - t$ (cf. Sect. 2.1). To construct a certificate α valid for «*some claim*» a party collects t_S correct *signature shares* $\sigma_j = \mathscr{S}_j(\text{«some claim»})$ from different parties, and combines them to a signature $\alpha = \mathscr{S}(\text{«some claim»}, \{\sigma_j\}_{j \in J})$, where $|J| \geq t_S$. Any party knowing the corresponding public verification key v can verify α using the algorithm $\mathscr{V}$. Depending on the context, we use different methods for creating certificates:

bilateral proofs: if P_i needs to certify knowledge of some value, or validity of some NP-statement (cf. examples (i) and (iii), respectively), we will use 2-party zero-knowledge proofs: P_i bilaterally proves a claim m in zero-knowledge to every P_j, who then, upon successful completion of the proof, sends to P_i a signature share $\sigma_j = \mathscr{S}_j(m)$ with a proof of correctness of the share, and P_i combines the correct shares to get a certificate α_i. We say then that "P_i *constructs certificate α_i for* «*some claim*» *by bilateral, zero-knowledge proofs*", denoted as $\alpha_i := \text{certify}_{\text{zkp}}(\text{«some claim»})$.

protocol-driven: For other claims, like (iv) and (ii), P_i also constructs a certificate α_i from a set of $n - t$ signature shares σ_j, but this time P_j sends σ_j not in response to a bilateral proof, but based on the current context of execution, as required by the protocol. In this case we just say "P_i *constructs certificate α_i for* «*some claim*»" and write $\alpha_i := \text{certify}(\text{«some claim»}).$ [1]

---

[1] Note that the signed messages can be different from the actual claim being certified, e.g, each P_j could provide a signature share for the message «*I have seen X_i*», and a complete signature on such a message can be interpreted as a certificate for (iv).

An adversary corrupting up to t players can never obtain sufficiently many signature shares: In the case of bilateral proofs an honest party never signs an incorrect claim, hence the adversary can collect at most $t < n - t$ shares. In the protocol-driven case, the soundness depends on the actual claim being certified, but it will be clear from the context. Note that the threshold $n - t$ implies that $n - 2t$ honest parties must sign to create a certificate, which ensures uniqueness.

4   The New Protocol

Our protocol needs that the encryption key of a public-key encryption scheme is publicly known, while the corresponding decryption key is shared among all the players. Given such a setup the evaluation of a circuit proceeds as follows. First the parties provide their inputs as ciphertexts of the encryption scheme. Then they cooperate to evaluate the circuit gate-by-gate: given encryptions of inputs of a gate, parties compute an encryption of the corresponding output of the gate, while maintaining privacy of the intermediate values. Finally, after an encryption of the output gate is computed, parties decrypt this encryption to learn the output. The robustness against corrupted parties is achieved with help of *certificates*, which are used to certify the correct execution of the protocol.

Intuitively, the efficiency gain stems from a combination of a ballanced distribution of work, with the so-called *circuit-randomization technique* due to Beaver [1]. In this technique the multiplication of two encrypted values is performed using a *pre-generated* random triple, which in our case consists of three ciphertexts (U, V, W) containing secret random plaintexts $u, v, w \in \mathbb{M}$, satisfying $u \cdot v = w$. Due to homomorphic encryption, given such a triple and two ciphertexts A, B containing plaintexts a, b, we compute a ciphertext C of $c = a \cdot b$ by *publicly decrypting* $A + U$ and $B + V$, and by using the following identity

$$a \cdot b = (a + u) \cdot (b + v) - (a + u) \cdot v - u \cdot (b + v) + w . \tag{1}$$

Main Protocol — A High-level Overview. The protocol proceeds in four stages, a *precomputation stage*, an *input stage*, an *evaluation stage*, and a *termination stage*. We briefly summarize the goal of each stage:

- *Precomputation stage*: Players generate random triples.
- *Input stage*: Each player provides an encryption of his input to every other player, and the players agree on a set of *input providers*.
- *Evaluation stage*: Players evaluate the circuit gate-by-gate, by executing concurrently subprotocols for every gate of the circuit.
- *Termination stage*: Executed concurrently to the evaluation stage, this stage ensures that every player eventually receives the output(s) and terminates.

Strictly speaking, the presented protocol is limited to the evaluation of *deterministic* circuits, but can be easily extended also to *randomized* circuits [20].

The Circuit. For the clarity of presentation we assume that every party provides exactly one input, and that the outputs are public, but this is without loss of generality [19]. The function to be computed is given as a circuit Circ over the plaintext space $\mathbb{M}$ of the homomorphic encryption scheme in use. The circuit is a set of labeled gates, where a label G uniquely identifies the gate. The full description of a gate is a tuple $(G, \ldots)$, where the parameters after the label depend on the type and the position of the gate. We denote by $\mathcal{G}$ the set of all gate labels of Circ, and we use $v : \mathcal{G} \rightarrow \mathbb{M} \cup \{\bot\}$ to refer to the values of gates, i.e., $v(G)$ denotes the value of gate G. Each gate has one of the following types:

input gate: (G), consisting only of its label $G = (P_i, \text{input})$, where $v(G)$ is equal to x_i, the input value provided by player P_i.
linear gate: $(G, \text{linear}, a_0, a_1, G_1, \ldots, a_l, G_l)$, where $l \geq 0$, $a_0, \ldots, a_l \in \mathbb{M}$ are constants, and $v(G) = a_0 + \sum_{j=1}^{l} a_j \cdot v(G_j)$.
multiplication gate: $(G, \text{mul}, G_1, G_2)$, where $v(G) = v(G_1) \cdot v(G_2)$.
output gate: (G, output, G_1), where $v(G) = v(G_1)$ is an output value of Circ.

Dictionary. Throughout the computation each party P_i maintains a *dictionary* $\Gamma_i : \mathcal{G} \rightarrow \mathbb{C} \cup \{\bot\}$, containing P_i's view on the intermediate values (encryptions) in the circuit. Initially $\Gamma_i(G) = \bot$ for all labels from $\mathcal{G}$. If $\Gamma_i(G) = X \neq \bot$, then from P_i's point of view evaluation of gate G was completed, and X is a ciphertext encrypting the value $v(G)$. We say then that P_i *has accepted X for G*. Honest parties will agree on accepted ciphertexts, allowing us to define a common map Γ. Furthermore, for all input gates $\Gamma(G) = X$ will be an encryption of the input that the party supplying input to that gate intended to deliver, except for at most t parties, where X might be an encryption of a default value. This is allowed by the security model.

Random Triples. Each party P_i maintains also a mapping Δ_i assigning to each multiplication gate a random triple generated during the precomputation stage, $\Delta_i : \mathcal{G} \rightarrow \mathbb{C} \times \mathbb{C} \times \mathbb{C} \cup \{\bot\}$. Initially $\Delta_i(G) = \bot$, for all gates G and all $j \in [n]$. If $\Delta_i(G) = (U, V, W) \neq \bot$, then P_i will use this triple for evaluating gate G. The honest parties P_i and P_j will have $\Delta_i = \Delta_j$.

5   Subprotocols Used by the Main Protocol

Below we present subprotocols of the main protocol. First we present a protocol SELECT, which is a basic subprotocol used both in the precomputation and input stages. Then we describe the subprotocols for the main stages.

Selecting Values. Protocol SELECT is used for selecting values provided by the players during the computation. It is parametrized by a condition φ (like e.g. «X_i is P_i's valid input»), which has to be satisfied for each input to the protocol, and certified by an appropriate certificate. We require that φ *implies uniqueness*, i.e., that every party can obtain a corresponding certificate valid for φ for at most one input value used in any execution of SELECT.

The protocol proceeds as follows (cf. Fig. 1): First P_i distributes its input (X_i, α_i) to all parties, and then constructs and distributes a *certificate of distribution* β_i, which

---

Protocol SELECT(φ), *code for* P_i: given input X_i with a certificate α_i valid for condition $\varphi(i)$ initialize sets $A_i, \mathcal{A}_i, C_i$ as empty, then execute the following rules concurrently:

DISTRIBUTION:

1. send (X_i, α_i) to all parties.
2. construct and send to all parties $\beta_i :=$ certify(«*we hold P_i's input X_i*»)

GRANT CERTIFICATE OF DISTRIBUTION: Upon first (X_j, α_j) from P_j with α_j valid for $\varphi(j)$: add j to A_i, add (X_j, α_j) to $\mathcal{A}_i$, and send $\sigma_i :=$ Sign$_i$(«*we hold P_j's input X_j*») to P_j.

ECHO CERTIFICATE OF DISTRIBUTION: Upon (X_j, β_j) with β_j valid for «*we hold P_j's input X_j*» and $j \notin C_i$: add j to C_i and send (X_j, β_j) to all parties.

SELECTION: If $|C_i| \geq n-t$, stop executing all above rules and proceed as follows:

1. send $(A_i, \mathcal{A}_i)$ to all parties.
2. collect a set $\{(A_j, \mathcal{A}_j)\}_{j \in J}$ of $(n - t)$ well-formed $(A_j, \mathcal{A}_j)$; let $B_i := \bigcup_{j \in J} A_j$ and $\mathcal{B}_i := \bigcup_{j \in J} \mathcal{A}_j$
3. enter n Byzantine Agreements (BAs) with inputs $v_1 \ldots v_n$, where $v_j = 1$ iff $j \in B_i$.
4. let $w_1, \ldots, w_n$ be the outputs of the BAs; let $\mathcal{W} := \{j \in [n] |\ w_j = 1\}$.
5. $\forall j \in B_i \cap \mathcal{W}$: send $(X_j, \alpha_j) \in \mathcal{B}_i$ to all parties.
6. collect and output $(\mathcal{W}, \{(X_j, \alpha_j)\}_{j \in \mathcal{W}})$.

Fig. 1. Protocol SELECT(φ)

proves that P_i has distributed (X_i, α_i) to at least $n - t$ parties. When a party collects $n - t$ certificates of distribution, she knows that at least $n - t$ parties have their certified inputs distributed to at least $n - t$ parties. So, at least $n - t$ parties had their certified inputs distributed to at least $(n-t)-t \geq t+1$ *honest* parties. Hence, if all honest parties echo the certified inputs they saw and collect $n - t$ echoes, then all honest parties will end up holding the certified input of the $n - t$ parties, which had their certified inputs distributed to at least $t + 1$ honest parties. These $n - t$ parties will eventually be the input providers. To determine who they are, n Byzantine Agreements are run.

Precomputation Stage. The goal of this stage is the generation of certified random triples. The corresponding protocol GEN-TRIPLES uses two subprotocols: protocol SELECT presented above, and protocol ONE-TRIPLE for generating a single random triple in a computation lead by one of the parties.[2] Given these two sub-protocols, we proceed as follows (Fig.5): first every party generates its own random triple using ONE-TRIPLE, and then uses this triple as input to SELECT, in which parties agree on at least $(n - t)$ triples.

In protocol ONE-TRIPLE we need to generate certified, encrypted random values unknown to any party, so first we present a sub-protocol RANDOM (Fig. 3), which achieves exactly that. Given (U, α) output by RANDOM, king P_k can extend it to a random triple using ONE-TRIPLE, see Fig. 4. Note that when computing a certificate β_i for the claim «*P_i knows v_i in V_i, and W_i is a randomization of $v_i \star U$*» the variables P_i, V_i, W_i, and U

---

[2] In the ONE-TRIPLE protocol one party, say P_k, plays the role of a leader (called *king*) who with help of other players (called *slaves*), generates P_k's own random triple $(U^{(k)}, V^{(k)}, W^{(k)})$ together with a certificate $\sigma^{(k)}$ certifying the triple's correctness.

To generate c_M random triples and reach agreement on them, parties proceed as follows:

1. Every party P_k, $k \in [n]$ starts as a king for $\ell = \lceil c_M/(n-t) \rceil$ instances of ONE-TRIPLE(id, k) protocol to generate ℓ certified random triples $(U^{(k,s)}, V^{(k,s)}, W^{(k,s)}; \sigma^{(k,s)})$, $s = 1, \ldots, \ell$ (all parties play roles of slaves to help the king in P_k's instances)

2. All parties start SELECT, where party P_i uses as its input the triples $(U^{(i,s)}, V^{(i,s)}, W^{(i,s)}; \sigma^{(i,s)})$ generated in the previous step. When SELECT terminates, parties have agreed on a set of at least $\ell(n-t) \geq c_M$ valid triples $\{(U^{(j)}, V^{(j)}, W^{(j)})\}_{j \in J}$.

3. Every party P_i initializes its mapping Δ_i, using the triples from the previous step in some pre-agreed order.

Fig. 2. Protocol GEN-TRIPLES for generating random triples

To generate for king P_k a certified random ciphertext (U, α), with α valid for «U: *1st part of triple* id(k)» parties proceed as follows:

GENERATION: code for every P_i:

1. pick random $u_i \in \mathbb{M}$ and compute $U_i := \mathcal{E}(u_i)$
2. construct $\beta_i = \mathsf{certify}_{\mathsf{zkp}}($«$P_i$ knows u_i in U_i»$)$
3. compute $\sigma_i := \mathsf{Sign}_i($«$U_i$: *component of 1st part of triple* id(k)»$)$
4. send (U_i, β_i, σ_i) to P_k:

CONSTRUCTION: code for P_k:

1. collect a set $S_{\mathsf{id}(k)} := \{(U_i, \beta_i, \sigma_i)\}_{i \in I_{\mathsf{id}(k)}}$, $|I_{\mathsf{id}(k)}| \geq t+1$, with each β_i valid for «P_i knows u_i in U_i», and each σ_i valid for «U_i : *component of 1st part of triple* id(k)».
2. send $S_{\mathsf{id}(k)}$ to all parties; each P_i computes $U := \bigoplus_{i \subset I_{\mathsf{id}(k)}} U_i$, and helps to construct α in the next step.
3. construct $\alpha := \mathsf{certify}($«$U$: *1st part of triple* id(k)»$)$.
4. output (U, α).

Fig. 3. Protocol RANDOM(id, k) for generating a certified random value for king P_k

are replaced by the actual values they stand for, but v_i stays as a literal, since it is just a name for the plaintext from V_i.

On the use of Byzantine Agreement. The protocol GEN-TRIPLES uses n BAs (as it invokes SELECT) and generates $n - t$ random triples. To implement multiplication of encrypted values via circuit randomization (cf. Fig. 6), we need one random triple per multiplication gate. A straightforward solution would be to use $\ell = \lceil c_M/(n-t) \rceil$ runs of GEN-TRIPLES, but this would lead to $\mathcal{O}(c_M)$ invocations of BA. To avoid this, we run ℓ invocations of ONE-TRIPLE in parallel, using only one invocation of SELECT. In particular, in the second step of GEN-TRIPLE each P_i uses all ℓ triples as its input to SELECT. Since SELECT returns a set of at least $(n-t)$ inputs, we obtain an agreement on c_M random triples with only n BAs, which is independent of the circuit size.

To generate for P_k a certified random triple $(U, V, W; \beta)$, with β valid for «(U, V, W): *correct triple* id(k)» parties proceed as follows:

REQUEST: code for P_k:

1. run RANDOM(id, k) to generate (U, α), with α valid for «U: *1st part of triple* id(k)», and send (U, α) to all parties.

REPLY: code for every P_i:

1. wait for (U, α) from P_k
2. compute $V_i := \mathscr{E}(v_i)$ and $W_i = \mathscr{R}(v_i \star U)$ for a random $v_i \in \mathbb{M}$
3. construct $\beta_i := \text{certify}_\text{zkp}$(«$P_i$ *knows* v_i *in* V_i, *and* W_i *is a randomization of* $v_i \star U$»)
4. compute $\sigma_i := \text{Sign}_i$(«$(V_i, W_i)$: *part of triple* id(k)»)
5. send $(V_i, W_i; \beta_i, \sigma_i)$ to P_k

CONSTRUCTION: code for P_k:

1. collect $T_{\text{id}(k)} := \{(V_i, W_i; \beta_i, \sigma_i)\}_{i \in I_{\text{id}(k)}}$, with each σ_i valid for «(V_i, W_i) : *part of triple* id(k)», and each β_i valid for «P_i *knows* v_i *in* V_i, *and* W_i *is a randomization of* $v_i \star U$» $|I_{\text{id}(k)}| \geq t + 1$.
2. send $T_{\text{id}(k)}$ to all parties; each P_i computes $V := \bigoplus_{i \in I_{\text{id}(k)}} V_i$, $W := \bigoplus_{i \in I_{\text{id}(k)}} W_i$, and helps to construct β in the next step.
3. construct $\beta := \text{certify}$(«(U, V, W): *correct triple* id(k)»).
4. output $(U, V, W; \beta)$.

Fig. 4. Protocol ONE-TRIPLE(id, k) for generating a random triple for king P_k

Input stage code for P_i: given an input $x_i \in \mathbb{M}$ do the following:

1. compute $X_i := \mathscr{E}(x_i)$ and construct $\alpha_i := \text{certify}_\text{zkp}$(«$X_i$ *is* P_i's *valid input*») (every party P_j helps to construct at most one α_i, for each $P_i \in \mathcal{P}$).
2. enter execution of SELECT protocol with input (X_i, α_i).
3. output value(s) returned by SELECT.

Fig. 5. The input stage code for P_i holding input $x_i \in \mathbb{M}$

Input Stage. The protocol is presented in Fig. 5. When providing (encrypted) inputs, the parties are required to prove plaintext knowledge for their encryptions, to ensure independence of the inputs. To cope with the inherent problems of the asynchronous setting we use protocol SELECT to agree on inputs from at least $(n - t)$ *input providers*, whose private inputs will be used in the actual computation. For the remaining inputs the default values will be used.

Computing Linear Gates. Due to the homomorphic property of encryption, linear gates are computed locally, without interaction: after P_i accepts encryptions of inputs to a gate $(G, \text{linear}, a_0, G_1, a_1, \ldots, G_l, a_l)$, i.e. when $\Gamma_i(G_u) \neq \perp$, for $u = 1 \ldots l$, then P_i

---

Party P_i evaluating a multiplication gate $(G, \mathsf{mul}, G_1, G_2)$:

1. wait until $A := \Gamma_i(G_1) \neq \bot$, $B := \Gamma_i(G_2) \neq \bot$, and $(U, V, W) := \Delta_i(G) \neq \bot$
2. compute $X := A \oplus U$ and $Y := B \oplus V$
3. compute decryption shares and corresponding validity proofs: $x_i := \mathscr{D}_i(X)$, $\beta_i :=$ $\mathsf{certify}_{\mathsf{zkp}}(\text{«}x_i \text{ is valid»})$ $y_i := \mathscr{D}_i(Y)$, $\gamma_i := \mathsf{certify}_{\mathsf{zkp}}(\text{«}y_i \text{ is valid»})$; send (x_i, β_i) and (y_i, γ_i) to all parties
4. collect sets $\mathcal{X} := \{(x_j, \beta_j)\}$ and $\mathcal{Y} := \{(y_j, \gamma_j)\}$, each containing t_D correct decryption shares, with corresponding validity proofs.
5. compute plaintexts $x := \mathscr{D}(X, \mathcal{X})$ and $y := \mathscr{D}(Y, \mathcal{Y})$.
6. compute $Z := \mathscr{E}(x \cdot y, r_0)$ for a public constant r_0, and set $\Gamma_i(G) := Z \ominus (x \star V) \ominus (y \star U) \oplus W$

Fig. 6. Code for P_i evaluating a multiplication gate

Party P_i evaluating an output gate $(G, \mathsf{output}, G_1)$:

1. wait until $\Gamma_i(G_1) = C \neq \bot$
2. compute a decryption share $c_i := \mathscr{D}_i(C)$ & a certificate $\delta_i := \mathsf{certify}_{\mathsf{zkp}}(\text{«}c_i \text{ is valid»})$; send (c_i, δ_i) to every P_j
3. collect a set $T = \{(c_j, \delta_j)\}$ of t_D decryption shares for C, with corresponding validity certificates δ_j
4. compute $c := \mathscr{D}(C, T)$
5. compute and send to all parties a signature share $\sigma_{i,G} = \mathscr{S}_i(\text{«}The \ value \ of \ G \ is \ c\text{»})$, together with a certificate of its correctness, $\xi_{i,G} := \mathsf{certify}_{\mathsf{zkp}}(\text{«}\sigma_{i,G} \text{ is correct»})$
6. collect a set $\{\sigma_{i,G}, \xi_{i,G}\}_{i \in I}$ of t_S certified signature shares and compute $\zeta_G = \mathscr{S}(x, \{\sigma_i\}_{i \in I})$ valid for «The value of G is c»
7. mark G as decrypted

Fig. 7. Code for player P_i evaluating an output gate

---

computes $\Gamma_i(G) := A_0 \oplus \left(\bigoplus_{u=1}^l (a_j \star \Gamma_i(G_u)) \right)$, where A_0 is a "dummy" encryption of a_0, computed using fixed, public random bits.

Computing Multiplication Gates. The multiplication protocol (Fig. 6) is based on a trick by Beaver [1]. Essentially, this trick reduces the problem of multiplication to two decryptions and a few linear operations (cf. eq. (1)).

Output Stage. When P_i completes the computation of a gate $(G, \mathsf{output}, G_1)$ (i.e. when $\Gamma_i(G) = C \neq \bot$), but the gate has not been decrypted yet, then P_i sends a decryption share c_i of C to all parties, along with a certificate for the correctness of the share. Every P_j collects sufficiently many certified decryption shares, and uses them to decrypt the output. Subsequently the parties construct a certificate ζ_G, which certifying that the decrypted output value is correct. With such a certificate any party P_i can convince any other party about the correctness of the output.

Termination Stage. Essentially, every P_i waits until he receives or computes the decrypted output value with a correctness certificate, and echoes this certified output to all parties before terminating (see [20] for details).

Summary. The main result of this paper is summarized in the theorem below. The analysis leading to this theorem is presented in the full version [20].

Theorem 1. *Assuming the cryptographic primitives from Sect. 2.1, there exists a protocol allowing n parties connected by an asynchronous network to securely evaluate any circuit in the presence of a poly-time adversary actively corrupting up to $t < n/3$ parties. The bit complexity of the protocol is $\mathcal{O}((c_I + c_M + c_O)n^2\kappa)$, where c_I, c_M, c_O denote the number of input, multiplication, and output gates, respectively, and κ is a security parameter.*

References

1. Beaver, D.: Efficient multiparty protocols using circuit randomization. In: Feigenbaum, J. (ed.) CRYPTO 1991. LNCS, vol. 576, pp. 420–432. Springer, Heidelberg (1992)
2. Bellare, M., Rogaway, P.: Random oracles are practical: A paradigm for designing efficient protocols. In: Proc. ACM CCS, pp. 62–73 (1993)
3. Bellare, M., Rogaway, P.: The exact security of digital signatures — How to sign with RSA and Rabin. In: Maurer, U.M. (ed.) EUROCRYPT 1996. LNCS, vol. 1070, pp. 399–416. Springer, Heidelberg (1996)
4. Ben-Or, M., Canetti, R., Goldreich, O.: Asynchronous secure computation. In: STOC, pp. 52–61 (1993)
5. Ben-Or, M., Goldwasser, S., Wigderson, A.: Completeness theorems for non-cryptographic fault-tolerant distributed computation. In: Proc. 20th STOC, pp. 1–10 (1988)
6. Ben-Or, M., Kelmer, B., Rabin, T.: Asynchronous secure computations with optimal resilience. In: Proc. 13th PODC, pp. 183–192 (1994)
7. Cachin, C., Kursawe, K., Shoup, V.: Random oracles in Constantinopole: Practical asynchronous Byzantine agreement using cryptography. In: Proc. 19th PODC, pp. 123–132 (2000)
8. Canetti, R.: Studies in Secure Multiparty Computation and Applications. PhD thesis, Weizmann Institute of Science, Rehovot 76100, Israel (June 1995)
9. Canetti, R.: Security and composition of multiparty cryptographic protocols. JoC 13(1), 143–202 (2000)
10. Canetti, R., Goldreich, O., Halevi, S.: The random oracle methodology, revisited. In: Proc. 30th STOC, pp. 209–218 (1998)
11. Canetti, R., Rabin, T.: Fast asynchronous byzantine agreement with optimal resilience. In: Proc. 25th STOC, pp. 42–51 (1993)
12. Chaum, D., Crépeau, C., Damgård, I.: Multiparty unconditionally secure protocols (extended abstract). In: Proc. 20th STOC, pp. 11–19 (1988)
13. Cramer, R., Damgård, I., Nielsen, J.B.: Multiparty computation from threshold homomorphic encryption. In: Pfitzmann, B. (ed.) EUROCRYPT 2001. LNCS, vol. 2045, pp. 280–300. Springer, Heidelberg (2001)
14. Damgård, I., Jurik, M.: A generalisation, a simplification and some applications of Paillier's probabilistic public-key system. In: Kim, K.-c. (ed.) PKC 2001. LNCS, vol. 1992, pp. 110–136. Springer, Heidelberg (2001)

15. Fiat, A., Shamir, A.: How to prove yourself: Practical solutions to identification and signature problems. In: Odlyzko, A.M. (ed.) CRYPTO 1986. LNCS, vol. 263, pp. 186–194. Springer, Heidelberg (1987)
16. Fouque, P.-A., Poupard, G., Stern, J.: Sharing decryption in the context of voting or lotteries. In: Proc. Financial Cryptography 2000 (2000)
17. Franklin, M., Haber, S.: Joint encryption and message-efficient secure computation. JoC 9(4), 217–232 (1996); Preliminary version in Proc. CRYPTO 1993
18. Goldreich, O., Micali, S., Wigderson, A.: How to play any mental game — a completeness theorem for protocols with honest majority. In: Proc. 19th STOC, pp. 218–229 (1987)
19. Hirt, M., Nielsen, J.B., Przydatek, B.: Cryptographic asynchronous multi-party computation with optimal resilience (extended abstract). In: Cramer, R.J.F. (ed.) EUROCRYPT 2005. LNCS, vol. 3494, pp. 322–340. Springer, Heidelberg (2005)
20. Hirt, M., Nielsen, J.B., Przydatek, B.: Asynchronous multi-party computation with quadratic communication (2008), eprint.iacr.org
21. Maurer, U., Renner, R., Holenstein, C.: Indifferentiability, impossibility results on reductions, and applications to the random oracle methodology. In: Naor, M. (ed.) TCC 2004. LNCS, vol. 2951, pp. 21–39. Springer, Heidelberg (2004)
22. Nielsen, J.B.: A threshold pseudorandom function construction and its applications. In: Yung, M. (ed.) CRYPTO 2002. LNCS, vol. 2442, pp. 401–416. Springer, Heidelberg (2002)
23. Paillier, P.: Public-key cryptosystems based on composite degree residuosity classes. In: Stern, J. (ed.) EUROCRYPT 1999. LNCS, vol. 1592, pp. 223–238. Springer, Heidelberg (1999)
24. Prabhu, B., Srinathan, K., Rangan, C.P.: Asynchronous unconditionally secure computation: An efficiency improvement. In: Menezes, A., Sarkar, P. (eds.) INDOCRYPT 2002. LNCS, vol. 2551, pp. 93–107. Springer, Heidelberg (2002)
25. Shoup, V.: Practical threshold signatures. In: Prenecl, B. (ed.) EUROCRYPT 2000. LNCS, vol. 1807, pp. 207–220. Springer, Heidelberg (2000)
26. Yao, A.C.: Protocols for secure computations. In: Proc. 23rd IEEE FOCS, pp. 160–164 (1982)

Improved Garbled Circuit: Free XOR Gates and Applications

Vladimir Kolesnikov[1] and Thomas Schneider[2,*]

[1] Bell Laboratories, 600 Mountain Ave. Murray Hill, NJ 07974, USA
kolesnikov@research.bell-labs.com
[2] Horst Görtz Institute for IT-Security, Ruhr-University Bochum, Germany
thomas.schneider@trust.rub.de

Abstract. We present a new garbled circuit construction for two-party secure function evaluation (SFE). In our one-round protocol, XOR gates are evaluated "for free", which results in the corresponding improvement over the best garbled circuit implementations (e.g. Fairplay [19]).

We build permutation networks [26] and Universal Circuits (UC) [25] almost exclusively of XOR gates; this results in a factor of up to 4 improvement (in both computation and communication) of their SFE. We also improve integer addition and equality testing by factor of up to 2.

We rely on the Random Oracle (RO) assumption. Our constructions are proven secure in the semi-honest model.

1 Introduction

Two-party general secure function evaluation (SFE) allows two parties to evaluate any function on their respective inputs x and y, while maintaining privacy of both x and y. SFE is (justifiably) a subject of immense amount of research, e.g. [27,28,17]. Efficient SFE algorithms enable a variety of electronic transactions, previously impossible due to mutual mistrust of participants. Examples include auctions [21,6,8,4], contract signing [7], distributed database mining [12,16], etc. As computation and communication resources have increased, SFE has become truly practical for common use. Fairplay [19] is a full-fledged implementation of generic two-party SFE with malicious players. It clearly demonstrates feasibility and efficiency of SFE of many useful functions, represented as circuits of up to $\approx 10^6$ gates. Today, generic SFE is a relatively mature technology, and even improvements by a small factor are non-trivial and are most welcome.

One area of SFE that especially benefits from our work is the SFE of *private functions* (PF-SFE). It is an extension of SFE where the evaluated function is known only by one party and needs to be kept secret (i.e. everything besides the size, the number of inputs and the number of outputs is hidden from the other party). Examples of real-life private functions include airport no-fly check function, credit evaluation function, background- and medical history checking

---

* The work was done while the author was visiting Bell Laboratories.

L. Aceto et al. (Eds.): ICALP 2008, Part II, LNCS 5126, pp. 486–498, 2008.

function, etc. Full or even partial revelation of these functions opens vulnerabilities in the corresponding process, exploitable by dishonest participants (e.g. credit applicants), and should be prevented. It is known that the problem of PF-SFE can be reduced to the "regular" SFE [24,23]. This is done by evaluating a *Universal Circuit* (UC) [25,15] instead of a circuit defining the evaluated function. UC can be thought of as a "program execution circuit", capable of simulating any circuit C of certain size, given the description of C as input. Therefore, disclosing the UC does not reveal anything about C, except its size. At the same time, the SFE computes output correctly and C remains private, since the player holding C simply treats description of C as additional (private) input to SFE. This reduction is the most common (and often the most efficient) way of securely evaluating private functions [24,23,15].

1.1 Related Work

General SFE has been a subject of immense amount of research, started by Yao [27,28], which resulted in significant advances in the field [9,21,17]. Fairplay [19] is a full practical implementation of general SFE based on garbled circuits.

Information-theoretic setting of SFE has also received a large amount of attention, e.g. [13,11]. However, due to the restrictions of the model, the resulting protocols are less efficient than those in the generous RO model. We apply some of the ideas of this setting, such as the efficient XOR gate construction (e.g. Construction 4 of [14]), in the RO setting, to obtain more efficient protocols.

1.2 Our Contributions

We present a new garbled circuit construction for two-party secure function evaluation (SFE) in the semi-honest model. In our one-round protocol, XOR gates are evaluated "for free" (that is, without the use of the associated garbled tables and the corresponding hashing or symmetric key operations). Our construction is as efficient as the best garbled circuit implementations (e.g. Fairplay [19]) in handling other gates.

We next show that free XOR gates bring significant benefit to many SFE settings. We show how to build permutation networks [26] and UC [25,15] almost exclusively of XOR gates; this results in a factor of up to 4 improvement (in both computation and communication) of their SFE. As discussed above, SFE of UC is the most efficient way of evaluating private functions; thus our work improves performance of PF-SFE almost fourfold. We note that other useful functions can benefit from free XOR gates. We show how to obtain a factor of up to 2 improvement of SFE of integer addition and equality testing.

We rely on the RO assumption; we discuss its (conservative) use in Sect. 3.1.

2 Setting and Preliminaries

We consider *acyclic* boolean circuits with k gates and arbitrary fan-out. That is, the (single) output of each gate can be used as input to an arbitrary number of gates. We assume that the gates $G_1, \ldots, G_k$ of the circuit are ordered topologically. This order (which is not necessarily unique) ensures that the i-th gate G_i has no inputs that are outputs of a successive gate G_j, where $j > i$. A topological order can always be obtained on acyclic circuits, with $O(k)$ computation.

We concentrate on the *semi-honest* model, where players follow the protocol, but try to learn information from the execution transcripts.

We use the following standard notation: $\in_R$ denotes uniform random sampling, $||$ denotes concatenation of bit strings. $\langle a, b \rangle$ is a vector with two components a and b, and its bit string representation is $a||b$. $W_c = g(W_a, W_b)$ denotes a 2-input gate G that computes function $g : \{0, 1\}^2 \to \{0, 1\}$ with input wires W_a and W_b and output wire W_c.

Let N be the security parameter. Let S be an infinite set and let $X = \{X_s\}_{s \in S}$ and $Y = \{Y_s\}_{s \in S}$ be distribution ensembles. We say that X and Y are computationally indistinguishable, denoted $X \overset{c}{\equiv} Y$, if for every non-uniform polynomial-time distinguisher D and all sufficiently large $s \in S$, $|Pr[D(X_s) = 1] - Pr[D(Y_s) = 1]| < 1/p(|s|)$ for every polynomial p.

Random Oracle. RO model is a useful abstraction, introduced and justified by [3]. RO is simply a randomly chosen function $\{0, 1\}^* \mapsto \{0, 1\}^N$ – a large object which cannot be fully stored or traversed by polytime players. RO model gives oracle access to such function to all players. In practice, ROs are modeled by hash functions, such as SHA. Although it was shown [5] that a protocol secure in the RO model may not be secure once RO is instantiated, "natural" RO protocols maintain their security in practice, and are widely used.

Oblivious Transfer (OT). The 1-out-of-2 OT is a two-party protocol. The *sender* P_1 has two secrets m_0, m_1, and the *receiver* P_2 has an selection bit $i \in \{0, 1\}$. At the end of the protocol, P_2 learns m_i, but nothing about m_{1-i}, and P_1 learns nothing about i. One-round OT is a widely studied primitive in the standard model [2,1], with improved implementations in the RO model [20,3].

Yao's Garbled Circuit (GC). The GC approach, excellently presented in [17], is the most efficient method of SFE of boolean circuits. Here we summarize its idea. Player P_1 first *garbles* circuit C: for each wire W_i, he randomly chooses two secrets, w_i^0 and w_i^1, where w_i^j is a *garbled value*, or *garbling*, of the W_i's value j. (Note: w_i^j does not reveal j.) Further, for each gate G_i, P_1 creates and sends to P_2 a *garbled table* T_i, with the following property: given a set of garblings of G_i's inputs, T_i allows to recover the garbling of the corresponding G_i's output, and nothing else. Then garblings of players' inputs are (obliviously) transferred to P_2. Now, P_2 can obtain the garbled output simply by evaluating the garbled circuit gate by gate, using the tables T_i. We call W_i's garbling w_i^j *active* if W_i assumes the value j when C is evaluated on the given input. Observe that for each wire, P_2 can obtain only its active garbling. The output wires of the circuit are not garbled (or their garblings are published), thus P_2 learns (only) the output of the circuit, and no internal wire values. P_1 learns the output from (semi-honest) P_2. (This step is trivial in the semi-honest model, and is usually not considered in the analysis.) Correctness of GC follows from method of construction of tables T_i. Neither party learns any additional information from the protocol execution.

3   Our Protocol

Overview. In our construction, we combine GC with the simple information-theoretic SFE implementation of XOR-gates (e.g., Construction 4 of [14]). In all GC implementations, XOR gates cost as much as AND or OR gates (i.e. in computation and communication required for creation, transfer and evaluation of the garbled tables). The XOR gates of Kolesnikov [14] are free of these costs. However, his construction imposes a restrictive global relationship on the wire secrets, which prevents its use in previous GC schemes. In this work, we show how to overcome this restriction.

First, we show an SFE implementation of the XOR gate G, derived from one of [14]. Let G have two input wires W_a, W_b and output wire W_c. Garble the wire values as follows. Randomly choose $w_a^0, w_b^0, R \in_R \{0,1\}^N$. Set $w_c^0 = w_a^0 \oplus w_b^0$, and $\forall i \in \{a, b, c\} : w_i^1 = w_i^0 \oplus R$. It is easy to see that the garbled gate output is simply obtained by XORing garbled gate inputs:

$$w_c^0 = w_a^0 \oplus w_b^0 = (w_a^0 \oplus R) \oplus (w_b^0 \oplus R) = w_a^1 \oplus w_b^1$$
$$w_c^1 = w_c^0 \oplus R = w_a^0 \oplus (w_b^0 \oplus R) = w_a^0 \oplus w_b^1 = (w_a^0 \oplus R) \oplus w_b^0 = w_a^1 \oplus w_b^0.$$ Further,
garblings w_i^j do not reveal the wire values they correspond to.

We can now pinpoint the restriction that the above XOR construction imposes on the garbled values – the garblings of the two values of each wire in the circuit must differ by the same value, i.e. $\forall i : w_i^1 = w_i^0 \oplus R$, for some global R. In contrast, in previous GC constructions, *all* garblings w_i^j were chosen independently at random, and proofs of security relied on that property.

Our main observation is that it is not necessary to select all garblings independently. In our construction (Sect. 3.1), we choose a random R once, and garble wire values, so that $\forall i : w_i^1 = w_i^0 \oplus R$.

3.1   Our Garbled Circuit Construction

Let C be a circuit. We first note that NOT gates can be implemented "for free" by simply eliminating them and inverting the correspondence of the wires' values and garblings. We thus do not further consider NOT gates.

We implement XOR gates as discussed above in Sect. 3. Further, we replace each XOR-gate with $n > 2$ inputs with $n - 1$ two-input XOR-gates.

We implement all other gates using standard garbled tables [19]. Namely, each gate with n inputs is assigned a table with 2^n randomly permuted entries. Each entry is an encrypted garbling of the output wire, and garblings of the input wires serve as keys to decrypt the "right" output value. For simplicity, we present our construction and proof for the case $n = 2$. The generalization to n-input gates ($n \geq 1$) is straightforward.

In Alg. 1 below, each garbling $w = \langle k, p \rangle$ consists of a key $k \in \{0,1\}^N$ and a permutation bit $p \in \{0,1\}$. The key is used for decryption of the table entries, and p is used to select the entry for decryption. The two garblings w_i^0, w_i^1 of each wire W_i are related as required by the XOR construction: for a chosen $R \in_R \{0,1\}^N$, $\forall i : w_i^1 = \langle k_i^1, p_i^1 \rangle = \langle k_i^0 \oplus R, p_i^0 \oplus 1 \rangle$, where $w_i^0 = \langle k_i^0, p_i^0 \rangle$. $H : \{0,1\}^* \mapsto \{0,1\}^{N+1}$ is a RO.

We now formalize the above intuition and present the GC construction (Alg. 1) and evaluation (Alg. 2). In SFE, Alg. 1 is run by P_1 and Alg. 2 is run by P_2.

Algorithm 1. *(Construction of a garbled circuit)*

1. *Randomly choose global key offset $R \in_R \{0,1\}^N$*
2. *For each input wire W_i of C*
   (a) *Randomly choose its garbled value $w_i^0 = \langle k_i^0, p_i^0 \rangle \in_R \{0,1\}^{N+1}$*
   (b) *Set the other garbled output value $w_i^1 = \langle k_i^1, p_i^1 \rangle = \langle k_i^0 \oplus R, p_i^0 \oplus 1 \rangle$*
3. *For each gate G_i of C in topological order*
   (a) *label $G(i)$ with its index: $label(G_i) = i$*
   (b) *If G_i is an XOR-gate $W_c = XOR(W_a, W_b)$ with garbled input values $w_a^0 = \langle k_a^0, p_a^0 \rangle, w_b^0 = \langle k_b^0, p_b^0 \rangle, w_a^1 = \langle k_a^1, p_a^1 \rangle, w_b^1 = \langle k_b^1, p_b^1 \rangle:$*
      i. *Set garbled output value $w_c^0 = \langle k_a^0 \oplus k_b^0, p_a \oplus p_b \rangle$*
      ii. *Set garbled output value $w_c^1 = \langle k_a^0 \oplus k_b^0 \oplus R, p_a \oplus p_b \oplus 1 \rangle$*
   (c) *If G_i is a 2-input gate $W_c = g_i(W_a, W_b)$ with garbled input values $w_a^0 = \langle k_a^0, p_a^0 \rangle, w_b^0 = \langle k_b^0, p_b^0 \rangle, w_a^1 = \langle k_a^1, p_a^1 \rangle, w_b^1 = \langle k_b^1, p_b^1 \rangle:$*
      i. *Randomly choose garbled output value $w_c^0 = \langle k_c^0, p_c^0 \rangle \in_R \{0,1\}^{N+1}$*
      ii. *Set garbled output value $w_c^1 = \langle k_c^1, p_c^1 \rangle = \langle k_c^0 \oplus R, p_c^0 \oplus 1 \rangle$*
      iii. *Create G_i's garbled table. For each of 2^2 possible combinations of G_i's input values $v_a, v_b \in \{0,1\}$, set*

$$e_{v_a, v_b} = H(k_a^{v_a} || k_b^{v_b} || i) \oplus w_c^{g_i(v_a, v_b)}$$

   Sort entries e in the table by the input pointers, i.e. place entry e_{v_a, v_b} in position $\langle p_a^{v_a}, p_b^{v_b} \rangle$
4. *For each circuit-output wire W_i (the output of gate G_j) with garblings $w_i^0 = \langle k_i^0, p_i^0 \rangle, w_i^1 = \langle k_i^1, p_i^1 \rangle:$*
   (a) *Create garbled output table for both possible wire values $v \in \{0,1\}$. Set*

$$e_v = H(k_i^v || \text{"out"} || j) \oplus v$$

   Sort entries e in the table by the input pointers, i.e. place entry e_v in position p_i^v. (There is no conflict, since $p_i^1 = p_i^0 \oplus 1$.)

Note, our encryption of table entries (Step 3(c)iii) is similar to that of Fairplay [19, Section 4.2]. Fairplay uses $e_{v_a, v_b} = H(k_a^{v_a} || i || p_a^{v_a} || p_b^{v_b}) \oplus H(k_b^{v_b} || i || p_a^{v_a} || p_b^{v_b}) \oplus w_c^{g_i(v_a, v_b)}$. This is a non-essential difference; we could use Fairplay's encryption.

Intuition for security. (A formal proof is given in Sect. 3.2.) Alg. 1 uses the output of the RO H as a one-time pad to encrypt the garbled output values in the garbled tables (Step 3(c)iii) and the garbled output tables (Step 4a). Note, any specific combination of H's inputs (keys and gate indices) is used for encryption of at most one table entry throughout our construction. (We assume that concatenation and string representation inside H is done "right".) Further, since the evaluator of the garbled circuit only knows one garbled value per wire, he can decrypt exactly one entry of G_i's garbled table. All other entries are encrypted with at least one key that cannot be guessed by a polytime evaluator. Therefore, one of the two of garbled values of every wire looks random to him.

We now give the corresponding GC evaluation algorithm, run by P_2. Recall, P_2 obtains all garbled tables and the garblings of P_1's input values from P_1. Garblings of input values held by P_2 are sent via OT.

Algorithm 2. *(Evaluation of a garbled circuit):*

1. *For each input wire W_i of C*
   (a) *Receive corresponding garbled value $w_i = \langle k_i, p_i \rangle$*
2. *For each gate G_i (in the topological order given by labels)*
   (a) *If G_i is an XOR-gate $W_c = XOR(W_a, W_b)$ with garbled input values*
      $w_a = \langle k_a, p_a \rangle, w_b = \langle k_b, p_b \rangle$
      i. *Compute garbled output value $w_c = \langle k_c, p_c \rangle = \langle k_a \oplus k_b, p_a \oplus p_b \rangle$*
   (b) *If G_i is a 2-input gate $W_c = g_i(W_a, W_b)$ with garbled input values $w_a = \langle k_a, p_a \rangle, w_b = \langle k_b, p_b \rangle$*
      i. *Decrypt garbled output value from garbled table entry e in position $\langle p_a, p_b \rangle$: $w_c = \langle k_c, p_c \rangle = H(k_a||k_b||i) \oplus e$*
3. *For each C's output wire W_i (output of gate G_j) with garbling $w_i = \langle k_i, p_i \rangle$*
   (a) *Decrypt output value f_i from garbled output table entry e in row p_i:*
      $f_i = H(k_i||\text{"out"}||j) \oplus e$

The GC construction and evaluation algorithms can be directly used to obtain the GC-based SFE protocol, in a standard manner. For completeness, we include the description of this protocol.

Protocol 1. *(Two-party SFE protocol):*

- **Inputs:** *P_1 has private input $x = \langle x_1, .., x_{u_1} \rangle \in \{0,1\}^{u_1}$ and P_2 has private input $y = \langle y_1, .., y_{u_2} \rangle \in \{0,1\}^{u_2}$.*
- **Auxiliary input:** *A boolean acyclic circuit C such that $\forall x \in \{0,1\}^{u_1}, y \in \{0,1\}^{u_2}$, it holds that $C(x,y) = f(x,y)$, where $f : \{0,1\}^{u_1} \times \{0,1\}^{u_2} \to \{0,1\}^v$. We require that C is such that if a circuit-output wire leaves some gate G, then gate G has no other wires leading from it into other gates (i.e., no circuit-output wire is also a gate-input wire). Likewise, a circuit-input wire that is also a circuit-output wire enters no gates. We also require that C is modified to contain no NOT-gates and all n-input XOR-gates with $n > 2$ replaced by 2-input XOR-gates as described in Section 3.1.*
- **The protocol:**
   1. *P_1 constructs the garbled circuit using Algorithm 1 and sends it (i.e. the garbled tables) to P_2.*
   2. *Let $W_1, .., W_{u_1}$ be the circuit input wires corresponding to x, and let $W_{u_1+1}, .., W_{u_1+u_2}$ be the circuit input wires corresponding to y. Then,*
      (a) *P_1 sends P_2 the garbled values $w_1^{x_1}, .., w_{u_1}^{x_{u_1}}$.*

(b) For every $i \in \{1, .., u_2\}$, P_1 and P_2 execute a 1-out-of-2 oblivious transfer protocol, where P_1's input is $(k^0_{u_1+i}, k^1_{u_1+i})$, and P_2's input is y_i. All u_2 OT instances can be run in parallel.

3. P_2 now has the garbled tables and the garblings of circuit's input wires. P_2 evaluates the garbled circuit, as described in Alg. 2, and outputs $f(x, y)$.

It is easy to verify protocol's correctness; we do not discuss it further.

On Our use of RO. In previous GC work, RO's use improves efficiency in the malicious model, but is not inherent. Here, while we rely on RO, we do so conservatively. First, we use *non-programmable* RO [22], i.e. we don't allow simulator to fake RO's answers. Second, (a variant of) *correlation-robust* functions [10], a weaker notion than RO, is sufficient for our purposes. (Recall, if h is correlation-robust and $R, t_1, .., t_n$ are random, $(h(t_1 \oplus R), .., h(t_n \oplus R))$ is pseudo-random, given $t_1, .., t_n$.)

Further, concrete security of our construction is comparable to that of standard GC with RO as the encryption function. This makes even constant-factor efficiency improvements, such as those suggested in this work, meaningful. For the lack of space, we omit the detailed analysis. We only note that the main feature of our protocol, the use of the global R, has very slight impact on security (e.g., our adversary can decrypt all garbled tables, once he breaks any one of them and learns R). Further, our use of RO is not vulnerable to birthday attacks in the semi-honest model. Indeed, the circuit is small, and P_2 w.h.p. will not see RO collisions.

3.2   Proof of Security

Our protocol is secure against semi-honest adversaries, who are not allowed to deviate from the protocol. Analogously to [19,18], (w.h.p.) malicious behavior of players can be prevented by using cut-and-choose method; we don't discuss malicious players further.

We prove security in the simulation paradigm. Intuitively, a protocol π is secure if whatever is seen by its party, can be computed only from that party's input and output. The view of a party P_i, $view^{\pi}_{P_i}(x, y)$, consists of the party's own input, randomness, and all messages that P_i receives in the execution of π. Thus, a protocol is secure, if there exist *simulators* S_1, S_2, such that $\{S_1(x, f(x,y))\} \overset{c}{\equiv} \{view^{\pi}_{P_1}(x, y)\}$ and $\{S_2(y, f(x, y))\} \overset{c}{\equiv} \{view^{\pi}_{P_2}(x, y)\}$.

Case 1 - P_1 is corrupted. P_1's view in Protocol 1 consists only of the view in the OT protocols in Step 2b. The following $S_1(x, f(x, y))$ simulates the view of P_1. Let S^{OT}_1 be the simulator that is guaranteed to exist for P_1 in the secure 1-out-of-2 OT protocol. S_1 constructs a garbled circuit using Alg. 1. Then S_1 feeds the constructed garblings of the input wires corresponding to y to S^{OT}_1, and obtains the simulated transcript of the OT, which he outputs. S additionally outputs x and the randomness used in construction of GC. It is not hard to see that the output of the simulator is indistinguishable from the view of P_1.

Case 2 - P_2 is corrupted. We construct a simulator S_2 that given input $(y, f(x, y))$ simulates the view of P_2. P_2 receives a garbled circuit (including garbled inputs), which S_2 must simulate. However, S_2 doesn't know P_1's input x. Thus, S_2 can not honestly generate the garbled circuit, since it doesn't know which of the input garblings corresponding to x to hand to P_2 in Step 2a of the protocol. Instead, S_2 generates a fake garbled circuit that always evaluates to $f(x, y)$, using a slightly modified Alg. 1. The only modification, in Step 4a, appropriately forges the output tables:

4. For each circuit-output wire W_i (the output of gate G_j) with garblings
   $w_i^0 = \langle k_i^0, p_i^0 \rangle, w_i^1 = \langle k_i^1, p_i^1 \rangle$:
   (a) Create **fake** garbled output table for both possible wire values $v \in \{0, 1\}$
       of the same encrypted output value. Set

$$e_v = H(k_i^v \| \text{``out''} \| j) \oplus \mathbf{f_i(x, y)}$$

   Sort entries e in the table by the input pointers, i.e. place entry e_v in
   position p_i^v.

Let S_2^{OT} be an OT simulator for P_2. S_2 outputs y, and the fake garbled circuit (i.e. its tables). Further, for each input wire W_i held by P_2, S_2 runs and outputs $S_2^{OT}(y_i, w_i^{y_i})$. Finally, S_2 simulates the received garblings of the input wires W_j held by P_1 simply by outputting w_j^0 (fake garblings corresponding to $x = 0..0$).

Theorem 1. *The output of S_2 is indistinguishable from the real view of P_2.*

Proof. (sketch) First, observe that S_2 feeds S_2^{OT} proper inputs (i.e. y and the corresponding honestly generated garblings). Thus, simulation of Step 2b of the protocol is indistinguishable from the real execution. The crux of the proof is in showing the indistinguishability of the fake and real circuits (which include the tables and the input garblings that P_2 sees). This is addressed next.

First, observe, pointers p_i^j are independent of the parties' inputs, and thus are easily simulated by S_2. For ease of presentation, we omit the details of pointer simulation from the proof.

We now show that no polytime procedure D can distinguish simulated and real garbled circuit transcripts with non-negligible probability. We proceed inductively, gate by gate in topological order, in proving this for each partial transcript τ_i, where τ_0 includes all active secrets on the input wires, and each τ_i additionally includes the garbled tables of first i gates.

Induction base. It is easy to see that the partial transcript τ_0 – active secrets on the input wires – is distributed identically in real and simulated cases. Indeed, these secrets are uniformly random in the domain. Moreover, clearly, no distinguisher D_0 can output with non-negligible probability the global key offset $\hat{R}$ used in the construction of the (either simulated or real) transcript.

For the induction step, suppose no polytime D_{i-1} can with non-negligible advantage distinguish the τ_{i-1} transcripts (i.e. those including the active secrets on the inputs and the first $i - 1$ garbled tables). Moreover, assume that no polytime D_{i-1} can output the global key offset $\hat{R}$ with non-negligible probability when given τ_{i-1}. We show that these properties hold also when additionally given the i-th garbled table.

Recall, the i-th garbled table contains (a permutation of) entries:

$$H(k_a||k_b||i) \oplus v_{00}$$
$$H(k_a||k_b \oplus \hat{R}||i) \oplus v_{01}$$
$$H(k_a \oplus \hat{R}||k_b||i) \oplus v_{10}$$
$$H(k_a \oplus \hat{R}||k_b \oplus \hat{R}||i) \oplus v_{11}$$

where $v_{00}, .., v_{11} \in \{k_c, k_c \oplus \hat{R}\}$ are the output secrets that correspond to the four possible gate input combinations. (Garbled output tables have one input and consist of two entries. The corresponding claims hold for these cases as well, via a natural modification of the following argument addressing two-input gates.)

Without loss of generality, suppose the active gate input secrets are k_a and k_b. By the induction assumption, no polytime D_{i-1} can compute both k_a and $k_a \oplus \hat{R}$, or both k_b and $k_b \oplus \hat{R}$ (otherwise D_{i-1} can output $\hat{R}$). Thus, D_{i-1} can call functions $H(k_a||k_b \oplus \hat{R}||i), H(k_a \oplus \hat{R}||k_b||i)$, or $H(k_a \oplus \hat{R}||k_b \oplus \hat{R}||i)$ only with negligible probability. Further, because of the inclusion of the gate index i, these function calls have not been made in the construction of (real or simulated) τ_i. Therefore, due to RO properties, except with negligible probability, all the inactive entries in the i-th table are distributed identically to random strings, from the point of view of D_{i-1}, and thus do not provide help to D_{i-1} in computing $\hat{R}$. Therefore, polytime D_i cannot output $\hat{R}$ or call any of $H(k_a||k_b \oplus \hat{R}||i), H(k_a \oplus \hat{R}||k_b||i)$, or $H(k_a \oplus \hat{R}||k_b \oplus \hat{R}||i)$, except with negligible probability. Therefore, no polytime D_i can distinguish the real and simulated transcripts τ_i with non-negligible probability.

This completes the induction and the proof of the theorem.   □

4   Application of Our SFE Constructions

We now present several motivating examples – practical functions whose SFE benefits from improvements of our construction. Universal circuit (UC) constructions [25,15] do not explicitly use many XOR gates. We show how to modify these circuits to mainly consists of XOR gates, achieving fourfold reduction of garbled circuit size. This construction may be of independent interest. Further, we show how to reduce in half the size of garbled circuits of commonly used blocks, such as integer addition and equality test.

Universal Circuits [25,15] and Permutation Networks [26]. The size of a UC mainly comes from programmable switching networks (such as permutation network [26]) connecting the simulated gates. In turn, these networks are constructed from two types of switching blocks shown in Fig. 1, as discussed in [26,25,15]. The Y-block can be programmed to output one of its two inputs. The X-block can be programmed to either pass or cross over its two inputs to the two outputs. A natural SFE implementation of the Y-block uses a 2-input garbled gate with a garbled table with $2^2 = 4$ encrypted table entries. Similarly, X-block

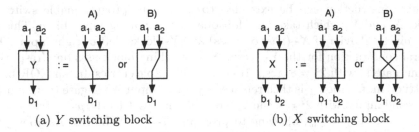

(a) Y switching block   (b) X switching block

Fig. 1. Switching blocks

is implemented by two 2-input garbled gates (one for each of its two outputs), resulting in a garbled table of $2 \cdot 2^2 = 8$ entries.

We show how to take advantage of free XOR gates and implement both X- and Y-gates with only two garbled table entries each. Since permutation network [26] consists only of X-gates, this results in 75% size reduction of its SFE. UC consists *almost exclusively* of X-gates. Valiant's UC [25] for a circuit of k gates has size $\sim 19k \log k$. The $\sim 19k \log k - k$ overhead gates are X-gates that come from switching networks. A recent UC construction [15] similarly consists almost exclusively of X-gates, and of very few Y-gates and simulated gates. Thus, UC enjoys almost 75% garbled table size reduction.

Let $f : \{0,1\} \mapsto \{0,1\}$ be a function (implemented with two garbled table entries). We implement X- and Y-blocks as follows (see Fig. 2). $Y(a_1, a_2) = b_1 = f(a_1 \oplus a_2) \oplus a_1$; $X(a_1, a_2) = (b_1, b_2)$, where $b_1 = f(a_1 \oplus a_2) \oplus a_1, b_2 = f(a_1 \oplus a_2) \oplus a_2$. It is easy to see that setting $f = f_0$ to the zero function results in Y choosing left input, and X passing the inputs. Further, setting $f = f_{id}$ to the identity function results in Y choosing the right input, and in X crossing its inputs:

$$f = f_0 : \qquad b_1 = 0 \oplus a_1 = a_1; \qquad\qquad b_2 = 0 \oplus a_2 = a_2.$$
$$f = f_{id} : \qquad b_1 = (a_1 \oplus a_2) \oplus a_1 = a_2; \qquad b_2 = (a_1 \oplus a_2) \oplus a_2 = a_1.$$

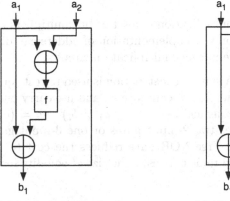

(a) Y switching block   (b) X switching block

Fig. 2. Efficient implementation of switching blocks

This construction can be extended to implement programmable switching blocks X and Y, which take an additional programming input bit p. This bit determines behavior of X- (pass or cross) and Y-blocks (left or right input). The natural construction for the Y- (resp. X-) switching block uses one (resp. two) 3-input gate(s) with $2^3 = 8$ (resp. 16) encrypted table entries. In our XOR-based construction, function f is then replaced by a two-input AND-gate (with p being the second input) with $2^2 = 4$ encrypted table entries. Clearly, $p = 0$ sets $f = f_0$, and $p = 1$ sets $f = f_{id}$, allowing to program X- and Y-blocks. As above, the size of Y- and X-blocks is reduced by 50% and 75% respectively.

Integer Adder and Multiplier. An adder for n-bit integers a, b is composed from a chain of n full adder (FA) blocks as shown in Fig. 3(b). (The last FA block can be replaced by a smaller half-adder block.) A FA block (see Fig. 3(a)) has as inputs a carry-in c_i from the previous FA block and the two input bits a_i and b_i. It outputs two bits: carry-out $c_{i+1} = (a_i \wedge b_i) \vee (a_i \wedge c_i) \vee (b_i \wedge c_i)$ and sum $s_i = a_i \oplus b_i \oplus c_i$. The straightforward implementation of a FA uses two 3-input gates with $2 \cdot 2^3 = 16$ encrypted table entries. We can compute s_i "for free" using free XOR-gates, and use only one 3-input gate with $2^3 = 8$ encrypted table entries to compute c_{i+1}. The size of a FA block, and hence that of an n-bit adder is reduced by 50%.

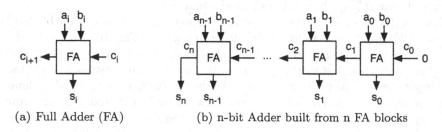

(a) Full Adder (FA)   (b) n-bit Adder built from n FA blocks

Fig. 3. Adder for two n-bit integers a and b

As circuits for integer multiplication consist of bit-multipliers (2-input AND-gates) and adders, the improved implementation of adders can directly be used to correspondingly improve integer-multiplication circuits.

Integer Equality Test. A similar construction is used to test equality of two n-bit integers a and b. Now, we do not compute s_i, and use carry bits as inequality flags. The carry-out bit is defined as $c_{i+1} = (a_i \neq b_i) \vee c_i = (a_i \oplus b_i) \vee c_i$. A simple implementation uses two 2-input gates or one 3-input gate (each costs 8 encrypted table entries). Free XOR gate reduces the cost to that of one 2-input OR gate (4 encrypted table entries). The size of equality test block is thus reduced by 50%.

Acknowledgments. We thank Yuval Ishai and anonymous referees for helpful comments. The second author thanks CACE project for funding his ICALP trip.

References

1. Aiello, W., Ishai, Y., Reingold, O.: Priced oblivious transfer: How to sell digital goods. In: Pfitzmann, B. (ed.) EUROCRYPT 2001. LNCS, vol. 2045, pp. 119–135. Springer, Heidelberg (2001)
2. Bellare, M., Micali, S.: Non-interactive oblivious transfer and applications. In: Brassard, G. (ed.) CRYPTO 1989. LNCS, vol. 435, pp. 547–557. Springer, Heidelberg (1990)
3. Bellare, M., Rogaway, P.: Random oracles are practical: A paradigm for designing efficient protocols. In: ACM CCS, pp. 62–73 (1993)
4. Blake, I.F., Kolesnikov, V.: Conditional encrypted mapping and comparing encrypted numbers. In: Di Crescenzo, G., Rubin, A. (eds.) FC 2006. LNCS, vol. 4107, pp. 206–220. Springer, Heidelberg (2006)
5. Canetti, R., Goldreich, O., Halevi, S.: The random oracle methodology, revisited. In: Proc. 30th ACM Symp. on Theory of Computing, pp. 209–218 (1998)
6. Crescenzo, G.D.: Private selective payment protocols. In: Frankel, Y. (ed.) FC 2000. LNCS, vol. 1962. Springer, Heidelberg (2001)
7. Even, S., Goldreich, O., Lempel, A.: A randomized protocol for signing contracts. Commun. ACM 28(6), 637–647 (1985)
8. Fischlin, M.: A cost-effective pay-per-multiplication comparison method for millionaires. In: Naccache, D. (ed.) CT-RSA 2001. LNCS, vol. 2020, pp. 457–471. Springer, Heidelberg (2001)
9. Goldreich, O., Vainish, R.: How to solve any protocol problem - an efficiency improvement. In: Pomerance, C. (ed.) CRYPTO 1987. LNCS, vol. 293, pp. 73–86. Springer, Heidelberg (1988)
10. Ishai, Y., Kilian, J., Nissim, K., Petrank, E.: Extending oblivious transfers efficiently. In: Boneh, D. (ed.) CRYPTO 2003. LNCS, vol. 2729. Springer, Heidelberg (2003)
11. Ishai, Y., Kushilevitz, E.: Perfect constant-round secure computation via perfect randomizing polynomials. In: Widmayer, P., Triguero, F., Morales, R., Hennessy, M., Eidenbenz, S., Conejo, R. (eds.) ICALP 2002. LNCS, vol. 2380, pp. 244–256. Springer, Heidelberg (2002)
12. Kantarcioglu, M., Clifton, C.: Privacy-preserving distributed mining of association rules on horizontally partitioned data. In: ACM SIGMOD Workshop on Research Issues on Data Mining and Knowledge Discovery (DMKD 2002) (2002)
13. Kilian, J.: Founding cryptography on oblivious transfer. In: Proc. 20th ACM Symp. on Theory of Computing, Chicago, pp. 20–31. ACM, New York (1988)
14. Kolesnikov, V.: Gate evaluation secret sharing and secure one-round two-party computation. In: Roy, B. (ed.) ASIACRYPT 2005. LNCS, vol. 3788, pp. 136–155. Springer, Heidelberg (2005)
15. Kolesnikov, V., Schneider, T.: A practical universal circuit construction and secure evaluation of private functions. In: Financial Cryptography and Data Security, FC 2008. LNCS. Springer, Heidelberg (2008)
16. Lindell, Y., Pinkas, B.: Privacy preserving data mining. In: Bellare, M. (ed.) CRYPTO 2000. LNCS, vol. 1880. Springer, Heidelberg (2000)
17. Lindell, Y., Pinkas, B.: A proof of Yao's protocol for secure two-party computation. Cryptology ePrint Archive, Report 2004/175 (2004)
18. Lindell, Y., Pinkas, B.: An efficient protocol for secure two-party computation in the presence of malicious adversaries. In: Naor, M. (ed.) EUROCRYPT 2007. LNCS, vol. 4515, pp. 52–78. Springer, Heidelberg (2007)

19. Malkhi, D., Nisan, N., Pinkas, B., Sella, Y.: Fairplay — a secure two-party computation system. In: USENIX (2004)
20. Naor, M., Pinkas, B.: Efficient oblivious transfer protocols. In: SODA 2001: Proceedings of the twelfth annual ACM-SIAM symposium on Discrete algorithms, Philadelphia, PA, USA. Society for Industrial and Applied Mathematics (2001)
21. Naor, M., Pinkas, B., Sumner, R.: Privacy preserving auctions and mechanism design. In: 1st ACM Conf. on Electronic Commerce (1999)
22. Nielsen, J.B.: Separating random oracle proofs from complexity theoretic proofs: The non-committing encryption case. In: Yung, M. (ed.) CRYPTO 2002. LNCS, vol. 2442, pp. 111–126. Springer, Heidelberg (2002)
23. Pinkas, B.: Cryptographic techniques for privacy-preserving data mining. SIGKDD Explor. Newsl. 4(2), 12–19 (2002)
24. Sander, T., Young, A., Yung, M.: Non-interactive cryptocomputing for NC^1. In: Proc. 40th FOCS, New York, pp. 554–566. IEEE, Los Alamitos (1999)
25. Valiant, L.G.: Universal circuits (preliminary report). In: Proc. 8th ACM Symp. on Theory of Computing, pp. 196–203. ACM Press, New York (1976)
26. Waksman, A.: A permutation network. J. ACM 15(1), 159–163 (1968)
27. Yao, A.C.: Protocols for secure computations. In: Proc. 23rd IEEE Symp. on Foundations of Comp. Science, Chicago, pp. 160–164. IEEE, Los Alamitos (1982)
28. Yao, A.C.: How to generate and exchange secrets. In: Proc. 27th IEEE Symp. on Foundations of Comp. Science, Toronto, pp. 162–167. IEEE, Los Alamitos (1986)

Improving the Round Complexity of VSS in Point-to-Point Networks

Jonathan Katz[1,*], Chiu-Yuen Koo[1,2], and Ranjit Kumaresan[1]

[1] Dept. of Computer Science, University of Maryland
{jkatz,cykoo,ranjit}@cs.umd.edu
[2] Google Labs, Mountain View, CA

Abstract. We revisit the following question: *what is the optimal round complexity of verifiable secret sharing (VSS)?* We focus here on the case of perfectly-secure VSS where the number of corrupted parties t satisfies $t < n/3$, with n being the total number of parties. Work of Gennaro et al. (STOC 2001) and Fitzi et al. (TCC 2006) shows that, assuming a broadcast channel, 3 rounds are necessary and sufficient for efficient VSS. The efficient 3-round protocol of Fitzi et al., however, treats the broadcast channel as being available "for free" and does not attempt to minimize its usage. This approach leads to relatively poor round complexity when protocols are compiled for a point-to-point network.

We show here a VSS protocol that is *simultaneously* optimal in terms of both the number of rounds and the number of invocations of broadcast. Our protocol also has a certain "2-level sharing" property that makes it useful for constructing protocols for general secure computation.

1 Introduction

The round complexity of cryptographic protocols has been the subject of intense study. Besides protocols for general secure computation, protocols for various specific functionalities of interest (e.g., broadcast, zero-knowledge proofs, etc.) have also been explored. Here, we revisit the case of *verifiable secret sharing*, whose definition we now recall informally. (Formal definitions appear in Section 2.) In secret sharing [2,19], there is a *dealer* who shares a secret among a group of n parties in a *sharing phase*. The requirements are that, for some parameter $t < n$, any set of t colluding parties gets no information about the dealer's secret at the end of the sharing phase, yet any set of $t+1$ parties can recover the dealer's secret in a later *reconstruction phase*. Secret sharing assumes the dealer is honest; *verifiable* secret sharing (VSS) [3] also requires that, no matter what a cheating dealer does (in conjunction with $t-1$ other colluding parties), there is *some* unique secret to which the dealer is "committed" by the end of the sharing phase. VSS serves as a fundamental building block in the design of protocols for general secure multi-party computation as well as other specialized goals (such

---

* Research supported in part by NSF awards #0310751 and #0447075 (CAREER), and US-Israel Binational Science Foundation grant #2004240.

L. Aceto et al. (Eds.): ICALP 2008, Part II, LNCS 5126, pp. 499–510, 2008.
© Springer-Verlag Berlin Heidelberg 2008

as Byzantine agreement); thus, it is of interest to understand the inherent round complexity for carrying out this task.

In this work we will always consider perfectly-secure VSS, where the protocol is required to be error-free and security should hold even against an all-powerful adversary. This is known to be possible if and only if $t < n/3$ [1,5]. Previous research investigating the round complexity of VSS, surveyed further below, has focused on optimizing the round complexity *assuming a broadcast channel is available "for free"*. (We remark that broadcast is essential for VSS, in a way we make precise below.) As argued previously [12], however, if the ultimate goal is to optimize the round complexity of protocols for point-to-point networks (where protocols are likely to be run), then it is preferable to minimize the *number of rounds in which broadcast is used* rather than to minimize the *total number of rounds*. This is due to the high overhead of emulating a broadcast channel over a point-to-point network: deterministic broadcast protocols require $\Omega(t)$ rounds [7]; known randomized protocols [6,8,11] require only $O(1)$ rounds in expectation, but the constant is rather high. (The most round-efficient protocol known [11,12] requires 23 rounds in expectation for $t < n/3$.[1]) Moreover, when using randomized broadcast protocols, if *more than one* invocation of broadcast is used then special care must be taken to deal with sequential composition of protocols without simultaneous termination (see [15,11,12]), leading to a substantial increase in the round complexity. As a consequence, a constant-round protocol that only uses a *single* round of broadcast is likely to yield a more round-efficient protocol in a point-to-point setting than any protocol that uses *two* rounds of broadcast (even if that protocol uses no additional rounds).

As a concrete example (taken from [12]) to illustrate the point, consider the VSS protocol of Micali and Rabin [16] and the 'round-optimal' VSS protocol of Fitzi et al. [9]. The former uses 16 rounds but only a single round of broadcast; the latter uses 3 rounds, two of which require broadcast. Compiling these protocols for a point-to-point network using the most round-efficient techniques known (see [12]), the Micali-Rabin protocol runs in an expected 31 rounds while the protocol of Fitzi et al. requires an expected 55 rounds!

In light of the above, when discussing the round complexity of protocols that assume a broadcast channel we keep track of both the number of rounds as well as the number of rounds in which broadcast is used. (In a given round when broadcast is used, each party may use the broadcast channel but a rushing adversary is still assumed. Existing broadcast protocols can be modified so that the round complexity is unchanged even if many parties broadcast in parallel.) We say a protocol has round complexity (r, r') if it uses r rounds in total, and $r' \leq r$ of these rounds invoke broadcast. The round complexity of VSS refers to the sharing phase only, since the reconstruction phase of most known protocols utilizes only a single round, without broadcast. (An exception is the protocol of [9], whose reconstruction phase uses a single round of broadcast.)

Our results and techniques. Gennaro et al. [10] show that three rounds are necessary for VSS, assuming a broadcast channel. We also observe that it is

---

[1] Actually, the VSS protocol given here can be used to improve this slightly.

impossible to construct a *strict* constant-round protocol for VSS without using a broadcast channel in at least one round: VSS implies broadcast using one additional round (the message to be broadcast can be treated as the input for VSS), and the results of Fischer and Lynch [7] rule out strict constant-round protocols for broadcast. Prior work [16,9,12,14] shows that optimal round complexity as well as optimal use of the broadcast channel could each be obtained *individually* for VSS, but it was unknown whether they could be obtained *simultaneously*. Here, we resolve this question and show a $(3,1)$-round VSS protocol that is optimal in both measures. (Our protocol has a 1-round reconstruction phase that does not use broadcast.) As a consequence, we obtain a VSS protocol with the best known round complexity in point-to-point networks. Our work also leads to an improvement in the round complexity of the most round-efficient broadcast protocols known [11].

A nice feature of our VSS protocol is that it also satisfies a certain "2-level sharing" property that is not achieved by the 3-round protocol from [9]. Roughly speaking, this means that the following conditions hold at the end of the sharing phase when the dealer's (effective) input is s:

1. There exists a polynomial $f(x)$ of degree at most t such that $f(0) = s$ and each honest party P_i holds the value $f(i)$.
2. For each party P_i, there exists a polynomial $f_i(x)$ of degree at most t such that $f_i(0) = f(i)$ and each honest party P_j holds the value $f_i(j)$.

VSS protocols with this property constitute a useful building block for protocols for general secure multi-party computation (see, e.g., [12,14]).

Our protocol is efficient, in that the computation and communication are polynomial in n. The communication complexity of our protocol is $\mathcal{O}(n^2 t)$ field elements, which matches the communication complexity of [9] but is worse than that of [10].

We now summarize the basic techniques used to prove our main result. As in [9], we begin by constructing a protocol for *weak* verifiable secret sharing (WSS) [18]. (In WSS, informally, if the dealer is dishonest then, in the reconstruction phase, each honest party recovers either the dealer's input or a special failure symbol.) Fitzi et al. show a $(3,2)$-round WSS protocol that essentially consists of the first three rounds of the 4-round VSS protocol from [10]. On a high level, their protocol works as follows: In the first round, the dealer distributes the shares of the secret using a random bivariate polynomial; in parallel, each pair of parties (P_i, P_j) exchanges a random pad $r_{i,j}$. In the second round, P_i and P_j check for an inconsistency between their shares by broadcasting their common shares masked with the random pad. In the third round, if there is a disagreement between P_i and P_j in round 2 (note that all parties agree whether there is disagreement since broadcast is used in round 2), then the dealer, P_i, and P_j all broadcast the share in question. This allows the rest of the parties to determine whether the dealer "agrees" with P_i or with P_j.

A $(5,1)$-round WSS protocol is implicitly given in [12].[2] There, rather than using the "random pad" technique, a different method is used to detect disagreement between P_i and P_j. While this saves one round of broadcast, it requires additional rounds of interaction.

To construct a $(3,1)$-round WSS protocol, we modify the $(3,2)$-round WSS protocol from [9] by using the random pad idea with the following twist: in the second round of the protocol, P_i and P_j check if there is any inconsistency between their shares by exchanging their common shares over a *point-to-point* link; they also send the random pad $r_{i,j}$ to the dealer. In the third round of the protocol, if there is a disagreement between P_i and P_j, then P_i and P_j each broadcast the shares they hold; otherwise, they broadcast the value of their common share masked with the random pad. The dealer will broadcast the corresponding share masked with the random pad (or the share itself if the random pads it received from P_i and P_j are different). Notice that secrecy of the share is preserved if P_i, P_j, and the dealer are all honest. On the other hand, if the dealer is malicious and there is a disagreement between honest parties P_i and P_j, then the dealer can only "agree" with at most one of P_i and P_j in round 3, but not both of them.

The above is the high-level idea of our WSS protocol. Using the same techniques as in [9], we can then immediately obtain a $(3,1)$-round VSS protocol. However, the VSS protocol constructed in this manner will not have the "2-level sharing" property; as a consequence, the resulting protocol cannot directly be plugged in to existing protocols for general secure multi-party computation.

To convert the VSS protocol into one with 2-level sharing we note that, by the end of the sharing phase, there is a set of honest parties (that we call a "core set") who already *do* have the required 2-level shares; thus, we only need to provide honest parties outside the core set with their required shares. We achieve this, as in [4], by having the dealer use a *symmetric* bivariate polynomial to share its input, and then modifying the protocol so that honest parties who are not in the core set can still generate appropriate shares by interpolating the shares of the parties in the core set. Of course, this process needs to be carefully designed so that no additional information is leaked to the adversary. We defer the details of this to a later section.

Other related work. Gennaro et al. [10] initiated a study of the exact round complexity of VSS. For $t < n/3$, they show an efficient (i.e., polynomial-time) $(4,3)$-round protocol, and an inefficient $(3,2)$-round protocol. (Recall that the round complexity of VSS is defined as the number of rounds in the sharing phase; unless otherwise stated, all protocols mentioned use only one round, without broadcast, in the reconstruction phase.) They also show that three rounds are necessary for VSS when $t < n/3$. For $t < n/4$, they show that two rounds are necessary and sufficient for efficient VSS. Settling the question of the absolute round complexity of efficient VSS for $t < n/3$, Fitzi et al. [9] show an efficient

---

[2] That work shows a 6-round VSS protocol that uses broadcast in the final two rounds. The first five rounds of that protocol suffice for WSS.

$(3, 2)$-round VSS protocol. The reconstruction phase of their protocol requires one round of broadcast as well.

As discussed extensively already, although the protocol by Fitzi et al. is optimal in terms of the total number of rounds, it is not optimal in terms of its usage of the broadcast channel. VSS protocols for $t < n/3$ using one round of broadcast are known, but these protocols are not optimal in terms of their overall round complexity. Micali and Rabin [16] give a $(16, 1)$-round VSS protocol, and recent work of the authors [12,14] improves this to give a $(7, 1)$-round protocol.

Our work, as well as all the work referenced above, focuses on VSS protocols with perfect security (i.e., *0-error VSS*). A natural relaxation is to consider *statistical* VSS where the security properties may fail with negligible probability. Surprisingly, recent work subsequent to our own [17] shows that the lower bound of Gennaro et al. [10] no longer holds in this setting, and that 2-round protocols are in fact possible.

Future directions. It would, of course, be nice to characterize the optimal round complexity of VSS in point-to-point networks. Though our work represents progress toward this goal, the question is complicated by the fact that one must consider the *distribution* of running times of any protocol (since strict constant-round protocols are ruled out). It will also be interesting to understand the round complexity of VSS when $t < n/2$; see [17] for an almost-tight characterization.

2 Model and Definitions

We consider the standard communication model where parties communicate in synchronous rounds using pairwise private and authenticated channels. We also assume a broadcast channel, with the understanding that it can be emulated in a point-to-point network using a broadcast protocol. A broadcast channel allows any party to send the same message to all other parties (and all parties to be assured they have received identical messages) in a single round. We stress that we do not assume *simultaneous* broadcast, but allow rushing here as well.

When we say a protocol tolerates t malicious parties, we always mean that it is secure against an adversary who may *adaptively* corrupt up to t parties during an execution of the protocol and coordinate the actions of these parties as they deviate from the protocol in an arbitrary manner. Parties not corrupted by the adversary are called *honest*. We always assume a *rushing* adversary; i.e., in any round the malicious parties receive the messages sent by the honest parties before deciding on their own messages.

2.1 VSS and Variants

We now present definitions of WSS, VSS, and VSS with 2-level sharing.

Definition 1 (Weak verifiable secret sharing). *A two-phase protocol for parties $\mathcal{P} = \{P_1, \ldots, P_n\}$, where a distinguished dealer $D \in \mathcal{P}$ holds initial input s, is a* WSS *protocol tolerating t malicious parties if the following conditions hold for any adversary controlling at most t parties:*

Privacy. *If the dealer is honest at the end of the first phase (the* sharing phase*), then at the end of this phase the joint view of the malicious parties is independent of the dealer's input* s*.*

Correctness. *Each honest party* P_i *outputs a value* s_i *at the end of the second phase (the* reconstruction phase*). If the dealer is honest then* $s_i = s$*.*

Weak commitment. *At the end of the sharing phase the joint view of the honest parties defines a value* s' *(which can be computed in polynomial time from this view) such that each honest party will output either* s' *or a default value* $\perp$ *at the end of the reconstruction phase.*   ◇

Definition 2 (Verifiable secret sharing). *A two-phase protocol for parties* $\mathcal{P}$*, where a distinguished dealer* $D \in \mathcal{P}$ *holds initial input* s*, is a* VSS *protocol tolerating* t malicious parties *if it satisfies the privacy and correctness requirements of WSS as well as the following (stronger)* commitment *requirement:*

Commitment. *At the end of the sharing phase the joint view of the honest parties defines a value* s' *(which can be computed in polynomial time from this view) such that all honest parties will output* s' *at the end of the reconstruction phase.*   ◇

Definition 3 (Verifiable secret sharing with 2-level sharing). *A two-phase protocol for parties* $\mathcal{P} = \{P_1, \ldots, P_n\}$*, where a distinguished dealer* $D \in \mathcal{P}$ *holds initial input* s*, is a* VSS *protocol with 2-level sharing tolerating* t *malicious parties if it satisfies the privacy and correctness requirements of VSS as well as the following requirement:*

Commitment with 2-level sharing. *At the end of the sharing phase each honest party* P_i *outputs* s_i *and* $s_{i,j}$ *for* $j \in \{1, \ldots, n\}$*, satisfying the following requirements:*

1. *There exists a polynomial* $p(x)$ *of degree at most* t *such that* $s_i = p(i)$ *for every honest party* P_i*, and furthermore all honest parties will output* $s' = p(0)$ *at the end of the reconstruction phase.*
2. *For each* $j \in \{1, \ldots, n\}$*, there exists a polynomial* $p_j(x)$ *of degree at most* t *such that (1)* $p_j(0) = p(j)$ *and (2)* $s_{i,j} = p_j(i)$ *for every honest party* P_i*.*   ◇

The above implies the commitment property of VSS, since the value $s' = p(0)$ that will be output in the reconstruction phase is defined by the view of the honest parties at the end of the sharing phase.

In our protocol descriptions, we implicitly assume all parties send properly-formatted messages at all times; this is without loss of generality, as we may interpret an improper or missing message as some default message. We assume the dealer's input s lies in a finite field $\mathbb{F}$ containing $\{0, 1, \ldots, n\}$ as a subset.

3   Weak Verifiable Secret Sharing

We show a $(3, 1)$-round WSS protocol tolerating $t < n/3$ malicious parties.

3.1   The Protocol

Sharing phase. The sharing phase consists of three rounds, with broadcast used in the last round.

Round 1: The dealer holds s. The following steps are carried out in parallel:

- The dealer chooses a random bivariate polynomial $F(x, y)$ of degree at most t in each variable such that $F(0,0) = s$. The dealer then sends to each party P_i the polynomials $f_i(x) := F(x, i)$ and $g_i(y) := F(i, y)$.
- Each party P_i picks a random pad $r_{i,j} \in \mathbb{F}$ for $j \in \{1, \ldots, n\}$, and sends $r_{i,j}$ to both P_j and the dealer D.

Round 2: For every ordered pair (i, j), parties P_i and P_j proceed as follows:

- Party P_i sends $a_{i,j} := f_i(j)$ to P_j.
- Party P_j sends $b_{j,i} := g_j(i)$ to P_i.
  (Note that, when everyone is honest, then $a_{i,j} = b_{j,i} = F(j, i)$.)
- Let $r'_{i,j}$ be the random pad that P_j received from P_i in the previous round. Then P_j sends $r'_{i,j}$ to D.

Round 3: For every ordered pair (i, j), parties P_i, P_j, and D do:

- (From the viewpoint of P_i:) If $b_{j,i} \neq f_i(j)$, then P_i broadcasts ("disagree", $f_i(j)$, $r_{i,j}$). Otherwise, P_i broadcasts ("agree", $f_i(j) + r_{i,j}$).
- (From the viewpoint of P_j:) If $a_{i,j} \neq g_j(i)$, then P_j broadcasts ("disagree", $g_j(i)$, $r'_{i,j}$). Otherwise, P_j broadcasts ("agree", $g_j(i) + r'_{i,j}$).
- (From the viewpoint of D:) If $r_{i,j} \neq r'_{i,j}$, then D broadcasts ("not equal", $F(j, i)$). Otherwise, D broadcasts ("equal", $F(j, i) + r_{i,j}$).

Local computation. An ordered pair of parties (P_i, P_j) is *conflicting* if, in round 3, party P_i broadcasts ("disagree", $f_i(j)$, $r_{i,j}$); party P_j broadcasts ("disagree", $g_j(i)$, $r'_{i,j}$); and $r_{i,j} = r'_{i,j}$. For a pair of conflicting parties (P_i, P_j), we say that P_i (resp., P_j) is *unhappy* if one of the following conditions hold:

- The dealer broadcasts ("not equal", $d_{i,j}$) and $d_{i,j} \neq f_i(j)$ (resp., $d_{i,j} \neq g_j(i)$).
- The dealer broadcasts ("equal", $d_{i,j}$) and $d_{i,j} \neq f_i(j) + r_{i,j}$ (resp., $d_{i,j} \neq g_j(i) + r'_{i,j}$).

Note that all parties agree on who is unhappy. If there are more than t unhappy parties, the dealer is disqualified and a default value is shared.

Reconstruction phase. The reconstruction phase is similar to the one in [9], except that we do not use broadcast.

1. Every party P_i that is not unhappy sends $f_i(x)$ and $g_i(y)$ to all other parties.
2. Let f^i_j, g^i_j denote the polynomials that P_j sent to P_i in the previous step. P_i then constructs a *consistency graph* G_i whose vertices correspond to the parties who are not unhappy:
   - Initially, there is an edge between P_j and P_k in G_i if and only if $f^i_j(k) = g^i_k(j)$ and $g^i_j(k) = f^i_k(j)$. (Note that we allow also the case $j = k$ here.)

- If there exists a vertex in G_i whose degree is less than $n - t$ (including self-loops), then that vertex is removed from G_i. This is repeated until no more vertices can be removed.

Let Core_i denote the parties whose corresponding vertices remain in G_i.

3. If $|\mathsf{Core}_i| < n - t$, then P_i outputs $\bot$. Otherwise, P_i reconstructs the polynomial $F'(x, y)$ defined by any $t + 1$ parties in Core_i, and outputs $s' := F'(0, 0)$.

We remark that, since we do not use broadcast in the reconstruction phase, it is possible that $\mathsf{Core}_i, \mathsf{Core}_j$ are different for different honest parties P_i, P_j.

3.2   Security of the Protocol

We state the following claims regarding the protocol of the previous section; all proofs appear in the full version of this work [13].

Lemma 1. *If the dealer is not corrupted by the end of the sharing phase, then privacy is preserved.*

Lemma 2. *If the dealer is not corrupted by the end of the sharing phase, then correctness holds.*

Lemma 3. *Weak commitment holds.*

Our WSS protocol also satisfies a weak variant of 2-level sharing that we state for future reference:

Lemma 4. *Say the dealer is not disqualified in an execution of the WSS protocol, and let $\mathcal{H}$ denote the set of all honest parties who are not unhappy. Then there is a bivariate polynomial $\hat{F}$ of degree at most t in each variable such that, at the end of the sharing phase, the polynomials f_i, g_i held by each $P_i \in \mathcal{H}$ satisfy $f_i(x) = \hat{F}(x, i)$ and $g_i(y) = \hat{F}(i, y)$.*

As a consequence, each $P_i \in \mathcal{H}$ can compute s_i and $s_{i,j}$ for $j \in \{1, \dots, n\}$ such that:

1. *There is a polynomial $p(x)$ of degree at most t with $s_i = p(i)$, and furthermore all honest parties output either $s' = p(0)$ or $\bot$ in the reconstruction phase.*
2. *For each $j \in \{1, \dots, n\}$, there exists a polynomial $p_j(x)$ of degree at most t such that (1) $p_j(0) = p(j)$ and (2) $s_{i,j} = p_j(i)$.*

4   Verifiable Secret Sharing

Before we describe our VSS protocol with 2-level sharing, we review the ideas used in [9] to transform their WSS protocol into a VSS protocol (that does not have 2-level sharing). At a high level, the sharing phase of the VSS protocol is more-or-less the same as the sharing phase of the underlying WSS protocol; the difference is that now, in the reconstruction phase, each party reveals the random pads they used in the sharing phase. A problem that arises is to ensure that a malicious party P_i reveals the "correct" random pads. This is enforced by having each player act as a dealer in its own execution of WSS, and "binding" the random pads of each party to this execution of WSS. In more detail: in parallel with the sharing phase of the larger VSS protocol, each party P_i also acts as a dealer and shares a random secret using the WSS protocol. Let $F_i^{pad}(x, y)$ be the corresponding bivariate polynomial chosen by P_i. Then P_i will use $r_{i,j} := F_i^{pad}(0, j)$ as the appropriate "random pad" in the larger VSS protocol. (The pads used by any player are now only $(t + 1)$-wise independent, but this suffices for secrecy.) These random pads are then revealed in the reconstruction phase by using the reconstruction phase of the underlying WSS protocol.

We can use the ideas outlined in the previous paragraph to obtain a $(3, 1)$-round VSS protocol, but the resulting protocol will not have 2-level sharing. Yet all is not lost. As observed already in Lemma 4, by the end of the sharing phase of the resulting VSS protocol the honest parties that are *not* unhappy *do* have the required 2-level shares. To achieve our desired result we must therefore only enable any *unhappy* honest party to construct *its* 2-level shares.

At a high level, we do this as follows: Suppose $\hat{F}(x, y)$ is the dealer's bivariate polynomial, defined by the end of the sharing phase of the VSS protocol, and let P_i be an honest party who is unhappy. We need to show how P_i constructs the polynomials $\hat{F}(x, i)$ and $\hat{F}(i, y)$ (which it will use to generate its 2-level shares exactly as in the proof of Lemma 4). Let P_j be a party such that:

- P_j is not unhappy (in the larger VSS protocol);
- P_j was not disqualified as a dealer it its own execution of WSS; and
- P_i is not unhappy in P_j's execution of WSS.

From the proof of Lemma 4, we know there is a bivariate polynomial $\hat{F}_j^{pad}(x, y)$ for which P_i holds the univariate polynomial $\hat{F}_j^{pad}(x, i)$. Furthermore, P_j has effectively broadcasted the polynomial $B_j(x) \overset{\text{def}}{=} \hat{F}(x, j) + \hat{F}_j^{pad}(0, x)$ in round 3, since it has broadcasted $\hat{F}(k, j) + \hat{F}_j^{pad}(0, k)$ for all k. Thus, party P_i can compute

$$\hat{F}(i, j) := B_j(i) - \hat{F}_j^{pad}(0, i) = \hat{F}(i, j)$$

for any party P_j satisfying the above conditions. If there are $t + 1$ parties satisfying the above conditions, then P_i can reconstruct the polynomial $\hat{F}(i, y)$.

Unfortunately, it is not clear how to extend the above approach to enable P_i to also reconstruct the polynomial $\hat{F}(x, i)$ in the case when $\hat{F}$ is an *arbitrary* bivariate polynomial. For this reason, we have the dealer use a *symmetric*[3] bivariate polynomial. Then $\hat{F}(x, i) = \hat{F}(i, x)$ and we are done.

4.1   The Protocol

We show a $(3, 1)$-round VSS protocol with 2-level sharing that tolerates $t < n/3$ malicious parties. Proofs of security are deferred to the appendix.

---

[3] A polynomial F is symmetric if, for all ℓ, m, the coefficient of the term $x^\ell y^m$ is equal to the coefficient of the term $x^m y^\ell$. If F is symmetric then $F(i, j) = F(j, i)$ for all i, j.

Sharing phase. The sharing phase consists of three rounds, with broadcast used in the last round.

Round 1: The dealer holds s. The following steps are carried out in parallel:

1. The dealer chooses a random *symmetric* bivariate polynomial $F(x, y)$ of degree t in each variable such that $F(0, 0) = s$. Then D sends to each party P_i the polynomial $f_i(x) := F(x, i)$. Note that $F(x, i) = F(i, x)$ since F is symmetric.

2. Each party P_i picks a random value $\hat{s}_i$ and executes the first round of the WSS protocol described in the previous section, acting as a dealer to share the "input" $\hat{s}_i$. We refer to this instance of the WSS protocol as WSS_i.

3. Let $F_i^{pad}(x, y)$ denote the bivariate polynomial used by P_i in WSS_i (i.e., $F_i^{pad}(0, 0) = \hat{s}_i$). Party P_i sends the polynomial $r_i(y) := F_i^{pad}(0, y)$ to the dealer D.

Round 2: Round 2 of WSS_i is run, for all i. Concurrently, each party P_j does the following:

1. For all i, send $a_{j,i} := f_j(i)$ to P_i.

2. Let $f_{i,j}^{pad}(x)$ be the x-polynomial that P_i sent to P_j in round 1 of WSS_i. (If P_i is honest then $f_{i,j}^{pad}(x) = F_i^{pad}(x, j)$.) Party P_j sends $r'_{i,j} := f_{i,j}^{pad}(0)$ to D.

Round 3: Round 3 of WSS_i is run, for all i. Concurrently, for every ordered pair (i, j):

1. (From the viewpoint of P_i:) If $a_{j,i} \neq f_i(j)$, then P_i broadcasts ("disagree", $f_i(j)$, $F_i^{pad}(0, j)$). Otherwise, P_i broadcasts ("agree", $f_i(j) + F_i^{pad}(0, j)$).

2. (From the viewpoint of P_j:) If $a_{i,j} \neq f_j(i)$, then P_j broadcasts ("disagree", $f_j(i)$, $f_{i,j}^{pad}(0)$). Otherwise, P_j broadcasts ("agree", $f_j(i) + f_{i,j}^{pad}(0)$).

3. (From the viewpoint of D:) If $r_i(j) \neq r'_{i,j}$, then D broadcasts ("not equal", $F(j, i)$). Otherwise, D broadcasts ("equal", $F(j, i) + r_i(j)$).

Local computation. Each party locally carries out the following steps:

1. An ordered pair of parties (P_i, P_j) is *conflicting* if, in round 3, party P_i broadcasts ("disagree", $f_i(j)$, $F_i^{pad}(0, j)$); party P_j broadcasts ("disagree", $f_j(i)$, $f_{i,j}^{pad}(0)$); and it holds that $F_i^{pad}(0, j) = f_{i,j}^{pad}(0)$. For a pair of conflicting parties (P_i, P_j), we say that P_i (resp., P_j) is *unhappy* if one of the following conditions hold:

   (a) D broadcasts ("not equal", $d_{i,j}$) and $d_{i,j} \neq f_i(j)$ (resp., $d_{i,j} \neq f_j(i)$).

   (b) D broadcasts ("equal", $d_{i,j}$) and $d_{i,j} \neq f_i(j) + F_i^{pad}(0, j)$ (resp., $d_{i,j} \neq f_j(i) + f_{i,j}^{pad}(0)$).

   Let Core denote the set of parties who are not unhappy with respect to the definition above. For every P_i who was not disqualified as the dealer in WSS_i, let Core_i denote the set of parties who are not unhappy with respect to WSS_i. (If P_i was disqualified in WSS_i, then set $\mathsf{Core}_i := \emptyset$.)

2. For all i, j, remove P_j from Core_i if either of the following hold for the ordered pair (i, j) in round 3:
   - P_i broadcasts ("agree", y) and P_j did not broadcast ("agree", y).
   - P_i broadcasts ("disagree", $\star$, w) and P_j broadcasts anything other than ("disagree", $\star$, w). (Here, $\star$ denotes an arbitrary value.)

3. Remove P_i from Core if $|\mathsf{Core} \cap \mathsf{Core}_i| < n - t$. (Thus, if P_i was disqualified in WSS_i then $P_i \notin \mathsf{Core}$.)
   Note that all parties have the same view regarding Core and the $\{\mathsf{Core}_i\}$.

4. If $|\mathsf{Core}| < n - t$, then the dealer is disqualified and a default value (and appropriate 2-level shares) are shared.

5. Each party P_i computes a polynomial $\hat{f}_i(x)$ of degree at most t:
   (a) If $P_i \in \mathsf{Core}$, then $\hat{f}_i(x)$ is the polynomial that P_i received from the dealer in round 1.
   (b) If $P_i \notin \mathsf{Core}$, then P_i computes $\hat{f}_i(x)$ in the following way:
       i. P_i first defines a set Core_i' as follows: A party P_j is in Core_i' if and only if all the following conditions hold:
          - $P_j \in \mathsf{Core}$ and $P_i \in \mathsf{Core}_j$.
          - Define $p_{j,k}$, for $k \in \{1, \ldots, n\}$, as follows: if, in step 1 of round 3 for the ordered pair (j, k), party P_j broadcasted ("agree", $y_{j,k}$), then set $p_{j,k} := y_{j,k}$. If P_j broadcasted ("disagree", $w_{j,k}, z_{j,k}$), then set $p_{j,k} := w_{j,k} + z_{j,k}$.

          We require that the $\{p_{j,k}\}$ are consistent with a polynomial $B_j(x)$ of degree at most t; i.e., $B_j(k) = p_{j,k}$ for all k. (If not, then P_j is not included in Core_i'.)
          Our proofs show that $|\mathsf{Core}_i'| \geq t + 1$ if the dealer is not disqualified.
       ii. For each $P_j \in \mathsf{Core}_i'$, set $p_j := p_{j,i} - f_{j,i}^{pad}(0)$. Let $\hat{f}_i$ be the polynomial of degree at most t such that $\hat{f}_i(j) = p_j$ for every $P_j \in \mathsf{Core}_i'$. (It will follow from our proof that such an $\hat{f}_i$ exists.)

6. Finally, P_i outputs $s_i := \hat{f}_i(0)$ and $s_{i,j} := \hat{f}_i(j)$ for all $j \in \{1, \ldots, n\}$.

Reconstruction phase. Each party P_i sends s_i to all other parties. Let $s_{j,i}'$ be the value that P_j sends to P_i. Using Reed-Solomon decoding, P_i computes a polynomial $f(x)$ of degree at most t such that $f(j) = s_{j,i}'$ for at least $2t + 1$ values of j. The final output of P_i is $f(0)$.

A proof of security appears in the full version of this work [13].

References

1. Ben-Or, M., Goldwasser, S., Wigderson, A.: Completeness theorems for non-cryptographic fault-tolerant distributed computation. In: 20th Annual ACM Symposium on Theory of Computing (STOC), pp. 1–10 (1988)
2. Blakley, G.R.: Safeguarding cryptographic keys. In: National Computer Conference, vol. 48, pp. 313–317. AFIPS Press (1979)

3. Chor, B., Goldwasser, S., Micali, S., Awerbuch, B.: Verifiable secret sharing and achieving simultaneity in the presence of faults. In: 26th Annual IEEE Symposium on Foundations of Computer Science (FOCS), pp. 383–395 (1985)
4. Cramer, R., Damgård, I., Maurer, U.: General secure multi-party computation from any linear secret sharing scheme. In: Preneel, B. (ed.) EUROCRYPT 2000. LNCS, vol. 1807, pp. 316–334. Springer, Heidelberg (2000)
5. Dolev, D., Dwork, C., Waarts, O., Yung, M.: Perfectly secure message transmission. J. ACM 40(1), 17–47 (1993)
6. Feldman, P., Micali, S.: An optimal probabilistic protocol for synchronous Byzantine agreement. SIAM J. Computing 26(4), 873–933 (1997)
7. Fischer, M.J., Lynch, N.A.: A lower bound for the time to assure interactive consistency. Information Processing Letters 14(4), 183–186 (1982)
8. Fitzi, M., Garay, J.: Efficient player-optimal protocols for strong and differential consensus. In: 22nd Annual ACM Symp. on Principles of Distributed Computing, pp. 211–220 (2003)
9. Fitzi, M., Garay, J.A., Gollakota, S., Rangan, C.P., Srinathan, K.: Round-optimal and efficient verifiable secret sharing. In: Halevi, S., Rabin, T. (eds.) TCC 2006. LNCS, vol. 3876, pp. 329–342. Springer, Heidelberg (2006)
10. Gennaro, R., Ishai, Y., Kushilevitz, E., Rabin, T.: The round complexity of verifiable secret sharing and secure multicast. In: 33rd Annual ACM Symposium on Theory of Computing (STOC), pp. 580–589 (2001)
11. Katz, J., Koo, C.-Y.: On expected constant-round protocols for Byzantine agreement. In: Dwork, C. (ed.) CRYPTO 2006. LNCS, vol. 4117, pp. 445–462. Springer, Heidelberg (2006)
12. Katz, J., Koo, C.-Y.: Round-efficient secure computation in point-to-point networks. In: Naor, M. (ed.) EUROCRYPT 2007. LNCS, vol. 4515, pp. 311–328. Springer, Heidelberg (2007)
13. Katz, J., Koo, C.-Y., Kumaresan, R.: Improving the round complexity of VSS in point-to-point networks, http://eprint.iacr.org/2007/358
14. Koo, C.: Studies on Fault-Tolerant Broadcast and Secure Computation. PhD thesis, University of Maryland (2007)
15. Lindell, Y., Lysyanskaya, A., Rabin, T.: Sequential composition of protocols without simultaneous termination. In: 21st Annual ACM Symposium on Principles of Distributed Computing, pp. 203–212 (2002)
16. Micali, S., Rabin, T.: Collective coin tossing without assumptions nor broadcasting. In: Menezes, A., Vanstone, S.A. (eds.) CRYPTO 1990. LNCS, vol. 537, pp. 253–266. Springer, Heidelberg (1991)
17. Patra, A., Choudhary, A., Ashwinkumar, B., Rangan, C.: Probabilistic verifiable secret sharing tolerating an adaptive adversary, http://eprint.iacr.org/2008/101
18. Rabin, T., Ben-Or, M.: Verifiable secret sharing and multiparty protocols with honest majority. In: 21st Annual ACM Symposium on Theory of Computing, pp. 73–85 (1989)
19. Shamir, A.: How to share a secret. Comm. ACM 22(11), 612–613 (1979)

How to Protect Yourself without Perfect Shredding*

Ran Canetti[1], Dror Eiger[2], Shafi Goldwasser[3], and Dah-Yoh Lim[4]

[1] IBM T. J. Watson Research Center
[2] Google, Inc. (work done at Weizmann Institute of Science)
[3] MIT and Weizmann Institute of Science
[4] MIT

Abstract. Erasing old data and keys is an important tool in cryptographic protocol design. It is useful in many settings, including proactive security, adaptive security, forward security, and intrusion resilience. Protocols for all these settings typically assume the ability to *perfectly erase* information. Unfortunately, as amply demonstrated in the systems literature, perfect erasures are hard to implement in practice.

We propose a model of *partial erasures* where erasure instructions leave almost all the data erased intact, thus giving the honest players only a limited capability for disposing of old data. Nonetheless, we provide a general compiler that transforms any secure protocol using perfect erasures into one that maintains the same security properties when only partial erasures are available. The key idea is a new redundant representation of secret data which can still be computed on, and yet is rendered useless when partially erased. We prove that any such a compiler must incur a cost in additional storage, and that our compiler is near optimal in terms of its storage overhead.

Keywords: mobile adversary, proactive security, adaptive security, forward security, intrusion resilience, universal hashing, partial erasures, secure multiparty computation, randomness extractors.

1 Introduction

As anyone who has ever tried to erase an old white board knows, it is often easier to erase a large amount of information imperfectly, than to erase a small amount of information perfectly.

In cryptographic protocol design, *perfect erasures*, namely the complete disposal of old and sensitive data and keys, is an important ability of honest players in fighting future break-ins, as this leaves no trace of sensitive data for the adversary to recover.

Examples where perfect erasures have been used extensively include the areas of *proactive security* [7,17,19,22,30,36], *forward security* [1,12,20] and *intrusion*

---

* Full version of paper available at http://eprint.iacr.org/ as [5]

L. Aceto et al. (Eds.): ICALP 2008, Part II, LNCS 5126, pp. 511–523, 2008.

resilience [27], and *adaptive security* [2,6,28,37]. Whereas erasures merely simplify the design of adaptively secure protocols, some form of erasures is provably necessary for achieving proactive security and even for defining the task of forward security as we explain below.

The goal of *Proactive Security* introduced in [36] is to achieve secure multiparty computations where some fraction of the parties are faulty. The identity of faulty parties are decided by a *mobile* adversary who can corrupt a different set of players in different *time periods* (here the protocols assume time is divided into well-defined intervals called time periods) subject to an upper bound on the total number of corrupted players per time period. At the heart of the solutions pursued in the literature are secret sharing methods in which in every time period, the old shares held by players are first replaced by new shares and then perfectly erased. It is easy to prove that secret sharing would be impossible to achieve without some form of erasures: otherwise a mobile adversary which is able to corrupt every single player in some time period or another, can eventually recover all old shares for some single time period and recover the secret. Forward security [1,12,20]. is an approach taken to tackle the *private key exposure* problem, so that exposure of long-term secret information does not compromise the security of previous sessions. Again, the lifetime of the system is divided into time periods. The receiver initially stores secret key SK_0 and this secret key "evolves" with time: at time period i, the receiver applies some function to the previous key SK_{i-1} to derive the current key SK_i and then key SK_{i-1} is perfectly erased. The public (encryption) key remains fixed throughout the lifetime of the scheme. A forward-secure encryption scheme guarantees that even if an adversary learns the secret key available at time i, SK_i, messages encrypted during all time periods prior to i remain secret. Intrusion Resilience is a strengthening of forward security [27] which can be viewed as combination of forward and backward security. Obviously, erasures are essential to define (and solve) both the forward security and intrusion resilience problems.

Finally, an example of a different flavor of the utility of erasures to guard against adversaries that can choose which future parties to corrupt as the protocol proceeds, based on information already gathered. Erasures are useful in this context since they limit the information the adversary sees upon corrupting a party. Protocols designed without erasures (although possible in this context), tend to be much more complex than those that rely on data erasures [2,6,28,37].

Unfortunately, perfect erasures of data are hard to achieve in practice and are thus problematic as a security assumption, as pointed out by Jarecki and Lysyanskaya [28] in their study of adaptive adversaries versus static adversaries in the context of threshold secret sharing.

Some of the difficulty in implementing perfect erasures is illustrated in the works of Hughes and Coughlin, Garfinkel, and Vaarala [18,24,25,39]. The root cause of this difficulty is that systems are actually designed to preserve data, rather than to erase it. Erasures present difficulties at both the hardware level (e.g. due to physical properties of the storage media) and at the software level (e.g. due to the complications with respect to system bookkeeping and backups).

At the hardware level, e.g. for hard drives, the Department of Defense recommends overwriting with various bit patterns [35]. This takes the order of days per 100GB, and is not fully effective because modern hard drives use block replacement and usually employ some form of error correction. For main memory, due to "ion migration", previous states of memory can be determined even after power off. At the software level, many operating systems detect and remap bad sectors of the hard drive on the fly, but original data can remain in the bad sectors and be recoverable.

1.1   This Paper

In light of the above difficulties, we propose to study protocols that can guarantee security even when only *imperfect* or *partial* erasures are available.

The first question to be addressed is how to *model* erasures that are only partially effective. One option is to simply assume that each erasure operation succeeds with some probability. However, such a modeling does not capture all the difficulties described above. In particular, it allows obtaining essentially perfect erasures by applying the erasure operation several times on a memory location; therefore such a model is unlikely to yield interesting or effective algorithms. In addition, such modeling does not take into account potential dependencies among information in neighboring locations.

The model of Partial Erasures. We thus take a much more conservative approach. Specifically, we model partial erasures by a length-shrinking function $h : \{0,1\}^m \mapsto \{0,1\}^{\lfloor \phi m \rfloor}$, that shrinks stored information by a given fraction $0 \le \phi \le 1$. We call ϕ the leakage fraction. When $\phi = 0$ then we get the perfect erasures case; when $\phi = 1$ nothing is ever erased. For the rest of this work we think of ϕ being a value close to 1 (namely, the size of what remains after data is partially-erased is close to the original size). Note that we do not require ϕ to be a constant – for instance, for reasonable settings of the parameters, it may be $\frac{1}{poly(\alpha)}$ close to 1, where α is a security parameter of the protocol in question.

The shrinking function may be chosen adversarially. In particular, it is not limited to outputting exact bits, and any length-shrinking function (efficiently computable or not) on the inputs is allowed. This modeling captures the fact that the remaining information may be a function of multiple neighboring bits rather than on a single bit. It also captures the fact that repeated erasures may not be more effective than a single one.

The function h is assumed to be a function only of the storage contents to be erased. Furthermore, for simplicity we assume that h is fixed in advance – our schemes remain secure without any modification even if the adversary chooses a new h_i prior to each new erasure. This choice seems to adequately capture erasures that are only partially successful due to the physical properties of the storage media[1]. However, this may not adequately capture situations where the failure to erase comes from interactions with an operating system, for

---

[1] Indeed, physical properties of the storage are mostly fixed at the factory; from then on the behavior of the hardware only depends on what is written.

instance memory swapping, and caching. In order to capture this sort of erasure failures, one might want to let h to be a function of some information other than the contents to be erased, or alternatively to be determined adaptively as the computation evolves.

We treat m, the input length of h, as a system parameter. (For instance, m might be determined by the physical properties of the storage media in use.) One can generalize the current model to consider the cases where h is applied to variable-length blocks, and where the block locations are variable.

Our Memory Model. We envision that processors participating in protocols can store data (secret and otherwise) in main memory as well as cache, hard drives, and CPU registers. We assume all types of storage are partially erasable, except a constant number of constant size CPU registers which are assumed to be perfectly erasable. We emphasize that the constant size of the registers ensures that we do not use this to effectively perfectly erase main memory and thus circumvent the lack of perfect erasures in main memory, since at no time can the registers hold any non-negligible part of the secret. We call this the *register model.*

We shall use these registers to perform intermediate local computations during our protocols. This will allow us to ignore the traces of these computations, which would otherwise be very messy to analyze.

Results and Techniques. Our main result is a compiler that on input any protocol that uses perfect erasures, outputs one that uses only partial erasures, and preserves both the functionality and the security properties of the original protocol. Our transformation only applies to the storage that needs to be erased.

The idea is to write secrets in an *encoded form* so that, on the one hand, the secret can be explicitly extracted from its encoded form, and on the other hand loss of even a small part of the encoded form results in loss of the secret.

Perhaps surprisingly, our encoding results in *expanding* the secret so that the encoded information is longer than the original. We will prove that expanding the secret is *essential* in this model (see more discussion below). This expansion of secrets seems a bit strange at first, since now there is more data to be erased (although only partially). However, we argue that it is often easier to erase a large amount of data imperfectly than to erase even one bit perfectly.

We describe the compiler in two steps. First we describe a special case where there is only a single secret to be erased. Next we describe the complete compiler.

Our technique for the case of a single secret is inspired by results in the *bounded storage model*, introduced by Maurer [32,33]. Work by Lu [29] casted results in the bounded storage model in terms of *extractors* [34], which are functions that when given a source with some randomness, purifies and outputs an almost random string.

At a high level, in order to make an n-bit secret s partially erasable, we choose random strings R, X and store $R, X, \mathrm{Ext}(R, X) \oplus s$, where Ext is a strong extractor that takes R as seed and X as input, and generates an n-bit output such that $(R, \mathrm{Ext}(R, X))$ is statistically close to uniform as long as the input X has sufficient min-entropy. To erase s, we apply the imperfect erasure operation on X. Both R and $\mathrm{Ext}(R, X) \oplus s$ are left intact.

For sake of analysis, assume that $|X| = m$, where m is the input length for the partial erasure function h. Recall that erasing X amounts to replacing X with a string $h(X)$ whose length is ϕm bits. Then, with high probability (except with probability at most $2^{-(1-\phi)m/2}$), X would have about $(1 - \phi)m/2$ min-entropy left given $h(X)$. This means that, as long as $(1 - \phi)m/2 > n$, the output of the extractor is roughly $2^{-(1-\phi)|X|/2}$-close to uniform even given the seed R and the partially erased source, $h(X)$. Consequently, s is effectively erased.

There is however a snag in the above description: in order to employ this scheme, one has to evaluate the extractor Ext without leaving any trace of the intermediate storage used during the evaluation. Recall that our model the size of the perfectly erasable memory is constant independently of n, the length of the secret. This means that Ext should be computable with constant amount of space, even when the output length tends to infinity. We identify several such extractors, including ϵ-almost universal hashing, strong extractors in NC^0, and Toeplitz Hashing [31]. It would seem superficially that locally computable strong extractors [40] can be used, but unfortunately they cannot (proof deferred to full version [5]).

Now let us move on to describe the general compiler. Suppose we want to compute some function g (represented as a circuit) on some secret s, only now, s is replaced by a representation that is partially erasable, and we would like to make sure that we can still compute $g(s)$. We are going to evaluate the circuit in a gate-by-gate manner where the gate inputs are in expanded form. The inputs are reconstructed in the registers, and the gate is evaluated to get an output, which is in turn expanded and stored in main memory. Even though some small (negligible) amount of information is leaked at each partial erasure, we show that as long as the number of erasure operations is sub-exponential, the overall amount of information gathered by the adversary on the erased data is negligible.

For maximum generality we formulate our results in the *Universally Composable (UC)* framework. In particular we use the notion of *UC emulation* [3], which is a very tight notion of correspondence between the emulated and emulating protocols. Our analysis applies to essentially any type of corruption – adaptive, proactive, passive, active, etc. That is, we show:

Theorem (informal): For any protocol Π_{org} that requires perfect erasures (for security), the protocol $\Pi_{new} = \text{COMPILER}(\Pi_{org})$ UC-emulates Π_{org}, and tolerates (maintains security even with) imperfect/partial erasures in the register model. For leakage fraction of ϕ, if Π_{org} uses n bits of storage then Π_{new} uses about $\frac{2}{1-\phi}n$ bits of storage.

Optimality of the scheme. One of the main cost parameters of such compilers is the *expansion factor*, the amount by which they increase the (erasable part of the) storage. That is, a compiler has expansion factor Ψ if whenever Π_{org} uses n bits of storage, Π_{new} uses at most Ψn bits of storage. It can be seen that our compiler has expansion factor $\Psi \leq \frac{2}{1-\phi} + \nu(n)$ where ν is a negligible function. In addition, in [5] we show that if ϵ-almost universal hashing is used and $\phi > 1/4$, then our compiler would have an expansion factor of about $\frac{c}{1-\phi}$, where $1 < c < 2$ is a constant.

We show that our construction is at most twice the optimal in this respect. That is, we show that *any* such compiler would necessarily have an expansion of roughly $\Psi \geq \frac{1}{1-\phi}$. This bound holds even for the simplest case of compiling even a single secret into one that is partially erasable. Roughly speaking, the argument is as follows. If we do not want to leak any information on a secret of n bits the function h must shrink the expanded version of s by at least n bits. In our model, h shrinks by $(1-\phi)\Psi n$ bits and therefore, $(1-\phi)\Psi n \geq n \Rightarrow \Psi \geq \frac{1}{1-\phi}$.

Some specific solutions. In addition to the general compiler, in [5] we describe some special-tailored solutions to two specific cases. One case is where the function to be evaluated is computable by NC^0 circuits. The second case is the case for all known proactive secret sharing schemes. These solutions are computationally more efficient since they do not require running the compiler on a gate by gate basis. In particular, in the case of proactive secret sharing we can apply our expanded representation directly to the secret and its shares and the instructions which modify the shares (to accordingly modify the new representations) and leave the rest of the protocol intact. Note that this greater efficiency also translates into tighter security – for instance if the original protocol assumed some timing guarantees, then the new protocol need not assume a timing that is much looser than the original.

Remark 1. As we elaborate in [5], a side benefit of using using our constructions is that it can be resistant to a practical class of physical attacks [21] that involves freezing RAM and recovering secrets from it.

Remark 2. Note that because we prove statistical security, our schemes are "everlastingly secure" in the sense that even if the adversary stores all the partially erased information, whatever happens in the future will not help him, e.g. even if it turns out that $P = NP$.

1.2   Related Work

The Bounded Storage Model (BSM). The Bounded Storage Model (BSM) proposed by Maurer [32,33], considers computationally unbounded but storage limited adversaries. This enables novel approaches to the secure communication problem as follows. The communicating parties begin with a short initial secret key k. In the first phase they use this key k and access to a long public random string R to derive a longer key X. The storage bounded adversary computes an arbitrary length-shrinking function on R. In the second phase, R "disappears", and the parties will use X as a one-time pad to communicate privately.

We will use the same kind of length-shrinking function to capture the act of partially erasing old shares of a secret.

However, conceptually the settings of the BSM and partial erasures are fundamentally different. In the BSM model possibility is proved by putting limitations on the **adversary** (storage), where as in our work possibility is proved *inspite* of putting limitation on the **honest parties** (erasing capability). Thus, although some of techniques are similar the setup is entirely different. From a technical point of view there are two differences we emphasize as well. Firstly, the extractors that we use must be computable with constant sized memory), whereas in the BSM the extractors are not necessarily computable with constant-sized memory. Secondly, in the BSM, it is assumed that the adversary's storage bound remains the same as time goes by, namely a constant fraction ϕ of the public randomness R. The same assumption is used in the bounded retrieval model [11,10,14,15,16]. For instance [16] constructs intrusion resilient secret sharing schemes by making the shares large and assuming that the adversary will never be able to retrieve any share completely. For partial erasures this bound is unreasonable, and we allow the adversary to get ϕ fraction of R_i for each erasure operation.

Exposure Resilient Functions. Exposure-Resilient Functions, or ERFs, were introduced by Canetti et al. [4,13]. An ℓ-ERF is a function with a random input, such that an adversary which learns all but ℓ bits of the input, cannot distinguish the output of the function from random.

At a high level the ERF objectives seem very similar to partial erasures. However, the settings are different. In particular, ERFs only deal with the leakage of exact bits whereas we deal with the leakage of general information. (We remark that this limitation of ERFs is inherent in their model: It is easy to see that there do not exist ERFs that resists arbitrary leakage functions).

Encryption as Deletion. As Di Crescenzo et al. [9] noted, one simple but inefficient way to implement erasable memory can be obtained by using the crypto-paging concept of Yee [41]. Assume that some amount of storage that is linear in the security parameter is available that is perfectly erasable, and some other poly storage is persistent. To make the persistent memory effectively erasable, pick an encryption scheme and keep the key in the erasable part. Always encrypt the contents to be kept on the persistent storage. Then erasing the key is as good as erasing the contents.

By combining these ideas with ours, it is possible to have an increase in storage that is linear in the security parameter, while using only a *constant* amount of perfectly erasable memory.

2   How to Make Secrets Partially Erasable

To change a protocol using perfect erasures to one that uses only partial erasures, the high level idea is that instead of having a piece of secret $s \in \{0,1\}^n$ directly in the system, we let the parties store it in expanded form. At the cost of more storage, this gives the ability to effectively erase a secret even when only partial erasures are available. In the end, the number of bits that have to be partially erased might be more than the number of bits that have to be perfectly erased. This is still reasonable because it is often much easier to partially erase a large number of bits than to perfectly erase a small number. Furthermore, we show in 2.1 that such expansion is inherent in the model.

We write $U_{\mathcal{X}}$ to denote an r.v. uniformly random on some set $\mathcal{X}$.

Definition 1 (Statistical Distance). *Suppose that A and B are two distributions over the same finite set Σ. The statistical distance between A and B is defined as $\Delta(A; B) := \frac{1}{2} \sum_{\sigma \in \Sigma} \left| \Pr_A[\sigma] - \Pr_B[\sigma] \right|$.*

Definition 2 (Statistical Distance from the Uniform). *Let $d(A) := \Delta (A; U_A)$. Also define $d(A|B) := \sum_b d(A|B = b) \cdot \Pr_B[b] = \sum_b \Pr_B[b] \frac{1}{2} \sum_a | \Pr_{A|B=b}[a] - \frac{1}{|A|} |$. We say that a random variable A is ϵ-close to the uniform given B to mean that $d(A|B) \leq \epsilon$.*

Due to lack of space we refer the reader to the introduction for the definitions of a *partial erasure function* $h : \{0,1\}^m \mapsto \{0,1\}^{\lfloor \phi m \rfloor}$ and a *leakage fraction* $0 \leq \phi \leq 1$.

Definition 3 (Partially Erasable (or Expanded) Form of a Secret). *Let $\mathsf{Exp}(\circ, \circ)$ be the "expansion" function taking the secret s to be expanded as the first input and randomness as the second, and Con be the "contraction" function taking the output of Exp as the input. Let $h_i^s := h(\mathsf{Exp}(s, \$_i))$, where $\$_i$ are independent randomness. We say that $(\mathsf{Exp}, \mathsf{Con})$ is (ℓ, α, ϕ)-partially erasable form of a secret if $\forall s \in \{0,1\}^n$, for any h with leakage fraction ϕ,*

1. *(Correctness) $\mathsf{Con}(\mathsf{Exp}(s, r)) = s$ for all $r \in \{0,1\}^{poly(n)}$.*
2. *(Secrecy) $\forall s' \in \{0,1\}^n, \Delta(h_1^s, ..., h_\ell^s; h_1^{s'}, ..., h_\ell^{s'}) \leq 2^{-\alpha}$.*
3. *(With Constant Memory) Both $\mathsf{Exp}, \mathsf{Con}$ are computable with constant memory.*

Remark 3. We require both Exp and Con to be computable with constant memory to ensure that intermediate computations can be kept in the registers which are perfectly erasable.

Remark 4. We require indistinguishability for many (ℓ above) erasures to account for the fact that many computations may be done during the protocol (directly or indirectly) on the secret, from which the adversary might gain more information on a secret. Generally, an adversary may have many partially erasable forms of the same secret (i.e adversary can see $h(\mathsf{Exp}(s, \$_i))$ s.t for each i adversary knows a 1-1 and onto correspondence q_i from $\mathsf{Exp}(s, \$_i)$ to s).

An example of an expanded form of a secret which can be partially erased and satisfies correctness and secrecy (in the above definition) would be to use a universal hash function family $\{H_R\}$ as follows: expand s to (v, R, k) s.t. $s = H_R(k) \oplus v$. By using the leftover hash lemma [26], for any constant ϕ such that $0 < \phi < 1$, for any arbitrary partial erasure function h with leakage fraction ϕ, for any universal hash function family $\{H_R\}$, $H_R(k)$ can be made negligibly close to uniform given R and $h(k)$ (so $H_R(k)$ is as good as a one time pad). Let us first focus on bounding $d(H_R(k)|R, h(k))$.

Theorem 1 (Security for a Single Erasure using Universal Hash). *Let* $\{H_R\}$ *be a universal family of hash functions. Let* $(R, h(k))$ *be a tuple such that* $R \in \{0,1\}^{n \times m}$, $k \in \{0,1\}^m$, *and* $h(k) \in \{0,1\}^{\phi m}$, *where* R *picks out a random function out of* $\{H_R\}$, *and* k *is random. Then,* $d(H_R(k)|R, h(k)) \leq 2^{-\frac{1}{3}(1-\phi)m+\frac{n}{3}+1}$.

The proof of this theorem and the next is deferred to the full version [5].

Theorem 2 (Security for Multiple Erasures using Universal Hash). *Let* $\{H_R\}$ *be a family of universal hash functions. Let* $(R_1, h(k_1)), ..., (R_\ell, h(k_\ell))$ *be* ℓ *tuples such that* $R_i \in \{0,1\}^{n \times m}$, $k_i \in \{0,1\}^m$, *and* $h(k_i) \in \{0,1\}^{\phi m}$, *where* R_i *picks out a random function out of* $\{H_R\}$, k_i *is random, and* q_i *are public 1-1 correspondences such that* $s = q_i(H_{R_i}(k_i))$. *Then, for any* $\beta > 0$, m *poly in* n, *and sufficiently large* n,

$$d(H_{R_i}(k_i)|R_1, h(k_1), ..., R_\ell, h(k_\ell)) \leq \sqrt{\frac{\ln 2}{2} \ell 2^{-\frac{1}{3}(1-\phi)m+\frac{(\beta+1)n}{3}-\frac{1}{3}}}.$$

Note that to get $2^{-(\alpha+1)}$ security when the adversary gets ℓ partially erased tuples, we need:

$$\sqrt{\frac{\ln 2}{2} \ell 2^{-\frac{1}{3}(1-\phi)m+\frac{(\beta+1)n}{3}-\frac{1}{3}}} \leq 2^{-(\alpha+1)}$$

$$\Leftrightarrow \ell \leq \frac{2^{\frac{4}{3}-2(\alpha+1)-\frac{1}{3}(1+\beta)n+\frac{1}{3}m(1-\phi)}}{\ln 2} \tag{1}$$

$$\Leftrightarrow m \geq \frac{3\log\left((\ln 2)\ell\right) - 4 + 6(\alpha+1) + (1+\beta)n}{(1-\phi)} \tag{2}$$

Let us make a few observations. Inequality 1 shows that if h has leakage fraction ϕ, how many times can you partially erase a secret (or computations on the secret) without leaking too much information. Rearranging, and fixing the other parameters, we can also see that the fraction that needs to be erased, $(1 - \phi)$, has to be at least logarithmic in ℓ. Inequality 2 on the other hand lower bounds m, which as we will see shortly, translates into a statement about the space efficiency of using universal hashing to get partially erasable forms.

Let us now consider two partially erasable forms based on universal hashing which satisfy correctness, secrecy and moreover can be computed with constant size memory. The expansions we consider can be thought of as having two parts, R, k, each serving different purposes. Furthermore, only one part, k, needs to be partially erased[2].

The first expanded form of s is random matrix $R \in \{0,1\}^{n \times m}$ and vector $k \in \{0,1\}^m$ subject to the constraint that $R \cdot k = s$. Only the vector k needs to be erased. However, this simple construction is highly randomness inefficient.

Our preferred partial erasable form will be to use Toeplitz hashing instead (whose universality is proven in [31]). A random Toeplitz matrix $R \in \{0,1\}^{n \times m}$

---

[2] Which makes our results stronger than required by the definition.

which selects a random hash function out of the Toeplitz family $\{H_R\}$, where $H_R : \{0,1\}^m \mapsto \{0,1\}^n$. In order to store an n-bit secret s in a partially erasable manner, we choose random Toplitz matric $R \in \{0,1\}^{n \times m}, k \in \{0,1\}^m$ (where R is fully specified by a random string of $n + m - 1$ bits), and store $R, k, R \cdot k \oplus s$ instead. Again, only the k part needs to be erased. In this case $\mathsf{Exp}(s,r)$ uses r to form a random Toeplitz R and a random k, and outputs $R, k, R \cdot k \oplus s$, and $\mathsf{Con}(R, k, x)$ recovers s by computing $R \cdot k \oplus x$.

It is easy to see how to implement $(\mathsf{Exp}, \mathsf{Con})$ corresponding to Toeplitz hash that is computable with constant memory, bit by bit. From the triangle inequality, for all $s, s' \in \{0,1\}^n$, $\Delta\big(H_R(k) \oplus s, R, h(k); H_R(k) \oplus s', R, h(k)\big) \leq 2d(H_R(k)|R, h(k))$. Combining these with theorem 2 proves that:

Corollary 1 (Toeplitz Hashing gives a Partially Erasable Form). *Toeplitz hashing yields a partially erasable form of a secret.*

2.1 Space Efficiency

Lower Bound. Say that an expansion function Exp is Ψ-expanding if for any r we have $|\mathsf{Exp}(s,r)| \leq \Psi|s|$. One parameter we would like to minimize is Ψ, the storage overhead, whose lower bound is given below (proof in [5]):

Theorem 3 (Lower Bound on the Storage Expansion Ψ). *For any Ψ-expanding, (ℓ, α, ϕ)-partially erasable expansion function Exp that is applied to inputs of length n we have:* $\Psi \geq \frac{1}{1-\phi}\left(1 - \frac{n+\alpha-1}{n\ell 2^{\alpha-1}}\right)$.

For typical settings of the parameters, where both α and ℓ are polynomial in n, we get that $\Psi \geq \frac{1}{1-\phi}(1 - \mathrm{neg}(\alpha))$.

Efficiency. Let us see how tight our construction is to the lower bound. If a completely random R is used and $H_R := R \cdot k$ (whose universality is proven in [8]), then the expansion factor Ψ of the storage would be (size of R + size of k) $\cdot \frac{1}{n}$, which is $(n+1)m \cdot \frac{1}{n} = (1 + \frac{1}{n})m$. Plugging this into inequality 2, we see that this bound is a (growing) factor of n away from the optimal.

If a Toeplitz matrix R is used instead, then the corresponding expanded form will be $R \in \{0,1\}^{n \times m}, k \in \{0,1\}^m$ and $x \in \{0,1\}^n$ such that $R \cdot k \oplus x = s$. In this case, R requires $n + m - 1$ bits to specify (since it is Toeplitz), and k requires m bits and x requires n bits respectively. So in this case, n bits get expanded into $2m + 2n - 1$ bits, and Ψ is $2\frac{m}{n} + 2 - \frac{1}{n}$. Plugging in inequality 2, we see that for $\alpha = O(n)$ and $\ell = 2^{O(n)}$, then this bound is a constant factor away from the optimal given in theorem 3. If ℓ is subexponential in n and α is sublinear in n, then the bound we get is about $\Psi \geq \frac{2}{1-\phi} + 2$, so it is essentially a factor of 2 away from the optimal bound[3].

---

[3] In the full version [5] we discuss two other partially erasable forms: 1. ϵ-almost universal hashing which, provided that $\phi > 1/4$, gives roughly $\Psi \geq \frac{c}{1-\phi}$ for some constant $1 < c < 2$, and 2. strong extractors computable with constant memory (e.g. those in NC^0), which, provided that the extractor is near optimal, would achieve the lower bound. Unfortunately we do not know of such extractors.

3   A General Construction

3.1   Computing on Partially Erasable Secrets at the Gate Level

Let $s \in \{0,1\}^n$ be the secret involved, and let $(\mathsf{Exp}, \mathsf{Con})$ be the partially erasable form of s. Consider any efficient computation on s, which can be modeled as a $\text{poly}(n)$-sized circuit. Without loss of generality, we consider gates with fan-in of two and fan-out of one, and consider each output bit separately as being computed by a polynomial-sized circuit.

To evaluate a gate, the two corresponding input bits are reconstructed from their expanded forms in the registers (using Con). The gate is evaluated, resulting in an output bit b in the registers. This output bit is expanded into the partially erasable form and output to main memory[4], by using Exp. This can be done with constant memory. Note that if we just store the values of the wires naïvely, i.e. by individually expanding the 1-bit value of each wire to a Ψn size secret, then the overhead of our scheme will not even be constant. So we must amortize the cost: group the wires of the circuit into groups of size t (i.e., there are t wires in each group), where t is such that secrets of size t are expanded into m-bit strings. Now, when we write the values of the wires in an expanded form, we expand all the t values into a single m-bit string. This will make sure that the overhead of the general compiler will still be the same as the overhead for the scheme described in section 2.

The above is an informal description of $\textsc{Compute-in-register}(g, \mathsf{Exp}(s, \$))$, which makes sure that the computation of $g(s)$ is done properly without leaking intermediate computation (through expressing them in expanded form). The proof of the following lemma is in [5].

Lemma 1. *Let $s \in \{0,1\}^n$ be any secret, g be the function to be computed on s, where each bit i of output of $g(s)$ is computed by a $\text{poly}(n)$-sized circuit C_{g_i}, consisting of gates $\{X_i^j\}_j$. Let v_i denote the number of partially erased intermediate computations while computing the i-th output bit ($\textsc{Compute-in-register}(g, \mathsf{Exp}(s, \$))$), where $\mathsf{Exp}(s, \$)$ is the expansion function of a (ℓ, α, ϕ)-partially erasable form and $v := \sum_i v_i \leq \ell$.*

Then, the adversary cannot distinguish the case of s versus any $s' \in \{0,1\}^n$ being partially erased, by more than $2^{-\alpha}$ probability, i.e.:

$$\Delta\big(h(\mathsf{Exp}(s, \$_1)), ..., h(\mathsf{Exp}(s, \$_v)); h(\mathsf{Exp}(s', \$_1')), ..., h(\mathsf{Exp}(s', \$_v'))\big) \leq 2^{-\alpha}.$$

Starting with any protocol that uses perfect erasures, we replace all computations on the secrets to use $\textsc{Compute-in-register}$ instead. The result that we get is (proof in [5]):

---

[4] Note that even if in practice, storage locations holding expanded forms of the intermediate computations may be overwritten, for analyzing security we can think of all the expanded forms as being written in a new memory location. Put another way, overwriting is just one form of the imperfect erasures we are capturing.

Theorem 4. *For any protocol Π_{org} that requires perfect erasures (for security), the protocol $\Pi_{new} = $ COMPILER(Π_{org}) UC-emulates Π_{org}, and tolerates (maintains security even with) imperfect/partial erasures in the register model.*

References

1. Anderson, R.: Two remarks on public key cryptology invited lecture. In: Acm-Ccs 1997 (1997)
2. Beaver, D.: Plug and play encryption. In: Kaliski Jr., B.S. (ed.) CRYPTO 1997. LNCS, vol. 1294. Springer, Heidelberg (1997)
3. Canetti, R.: Universally composable security: A new paradigm for cryptographic protocols. In: Proc. 42nd IEEE Symp. on Foundations of Comp. Science, pp. 136–145 (2001)
4. Canetti, R., Dodis, Y., Halevi, S., Kushilevitz, E., Sahai, A.: Exposure-resilient functions and all-or-nothing transforms. In: Preneel, B. (ed.) EUROCRYPT 2000. LNCS, vol. 1807, pp. 453–469. Springer, Heidelberg (2000)
5. Canetti, R., Eiger, D., Goldwasser, S., Lim, D.-Y.: How to protect yourself without perfect shredding (full version) (2008), http://eprint.iacr.org/
6. Canetti, R., Feige, U., Goldreich, O., Naor, M.: Adaptively secure computation (1995)
7. Canetti, R., Gennaro, R., Herzberg, A., Naor, D.: Proactive security: Long-term protection against break-ins. In: CryptoBytes (1) (1999)
8. Carter, J.L., Wegman, M.N.: Universal classes of hash functions. JCSS 18 (1979)
9. Di Crescenzo, G., Ferguson, N., Impagliazzo, R., Jakobsson, M.: How to forget a secret. In: Meinel, C., Tison, S. (eds.) STACS 1999. LNCS, vol. 1563, pp. 500–509. Springer, Heidelberg (1999)
10. Di Crescenzo, G., Lipton, R.J., Walfish, S.: Perfectly secure password protocols in the bounded retrieval model. In: Theory of Cryptography Conference, pp. 225–244 (2006)
11. Dagon, D., Lee, W., Lipton, R.J.: Protecting secret data from insider attacks. In: Financial Cryptography, pp. 16–30 (2005)
12. Diffie, W., Van-Oorschot, P.C., Weiner, M.J.: Authentication and authenticated key exchanges. In: Designs, Codes, and Cryptography, pp. 107–125 (1992)
13. Dodis, Y.: Exposure-Resilient Cryptography. PhD thesis. MIT, Cambridge (2000)
14. Dziembowski, S.: Intrusion-resilience via the bounded-storage model. In: Theory of Cryptography Conference, pp. 207–224 (2006)
15. Dziembowski, S.: On forward-secure storage. In: Dwork, C. (ed.) CRYPTO 2006. LNCS, vol. 4117, pp. 251–270. Springer, Heidelberg (2006)
16. Dziembowski, S., Pietrzak, K.: Intrusion-resilient secret sharing. In: FOCS 2007, Washington, DC, USA, pp. 227–237. IEEE Computer Society, Los Alamitos (2007)
17. Frankel, Y., Gemmel, P., MacKenzie, P.D., Yung, M.: Proactive rsa. In: Kaliski Jr., B.S. (ed.) CRYPTO 1997. LNCS, vol. 1294, pp. 440–454. Springer, Heidelberg (1997)
18. Garfinkel, S.L.: Design Principles and Patterns for Computer Systems That Are Simultaneously Secure and Usable. PhD thesis. MIT, Cambridge (2005)
19. Gennaro, R., Jarecki, S., Krawczyk, H., Rabin, T.: Robust threshold Dss signatures. In: Maurer, U.M. (ed.) EUROCRYPT 1996. LNCS, vol. 1070, pp. 354–371. Springer, Heidelberg (1996)

20. Günther, C.G.: An identity-based key-exchange protocol. In: Proc. EUROCRYPT 1989, pp. 29–37 (1989)
21. Halderman, J.A., Schoen, S.D., Heninger, N., Clarkson, W., Paul, W., Calandrino, J.A., Feldman, A.J., Appelbaum, J., Felten, E.W.: Lest We Remember: Cold Boot Attacks on Encryption Keys (April 2008), http://citp.princeton.edu/memory/
22. Herzberg, A., Jakobsson, M., Jarecki, S., Krawczyk, H., Yung, M.: Proactive public key and signature systems. In: ACM Conference on Computers and Communication Security (1997)
23. Herzberg, A., Jarecki, S., Krawczyk, H., Yung, M.: Proactive secret sharing, or: How to cope with perpetual leakage. In: Coppersmith, D. (ed.) CRYPTO 1995. LNCS, vol. 963, pp. 339–352. Springer, Heidelberg (1995)
24. Hughes, G., Coughlin, T.: Tutorial on hard drive sanitation (2006), http://www.tomcoughlin.com/
25. Hughes, G., Coughlin, T.: Secure erase of disk drive data, pp. 22–25 (2002)
26. Impagliazzo, R., Levin, L.A., Luby, M.: Pseudo-random generation from one-way functions. In: STOC 1989, pp. 12–24 (1989)
27. Itkis, G., Reyzin, L.: Sibir: Signer-base intrusion-resilient signatures. In: Yung, M. (ed.) CRYPTO 2002. LNCS, vol. 2442, pp. 499–514. Springer, Heidelberg (2002)
28. Jarecki, S., Lysyanskaya, A.: Adaptively secure threshold cryptography: Introducing concurrency, removing erasures. In: Preneel, B. (ed.) EUROCRYPT 2000. LNCS, vol. 1807, pp. 221–243. Springer, Heidelberg (2000)
29. Lu, C.-J.: Encryption against storage-bounded adversaries from on-line strong extractors. In: Proc. CRYPTO 2002, pp. 257–271 (2002)
30. Lysyanskaya, A.: Efficient threshold and proactive cryptography secure against the adaptive adversary (extended abstract) (1999)
31. Mansour, Y., Nisan, N., Tiwari, P.: The computational complexity of universal hashing. In: Proc. 22nd ACM Symp. on Theory of Computing (2002)
32. Maurer, U.: A provably-secure strongly-randomized cipher. In: Damgård, I.B. (ed.) EUROCRYPT 1990. LNCS, vol. 473, pp. 361–373. Springer, Heidelberg (1991)
33. Maurer, U.: Conditionally-perfect secrecy and a provably-secure randomized cipher, pp. 53–66 (1992)
34. Nisan, N., Zuckerman, D.: Randomness is linear in space. Journal of Computer and System Sciences 52(1), 43–52 (1996)
35. Department of Defense. DoD 5220.22-M: National Industrial Security Program Operating Manual (1997)
36. Ostrovsky, R., Yung, M.: How to withstand mobile virus attacks, pp. 51–61 (1991)
37. Damgård, I., Nielsen, J.: Improved non-committing encryption schemes based on a general complexity assumption. In: Bellare, M. (ed.) CRYPTO 2000. LNCS, vol. 1880, Springer, Heidelberg (2000)
38. Shannon, C.E.: Communication theory of secrecy systems. Bell System Technical Journal, 656–715
39. Vaarala, S.: T-110.5210 cryptosystems lecture notes (2006)
40. Vadhan, S.P.: Constructing locally computable extractors and cryptosystems in the bounded-storage model. J. Cryptol. 17(1), 43–77 (2004)
41. Yee, B.: Using secure coprocessors. PhD thesis (May 1994)

Universally Composable Undeniable Signature

Kaoru Kurosawa[1] and Jun Furukawa[2]

[1] Ibaraki University, Japan
kurosawa@mx.ibaraki.ac.jp
[2] NEC Corporation, Japan
j-furukawa@ay.jp.nec.com

Abstract. How to define the security of undeniable signature schemes is a challenging task. This paper presents two security definitions of undeniable signature schemes which are more useful or natural than the existing definition. It then proves their equivalence.

We first define the UC-security, where UC means universal composability. We next show that there exists a UC-secure undeniable signature scheme which does not satisfy the standard definition of security that has been believed to be adequate so far. More precisely, it does not satisfy the invisibility defined by [10]. We then show a more adequate definition of invisibility which captures a wider class of (naturally secure) undeniable signature schemes.

We finally prove that the UC-security against non-adaptive adversaries is equivalent to this definition of invisibility and the strong unforgeability in $\mathcal{F}_{ZK}$-hybrid model, where $\mathcal{F}_{ZK}$ is the ideal ZK functionality. Our result of equivalence implies that all the known proven secure undeniable signature schemes (including Chaum's scheme) are UC-secure if the confirmation/disavowal protocols are both UC zero-knowledge.

Keywords: Universal composability, undeniable signature scheme.

1 Introduction

The concept of undeniable signature schemes was introduced by Chaum and van Antwerpen [9]. In an undeniable signature scheme, the signer issues an undeniable signature σ which is not publicly verifiable. She then proves the validity or invalidity of σ to the verifier in zero-knowledge (ZK) by running a confirmation protocol or disavowal protocol. Undeniable signature schemes have found various applications in cryptography such as in licensing software, electronic cash, electronic voting and auction. Then there have been a wide range of research covering a variety of different schemes for undeniable signatures over the past 15 years [1,2,8,10,11,12,13,17,19,20,21].

Recently, the security of Chaum's undeniable signature scheme is proved formally in the random oracle model under the decisional Diffie-Hellman (DDH) assumption by [22]. In the standard model, Laguillaumie and Vergnaud showed an undeniable signature scheme which is secure under a decisional variant of the strong Diffie-Hellman (DH) assumption [18]. Kurosawa and Takagi showed an

L. Aceto et al. (Eds.): ICALP 2008, Part II, LNCS 5126, pp. 524–535, 2008.

undeniable signature scheme which is secure under the strong RSA assumption and the decisional Nth residuosity assumption [16].

However, how to define the security of undeniable signature schemes is a challenging task. For example, it is not known if the security of these schemes is maintained under a general protocol composition. This concern is serious because undeniable signatures are often used as a building block in a more complicated protocol as shown above.

This paper presents two security definitions of undeniable signature schemes which are more useful or natural than the existing definition. It then proves their equivalence.

We first present an ideal functionality of undeniable signature schemes Σ in the universally composable (UC) framework [3,4]. We next show that there exists a UC-secure undeniable signature scheme which does not satisfy the standard definition of security that has been believed to be adequate so far. More precisely, it does not satisfy the invisibility defined by [10]. The invisibility means that, for a message m, the receiver cannot tell if σ is a valid signature or a simulated signature. We then show a more adequate definition of invisibility which captures a wider class of (naturally secure) undeniable signature schemes.

We finally prove that the UC-security against non-adaptive adversaries is equivalent to this definition of invisibility and the strong unforgeability in $\mathcal{F}_{ZK}$-hybrid model where $\mathcal{F}_{ZK}$ is the ideal ZK functionality. For adaptive adversaries, we show that it is impossible to construct a UC-secure undeniable signature scheme even in the $\mathcal{F}_{ZK}$-hybrid model.

Our result of equivalence implies that all the known proven secure undeniable signature schemes (including Chaum's scheme) [22,18,16] are UC-secure against non-adaptive adversaries if the confirmation protocol and the disavowal protocol are UC zero-knowledge. Hence the security of these schemes is maintained under a general protocol composition against non-adaptive adversaries.

2   Preliminaries

2.1   Undeniable Signature Scheme

According to [10], an undeniable signature scheme is denoted by

$$\Sigma = (\mathsf{G}_{sign}, \mathsf{Sign}, \mathsf{Check}, \mathsf{Sim}, \pi_{con}, \pi_{dis}).$$

It consists of a key generation algorithm G_{sign}, a signing algorithm Sign, a validity check algorithm Check, a signature simulator Sim, a confirmation protocol π_{con} and a disavowal protocol π_{dis}.

The key generation algorithm G_{sign} is a PPT (probabilistic polynomial-time) algorithm which outputs (vk, sk), where vk is a verification key and sk is the signing key. [1] The message space $\mathcal{M}$ is specified by vk.

The signing algorithm Sign is a PPT algorithm which generates a signature σ on input a message $m \in \mathcal{M}$ and the signing key sk.

---

[1] We assume that sk is uniquely determined by vk.

We say that (m, σ) is valid if σ is an output of $\mathsf{Sign}(sk, m)$ for some random string r. Otherwise, we say that (m, σ) is invalid. The validity check algorithm Check is a deterministic polynomial time algorithm such that

$$\mathsf{Check}((vk, m, \sigma), sk) = \begin{cases} 1 \; if \; (m, \sigma) = \mathsf{valid} \\ 0 \; if \; (m, \sigma) = \mathsf{invalid} \end{cases}$$

The signature simulator Sim is a PPT algorithm which outputs a simulated signature such that $\sigma' = \mathsf{Sim}(vk, m)$.

An undeniable signature scheme must satisfy unforgeability and invisibility. Invisibility means that for a message m, the receiver cannot tell if σ is a valid signature or a simulated signature.

This implies that the receiver cannot verify the validity of (m, σ) by himself. Instead, the cooperation of the signer is needed to verify the validity and invalidity of (m, σ) by running a confirmation protocol π_{con} and a disavowal protocol π_{dis} with the receiver respectively. π_{con} is a zero-knowledge interactive proof system (ZKIP) on a language $L_0 = \{(vk, m, \sigma) \mid (m, \sigma) \text{ is valid}\}$, and π_{dis} is a ZKIP on a language $L_1 = \{(vk, m, \sigma) \mid (m, \sigma) \text{ is invalid}\}$. Each ZKIP must satisfy completeness, soundness and zero-knowledgeness.

2.2   Security of Undeniable Signature

Unforgeability. The unforgeability is defined as follows. Consider the following game between a challenger and an adversary A.

1. The challenger generates a key pair (vk, sk) randomly, and gives the verification key vk to A.
2. For $i = 1, 2, \ldots, q_s$ for some q_s, A queries a message m_i to the signing oracle adaptively and receives a signature σ_i.
3. Eventually, A outputs a forgery (m^*, σ^*).

We allow the adversary A to query (m_j, σ_j) to the confirmation/disavowal oracle adaptively at step 2, where the confirmation/disavowal oracle responds as follows.

- If (m_j, σ_j) is a valid pair, then the oracle returns a bit $\mu = 1$ and proceeds with the execution of the confirmation protocol π_{con} with A.
- Otherwise, the oracle returns a bit $\mu = 0$ and executes the disavowal protocol π_{dis} with A accordingly.

We say that A succeeds in strong forgery if (m^*, σ^*) is valid and (m^*, σ^*) is not among the pairs (m_i, σ_i) generated during the signing oracle queries. [2]

Definition 1. *We say that Σ is strongly unforgeable if* $\Pr[A$ *succeeds in strong forgery* $]$ *is negligible for any PPT adversary A in the above game.*

---

[2] We say that A succeeds in weak forgery if (m^*, σ^*) is valid and m^* has never been queried to the signing oracle. Weak unforgeability and strong one are equivalent if the signing algorithm is deterministic, and there exists a unique signature for each message that is verified correctly.

Invisibility. Damgård and Pedersen defined the invisibility by using the following game between a challenger and an adversary A [10].

1. The challenger generates a key pair (vk, sk) randomly, and gives the verification key vk to A.
2. A is permitted to issue a series of signing queries m_i to the signing oracle adaptively and receives a signature σ_i.
3. At some point, A chooses a message m^* and sends it to the challenger.
4. The challenger chooses a random bit b.
   If $b = 1$, then he computes a real signature $\sigma^* = \mathsf{Sign}(sk, m^*)$.
   Otherwise, he computes a fake signature $\sigma^* = \mathsf{Sim}(vk, m^*)$.
   He then returns σ^* to A.
5. A performs some signing queries again
6. At the end of this attack game, A outputs a guess b'.

We allow the adversary A to query (m_j, σ_j) to the confirmation/disavowal oracle adaptively at step 2 and at step 5.

However, A is not allowed to query the challenge (m^*, σ^*) to the confirmation/disavowal oracle at step 5. Also A is not allowed to query m^* to the signing oracle.

Definition 2. *We say that Σ is invisible if for any PPT adversary A, $|\Pr[b = b'] - 1/2|$ is negligible in the above game.*

2.3 Universal Composability

The security of a protocol $\pi = (P_1, \cdots, P_n)$ is maintained under a general protocol composition if π is secure in the universally composable (UC) security framework. See [3,4,5] for the details.

3 UC Undeniable Signature

3.1 Ideal Functionality

Suppose that there exists a trusted third party (TTP) who has magical ink such that anything written by it is not visible. Only TTP can see it by using a special pair of glasses. Then the ideal functionality of undeniable signature schemes can be illustrated as follows.

1. A signer, Alice, first receives a registered number vk from TTP.
2. Upon signing request on a message m from Alice, TTP makes a signature σ on m (on behalf of vk) by using the magical ink.
3. Upon verification request on (m, σ, Bob) from Alice, TTP checks if σ is a correct signature (on behalf of vk) by using the special pair of glasses. Then it tells Bob if (m, σ) is valid or not.

We now present the ideal functionality $\mathcal{F}_{usig}$ of undeniable signature schemes in the UC framework.

Key Generation: 1. Upon receiving a value (KeyGen, sid) from some party P, verify that $sid = (P, sid')$ for some sid'. If not, then ignore the request. Else, hand (KeyGen, sid) to the adversary.
   2. Upon receiving (Keys, sid, vk, Sim) from the adversary, output (VerifyKey, sid, vk, Sim) to P, where vk is a verification key and Sim is a PPT algorithm.

Signature Generation: Upon receiving a value (Sign, sid, m) from P, verify that $sid = (P, sid')$ for some sid'. If not, then ignore the request. Else do:
   1. If $(m, \sigma, 1)$ is recorded, then output (Signature, sid, m, σ) to P. [3]
   2. Else, if P is not corrupted, generate $\sigma = \mathsf{Sim}(vk, m)$ randomly such that no entry $(m, \sigma, 0)$ is recorded. Then output (Signature, sid, m, σ) to P and the adversary.
   3. Else send (Sign, sid, m) to the adversary.
      Upon receiving (Signature, sid, m, σ) from the adversary, verify that no entry $(m, \sigma, vk, 0)$ is recorded. If it is, then output an error message to P and halt. Else, output (Signature, sid, m, σ) to P, and record the entry $(m, \sigma, vk, 1)$.

Verification: Upon receiving a value (Verify, sid, m, σ, V) from P, where V is a verifier, verify that $sid = (P, sid')$ for some sid'. If not, then ignore the request. Else do:
   1. If $(m, \sigma, flag')$ is recorded, then set $flag = flag'$.
   2. Else, if P is not corrupted, then set $flag = 0$ and record $(m, \sigma, 0)$. (This condition guarantees strong unforgeability: if the signer is not corrupted, and never signed m, then the verification fails.)
   3. Else, hand (Verify, sid, m, σ, V) to the adversary.
      Upon receiving (AdVerified, sid, m, σ, ϕ) from the adversary, let $flag = \phi$ and record (m, σ, ϕ).
   Finally output (Verified, $sid, (m, \sigma), flag$) to V and the adversary.

3.2   Remarks

The main differences between $\mathcal{F}_{usig}$ and $\mathcal{F}_{sig}$ are as follows, where $\mathcal{F}_{sig}$ is the signature functionality given by [4].

- At key generation, $\mathcal{F}_{sig}$ receives vk from the adversary, and hands it to P. On the other hand, $\mathcal{F}_{usig}$ receives (vk, Sim) from the adversary, and hands it to P.
- At signature generation, $\mathcal{F}_{sig}$ receives σ from the adversary, and hands it to P. On the other hand, $\mathcal{F}_{usig}$ computes $\sigma = \mathsf{Sim}(vk, m)$, and hands it to P. This is because σ must be invisible in undeniable signature schemes.

---

[3] Ignore this step if the signing algorithm is probabilistic.

– The *signer* (P) issues Verify command to $\mathcal{F}_{usig}$ while the *verifier* (V) issues it to $\mathcal{F}_{sig}$. This is because V should not be able to verify the validity of (m, σ) without the cooperation of P in undeniable signature schemes.

The adversary returns (Keys, sid, vk, Sim) to $\mathcal{F}_{usig}$ at key generation. Hence Sim depends on vk. This means that we can write $\sigma = \mathsf{Sim}(m)$ instead of $\sigma = \mathsf{Sim}(vk, m)$ at signature generation.

4   Subtlety on Invisibility and New Definition

4.1   Problem of Previous Definition

The standard definition of invisibility (Def. 2) was given by Damgård and Pedersen [10], where Sim is a part of Σ. However, we show that there exists a UC-secure (and naturally secure) undeniable signature scheme which does not satisfy this definition of invisibility (Def. 2).

Let Σ be an undeniable signature scheme which satisfies the strong unforgeability and the invisibility defined by Def. 1 and Def. 2. Let Σ' be a strongly unforgeable (usual) signature scheme. Then consider an undeniable signature scheme Ω based on Σ and Σ' as follows.

– The public-key of Ω is (vk, vk'), where vk is a public-key of Σ, and vk' is a public-key of Σ'.
– The undeniable signature $\tilde{\sigma}$ on a message m is (σ, sk', σ'), where σ is an undeniable signature of Σ on m, sk' is a secret-key of Σ' and σ' is a (usual) signature of Σ' on m.

This undeniable signature scheme Ω does not satisfy the invisibility defined by Def. 2 because any PPT Sim() cannot compute sk'.

However, we can show that Ω is UC-secure. Intuitively, it is strongly unforgeable because Σ is strongly unforgeable. It is *naturally* invisible because σ is invisible, and everyone can compute σ' for any message by using sk' once he obtains sk' (for example, by known message attack). Indeed, our ideal process adversary S has only to return Sim which includes sk' at Key Generation.

4.2   New Definition of Invisibility

The above difference comes from the fact that Sim is independent of vk in the previous definition while it is not in the UC framework. Indeed, the adversary returns (vk, Sim) to $\mathcal{F}_{usig}$ in the UC framework.

We now show a new definition of invisibility. We delete Sim from Σ, and let Sim be a part of a public-key. That is, we define an undeniable signature scheme as

$$\Sigma = (\mathsf{G}_{sign}, \mathsf{Sign}, \mathsf{Check}, \pi_{con}, \pi_{dis})$$

such that

- the key generation algorithm G_{sign} outputs (vk, sk) and Sim. The signer makes (vk, Sim) public, and keeps sk secret.

The other parts of Σ remain the same. Accordingly, we need to modify step 1 of the attack game of invisibility shown in Sec.2.2 as follows.

1. The challenger generates (vk, sk) and Sim by running G_{sign}, and gives (vk, Sim) to A.

We then define invisibility as follows.

Definition 3. *We say that Σ is invisible if for any PPT adversary A, $|\Pr[b = b'] - 1/2|$ is negligible in the modified attack game.*

Now Ω is invisible under our new definition. More generally, it is easy to see that our new definition captures a wider class of (naturally secure) undeniable signature schemes.

4.3 New Definition of Unforgeability

We also need to modify step 1 of the attack game of unforgeability shown in Sec.2.2 as follows.

1. The challenger generates (vk, sk) and Sim by running G_{sign}, and gives (vk, Sim) to A.

We then define strong unforgeability as follows.

Definition 4. *We say that Σ is strongly unforgeable if $\Pr[A$ succeeds in strong forgery $]$ is negligible for any PPT adversary A in the modified attack game.*

4.4 Translation to Protocol

Under our new definition of Sec.4.2 and Sec.4.3, we show how to translate an undeniable signature scheme $\Sigma = (G_{sign}, Sign, Check, \pi_{con}, \pi_{dis})$ into a protocol π_Σ in $\mathcal{F}_{ZK}$-hybrid model, where $\mathcal{F}_{ZK}$ is the ZK functionality on the binary relation Check.

1. When party P receives an input $(KeyGen, sid)$, it verifies that $sid = (P, sid')$ for some sid'. If not, it ignores the input. Else it generates (vk, sk) and Sim by running G_{sign}, and outputs $(VerifyKey, sid, vk, Sim)$.
2. When P receives an input $(Sign, sid, m)$ with $sid = (P, sid')$, it sets $\sigma = Sign(sk, m)$ and outputs $(Signature, sid, m, \sigma)$.
3. When P receives an input $(Verify, sid, m, \sigma, V)$, do:
   (a) P sends $((vk, m, \sigma), sk)$ to $\mathcal{F}_{ZK}$.
   (b) $\mathcal{F}_{ZK}$ then sends $(Verified, sid, P, (vk, m, \sigma), f)$ to V and the adversary, where $f = Check((vk, m, \sigma), sk)$.
   (c) Finally V outputs $(Verified, sid, (m, \sigma), f)$.

When a party is corrupted, it reveals its internal state, which includes all past signing and verification requests and answers, and for P also the state of the signing algorithm, including the signing key and the randomness used to sign past messages.

Definition 5. *We say that an undeniable signature scheme Σ is UC-secure if π_Σ securely realizes $\mathcal{F}_{usig}$.*

5   Equivalence

In this section, we prove that our UC-security notion of undeniable signature schemes is equivalent to our new definition of strong unforgeability and our new definition of invisibility.

Theorem 1. *Σ satisfies strong unforgeability and invisibility if Σ is UC-secure against non-adaptive adversaries in the $\mathcal{F}_{ZK}$-hybrid model.*

A proof is given in [14].

Corollary 1. *Σ satisfies weak unforgeability and invisibility if Σ is UC-secure against non-adaptive adversaries.*

Theorem 2. *Σ is UC-secure against non-adaptive adversaries if Σ satisfies strong unforgeability and invisibility in the $\mathcal{F}_{ZK}$-hybrid model.*

5.1   Proof of Theorem 2

Assume that π_Σ does not securely realize $\mathcal{F}_{usig}$ against non-adaptive adversaries. We show that Σ does not satisfy strong unforgeability or invisibility. Assume that Σ is invisible (otherwise the theorem is proven). Then there exists a PPT algorithm Sim which satisfies the definition of the invisibility. Our goal is to construct a forger G.

Using the equivalent notion of security against the (non-adaptive) dummy adversary D, [4] we have that for any ideal process adversary S, there exists an environment $\mathcal{Z}$ that can tell whether it is interacting with $\mathcal{F}_{usig}$ and S, or with π_Σ and the non-adaptive dummy adversary D. (Remember that non-adaptive adversaries corrupt parties at the beginning of executions only.)

We consider a particular S as shown below. For this particular S, there exists an environment $\mathcal{Z}_S$ that can distinguish the real world and the ideal world. We will use this $\mathcal{Z}_S$ to construct a forger G on Σ.

First our particular S behaves as follows.

---

[4] The dummy adversary D only delivers to parties messages generated by the environment $\mathcal{Z}$, and delivers to $\mathcal{Z}$ all messages generated by the parties. Instead of quantifying over all possible adversary A, it suffices to require that the ideal protocol adversary S be able to simulate, for any environment $\mathcal{Z}$, the behavior of the dummy adversary D. [5]

- Suppose that there are no party corruption instructions by $\mathcal{Z}$. In this case, $\mathcal{S}$ provides $\mathcal{F}_{usig}$ with vk and Sim at key generation. $\mathcal{S}$ outputs nothing other than this.

- Suppose that $\mathcal{Z}$ instructs $\mathcal{S}$ to corrupt P at the beginning. In this case, $\mathcal{F}_{usig}$ forwards all commands of $\mathcal{Z}$ (to P) to $\mathcal{S}$. Then $\mathcal{S}$ behaves in the same way as the real signer of π_Σ does. That is:

1. At key generation, $\mathcal{S}$ generates (vk, sk) randomly and returns vk and Sim to $\mathcal{F}_{usig}$.
2. At signature generation, $\mathcal{S}$ computes $\sigma = \mathsf{Sign}(sk, m)$ and returns σ to $\mathcal{F}_{usig}$.
3. At signature verification, $\mathcal{S}$ computes $\phi = \mathsf{Check}((vk, m, \sigma), sk))$ and returns $(\mathtt{AdVerified}, sid, m, \sigma, \phi)$ to $\mathcal{F}_{usig}$.

Lemma 1. $\mathcal{Z}_S$ *does not corrupt P with nonnegligible probability.*

Proof. If $\mathcal{Z}_S$ always corrupts P (at the beginning), then such $\mathcal{Z}_S$ cannot distinguish the real world and the ideal world because our $\mathcal{S}$ behaves in the same way as the real signer. Hence $\mathcal{Z}_S$ does not corrupt P with nonnegligible probability. $\square$

---

1. G is given (vk, Sim) as an input. G then runs $\mathcal{Z}_S$.
2. If $\mathcal{Z}_S$ corrupts some party P at the beginning, then G outputs failure.
   If $\mathcal{Z}_S$ activates P with input $(\mathtt{KeyGen}, sid)$ with $sid = (P, sid')$ for some sid',
   then G returns (vk, Sim) to $\mathcal{Z}_S$.
3. When $\mathcal{Z}_S$ activates P with input $(\mathtt{Sign}, sid, m)$,
   then G asks its signing oracle for a signature σ on m, and returns σ to $\mathcal{Z}_S$.
4. When $\mathcal{Z}_S$ activates P with input $(\mathtt{Verify}, sid, m, \sigma, V)$ for some party V,
   then G queries (m, σ) to its confirmation/disavowal oracle,
   and returns the answer to $\mathcal{Z}_S$ through V.
5. If the answer is valid, and (m, σ) is not a pair generated at step 3,
   then G outputs (m, σ) as a strong forgery and stops.

Fig. 1. Forger G

Next let **FORGE** denote the event that $\mathcal{Z}_S$ activates P with input $(\mathtt{Verify}, sid, m, \sigma, V)$ such that (m, σ) is a strong forgery.

Lemma 2. *Suppose that Σ satisfies the invisibility. Also suppose that $\mathcal{Z}_S$ does not corrupt P, and can distinguish the real world from the ideal world. Then* **FORGE** *happens in the real world with nonnegligible probability.*

Now we present our forger G in Fig.1. Suppose that $\mathcal{Z}_S$ does not corrupt P. Then G simulates the real world for $\mathcal{Z}_S$ until step 5. Therefore the view of $\mathcal{Z}_S$ of Fig.1 is the same as the view of $\mathcal{Z}_S$ in the real world until step 5. Hence from Lemma 1 and Lemma 2, it is clear that G succeeds in strong forgery with nonnegligible probability if Σ is not UC-secure and satisfies the invisibility. This completes the proof of Theorem 2.

(Proof of Lemma 2)
It is clear that **FORGE** never happens in the ideal world. We prove that $\mathcal{Z}_S$ cannot distinguish the real world and the ideal world if **FORGE** never happens in the real world.

Suppose that **FORGE** never happens in the real world. Then the view of $\mathcal{Z}_S$ in the real world is identical to the view of $\mathcal{Z}_S$ shown in Fig.2.

---

1. When party P receives an input (KeyGen, sid),
   it verifies that $sid = (P, sid')$ for some sid'. If not, it ignores the input.
   Else it generates (vk, sk) and Sim by running G_{sign},
   and outputs (VerifyKey, $sid, vk,$ Sim).
2. When P receives an input (Sign, sid, m) with $sid = (P, sid')$,
   it sets $\sigma = \mathsf{Sign}(sk, m)$ and outputs (Signature, sid, m, σ).
   P records (m, σ).
3. When P receives an input (Verify, sid, m, σ, V), do:
   If (m, σ) is recorded, then P outputs valid. Otherwise P outputs invalid.

Fig. 2. FORGE never happens

We consider a series of games on $\mathcal{Z}_S$ as follows. **Game**$_0$ is the same as Fig.2 except for that σ_i are all simulated signatures. Assume that $\mathcal{Z}_S$ activates P with input (Sign, sid, m_i) and the signing oracle returns σ_i for $i = 1, \cdots, q_s$. For $j - 1, \cdots, q_s$, **Game**$_j$ is the same as Fig.2 except for that σ_i is a real signature for $i = 1, \cdots, j$, and σ_i is a simulated signature for $i = j + 1, \cdots, q_s$. Note that **Game**$_{q_s}$ is the same as Fig.2, where σ_i are all real signatures.

From a view point of $\mathcal{Z}_S$, it is clear that **Game**$_0$ is the ideal world and **Game**$_{q_s}$ is the real world. Therefore from our assumption, $\mathcal{Z}_S$ can distinguish **Game**$_0$ and **Game**$_{q_s}$. Then it is easy to show that there exists J such that $\mathcal{Z}_S$ can distinguish **Game**$_{J-1}$ and **Game**$_J$.

Now we construct an adversary A who can break the invisibility by using $\mathcal{Z}_S$ as follows. A engages in the attack game on the invisibility. First, A is given (vk, Sim) by the challenger. It then runs $\mathcal{Z}_S$. When $\mathcal{Z}_S$ invokes some uncorrupted P, A returns (vk, Sim) to $\mathcal{Z}_S$.

Suppose that $\mathcal{Z}_S$ activates P with input (Sign, sid, m_i).

- If $i < J$, then A queries m_i to his own signing oracle and receives a real signature σ_i. A records (m_i, σ_i).
- If $i > J$, then A computes a simulated signature $\sigma_i = \mathsf{Sim}(vk, m_i)$.
- If $i = J$, then A sends m_J to the challenger as a challenge message, and receives σ_J from the challenger.

A then returns the above σ_i to $\mathcal{Z}_S$.

Suppose that $\mathcal{Z}_S$ activates P with input (Verify, sid, m, σ, V) for some party V. If (m, σ) is recorded, then A returns valid. Otherwise, A returns invalid. (Remember that **FORGE** never happens.)

Let b' be the final output of $\mathcal{Z}_S$. A outputs this b'.

It is clear that the view of $\mathcal{Z}_S$ is exactly the same as that of **Game**$_{J-1}$ and **Game**$_J$ according to the challenge bit b of the challenger. Therefore from the definition of J, $|\Pr(b' = b) - 1/2|$ is nonnegligible. This means that A wins the attack game on the invisibility. However, this is a contradiction.

This completes the proof of Lemma 2.

5.2 Application

From the result of [22], it is easy to see that Chaum's undeniable signature scheme is (strongly) unforgeable under CDH assumption, and invisible under DDH assumption in the random oracle model even under our definitions. Hence we have the following corollary from Theorem 2.

Corollary 2. *Chaum's undeniable signature scheme is UC-secure against nonadaptive adversaries under the DDH assumption in the random oracle model if it uses a confirmation protocol and a disavowal protocol which are UC zeroknowledge.*

6   Impossibility Result

In this section, we show that it is impossible to construct an undeniable signature scheme which satisfies our UC-security against adaptive adversaries.

Theorem 3. *There exists no undeniable signature scheme Σ which is UC-secure against adaptive adversaries even in the $\mathcal{F}_{ZK}$-hybrid model.*

A proof is given in [14].

References

1. Boyar, J., Chaum, D., Damgård, I., Pedersen, T.: Convertible undeniable signatures. In: Menezes, A., Vanstone, S.A. (eds.) CRYPTO 1990. LNCS, vol. 537, pp. 189–208. Springer, Heidelberg (1991)
2. Biehl, I., Paulus, S., Takagi, T.: Efficient undeniable signature schemes based on ideal arithmetic in quadratic orders. Designs, Codes and Cryptography 31(2), 99–123 (2004)
3. Canetti, R.: Universally Composable Security: A New Paradigm for Cryptographic Protocols, Revision 1 of ECCC Report TR01-016 (2001)
4. Canetti, R.: Universally Composable Signatures, Certification and Authentication, IACR ePrint 2003/239
5. Canetti, R.: Universally Composable Security: A New Paradigm for Cryptographic Protocols, IACR ePrint 2000/067 (2005)

6. Canetti, R., Fischlin, M.: Universally Composable Commitments. In: Kilian, J. (ed.) CRYPTO 2001. LNCS, vol. 2139, pp. 19–40. Springer, Heidelberg (2001)
7. Canetti, R., Lindell, Y., Ostrovsky, R., Sahai, A.: Universally composable two-party and multi-party secure computation. In: STOC 2002, pp. 494–503 (2002)
8. Chaum, D.: Zero-knowledge undeniable signatures. In: Damgård, I.B. (ed.) EUROCRYPT 1990. LNCS, vol. 473, pp. 458–464. Springer, Heidelberg (1991)
9. Chaum, D., van Antwerpen, H.: Undeniable signatures. In: Brassard, G. (ed.) CRYPTO 1989. LNCS, vol. 435, pp. 212–216. Springer, Heidelberg (1990)
10. Damgård, I., Pedersen, T.: New convertible undeniable signature schemes. In: Maurer, U.M. (ed.) EUROCRYPT 1996. LNCS, vol. 1070, pp. 372–386. Springer, Heidelberg (1996)
11. Galbraith, S., Mao, W.: Invisibility and anonymity of undeniable and confirmer signatures. In: Joye, M. (ed.) CT-RSA 2003. LNCS, vol. 2612, pp. 80–97. Springer, Heidelberg (2003)
12. Galbraith, S., Mao, W., Paterson, K.G.: RSA-based undeniable signatures for general moduli. In: Preneel, B. (ed.) CT-RSA 2002. LNCS, vol. 2271, pp. 200–217. Springer, Heidelberg (2002)
13. Gennaro, R., Rabin, T., Krawczyk, H.: RSA-based undeniable signatures. Journal of Cryptology 13(4), 397–416 (2000)
14. Kurosawa, K., Furukawa, J.: Universally Composable Undeniable Signature, Cryptology ePrint Archive, Report 2008/094 (2008), http://eprint.iacr.org/
15. Kurosawa, K., Heng, S.: Relations among security notions for undeniable signature schemes. In: De Prisco, R., Yung, M. (eds.) SCN 2006. LNCS, vol. 4116, pp. 34–48. Springer, Heidelberg (2006)
16. Kurosawa, K., Takagi, T.: New Approach for Selectively Convertible Undeniable Signature Schemes. In: Lai, X., Chen, K. (eds.) ASIACRYPT 2006. LNCS, vol. 4284, pp. 428–443. Springer, Heidelberg (2006)
17. Libert, B., Quisquater, J.-J.: Identity based undeniable signatures. In: Okamoto, T. (ed.) CT-RSA 2004. LNCS, vol. 2964, pp. 112–125. Springer, Heidelberg (2004)
18. Laguillaumie, F., Vergnaud, D.: Short undeniable signatures without random oracles: The Missing Link. In: Maitra, S., Veni Madhavan, C.E., Venkatesan, R. (eds.) INDOCRYPT 2005. LNCS, vol. 3797, pp. 283–296. Springer, Heidelberg (2005)
19. Michels, M., Stadler, M.: Efficient convertible undeniable signature schemes. In: SAC 1997, pp. 231–244. Springer, Heidelberg (1997)
20. Monnerat, J., Vaudenay, S.: Undeniable signatures based on characters: how to sign with one bit. In: Bao, F., Deng, R., Zhou, J. (eds.) PKC 2004. LNCS, vol. 2947, pp. 361–396. Springer, Heidelberg (2004)
21. Monnerat, J., Vaudenay, S.: Generic homomorphic undeniable signatures. In: Lee, P.J. (ed.) ASIACRYPT 2004. LNCS, vol. 3329, pp. 354–371. Springer, Heidelberg (2004)
22. Ogata, W., Kurosawa, K., Heng, S.: The security of the FDH variant of Chaum's undeniable signature scheme. IEEE Transactions on Information Theory 52(5), 2006–2017 (2006)

Interactive PCP

Yael Tauman Kalai* and Ran Raz**

Abstract. A central line of research in the area of PCPs is devoted to constructing short PCPs. In this paper, we show that if we allow an additional interactive verification phase, with very low communication complexity, then for some NP languages, one can construct PCPs that are significantly shorter than the known PCPs (without the additional interactive phase) for these languages. We give many cryptographical applications and motivations for our results and for the study of the new model in general.

More specifically, we study a new model of proofs: **interactive-PCP**. Roughly speaking, an interactive-PCP (say, for the membership $x \in L$) is a proof-string that can be verified by reading only one of its bits, with the help of an interactive-proof with very small communication complexity. We show that for membership in some NP languages L, there are interactive-PCPs that are significantly shorter than the known (non-interactive) PCPs for these languages.

Our main result is that for any constant depth Boolean formula $\Phi(z_1, \ldots, z_k)$ of size n (over the gates $\wedge, \vee, \oplus, \neg$), a prover, Alice, can publish a proof-string for the satisfiability of Φ, where the size of the proof-string is $\mathrm{poly}(k)$. Later on, any user who wishes to verify the published proof-string needs to interact with Alice via a short interactive protocol of communication complexity $\mathrm{poly}(\log n)$, while accessing the proof-string at a single location.

Note that the size of the published proof-string is $\mathrm{poly}(k)$, rather than $\mathrm{poly}(n)$, i.e., the size is polynomial in the size of the witness, rather than polynomial in the size of the instance. This compares to the known (non-interactive) PCPs that are of size polynomial in the size of the instance. By reductions, this result extends to many other central NP languages (e.g., SAT, k-clique, Vertex-Cover, etc.).

More generally, we show that the satisfiability of $\bigwedge_{i=1}^{n}[\Phi_i(z_1, \ldots, z_k) = 0]$, where each $\Phi_i(z_1, \ldots, z_k)$ is an arithmetic formula of size n (say, over $\mathbb{GF}[2]$) that computes a polynomial of degree d, can be proved by a published proof-string of size $\mathrm{poly}(k, d)$. Later on, any user who wishes to verify the published proof-string needs to interact with the prover via an interactive protocol of communication complexity $\mathrm{poly}(d, \log n)$, while accessing the proof-string at a single location.

We give many applications and motivations for our results and for the study of the notion of interactive PCP in general. In particular, we have the following applications:

---

* Georgia Institute of Technology. Supported in part by NSF CyberTrust grant CNS-0430450. Part of this work was done when the author visited the Weizmann Institute.
** Weizmann Institute of Science. Supported by Binational Science Foundation (BSF), Israel Science Foundation (ISF) and Minerva Foundation.

L. Aceto et al. (Eds.): ICALP 2008, Part II, LNCS 5126, pp. 536–547, 2008.
© Springer-Verlag Berlin Heidelberg 2008

Succinct zero knowledge proofs: We show that any interactive PCP, with certain properties, can be converted into a zero-knowledge interactive proof. We use this to construct zero-knowledge proofs of communication complexity polynomial in the size of the witness, rather than polynomial in the size of the instance, for many NP languages.

Succinct probabilistically checkable arguments: In a subsequent paper, we study the new notion of *probabilistically checkable argument*, and show that any interactive PCP, with certain properties, translates into a probabilistically checkable argument [18]. We use this to construct probabilistically checkable arguments of size polynomial in the size of the witness, rather than polynomial in the size of the instance, for many NP languages.

Commit-Reveal schemes: We show that Alice can commit to a string w of k bits, by a message of size poly(k), and later on, for any predicate Φ of size n, whose satisfiability can be proved by an efficient enough interactive PCP with certain properties, Alice can prove the statement $\Phi(w) = 1$, by a zero-knowledge interactive proof with communication complexity poly($\log n$). (Surprisingly, the communication complexity may be significantly smaller than k and n).

1 Introduction

Different interpretations and views of the notion of *proof* have played a central role in the development of complexity theory. Many of the most exciting ideas in complexity theory were originated by defining and studying new models of proofs. Three of the most successful models that were suggested are: Interactive Proofs [6,13,20,26], where the proof is interactive, Probabilistically Checkable Proofs (PCP) [1,2,9], where the verifier is only allowed to read a small number of symbols of the proof, and Multi-prover Interactive Proofs [4,5], where the verifier interacts with several (say, two) different provers that are not allowed to communicate between them.

A two provers interactive proof can be viewed as a proof where the verifier is given access to two independent interactive proofs. In light of the great success of this model, it seems very interesting to study the case where the verifier is given access to two independent proofs of a different nature. For example, in [25] the case where the verifier is given access to both, a quantum proof on one hand, and a classical PCP on the other hand, was studied.

Since interactive proofs and PCPs are among the most exciting models of proofs that were suggested, it seems very interesting to ask what happens when the two models are combined. In this paper, we study the new model where the verifier is given access to both, a PCP on one hand, and an interactive proof on the other hand. We think of this model also as an Interactively Verifiable PCP (or, in short, Interactive PCP), that is, a PCP that is verified by an interaction between a prover and the verifier. We show that in some cases this model has advantages over both PCPs and interactive proofs (given separately). More specifically, we show that the membership in many NP languages can be proved by a combination of a PCP and an interactive proof, where the PCP is much shorter than the best PCPs known for these languages, and the interactive proof is of a much lower communication complexity than the best interactive proofs known for these languages.

For example, one of our main results shows that for any constant depth Boolean formula $\Phi(z_1, \ldots, z_k)$ of size n (over the gates $\wedge, \vee, \oplus, \neg$), a prover, Alice, can publish a proof-string (PCP) for the satisfiability of Φ, where the size of the proof-string is poly(k) (rather than poly(n), i.e., the size of the proof-string is polynomial in the size of the witness, rather than polynomial in the size of the instance). Later on, any user who wishes to verify the published proof-string needs to interact with Alice via an interactive protocol of communication complexity poly($\log n$), while accessing the proof-string at a single location.

By reductions, the result extends to many other central NP languages, e.g., SAT, k-clique, Vertex-Cover, etc. Moreover, a subsequent theorem of Goldwasser, Kalai, and Rothblum [12], in the context of computation delegation, improves the above mentioned result so that it holds for any Boolean formula $\Phi(z_1, \ldots, z_k)$ of size n (rather than only for constant depth formulas). More generally, they show that the same result holds for any Boolean circuit $\Phi(z_1, \ldots, z_k)$ of size n and depth d, where now the size of the proof-string is poly(k, d), and the interactive verification phase is of communication complexity poly($d, \log n$).

We give many motivations and applications for these results, and for the study of the new model in general. Most of these applications are general reductions that convert any interactive PCP, with certain properties, into another object (i.e., the application). The motivations and applications are described in details in Subsection 1.4.

1.1   Interactive PCP

In this paper, we study a new model of proofs: **interactive PCP**, and show that for membership in some NP languages L, there are interactive-PCPs that are significantly shorter than the known (non-interactive) PCPs for these languages.

An interactive PCP (say, for the membership $x \in L$) is a combination of a PCP and a short interactive proof. Roughly speaking, an interactive PCP is a proof that can be verified by reading only a small number of its bits, with the help of a short interactive proof.

More precisely, let L be an NP language, defined by $L = \{x : \exists w \ s.t. \ (x, w) \in R_L\}$. Let p, q, l, c, s be parameters as follows: p, q, l are integers and c, s are reals, s.t. $0 \le s < c \le 1$. (Informally, p is the **size of the PCP**, q is the **number of queries** allowed to the PCP, l is the **communication complexity** of the interactive proof, c is the **completeness** parameter and s is the **soundness** parameter). We think of the parameters p, q, l, c, s as functions of the instance size n. An interactive PCP with parameters (p, q, l, c, s) for membership in L is an interactive protocol between an (efficient[1]) prover P and an (efficient) verifier V, as follows:

We assume that both the prover and the verifier know L and get as input an instance x of size n, and the prover gets an additional input w (supposed to be a

---

[1] We could also consider a model with a not necessarily efficient prover.

witness for the membership $x \in L$). In the first round of the protocol, the prover generates a string π of p bits. (We think of π as an encoding of the witness w). The verifier is still not allowed to access π. The prover and the verifier then apply an interactive protocol, where the total number of bits communicated is l. During the protocol, the verifier is allowed to access at most q bits of the string π. After the interaction, the verifier decides whether to accept or reject the statement $x \in L$. We require the following (standard) completeness and soundness properties: There exists an (efficient) verifier V such that:

Completeness: There exists an (efficient) prover P, such that: for every $x \in L$ and any witness w (given to the prover P as an input), if $(x, w) \in R_L$ then the verifier accepts with probability at least c.

Soundness: For any $x \notin L$ and any (not necessarily efficient) prover $\tilde{P}$, and any w (given to the prover $\tilde{P}$ as an input), the verifier accepts with probability at most s.

Note that in the above definition we allow π to depend on L, x and w. However, in our results we use π that depends only on w, and is of size polynomial in the size of w. We hence think of π as an encoding of the witness w (and this encoding will always be efficient in our results). The fact that π depends only on w (and not on x) is important for many of our applications.

Note also that the notion of interactive PCP is very related to the notion of multi-prover interactive proof [5]. For example, an interactive PCP with $q = 1$ can be viewed as a two provers interactive proof, where the interaction with the first prover is of only one round and is of question size $\log p$ and answer size 1, and the interaction with the second prover is of communication complexity l, (and where both provers are efficient).

1.2 Our Results

We show that the membership in some NP languages, with small witness size, can be proved by *short* interactive PCPs with $q = 1$. We have two main results.

I) Let $\Phi(z_1, \ldots, z_k)$ be a constant depth Boolean formula of size n (over the gates $\wedge, \vee, \oplus, \neg$). For any constant $\epsilon > 0$, the satisfiability of Φ can be proved by an interactive PCP with the following parameters. Size of the PCP: $p = \text{poly}(k)$. Number of queries: $q = 1$. Communication complexity of the interactive proof: $l = \text{poly}(\log n)$. Completeness: $c = 1 - \epsilon$. Soundness: $s = 1/2 + \epsilon$.

Moreover, the string π (generated by the prover in the first round of the protocol) depends only on the witness $w_1, \ldots, w_k$, and not on the instance Φ.

II) Let $\Phi_1(z_1, \ldots, z_k), \ldots, \Phi_n(z_1, \ldots, z_k)$ be arithmetic formulas of size n (say, over $\mathbb{GF}[2]$) that compute polynomials of degree d. For any constant $\epsilon > 0$, the satisfiability of the formula $\bigwedge_{i=1}^{n}[\Phi_i(z_1, \ldots, z_k) = 0]$ can be proved by an interactive PCP with the following parameters. Size of the PCP: $p = \text{poly}(k, d)$. Number of queries: $q = 1$. Communication complexity of the interactive proof: $l = \text{poly}(d, \log n)$. Completeness: $c = 1$. Soundness: $s = 1/2 + \epsilon$.

Moreover, the string π (generated by the prover in the first round of the protocol) depends only on the witness $w_1, \ldots, w_k$ (and on the parameter d), and not on $\Phi_1, \ldots, \Phi_n$. The result works over any other finite field.

In both results, we could actually take ϵ to be poly-logarithmically small. Also, the constant $1/2$, in the soundness parameter of both results, appears only because the string π is a string of bits. We could actually take π to be a string of symbols in $\{1, \ldots, 2^k\}$ and obtain soundness $2^{-k} + \epsilon$.

An additional property of our constructions is that we can assume that the verifier queries the string π before the interaction with the prover starts. This is the case, because the queries to π are non-adaptive (i.e., they do not depend on the values returned on previous queries) and do not depend on the interaction with the prover.

Note that many of the central NP languages can be reduced to the satisfiability of a constant depth formula, without increasing the witness size (e.g., SAT, k-clique, Vertex-Cover, etc.). We hence obtain short interactive PCPs for many other NP languages. Moreover, many NP languages can be reduced to the satisfiability of a formula of the form $\bigwedge_{i=1}^{n}[\Phi_i(z_1, \ldots, z_k) = 0]$ (without increasing the witness size), where $\Phi_1, \ldots, \Phi_n$ are arithmetic formulas of small degree. In these cases, by the second result, perfect completeness can be obtained.

1.3   Subsequent Result

In a subsequent work [12], Goldwasser et al. improved our results as follows.

Let $\Phi(z_1, \ldots, z_k)$ be any Boolean circuit of size n and depth d. For any constant $\epsilon > 0$, the satisfiability of Φ can be proved by an interactive PCP with the following parameters. Size of the PCP: $p = \text{poly}(k, d)$. Number of queries: $q = 1$. Communication complexity of the interactive proof: $l = \text{poly}(d, \log n)$. Completeness: $c = 1$. Soundness: $s = 1/2 + \epsilon$.

Moreover, the string π (generated by the prover in the first round of the protocol) depends only on the witness $w_1, \ldots, w_k$, and not on the instance Φ.

Thus, in particular, Goldwasser et al. improve our first result so that it holds for any formula (rather than for a constant-depth formula) and with perfect completeness. Consequently, their result improves all the applications.

1.4   Motivations and Applications

Below we give several applications and motivations for our results, and for the study of the model of interactive PCP in general. In most of these applications what we actually have is a general reduction that converts any interactive PCP, with certain properties, into another object (i.e., the application). For simplicity, we concentrate on the applications of our first main result and the improved result given in [12].

Motivation: Succinct PCPs with interaction: The PCP theorem states that the satisfiability of a formula $\Phi(z_1, \ldots, z_k)$ of size n can be proved by a proof of size $\text{poly}(n)$ that can be verified by reading only a constant number of its bits [4,9,2,1].

A central line of research in the area of PCPs is devoted to constructing short PCPs. An extremely interesting question is: Do there exist PCPs of size polynomial in the size of the witness, rather than polynomial in the size of the instance (think of the instance as significantly larger than the witness) ? For example, does the satisfiability of a formula $\Phi(z_1, \ldots, z_k)$ of size n can be proved by a PCP of size poly(k), rather than poly(n) (think of n as significantly larger than k) ? A positive answer for this question would have important applications in complexity theory and cryptography (see for example [15]). However, a very interesting recent result by Fortnow and Santhanam shows that this is very unlikely, as it implies that $NP \subseteq coNP/poly$ [11].

Our main results imply that for any constant depth Boolean formula $\Phi(z_1, \ldots, z_k)$ of size n, Alice can publish on the internet a "succinct" proof for the satisfiability of Φ, where the size of the proof is poly(k) (rather than poly(n), i.e., the size of the proof is polynomial in the size of the witness, rather than polynomial in the size of the instance). Later on, any user who wishes to verify the published proof needs to interact with Alice via an interactive protocol of communication complexity poly$(\log n)$, while accessing the published proof at a single location. By reductions, the result extends to many other central NP languages (e.g., SAT, k-clique, Vertex-Cover, etc.).

Using the above mentioned improvement of [12] the same holds for any formula $\Phi(z_1, \ldots, z_k)$ of size n, (rather than a constant depth formula). Moreover, the result holds for any Boolean circuit $\Phi(z_1, \ldots, z_k)$ of size n and depth d, where now the size of the published proof is poly(k, d), and the interactive verification protocol is of communication complexity poly$(d, \log n)$.

Application: Succinct probabilistically checkable arguments: In a very recent subsequent work [18], we give the following application of interactive PCPs. We study the new notion of *probabilistically checkable argument* (PCA) and we give a general way to construct a PCA from an interactive PCP. We use this to construct short PCAs for some languages in NP.

A probabilistically checkable argument (PCA) is a relaxation of the notion of probabilistically checkable proof (PCP). It is defined analogously to PCP, except that the soundness property is required to hold only *computationally*, rather than information theoretically. We consider the model where each verifier is associated with a public key, and each PCA is verifier-dependent, that is, it depends on the verifier's public key. (The key does not need to be certified, and we can assume that the verifier simply publishes it on his web-page). We show that for membership in some languages L, there are PCAs that are significantly shorter than the known PCPs for these languages.

More precisely, our reduction in [18], combined with the above mentioned result of [12], gives the following result: the satisfiability of a Boolean formula $\Phi(z_1, \ldots, z_k)$ of size n can be proved by a PCA of size poly(k). That is, the size of the PCA is polynomial in the size of the witness (as opposed to known PCPs, that are of size polynomial in the size of the instance). The number of queries to the PCA is poly-logarithmic in n. As before, by reductions, the result extends to many other central NP languages (e.g., SAT, k-clique, Vertex-Cover, etc.). Moreover, the result holds for any Boolean circuit $\Phi(z_1, \ldots, z_k)$ of size n and depth d, where now the size of the PCA is $\text{poly}(k, d)$ and the number of queries to the PCA is $\text{poly}(d, \log n)$. The soundness property relies on exponential hardness assumptions.

Application: Succinct zero-knowledge proofs: The notion of *zero-knowledge proof*, first introduced by Goldwasser, Micali and Rackoff [13], has become one of the central notions of modern cryptography. Goldreich, Micali and Wigderson showed that for any language $L \in \text{NP}$, the membership $x \in L$ can be proved by an interactive zero-knowledge proof of polynomial communication complexity [14]. An extremely interesting open problem in cryptography is: Can we significantly reduce the communication complexity of zero-knowledge protocols ? Kilian and Micali, independently, showed that for any language $L \in \text{NP}$, the membership $x \in L$ can be proved by a succinct interactive zero-knowledge *argument* of poly-logarithmic communication complexity [19,22]. Note, however, that the succinct zero-knowledge protocols of [19,22] are *arguments*, rather than *proofs*, that is, their soundness property holds *computationally*. These works left open the problem of constructing "short" zero-knowledge *proofs* for NP.

As an application of our results we show that the satisfiability of a constant depth Boolean formula $\Phi(z_1, \ldots, z_k)$ of size n (over the gates $\wedge, \vee, \oplus, \neg$) can be proved by an interactive zero-knowledge proof of communication complexity $\text{poly}(k)$ (rather than $\text{poly}(n)$). That is, we obtain zero-knowledge proofs of communication complexity polynomial in the size of the witness, rather than polynomial in the size of the instance. As before, the result extends to many other central NP languages.

We note that a similar result, for the case of constant depth formulas, was proved independently (and roughly at the same time) by Ishai, Kushilevitz, Ostrovsky and Sahai [16], using different methods.

Once again, using the above mentioned improvement of [12] the same holds for any formula $\Phi(z_1, \ldots, z_k)$ of size n (rather than a constant depth formula). Moreover, the result holds for any Boolean circuit $\Phi(z_1, \ldots, z_k)$ of size n and depth d, where now the communication complexity of the zero-knowledge proof is $\text{poly}(k, d)$.

We note that for this application we do not use the full power of interactive-PCP, and use mainly the fact that the interactive phase is of very low communication complexity. The results are proved by a general reduction that converts any interactive PCP, with certain properties, into a zero-knowledge interactive proof.

Application: Succinct Commit-Reveal schemes: The zero-knowledge proofs that we construct have an additional property that makes them very useful for many applications. They consist of two phases: The first phase is non-interactive, and depends only on the witness $w = (w_1, \ldots, w_k)$ (and is independent of the instance Φ). In this phase the prover sends to the verifier a certain (non-interactive) commitment to her witness at hand. The second phase is interactive and is very short. It is of communication complexity poly-logarithmic in n. In this phase the prover and verifier engage in a zero-knowledge proof that indeed the string that the prover committed to (in the first phase) is a valid witness for Φ.

This additional property has the following immediate application: Given a string of bits, $w = (w_1, \ldots, w_k)$, a user, Alice, can publish a commitment to w. The commitment is of size $\operatorname{poly}(t, k)$, where t is the security parameter, and is non-interactive. Later on, for any constant depth Boolean formula $\Phi(z_1, \ldots, z_k)$ of size n (over the gates $\wedge, \vee, \oplus, \neg$), such that $\Phi(w) = 1$, Alice can prove the statement $\Phi(w) = 1$, by an interactive zero-knowledge *proof* (rather than argument) with communication complexity $\operatorname{poly}(t, \log n)$. Verifying this proof requires accessing the published commitment in only $\operatorname{poly}(t)$ locations. In other words, after publishing the commitment to w, Alice can prove the statement $\Phi(w) = 1$, by a zero-knowledge proof with communication complexity poly-logarithmic in n.

Note that the proof of the statement $\Phi(w) = 1$ is of communication complexity poly-logarithmic in n, and may be significantly shorter than the length of w. This is interesting even if we discard the requirement of the proof being zero-knowledge.

Once again, using the above mentioned improvement of [12] the same holds for any formula $\Phi(z_1, \ldots, z_k)$ of size n (rather than a constant depth formula). Moreover, the result holds for any Boolean circuit $\Phi(z_1, \ldots, z_k)$ of size n and depth d, where now the commitment is of size $\operatorname{poly}(t, k, d)$, and the communication complexity of the zero-knowledge proof phase is $\operatorname{poly}(d, t, \log n)$.

Application: How to commit to a formula: Below, we give several examples for situations where the commit-reveal protocol may be useful. Many of these motivations and applications are taken from [17] (where succinct non-interactive argument systems with similar properties were given). Note, however, that the situation here is completely different than the one in [17], as here we give in the proof phase *interactive proofs* and there we gave *non-interactive arguments*. Nevertheless, many of the motivations and applications are similar.

For simplicity, we describe all these applications using the above mentioned improved result of [12]. That is, we present the applications for any Boolean formula[2], rather than only for constant depth formulas. We note that one could get results for formulas (rather than for constant depth formulas) also by using our second main result. These results, however, are not as strong as the ones obtained by using the improved result of [12].

In all the applications that we describe below, except for the first one, the main idea is that the string w can itself be a description of a formula Λ. Thus, the commitment that Alice publishes is just a commitment to the formula Λ. We then take the formula Φ to be a *(universal)* formula that runs the formula Λ on N different inputs, $x_1, \ldots, x_N$, and checks that the N outputs are $z_1, \ldots, z_N$. Thus, Alice publishes a commitment to Λ, and later on proves the statement $\bigwedge_{i=1}^{N} [\Lambda(x_i) = z_i]$, by a very short interactive zero-knowledge proof. The commitment is of size

---

[2] Using [12], one can also obtain similar results for circuits, rather than formulas, where the depth of the circuit is also taken into account. For simplicity, we present the applications here only for formulas.

poly$(t, |\Lambda|)$ (where t is the security parameter), and the communication complexity of the zero-knowledge proof is poly$(t, \log |\Lambda|, \log N)$. Note that the communication complexity is logarithmic in both $|\Lambda|$ and N. The main drawback of our protocol, in these contexts, is that it works only for Boolean formulas, and not for general Boolean circuits (with unbounded depth).

I) One of the main tasks of cryptography is to protect honest parties from malicious parties in interactive protocols. Assume that in an interaction between Alice and Bob, Alice is supposed to follow a certain protocol Φ. That is, on an input x, she is supposed to output $\Phi(x, w)$, where w is her secret key. How can Bob make sure that Alice really follows the protocol Φ ? A standard solution, effective for many applications, is to add a commitment phase and a proof phase as follows: Before the interactive protocol starts, Alice is required to commit to her secret key w. After the interactive protocol ends, Alice is required to prove that she actually acted according to Φ, that is, on inputs $x_1, \ldots, x_N$, her outputs were $\Phi(x_1, w), \ldots, \Phi(x_N, w)$. In other words, Alice is required to prove the N statements $\Phi(x_i, w) = z_i$. Typically, we want the proof to be zero-knowledge, since Alice doesn't want to reveal her secret key.

Thus, Alice has to prove in zero-knowledge the statement $\bigwedge_{i=1}^{N} [\Phi(x_i, w) = z_i]$. The only known way to do this is by a proof of length $N \cdot q$, where q is the size of proof needed for proving a single statement of the form $\Phi(x_i, w) = z_i$. Note that $N \cdot q$ may be significantly larger than the total size of all other messages communicated between Alice and Bob.

Our results imply that if Φ is a Boolean formula, there is a much more efficient way. Alice will commit to her secret key w. Then, Alice can prove to Bob the statement $\bigwedge_{i=1}^{N} [\Phi(x_i, w) = z_i]$, by a zero-knowledge proof of size poly$(t, \log |\Phi|, \log N)$, (where t is the security parameter). That is, the proof is of size polylogarithmic in $|\Phi|$ and N.

II) Alice claims that she found a short formula for factoring integers, but of course she doesn't want to reveal it. Bob sends Alice N integers $x_1, \ldots, x_N$ and indeed Alice menages to factor all of them correctly. But how can Bob be convinced that Alice really applied her formula, and not, say, her quantum computer ? We suggest that Alice commits to her formula Λ, and then proves that she actually used her formula Λ to factor $x_1, \ldots, x_N$. The commitment is of size poly$(t, |\Lambda|)$ (where t is the security parameter), and the communication complexity of the zero-knowledge proof is poly$(t, \log |\Lambda|, \log N)$.

III) We want to run a chess contest between formulas. Obviously, the parties don't want to reveal their formulas (e.g., because they don't want their rivals to plan their next moves according to it). Of course we can just ask the parties to send their next move at each step. But how can we make sure that the parties actually use their formulas, and don't have teams of grand-masters working for them ? We suggest that each party commits to her formula Λ before the contest starts. After the contest ends, each party will prove that she actually played according to the formula Λ that she committed to. As before the commitment is of size poly$(t, |\Lambda|)$ and the communication complexity of the zero-knowledge proof is poly$(t, \log |\Lambda|, \log N)$ (where N is the number of moves played).

IV) Assume that both Alice and Bob have access to a very large database $[(x_1, z_1), \ldots, (x_N, z_N)]$. Their goal is to learn a small Boolean formula Λ that explains the database. That is, the goal is to learn Λ such that $\Lambda(x_1) = z_1, \ldots, \Lambda(x_N) = z_N$. Alice claims that she managed to learn such a formula Λ, but she doesn't want to reveal it. It follows from our result that Alice can prove to Bob the existence of such a Λ by a zero-knowledge proof with communication complexity $\text{poly}(t, |\Lambda|, \log N)$, where t is the security parameter.

1.5   Techniques

Our proofs combine many techniques that were previously used in constructing PCPs, and in computational complexity in general, together with some new techniques. Our main technical contribution is a new *sum-check* procedure, which is, in many cases, more efficient than the standard one. More precisely, the standard sum-check procedure requires a prover that runs in time polynomial in the size of the space (on which the sum-check is performed). Our new procedure enables to perform sum-check, with a polynomial-time prover, in many cases where the space is of super-polynomial size.

The first result is proved by a reduction to the second result. This is done by approximating a constant depth formula by a (family of) polynomials of degree $d = \text{poly}(\log n)$ (a well known method in complexity theory, first used by Razborov and Smolensky [24,27] for proving lower bounds for constant depth formulas). It is well known that the approximation can be done by a relatively small family of polynomials. That is, the approximation can be done using a relatively small number of random bits (see for example the survey paper of [7]).

Suppose that we have a small family of polynomials of degree $d = \text{poly}(\log n)$ that approximate the constant depth formula. After the prover generates the string π, the verifier chooses randomly a polynomial Φ from the family, and the prover is required to prove that $\Phi(w_1, \ldots, w_k) = 1$ for the witness $w = (w_1, \ldots, w_k)$ encoded by π. This proves that the constant depth formula is satisfied by the witness w. We loose the perfect completeness because the low degree polynomials only approximate the constant depth formula and are not equivalent to it.

For the proof of the second result we use methods that were previously used for constructing PCPs and interactive proofs, together with some new techniques. The proof has two parts: The first part shows the needed result but with q larger than 1, that is, more than one query to the PCP. The second part shows that in any interactive PCP, the number of queries, q, can be reduced to 1, using a short additional interaction with the prover.

For the proof of the first part, we take π to be the *low degree extension* of the witness w. The verifier checks that π is indeed a low degree polynomial using a *low degree test*, as is frequently done in constructions of PCPs (e.g., [4,2,1,23,3,21]). Given an arithmetic formula Φ of size n and degree d, the verifier can verify that $\Phi(w_1, \ldots, w_k) = 0$ by an interactive *sum-check procedure*, as is frequently done in constructions of PCPs and interactive proofs (e.g., [20,26]). However, we need to apply the sum-check procedure on a space of size $> k^d$, which is, in most interesting cases, super-polynomial in n. This seems to require a prover that runs in super-polynomial time. Nevertheless, we show that we can use our new sum-check procedure, that can be performed by an efficient prover, even when the space is of super-polynomial size.

Note that the prover needs to prove that $\Phi_i(w_1, \ldots, w_k) = 0$ for every polynomial $\Phi_i \in \{\Phi_1, \ldots, \Phi_n\}$. Since this would require too much communication, we take Φ to be a (pseudo-random) linear combination of $\Phi_1, \ldots, \Phi_n$ (using any linear error correcting code). The combination is chosen by the verifier, and the prover is only required to prove that $\Phi(w_1, \ldots, w_k) = 0$.

As mentioned above, the second part of the proof of the second result is a general theorem that shows that in any interactive PCP the number of queries q can be reduced to 1 (using some additional interaction with the prover). This is done as follows. First, we can assume w.l.o.g. that all the queries to π are made after the interactive protocol ends. This is because rather than querying π, the verifier can simply ask the prover for the answers, and after the interactive protocol ends make the actual queries and compare the answers. Second, we can assume w.l.o.g. that the string π, generated by the (honest) prover, is a multivariate polynomial of low degree. Otherwise, we just ask the prover to generate the low degree extension of π, rather than π itself.

We can now apply a method that is based on methods that are frequently used in constructions of PCPs (e.g., [10,1,8,25]). If the verifier wants to query π in q points, the verifier takes a curve γ of degree $q + 1$ that passes through all these points and an additional random point. If π is a low degree polynomial, the restriction of π to γ is also a low degree polynomial. The verifier asks the prover to send this low degree polynomial, and verifies the answer by checking it in a single point, using a single query to π. The verifier still needs to check that π is indeed a low degree polynomial. This is done, once again, using a low degree test. The verifier asks the prover for the restriction of π to a low dimensional subspace ν, and verifies the answer by checking it in a single point, using a single query to π. This, however, requires an additional query to π, while we only allow one query altogether. We hence need to combine the two tests. That is, we need to combine the low degree test and the test that the prover's answer for the curve γ is correct. To do this, the verifier actually asks the prover to send the restriction of π to a manifold spanned by both γ and ν. The verifier verifies the answer by checking it in a single point, using a single query to π.

This shows that the verifier can make only one query to π. However, the point queried in π contains a field element and not a single bit. To reduce the answer size to a single bit, the prover is required to generate in π the error correcting code of each field element, rather than the element itself. The verifier can now verify an element by querying only one bit in its error correcting code.

References

1. Arora, S., Lund, C., Motwani, R., Sudan, M., Szegedy, M.: Proof Verification and Hardness of Approximation Problems. J. ACM 45(3), 501–555 (1998)
2. Arora, S., Safra, S.: Probabilistic Checking of Proofs: A New Characterization of NP. J. ACM 45(1), 70–122 (1998)

3. Arora, S., Sudan, M.: Improved Low-Degree Testing and its Applications. Combinatorica 23(3), 365–426 (2003)
4. Babai, L., Fortnow, L., Lund, C.: Non-Deterministic Exponential Time has Two-Prover Interactive Protocols. Computational Complexity 1, 3–40 (1991)
5. Ben-Or, M., Goldwasser, S., Kilian, J., Wigderson, A.: Multi-Prover Interactive Proofs: How to Remove Intractability Assumptions. In: STOC 1988, pp. 113–131 (1988)
6. Babai, L., Moran, S.: Arthur-Merlin Games: A Randomized Proof System, and a Hierarchy of Complexity Classes. J. Comput. Syst. Sci. 36(2), 254–276 (1988)
7. Beigel, R.: The Polynomial Method in Circuit Complexity. In: Structure in Complexity Theory Conference, pp. 82–95 (1993)
8. Dinur, I., Fischer, E., Kindler, G., Raz, R., Safra, S.: PCP Characterizations of NP: Towards a Polynomially-Small Error-Probability. In: STOC 1999, pp. 29–40 (1999)
9. Feige, U., Goldwasser, S., Lovasz, L., Safra, S., Szegedy, M.: Interactive Proofs and the Hardness of Approximating Cliques. J. ACM 43(2), 268–292 (1996)
10. Feige, U., Lovasz, L.: Two-Prover One-Round Proof Systems: Their Power and Their Problems (Extended Abstract). In: STOC 1992, pp. 733–744 (1992)
11. Fortnow, L., Santhanam, R.: Infeasibility of Instance Compression and Succinct PCPs for NP. In: STOC 2008 (2008)
12. Goldwasser, S., Kalai, Y.T., Rothblum, G.: Delegating Computation: Interactive Proofs for Mortals. In: STOC 2008 (2008)
13. Goldwasser, S., Micali, S., Rackoff, C.: The Knowledge Complexity of Interactive Proof Systems. SIAM Journal on Computing 18(1), 186–208 (1989)
14. Goldreich, O., Micali, S., Wigderson, A.: Proofs that Yield Nothing But Their Validity or All Languages in NP Have Zero-Knowledge Proof Systems. J. ACM 38(3), 691–729 (1991)
15. Harnik, H., Naor, M.: On the Compressibility of NP instances and Cryptographic Applications. In: FOCS, pp. 719–728 (2006)
16. Ishai, Y., Kushilevitz, E., Ostrovsky, R., Sahai, A.: Zero-Knowledge from Secure Muliparty Computation. In: STOC 2007, pp. 21–30 (2007)
17. Kalai, Y.T., Raz, R.: Succinct Non-Interactive Zero-Knowledge Proofs with Preprocessing for LOGSNP. In: FOCS 2006, pp. 355–366 (2006)
18. Kalai, Y.T., Raz, R.: Probabilistically Checkable Arguments
19. Kilian, J.: A note on efficient zero-knowledge proofs and arguments. In: STOC 1992, pp. 723–732 (1992)
20. Lund, C., Fortnow, L., Karloff, H.J., Nisan, N.: Algebraic Methods for Interactive Proof Systems. J. ACM 39(4), 859–868 (1992)
21. Moshkovitz, D., Raz, R.: Sub-Constant Error Low Degree Test of Almost Linear Size. In: STOC 2006, pp. 21–30 (2006)
22. Micali, S.: CS Proofs (Extended Abstracts). In: FOCS 1994, pp. 436–453 (1994)
23. Raz, R., Safra, S.: A Sub-Constant Error-Probability Low-Degree Test, and a Sub-Constant Error-Probability PCP Characterization of NP. In: STOC 1997, pp. 475–484 (1997)
24. Razborov, A.: Lower Bounds for the Size of Circuits of Bounded Depth with Basis $\{\wedge, \oplus\}$. Math. Notes of the Academy of Science of the USSR 41(4), 333–338 (1987)
25. Raz, R.: Quantum Information and the PCP Theorem. In: FOCS 2005, pp. 459–468 (2005)
26. Shamir, A.: IP=PSPACE. J. ACM 39(4), 869–877 (1992)
27. Smolensky, R.: Algebraic Methods in the Theory of Lower Bounds for Boolean Circuit Complexity. In: STOC 1987, pp. 77–82 (1987)

Constant-Round Concurrent Non-malleable Zero Knowledge in the Bare Public-Key Model

Rafail Ostrovsky[1], Giuseppe Persiano[2], and Ivan Visconti[2]

[1] UCLA, Los Angeles, CA 90095, USA
rafail@cs.ucla.edu
[2] Dipartimento di Informatica ed Applicazioni, Università di Salerno,
84084 Fisciano (SA), Italy
{giuper,visconti}@dia.unisa.it

Abstract. One of the central questions in Cryptography is the design of round-efficient protocols that are secure under concurrent man-in-the-middle attacks. In this paper we present the first *constant-round concurrent non-malleable zero-knowledge* argument system for NP in the Bare Public-Key model [Canetti et al., STOC 2000], resolving one of the major open problems in this area. To achieve our result, we introduce and study the notion of non-malleable witness indistinguishability, which is of independent interest. Previous results either achieved relaxed forms of concurrency/security or needed stronger setup assumptions or required a non-constant round complexity.

Keywords: non-malleable zero knowledge, witness indistinguishability.

1 Introduction

In [1] Dolev, Dwork and Naor proposed the notion of a non-malleable zero-knowledge (NMZK, for short) proof system where security must be preserved even under a man-in-the-middle attack. This strong attack allows the adversary to act as a prover in a proof and as a verifier in another proof with full control over the scheduling of the messages. The notion of NMZK is proved to be extremely important in cryptography, since it captures the notion of *proof independence*, and led to multiple applications. Feasibility results for NMZK have been shown by using either black-box techniques and a super-constant number of rounds by Dolev et al. [1] or by using non-black-box techniques and obtaining computational soundness in a constant number of rounds by Barak [2] and Pass and Rosen [3]. Another strong security notion for proof systems is that of *concurrent* zero knowledge, introduced by Dwork, Naor and Sahai [4], where security has to work against adversaries that are involved in many concurrent executions of a proof system.

In this paper we consider an adversary $\mathcal{A}$ mounting a *concurrent* man-in-the-middle attack in which $\mathcal{A}$ acts as a verifier interacting with a honest prover in polynomially many *left* proofs and acts as a prover interacting with honest verifiers in polynomially many *right* proofs. The problem of designing protocols that

L. Aceto et al. (Eds.): ICALP 2008, Part II, LNCS 5126, pp. 548–559, 2008.

combine concurrent security with security against man-in-the-middle adversaries has received a lot of attention; several questions still remain open, though. In particular, constant-round concurrent non-malleable zero-knowledge (cNMZK, for short) proof systems have been shown to exist by assuming the existence of trusted third parties or a trusted common reference string or using relaxed security notions or relaxed concurrency. A construction with poly-logarithmic round complexity for concurrent NMZK in the plain model has been given by Barak, Prabhakaran, and Sahai [5]. The possibility of constructing constant round cNMZK proof systems in the plain model or under weaker setup assumptions is still an open problem.

Witness indistinguishability. A weaker but still useful security notion for proof systems is that of witness indistinguishability [6], where it is required that the adversarial verifier does not distinguish the witness used by the prover. Despite the tremendous applicability of witness indistinguishability, while a lot of attention has been given to zero knowledge with respect to man-in-the-middle attacks, very little attention has been given to witness indistinguishability with respect to concurrent man-in-the-middle attacks.

1.1   Our Results

In this paper we study concurrent man-in-the-middle attacks with respect to proof systems and show the following two results.

We first show the definition and construction of a new concurrent non-malleable primitive that extends the notion of witness indistinguishability to the setting in which the adversary is a concurrent man-in-the-middle. For defining this new primitive, we focus on a specific class of argument systems referred to as *commit-and-prove*[1] functionality introduced in [7]. We then construct a *constant-round* concurrent non-malleable witness indistinguishable (cNMWI, for short) argument of knowledge (under Def. 2) for all NP in the plain model (see Theorem 1). This construction relies upon the work by Pass and Rosen [8] where constant-round concurrent non-malleable (NM, for short) commitments have been achieved. In a next work we also show that the notions of NMWI and NMZK argument systems are incomparable, this is surprising since all previously introduced notions of witness indistinguishability were implied by the corresponding notions of zero knowledge.

Second, we show the construction of a *a constant-round cNMZK argument system* under standard complexity theoretic assumptions and security notions in the Bare Public-Key model, a set-up assumption introduced in [9] that does not require any trusted third party. So far this has been achieved only under stronger setup assumptions. Previously, constant-round concurrent zero knowledge has been obtained in the BPK model in [9] (in [10] with a concurrent soundness guarantee, and in [11, 12] under standard assumptions). Given our results, the

---

[1] We restrict our study to this class of argument systems as: 1) they allow us to define the notion of witness encoded in a proof; 2) they suffice for our constructions and applications. It is possible however to generalize this notion.

BPK model is, at the best of our knowledge, the weakest model in which *constant-round* cNMZK has been achieved.

Corruption model and adaptive inputs. In all our results we consider the static corruption model where the adversary has to choose the corrupted parties before the protocols start. Following the previous work on NMZK, in the proof of our concurrent NMZK argument of knowledge in the BPK model we assume that the inputs (i.e., statements) for honest parties are fixed according to some predetermined distribution while the adversary can choose its inputs adaptively. Instead, for our cNMWI argument of knowledge in the plain model, following [13] we also allow the adversary to choose the inputs of the prover by giving it both the statements and the witnesses.

Work related to witness indistinguishability and cNMZK zero knowledge in the plain model. Recently and independently from our work Micali, Pass and Rosen [14] presented an extension of the notion of witness indistinguishability for achieving a relaxed notion of secure computation that does not resort to the simulation paradigm. Their techniques are similar to ours but in this work, in contrast to [14], we achieve arguments of knowledge and focus on the use of these strong notions of witness indistinguishability for achieving a notion of security based on simulation (i.e., concurrent NMZK). Moreover, achieving input-indistinguishability involves significantly more complicated protocols; furthermore, it is not clear how easy this notion is to work with when used as a "sub-protocol". The power of our simple and specific definition of non-malleable witness indistinguishability is that it can be achieved essentially directly by relying on the non-malleable commitment protocol of [8] and it is easy to work with.

We observe that in the plain model constant-round (non-concurrent) NMZK has been recently obtained [2, 3] whereas obtaining constant-round concurrent zero knowledge in the plain model has been open for quite some time. The only constant-round concurrent zero-knowledge arguments known in the plain model impose a bound on the number of concurrent executions that the adversary can perform [15]. If we do not insist on constant-round protocols, non-malleability and security in a concurrent setting have been achieved by [5] which present a protocol with logarithmic round complexity.

2   Non-malleable Witness Indistinguishability

For lack of space, the definition of standard tools and the ones about non-malleability can be found in the full version of this work [16, 17].

We now start by discussing and defining the new non-malleable notion of proof systems. In our definition of NM witness indistinguishability we shall require that the witness *encoded in the proof* given by the man-in-the-middle adversary $\mathcal{A}$ is independent from the witness used by the honest prover in the left proof. Notice that $\mathcal{A}$ might be unaware of the witness it has used in the right proof. More specifically, we focus on a specific class of argument systems referred to as *commit-and-prove* argument systems (previously considered in [7]). Informally, the transcript of a commit-and-prove argument encodes in an unambiguous way the witness used by the prover (even though it might not be efficiently extracted from the transcript). In a NMWI commit-and-prove argument we require the witness encoded in the proof produced by the man-in-the-middle adversary to be independent of the witness used (by the honest prover) in the proof in which the adversary acts as a verifier.

For general argument systems it is not clear whether the notion of witness encoded is well defined as there could be more than one. Therefore, we focus on commit-and-prove argument systems for which the notion of the witness encoded is well defined and commit-and-prove arguments actually suffice for proving our next result (i.e., cNMZK n the BPK model).

Commit-and-prove argument systems. A commit-and-prove argument system $\Pi = \langle P, V \rangle$ for a language L is a two-stage protocol. On input x, in the first stage the prover and the verifier execute a commitment protocol by which the prover commits to a string w. In the second stage, the prover proves to the verifier that the committed string w is a valid witness for "$x \in L$". We study commit-and-prove argument systems in which the commitment scheme used in the first stage is non-interactive and statistically binding, therefore the notion of *witness encoded in the proof* is well defined and it corresponds to the string committed to by the *first* prover-to-verifier message. If the proof is not accepted by the verifier, we consider the witness to be encoded in the proof to be the string $\perp$. We shall require that in a NMWI commit-and-prove argument system the man-in-the-middle adversary encodes in the right proof a witness that is independent from the one that the honest prover has used in the left proof.

Tag-based NMWI commit-and-prove arguments. We consider a man-in-the-middle adversary $\mathcal{A}$ interacting in the left proof with tag tag with the honest prover P that is running on input instance x and witness w. In the right proof, $\mathcal{A}$ is interacting with the honest verifier V on common input $\tilde{x}$ and tag $\tilde{\mathsf{tag}}$ of its choice. We denote by z the auxiliary information available to $\mathcal{A}$.

The notion of tag-based NM witness indistinguishability is defined in terms of the random variable $\mathsf{wmim}^{\mathcal{A}}(\mathsf{tag}, x, w, z)$ that is the distribution of the output of the following process: a transcript trans of an interaction of $\mathcal{A}$, including the left and the right proof, is picked according to distribution $\mathsf{View}^{P}_{\mathcal{A}}(\mathsf{tag}, x, w, z)$ (i.e., the view of $\mathcal{A}$ when running with z as auxiliary input and playing with P that runs on input (x, w) and tag tag) and the output of a procedure wit applied to trans is returned. The procedure wit returns $\perp$ if the right proof is not accepting (i.e., V outputs 0) or tag is the tag of the right proof. Otherwise it returns the witness encoded in the right proof.

Definition 1 (tag-based NMWI argument). *A family of commit-and-prove argument systems* $\Pi = \{\langle P_{\mathsf{tag}}, V_{\mathsf{tag}} \rangle\}_{\mathsf{tag}}$ *for an* NP*-language L is a tag-based non-malleable witness indistinguishable (tag-based NMWI, in short) argument with tags of length ℓ if, for all probabilistic polynomial-time man-in-the-middle* *adversaries $\mathcal{A}$, for all probabilistic polynomial-time algorithms D, there exists a negligible function ν such that for all $x \in L$, for all tags $\mathsf{tag} \in \{0,1\}^{\ell}$, for all pairs (w, w') of witnesses for x, and for all auxiliary information z it holds that*

$$|\mathrm{Prob}[\, D(x, w, w', \mathsf{wmim}^{\mathcal{A}}(\mathsf{tag}, x, w, z), z) = 1\,]-$$

$$\mathrm{Prob}[\, D(x, w, w', \mathsf{wmim}^{\mathcal{A}}(\mathsf{tag}, x, w', z), z) = 1\,]| < \nu(|x|).$$

A NMWI argument system is an argument of knowledge when for any prover that proves a given statement with probability p, there exists an efficient extractor that outputs a valid witness with essentially the same probability p (see the definition of [18]).

Comparison with NMZK and NM commitments. We stress here that NMZK requires the existence of a simulator while NM witness indistinguishability does not. Instead, NM witness indistinguishability crucially considers the possible witnesses that are encoded in the proofs given by the man-in-the-middle while NMZK requirements are satisfied when a valid witness is given in output by the simulator-extractor. The notion of NM witness indistinguishability is similar to the notion of NM commitment with respect to commitment [1,3]. Indeed, both notions concern the security of a primitive against man-in-the-middle attacks by considering a string that is encoded in the messages sent by the adversary. This string is a committed message in case of NM commitments while it is an encoded witness in case of NM witness indistinguishability.

2.1   Concurrent and Simulation-Based NMWI Arguments

We extend the notion of non-malleable witness indistinguishability to the concurrent setting by considering a concurrent man-in-the-middle adversary $\mathcal{A}$ that opens $m = \mathsf{poly}(k)$ left and right proofs each with a common input of length $n = \mathsf{poly}(k)$. Here k refers to the security parameter. $\mathcal{A}$ interacts in the i-th left proof with an instance of the honest prover P on common input "$x_i \in L$" and private prover's input $w_i \in W(x_i)$. In the j-th right proof $\mathcal{A}$ is interacting with the honest verifier V on common input $\tilde{x}_j$ of its choice.

To define concurrent non-malleable witness indistinguishability, we extend $\mathsf{wmim}^{\mathcal{A}}(X, W, z)$ to sequences of inputs and witnesses in the following way. The distribution $\mathsf{wmim}^{\mathcal{A}}(X, W, z)$ is the distribution of the output of the following procedure. First a transcript trans is sampled according to the view $\mathsf{View}_{\mathcal{A}}^{P}(X, W, z)$ of $\mathcal{A}$. Then the output of the following extension of the procedure wit applied to trans is returned. Procedure wit returns a sequence $(\tilde{w}_1, \cdots, \tilde{w}_m)$ where m is the number of right proofs and it holds that: if the j-th right proof is non-accepting or has the same common input as one of the left proofs then $\tilde{w}_j = \perp$; otherwise, $\tilde{w}_j$ is the witness encoded in the j-th right proof.

As done for non-malleable witness indistinguishability, we can obtain a tag-based definition of concurrent non-malleable witness indistinguishability and we define $\mathsf{wmim}^{\mathcal{A}}(T, X, W, z)$ so to take into account the tags and not the inputs of the right proofs. We stress again that $\mathcal{A}$ is allowed to choose the inputs and the tags for the right proofs.

Definition 2 (tag-based cNMWI argument). *A family of commit-and-prove argument systems* $\Pi = \{\langle P_{\text{tag}}, V_{\text{tag}}\rangle\}_{\text{tag}}$ *for the language L is a tag-based concurrent non-malleable witness indistinguishable argument (a tag-based cNMWI) with tags of length ℓ if, for all probabilistic polynomial-time concurrent man-in-the-middle adversaries $\mathcal{A}$, for all $m = \text{poly}(k)$, for all $n = \text{poly}(k)$ and for all probabilistic polynomial-time algorithms D, there exists a negligible function ν such that for all k, for all sequences X of m elements of L of length n, for all sequences T of tags of length ℓ, for all sequences W and W' of witnesses for X, and for all auxiliary information z it holds that*

$$|\text{Prob}[\, D(X, W, W', \text{wmim}^{\mathcal{A}}(T, X, W, z), z) = 1\,]-$$

$$\text{Prob}[\, D(X, W, W', \text{wmim}^{\mathcal{A}}(T, X, W', z), z) = 1\,]| < \nu(k).$$

We stress that the two above definitions can be adapted by requiring that each statement to be proved is adaptively chosen by the adversary (that will also provide valid witnesses to the provers) before the corresponding proof starts, as discussed in [19]. Our constructions will enjoy this extra property.

We will also consider a relaxed notion of concurrent non-malleable witness indistinguishability where the adversary is allowed to run only one left proof. We denote this restricted notion of concurrent NM witness indistinguishability as *one-left many-right* concurrent NM witness indistinguishability.

Simulation-based cNMWI Arguments. We also give a simulation-based definition of non-malleable witness indistinguishability. We consider only the tag-based case. Let $\mathcal{A}$ be a concurrent man-in-the-middle adversary and consider the following two executions. The first execution is the *man-in-the-middle* execution where the concurrent man-in-the-middle adversary $\mathcal{A}$ interacts with several copies of the honest prover in the left proofs and with several copies of the honest verifier in the right proofs. For this execution we define distribution $\text{wmim}^{\mathcal{A}}(T, X, W, z)$ as done previously. Also, we stress that $\mathcal{A}$ can choose the inputs for the right proofs as well as the tags. In the second execution, called the *stand-alone* execution, we consider a simulator S that, without receiving any witness for the inputs X of the left instances and without interacting with a honest prover, manages to output the transcripts of the left and the right proofs. We denote by $\text{wsta}^{S}(T, X, z)$ the random variable that describes output of the following procedure. First a transcript **trans** is sampled according to the distribution of the output of $S(T, X, z)$. Then the procedure wit is applied to **trans** and the output is returned.

Definition 3 (tag-based SBcNMWI argument). *A family of commit-and-prove argument system* $\Pi = \{\langle P_{\text{tag}}, V_{\text{tag}}\rangle\}_{\text{tag}}$ *is a tag-based simulation-based concurrent non-malleable witness indistinguishable (tag-based SBcNMWI, in short) argument for the language L, if for all polynomials $m = \text{poly}(k)$ and $n = \text{poly}(k)$, for all probabilistic polynomial-time concurrent man-in-the-middle adversaries $\mathcal{A}$, there exists a simulator S running in expected polynomial time, such that the following distributions are computationally indistinguishable:*

$$\{\text{wmim}^{\mathcal{A}}(T, X, W, z)\}_{T \in \{0,1\}^{ml}, X \in L_n^m, W \in W(X), z \in \{0,1\}^*} \text{ and}$$

$$\{\mathsf{wsta}^S(T, X, z)\}_{T \in \{0,1\}^{ml}, X \in L_n^m, z \in \{0,1\}^*}.$$

The notion of a simulation-based non-malleable witness indistinguishable commit-and-prove *argument of knowledge* can be obtained by further requiring that S is able to extract witnesses from the right proofs whenever they use tags different from the left proofs.

The notion of one-left many-right SBcNMWI argument can be obtained by restricting the adversary to be involved only in one left proof.

Theorem 1. *Assume that there exists a family of claw-free permutations. Then there exists a constant-round tag-based cNMWI commit-and-prove argument of knowledge for all* NP *in the plain model.*

The proof of this theorem is obtained by first noticing that a variation of the commitment scheme of [3] actually allows one to obtain a one-left many-right SBcNMWI argument of knowledge, then by noticing that any one-left many-right SBcNMWI argument of knowledge is a one-left many-right cNMWI argument of knowledge, and finally by noticing that any one-left many-right cNMWI argument of knowledge is a many-left many-right cNMWI argument of knowledge (see the full version of this work [16,17] for the protocol and the security proof.)

We finally stress that the above theorem still holds in case the adversary chooses the inputs of the honest prover, by feeding it also valid witnesses.

3   cNMZK in the BPK Model

In the BPK model [9], each verifier registers some public information (called the *public key*) in a public file during a preprocessing stage. Each public key is associated with some secret information (called the *secret key*) that is known only to the owner of the public key. After the preprocessing is completed, parties engage in the proof stage where proofs are run.

We will define and construct in the BPK model constant-round arguments for any NP-language that are secure with respect to a BPK concurrent man-in-the-middle adversary $\mathcal{A}$ which during the preprocessing stage has complete control over the public file where keys are registered (that is, $\mathcal{A}$ can modify, omit and, add new adaptively chosen keys to the public file) and, once the preprocessing stage is completed, $\mathcal{A}$ acts as a concurrent man-in-the-middle adversary. We stress that no form of key-authentication is required thus making the BPK model a setting very close to the plain model.

The BPK *model for interactive argument systems.* We now review the definition of an interactive argument system in the BPK model that were previously given in [20] and the extension to the concurrent man-in-the-middle attack case.

Formally, a BPK *pair* is a pair $\langle P, V \rangle$ where P is a probabilistic polynomial-time algorithm and V is a pair $V = (V_0, V_1)$ of probabilistic polynomial-time algorithms. The interaction between provers and verifiers takes place in two stages. In the first stage, called the *set-up* stage, verifiers run algorithm V_0, on input a security parameter 1^k, to obtain a pair (pk, sk) consisting of a public and a secret key. Each verifier publishes his public key pk in a public file F. The second stage, called the *proof* stage, consists of polynomially (in the security parameter) many proofs. In each of them a prover interacts with a verifier; specifically, the prover runs algorithm P on input x (of length polynomial in the security parameter), some auxiliary information w (typically w is a witness for x to be member of some fixed language L) and the public key pk chosen by the verifier. The verifier instead runs algorithm V_1 on input x and sk.

A BPK pair $\langle P, V \rangle$ is *complete* for the language L if in any interaction on common input $x \in L$ and pk constructed by V_0, where P receives as additional input $w \in W(x)$, and V_1 secret key sk associated with pk, V_1 accepts except with negligibly probability.

The definitions of argument systems in the BPK model can be found in [9], in particular in [20, 21] the notions of concurrent zero-knowledge and concurrent soundness have been defined. We will focus on cNMZK arguments of knowledge in the BPK model that imply both concurrent zero knowledge and concurrent soundness. Indeed, concurrent zero-knowledge corresponds to a special case where the man-in-the-middle does not run any right proof. Instead, concurrent soundness corresponds to the special case where the man-in-the-middle does not run any left proof and is implied by the fact that we require that a legal NP witness is obtained for any accepting proof given by the adversary (i.e. proofs where V outputs 1).

We next define *cNMZK argument of knowledge* in the BPK model.

A BPK concurrent man-in-the-middle adversary $\mathcal{A} = (\mathcal{A}_0, \mathcal{A}_1)$ is a pair of probabilistic algorithms. $\mathcal{A}_0$ on input an auxiliary information z receives the public file F containing the public keys as computed by the honest verifiers and outputs a modified public file F'. In computing F', $\mathcal{A}_0$ is allowed to add new adaptively chosen keys and to remove some of the keys of the honest verifiers. $\mathcal{A}_0$ also outputs some secret auxiliary information Z relative to F'. Once F' is made public by $\mathcal{A}_0$, it cannot be changed and the control passes to $\mathcal{A}_1$ that runs on input F' and Z. In the proof stage, $\mathcal{A}_1$ behaves like a concurrent man-in-the-middle adversary with the only restriction that he can start right proofs in which he plays as a prover with honest verifiers only with respect to entries of F' that were chosen by the honest verifiers and not modified by $\mathcal{A}_0$.

We define the view $\mathsf{BView}_\mathcal{A}(X, W, z)$ of a BPK concurrent man-in-the-middle adversary $\mathcal{A} = (\mathcal{A}_0, \mathcal{A}_1)$ with respect to the vector X of left inputs with witnesses W as consisting of the initial public file received by $\mathcal{A}_0$, of all messages received by $\mathcal{A}_1$ in the proof stage both in the left proofs run on input X and right proofs run on inputs adaptively chosen by $\mathcal{A}_1$, along with the sequence of internal states of $\mathcal{A}_0$ and $\mathcal{A}_1$ and coin tosses, and the output of the honest verifiers.

Definition 4. *(cNMZK arguments of knowledge in the BPK) A BPK pair $\Pi = \langle P, V \rangle$ complete for the language L is a BPK cNMZK argument of knowledge if for every probabilistic polynomial-time BPK concurrent man-in-the-middle adversary $\mathcal{A}$, there exists a probabilistic algorithm S running in expected polynomial time such that, for all $m = \mathsf{poly}(k)$ and $n = \mathsf{poly}(k)$, by denoting with $S(X, z) = (S_0(X, z), S_1(X, z))$ the output of S on input (X, z), we have*

1. $\{S_0(X, z)\}_{X \in L_n^m, z \in \{0,1\}^*}$ and $\{\text{BView}_{\mathcal{A}}(X, W, z)\}_{X \in L_n^m, W \in W(X), z \in \{0,1\}^*}$ are computationally indistinguishable.

2. Writing the second component of S's output as $S_1(X, z) = (\tilde{w}_1, \ldots, \tilde{w}_m)$, we have that, for all accepting right proofs j of $S_0(X, z)$ with common input $\tilde{x}_j \notin X$, $\tilde{w}_j \in W(\tilde{x}_j)$ except with negligible probability.

We stress that the adversary can always see the output of the verifier. This is an important issue for proof systems in which the internal state of the verifier is needed to decide whether a proof is accepted or not.

As a concurrent verifier and a concurrent prover are both special cases of a concurrent man-in-the-middle adversary, then it is obvious that a cNMZK argument of knowledge in the BPK model is both concurrent zero-knowledge and concurrently sound.

3.1   The Constant-Round Protocol

The main idea is to use the FLS paradigm by having the prover prove knowledge of either a legal witness of the input statement or of the secret key of the verifier. The goal is to design a simulator that runs the honest verifier algorithm and plays the role of the prover by first extracting the secret keys used by the adversary and then by using them as witnesses running in a straight-line fashion the honest prover algorithm. In order to make this possible, we have the verifier first prove knowledge of his secret key so that the simulator will first extract the secret keys of the adversary. To withstand concurrent man-in-the-middle attack, we employ the cNMWI argument of knowledge we have developed in the previous section along with the two-key technique by [6].

More in details, in the preprocessing stage, each verifier computes a pair of public keys along with the corresponding secret keys. He then randomly chooses one of the two secret keys and discards the other one. This step can be implement by using a one-way function f in the following way: randomly pick two messages $\mathsf{sk}_0, \mathsf{sk}_1$ in the domain of f; compute public keys $\mathsf{pk}_0 = f(\mathsf{sk}_0), \mathsf{pk}_1 = f(\mathsf{sk}_1)$; randomly select $b \leftarrow \{0, 1\}$; set $\mathsf{sk} = (b, \mathsf{sk}_b)$ and $\mathsf{pk} = (\mathsf{pk}_0, \mathsf{pk}_1)$.

The actual argument on input x consists of a sequential composition of two instances of the tag-based constant-round cNMWI commit-and-prove argument of knowledge we have constructed. First the verifier proves knowledge of one of the two secret keys associated to his entry in the public file (this is obviously done by NP-reducing this instance to the NP-complete language used by the sub-protocol). This subprotocol is run using $x \circ 0$ as tag. Obviously the honest verifier uses his knowledge of one of two secret keys to successfully complete this subprotocol. In the second execution the prover proves knowledge of either w such that $R(x, w) = 1$ or of one of the two secret keys associated with the two public keys of the verifier. The tag used in this subprotocol is $x \circ 1$. Obviously the honest prover uses knowledge of a witness w for $R(x, \cdot)$ to complete the protocol.

Let us explain how we plan to perform simulation of the protocol. Simulation is easy for right proofs where the simulator plays the role of the honest verifier. Indeed right proofs are executed relatively to entry of the public file that have been constructed by the simulator itself and thus it knows one of the secret keys to perform the first subprotocol of a right proof. Simulating the second subprotocol of right proofs and the first subprotocol of the left proofs is trivial as the simulator can simply play the honest verifier algorithm of the subprotocol. In order to simulate the second subprotocol of left proofs instead the simulator needs to know either a witness for "$x \in L$" or one of the secret keys associated with the corresponding entries of the public file that are *used* by the adversary. However, the adversary has just proved knowledge of at least one of the two keys in the first subprotocol of the same proof. Therefore we plan on extracting one of these keys from the adversary and then use it to perform the second subprotocol. The use of rewinds is dangerous in concurrent setting but not in the BPK model as shown in [9]. Indeed the number of extraction procedures that have to be successfully run is independent of the number of concurrent proofs, since it is bounded by the size of the public file. Once the simulator knows at least one secret key for each of the entries of the public file used by the adversary, the simulation is straight-line.

Let us now explain why we can also extract valid witness for all theorems proved by the adversary. We know that in all succeeding proofs for $x \in L$ given by the adversary, there is a cNMWI argument of knowledge for proving that $x \in L$ or that the adversary knows one of the two secret keys of the verifier. During the simulated game we can run the extractor for all these proofs in order to obtain the valid witnesses thus satisfying definition 4. If instead we extract as witnesses the secret keys of the verifier, we distinguish two cases. In the former case we extract a secret key that was not used by the simulator; we show how to reduce this case to an adversary that inverts the one-way function used for generating the public keys. In the latter case we always extract the same secret keys used by the simulator; this last case means that the adversary succeeded in encoding in the cNMWI arguments of knowledge that it proved, the same witness encoded by the simulator in the cNMWI arguments of knowledge where the adversary played as verifier. This last case contradicts the NM witness indistinguishability of the cNMWI arguments of knowledge.

The protocol in details. Let L be an NP-language with polynomial-time relation R and let f be a one-way function. Associated with L and f, we consider two auxiliary NP-languages L_1 and L_2 with polynomial-time relations R_1 and R_2 defined as follows:

- $(\mathsf{pk}_0, \mathsf{pk}_1) \in L_1$ iff there exist b and sk such that $\mathsf{pk}_b = f(\mathsf{sk})$;
- $(x, \mathsf{pk}_0, \mathsf{pk}_1) \in L_2$ iff $x \in L$ or $(\mathsf{pk}_0, \mathsf{pk}_1) \in L_1$.

In the description of our BPK cNMZK argument of knowledge (P, V) for any NP-language L we will use a tag-based cNMWI argument of knowledge $\Pi = \{\langle \mathcal{P}_{\mathsf{tag}}, \mathcal{V}_{\mathsf{tag}} \rangle\}_{\mathsf{tag}}$ for an NP-complete language Λ. When we say that we execute Π for proving that $\tau \in L_1$ (or $\sigma \in L_2$) we actually mean that τ (or σ) is reduced to an instance of Λ and $\mathcal{P}_{\mathsf{tag}}$ and $\mathcal{V}_{\mathsf{tag}}$ are executed on input this instance. We also remark that known reductions have the property that, if a witness for $\tau \in L_1$ (or for $\sigma \in L_2$) is known then a witness for the new instance can be constructed in polynomial time. (The protocol is formally described in Fig. 1.)

Input: security parameter 1^k.

PREPROCESSING STAGE:
Entry l of the public file is constructed by V_0 as follows:
  pick $sk_0^l, sk_1^l \leftarrow \{0,1\}^k$, compute $pk_0^l = f(sk_0^l)$ and $pk_1^l = f(sk_1^l)$,
  randomly pick $b^l \leftarrow \{0,1\}$, set $pk^l = (pk_0^l, pk_1^l)$ and $sk^l = (b_l, sk_{b_l}^l)$.
  output: (pk, sk).

PROOF STAGE:
Sub-protocol: tag-based cNMWI argument of knowledge $\Pi =$ $\{\langle \mathcal{P}_{tag}, \mathcal{V}_{tag} \rangle\}_{tag}$ for a NP-complete language Λ.
Common input: the public file F, entry $pk^l = (pk_0^l, pk_1^l)$ of F, $n = \mathsf{poly}(k)$-bit string $x \in L$.
P's private input: a witness w for $x \in L$.
V_1's private input: secret key $sk^l = (b_l, sk_{b_l}^l)$.

$V_1 \longrightarrow P$: V_1 and P engage in an execution of Π with tag $x \circ 0$ where V_1 runs $\mathcal{P}_{x \circ 0}$ to prove to P (running $\mathcal{V}_{x \circ 0}$) knowledge of a witness (b_l, sk^l) for $\sigma = (pk_0^l, pk_1^l) \in L_1$.

$P \longrightarrow V_1$: P and V_1 engage in an execution of Π with tag $x \circ 1$ where P runs $\mathcal{P}_{x \circ 1}$ to prove to V_1 (running $\mathcal{V}_{x \circ 1}$) knowledge of a witness for $\tau = (x, pk_0^l, pk_1^l) \in L_2$.

Fig. 1. The constant-round BPK cNMZK argument of knowledge $\langle P, V \rangle$ for NP

Lemma 1. *If f is a one-way function and Π is a cNMWI argument of knowledge then the protocol of Fig. 1 is a cNMZK argument of knowledge in the* BPK *model for any* NP *language.*

For lack of space, the formal proof can be found in [17, 16].

Theorem 2. *If a family of claw-free permutations exists, then in the* BPK *model there exists a constant-round cNMZK argument of knowledge for all* NP.

The proof follows by Theorem 1, and by the observation that claw-free permutations imply the existence of one-way functions.

Acknowledgments. We thank the anonymous reviewers for their suggestions. The work of the first author has been supported in part by Intel equipment grant, NSF Cybertrust grant No. 0430254, Xerox Innovation group Award and IBM Faculty Award. The work of the authors has been supported in part by the European Commission through the IST program under Contract IST-2002-507932 ECRYPT and the one of the last two authors through the FP6 program under contract FP6-1596 AEOLUS.

References

1. Dolev, D., Dwork, C., Naor, M.: Nonmalleable cryptography. Siam J. on Computing 30, 391–437 (2000)
2. Barak, B.: Constant-round coin-tossing with a man in the middle or realizing the shared random string model. In: Proc. of FOCS, pp. 345–355 (2002)
3. Pass, R., Rosen, A.: New and Improved Constructions of Non-Malleable Cryptographic Protocols. In: Proc. of STOC, pp. 533–542 (2005)
4. Dwork, C., Naor, M., Sahai, A.: Concurrent zero-knowledge. In: Proc. of STOC, pp. 409–418 (1998)
5. Barak, B., Prabhakaran, M., Sahai, A.: Concurrent non-malleable zero knowledge. In: Proc. of FOCS, pp. 345–354 (2006)
6. Feige, U., Shamir, A.: Witness indistinguishable and witness hiding protocols. In: Proc. of STOC, pp. 416–426 (1990)
7. Kilian, J.: Uses of randomness in Algorithms and Protocols. MIT Press, Cambridge (1990)
8. Pass, R., Rosen, A.: Concurrent non-malleable commitments. In: Proc. of FOCS, pp. 563–572 (2005)
9. Canetti, R., Goldreich, O., Goldwasser, S., Micali, S.: Resettable zero-knowledge. In: Proc. of STOC, pp. 235–244 (2000)
10. Di Crescenzo, G., Persiano, G., Visconti, I.: Constant-round resettable zero knowledge with concurrent soundness in the bare public-key model. In: Franklin, M. (ed.) CRYPTO 2004. LNCS, vol. 3152, pp. 237–253. Springer, Heidelberg (2004)
11. Di Crescenzo, G., Visconti, I.: Concurrent zero knowledge in the public-key model. In: Caires, L., Italiano, G.F., Monteiro, L., Palamidessi, C., Yung, M. (eds.) ICALP 2005. LNCS, vol. 3580, pp. 816–827. Springer, Heidelberg (2005)
12. Visconti, I.: Efficient zero knowledge on the internet. In: Bugliesi, M., Preneel, B., Sassone, V., Wegener, I. (eds.) ICALP 2006. LNCS, vol. 4052, pp. 22–33. Springer, Heidelberg (2006)
13. Feige, U., Lapidot, D., Shamir, A.: Multiple NonInteractive Zero Knowledge Proofs under General Assumptions. SIAM Journal on Computing 29, 1–28 (1999)
14. Micali, S., Pass, R., Rosen, A.: Input-indistinguishable computation. In: Proc. of FOCS, pp. 136–145 (2006)
15. Barak, B.: How to go beyond the black-box simulation barrier. In: Proc. of FOCS, pp. 106–115 (2001)
16. Ostrovsky, R., Persiano, G., Visconti, I.: Constant-round concurrent nmwi and its relation to nmzk. Technical Report ECCC Report TR06-095, ECCC (2006)
17. Ostrovsky, R., Persiano, G., Visconti, I.: Constant-round concurrent nmwi and its relation to nmzk. Technical Report 2006-256, Cryptology ePrint Archives (2006)
18. Bellare, M., Goldreich, O.: On defining proofs of knowledge. In: Brickell, E.F. (ed.) CRYPTO 1992. LNCS, vol. 740, pp. 390–420. Springer, Heidelberg (1993)
19. Sahai, A.: Non-malleable non-interactive zero knowledge and adaptive chosen-ciphertext security. In: Proc. of FOCS, pp. 543–553 (1999)
20. Micali, S., Reyzin, L.: Soundness in the public-key model. In: Kilian, J. (ed.) CRYPTO 2001. LNCS, vol. 2139, pp. 542–565. Springer, Heidelberg (2001)
21. Reyzin, L.: Zero-Knowledge with Public Keys, Ph.D. Thesis. MIT Press, Cambridge (2001)

Delegating Capabilities in Predicate Encryption Systems

Elaine Shi[1] and Brent Waters[2]

[1] Carnegie Mellon University
[2] SRI International

Abstract. In predicate encryption systems, given a capability, one can evaluate one or more predicates on the plaintext encrypted, while all other information about the plaintext remains hidden. We consider the role of delegation in such predicate encryption systems. Suppose Alice has a capability, and she wishes to delegate to Bob a more restrictive capability allowing the decryption of a subset of the information Alice can learn about the plaintext encrypted. We formally define delegation in predicate encryption systems, propose a new security definition for delegation, and give an efficient construction supporting conjunctive queries. The security of our construction can be reduced to the general 3-party Bilinear Diffie-Hellman assumption, and the Bilinear Decisional Diffie-Hellman assumption in composite order bilinear groups.

1 Introduction

In traditional public key encryption a user creates a public and private key pair where the private key is used to decrypt all messages encrypted under that public key. Traditional public key encryption allows "all-or-nothing" access to the encrypted data: the private key owner can decrypt everything; and any party without the private key learns nothing about the data encrypted. Recently, cryptographers have proposed a new notion of encryption called *predicate encryption* [5,9,8,21,1,7,17] (also referred to as *searching on encrypted data*). In predicate encryption, the private key owner can compute a capability that allows one to evaluate predicates on the encrypted data. Capabilities can be regarded as partial decryption keys that release partial information about the plaintext encrypted in a controlled manner.

For example, imagine a network audit log collection effort similar to the one mentioned in the recent work[21]. Suppose different Internet Service Providers (ISPs) contribute network audit logs to an untrusted repository. The audit logs will later be used to study network intrusions and worms. Due to privacy concerns, the ISPs encrypt their audit logs before submitting them to the repository, and only a trusted authority has the private key to search the logs. Now suppose there has been an outbreak of a new network worm. An auditor (e.g., a research institute) has been asked to study the behavior of the worm and propose countermeasures. The auditor can now request the authority for a capability that allows

L. Aceto et al. (Eds.): ICALP 2008, Part II, LNCS 5126, pp. 560–578, 2008.

the decryption of suspicious log entries, e.g., flows satisfying the following characteristic: (PORT $\in [p_1, p_2]$) $\land$ (TIME $\in$ LAST MONTH). Meanwhile, the privacy of all other log entries are still preserved.

In predicate encryption, it is often important for a user holding a capability (or a set of capabilities) to generate another capability that is more restrictive than the ones she currently holds. For example, suppose that Carnegie Mellon University has the capability to decrypt all log entries satisfying characteristics of the SQL Slammer worm. Now the university may ask a specific group of researchers to study the SQL Slammer worm originating from an IP address range. To do this, the head of the university can create a more restrictive capability that can decrypt all log entries having the worm characteristic, and originating from this IP range. We say that a predicate encryption system allows for delegation if a user can create capabilities more restrictive than the one she currently owns and if she can do this operation *autonomously*; that is, without interacting with an authority.

In this paper, we study delegation in predicate encryption systems. We propose new security definitions of delegation, and a delegateable predicate encryption scheme supporting conjunctive queries. In the remainder of this section, we first give an overview of related work, and then explain our approach and contributions.

1.1   Related Work

From traditional public-key encryption to predicate encryption. While traditional public-key encryption is sufficient for applications where there is a one to one association between a particular user and a public key, several applications will demand a finer-grained and more expressive decryption capabilities. Shamir [20] provided the first vision for finer-grained encryption systems by introducing the concept of Identity-Based Encryption (IBE). In an IBE system, a party encrypts a message under a particular public key and associates the ciphertext with a given string or "identity". A user can obtain a private key, that is derived from a master secret key, for a particular identity and can use it to decrypt any ciphertext that was encrypted under his identity.

Since the realization of the first Identity-Based Encryption schemes by Boneh and Franklin [6] and Cocks [13], there have been a number of new crypto-systems that provided increasing functionality and expressiveness of decryption capabilities. In Attribute-Based Encryption systems [2,12,15,18,19] a user can receive a private capability that represents complex access control policies over the attributes of an encrypted record. Other encryption systems, including keyword search (or anonymous IBE) [1,5,7,8,9,17,21] systems, allow for a capability holder to evaluate a predicate on the the encrypted data itself and learn nothing more. We henceforth refer to such encryption systems as predicate encryption. Predicate encryption represents a significant breakthrough in the sense that access to the encrypted data is no longer "all-or-nothing"; a user with a predicate capability is able to learn partial information about encrypted data.

Delegation. The concept of delegation was first introduced in this context by Horwitz and Lynn [16] in the form of Hierarchical Identity-Based Encryption (HIBE) [4,14,16]. In an HIBE scheme both private keys and ciphertexts are associated with ordered lists of identities. A user with a given hierarchical identity is able to decrypt any ciphertext where his identity is a prefix of the ciphertext's identity; moreover, a user is able to delegate by creating any other private key with for which his identity is a prefix. For example, a user in charge of the UC Davis domain with a private key for EDU:UCDAVIS can delegate to the computer science department a private key for EDU:UCDAVIS.CS. Since then, the introduction of HIBE the principle of delegation has been applied to other access control systems such as attribute-based encryption systems [15].

1.2   Delegation in Predicate Encryption

In this paper, we examine the problem of delegating capabilities in the more general context of predicate encryption systems [1,5,8,9,17,21,23]. Apart from the aforementioned network audit log example, delegation in predicate encryption can also be useful in other scenarios. For example, suppose Alice has the capability to decrypt all email labeled with "TO:ALICE@YAHOO.COM". If Alice plans to go on vacation over the next two weeks she might want to delegate to her assistant the ability to read all of her incoming emails, but only over this period. To do this, Alice can create a more restrictive capability that can decrypt all such messages that are sent the next two weeks. In another example, suppose Alice's email gateway has the capability to decrypt certain labels of the email and makes forwarding decisions accordingly. For example, emails label as "urgent" by her boss should be sent to her pager; emails from her family should be forwarded to her home computer, etc. The email gateway might want to install similar filtering capabilities on an upstream gateway for cost saving reasons, however, this gateway might be a less trusted device; and Alice may only wish to have the upstream gateway classify emails as "urgent" and "non-urgent" and give preference in forwarding the urgent emails.

Delegation in predicate encryption poses a unique set of challenges; and is typically harder to realize than delegation in Identity-Based Encryption (IBE). This is due to the fact that in an IBE system, a user is able to access an encrypted message if and only if his private key identity matches the ciphertext identity, but the ciphertext identity itself is not hidden. In contrast, predicate encryption systems such as anonymous IBE hides the "identity" of the ciphertext itself. In fact, one can equivalently regard the "identity" as part of the data to be encrypted; and the query predicates are directly evaluated over the encrypted data itself. In practice, this means that it is typically much more difficult to realize delegation in predicate encryption systems. For instance, in anonymous HIBE systems one needs to be careful that the delegation components themselves cannot be used to answer queries.

Another difficulty in building delegation into encryption systems is that previous definitions for security of HIBE appear to be incomplete. In the existing definitions of HIBE security, the attacker plays a game where he receives all

all of his private key queries *directly* from the HIBE authority; however this does not accurately model an adversary's view in a real system. In a real system an adversary might get the private key EDU:UCDAVIS.CS directly from an authority or it might choose to get it from a user with the key EDU:UCDAVIS. In general, private keys received directly from the authority and delegated private keys may have different distribution or forms. For example, in the Gentry and Silverberg [14] and Boneh and Boyen HIBE [3] schemes if a HIBE private key of depth ℓ is received directly from an authority, the authority will create ℓ newly random elements of $\mathbb{Z}_p^*$ in creating the key; however, if the key is generated by another user only one new degree of randomness will be added and the rest will be in common with the previous key. As a result, in the security game, we should not assume in general that delegated keys have the same distribution as keys directly computed by the authority.

Our Approach. We first set out to create a general framework and definitions for delegation in predicate encryption systems. In order to do this we create a general definition that accounts for how predicate capabilities were created. In particular, our definition allows for the adversary to make queries both for capabilities that are created by an authority and for capabilities delegated by users. The adversary may then ask for some subset of these capabilities to be revealed to him.

Using our new definition we set out to realize delegation in an expressive predicate encryption system by extending the Hidden Vector Encryption (HVE) system of Boneh and Waters [8] to allow for delegation. In order to realize security under our new definition we apply two new techniques.

First, we need to make sure that the additional delegation components do not compromise the security of our scheme. We enforce this by "tying" the delegation components of a key to the restrictions of the original key itself. Second, we have the challenge that in the previous HVE techniques of Boneh and Waters [8], the simulator typically creates key that are "completely random" in the sense that they have the same distribution as those coming directly from the authority; however our security definition demands that the keys reflect the distribution of delegation steps specified by the adversary. In order to overcome this we modify the basic scheme such that the distribution of the keys is hidden from a computationally bounded adversary. We show that no adversary can tell whether any key was delegated as he specified or came directly from the authority. After applying this hybrid step we can then proceed to use a simulation that is similar to the previous ones. We believe that our approach is novel in that it is the first instance of a computational game over the private keys in a capability oriented crypto-system.

Finally, we provide a more efficient realization of Anonymous HIBE, which can be seen as a special case of our delegateable HVE scheme. Our Anonymous HIBE scheme has the property that private keys are $O(D)$ in size for a system that allows hierarchies of depth D. Our private key space efficiency can be viewed as a direct result of our corrected definition as the previous scheme of Boyen and Waters required $O(D^2)$ to make all delegated keys have the same distribution as those that came directly from the authority.

2 Definitions

In this section, we introduce the notion of delegation in predicate encryption systems and provide a formal definition of security.

In a predicate encryption system, some user, Alice, creates a public key and a corresponding master key. Using her master key, Alice can compute and hand out a token to Bob, such that Bob is able to evaluate some function[1], f, on the plaintext that has been encrypted. Meanwhile, Bob cannot learn any more information about the plaintext, apart from the output of the function f.

In this paper, we consider the role of delegation in predicate encryption systems. Suppose Alice (the master key owner) has given Bob tokens to evaluate a set of functions $f_1, f_2, \ldots, f_m$ over ciphertexts. Now Bob wishes to delegate to Charles the ability to evaluate the functions $\{f_1 + f_2, f_3, f_4\}$ over the ciphertext. Charles should not be able to learn more information about the plaintext apart from the output of the functions $\{f_1 + f_2, f_3, f_4\}$. For example, although Charles can evaluate $f_1 + f_2$, he should not be able to learn f_1 or f_2 separately. In general, Bob may be interested in delegating any set of functions that is more *restrictive* than what he is able to evaluate with his tokens. Delegation can also happen more than a single level. For example, after obtaining a token from Bob for functions $\{f_1 + f_2, f_3, f_4\}$, Charles may now decide to delegate to his friend David a token to evaluate $f_3 \cdot f_4$.

2.1 Definition

We now formally define delegation in predicate encryption systems that captures the above notion.

Let $X = (x_1, x_2, \ldots, x_\ell) \in \{0, 1\}^\ell$ denote a plaintext. Without loss of generality, assume that we would like to evaluate from the ciphertext boolean functions (a.k.a. predicates) on X. Functions that output multiple bits can be regarded as concatenation of boolean functions. Let $\mathcal{F}$ denote the set of all boolean functions from $\{0, 1\}^\ell$ to $\{0, 1\}$, i.e., $\mathcal{F} := \{f \mid f : \{0, 1\}^\ell \to \{0, 1\}\}$.

A token allows one to evaluate from the ciphertext a set of functions on X. Let $\mathcal{G} = \{g_1, g_2, \ldots, g_m\} \subseteq \mathcal{F}$ denote a collection of functions (also referred to as a *function family*). We use the notation closure($\mathcal{G}$) to denote the set of functions mapping $\{0, 1\}^\ell$ to $\{0, 1\}$ that can be evaluated from $\{g_1(X), g_2(X), \ldots, g_m(X)\}$, i.e.,

$$\mathsf{closure}(\mathcal{G}) = \left\{ f' : \{0,1\}^\ell \to \{0,1\} \;\middle|\; f'(X) = h(g_1(X), g_2(X), \ldots, g_m(X)), \quad \text{where } h : \{0,1\}^m \to \{0,1\} \right\}$$

Given a token to evaluate a function family $\mathcal{G} \subseteq \mathcal{F}$ from a ciphertext, we have sufficient information to evaluate any function in closure($\mathcal{G}$) (assuming unrestricted computational power). A party with a token for a function family $\mathcal{G}$ may be interested in delegating to a friend the ability to evaluate a subset of closure($\mathcal{G}$). In other words, any subset of closure($\mathcal{G}$) can be used to define a token more *restrictive* than a token for the function family $\mathcal{G}$.

---

[1] Although we focus on functions that are predicates in our solutions, we use the more general term of functions in this discussion and our formal definitions.

A Delegateable Predicate Encryption (DPE) scheme consists of the following (possibly randomized) algorithms.

$Setup(1^\lambda)$. The *Setup* algorithm takes as input a security parameter 1^λ, and outputs a public key PK and a master secret key MSK.

$Encrypt(PK, X)$. The *Encrypt* algorithm takes as input a public key PK and a plaintext $X = (x_1, x_2, \ldots, x_\ell) \in \{0, 1\}^\ell$; and outputs a ciphertext CT.

$GenToken(PK, MSK, \mathcal{G})$. The *GenToken* algorithm takes as input a public key PK, master secret key MSK, and a set of boolean functions $\mathcal{G} \subseteq \mathcal{F}$. It outputs a token for evaluating the set of functions $\mathcal{G}$ from a ciphertext.

$Query(PK, TK_{\mathcal{G}}, CT, f)$. The *Query* algorithm takes as input a public key PK, a token $TK_{\mathcal{G}}$ for the function family $\mathcal{G}$, a function $f \in \mathcal{G}$, and a ciphertext CT. Suppose CT is an encryption of the plaintext X; the algorithm outputs $f(X)$.

$Delegate(PK, TK_{\mathcal{G}}, \mathcal{G}')$. The *Delegate* algorithm takes as input a public key PK, a token for the function family $\mathcal{G} \subseteq \mathcal{F}$, and $\mathcal{G}' \subseteq \text{closure}(\mathcal{G})$. It computes a token for evaluating the function family $\mathcal{G}'$ on a ciphertext.

Remark 1. We note that the above definition captures delegation in predicate encryption systems in the broadest sense. In a predicate encryption system, we would like to maximize the expressiveness of delegation; however, one should not be able to delegate beyond what she can learn with her own tokens. Otherwise, the security of predicate encryption would be broken.

Since we care about being able to perform expressive delegations, we can judge a system by its expressiveness, e.g., what types of functions one can evaluate over the ciphertext, and what types of delegations one can perform. Our vision is to design a predicate encryption system that supports a rich set of queries and delegations. As an initial step, we restrict ourselves to some special classes of functions. At the time of writing this paper, the most expressive predicate encryption system (without delegation) we know of supports conjunctive queries [8]. However, soon after this writing, Katz, Sahai and Waters propose a novel predicate encryption system supporting inner product queries [17].

2.2   Security

To define the security for delegation in predicate encryption, we describe a query security game between a challenger and an adversary. This game formally captures the notion that the tokens reveal no unintended information about the plaintext. In this game, the adversary asks the challenger for a number of tokens. For each queried token, the adversary gets to specify its path of derivation: whether the token is directly generated by the root authority, or delegated from another token. If the token is delegated, the adversary also gets to specify from which token it is delegated. The game proceeds as follows:

Setup. The challenger runs the *Setup* algorithm, and gives the adversary the public key PK.

Query 1. The adversary adaptively makes a polynomial number of queries of the following types:

- *Create token.* The adversary asks the challenger to create a token for a set functions $\mathcal{G} \subseteq \mathcal{F}$. The challenger creates a token for $\mathcal{G}$ without giving it the adversary.
- *Create delegated token.* The adversary specifies a token for function family $\mathcal{G}$ that has already been created, and asks the challenger to perform a delegation operation to create a child token for $\mathcal{G}' \subseteq \mathsf{closure}(\mathcal{G})$. The challenger computes the child token without releasing it to the adversary.
- *Reveal token.* The adversary asks the challenger to reveal an already created token for function family $\mathcal{G}$.

Note that when token creation requests are made, the adversary does not automatically see the created token. The adversary only sees a token when it makes a reveal token query.

Challenge. The adversary outputs two strings $X_0^*, X_1^* \in \{0,1\}^\ell$ subject to the following constraint:

For any token revealed to the adversary in the **Query 1** stage, let $\mathcal{G}$ denote the function family corresponding to this token. For all $f \in \mathcal{G}, f(X_0^*) = f(X_1^*)$.

Next, the challenger flips a random coin b, and encrypts X_b^*. It returns the ciphertext to the adversary.

Query 2. Repeat the **Query 1** stage. All tokens revealed in this stage must satisfy the same condition as above.

Guess. The adversary outputs a guess b' of b. The advantage of an adversary $\mathcal{A}$ in the above game is defined to be $\mathsf{Adv}_\mathcal{A} = |\Pr[b = b'] - 1/2|$.

Definition 1. *We say that a delegateable predicate encryption system is* secure *if for all polynomial-time adversaries $\mathcal{A}$ attacking the system, its advantage $\mathsf{Adv}_\mathcal{A}$ is a negligible function of λ.*

Selective Security. We also define a weaker security notion called *selective security.* In the selective security game, instead of submitting two strings X_0^*, X_1^* in the **Challenge** stage, the adversary first commits to two strings at the beginning of the security game. The rest of the security game proceeds exactly as before. The selective security model has appeared in various constructions in the literature [10,11,3,8,9,21], since it is often easier to prove security in the selective model.

We say that a delegateable predicate encryption system is *selectively secure* if all polynomial time adversaries $\mathcal{A}$ have negligible advantage in the selective security game.

Remark 2. We note that our security definition is complete in the sense that in the query phase, the adversary gets to specify, for each queried token, its path of derivation: whether the token is generated by the root authority, or from whom the token has been delegated. Previously when researchers studied delegation in identity-based encryption systems, (e.g., Hierarchical Identity-Based Encryption

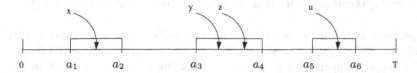

Fig. 1. A simple example of predicate encryption similar to the one described in BW06 [8]

(HIBE) [4], Anonymous Hierarchical Identity-Based Encryption (AHIBE) [9]), the security game was under-specified: the adversary does not get to specify from whom each queried token is delegated. One way to interpret this is to assume that all tokens are generated from the same probability distribution. For example, the AHIBE [9] work uses this approach. While this allows us to prove the security of these systems, it is in fact an overkill. This motivates our new security definition for delegation. Under the new security definition, the delegated token need not be picked from the same probability distribution as the non-delegated tokens. In fact, we show that the ability to capture such nuances in our security definition allows us to construct a simpler AHIBE scheme with smaller private key size.

2.3 A Simple Example

To help understand the above definition, we give a simple example similar to that in the BW06 paper [8]. As shown by Figure 1, suppose the point X encrypted takes on integer values between 0 and T. Given $a, b \in [0, T]$, let $f_{a,b}$ denote the function that decides whether $X \in [a, b]$:

$$f_{a,b}(X) = \begin{cases} 1 & X \in [a, b] \\ 0 & \text{o.w.} \end{cases}$$

In Figure 1, we mark three disjoint segments $[a_1, a_2]$, $[a_3, a_4]$ and $[a_5, a_6]$; and four points x, y, z, u. Suppose Alice has a token for functions $\{f_{a_1,a_2}, f_{a_3,a_4}, f_{a_5,a_6}\}$. This allows her to evaluate the following three predicates: whether $a_1 \leq X \leq a_2$, $a_3 \leq X \leq a_4$, and $a_5 \leq X \leq a_6$. Alice can now distinguish between ciphertexts $Encrypt(\mathsf{PK}, x)$ and $Encrypt(\mathsf{PK}, y)$; but she cannot distinguish between ciphertexts $Encrypt(\mathsf{PK}, y)$ and $Encrypt(\mathsf{PK}, z)$.

Suppose now Alice performs a delegation, and computes a child token for the function $g(X) = f_{a_1,a_2}(X) \vee f_{a_3,a_4}(X)$. Suppose that Bob receives this delegated token from Alice. Now Bob is able to decide whether $(a_1 \leq X \leq a_2) \vee (a_3 \leq X \leq a_4)$; this is a subset of information allowed by Alice's token. Given this new token, Bob can decide whether X falls inside these two ranges, but he cannot decide between the cases whether $X \in [a_1, a_2]$ or $X \in [a_3, a_4]$. For example, Bob can distinguish between the ciphertexts $Encrypt(\mathsf{PK}, x)$ and $Encrypt(\mathsf{PK}, u)$, but he cannot distinguish between the ciphertexts $Encrypt(\mathsf{PK}, x)$ and $Encrypt(\mathsf{PK}, y)$.

3   Delegateable Hidden Vector Encryption (dHVE)

We propose a primitive called delegateable hidden vector encryption (dHVE), where we add delegation to the HVE construction proposed in BW06 [8]. This is an interesting special case to the general definition given in Section 2.1, and represents an initial step towards our bigger vision of enabling expressive queries and delegations in predicate encryption systems.

3.1   Delegateable HVE Overview (dHVE)

In our dHVE system, plaintexts consists of multiple "fields". For example, a plaintext can be the tuple (IP, PORT, TIME, LENGTH). A token corresponds to a conjunction of a subset of these fields: we can fix a field to a specific value, make a field "delegateable", or choose not to include a field in a query. For example, the query (IP = ?) $\land$ (PORT = 80) $\land$ (TIME = 02/10/08) fixes the values of the PORT and TIME fields, and makes the IP field delegateable. The LENGTH field is not included in the query. A party in possession of this token can fill in any appropriate value for the delegateable field IP, however, she cannot change the values of the fixed field or delete them from the query, nor can she add in the missing field LENGTH to the query. We now give formal definitions for the above notions.

Let Σ denote a finite alphabet and let $?, \perp$ denote two special symbols not in Σ. Define $\Sigma_{?,\perp} := \Sigma \cup \{?, \perp\}$. The symbol ? denotes a delegateable field, i.e., a field where one is allowed to fill in an arbitrary value and perform delegation. The symbol $\perp$ denotes a "don't care" field, i.e., a field not involved in some query. Typically, if a query predicate does not concern a specific field, we call this field a "don't care" field. In the aforementioned example, (IP = ?)$\land$(PORT = 80) $\land$ (TIME = 02/10/08), the IP field is a delegateable field, LENGTH is a "don't care" field, and the remaining are fixed fields.

Plaintext. In dHVE, our plaintext is composed of a message $\mathsf{M} \in \{0,1\}^*$ and ℓ fields, denoted by $X = (x_1, x_2, \ldots, x_\ell) \in \Sigma^\ell$. The *Encrypt* algorithm takes as input a public key PK, a pair $(X, \mathsf{M}) \in \{0,1\}^* \times \Sigma^\ell$, and outputs a ciphertext CT.

Tokens. In dHVE, a token allows one to evaluate a special class of boolean functions on the fields $X \in \Sigma^\ell$. We use a vector $\sigma = (\sigma_1, \sigma_2, \ldots, \sigma_\ell) \in (\Sigma_{?,\perp})^\ell$ to specify a set of functions being queried. Given σ, let $\mathcal{W}(\sigma)$ denote the indices of all delegateable fields, let $\mathcal{D}(\sigma)$ denote the indices of all "don't care" fields, and let $\mathcal{S}(\sigma)$ denote the indices of the remaining fixed fields. In the following, we use the notation $[\ell]$ to denote the set $\{1, 2, \ldots, \ell\}$.

$$\mathcal{W}(\sigma) := \{i \mid \sigma_i = ?\}, \qquad \mathcal{D}(\sigma) := \{i \mid \sigma_i = \perp\}$$
$$\mathcal{S}(\sigma) := \{i \mid \sigma_i \in \Sigma\} = [\ell] \backslash (\mathcal{W}(\sigma) \cup \mathcal{D}(\sigma))$$

Let $\sigma = (\sigma_1, \sigma_2, \ldots, \sigma_\ell) \in (\Sigma_?, \perp)^\ell$, σ specifies the following function family C_σ on the point $X = (x_1, \ldots, x_\ell)$ encrypted:

$$C_\sigma := \left\{ \left(\bigwedge_{i \in W'} (x_i = a_i) \right) \wedge \left(\bigwedge_{j \in S(\sigma)} (x_j = \sigma_j) \right) \mid W' \subseteq \mathcal{W}(\sigma), \forall i \in W', a_i \in \Sigma \right\} \tag{1}$$

In other words, given a token for σ, the family C_σ denotes the set of functions we can evaluate from a ciphertext. For the delegateable fields, we can fill in any appropriate value, but we cannot change or delete any of the fixed fields or add a "don't care" field to the query. In addition, if any function in C_σ evaluates to 1, one would also be able to decrypt the payload message M.

Remark 3. The family C_σ is a set of conjunctive equality tests, where we can fill in every delegateable field in σ with a value in Σ or "don't care". In particular, we fill in fields in W' with appropriate values in σ, and for the remaining delegateable fields $\mathcal{W}(\sigma) - W'$, we fill them with "don't care". If σ has no delegateable field, then the set C_σ contains a single function. This is exactly the case considered by the original HVE construction, where each token allows one to evaluate a single function from a ciphertext.

Delegation. In dHVE, Alice, who has a token for σ, can delegate to Bob a subset of the functions she can evaluate: 1) Alice can fill in delegateable fields (i.e., $\mathcal{W}(\sigma)$) with a value in Σ or with the "don't care" symbol $\perp$; 2) Alice can also leave a delegateable field unchanged (with the ? symbol). In this case, Bob will be able to perform further delegation on that field.

Definition 2. *Let $\sigma = (\sigma_1, \sigma_2, \ldots, \sigma_\ell), \sigma' = (\sigma'_1, \sigma'_2, \ldots, \sigma'_\ell) \in \Sigma^\ell_{?, \perp}$. We say that $\sigma' \prec \sigma$, if for all $i \in S(\sigma) \cup \mathcal{D}(\sigma)$, $\sigma'_i = \sigma_i$.*

Note that $\sigma' \prec \sigma$ means that from TK_σ we can perform a delegation operation and compute $\mathsf{TK}_{\sigma'}$. In addition, if $\sigma' \prec \sigma$, then $C_{\sigma'} \subseteq C_\sigma$, i.e., $\mathsf{TK}_{\sigma'}$ allows one to evaluate a subset of the functions allowed by TK_σ.

In summary, we introduce delegateable fields to the original HVE construction. We use the notation $\sigma \in \Sigma^\ell_{?, \perp}$ to specify a function family. Given TK_σ, one can perform a set of conjunctive equality tests (defined by Equation (1)) from the ciphertext. One may also fill in the delegateable fields in σ with any value in $\Sigma \cup \{\perp\}$ and compute a child token for the resulting vector. The child token allows one to evaluate a subset of the functions allowed by the parent token.

Example. Suppose the trusted authority T issues to A a token for $\sigma_A = (\mathcal{I}_1, \mathcal{I}_2, ?, ?, \perp, \perp, \ldots, \perp)$. This token allows A to evaluate the following functions from the ciphertext:

- $(x_1 = \mathcal{I}_1) \wedge (x_2 = \mathcal{I}_2)$
- $\forall \mathcal{I}_3 \in \Sigma$: $(x_1 = \mathcal{I}_1) \wedge (x_2 = \mathcal{I}_2) \wedge (x_3 = \mathcal{I}_3)$
- $\forall \mathcal{I}_4 \in \Sigma$: $(x_1 = \mathcal{I}_1) \wedge (x_2 = \mathcal{I}_2) \wedge (x_4 = \mathcal{I}_4)$
- $\forall \mathcal{I}_3, \mathcal{I}_4 \in \Sigma$: $(x_1 = \mathcal{I}_1) \wedge (x_2 = \mathcal{I}_2) \wedge (x_3 = \mathcal{I}_3) \wedge (x_4 = \mathcal{I}_4)$

Later, suppose A delegates to B the following token: $\sigma_B = (\mathcal{I}_1, \mathcal{I}_2, \mathcal{I}_3, ?, \perp, \perp, \ldots, \perp)$, where $\mathcal{I}_3 \in \Sigma$. Note that this allow B to evaluate the following functions:

- $(x_1 = \mathcal{I}_1) \wedge (x_2 = \mathcal{I}_2) \wedge (x_3 = \mathcal{I}_3)$
- $\forall \mathcal{I}_4 \in \Sigma : \quad (x_1 = \mathcal{I}_1) \wedge (x_2 = \mathcal{I}_2) \wedge (x_3 = \mathcal{I}_3) \wedge (x_4 = \mathcal{I}_4)$

Clearly, a token for σ_B releases a subset of information allowed by σ_A. Meanwhile, B is able to further delegate on the x_4 field.

3.2 dHVE Definition

We now give a formal definition of dHVE.

$Setup(1^\lambda)$. The *Setup* algorithm takes as input a security parameter 1^λ, and outputs a public key PK and a master secret key MSK.

$Encrypt(\mathsf{PK}, X, \mathsf{M})$. The *Encrypt* algorithm takes as input a public key PK, a pair $(X, \mathsf{M}) \in \Sigma^\ell \times \{0,1\}^*$; and outputs a ciphertext CT.

$GenToken(\mathsf{PK}, \mathsf{MSK}, \sigma)$. The *GenToken* algorithm takes as input a public key PK, master secret key MSK, and a vector $\sigma \in (\Sigma_{?,\perp})^\ell$. It outputs a token for evaluating the set of conjunctive queries C_σ from a ciphertext.

$Delegate(\mathsf{PK}, \mathsf{TK}_\sigma, \sigma')$. The *Delegate* algorithm takes as input a public key PK, a token TK_σ for the vector σ, and another vector $\sigma' \prec \sigma$. It outputs a delegated token $\mathsf{TK}_{\sigma'}$ for the new vector σ'.

$Query(\mathsf{PK}, \mathsf{TK}_\sigma, \mathsf{CT}, \sigma')$. The *Query* algorithm takes as input a public key PK, a token TK_σ for the vector σ, a ciphertext CT, and a new vector σ' satisfying the following conditions: (1) $\sigma' \prec \sigma$; (2) σ' does not contain delegatable fields, that is, such a σ' specifies a single conjunctive query (denoted $f_{\sigma'}$) over the point X encrypted. The algorithm outputs $f_{\sigma'}(X)$; in addition, if $f_{\sigma'}(X) = 1$, it also outputs the message M.

Remark 4. We note that in comparison to the general definition given in Section 2, in dHVE, we add a payload message $\mathsf{M} \in \{0,1\}^*$ to the plaintext. Meanwhile, the conjunctive queries in dHVE are functions on the attributes $X \in \Sigma^\ell$, but not the payload M. Additionally, if a query matches a point X encrypted, one can successfully decrypt the payload message using the corresponding token. It is not hard to show that the above formalization for dHVE is captured by the general definition given in Section 2: We can regard (M, X) as an entire bit-string, and decrypting the payload M can be regarded as evaluating a concatenation of bits from the bit-string (M, X). We choose to define dHVE with a payload message to be consistent with the HVE definition in BW06 [8].

Selective security of dHVE. We will prove the selective security of our dHVE construction. We give the formal selective security definition below. The full security definition for dHVE can be found in the Appendix.

- **Init.** The adversary commits to two strings $X_0^*, X_1^* \in \Sigma^\ell$.
- **Setup.** The challenger runs the *Setup* algorithm and gives the adversary the public key PK.

- **Query 1.** The adversary adaptively makes a polynomial number of "create token", "create delegated token" or "reveal token" queries. The queries must satisfy the following constraint: For any token σ revealed to the adversary, let $\mathcal{C}_\sigma$ denote the set of conjunctive queries corresponding to this token.

$$\forall \mathsf{TK}_\sigma \text{ revealed,} \quad \forall f \in \mathcal{C}_\sigma : \quad f(X_0^*) = f(X_1^*) \tag{2}$$

- **Challenge.** The adversary outputs two equal length messages M_0 and M_1 subject to the following constraint:
  For any token σ revealed to the adversary in the **Query 1** stage, let $\mathcal{C}_\sigma$ denote the set of conjunctive queries corresponding to this token.

$$\forall \mathsf{TK}_\sigma \text{ revealed} : \quad \text{if } \exists f \in \mathcal{C}_\sigma, f(X_0^*) = f(X_1^*) = 1, \text{then } \mathsf{M}_0 = \mathsf{M}_1 \tag{3}$$

  The challenger flips a random coin b and returns an encryption of (M_b, X_b) to the adversary.
- **Query 2.** Repeat the **Query 1** stage. All tokens revealed in this stage must satisfy constraints (2) and (3).
- **Guess.** The adversary outputs a guess b' of b.

The advantage of an adversary $\mathcal{A}$ in the above game is defined to be $\mathsf{Adv}_\mathcal{A} = |\Pr[b = b'] - 1/2|$. We say that a dHVE construction is *selectively secure* if for all polynomial time adversaries, its advantage in the above game is a negligible function of λ.

Observation 1. *Anonymous Hierarchical Identity-Based Encryption (AHIBE) is a special case of the above defined dHVE scheme.*

AHIBE is very similar to the dHVE definition given above. The only difference is that in AHIBE, the function family queried is $\mathcal{C}_\sigma$, where σ has the special structure such that $\mathcal{S}(\sigma) = [d]$ where $d \in [\ell]$, $\mathcal{W}(\sigma) = [d+1, \ell]$, and $\mathcal{D}(\sigma) = \emptyset$. In fact, we show that the new security definition and the techniques we use to construct dHVE can be directly applied to give *an AHIBE scheme with shorter private key size*. While the previous AHIBE scheme by Boyen and Waters require $O(D^2)$ private key size, our new construction has $O(D)$ private key size, where D is the maximum depth of the hierarchy. We refer readers to the Appendix for details of the construction.

4   Background on Pairings and Complexity Assumptions

Our construction relies on bilinear groups of composite order $n = pqr$, where p, q and r are distinct large primes. We assume that the reader is familiar with bilinear groups. More background on composite order bilinear groups can be found in the Appendix.

Our construction relies on two complexity assumptions: the bilinear Diffie-Hellman assumption (BDH) and the generalized composite 3-party Diffie-Hellman assumption (C3DH). *Although our construction only requires bilinear groups whose order is the product of three primes* $n = pqr$, *we state our assumptions more generally for bilinear groups of order* n *where* n *is the product of three or more primes.*

We begin by defining some notation. We use the notation GG to denote the *group generator* algorithm that takes as input a security parameter $\lambda \in \mathbb{Z}^{>0}$, a number $k \in \mathbb{Z}^{>0}$, and outputs a tuple $(p, q, r_1, r_2, \ldots, r_k, \mathbb{G}, \mathbb{G}_T, e)$ where $p, q, r_1, r_2, \ldots, r_k$ are $k + 2$ distinct primes, $\mathbb{G}$ and $\mathbb{G}_T$ are two cyclic groups of order $n = pq \prod_{i=1}^{k} r_i$, and $e : \mathbb{G}^2 \to \mathbb{G}_T$ is the bilinear mapping function. We use the notation $\mathbb{G}_p, \mathbb{G}_q, \mathbb{G}_{r_1}, \ldots, \mathbb{G}_{r_k}$ to denote the respective subgroups of order $p, q, r_1, \ldots, r_k$ of $\mathbb{G}$. Similarly, we use the notation $\mathbb{G}_{T,p}, \mathbb{G}_{T,q}, \mathbb{G}_{T,r_1}, \ldots, \mathbb{G}_{T,r_k}$ to denote the respective subgroups of order $p, q, r_1, \ldots, r_k$ of $\mathbb{G}_T$.

The bilinear Diffie-Hellman assumption. We review the standard Bilinear Diffie-Hellman assumption, but in groups of composite order. For a given group generator GG define the following distribution $P(\lambda)$:

$$(p, q, r_1, \ldots, r_k, \mathbb{G}, \mathbb{G}_T, e) \xleftarrow{R} GG(\lambda, k), \quad n \leftarrow pq \prod_{i=1}^{k} r_i,$$

$$g_p \xleftarrow{R} \mathbb{G}_p, \quad g_q \xleftarrow{R} \mathbb{G}_q, \quad h_1 \xleftarrow{R} \mathbb{G}_{r_1}, \quad \ldots, \quad h_k \xleftarrow{R} \mathbb{G}_{r_k}$$

$$a, b, c \xleftarrow{R} \mathbb{Z}_n$$

$$\bar{Z} \leftarrow ((n, \mathbb{G}, \mathbb{G}_T, e), \quad g_q, \ g_p, \ h_1, \ h_2, \ \ldots, \ h_k, \ g_p^a, \ g_p^b, \ g_p^c)$$

$$T \leftarrow e(g_p, g_p)^{abc}$$

Output $(\bar{Z}, T)$

Define algorithm $\mathcal{A}$'s advantage in solving the composite bilinear Diffie-Hellman problem as $\mathsf{cBDH}\,\mathsf{Adv}_{GG,\mathcal{A}}(\lambda) := \left| \Pr[\mathcal{A}(\bar{Z}, T) = 1] - \Pr[\mathcal{A}(\bar{Z}, R) = 1] \right|$, where $(\bar{Z}, T) \xleftarrow{R} P(\lambda)$ and $R \xleftarrow{R} \mathbb{G}_{T,p}$. We say that GG satisfies the composite bilinear Diffie-Hellman assumption (cBDH) if for any polynomial time algorithm $\mathcal{A}$, $\mathsf{cBDH}\,\mathsf{Adv}_{GG,\mathcal{A}}(\lambda)$ is a negligible function of λ.

The generalized composite 3-party Diffie-Hellman assumption. We also rely on the composite 3-party Diffie-Hellman assumption first introduced by Boneh and Waters [8]. For a given group generator GG define the following distribution $P(\lambda)$:

$$(p, q, r_1, \ldots, r_k, \mathbb{G}, \mathbb{G}_T, e) \xleftarrow{R} GG(\lambda, k), \quad n \leftarrow pq \prod_{i=1}^{k} r_i,$$

$$g_p \xleftarrow{R} \mathbb{G}_p, \quad g_q \xleftarrow{R} \mathbb{G}_q, \quad h_1 \xleftarrow{R} \mathbb{G}_{r_1}, \quad \ldots, \quad h_k \xleftarrow{R} \mathbb{G}_{r_k}$$

$$R_1, R_2, R_3 \xleftarrow{R} \mathbb{G}_q, \quad a, b, c \xleftarrow{R} \mathbb{Z}_n$$

$$\bar{Z} \leftarrow ((n, \mathbb{G}, \mathbb{G}_T, e), \quad g_q, \ g_p, \ h_1, \ h_2, \ \ldots, \ h_k, \ g_p^a, \ g_p^b, \ g_p^{ab} \cdot R_1, \ g_p^{abc} \cdot R_2)$$

$$T \leftarrow g_p^c \cdot R_3$$

Output $(\bar{Z}, T)$

Define algorithm $\mathcal{A}$'s advantage in solving the generalized composite 3-party Diffie-Hellman problem for GG as $\mathsf{C3DH}\,\mathsf{Adv}_{GG,\mathcal{A}}(\lambda) := \left| \Pr[\mathcal{A}(\bar{Z}, T) = 1] - \Pr[\mathcal{A}(\bar{Z}, R) = 1] \right|$, where $(\bar{Z}, T) \xleftarrow{R} P(\lambda)$ and $R \xleftarrow{R} \mathbb{G}$. We say that GG satisfies the composite 3-party Diffie-Hellman assumption (C3DH) if for any polynomial time algorithm $\mathcal{A}$, its advantage $\mathsf{C3DH}\,\mathsf{Adv}_{GG,\mathcal{A}}(\lambda)$ is a negligible function of λ.

The assumption is formed around the intuition that it is hard to test for Diffie-Hellman tuples in the subgroup $\mathbb{G}_p$ if the elements have a random $\mathbb{G}_q$ subgroup component.

Remark 5. Consider bilinear groups of order $n = pqr$, where p, q and r are three distinct primes. In the above generalized composite 3-party Diffie-Hellman assumption, whether to call a prime p, q or r is merely a nominal issue. So equivalently, we may assume that it is hard to test for Diffie-Hellman tuples in the subgroup $\mathbb{G}_p$, if each element is multiplied by a random element from $\mathbb{G}_r$ instead of $\mathbb{G}_q$.

5   dHVE Construction

We construct a dHVE scheme by extending the HVE construction by Boneh and Waters [8] (also referred to as the BW06 scheme). One of the challenges that we have to overcome is how to add delegation in anonymous IBE systems. We note that delegation is easier in non-anonymous IBE systems, such as in HIBE [4]. In the HIBE construction [4], the public key contains an element corresponding to each attribute, and the delegation algorithm can use these elements in the public key to rerandomize the tokens. In anonymous systems, however, as the encryption now has to hide the attributes as well, we have extra constraints on what information we can release in the public key. This makes delegation harder in anonymous settings.

5.1   Construction

In our construction, the public key and the ciphertext are constructed in a way similar to the BW06 scheme. However, we use a new trick to reduce the number of group elements in the ciphertext asymptotically by one half. Our token consists of two parts, a decryption key part denoted as DK and a delegation component denoted as DL. The decryption key part DK is similar to that in the BW06 scheme. The delegation component DL is more difficult to construct, since we need to make sure that the delegation component itself does not leak unintended information about the plaintext encrypted.

We will use $\Sigma = \mathbb{Z}_m$ for some integer m. Recall that $\Sigma_{?,\perp} := \Sigma \cup \{?, \perp\}$, where ? denotes a delegateable field, and $\perp$ denotes a "don't care" field.

Setup(1^λ). The setup algorithm first chooses random large primes $p, q, r > m$ and creates a bilinear group $\mathbb{G}$ of composite order $n = pqr$, as specified in Section 4. Next, it picks random elements

$$(u_1, h_1), \ldots, (u_\ell, h_\ell) \in \mathbb{G}_p^2, \quad g, v, w, \overline{w} \in \mathbb{G}_p, \quad g_q \in \mathbb{G}_q, \quad g_r \in \mathbb{G}_r$$

and an exponent $\alpha \in \mathbb{Z}_p$. It keeps all these as the secret key MSK.
It then chooses $2\ell + 3$ random blinding factors in $\mathbb{G}_q$:

$$(R_{u,1}, R_{h,1}), \ldots, (R_{u,\ell}, R_{h,\ell}) \in \mathbb{G}_q \text{ and } R_v, R_w, \overline{R}_w \in \mathbb{G}_q.$$

For the public key, PK, it publishes the description of the group $\mathbb{G}$ and the values

$$g_q, \quad g_r, \quad V = vR_v, \quad W = wR_w, \quad \overline{W} = \overline{w}R_w, \quad A = \mathbf{e}(g, v)^\alpha, \quad \begin{pmatrix} U_1 = u_1 R_{u,1}, & H_1 = h_1 R_{h,1} \\ \cdots \\ U_\ell = u_\ell R_{u,\ell}, & H_\ell = h_\ell R_{h,\ell} \end{pmatrix}$$

The message space $\mathcal{M}$ is set to be a subset of $\mathbb{G}_T$ of size less than $n^{1/4}$.

$Encrypt(\text{PK}, X \in \Sigma^\ell, M \in \mathcal{M} \subseteq \mathbb{G}_T)$. Assume that $\Sigma \subseteq \mathbb{Z}_m$. Let $X = (x_1, \ldots, x_\ell) \in \mathbb{Z}_m^\ell$. The encryption algorithm first chooses a random $\rho \in \mathbb{Z}_n$ and random $Z, Z_0, Z_\phi, Z_1, Z_2, \ldots, Z_\ell \in \mathbb{G}_q$. (The algorithm picks random elements in $\mathbb{G}_q$ by raising g_q to random exponents from $\mathbb{Z}_n$.) Then, the encryption algorithm outputs the ciphertext:

$$\text{CT} = \left(\tilde{C} = MA^\rho, \quad C = V^\rho Z, \quad C_0 = W^\rho Z_0, \quad C_\phi = \overline{W}^\rho Z_\phi, \quad \begin{pmatrix} C_1 = (U_1^{x_1} H_1)^\rho Z_1, \\ C_2 = (U_2^{x_2} H_2)^\rho Z_2, \\ \cdots\cdots \\ C_\ell = (U_\ell^{x_\ell} H_\ell)^\rho Z_\ell \end{pmatrix} \right)$$

Remark 6. We note that the ciphertext size is cut down by roughly a half when compared to the BW06 construction [8]. Therefore, our construction immediately implies an HVE scheme with asymptotically half the ciphertext size as the original BW06 construction.

$GenToken(\text{PK}, \text{MSK}, \quad \sigma \in \Sigma_{?,\perp}^\ell)$. The token generation algorithm will take as input the master secret key MSK and an ℓ-tuple $\sigma = (\sigma_1, \ldots, \sigma_\ell) \in \Sigma_{?,\perp}^\ell$. The token for σ consists of two parts: (1) a decryption key component denoted as DK, and (2) a delegation component denoted DL.

- The decryption key component DK is composed in a similar way to the original HVE construction [8]. Recall that $\mathcal{S}(\sigma)$ denotes the indices of the fixed fields, i.e., indices j such that $\sigma_j \in \Sigma$. Randomly select $\gamma, \overline{\gamma} \in \mathbb{Z}_p$ and $t_j \in \mathbb{Z}_p$ for all $j \in \mathcal{S}(\sigma)$. Pick random $Y, Y_0, Y_\phi \in \mathbb{G}_r$ and $Y_j \in \mathbb{G}_r$ for all $j \in \mathcal{S}(\sigma)$. Observe that picking random elements from the subgroup $\mathbb{G}_r$ can be done by raising g_r to random exponents in $\mathbb{Z}_n$. Next, output the following decryption key component:

$$\text{DK} = \left(K = g^\alpha w^\gamma \overline{w}^{\overline{\gamma}} \prod_{j \in \mathcal{S}(\sigma)} (u_j^{\sigma_j} h_j)^{t_j} Y, \quad K_0 = v^\gamma Y_0, \quad K_\phi = v^{\overline{\gamma}} Y_\phi, \quad \forall j \in \mathcal{S}(\sigma): \ K_j = v^{t_j} Y_j \right)$$

- The delegation component DL is constructed as below. Recall that $\mathcal{W}(\sigma)$ denotes the set of all indices i where $\sigma_i = ?$. Randomly select $Y_{i,u}, Y_{i,h} \in \mathbb{G}_r$. For each $i \in \mathcal{W}(\sigma)$, for each $j \in \mathcal{S}(\sigma) \cup \{i\}$, randomly select $s_{i,j} \in \mathbb{Z}_p$, $Y_{i,j} \in \mathbb{G}_r$. For each $i \in \mathcal{W}(\sigma)$, randomly select $\gamma_i, \overline{\gamma}_i \in \mathbb{Z}_p$, $Y_{i,h}, Y_{i,u}, Y_{i,0}, Y_{i,\phi} \in \mathbb{G}_r$. Next, output the following delegation component DL_i for coordinate i.

$$\forall i \in \mathcal{W}(\sigma): \quad \text{DL}_i = \begin{pmatrix} L_{i,h} = h_i^{s_{i,i}} w^{\gamma_i} \overline{w}^{\overline{\gamma}_i} \prod_{j \in \mathcal{S}(\sigma)} (u_j^{\sigma_j} h_j)^{s_{i,j}} Y_{i,h}, & L_{i,u} = u_i^{s_{i,i}} Y_{i,u} \\ L_{i,0} = v^{\gamma_i} Y_{i,0}, \quad L_{i,\phi} = v^{\overline{\gamma}_i} Y_{i,\phi}, \quad \forall j \in \mathcal{S}(\sigma) \cup \{i\}: \ L_{i,j} = v^{s_{i,j}} Y_{i,j} \end{pmatrix}$$

Remark 7. Later, suppose we want to delegate on the kth field by fixing it to $\mathcal{I} \in \Sigma$. To do so, we will multiply $L_{k,u}^{\mathcal{I}}$ to $L_{k,h}$, resulting in something similar to the decryption key DK (except without the g^α term). Observe that the $L_{i,h}$ terms encode all the fixed fields (i.e., $\mathcal{S}(\sigma)$). This effectively restricts the use of the delegation components, such that they can only be added on top of the fixed fields, partly ensuring that the delegation components do not leak unintended information.

Delegate(PK, σ, σ'). Given a token for $\sigma \in \Sigma_{?,\perp}^{\ell}$, the *Delegate* algorithm computes a token for $\sigma' \prec \sigma$. Without loss of generality, we assume that σ' fixes only one delegateable field of σ to a symbol in Σ or to $\perp$. Clearly, if we have an algorithm to perform delegation on one field, then we can perform delegation on multiple fields. This can be achieved by fixing the multiple delegateable fields one by one.

We now describe how to compute $\mathsf{TK}_{\sigma'}$ from TK_{σ}. Suppose σ' fixes the kth coordinate of σ. We consider the following two types of delegation: 1) the kth coordinate is fixed to some value in the alphabet Σ, and 2) the kth coordinate is set to $\perp$, i.e., it becomes a "don't care" field.

Type 1: σ' fixes the kth coordinate of σ to $\mathcal{I} \in \Sigma$, and the remaining coordinates of σ remain unchanged. In this case, $\mathcal{S}(\sigma') = \mathcal{S}(\sigma) \cup \{k\}$, and $\mathcal{W}(\sigma') = \mathcal{W}(\sigma) \backslash \{k\}$. (Recall that $\mathcal{S}(\sigma)$ denotes the set of indices j where $\sigma_j \in \Sigma$, and $\mathcal{W}(\sigma)$ denotes the set of delegateable fields of σ.)

Step 1: Let (DK, DL) denote the parent token. Pick a random exponent $\mu \in \mathbb{Z}_n$ and rerandomize the delegation component DL by raising every element in DL to μ. Denote the rerandomized delegation component as:

$$\forall i \in \mathcal{W}(\sigma): \quad \widehat{\mathsf{DL}}_i = \begin{pmatrix} \widehat{L}_{i,h} = L_{i,h}^{\mu}, & \widehat{L}_{i,u} = L_{i,u}^{\mu}, \\ \widehat{L}_{i,0} = L_{i,0}^{\mu}, & \widehat{L}_{i,\phi} = L_{i,\phi}^{\mu}, & \forall j \in \mathcal{S}(\sigma) \cup \{i\}: \ \widehat{L}_{i,j} = L_{i,j}^{\mu} \end{pmatrix}$$

In addition, compute a partial decryption key component with the kth coordinate fixed to $\mathcal{I}$:

$$\mathsf{pDK} = \left(T = \widehat{L}_{k,u}^{\mathcal{I}} \widehat{L}_{k,h}, \quad T_0 = \widehat{L}_{k,0}, \quad T_{\phi} = \widehat{L}_{k,\phi}, \quad \forall j \in \mathcal{S}(\sigma'): \ T_j = \widehat{L}_{k,j} \right)$$

The partial decryption key pDK is formed similarly to the decryption key DK, except that pDK does not contain the term g^{α}.

Step 2: Compute $|\mathcal{W}(\sigma')|$ rerandomized versions of the above. For all $i \in \mathcal{W}(\sigma')$, randomly select $\tau_i \in Z_n$, and compute:

$$\mathsf{pDK}_i = \left(\Gamma_i = T^{\tau_i}, \quad \Gamma_{i,0} = T_0^{\tau_i}, \quad \Gamma_{i,\phi} = T_{\phi}^{\tau_i}, \quad \forall j \in \mathcal{S}(\sigma'): \ \Gamma_{i,j} = T_j^{\tau_i} \right)$$

Step 3: We are now ready to compute the decryption key component DK' of the child token. DK' is computed from two things: 1) DK, the decryption key component of the parent token and 2) pDK, the partial decryption key computed in Step 1. In particular, pDK is the partial decryption key with the kth field fixed; however, as pDK does not contain the g^{α} term, we need to multiply appropriate components of pDK to those in DK.

To compute DK', first, randomly select $Y', Y_0', Y_{\phi}' \in \mathbb{G}_r$. For all $j \in \mathcal{S}(\sigma')$, randomly select $Y_j' \in \mathbb{G}_r$. Now output the following DK':

$$\mathsf{DK}' = (K' = KTY', \quad K_0' = K_0 T_0 Y_0', \quad K_{\phi}' = K_{\phi} T_{\phi} Y_{\phi}', \quad K_k' = T_k Y_k', \quad \forall j \in \mathcal{S}(\sigma): \ K_j' = K_j T_j Y_j')$$

Step 4: We now explain how to compute the delegation component DL' of the child token. DL' is composed of a portion DL'_i for each $i \in \mathcal{W}(\sigma')$. Moreover, each DL'_i is computed from two things: 1) $\widehat{DL}_i$ as computed in Step 1 and 2) pDK_i as computed in Step 2.

Follow the steps below to compute DL'. For each $i \in \mathcal{W}(\sigma')$, randomly select $Y'_{i,h}, Y'_{i,u}, Y'_{i,0}, Y'_{i,\phi}$ from $\mathbb{G}_r$. For each $i \in \mathcal{W}(\sigma')$, for each $j \in \mathcal{S}(\sigma) \cup \{i, k\}$, pick at random $Y'_{i,j}$ from $\mathbb{G}_r$. Compute the delegation component DL' of the child token as below:

$$\forall i \in \mathcal{W}(\sigma'): \quad DL'_i = \begin{pmatrix} L'_{i,h} = \widehat{L}_{i,h}\Gamma_i Y'_{i,h}, & L'_{i,u} = \widehat{L}_{i,u} Y'_{i,u} \\ L'_{i,0} = \widehat{L}_{i,0}\Gamma_{i,0} Y'_{i,0}, & L'_{i,\phi} = \widehat{L}_{i,\phi}\Gamma_{i,\phi} Y'_{i,\phi}, \\ L'_{i,i} = \widehat{L}_{i,i} Y'_{i,i}, & L'_{i,k} = \Gamma_{i,k} Y'_{i,k}, \quad \forall j \in \mathcal{S}(\sigma): \; L'_{i,j} = \widehat{L}_{i,j}\Gamma_{i,j} Y'_{i,j} \end{pmatrix}$$

Type 2: We now go on to explain how to perform a Type 2 delegation. Suppose σ' fixes the kth coordinate of σ to $\perp$. In this case, $\mathcal{S}(\sigma') = \mathcal{S}(\sigma)$, and $\mathcal{W}(\sigma') = \mathcal{W}(\sigma)\backslash\{k\}$. The child token is formed by removing the part DL_k from the parent token:

$$TK_{\sigma'} = (DK, \quad DL\backslash\{DL_k\})$$

Remark 8. It is not hard to verify that delegated tokens have the correct form, except that their exponents are no longer distributed independently at random, but are correlated with the parent tokens. In the proof in the Appendix, we show that Type 1 delegated tokens "appear" (in a computational sense) as if there were generated directly by calling the *GenToken* algorithm, that is, with exponents completely at random. This constitutes an important idea in our security proof.

Query$(PK, TK_\sigma, CT, \sigma')$. A token for $\sigma \in \Sigma_{?,\perp}^\ell$ allows one to evaluate a set of functions $\mathcal{C}_\sigma$ defined by Equation (1) from the ciphertext. Let $\sigma' \prec \sigma$ and assume σ' has no delegateable fields. Then σ' represents a single function $f_{\sigma'}$ (a conjunctive equality test), and the *Query* algorithm allows us to evaluate $f_{\sigma'}$ over the ciphertext.

To evaluate $f_{\sigma'}$ from the ciphertext using TK_σ, first call the *Delegate* algorithm to compute a decryption key for σ'. Write this decryption key in the form $DK = (K, \quad K_0, \quad K_\phi, \quad \forall j \in \mathcal{S}(\sigma'): K_j)$. Furthermore, parse the ciphertext as $CT = \left(\widetilde{C}, \quad C, \quad C_0, \quad C_\phi, \quad \forall j \in \ell: C_j\right)$.

Now use the same algorithm as the original HVE construction to perform the query. First, compute

$$M \leftarrow \widetilde{C} \cdot e(C, K)^{-1} \cdot e(C_0, K_0) e(C_\phi, K_\phi) \prod_{j \in \mathcal{S}(\sigma')} e(C_j, K_j) \qquad (4)$$

If $M \notin \mathcal{M}$, output 0, indicating that $f_{\sigma'}$ is not satisfied. Otherwise, output 1, indicating that $f_{\sigma'}$ is satisfied and also output M. We explain why the *Query* algorithm is correct in the Appendix.

5.2   Security of Our Construction

Theorem 1. *Assuming that the bilinear Diffie-Hellman assumption and the generalized composite 3-party Diffie-Hellman assumptions hold in* $\mathbb{G}$, *then the above dHVE construction is selectively secure.*

We now explain the main techniques used in the proof; however, we defer the detailed proof to the Appendix. In our main construction, delegated tokens have certain correlations with their parent tokens. As a result, the distribution of delegated tokens differ from tokens generated freshly at random by calling the *GenToken* algorithm. A major technique used in the proof is *"token indistinguishability"*: although delegated tokens have correlations with their parent tokens, they are in fact computationally indistinguishable from tokens freshly generated through the *GenToken* algorithm. (Strictly speaking, Type 1 delegated tokens are computationally indistinguishable from freshly generated tokens.) This greatly simplifies our simulation, since now the simulator can pretend that all Type 1 tokens queried by the adversary are freshly generated, without having to worry about their correlation with parent tokens. Intuitively, the above notion of token indistinguishability relies on the C3DH assumption: if we use a random hiding factor from $\mathbb{G}_r$ to randomize each term in the token, then DDH becomes hard for the subgroup $\mathbb{G}_p$.

Acknowledgement

We would like to thank John Bethencourt and Jason Franklin for helpful suggestions and comments. We also would like to thank the anonymous reviewers for their helpful reviews.

References

1. Abdalla, M., Bellare, M., Catalano, D., Kiltz, E., Kohno, T., Lange, T., Malone-Lee, J., Neven, G., Paillier, P., Shi, H.: Searchable encryption revisited: Consistency properties, relation to anonymous IBE, and extensions. In: Shoup, V. (ed.) CRYPTO 2005. LNCS, vol. 3621. Springer, Heidelberg (2005)
2. Bethencourt, J., Sahai, A., Waters, B.: Ciphertext-policy attribute-based encryption. In: Proceedings of the 2007 IEEE Symposium on Security and Privacy (2007)
3. Boneh, D., Boyen, X.: Efficient selective-ID secure identity based encryption without random oracles. In: Cachin, C., Camenisch, J.L. (eds.) EUROCRYPT 2004. LNCS, vol. 3027. Springer, Heidelberg (2004)
4. Boneh, D., Boyen, X., Goh, E.-J.: Hierarchical identity based encryption with constant size ciphertext. In: Cramer, R.J.F. (ed.) EUROCRYPT 2005. LNCS, vol. 3494, pp. 440–456. Springer, Heidelberg (2005)
5. Boneh, D., Di Crescenzo, G., Ostrovsky, R., Persiano, G.: Public key encryption with keyword search. In: Cachin, C., Camenisch, J.L. (eds.) EUROCRYPT 2004. LNCS, vol. 3027, pp. 506–522. Springer, Heidelberg (2004)
6. Boneh, D., Franklin, M.: Identity-based encryption from the Weil pairing. In: Kilian, J. (ed.) CRYPTO 2001. LNCS, vol. 2139, pp. 213–229. Springer, Heidelberg (2001)

7. Boneh, D., Gentry, C., Hamburg, M.: Space-efficient identity based encryption without pairings. In: Proceedings of FOCS (2007)
8. Boneh, D., Waters, B.: A fully collusion resistant broadcast trace and revoke system with public traceability. In: ACM Conference on Computer and Communication Security (CCS) (2006)
9. Boyen, X., Waters, B.: Anonymous hierarchical identity-based encryption (without random oracles). In: Dwork, C. (ed.) CRYPTO 2006. LNCS, vol. 4117. Springer, Heidelberg (2006)
10. Canetti, R., Halevi, S., Katz, J.: A forward-secure public-key encryption scheme. In: EUROCRYPT, pp. 255–271 (2003)
11. Canetti, R., Halevi, S., Katz, J.: Chosen-ciphertext security from identity-based encryption. In: Cachin, C., Camenisch, J.L. (eds.) EUROCRYPT 2004. LNCS, vol. 3027, pp. 207–222. Springer, Heidelberg (2004)
12. Chase, M.: Multi-authority attribute based encryption. In: TCC, pp. 515–534 (2007)
13. Cocks, C.: An identity based encryption scheme based on quadratic residues. In: Proceedings of the 8th IMA International Conference on Cryptography and Coding, London, UK, pp. 360–363. Springer, Heidelberg (2001)
14. Gentry, C., Silverberg, A.: Hierarchical id-based cryptography. In: Zheng, Y. (ed.) ASIACRYPT 2002. LNCS, vol. 2501. Springer, Heidelberg (2002)
15. Goyal, V., Pandey, O., Sahai, A., Waters, B.: Attribute-based encryption for fine-grained access control of encrypted data. In: ACM conference on Computer and communications security (CCS) (2006)
16. Horwitz, J., Lynn, B.: Towards hierarchical identity-based encryption. In: Knudsen, L.R. (ed.) EUROCRYPT 2002. LNCS, vol. 2332. Springer, Heidelberg (2002)
17. Katz, J., Sahai, A., Waters, B.: Predicate encryption supporting disjunctions, polynomial equations, and inner products. In: Eurocrypt (to appear, 2008)
18. Pirretti, M., Traynor, P., McDaniel, P., Waters, B.: Secure attribute-based systems. In: CCS 2006: Proceedings of the 13th ACM conference on Computer and communications security (2006)
19. Sahai, A., Waters, B.: Fuzzy identity-based encryption. In: Cramer, R.J.F. (ed.) EUROCRYPT 2005. LNCS, vol. 3494, pp. 457–473. Springer, Heidelberg (2005)
20. Shamir, A.: Identity-based cryptosystems and signature schemes. In: Proceedings of Crypto (1984)
21. Shi, E., Bethencourt, J., Chan, T.-H.H., Song, D., Perrig, A.: Multi-dimension range query over encrypted data. In: IEEE Symposium on Security and Privacy (May 2007)
22. Shi, E., Waters, B.: Delegating capabilities in predicate encryption systems. In: Aceto, L., Damgaard, I., Goldberg, L.A., Halldorsson, M.M., Ingolfsdottir, A., Walukiewicz, I. (eds.) ICALP 2008. LNCS, vol. 5125. Springer, Heidelberg (2008), http://sparrow.ece.cmu.edu/~elaine/docs/delegation.pdf
23. Song, D.X., Wagner, D., Perrig, A.: Practical techniques for searches on encrypted data. In: IEEE Symposium on Security and Privacy (2000)

Appendix

Due to limit of space, please refer to the online full version of this paper for the appendix [22].

Bounded Ciphertext Policy Attribute Based Encryption

Vipul Goyal*, Abhishek Jain*, Omkant Pandey*, and Amit Sahai*

Department of Computer Science, UCLA
{vipul,abhishek,omkant,sahai}@cs.ucla.edu

Abstract. In a ciphertext policy attribute based encryption system, a user's private key is associated with a set of attributes (describing the user) and an encrypted ciphertext will specify an access policy over attributes. A user will be able to decrypt if and only if his attributes satisfy the ciphertext's policy.

In this work, we present the first construction of a ciphertext-policy attribute based encryption scheme having a security proof based on a number theoretic assumption and supporting advanced access structures. Previous CP-ABE systems could either support only very limited access structures or had a proof of security only in the generic group model. Our construction can support access structures which can be represented by a bounded size access tree with threshold gates as its nodes. The bound on the size of the access trees is chosen at the time of the system setup. Our security proof is based on the standard Decisional Bilinear Diffie-Hellman assumption.

1 Introduction

In many access control systems, every piece of data may legally be accessed by several different users. Such a system is typically implemented by employing a trusted server which stores all the data in clear. A user would log into the server and then the server would decide what data the user is permitted to access. However such a solution comes with a cost: what if the server is compromised? An attacker who is successful in breaking into the server can see all the sensitive data in clear.

One natural solution to the above problem is to keep the data on the server encrypted with the private keys of the users who are permitted to access it. However handling a complex access control policy using traditional public key encryption systems can be difficult. This is because the access policy might be described in terms of the *properties or attributes* that a valid user should have rather than in terms of the actual identities of the users. Thus, a priori, one may

---

* This research was supported in part from grants from the NSF ITR and Cybertrust programs (including grants 0627781, 0456717, and 0205594), a subgrant from SRI as part of the Army Cyber-TA program, an equipment grant from Intel, an Alfred P. Sloan Foundation Fellowship, and an Okawa Foundation Research Grant.

L. Aceto et al. (Eds.): ICALP 2008, Part II, LNCS 5126, pp. 579–591, 2008.

not even know the exact list of users authorized to access a particular piece of data.

The concept of attribute based encryption (ABE) was introduced by Sahai and Waters [1] as a step towards developing encryption systems with high expressiveness. Goyal et al [2] further developed this idea and introduced two variants of ABE namely ciphertext-policy attribute based encryption (CP-ABE) and key-policy attribute based encryption (KP-ABE). In a CP-ABE system, a user's private key is associated with a set of attributes (describing the *properties* that the user has) and an encrypted ciphertext will specify an access policy over attributes. A user will be able to decrypt if and only if his attributes satisfy the ciphertext's policy. While a construction of KP-ABE was offered by [2], constructing CP-ABE was left as an important open problem.

Subsequently to Goyal et al [2], Bethencourt et al [3] gave the first construction of a CP-ABE system. Their construction however only had a security argument in the generic group model. Cheung and Newport [4] recently gave a CP-ABE construction supporting limited type of access structures which could be represented by **AND** of different attributes. Cheung and Newport also discussed the possibility of supporting more general access structures with threshold gates. However as they discuss, a security proof of this generalization would involve overcoming several subtleties. In sum, obtaining a CP-ABE scheme for more advanced access structures based on any (even relatively non-standard) number theoretic assumption has proven to be surprisingly elusive.

Our Results. We present the first construction of a ciphertext-policy attribute based encryption scheme having a security proof based on a standard number theoretic assumption and supporting advanced access structures. Our construction can support access structures which can be represented by a bounded size access tree with threshold gates as its nodes. The bound on the size of the access trees is chosen at the time of the system setup and is represented by a tuple (d, num) where d represents the maximum depth of the access tree and num represents the maximum number of children each non-leaf node of the tree might have. We stress that any access tree satisfying these upper bounds on the size can be dynamically chosen by the encrypter. Our construction has a security proof based on the standard Decisional Bilinear Diffie-Hellman (BDH) assumption. We note that previous CP-ABE systems could either support only very limited access structures [4] or had a proof of security only in the generic group model [3] (rather than based on a number theoretic assumption). Further, we show how to extend our constructions to support non-monotonic access policies. Finally, we observe that our constructions for non-monotonic access policies can in fact support any access formula with bounded polynomial size.

Our Techniques. Our construction can be seen as a way to reinterpret Key-Policy ABE schemes (e.g. [2]) with a fixed "universal" tree access structure as a CP-ABE scheme. Such a reinterpretation presents some problems because in a KP-ABE scheme, the key material for each attribute is "embedded" into the access structure in a unique way depending on where it occurs in the access policy. To overcome this difficulty, we introduce many "copies" of each attribute for every position in the access structure tree where it can occur. This causes a significant increase in private key size, but does not significantly affect ciphertext size. However, since the actual access structure to be used for a particular ciphertext must be embedded into the fixed "universal" tree access structure in the KP-ABE scheme, this causes a blowup in ciphertext size. This effect can be moderated by having multiple parallel CP-ABE schemes with different sized "universal" tree access structures underlying the scheme, which allows for a trade-off between ciphertext size and the size of the public parameters and private keys.

Note that in general a Boolean formula of size n can be represented by a balanced formula of size $O(n^{2/\log(3/2)})$ (roughly $O(n^{3.42})$). Thus, in general our methodology would yield a ciphertext blowup of $O(n^{3.42})$ group elements. As such, our result can be seen as a "feasibility result" for CP-ABE for general Boolean formulas of bounded size. We leave constructing more efficient CP-ABE schemes based on number-theoretic assumptions as an important open question.

2   Background

We first give formal definitions for the security of Bounded Ciphertext Policy Attribute Based Encryption (BCP-ABE). Then we give background information on bilinear maps and our cryptographic assumption. Like the work of Goyal et al. [2], we describe our constructions for access trees. Roughly speaking, given a set of attributes $\{P_1, P_2, \ldots, P_n\}$, an access tree is an access structure $\mathcal{T} \subseteq 2^{\{P_1, P_2, \ldots, P_n\}}$, where each node in the tree represents a threshold gate (see Section 3 for a detailed description). We note that contrary to the work of Goyal et al., in our definitions, users will be identified with a set of attributes while access trees will be used to specify policies for encrypting data.

A Bounded Ciphertext Policy Attribute Based Encryption scheme consists of four algorithms.

Setup (d, num). This is a randomized algorithm that takes as input the implicit security parameter and a pair of system parameters (d, num). These parameters will be used to restrict the access trees under which messages can be encrypted in our system. It outputs the public parameters PK and a master key MK.

Key Generation (γ, MK). This is a randomized algorithm that takes as input
– the master key MK and a set of attributes γ. It outputs a decryption key D corresponding to the attributes in γ.

Encryption $(M, PK, \mathcal{T}')$. This is a randomized algorithm that takes as input –
the public parameters PK, a message M, and an access tree $\mathcal{T}'$ over the universe of attributes, with depth $d' \leq d$, and where each non-leaf node x has at most num child nodes. The algorithm will encrypt M and output the ciphertext E. We will assume that the ciphertext implicitly contains $\mathcal{T}'$.

Decryption (E, D). This algorithm takes as input – the ciphertext E that was encrypted under the access tree $\mathcal{T}'$, and the decryption key D for an attribute set γ. If the set γ of attributes satisfies the access tree $\mathcal{T}'$ (i.e. $\gamma \in \mathcal{T}'$), then the algorithm will decrypt the ciphertext and return a message M.

We now discuss the security of a bounded ciphertext-policy ABE scheme. We define a selective-tree model for proving the security of the scheme under the chosen plaintext attack. This model can be seen as analogous to the selective-ID model [5,6,7] used in identity-based encryption (IBE) schemes [8,9,10].

Selective-Tree Model for BCP-ABE. Let $\mathcal{U}$ be the universe of attributes fixed by the security parameter. The system parameters d, num are also defined.

Init. The adversary declares the access tree $\mathcal{T}'$, that he wishes to be challenged upon.

Setup. The challenger runs the Setup algorithm of ABE and gives the public parameters to the adversary.

Phase 1. The adversary is allowed to issue queries for private keys for many attribute sets γ_j, where γ_j does not satisfy the access tree $\mathcal{T}'$ for all j.

Challenge. The adversary submits two equal length messages M_0 and M_1. The challenger flips a random coin b, and encrypts M_b with $\mathcal{T}'$. The ciphertext is passed on to the adversary.

Phase 2. Phase 1 is repeated.

Guess. The adversary outputs a guess b' of b.

The advantage of an adversary $\mathcal{A}$ in this game is defined as $\Pr[b' = b] - \frac{1}{2}$. We note that the model can easily be extended to handle chosen-ciphertext attacks by allowing for decryption queries in Phase 1 and Phase 2.

Definition 1. *A bounded ciphertext-policy attribute-based encryption scheme (BCP-ABE) is secure in the Selective-Tree model of security if all polynomial time adversaries have at most a negligible advantage in the Selective-Tree game.*

2.1 Bilinear Maps

We present a few facts related to groups with efficiently computable bilinear maps. Let $\mathbb{G}_1$ and $\mathbb{G}_2$ be two multiplicative cyclic groups of prime order p. Let g be a generator of $\mathbb{G}_1$ and e be a bilinear map, $e : \mathbb{G}_1 \times \mathbb{G}_1 \to \mathbb{G}_2$. The bilinear map e has the following properties:

1. Bilinearity: for all $u, v \in \mathbb{G}_1$ and $a, b \in \mathbb{Z}_p$, we have $e(u^a, v^b) = e(u, v)^{ab}$.
2. Non-degeneracy: $e(g, g) \neq 1$.

We say that $\mathbb{G}_1$ is a bilinear group if the group operation in $\mathbb{G}_1$ and the bilinear map $e : \mathbb{G}_1 \times \mathbb{G}_1 \to \mathbb{G}_2$ are both efficiently computable. Notice that the map e is symmetric since $e(g^a, g^b) = e(g, g)^{ab} = e(g^b, g^a)$.

2.2 The Decisional Bilinear Diffie-Hellman (BDH) Assumption

Let $a, b, c, z \in \mathbb{Z}_p$ be chosen at random and g be a generator of $\mathbb{G}_1$. The decisional BDH assumption [7,1] is that no probabilistic polynomial-time algorithm

$\mathcal{B}$ can distinguish the tuple $(A = g^a, B = g^b, C = g^c, e(g, g)^{abc})$ from the tuple $(A = g^a, B = g^b, C = g^c, e(g, g)^z)$ with more than a negligible advantage. The advantage of $\mathcal{B}$ is

$$\left| \Pr[\mathcal{B}(A, B, C, e(g, g)^{abc}) = 0] - \Pr[\mathcal{B}(A, B, C, e(g, g)^z)] = 0 \right|$$

where the probability is taken over the random choice of the generator g, the random choice of a, b, c, z in $\mathbb{Z}_p$, and the random bits consumed by $\mathcal{B}$.

3   Access Trees

In our constructions, user decryption keys will be identified with a set γ of attributes. A party who wishes to encrypt a message will specify through an access tree structure a policy that private keys must satisfy in order to decrypt. We now proceed to explain the access trees used in our constructions.

Access Tree. Let T be a tree representing an access structure. Each non-leaf node of the tree represents a threshold gate, described by its children and a threshold value. If num_x is the number of children of a node x and k_x is its threshold value, then $0 < k_x \leq num_x$. For ease of presentation, we use the term *cardinality* to refer to the number of children of a node. Each leaf node x of the tree is described by an attribute and a threshold value $k_x = 1$.

We fix the root of an access tree to be at level 0. Let Φ_T denote the set of all the non-leaf nodes in the tree T. Further, let Ψ_T be the set of all the non-leaf nodes at depth $d - 1$, where d is the depth of T. To facilitate working with the access trees, we define a few functions. We denote the parent of the node x in the tree by parent(x). The access tree T also defines an ordering between the children of every node, that is, the children of a node x are numbered from 1 to num_x. The function index(x) returns such a number associated with a node x, where the index values are uniquely assigned to nodes in an arbitrary manner for a given access structure. For simplicity, we provision that index$(x) = $ att(x), when x is a leaf node and att(x) is the attribute associated with it.

Satisfying an Access Tree. Let T be an access tree with root r. Denote by T_x the subtree of T rooted at the node x. Hence T is the same as T_r. If a set of attributes γ satisfies the access tree T_x, we denote it as $T_x(\gamma) = 1$. We compute $T_x(\gamma)$ recursively as follows. If x is a non-leaf node, evaluate $T_z(\gamma)$ for all children z of node x. $T_x(\gamma)$ returns 1 if and only if at least k_x children return 1. If x is a leaf node, then $T_x(\gamma)$ returns 1 iff att$(x) \in \gamma$.

Universal Access Tree. Given a pair of integer values (d, num), define a complete num-ary tree T of depth d, where each non-leaf node has a threshold value of num. The leaf nodes in T are empty, i.e., no attributes are assigned to the leaf nodes. Next, $num - 1$ new leaf nodes are attached to each non-leaf node x, thus increasing the cardinality of x to $2 \cdot num - 1$ while the threshold value num is left intact. Choose an arbitrary assignment of *dummy* attributes (explained later in Section 4) to these newly added leaf nodes[1] for each x. The resultant tree $\mathcal{T}$ is called a (d, num)-universal access tree (or simply the universal access tree when d, num are fixed by the system).

Bounded Access Tree. We say that $\mathcal{T}'$ is a (d, num)-bounded access tree if it has depth $d' \leq d$, and each non-leaf node in $\mathcal{T}'$ exhibits a cardinality at most num.

Normal Form. Consider a (d, num)-bounded access tree $\mathcal{T}'$. We say that $\mathcal{T}'$ exhibits the (d, num)-normal form if (a) it has depth $d' = d$, and (b) all the leaves in $\mathcal{T}'$ are at depth d. Any (d, num)-bounded access tree $\mathcal{T}'$ can be converted to the (d, num)-normal form (or simply the normal form when d, num are fixed by the system) in the following way in a top down manner, starting from the root node r'. Consider a node x at level l_x in $\mathcal{T}'$. If the depth d_x of the subtree $\mathcal{T}'_x$ is less than $(d - l_x)$, then insert a vertical chain of $(d - l_x - d_x)$ nodes (where each node has cardinality 1 and threshold 1) between x and parent(x). Repeat the procedure recursively for each child of x. Note that conversion to the normal form does not affect the satisfying logic of an access tree.

Map between Access Trees. Consider a (d, num)-universal access tree $\mathcal{T}$ and another tree $\mathcal{T}'$ that exhibits the (d, num)-normal form. A map between the nodes of $\mathcal{T}'$ and $\mathcal{T}$ is defined in the following way in a top-down manner. First, the root of $\mathcal{T}'$ is mapped to the root of $\mathcal{T}$. Now suppose that x' in $\mathcal{T}'$ is mapped to x in $\mathcal{T}$. Let $z'_1, \ldots, z'_{num_{x'}}$ be the child nodes of x', ordered according to their index values. Then, for each child z'_i ($i \in [1, num_{x'}]$) of x' in $\mathcal{T}'$, set the corresponding child z_i (i.e. with index value index(z'_i)) of x in $\mathcal{T}$ as the map of z'. This procedure is performed recursively, until each node in $\mathcal{T}'$ is mapped to a corresponding node in $\mathcal{T}$. To capture the above node mapping procedure, we define a public function map$(\cdot)$ that takes a node (or a set of nodes) in $\mathcal{T}'$ as input and returns the corresponding node (or a set of nodes) in $\mathcal{T}$.

4   Small Universe Construction

Before we explain the details of our construction, we first present an overview highlighting the main intuitions behind our approach.

4.1   Overview of Our Construction

Fixed Tree Structure. We first give the outline of a basic target system. For simplicity, let us consider a very simple and basic access tree $\mathcal{T}$. The tree $\mathcal{T}$ has depth, say, d; and all leaf nodes in the tree are at depth d. Each non-leaf node x

---

[1] From now onwards, by *dummy nodes*, we shall refer to the leaf nodes with dummy attributes associated with them.

is a "k_x-out-of-num_x" threshold gate where k_x and num_x are fixed beforehand. Thus the "structure" of the access tree is *fixed*. However, the leaf nodes of $\mathcal{T}$ are "empty", i.e., no attributes are associated with them. At the time of encryption, an encrypter will assign attributes to each leaf-node, in order to define the access structure completely. That is, once the encrypter assigns an attribute to each leaf node in $\mathcal{T}$, it fixes the set of "authorized sets of attributes". A user having keys corresponding to an authorized set will be able to decrypt a message encrypted under the above access structure.

We now explain how such a system can be constructed. Recall that $\Psi_\mathcal{T}$ denotes the set of non-leaf nodes at depth $d - 1$. Each leaf child z of an $x \in \Psi_\mathcal{T}$ will be assigned an attribute j.[2] However, the same j may be assigned to z_1 as well as z_2 where z_1 is a child node of $x_1 \in \Psi_\mathcal{T}$, and z_2 is a child node of $x_2 \in \Psi_\mathcal{T}$. Thus, any given attribute j may have at most $|\Psi_\mathcal{T}|$ distinct parent nodes. Intuitively, these are all the distinct positions under which j can appear as an attribute of a leaf in the tree $\mathcal{T}$. Now, during system setup, we will publish a unique public parameter corresponding to each such appearance of j, for each attribute j. Next, consider a user A with an attribute set γ. Imagine an access tree $\mathcal{T}'$ that has the same "structure" as $\mathcal{T}$, except that each node $x \in \Psi_{\mathcal{T}'}$ has cardinality $|\gamma|$ instead of num_x (while the threshold is still k_x). Additionally, each attribute $j \in \gamma$ is attached to a distinct leaf child of each node $x \in \Psi_{\mathcal{T}'}$. A will be assigned a private key that is computed for such an access tree $\mathcal{T}'$ as in the KP-ABE construction of Goyal et al [2]. Now, suppose that an encrypter $\mathcal{E}$ has chosen an assignment of attributes to the leaf nodes in $\mathcal{T}$ to define it completely. Let $f(j, x)$ be a function that outputs 1 if an attribute j is associated with a leaf child of x and 0 otherwise. Then, $\mathcal{E}$ will compute (using the public parameters published during system setup) and release a ciphertext component $E_{j,x}$ corresponding to an attribute j attached to a leaf child of $x \in \Psi_\mathcal{T}$ (i.e., iff $f(j, x) = 1$). A receiver who possesses an authorized set of attributes for the above tree can choose from his private key - the components $D_{j,x}$, such that $f(j, x) = 1$; and use them with corresponding $E_{j,x}$ during the decryption process.

Varying the Thresholds. The above system, although dynamic, may be very limited in its expressibility for some applications. In order to make it more expressible, we can further extend the above system as follows. At the time of encryption, an encrypter is now given the flexibility of choosing the threshold value between 1 and some maximum fixed value, (say) num for each node x in the access tree.

As a first step, we will construct $\mathcal{T}$ as a complete num-ary tree of depth d, where each non-leaf node is a "num-out-of-num" threshold gate. As earlier, the leaf nodes in the tree are empty. Next, we introduce a $(num - 1)$-sized set of special attributes called *dummy* attributes [1] that are different from the usual

---

[2] For simplicity of exposition, we provision that an encrypter cannot assign the same attribute j to two child nodes z_1, z_2 of a given x. We note that this restriction can be removed by some simple modifications. Details will be given in the full version [11].

attributes (which we will henceforth refer to as *real* attributes[3]). Now, attach $(num - 1)$ leaf nodes to each $x \in \varPhi_\mathcal{T}$, and assign a dummy attribute to each such newly-added leaf node (henceforth referred to as dummy nodes).

Note that a dummy attribute j may have atmost $|\varPhi_\mathcal{T}|$ parent nodes. Intuitively, these are all the distinct positions where j can appear as an attribute of a dummy leaf in $\mathcal{T}$. Therefore, during the system setup, for each dummy attribute j, we will publish a unique public parameter corresponding to *each appearance* of j (in addition to the public parameters corresponding to the real attributes as in the previous description). Next, consider a user A with an attribute set γ. Imagine an access tree $\mathcal{T}'$ that is similar to $\mathcal{T}$, except that each node $x \in \varPsi_{\mathcal{T}'}$ has $|\gamma|$ leaf child (dummy nodes not inclusive) instead of num (while the threshold is still num). Additionally, each attribute $j \in \gamma$ is attached to a distinct leaf child of each node $x \in \varPsi_{\mathcal{T}'}$. A will be assigned a private key that is computed for such an access tree $\mathcal{T}'$ as in the KP-ABE construction of Goyal et al [2] (the difference from the previous description for fixed trees is that here A will additionally receive key-components corresponding to dummy attributes). Now, at the time of encryption, an encrypter E will first choose a threshold value $k_x \leq num$ for each $x \in \varPhi_\mathcal{T}$. Next, E will choose an assignment of real attributes to the leaf nodes in $\mathcal{T}$, and an arbitrary $(num - k_x)$-sized subset ω_x of dummy child nodes of each $x \in \varPhi_\mathcal{T}$. Finally, E will release the ciphertext components as in the previous description. E will additionally release a ciphertext component corresponding to each dummy node in ω_x, for each $x \in \varPhi_\mathcal{T}$. Now, consider a receiver with an attribute set γ. For any $x \in \varPhi_\mathcal{T}$, if k_x children of x can be satisfied with γ, then the receiver can use the key-components (from his private key) corresponding to the dummy attributes in order to satisfy each dummy leaf $z \in \omega_x$; thus satisfying the num-out-of-num threshold gate x.

Varying Tree Depth and Node Cardinality. Finally, we note that the above system can be further extended to allow an encrypter to choose the depth of the access tree and also the cardinality of each node; thus further increasing the expressibility of the system. To do this, we will assume an upper bound on the maximum tree depth d and the maximum node cardinality num, fixed beforehand. We will then make use of the techniques presented in the latter part of Section 3 to achieve the desired features. Details are given in the construction.

4.2   The Construction

Let $\mathbb{G}_1$ be a bilinear group of prime order p, and let g be a generator of $\mathbb{G}_1$. In addition, let $e : \mathbb{G}_1 \times \mathbb{G}_1 \to \mathbb{G}_2$ denote the bilinear map. A security parameter, κ, will determine the size of the groups. We also define the Lagrange coefficient $\Delta_{i,S}$ for $i \in \mathbb{Z}_p$ and a set, S, of elements in $\mathbb{Z}_p$: $\Delta_{i,S}(x) = \prod_{j \in S, j \neq i} \frac{x-j}{i-j}$. We will associate each attribute with a unique element in $\mathbb{Z}_p^*$. Our construction follows.

---

[3] As the name suggests, we will identify users with a set of "real" attributes, while the dummy attributes will be used for technical purposes, i.e., varying the threshold of the nodes when needed.

Setup (d, num). This algorithm takes as input two system parameters, namely, (a) the maximum tree depth d, and (b) the maximum node cardinality num. The algorithm proceeds as follows. Define the universe of real attributes $\mathcal{U} = \{1, \ldots, n\}$, and a $(num - 1)$-sized universe of dummy attributes[4] $\mathcal{U}^* = \{n + 1, \ldots, n+num-1\}$. Next, define a (d, num)-universal access tree $\mathcal{T}$ as explained in the section 3. In the sequel, $d, num, \mathcal{U}, \mathcal{U}^*, \mathcal{T}$ will all be assumed as implicit inputs to all the procedures.

Now, for each real attribute $j \in \mathcal{U}$, choose a set of $|\Psi_{\mathcal{T}}|$ numbers $\{t_{j,x}\}_{x \in \Psi_{\mathcal{T}}}$ uniformly at random from $\mathbb{Z}_p$. Further, for each dummy attribute $j \in \mathcal{U}^*$, choose a set of $|\Phi_{\mathcal{T}}|$ numbers $\{t^*_{j,x}\}_{x \in \Phi_{\mathcal{T}}}$ uniformly at random from $\mathbb{Z}_p$. Finally, choose y uniformly at random in $\mathbb{Z}_p$. The public parameters PK are:

$$Y = e(g,g)^y, \ \{T_{j,x} = g^{t_{j,x}}\}_{j \in \mathcal{U}, x \in \Psi_{\mathcal{T}}}, \ \{T^*_{j,x} = g^{t^*_{j,x}}\}_{j \in \mathcal{U}^*, x \in \Phi_{\mathcal{T}}}$$

The master key MK is:

$$y, \ \{t_{j,x}\}_{j \in \mathcal{U}, x \in \Psi_{\mathcal{T}}}, \ \{t^*_{j,x}\}_{j \in \mathcal{U}^*, x \in \Phi_{\mathcal{T}}}$$

Key Generation (γ, MK). Consider a user A with an attribute set γ. The key generation algorithm outputs a private key D that enables A to decrypt a message encrypted under a (d, num)-bounded access tree $\mathcal{T}'$ iff $\mathcal{T}'(\gamma) = 1$.

The algorithm proceeds as follows. For each user, choose a random polynomial q_x for each non-leaf node x in the universal access tree $\mathcal{T}$. These polynomials are chosen in the following way in a top-down manner, starting from the root node r. For each x, set the degree c_x of the polynomial q_x to be one less than the threshold value, i.e., $c_x = num - 1$. Now, for the root node r, set $q_r(0) = y$ and choose c_r other points of the polynomial q_r randomly to define it completely. For any other non-leaf node x, set $q_x(0) = q_{\text{parent}(x)}(\text{index}(x))$ and choose c_x other points randomly to completely define q_x. Once the polynomials have been decided, give the following secret values to the user:

$$\{D_{j,x} = g^{\frac{q_x(j)}{t_{j,x}}}\}_{j \in \gamma, x \in \Psi_{\mathcal{T}}}, \ \{D^*_{j,x} = g^{\frac{q_x(j)}{t^*_{j,x}}}\}_{j \in \mathcal{U}^*, x \in \Phi_{\mathcal{T}}}$$

The set of above secret values is the decryption key D.

Encryption $(M, \text{PK}, \mathcal{T}')$. To encrypt a message $M \in \mathbb{G}_2$, the encrypter $\mathcal{E}$ first chooses a (d, num)-bounded access tree $\mathcal{T}'$. $\mathcal{E}$ then chooses an assignment of real attributes to the leaf nodes in $\mathcal{T}'$.

Now, to be able to encrypt the message M with the access tree $\mathcal{T}'$, the encrypter first converts it to the normal form (if required). Next, $\mathcal{E}$ defines a map between the nodes in $\mathcal{T}'$ and the universal access tree $\mathcal{T}$ as explained in section 3. Finally, for each non-leaf node x in $\mathcal{T}'$, $\mathcal{E}$ chooses an arbitrary $(num - k_x)$-sized set ω_x of dummy child nodes of $\text{map}(x)$ in $\mathcal{T}$.

---

[4] Recall the distinction between real attributes and dummy attributes that was introduced in the Overview section.

Let $f(j, x)$ be a boolean function such that $f(j, x) = 1$ if a real attribute $j \in \mathcal{U}$ is associated with a leaf child of node $x \in \Psi_{T'}$ and 0 otherwise. Now, choose a random value $s \in \mathbb{Z}_p$ and publish the ciphertext E as:

$$\langle T', E' = M \cdot Y^s, \{E_{j,x} = T_{j,\mathrm{map}(x)}^s\}_{j \in \mathcal{U}, x \in \Psi_{T'}: f(j,x)=1}, \{E_{j,x}^* = T_{j,\mathrm{map}(x)}^{*s}\}_{j = \mathrm{att}(z): z \in \omega_x, x \in \Phi_{T'}} \rangle$$

Decryption (E, D). We specify our decryption procedure as a recursive algorithm. For ease of exposition, we present the simplest form of the decryption algorithm here. The performance of the decryption procedure can potentially be improved by using the techniques explained in [2].

We define a recursive algorithm $\mathrm{DecryptNode}(E, D, x)$ that takes as input the ciphertext E, the private key D, and a node x in T'. It outputs a group element of $\mathbb{G}_2$ or $\perp$. First, we consider the case when x is a leaf node. Let $j = \mathrm{att}(x)$ and w be the parent of x. Then, we have:

$$\mathrm{DecryptNode}(E, D, x) = \begin{cases} e(D_{j,\mathrm{map}(w)}, E_{j,w}) = e(g^{\frac{q_{\mathrm{map}(w)}(j)}{t_{j,\mathrm{map}(w)}}}, g^{s \cdot t_{j,\mathrm{map}(w)}}) & \text{if } j \in \gamma \\ \perp & \text{otherwise} \end{cases}$$

which reduces to $e(g, g)^{s \cdot q_{\mathrm{map}(w)}(j)}$ when $j \in \gamma$. We now consider the recursive case when x is a non-leaf node in T'. The algorithm proceeds as follows: For all nodes z that are children of x, it calls $\mathrm{DecryptNode}(E, D, z)$ and stores the output as F_z. Additionally, for each dummy node $z \in \omega_x$ (where ω_x is a select set of dummy nodes of $\mathrm{map}(x)$ in T chosen by the encrypter), it invokes a function $\mathrm{DecryptDummy}(E, D, z)$ that is defined below, and stores the output as F_z. Let j be the dummy attribute associated with z. Then, we have:

$$\mathrm{DecryptDummy}(E, D, z) = e(D_{j,\mathrm{map}(x)}^*, E_{j,x}^*) = e(g^{\frac{q_{\mathrm{map}(x)}(j)}{t_{j,\mathrm{map}(x)}^*}}, g^{s \cdot t_{j,\mathrm{map}(x)}^*}),$$

which reduces to $e(g, g)^{s \cdot q_{\mathrm{map}(x)}(j)}$. Let Ω_x be an arbitrary k_x-sized set of child nodes z such that $F_z \neq \perp$. Further, let S_x be the union of the sets Ω_x and ω_x. Thus we have that $|S_x| = num$. Let $\hat{g} = e(g, g)$. If no k_x-sized set Ω_x exists, then the node x was not satisfied and the function returns $\perp$. Otherwise, we compute:

$$F_x = \prod_{z \in S_x} F_z^{\Delta_{i, S_x'}(0)}, \quad \text{where} \quad \begin{matrix} i = \mathrm{att}(z) \text{ if } z \text{ is a leaf node} \\ i = \mathrm{index}(\mathrm{map}(z)) \text{ otherwise} \\ S_x' = \{i : z \in S_x\} \end{matrix}$$

$$= \prod_{z \in \Omega_x} F_z^{\Delta_{i, S_x'}(0)} \prod_{z \in \omega_x} F_z^{\Delta_{i, S_x'}(0)}$$

$$= \begin{cases} \prod_{z \in \Omega_x} (\hat{g}^{s \cdot q_{\mathrm{map}(x)}(i)})^{\Delta_{i, S_x'}(0)} \prod_{z \in \omega_x} (\hat{g}^{s \cdot q_{\mathrm{map}(x)}(i)})^{\Delta_{i, S_x'}(0)} & \text{if } x \in \Psi_{T'} \\ \prod_{z \in \Omega_x} (\hat{g}^{s \cdot q_{\mathrm{map}(z)}(0)})^{\Delta_{i, S_x'}(0)} \prod_{z \in \omega_x} (\hat{g} t^{s \cdot q_{\mathrm{map}(x)}(i)})^{\Delta_{i, S_x'}(0)} & \text{else} \end{cases}$$

$$= \begin{cases} \prod_{z \in S_x} \hat{g}^{s \cdot q_{\mathrm{map}(x)}(i) \cdot \Delta_{i, S_x'}(0)} & \text{if } x \in \Psi_{T'} \\ \prod_{z \in \Omega_x} (\hat{g}^{s \cdot q_{\mathrm{map}(\mathrm{parent}(z))}(\mathrm{index}(\mathrm{map}(z)))})^{\Delta_{i, S_x'}(0)} \prod_{z \in \omega_x} (\hat{g}^{s \cdot q_{\mathrm{map}(x)}(i)})^{\Delta_{i, S_x'}(0)} & \text{else} \end{cases}$$

$$= \prod_{z \in S_x} \hat{g}^{s \cdot q_{\mathrm{map}(x)}(i) \cdot \Delta_{i, S_x'}(0)}$$

$$= \hat{g}^{s \cdot q_{\mathrm{map}(x)}(0)} = e(g, g)^{s \cdot q_{\mathrm{map}(x)}(0)} \quad \text{(using polynomial interpolation)}$$

and return the result.

Now that we have defined DecryptNode, the decryption algorithm simply invokes it on the root r' of T'. We observe that DecryptNode$(E, D, r') = e(g, g)^{sy}$ iff $T'(\gamma) = 1$ (note that $F_{r'} = e(g, g)^{s \cdot q_{\text{map}(r')}(0)} = e(g, g)^{s \cdot q_r(0)} = e(g, g)^{sy}$, where r is the root of the universal tree T). Since $E' = M \cdot e(g, g)^{sy}$, the decryption algorithm simply divides out $e(g, g)^{sy}$ and recovers M.

Theorem 1. *If an adversary can break our scheme in the Selective-Tree model, then a simulator can be constructed to play the Decisional BDH game with a non-negligible advantage.*

PROOF: See full version [11].

5 Non-monotonic Access Trees

One limitation of our original construction is that it does not support *negative* constraints in a ciphertext's access formula. With some minor modifications to our small universe construction, we can allow an encrypter to use non-monotonic ciphertext policies. Below we highlight the necessary modifications in the small universe case.

We introduce explicit attributes that indicate the *negative* of attributes in the system. A user will be assigned a negative attribute for each attribute *not* present in his attribute set. In this manner, each user will have $|\mathcal{U}|$ number of attributes. It is known that by applying DeMorgan's law, we can transform a non-monotonic access tree T'' into T' so that T' represents the same access scheme as T'', but has **NOTs** only at the leaves, where the attributes are. Further, we can replace an attribute j with its corresponding negative attribute $\bar{j}$ if the above transformation results in a **NOT** gate at the leaf to which j is associated. Now consider a (d, num)-bounded non-monotonic access tree T'' chosen by an encrypter. Using the above mechanism, the encrypter first transforms it to T' such that the interior gates of T' consist only of positive threshold gates, while both positive and negative attributes may be associated with the leaf nodes. Then, the encryption and decryption procedures follow as in the original construction.

Supporting any Access Formula of Bounded Polynomial Size. It is known that any access formula can be represented by a non-monotonic NC^1 circuit [12]. It is intuitive to see that any circuit of logarithmic depth can be converted to a tree with logarithmic depth. To this end, we note that our modified construction for non-monotonic access trees can support any access formula of bounded polynomial size.

6 Discussion and Extensions

We discuss various extensions to our scheme.

Large Universe Case. In our previous construction, the size of public parameters corresponding to the real attributes grows linearly with the size of the universe of real attributes. Combining the tricks presented in section 4 with those in [2], we construct another scheme that allows us to use arbitrary strings as attributes in the system, yet the public parameters corresponding to the real attributes grow only linearly in a parameter n which we fix as the maximum number of leaf child nodes of a node in an access tree we can encrypt under. Details will be given in the full version [11].

Non-Monotonic Access Policies in the Large Universe Case. We note that the solution for supporting non-monotonic access policies in the small universe construction is inapplicable in the large universe case. This is because the attributes in the system may not be fixed at the time of key generation. However, we can leverage the techniques from [13] in order to support non-monotonic access policies in the large universe case. Details will be give in the full version [11].

Delegation of Private Keys. Similar to the system of Goyal et al. [2], our constructions come with the added capability of delegation of private keys. Details will be given in the full version [11].

References

1. Sahai, A., Waters, B.: Fuzzy Identity Based Encryption. In: Cramer, R.J.F. (ed.) EUROCRYPT 2005. LNCS, vol. 3494, pp. 457–473. Springer, Heidelberg (2005)
2. Goyal, V., Pandey, O., Sahai, A., Waters, B.: Attribute Based Encryption for Fine-Grained Access Conrol of Encrypted Data. In: ACM conference on Computer and Communications Security (ACM CCS) (2006)
3. Bethencourt, J., Sahai, A., Waters, B.: Ciphertext-policy attribute-based encryption. In: IEEE Symposium on Security and Privacy, pp. 321–334. IEEE Computer Society, Los Alamitos (2007)
4. Cheung, L., Newport, C.: Provably Secure Ciphertext Policy ABE. In: ACM conference on Computer and Communications Security (ACM CCS) (2007)
5. Canetti, R., Halevi, S., Katz, J.: A Forward-Secure Public-Key Encryption Scheme. In: EUROCRYPT 2003. LNCS, vol. 2656. Springer, Heidelberg (2003)
6. Canetti, R., Halevi, S., Katz, J.: Chosen Ciphertext Security from Identity Based Encryption. In: Cachin, C., Camenisch, J.L. (eds.) EUROCRYPT 2004. LNCS, vol. 3027, pp. 207–222. Springer, Heidelberg (2004)
7. Boneh, D., Boyen, X.: Efficient Selective-ID Secure Identity Based Encryption Without Random Oracles. In: Cachin, C., Camenisch, J.L. (eds.) EUROCRYPT 2004. LNCS, vol. 3027, pp. 223–238. Springer, Heidelberg (2004)
8. Shamir, A.: Identity Based Cryptosystems and Signature Schemes. In: Blakely, G.R., Chaum, D. (eds.) CRYPTO 1984. LNCS, vol. 196, pp. 37–53. Springer, Heidelberg (1985)
9. Boneh, D., Franklin, M.: Identity Based Encryption from the Weil Pairing. In: Kilian, J. (ed.) CRYPTO 2001. LNCS, vol. 2139, pp. 213–229. Springer, Heidelberg (2001)
10. Cocks, C.: An identity based encryption scheme based on quadratic residues. In: IMA Int. Conf., pp. 360–363 (2001)

11. Goyal, V., Jain, A., Pandey, O., Sahai, A.: Bounded Ciphertext Policy Attribute Based Encryption, http://eprint.iacr.org/2008/
12. Brent, R.P.: The parallel evaluation of general arithmetic expressions. Journal of ACM 21, 201–206 (1974)
13. Ostrovsky, R., Sahai, A., Waters, B.: Attribute Based Encryption with Non-Monotonic Access Structures. In: ACM conference on Computer and Communications Security (ACM CCS) (2007)

Making Classical Honest Verifier Zero Knowledge Protocols Secure against Quantum Attacks

Sean Hallgren[1], Alexandra Kolla[2], Pranab Sen[3], and Shengyu Zhang[4]

[1] Pennsylvania State University, University Park, PA, U.S.A.
[2] U C Berkeley, Berkeley, CA, U.S.A.
akolla@cs.berkeley.edu
[3] Tata Institute of Fundamental Research, Mumbai, India
pgdsen@tcs.tifr.res.in
[4] California Institute of Technology, Pasadena, CA, U.S.A.
shengyu@caltech.edu

Abstract. We show that any problem that has a classical zero-knowledge protocol against the honest verifier also has, under a reasonable condition, a classical zero-knowledge protocol which is secure against all classical and quantum polynomial time verifiers, even cheating ones. Here we refer to the generalized notion of zero-knowledge with classical and quantum auxiliary inputs respectively.

Our condition on the original protocol is that, for positive instances of the problem, the simulated message transcript should be quantum computationally indistinguishable from the actual message transcript. This is a natural strengthening of the notion of honest verifier computational zero-knowledge, and includes in particular, the complexity class of honest verifier statistical zero-knowledge. Our result answers an open question of Watrous [Wat06], and generalizes classical results by Goldreich, Sahai and Vadhan [GSV98], and Vadhan [Vad06] who showed that honest verifier statistical, respectively computational, zero knowledge is equal to general statistical, respectively computational, zero knowledge.

1 Introduction

Zero knowledge protocols are a central concept in cryptography. These protocols allow a prover to convince a verifier about the truth of a statement without revealing any additional information about the statement, even if the verifier *cheats* by deviating from the prescribed protocol. For a nice overview of definitions and facts about zero-knowledge we refer the reader to [Gol01]. In practice, zero-knowledge protocols are used as primitives in larger cryptographic protocols in order to limit the power of malicious parties to disrupt the security of the larger protocol. For example, at the start of a secure online transaction Alice may be required to prove her identity to Bob. She does this by demonstrating that she knows a particular secret which only she is supposed to know. However, Alice wants to prevent the possibility of Bob committing identity theft, that is, Bob should not be able to masquerade as Alice later on. Thus, Bob should gain no information about Alice's secret even if he acts maliciously during the identity verification protocol.

L. Aceto et al. (Eds.): ICALP 2008, Part II, LNCS 5126, pp. 592–603, 2008.
© Springer-Verlag Berlin Heidelberg 2008

With the advent of quantum computation an important question rears its head: what happens to classical zero-knowledge protocols when the cheating verifier has access to a quantum computer? Note that even if the verifier cheats quantumly, the messages exchanged with the prover and the prover itself continue to be classical. Thus, the prover does not know if it is interacting with a classical or quantum verifier. One may expect that quantum computers can break some classical zero-knowledge protocols, i.e. a quantum verifier interacting with the prover may be able to extract information about from the message transcript (sequence of all messages exchanged) that a classical verifier cannot. As one example, the Feige-Fiat-Shamir [FFS88] zero-knowledge protocol for identity verification can be broken by a quantum computer simply because it relies on the hardness of factoring for security.

Watrous [Wat06] recently showed that two well-known classical protocols continue to be zero-knowledge against cheating quantum verifiers. In particular, he showed that the graph isomorphism protocol of Goldreich, Micali and Wigderson [GMW91] is secure, and also that the graph 3-coloring protocol in [GMW91] is secure if one can find classical commitment schemes that are concealing against quantum computers. However, the general question of which classical zero-knowledge protocols continue to be secure against cheating quantum verifiers was left open by Watrous.

In this paper, we answer this question for a large family of classical protocols. We show that all protocols that are honest verifier zero-knowledge (**HVZK**) and satisfy some reasonable assumption on their simulated transcripts can be made secure against all efficient classical and quantum machines. More specifically, any protocol which is honest verifier statistical zero-knowledge (**HVSZK**) can be transformed to be statistical zero-knowledge against all classical and quantum verifiers (**SZKQ**). Also, any protocol which is honest verifier computational zero-knowledge and has classical message transcripts of the interaction between the prover and the honest verifier that yield no information to an efficient quantum machine (**HVCZK$_Q$**), can be transformed to be computational zero knowledge against all classical and quantum verifiers (**CZKQ**). Note that classically it was shown that any language in **HVCZK** also has a protocol which is zero-knowledge against any cheating verifier (the class **CZK**).

As in the classical case, by starting with fairly weak assumption on protocols, we show that a much stronger protocol exists. Note that being zero-knowledge against quantum verifiers does not imply being zero-knowledge against classical verifiers owing to a technical requirement in the definition of zero-knowledge to be elucidated later. The significance of our result is that we give a single classical protocol zero-knowledge against both types of verifiers. Our work substantially generalizes Watrous' results [Wat06].

Formally, a protocol is said to be zero-knowledge if for every non-uniform polynomial time verifier there is a non-uniform polynomial time simulator that can produce, for inputs in the language, a simulated *view* of the verifier that is indistinguishable to the verifier's view in an actual interaction with the prover. The view of the verifier consists of the message transcript together with the internal state of the verifier, and represents what the verifier can 'learn' from interacting with the prover. The existence of a polynomial time simulator for every polynomial time verifier captures the intuition that the verifier learns nothing that it could not have learned on its own from the input, even by being malicious. For a classical verifier the simulator is required to be classical. For a quantum verifier the simulator is quantum. Thus, zero-knowledge against quantum verifiers does not immediately imply zero-knowledge against classical verifiers.

Constructing a simulator appears to be counterintuitive since it seems to replace the role of the prover who is usually assumed to be computationally unbounded whereas the simulator is polynomial time. The difference between the prover and the simulator is that the prover has to respond to verifiers queries in an 'online' fashion, that is immediately, whereas the simulator can work 'offline' and generates the messages 'out of turn', as well as 'rewind'. By rewinding, we mean a simulator runs parts of the verifier during the simulation and produces a fragment of the conversation that has some desired property with a certain probability. If the simulator fails then it rewinds, that is it just runs the part of the verifier again from scratch. In the quantum case one would have a quantum simulator using the quantum verifier to produce such a fragment of the conversation and attempting to rewind if it fails.

Protocols that are classically zero-knowledge are not necessarily zero-knowledge against quantum verifiers. In the case of the two problems graph isomorphism and graph 3-coloring that Watrous [Wat06] studied, the essential difference between classical and quantum simulators comes from one additional requirement of zero-knowledge protocols. In order for zero-knowledge protocols to sequentially compose, which is essential to achieve reasonable error parameters as well as ensure the security of the protocol when used as part of a larger cryptographic system, the simulator must still work when the simulators and verifiers are given an arbitrary *auxiliary* state. This is a natural requirement if one considers that, for example, perhaps the verifier has interacted with the prover already to compute some intermediate information modeled by the auxiliary state, and now during the next interaction it gains even more information. In the quantum case the auxiliary state is an unknown *quantum* state. But unknown quantum states cannot be copied, and measurements of unknown quantum states are irreversible operations in general, and as pointed out by Watrous [Wat06], even determining if the simulator was successful in producing a fragment of the conversation with the desired property may destroy the state. Therefore the simulator cannot trivially rewind since it cannot feed the auxiliary state into the verifier a second time if the state was destroyed during the first attempt at simulation. Nevertheless, Watrous [Wat06] showed that it is possible to quantumly rewind in a clever way in the case of Goldreich, Micali and Wigderson's [GMW91] classical zero-knowledge protocols for graph isomorphism and graph 3-coloring.

When searching for more classical zero-knowledge protocols that are secure against quantum cheating verifiers we come across new difficulties not encountered by Watrous [Wat06]. One restriction of the protocols he analyzes is that they are three-round public coin protocols where the second message is $O(\log n)$ uniformly random bits from the verifier. This leaves out many languages in **SZK** and **CZK** including the complete problems *statistical difference* [SV03] and *entropy difference* [GV97] for **SZK**. In a different vein [Wat02, Wat06], Watrous showed that every problem in **SZK** has a *quantum* protocol that is statistical zero-knowledge against any cheating non-uniform polynomial time quantum verifier. Very recently, Kobayashi [Kob08] extended Watrous' result to the case of quantum protocols that are quantum computationally zero knowledge. However, it is preferable that the prescribed protocols themselves are classical since they can be implemented using current technology yet remain secure against all potential quantum attacks in the future. In this paper, we show that a large class of polynomial round, polynomial verifier message length classical zero-knowledge protocols can be made secure against cheating quantum verifiers.

Classically, the construction of zero-knowledge protocols has been greatly simplified by showing that **HVSZK** or **HVCZK** is equal to **SZK** or **CZK** [GSV98, Vad06]. Concretely, if one can design a protocol for a given language that is zero-knowledge against (only) the honest verifier, which is typically much easier, then there is also a protocol for the language that is zero-knowledge against an arbitrary cheating verifier. We follow this approach: we show that if one can find a classical protocol zero-knowledge for just the honest (classical!) verifier such that the actual and simulated message transcripts with respect to the honest verifier are indistinguishable by polynomial sized quantum circuits, then there is also a classical protocol that is zero-knowledge against all classical and quantum cheating verifiers. More precisely, our result can be stated as:

Result 1

1. **SZK** = **HVSZK** = **SZKQ**, *where* **SZKQ** *is the class of languages with a classical protocol that is statistical zero knowledge against all classical and quantum verifiers.*

2. **HVCZK$_Q$** = **CZKQ** = **CZK$_Q$**, *Where* **HVCZK$_Q$** *(resp.* **CZK$_Q$**) *is the class of languages with a classical protocol that is honest verifier computational zero-knowledge (resp. computational zero-knowledge) and for YES instances, the classical message transcripts of the interaction between the prover and the honest verifier are quantum computationally indistinguishable from the simulated message transcripts. Similarly,* **CZKQ** *is the class of languages with a classical protocol that is computational zero knowledge against all classical and quantum verifiers.*

We note that the classical results **HVSZK** = **SZK** and **HVCZK** = **CZK** are known and can be found in Goldreich, Sahai and Vadhan [GSV98] and Vadhan [Vad06]. Also, observe that **HVSZK** ⊆ **HVCZK$_Q$** ⊆ **HVCZK**.

Finally, we would like to remark that the definition of zero knowledge in quantum computation in the literature assumes that we can do error-free computation. Constructing a simulator for a cheating verifier typically involves a polynomial multiplicative factor overhead. Thus in reality, it may happen that a simulator fails to successfully simulate the cheating verifier's view because of additional noise incurred by the overheads. However, if we take the view that noise rates in hardware can be decreased by polynomial factors with polynomial effort, the current definition of zero knowledge in quantum computation is justified.

1.1 Overview of Our Proof: Ideas and Difficulties

Damgård, Goldreich and Wigderson [DGW94] gave a method, hereafter called DGW, for transforming any classical constant round public coin honest verifier zero knowledge protocol into another classical constant round public coin protocol that is zero knowledge against all classical verifiers. We first observe that Watrous' quantum rewinding trick [Wat06] can be used to show that the new protocol resulting from DGW is secure against all quantum verifiers also. This allows us to handle protocols with verifier messages of polynomial length. The shortcoming is that, as in the classical case, the quantum simulator succeeds in almost correctly simulating the prover-verifier interaction with non-negligible probability only if the original protocol has a constant number of rounds. This arises from the fact that the classical and quantum simulators from DGW 'rewind from scratch', that is, they attempt to simulate all the rounds of the protocol in one shot, and if they fail, they rewind the verifier to the beginning of the protocol. The success probability of one attempt at simulation drops exponentially in the number of rounds, and hence, we can only handle a constant number of rounds using the DGW transformation.

Building on Damgård et al.'s work, Goldreich, Sahai and Vadhan [GSV98] gave a method, hereafter called GSV, for transforming any classical public-coin **HVZK** protocol into another public-coin protocol **ZK** against all classical verifiers. Their transformation handles protocols with a polynomial number of rounds. However, one cannot apply Watrous' quantum rewinding technique [Wat06] to the new protocol resulting from GSV for the following technical reason: the simulator for the new protocol rewinds the new verifier polynomial number of times for each round. In order to do the same thing quantumly using Watrous' rewinding lemma, one needs that for most messages of the verifier in the original protocol, the success probability of the simulation attempt conditioned on the old verifier's message be independent of the quantum auxiliary state. Unfortunately this cannot be ensured for any message of the verifier in the original protocol, and hence, we are unable to show that GSV makes the protocol secure against cheating quantum verifiers.

Our crucial observation is that if the honest-verifier simulator for the original classical public coin **ZK** protocol uses its internal randomness in a *stage-by-stage* fashion, where each stage consists of a constant number of rounds, then applying DGW gives a new protocol which is zero-knowledge against all classical and quantum verifiers. This is still the case even the original protocol has a polynomial number of rounds. This is because now the classical or quantum simulator for the new protocol can rewind the verifier polynomial number of times within each stage, where each iteration preserves the simulated message transcript of the earlier rounds and uses fresh random coins to attempt to simulate the current round. Since the success probability of one simulation attempt for a stage is inverse polynomial as it has a constant number of rounds, polynomially many rewinding steps will result in a successful simulation of the current stage with very high probability. This leads us to the question of which problems possess zero-knowledge protocols with stage-by-stage honest-verifier simulators.

Our next observation is that the standard technique of converting any public coin interactive protocol into a zero-knowledge protocol [IY88, BGG+90] based on bit commitments actually gives rise to a new protocol with a stage-by-stage honest verifier simulator. Note that any interactive protocol can be converted into a public coin protocol [GS89] where the messages of the verifier are uniformly distributed random strings independent of the previous messages of the protocol, and the final decision of the verifier to accept or reject is a deterministic function of the message transcript and the input. The only caveat is that the existence of bit commitment schemes seems to be conditional on the existence of one-way functions. However, the recent work of Vadhan [Vad06], Nguyen and Vadhan [NV06] and Ong and Vadhan [OV08] gives a way of replacing standard bit commitments by instance-dependent bit commitments, which exist unconditionally as shown by them. An instance-dependent bit commitment scheme is a protocol which depends on the input instance to the problem such that the protocol is hiding on the bit to be committed for positive instances of the problem and binding on the bit for negative instances of the problem. Since the hiding and binding properties are not required to hold simultaneously, the need for unproven assumptions like the existence of one-way functions is avoided. Ong and Vadhan [OV08] show that every problem with an honest verifier zero-knowledge protocol gives rise to a public coin constant round instance dependent bit commitment scheme which is statistically binding on the negative instances. For positive instances, the hiding property of the commitment scheme is statistical if the original protocol is **HVSZK**, and computational against polynomial sized classical circuits if the original protocol is **HVCZK**. We can show that their proofs can be modified to ensure that the hiding property is computational against polynomial sized quantum circuits if the original classical protocol is in **HVCZK$_\mathbf{Q}$**. Replacing the bit commitments in the standard compilation of interactive proofs to zero-knowledge by instance dependent commitments gives us a zero-knowledge protocol with an honest-verifier simulator that uses its internal randomness in a stage-by-stage fashion, where each stage consists of a constant number of rounds. Applying the DGW transformation to such a protocol gives rise to a new public coin classical protocol zero-knowledge against all non-uniform polynomial time classical and quantum verifiers. That fact follows since the success probability of correctly simulating a stage in the new protocol continues to be inverse polynomial and also the simulator for the new protocol can rewind in a stage-by-stage fashion.

2 Preliminaries

2.1 The DGW Transformation

We denote a classical N-round public coin interactive protocol by the notation (P, V) : $(\alpha_1, \beta_1, ..., \alpha_N, \beta_N)$, which means that in the round i, the (honest) classical verifier V sends a uniformly random string α_i and the (honest) classical prover P responds with a string β_i, which in general is a function of the previous transcript and the prover's randomness. Without loss of generality, each α_i has the same length s. Let $t < s$ be a positive integer. Damgård, Goldreich and Wigderson [DGW94] describe a family $\mathcal{F}_{s,t}$ of nearly s-wise independent hash functions from $\{0, 1\}^s$ to $\{0, 1\}^t$. Every function $f \in \mathcal{F}_{s,t}$ has a description of length s^2 bits and for all $y \in \{0, 1\}^t$, $1 \le |f^{-1}(y)| \le (s-1)2^{s-t} + 1$, where $f^{-1}(y) := \{x \in \{0, 1\}^s : f(x) = y\}$. Computing $f^{-1}(y)$ can be done in randomized time polynomial in s and 2^{s-t}. In DGW, $s - t$ is taken to be logarithmic in the input length, so 2^{s-t} will be a polynomial in the input length. Using this family $\mathcal{F}_{s,t}$, Damgård et al. describe a process to transform a random message $\alpha \in_R \{0, 1\}^s$ from the verifier in the original protocol, giving rise to a new protocol with twice as many messages.

1. The verifier chooses f uniformly in $\mathcal{F}_{s,t}$ and sends it to the prover.
2. The prover chooses y uniformly in $\{0, 1\}^t$ and sends it to the verifier.
3. The verifier chooses α uniformly in $f^{-1}(y)$ and sends it to the prover.

As described, the second message of the verifier in the DGW transformation is not public coin. However, it can be made public coin by letting the verifier send a random $r \in ((s-1)2^{s-t}+1)!$, which the prover interprets as the $(r \bmod |f^{-1}(y)|)$th element of $f^{-1}(y)$. Note that since $(s-1)2^{s-t}+1$ is polynomial in the input size, r can be described using polynomially many bits. Henceforth, we shall assume that the new protocol arising from the application of DGW is public coin but we shall continue to use the description of DGW given above for simplicity.

Applying DGW to an N-round public coin protocol $(\alpha_1, \beta_1, ..., \alpha_N, \beta_N)$ gives a new public coin protocol $(f_1, y_1, \alpha_1, \beta_1, ..., f_N, y_N, \alpha_N, \beta_N)$ where each β_i is obtained in the same way as the original prover does on seeing the previous $(\alpha_1, ..., \alpha_i)$. The DGW transformation satisfies the following soundness and completeness property which we will crucially use [DGW94].

Fact 1. *Suppose the original N-round public coin protocol has perfect completeness and soundness error ϵ_0, then the DGW transformation gives a new public coin protocol with perfect completeness and soundness error $\epsilon_1 = \epsilon_0 + N(2s2^{(t-s)/4} + 2^{-s})$.*

The zero knowledge properties of DGW will be the main topic of discussion in the later sections of this paper.

2.2 Stage-by-Stage Simulator

We now give the formal definition of the important notion of an interactive protocol possessing a 'stage-by-stage' honest-verifier simulator, which is central to our work.

Definition 1. *Suppose (P, V) is a classical public coin protocol with N stages, each stage i containing constant number c of rounds $(\alpha_{i1}, \beta_{i1}, ..., \alpha_{ic}, \beta_{ic})$, where α_{ij}, β_{ij} are verifier's, respectively prover's messages and all α_{ij}s are of the same length. We say that an honest-verifier simulator M is* stage-by-stage *if its internal random string r can be decomposed as $r = r_1 \circ \cdots \circ r_N$, $r_1, ..., r_N$ uniform and independent random variables, such that in each stage i, the simulated messages $(\hat{\beta}_{i1}, ..., \hat{\beta}_{ic})$ are functions of $r_1, ..., r_i$ and the input alone, and $(\hat{\alpha}_{i1}, ..., \hat{\alpha}_{ic})$ is a function of r_i alone.*

A public coin constant round protocol can be trivially considered to be a stage-by-stage with only one stage. Note that we do not assume anything about how the simulator uses its randomness in each stage; it can be used arbitrarily. But since each stage only contains a constant number of rounds, rewinding to the beginning of the stage is affordable while simulating the new protocol arising from the application of DGW.

2.3 Instance-Dependent Bit Commitments

We recall the definition of *instance-dependent bit commitment* protocols [OV08] which will be used in our construction of interactive protocols with honest-verifier stage-by-stage simulators. Below, by an *exponentially small* function $\epsilon(n)$ we mean a function of a positive parameter n that grows smaller than 2^{-n^c} for some fixed $c > 0$. By the *total variation distance*, also known as *statistical distance*, between two probability distributions P, Q on the same sample space, we mean the ℓ_1-distance $\|P - Q\| = \sum_i |P(i) - Q(i)|$.

Definition 2. *For a promise problem $\Pi = (\Pi_Y, \Pi_N)$, a classical public coin constant round instance-dependent bit commitment scheme consists of a classical public coin interactive protocol Com_x for every $x \in \Pi_Y \cup \Pi_N$ between two parties called* sender S_x *and receiver R_x, with the following properties:*

1. *Protocol Com_x has two stages, a* commit *stage and a* reveal *stage;*
2. *At the beginning of the commit stage, S_x gets a private input $b \in \{0, 1\}$ which represents the bit he has to commit to. The commit stage proceeds for a constant number of rounds, and its transcript $c_{x;b}$ is defined to be the* commitment to the *bit b;*
3. *Later on, in the reveal stage, S_x reveals the bit b and sends another string $d_{x;b}$ called the* decommitment *string for b. The receiver R_x accepts or rejects deterministically based on $c_{x;b}$, b and $d_{x;b}$.*
4. *Sender S_x and receiver R_x can be implemented in randomized time polynomial in $|x|$;*
5. *For all $x \in \Pi_Y \cup \Pi_N$, for all $b \in \{0, 1\}$, R_x accepts with probability 1 if both S_x and R_x follow the prescribed protocol;*

The scheme Com_x is said to be exponentially binding statistically *for all $x \in \Pi_N$, if for any sender S_x^*, there exists an exponentially small function $\epsilon(\cdot)$ such that if c_x^* denotes the commitment obtained by the interaction of S_x^* and the honest R_x, the probability that there exist decommitment strings $d_{x;0}^*$, $d_{x;1}^*$ in the reveal stage so that R_x accepts on c_x^*, 0, $d_{x;0}^*$ as well as c_x^*, 1, $d_{x;1}^*$ is less than $\epsilon(|x|)$. The binding property is required to hold for malicious senders too who do not follow the prescribed protocol. In addition, the scheme Com_x is said to be* exponentially hiding statistically *for all $x \in \Pi_Y$ if the views of the honest receiver R_x when $b = 0$ and $b = 1$ have exponentially small total variation distance. Similarly, if the two views are negligibly distinguishable by polynomial sized classical or quantum circuits, the scheme Com_x is said to be* computationally, *respectively* quantum computationally, *hiding.*

Remark: Observe that we only require the hiding property to hold for the honest receiver R_x in the above definition. The reason for this is as follows. As mentioned earlier in the introduction, our initial aim is only to get a protocol with a stage-by-stage honest verifier simulator. We will then make that protocol resilient against all malicious verifiers by applying the DGW transformation. The hiding property of the commitment scheme against the honest receiver translates to zero knowledge against the honest verifier in Proposition 1, where we show how to achieve our initial aim.

3   Applying DGW to Protocols with Stage-by-Stage Simulators

In this section, we will show that applying the DGW transformation to a classical public coin interactive protocol with a stage-by-stage honest verifier simulator results in a classical public coin protocol zero-knowledge against all non-uniform polynomial time classical and quantum verifiers.

Lemma 1. *If a classical public-coin protocol P has a stage-by-stage honest-verifier simulator M such that the simulated transcript is quantum computationally indistinguishable from the actual prover honest-verifier interaction, then applying DGW to it gives a new classical public coin protocol P' with inverse polynomially larger soundess error that is computationally zero-knowledge against all non-uniform polynomial time classical and quantum verifiers. If in addition P is statistical zero knowledge against the honest verifier, P' is statistically zero knowledge against all non-uniform polynomial time classical and quantum verifiers.*

Proof. **(Sketch)** The claim about soundness error follows from Fact 1 with an appropriate setting of the parameters of the DGW transformation. The zero-knowledge property crucially relies on the stage-by-stage assumption and the zero-knowledge property of DGW. Below we sketch the main points of difference from the standard classical setting.

First, the classical proof attempts to simulate all the rounds of the protocol failing which it rewinds from scratch. Here, we do a stage-by-stage simulation, that is, we try to simulate all the rounds of one stage failing which we rewind to the beginning of the stage only. The stage-by-stage property of the honest-verifier simulator M allows us to do this, since rewinding to the beginning of stage i just means tossing a fresh coin r_i without disturbing the earlier coin tosses $r_1, \ldots, r_{i-1}$. Since each stage consists of only a constant number of rounds, the success probability of one attempt at simulating DGW on a stage is inverse polynomial. Thus polynomially many rewinding steps for a stage suffices to simulate the stage successfully with very high probability. After successfully simulating a stage, we can proceed to simulating the next stage, and so on for polynomially many stages.

The second point of difference is that in the proof of security against quantum verifiers, we use Watrous' rewinding technique [Wat06] at the end of a stage. The reason this is possible is because the DGW transformation ensures that the success probability of one attempt at simulation of a stage is independent of the quantum auxiliary input. Combined with the observation above that the probability of successfully simulating a stage is inverse polynomial, this allows us to rewind a stage polynomially many times quantumly without disturbing previous stages and ensure a successful simulation with very high probability. □

A more formal proof of the classical and quantum parts of the above lemma is given in the appendix.

4 Designing Protocols with Stage-by-Stage Simulators

In this section, we indicate how to design a classical public coin interactive protocol for any promise problem in **HVSZK** and **HVCZK$_Q$** with perfect completeness, exponentially small soundness and possessing a stage-by-stage honest-verifier simulator. For problems in **HVSZK** the simulated transcript will be exponentially close in total variation distance to the actual transcript, and for problems in **HVCZK$_Q$** the two transcripts will be negligibly distinguishable against polynomial sized quantum circuits.

The following statement follows by modifying the arguments of Vadhan [Vad06]. But first, we have to define the notion of a *quantumly secure false entropy generator* which is the natural quantum generalization of a so-called false entropy generator [HILL99].

Definition 3. *Let $I \subseteq \{0,1\}^*$, and $m(\cdot)$ be a polynomial function. For $x \in I$, a family D_x of probability distributions on $\{0,1\}^{m(|x|)}$ is said to be P-sampleable if there exists a probabilistic polynomial time algorithm whose output is distributed according to D_x on input x. A P-sampleable family D_x is said to be a* quantumly secure false entropy generator *if there exists a family F_x of probability distributions on $\{0,1\}^{m(|x|)}$ that is negligibly distinguishable from D_x by polynomial sized quantum circuits such that $H(F_x) \geq H(D_x) + 1$, where $H(\cdot)$ is the Shannon entropy of a probability distribution.*

Lemma 2. *Suppose $\Pi = (\Pi_Y, \Pi_N)$ is a promise problem in $\mathbf{HVCZK_Q}$. Then there is a family $\{D_x\}_{x \in \Pi_Y \cup \Pi_N}$ of P-sampleable probability distributions on $\{0,1\}^{m(|x|)}$, and a subset $I \subseteq \Pi_Y$ such that $\{D_y\}_{y \in I}$ is a quantumly secure false entropy generator. Also, $(\Pi_Y \setminus I, \Pi_N) \in \mathbf{HVSZK}$.*

Proof. **(Sketch)** The proof follows by observing that the arguments of [Vad06] go through equally well for quantum indistinguishability as for classical indistinguishability. Essentially, this is because the proof of [Vad06] uses reducibility arguments where the computational hardness of a primitive is used as a black box. A more detailed proof is left for the full version of the paper. $\square$

We need the following result about the existence of classical public coin constant round instance dependent bit commitment protocols for problems in $\mathbf{HVSZK}$ by Ong and Vadhan [OV08].

Fact 2. *Every promise problem in $\mathbf{HVSZK}$ gives rise to a classical constant round public coin instance dependent bit commitment scheme that is exponentially hiding on the positive instances and exponentially binding on the negative instances statistically.*

Remark: In fact for our purposes, we do not really require the full strength of the above fact. A weaker primitive of classical constant round public coin instance-dependent *two-phase* bit commitment scheme that is statistically hiding on the positive instances and statistically 1-*out-of*-2 binding on the negative instances suffices for us. Such schemes were first constructed by Nguyen and Vadhan [NV06]. However, our construction of an interactive protocol with a stage-by-stage honest-verifier simulator is more complicated if we use 1-out-of-2 binding schemes. Hence, we use the stronger scheme of the above fact in our proof.

Finally, we need the following statement which follows by modifying the arguments of Håstad, Impagliazzo, Levin and Luby [HILL99], and Naor [Nao91].

Lemma 3. *Let $I \subseteq J \subseteq \{0,1\}^*$. Suppose D_x, $x \in J$ is a P-sampleable family of probability distributions on $\{0,1\}^{m(x)}$. Also, suppose D_x, $x \in I$ is a quantumly secure false entropy generator. Then there is a classical constant round public coin instance-dependent bit commitment scheme for all $x \in J$ which is exponentially binding statistically for all $x \in J$ and quantum computationally hiding for all $x \in I$.*

Proof. (**Sketch**) Same reasoning as in the proof of Lemma 2.   □

By combining Lemmas 2 and 3, and Fact 2, and using the techniques of Vadhan [Vad06], we can conclude the following quantum analogue of results of Ong and Vadhan [OV08].

Lemma 4. *Every promise problem in* $\mathbf{HVCZK_Q}$ *gives rise to a classical constant round public coin instance dependent bit commitment scheme that is quantum computationally hiding on the positive instances and exponentially binding statistically on the negative instances.*

We are now finally in a position to show that every problem in $\mathbf{HVCZK_Q}$ has a classical public coin interactive protocol with a stage-by-stage honest verifier simulator. For the classical counterparts of the proposition below, we refer the reader to Ong and Vadhan [OV08].

Proposition 1. *Every promise problem* $\Pi = (\Pi_Y, \Pi_N)$ *in* $\mathbf{HVCZK_Q}$ *has a classical public coin interactive protocol with perfect completeness, exponentially small soundness and a stage-by-stage honest-verifier simulator that produces simulated transcripts that are negligibly quantum computationally distinguishable from the actual prover honest-verifier interaction transcripts. Furthermore if* $\Pi \in \mathbf{HVSZK}$, *then the resulting protocol is constant round and the simulated transcripts are exponentially close in total variation distance from the actual transcripts.*

A proof sketch can be found in the appendix.

Combining Lemma 1 together with Proposition 1, we prove the main theorem of the paper.

Theorem 1. $\mathbf{HVCZK_Q} \subseteq \mathbf{CZKQ}$ *and* $\mathbf{HVSZK} \subseteq \mathbf{SZKQ}$.

Acknowledgments

We are grateful to an anonymous referee of an earlier version of this paper for detecting a subtle bug in that version, and also for informing us about the recent work of Ong and Vadhan [OV08] on zero-knowledge. We thank Shien Jin Ong and Salil Vadhan for clarifying many doubts about instance-dependent bit commitments and zero-knowledge. We also thank the anonymous referees of this version for helpful comments. S.Z. thanks Iordanis Kerenidis, Manoj Prabhakaran, Ben Reichardt, Amit Sahai, Yaoyun Shi, Robert Spalek, Shanghua Teng, Umesh Vazirani and Andy Yao for listening to the progress of the work, clarifying things and giving interesting comments. P.S. thanks Jaikumar Radhakrishnan and Thomas Vidick for helpful feedback. All authors thank Wei Huang and Martin Rötteler for discussions at an early stage of the work.

References

[BGG+90]   Ben-Or, M., Goldreich, O., Goldwasser, S., Håstad, J., Kilian, J., Micali, S., Rogaway, P.: Every provable is provable in zero-knowledge. In: Goldwasser, S. (ed.) CRYPTO 1988. LNCS, vol. 403, pp. 37–56. Springer, Heidelberg (1990)

[DGW94]   Damgård, I., Goldreich, O., Wigderson, A.: Hashing functions can simplify zero-knowledge protocol design (too). Technical Report RS-94-39, BRICS (1994)

[FFS88]   Feige, U., Fiat, A., Shamir, A.: Zero-knowledge proofs of identity. Journal of Cryptology 1(2), 77–94 (1988)

[GMW91]   Goldreich, O., Micali, S., Widgerson, A.: Proofs that yield nothing but their validity or all languages in NP have zero-knowledge proof systems. Journal of the ACM 38(1), 691–729 (1991)

[Gol01]   Goldreich, O.: Foundations of cryptography, vol. 1. Cambridge University Press, Cambridge (2001)

[GS89]   Goldwasser, S., Sipser, M.: Private coins versus public coins in interactive proof systems. Advances in Computing Research, vol. 5, pp. 73–90. JAC Press, Inc. (1989)

[GSV98]   Goldreich, O., Sahai, A., Vadhan, S.: Honest-verifier statistical zero-knowledge equals general statistical zero-knowledge. In: Proceedings of the 30th Annual ACM Symposium on Theory of Computing, pp. 399–408 (1998)

[GV97]   Goldreich, O., Vadhan, S.: Comparing entropies in statistical zero knowledge with applications to the structure of SZK. In: Proceedings of the 14th Annual IEEE Symposium on Foundations of Computer Science, pp. 448–457 (1997)

[HILL99]   Håstad, J., Impagliazzo, R., Levin, L., Luby, M.: A pseudorandom generator from any one-way function. SIAM Journal on Computing 28(4), 1364–1396 (1999)

[IY88]   Impagliazzo, R., Yung, M.: Direct zero-knowledge computations. In: Pomerance, C. (ed.) CRYPTO 1987. LNCS, vol. 293, pp. 40–51. Springer, Heidelberg (1988)

[Kob08]   Kobayashi, H.: General properties of quantum zero-knowledge proofs. In: Proceedings of the 5th Theory of Cryptography Conference, pp. 107–124 (2008), Also quant-ph/0705.1129

[Nao91]   Naor, M.: Bit commitment using pseudorandom generator. Journal of Cryptology 4, 151–158 (1991)

[NV06]   Nguyen, M.-H., Vadhan, S.: Zero knowledge with efficient provers. In: Proceedings of the 38th Annual ACM Symposium on Theory of Computing, pp. 287–295 (2006)

[OV08]   Ong, S., Vadhan, S.: An equivalence between zero knowledge and commitments. In: Proceedings of the 5th Theory of Cryptography Conference (to appear, 2008)

[SV03]   Sahai, A., Vadhan, S.: A complete promise problem for statistical zero-knowledge. Journal of the ACM 50(2), 196–249 (2003)

[Vad06]   Vadhan, S.: An unconditional study of computational zero knowledge. SIAM Journal on Computing 36(4), 1160–1214 (2006)

[Wat02]   Watrous, J.: Limits on the power of quantum statistical zero-knowledge. In: Proceedings of the 43rd Annual IEEE Symposium on Foundations of Computer Science, pp. 459–468 (2002)

[Wat06]   Watrous, J.: Zero-knowledge against quantum attacks. In: Proceedings of the 38th Annual ACM Symposium on Theory of Computing, pp. 296–305 (2006)

Composable Security in the Bounded-Quantum-Storage Model

Stephanie Wehner[1] and Jürg Wullschleger[2]

[1] California Institute of Technology, IQI, Pasadena CA 91125, USA
[2] University of Bristol, University Walk, Bristol BS8 1TW, United Kingdom

Abstract. We give a new, simulation-based, definition for security in the bounded-quantum-storage model, and show that this definition allows for sequential composition of protocols. Damgård *et al.* (FOCS '05, CRYPTO '07) showed how to securely implement bit commitment and oblivious transfer in the bounded-quantum-storage model, where the adversary is only allowed to store a limited number of qubits. However, their security definitions did only apply to the standalone setting, and it was not clear if their protocols could be composed. Indeed, we show that these protocols are *not* composable in our framework without a small refinement. We then prove the security of their randomized oblivious transfer protocol with our refinement. Secure implementations of oblivious transfer and bit commitment follow easily by a (classical) reduction to randomized oblivious transfer.

1 Introduction

Secure two-party computation [1] allows two mutually distrustful players to jointly compute the value of a function without revealing more information about their inputs than can be inferred from the function value itself. The primitive known as oblivious transfer (OT) [2,3,4] is thereby of particular importance: *any* two-party computation can be implemented, if this primitive is available [5,6]. Another important primitive in this context is *bit commitment* (BC) [7]. But since bit commitment can be implemented from oblivious transfer, a direct implementation of bit commitment is only important if we cannot implement oblivious transfer itself, or if we want to improve efficiency. In oblivious transfer, the sender (Alice) chooses two bits x_0 and x_1, the receiver (Bob) chooses a bit c. The protocol of oblivious transfer allows Bob to retrieve x_c in such a way that Alice cannot gain any information about c. At the same time, Alice can be ensured that Bob only retrieves x_c, but no information about x_{1-c}.

Unfortunately, BC and OT are impossible to implement securely without any additional assumptions, even in the quantum model [8,9]. This result holds even in the presence of the so-called superselection rules [10]. Exact trade-offs on how well we can implement BC in the quantum world can be found in [11]. To circumvent this problem (classically and quantumly), we thus need to assume that the adversary is limited. In the classical case, one such limiting assumption is that the adversary is *computationally bounded*. In the quantum model, it is

L. Aceto et al. (Eds.): ICALP 2008, Part II, LNCS 5126, pp. 604–615, 2008.

also possible to securely implement both protocols provided that an adversary cannot measure more than a fixed number of qubits simultaneously [12]. String commitments can be obtained with very weak security parameters [13].

The Bounded-Quantum-Storage Model. In the quantum case, it is *very* difficult to store states even for a very short period of time. This leads to the protocol presented in [14,15], which show how to implement BC and OT if the adversary is not able to store *any* qubits at all. In [16,17], these ideas have been generalized to the *bounded-quantum-storage model*, where the adversary is computationally unbounded and allowed to have an unlimited amount of *classical* memory. However, he is only allowed a limited amount of *quantum* memory. The honest players do not require any quantum storage at all, making the protocols implementable using present day technology.

Security Definitions and Composability. As cryptographic protocols are almost never executed on their own, it is important that they remain secure when they are composed. [18,19,20] introduced *simulation-based* security definitions and showed that they can be composed *sequentially*, i.e, at any point in time at most one protocol is running. A stronger security definition called *universal composability* has been introduced in [21,22,23]. It guarantees that protocols can be securely composed in an arbitrary way (also concurrently) in any environment.

Based on earlier an earlier definition of security in the quantum setting [24], a simulation-based security definition has been presented in [25], however no composability theorem was proven. Universal composability in the quantum world has been introduced in [26], and independently in [27]. In [28], it has been shown that classical protocols are universally composable using their *classical* definitions, are secure against *quantum* adversaries.

1.1   Contribution

In [17], protocols for OT and BC have been presented and shown to be secure against adversaries who have bounded quantum storage. However, the proofs only guarantee security in a standalone setting, and it was not clear whether these protocols remain secure when they are composed with other protocols. Indeed, the following simple example shows that in some situations, the protocols presented in [16,17] do not guarantee security in a strong sense. (However, Fehr and Schaffner [29] recently showed that the original definitions still allow for some weak form of composability.) Suppose the adversary receives a large number of halves of EPR-pairs from the environment as his auxiliary input. He can then effectively enlarge his own quantum memory by teleporting quantum states to the environment, which has unlimited memory. The classical communication needed to teleport can be part of the adversary's classical storage that he later outputs. In the case of the protocol presented in [17] (where the security depends on the fact that the adversary does not know in which basis to measure before his quantum memory bound is applied) this allows the environment to distinguish easily between the real and the ideal setting.

We present a formal model for secure two-party computation in the bounded-quantum-storage model and show that our model implies that secure protocols are sequentially composable. Then, we slightly modify the protocol for randomized OT presented [17] by introducing a second memory bound and prove the security of the protocol in our model.

In the full version of this work, we give well-known *classical* reductions of BC and OT to randomized OT. An important consequence is that *any* secure function evaluation can be achieved in the bounded-quantum-storage model. This follows from the fact that the proof of [28] carries over to our model, which means that any classical protocol that is secure in the classical universal composability model is also secure in our model. Therefore, we can use the protocol from [30] (based on [6]) to implement any secure function evaluation [1].

2   Preliminaries

We use the term *computational basis* to refer to the basis given by $\{|0\rangle, |1\rangle\}$. We write $+$ for the computational basis, and let $|0\rangle_+ = |0\rangle$ and $|1\rangle_+ = |1\rangle$. The *Hadamard basis* is denoted by $\times$, and given by $\{|0\rangle_\times, |1\rangle_\times\}$, where $|0\rangle_\times = (|0\rangle + |1\rangle)/\sqrt{2}$ and $|1\rangle_\times = (|0\rangle - |1\rangle)/\sqrt{2}$. For a string $x \in \{0,1\}^n$ encoded in bases $b \in \{+, \times\}^n$, we write $|x\rangle_b = |x_1\rangle_{b_1}, \ldots, |x_n\rangle_{b_n}$. We also use 0 to denote $+$, and 1 to denote $\times$. Finally, we use $x_{|c}$ to denote the sub-string of an encoded string x consisting of all x_i where $b_i = c$.

We use the font $\mathcal{A}$ to label a quantum register, corresponding to a Hilbert space $\mathcal{A}$. A *quantum channel* from $\mathcal{A}$ to $\mathcal{B}$ is a completely positive trace preserving (CPTP) map $\Lambda : \mathcal{A} \to \mathcal{B}$. We also call a map from $\mathcal{A}$ to itself a *quantum operation*. Any quantum operation on the register $\mathcal{A}$ can be phrased as a unitary operation on $\mathcal{A}$ and an additional ancilla register $\mathcal{A}'$, where we trace out $\mathcal{A}'$ to obtain the actions of the quantum operation on register $\mathcal{A}$ [32]. We use $\mathbb{S}(\mathcal{A})$ to refer to the set of all quantum states in $\mathcal{A}$, and $\mathbb{T}(\mathcal{A})$ to refer to the set of all Hermitian matrices in $\mathcal{A}$. We use $\mathbf{U}$ to refer to a quantum operation, upper case letters X to refer to classical random variables, the font $\mathbb{S}$ for a set, and the font A to refer to a player in the protocol.

Our ability to distinguish two quantum states $\rho, \rho' \in \mathbb{S}(\mathcal{H})$ is determined by their *trace distance* defined as $D(\rho, \rho') := \frac{1}{2} \operatorname{Tr} |\rho - \rho'|$, where $|A| = \sqrt{A^\dagger A}$. The triangle inequality holds. I.e., for all ρ, ρ' and ρ'', we have $D(\rho, \rho'') \leq D(\rho, \rho') + D(\rho', \rho'')$. We also write $\rho \equiv_\varepsilon \rho'$, if $D(\rho, \rho') \leq \varepsilon$. For all practical purposes, $\rho \equiv_\varepsilon \rho'$ means that the state ρ' behaves like the state ρ, except with probability ε [33]. For any quantum channel Λ, we have $D(\Lambda(\rho), \Lambda(\rho')) \leq D(\rho, \rho')$. Let $\rho_{AB} \in \mathbb{S}(\mathcal{A} \otimes \mathcal{B})$ be classical on $\mathcal{A}$, i.e. $\rho_{AB} = \sum_{x \in \mathcal{X}} P_X(x)|x\rangle\langle x| \otimes \rho_x$ for some distribution P_X over a finite set $\mathcal{X}$. We say that A *is ε-close to uniform with respect to B*, if $D(\rho_{AB}, \mathbb{I}_A/d \otimes \rho_B) \leq \varepsilon$, where $d = \dim(\mathcal{H}_A)$.

For random variables X and Y with joint distribution P_{XY}, the *smooth conditional min-entropy* [34] can be expressed in terms of an optimization over events

---

[1] Note that because our implementation of OT is physical, the results presented in [31] cannot be applied, as explained in [30] on page 11.

$\mathcal{E}$ occurring with probability at least $1 - \varepsilon$. Let $P_{X\mathcal{E}|Y=y}(x)$ be the probability that $\{X = x\}$ *and* $\mathcal{E}$ occur conditioned on $Y = y$. We have $H_{\min}^{\varepsilon}(X|Y) = \max_{\mathcal{E}:\Pr(\mathcal{E})\geq 1-\varepsilon} \min_y \min_x (-\log P_{X\mathcal{E}|Y=y}(x))$. The smooth min-entropy allows us to use the following chain rule.

Lemma 1 (Chain Rule [34]). *For all random variables X, Y, Z and for all $\varepsilon, \varepsilon' > 0$, $H_{\min}^{\varepsilon+\varepsilon'}(X|YZ) \geq H_{\min}^{\varepsilon}(XY \mid Z) - \log|\mathbb{Y}| - \log(1/\varepsilon')$.*

We also need the monotonicity of the smooth min-entropy, $H_{\min}^{\varepsilon}(XY \mid Z) \geq H_{\min}^{\varepsilon}(X \mid Z)$. A function $h : \mathbb{S} \times \mathbb{X} \to \{0,1\}^{\ell}$ is called a *two-universal hash function* [35], if for all $x_0 \neq x_1 \in \mathbb{X}$, we have $\Pr[h(S,x_0) = h(S,x_1)] \leq 2^{-\ell}$ if S is uniform over $\mathbb{S}$. We thereby say that a random variable S is *uniform over* a set $\mathbb{S}$ if S is chosen from $\mathbb{S}$ according to the uniform distribution. The following theorem is from [17], stated slightly differently than in [33,36].

Theorem 1 (Privacy Amplification [33,36]). *Let X and Z be (classical) random variables distributed over $\mathbb{X}$ and $\mathbb{Z}$, and let Q be a random state of q qubits. Let $h : \mathbb{S} \times \mathbb{X} \to \{0,1\}^{\ell}$ be a two-universal hash function and let S be uniform over $\mathbb{S}$ and independent from X and Z. If $\ell \leq H_{\min}^{\varepsilon'}(X \mid Z) - q - 2\log(1/\varepsilon)$, then $h(S,X)$ is $(\varepsilon + 2\varepsilon')$-close to uniform with respect to (S, Z, Q).*

The following lemma that we prove in the full version follows from the uncertainty relation presented in [17].

Lemma 2. *Let $X \in \{0,1\}^n$ be a uniform random string, let $B \in \{+, \times\}^n$ be a uniform random basis. Let $|X\rangle_B = (|X_1\rangle_{B_1}, \ldots, |X_n\rangle_{B_n})$ be a state of n qubits, and let K be the outcome of an arbitrary measurement of $|X\rangle_B$, which does not depend on X and B. Then, for any ε, we have $H_{\min}^{\varepsilon}(X|BK) \geq \frac{n}{2} - 10\sqrt[3]{n^2 \log \frac{1}{\varepsilon}}$, which is positive if $n > 8000\log(1/\varepsilon)$.*

3   Security in the Bounded-Quantum-Storage Model

We now give a definition of offline-security in the bounded-quantum-storage model, and show that it allows protocols to be composed *sequentially* (at any given time only one sub-protocol is executed). More detail can be found in the long version of our paper. Our definitions are closely related to [25].

We look at the following setting: Two *players*, A and B, execute a *protocol* $\mathbf{P} = (\mathbf{P_A}, \mathbf{P_B})$, where $\mathbf{P_A}$ is the program executed by A and $\mathbf{P_B}$ the program executed by B. Before the first round, each program receives an input (that might be entangled with the input of the other player) and stores it. In each round, each program may first send/receive messages to/from a given functionality $\mathbf{G}$, then apply a quantum operation to its current internal storage (including the message space), and finally send/receive further messages at the end of each round. $\mathbf{G}$ defines the communication resources available between the players, modeled as an interactive quantum functionality. It may contain a classical and/or a quantum communication channel, or other functionalities such as oblivious transfer or bit commitment. Finally, in the last step of the protocol each program outputs an output value. The execution of **P** using **G** (denoted by **P(G)**) is a quantum channel, which takes the input of both parties to the output of both parties.

Players may be *honest*, which means that they follow the protocol, or they may be *corrupted*. All corrupted players belong to the *adversary*, $\mathbb{A} \subset \{A, B\}$. Note that we can ignore the case where both players are corrupted. To simplify the proofs, we assume the set $\mathbb{A}$ to be *static*, i.e., it is already fixed before the protocol starts. We take the adversary to be *active*, i.e., he may not follow the protocol. The adversary $\mathbb{A} = \{p\}$ may replace his part of the protocol $\mathbf{P}_p$ by another program $\mathbf{A}_p$. Opposed to $\mathbf{P}_p$, $\mathbf{A}_p$ receives some *auxiliary (quantum) input* at the start of the protocol that may also be entangled with the environment. This input can be given to the adversary from the environment, but also come from the output of an honest player from a previous run of the protocol. At the end of the protocol, the adversary may return a (quantum) output to the environment. There is no communication between the adversary and the environment between the beginning and the end of the protocol. After receiving the (quantum) output, the environment tries to distinguish the protocol from the ideal setting based on its knowledge of its own input and output to and from the adversary.

We do not restrict the computational power of $\mathbf{A}_p$ in any way, however we do limit its internal quantum storage to a certain *memory-bound* of m qubits. We call such an $\mathbf{A}_p$ m-*bounded*. $\mathbf{A}_p$ is allowed to perform arbitrary quantum operations in each round of the protocol. However after receiving his input, and after every round, all of his internal memory is measured, except for m qubits. He may, however, store an unlimited amount of classical information.

The *ideal functionality*, denoted by **F**, defines what functionality we expect the protocol to implement. In this paper, we only consider *non-interactive* functionalities, i.e., both players can send it input only once at the beginning, and obtain the output only once at the end. These functionalities have the form of a quantum channel. To make the definitions more flexible, we allow **F** to look differently depending on whether both players are honest, or either A or B belongs to the adversary. So the ideal functionality is in fact a *collection of functionalities*, $\mathbf{F} = (\mathbf{F}_\emptyset, \mathbf{F}_{\{A\}}, \mathbf{F}_{\{B\}})$. $\mathbf{F}_\emptyset$ denotes the functionality for the case when both players are honest, and $\mathbf{F}_{\{A\}}$ and $\mathbf{F}_{\{B\}}$ for the cases when A or B respectively are dishonest. As a honest player does not know whether the other player is also honest or not, we require that $\mathbf{F}_{\{A\}}$ ($\mathbf{F}_{\{B\}}$) and $\mathbf{F}_{\{\emptyset\}}$ must look the same from him. We also require that $\mathbf{F}_{\{A\}}$ and $\mathbf{F}_{\{B\}}$ allow the adversary to play honestly, i.e., they must be at least as good for the adversary as the functionality $\mathbf{F}_\emptyset$.

As we formally define in the long version, we say that a protocol **P** having access to the functionality **G** implements a functionality **F**, if the following conditions are satisfied: First of all, we require that output of the protocol is ε-close to that of **F**, if both players are honest. Second, for $\mathbb{A} = \{p\}$, we require that the adversary attacking the protocol has basically no advantage over attacking **F** directly. We thus require that for every m-bounded program $\mathbf{A}_p$, there exists an s-bounded program $\mathbf{S}_p$ (called the *simulator*), such that the overall outputs of both situations are ε-close, for all inputs. For simplicity, we do not make any restrictions on the efficiency of the simulators[2]. Also, we do not require him to use the adversary $\mathbf{A}_p$ as a black-box: $\mathbf{S}_p$ may be constructed from scratch, under full knowledge of the behavior of $\mathbf{A}_p$.

It is important to note that we allow the simulator to execute some or all actions of $\mathbf{A}_p$ in a single round. This will allow the simulator to execute $\mathbf{A}_p$ *without* a memory bound being applied: Recall, that a memory bound is applied only after each round. This model is motivated by the physically realistic assumption that such memory bounds are introduced by adding specific waiting times after each round. Since the adversary is computationally unbounded, he would essentially also be able to perform any computation before the memory bound is applied and hence the simulator does not gain any more powers than the adversary. In particular, this does not give the simulator any memory.

However, in order to make protocols composable with other protocols in our model, we do require the simulator to be memory-bounded as well. The amount of memory required by the simulator gives a bound on the *virtual* memory the adversary seems to have by attacking the real protocol instead of the ideal one. Ideally, we would like $\mathbf{S}_p$ to use the same amount of memory as $\mathbf{A}_p$.

An important property of our definition is that it allows protocols to be composed. The following theorem shows that in a secure protocol that is based on an ideal, non-interactive functionality $\mathbf{G}$ and some other functionalities $\mathbf{G}'$, we can replace $\mathbf{G}$ with a secure implementation of $\mathbf{G}$, without making the protocol insecure. We thereby denote the concatenation of the functionalities $\mathbf{G}$ and $\mathbf{G}'$ by $\mathbf{G}\|\mathbf{G}'$. The theorem requires that $\mathbf{G}$ is called sequentially, i.e., that no other sub-protocols are running parallel to $\mathbf{G}$. The proof uses the same idea as in the classical case [20].

Theorem 2 (Sequential Composition Theorem). *Let $\mathbf{F}$ and $\mathbf{G}$ be non-interactive, and $\mathbf{G}'$ and $\mathbf{H}$ arbitrary functionalities. Let $\mathbf{P}(\mathbf{G}\|\mathbf{G}')$ be a protocol that calls $\mathbf{G}$ sequentially and that implements $\mathbf{F}$ with error at most ε_1 secure against m_1-bounded adversaries using s_1-bounded simulators, and let $\mathbf{Q}(\mathbf{H})$ be a protocol that implements $\mathbf{G}$ with error at most ε_2 secure against m_2-bounded adversaries using s_2-bounded simulators, where $m_2 \geq s_1$. Then $\mathbf{P}(\mathbf{Q}(\mathbf{H})\|\mathbf{G}')$ implements $\mathbf{F}$ with error at most $\varepsilon_1 + \varepsilon_2$, secure against $\min(m_1, m_2)$-bounded adversaries using s_2-bounded simulators.*

4 Randomized Oblivious Transfer

We now apply our framework to the randomized OT protocol presented in [16]. In particular, we prove security with respect to the following definition of randomized oblivious transfer. We show in the long version how to obtain the standard notion of OT from randomized OT. Note that in our version of randomized OT, also the choice bit c of the receiver is randomized.

Definition 1 (Randomized oblivious transfer). $\binom{2}{1}$-*ROT$^\ell$ (or, if ℓ is clear from the context, ROT) is defined as* $ROT = (ROT_\emptyset, ROT_{\{A\}}, ROT_{\{B\}})$, *where*

---

[2] Recall the adversary is computationally unbounded as well.

- $ROT_\emptyset$: *The functionality chooses uniformly at random the value* $(x_0, x_1) \in_R$ $\{0, 1\}^{2\ell}$ *and* $c \in_R \{0, 1\}$. *It sends* (x_0, x_1) *to* A *and* (c, y) *to* B *where* $y = x_c$.
- $ROT_{\{A\}}$: *The functionality receives* $(x_0, x_1) \in \{0, 1\}^{2\ell}$ *from* A. *Then, it chooses* $c \in_R \{0, 1\}$ *uniformly at random and sends* (c, y) *to* B, *where* $y = x_c$.
- $ROT_{\{B\}}$: *The functionality receives* $(c, y) \in \{0, 1\} \times \{0, 1\}^\ell$ *from* B. *Then, it sets* $x_c = y$, *chooses* $x_{1-c} \in_R \{0, 1\}^\ell$ *uniformly at random, and sends* (x_0, x_1) *to* A.

The protocol BQS-OT = $(\text{BQS-OT}_A, \text{BQS-OT}_B)$ uses a noiseless unidirectional quantum channel Q-Comm, and a noiseless unidirectional classical channel Comm, both from the sender to the receiver. Let $h : \mathcal{R} \times \{0, 1\}^n \to \{0, 1\}^\ell$ be a two-universal hash function. A memory bound is applied before step 1, and between step 2 and 3. The sender (A) and receiver (B) execute:

Protocol 1: BQS-OT$_A$

1. Choose $x \in_R \{0, 1\}^n$ and $b \in_R \{0, 1\}^n$ uniformly at random.
2. Send $|x\rangle_b := (|x_1\rangle_{b_1}, \ldots, |x_n\rangle_{b_n})$ to Q-Comm, where $|x_i\rangle_{b_i}$ is x_i encoded in the basis b_i.
3. Choose $r_0, r_1 \in_R \mathcal{R}$ uniformly at random and send (b, r_0, r_1) to Comm.
4. Output $(s_0, s_1) := (h(r_0, x_{|0}), h(r_1, x_{|1}))$, where $x_{|j}$ is the string of all x_i where $b_i = j$.

Protocol 2: BQS-OT$_B$

1. Choose $c \in_R \{0, 1\}$ uniformly at random.
2. Receive the qubits $(q_1, \ldots, q_n)$ from Q-Comm and measure them in the basis c, which gives output $x' \in \{0, 1\}^n$.
3. Receive (b, r_0, r_1) from Comm.
4. Output $(c, y) := (c, h(r_c, x'_{|c}))$, where $x'_{|c}$ is the string of all x'_i where $b_i = c$.

Security against the sender. We first consider the case when the sender, A, is dishonest. This case turns out to be quite straightforward and closely follows the proof given in [17]. We use the following letters to refer to the different classical and quantum registers available to the adversary: Let $\mathcal{Q}$ denote the quantum register. Note that since we assume that our adversary's memory is m-bounded, the size of $\mathcal{Q}$ does not exceed m. Let $\mathcal{M}_Q$ and $\mathcal{M}_K$ denote the quantum and classical registers, that hold the messages sent to the receiver. Let $\mathcal{K}$ denote the classical input register of the adversary. Finally, let $\mathcal{A}$ denote an auxiliary quantum register. Recall from Section 2, that any quantum operation on $\mathcal{Q}$ and $\mathcal{M}_Q$ can be implemented by a unitary followed by a measurement on an additional register $\mathcal{A}$. Wlog we let $\mathcal{A}$ and $\mathcal{M}_Q$ be measured in the computational basis to enforce a memory bound, and $\mathcal{Q}$ be the sole quantum memory.

To model quantum and classical input that a malicious A may receive, we let $\mathcal{Q}$ start out in any state ρ_{in}, unknown to the simulator. Likewise, $\mathcal{K}$ may contain some classical input k_{in} of A. Wlog we assume that all other registers start out in a fixed state of $|0\rangle$. We can then describe the actions of A by a single unitary

$\mathbf{A_A}$ defined by

$$\mathbf{A_A}(\underbrace{\rho_{\text{in}}}_{Q} \otimes \underbrace{|0\rangle\langle 0|}_{A} \otimes \underbrace{k_{\text{in}}}_{K} \otimes \underbrace{|0\rangle\langle 0|}_{M_Q} \otimes \underbrace{|0\rangle\langle 0|}_{M_K})\mathbf{A_A^\dagger} = \underbrace{\rho_{\text{out}}}_{Q,A} \otimes \underbrace{k_{\text{in}}}_{K} \otimes \underbrace{\rho_{x_b}}_{M_Q} \otimes \underbrace{|b r_0 r_1\rangle\langle b r_0 r_1|}_{M_K}.$$

Note that without loss of generality $\mathbf{A_A}$ leaves $\mathcal{K}$ unmodified: since $\mathcal{K}$ is classical we can always copy its contents to $\mathcal{A}$ and let all classical output be part of $\mathcal{A}$. To enforce the memory bound, assume wlog that $\mathcal{A}$ and $\mathcal{M}_Q$ are now measured completely in the computational basis. We now show that for any adversary $\mathbf{A_A}$ there exists an appropriate simulator $\mathbf{S_A}$.

Lemma 3. *Protocol BQS-OT is secure against dishonest A.*

Proof. Let $\mathbf{S_A}$ be defined as follows: $\mathbf{S_A}$ runs $\mathbf{A_A}$. Note that $\mathbf{S_A}$ can effectively skip the wait time required for the memory bound to take effect, since he can execute $\mathbf{A_A}$ in one round before his memory bound is applied, where we refer to Section 3 for an important discussion and justification of this procedure. The simulator then measures register $\mathcal{M}_Q$ in the basis determined by $\mathcal{M}_K$. This allows him to compute $s_0 = h(r_0, x_{|0})$ and $s_1 = h(r_1, x_{|1})$. $\mathbf{S_A}$ then sends s_0 and s_1 to $\text{ROT}_{\{A\}}$. It is clear that since the simulator based his measurement on $\mathcal{M}_K$, s_0 and s_1 are consistent with the run of the protocol. Furthermore, note that $\mathbf{S_A}$ did not need to touch register $\mathcal{Q}$ at all. We can thus immediately conclude that the environment can tell no difference between the real protocol and the ideal setting. □

Security against the receiver. The proof of security against a dishonest receiver requires a more careful treatment of the quantum input given to the adversary. The main idea behind our proof is that the memory bound in fact *fixes* a classical bit c. Our main challenge is to find a c that the simulator can calculate and that is consistent with the adversary and his input, while keeping the output state of the adversary intact. To do so, we use a generalization of the *min-entropy splitting lemma* in [17], which in turn is based on an earlier version of [37]. It states that if two random variables X_0 and X_1 together have high min-entropy, then we can define a random variable C, such that X_{1-C} has at least half of the original min-entropy. To find C, one must know the distributions of X_0 and X_1. In the following generalization, we do *not* exactly know the distribution of X_0 and X_1, since we assume that its distribution also depends on an unknown random variable J, distributed over a domain of the size 2^β. $\beta = 0$ gives the min-entropy splitting lemma in [17].

Lemma 4 (Generalized Min-Entropy Splitting Lemma). *Let $\varepsilon \geq 0$, and $0 < \beta < \alpha$. Let J be a random variable over $\{0, \ldots, 2^\beta - 1\}$, and let X_0, X_1 and K be random variables such that $H_{\min}^\varepsilon(X_0 X_1 \mid KJ) \geq \alpha$. Let $f(x_1, k) = 1$, if there exists a $j \in \{0, \ldots, 2^\beta - 1\}$ such that $P_{X_1 | KJ}(x_1, k, j) \geq 2^{-(\alpha-\beta)/2}$, and 0 otherwise, and let $C := f(X_1, K)$. We have $H_{\min}^\varepsilon(X_{1-C} C \mid KJ) \geq \frac{\alpha - \beta}{2}$.*

Proof. Let S_k^j be the set of values x_1 for which $P_{X_1 | KJ}(x_1, k, j) \geq 2^{-(\alpha-\beta)/2}$. We have $|S_k^j| \leq 2^{(\alpha-\beta)/2}$, since all values in S_k^j have a probability that is at least $2^{-(\alpha-\beta)/2}$. Let $S_k := \bigcup_j S_k^j$. We have $|S_k| \leq 2^\beta \cdot 2^{(\alpha-\beta)/2} = 2^{(\alpha+\beta)/2}$.

Let $K = k$ and $J = j$. Because $C = 0$ implies that $X_1 \notin S_k$, and thus also that $X_1 \notin S_k^j$, we have $P_{X_1 C | K J}(x_1, 0, k, j) < 2^{-(\alpha - \beta)/2}$. It follows from the assumption that there exists an event $\mathcal{E}$ with probability $1 - \varepsilon$ such that for all x_0, x_1, k and j, we have $P_{X_0 X_1 \mathcal{E} | K J}(x_0, x_1, k, j) \leq 2^{-\alpha}$. Hence $P_{X_0 C \mathcal{E} | K J}(x_0, 1, k, j) = \sum_{x_1 \in S_k} P_{X_0 X_1 \mathcal{E} | K J}(x_0, x_1, k, j) \leq 2^{(\alpha + \beta)/2} \cdot 2^{-\alpha} = 2^{-(\alpha - \beta)/2}$. $\square$

We now describe the actions of the adversary. Let $\mathcal{M}$ denote the register holding the quantum message he receives from the sender in step 2. Let his registers $\mathcal{Q}$, $\mathcal{A}$ and $\mathcal{K}$ be initialized as above. We can now describe the actions of the adversary by two unitaries, where a memory bound is applied after the first. The action of the adversary following step 2 can be described as a unitary $\mathbf{A}_B^{(1)}$ as before. Note we can again assume that $\mathbf{A}_B^{(1)}$ leaves $\mathcal{K}$ unmodified. To enforce the memory bound, we now let register $\mathcal{M}$ and $\mathcal{A}$ be measured in the computational basis. We use $\rho_{\text{out}} \in \mathcal{Q}$ to denote the adversary's quantum output, and $k_{\text{out}} \in \mathcal{M} \otimes \mathcal{A}$ to denote his classical output. After the memory bound is applied, the receiver obtains additional information from the sender. The actions of the adversary after step 3 can then be described by a unitary $\mathbf{A}_B^{(2)}$ followed by a measurement of quantum registers $\mathcal{M}$ and $\mathcal{A}$ in the computational basis.

First, we analyze the case where the adversary's auxiliary quantum input is a pure state of β qubits. Note that this means that the adversary cannot be entangled with the environment. Then we extend it, by allowing the adversary some arbitrary mixed quantum auxiliary input.

Lemma 5. *Protocol BQS-OT is secure against dishonest B with an error of at most 5ε, if he receives a pure state quantum (auxiliary) input, and his quantum memory is bounded before step 1 by β qubits, and between step 2 and 3 by m qubits, for*

$$8\ell + 2\beta + 4m \leq n - 20\sqrt[3]{n^2 \log \frac{1}{\varepsilon}} - 12 \log \frac{1}{\varepsilon} - 4.$$

Proof. Let K_{in} be the classical auxiliary input the adversary receives, and let $|j\rangle$ for $j \in \{0, \ldots, 2^\beta - 1\}$ be a basis for the quantum auxiliary input. Any fixed auxiliary input $|j\rangle$ and k_{in} fixes a distribution $P_{X_0 X_1 K | J = j}$, where K is the classical value the adversary has after second memory bound. The choice of input state $|\Psi_{\text{in}}\rangle$ thus defines the distribution of J. First of all, the simulator simulates the actions of the sender following steps 1 and 2, using a random string X and a random basis B. The simulator then applies $\mathbf{A}_B^{(1)}$, which gives him some classical output K_{out}, and a quantum state ρ_{out}. It follows from the uncertainty relation of Lemma 2 that $H_{\min}^\varepsilon(X \mid BK_{\text{out}}K_{\text{in}}) \geq \alpha$ for $\alpha := n/2 - 10\sqrt[3]{n^2 \log(1/\varepsilon)}$. Let $(X_0, X_1) := X$, where $X_0 := X_{|0}$ and $X_1 := X_{|1}$ are the substrings of X defined in the same way as in the protocol.

It follows from Lemma 4 and the fact that the simulator holds a description of $\mathbf{A}^{(1)}$, $K = (B, K_{\text{out}}, K_{\text{in}})$ and X_0, X_1 that he can calculate the value $C := f(X_1, K)$. This means that the simulator can construct a linear transformation $\mathbf{S}_B$ acting on registers $\mathcal{Q}, \mathcal{M}, \mathcal{A}, \mathcal{K}, \mathcal{X}, \mathcal{B}, \mathcal{R}$, and $\mathcal{C}$ combining the actions of $\mathbf{A}_A^{(1)}$ and the choice of c using the function f as defined in the min-entropy splitting

Lemma 4. We have

$$
\mathbf{S_B}\Big(\sum_j \alpha_j \underbrace{|j\rangle}_{\mathcal{Q}} \otimes \underbrace{|x_b\rangle}_{\mathcal{M}} \otimes \underbrace{|0\rangle}_{\mathcal{A}} \otimes \underbrace{|k_{\text{in}}\rangle}_{\mathcal{K}} \otimes \underbrace{|x\rangle}_{\mathcal{X}} \otimes \underbrace{|b\rangle}_{\mathcal{B}} \otimes \underbrace{|r_0, r_1\rangle}_{\mathcal{R}} \otimes \underbrace{|0\rangle}_{\mathcal{C}} \otimes \underbrace{|0\rangle}_{\mathcal{Y}}\Big) =
$$

$$
\sum_{q, m_1, a_1} \alpha_{q, m_1, a_1} \underbrace{|q\rangle}_{\mathcal{Q}} \otimes \underbrace{|m_1\rangle}_{\mathcal{M}} \otimes \underbrace{|a_1\rangle}_{\mathcal{A}} \otimes \underbrace{|k_{\text{in}}\rangle}_{\mathcal{K}} \otimes \underbrace{|x\rangle}_{\mathcal{X}} \otimes \underbrace{|b\rangle}_{\mathcal{B}} \otimes \underbrace{|r_0, r_1\rangle}_{\mathcal{R}} \otimes \underbrace{|c\rangle}_{\mathcal{C}} \otimes \underbrace{|s_0, s_1\rangle}_{\mathcal{Y}}
$$

for any pure state input $|\Psi_{\text{in}}\rangle = \sum_j \alpha_j |j\rangle$. Wlog, all registers except $\mathcal{Q}$ are now measured in the computational basis as the memory bound takes effect. It is an important consequence of our generalized min-entropy splitting lemma that the simulator can measure register $\mathcal{C}$ in the computational basis to extract c without causing any disturbance to the quantum output: Note that the definition of f did not take j into account explicitly and hence $\mathcal{C}$ is not entangled with the quantum output. From Lemma 4 we thus have that $H^\varepsilon_{\min}(X_{1-C}C \mid K) \geq \frac{\alpha - \beta}{2}$. The simulator now chooses R_0 and R_1 uniformly at random and calculates $S_0 = h(R_0, X_0)$ and $S_1 = h(R_1, X_1)$. Since R_0 and R_1 are independent of X_0, X_1 and C, we have $H^\varepsilon_{\min}(X_{1-C}C \mid K) = H^\varepsilon_{\min}(X_{1-C}C \mid R_C K)$. Using the chain rule from Lemma 1 and the monotonicity of $H^\varepsilon_{\min}$, we obtain $H^{2\varepsilon}_{\min}(X_{1-C} \mid CR_C KS_C) \geq \frac{\alpha - \beta}{2} - \ell - 1 - \log \frac{1}{\varepsilon}$. By using the privacy amplification Theorem 1, we get that S_{1-C} is 5ε close to uniform with respect to $(R_0, R_1, C, S_C, B, K_{\text{out}}, K_{\text{in}})$ and ρ_{out} if $\ell \leq \frac{\alpha - \beta}{2} - \ell - 1 - \log \frac{1}{\varepsilon} - m - 2\log \frac{1}{\varepsilon}$. By replacing α and rearranging the terms we get the claimed equation.

The simulator now sets $Y := S_C$, and sends (C, Y) to $\mathsf{ROT}_{\{B\}}$. To complete the simulation, he runs $\mathbf{A}^{(2)}_\mathsf{A}$ as the adversary would have. Note that the simulator did not require any more memory than the adversary itself, i.e., we can take $\mathbf{S_B}$ to be m-bounded as well. Clearly, the simulator determined C solely from the classical output of the adversary and thus the adversary's output state in the simulated run is equal to the original output state of the adversary $\rho_{\text{out}} \otimes k_{\text{out}}$. Since the only difference between the simulation and the real execution is that in the simulation, S_{1-C} is chosen completely at random, the simulation is 5ε-close to the output of the real protocol. □

It remains to address the case where the receiver gets a mixed state quantum input. This is the case where the adversary receives a state that is entangled with the environment. Note that this means that we must decrease the size of the adversary's memory: If he could receive an entangled state of β qubits as input, he could use it to increase his memory to $m + \beta$ qubits by teleporting β qubits to the environment, and storing the remaining m. Hence, we now have to take the adversary to be m'-bounded, where $m' := m - \beta$. Luckily, using a a similar argument as in [38], we can now extend the argument given above: Note that for any pure state input $|\Psi\rangle = |\Psi_{\text{in}}\rangle \otimes k_{\text{in}}$, the output of the simulated adversary is *exactly* $\Lambda(|\Psi\rangle\langle\Psi|)$, where Λ is the adversary's channel. Since $\{|\Psi\rangle\langle\Psi| \mid |\Psi\rangle \in \mathcal{Q} \otimes \mathcal{K}, \||\Psi\rangle\| = 1\}$ spans all of $\mathbb{T}(\mathcal{Q} \otimes \mathcal{K})$ and the map given by the simulation procedure is the same as Λ on all inputs, we can conclude that the complete map is equal to Λ. Note that the simulator does not need to consider the β qubits

614   S. Wehner and J. Wullschleger

that the adversary might have teleported to the environment: we can essentially view it as part of the original adversary's quantum memory, and the simulator bases his decision solely on the classical output of the adversary. Hence,

Lemma 6. *Protocol BQS-OT is secure against dishonest B with an error of at most 5ε, if he receives a quantum (auxiliary) input, and his quantum memory is bounded before step 1 by β qubits and between step 2 and 3, by m qubits, for*
$$8\ell + 6\beta + 4m \leq n - 20\sqrt[3]{n^2 \log \tfrac{1}{\varepsilon}} - 12\log \tfrac{1}{\varepsilon} - 4.$$

Theorem 3. *Protocol BQS-OT(Q-Comm$\|$Comm) implements $\binom{2}{1}$-ROT$^\ell$ with an error of at most 5ε, secure against m-bounded adversaries using m-bounded simulators, if $8\ell + 10m \leq n - 20\sqrt[3]{n^2 \log \tfrac{1}{\varepsilon}} - 12\log \tfrac{1}{\varepsilon} - 4$.*

Acknowledgments

We thank S. Desrosiers and C. Schaffner for useful comments, and D. Unruh for a kind explanation of his work. SW is supported by NSF grant PHY-0456720. JW is supported by the EPSRC. Part of this work was done while SW was a PhD student at CWI, Amsterdam, and JW was a PhD student at ETH Zürich, and during a 3 month visit at McGill University, Montreal, Quebec.

References

1. Yao, A.C.: Protocols for secure computations. In: 23rd IEEE FOCS, pp. 160–164 (1982)
2. Wiesner, S.: Conjugate coding. SIGACT News 15(1), 78–88 (1983)
3. Rabin, M.O.: How to exchange secrets by oblivious transfer. Technical Report TR-81, Harvard Aiken Computation Laboratory (1981)
4. Even, S., Goldreich, O., Lempel, A.: A randomized protocol for signing contracts. Commun. ACM 28(6), 637–647 (1985)
5. Kilian, J.: Founding cryptography on oblivious transfer. In: Proceedings of the 20th STOC, pp. 20–31 (1988)
6. Crépeau, C., van de Graaf, J., Tapp, A.: Committed oblivious transfer and private multi-party computation. In: Coppersmith, D. (ed.) CRYPTO 1995. LNCS, vol. 963, pp. 110–123. Springer, Heidelberg (1995)
7. Blum, M.: Coin flipping by telephone a protocol for solving impossible problems. SIGACT News 15(1), 23–27 (1983)
8. Mayers, D.: Unconditionally secure quantum bit commitment is impossible. Physical Review Letters 78, 3414–3417 (1997)
9. Lo, H.K., Chau, H.F.: Is quantum bit commitment really possible? Physical Review Letters 78, 3410–3413 (1997)
10. Kitaev, A., Mayers, D., Preskill, J.: Superselection rules and quantum protocols. Physical Review A 69, 052326 (2004)
11. Spekkens, R., Rudolph, T.: Degrees of concealment and bindingness in quantum bit commitment protocols. Physical Review A 65, 012310 (2002)
12. Salvail, L.: Quantum bit commitment from a physical assumption. In: Krawczyk, H. (ed.) CRYPTO 1998. LNCS, vol. 1462, pp. 338–353. Springer, Heidelberg (1998)
13. Buhrman, H., Christandl, M., Hayden, P., Lo, H.K., Wehner, S.: Security of quantum bit string commitment depends on the information measure. Physical Review Letters 97, 250501 (2006)

14. Bennett, C.H., Brassard, G., Crépeau, C., Skubiszewska, H.: Practical quantum oblivious transfer. In: Feigenbaum, J. (ed.) CRYPTO 1991. LNCS, vol. 576, pp. 351–366. Springer, Heidelberg (1992)
15. Crépeau, C.: Quantum oblivious transfer. J. of Mod. Opt. 41(12), 2455–2466 (1994)
16. Damgård, I., Fehr, S., Salvail, L., Schaffner, C.: Cryptography in the Bounded Quantum-Storage Model. In: 46th IEEE FOCS, pp. 449–458 (2005)
17. Damgård, I., Fehr, S., Renner, R., Salvail, L., Schaffner, C.: A tight high-order entropic uncertainty relation with applications in the bounded quantum-storage model. In: Menezes, A. (ed.) CRYPTO 2007. LNCS, vol. 4622. Springer, Heidelberg (2007)
18. Micali, S., Rogaway, P.: Secure computation. In: Feigenbaum, J. (ed.) CRYPTO 1991. LNCS, vol. 576, pp. 392–404. Springer, Heidelberg (1992)
19. Beaver, D.: Foundations of secure interactive computing. In: Feigenbaum, J. (ed.) CRYPTO 1991. LNCS, vol. 576, pp. 377–391. Springer, Heidelberg (1992)
20. Canetti, R.: Security and composition of multiparty cryptographic protocols. Journal of Cryptology 13(1), 143–202 (2000)
21. Canetti, R.: Universally composable security: A new paradigm for cryptographic protocols. In: 42th IEEE FOCS, pp. 136–145 (2001)
22. Pfitzmann, B., Waidner, M.: A model for asynchronous reactive systems and its application to secure message transmission. In: IEEE SP, p. 184 (2001)
23. Backes, M., Pfitzmann, B., Waidner, M.: A universally composable cryptographic library (2003), http://eprint.iacr.org/2003/015
24. van de Graaf, J.: Towards a formal definition of security for quantum protocols. Ph.D. thesis (1998), http://www.cs.mcgill.ca/~crepeau/PS/these-jeroen.ps
25. Smith, A.: Multi-party quantum computation. Masters Thesis (2001), quant-ph/0111030
26. Ben-Or, M., Mayers, D.: General security definition and composability for quantum and classical protocols (2004), quant-ph/0409062
27. Unruh, D.: Simulatable security for quantum protocols (2004), quant-ph/0409125
28. Unruh, D.: Formal security in quantum cryptology. Student research project, Institut für Algorithmen und Kognitive Systeme. University of Karlsruhe (2002)
29. Fehr, S., Schaffner, C.: Composing quantum protocols in a classical environment (2008), arxiv:0804.1059
30. Estren, G.: Universally composable committed oblivious transfer and multi-party computation assuming only basic black-box. M.Sc. thesis, School of Computer Science. McGill University (2004)
31. Canetti, R., Lindell, Y., Ostrovsky, R., Sahai, A.: Universally composable two-party and multi-party secure computation. In: 34th STOC, pp. 494–503 (2002)
32. Hayashi, M.: Quantum Information: An introduction. Springer, Heidelberg (2006)
33. Renner, R., König, R.: Universally composable privacy amplification against quantum adversaries. In: Kilian, J. (ed.) TCC 2005. LNCS, vol. 3378, pp. 407–425. Springer, Heidelberg (2005)
34. Renner, R., Wolf, S.: Simple and tight bounds for information reconciliation and privacy amplification. In: Roy, B. (ed.) ASIACRYPT 2005. LNCS, vol. 3788, pp. 199–216. Springer, Heidelberg (2005)
35. Carter, J.L., Wegman, M.N.: Universal classes of hash functions. Journal of Computer and System Sciences 18, 143–154 (1979)
36. Renner, R.: Security of Quantum Key Distribution. PhD thesis, ETH Zurich, Switzerland (2005), http://arxiv.org/abs/quant-ph/0512258
37. Wullschleger, J.: Oblivious-transfer amplification. In: Naor, M. (ed.) EUROCRYPT 2007. LNCS, vol. 4515. Springer, Heidelberg (2007)
38. Watrous, J.: Zero-knowledge against quantum attacks (2005), quant-ph/0511020

On the Strength of the Concatenated Hash Combiner When All the Hash Functions Are Weak

Jonathan J. Hoch and Adi Shamir

Department of Computer Science and Applied Mathematics,
The Weizmann Institute of Science, Israel
{yaakov.hoch,adi.shamir}@weizmann.ac.il

Abstract. At Crypto 2004 Joux showed a novel attack against the concatenated hash combiner instantiated with Merkle-Damgård iterated hash functions. His method of producing multicollisions in the Merkle-Damgård design was the first in a recent line of generic attacks against the Merkle-Damgård construction. In the same paper, Joux raised an open question concerning the strength of the concatenated hash combiner and asked whether his attack can be improved when the attacker can efficiently find collisions in both underlying compression functions. We solve this open problem by showing that even in the powerful adversarial scenario first introduced by Liskov (SAC 2006) in which the underlying compression functions can be fully inverted (which implies that collisions can be easily generated), collisions in the concatenated hash cannot be created using fewer than $2^{n/2}$ queries. We then expand this result to include the double pipe hash construction of Lucks from Asiacrypt 2005. One of the intermediate results is of interest on its own and provides the first streamable construction provably indifferentiable from a random oracle in this model.

Keywords: hash functions, cryptographic combiners, indifferentiability.

1 Introduction

Cryptanalysis of hash functions has been a very active area of research in the past few years. A flurry of attacks have been found against various hash functions including SHA-1 and the MD variants (see [10,16,17,18,19]). Besides these attacks on specific hash functions, a number of novel generic attacks against the Merkle-Damgård [5,14] iterated construction have been published as well. These include among others Joux's multicollision attack [7], Kelsey and Schneier's expandable message attack [9] and Kelsey and Kohno's herding attack [8]. Joux's multicollision attack demonstrates how to find collisions in a concatenated hash construction $H(M) = F(M) \| G(M)$ when at least one of the underlying hash functions is iterated.

In the classic combiner scenario we have two instantiations, I_1 and I_2, of some cryptographic primitive, e.g., two encryption schemes or two hash functions. The

L. Aceto et al. (Eds.): ICALP 2008, Part II, LNCS 5126, pp. 616–630, 2008.

goal is to build a new combined instantiation I of the primitive, which remains secure even when one of the underlying primitives is broken, as long as the other remains secure. In contrast to this classical approach, we will show that certain hash combiners retain a provable level of security even if all of the underlying hash functions are compromised, provided that the two primitives are sufficiently random and sufficiently different in a sense which will be made precise later.

1.1 Related Work

Joux's innovative attack focused attention on the security properties of hash combiners as his attack shows that the trivial combiner does not improve over the security of the underlying hash functions. A line of research concerning hash combiners has followed, demonstrating that security amplifying combiners exist [6] and on the other hand proving that any provably secure black-box combiner must preserve the total length of the underlying hash functions [1,15]. Other responses to Joux's paper include Lucks' [12] proposal of the wide/double piped constructions whose aim was to overcome the multicollision attack by using a larger internal state. Lucks' proposal is provably secure in the random oracle model against multicollisions. Maurer et al. [13] introduced the notion of indifferentiability. Similar to the concept of indistinguishability, this notion describes a situation in which two systems are indistinguishable despite having extra access to the internal structure of the systems. Inspired by the generic attacks against the Merkle-Damgård iterated construction, Coron et al. [3] operated within the indifferentiability framework to show how iterated hash functions can be proved indifferentiable from random oracles in the ideal cipher model.[1] Liskov further pursued this approach in [11] by introducing *weak compression functions*. A weak compression function behaves like a random oracle except that the adversary is given access to corresponding inversion oracles. Liskov presented a new hash construction, the *zipper hash*, composed of a pair of weak compression functions and using the framework of Coron et al. proved it indifferentiable from a random oracle. In Joux's attack he did not assume that the attacker can find collisions in the underlying compression functions faster than the birthday paradox bound. Joux then posed the question whether the ability to find collisions efficiently in both the underlying compressions functions can help the attacker improve the complexity of his attack.

1.2 Our Results

In this paper we prove that even in a very strong attack scenario in which the attacker can find not only collisions but even invert in unit time *all* the compression functions on inputs of his choice, the best attack against the concatenated construction is Joux's multicollision attack with complexity $\mathcal{O}\left(2^{n/2}\right)$. Furthermore, as an intermediate result we show a streamable[2] hash construction, provably

---

[1] The underlying compression function is modelled as an ideal cipher.

[2] A hash construction in which each block of the message can be processed once and then be forgotten. This is an essential requirement in applications where the hash is computed on the fly from a data stream.

indifferentiable from a random oracle in the model of weak compression functions, which has the same rate as the non-streamable zipper hash of Liskov [11]. This result is then extended to prove that the double pipe hash construction of Lucks [12] is also indifferentiable from a random oracle in the same model. We stress that the model of weak compression functions captures all black-box generic attacks arising from collision or preimage finding attacks against the underlying compression functions.

1.3 Paper Organization

Section 2 describes the model of weak compression functions and gives our notation for the rest of the paper. Section 3 proves the main result of the paper, namely that in the model of weak compression functions, finding collisions in the concatenated hash combiner requires $\mathcal{O}\left(2^{n/2}\right)$ operations. Finally, Section 4 proves the indifferentiability of Lucks' double pipe hash construction.

2 The Model

We first give a short description of the iterated hash construction. An iterated hash function $F^f : \{0,1\}^* \to \{0,1\}^n$ is built by iterating a basic compression function $f : \{0,1\}^n \times \{0,1\}^m \to \{0,1\}^n$ as follows:

- Split a message M into k, m-bit blocks $x_1, \ldots, x_k$.
- Set $h_0 = IV$ where IV is the initialization vector.
- For each message block x_i compute $h_i = f(h_{i-1}, x_i)$.
- Output $F^f(M) = h_k$.

The classical Merkle-Damgård construction also contains padding and length encoding which we will ignore for the sake of simplicity since they do not affect our results.

Following Joux's open question, we will try to model a situation in which the attacker can efficiently find collisions in either compression function, but do not assume any other special properties of these colliding pairs. In fact we will give our adversary even stronger oracle access and allow him to find in unit time random preimages of two different types as well. Formally, let f and g be compression functions from $m + n$ bits to n bits, and let F and G be the corresponding hash functions built by instantiating the Merkle-Damgård paradigm with f and g respectively. We will model f and g as random functions provided as black box oracles with additional respective *inversion* oracles.

We define the following oracles:

- $f^*(x, ?, z) \to (x, y, z)$ where y is chosen uniformly such that $f(x, y) = z$, or $\perp$ if no such y exists.
- $f^{-1}(?, y, z) \to (x, y, z)$ where x is chosen uniformly such that $f(x, y) = z$, or $\perp$ if no such x exists.

- $g^*(x,?,z) \to (x,y,z)$ where y is chosen uniformly such that $g(x,y) = z$, or $\perp$ if no such y exists.
- $g^{-1}(?,y,z) \to (x,y,z)$ where x is chosen uniformly such that $g(x,y) = z$, or $\perp$ if no such x exists.

f and g queries will be called *forward* queries, g^{-1} and f^{-1} queries will be called *backward* queries and f^* and g^* queries will be called *bridging* queries.[3] The slightly more complicated case in which these inverses are not uniformly distributed will be discussed at the end of this section. One should notice that while weak compression functions are indeed weak in the sense that they allow trivial collision and preimage attacks, there are some operations in which they do not assist at all. For example, given two chaining values x_1 and x_2 finding a message block y such that $f(x_1,y,?) = f(x_2,y,?)$ still requires $\mathcal{O}\left(2^{n/2}\right)$ queries.

We now introduce a slight modification due to Liskov [11] of the framework of Coron *et al.* [3] and Maurer *et al.* [13]. This framework will enable us to prove that certain hash functions based on weak compression functions are indifferentiable from random oracles. Let Γ be an oracle encapsulating $f, f^{-1}, f^*, g, g^{-1}$ and g^*.

Definition 1 (indifferentiability). *A construction C is (q, ϵ)-indifferentiable in the presence of Γ from a random oracle RO if there exists a polynomial time simulator S, such that for every distinguisher D which uses at most q oracle queries (to either of the oracles),*

$$\left| Pr[D^{C,\Gamma} = 1] - Pr[D^{RO,S^{RO}} = 1] \right| < \epsilon$$

Notice that this definition is slightly different from the usual notion of indistinguishability in that the simulator, besides simulating the behavior of Γ, must also remain consistent with the random oracle RO. The following example illustrates the problem. Let C be an iterated hash function built from a compression function f and assume that f is a random oracle. The pair (C, f) is differentiable from (RO, S^{RO}) for any simulator S. The distinguisher D, when presented with a pair (A, B), performs the following queries $h_1 = A(m_1)$, $h_2 = B(h_1, m_2)$, $h = A(m_1 m_2)$. If $h = h_2$ the distinguisher returns 1 and otherwise 0. When D is presented with the pair (C, f), the equality will always hold and $Pr[D^{C,f} = 1] = 1$. On the other hand, for any simulator S, the probability over the random coins of S and the random oracle that $S^{RO}(m_2) = RO(m_1 m_2)$ is negligible. In this example, the distinguisher worked since the simulator could not maintain the required consistency with RO. So we see that S does not only need to simulate Γ *per se* but also needs to maintain the relation of S relative to the RO, simulating the relationship between Γ and C as well. Maurer *et al.* [13] proved that this definition of indifferentiability will allow us to use the construction C in place of a random oracle in any cryptography protocol and retain the same level of provable security.

Another subtle issue is the fact that in our case Γ includes inversion oracles. Notice that when f is a random function, a fixed fraction of the queries

---

[3] Liskov in [11] used the term *squeezing* queries.

$f^{-1}(?, y, z)$ do not have answers, while other queries might have multiple possible answers. We have defined f^{-1} and f^* to return an answer uniformly distributed the possible answers, and thus the simulator S must reproduce the same distribution of the number of inverses which is known to be Poisson.[4] If we would like to model inversion oracles with a non-uniform distribution, the simulator will need to model this distribution as well.

3   A Lower Bound

Using techniques similar to those introduced by Coron et al. we will show that the construction $C(M) = F(M) \oplus G(M)$ is indifferentiable from a random oracle RO when less than $\mathcal{O}\left(2^{n/2}\right)$ queries are performed. Since finding collisions in $H(M) = F(M) \| G(M)$ implies finding collisions in $C(M)$ as well, the indifferentiability of $C(M)$ will give us a lower bound on the number of queries required to find a collision in $H(M)$ with non-negligible probability. Notice that the same proof can be used for any construction of the form $H(M) = \alpha(F(M), G(M))$ for any n-bit function α which is uniquely invertible when its output and any one of its input parameters in known.

Let Γ be an oracle implementing $f, g, f^{-1}, f^*, g^{-1}$ and g^*. Let RO be a random oracle and let S^{RO} be an oracle Turing machine with the same black-box interface as Γ. In order to prove the indifferentiability result, we will give a hybrid argument and show that any distinguisher D cannot differentiate between interacting with the pair (C, Γ) and the pair (RO, S^{RO}).

3.1   The Simulator S

We want the simulator S^{RO} to simulate Γ such that for any distinguisher D, which performs $q \ll 2^{\frac{n}{2}}$ queries [5], $|Pr[D^{C,\Gamma} = 1] - Pr[D^{RO,S} = 1]|$ is negligible. Obviously we would like the simulator S to produce random responses to the simulated queries while maintaining consistency. The naive approach would be to keep a list of all answers given so far and each time S receives a new query, it will return a random value consistent with the values returned so far. Notice that there are two types of consistency involved: self consistency and consistency with the random oracle RO. Handling the self consistency can be done efficiently with the list of answers, however consistency with the random oracle is a bit more tricky. The following definition will capture the essence of maintaining consistency with the random oracle.

---

[4] Note that Liskov in [11] neglected to handle this problem, and therefore his simulator suffers from the fact that a distinguisher can query f^{-1} on a large number of random inputs and the simulator will always return an inverse whereas a true random function will only have inverses for $1 - 1/e$ fraction of the inputs.

[5] We will charge queries to C or RO differently than queries to Γ or S. An l block message query to C or RO will cost l queries. The reason for this different cost will become clear in the remainder of the proof.

Definition 2 (Chains). *A chain is a triplet* (M, h_f, h_g), *where* M *is a* k *block message and* h_f, h_g *are hash values. In addition we require that*

$$f(f(...f(IV_f, m_1), m_2), ..), m_k) = h_f$$

$$g(g(...g(IV_g, m_1), m_2), ...), m_k) = h_g$$

and all the intermediate links are defined in the list of known values (i.e., have been queried previously).

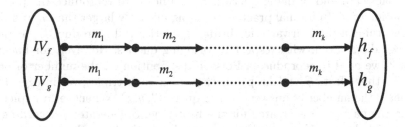

Fig. 1. Chains in the concatenated hash combiner

The chains create a tree structure, with the triplet $(\perp, IV_f, IV_g)$ at the root. An edge between (M, h_f, h_g) and $(M \| m_{k+1}, h'_f, h'_g)$ corresponds to a pair of queries, linking h_f to h'_f and h_g to h'_g with the same message block m_{k+1}. Each node/chain in the tree corresponds to a constraint $h_f \oplus h_g = RO(M)$. The fact that with overwhelming probability the chains form a tree rather than a general graph structure will be proven later. To maintain consistency with the random oracle RO, our naive S will examine each new query and check if answering it will create a chain. If the response creates a chain, S will return a value consistent with RO. As stated, however, this task may require exponential time. Let us assume that the adversary uses a small number of calls to f and f^* in order to create a exponential size multicollision in F. When receiving a new g query, S must check exponentially many possible messages for G as there are that many messages with known chaining values for F. To overcome this problem the simulator will maintain three data structures in order to perform its operation. The first two structures T_f and T_g will contain explicit lists of the triplet answers given by S so far. The third structure will hold the tree of chains created so far. Notice that while the chain tree is implied from the first two lists, keeping it explicitly allows the simulator to run in polynomial time.

We will show how S updates these structures after each query and uses them in order to give consistent answers. For each forward query to f or g, S checks whether the value is already defined in the corresponding data structure of triplets, and if so returns the same value; if not, it returns a random value. To check if the value is defined, S checks if the query appears in its list of responses and additionally checks if the query completes a chain, i.e., extends the chain tree. If the query completes a chain with message M, S queries $RO(M)$ and uses the answer to give a consistent answer to the query. Notice that although chains might be created by bridging or backward queries, we will show that this will only happen with negligible probability and thus we can ignore these possibilities. In fact, we will show that with very high probability the chain tree does not contain any hash value more than once. I.e., the combined list of all x's and z's in the chain tree does not contain duplicates. Our main lemma will show that with high probability, the above holds and chains are only created though forward queries. This in turn will imply that the answers S gives are consistent with the random oracle RO.

For backward and bridging queries, S also needs to reproduce the preimage distribution of Γ. In normal practice, m is significantly larger than n and therefore, returning a random value for bridging queries will reproduce the expected preimage distribution with respect to bridging queries. However, for backward queries[6], we need to reproduce a Poisson distribution on the number of preimages. To this effect, S will keep together with each triplet, an integer j that represents the number of answers to the query $(?, y, z)$. Whenever a triplet containing the pair (y, z) is created for the first time, S generates j according to a Poisson distribution. If on a backward query $j = 0$, S returns the triplet $(\perp, y, z)$. For forward and bridging queries, j is generated according to a Poisson distribution conditioned on the output being non-zero. In future backward queries, S will return a uniform answer from the j possible answers. If one of the j possible answers is not defined yet, S will simply return a random value.

The simulator S formally acts as follows:

Forward queries
On input $(x, y, ?)$:

1. Check if there exists a triplet (x', y', z') in the same[7] list and return that triplet if it exists.
2. If no such triplet exists, generate an integer j with Poisson distribution conditioned on being non-zero.
3. Check whether the query extends the chain tree.
4. If it does, query $RO(M)$ where M is the message corresponding to the new chain, and return the answer compatible with $RO(M)$.
5. Update the chain tree.
6. If no such chain is found, return a uniformly distributed answer.
7. In any case update the list of triplets with the answer and memorize the generated j.

Backward queries
On input $(?, y, z)$:

1. Check if there exists a triplet (x', y', z') in the same list with $(y, z) = (y', z')$.

---

[6] The same special treatment given to backward queries can be given to bridging queries as well when m is not significantly larger than n.

[7] I.e., T_f for f queries and T_G for g queries.

2. If no such triplet exists, generate an integer j with Poisson distribution.
3. Choose uniformly from the j possible answers (some may not be defined yet).
4. If the chosen answer is not defined, generate a uniform answer x.
5. If $j = 0$, set $x = \perp$.
6. In any case (even if $j = 0$) update the list of triplets with the answer and memorize the generated j.

Bridging queries

On input $(x, ?, z)$:

1. Generate a random y.
2. Generate an integer j with Poisson distribution conditioned on being non-zero.
3. Update the list of triplets with the answer and memorize the generated j.

3.2   The Indifferentiability Proof

Our hybrid argument will have five settings. In the first setting, we simply have the pair (RO, S^{RO}). In the second setting, we have the pair (R^{RO}, S^{RO}) where R simply relays the queries it receives to RO and answers with the responses it gets from RO. Since the view of any distinguisher D is identical with both pairs, we clearly have that

$$Pr[D^{RO, S^{RO}} = 1] = Pr[D^{R^{RO}, S^{RO}} = 1]$$

In the third setting, we have the pair $(R^{RO}, S_1{}^{RO})$ in which we slightly change the simulator S to S_1 such that when certain *unexpected events* occur, S_1 explicitly fails. Whenever an *unexpected event* occurs, S_1 fails explicitly, otherwise S_1 behaves exactly as S does.

Definition 3 (Unexpected events). *Let an* unexpected event *be the event that during an S query one of the following occurs:*

U_1 *During a forward query the answer triplet (x, y, z) is such that there exists a triplet (x', y', z')[8] in one of the lists, $(x, y) \neq (x', y')$ and either $z = z'$, $z = x'$ or $z = IV$.*

U_2 *During a backward query the answer triplet (x, y, z) is such that there exists a triplet (x', y', z') in one of the lists, $(y, z) \neq (y', z')$ and either $x = x'$, $x = z'$ or $x = IV$.*

U_3 *During a bridging query the answer triplet (x, y, z) is such that there exists a triplet (x', y', z') in one of the lists and $y = y'$.*

Lemma 1. *For any distinguisher D, the probability over the random coins of S and the random oracle RO that one of the unexpected events occurs is $\mathcal{O}\left(\frac{q^2}{2^n}\right)$.*

---

[8] Throughout this definition, answer triplets of the form $(\perp, y', z')$ are also considered.

Proof. We will prove that the probability of an unexpected event occurring at query number i, conditioned on the event that so far no unexpected events have occurred is $\mathcal{O}\left(\frac{q}{2^n}\right)$. Using the union bound over all q queries we then get that the probability of the unexpected event is $\mathcal{O}\left(\frac{q^2}{2^n}\right)$.

We will examine each of the three possible unexpected events and bound their probability. We first analyze what happens if during the query, no chain is completed. In this case, for forward queries the answer triplet has a uniformly distributed z and therefore the probability of $z = z'$ or $z = x'$ for any of the existing x', z' is bounded by $\frac{2q}{2^n}$. For bridging queries, the answer triplet has a uniformly distributed y and therefore the probability of $y = y'$ for any of the existing y' list is bounded by $\frac{q}{2^m}$. For backward queries the answer triplet has a uniformly distributed x and therefore the probability of $x = x'$ or $x = z'$ for any of the existing x', z' in the respective list is bounded by $\frac{2q}{2^n}$.

We now examine the case in which the query completes a chain. For a backward and bridging queries the simulator's answer does not depend on the fact that a chain has been completed and therefore the probability of an unexpected event is the same as before. For forward queries, the response of the simulator is fully determined by $RO(M)$. However, the value of $RO(M)$ is uniformly distributed and hence so is the simulator's answer. Therefore, also in this case the probability of U_1 occurring is at most $\frac{2q}{2^n}$. Concluding, we have that the probability of an unexpected event occurring conditioned that no such events have happened so far is $\mathcal{O}\left(\frac{q}{2^n}\right)$. A union bound over all the queries gives us the bound $\mathcal{O}\left(\frac{q^2}{2^n}\right)$ as required.[9] □

Lemma 2

1. *If we condition on the event that no unexpected events occur, then for every distinguisher the view when interacting with the pair (R^{RO}, S^{RO}) is identical to view when interacting with the pair (R^{RO}, S_1^{RO})*
2. $\left| Pr[D^{R^{RO}, S^{RO}} = 1] - Pr[D^{R^{RO}, S_1^{RO}} = 1] \right| = \mathcal{O}\left(\frac{q^2}{2^n}\right)$

Proof. Unless an unexpected event occurs, S_1 behaves exactly the same as S. This proves the first part of the lemma. Putting this result together with the fact that the probability of an unexpected event is bounded by $\mathcal{O}\left(\frac{q^2}{2^n}\right)$, proves the second part as well. □

Now we turn our attention to the fourth setting, in which we examine the pair $(R_1^{S_1}, S_1^{RO})$, where R_1 answers its RO queries on input M by using S_1 to calculate $C(M)$. I.e., R_1 queries S on all the required f and g queries. Notice that a single query to R_1 with an l-block message will result in l queries to S_1, for this reason R_1 queries cost l times more than a S_1 query.

Lemma 3. *If no unexpected events occur, then chains are only created by forward queries.*

---

[9] Note that even though using the union bound is usually not tight, in this case we get the birthday bound which is indeed tight.

Proof. Notice that when a chain is created, the message M is already determined. Without loss of generality, let the query which completes the chain be a f, f^{-1} or f^* query. In this case, all the g triplets in the chain have already been made and in particular, M is defined. Now, if a chain were created using a bridging query f^*, then the answer triplet (x, y, z) is such that $y \in M$ (as it completes a chain) and in particular y appears in a triplet in T_g, implying that the unexpected event U_2 occurred. If the chain were created using a backward query f^{-1}, then as the answer query (x, y, z) completed a chain, we know that x appears in a triplet in the T_f list or $x = IV$. Since (x, y, z) did not appear in T_f prior to the query (otherwise the chain would have been completed before) this implies that the unexpected event U_3 has occurred. Therefore, if no unexpected events occur all chains are created by forward queries. □

Corollary 1. *If no unexpected events occur, the chain data structure is a tree containing all chains.*

Proof. If a forward call creates a cycle in the chain data structure, then unexpected event U_1 occurs. Hence, the chain data structure is a tree. Notice that if more that one chain is created during a forward call, then unexpected event U_1 has occurred previously (as there are two identical nodes in the chain tree). Therefore, at most a single chain is created during each forward call and the simulator tracks them correctly. □

Lemma 4. *Unless an unexpected event occurs, then for every distinguisher the view when interacting with the pair $(R^{RO}, S_1{}^{RO})$ is indifferentiable from the view when interacting with the pair $(R_1{}^{S_1}, S_1{}^{RO})$.*

Proof. The proof will demonstrate the following three points:

1. Unless an unexpected event occurs when interacting with the pair $(R^{RO}, S_1{}^{RO})$, the answers given by S_1 are consistent with those given by R^{RO}.
2. Unless an unexpected event occurs when interacting with the pair $(R_1{}^{S_1}, S_1{}^{RO})$, the answers given by S_1 are consistent with those given by $R_1{}^{S_1}$.
3. Unless an unexpected event occurs when interacting either with the pair $(R^{RO}, S_1{}^{RO})$ or with the pair $(R_1{}^{S_1}, S_1{}^{RO})$, the answers given by R^{RO} are exactly the same as those given by $R_1{}^{S_1}$.

Proof of point 1. Notice that from Lemma 3 we know that chains are only completed by forward queries. This implies that the simulator's answers are consistent with the value $RO(M)$ for any message M. Since $R^{RO}(M)$ simply replies with $RO(M)$, the answers given by both oracles are consistent.

Proof of point 2. The proof is similar to the proof of the previous point. The simulator's answers are always consistent with the value $RO(M)$ for any message M and $R_1{}^{S_1}(M) = RO(M)$ since the behavior of S_1 ensures this result.

Proof of point 3. This point is obvious since $R^{RO}(M) = RO(M)$ and also $R_1{}^{S_1}(M) = RO(M)$.

It now follows that unless an unexpected event occurs, the views generated by any distinguisher's interaction with the pairs $(R^{RO}, S_1{}^{RO})$ and $(R_1{}^{S_1}, S_1{}^{RO})$ are indifferentiable. □

We are now ready for the proof of our main theorem:

Theorem 1. *The construction $C(M) = F(M) \oplus G(M)$, where F, G are iterated hash functions based on the compression function f and g respectively, is indifferentiable in $q \ll 2^{n/2}$ queries from a random oracle even in the presence of f^{-1}, f^*, g^{-1} and g^* oracles.*

Proof. Let S be the simulator defined above and let Γ be an oracle encapsulating f, g, f^{-1}, f^*, g^{-1} and g^*. We will prove that for any distinguisher D

$$|Pr[D^{C,\Gamma} = 1] - Pr[D^{RO,S^{RO}} = 1]| = \mathcal{O}\left(\frac{q^2}{2^n}\right)$$

The lemmas so far have shown that $|Pr[D^{R_1^{S_1},S_1} = 1] - Pr[D^{RO,S^{RO}} = 1]| = \mathcal{O}\left(\frac{q^2}{2^n}\right)$ and that S can be implemented in time polynomial in the number of queries q. It remains to show that for any possible distinguisher the pairs $(R_1^{S_1}, S_1{}^{RO})$ and (C, Γ) are indifferentiable. Notice however that unless an unexpected event occurs, S exactly simulates Γ and $R_1{}^S$ exactly computes C. This completes the proof. □

We have shown that the construction $C(M) = F(M) \oplus G(M)$ (or any n-bit function of F and G which is uniquely invertible when its output and any one of its input parameters is known) is indifferentiable in $q \ll 2^{n/2}$ queries from a random oracle even in the presence of f^{-1}, f^*, g^{-1} and g^* oracles and hence finding collisions in $H(M) = F(M)\|G(M)$ requires $\mathcal{O}\left(2^{n/2}\right)$ queries, matching the known upper bound of Joux. Notice that the construction $C(M) = F(M) \oplus G(M)$ requires the same amount of underlying function calls as the zipper hash of Liskov, albeit having a larger internal state, while having the advantage of being streamable.

3.3   Comments

Note that even though we have proved a lower bound on the number of calls to the compression functions and hence on the running time of a collision finding attack, this does not give a corresponding lower bound on the amount of memory required for the attack. In fact we can use Pollard's rho algorithm to find such a collision using only a linear amount of memory. Let $M_1^{0,1} M_2^{0,1}...M_n^{0,1}$ and $N_1^{0,1} N_2^{0,1}...N_n^{0,1}$ be Joux multicollisions for F and G respectively. We define two functions r_1, r_2 s.t. $r_1(x) = F(N_1^{x_1} N_2^{x_2}...N_n^{x_n})$ and $r_2(x) = G(M_1^{x_1} M_2^{x_2}...M_n^{x_n})$. We now use the rho algorithm to find a cycle in the path generated by iteratively alternating between applications of r_1 and r_2. The memory complexity is $O(n)$ while the time complexity is $O(n2^{\frac{n}{2}})$.

4   Application to Lucks' Double Pipe Proposal

The same proof framework can be used to prove other indifferentiability results. For example, the double pipe hash from [12] can also be proved indifferentiable from a random oracle in the model of weak compression functions. Given a compression function f, the double pipe hash has a 2n bit internal state (r, s) and is defined as follows:

- Split a message M into k blocks each of size $(m - n)$ bits, $x_1, \ldots, x_k$.
- Set $r_0 = IV_1$, $s_0 = IV_2$ where IV_1 and IV_2 are the initialization vectors.
- For each message block x_i compute $r_i = f(r_{i-1}, s_{i-1} \| x_i)$ and $s_i = f(s_{i-1}, r_{i-1} \| x_i)$.
- Output $DP^f(M) = f(IV_3, r_k \| s_k \| o^{m-2n})$.

The double pipe hash is schematically described in Figure 2.

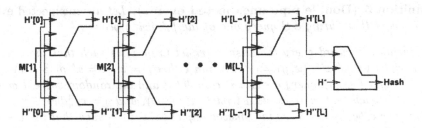

Fig. 2. Lucks' double pipe hash (taken from [12])

Note that Lucks proved that the double pipe hash is not vulnerable to multicollision or multi-(second)-preimage attacks when the underlying compression function is modeled as a random oracle (or ideal cipher) which has no weaknesses, while Liskov [11] claimed (without proof) that the construction is indifferentiable from a random oracle if the two pipes use two unrelated weak compression functions f and g. We will prove that the original construction is indifferentiable from a random oracle even when the same function is used in both pipes, and it is weak in the sense that the attacker is given both inversion and bridging oracles. Our proof will also hold if the final hash is replaced by a xor operation, or any function which is uniquely invertible when its output and any one of its input parameters are known.

The proof outline is identical to the one presented in Section 3; we will therefore only give the main lemmas required. We start by giving an adequate definition of *chains* in the double pipe hash, that following the example in section 3 captures the essence of consistency between the simulator S and the random oracle RO.

Definition 4 (Double pipe hash chains). *A (double pipe hash) chain is a triplet M, h_1, h_2, where M is a k block message and h_1 and h_2 are hash values. In addition we require that*

$$f(f(...f(IV_1, s_1 \| m_1), s_2 \| m_2), ..), s_k \| m_k) = h_1$$

$$f(f(...f(IV_2, r_1 \| m_1), r_2 \| m_2), ...), r_k \| m_k) = h_2$$

where r_i is the chaining value of the upper pipe after the first i blocks and s_i is the chaining value of the lower pipe after the same i blocks. We additionally require that all the intermediate links are defined in the list of known values (i.e., have been queried previously).

The simulator will be identical to the one introduced in Section 3 with the following changes: We will change the simulator's behavior when a chain is completed with message M. Without loss of generality, assume that the query which completed the chain is in the lower pipe. The simulator computes the value $z = RO(M)$, generates a random value d, sets the triplet $(IV_3, r_k \| d \| 0^{m-2n}, z)$ and returns d as the response to the query.

The unexpected events will now become:

Definition 5 (Double pipe unexpected events). *Let an unexpected event be the event that during a S query one of the following occurs:*

V_1 *During a forward query the answer triplet (x, y, z) is such that there exists a triplet (x', y', z'), $(x, y) \neq (x', y')$ and either $z = z'$, $z = x'$ or $z = IV$.*

V_1^* *During a forward query a chain is completed and the random value d generated is such that there exists a triplet (x', y', z'), and $y = r_k \| d \| 0^{m-2n}$.*

V_2 *During a backward query the answer triplet (x, y, z) is such that there exists a triplet (x', y', z'), $(y, z) \neq (y', z')$ and either $x = x'$, $x = z'$ or $x = IV$.*

V_3 *During a bridging query the answer triplet (x, y, z) is such that there exists a triplet (x', y', z') and $y = y'$.*

Lemma 5. *The probability over the random coins of S and the random oracle RO, that an unexpected event occurs is $\mathcal{O}\left(\frac{q^2}{2^n}\right)$.*

Proof. As before, we will prove that the probability of an unexpected event occurring, conditioned on the event that so far no unexpected events have occurred is $\mathcal{O}\left(\frac{q}{2^n}\right)$. The proof for events V_1, V_2 and V_3 is identical to the proof for the corresponding events U_1, U_2 and U_3. It remains to bound the probability of V_1^*. However, since d is completely uniform, we have that the probability is at most $\frac{q}{2^n}$. Using a union bound over all queries gives us the required bound of $\mathcal{O}\left(\frac{q^2}{2^n}\right)$. □

As in the proof in Section 3, the main lemma will show that unless an unexpected event occurs, chains are only created during forward queries.

Lemma 6. *If no unexpected events occur, then chains are only created by forward queries.*

Proof. As before, notice that when a chain is created, the message M is already determined. Now, if a chain were created using a bridging query f^*, then the answer triplet (x, y, z) is such that $y \in M$ (as it completes a chain) and in particular y appears in an existing triplet, implying that the unexpected event V_2 occurred. If the chain were created using a backward query f^{-1}, then as the answer query (x, y, z) completed a chain, we know that x appears in an existing triplet or $x = IV$. Since (x, y, z) did not appear in the list of triplets prior to the query (otherwise the chain would have been completed before) this implies that the unexpected event V_3 has occurred. Therefore, if no unexpected events occur all chains are created by forward queries. $\square$

Theorem 2. *The double pipe hash construction is indifferentiable from a random oracle in the model of weak compression functions.*

Proof. The sequence of hybrids is the same as in the proof in Section 3 and culminates with the required result. $\square$

5 Conclusion

While the results of Joux [7], Kelsey and Schneier [9] and Kelsey and Kohno [8] have shown that there are a number of surprising attacks when the attacker is allowed more than $2^{n/2}$ time, we have shown that there is a surprising amount of 'life' below the $2^{n/2}$ barrier: Even an adversary with the power to invert compression functions on inputs of his choice in unit time is still unable to differentiate between a variety of hash constructions and a random oracle. It seems that there are two main issues at the heart of our results. The first is the assumed randomness of the compression function, which implies that with less than $2^{n/2}$ queries it is not feasible to use in an effective way the given inversion oracles. The second issue is the fact that during the simulation the simulator needs to maintain consistency with the random oracle. In order to do this, the simulator must somehow 'know' when the queries given so far define some final hash value. In all the examples we gave as well as in the zipper hash[11] of Liskov, the construction of the combined hash function is such that with overwhelming probability the simulator can always tell when a query determines the output of the hash.

References

1. Boneh, D., Boyen, X.: On the impossibility of efficiently combining collision resistant hash functions. In: Dwork, C. (ed.) CRYPTO 2006. LNCS, vol. 4117, pp. 570–583. Springer, Heidelberg (2006), http://www.cs.stanford.edu/~xb/crypto06b/
2. Brassard, G. (ed.): CRYPTO 1989. LNCS, vol. 435. Springer, Heidelberg (1990)
3. Coron, J.S., Dodis, Y., Malinaud, C., Puniya, P.: Merkle-damgård revisited: How to construct a hash function. In: Shoup, V. (ed.) CRYPTO 2005. LNCS, vol. 3621, pp. 430–448. Springer, Heidelberg (2005)
4. Cramer, R. (ed): Advances in Cryptology - EUROCRYPT 2005, 24th Annual International Conference on the Theory and Applications of Cryptographic Techniques, Aarhus, Denmark, May 22-26 (2005)Proceedings. Cramer, R. (ed): EUROCRYPT 2005. LNCS. vol. 3494. Springer (2005)
5. Damgård, I.: A Design Principle for Hash Functions. In: [2], pp. 416–427

6. Fischlin, M., Lehmann, A.: Security-amplifying combiners for collision-resistant hash functions. In: Menezes, A. (ed.) CRYPTO 2007. LNCS, vol. 4622, pp. 224–243. Springer, Heidelberg (2007)
7. Joux, A.: Multicollisions in Iterated Hash Functions. Application to Cascaded Constructions. In: Franklin, M. (ed.) CRYPTO 2004. LNCS, vol. 3152, pp. 306–316. Springer, Heidelberg (2004)
8. Kelsey, J., Kohno, T.: Herding Hash Functions and the Nostradamus Attack. In: Vaudenay, S. (ed.) EUROCRYPT 2006. LNCS, vol. 4004, pp. 183–200. Springer, Heidelberg (2006)
9. Kelsey, J., Schneier, B.: Second Preimages on n-Bit Hash Functions for Much Less than 2^n Work. In: [4], pp. 474–490
10. Klima, V.: Tunnels in Hash Functions: MD5 Collisions Within a Minute. Cryptology ePrint Archive, Report 2006/105 (2006), http://eprint.iacr.org/
11. Liskov, M.: Constructing an ideal hash function from weak ideal compression functions. In: Selected Areas in Cryptography, pp. 358–375 (2006)
12. Lucks, S.: A Failure-Friendly Design Principle for Hash Functions.. In: Roy, B. (ed.) ASIACRYPT 2005. LNCS, vol. 3788, pp. 474–494. Springer, Heidelberg (2005)
13. Maurer, U.M., Renner, R., Holenstein, C.: Indifferentiability, impossibility results on reductions, and applications to the random oracle methodology. In: Naor, M. (ed.) TCC 2004. LNCS, vol. 2951, pp. 21–39. Springer, Heidelberg (2004)
14. Merkle, R.C.: One Way Hash Functions and DES. In: [2], pp. 428–446
15. Pietrzak, K.: Non-trivial black-box combiners for collision-resistant hash-functions don't exist. In: Naor, M. (ed.) EUROCRYPT 2007. LNCS, vol. 4515, pp. 23–33. Springer, Heidelberg (2007)
16. Wang, X., Lai, X., Feng, D., Chen, H., Yu, X.: Cryptanalysis of the Hash Functions MD4 and RIPEMD. In: [4], pp. 1–18
17. Wang, X., Yu, H.: How to Break MD5 and Other Hash Functions. In: [4], pp. 19–35
18. Wang, X., Yu, H., Yin, Y.L.: Efficient Collision Search Attacks on SHA-0. In: Shoup, V. (ed.) CRYPTO 2005. LNCS, vol. 3621, pp. 1–16. Springer, Heidelberg (2005)
19. Yu, H., Wang, G., Zhang, G., Wang, X.: The Second-Preimage Attack on MD4. In: Desmedt, Y.G., Wang, H., Mu, Y., Li, Y. (eds.) CANS 2005. LNCS, vol. 3810, pp. 1–12. Springer, Heidelberg (2005)

History-Independent Cuckoo Hashing[*]

Moni Naor[1],[**], Gil Segev[1],[**], and Udi Wieder[2]

[1] Department of Computer Science and Applied Mathematics,
Weizmann Institute of Science, Rehovot 76100, Israel
{moni.naor,gil.segev}@weizmann.ac.il
[2] Microsoft Research, Silicon Valley Campus,
1065 La Avenida, Mountain View, CA 94043
uwieder@microsoft.com

Abstract. Cuckoo hashing is an efficient and practical dynamic dictionary. It provides expected amortized constant update time, worst case constant lookup time, and good memory utilization. Various experiments demonstrated that cuckoo hashing is highly suitable for modern computer architectures and distributed settings, and offers significant improvements compared to other schemes.

In this work we construct a practical *history-independent* dynamic dictionary based on cuckoo hashing. In a history-independent data structure, the memory representation at any point in time yields no information on the specific sequence of insertions and deletions that led to its current content, other than the content itself. Such a property is significant when preventing unintended leakage of information, and was also found useful in several algorithmic settings.

Our construction enjoys most of the attractive properties of cuckoo hashing. In particular, no dynamic memory allocation is required, updates are performed in expected amortized constant time, and membership queries are performed in worst case constant time. Moreover, with high probability, the lookup procedure queries only two memory entries which are independent and can be queried in parallel. The approach underlying our construction is to enforce a canonical memory representation on cuckoo hashing. That is, up to the initial randomness, each set of elements has a unique memory representation.

1 Introduction

Over the past decade an additional aspect in the design of data structures has emerged due to security and privacy considerations: a data structure may give away much more information than it was intended to. Computer folklore is rich with tales of such cases, for example, files containing information whose creators assumed had been erased, only to be revealed later in embarrassing circumstances[1].

---

[*] Due to space limitations we refer the reader to a longer version available at http://www.wisdom.weizmann.ac.il/~naor.

[**] Most of the work was done at Microsoft Research, Silicon Valley Campus.

[1] See [5] for some amusing anecdotes of this nature.

L. Aceto et al. (Eds.): ICALP 2008, Part II, LNCS 5126, pp. 631–642, 2008.
© Springer-Verlag Berlin Heidelberg 2008

When designing a data structure whose internal representation may be revealed, a highly desirable goal is to ensure that an adversary will not be able to infer information that is not available through the legitimate interface. Informally, a data structure is *history independent* if its memory representation does not reveal any information about the sequence of operations that led to its current content, other than the content itself.

In this paper we design a practical history-independent data structure. We focus on the *dictionary* data structure, which is used for maintaining a set under insertions and deletions of elements, while supporting membership queries. Our construction is inspired by the highly practical cuckoo hashing, introduced by Pagh and Rudler [25], and guarantees history independence while enjoying most of the attractive features of cuckoo hashing. In what follows we briefly discuss the notion of history independence and several of its applications, and the main properties of cuckoo hashing.

Notions of history independence. Naor and Teague [23], following Micciancio [18], formalized two notions of history independence: a data structure is weakly history independent if any two sequences of operations that lead to the same content induce the same distribution on the memory representation. This notion assumes that the adversary gaining control is a one-time event, but in fact, in many realistic scenarios the adversary may obtain the memory representation at several points in time. A data structure is strongly history independent if for any two sequences of operations, the distributions of the memory representation at all time-points that yield the same content are identical. Our constructions in this paper are strongly history independent. An alternative characterization of strong history independence was provided by Hartline et al. [14]. Roughly speaking, they showed that strong history independence is equivalent to having a canonical representation up to the choice of initial randomness. More formal definitions of the two notions are provided in Section 2.

Applications of history independence. History-independent data structures were introduced in a cryptographic setting. Micciancio showed that *oblivious trees*[2] can be used to guarantee privacy in the incremental signature scheme of Bellare, Goldreich and Goldwasser [2,3]. An incremental signature scheme is private if the signatures it outputs do not give any information on the sequence of edit operations that have been applied to produce the final document.

An additional cryptographic application includes, for example, designing vote storage mechanisms (see [4,21,22]). As the order in which votes are cast is public, a vote storage mechanism must be history independent in order to guarantee the privacy of the election process.

History-independent data structures are valuable beyond the cryptographic setting as well. Consider, for example, the task of reconciling two dynamic sets. We consider two parties each of which receives a sequence of insert and delete operations, and their goal is to determine the elements in the symmetric difference between their sets. Now, suppose that each party processes its sequence of

---

[2] These are trees whose shape does not leak information.

operations using a data structure in which each set of elements has a canonical representation. Moreover, suppose that the update operations are efficient and change only a very small fraction of the memory representation. In such a case, if the size of the symmetric difference is rather small, the memory representations of the data structures will be rather close, and this can enable an efficient reconciliation algorithm.

Cuckoo hashing. Pagh and Rudler [25] constructed an efficient hashing scheme, referred to as cuckoo hashing. It provides worst case constant lookup time, expected amortized constant update time, and uses roughly $2n$ words for storing n elements. Additional attractive features of cuckoo hashing are that no dynamic memory allocation is performed (except for when the tables have to be resized), and the lookup procedure queries only two memory entries which are independent and can be queried in parallel. These properties offer significant improvements compared to other hashing schemes, and experiments have shown that cuckoo hashing and its variants are highly suitable for modern computer architectures and distributed settings. Cuckoo hashing was found competitive with the best known dictionaries having an average case (but no non-trivial worst case) guarantee on lookup time (see, for example, [10,25,27,29]).

1.1   Related Work

Micciancio [18] formalized the problem of designing oblivious data structures. He considered a rather weak notion of history independence, and devised a variant of 2–3 trees whose shape does not leak information. This notion was strengthened by Naor and Teague [23] to consider data structures whose memory representation does not leak information. Their main contributions are two history-independent data structures. The first is strongly history independent, and supports only insertions and membership queries which are performed in expected amortized constant time. Roughly speaking, the data structure includes a logarithmic number of pair-wise independent hash functions, which determine a probe sequence for each element. Whenever a new element is inserted and the next entry in its probe sequence is already occupied, a "priority function" is used to determine which element will be stored in this entry and which element will be moved to the next entry in its probe sequence. The second data structure is a weakly history-independent data structure supporting insertions, deletions and membership queries. Insertions and deletions are performed in expected amortized constant time, and membership queries in worst case constant time. Roughly speaking, this data structure is a history-independent variant of the perfect hash table of Fredman, Komlós and Szemerédi [12] and its dynamic extension due to Dietzfelbinger et al. [7].

Buchbinder and Petrank [6] provided a separation between the two notions of history independence for comparison based algorithms. They established lower bounds for obtaining strong history independence for a large class of data structures, including the heap and the queue data structures. They also demonstrated that the heap and queue data structures can be made weakly history independent without incurring any additional (asymptotic) cost.

Blelloch and Golovin [5] constructed two strongly history-independent data structures based on linear probing. Their first construction supports insertions, deletions and membership queries in expected constant time. This essentially extends the construction of Naor and Teague [23] that did not support deletions. While the running time in the worst case may be large, the expected update time and lookup time is tied to that of linear probing and thus is $O(1/(1 - \alpha)^3)$ where α is the memory utilization of the data structure (i.e., the ratio between the number of items and the number of slots). Their second construction supports membership queries in worst case constant time while maintaining an expected constant time bound on insertions and deletions. However, the memory utilization of their second construction is only about 9%. In addition, it deploys a two-level encoding, which may involve hidden constant factors that affect the practicality of the scheme. Furthermore, the worst case guarantees rely on an exponential amount of randomness and serves as a basis for a different hash table with more relaxed guarantees. The goal of our work is to design a hash table with better memory utilization and smaller hidden constants in the running time, even in the worst case.

1.2   Our Contributions

We construct a practical history-independent data structure that supports insertions, deletions, and membership queries. Our construction is based on cuckoo hashing, and shares most of its properties. Our construction provides the following performance guarantees (where the probability is taken only over the randomness used during the initialization phase of the data structure):

1. Insertions and deletions are performed in expected amortized constant time. Moreover, with high probability, insertion and deletions are performed in time $O(\log n)$ in the worst case.
2. Membership queries are performed in worst case constant time. Moveover, with high probability, the lookup procedure queries only two memory entries which are independent and can be queried in parallel.
3. The memory utilization of the data structure is roughly 50% when supporting only insertions and membership queries. When supporting deletions the data structure allocates an additional pointer for each entry. Thus, the memory utilization in this case is roughly 25%, under the conservative assumption that the size of a pointer is not larger than that of a key.

We obtain the same bounds as the second construction of Blelloch and Golovin [5] (see Section 1.1). The main advantages of our construction are its simplicity and practicality: membership queries would mostly require only two independent memory probes, and updates are performed in a way which is almost similar to cuckoo hashing and thus is very fast. A major advantage of our scheme is that it *does not use rehashing*. Rehashing is a mechanism for dealing with a badly behaved hash function by choosing a new one; using such a strategy in a strongly history-independent environment causes many problems (see below).

Furthermore, our data structure enjoys a better memory utilization, even when supporting deletions. We expect that in any practical scenario, whenever cuckoo hashing is preferred over linear probing, our construction should be preferred over those of Blelloch and Golovin.

1.3   Overview of the Construction

In order to describe our construction we first provide a high-level overview of cuckoo hashing. Then, we discuss our approach which is based on the underlying properties of cuckoo hashing.

Cuckoo hashing. Cuckoo hashing uses two tables T_0 and T_1, each consisting of $r \geq (1 + \epsilon)n$ words for some constant $\epsilon > 0$, and two hash functions $h_0, h_1 : \mathcal{U} \to \{0, \ldots, r - 1\}$. An element $x \in \mathcal{U}$ is stored either in entry $h_0(x)$ of table T_0 or in entry $h_1(x)$ of table T_1, but never in both. The lookup procedure is straightforward: when given an element $x \in \mathcal{U}$, query the two possible memory entries in which x may be stored. The deletion procedure deletes x from the entry in which it is stored. As for insertions, Pagh and Rudler [25] demonstrated that the "cuckoo approach", kicking other elements away until every element has its own "nest", leads to a highly efficient insertion procedure when the functions h_0 and h_1 are assumed to sample an element in $[r]$ uniformly and independently. More specifically, in order to insert an element $x \in \mathcal{U}$ we first query entry $T_0[h_0(x)]$. If this entry is not occupied, we store x in that entry. Otherwise, we store x at that entry anyway, thus making the previous occupant "nestless". This element is then inserted to T_1 in the same manner, and so forth iteratively. We refer the reader to [25] for a more comprehensive description of cuckoo hashing.

Our approach. Cuckoo hashing is not history independent. The table in which an element is stored depends upon the elements inserted previously. Our approach is to enforce a canonical memory representation on cuckoo hashing. That is, up to the initial choice of the two hash functions, each set of elements has only one possible representation. As in cuckoo hashing, our construction uses two hash tables T_0 and T_1, each consisting of $r \geq (1 + \epsilon)n$ entries for some constant $\epsilon > 0$, and two hash functions $h_0, h_1 : \mathcal{U} \to \{0, \ldots, r - 1\}$. An element $x \in \mathcal{U}$ is stored either in cell $h_0(x)$ of table T_0 or in cell $h_1(x)$ of table T_1.

Definition 1.1. *Given a set $S \subseteq \mathcal{U}$ and two hash functions h_0 and h_1, the cuckoo graph is the bipartite graph $G = (L, R, E)$ where $L = R = \{0, \ldots, r - 1\}$, and $E = \{(h_0(x), h_1(x)) : x \in S\}$.*

The cuckoo graph plays a central role in our analysis. It is easy to see that a set S can be successfully stored using the hash functions h_0 and h_1 if and only if no connected component in G has more edges then nodes. In other words, every component contains at most one cycle (i.e., unicyclic). The analysis of the insertion and deletion procedures are based on bounds on the size of a connected component. The following lemma is well known in random graph theory (see, for example, [15, Section 5.2]), and implies that the expected size of each component is constant, and with high probability it is $O(\log n)$:

Lemma 1.1. *Assume $r \geq (1+\epsilon)n$ and the two hash functions are truly random. Let v be some node and denote by C the connected component of v. Then there exists some constant $\beta = \beta(\epsilon) \in (0,1)$ such that for any integer $k > 0$ it holds that $\Pr[||C| > k] \leq \beta^k$.*

In order to describe the canonical representation that our construction enforces it is sufficient to describe the canonical representation of each connected component in the graph. Let C be a connected component, and denote by S be the set of elements that are mapped to C. In case C is acyclic, we enforce the following canonical representation: the minimal element in S (according to some fixed ordering of $\mathcal{U}$) is stored in *both tables*, and this yields only one possible way of storing the remaining elements. In case C is unicyclic, we enforce the following canonical representation: the minimal element *on the cycle* is stored in table T_0, and this yields only one possible way of storing the remaining elements. The most challenging aspect of our work is dealing with the unlikely event in which a connected component contains more than one cycle.

Rehashing and history independence. It is known [17] that even if h_0 and h_1 are completely random functions, with probability $\Omega(1/n)$ there will be a connected component with more than one cycle. In this case the given set cannot be stored using h_0 and h_1. The standard solution for this scenario is to choose new functions and rehash the entire data. In the setting of strongly history-independent data structures, however, rehashing is particular problematic and affects the practical performance of the data structure. Consider, for example, a scenario in which a set is stored using h_0 and h_1, but when inserting an additional element x it is required to choose new hash functions h'_0 and h'_1, and rehash the entire data. If the new element x is now deleted, then in order to maintain history independence we must "roll back" to the previous hash functions h_0 and h_1, and once again rehash the entire data. This has two undesirable properties: First, when rehashing we cannot erase the description of any previous pair of hash functions, as we may be forced to roll back to this pair later on. When dealing with strongly history-independent data structures, a canonical representation for each set of elements must be determined at the initialization phase of the data structure. Therefore, all the hash functions must be chosen in advance, and this may lead to a high storage overhead (as is the case in [5]). Secondly, if an element that causes a rehash is inserted and deleted multiple times, each time an entire rehash must be performed.

Avoiding rehashing by stashing elements. Kirsch et al. [16] suggested a practical augmentation to cuckoo hashing in order to avoid rehashing: exploiting a secondary data structure for storing elements that create cycles, starting from the second cycle of each component. That is, whenever an element is inserted to a unicyclic component and creates an additional cycle in this component, the element is *stashed* in the secondary data structure. In our case, the choice of the stashed element must be history independent in order to guarantee that the whole data structure is history independent. Kirsch et al. prove the following bound on the number of stashed elements in the secondary data structure:

Lemma 1.2. *Assume* $r \geq (1+\epsilon)n$ *and the two hash functions are truly random. The probability that the secondary data structure has more than s elements is* $O(r^{-s})$.

The secondary data structure in our construction can be any strongly history-independent data structure (such as a sorted list). This approach essentially reduces the task of storing n elements in a history independent manner to that of storing only a few elements in a history-independent manner. In addition, it enables us to avoid rehashing and to increase the practicality of our scheme.

1.4   Paper Organization

The remainder of this paper is organized as follows. In Section 2 we overview the notion of history independence. In Section 3 we describe our data structure. In Section 4 we propose several possible instantiations for the secondary data structure used in our construction, and in Section 5 we provide several concluding remarks.

2   Preliminaries

A data structure is defined by a list of operations. We say that two sequences of operations, S_1 and S_2, yield the same content if for all suffixes T, the results returned by T when the prefix is S_1 are identical to those returned by T when the prefix is S_2.

Definition 2.1 (Weak History Independence). *A data structure implementation is weakly history independent if any two sequences of operations that yield the same content induce the same distribution on the memory representation.*

We consider a stronger notion of history independence that deals with cases in which an adversary may obtain the memory representation at several points in time. In this case it is required that for any two sequences of operations, the distributions of the memory representation at all time-points that yield the same content are identical.

Definition 2.2 (Strong History Independence). *Let S_1 and S_2 be sequences of operations, and let $P_1 = \{i_1^1, \ldots, i_\ell^1\}$ and $P_2 = \{i_1^2, \ldots, i_\ell^2\}$ be two lists such that for all $b \in \{1, 2\}$ and $1 \leq j \leq \ell$ it holds that $1 \leq i_j^b \leq |S_b|$, and the content of the data structure following the i_j^1 prefix of S_1 and the i_j^2 prefix of S_2 are identical. A data structure implementation is strongly history independent if for any such sequences the distributions of the memory representation at the points of P_1 and at the corresponding points of P_2 are identical.*

Note that Definition 2.2 implies, in particular, that any data structure in which the memory representation of each state is fully determined given the randomness used during the initialization phase is strongly history independent. Our construction in this paper enjoys such a canonical representation, and hence is strongly history independent.

3   The Data Structure

Our data structure uses two tables T_0 and T_1, and a secondary data structure. Each table consists of $r \geq (1 + \epsilon)n$ entries for some constant $\epsilon > 0$. In the insert-only variant each entry stores at most one element. In the variant which supports deletions each entry stores at most one element and a pointer to another element. The secondary data structure can be chosen to be any strongly history-independent data structure (we refer the reader to Section 4 for several possible instantiations of the secondary data structure).

Elements are inserted into the data structure using two hash functions $h_0, h_1 :$ $\mathcal{U} \rightarrow \{0, \ldots, r - 1\}$, which are independently chosen at the initialization phase. An element $x \in \mathcal{U}$ can be stored in three possible locations: entry $h_0(x)$ of table T_0, entry $h_1(x)$ of table T_1, or stashed in the secondary data structure. The lookup procedure is straightforward: when given an element $x \in \mathcal{U}$, query the two tables and perform a lookup in the secondary data structure.

In the remainder of this section we first describe the canonical representation of the data structure and some of its useful properties. Then, we provide a high-level description of the insertion and deletion procedures. Due to space limitations the proof of history independence and the efficiency analysis (proving the performance guarantees claimed in Section 1.2) are omitted from this version.

The canonical representation. As mentioned in Section 1.3, it is sufficient to consider a single connected component in the cuckoo graph. Let C be a connected component, and denote by S the set of elements that are mapped to C. We distinguish between the following cases:

- C is a tree. In this case the minimal element in S is stored in both tables, and this yields only one possible way of storing the remaining elements.
- C is unicyclic. In this case the minimal element *on the cycle* is stored in table T_0, and this yields only one possible way of storing the remaining elements.
- C contains at least two cycles. In this case we iteratively put in the secondary data structure the largest element that lies in a cycle, until C contains only one cycle. The elements which remain in the component are arranged according to the previous case. We note that this case is rather unlikely, and occurs with only a polynomially small probability.
- When supporting deletions each table entry includes additional space for one pointer. These pointers form a cyclic sorted list of the elements of the component (not including stashed elements). When deletions are not supported, there is no need to allocate or maintain the additional pointers.

When describing the insertion and deletion procedures it will be convenient to consider the cuckoo graph as a *directed* graph. Given an element x, we orient the edge so that x is stored at its tail. In other words, if x is stored in table T_b for some $b \in \{0, 1\}$, we orient its corresponding edge in the graph from $T_b[h_b(x)]$ to $T_{1-b}[h_{1-b}(x)]$. An exception is made for the minimal element of an acyclic component, since such an element is stored in both tables. In such a case we orient the corresponding edge in both directions. The following claims state straightforward properties of the directed graph:

Claim 3.1. *Let $x_1 \to \cdots \to x_k$ be any directed path. Then, given the element x_1 it is possible to retrieve all the elements on this path using k probes to memory. Furthermore, if x_{min} is a minimal element in an acyclic component C, then for any element x stored in C there exists a directed path from x to x_{min}.*

Claim 3.2. *Let C be a unicyclic component, and let x^* be any element on its cycle. Then for any element x stored in C there exists a simple directed path from x to x^*.*

The insertion procedure. Given an element to insert x, the goal of the insertion procedure is to insert x while maintaining the canonical representation. Note that one only has to consider the representation of the connected component of the cuckoo graph in which x resides. Furthermore, Lemma 1.1 implies that the size of the component is $O(1)$ on expectation, thus an algorithm which is linear in the size of the component would have a constant expected running time. In the following we show that the canonical memory representation could be preserved without using the additional pointers. The additional pointers are only needed for supporting deletions. If the additional pointers are maintained, then once the element is inserted the pointers need to be updated so that the element is in its proper position in the cyclic linked list. This could be done in a straightforward manner in time linear in the size of the component.

Given an element $x \in U$ there are four possible cases to consider. The first and simplest case is when both $T_0[h_0(x)]$ and $T_1[h_1(x)]$ are unoccupied, and we store x in both entries. The second and third cases are when one of the entries is occupied and the other is not occupied. In these cases x does not create a new cycle in the graph. Thus, unless x is the new minimal element in an acyclic component it is simply put in the empty slot. If x is the new minimal element in an acyclic component, it is put in both tables and the appropriate elements are pushed to their alternative location, effectively removing the previous minimum element from one of the tables. The fourth case, in which both entries are occupied involves slightly more details, but is otherwise straightforward. In this case x either merges two connected components, or creates a new cycle in a component. The latter case may also trigger the low probability event of stashing an element in the secondary data structure. Due to space limitations, a formal description of the procedure is provided in the longer version of the paper.

The deletion procedure. The deletion procedure takes advantage of the additional pointer stored in each entry. Recall that these pointers form a cyclic list of all the elements of a connected component. Note that since the expected size of a connected component is constant, and the expected size of the secondary data structure is constant as well, a straightforward way of deleting an element is to retrieve all the elements in its connected component, reinsert them without the deleted element, and then reinsert all the elements that are stashed in the secondary data structure. This would result in expected amortized constant deletion time. In practice, however, it is desirable to minimize the amount of memory manipulations. In the longer version of the paper we detail a more refined procedure, which although shares the same asymptotic performance, is more sensible in practice.

4   The Secondary Data Structure

In this section we propose several possible instantiations for the secondary data structure. As discussed in Section 1.3, the secondary data structure can be any strongly history-independent data structure. Recall that Lemma 1.2 implies in particular that the expected number of stashed elements is constant, and with overwhelming probability there are no more than $\log n$ stashed elements. Thus, the secondary data structure is essentially required to store only a very small number of elements. Furthermore, since the secondary data structure is probed every time a lookup is performed, it is likely to reside most of the time in the cache, and thus impose a minimal cost.

The practical choice. The most practical approach is instantiating the secondary data structure with a sorted list. A sorted list is probably the simplest data structure which is strongly history independent. When a sorted list contains at most s elements, insertions and deletions are performed in time $O(s)$ in the worst case, and lookups are performed in time $O(\log s)$ in the worst case. In turn, instantiated with a sorted list, our data structure supports insertions, deletions, and membership queries in expected constant time. Moreover, Lemma 1.2 implies that the probability that a lookup requires more than k probes is at most $O(n^{-2^k})$.

Constant worst case lookup time. We now propose two instantiations that guarantee constant lookup time in the *worst case*. We note that these instantiations result in a rather theoretical impact, and in practice we expect a sorted list to perform much better.

One possibility is using the strongly history-independent data structure of Blelloch and Golovin [5], and in this case our data structure supports insertions and deletions in expected constant time, and membership queries in worst case constant time. Another possibility is using any deterministic perfect hash table with constant lookup time. On every insertion and deletion we reconstruct the hash table, and since its construction is deterministic, the resulting data structure is strongly history independent. The repeated reconstruction allows us to use a static hash table (instead of a dynamic hash table), and in this case the construction time of the table determines the insertion and deletion time. Perfect hash tables with such properties were suggested by Alon and Naor [1], Miltersen [19], and Hagerup, Miltersen and Pagh [13]. Asymptotically, the construction of Hagerup et al. is the most efficient one, and provides an $O(s \log s)$ construction time on s elements. Instantiated with their construction, our data structure supports insertions and deletion in expected constant time, and membership queries in worst case constant time.

5   Concluding Remarks

On using $O(\log n)$-wise independent hash functions. One possible drawback of our construction, from a purely theoretical point of view, is that we assume the availability of truly random hash functions, while the constructions of Blelloch and Golovin assume $O(\log n)$-wise independent hash functions (when guaranteeing worst case constant lookup time) or 5-wise independent hash functions (when guaranteeing expected constant lookup time). Nevertheless, simulations (see, for example, [25]) give a strong evidence that simple heuristics work for the choice of the hash functions as far as cuckoo hashing is concerned (Mitzenmacher and Vadhan [20] provide some theoretical justification). Thus we expect our scheme to be efficient in practice.

Our construction can be instantiated with $O(\log n)$-wise independent hash functions, and still provide the same performance guarantees for insertions, deletions, and membership queries. However, in this case the bound on the number of stashed elements is slightly weaker than that stated in Lemma 1.2. Nevertheless, rather standard probabilistic arguments can be applied to argue that (1) the expected number of stashed elements is constant, and (2) the expected size of a connected component in the cuckoo graph is constant.

Alternatively, our construction can be instantiated with the highly efficient hash functions of Dietzfelbinger and Woelfel [9] (improving the constructions of Siegel [28] and Ostlin and Pagh [24]). These hash functions are almost n^δ-wise independent with high probability (for some constant $0 < \delta < 1$), can be evaluated in constant time, and each function can be described using only $O(n)$ memory words. One possible drawback of this approach is that the distance to n^δ-independence is only polynomially small.

Memory utilization. Our construction achieves memory utilization of essentially 50% (as in cuckoo hashing), and of 25% when supporting deletions. More efficient variants of cuckoo hashing [8,11,26] circumvent the 50% barrier and achieve better memory utilization by either using more than two hash functions, or storing more than one element in each entry. It would be interesting to transform these variants to history-independent data structures while essentially preserving their efficiency.

References

1. Alon, N., Naor, M.: Derandomization, witnesses for Boolean matrix multiplication and construction of perfect hash functions. Algorithmica 16(4-5), 434–449 (1996)
2. Bellare, M., Goldreich, O., Goldwasser, S.: Incremental Cryptography: The Case of Hashing and Signing. In: Desmedt, Y.G. (ed.) CRYPTO 1994. LNCS, vol. 839, pp. 216–233. Springer, Heidelberg (1994)
3. Bellare, M., Goldreich, O., Goldwasser, S.: Incremental cryptography and application to virus protection. In: 27th STOC, pp. 45–56 (1995)
4. Bethencourt, J., Boneh, D., Waters, B.: Cryptographic methods for storing ballots on a voting machine. In: 14th NDSS, pp. 209–222 (2007)

5. Blelloch, G.E., Golovin, D.: Strongly history-independent hashing with applications. In: 48th FOCS, pp. 272–282 (2007)
6. Buchbinder, N., Petrank, E.: Lower and upper bounds on obtaining history-independence. Inf. Comput. 204(2), 291–337 (2006)
7. Dietzfelbinger, M., Karlin, A.R., Mehlhorn, K., auf der Heide, F.M., Rohnert, H., Tarjan, R.E.: Dynamic perfect hashing: Upper and lower bounds. SIAM J. Comput. 23(4), 738–761 (1994)
8. Dietzfelbinger, M., Weidling, C.: Balanced allocation and dictionaries with tightly packed constant size bins. Theor. Comput. Sci. 380(1-2), 47–68 (2007)
9. Dietzfelbinger, M., Woelfel, P.: Almost random graphs with simple hash functions. In: 35th STOC, pp. 629–638 (2003)
10. Erlingsson, Ú., Manasse, M., McSherry, F.: A cool and practical alternative to traditional hash tables. In: 7th Workshop on Distributed Data and Structures (2006)
11. Fotakis, D., Pagh, R., Sanders, P., Spirakis, P.G.: Space efficient hash tables with worst case constant access time. Theor. Comput. Sys. 38(2), 229–248 (2005)
12. Fredman, M.L., Komlós, J., Szemerédi, E.: Storing a sparse table with $O(1)$ worst case access time. J. ACM 31(3), 538–544 (1984)
13. Hagerup, T., Miltersen, P.B., Pagh, R.: Deterministic dictionaries. J. Algorithms 41(1), 69–85 (2001)
14. Hartline, J.D., Hong, E.S., Mohr, A.E., Pentney, W.R., Rocke, E.: Characterizing history independent data structures. Algorithmica 42(1), 57–74 (2005)
15. Janson, S., Łuczak, T., Ruciński, A.: Random Graphs. Wiley-Interscience, Chichester (2000)
16. Kirsch, A., Mitzenmacher, M., Wieder, U.: More robust hashing: Cuckoo hashing with a stash (manuscript, 2008)
17. Kutzelnigg, R.: Bipartite random graphs and cuckoo hashing. In: 4th Colloquium on Mathematics and Computer Science, pp. 403–406 (2006)
18. Micciancio, D.: Oblivious data structures: Applications to cryptography. In: 29th STOC, pp. 456–464 (1997)
19. Miltersen, P.B.: Error correcting codes, perfect hashing circuits, and deterministic dynamic dictionaries. In: 9th SODA, pp. 556–563 (1998)
20. Mitzenmacher, M., Vadhan, S.: Why simple hash functions work: Exploiting the entropy in a data stream. In: 19th SODA, pp. 746–755 (2008)
21. Molnar, D., Kohno, T., Sastry, N., Wagner, D.: Tamper-evident, history-independent, subliminal-free data structures on PROM storage -or- How to store ballots on a voting machine. In: IEEE S&P, pp. 365–370 (2006)
22. Moran, T., Naor, M., Segev, G.: Deterministic history-independent strategies for storing information on write-once memories. In: 34th ICALP, pp. 303–315 (2007)
23. Naor, M., Teague, V.: Anti-persistence: History independent data structures. In: 33rd STOC, pp. 492–501 (2001)
24. Ostlin, A., Pagh, R.: Uniform hashing in constant time and linear space. In: 35th STOC, pp. 622–628 (2003)
25. Pagh, R., Rodler, F.F.: Cuckoo hashing. J. of Algorithms 51(2), 122–144 (2004)
26. Panigrahy, R.: Efficient hashing with lookups in two memory accesses. In: 16th SODA, pp. 830–839 (2005)
27. Ross, K.A.: Efficient hash probes on modern processors. In: 23nd International Conference on Data Engineering, pp. 1297–1301 (2007)
28. Siegel, A.: On universal classes of fast high performance hash functions, their time-space tradeoff, and their applications. In: 30th FOCS, pp. 20–25 (1989)
29. Zukowski, M., Héman, S., Boncz, P.A.: Architecture conscious hashing. In: 2nd International Workshop on Data Management on New Hardware, vol. 6 (2006)

Building a Collision-Resistant Compression Function from Non-compressing Primitives

(Extended Abstract)

Thomas Shrimpton[1] and Martijn Stam[2]

[1] University of Lugano and Portland State University
thomas.shrimpton@unisi.ch
[2] EPFL
martijn.stam@epfl.ch

Abstract. We consider how to build an efficient compression function from a small number of random, non-compressing primitives. Our main goal is to achieve a level of collision resistance as close as possible to the optimal birthday bound. We present a $2n$-to-n bit compression function based on three independent n-to-n bit random functions, each called only once. We show that if the three random functions are treated as black boxes then finding collisions requires $\Theta(2^{n/2}/n^c)$ queries for $c \approx 1$. This result remains valid if two of the three random functions are replaced by a fixed-key ideal cipher in Davies-Meyer mode (i.e., $E_K(x) \oplus x$ for permutation E_K). We also give a heuristic, backed by experimental results, suggesting that the security loss is at most four bits for block sizes up to 256 bits. We believe this is the best result to date on the matter of building a collision-resistant compression function from non-compressing functions. It also relates to an open question from Black et al. (Eurocrypt'05), who showed that compression functions that invoke a single non-compressing random function cannot suffice.

Keywords: Hash Functions, Random Oracle Model, Compression Functions, Collision Resistance.

1 Introduction

The design of hash functions usually proceeds in two stages. First one designs a compression function with fixed domain, typically bitstrings of some small length. One then applies a domain extension method, such as the Merkle-Damgård transform [14,6], to the compression function in order to construct a hash function for messages of arbitrary length. The first part has our interest; in particular, the central problem considered by this paper is the following one:

Given a (small) number of independent n-to-n bit random (one-way) functions, construct a $2n$-to-n bit compression function with provable collision resistance as close as possible to the optimal $2^{n/2}$ birthday bound.

This problem is related to one recently considered by Maurer and Tessaro [12], who consider the problem of constructing a function $C : \{0,1\}^{m(n)} \rightarrow \{0,1\}^{l(n)}$ given a n-to-n bit random one-way function f. Setting $m(n) = 2n$ and $l(n) = n$ gives rise to a

L. Aceto et al. (Eds.): ICALP 2008, Part II, LNCS 5126, pp. 643–654, 2008.
© Springer-Verlag Berlin Heidelberg 2008

$2n$-to-n bit compression function. Maurer and Tessaro cast security in the indifferentia-bility framework, in which case domain extension is actually far more challenging than range extension. For all $\epsilon > 0$ they give a construction indifferentiable from a random function against adversaries making $\Theta(2^{n(1-\epsilon)})$ queries. Indifferentiability is a much stronger requirement than just collision resistance, indeed for $\epsilon \geq 1/2$ their functions have better proven security than ours. Unfortunately, the construction by Maurer and Tessaro is not very efficient, to get the required collision resistance for a $2n$-to-n bit compression function requires 99 calls to the underlying primitive f.

This raises the question whether more efficient constructions are possible, at least when one focuses primarily on collision resistance. Ideally we would like to make do with just one random function, and to invoke it once for each n-bit block of message digested; such a compression function would be called rate-1. Unfortunately, Black et al. [4] have given a negative result that all but rules this out. One can also show that rate-1/2 compression functions (with optimal collision resistance) are unlikely to exist. Thus the best one can hope for will be a rate-1/3 compression function.

We show that this hope *can* be realized. In particular, we present a compression function that calls random n-to-n bit functions f_1, f_2, f_3, and that has almost optimal collision resistance (here n is a parameter that can be chosen freely). When we consider the construction for increasing n, we show that any adversary making $\Theta(2^{n/2}/n^c)$ total oracle queries, for $c > 1$, has a vanishing probability of finding a collision in H. On the other hand, for $c < 1$ we provide compelling arguments that an adversary exists that will find a collision with high probability. Thus it is fair to say that finding collisions takes around $\Theta(2^{n/2}/n)$ queries. Experimental results, detailed in the full version [21], suggest that for n of cryptographic relevance (up to 256 bits), the loss is at most 4 bits of collision resistance. Of course, the standard caveats apply to these results when one instantiates the random functions in practice.

Central to our proof is the concept of the *yield* of a set of queries. This is the number of compression function evaluations an adversary can make given a certain number of queries to the underlying primitives. Somewhat surprisingly, we can show that for our compression function an adversary can do not much better than simply optimizing his yield and hoping for a collision via the birthday bound. The yield can also be used to get (crude) negative results, for instance for any rate-1/2 compression function there exists a (greedy) adversary with yield $2^{n/2}$ using only $2^{n/4}$ queries. Since we expect good hash functions to behave randomly, this indicates it is unlikely to find a rate-1/2 compression function with good collision resistance. This use of the yield to obtain impossibility results has recently been extensively generalized by Rogaway and Steinberger [20].

The construction itself is as follows: $H^{f_1, f_2, f_3}(V, M) = f_3(f_1(M) \oplus f_2(V)) \oplus f_1(M)$. A picture of the compression function is given in Figure 1. The compres-sion function can easily be transformed into a hash function with arbitrary domain while preserving the collision resistance (e.g., using the Merkle-Damgård transform). As a note of warning, we do not claim any "beyond-birthday" properties one might hope for from a hash function, such as resistance against multi-collisions and optimal preimage resistance. Indeed, preimages can typically be found in $O(2^{2n/3})$ queries, rather than the desired $\Omega(2^n)$. Rogaway and Steinberger [20] show that this reduced

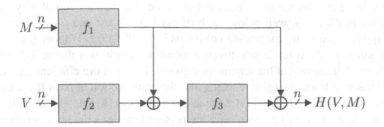

Fig. 1. The triple-function compression function. The functions f_1, f_2, f_3 are random n-to-n bit functions.

preimage-resistance is to a large extent inherent to rate-1/3 schemes and not an artefact of our particular scheme.

In the full version we also investigate what happens when one wants to instantiate f_1, f_2 and f_3 with a blockcipher with its key fixed (thus supplying the adversary with inverse oracles for each). When f_1 or f_2 are instantiated directly with a fixed-key ideal cipher (i.e., random two-way permutations instead of random functions), the construction breaks down badly. However, we show that by using a Davies-Meyer like construction in the place of f_1 and f_2 we can also deal with a fixed-key ideal cipher in those places without loss of security. Using a fixed-key ideal cipher for f_3 does not seem to affect collision resistance, yet we do not have a full proof for this.

The problem of building compression functions from non-compressing primitives has its historical roots in blockcipher-based hashing, although a blockcipher can already be thought of as compressing its key and plaintext into a ciphertext. That said, there seems to be no intrinsic theoretical reason to restrict designs to using blockciphers (or permutations), hence our focus on simple random one-way functions (cf. [3,12]).

Interestingly, the problem of building a compression function using random functions (as opposed to a fixed-key ideal cipher) has an unexpected link with that of constructing *double-length* hash functions. Specifically: given a $2n$-to-n bit compression function create a $4n$-to-$2n$ bit compression function with collision resistance close to the optimal 2^n. In the full version of this paper, we show that range extenders could be used as an alternative means to turn a number of non-compressing random functions into a compressing one, and vice versa. In particular, one can use our method to get a rate-1/3 $4n$-to-$2n$ bit double-length compression function with close to optimal collision resistance (in the output size) based on a set of $2n$-to-n bit random functions.

RELATED WORK. Bellare and Micciancio [1] introduce incremental hash functions, which in principle could be built upon a non-compressing primitive. Crucial differences with our work are that they build an entire hash function, not just a compression function, and that the collision resistance of their schemes is not based on query complexity (usually just n queries suffice for a collision), but on the presumed computational hardness of combining the query answers into an actual collision.

Bernstein [3] bases his Rumba20 compression function on these ideas. He xor's together the output of four pseudorandom generators, components of the Salsa20 streamcipher. By modelling the underlying primitives as (independent) random one-way functions, he shows an *upper* bound (and estimate) on the full complexity for collision finding of $O(2^{n/3})$, well below the birthday bound (and our *lower* bound). Note that the query complexity for finding collisions is $\Theta(2^{n/8})$ (cf. Wagner [22]). The expanding nature of Bernstein's primitives somewhat complicates theoretical efficiency comparisons, but the rate of his scheme is arguably 1/2, so more efficient than ours.

There has also been extensive research into the construction of hash functions based on blockciphers. Davies-Meyer [13], Matyas-Meyer-Oseas [11], and Miyaguchi-Preneel [15,18] are all well-known $2n$-to-n bit compression functions based on a single call to a blockcipher with n-bit key operating on n-bit blocks. These type of *rate-1* constructions were later systematically studied by Preneel et al. [18] and Black et al. [5], who identified twelve distinct constructions that provide optimal collision resistance when the blockcipher is modelled as an ideal cipher. Black et al. [5] showed that an additional eight constructions do not yield collision resistant compression functions, yet still lead to collision resistant hash functions when properly MD-iterated.

Most of the work related to blockcipher-based compression functions allows per-round rekeying [5,9,16,18]. This significantly eases design and proof. Unfortunately, rekeying has the drawback of entailing a significant computational cost: Gladman's implementation survey [8] shows that AES key scheduling would account for nearly 50% of the overall runtime. Thus a lower rate fixed-key solution might actually be more efficient. Fixing the key would make use of the blockcipher in a more natural way, namely setting up a key once and then processing relatively large amounts of data with it (the way that blockciphers are used for encryption). A fixed-key blockcipher would be modelled as a random two-way permutation.

Preneel et al. [17] propose a family of fixed-key constructions, but no formal security proof is given. The rate of their scheme is always strictly smaller then 1/2 and typically between 1/4 and 1/8. Recently (but subsequent to the initial presentation of our work), Rogaway and Steinberger [19] have constructed a rate-1/3 scheme based on a fixed-key ideal cipher that achieves security comparable to ours.

2   Preliminaries

GENERAL NOTATION. For a positive integer n, we write $\{0, 1\}^n$ for the set of all bitstrings of length n. When X and Y are strings we write $X \,\|\, Y$ to mean their concatenation and $X \oplus Y$ to mean their bitwise exclusive-or (xor). Unless specified otherwise, we will consider bitstrings as elements in the group $(\{0, 1\}^n, \oplus)$.

For positive integers m and n, we let $\mathrm{Func}(m, n)$ denote the set of all functions mapping $\{0, 1\}^m$ into $\{0, 1\}^n$. We write $f \xleftarrow{\$} \mathrm{Func}(m, n)$ to denote random sampling from the set $\mathrm{Func}(m, n)$ and assignment to f. Unless otherwise specified, all finite sets are equipped with a uniform distribution.

DISTRIBUTIONS AND TENSORS. With $(\{0, 1\}^n)^q$ we denote the set of q-element vectors, or *q-vectors*, in which each element is an n-bit string. When $\mathbf{a} \in (\{0, 1\}^n)^q$ and $\mathbf{b} \in (\{0, 1\}^n)^q$, we will write $\mathbf{a} = (a_1, \ldots, a_q)$ and $\mathbf{b} = (b_1, \ldots, b_q)$ when we wish to stress its components.

Fix a value q, and let $Q = q^2$. We define $\mathbf{a} \otimes \mathbf{b} \in (\{0,1\}^n)^Q$ as the tensor product under exclusive-or, where we identify $(\{0,1\}^n)^{q \times q}$ with $(\{0,1\}^n)^{q \cdot q} = (\{0,1\}^n)^Q$. More concretely $(\mathbf{a} \otimes \mathbf{b})_{i,j} = a_i \oplus b_j$ for i and j in $[1, \ldots, q]$. (Whenever possible we will try to use dummy i to refer to elements of $\mathbf{a}$ and dummy j to refer to those of $\mathbf{b}$.) If A and B are both distributions over $(\{0,1\}^n)^q$, this tensor operation induces a distribution over $(\{0,1\}^n)^Q$, which we will denote by the symbol $A \oplus B$. Unless otherwise specified, we will assume throughout that A and B are two distributions induced by sampling from $(\{0,1\}^n)^q$ *without* replacement. We will use U to denote the uniform distribution over $(\{0,1\}^n)^Q$ (where n and Q will often follow from the context). Thus U corresponds to sampling Q strings from $\{0,1\}^n$ uniformly and independently *with* replacement.

If in a random sample some value appears *exactly* k times, we say there is a *k-way collision* in that sample. Let $M_U(k)$ be the random variable describing the number of k-way collisions when the samples are drawn according to the distribution U. Similarly, let $M_{A \oplus B}(k)$ be the random variable describing the number of k-way collisions when the samples are drawn according to the distribution $A \oplus B$.

COMPRESSION FUNCTION SECURITY. When algorithms are provided with oracles, we write them as superscripts. A *compression function* is a mapping from $\{0,1\}^n \times \{0,1\}^m$ to $\{0,1\}^n$ for some $m, n > 0$. For us, a compression function H must be given by a program that, given (V, M), computes $H^{\cdots}(V, M)$ via access to a finite number of specified oracles. A *collision-finding adversary* is an algorithm with access to one or more oracles, whose goal it is to find collisions in some specified compression function.

Definition 1. *Let $n, \ell > 0$ be integer parameters. Let $H: \{0,1\}^n \times \{0,1\}^n \to \{0,1\}^n$ be a compression function taking ℓ oracles. Let A be a collision-finding adversary for H that takes ℓ oracles. The collision-finding advantage of A is defined to be*

$$\mathbf{Adv}_{H(n)}^{\text{coll}}(A) = \Pr\left[f_1, \ldots, f_\ell \xleftarrow{\$} \text{Func}(n, n), (V, M), (V', M') \leftarrow A^{f_1(\cdot), \ldots, f_\ell(\cdot)} : \right.$$
$$\left. (V, M) \neq (V', M') \text{ and } H^{f_1, \ldots, f_\ell}(V, M) = H^{f_1, \ldots, f_\ell}(V', M') \right]$$

Furthermore, for $q > 0$, we define $\mathbf{Adv}_{H(n)}^{\text{coll}}(q)$ as the maximum of $\mathbf{Adv}_{H(n)}^{\text{coll}}(A)$ over all adversaries A making at most q queries.

When the compression function H is defined for arbitrary positive integers n, we will be able to make asymptotic statements about collision resistance.

We consider information-theoretic adversaries in order to make a strong statement about collision resistance. That is, our adversaries are computationally unbounded and their complexity is measured only by the number of queries made to the oracles for the non-compressing primitives. Without loss of generality, we assume that adversaries do not repeat queries to oracles and that they do not query an oracle outside of its specified domain.

COLLISIONS. In bounding the probability of finding a collision in our compression function, we will need the distribution of k-way collisions when the samples from $\{0,1\}^n$ are distributed according to $A \oplus B$. For the sequel, to bound the collision finding probability either upwards or downwards, the following lemma suffices (proven in the full version).

Lemma 2. *For some positive integers q, n, let $Q = q^2$ and $N = 2^n$, and let $\lambda = Q/N$. Let q-vectors $\mathbf{a}$ and $\mathbf{b}$ have elements drawn according to A and B, respectively (i.e., uniformly from $\{0, 1\}^n$ without replacement). Then*

1. *When q, n tend to infinity such that $\lambda \to 0$, $\mathbb{E}[M_{A \oplus B}(k)]$ tends to $N e^{-\lambda} \frac{\lambda^k}{k!}$.*
2. *For all $k > 0$ we have $\sum_{\kappa \geq k} \Pr[M_{A \oplus B}(\kappa) > 0] \leq \frac{(q!)^2 2^n (2^n - k)!}{((q-k)!)^2 k! (2^n)!}$.*

For notational convenience, the righthand side of the inequality will be denoted by $P_{q,k,n}$.

3   The Rate-1/3 Compression Function

We want to show that H (see Figure 1) is collision-resistant, so that it can be iterated to create a collision-resistant hash function [14,6,2,7]. In the sequel, we will model f_1, f_2, and f_3 as three independent, uniform elements of $\mathrm{Func}(n, n)$, and simply refer to our construction as H^{f_1, f_2, f_3} (or just H when it is not necessary to make the component functions explicit). We proceed to bound the probability that a computationally unbounded adversary can find a collision as a function of the number of its oracle queries. In particular, we will prove the following statement.

Theorem 3. *Fix $n > 0$ and let H^{f_1, f_2, f_3} be as previously defined. Then for all $k, q \geq 0$ we have $\mathbf{Adv}_{H(n)}^{\mathrm{coll}}(q) \leq \frac{q^2}{2^n} + \frac{(kq)^2}{2^n} + P_{q,k,n}$.*

Section 4 is dedicated to a proof of this theorem. What is more the following lemma provides asymptotic upper and lower bounds on the number of queries required to find collisions. The first item is a corollary of Theorem 3 and we will give a short proof sketch. The second item is proven after the proof of Theorem 3. Lemma 4 can be loosely rephrased by saying that to find collisions with any constant (non-zero) probability $\tilde{\Theta}(2^{n/2})$ queries are necessary and sufficient.

Lemma 4.   1. *For any $c > 1$ and all adversaries making at most $O(2^{n/2}/n^c)$ queries it holds that $\lim_{n \to \infty} \mathbf{Adv}_{H(n)}^{\mathrm{coll}}(\mathcal{A}) = 0$.*
2. *For any $c < 1$ there exists an $\epsilon > 0$ and an adversary asking $\Theta(2^{n/2}/n^c)$ queries for which, under a uniformity assumption, $\mathbf{Adv}_{H(n)}^{\mathrm{coll}}(\mathcal{A}) > \epsilon$ for all sufficiently large n.*

Proof: (Sketch, item 1.) Let d be such that $c > d > 1$ and consider the upper bound from Theorem 3 with $k = n^d$ and $q = 2^{n/2}/n^c$. Then as $n \to \infty$ all three terms tend to zero. Details of the asymptotic analysis can be found in the full version.   *Q.E.D.*

In the remainder of this section we build some intuition about the compression function and the necessary requirements on f_1, f_2, f_3 when instantiated in practice. For a classical birthday attack, an attacker would need to evaluate the compression function H on roughly $2^{n/2}$ inputs to succeed. Clearly, one can obtain this many evaluations by querying each of the f_1, f_2, f_3 oracles on this many points. However, the structure of the compression function may make things easier for the adversary. In particular, asking q queries to each of the oracles can provide more than q evaluations of H, due to internal xor-collisions at the input to f_3. In the next section we will introduce the *yield* of a query set, and use it to measure the number of H evaluations an adversary can make given q queries to each of the oracles.

We note that in the construction of H any bijection can be applied to the inputs and the output without affecting the collision resistance as bounded by Theorem 3. This freedom might yield a possible avenue to strengthen the hash function when iterating the compression function, without aversely affecting the security of the compression function itself.

PRACTICAL CONSIDERATIONS. Firstly, a collision in either f_1 or f_2 easily leads to a collision on the full compression function. In fact, it is even worse, since a single colliding pair M, M' for f_1 can be used for any chaining value V. That is, if $f_1(M) = f_1(M')$, then for all V it holds that $H(V, M) = H(V, M')$. The precise ramifications of such an attack are unclear, although it for instance allows finding k-way collisions in a standard (strengthened) MD-iterate in query complexity $\Theta(2^{n/2})$, which is an improvement over Joux' [10] $\Theta(2^{n/2} \log k)$. We will not delve into this in great detail; actual attacks based on this property will also crucially rely on the iteration method used (and if multi-collisions are an issue one should not rely on the standard MD-transform).

Secondly, from the proof it will be clear that our construction is equally secure if the random functions are replaced by random one-way permutations. However, if either f_1 or f_2 are invertible, an adversary can find collisions in H by making $O(2^{n/4})$ oracle queries. Say that f_2 is invertible. Then the adversary makes $2^{n/4}$ queries to both f_1 and f_3. With reasonable probability this will result in an internal xor-collision $f_1(M) \oplus f_3(Z) = f_1(M') \oplus f_3(Z')$. Inverting f_2 on $Z \oplus f_1(M)$, resp. $Z' \oplus f_1(M')$ will give a collision for H. Similarly if f_1 is invertible, call f_2 and f_3 each $2^{n/4}$ times to find an internal xor-collision $f_2(V) \oplus f_3(Z) \oplus Z = f_2(V') \oplus f_3(Z') \oplus Z'$. Now inverting f_1 on $Z \oplus f_2(V)$ and $Z' \oplus f_2(V')$ will complete the collision. If both f_1 and f_2 are invertible, only two calls to f_3 are needed to find a collision. Thus we will need (at least) for f_1 and f_2 to be collision-resistant and one-way. In particular, this rules out the straightforward blockcipher implementation $f_i(M) = E_{K_i}(M)$ for fixed (distinct) keys K_i, $i \in \{1, 2\}$, as this violates the one-way requirement. Nonetheless, one could instantiate the functions f_1, f_2 with a simple blockcipher-based function. In the full version we show that, for example, instantiation as $f_i(X) = E_{K_i}(X) \oplus X$, ($i \in \{1, 2\}$, where K_1 and K_2 are fixed and publicly known keys) gives a collision-resistant compression function in a combined model using a random oracle for f_3 and an ideal cipher for E.

Instantiating f_3 with $f_3(X) = E_{K_3}(X) \oplus X$ is pointless: it essentially results in the original rate-1/3 scheme with the inputs swapped and f_3 replaced with E_{K_3}. Luckily, neither invertibility of f_3 nor collisions in f_3 appear to be useful for finding collisions in H. It would be interesting to see whether our construction can be proven secure (with similar collision resistance), in the ideal cipher model for f_3. Note that invertibility of

f_3 is useful for finding preimages, allowing a meet-in-the-middle attack using only $\Theta(2^{n/2})$ (as shown later).

4   Proof of Theorem 3 and Lemma 4

In this section we will prove Theorem 3 and the second item of Lemma 4. Following the proof, we will also give some intuition why one expects that $\tilde{\Theta}(2^{2n/3})$ queries are necessary and sufficient to find preimages. This provides a fairly complete asymptotic characterization of the newly proposed construction.

SETTING UP THE PROOF. We will distinguish between three ways for an adversary to find a collision in H. It can try to find a collision in f_1 or f_2, since either would lead to a collision in H, as already shown above. Failing that, it can try to find a collision in the final output. This leads to the following upper bound

$$\mathbf{Adv}_{H(n)}^{\mathrm{coll}}(\mathcal{A}) \leq \Pr[\mathcal{A} \text{ finds collision in } f_1] + \Pr[\mathcal{A} \text{ finds collision in } f_2]$$
$$+ \Pr[\mathcal{A} \text{ finds collision in } H | \text{no collisions in } f_1 \text{ or } f_2] .$$

The probabilities of finding a collision in f_1 or f_2 are ordinary collision-finding problems and hence well understood. For $q \leq \sqrt{N}$ these probabilities roughly sum up to (and are upper bounded by) $\frac{q^2}{N}$, the first term in the upper bound of Theorem 3. Needless to say, if f_1 and f_2 are (random) permutations, no collisions exist and both probabilities are always zero. In any case, we can concentrate on the probability of $\mathcal{A}$ finding a collision in H when f_1 and f_2 are collision free. Henceforth we will therefore model f_1 and f_2 as random (one-way) permutations, but f_3 still as a random (one-way) function.

We also make the following standard assumptions, all without loss of generality. Firstly, we assume that the adversary makes exactly q queries to each of the three oracles, f_1, f_2, f_3: for any adversary that makes q_i queries to f_i there is an adversary that makes $q = \max(q_1, q_2, q_3)$ queries to *each* of the oracles with identical success probability. Secondly, we will assume that adversaries actually compute $H^{f_1,f_2,f_3}(V, M)$ and $H^{f_1,f_2,f_3}(V', M')$ before outputting their candidate collisions. In particular, this means that all necessary queries to f_1, f_2 and f_3 are made before halting.

REMOVING THE ADVERSARY. Normally we would imagine that the adversary makes queries to all three oracles in some adaptive, probabilistic manner. But here we cannot only argue away the adversary's adaptivity, but we can remove the adversary altogether. Recall that f_1, f_2, f_3 are independent oracles, and let us focus for a moment on the adversary's queries to f_1 and f_2.

Knowing that $\mathcal{A}$ makes q queries to each, we can imagine preparing the answers in advance. That is, before the adversary starts querying the oracles, we make two lists, each of q random elements, and when the adversary makes a query to one of the two oracles f_1 or f_2, we supply it with the next element of the respective list. Because the inputs to f_1 and f_2 are not used elsewhere in the compression function, the actual correspondence between query and response is irrelevant. Consequently, we might as well have provided the two lists to the adversary *before* any queries to f_1 or f_2, and even at the very beginning of the collision-finding game, in advance of any f_3 queries, given f_3's independence of f_1 and f_2.

DEFINING THE YIELD. A netral quantity in bounding an adversary's succes is what we call the *yield*. Formally, given a vector $\mathbf{c} = (c_1, \ldots, c_Q) \in (\{0,1\}^n)^Q$, define

$$\text{yield}(\mathbf{c}) = \max_{\substack{G \subseteq \{0,1\}^n \\ |G|=q}} \sum_{g \in G} \sum_{i=1}^{Q} [c_i = g]$$

where $[\text{true}] = 1$ and $[\text{false}] = 0$. Thus the yield counts the total number of occurences of the q most frequent elements in a vector. We also define the yield over the tensor of two q-vectors. Given vectors $\mathbf{a} = (a_1, \ldots, a_q)$ and $\mathbf{b} = (b_1, \ldots, b_q)$ in $(\{0,1\}^n)^q$, we will define the yield of tensor $\mathbf{a} \otimes \mathbf{b}$ to be

$$\text{yield}(\mathbf{a} \otimes \mathbf{b}) = \max_{\substack{G \subseteq \{0,1\}^n \\ |G|=q}} \sum_{g \in G} \sum_{i=1}^{q} \sum_{j=1}^{q} [a_i \oplus b_j = g] .$$

Let us give some intuition for this latter definition, in particular. Recall that we will give the response lists of f_1 and f_2, call these $\mathbf{a} = (a_1, \ldots, a_q)$ and $\mathbf{b} = (b_1, \ldots, b_q)$ (resp.), to the adversary prior to its making any f_3 queries. These q queries to f_3 can be made according to any strategy. One such strategy, already mentioned in the previous section, is to query the f_3 oracle on those values for which it knows the greatest total number of xor-preimages, thereby maximizing the number of inputs for which it can evaluate the compression function. It is precisely this number that the yield of $\mathbf{a} \otimes \mathbf{b}$ represents: the number of compression function outputs that the adversary can evaluate by asking q queries to f_3.

CONNECTING THE PIECES. We still need to show how the yield relates to the collision-finding probability of the adversary. Let d_r, for $r = 1, \ldots, q$, denote the number of pairs (i, j) such that $a_i \oplus b_j$ equals the input of the r-th query to f_3. Suppose that after $r-1$ queries to f_3, the adversary still has not found a collision. Then it will be able to output preimages for $\sum_{s=1}^{r-1} d_s$ hash values (or, equivalently, it will be able to output the hash value for that many preimages). With its rth call to f_3 it will be able to evaluate d_r new hash values, and the probability that one collides with one of the older values is therefore upper bounded by $d_r \sum_{s=1}^{r-1} d_s / 2^n$. The upper bound is not always tight. Consequently, picking the elements corresponding to the maximal yield is sometimes not the optimal strategy for finding a collision; picking elements that are slightly less common might actually increase the chances of finding a collision, nonetheless the same upper bound will apply.

Summing over all queries to f_3 leads us to the following upper bound

$$\Pr[\mathcal{A} \text{ finds collision in } H \mid \text{ no collisions in } f_1 \text{ or } f_2] \leq \sum_{r=1}^{q} \sum_{s=1}^{r-1} d_r d_s / 2^n .$$

What can we say about this value? Firstly, the possible values of d_r are determined by $\mathbf{a}$ and $\mathbf{b}$ and the maximum $\sum_{s=1}^{q} d_s = \text{yield}(\mathbf{a} \otimes \mathbf{b})$. Suppose we allow the adversary to partition $\text{yield}(\mathbf{a} \otimes \mathbf{b})$ arbitrarily in q (real) parts d_r. The optimal way, in the sense of maximizing the sum above, is then to choose $d_r = \text{yield}(\mathbf{a} \otimes \mathbf{b})/q$ for all $r = 1, \ldots, q$ (optimality of this choice can be shown by induction). In that case we have

$$\Pr[\mathcal{A} \text{ finds collision in } H | \text{ no collisions in } f_1 \text{ or } f_2] \leq \sum_{r=1}^{q} \sum_{s=1}^{r-1} d_r d_s / 2^n$$

$$\leq \sum_{r=1}^{q} \sum_{s=1}^{r-1} (\text{yield}(\mathbf{a} \otimes \mathbf{b})/q)^2 / 2^n$$

$$\leq \text{yield}(\mathbf{a} \otimes \mathbf{b})^2 / 2^{n+1} .$$

Moreover, if we would assume that the compression function outcomes for an adversary optimizing its yield are uniformly distributed, the probability that a collision occurs will satisfy the birthday bound, that is $\mathbf{Adv}_{H(n)}^{\text{coll}}(\mathcal{A}) \approx 0.63 \cdot \text{yield}(\mathbf{a} \otimes \mathbf{b})^2 / 2^{n+1}$ giving us nearly matching upper and lower bounds. Our task then is to put bounds on the expected value of $\text{yield}(\mathbf{a} \otimes \mathbf{b})$, or rather its square, where the elements in $\mathbf{a}$ and $\mathbf{b}$ are chosen independently, uniformly at random from $\{0,1\}^n$ (without replacement).

To upper bound the yield we recall that it is the sum of the frequencies of the q most frequent elements in $\mathbf{a} \otimes \mathbf{b}$. As such, the trivial upper bound on the yield is the cardinality of $\mathbf{a} \otimes \mathbf{b}$, that is $Q = q^2$. Moreover, if all collisions in $\mathbf{a} \otimes \mathbf{b}$ are less than k-way, then the yield is (strictly) smaller than kq. We can combine the two bounds as well. Let p be an upper bound on the probability that at least one collision that is at least k-way occurs in $\mathbf{a} \otimes \mathbf{b}$. Then conditioning on this event and employing the above observations yields that

$$\Pr[\mathcal{A} \text{ finds collision in } H | \text{ no collisions in } f_1 \text{ or } f_2] \leq (kq)^2 / 2^n + p .$$

By Lemma 2 we can use $P_{q,k,n}$ as our upper bound p. This concludes the proof of the upper bound of an adversary's advantage.

To lower bound the adversary's advantage, we need to lower bound the expected yield. We will only do this asymptotically, proving the second item of Lemma 4. Given $c < 1$, let d be such that $c < d < 1$ and consider the expected number of k-way collisions for $k = n^d$. By Lemma 2, this number is given by λ_k. In the full version, we show that asymptotically the expected number of k-way collisions exceeds q. Consequently the yield is at least $kq = 2^{n/2} n^{d-c}$ and the probability of a collision tends to one.

A NOTE ON PREIMAGE RESISTANCE. Although our goal is to demonstrate a construction that yields a compression function with good collision resistance, other useful properties should also be mentioned. Ideally, finding a preimage takes expected time 2^n for an n-bit primitive. To get an idea of the preimage resistance of the current proposal, we can look at the value of q for which the yield is around 2^n. If $q > 2^{n/2}$, a lower bound (and reasonable estimate) for the yield is $q^3 / 2^n$. Since $q^3 / 2^n \approx 2^n$ for $q \approx 2^{\frac{2}{3}n}$ it follows that our construction is not as preimage resistant as one might wish for. However, Rogaway and Steinberger [20] recently showed that this reduced preimage resistance is all but inevitable. Indeed, for any rate-1/3 scheme there exists an adversary whose yield is at least 2^n after $2^{2n/3}$ queries, which will likely lead to a preimage.

When f_3 is a random two-way permutation instead of a random one-way function, finding preimages becomes easier due to a meet-in-the-middle attack pointed out to us by Antoine Joux. Given a target h, the adversary queries f_1 on $q = 2^{n/2}$ arbitrary values leading to $f_1(V_1), \ldots, f_1(V_q)$ and subsequently queries f_3^{-3} on the values $h \oplus f_1(V_i)$ for $i = 1, \ldots, q$. After querying f_2 on q arbitrary values (leading to $f_2(M_1), \ldots, f_2(M_q)$) a preimage is obtained if $f_2(M_i) = f_3^{-1}(h \oplus f_1(V_j))$ which will occur with probability $q^2/2^n \approx 1$.

POISSON HEURISTIC TO APPROXIMATE THE YIELD. In the full version we develop an alternative characterization of yield$(\mathbf{a} \otimes \mathbf{b})$ and recast the problem of finding the expected value yield$_n^{A \oplus B}(q)$ into that of determining a certain property of the tail of a Poisson distribution. Experimental results give concrete estimates of the collision resistance of our proposal in practice, and it turns out that for n up to 256, the loss in collision resistance is at most four bits.

5   Conclusion

In this paper we have proposed a rate-1/3 $2n$-to-n bit compression function based on three random n-to-n bit functions. If the three underlying functions are modelled as random oracles, finding collisions requires roughly $2^{n/2}/n$ queries. Preimage resistance is loosely estimated to be around $2^{2n/3}$. Since the attacks based on optimizing the yield are inherently time and space consuming, it is unclear whether in practice algorithms can be found with a time complexity matching these query complexities (meaning our scheme will be harder to break).

Acknowledgements. We would like to thank Antoine Joux, Thomas Ristenpart and David Wagner for clarifying discussions. We also thank the anonymous reviewers for their feedback. Much of this paper was developed while the first author was visiting LACAL at EPFL, and he thanks them for their hospitality. Also, he was supported in part by NSF grant CNS-0627752.

References

1. Bellare, M., Micciancio, D.: A new paradigm for collision-free hashing: incrementality at reduced cost. In: Fumy, W. (ed.) EUROCRYPT 1997. LNCS, vol. 1233, pp. 163–192. Springer, Heidelberg (1997)
2. Bellare, M., Ristenpart, T.: Multi-property-preserving hash domain extension and the EMD transform. In: Lai, X., Chen, K. (eds.) ASIACRYPT 2006. LNCS, vol. 4284, pp. 299–314. Springer, Heidelberg (2006)
3. Bernstein, D.: The Rumba20 compression function (2007),
   http://cr.yp.to/rumba20.html
4. Black, J., Cochran, M., Shrimpton, T.: On the impossibility of highly efficient blockcipher-based hash functions. In: Cramer, R.J.F. (ed.) EUROCRYPT 2005. LNCS, vol. 3494, pp. 526–541. Springer, Heidelberg (2005)
5. Black, J., Rogaway, P., Shrimpton, T.: Black-box analysis of the block-cipher-based hash-function constructions from PGV. In: Yung, M. (ed.) CRYPTO 2002. LNCS, vol. 2442. Springer, Heidelberg (2002)

6. Damgård, I.: A design principle for hash functions. In: Brassard, G. (ed.) CRYPTO 1989. LNCS, vol. 435. Springer, Heidelberg (1990)
7. Gauravaram, P., Millan, W., Dawson, E., Viswanathan, K.: Constructing secure hash functions by enhancing Merkle-Damgård construction. In: Batten, L.M., Safavi-Naini, R. (eds.) ACISP 2006. LNCS, vol. 4058, pp. 407–420. Springer, Heidelberg (2006)
8. Gladman, B.: Implementation experience with AES candidate algorithms. In: Second AES Conference (1999)
9. Hirose, S.: Provably secure double-block-length hash functions in a black-box model. In: Park, C.-s., Chee, S. (eds.) ICISC 2004. LNCS, vol. 3506, pp. 330–342. Springer, Heidelberg (2005)
10. Joux, A.: Multicollisions in iterated hash functions. Application to cascaded constructions. In: Franklin, M.K. (ed.) Advances in Cryptology – CRYPTO 2004. LNCS, vol. 3621, pp. 306–316. Springer, Heidelberg (2004)
11. Matyas, S., Meyer, C., Oseas, J.: Generating strong one-way functions with cryptographic algorithms. IBM Technical Disclosure Bulletin 27(10a), 5658–5659 (1985)
12. Maurer, U., Tessaro, S.: Domain extension of public random functions: Beyond the birthday barrier. In: Menezes, A. (ed.) CRYPTO 2007. LNCS, vol. 4622, pp. 187–204. Springer, Heidelberg (2007)
13. Menezes, A., van Oorschot, P., Vanstone, S.: Handbook of Applied Cryptography. CRC Press, Boca Raton (1996)
14. Merkle, R.: One way hash functions and DES. In: Brassard, G. (ed.) Advances in Cryptology – CRYPTO 1989. LNCS, vol. 435, pp. 428–466. Springer, Heidelberg (1990)
15. Miyaguchi, S., Iwata, M., Ohta, K.: New 128-bit hash function. In: Proceedings 4th International Joint Workshop on Computer Communications, pp. 279–288 (1989)
16. Peyrin, T., Gilbert, H., Muller, F., Robshaw, M.: Combining compression functions and block cipher-based hash functions. In: Lai, X., Chen, K. (eds.) ASIACRYPT 2006. LNCS, vol. 4284, pp. 315–331. Springer, Heidelberg (2006)
17. Preneel, B., Govaerts, R., Vandewalle, J.: On the power of memory in the design of collision resistant hash functions. In: Seberry, J., Zheng, Y. (eds.) AUSCRYPT 1992. LNCS, vol. 718, pp. 105–121. Springer, Heidelberg (1993)
18. Preneel, B., Govaerts, R., Vandewalle, J.: Hash functions based on block ciphers: A synthetic approach. In: Stinson, D.R. (ed.) CRYPTO 1993. LNCS, vol. 773, pp. 368–378. Springer, Heidelberg (1994)
19. Rogaway, P., Steinberger, J.: How to build a permutation-based hash function (manuscript, 2008)
20. Rogaway, P., Steinberger, J.: Security/efficiency tradeoffs for permutation-based hashing. In: Smart, N. (ed.) EUROCRYPT 2008. LNCS, vol. 4965, pp. 220–236. Springer, Heidelberg (2008)
21. Shrimpton, T., Stam, M.: Building a collision-resistant compression function from non-compressing primitives. Technical Report 409, IACR e-print (2007)
22. Wagner, D.: A generalized birthday problem. In: Yung, M. (ed.) CRYPTO 2002. LNCS, vol. 2442, pp. 288–303. Springer, Heidelberg (2002)

Robust Multi-property Combiners for Hash Functions Revisited

Marc Fischlin[1], Anja Lehmann[1], and Krzysztof Pietrzak[2]

[1] Darmstadt University of Technology, Germany
www.minicrypt.de
[2] CWI, Amsterdam, Netherlands

Abstract. A robust multi-property combiner for a set of security properties merges two hash functions such that the resulting function satisfies each of the properties which at least one of the two starting functions has. Fischlin and Lehmann (TCC 2008) recently constructed a combiner which simultaneously preserves collision-resistance, target collision-resistance, message authentication, pseudorandomness and indifferentiability from a random oracle (IRO). Their combiner produces outputs of $5n$ bits, where n denotes the output length of the underlying hash functions.

In this paper we propose improved combiners with shorter outputs. By sacrificing the indifferentiability from random oracles we obtain a combiner which preserves all of the other aforementioned properties but with output length $2n$ only. This matches a lower bound for black-box combiners for collision-resistance as the only property, showing that the other properties can be achieved without penalizing the length of the hash values. We then propose a combiner which also preserves the IRO property, slightly increasing the output length to $2n + \omega(\log n)$. Finally, we show that a twist on our combiners also makes them robust for one-wayness (but at the price of a fixed input length).

1 Introduction

The concept of hash function combiners has been introduced by Herzberg [4] as an approach to create hash functions which are more resistant to cryptanalytic results. A combiner for some security property is a combination of two candidate hash functions such that the resulting function satisfies the property as long as at least one of the candidates has this property. For example, the "concatenation combiner" $C_{\parallel}^{H_0,H_1}(M) = H_0(M) \| H_1(M)$ preserves the property of being collision-resistant (CR) and target collision-resistant (TCR), because a collision $M \neq M'$ for the combiner is always also a collision for both components H_0 or H_1. Thus if either of the hash function H_0 or H_1 is collision-resistant, then so is the combined function.

Nowadays hash functions are often deployed in many facets, e.g., as pseudorandom functions in TLS or message authentication codes in IPSec, and thus provide numerous properties beyond collision-resistance. While the concatenation combiner preserves the MAC property, the PRF property is in general not

L. Aceto et al. (Eds.): ICALP 2008, Part II, LNCS 5126, pp. 655–666, 2008.

conserved. In contrast, the "XOR combiner" $C_{\oplus}^{H_0,H_1}(M) = H_0(M) \oplus M_1(M)$ is robust with respect to PRF, and also for indistinguishability from a random oracle (IRO), but neither preserves the CR nor the TCR property.

Ideally, one would like to have a single combiner preserving many properties simultaneously. To this end, Fischlin and Lehmann [3] have introduced the notion of robust multi-property combiners for a set of security properties PROP. According to their strongest notion such a combiner satisfies the property P ∈ PROP if P is satisfied by at least one of the two candidate hash functions. Their combiner, denoted here as C_{5P}, preserves all of the discussed properties, i.e., (target) collision-resistance (TCR, CR), pseudorandomness (PRF), message authentication (MAC) and indifferentiability from a random oracle[1] (IRO).

Paying tribute to the fact that several properties are conserved, the C_{5P} combiner requires an output length of $5n$ bits for the underlying hash functions with n-bit outputs, as opposed to $2n$ bits for the concatenating combiner for collision resistance or the n bits for the XOR combiner for pseudorandomness. Note that the lower bounds for black-box combiners preserving collision-resistance only [2,7] suggest that $2n$ is essentially the best we can hope for if collision-resistance is among the multiple properties. This raises the question if the lower bound can be matched.

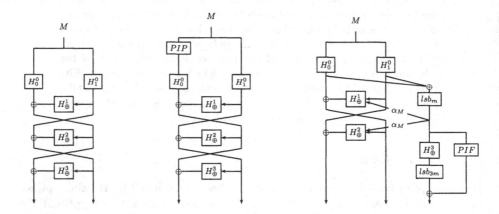

Fig. 1. Illustration of the basic construction C_{4P} (left) preserving CR, PRF, TCR and MAC. Here $H_b^i(\cdot)$ denotes $H_b(\langle i \rangle_2 \| \cdot)$ where $\langle i \rangle_2$ is the binary representation of the integer i with two bits. $H_{\oplus}^i(\cdot)$ denotes $H_0^i(\cdot) \oplus H_1^i(\cdot)$. By applying a pairwise independent permutation to the input of H_0^0 we get our construction $C_{4P\&OW}$ (center), which also preserves OW. Because of the PIP, the input length of the construction must now be fixed. The combiner $C_{4P\&IRO}$ (right) also preserves the IRO property (besides the four properties preserved by C_{4P}), at the prize of an increased output length. This is achieved by adding a third branch to the basic construction which is based on a "signature" value α_M depending on an input M and a pairwise independent function.

---

[1] Indifferentiability from a random oracle is sometimes also referred to as "being a pseudorandom oracle".

The Combiner C_{4P}. In this paper we first propose a combiner C_{4P} with optimal output length of $2n$ bits and which preserves all the properties of the C_{5P}
combiner from [3], except for indifferentiability from random oracles. The basic
idea of this construction is to use the concatenation combiner $C_{\|}$, and to apply a
three-round Feistel permutation to its output, where the round functions of the
Feistel network are constructed by using the XOR-combiner $C_{\oplus}$ (cf. Figure 1).
The round functions are made somewhat independent by prepending the round
number to the input.

The rationale here is that applying the Feistel (or any other) permutation to
the output of $C_{\|}$ still preserves the CR, TCR and MAC properties, e.g., collisions
for $C_{\|}$ are pulled through the downstream permutation and can be traced back
to collisions for $C_{\|}$. At the same time, one achieves robustness for the PRF
property. The latter can be seen as follows: if either H_0 or H_1 is pseudorandom,
then the round functions in the Feistel network are pseudorandom as $H_{\oplus}$ is a
secure combiner for pseudorandom functions. The Luby-Rackoff [5] result now
states that a three-round Feistel-network, instantiated with quasi independent
pseudorandom functions, is a pseudorandom permutation. We note that the
formal argument also needs to take into account that finding collisions in the
keyed version of the initial $C_{\|}$ computation is infeasible.

Preserving IRO. In Section 4.2 we modify the C_{4P} construction such that it
also preserves indifferentiability from a random oracle. The obstruction of the
IRO robustness in the C_{4P} combiner stems from the invertibility of the Feistel
permutation: an adversary trying to distinguish the output of the combiner from
a random function (given access to the underlying hash function components, as
opposed to the case of pseudorandom functions for example) can partly "reverse
engineer" images under the combiner. Hence, we introduce a "signature" value
α_M (depending on the input message M), entering the round functions in the
Feistel network and basically allowing combiner computations in the forward
direction only.

The description of our enhanced combiner $C_{4P\&IRO}$ is given in Figure 1. The
signature α_M is taken as (a prefix of) the XOR of the output halves of the $C_{\|}$
combiner and is used as additional input parameter in the Feistel round functions, allowing us to also save one round of the Feistel structure. Note that this
essentially means that different Feistel permutations may be used for different
inputs M, M', because the signatures $\alpha_M, \alpha_{M'}$ may be distinct. In order to apply again the argument that the Feistel permutation does not interfer with the
CR, TCR and MAC robustness of the concatenating combiner, we therefore also
need to ensure that finding "bad" pairs α_M and $\alpha_{M'}$ is infeasible. To this end we
introduce another output branch which basically guarantees collision resistance
of the signatures. This additional output is of length $3m$ for some $m = \omega(\log n)$,
yielding an overall output length of $2n + \omega(\log n)$.

Preserving One-Wayness. Even though both our solutions are robust for an
important set of properties they are still not known to be good combiners for
one-wayness. Our results so far merely show that they are one-way functions

making for example the potentially stronger assumption that one of the two hash functions is collision-resistance. In Section 5 we therefore show how to augment our constructions such they also preserves the one-wayness property.

The idea is that applying a pairwise-independent permutation (PIP) to the input of H_0 (or H_1) in the concatenation combiner $\mathsf{C}_\|$ makes this combiner also robust for one-wayness. Then we can use this modified concatenation combiner in the initial stages of our previous constructions, noting again the subsequent Feistel permutations do not interfer with this property either. Yet, as the description length of a PIP is linear in its input length, the input length of the derived combiners must be fixed, too, giving one-wayness as an additional property.

2   Preliminaries

2.1   Hash Functions and Their Properties

A hash function $\mathcal{H} = (\mathsf{HKGen}, \mathsf{H})$ is a pair of efficient algorithms such that HKGen for input 1^n returns (the description of) a hash function H, and H for input H and $M \in \{0,1\}^*$ deterministically outputs the hash value $H(M) \in \{0,1\}^n$. Often, the hash function is based on a public initial value IV which is replaced by a secret key K when the hash function is used as a pseudorandom function or a MAC.[2] For such a keyed hash function we write $H(K, M)$. If the key generation algorithm is clear from the context, we simply identify the hash function with its digest values $H(\cdot)$.

The following are security properties that are often required by cryptographic applications from hash functions.

collision resistance (CR): The hash function is called *collision-resistant* if for any efficient algorithm $\mathcal{A}$ the probability that for $H \leftarrow \mathsf{HKGen}(1^n)$ and $(M, M') \leftarrow \mathcal{A}(H)$ we have $M \neq M'$ but $\mathsf{H}(H, M) = \mathsf{H}(H, M')$, is negligible (as a function of n).

target collision-resistance (TCR): A hash function is called *target collision-resistant* if any adversary $\mathcal{A}$ consisting of two efficient algorithms $(\mathcal{A}^1, \mathcal{A}^2)$ has negligible success probability of winning the following experiment. Let $\mathcal{A}^1(1^n)$ first generate the target message M and possibly some additional state information st. Then, a hash function $H \leftarrow \mathsf{HKGen}(1^n)$ is chosen and $\mathcal{A}^2$ on input (H, M, st) tries to compute a colliding message M'. The adversary $\mathcal{A}$ wins if $M \neq M'$ but $H(M) = H(M')$.

pseudorandomness (PRF): A keyed hash function $H(K, \cdot)$ where the key generation algorithm outputs a public part (H, IV) and IV is replaced by a secret key K, is called *pseudorandom* if for any efficient adversary $\mathcal{D}$ the advantage $\Pr\left[\mathcal{D}^{H(K, \cdot)}(H) = 1\right] - \Pr\left[\mathcal{D}^f(H) = 1\right]$ is negligible, where the probability in the first case is over $\mathcal{D}$'s coin tosses, the choice of $H \leftarrow \mathsf{HKGen}(1^n)$ and

---

[2] Here IV is understood as a general parameter, possibly consisting of pairs of keys as in NMAC or HMAC, and not only as a single-valued paramter as in the Merkle-Damgård design principle.

the key K, and in the second case over $\mathcal{D}$'s coin tosses, the choice of $H \leftarrow$ HKGen(1^n), and the choice of the random function $f : \{0,1\}^* \rightarrow \{0,1\}^n$.

message authentication (MAC): We say that a keyed hash function (as defined for PRFs) is a *secure MAC* if for any efficient adversary $\mathcal{A}$ the probability that for $H \leftarrow$ HKGen(1^n) and random key K and $(M, \tau) \leftarrow \mathcal{A}^{H(K, \cdot)}(H)$ we have $\tau = H(K, M)$ and M has never been queried to oracle $H(K, \cdot)$, is negligible.

indifferentiability from random oracles (IRO): Indifferentiability [6] is a generalization of indistinguishability allowing to consider random oracles that are used as a public component. More formally, a hash function H^f based on a random oracle f is *indifferentiable* from a random oracle $\mathcal{F}$ if for any efficient adversary $\mathcal{D}$ there exists an efficient algorithm $\mathcal{S}$ such that the advantage $\Pr\left[\mathcal{D}^{H^f, f}(H) = 1\right] - \Pr\left[\mathcal{D}^{\mathcal{F}, \mathcal{S}^{\mathcal{F}}(H)}(H) = 1\right]$ is negligible in n, where the probability in the first case is over $\mathcal{D}$'s coin tosses, $H \leftarrow$ HKGen(1^n) and the choice of the random function f, and in the second case over the coin tosses of $\mathcal{D}$ and $\mathcal{S}$, and $H \leftarrow$ HKGen(1^n) and over the choice of $\mathcal{F}$.

one-wayness (OW): A hash function is called *one-way* if for any efficient algorithm $\mathcal{A}$ the probability that for $H \leftarrow$ HKGen(1^n) and for random M (chosen from some domain which is clear from the context) the probability that $\mathcal{A}(H, H(M))$ returns M' with $H(M') = H(M)$, is negligible.

2.2 Robust Multi-property Combiners

We now give a formal definition of robust multi-property combiners. A hash function combiner $\mathcal{C} = (\text{CKGen}, \text{C})$ for some security property P is a pair of algorithms which, when instantiated with two hash functions $\mathcal{H}_0, \mathcal{H}_1$, itself implements a hash function, such that the combined function satisfies P if at least one of the two candidates satisfies P. The concept of combiners for multiple properties PROP $= \{\text{P}_1, \text{P}_2, \ldots, \text{P}_N\}$ has been introduced in [3] and distinguishes between different levels of robustness. In the weakest case the combiner inherits a set of multiple properties if one of the hash functions is strong and has all the properties (weakly robust), whereas the strongest notion only requires that each property individually is provided by at least one of the two candidates (strongly robust). In between, there are mildly robust combiners for which one property may support the implementation of another property. In this paper we only consider strongly robust multi-property combiners. We denote by PROP($\mathcal{H}$) $\subseteq$ PROP for a set PROP $= \{\text{P}_1, \text{P}_2, \ldots, \text{P}_N\}$ the properties which hash function $\mathcal{H}$ has.

Definition 1 (Multi-Property Robustness). *A hash function combiner* $\mathcal{C} =$ (CKGen, C) *is* strongly multi-property-robust *(sMPR) for a set* PROP $=$ $\{\text{P}_1, \text{P}_2, \ldots, \text{P}_N\}$ *of properties, if for any hash functions* $\mathcal{H}_0, \mathcal{H}_1$ *we have* $\text{P}_i \in$ PROP($\mathcal{H}_0$) $\cup$ PROP($\mathcal{H}_1$) $\implies \text{P}_i \in$ PROP($\mathcal{C}^{\mathcal{H}_0, \mathcal{H}_1}$).

In our construction the key-generation procedure CKGen of the combiner calls the key-generation procedure HKGen of $\mathcal{H}_0$ and $\mathcal{H}_1$ (and possibly samples some other random variables like pairwise independent permutations), and then uses the two sampled functions H_0 and H_1 in the evaluation procedure C^{H_0,H_1} as "black-boxes". For the IRO property we assume that the evaluation procedure is given access to the oracles directly. The security property then requires that C^{H_0,H_1} is indifferentiable from a random oracle if H_0 or H_1 is a random oracle, and the other oracle is arbitrary.

3   The C_{4P} Combiner for CR, PRF, TCR and MAC

In this section we introduce the construction of our basic combiner C_{4P} as illustrated in Figure 1. Recall that the idea of this combiner is to apply a Feistel permutation (with quasi independent round functions given by the XOR combiner) to the concatenating combiner to ensure CR, PRF, TCR and MAC robustness.

3.1   Our Construction

The three-round Feistel permutation P^3 over $\{0,1\}^{2n}$ is given by the round functions $H_\oplus^i(\cdot) = H_0^i(\cdot) \oplus H_1^i(\cdot)$ for $i = 1, 2, 3$, with $H_b^i(\cdot)$ denoting the function $H_b(\langle i \rangle_2 \| \cdot)$ where $\langle i \rangle_2$ is the binary representation of the integer i with two bits. In the i-th round the input (L_i, R_i) is mapped to the output $(R_i, L_i \oplus H_\oplus^i(R_i))$. We occassionally denote this Feistel permutation more explicitly by $\psi[H_\oplus^1, H_\oplus^2, H_\oplus^3](\cdot)$.

Our combiner, instantiated with hash functions $\mathcal{H}_0, \mathcal{H}_1$, is a pair of efficient algorithms $C_{4P} = (\mathsf{CKGen}_{4P}, \mathsf{C}_{4P})$ where the key generation algorithm $\mathsf{CKGen}_{4P}(1^n)$ samples $H_0 \leftarrow \mathsf{HKGen}_0(1^n)$ and $H_1 \leftarrow \mathsf{HKGen}_1(1^n)$. The evaluation algorithm $\mathsf{C}_{4P}^{H_0,H_1}$ for parameters H_0, H_1 and input message M outputs

$$\mathsf{C}_{4P}^{H_0,H_1}(M) = P^3(H_0^0(M) \| H_1^0(M)).$$

3.2   Multi-property Robustness

We next show that the construction satisfies the strongest notion for robust multi-property combiners:

Theorem 1. C_{4P} is a strongly robust multi-property combiner for PROP = $\{CR, PRF, TCR, MAC\}$.

Recall that a strong robust multi-property combiner inherits all properties that are provided by at least one of the underlying hash functions. Thus, we have to prove that each property CR, PRF, TCR and MAC is preserved independently. As explained in the introduction the Feistel permutation clealy preserves some the properties from the initial $C_\|$ combiner due to its invertibility, making C_{4P} is CR-, TCR- and MAC-robust (proof omitted from this version).

Lemma 1. The combiner C_{4P} is PRF-robust.

Proof. As the XOR combiner is a good combiner for pseudorandom functions (PRFs), the round functions in the Feistel network $P^3 = \psi[H_\oplus^1, H_\oplus^2, H_\oplus^3]$ are instantiated with PRFs, as long as at least $\mathcal{H}_0$ or $\mathcal{H}_1$ is a PRF. Prepending the unique prefix $\langle i \rangle_2$ for $i = 1, 2, 3$ to the input of $H_\oplus^i(\cdot) = H_\oplus(\langle i \rangle_2 \| \cdot)$ in each round ensures that the functions in different rounds are never invoked on the same input, which means they are indistinguishable from three independent random functions. We can now apply the result due to Luby-Rackoff [5] which states that a three-round Feistel-network instantiated with independent pseudorandom functions is a pseudorandom permutation (PRP). Further, if either $\mathcal{H}_0$ or $\mathcal{H}_1$ is a PRF, then the initial concatenation combiner $C_\|^{H_0, H_1}$ is weakly collision resistant[3], i.e., the probability that the adversary will invoke the combiner on distinct inputs M, M' where $H_0^0(M) \| H_1^0(M) = H_0^0(M') \| H_1^0(M')$, is negligible. Thus, with overwhelming probability, all the adversary sees is the output of a PRP on distinct inputs. This distribution is indistinguishable from uniformly random (this follows from the PRP/PRF switching lemma [1]), thus C_{4P} is PRF robust. More precisely, if the distinguishing advantage of either $\mathcal{H}_0$ or $\mathcal{H}_1$ is at most ϵ, then the advantage for our combiner is at most $O(q^2 \cdot 2^{-n}) + \epsilon$, taking into account the probability $O(q^2/2^n)$ for weak collisions in case of a truly random function. $\square$

4   Preserving Indifferentiability: The $C_{4P\&IRO}$ Combiner

To be IRO-robust, a combiner C^{H_0, H_1} has to be indifferentiable from a random oracle for any efficient adversary $\mathcal{D}$, if H_b is a random oracle for some $b \in \{0, 1\}$. Thereby the adversary $\mathcal{D}$ has oracle access either to the combiner C^{H_0, H_1} and the random oracle H_b, or to $\mathcal{F}$ and a simulator $S^{\mathcal{F}}$. The simulator's goal is to mimic H_b such that $\mathcal{D}$ cannot have a significant advantage on deciding whether its interacting with C^{H_0, H_1} and H_b, or with $\mathcal{F}$ and $S^{\mathcal{F}}$.

The reason why the previous combiner $C_{4P}^{H_0, H_1}$ fails in preserving the IRO-property in a robust fashion is basically the invertibility of the combiner. This circumvents the usual strategy of the simulator, which is to check if a query is a potential attempt of $\mathcal{D}$ to simulate the construction of the combiner and then to precompute further answers that are consistent with the information $\mathcal{D}$ can get from $\mathcal{F}$. However, for $C_{4P}^{H_0, H_1}$ the simulator may be unable to precompute those consistent values, because an adversary $\mathcal{D}$ can compute the permutation part of the combiner backwards such that $S^{\mathcal{F}}$ has to commit to its round values used in the permutation P^3 before knowing the initial input M.

Thus, in order to guarantee the IRO property, we modify the $C_{4P}^{H_0, H_1}$ combiner such that the adversary is forced to query the message M before he can create meaningful queries aiming to imitate the construction. By this the simulator becomes able to switch to the common strategy of preparing consistant answers in advance. As explained in the introduction, adding a signature value α_M into the computation does the job.

---

[3] Weak collision resistance is defined similarly to collision resistance, except that here the function is keyed and the adversary only gets black-box access to the function.

4.1   The Combiner $\mathcal{C}_{4P\&IRO}$

In this section we consider the modified combiner $\mathcal{C}_{4P\&IRO}$ as illustrated in Figure 1. The combiner $\mathcal{C}_{4P\&IRO} = (\mathsf{CKGen}_{4P\&IRO}, \mathsf{C}_{4P\&IRO})$ is defined as follows: $\mathsf{CKGen}_{4P\&IRO}$ first samples $H_0 \leftarrow \mathsf{HKGen}_0(1^n), H_1 \leftarrow \mathsf{HKGen}_1(1^n)$ and a pairwise independent function $h : \{0,1\}^m \rightarrow \{0,1\}^{3m}$ for some $m \leq n/3$ (the larger m, the better the security level, but the longer the output, too):

Definition 2. *A familiy of functions $h : A \rightarrow B$ from domain A to range B is called pairwise independent iff for all $x \neq x' \in A$ and $z \neq z' \in B$ we have $\Pr[h(x) = z \wedge h(x') = z'] = \frac{1}{|B| \cdot (|B|-1)}$ (over the choice of h).*

For simplicity we often call h a pairwise independent function if it is chosen at random from such a family. The function is called a *pairwise independent permutation* if, in addition, $A = B$. An example of such a permutation are the functions $h_{(a,b)}(x) = ax + b$ for random $a \neq 0$ and b from the field $A = \mathrm{GF}(2^n)$. Throughout this work all pairwise independent functions and permutations are understood to be efficiently computable.

The evaluation algorithm $\mathsf{C}_{4P\&IRO}^{H_0,H_1,h}(M)$ first computes the concatenation combiner $H_0^0(M) \| H_1^0(M)$ and a signature α_M depending on M as $\alpha_M = lsb_m (H_\oplus^0(M))$ where $H_\oplus^0(M) = H_0^0(M) \oplus H_1^0(M)$ and $lsb_a(x)$ denotes the first a bits of x. The value α_M is used as additional prefix in the round functions of the two-round Feistel permutation $P_\alpha^2(\cdot) = \psi[H_\oplus^1(\alpha_M \| \cdot), H_\oplus^2(\alpha_M \| \cdot)]$. Applying P_α^2 on $H_0^0(M) \| H_1^0(M)$ then gives the first part of the combiners output. The second part is determined only by α_M, which is necessary to guarantee robustness for CR, TCR and MAC. To avoid that the output of the second branch leaks information about α_M when the combiner is instantiated with pseudorandom functions (which would allow inverting the Feistel network and thereby preventing the IRO-property again) we hide α_M by applying the function $H_\oplus^3$.

Overall, the combiner computes for input message M and its corresponding signature α_M the following output:

$$\mathsf{C}_{4P\&IRO}^{H_0,H_1,h}(M) = P_\alpha^2(H_0^0(M) \| H_1^0(M)) \, \| \, lsb_{3m}(H_\oplus^3(\alpha_M)) \oplus h(\alpha_M).$$

4.2 $\mathcal{C}_{4P\&IRO}$ Preserves the IRO Property

We show that this combiner is indifferentiable from a random oracle when instantiated with two functions H_0, H_1, where one of them is a random oracle (we refer to it as $H_b, b \in \{0,1\}$), and the other function $H_{\bar{b}}$ is arbitrary. Like the random oracle H_b, also $H_{\bar{b}}$ is given as an oracle and accessible by all parties. The pairwise independent function h that comes up in this construction is only needed to prove that $\mathcal{C}_{4P\&IRO}$ still preserves the CR and TCR properties; for the IRO property this function can be arbitrary.

Lemma 2. *The combiner $\mathsf{C}_{4P\&IRO}$ is IRO-robust.*

Remark 1. Note that the security of $\mathsf{C}_{4P\&IRO}$ as a random oracle combiner depends on m, and thus on the output length, which is $2n + 3m$. This can be slightly improved to $2n + 2m + m'$ for some $m' < m$ (by simply replacing $3m$ with $2m + m'$ in Figure 1), though m' should not be too small, as $C_{4P\&IRO}$ is a good combiner for the CR and TCR with probability $2^{-m'}$ (this probability is over the choice of the PIF, as we explain later in Section 4.3).

Proof. For the proof we assume that $b = 0$, i.e., the hash function $H_0 : \{0,1\}^* \to \{0,1\}^n$ is a random oracle. The case $b = 1$ is proved analogously. The adversary D has then access either to the combiner $C_{4P\&IRO}$ and H_0 or to a random oracle $\mathcal{F} : \{0,1\}^* \to \{0,1\}^{2n+3m}$ and a simulator $S^\mathcal{F}$. Our combiner is indifferentiable from a random oracle $\mathcal{F}$ if there exists a simulator $S^\mathcal{F}$, such that the adversary D can distinguish between $C_{4P\&IRO}, H_0$ and $\mathcal{F}, S^\mathcal{F}$ only with negligible probability.

The simulator keeps as state the function table of a (partially defined) function $\hat{H}_0 : \{0,1\}^* :\to \{0,1\}^n$, which initially is empty, i.e., $\hat{H}_0(X) = \bot$ for all X. We define $\hat{H}_0^i(M) = \hat{H}_0(\langle i \rangle_2 \| M)$ to mimic the notion used in Figure 1. The goal of $S^\mathcal{F}$ is to define $\hat{H}_0$ in such a way that, from D's point of view, $(\mathcal{F}, \hat{H}_0)$ look like $(C_{4P\&IRO}^{H_0,H_1,h}, H_0)$, i.e., the output of $\hat{H}_0$ has to be random and consistent to what the distinguisher can obtain from $\mathcal{F}$. Therefore, our simulator $S^\mathcal{F}$ parses each query X it is invoked on as $X = \langle i \rangle_2 \| M$ and proceeds as follows:

Simulator $S_{H_1,f}^\mathcal{F}(X)$:
on query X check if some entry $Y \leftarrow \hat{H}_0(X)$ already exists
    if $Y = \bot$ //no entry so far
        if $X = \langle 0 \rangle_2 \| M$ for some M
            set $\hat{H}_0^0(X) = y_0$ where y_0 is randomly chosen from $\{0,1\}^n$
            get $y_1 \leftarrow H_1^0(M)$ and compute $\alpha_M = lsb_m(y_0 \oplus y_1)$
            get $U \leftarrow \mathcal{F}(M)$ for query M and parse U as $U_1 \| U_2 \| U_3$
            where $|U_1| = |U_2| = n$ and $|U_3| = 3m$.
            set $\hat{H}_0^1(\alpha_M \| y_1) = U_2 \oplus y_0 \oplus H_1^1(\alpha_M \| y_1)$
            set $\hat{H}_0^2(\alpha_M \| U_2) = U_1 \oplus y_1 \oplus H_1^2(\alpha_M \| U_2)$
            set $\hat{H}_0^3(\alpha_M) = (U_3 \| z) \oplus (h(\alpha_M) \| 0^{n-3m}) \oplus H_1^3(\alpha_M)$
            where z is randomly chosen from $\{0,1\}^{n-3m}$
        if $X \neq \langle 0 \rangle_2 \| M$, choose a random $Y \in \{0,1\}^n$
            and save the value by setting $\hat{H}_0(X) = Y$
output $Y \leftarrow \hat{H}_0(X)$

The interesting queries are the queries of the form $X = \langle 0 \rangle_2 \| M$ which could be an attempt of D to simulate the construction of the combiner, such that the simulator has to compute in addition consistent answers to potential subsequent queries of D. The simulator starts by sampling a random $y_0 \in \{0,1\}^n$ and sets $\hat{H}_0^0(M) = y_0$. To define the "signature" α_M of M, $S^\mathcal{F}$ queries its oracle H_1 on $\langle 0 \rangle_2 \| M$ and uses the answer $y_1 = H_1^0(M)$ to compute $\alpha_M = lsb_m(y_0 \oplus y_1)$.

The simulator then defines the outputs of $\hat{H}_0^1, \hat{H}_0^2$ and $\hat{H}_0^3$ such that $C_{4P\&IRO}^{\hat{H}_0,H_1,h}(M) = \mathcal{F}(M)$. Therefore $S^\mathcal{F}$ invokes its random oracle $\mathcal{F}$ on input M and computes the corresponding outputs of $\hat{H}_0$ by retracing the combiners construction as defined in the simulators description. Note that this is possible in a unique way, except for the $n - 3m$ last bits of $\hat{H}_0^3(\alpha_M)$, which must be chosen uniformly at random. We say the simulator "loses" if, for some $i \in \{1, 2, 3\}$, the function $\hat{H}_0^i$ is already defined on any input of the form $\alpha_M \| *$, such that $\mathcal{S}^{\mathcal{F}}$ cannot define all $\hat{H}_0^i$ values in order to provide consistent outputs.

As $\alpha_M \in \{0,1\}^m$ is uniformly random, the probability that the simulator loses in the q-th query is at most $3q \cdot 2^{-m}$ (as each $\hat{H}_0^i$ for $i \in \{1,2,3\}$ is defined on at most $q - 1$ inputs). Let $\mathcal{E}$ denote the event that the simulator loses in any of its q queries, then the overall probability that $\mathcal{E}$ happens is at most $3q^2 \cdot 2^{-m}$. If $\mathcal{E}$ does not occur, the replies of $\mathcal{S}^{\mathcal{F}}$ are consistent with $\mathcal{F}$ and random, since $\mathcal{S}^{\mathcal{F}}$ answers are determined by its random choices and the replies of $\mathcal{F}$. Hence, the advantage of the adversary $\mathcal{D}$ in distinguishing $(\mathsf{C}_{4\mathsf{P\&IRO}}^{H_0,H_1,h}, H_0)$ from $(\mathcal{F}, \mathcal{S}^{\mathcal{F}})$ is at most the probability that event $\mathcal{E}$ happens, which is by $\Pr[\mathcal{E}] = 3q^2 \cdot 2^{-m}$ negligible. □

4.3 $\mathcal{C}_{4\mathsf{P\&IRO}}$ Is Robust for CR, TCR, MAC, PRF

We now prove that, like the $\mathcal{C}_{4\mathsf{P}}$ combiner, $\mathcal{C}_{4\mathsf{P\&IRO}}$ also preserves the CR, TCR, MAC and PRF property in a robust manner. The proofs for TCR, MAC and PRF are similar to the proofs for $\mathcal{C}_{4\mathsf{P}}$ and appear in the full version.

Lemma 3. *The combiner* $\mathcal{C}_{4\mathsf{P\&IRO}}$ *is CR- and TCR-robust.*

Proof. We will prove that for any H_0, H_1, with probability $1 - 2^{-m}$ over the choice of the pairwise independent function h, any collision for $\mathsf{C}_{4\mathsf{P\&IRO}}^{H_0,H_1,f}$ is simultaneously a collision for H_0^0 and H_1^0. To this end, let $M \neq M'$ be a collision for $\mathsf{C}_{4\mathsf{P\&IRO}}^{H_0,H_1,h}$ and let α_M and $\alpha_{M'}$ denote their signatures. Let $Y \| Y' = \mathsf{C}_{4\mathsf{P\&IRO}}^{H_0,H_1,h}(M)$ where $Y \in \{0,1\}^{2n}$ and $Y' \in \{0,1\}^{3m}$.

If $\alpha_M = \alpha_{M'}$, then M, M' must be a collision for H_0^0 and H_1^0, as we have

$$H_0^0(M) \| H_1^0(M) = P_\alpha^{2^{-1}}(Y) = P_{\alpha'}^{2^{-1}}(Y) = H_0^0(M') \| H_1^0(M') \qquad (1)$$

and the Feistel permutations $P_\alpha^2, P_{\alpha'}^2$ are identical if $\alpha_M = \alpha_{M'}$.

For M, M' where $\alpha_M \neq \alpha_{M'}$, a collision $\mathsf{C}_{4\mathsf{P\&IRO}}^{H_0,H_1,h}(M) = \mathsf{C}_{4\mathsf{P\&IRO}}^{H_0,H_1,h}(M')$ does not imply (1), and thus will in general not be a collision for H_0 and H_1. Yet, as with probability $1 - 2^{-m}$ over the choice of the pairwise independent function $h : \{0,1\}^m \to \{0,1\}^{3m}$, there does not exist a collision M, M' for $\mathsf{C}_{4\mathsf{P\&IRO}}^{H_0,H_1,h}$ where $\alpha_M \neq \alpha_{M'}$. Note that for this it is sufficient to prove that for any two potential signatures $\alpha \neq \alpha' \in \{0,1\}^m$, we have

$$lsb_{3m}(H_\oplus^3(\alpha)) \oplus h(\alpha) \neq lsb_{3m}(H_\oplus^3(\alpha')) \oplus h(\alpha') \qquad (2)$$

as this implies that the final outputs are distinct for any two messages with different signatures. As h is pairwise independent, for any particular $\alpha \neq \alpha'$, equation (2) holds with probability $1 - 2^{-3m}$. Taking the union bound over all $2^m(2^m - 1)/2 < 2^{2m}$ distinct values $\alpha \neq \alpha'$, we get that the probability that there exists some $\alpha \neq \alpha'$ not satisfying (2) is at most $2^{2m}/2^{3m} = 2^{-m}$.

The proof of TCR-robustness follows a similiar argument and appears in the full version. □

5   Preserving One-Wayness and the $\mathcal{C}_{4P\&OW}$ Combiner

In this section we first propose a combiner which preserves collision resistance and one-wayness simultaneously. We remark that this is a strong combiner in the sense that it preserves each property as long as one of the hash functions has the property in question. This additional guarantee here comes at the price of restricted input length. At the end of this section we briefly discuss how to plug in this combiner into our combiners $\mathcal{C}_{4P}$ and $\mathcal{C}_{4P\&IRO}$ to get our construction $\mathcal{C}_{4P\&OW}$ and $\mathcal{C}_{6P}$, respectively.

We remark that the concatenating combiner $C_{\parallel}^{H_0,H_1}(M) = H_0(M)\|H_1(M)$ for collision resistance is readily verified to be an insufficient combiner for one-wayness. Similarly, the "input-splitting" combiner defined by $C_{OW}^{H_0,H_1}(M_0\|M_1) = H_0(M_0)\|H_1(M_1)$ preserves one-wayness but clearly not collision resistance. Below we construct a combiner which draws from the best of both constructions: it uses the same input M for both hash functions but re-randomizes the part for H_0 via a pairwise-independent permutation π.

5.1   A Combiner for CR and OW

We define the combiner $\mathcal{C}_{\parallel\&OW}$ for preserving collision-resistance and one-wayness in a robust manner as follows. The key generation algorithm $\mathsf{CKGen}_{\parallel\&OW}(1^n)$ generates $H_0 \leftarrow \mathsf{HKGen}_0(1^n)$ and $H_1 \leftarrow \mathsf{HKGen}_1(1^n)$ and picks a pairwise independent permutation $\pi : \{0,1\}^{5n} \to \{0,1\}^{5n}$. It outputs (H_0, H_1, π). The evaluation algorithm $C_{\parallel\&OW}^{H_0,H_1,\pi}$ on input $M \in \{0,1\}^{5n}$ returns $H_0(\pi(M))\|H_1(M)$. Note that we fix the input length to $5n$ bits. This value can be replaced by any length kn with $k \geq 5$, as we only use the fact that the input length is *at least* $5n$ bits (but then π must also be over $\{0,1\}^{kn}$). Getting shorter output length is possible, at the prize of worse security bounds. By the following theorem $\mathcal{C}_{\parallel\&OW}$ preserves the properties of $\mathcal{C}_{\parallel}$ and $\mathcal{C}_{OW}$ simultaneously.

Theorem 2. *The combiner $\mathcal{C}_{\parallel\&OW}$ is a strongly robust multi-property combiner for* PROP $= \{CR, TCR, MAC, OW\}$.

The proof is omitted due to space constraints and appears in the full version of the paper.

5.2   Combining Things

We can now plug in the combiner $\mathcal{C}_{\parallel\&OW}$ into the initial computation of the combiner $\mathcal{C}_{4P}$, obtaining our construction $\mathcal{C}_{4P\&OW}$. That is, we replace the initial computation $H_0^0(M)\|H_1^0(M)$ in our original combiner by $H_0^0(\pi(M))\|H_1^0(M)$ for messages of $5n$ bits. Note that if $H_b(\cdot)$ is one way on inputs of length $5n + 2$, then also $H_b^0(\cdot)$ is one-way on inputs of length $5n$, and we only lose a factor of 4 in the security.

Theorem 3. *The combiner $\mathcal{C}_{4P\&OW}$ is a strongly robust multi-property combiner for* PROP $= \{CR, PRF, TCR, MAC, OW\}$.

When we apply the modifications from Section 5 and the combiner $C_{4P\&IRO}$ from Section 4 together, we get our construction C_{6P}. This construction is defined like $C_{4P\&IRO}$, where one additionally applies a pairwise-independent permutation over $\{0,1\}^{kn}$ (with $k \geq 5$) to the input of H_0^0.

Theorem 4. *The combiner C_{6P} is a strongly robust multi-property combiner for* PROP $= \{CR, TCR, PRF, MAC, OW, IRO\}$.

The proofs of theorem 3 and 4 are given in the full version of this paper.

Acknowledgments

We thank Thomas Holenstein for a helpful conversation on one-way functions. We also thank the anonymous reviewers for valuable comments. The first two authors are supported by the Emmy Noether Program Fi 940/2-1 of the German Research Foundation (DFG).

References

1. Bellare, M., Rogaway, P.: The security of triple encryption and a framework for code-based game-playing proofs. In: Vaudenay, S. (ed.) EUROCRYPT 2006. LNCS, vol. 4004, pp. 409–426. Springer, Heidelberg (2006)
2. Boneh, D., Boyen, X.: On the impossibility of efficiently combining collision resistant hash functions. In: Dwork, C. (ed.) CRYPTO 2006. LNCS, vol. 4117, pp. 570–583. Springer, Heidelberg (2006)
3. Fischlin, M., Lehmann, A.: Multi-property preserving combiners for hash functions. In: Canetti, R. (ed.) TCC 2008. LNCS, vol. 4948, pp. 375–392. Springer, Heidelberg (2008)
4. Herzberg, A.: On tolerant cryptographic constructions. In: Menezes, A. (ed.) CT-RSA 2005. LNCS, vol. 3376, pp. 172–190. Springer, Heidelberg (2005)
5. Luby, M., Rackoff, C.: How to construct pseudorandom permutations from pseudorandom functions. SIAM Journal on Computing 17(2), 373–386 (1988)
6. Maurer, U., Renner, R., Holenstein, C.: Indifferentiability, impossibility results on reductions, and applications to the random oracle methodology. In: Naor, M. (ed.) TCC 2004. LNCS, vol. 2951, pp. 21–39. Springer, Heidelberg (2004)
7. Pietrzak, K.: Non-trivial black-box combiners for collision-resistant hash-functions don't exist. In: Naor, M. (ed.) EUROCRYPT 2007. LNCS, vol. 4515. Springer, Heidelberg (2007)

Homomorphic Encryption with CCA Security

Manoj Prabhakaran[*] and Mike Rosulek[*]

University of Illinois, Urbana-Champaign
{mmp,rosulek}@uiuc.edu

Abstract. We address the problem of constructing public-key encryption schemes that meaningfully combine useful *computability features* with *non-malleability*. In particular, we investigate schemes in which anyone can change an encryption of an unknown message m into an encryption of $T(m)$ (as a *feature*), for a specific set of allowed functions T, but the scheme is "non-malleable" with respect to all other operations. We formulate precise definitions that capture these intuitive requirements and also show relationships among our new definitions and other more standard ones (IND-CCA, gCCA, and RCCA). We further justify our definitions by showing their equivalence to a natural formulation of security in the Universally Composable framework. We also consider extending the definitions to features which combine *multiple* ciphertexts, and show that a natural definition is unattainable for a useful class of features. Finally, we describe a new family of encryption schemes that satisfy our definitions for a wide variety of allowed transformations T, and which are secure under the standard Decisional Diffie-Hellman (DDH) assumption.

1 Introduction

A recurring theme in cryptography is the tension between achieving powerful functionality and making strong security guarantees. In the case of encryption, IND-CCA security is well-accepted as a sufficiently strong security guarantee. On the other hand, for encryption to be useful in sophisticated applications (such as voting or mix-nets), the scheme should have features which allow *computation* on encrypted messages (e.g., features like rerandomizability [20,21], proxy re-encryption [5,9], searchability [36,11] and different kinds of homomorphism properties [18,27]). However, IND-CCA security rules out *any* such feature which operates on encrypted messages, while the other extreme of IND-CPA security does not exclude the possibility that a scheme may have additional "unforeseen features" that an adversary can exploit when the scheme is used in a larger application. Is it possible to express (and achieve via a construction) a security requirement capturing *the best of both worlds*: to be malleable enough to allow rich features, but non-malleable enough to rule out "everything else?"

In this work we address this question in the context of *homomorphic* public-key encryption schemes — those which allow anyone to change encryptions of

---

[*] Partially supported by NSF grant CNS 07-47027.

L. Aceto et al. (Eds.): ICALP 2008, Part II, LNCS 5126, pp. 667–678, 2008.
© Springer-Verlag Berlin Heidelberg 2008

unknown messages $m_1, \ldots, m_k$ into an encryption of $T(m_1, \ldots, m_k)$, for some allowed set of functions T. Such schemes have been extensively studied for a long time and have a wide variety of applications (cf. [4,12,14,15,20,23,24,25,33,34]). Homomorphic encryption schemes have additional utility in that ciphertexts hide not only the underlying plaintext, but also the way in which the ciphertext was derived (i.e., as a regular encryption, or via some homomorphic operation applied to some other ciphertexts). We explicitly formalize this requirement, which we call *unlinkability*.

Challenges and Related Work. The first challenge is formally defining (in a convincing way) the intuitive requirement that a scheme "allow particular features but forbid all others." Security notions for regular encryption developed and matured over many years [19,26,32,3,17,7], while arguably security definitions for homomorphic encryptions have lagged behind — to date, homomorphic encryptions are almost exclusively held to the weak standard of IND-CPA security. In some applications (e.g., [14]) CPA security is indeed sufficient, but for others (e.g. [16]) it is not. Very little work has addressed the possibility of homomorphic encryption schemes having "unforeseen features" beyond the prescribed operations; one exception is Wikström [37], who addresses this question in a simpler setting for El Gamal.

Benignly-malleable (gCCA) security [35,1] was proposed as a relaxation of CCA security, and was further relaxed in the definition of Replayable-CCA (RCCA) security [10]. RCCA security allows a scheme to have homomorphic operations which preserve the underlying plaintext, but enforces non-malleability "everywhere else." However, relaxing CCA security in the same way does not yield an acceptable level of security when applied to more expressive homomorphic operations (see Section 3.1); a new approach to defining security is needed.

The second challenge is achieving the desired security with a construction based on standard assumptions — i.e., an encryption scheme that has a particular set of (unlinkable) homomorphic operations, but is non-malleable with respect to all other operations. Note that even if the set of allowed operations is very simple, supporting it can be very involved. Indeed, the problem of unlinkable (rerandomizable) RCCA encryption considered in a recent series of works [10,21,30] corresponds to arguably the simplest special case of our definitions.

Our Results. We give several new security definitions to precisely capture the desired requirements in the case of *unary* homomorphic operations (those which transform a single encryption of m to an encryption of $T(m)$, for a particular set of functions T). We provide two new indistinguishability-based security definitions: one formalizing the unlinkability requirement and one formalizing the intuition of "non-malleability except for certain prescribed operations." To justify this last security definition, we show that it subsumes the standard IND-CCA, gCCA, and RCCA security definitions (Theorem 1). We further show that our two new security requirements imply a natural definition of security in the

Universal Composition framework (Theorem 2). Using the UC framework to define security of encryption schemes was already considered in [7,10,28,8].

We also consider extending our definitions to the case of *binary* homomorphic operations (those which combine pairs of ciphertexts). We show that the natural generalization of our UC security definition to this scenario is unachievable for a large class of useful homomorphic operations (Theorem 5).

Finally, we describe a family of encryption schemes which achieves our definitions for a wide range of allowed (unary) homomorphism operations. The construction, which is a careful generalization of the rerandomizable RCCA-secure scheme of [30], is secure under the standard DDH assumption, and supports the group operation as a homomorphic feature (as well as several related operations).

We refer the reader to the full version of this work [31] for the technical details omitted in this extended abstract.

2   Homomorphic Encryption Preliminaries

Let $\mathcal{M}$ be a space of plaintext messages, let $\bot$ be a special error indicator symbol not in $\mathcal{M}$, and let $\mathcal{T}$ be a "transformation space" — i.e., a set of polynomial-time computable functions from $(\mathcal{M} \cup \{\bot\})^k$ to $\mathcal{M} \cup \{\bot\}$. We call the elements of $\mathcal{T}$ the *allowable transformations*.

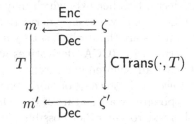

Fig. 1. Syntax and correctness of a homomorphic encryption scheme

An encryption scheme consists of three polynomial-time (polynomial in the implicit security parameter) algorithms, KeyGen, Enc and Dec.

A $\mathcal{T}$-*homomorphic encryption scheme* comes with an additional algorithm CTrans, the homomorphic operation feature: a randomized algorithm which takes k ciphertexts and (the description of) a *transformation* from $\mathcal{T}$, and outputs another ciphertext.[1] We mostly restrict attention to the case where $k = 1$; i.e., when the homomorphic operation is *unary*.

Correctness Properties. Below we give the correctness properties for unary homomorphic encryption. These requirements can be slightly relaxed (e.g., to hold only with overwhelming probability over key generation), but our construction achieves these simpler requirements.

For all key pairs (PK, SK) in the support of KeyGen, we require the following:

1. For every plaintext msg $\in \mathcal{M}$, we require $\mathsf{Dec}_{SK}(\mathsf{Enc}_{PK}(\mathsf{msg})) = \mathsf{msg}$, with probability 1 over the randomness of Enc.
2. For every purported ciphertext ζ and every $T \in \mathcal{T}$, we require $\mathsf{Dec}_{SK}(\mathsf{CTrans}(\zeta, T)) = T(\mathsf{Dec}_{SK}(\zeta))$, with probability 1 over the randomness of CTrans.

---

[1] Allowing CTrans to take the public key as additional input would also be a meaningful relaxation, but may not be suitable in some applications.

3   Defining Security

In this section we present our formal security definitions. The first two are traditional indistinguishability-based definitions, while the third is a definition in the Universal Composition framework.

3.1   Homomorphic-CCA (HCCA) Security

Our first indistinguishability-based security definition formalizes the intuitive notions of message privacy and "non-malleability other than certain operations."

Existing non-malleability definitions such as IND-CCA, benignly-malleable (a.k.a. gCCA) security [35,1] and Replayable CCA (RCCA) security [10] share a similar structure, in which an experimenter encrypts one of two adversarially chosen plaintexts and provides a decryption oracle to the adversary, whose task it is to guess which plaintext was encrypted. Since the adversary could simply ask to decrypt the challenge ciphertext itself, the decryption oracles are guarded to not decrypt ciphertexts which may be "derivatives" of the challenge ciphertext. In CCA security, the only derivative is the challenge ciphertext itself; in gCCA, derivatives are those which satisfy a particular binary relation with the challenge ciphertext; in RCCA, derivatives are those which decrypt to either of the two adversarially-chosen plaintexts.

However, in the case of more general homomorphic encryption, it may be legal (i.e., possible via a feature of the scheme) to change the underlying plaintext of a ciphertext to any other possible plaintext. Indeed, in some instantiations of our construction, *every ciphertext* in the support of the Enc operation is a possible derivative of every other such ciphertext. Following the IND-CCA paradigm here would weaken it essentially to IND-CCA1 (i.e., "lunchtime attack") security.

Our approach to identifying "derivative" ciphertexts is completely different, and as a result our definition initially appears incomparable to these other standard definitions. However, Theorem 1 demonstrates that our new definition gives a generic notion of non-malleability which subsumes these existing definitions.

The formal definition, which we call *Homomorphic-CCA (HCCA) security*, appears below. Informally, in the security experiment we identify derivative ciphertexts not for regular encryptions, but for special "rigged" ciphertexts that carry no message. In other words, there should be a procedure RigEnc_{PK} which outputs a rigged ciphertext ζ and some auxiliary information S, such that ζ is indistinguishable from a normal ciphertext. There should also be a corresponding procedure $\mathsf{RigExtract}_{SK}$ which, when given another ciphertext ζ' and auxiliary information S, determines whether ζ' was obtained by applying a transformation to ζ, and outputs the transformation.

Intuitively, the transformations output by $\mathsf{RigExtract}$ constitute all the ways a ciphertext's message can depend on another ciphertext in the scheme, so restricting the range of $\mathsf{RigExtract}$ restricts the malleability of the scheme.

Definition 1. *A homomorphic encryption scheme is* Homomorphic-CCA (HCCA) *secure with respect to* T *if there are PPT algorithms* RigEnc *and*

RigExtract, *where the range of* RigExtract *is* $\mathcal{T} \cup \{\bot\}$, *and such that for all PPT adversaries* $\mathcal{A}$, *the advantage of* $\mathcal{A}$ *in the following IND-HCCA experiment is negligible:*

1. **Setup:** *Pick* $(PK, SK) \leftarrow$ KeyGen *and give* PK *to* $\mathcal{A}$.
2. **Phase I:** $\mathcal{A}$ *gets access to the* $\text{Dec}_{SK}(\cdot)$ *oracle and the following two "guarded"* RigEnc *and* RigExtract *oracles:*

$$\text{GRigEnc}_{PK}() = \zeta_i, \text{ where } (\zeta_i, S_i) \leftarrow \text{RigEnc}_{PK}, \text{ when called for the ith time}$$
$$\text{GRigExtract}_{SK}(\zeta, i) = \text{RigExtract}_{SK}(\zeta, S_i)$$

3. **Challenge:** $\mathcal{A}$ *outputs a plaintext* msg*. *We privately flip a coin* $b \leftarrow \{0, 1\}$. *If* $b = 0$, *we compute* $\zeta^* \leftarrow \text{Enc}_{PK}(\text{msg}^*)$. *If* $b = 1$, *we compute* $(\zeta^*, S^*) \leftarrow \text{RigEnc}_{PK}$. *In both cases, we give* ζ^* *to* $\mathcal{A}$.
4. **Phase II:** $\mathcal{A}$ *gets access to the same* GRigEnc *and* GRigExtract *oracles as in Phase I, as well as a "rigged" version of the decryption oracle* RigDec. *When* $b = 0$, RigDec *is simply the normal decryption oracle* $\text{Dec}_{SK}(\cdot)$. *When* $b = 1$, RigDec *is implemented as follows:*

$$\text{RigDec}_{SK}(\zeta) = \begin{cases} T(\text{msg}^*) & \text{if } \bot \neq T \leftarrow \text{RigExtract}_{SK}(\zeta, S^*) \\ \text{Dec}_{SK}(\zeta) & \text{otherwise} \end{cases}.$$

5. **Output:** $\mathcal{A}$ *outputs a bit* b'. *The advantage of* $\mathcal{A}$ *is* $\Pr[b' = b] - \frac{1}{2}$.

We immediately observe that in order to achieve HCCA security, $\mathcal{T}$ must be closed under composition (or at least approximately so). $\mathcal{T}$ must also contain the identity function, since the adversary can simply submit the challenge ciphertext ζ^* to the RigDec oracle.

3.2   Unlinkability

There indeed is some tension between the HCCA definition given above and the intuitive notion of unlinkability that we desire. HCCA security implies that it is possible to track transformations applied to rigged ciphertexts, while unlinkability demands that ciphertexts not leak whether they were generated via a transformation. To reconcile this, we require unlinkability only on *ciphertexts that succesfully decrypt under a private key chosen by the challenger.* This excludes linkability via the RigEnc and RigExtract procedures, since tracking ciphertexts using RigExtract in general requires the tracking party to know the private key.

Our formal definition of unlinkability is given below. We highlight that the adversary has access to a decryption oracle in the experiment, making it meaningful for modeling chosen-ciphertext attacks.

Definition 2. *A homomorphic encryption scheme is* unlinkably homomorphic *with respect to* $\mathcal{T}$ *if for all PPT adversaries* $\mathcal{A}$, *the advantage of* $\mathcal{A}$ *in the following experiment is negligible:*

1. **Setup:** *Pick* $(PK, SK) \leftarrow$ KeyGen *and give* PK *to* $\mathcal{A}$.
2. **Phase I:** $\mathcal{A}$ *is given access to the decryption oracle* $\mathsf{Dec}_{SK}(\cdot)$.
3. **Challenge:** *Flip a coin* $b \leftarrow \{0, 1\}$. *Receive from* $\mathcal{A}$ *a ciphertext* ζ *and transformation* $T \in \mathcal{T}$. *If* $\mathsf{Dec}_{SK}(\zeta) = \bot$, *abort; else give* ζ^* *to* $\mathcal{A}$, *where:*

$$\zeta^* \leftarrow \begin{cases} \mathsf{Enc}_{PK}(T(\mathsf{Dec}_{SK}(\zeta))) & \text{if } b = 0 \\ \mathsf{CTrans}(\zeta, T) & \text{if } b = 1 \end{cases}$$

4. **Phase II:** $\mathcal{A}$ *is given access to the decryption oracle* $\mathsf{Dec}_{SK}(\cdot)$.
5. **Output:** $\mathcal{A}$ *outputs a bit* b'. *The* advantage *of* $\mathcal{A}$ *is* $\Pr[b' = b] - \frac{1}{2}$.

Note that unlinkability is a security guarantee involving maliciously crafted ciphertexts, and is not necessarily implied by guaranteeing that the distributions $\mathsf{Enc}_{PK}(T(\mathsf{msg}))$ and $\mathsf{CTrans}(\mathsf{Enc}_{PK}(\mathsf{msg}), T)$, which involve only honestly-generated ciphertexts, are indistinguishable.

We have defined unlinkability with the goal that a scheme can be both $\mathcal{T}$-unlinkably homomorphic, and $\mathcal{T}$-HCCA-secure (for the same $\mathcal{T}$). Indeed, it is easy to see that if a scheme is $\mathcal{T}$-unlinkably homomorphic and $\mathcal{T}'$-HCCA-secure, then $\mathcal{T} \subseteq \mathcal{T}'$. For simplicity, and to highlight the compatibility and sharp tradeoff between these two definitions, we only focus on schemes which satisfy them both with respect to the same transformation space $\mathcal{T}$.

3.3   UC Definition: Homomorphic Message Posting

We also define the "Homomorphic Message Posting" functionality $\mathcal{F}_{\mathrm{HMP}}^{\mathcal{T}}$ in the framework of Universally Composable security [7,29] as a natural security definition encompassing both unlinkability and our desired notion of non-malleability. We give an informal overview below (the more technical formal definition appears in the full version).

$\mathcal{F}_{\mathrm{HMP}}^{\mathcal{T}}$ allows parties to post private messages for other parties, as on a bulletin board, represented by abstract *handles* which reveal no information about the message (they are generated by the adversary without knowledge of the message). Only the designated receiver is allowed to obtain the corresponding message for a handle. To model the homomorphic feature, the functionality allows parties to post messages derived from other handles, as follows: When a party provides a previously posted handle and a transformation $T \in \mathcal{T}$, the functionality retrieves the message m corresponding to the handle and then acts as if the party had actually posted $T(m)$. The requesting party does not need to know, nor is it told, the underlying message m of the existing handle.

$\mathcal{F}_{\mathrm{HMP}}^{\mathcal{T}}$ models the non-malleability we require, since the *only* way a posted message can influence a subsequent message is via an allowed transformation.

The functionality also models unlinkability by internally behaving identically (in particular, in its interaction with the adversary) for the two different kinds of posts. The only exception is that corrupt parties may generate "dummy" handles which look like normal handles but do not contain any message. When a party derives a new handle from such a dummy handle, the adversary learns the transformation. This apparent slight weakness is natural[2] and it mirrors the tradeoff between our indistinguishability definitions. In our security proofs, this additional dummy handle feature is crucial.

Homomorphic Encryption Schemes and Protocols for $\mathcal{F}_{HMP}^{\mathcal{T}}$. The UC framework defines when a protocol is said to *securely realize* the functionality $\mathcal{F}_{HMP}^{\mathcal{T}}$: for every PPT adversary in the real world interaction (using the protocol), there exists a PPT simulator in the ideal world interaction with $\mathcal{F}_{HMP}^{\mathcal{T}}$, such that no PPT environment can distinguish between the two interactions.

We associate homomorphic encryption schemes with candidate protocols for $\mathcal{F}_{HMP}^{\mathcal{T}}$ in the following natural way (for simplicity assume all communication is on an authenticated broadcast channel). To setup an instance of $\mathcal{F}_{HMP}^{\mathcal{T}}$, a party generates a key pair and broadcasts the public key. To post a message, a party encrypts it under the public key and broadcasts the resulting ciphertext. The "derived post" feature is implemented via the CTrans procedure. To retrieve a message from a handle, the receiver decrypts it using the private key.

4 Relationships among Security Definitions

To justify our new security definitions, we prove some relationships among them and among the more established definitions of IND-CCA, gCCA [1,35], and RCCA [10] security. We defer the proofs to the full version.

Theorem 1. *CCA, gCCA, and RCCA security can be obtained as special cases of the HCCA definition, by appropriately restricting* RigEnc *and* RigExtract.

Theorem 2. *Every $\mathcal{T}$-homomorphic encryption scheme which is HCCA-secure, unlinkably homomorphic (with respect to $\mathcal{T}$) and satisfies the correctness properties, is a secure realization of $\mathcal{F}_{HMP}^{\mathcal{T}}$ in the standard UC model, where the adversary is allowed to make only static (non-adaptive) corruptions.*

Proof (sketch). To show that the UC security definition is satisfied, we must construct a simulator for every real-world adversary. The simulator uses RigEnc to simulate ciphertexts between uncorrupted parties (for which the simulator does not know the corresponding plaintext), and it uses RigExtract to detect when the adversary has applied a transformation to them.

In general, one cannot easily modify a $\mathcal{T}_1$-unlinkable-HCCA-secure scheme into a $\mathcal{T}_2$-unlinkable-HCCA-secure scheme, even if $\mathcal{T}_2 \subseteq \mathcal{T}_1$. The problem of "disabling" the transformations in $\mathcal{T}_1 \setminus \mathcal{T}_2$ while at the same time maintaining those in $\mathcal{T}_2$ appears just as challenging as constructing a $\mathcal{T}_2$-unlinkable-HCCA scheme from scratch. However, such a generic reduction is possible for the special case of unlinkable (a.k.a. rerandomizable) RCCA security, where the only allowed transformation is the identity function:

---

[2] For example, an adversary may broadcast a single encryption under a public key that he keeps hidden. The ciphertext will be meaningless to the recipient, but if the adversary later encounters another ciphertext that decrypts under this same key, he can deduce that the it was derived from his previous ciphertext.

Theorem 3. *Given a $\mathcal{T}$-unlinkable-HCCA-secure scheme and a (not necessarily unlinkable) RCCA-secure scheme, it is possible to construct an unlinkable-RCCA-secure scheme.*

Proof (sketch). To encrypt a message in the new scheme, first encrypt it in the RCCA-secure scheme, then encrypt the resulting ciphertext in the unlinkable-HCCA-secure scheme. The other operations of the scheme are defined appropriately, and the resulting scheme achieves unlinkable RCCA security.

Note that RCCA security without unlinkability is a weaker requirement than CCA security [10]. Thus, for example, an unlinkable HCCA-secure scheme along with a plain CCA-secure encryption scheme will yield an unlinkable RCCA-secure encryption scheme.

5   Achieving Unlinkable HCCA Security

Our main result is a family of construtions which achieves both HCCA security and unlinkable homomorphism, with respect to a wide range of message transformations, under the standard DDH assumption in two related groups.

Our construction is based on the rerandomizable RCCA scheme of Prabhakaran and Rosulek [30]. Recall that rerandomizable RCCA security is a special case of unlinkable HCCA security where the only allowed transformation is the identity function. Indeed, for the appropriate choice of parameters, our construction coincides with the one presented there.

Requirements. As in [30], our construction requires two (multiplicative) cyclic groups with a specific relationship: $\mathbb{G}$ of prime order p, and $\widehat{\mathbb{G}}$ of prime order q, where $\widehat{\mathbb{G}}$ is a subgroup of $\mathbb{Z}_p^*$. We require the DDH assumption to hold in both groups (with respect to the same security parameter). Given a sequence of primes $q, 2q + 1, 4q + 3$ (a *Cunningham chain* of the first kind of length 3 [2]), the two quadratic-residue groups $\widehat{\mathbb{G}} = \mathbb{QR}_{2q+1}^*$ and $\mathbb{G} = \mathbb{QR}_{4q+3}^*$, in which the DDH assumption is believed to hold, represent a suitable choice.

Features. Our construction uses $\mathbb{G}^n$ as its message space, where n is a parameter of the construction. We write the group operation in $\mathbb{G}$ as multiplication. For $\boldsymbol{\tau} = (\tau_1, \ldots, \tau_n) \in \mathbb{G}^n$, define $T_{\boldsymbol{\tau}}$ to be the "componentwise multiplication by $\boldsymbol{\tau}$" transformation: $(m_1, \ldots, m_n) \mapsto (\tau_1 m_1, \ldots, \tau_n m_n)$. We also let $T_{\boldsymbol{\tau}}(\bot) = \bot$ for simplicity.

Given n we can construct a scheme whose message space is $\mathbb{G}^n$, and whose set of allowable transformations is $\mathcal{T}_f = \{T_{\boldsymbol{\tau}} \mid \boldsymbol{\tau} \in \mathbb{G}^n \text{ and } f \circ T_{\boldsymbol{\tau}} \equiv f\}$, provided that f satisfies a certain technical property. In other words, the allowed transformations are the subspace of componentwise-multiplication transformations which are invariant with respect to the f function. By setting f appropriately, we can obtain the following notable classes $\mathcal{T}_f$:

- The identity function alone (i.e., rerandomizable RCCA security)
- All transformations $T_{\boldsymbol{\tau}}$. That is, all component-wise multiplications in $\mathbb{G}^n$.
- All "scalar multiplications" of tuples in $\mathbb{G}^n$ by coefficients in $\mathbb{G}$.

High-level Overview. The full and lengthy details of the construction are deferred to the full version. Some important ideas behind the scheme are outlined here:

The Cramer-Shoup CCA-secure scheme [13] achieves non-malleability via a ciphertext component of the form $(DE^\mu)^x$, where D, E are parts of the public key, x is a random value used in encryption, and μ is a hash of the ciphertext's prefix. The rerandomizable RCCA scheme of [30] uses the same paradigm, except that the value μ is a direct encoding of the plaintext (in rerandomizable RCCA, ciphertexts are malleable but only in ways which preserve the plaintext). In our HCCA-secure scheme, μ is a hash of $f(m_1, \ldots, m_n)$. Intuitively, the scheme can therefore only be malleable in ways which preserve $f(m_1, \ldots, m_n)$. Finally, we use the same "double-strand" technique of [30] to achieve a CTrans operation which is unlinkable and perfectly rerandomizable. We also expand the public key size to allow n group elements to be encrypted, instead of just one.

Theorem 4. *The construction satisfies the correctness requirements, HCCA security, and unlinkable homomorphism properties with respect to T_f, for any suitable f, under the DDH assumption in the two cyclic groups.*

6   Beyond Unary Transformations

Many interesting applications of homomorphic encryptions involve (at least) *binary* operations — those which accept encryptions of plaintexts m_0 and m_1 and output a ciphertext encoding $T(m_0, m_1)$. A common example is ElGamal encryption, where T may be the group operation of the underlying cyclic group. In this section, we examine the possibility of extending our results to schemes with *binary* transformations. We show some simple positive results, and also an impossibility for the specific case of group operations.

6.1   Extending Definitions

It is straight-forward to extend the UC definition of $\mathcal{F}_{\text{HMP}}^T$ to handle binary transformations. We define $\mathcal{F}_{2\text{-HMP}}^T$ to act like $\mathcal{F}_{\text{HMP}}^T$, except that T is a set of allowed *binary* transformations. Honest parties may then generate a derived post by giving *two* handles an an allowed transformation T. Naturally, the functionality internally acts as if the party had requested a post of $T(m_1, m_2)$, where m_1, m_2 are the messages corresponding to the given handles.

It is not clear what is the most appropriate behavior for $\mathcal{F}_{2\text{-HMP}}^T$ when one or both of the given handles is a dummy handle. Our impossibility results in the next section do not depend on any particular behavior of $\mathcal{F}_{2\text{-HMP}}^T$ in such a case, so we opt to make the definition as weak as possible. When such a request is made, we let the adversary learn the transformation and the two handles.

Defining an analog of our indistinguishability definition, however, appears to be much more difficult task. Indeed, it is not clear how to appropriately handle the case where the adversary applies a transformation to a pair of ciphertexts where one or both were generated via (independent calls to) RigEnc.

Below we show that it is impossible to securely realize $\mathcal{F}^{\mathcal{T}}_{2\text{-HMP}}$ (i.e., achieve unlinkability and an HCCA-like definition for binary transformations) for a large class of useful transformations $\mathcal{T}$. Still, one may be willing to relax the unlinkability requirement (e.g., as in [34]) and still demand some non-malleability. Thus we leave it as an important open problem to give a meaningful generalization of HCCA security, for transformations that combine multiple ciphertexts.

6.2   A Simple Positive Result

$\mathcal{F}^{\mathcal{T}}_{2\text{-HMP}}$ can be achieved for some simple transformation spaces, by appropriately composing an HCCA-secure, unlinkably homomorphic (unary) scheme Π. To see this, construct a new scheme whose ciphertexts are k-tuples of completely independent ciphertexts from Π. The (binary) transformation operations in the new scheme have the following form: Given two tuples of k ciphertexts each, choose k components from among these $2k$ ciphertexts, apply an allowed transformation to each separately, and output those k ciphertexts in a tuple.

The resulting scheme's transformations simply "mix and match" the independent components of the two tuples of ciphertexts to form a new tuple. The unlinkability of the transformations applied to each individual component implies the unlinkability of the new scheme. It is easy to see that such a scheme securely realizes of $\mathcal{F}^{\mathcal{T}}_{2\text{-HMP}}$ with respect to the appropriate transformation space.

6.3   Negative Results

The positive result presented above appears to be much less sophisticated than, say, a scheme that is homomorphic with respect to a group operation. Indeed, this limitation turns out to be inherent in securely realizing $\mathcal{F}^{\mathcal{T}}_{2\text{-HMP}}$.

Theorem 5. *There is no secure realization of $\mathcal{F}^{\mathcal{T}}_{2\text{-HMP}}$ via a homomorphic encryption scheme, when $\mathcal{T}$ contains a quasigroup operation[3] on the message space.*

Proof (sketch). The main observation is that each handle (ciphertext) must have a bounded length independent of its "history" (i.e., whether it was generated via the homomorphic operation and if so, which operations applied to which existing handles), and thus can only encode a bounded amount of information about its history. We show that any simulator for $\mathcal{F}^{\mathcal{T}}_{2\text{-HMP}}$ must be able to extract a reasonble history from any handle output by the adversary.

However, when a quasigroup operation is an allowed transformation, there can be far more possible histories than can be encoded in a single handle. We use this fact to construct an environment and adversary which can distinguish between the real world and the ideal world with any simulator, contradicting the security definition.

---

[3] A quasigroup operation $\star$ on a set X is an operation such that fixing any two values in the equation $x \star y = z$ *uniquely* determines the third value.

Acknowledgments

We thank Rui Xue for helpful discussions regarding Theorem 3, as well as the feedback from the anonymous referees.

References

1. An, J.H., Dodis, Y., Rabin, T.: On the security of joint signature and encryption. In: Knudsen, L.R. (ed.) EUROCRYPT 2002. LNCS, vol. 2332, pp. 83–107. Springer, Heidelberg (2002)
2. Andersen, J.K., Weisstein, E.W.: Cunningham chain. From MathWorld–A Wolfram Web Resource (2005), http://mathworld.wolfram.com/CunninghamChain.html
3. Bellare, M., Sahai, A.: Non-malleable encryption: Equivalence between two notions, and an indistinguishability-based characterization. In: Wiener, M.J. (ed.) CRYPTO 1999. LNCS, vol. 1666, pp. 519–536. Springer, Heidelberg (1999)
4. Benaloh, J.: Verifiable Secret-Ballot Elections. PhD thesis, Department of Computer Science. Yale University (1987)
5. Blaze, M., Bleumer, G., Strauss, M.: Divertible protocols and atomic proxy cryptography. In: Nyberg, K. (ed.) EUROCRYPT 1998. LNCS, vol. 1403, pp. 127–144. Springer, Heidelberg (1998)
6. Boneh, D. (ed.): CRYPTO 2003. LNCS, vol. 2729. Springer, Heidelberg (2003)
7. Canetti, R.: Universally composable security: A new paradigm for cryptographic protocols. Cryptology ePrint Archive, Report 2000/067 (2005)
8. Canetti, R., Herzog, J.: Universally composable symbolic analysis of mutual authentication and key-exchange protocols. In: Halevi, Rabin (eds.) [22], pp. 380–403
9. Canetti, R., Hohenberger, S.: Chosen-ciphertext secure proxy re-encryption. In: ACM Computer and Communication Security (CCS) (2007)
10. Canetti, R., Krawczyk, H., Nielsen, J.B.: Relaxing chosen-ciphertext security. In: Boneh (ed.) [6], pp. 565–582
11. Chor, B., Gilboa, N., Naor, M.: Private information retrieval by keywords. TR CS0917, Department of Computer Science, Technion (1997)
12. Cramer, R., Franklin, M.K., Schoenmakers, B., Yung, M.: Multi-autority secret-ballot elections with linear work. In: Maurer, U.M. (ed.) EUROCRYPT 1996. LNCS, vol. 1070, pp. 72–83. Springer, Heidelberg (1996)
13. Cramer, R., Shoup, V.: A practical public key cryptosystem provably secure against adaptive chosen ciphertext attack. In: Krawczyk, H. (ed.) CRYPTO 1998. LNCS, vol. 1462. Springer, Heidelberg (1998)
14. Damgård, I., Fazio, N., Nicolosi, A.: Non-interactive zero-knowledge from homomorphic encryption. In: Halevi, Rabin (eds.) [22], pp. 41–59
15. Damgård, I., Nielsen, J.B.: Universally composable efficient multiparty computation from threshold homomorphic encryption. In: Boneh (ed.) [6], pp. 247–264
16. Danezis, G.: Breaking four mix-related schemes based on universal re-encryption. In: Proc. Information Security Conference. Springer, Heidelberg (2006)
17. Dolev, D., Dwork, C., Naor, M.: Nonmalleable cryptography. SIAM J. Comput. 30(2), 391–437 (electronic) (2000); Preliminary version in STOC (1991)
18. Gamal, T.E.: A public key cryptosystem and a signature scheme based on discrete logarithms. In: Blakely, G.R., Chaum, D. (eds.) CRYPTO 1984. LNCS, vol. 196, pp. 10–18. Springer, Heidelberg (1985)

19. Goldwasser, S., Micali, S.: Probabilistic encryption. J. Comput. Syst. Sci. 28(2), 270–299 (1984); Preliminary version appeared in STOC 1982
20. Golle, P., Jakobsson, M., Juels, A., Syverson, P.: Universal re-encryption for mixnets. In: Proceedings of the 2004 RSA Conference, Cryptographer's track, San Francisco, USA (February 2004)
21. Groth, J.: Rerandomizable and replayable adaptive chosen ciphertext attack secure cryptosystems. In: Naor, M. (ed.) TCC 2004. LNCS, vol. 2951, pp. 152–170. Springer, Heidelberg (2004)
22. Halevi, S., Rabin, T. (eds.): TCC 2006. LNCS, vol. 3876. Springer, Heidelberg (2006)
23. Hirt, M., Sako, K.: Efficient receipt-free voting based on homomorphic encryption. In: Preneel, B. (ed.) EUROCRYPT 2000. LNCS, vol. 1807, pp. 539–556. Springer, Heidelberg (2000)
24. Ishai, Y., Kushilevitz, E., Ostrovsky, R.: Sufficient conditions for collision-resistant hashing. In: Kilian, J. (ed.) TCC 2005. LNCS, vol. 3378, pp. 445–456. Springer, Heidelberg (2005)
25. Jurik, M.J.: Extensions to the Paillier Cryptosystem with Applications to Cryptological Protocols. PhD thesis, BRICS (2003)
26. Naor, M., Yung, M.: Public-key cryptosystems provably secure against chosen ciphertext attacks. In: STOC, pp. 427–437. ACM, New York (1990)
27. Paillier, P.: Public-key cryptosystems based on composite degree residuosity classes. In: Stern, J. (ed.) EUROCRYPT 1999. LNCS, vol. 1592, pp. 223–238. Springer, Heidelberg (1999)
28. Patil, A.: On symbolic analysis of cryptographic protocols. Master's thesis, Massachusetts Institute of Technology (2005)
29. Pfitzmann, B., Waidner, M.: Composition and integrity preservation of secure reactive systems. In: ACM Conference on Computer and Communications Security, pp. 245–254 (2000)
30. Prabhakaran, M., Rosulek, M.: Rerandomizable RCCA encryption. In: Menezes, A. (ed.) CRYPTO 2007. LNCS, vol. 4622, Springer, Heidelberg (to appear, 2007)
31. Prabhakaran, M., Rosulek, M.: Homomorphic encryption with chosen-ciphertext security. Cryptology ePrint Archive, Report 2008/079 (2008), http://eprint.iacr.org/2008/079
32. Rackoff, C., Simon, D.R.: Non-interactive zero-knowledge proof of knowledge and chosen ciphertext attack. In: Feigenbaum, J. (ed.) CRYPTO 1991. LNCS, vol. 576, pp. 433–444. Springer, Heidelberg (1992)
33. Sako, K., Kilian, J.: Secure voting using partially compatible homomorphisms. In: Desmedt, Y.G. (ed.) CRYPTO 1994. LNCS, vol. 839, pp. 411–424. Springer, Heidelberg (1994)
34. Sander, T., Young, A., Yung, M.: Non-interactive cryptocomputing for NC^1. In: FOCS, pp. 554–567 (1999)
35. Shoup, V.: A proposal for an ISO standard for public key encryption. Cryptology ePrint Archive, Report 2001/112 (2001), http://eprint.iacr.org/
36. Song, D.X., Wagner, D., Perrig, A.: Practical techniques for searches on encrypted data. In: IEEE Symposium on Security and Privacy, pp. 44–55 (2000)
37. Wikström, D.: A note on the malleability of the El Gamal cryptosystem. In: Menezes, A., Sarkar, P. (eds.) INDOCRYPT 2002. LNCS, vol. 2551, pp. 176–184. Springer, Heidelberg (2002)

How to Encrypt with the LPN Problem

Henri Gilbert, Matthew J.B. Robshaw, and Yannick Seurin

Orange Labs, 38–40 rue du General Leclerc, Issy les Moulineaux, France
{henri.gilbert,matt.robshaw,yannick.seurin}@orange-ftgroup.com

Abstract. We present a probabilistic private-key encryption scheme named LPN-C whose security can be reduced to the hardness of the Learning from Parity with Noise (LPN) problem. The proposed protocol involves only basic operations in GF(2) and an error-correcting code. We show that it achieves indistinguishability under adaptive chosen plaintext attacks (IND-P2-C0). Appending a secure MAC renders the scheme secure under adaptive chosen ciphertext attacks. This scheme enriches the range of available cryptographic primitives whose security relies on the hardness of the LPN problem.

Keywords: symmetric encryption, LPN, error-correcting code.

1 Introduction

The connections between cryptography and learning theory are well known since the celebrated paper by Impagliazzo and Levin [18]. They showed that these two areas are in a sense complementary since the possibility of cryptography rules out the possibility of efficient learning and *vice-versa*. Since then, a lot of work has dealt with building cryptographic primitives based on presumably hard learning problems. Perhaps the most well-known of these problems among the cryptographic community is the so called *Learning from Parity with Noise* (LPN) problem, which can be described as learning an unknown k-bit vector x given noisy versions of its scalar product $a \cdot x$ with random vectors a. The prominent lightweight authentication protocol HB$^+$ recently proposed by Juels and Weis [19], and its variants [7,9,12,26], are based on this problem.

Our work is concerned with encryption schemes in the symmetric setting, where a sender and a receiver share a secret key. Up to now, most of the work in this field has concentrated on studying various operating modes to use with a secure block cipher [2]. Departing from this approach, we will construct a symmetric encryption scheme that does not appeal to any assumption regarding the pseudorandomness of a block cipher, and whose security can *directly* be reduced to some hard problem, namely here the LPN problem. In a nutshell, our scheme, named LPN-C, uses a shared secret matrix M and random vectors a to compute "noisy" masking vectors $b = a \cdot M \oplus \nu$. The vector b is then used to mask the plaintext, preliminary encoded with an error-correcting code. The receiver, knowing M, can remove the mask $a \cdot M$, and then the noise with the error-correcting code. At the same time the noise ν prevents an attacker from "learning" the secret matrix M.

L. Aceto et al. (Eds.): ICALP 2008, Part II, LNCS 5126, pp. 679–690, 2008.
© Springer-Verlag Berlin Heidelberg 2008

Related Work. We briefly review the related work building cryptographic primitives based on hard learning problems. We have already cited the authentication protocol HB$^+$ [19], which was itself derived from a simpler protocol named HB by Hopper and Blum [17]. Both protocols possess a proof of security in a certain attack model relying on the LPN problem [19,20,21]. Gilbert, Robshaw, and Sibert [13] then showed a simple man-in-the-middle attack against HB$^+$. This triggered many trials to modify and protect HB$^+$ against man-in-the-middle attacks [7,9,26] but these three proposals were recently broken [11]. The subsequent proposal HB$^\#$ [12] is the only one to be provably secure against (some) man-in-the-middle attacks.

Former proposals were made by Blum *et al.* [5], who described a pseudorandom number generator (PRNG), a one-way function, and a private-key cryptosystem (encrypting only one bit at a time, thus much less efficient than the proposal in this paper) based on very general hard-to-learn class of functions. They also proposed a PRNG explicitly based on the LPN problem (rather than on general class of functions) derived from an older proposal of one-way function based on the hardness of decoding a random linear code [14]. More recently, Regev [28] proposed a public-key cryptosystem based on the so-called LWE (Learning with Error) problem, a generalization of the LPN problem to fields $GF(p)$, $p > 2$ (and proved that an efficient algorithm for the LWE problem would imply an efficient *quantum* algorithm for worst-case lattice problems).

LPN-C carries some similarity with a scheme by Rao and Nam [27], which may be seen as a secret-key variant of the McEliece cryptosystem, and with the trapdoor cipher *TCHo* [1], by Aumasson *et al.* In the later, the additional noise added to $C(x) \oplus \nu$ is introduced via an LFSR whose feedback polynomial has a low-weight multiple used as the trapdoor.

Organisation. Our paper is organised as follows. First we give some basic definitions and facts about the LPN problem and private-key encryption. Then we describe the encryption scheme LPN-C. In Section 4 we analyse the security of the scheme, in particular we establish that it is secure in the sense IND-P2-C0. In Section 5 we give some practical parameter values and explore some possible variants of the scheme. Finally, we draw our conclusions and suggest some potential future work.

2   Preliminaries

Basic Notation. In the sequel, the security parameter will be denoted by k, and we will say that a function of k (from positive integers to positive real numbers) is *negligible* if it approaches zero faster than any inverse polynomial, and *noticeable* if it is larger than some inverse polynomial (for infinitely many values of k). An algorithm will be *efficient* if it runs in time polynomial in k and possibly the size of its inputs. PPT will stand for *Probabilistic Polynomial-Time* Turing machine.

We use bold type x to indicate a row vector while scalars x are written in normal text. The i-th bit of x is denoted $x[i]$. The bitwise addition of two vectors will be denoted $\oplus$ just as for scalars, the scalar product of a and b will be denoted $a \cdot b$, and their concatenation $a\|b$. We denote the *Hamming weight* of x by $\mathrm{Hwt}(x)$.

Given a finite set S and a probability distribution Δ on S, $s \leftarrow \Delta$ denotes the drawing of an element of S according to Δ and $s \xleftarrow{\$} S$ denotes the random drawing of an element of S endowed with the uniform probability distribution. Ber_η will denote the Bernoulli distribution of parameter $\eta \in]0, \frac{1}{2}[$, *i.e.* a bit $\nu \leftarrow \mathrm{Ber}_\eta$ is such that $\Pr[\nu = 1] = \eta$ and $\Pr[\nu = 0] = 1 - \eta$. We also define the corresponding vectorial distribution $\mathrm{Ber}_{n,\eta}$: an n-bit vector $\nu \leftarrow \mathrm{Ber}_{n,\eta}$ is such that each bit of ν is independently drawn according to Ber_η. Finally, we will need to define the two following oracles: we will let U_n denote the oracle returning independent uniformly random n-bit strings, and for a fixed k-bit string s, $\Pi_{s,\eta}$ will be the oracle returning independent $(k + 1)$-bit strings according to the distribution (to which we will informally refer to as *an LPN distribution*):

$$\{a \xleftarrow{\$} \{0,1\}^k; \nu \leftarrow \mathrm{Ber}_\eta \; : \; (a, a \cdot s \oplus \nu)\} \; .$$

The LPN Problem. The LPN problem is the problem of retrieving s given access to the oracle $\Pi_{s,\eta}$. For a fixed value of k, we will say that an algorithm $\mathcal{A}$ (T, q, δ)-solves the LPN problem with noise parameter η if $\mathcal{A}$ runs in time at most T, makes at most q oracle queries, and

$$\Pr\left[s \xleftarrow{\$} \{0,1\}^k \; : \; \mathcal{A}^{\Pi_{s,\eta}}(1^k) = s\right] \geq \delta \; .$$

By saying that the LPN problem is hard, we mean that any efficient adversary solves it with only negligible probability. There is a significant amount of literature dealing with the hardness of the LPN problem. It is closely related to the problem of decoding a random linear code [4] and is NP-Hard. It is NP-Hard to even find a vector x satisfying more than half of the equations outputted by $\Pi_{s,\eta}$ [16]. The average-case hardness has also been intensively investigated [5,6,17]. The current best known algorithms to solve it are the BKW algorithm due to Blum, Kalai, and Wasserman [6] and its improved variants by Fossorier *et al.* [10] and Levieil and Fouque [23]. They all require $2^{\Theta(k/\log k)}$ oracle queries and running time.

Private-key Encryption. We briefly recall the basic definitions dealing with the semantics of probabilistic private-key encryption.

Definition 1 (Private-key cryptosystem). *A probabilistic private-key encryption scheme is a triple of algorithms* $\Gamma = (\mathcal{G}, \mathcal{E}, \mathcal{D})$ *such that:*

- *the key generation algorithm* $\mathcal{G}$*, on input the security parameter* k*, returns a random secret key* $K \in \mathcal{K}(k)$*:* $K \xleftarrow{\$} \mathcal{G}(1^k)$*;*
- *the encryption algorithm* $\mathcal{E}$ *is a PPT algorithm that takes as input a secret key* K *and a plaintext* $X \in \{0,1\}^*$ *and returns a ciphertext* Y*:* $Y \leftarrow \mathcal{E}_K(X)$*;*

– *the decryption algorithm $\mathcal{D}$ is a deterministic, polynomial-time algorithm that takes as input a secret key K and a string Y and returns either the corresponding plaintext X or a special symbol $\perp$: $\mathcal{D}_K(Y) \in \{0,1\}^* \cup \{\perp\}$.*

It is usually required that $\mathcal{D}_K(\mathcal{E}_K(X)) = X$ for all $X \in \{0,1\}^*$. One can slightly relax this condition, and only require that $\mathcal{D}_K(\mathcal{E}_K(X)) = X$ except with negligible probability.

3   Description of LPN-C

Let $C : \{0,1\}^r \rightarrow \{0,1\}^m$ be an $[m, r, d]$ error-correcting code (*i.e.* of length m, dimension r, and minimal distance d) with correction capacity $t = \lfloor \frac{d-1}{2} \rfloor$. This error-correcting code is assumed to be publicly known. Let M be a secret $k \times m$ matrix (constituting the secret key of the cryptosystem). To encrypt an r-bit vector x, the sender draws a k-bit random vector a and computes

$$y = C(x) \oplus a \cdot M \oplus \nu \ ,$$

where $\nu \leftarrow \text{Ber}_{m,\eta}$ is an m-bit noise vector such that each of its bits is (independently) 1 with probability η and 0 with probability $1 - \eta$. The ciphertext is the pair (a, y).

Upon reception of this pair, the receiver decrypts by computing $y \oplus a \cdot M = C(x) \oplus \nu$, and decoding the resulting value. If decoding is not possible (which may happen when the code is not *perfect*), then the decryption algorithm returns $\perp$. When the message is not r-bit long, it is padded till its length is the next multiple of r and encrypted blockwise. The steps for LPN-C are given in Fig. 1.

As can be seen from its description, LPN-C encryption involves only basic operations (at least when a simple linear code is used) reduced to scalar products and exclusive-or's. The decryption requires to implement the decoding procedure, which implies more work on the receiver side, though there are error-correcting codes with very efficient decoding algorithms [25].

Parameters	Security parameter k Polynomials (in k) m, r, d with $m > r$ Noise level $\eta \in]0, \frac{1}{2}[$
Public Components	An $[m, r, d]$ error-correcting code $C : \{0,1\}^r \rightarrow \{0,1\}^m$ and the corresponding decoding algorithm C^{-1}
Secret Key Generation	On input 1^k, output a random $k \times m$ binary matrix M
Encryption Algorithm	On input an r-bit vector x, draw a random k-bit vector a and a noise vector ν, compute $y = C(x) \oplus a \cdot M \oplus \nu$, and output (a, y)
Decryption Algorithm	On input (a, y), compute $y \oplus a \cdot M$, decode the resulting value by running C^{-1} and return the corresponding output or $\perp$ if unable to decode

Fig. 1. Description of LPN-C

Decryption Failures. Decryption failures happen when the Hamming weight of the noise vector ν is greater than the correction capacity t of the error-correcting code, $\mathrm{Hwt}(\nu) > t$. When the noise vector is randomly drawn, the probability of decryption failure is given by

$$P_{\mathrm{DF}} = \sum_{i=t+1}^{m} \binom{m}{i} \eta^i (1-\eta)^{m-i} .$$

In order to eliminate such decryption failures, the Hamming weight of the noise vector can be tested before being used. If $\mathrm{Hwt}(\nu) > t$, the sender draws a new noise vector according to $\mathrm{Ber}_{m,\eta}$. When the parameters are chosen such that $\eta m < t$, then this happens only with negligible probability and the encryption algorithm remains efficient.

4 Security Proofs

4.1 Security Model

The security notions for probabilistic private-key encryption have been formalized by Bellare et al. [2] and thoroughly studied by Katz and Yung in [22]. The two main security goals for symmetric encryption are indistinguishability (IND) and non-malleability (NM). Indistinguishability deals with the secrecy afforded by the scheme: an adversary must be unable to distinguish the encryption of two (adversarially chosen) plaintexts. This definition was introduced in the context of public-key encryption as a more practical equivalent to semantic security [15]. Non-malleability was introduced (again in the context of public-key encryption) by Dolev, Dwork, and Naor [8] and deals with ciphertext modification: given a challenge ciphertext Y, an adversary must be unable to generate a different ciphertext Y' so that the respective plaintexts are meaningfully related.

Adversaries run in two phases (they are denoted as a pair of algorithms $\mathcal{A} = (\mathcal{A}_1, \mathcal{A}_2)$) and are classified according to the oracles (encryption and/or decryption) they are allowed to access in each phase. At the end of the first phase, $\mathcal{A}_1$ outputs a distribution on the space of the plaintexts (i.e. a pair of plaintexts (x_1, x_2) of probability $1/2$ each in the case of IND or a more complex distribution in the case of NM). Then, a ciphertext is selected at random according to the distribution and transmitted to $\mathcal{A}_2$ (this represents $\mathcal{A}$'s challenge) and the success of $\mathcal{A}$ is determined according to the security goal (e.g. in the case of IND, determine whether x_1 or x_2 was encrypted). The adversary is denoted PX-CY, where P stands for the encryption oracle and C for the decryption oracle, and where $X, Y \in \{0, 1, 2\}$ indicates when $\mathcal{A}$ is allowed to access the oracle:

- 0: $\mathcal{A}$ never accesses the oracle
- 1: $\mathcal{A}$ can only access the oracle during phase 1, hence before seeing the challenge (also termed *non-adaptive*)
- 2: $\mathcal{A}$ can access the oracle during phases 1 and 2 (also termed *adaptive*)

We only give the formal definition of indistinguishability since this is the security goal we will be primarily interested in. A formal definition of non-malleability can be found in [22].

Definition 2 (IND-PX-CY). *Let $\Gamma = (\mathcal{G}, \mathcal{E}, \mathcal{D})$ be an encryption scheme and let $\mathcal{A} = (\mathcal{A}_1, \mathcal{A}_2)$ be an adversary. For $X, Y \in \{0, 1, 2\}$ and a security parameter k, the advantage of $\mathcal{A}$ in breaking the indistinguishability of Γ is defined as:*

$$\mathrm{Adv}_{\mathcal{A},\Gamma}^{\mathrm{IND\text{-}PX\text{-}CY}}(k) \stackrel{\mathrm{def}}{=} \left| \Pr\left[K \stackrel{\$}{\leftarrow} \mathcal{G}(1^k); (\boldsymbol{x_0}, \boldsymbol{x_1}, s) \leftarrow \mathcal{A}_1^{\mathcal{O}_1, \mathcal{O}_1'}(1^k); \right.\right.$$

$$\left.\left. b \stackrel{\$}{\leftarrow} \{0, 1\}; \boldsymbol{y} \leftarrow \mathcal{E}_K(\boldsymbol{x_b}) : \mathcal{A}_2^{\mathcal{O}_2, \mathcal{O}_2'}(1^k, s, \boldsymbol{y}) = b \right] - \frac{1}{2} \right|$$

where $(\mathcal{O}_1, \mathcal{O}_2)$ is $(\emptyset, \emptyset)$, $(\mathcal{E}_K(\cdot), \emptyset)$, $(\mathcal{E}_K(\cdot), \mathcal{E}_K(\cdot))$ when X is resp. 0, 1, 2 and $(\mathcal{O}_1', \mathcal{O}_2')$ is $(\emptyset, \emptyset)$, $(\mathcal{D}_K(\cdot), \emptyset)$, $(\mathcal{D}_K(\cdot), \mathcal{D}_K(\cdot))$ when Y is resp. 0, 1, 2, and s is some state information. Note that the plaintexts returned by $\mathcal{A}_1$ must respect $|\boldsymbol{x_0}| = |\boldsymbol{x_1}|$ and that when $Y = 2$, $\mathcal{A}_2$ is not allowed to query $\mathcal{D}_K(\boldsymbol{y})$.
We say that Γ is secure in the sense IND-PX-CY if $\mathrm{Adv}_{\mathcal{A},\Gamma}^{\mathrm{IND\text{-}PX\text{-}CY}}(k)$ is negligible for any PPT adversary $\mathcal{A}$.

Important relationships between the different security properties have been proved by Katz and Yung [22]. The most meaningful for us are:

– non-adaptive CPA-security implies adaptive CPA-security:

$$\mathrm{IND\text{-}P1\text{-}CY} \Rightarrow \mathrm{IND\text{-}P2\text{-}CY} \quad \text{and} \quad \mathrm{NM\text{-}P1\text{-}CY} \Rightarrow \mathrm{NM\text{-}P2\text{-}CY}$$

– IND and NM are equivalent in the case of P2-C2 attacks (but unrelated for other attacks): IND-P2-C2 ⇔ NM-P2-C2.

4.2   Proof of Indistinguishability Under Chosen Plaintext Attacks

We now prove that LPN-C is secure in the sense IND-P2-C0, by reducing its security to the LPN problem. First, we will recall the following useful lemma which was proved in [20] following [28], and which states that the hardness of the LPN problem implies that the two oracles U_{k+1} and $\Pi_{s,\eta}$ are indistinguishable.

Lemma 1 ([20], Lemma 1). *Assume there exists an algorithm $\mathcal{M}$ making q oracle queries, running in time T, and such that*

$$\left| \Pr\left[s \stackrel{\$}{\leftarrow} \{0, 1\}^k : \mathcal{M}^{\Pi_{s,\eta}}(1^k) = 1 \right] - \Pr\left[\mathcal{M}^{U_{k+1}}(1^k) = 1 \right] \right| \geq \delta .$$

Then there is an algorithm $\mathcal{A}$ making $q' = \mathcal{O}(q \cdot \delta^{-2} \log k)$ oracle queries, running in time $T' = \mathcal{O}(T \cdot k \delta^{-2} \log k)$, and such that

$$\Pr\left[s \stackrel{\$}{\leftarrow} \{0, 1\}^k : \mathcal{A}^{\Pi_{s,\eta}}(1^k) = s \right] \geq \frac{\delta}{4} .$$

A full proof of this result can be found in [20]. We will reduce the security of LPN-C to the problem of distinguishing U_{k+1} and $\Pi_{s,\eta}$ rather than directly to the LPN problem.

Theorem 1. *Assume there is an adversary $\mathcal{A}$, running in time T, and attacking LPN-C with parameters (k, m, r, d, η) in the sense of IND-P2-C0 with advantage δ by making at most q queries to the encryption oracle. Then there is an algorithm $\mathcal{M}$ making $\mathcal{O}(q)$ oracle queries, running in time $\mathcal{O}(T)$, and such that*

$$\left| \Pr\left[s \xleftarrow{\$} \{0,1\}^k : \mathcal{M}^{\Pi_{s,\eta}}(1^k) = 1 \right] - \Pr\left[\mathcal{M}^{U_{k+1}}(1^k) = 1 \right] \right| \geq \frac{\delta}{m} .$$

Proof. As already pointed out, non-adaptive CPA-security (P1) implies adaptive CPA-security (P2), hence we may restrict ourselves to adversaries accessing the encryption oracle only during the first phase of the attack (before seeing the challenge ciphertext).

The proof proceeds by a hybrid argument. We will first define the following hybrid distributions on $\{0,1\}^{k+m}$. For $j \in [0..m]$, let M' denote a $k \times (m - j)$ binary matrix. We define the probability distribution $\mathcal{P}_{j,M',\eta}$ as

$$\{ a \xleftarrow{\$} \{0,1\}^k; r \xleftarrow{\$} \{0,1\}^j; \nu \leftarrow \mathrm{Ber}_{(m-j),\eta} : a\|r\|(a \cdot M' \oplus \nu) \} .$$

Hence the returned vector $a\|b$ is such that the first j bits of b are uniformly random, whereas the last $(m - j)$ bits are distributed according to $(m - j)$ independent LPN distributions related to the respective columns of M'. Note that $\mathcal{P}_{m,M',\eta} = U_{k+m}$.

We will also define the following hybrid encryption oracles $\mathcal{E}'_{j,M',\eta}$ associated with the secret matrix M' and noise parameter η: on input the r-bit plaintext x, the encryption oracle encodes it to $C(x)$, draws a random $(k + m)$-bit vector $a\|b$ distributed according to $\mathcal{P}_{j,M',\eta}$, and returns $(a, C(x) \oplus b)$.

We now describe how the distinguisher $\mathcal{M}$ proceeds. Recall that $\mathcal{M}$ has access to an oracle and wants to distinguish whether this is U_{k+1} or $\Pi_{s,\eta}$. On input the security parameter 1^k, $\mathcal{M}$ draws a random $j \in [1..m]$. If $j < m$, it also draws a random $k \times (m - j)$ binary matrix M'. It then launches the first phase $\mathcal{A}_1$ of the adversary $\mathcal{A}$. Each time $\mathcal{A}_1$ asks for the encryption of some x, $\mathcal{M}$ obtains a sample (a, z) from its oracle, draws a random $(j - 1)$-bit vector $r \xleftarrow{\$} \{0,1\}^{j-1}$, and draws a $(m - j)$-bit noise vector ν distributed according to $\mathrm{Ber}_{(m-j),\eta}$. It then forms the masking vector $b = r\|z\|(a \cdot M' \oplus \nu)$ and returns $(a, C(x) \oplus b)$.

The adversary $\mathcal{A}_1$ then returns two plaintexts x_1 and x_2. The distinguisher $\mathcal{M}$ selects a uniformly random $\alpha \in \{1, 2\}$ and returns to $\mathcal{A}_2$ the ciphertext corresponding to x_α encrypted exactly as described just before. If the answer of $\mathcal{A}_2$ is correct, then $\mathcal{M}$ returns 1, otherwise it returns 0.

It is straightforward to verify that when $\mathcal{M}$'s oracle is U_{k+1}, $\mathcal{M}$ simulates the encryption oracle $\mathcal{E}'_{j,M',\eta}$, whereas when $\mathcal{M}$'s oracle is $\Pi_{s,\eta}$, then $\mathcal{M}$ simulates the encryption oracle $\mathcal{E}'_{j-1,M'',\eta}$ where $M'' = s\|M'$ is the matrix obtained as the concatenation of s and M'. Hence the advantage of the distinguisher can be expressed as

$$\text{Adv} = \left| \Pr\left[s \xleftarrow{\$} \{0,1\}^k : \mathcal{M}^{\Pi_{s,\eta}}(1^k) = 1 \right] - \Pr\left[\mathcal{M}^{U_{k+1}}(1^k) = 1 \right] \right|$$

$$= \frac{1}{m} \left| \sum_{j=0}^{m-1} \Pr\left[\mathcal{A}^{\mathcal{E}'_{j,M',\eta}} \text{ succeeds} \right] - \sum_{j=1}^{m} \Pr\left[\mathcal{A}^{\mathcal{E}'_{j,M',\eta}} \text{ succeeds} \right] \right|$$

$$= \frac{1}{m} \left| \Pr\left[\mathcal{A}^{\mathcal{E}'_{0,M',\eta}} \text{ succeeds} \right] - \Pr\left[\mathcal{A}^{\mathcal{E}'_{m,M',\eta}} \text{ succeeds} \right] \right| .$$

Note that the encryption oracle $\mathcal{E}'_{0,M',\eta}$ is exactly the real LPN-C encryption oracle. On the other hand the encryption oracle $\mathcal{E}'_{m,M',\eta}$ encrypts all plaintexts by blinding them with uniformly random vectors b so that in this case the adversary $\mathcal{A}$ cannot do better (or worse) than guessing α at random and has a success probability of $1/2$. Hence

$$\left| \Pr\left[\mathcal{A}^{\mathcal{E}'_{0,M',\eta}} \text{ succeeds} \right] - \Pr\left[\mathcal{A}^{\mathcal{E}'_{m,M',\eta}} \text{ succeeds} \right] \right|$$

is exactly the advantage of the adversary which is greater than δ by hypothesis. The theorem follows.   □

Remark 1. Note that when the error-correcting code is linear, the scheme is clearly malleable, even when the adversary has no access at all to the encryption nor the decryption oracle (the scheme is not NM-P0-C0). Indeed, an adversary receiving a ciphertext (a, y) corresponding to some plaintext x, can forge a new ciphertext corresponding to some other plaintext $x \oplus x'$ simply by modifying the ciphertext to $(a, y \oplus C(x'))$. The same kind of attacks, though more elaborate, would probably apply for non-linear error-correcting codes. Since NM-P2-C2 is equivalent to IND-P2-C2, the scheme cannot be IND-P2-C2 either. We investigate the security with respect to IND-P2-C1 attacks in the next subsection.

4.3   An IND-P0-C1 Attack

Here we show that the scheme is insecure (*i.e.* distinguishable) when the attacker has (non-adaptive) access to the decryption oracle. The idea is to query the decryption oracle many times with the same vector a in order to get many approximate equations on $a \cdot M$. Consider an adversary querying the decryption oracle with ciphertexts (a, y_i) for a fixed a and random y_i's. Each time $y_i \oplus a \cdot M$ is at Hamming distance less than t from a codeword, the decryption oracle will return x_i such that $\text{Hwt}(C(x_i) \oplus y_i \oplus a \cdot M) \leq t$. This will give an approximation for each bit of $a \cdot M$ with noise parameter less than $\frac{t}{m}$.

Indeed, let us fix some bit position j, and evaluate the probability p that, given that the decryption oracle returned the plaintext x_i, the j-th bit of $a \cdot M$ is *not* equal to the j-th bit of $C(x_i) \oplus y_i$:

$$p = \Pr_{y_i \xleftarrow{\$} \{0,1\}^m} \left[(a \cdot M)[j] \neq (C(x_i) \oplus y_i)[j] \,\middle|\, \mathcal{D}_K(a, y_i) = x_i \right] .$$

Obviously, the sum over j of this quantity is equal to the expected value of the number of errors, hence is less than t. Consequently the error probability is less than t/m. Assume the vector a was chosen to have only one non-null coordinate (say, the l-th one). Then this will enable to retrieve with high confidence the bit in position (l, j) of the secret matrix M with a few attempts (according to the Chernoff bound, since the repeated experiments use independent y_i's). Repeating the procedure $k \cdot m$ times will enable the adversary to retrieve the matrix M, which completely compromises the security of the scheme.

Note that for this reasoning to be correct, the probability that the decryption oracle does not return $\perp$ must be noticeable. Otherwise the adversary will have to make an exponential number of attempts to get enough equations. Clearly

$$\Pr_{y_i \xleftarrow{\$} \{0,1\}^m} \left[\mathcal{D}_K(a, y_i) \neq \perp \right] = 2^r \sum_{i=0}^t \frac{\binom{m}{i}}{2^m} \simeq 2^{-(1-\frac{r}{m}-H(\frac{t}{m}))m} \ ,$$

where H is the entropy function $H(x) = -x \log_2(x) - (1-x) \log_2(1-x)$. The concrete value of this probability will depend on the error-correcting code which is used. If it is good enough this value will not be too small.

At the same time this suggests a method to thwart the attack. Assume that LPN-C is modified in the following way: an additional parameter t' such that $\eta m < t' < t$ is chosen. When the number of errors in $y \oplus a \cdot M$ is greater than t' (i.e. $y \oplus a \cdot M$ is at Hamming distance greater than t' from any codeword), the decryption algorithm returns $\perp$. If t' is such that $2^{-(1-\frac{r}{m}-H(\frac{t'}{m}))m}$ is negligible, then the previous attack is not possible anymore. At the same time, this implies to drastically reduce the noise parameter η and the LPN problem becomes easier. The scheme also remains malleable, as the attack in Remark 1 remains applicable (hence the scheme cannot be IND-P2-C2 either). However, it could be that such a modified scheme is IND-P2-C1. This remains an open problem.

4.4   Achieving P2-C2 Security

The most straightforward way to get an encryption scheme secure against chosen-ciphertext attacks from an encryption scheme secure against chosen-plaintext attacks is to add message authenticity, e.g. by using a Message Authentication Code (MAC). This idea was suggested in [8,22] and was carefully studied by Bellare and Namprempre [3]. They explored the three paradigms Encrypt-and-MAC, MAC-then-Encrypt and Encrypt-then-MAC and showed that the later one was the most secure way to proceed. More precisely, assume that the sender and the receiver share an additional secret key K_m for the goal of message authentication, and let $\text{MAC}_{K_m}(\cdot)$ be a secure[1] MAC. LPN-C is modified as follows: let $A = (a_1, \ldots, a_n)$ be the vectors used to encrypt in LPN-C, and $Y = (y_1, \ldots, y_n)$ be the ciphertexts to transmit. A MAC of the ciphertext is added

---

[1] that is, strongly unforgeable under chosen plaintext attacks; see [3] for a precise definition.

to the transmission and computed as $\tau = \text{MAC}_{K_m}(A\|Y)$. The decryption algorithm is modified to return $\perp$ each time the MAC is not valid.

Given that the original scheme is IND-P2-C0, generic results of [3] imply that the enhanced scheme is IND/NM-P2-C2. This generic method has the drawback to rely on an additional assumption, namely the unforgeability of the MAC. We go one step further and propose a way to build a MAC only relying on the LPN problem and a one-way function.

Let M_2 be a secret $l \times l'$ binary matrix, where l and l' are polynomials in k. Let $H : \{0,1\}^* \to \{0,1\}^l$ be a one-way function. For $X \in \{0,1\}^*$ define

$$\text{MAC}_{M_2}(X) = H(X) \cdot M_2 \oplus \nu'$$

where $\nu' \leftarrow \text{Ber}_{l',\eta}$. We sketch the proof of the security of this MAC in the Random Oracle model in the full version of this paper.

5   Concrete Parameters for LPN-C

We now discuss some example parameters for LPN-C as well as some possible practical variants. We will define the expansion factor of the scheme as $\sigma = \frac{|\text{ciphertext}|}{|\text{plaintext}|} = \frac{m+k}{r}$, and the secret key size $|K| = k \cdot m$. There are various trade-offs possible when fixing the values of the parameters (k, η, m, r, d). First, the hardness of the LPN problem depends on k and η (it increases with k and η). However an increase to k implies a higher expansion factor and a bigger key size, whereas an increase to η implies to use a code with a bigger correction capacity and minimal distance, hence a bigger factor $\frac{m}{r}$. Depending on how the noise vectors ν are generated, decryption failures may also be an issue.

Example values for k and η were given by Levieil and Fouque [23]. If one is seeking 80-bit security, suitable parameters are $(k = 512, \eta = 0.125)$, or $(k = 768, \eta = 0.05)$. Example parameters for LPN-C are given below, where we used the list of Best Known Linear Codes available in MAGMA 2.13 [24].

LPN-C					expansion	storage	storage	decryption
k	η	m	r	d	factor σ	$|K|$ (bits)	(Toeplitz)	failure P_{DF}
512	0.125	80	27	21	21.9	40,960	591	0.42
512	0.125	160	42	42	16	81,920	671	0.44
768	0.05	80	53	9	16	61,440	847	0.37
768	0.05	160	99	17	9.4	122,880	927	0.41
768	0.05	160	75	25	12.4	122,880	927	0.06

Possible Variants. A first possibility is to increase the size of the secret matrix M in order to decrease the expansion factor σ. Indeed, assume that M is now a $k \times (n \cdot m)$ binary matrix for some integer $n > 1$. Then it becomes possible to encrypt n blocks of r bits with the same random vector a. The expansion factor becomes $\sigma = \frac{n \cdot m + k}{n \cdot r}$. Asymptotically when n increases, the expansion factor of the scheme tends to the one of the error-correcting code $\frac{m}{r}$.

Another possibility would be to pre-share the vectors a_i's, or to generate them from a small seed an a pseudorandom number generator. The expansion factor would then fall to $\sigma = \frac{m}{r}$, but synchronization issues could arise.

Finally, we mention the possibility (already used in HB$^\#$ [12]) to use Toeplitz matrices in order to decrease the size of the secret key. A $(k \times m)$-binary *Toeplitz* matrix M is a matrix for which the entries on every upper-left to lower-right diagonal have the same value. The entire matrix is specified by the top row and the first column. Thus a Toeplitz matrix can be stored in $k + m - 1$ bits rather than the km bits required for a truly random matrix. However, the security implications of such a design choice remain to be studied.

6 Conclusions

We have presented LPN-C, a novel symmetric encryption scheme whose security can be reduced to the LPN problem. Due to the low-cost computations (essentially of bitwise nature) required on the sender side, this encryption scheme could be suitable for environments with restricted computation power, typically RFIDs. Moreover, due to some similarities it could be possible to combine it with one of the authentication protocols HB$^+$ or HB$^\#$.

Among open problems we highlight the design of an efficient MAC directly from the LPN problem without any other assumption, as well as an understanding of the impact of the use of Toeplitz matrices in LPN-C (and HB$^\#$).

References

1. Aumasson, J.-P., Finiasz, M., Meier, W., Vaudenay, S.: TCHo: A Hardware-Oriented Trapdoor Cipher. In: Pieprzyk, J., Ghodosi, H., Dawson, E. (eds.) ACISP 2007. LNCS, vol. 4586, pp. 184–199. Springer, Heidelberg (2007)
2. Bellare, M., Desai, A., Jokipii, E., Rogaway, P.: A Concrete Security Treatment of Symmetric Encryption: Analysis of the DES Modes of Operation. In: Proceedings of FOCS 1997, pp. 394–403 (1997)
3. Bellare, M., Namprempre, C.: Authenticated Encryption: Relations Among Notions and Analysis of the Generic Composition Paradigm. In: Okamoto, T. (ed.) ASIACRYPT 2000. LNCS, vol. 1976, pp. 531–545. Springer, Heidelberg (2000)
4. Berlekamp, E.R., McEliece, R.J., van Tilborg, H.C.A.: On the Inherent Intractability of Certain Coding Problems. IEEE Trans. Info. Theory 24, 384–386 (1978)
5. Blum, A., Furst, M., Kearns, M., Lipton, R.: Cryptographic Primitives Based on Hard Learning Problems. In: Stinson, D.R. (ed.) CRYPTO 1993. LNCS, vol. 773, pp. 278–291. Springer, Heidelberg (1994)
6. Blum, A., Kalai, A., Wasserman, H.: Noise-Tolerant Learning, the Parity Problem, and the Statistical Query Model. J. ACM 50(4), 506–519 (2003); Preliminary version. In: Proceedings of STOC 2000
7. Bringer, J., Chabanne, H., Dottax, E.: HB^{++}: A Lightweight Authentication Protocol Secure Against Some Attacks. In: Proceedings of SecPerU 2006, pp. 28–33. IEEE Computer Society Press, Los Alamitos (2006)
8. Dolev, D., Dwork, C., Naor, M.: Nonmalleable Cryptography. SIAM Journal of Computing 30(2), 391–437 (2000)

9. Duc, D.N., Kim, K.: Securing HB$^+$ Against GRS Man-in-the-Middle Attack. In: Institute of Electronics, Information and Communication Engineers, Symposium on Cryptography and Information Security, January, pp. 23–26 (2007)
10. Fossorier, M.P.C., Mihaljevic, M.J., Imai, H., Cui, Y., Matsuura, K.: A Novel Algorithm for Solving the LPN Problem and its Application to Security Evaluation of the HB Protocol for RFID Authentication., http://eprint.iacr.org/2006/197.pdf
11. Gilbert, H., Robshaw, M.J.B., Seurin, Y.: Good Variants of HB$^+$ are Hard to Find. In: Proceedings of Financial Crypto 2008 (to appear, 2008)
12. Gilbert, H., Robshaw, M.J.B., Seurin, Y.: HB$^\#$: Increasing the Security and Efficiency of HB$^+$. In: Smart, N. (ed.) EUROCRYPT 2008. LNCS, vol. 4965, pp. 361–378. Springer, Heidelberg (2008)
13. Gilbert, H., Robshaw, M.J.B., Sibert, H.: An Active Attack Against HB$^+$: A Provably Secure Lightweight Authentication Protocol. IEE Electronics Letters 41(21), 1169–1170 (2005)
14. Goldreich, O., Krawczyk, H., Luby, M.: On the Existence of Pseudorandom Generators. In: Proceedings of FOCS 1988, pp. 12–21 (1988)
15. Goldwasser, S., Micali, S.: Probabilistic Encryption. Journal of Computer and System Science 28(2), 270–299 (1984)
16. Håstad, J.: Some Optimal Inapproximability Results. J. ACM 48(4), 798–859 (2001)
17. Hopper, N., Blum, M.: Secure Human Identification Protocols. In: Boyd, C. (ed.) ASIACRYPT 2001. LNCS, vol. 2248, pp. 52–66. Springer, Heidelberg (2001)
18. Impagliazzo, R., Levin, L.A.: No Better Ways to Generate Hard NP Instances than Picking Uniformly at Random. In: Proceedings of FOCS 1990, pp. 812–821 (1990)
19. Juels, A., Weis, S.A.: Authenticating Pervasive Devices With Human Protocols. In: Shoup, V. (ed.) CRYPTO 2005. LNCS, vol. 3621, pp. 293–308. Springer, Heidelberg (2005)
20. Katz, J., Shin, J.: Parallel and Concurrent Security of the HB and HB$^+$ Protocols. In: Vaudenay, S. (ed.) EUROCRYPT 2006. LNCS, vol. 4004, pp. 73–87. Springer, Heidelberg (2006)
21. Katz, J., Smith, A.: Analysing the HB and HB$^+$ Protocols in the "Large Error" Case, http://eprint.iacr.org/2006/326.pdf
22. Katz, J., Yung, M.: Complete Characterization of Security Notions for Probabilistic Private-Key Encryption. Journal of Cryptology 19(1), 67–95 (2006); Preliminary version. In: Proceedings of STOC 2000
23. Levieil, E., Fouque, P.-A.: An Improved LPN Algorithm. In: De Prisco, R., Yung, M. (eds.) SCN 2006. LNCS, vol. 4116, pp. 348–359. Springer, Heidelberg (2006)
24. MAGMA Computational Algebra System, http://magma.maths.usyd.edu.au/magma
25. MacWilliams, F.J., Sloane, N.J.A.: The Theory of Error-Correcting Codes. North-Holland Mathematical Library (1983)
26. Munilla, J., Peinado, A.: HB-MP: A Further Step in the HB-family of Lightweight Authentication Protocols. Computer Networks 51, 2262–2267 (2007)
27. Rao, T.R.N., Nam, K.H.: Private-Key Algebraic-Code Encryptions. IEEE Transactions on Information Theory 35(4), 829–833 (1989)
28. Regev, O.: On Lattices, Learning with Errors, Random Linear Codes, and Cryptography. In: Proceedings of STOC 2005, pp. 84–93 (2005)

Could SFLASH be Repaired?[*]

Jintai Ding[1], Vivien Dubois[2], Bo-Yin Yang[3,*],
Owen Chia-Hsin Chen[3], and Chen-Mou Cheng[4]

[1] Dept. of Mathematics and Computer Sciences, University of Cincinnati
[2] CELAR, France
[3] Institute of Information Sciences, Academia Sinica, Taiwan
[4] Dept. of Electrical Engineering, National Taiwan University

Abstract. The SFLASH signature scheme stood for a decade as the most successful cryptosystem based on multivariate polynomials, before an efficient attack was finally found in 2007. In this paper, we review its recent cryptanalysis and we notice that its weaknesses can all be linked to the fact that the cryptosystem is built on the structure of a large field. As the attack demonstrates, this richer structure can be accessed by an attacker by using the specific symmetry of the core function being used. Then, we investigate the effect of restricting this large field to a purely linear subset and we find that the symmetries exploited by the attack are no longer present. At a purely defensive level, this defines a countermeasure which can be used at a moderate overhead. On the theoretical side, this informs us of limitations of the recent attack and raises interesting remarks about the design itself of multivariate schemes.

Keywords: multivariate cryptography, signature, SFLASH, differential.

1   Introduction

Multivariate schemes are asymmetric primitives based on hard computational problems involving multivariate polynomials. Reference problems are for instance solving a system of multivariate polynomial equations, or deciding whether two sequences of multivariate polynomials are isomorphic. The research for such schemes originates from Matsumoto and Imai's work in the early 80s, but has really been active for a decade. The practical interest for considering such schemes, besides the obvious diversification effort, comes from their usual high performances which make them well-suited for implementation on small devices. On the other side, the area is young and much cryptanalytic effort is still to be done to understand well what their security might rely on.

Multivariate schemes are all based on a construction method inspired from McEliece [12]: an easy-to-invert multivariate vectorial function is transformed into a random-looking one by applying secret linear bijections on both variables and coordinates. Of course, such a linear hiding has the nice feature to be very

---

[*] Correspondence to BY at by@moscito.org; a full version of this extended abstract available from the authors, also cf. ePrint at http://eprint.iacr.org/2007/366.

L. Aceto et al. (Eds.): ICALP 2008, Part II, LNCS 5126, pp. 691–701, 2008.

easy to undo by the legitimate user, but it also has the drawback of leaking the invariant properties of the internal function. Whenever such invariant properties can be used in order to devise a cryptanalytic attack (e.g. elimination properties enhancing Gröbner basis computation), one uses additional transformations to destroy them.

SFLASH is a signature scheme proposed by Patarin, Goubin and Courtois [17], following a design they had introduced at Asiacrypt'98 [15]. The easy-to-invert internal function of SFLASH is defined from a single variable polynomial over some field extension $\mathbb{F}_{q^n}$ and turned into a function from $(\mathbb{F}_q)^n$ to itself by using the linear structure of $\mathbb{F}_{q^n}$ over $\mathbb{F}_q$. To allow efficient inversion, this function has a specific shape as a polynomial over $\mathbb{F}_{q^n}$, namely this is a *monomial* which is inverted by raising to the inverse exponent, like in RSA. The basic McEliece-type hiding, *i.e.* using two linear bijections, of such a function was the initial proposal – known as the C* cryptosystem – of Matsumoto and Imai [11], but it was later seen by Patarin [14] that the hidden monomial structure implies some algebraic properties of the public function which can be exploited for an attack. However, Patarin, Goubin and Courtois later showed [15] that algebraic attacks can be very easily avoided by using an additional transformation initially used by Shamir [16] which consists in simply *deleting a few coordinates* of the public function. Schemes obtained from the application of *minus* to C* are termed C*⁻ schemes; they are suitable for signature. SFLASH is a C*⁻ scheme chosen as a candidate for the selection organized by the NESSIE European consortium [1], and accepted in 2003 [13].

Recently, Dubois, Fouque, Shamir and Stern discovered a new property of C* monomials which is almost not affected by the *minus* transformation, and which can be used to recover missing coordinates of the public function [4,3]. As a consequence, all practical parameters choices for C*⁻ schemes, including those of SFLASH, were shown insecure. The attack found by Dubois *et al.* is the most effective development of a new kind of cryptanalysis which targets geometrical properties of multivariate functions. Consequences of this attack are of course a reevaluation of related cryptosystems and a more careful study of the properties of the internal functions being used. However it seems that the mere design principle of multivariate schemes is here in question : can we effectively hide a particular function such as a C* monomial using linear maps ?

Our results. In this paper, we review the recent cryptanalysis of SFLASH and we notice that its weaknesses can all be linked to the fact that the cryptosystem is built on the structure of a large field. As the attack demonstrates, this richer structure can be accessed by an attacker by using the specific symmetry of the internal C* function that can be perceived from even a small number of public polynomials. Then, we study the effect of restricting this large field to a purely linear subset, and we find that the symmetries exploited by the attack are no longer present. We provide mathematical proofs for the target cases explaining this phenomenon in detail. As we will see, this result conveys additional perspective on the general design of multivariate schemes.

Organization of the Paper. In Section 2, we give a brief introduction to SFLASH. In Section 3, we review its recent cryptanalysis [4,3]. In Section 4, we show that the geometrical properties which are exploited by the attack do not hold when restricting the internal function to a proper subspace of the large field. In Section 5, we define a modified family of schemes which resists the attack. We discuss our results in Section 6.

2   The SFLASH Scheme

2.1   The C* Scheme

The C* scheme was proposed by Matsumoto and Imai in 1988. It uses a *monomial* over $\mathbb{F}_{q^n} : F(x) = x^{1+q^\theta}, x \in \mathbb{F}_{q^n}$, where x can be identified with an n coordinates vector over $\mathbb{F}_q$ by fixing some basis of $\mathbb{F}_{q^n}$. The exponent $1+q^\theta$ is chosen invertible modulo $q^n - 1$ and raising to its inverse is inverting F. Since $1 + q^\theta$ has q-weight 2, F corresponds to a multivariate function from $(\mathbb{F}_q)^n$ into itself of degree 2. On the other hand, the inverse of $1 + q^\theta$ has very high q-weight $\mathcal{O}(n)$ for prescribed values of θ [11], and the inverse of F then corresponds to a multivariate function from $(\mathbb{F}_q)^n$ into itself with very high degree $\mathcal{O}(n)$. A C* scheme is built by transforming F with randomly chosen linear bijections S and $T : \boldsymbol{P} = T \circ F \circ S$. The resulting function $\boldsymbol{P}$ has the same *multivariate* properties as F, but the twisting provided by S and T hides the *single variable* representation which allows fast inversion. Unfortunately, Patarin showed in 1995 [14] that although the plaintext x is a high degree function in term of the ciphertext y, the pairs (x, y) satisfy many low degree algebraic relations, whose degree is independent of the security parameter n. This implies vulnerability to algebraic attacks.

2.2   SFLASH

To avoid an attacker to possibly reconstruct existing algebraic relations on the pairs (x, y), a simple idea is not to provide the entire description of how these variables are related. The most easy way to realize this was used by Shamir in 1993 [16] and consists in simply removing a few coordinate-polynomials of the public key, say the last r ones where r is an additional parameter. Furthermore, Patarin, Goubin and Courtois showed in 1998 [15] that for a C* scheme, the degree of algebraic relations between x and the partial y is quickly growing with the parameter r. Of course, the resulting scheme is no longer bijective but it can still be used for signature at no performance loss. These schemes were introduced as C*− by Patarin, Goubin and Courtois [15]. A public key consists of the $n - r$ first coordinates of an initial C* public key $\boldsymbol{P} = T \circ F \circ S$ with T and S as the secret key. A rationale for the parameter r is provided in [15]; choosing r with $q^r \geq 2^{80}$ is then required for a 2^{80} security level. Besides, no algebraic attack is expected to succeed when r is not too small in regards to n, the initial number of polynomials. SFLASH is a C*− scheme chosen by Patarin *et al.* for the NESSIE selection. For the recommended parameters $q = 2^7$, $n = 37$, $\theta = 11$ and $r = 11$, the signature length is 239 bits and the public key size is 15 Kbytes.

3 The Symmetry in SFLASH

The design of SFLASH was aimed at resisting algebraic attacks and stood challenging for almost ten years. However, in the last four years, a new kind of cryptanalysis for multivariate schemes has been developed based on geometrical properties of the so-called differential [8,5,6]. As defined in the initial paper by Fouque, Granboulan and Stern [8], the differential transforms a *quadratic* function $P(x)$ into its *bilinear symmetric* associate, denoted $DP(a, b)$. The differential of P can be obtained by substituting monomials $x_i x_j$ by $a_i b_j + a_j b_i$ in the expression of P (if P is not homogeneous, terms of degree 1 and 0 are discarded). The interest of doing so is that DP is linear separately in a and b and its properties relatively to these variables can then be described in terms of linear algebra. Furthermore, when considering a multivariate scheme $P = T \circ F \circ S$, these properties are isomorphic to those of F since S and T are linear bijections.

Recently, Dubois, Fouque, Shamir and Stern showed a very efficient cryptanalysis of C^{*-} schemes based on a class of geometrical invariants of the differential of C^* [4,3]. We summarize it below.

3.1 Skew-Symmetric Maps with Respect to the Differential

The differential of the internal C^* function is $DF(a, b) = a b^{q^\theta} + a^{q^\theta} b$ for $a, b \in \mathbb{F}_{q^n}$. When a and b are identified with n coordinates vectors over $\mathbb{F}_q$, DF is a bilinear symmetric function from $(\mathbb{F}_q)^n \times (\mathbb{F}_q)^n$ to $(\mathbb{F}_q)^n$. Each of the n coordinates of DF is a multivariate polynomial in the coordinates $a_1, \ldots, a_n$ and $b_1, \ldots, b_n$ of a and b respectively, which is linear separately in a and b, and where a and b play symmetric roles. Each such polynomial is written on the basis of terms $a_i b_j + a_j b_i$ so it has $n(n-1)/2$ coefficients. Now, it is observed in [4] that linear maps consisting of *multiplications* by some element ξ of $\mathbb{F}_{q^n}$ have a specific action on DF. Indeed, we have

$$DF(\xi.a, b) + DF(a, \xi.b) = (\xi + \xi^{q^\theta}).DF(a, b) \tag{1}$$

For the particular elements ξ such that $\xi + \xi^{q^\theta} = 0$ (at least 1 is solution), the associated multiplication maps M_ξ satisfy

$$DF(M_\xi(a), b) + DF(a, M_\xi(b)) = 0$$

that is, they are the *skew-symmetric* maps with respect to DF. The existence of non-trivial (*i.e.* not colinear to the identity) such maps is of course very unusual and even for a C^* monomial it does not happen for all parameters. However, even when it does not happen, the initial identity can also be interpreted as a skew-symmetry property. Let us indeed define for any linear map M, the skew-symmetric action of M over DF as the bilinear and symmetric function

$$\Sigma[M](a, b) = DF(M(a), b) + DF(a, M(b))$$

Our basic identity infers that in the special case of multiplication maps,

$$\Sigma[M_\xi](a, b) = M_\zeta \circ DF(a, b)$$

where M_ζ is the multiplication by $\xi + \xi^{q^\theta}$. As a consequence, for any element ξ of $\mathbb{F}_{q^n}$, the coordinates of the bilinear and symmetric function $\Sigma[M_\xi](a,b)$ are linear combinations of the coordinates of DF. Therefore, expressed in geometrical terms, multiplication maps have the specific property to leave unchanged under skew-symmetric action the subspace spanned by the coordinates of DF. Note that this property is very strong because the subspace spanned by the n coordinates of DF has dimension at most n while for a random linear map M, the coordinates of $\Sigma[M]$ might be any polynomials in the whole space of bilinear symmetric polynomials of dimension $n(n-1)/2$ and are very unlikely to all be confined in the tiny subspace spanned by the coordinates of DF.

The public key P of a C* scheme inherits of the above properties; the only difference is that the linear maps that play with regards to P the role of multiplications with regards to F are the conjugates $S^{-1} \circ M_\xi \circ S$. Now, a crucial point is : although the latter maps depend on the secret bijection S, they can be computed from their characteristic property with regards to the public key P. For instance, considering the simple skew-symmetry condition, $DP(M(a),b) + DP(a,M(b)) = 0$, we see that this equation is linear in M. It can be seen [4] that each coordinate of DP provides us with $n(n-1)/2$ linear conditions on the n^2 coefficients of M. Then, even a marginal number of coordinates of the public key allows to solve the space of skew-symmetric maps. Solving the more general skew-symmetry condition follows similar principles although more theory is involved; we refer the reader to the original paper [3] for the details.

3.2   Consequences

The properties described above allow an attacker to compute from a C*⁻ public key conjugates $S^{-1} \circ M_\xi \circ S$ of multiplications maps M_ξ. This of course is very annoying because these maps depend on the secret bijection S and were initially considered as secret information. Furthermore, it is shown in [4] that the nature of these maps is an additional problem. We do not consider these aspects here and focus on the initial breach *i.e.* the existence of linear maps which can be computed from the public key although they contain secret information. In the sequel, we investigate the possibility to destroy the skew-symmetry property of C*⁻ schemes.

4   Breaking the Symmetry

As we have seen, for C*⁻ schemes, the linear maps which are associated to the skew-symmetry property are connected to the internal field structure, namely they are multiplications by elements of $\mathbb{F}_{q^n}$. In principle, this means that the existence of these maps is tied to the internal field structure. A natural question is: would skew-symmetric maps exist if the internal field structure were truncated, *i.e.* restricted to a subspace of it?

4.1 Projection Breaks the Skew-Symmetry Property of C^{*-} Schemes

Suppose we consider the internal function F restricted to some proper subspace H of $\mathbb{F}_{q^n}$. We denote F_H this restriction. The skew-symmetric maps with respect to the differential DF_H of F_H are by definition the linear maps M_H from H to itself which satisfy :

$$DF_H(M_H(h), k) + DF_H(h, M_H(k)) = 0 , \quad h, k \in H \qquad (2)$$

We expect the solutions M_H to this condition to be the restrictions to H of the skew-symmetric maps w.r.t DF *which map H to itself*. When H is an arbitrary subspace, we do not expect non-trivial multiplications M_ξ to map H into itself. Then, the only solutions to our condition should be the scalar multiples of the Identity: $M_H = \lambda . Id_H, \lambda \in \mathbb{F}_q$. Let us now show that our expectation is correct using mathematical arguments. First, we characterize the linear maps M_H which are skew-symmetric with respect to DF_H by transforming the above condition (2) in a condition with respect to DF. That is, we embed the above condition over H in a condition over $\mathbb{F}_{q^n}$. We can embed M_H into a linear map $\bar{M}_H$ which is M_H over H and zero elsewhere. The same way, we can embed the Identity over H into the projection map to H, denoted π_H. Then, (2) is equivalent to:

$$DF(\bar{M}_H(a), \pi_H(b)) + DF(\pi_H(a), \bar{M}_H(b)) = 0 , \quad a, b \in \mathbb{F}_{q^n}$$

Therefore, the linear maps $\bar{M}_H$ are special solutions to the condition

$$DF(M(a), \pi_H(b)) + DF(\pi_H(a), M(b)) = 0 , \quad a, b \in \mathbb{F}_{q^n} \qquad (3)$$

They are those solutions M left unchanged by composition with π_H :

$$M = M \circ \pi_H = \pi_H \circ M$$

Our method to determine the linear maps $\bar{M}_H$ is then clear : we first find the solutions M to the condition (3), and then find those which are left unchanged by composition with π_H.

The Solutions to Condition 3. As we can see, obvious solutions to Condition 3 are the maps $M_\xi \circ \pi_H$ where M_ξ is skew-symmetric with respect to DF. Since our condition is greatly overdetermined, we do not expect any other solutions. This is confirmed experimentally. In the most simple case when H is a hyperplane, we can give it a mathematical proof.

Lemma 1. *Let H be a hyperplane of $\mathbb{F}_{q^n}$ and DF be the differential of a bijective C^* monomial. The linear maps M which satisfy the condition*

$$DF(M(a), \pi_H(b)) + DF(\pi_H(a), M(b)) = 0 , \quad a, b \in \mathbb{F}_{q^n}$$

are of the form $M_\xi \circ \pi_H$ where M_ξ is skew-symmetric with respect to DF.

Proof. The idea of the proof is to replace M and π_H by their expressions as sums of q-powerings, and to express our condition as the vanishing of a polynomial in a, b over $\mathbb{F}_{q^n}$. We have $M(a) = \sum_{i=0}^{n-1} \mu_i a^{q^i}$ and π_H can be expressed as the projection orthogonally to some element u, where the orthogonality is defined relatively to the *trace* product (see [10] for a definition). Recalling $tr(a) = \sum_{i=0}^{n-1} a^{q^i}$ and that $tr(a)$ is an element of $\mathbb{F}_q$, we have $\pi_H(a) = a - tr(au)u$. To simplify, we consider in the sequel $u = 1$. We can rewrite our condition : $A(a, b) - B(a, b) = 0$, where

$$A(a, b) = DF(M(a), b) + DF(a, M(b))$$
$$B(a, b) = tr(a)DF(M(b), 1) + tr(b)DF(M(a), 1)$$

Both expressions are written on the basis of symmetric terms of the form $a^{q^i} b^{q^j} + a^{q^j} b^{q^i}$ and their respective coefficients are :

$$A(a, b) : \text{coefficient}\{i, 0\} = \mu_{i-\theta}^{q^\theta} \ ; \ \text{coefficient}\{i, \theta\} = \mu_i$$
$$B(a, b) : \text{coefficient}\{i, j\} = \mu_i + \mu_j + (\mu_{i-\theta} + \mu_{j-\theta})^{q^\theta}$$

From these expressions, we easily resolve $\mu_0 = 0$ and $\mu_i = \xi$ for all $i \neq 0$ where ξ satisfies $\xi^{q^\theta} + \xi = 0$ (see the full version for the details). Therefore, $M(a) = \xi(a - tr(a)) = M_\xi \circ \pi_H(a)$ where M_ξ is skew-symmetric with respect to DF (which is obtained from $\xi^{q^\theta} + \xi = 0$). □

Solutions Which are Left Unchanged by Composition with the Projection. As we have shown, the linear maps $\bar{M}_H$ which correspond to the skew-symmetric maps with respect to DF_H, are the solutions to Condition 3 which are left unchanged by composition with π_H. As argued in the previous section, the solutions to this condition are $M_\xi \circ \pi_H$ where M_ξ is multiplication by some element ξ. These maps are unchanged by composition with π_H if and only if M_ξ commutes with π_H, *i.e.* if and only if M_ξ maps H to itself. Then, since for any ξ, M_ξ is bijective, we have $\xi.H = H$. Our goal is to show that, except for specific choices of H which are very sparse, the only ξ satisfying this property are the scalar multiples of 1. As a first step, we notice that these elements ξ form a multiplicative group, independently of the choice of H. Therefore, they actually form a subfield of $\mathbb{F}_{q^n}$ and H is a linear space over this subfield. Finally, the subspaces H for which our property is satisfied by non-trivial elements ξ are subspaces over intermediate subfields of $\mathbb{F}_{q^n}$. As a second step, we upperbound the probability that a random subspace H of a prescribed dimension s is a subspace over an intermediate subfield of $\mathbb{F}_{q^n}$. (In this case, we say that H is degenerate). We show that this probability is negligible in terms of q and n.

Lemma 2. *Degenerate subspaces of $\mathbb{F}_{q^n}$ only exist at dimensions s not coprime with n. Degenerate hyperplanes never exist. The proportion of degenerate subspaces in $\mathbb{F}_{q^n}$ of a prescribed dimension is always $\mathcal{O}(q^{-n})$.*

Proof. When H is a subspace over $\mathbb{F}_{q^r}$, its dimension over $\mathbb{F}_q$ is a multiple of r. Since r must itself be a divisor of n, degenerate subspaces only exist at dimensions s not coprime with n. For instance, we deduce that degenerate hyperplanes never exist since $n - 1$ is always coprime with n. Let r be a common divisor of s and n. It can be shown that the number of subspaces of dimension s in a vector space of dimension n is of the order of $q^{s(n-s)}$ [9]. Then, the number of $\mathbb{F}_{q^r}$-subspaces of dimension s/r in $\mathbb{F}_{q^n}$ is of the order of $q^{s(n-s)/r}$. The number of degenerate subspaces of dimension s in $\mathbb{F}_{q^n}$ is dominated by the latter quantity considered for the smallest common factor r of n and s. Since the smallest possible value of r is 2, the proportion of degenerate subspaces of dimension s in $\mathbb{F}_{q^n}$ is at most of the order of $q^{-s(n-s)/2}$. Since $s(n - s)$ is minimal for $s = 2$ (2 is a common factor of s and n), the searched proportion is dominated by $q^{-(n-2)}$ and therefore q^{-n} asymptotically. $\qquad\square$

Application to the General Skew-Symmetry Property of C*⁻ Schemes. In the preceding paragraphs, we have shown that restricting the internal function F to some proper subspace H of $\mathbb{F}_{q^n}$ destroys the simple skew-symmetry property (2). In this paragraph, we consider the general skew-symmetry property of C*⁻ schemes. This property expresses that there exists non-trivial linear maps which leave the space spanned by the coordinates of DF unchanged under skew-symmetric action. The linear maps satisfying this condition are the whole space of multiplications. Using similar techniques as before, we can show that this property considered for the restricted function F_H admits only trivial solutions.

4.2   Experimental Verifications

We checked experimentally, for various C* parameters n and θ, the effect of restricting the internal function to a randomly chosen subspace H of various dimensions s. For instance, for parameters $n = 36$ and $\theta = 4$, we obtain the table below for the solution space of the general skew-symmetry condition as the number of coordinate-wise conditions grows.

5   Projected C*⁻ Schemes

Based on the previous results, we are led to define a new family of schemes that we call *projected C^{*-}* schemes. As we will see, these schemes actually consists in hiding a C* monomial using non-bijective linear maps. We next define the (ad-hoc) computational problems on which the security of these schemes is based. Finally, we discuss possible choices of parameters and suggest one concrete choice with performances comparable to SFLASH.

Description. A projected C*⁻ scheme is defined as follows. Start from a C* scheme $F(x) = x^{1+q^\theta}$ with secret linear maps S and T. Let r and s be two integers between 0 and n. Let T^- be the projection of T on the last r coordinates and S^- be the restriction of S on the last s coordinates. Compute $\hat{P} = T^- \circ F \circ S^-$. The generated function $\hat{P}$ is used as the public key and the secret linear bijections S and T are used as the secret key. Note that $\hat{P}$ is a quadratic function from $(\mathbb{F}_q)^{n-s}$ to $(\mathbb{F}_q)^{n-r}$. To find a preimage by the public function of

# conditions	$s = 0$	$s = 1$	$s = 2$	$s = 3$	$s = 4$	$s = 9$	$s = 18$
1	1296	1225	1156	1089	1024	769	324
2	708	669	632	598	564	414	207
3	168	145	124	109	104	99	90
4	36	1	1	1	1	1	1
6	36	1	1	1	1	1	1
⋮	⋮	⋮	⋮	⋮	⋮	⋮	⋮

a given message m, the legitimate user first pads m with a random vector m' of $(\mathbb{F}_q)^r$ and compute the preimage of (m, m') by $T \circ F \circ S$. If this element has its last s coordinates to 0, then its $n - s$ first coordinates are a valid signature for m. Otherwise, he discards this element and tries with an other random padding m'. When $r > s$, the process ends with probability 1 and costs on average q^s inversions of F. In practice, r is chosen a significant fraction of n to make the public key resistant to algebraic attacks; s can be chosen as small as 1 to destroy symmetries arising from the internal field structure. As for C^{*-} schemes, the significant value of r makes projected C^{*-} schemes only suitable for signature, since reviewing all possible paddings m' is not efficient. Finally, we mention that projection already appeared in the literature as a possible modifier [18] but was never considered as a useful measure let alone a defensive measure.

Possible Angles of Analysis. As usual for multivariate schemes, the security relies on several ad-hoc computational problems. The first problem is solving the public system of quadratic equations. Since s is chosen small, this is about as hard as solving the initial C^{*-} system. The second problem is recovering the functional decomposition of the public key or at least some information on the secret maps S^-, T^-. There is no efficient strategy to solve this problem in general [7], and the attack by Dubois et al. which falls into this category for C^{*-} schemes is here prevented by the projection. Remains the strategy consisting in recovering the public key into a valid C^{*-} public key. Showing this to be possible is actually the new challenge opened by the new family of schemes.

Parameters. n, θ, r are chosen following the rationales for C^{*-} schemes. We choose $s = 1$ as it induces the minimal factor q on the secret operations. The value of q can be chosen small but, at constant blocksize, this requires a larger value of n and therefore a larger public key. As a possible trade-off, we propose pFLASH with $q = 2^4$, $n = 74$, $\theta = 11$, $r = 22$ and $s = 1$. Our tests have pFLASH signing at $\lesssim 1$ million K8/C2 cycles, in line with expectations of $\sim 16\times$ time of SFLASH [2]; private key size is $2\times$ at 5.4kB. These are still attractive features for small device implementation.

6   Conclusion

In this paper, we provide additional insight on the recent cryptanalysis of SFLASH by exhibiting a simple modification which provably avoids the attack.

Our study shows that the attack against SFLASH has deeper roots than the mere fact that it is based on a C* monomial : the attack is made possible because the large field structure is embedded in the public key and is stopped when it is no more the case. Then, we realize that, indeed, one might not hope to hide effectively a particular function defined on a large field using linear bijections; this might at most be achievable in some security range using compressive linear maps. But then, is it still possible to build a practical cryptosystem in this setting ? At the present state, we can still define a modified family of C* -based schemes which is of practical interest. Analysis of this most simple case would probably yield additional understanding of the ways to distinguish a specifically-built multivariate function and would provide further insight on the very possibility to obfuscate such a function using linear maps.

Acknowledgment

JD and BY are grateful to the Alexander van Humboldt Foundation, the Taft Fund, and TWISC [National Science Council project NSC 96-2219-E-011-008 / NSC 96-2219-E-001-001] without whose valuable support much of this work would've not happened. BY, CC, OC would also like to thank NSC for partial sponsorship via project NSC 96-2221-E-001-031-MY3.

References

1. European project IST-1999-12324 on New European Schemes for Signature, Integrity and Encryption, http://www.cryptonessie.org
2. Daniel, J.: Bernstein. eBATs benchmark results, http://ebats.cr.yp.to
3. Dubois, V., Fouque, P.-A., Shamir, A., Stern, J.: Practical Cryptanalysis of SFLASH. In: Menezes, A. (ed.) CRYPTO 2007. LNCS, vol. 4622, pp. 1–12. Springer, Heidelberg (2007)
4. Dubois, V., Fouque, P.-A., Stern, J.: Cryptanalysis of SFLASH with Slightly Modified Parameters. In: Naor, M. (ed.) EUROCRYPT 2007. LNCS, vol. 4515, pp. 264–275. Springer, Heidelberg (2007)
5. Dubois, V., Granboulan, L., Stern, J.: An Efficient Provable Distinguisher for HFE. In: Bugliesi, M., Preneel, B., Sassone, V., Wegener, I. (eds.) ICALP 2006. LNCS, vol. 4052, pp. 156–167. Springer, Heidelberg (2006)
6. Dubois, V., Granboulan, L., Stern, J.: Cryptanalysis of HFE with Internal Perturbation. In: Okamoto, T., Wang, X. (eds.) PKC 2007. LNCS, vol. 4450, pp. 249–265. Springer, Heidelberg (2007)
7. Faugère, J.-C., Perret, L.: Polynomial Equivalence Problems: Algorithmic and Theoretical Aspects. In: Vaudenay, S. (ed.) EUROCRYPT 2006. LNCS, vol. 4004, pp. 30–47. Springer, Heidelberg (2006)
8. Fouque, P.-A., Granboulan, L., Stern, J.: Differential Cryptanalysis for Multivariate Schemes.. In: Cramer, R.J.F. (ed.) EUROCRYPT 2005. LNCS, vol. 3494, pp. 341–353. Springer, Heidelberg (2005)
9. Goldman, J., Rota, G.-C.: The Number of Subspaces of a Vector Space. In: Tutte, W.T. (ed.) Recent Progress in Combinatorics, pp. 75–83. Academic Press, London (1969)

10. Lidl, R., Niederreiter, H.: Finite Fields. Encyclopedia of Mathematics and its applications, vol. 20. Cambridge University Press, Cambridge (1997)
11. Matsumoto, T., Imai, H.: Public Quadratic Polynominal-Tuples for Efficient Signature-Verification and Message-Encryption. In: Günther, C.G. (ed.) EUROCRYPT 1988. LNCS, vol. 330, pp. 419–453. Springer, Heidelberg (1988)
12. McEliece, R.J.: A Public-Key Cryptosystem based on Algebraic Coding Theory. In: JPL DSN Progress Report, pp. 114–116. California Inst. Technol., Pasadena (1978)
13. NESSIE, New European Schemes for Signatures, Integrity, and Encryption. Portfolio of Recommended Cryptographic Primitives, http://www.nessie.eu.org
14. Patarin, J.: Cryptanalysis of the Matsumoto and Imai Public Key Scheme of Eurocrypt 1988. In: Coppersmith, D. (ed.) CRYPTO 1995. LNCS, vol. 963, pp. 248–261. Springer, Heidelberg (1995)
15. Patarin, J., Goubin, L., Courtois, N.: C^{*}_{-+} and HM: Variations Around Two Schemes of T. Matsumoto and H. Imai. In: Ohta, K., Pei, D. (eds.) ASIACRYPT 1998. LNCS, vol. 1514, pp. 35–49. Springer, Heidelberg (1998)
16. Shamir, A.: Efficient Signature Schemes Based on Birational Permutations. In: Stinson, D.R. (ed.) CRYPTO 1993. LNCS, vol. 773, pp. 1–12. Springer, Heidelberg (1994)
17. Specifications of SFLASH. Final Report NESSIE, pp. 669–677 (2004)
18. Wolf, C., Preneel, B.: Taxonomy of Public Key Schemes based on the problem of Multivariate Quadratic equations. ePrint Archive Report 2005/077, http://eprint.iacr.org/2005/077

Password Mistyping in Two-Factor-Authenticated Key Exchange

Vladimir Kolesnikov[1] and Charles Rackoff[2]

[1] Bell Labs, Murray Hill, NJ 07974,USA
kolesnikov@research.bell-labs.com
[2] Dept. Computer Science, University of Toronto, Canada
rackoff@cs.utoronto.ca

Abstract. We study the problem of Key Exchange (KE), where authentication is two-factor and based on both electronically stored long keys and human-supplied credentials (passwords or biometrics). The latter credential has low entropy and may be *adversarily* mistyped. Our main contribution is the first formal treatment of mistyping in this setting.

Ensuring security in presence of mistyping is subtle. We show mistyping-related limitations of previous KE definitions and constructions (of Boyen et al. [6,7,10] and Kolesnikov and Rackoff [16]).

We concentrate on the practical two-factor authenticated KE setting where *servers* exchange keys with *clients*, who use short passwords (memorized) and long cryptographic keys (stored on a card). Our work is thus a natural generalization of Halevi-Krawczyk [15] and Kolesnikov-Rackoff [16]. We discuss the challenges that arise due to mistyping. We propose the first KE definitions in this setting, and formally discuss their guarantees. We present efficient KE protocols and prove their security.

1 Introduction

The problem of securing communication over an insecure network is generally solved using *key exchange* (KE). KE provides partners with matching randomly chosen keys, which are used for securing their conversation. Of course, no adversary *Adv* should be able to mismatch players. Therefore, players must possess secrets with which they can authenticate themselves. The kind of secrets that are available to players determines the setting of KE. In the simplest KE setting players have a long shared random string. KE is more complicated if parties establish key pairs with the public keys securely published. Using weak and/or fuzzy credentials, such as passwords or biometrics, further complicates the design of KE. Finally, using a combination of credentials may make certain aspects of KE easier (such as incorporating password authentication), but increases the overall complexity of the solution, as discussed in [16].

Our Setting. Two-factor authentication is critical and is used extensively in secure applications such as banking, VPN, etc. Stored long keys protect against online adversaries, but are vulnerable against theft. The extra layer of security is achieved with additional use of a theft-resistant credential, e.g. a short password

L. Aceto et al. (Eds.): ICALP 2008, Part II, LNCS 5126, pp. 702–714, 2008.

or a biometric. Unfortunately, neither password nor biometric can be expected
to be read reliably into the computer.

We give foundation to this setting by generalizing the work of Halevi-
Krawczyk (HK) [15] and Kolesnikov-Rackoff (KR) [16]. Recall, they address the
client-server setting where both long key and a short password are used for KE.
The servers are incorruptible, but client's card or password can be compromised.

Motivated by real scenarios, we study the effects of password mistyping. Mistyping need not be random, but may be skewed by the adversary, e.g. by technical
means or social engineering manipulation. We thus consider security against
adversaries who can *arbitrarily* affect user's mistyping. This consideration is
especially relevant in case biometric credentials are used for authentication, since,
due to technology limitations, biometric readings are *expected* to be misread.

Mistyping opens subtle vulnerabilities and raises complex definitional issues.
In the sequel, we use terms "password" and "mistype", although our work applies
to passwords, biometrics, and other short noisy credentials, as noted in Sect. 5.

1.1 Our Contributions and Outline of Work

Our main contribution is the first formal treatment of mistyping of passwords
in KE that uses a combination of credentials.

We discuss recent definitions that consider mistyping-related settings and
issues – robust fuzzy extractors of [6,7,10]. We point out a limitation of the
definitions of [6,7,10] with respect to robust handling of biometric misread-
ing/mistyping and discuss possible remedies. We demonstrate and correct a vulnerability of the definition and protocol of [16], which can only be exploited when
users mistype. These observations further emphasize the subtleties of mistyping
and the need for its formal treatment and deeper understanding.

In Sect. 3, we introduce our setting and the framework of [16] which we build
upon. Then, with simple protocols we illustrate mistyping-related issues, discuss
natural definitional approaches to handling mistyping and their shortcomings.
Most of the mistyping-related subtleties we uncover arise due to the simultaneous
use of both long keys and passwords. In Sect. 4, we formalize our discussion in
a definition, and formally argue that it prevents attacks that exploit mistyping.

In Sect. 5 we discuss applications of our work in biometric authentication.

In Sect. 6 we give efficient protocols; we prove their security in the full version.

1.2 Related Work

The problem of key exchange has deservedly received a vast amount of attention.
Password KE was first considered by Bellovin and Merritt [4]. Foundations –
formal definitions and protocols – were laid in [3,8,13,9], and other works.

The use of combined keys in authentication, where the client has a password
and the public key of the server, was introduced by Gong et al. [14] and first
formalized by Halevi and Krawczyk [15]. Kolesnikov and Rackoff [16] extended
this setting by allowing the client to also share a long key with the server, and
gave first definitions of KE in their (and thus in the Gong et al. and HK) setting.

Password Mistyping in KE. Despite the large research effort, the definitional issues of KE password mistyping are formally approached only in the UC definition of Canetti et al. [9]. In their password-only setting, mistyping is modelled by Environment $\mathcal{Z}$ providing players' inputs. Additional use of long key makes our setting significantly different (and more subtle with respect to mistyping) from that of [9]. Mistyping was also considered in different settings: related-key attacks on blockciphers [2] and signing authority delegation [17].

Biometric authentication and fuzzy extractors. A growing body of work, e.g. [5,10,11,12], addresses the use of biometrics in cryptography. Boyen et al. [6,7,10] consider its application to KE. They introduce the notion of *robust fuzzy extractor* (RFE), and give generic constructions of biometric-based KE from RFE. While their setting is similar to ours, the problems solved by [6,7,10] are different. They give KE protocols that accept "close enough" secrets, thus enabling security and privacy of biometric authentication. They do not aim to give a formal KE definition that handles biometric/password misreading. Moreover, as shown in Sect. 2, their notion of RFE is insufficiently strong to guarantee security of their generic KE protocol in many practical settings. (However, instantiating their KE protocol with their RFE construction is secure, since the latter satisfies stronger requirements than required by the definition.)

2   Mistyping-Related Limitations in Previous Work

On robust fuzzy extractor (RFE) definition and KE protocol [7,6,10]. We first clarify underlying biometric technology limitations and assumptions. Biometrics are "fuzzy", i.e. each scan is likely to be different from, but "close" to the "true" scan. Error-correction [12] is then used to extract non-fuzzy keys usable in cryptography. However, error-correction cannot correct many misreading errors (up to 10%), since this would imply high false acceptance rate[1]. Thus misreading beyond error-correction ball occurs often, and must be considered.

We note a limitation of RFE definition [7,6,10], prohibiting its use with the generic KE construction (Sect. 3.3 of [7]) in many scenarios. Roughly, definition's domains of correctness and security guarantees coincide. That is, extracted randomness is only guaranteed to be good if the scan is within the *error-correction* distance t from the original. There are no guarantees on the randomness if this condition does not hold. This is, perhaps, due to the papers' implicit assumption that "natural" misreadings are almost always "close" and are corrected (i.e. FRR is negligible). However, as discussed above, this assumption often does not hold. Strengthening the randomness guarantees of RFE would increase its usability.

More specifically, a RFE (Gen, Rep) may exhibit the following vulnerability. Given the public helper string P, if the biometric w_0 is misread in a special way w' outside the error-correction ball, the extracted randomness $Rep(w', P)$ is predictable. Even more subtly, $Rep(w', P)$ and $Rep(w_0, P)$ could be related,

---

[1] In balanced optimized real-life systems, which compare scans directly, False Reject Rate (FRR) is usually 1..10%. Notably, NIST reports FRR of fingerprints 0.1..2%, iris 0.2..1% and face 10%. See [1] for comprehensive overview and references.

but unequal. Clearly, KE protocols, including one of Sect. 3.3 of [7,6], constructed from such RFE would not be secure. One solution is to require, for w' outside the error-correction ball, that either $Rep(w', P) = \bot$ (property of RFE construction of [6,7]) or that $Rep(w', P)$ is either equal to or independent from $Rep(w_0, P)$.

Finally, although [6,7,10] consider adversarial substitution of P with P', they guarantee $Rep(w', P') = \bot$ only for w' in the error-correction ball. This vulnerability also can be resolved by separating the error-correction and security domains. We defer detailed definition, analysis and constructions as future work.

On the definition and construction of [16]. We present the following practical outside-of-the-model mistyping attack on the protocol (and thus also on the definition) of Kolesnikov and Rackoff [16]. Specifically, resistance to Denial of Access (DoA) attacks of the protocol of [16] is compromised if the honest client ever mistypes. Indeed, since their protocol is not challenge-response, client C's message can be replayed. This is not a problem if C always types the correct password (session keys of C and server S will be independent). However, if the password was mistyped, both the original and replayed message will cause S to register password failure, violating the intent of the DoA resistance. We stress that the KR protocol is otherwise secure against mistyping (and we prove it in Sect. 6). Our definitions and protocols address and correct the above insecurity.

Above limitations show subtleties of mistyping and the need to address them.

3   Pre-definition Discussion

Our main contribution is a formal treatment of mistyping in the combined keys KE setting of Kolesnikov and Rackoff [16]. The KR setting is a generalization of the Halevi-Krawczyk setting [15], in which clients have a password and the public key of S. In KR setting, clients carry stealable cards capable of storing cryptographic keys – public key of S and long key ℓ shared by C and S. Addition of the cards allows better functionality and security than that of HK. KR definitions and protocol guarantee and achieve strong security when C's card is secure, and weaker, password-grade, security, when the card is compromised.

We stress that the definition of KR does not handle mistyping. That is, it is possible to construct KR-secure protocols that "break" if the client ever mistypes his password. Sect. 3.3 of [16] provides an example and a short informal discussion on mistyping, and leaves the problem open. In Sect. 3.2, we expand this discussion, present more subtle mistyping threats, and discuss approaches to handling them. This leads to the presentation of our definitions in Sect. 4.

Notation. We concentrate on the two-factor authentication setting, where a client (denoted C) exchanges keys with a server (S). Both long and short keys are used for KE. Let P be a player. We denote by P_i the i-th instance of P. We write P_i^Q to emphasize that P_i intends to do KE with (some instance of) player Q. Denote the adversary by Adv. Sometimes we distinguish the game and real-life adversary, and denote the latter Adv_{Real}. Denote C's password by pwd

and long key by ℓ. S's public/ private keys are pk_S and sk_S. Password failure and the associated control symbol output by S is denoted by P⊥.

On the Style of Definition. We chose the game (Bellare-Pointcheval-Rogaway [3]) style, since this allowed using the intuitive definition of KR (only existing two-factor-authentication KE definition). Extending KR allowed reduction of security claims of our definition/setting to those of KR. Further, the stronger and arguably more intuitive UC model unfortunately is sometimes too strict, ruling out some efficient protocols which appear to be good enough in practice.

Proposing a simulation-based (especially, UC) definition, and exploring the relationship between it and our definition would add confidence in both our and the UC treatment of the problem. We thus leave as an important next step the design, detailed analysis and comparison of a corresponding UC definition. We expect that our discussions of ideas and obstacles would aid in this future work.

3.1   Review of the Framework of [16]

Our definition is an extension of the KE definition of KR (Def. 2 of [16]).

Recall, KR (and thus our) definition follows the common game-based paradigm. The real world and real adversary Adv_{Real} are abstracted as a game, played by the game adversary Adv. Game includes clients and servers – Interactive Turing Machines (ITM) running the KE protocol Π, communicating via channels controlled by Adv. Game rules mimic reality, and are designed so that Adv's wins correspond to real-life breaks. Π is defined secure if no polytime Adv is able to win above certain "allowed" probability. Definition is thus reduced to the design of the game. KR break down the real world into five intuitive games (KE$_1$, KE$_2$, KE$_3$, DOA and SID), which mimic possible real-life attack scenarios.

Game KE$_1$ is the core of the definition; it addresses password security when the long key is compromised. The difficulty of KE$_1$ design is in balancing the power given to Adv, since Adv_{Real}'s non-negligible advantage must be accounted exactly. It is achieved by "charging" Adv for each active attack (i.e. P⊥ output by S). The allowed Adv win probability is a function of the number of charges.

KE$_2$ models Adv_{Real} posing as S to C. KE$_3$ models KE with uncompromised card. In both cases, Adv is allowed only negligible success, which is easy to model. DOA models a "denial of access" attack formalized by KR, which requires that Adv is not able to cut C's access to S by exhausting allowed password failures. Finally, SID is a game preventing technicality-based insecure protocols.

We stress that a good model need not mimic the world *exactly*. E.g., Adv's ability to mistype or to know whether S failed may be different from Adv_{Real}'s, as long as Adv can win in *some* way (only) against bad protocols.

Mistyping in KR definitions. In KR games, client ITMs are always instantiated with correct password, which limits Adv's ability to emulate mistyping. Many real-life attacks that exploit mistyping cannot be carried in the game, allowing vulnerable protocol to withstand Adv's attacks and be defined secure. In Sect. 3.2, we discuss vulnerabilities, some natural "fixes" and their limitations.

3.2 Natural Definitional Approaches to Mistyping (That Don't Work)

To better expose subtle definitional issues and the limitations of some natural approaches, we build presentation incrementally. We propose several mistyping-vulnerable protocols, each progressively more "tricky", and show that they are KR-secure. We then discuss corresponding natural "fixes" of the KR definition – ways of allowing Adv to modify or substitute client's password, so as to mimic real-life mistyping and allow Adv to carry the real-world attacks. We show that ultimately they are insufficient and conclude that, for technical reasons, direct mimicking of mistyping in the games does not result in a good model. For readability, we keep discussion brief and informal (but readily formalizeable).

Mistyping vulnerabilities by example. Let Π be a KR-secure KE protocol. Π_1, Π_2, Π_3 below are KR-secure, but fail in progressively more subtle ways.

Π_1 *(S leaks long key upon mistyping).* Let Π_1 be a protocol as Π, except that in Π_1 S reveals the long key ℓ in a message, once password failure P$\perp$ occurred.

Clearly, Π_1 is "bad". But, it is easy to see that Π_1 is secure by KR definition. Since instances of C never mistype in the game, KR Adv cannot cause P$\perp$ without possession of ℓ. Thus, Adv cannot gain from S revealing ℓ, and Π_1 is KR-secure.

Π_2 *(S leaks password upon repeated mistyping).* Let pwd be C's password. Let Π_2 be a protocol as Π, except that in Π_2, S reveals pwd once $pwd + 1$ was tried twice. (Limited global state can be communicated among instances of S with the help of Adv, thus allowing Π_2 [16]; see full version for detailed discussion.)

At the first glance, it may appear that Π_2 is "good". Indeed, the advantage Adv gets from causing the leak is canceled by the effort to obtain it – a redundant password attempt for each attempt of causing the leak (this is the reason why Π_2 is KR-secure). However, this leak can be caused by real-life honest C mistakenly entering $pwd + 1$ twice. This is not an unusual situation, and the resulting password compromise is clearly unacceptable.

Π_3 *(S leaks a small hint about a password upon repeated mistyping).* Let pwd be C's password. Let Π_3 be a protocol as Π, except that in Π_3, S reveals whether $pwd = 0$ once $pwd + 1$ was tried 4 times. Π_3 is bad for the same reason as Π_2.

Definitional approaches. We consider strengthening Adv of KR by mimicking powers of real-life adversary. Our goal is to disallow above "bad" protocols.

Allowing Adv to specify the password of C's instances disqualifies Π_1. Indeed, Adv wins the game where he is not given ℓ, as follows. He instantiates C with a wrong password, causing P$\perp$ and leak of ℓ, which Adv uses to win.

To disqualify Π_2, Adv needs more than simple substitution of C's password. Adv needs the power to specify a "mistyping function" applied to the password given to C (idea also considered in [17]). That is, Adv specifies a map $F : D \mapsto D$, and C is instantiated with password $F(p)$. (Not every map F is allowed [16].)

While Π_3 is bad for the same reason as Π_2 (real-life C's mistyping leaks a password hint), it is harder to disqualify Π_3 due to the small size of the leak. It turns out that Π_3 is an important example, showing that allowing Adv to influence C's input is insufficient. We continue this discussion below in Sect. 4.

4   Mistyping-Secure KE Definition

Π_3, the last example of Sect. 3.2 is a (otherwise secure) protocol where S leaks a small password hint after four certain *repeated* mistypings. A repeated mistyping does not help Adv (he is checking already checked password). Since in KR definition, Adv is charged for each (even repeated) mistyping, the cost of mistyping outweighs the benefit of the leak, and Adv is not able to exploit the vulnerability.

This leads to our main idea – to allow Adv to run mistyped KE executions "for free". This way, Adv will be able to win whenever a non-negligible amount of information is leaked due to mistyping. It turns out that this additional power, applied properly, results in a good (i.e. sufficiently, but not too strong, and easy to use) definition, presented in this section.

Our extension of KR definition. We would like to give Adv the ability to observe and actively participate "for free" in mistyped KE sessions. This is not possible with the approaches we previously discussed, including that of [16]. This is because there Adv always learns whether S accepted the password, allowing Adv to verify a password guess, for which Adv must be charged. Our idea is to withhold failure information from Adv (and not charge him in case of P $\perp$) by default, thus allowing "free" mistypings. If Adv wants to obtain failure information, it is given to him upon special "check" request. Since this gives him information about the password, he is charged one attempt, if the check reveals P$\perp$. Note, this cost structure is a simple generalization of the one used in [16]. This amendment of KR is sufficient to handle mistyping.

Another advantage of this approach is allowing to mimic mistyping without Adv creating instances with substituted password. Indeed, Adv can make a password guess, and, based on it, emulate any mistyping sequence of C. As shown in Sect. 4.1, this guarantees security, since a "free" mistyping-dependent leak would confirm Adv's guess, allowing him to win. On the other hand, C's input substitution, especially using a mistyping map, is technically complex, and makes the definition less usable, since proofs would have to consider all such maps.

We now present our definition. Let n be a security parameter, and $D = \{0, 1\}^m$ is the password domain. (In general, m can be a function of n; interesting cases are when m is constant or logarithmic in n.) All players (Adv, C, S) are p.p.t. machines. As does [16], we use session IDs (SID) to partner instances of players, and impose the following correctness requirement. In the absence of adversary, all sessions terminate and intended parties output same sid and key.

Definition 1. *We say that an instance C_i^S of a client C and an instance S_j^C of a server S are* partners, *if they have output the same session id sid.*

We start by presenting KE games, which model attacks of a real-life adversary Adv_{Real}. The first game models the setting where Adv_{Real} obtained C's long key, is attacking a server, and is allowed a limited number of password tries.

Game KE$_1$. *Adv deterministically chooses active attack threshold $q \in 1..|D|$ (based on security parameter n) and creates an (honest) server S. Adv chooses*

S's name; then S's public/private keys are set up, and the public key revealed to Adv. Adv then runs players by executing steps 1-7 multiple times, in any order:

1. Adv creates an honest client C. Adv is allowed to pick any unused name for the client; the client C is registered with S, and long key ℓ and password pwd are set up and associated with C. Only one honest client can be created. Adv is given the long key ℓ, but not pwd.

2. Adv creates a corrupt client B^i. Adv is allowed to initialize him in any way, choosing any unused name, long key and password for him.

3. Adv creates an instance C_i of the honest client C. C_i is given (secretly from Adv) as input: his name C, the partner server's name S, the public key of S, the long key and the password of C.

4. Adv creates an instance S_j of the honest server S. S_j is given (secretly from Adv) as input: his name S, the private key of S, his partner's name (C or B^i) and that client's long key and password.

5. Adv delivers a message m to an honest party instance. The instance immediately responds with a reply (by giving it to Adv) and/or, terminates and outputs the result (a sid and either the session key, the failure symbol $\perp$, or, in case of the server instance, the password failure symbol $P\!\perp$) according to the protocol. Adv learns only the sid part of the output.

6. Adv "checks" any completed honest instance – then he is notified whether the instance output $P\!\perp$, $\perp$, or a session key. Adv gets charged one attempt, if he checked S^C and it output $P\!\perp$.
   When Adv accumulates q charges, he becomes restricted – he can neither deliver messages to any instances S_j^C nor check any instances.

7. Adv "opens" any successfully completed and checked honest instance – then he is given the session key output of that instance.

Then Adv asks for a challenge on an instance S_j^C of the server S. S_j^C, who has been instantiated to talk to the honest client C, must have completed, been checked by Adv, and output a session key. The challenge is, equiprobably, either the key output by S_j^C or a random string of the same length. Adv must not have opened S_j^C or a partner of S_j^C, and is not allowed to do it in the future.

Then Adv continues to run the game as before (execute steps 2-7). Finally, Adv outputs a single bit b which denotes Adv's guess at whether the challenge string was random. Adv wins if he makes a correct guess, and loses otherwise. Adv cannot "withdraw" from a challenge, and must produce his guess.

Note that we handle sid differently from [16]. Here we insist that parties always output sid, while previously sid was only output if a party did not fail. We need this change, since KE_1's interface needs to be the same for cases when an instance failed and did not fail. Outputting a sid only if KE succeeded (and letting it known to Adv for free) helps Adv determine whether $P\!\perp$ occurred.

In all other KE games (KE_2, KE_3, SID and DOA) below, password mistyping and even the knowledge of pwd should not help Adv. We thus choose to reveal the password to Adv and remove restrictions on the number of $P\!\perp$'s (thus removing the definition of q). We also allow Adv to specify C_i's password at its instantiations. These games are presented by modifying the above KE_1. All of the above four modifications are included in all games below.

KE_2 models the setting where Adv stole C's pwd and ℓ, but is attacking C.

Game KE_2 *is derived from KE_1 as noted in the previous paragraphs; further, Adv is given ℓ and must challenge an honest client instance C_i^S.*

KE_3 models the setting where Adv only stole C's pwd, and is attacking S.

Game KE_3 *derived from KE_1 as noted above, but Adv is not given ℓ.*

SID enforces non-triviality, preventing improper partnering (e.g. players unnecessarily outputting same sid). Recall, Adv is not allowed to challenge parties whose partner has been opened; SID ensures that Adv is not unfairly restricted.

Game SID *is derived from KE_1 as noted above; further, Adv does not ask for (nor answers) the challenge. Adv wins if any two honest partners output different session keys.*

Note, SID allows for one (or both) of the partners to output a failure symbol. Adv only wins if two successfully completed parties output different session keys.

Finally, game DOA models resistance to the Denial of Access (DoA) attacks. This game prevents vulnerabilities due to mistyping (see Sect. 4.1).

Game DOA *is derived from KE_1 as noted above; further, Adv does not ask for (nor answers) the challenge. Adv wins if the number of PL's is greater than the number of client instances where he substituted the password.*

Definition 2. *We say that a key exchange protocol Π is secure in the Combined Keys model with mistyping, if for every polytime adversaries Adv_1, Adv_2, Adv_3, Adv_{sid} and Adv_{doa} playing games KE_1, KE_2, KE_3, SID and DOA, their probabilities of winning (over the randomness used by the adversaries, all players and generation algorithms) is at most only negligibly (in n) better than:*

- *$1/2 + \frac{q}{2|D|}$, for KE_1,*
- *$1/2$, for KE_2 and KE_3,*
- *0, for SID and DOA.*

The definition for the HK setting (where C does not have ℓ) is extracted from Def. 2 by removing all uses of ℓ and the games where Adv doesn't know ℓ.

4.1   Why This Is a Good Definition

First, since Adv is not weaker than Adv of [16], Def. 2 enforces basic security properties of the protocols. We additionally need to argue that the definition is not too strict and that it prevents mistyping-caused leaks in protocols. The former property is intuitive, and we support it by proposing an efficient protocol and proving its security w.r.t. Def. 2 (Sect. 6). The latter property, on the other hand, requires significantly more careful consideration, presented in this section.

Note, KE_1 is the only game where we need to be careful with not giving Adv too much power w.r.t. mistyping. In other games, unlimited ability of Adv to substitute C's input should not help him win against a secure protocol. At the same time, such Adv directly models real-life adversary. Therefore, this simple allowance resolves mistyping problems w.r.t. other games we consider.

KE_1 is the core of the definition, and most of the definitional subtleties appear in KE_1. We start with the discussion of the details and ideas about this game.

Why KE_1 is a good model. Often, when a definition is proposed, a proof is provided, demonstrating the relationship between the new and previous definitions. This adds confidence in the proposed definition. We introduce the first definition in our setting; thus there is no previous definition to relate it to.

Our approach. Instead, we prove that if a protocol Π is secure by Def. 2, Adv of the game KE_1 cannot tell the difference between the following two executions, if he is not allowed to see the outputs of S. In one execution, selected (by Adv) client instances are instantiated with a mistyping sequence Adv chooses, and in the other they are instantiated with the password pwd of C. We stress that Adv is active during these executions; he can perform (almost) all the actions Adv of KE_1 can. This provides an informal "reduction" to the definition of [16], in the following sense. Assume the definition of [16] is "good", i.e. accurately identifies insecure protocols in its "no-mistyping" model. Then Def. 2 is "good" in the general setting, where clients are allowed to mistype.

Indeed, suppose Π is "bad". Due to the indistinguishability of the above executions, anything that Π leaks due to mistyping can also be seen and exploited without mistyping by Adv of KE_1 of [16]. Then Π will be insecure by definition of [16], since, by assumption, it is a good definition. Since KE_1 Adv of Def. 2 is at least as strong as that of [16], Π will also be insecure by Def. 2. From another angle, if active Adv cannot distinguish the above executions, then he is not learning anything from the mistypings, other than what may be inferred from the corresponding sequence of $P\bot$'s, but the latter is unavoidable anyway.

This reduction is informal, and serves only as evidence that our definition is good. By the nature of definitional work, it is not possible to "prove" definitions.

Formal theorem statement and proof of indistinguishability of the above executions is in full version. Proof idea is that some passwords used in the mistyped execution must be unequal to C's pwd. Ability to distinguish executions gives a free hint of what pwd is not, allowing corresponding KE_1 Adv to win.

On DoA protection. As mentioned in Sect. 2, the definition of [16] does not model (and fails to guarantee) DoA resistance when honest users mistype. We need that a replayed client's flow must not cause S output $P\bot$. Therefore, C must send at least one message that is dependent on S's message. Thus, the one-round, two-independent-flow protocols are not possible if DoA is desired.

We change the DOA game accordingly. Adv knows pwd, and is now allowed to instantiate clients with passwords of his choice. Adv wins DOA, if the number of $P\bot$ is greater than the number of client instances with substituted password.

5   Application to Biometric Authentication

We note that our definitions and protocols are directly applicable to biometric-based authentication. For example, fuzzy extractors [11] can be naturally used in our two-factor authentication setting, as follows. The storage card now additionally contains the public data pub_C of C's biometric b_C. The (potentially short) randomness extracted from b_C plays the role of the password. To authenticate, C first reconstructs the password using extractor's recovery procedure $Rec(pub_C, b'_C)$, and then uses it as prescribed by a KE protocol. Misreading b'_C of b_C can cause variety in the output of Rec and thus effect mistypings in the protocol. Still, our definitions (in-particular, mistyping-security property) and properties of fuzzy extractors guarantee security of this construction, even if Adv captured the card with the long key and pub_C. (In the HK setting, where C only has pk_S, we also can use our definition and above protocol – but pub_C is now sent by S to C authenticated by S's signature, as part of the protocol.)

However, we note that our definitions do not handle the general case, where b_C is used directly as input to C. That is, S knows "acceptance set" of C (AS_C), and accepts if C's submitted password/biometric $b_C \in AS_C$. We anticipate that a natural extension of our definition would handle this case. In particular, the correctness requirement should be amended w.r.t. AS_C, and Adv's allowed success rate may be dependent on AS as well. We leave this definition as future work, to be performed either as extension of our definition, or in the UC framework.

6   Mistyping-Secure KE Protocols

WLOG, assume protocol messages are formed properly (i.e. values drawn from appropriate domains, etc.). Let n be a security parameter, $E = (Gen, Enc, Dec)$ be a CCA2 secure public key encryption scheme, $F : \{0,1\}^n \times \{0,1\}^n \mapsto \{0,1\}^n$ be a PRFG, and $MAC : \{0,1\}^n \times \{0,1\}^* \mapsto \{0,1\}^n$ be a message authentication code. Let $N_C \in \{0,1\}^n$ be the name of client C. (Shorter names may be used.)

Although KR definitions do not handle mistyping, their protocol resists all mistyping-related attacks, except for (perhaps, unimportant in some settings) DoA resistance. We first prove this fact. Constr. 1 is the protocol of [16], only with updated handling of sid, to satisfy the syntactic requirements of Def. 2.

Construction 1. *(KE with mistyping, no DoA resistance [16])*

S^C	C^S
choose $r \in_R \{0,1\}^n$	choose $k \in_R \{0,1\}^n$,
	set $\alpha = Enc_{pk_S}(N_C, pwd, k)$

$$r \to \cdots \leftarrow \alpha, MAC_\ell(\alpha)$$

S^C	C^S
set $sid = (r, \alpha)$,	set $sid = (r, \alpha)$,
verify $MAC_\ell(\alpha)$ and N_C;	output
if fail, output $(sid, \perp)$, halt	$(sid, K = F_k(r))$
verify pwd;	
if fail, output $(sid, P\perp)$, halt	
else output $(sid, K = F_k(r))$	

Theorem 1. *Constr. 1 satisfies Def. 2, except for the success rate in game DoA.*

We now present a fully secure protocol in our model, derived from Constr. 1.

Construction 2. *is a challenge-response version of Constr. 1, where C^S replies with $(\alpha, MAC_\ell(r, \alpha))$ to message r.*

Theorem 2. *Constr. 2 is secure by Def. 2.*

We note that Constr. 2 can be modified to allow S to send confirmation to C whether he accepted, failed or password-failed. See full version for details.

Proofs of security of Theorems 1 and 2 are presented in the full version.

Achnowledgements. We thank Shai Halevi, Hugo Krawczyk, and anonymous referees for valuable comments.

References

1. http://en.wikipedia.org/wiki/Biometrics#Performance, Retrieved 02/10/08
2. Bellare, M., Kohno, T.: A theoretical treatment of related-key attacks: Rka-prps, rka-prfs, and applications. In: Biham, E. (ed.) EUROCRYPT 2003. LNCS, vol. 2656, pp. 491–506. Springer, Heidelberg (2003)
3. Bellare, M., Pointcheval, D., Rogaway, P.: Authenticated key exchange secure against dictionary attacks. In: Preneel, B. (ed.) EUROCRYPT 2000. LNCS, vol. 1807, pp. 139–155. Springer, Heidelberg (2000)
4. Bellovin, S.M., Merritt, M.: Encrypted key exchange: Password-based protocols secureagainst dictionary attacks. In: SP 1992: Proceedings of the 1992 IEEE Symposium on Security and Privacy, Washington, DC, USA, p. 72. IEEE Computer Society, Los Alamitos (1992)
5. Boyen, X.: Reusable cryptographic fuzzy extractors. In: CCS, pp. 82–91. ACM Press, New York (2004)
6. Boyen, X., Dodis, Y., Katz, J., Ostrovsky, R., Smith, A.: Secure remote authentication using biometric data (revised version),
   http://www.cs.stanford.edu/~xb/eurocrypt05b/
7. Boyen, X., Dodis, Y., Katz, J., Ostrovsky, R., Smith, A.: Secure remote authentication using biometric data. In: Cramer, R.J.F. (ed.) EUROCRYPT 2005. LNCS, vol. 3494, pp. 147–163. Springer, Heidelberg (2005)
8. Boyko, V., MacKenzie, P., Patel, S.: Provably Secure Password-Authenticated Key Exchange Using Diffie-hellman. In: Preneel, B. (ed.) EUROCRYPT 2000. LNCS, vol. 1807, pp. 156–171. Springer, Heidelberg (2000)
9. Canetti, R., Halevi, S., Katz, J., Lindell, Y., MacKenzie, P.D.: Universally composable password-based key exchange. In: Cramer, R.J.F. (ed.) EUROCRYPT 2005. LNCS, vol. 3494, pp. 404–421. Springer, Heidelberg (2005)
10. Dodis, Y., Katz, J., Reyzin, L., Smith, A.: Robust fuzzy extractors and authenticated key agreement from close secrets. In: Dwork, C. (ed.) CRYPTO 2006. LNCS, vol. 4117, pp. 147–163. Springer, Heidelberg (2006)
11. Dodis, Y., Ostrovsky, R., Reyzin, L., Smith, A.: Fuzzy extractors: How to generate strong keys from biometrics and other noisy data. Cryptology ePrint Archive, Report 2003/235 (2003), http://eprint.iacr.org/

12. Dodis, Y., Reyzin, L., Smith, A.: Fuzzy extractors: How to generate strong keys from biometrics and other noisy data. In: Cachin, C., Camenisch, J.L. (eds.) EUROCRYPT 2004. LNCS, vol. 3027, pp. 523–540. Springer, Heidelberg (2004)
13. Goldreich, O., Lindell, Y.: Session-key generation using human passwords only. In: Kilian, J. (ed.) CRYPTO 2001. LNCS, vol. 2139, pp. 408–432. Springer, Heidelberg (2001)
14. Li Gong, T., Lomas, M.A., Needham, R.M., Saltzer, J.H.: Protecting poorly chosen secrets from guessing attacks. IEEE Journal on Selected Areas in Communications 11(5), 648–656 (1993)
15. Halevi, S., Krawczyk, H.: Public-key cryptography and password protocols. ACM Trans. Inf. Syst. Secur. 2(3), 230–268 (1999)
16. Kolesnikov, V., Rackoff, C.: Key exchange using passwords and long keys. In: Halevi, S., Rabin, T. (eds.) TCC 2006. LNCS, vol. 3876, pp. 100–119. Springer, Heidelberg (2006)
17. MacKenzie, P., Reiter, M.: Delegation of cryptographic servers for capture-resilient devices. Distributed Computing 16(4), 307–327 (2003)

Affiliation-Hiding Envelope and Authentication Schemes with Efficient Support for Multiple Credentials

Stanisław Jarecki and Xiaomin Liu

University of California, Irvine*
{stasio,xiaominl}@ics.uci.edu

Abstract. We present an efficient implementation of affiliation-hiding envelope and authentication schemes. An envelope scheme enables secure message transmission between two parties s.t. the message can be decrypted only by a receiver who holds a credential from (i.e. is *affiliated with*) an entity specified by the sender's authorization policy. An envelope scheme is affiliation-hiding if it hides the receiver's affiliation, and if the sender's policy is revealed only to receivers who satisfy it. Similarly, an authentication scheme is affiliation-hiding if it reveals information about affiliations and the authentication policy of a participating party only to counterparties that satisfy this policy.

The novelty of our affiliation-hiding envelope scheme is that it remains practical in the *multi-affiliation setting* without relying on groups with bilinear maps. Namely, it requires $O(n)$ modular exponentiations and communicates $O(n)$ group elements, even if each party has n credentials, and each party's authentication policy specifies n admissible affiliations. Moreover, our affiliation-hiding envelope is chosen-ciphertext secure, which leads to a provably secure affiliation-hiding authentication scheme with same $O(n)$ efficiency in the multi-affiliation setting.

1 Introduction

Privacy Protection in Cryptographic Protocols. As the world becomes increasingly dependent on electronic communications, and as such communications fall prey to various forms of surveillance, it is important to investigate whether cryptographic protocols which enable secure electronic communication can have *privacy-protecting* variants that are efficient enough for practical usage. For example, group signature and privacy escrow schemes [10,19], provide privacy-protecting alternatives to standard signature and authentication schemes, where the verifier learns that the prover holds credentials that prove its membership in some group, but does not learn the identity of the prover within that group. In another example, key-private encryption or broadcast encryption [3,2] enable sending encrypted message to, respectively, a single receiver or to any member of a group, s.t. the ciphertext hides the encryptor's policy (i.e. the identity of intended recipients) from everyone except of the authorized recipients themselves.

Affiliation-Hiding Envelopes and Authentication. Affiliation-hiding envelopes address the same privacy issue as key-private broadcast encryption but with regards to *interactive* protocols for secure message transmission. An envelope scheme is an interactive

---

* Research supported by NSF CyberTrust Grant #0430622.

L. Aceto et al. (Eds.): ICALP 2008, Part II, LNCS 5126, pp. 715–726, 2008.

protocol between sender S and receiver R, where R receives S's message only if R satisfies S's authorization policy. We call receiver R certified by a Certification Authority (CA) a *member* of a *group* administered by this CA, and we say that R is *affiliated* with this group. Since in a PKI setting one can hold certificates from many CA's, we let $\mathsf{Afl}(R)$ be a *set* of groups R is affiliated with, and we assume that the sender's authorization policy is also expressed as a *set* of groups, denoted $\mathsf{Pol}(S)$. Using this notation, an envelope scheme should ensure that R learns nothing about sender S's message if $\mathsf{Afl}(R) \cap \mathsf{Pol}(S) = \emptyset$. An *affiliation-hiding* envelope scheme, introduced as a "Hidden Credentials" scheme by Holt et al. [12], must satisfy two additional privacy properties: (1) Receiver's affiliations $\mathsf{Afl}(R)$ are hidden from all parties, including a potentially malicious sender; and (2) Sender's policy $\mathsf{Pol}(S)$ is hidden from any non-authorized receiver R^*, i.e. any R^* s.t. $\mathsf{Afl}(R^*) \cap \mathsf{Pol}(S) = \emptyset$.

Affiliation-hiding authentication schemes, introduced as "Secret Handshakes" by Balfanz et al. [1], provide similar privacy property to authentication schemes: If each player U is affiliated with a set of groups $\mathsf{Afl}(U)$, and its authentication policy consists of a set of groups $\mathsf{Pol}(U)$, an authentication protocol between A and B should succeed only if $\mathsf{Afl}(A) \cap \mathsf{Pol}(B) \neq \emptyset$ and $\mathsf{Afl}(B) \cap \mathsf{Pol}(A) \neq \emptyset$. Moreover, such protocol is *affiliation-hiding* if sets $\mathsf{Afl}(A)$ and $\mathsf{Pol}(A)$ are hidden from any B^* who does not satisfy A's authentication policy, i.e. B^* s.t. $\mathsf{Afl}(B^*) \cap \mathsf{Pol}(A) = \emptyset$, and if sets $\mathsf{Afl}(B)$ and $\mathsf{Pol}(B)$ are similarly hidden from any A^* s.t. $\mathsf{Afl}(A^*) \cap \mathsf{Pol}(B) = \emptyset$.

Affiliation-Hiding vs. Unlinkability. Note that the property of affiliation-hiding is orthogonal to the property of *unlinkability* offered by group signatures and identity escrow schemes. Group signatures protect the privacy of a user within the group she is affiliated with, but they reveal this user's affiliation to any observer. In contrast, an affiliation-hiding authentication ensures that user's affiliation is revealed only to the entities that satisfy this user's authentication policy. Conversely, affiliation-hiding envelopes and authentication schemes protect the privacy of users' affiliations and authentication/authorization policies, but they might not protect the user from being *linkable*, i.e. it might be easy for the adversary to detect that two instances of the authentication or envelope scheme were executed by the same player. Indeed, the affiliation-hiding schemes we present here are linkable in this sense.

Prior Work on Affiliation-Hiding Envelopes and Authentication. Several solutions to affiliation-hiding envelope or authentication have been proposed under various assumptions. Affiliation-hiding envelope was given based on a Bilinear Diffie-Hellman (BDH) problem [12], while affiliation-hiding authentication was given based on BDH [1], a Computational Diffie-Hellman (DH) [9], and the RSA problem [13,14]. However, all these schemes consider a simplified setting where both the affiliation list and the policy of every player consists of a single group, and the straightforward extensions of these schemes to the *multi-affiliation* setting have $O(n^2)$ complexity, where n is the upper bound on the size of policies and certificate lists. The only previous work which achieves $O(n)$ complexity in the multi-affiliation is an affiliation-hiding envelope due to Bradshaw et al. [7]. However, this scheme relies on groups with bilinear maps, which, as we show, is not necessary. Moreover, it is not clear if this scheme offers chosen ciphertext security: [7] refers to the Fujisaki-Okamoto [11] method which requires ciphertext re-encryption during decryption, and it's not clear how this would work if encryption is simultaneously performed under n keys only some of which might be known by the receiver. Indeed, even defining CCA security for an envelope scheme is not trivial, and such definitions are missing from [7]. While CCA security is important in many applications of encryption, it is especially important in our context, because CCA encryption leads to provably secure encryption-based authentication [8], and the same holds for CCA-secure *affiliation-hiding* envelope and *affiliation-hiding* authentication.

Our Contributions. First, we define CCA security and privacy of an affiliation-hiding envelope scheme in the multi-affiliation setting. Second, we show such scheme with $O(n)$ efficiency, secure in the Random Oracle Model under DDH and GapDH assumptions on *any* multiplicative group, i.e. not necessarily one with a bilinear map. Note that DDH and GapDH assumptions are potentially weaker than BDH and that they can be plausibly posited on groups with smaller orders. In exact costs, the scheme of [7] requires $2n$ bilinear maps and n exponentiations, while our scheme requires about $2.5n$ exponentiations. Finally, in the full version of this paper [16] we show that a CCA-secure affiliation-hiding envelope scheme implies an affiliation-hiding authentication protocol. The resulting authentication protocol, included here, retains the same $O(n)$ complexity in the multi-affiliation setting as the underlying envelope scheme.

Other Related Work. While all the above affiliation-hiding schemes are linkable, and so is the scheme we present here, there are works which extend affiliation-hiding authentication schemes (and envelopes) to unlinkable schemes. The initial proposal by Tsudik and Xu [20] worked only if two communicating users assume the same revocation epoch. Jarecki and Liu [15] showed a scheme which tolerates up to a constant Δ lag in the revocation epochs at the $O(\Delta)$ cost to the protocol. Even though this scheme scales well in the multi-affiliation setting, the resulting protocol imposes a non-standard constraint that two players fail to communicate if their assumed revocation epochs are farther than Δ apart. In recent work same authors proposed a scheme that makes no assumptions on synchrony in revocation lists [17], but this scheme becomes $O(n^2)$ in the multi-affiliation setting, unlike the (linkable) affiliation-hiding scheme presented here. We note that affiliation-hiding authentication was also extended to a *group* key agreement protocol [13], but this protocol works only in the single-affiliation setting.

Organization. In Section 2 we intuitively explain our technical challenges and our solutions. Section 3 consists of preliminaries. In Section 4 we formally define affiliation-hiding envelopes. In Section 5 we show our construction of such envelope scheme and a sketch of a security argument for it. Finally in Section 6 we show a generic construction of an affiliation-hiding authentication from an affiliation-hiding envelope. For lack of space, we have relegated all security proofs to the full version of this paper [16].

2   Technical Roadmap

First note that given CCA-secure and affiliation-hiding envelope scheme that is efficient in the multi-affiliation setting, construction of an affiliation-hiding authentication efficient in this setting is immediate: Each party chooses a nonce and encrypts it for the other using the envelope scheme, where in each case the sender uses the keys corresponding to his authentication policy while the receiver attempts to decrypt using the keys and certificates corresponding to his set of affiliations. The resulting key is then a hash of the two nonces. The computational and communication costs of this scheme are just twice the costs of the underlying envelope. The affiliation-privacy of this scheme follows from affiliation-privacy of the envelope, while security follows from the chosen ciphertext security of the envelope scheme by a straightforward extension of a theorem shown in [15]. (Note, however, that if the envelope is not CCA secure then subtle attacks on this simple authentication scheme are possible, e.g. if the adversary modifies one of the ciphertexts without changing the plaintext.)

Thus the technical challenge is in building a CCA-secure affiliation-hiding envelope efficient in the multi-affiliation setting. Note that an envelope scheme can be built from CMA-secure signatures where the sender $U_{i'}$ encrypts its message under the group public key PK_j, and the receiver U_i can decrypt it only if he possesses a valid signature $\sigma_{i,j}$ issued under the key PK_j on some (fixed) message. Such envelope scheme can be affiliation-hiding because the authentication (of U_i as member in G_j) is done *implicitly*, i.e. by the fact that U_i can decrypt $U_{i'}$'s message. Note that existing efficient secret handshake schemes, e.g. [1,9,13], are constructed using such implicitly-authenticated envelope, each created from a different signature scheme. In the case of the DL-based scheme of [9], the envelope scheme works by splitting the signature $\sigma_{i,j}$ into two parts, the "certificate" $cert_{i,j}$ and the "secret key" $sk_{i,j}$. The receiver sends its certificate $cert_{i,j}$ to the sender, and the sender can derive an encryption public key $pk_{i,j}$, from $cert_{i,j}$ and the group public key PK_j, s.t. $pk_{i,j}$ is a public key encryption key corresponding to the private decryption key $sk_{i,j}$. For example, if (s, r) is a Schnorr signature on an empty message under the group public key $PK_j = y$, i.e. $g^s = r \cdot y^{H(r,y)}$, and if we set $cert_{i,j} = r$ and $sk_{i,j} = s$, then the ElGamal encryption key corresponding to $sk_{i,j}$, i.e. $pk_{i,j} = g^s$, can be indeed computed from $cert_{i,j}$ and PK_j. By the strong CMA unforgeability property of Schnorr signatures user U_i can be revoked from G_j by placing $= r$ on a revocation list. Moreover, neither $cert_{i,j} = r$ nor the ElGamal encryption under $pk_{i,j} = g^s$ can be linked to the group key y, and thus the scheme is affiliation-hiding (but not unlinkable).

However, this affiliation-hiding envelope leads to an $O(n^2)$ protocol in a setting where the sender encrypts under n group public keys $y_1, ..., y_n$ and the receiver has n certificate and secret key tuples $(r_1, s_1), ..., (r_n, s_n)$ to decrypt with. In order to get this cost down to $O(n)$ we make the following two moves: First, we create a special-purpose discrete-log based certification scheme which allows each user U_i to control the secret key part $sk_{i,j}$ for all signatures $\sigma_{i,j}$ it holds (i.e. for each group G_j it is affiliated with), so that it can set all these $sk_{i,j}$ values to a single common value s_i. In this way the receiver can use only one key $sk_i = s_i$ in its attempts to decrypt all the ciphertexts sent as part of this envelope scheme. Here is a CMA-secure signature scheme we create to meet this property: A signature on message s_i under key $y_j = g^{x_j}$ is a non-interactive (in ROM) zero-knowledge proof $\Pi_{r_{i,j}}$ of knowledge of a discrete logarithm $DL_g(r_{i,j})$, where $r_{i,j} = (y_j)^{s_i} = g^{x_j \cdot s_i}$. Setting $cert_{i,j} = (r_{i,j}, \Pi_{r_{i,j}})$ and $sk_i = s_i$, the ElGamal encryption key $pk_{i,j}$ is just the pair $(y_j, r_{i,j})$ because the decryption key sk_i is a discrete logarithm between these two values. Note that $cert_{i,j}$ is independent of key y_j. It is also plausible (and we prove that it indeed holds) that the ElGamal encryption of m under $pk_{i,j}$, a pair $((y_j)^t, (r_{i,j})^t \cdot m)$, cannot be linked to key y_j under the DDH assumption.

Secondly, in order for the sender not to have to create n^2 ElGamal ciphertexts, for each $r_{i,j}$ sent by the receiver and each group key y_l in sender's authentication policy, the encryptor *batches* these n^2 encryptions by using a single randomness t in all these instances. (We note that a similar idea of re-using randomness in several instances of ElGamal encryption was investigated also in [4] and [5].) We show this affiliation-hiding envelope scheme in Figure 2. Following the methodology of Bellare, Kohno, and Shoup [5] we use CCA-secure symmetric encryption and a hash function modeled as a random oracle to convert ElGamal into a CCA encryption scheme *while* re-using the same randomness in all the ciphertexts. One of the interesting parts of the security proof of the resulting scheme is that the reduction, to solve its Diffie-Hellman challenge from the attacker's computation of (one of) ElGamal "temporary key" values $r_{i,j}^t$, needs to know the discrete logarithm $DL(g, r_{i,j})$ for *the* value for which the adversary computes $(r_{i,j})^t$. Seemingly, this requires simultaneous extraction of the discrete logarithms of up to n values $r_{i,1}, ..., r_{i,n}$ sent by the malicious receiver. However, our reduction instead uses the standard forking-lemma to efficiently extract $DL(g, r_{i,j})$ for a *random* index j, and with probability $1/n$ that index corresponds to the correct $r_{i,j}$ value.

3   Cryptographic Assumptions and Tools

We state the security assumptions required by our construction, as well as two tools we employ, namely certain zero-knowledge proofs and a symmetric encryption scheme. From now on let g be a generator of a multiplicative group of order q, denoted $\langle g \rangle = \mathbb{G}$.

DDH Assumption on $\mathbb{G}$: We say that DDH is (T, ϵ)-hard in $\mathbb{G}$, if any T-time algorithm $\mathcal{A}$ has at most $1/2 + \epsilon$ advantage in distinguishing distributions $\{(g, g^a, g^b, g^{ab})\}_{a,b \leftarrow \mathbb{Z}_q}$ and $\{(g, g^a, g^b, g^c)\}_{a,b,c \leftarrow \mathbb{Z}_q}$.

GapDH Assumption on $\mathbb{G}$: Informally, GapDH assumption holds if the CDH problem in group $\mathbb{G}$ is hard even given access to a DDH oracle in this group. A DDH oracle in $\mathbb{G}$ is an algorithm on input $(g^\alpha, g^\beta, g^\gamma)$ outputs 1 if $\gamma = \alpha\beta$, and 0 otherwise. We say GapDH problem is (T, ϵ, q_{ddh})-hard in $\mathbb{G}$ if any T-time algorithm $\mathcal{A}$ making at most q_{ddh} DDH oracle queries, succeeds with probability at most ϵ in computing g^{ab} given (g, g^a, g^b), for random a, b in $\mathbb{Z}_q$.

Non-Interactive Zero Knowledge Proof of Knowledge of Discrete Logarithm: We use a standard NIZK proof of knowledge (in ROM) of discrete logarithm of value $y = g^x$, denoted NIZK-DL$^{H_1}(y)$, which is a pair (a, z), where $a = g^\alpha$ for random α in $\mathbb{Z}_q$, $z = \alpha + e \cdot x \pmod{q}$, and $e = H_1(a, y)$, where $H_1 : \{0, 1\}^* \to \mathbb{Z}_q$ is a hash function. Verifier accepts the proof $\Pi = (a, z)$ if $g^z = a \cdot y^e$, for $e = H_1(a, y)$. In our security proofs we use a standard simulator of this NIZK which on input y picks random z and e in $\mathbb{Z}_q$, computes $a = g^z/y^e$, "defines" $H_1(a, y)$ as e, and outputs (a, z).

A symmetric key encryption scheme: We use a symmetric encryption scheme $\Pi^{sym} = (\text{KGen}, \text{SEnc}, \text{SDec})$, which is chosen-plaintext secure (IND-CPA) and satisfies the integrity of ciphertext property (INT-CTXT). As shown in [6], IND-CPA and INT-CTXT imply IND-CCA security of encryption. We skip the standard definition of IND-CPA security of symmetric encryption, but we state the ciphertext integrity property [18,6]: A symmetric key encryption Π^{sym} has (T, ϵ, q_E, q_D)-*integrity* *of ciphertexts (INT-CTXT)* if for any T-time adversary $\mathcal{A}$ with at most q_E queries to $\mathsf{SEnc}(K, \cdot)$ and at most q_D queries to $\mathsf{SDec}(K, \cdot)$, $\mathcal{A}$ outputs a valid new ciphertext C, s.t. $\mathsf{SDec}(K, C) \neq \perp$, with probability at most ϵ, where we call C "new" if it was not returned by any call to $\mathsf{SEnc}(K, \cdot)$.

4   Affiliation-Hiding Envelope Schemes

An (implicitly authenticated) envelope scheme consists of a tuple of efficient probabilistic algorithms (Setup, GInit, UInit, MEnc, Dec, Check), and a 2-party protocol Add. The principals in the scheme are n' users $U_1, .., U_{n'}$ and n groups $G_1, ..., G_n$ administered by n respective group authorities $GA_1, ..., GA_n$. Each user U_i has both a policy $\mathsf{Pol}(U_i)$ and a set of affiliations $\mathsf{Afl}(U_i)$, both of which are subsets of groups $G_1, ..., G_n$. (However, our security and privacy definitions are all stated for the worst case, where the policy or affiliation set of an attacked entity is made of *all* n groups.) The syntax of an envelope scheme given below is custom-made to model two-round envelope protocols in which the sender and the receiver proceed in a way described in Figure. 1 below. We use $\mathcal{M}$ to denote the message space.

– Setup(1^λ), on security parameter λ, generates common parameters σ.
– GInit(σ), executed by GA_j creates public key pair (SK_j, PK_j) for group G_j.
– UInit(σ), executed by user U_i creates U_i's secret key sk_i.
– Add is an interactive procedure between user U_i and GA_j on U_i's private input sk_i, on GA_j's private input SK_j, and on public input PK_j. At the end of the interaction, U_i gets his public key $pk_{i,j}$ and a certificate $cert_{i,j}$ of membership for group G_j. (We additionally require that the resulting keys $pk_{i,j}$ come from a publicly samplable space.) We say that $(pk_{i,j}, cert_{i,j})$ is issued *under* key PK_j. To revoke key $pk_{i,j}$, it is simply added to G_j's revocation list Rev_j. We introduce the following notation: $\mathbf{pk}_i = \{pk_{i,j}\}_{G_j \in \mathsf{Afl}(U_i)}$; $\mathbf{cert}_i = \{cert_{i,j}\}_{G_j \in \mathsf{Afl}(U_i)}$; $\mathbf{PK}_i = \{PK_l\}_{G_l \in \mathsf{Pol}(U_i)}$; $\mathbf{Rev}_i = \bigcup_{G_l \in \mathsf{Pol}(U_i)} Rev_l$.
– MEnc($\mathbf{PK}_{i'}, \mathbf{pk}_i, m$) is executed by sender $U_{i'}$ on inputs $\mathbf{PK}_{i'}$, message $m \in \mathcal{M}$, and vector $\mathbf{pk}_i$ supplied by receiver U_i. The algorithm's output, denoted $\mathbf{e}$, encodes a set of ciphertexts $\mathsf{Rep}(\mathbf{e}) = \{e_{l,j} \mid (PK_l, pk_{i,j}) \in \mathbf{PK}_{i'} \times \mathbf{pk}_i\}$.
– Dec(sk_i, e) is executed by receiver U_i on private input sk_i and a ciphertext e, an element in $\mathsf{Rep}(\mathbf{e})$. The output is a message $m \in \mathcal{M}$ or a failure symbol $\perp$.
  We define MDec($sk_i, \mathbf{e}$) as a procedure that computes Dec($sk_i, e_{l,j}$) for every $e_{l,j} \in \mathsf{Rep}(\mathbf{e})$ (processed in some canonic order), and outputs the message output by the first successful instance of Dec($sk_i, e_{l,j}$), or $\perp$ if all these Dec instances fail.
– Check($\sigma, \mathbf{pk}_i, \mathbf{cert}_i$) is executed by sender $U_{i'}$ to test the validity of certificates $\mathbf{cert}_i$ supplied by the receiver U_i in the envelope procedure. It outputs 1 or 0.

Using these procedures, an envelope protocol between U_i and $U_{i'}$ proceeds as follows:

4.1   Security and Privacy Properties of an Affiliation-Hiding Envelope Scheme

Completeness. Let σ, pairs (PK_l, SK_l) for every group G_l, and $(pk_{i,j}, sk_{i,j})$ for every user U_i and every group $G_j \in \mathsf{Afl}(U_i)$ be properly generated by procedures Setup,

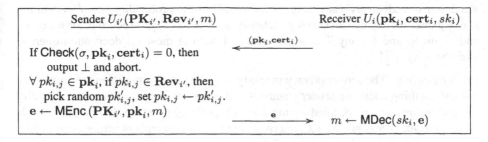

Fig. 1. Envelope Scheme in Multi-Affiliation Setting

GInit, and Add respectively. We say that an Envelope Scheme is ϵ-complete, if for any receiver U_i with input $(\mathbf{pk}_i, sk_i)$, and any sender $U_{i'}$ with input $(\mathbf{PK}_{i'}, \mathbf{Rev}_{i'}, m)$, if $\exists\, G_l$, s.t. $pk_{i,l} \in \mathbf{pk}_i$, and $PK_l \in \mathbf{PK}_{i'}$, then

$$Pr[\mathsf{MDec}(sk_i, \mathsf{MEnc}(\mathbf{PK}_{i'}, \mathbf{pk}_i, m)) = m] \geq 1 - \epsilon.$$

Chosen Ciphertext Security. The (CCA) security property for an envelope scheme is equivalent to CCA security for broadcast encryption. Namely, a malicious receiver who is not a valid member of any group specified by the sender's policy cannot tell anything about the encrypted message, even if he is given a CCA access to some valid group members acting as receivers, except that they cannot be queried on the challenge ciphertext. Formally, we define the security of an envelope scheme via the following game between an adversary $\mathcal{A}$ and a challenger.

- **Init:** Adversary $\mathcal{A}$ gets $\{PK_j\}_{j \in \{1,\ldots,n\}}$, and $\{(\mathbf{pk}_i, \mathbf{cert}_i)\}_{i \in \{1,\ldots,n'\}}$.
- **Join, Corruption and Decryption Query Phase I (JCD-I):**
  On a join request $\mathsf{JR}(j)$, the challenger performs Add between $G\Lambda_j$ and $\mathcal{A}$. The public key which $\mathcal{A}$ receives in this instance is then revoked (i.e. added to Rev_j).
  On corruption request $\mathsf{CR}(i)$, the challenger sends sk_i to $\mathcal{A}$, and all public keys $pk_{i,j}$ for $G_j \in \mathsf{Afl}(U_i)$ are added to respective revocation list Rev_j.
  On decryption query $\mathsf{DQ}(i,e)$, challenger sends $\mathsf{Dec}(sk_i,e)$ to $\mathcal{A}$.
- **Challenge:** $\mathcal{A}$ sends $(\mathbf{pk}^*, \mathbf{cert}^*, m_0, m_1)$. If $\mathsf{Check}(\sigma, \mathbf{pk}^*, \mathbf{cert}^*) = 0$, then the game aborts. Let $\mathbf{Rev} = \bigcup_{l=1}^{n} Rev_l$. For each pk_j in $\mathbf{pk}^*$, if $pk_j \in \mathbf{Rev}$, challenger picks random pk'_j and sets $pk_j \leftarrow pk'_j$. The challenger then picks random bit b, and replies to $\mathcal{A}$ with $e^* \leftarrow \mathsf{MEnc}(\mathbf{PK}, \mathbf{pk}^*, m_b)$ for $\mathbf{PK} = \{PK_l\}_{l=1,\ldots,n}$.
- **Join, Corruption and Decryption Query Phase II (JCD-II):** This is the same as in Join, Corruption and Decryption Query Phase I, except that:
  (a) the challenger rejects $\mathsf{CR}(i)$ if $pk_{i,j} \in \mathbf{pk}^*$ for some group G_j, i.e. if the encryption challenge included one of U_i's public keys; and
  (b) the challenger rejects $\mathsf{DQ}(i,e)$ if $\exists\, j, l$, s.t. $pk_{i,j} \in \mathbf{pk}^*$, $PK_l \in \mathbf{PK}$, $G_l = G_j$ and $e = e^*_{l,j}$ for some $e^*_{l,j} \in \mathsf{Rep}(e^*)$. Intuitively, $e^*_{l,j}$ is an encryption of m_b under PK_l and $pk_{i,j}$. Since $cert_{i,j}$ in $\mathbf{cert}^*$ that corresponds to $pk_{i,j}$ in $\mathbf{pk}^*$ is valid, U_i is a valid member of G_j, and therefore $pk_{i,j}$ was issued under $PK_j = PK_l$, so $\mathsf{Dec}(sk_i, e^*_{l,j})$ would reveal m_b.
- **Guess:** Adversary outputs a bit b' as her guess of b.

We define the adversary's advantage Adv-Sec($\mathcal{A}$) as the probability that $b' = b$ in the above game. We say that an envelope scheme is $(n, n', T, \epsilon, q_D)$-secure if for n groups and n' users, and for any T-time adversary $\mathcal{A}$ with at most q_D decryption queries, $|\mathsf{Adv\text{-}Sec}(\mathcal{A}) - \frac{1}{2}| \leq \epsilon$.

Sender Privacy. The sender privacy property says that even a malicious receiver cannot tell anything about the sender's authentication policy, if this receiver is *not* a valid member of *any* group specified by the sender's policy. Moreover, this property holds even if the attacker is given a CCA access to *valid* group members acting as receivers, except that they cannot be queried on the challenge ciphertext. Note that here we define sender privacy only against *outsider* attacks, i.e. against adversaries who are *not* valid members of *any* group in the sender's authentication policy. In [16] we also consider a stronger notion of sender privacy, namely against *insider* attacks.

Formally, we define sender privacy via an interactive game between an adversary $\mathcal{A}$ and a challenger. As in the game in the security definition above, this game has five phases: Init, JCD-I, Challenge, JCD-II, and Guess. All phases are the same as in the security game, except the Challenge phase:

- **Challenge:** Adversary sends $(\mathbf{pk}^*, \mathbf{cert}^*, m)$. If $\mathsf{Check}(\sigma, \mathbf{pk}^*, \mathbf{cert}^*) = 0$, then the game aborts. Let $\mathbf{Rev} = \bigcup_{l=1}^{n} Rev_l$. For each pk_j in $\mathbf{pk}^*$, if $pk_j \in \mathbf{Rev}$, challenger picks random pk_j' and sets $pk_j \leftarrow pk_j'$. The challenger then picks a random bit b and sends $\mathbf{e}^*$ to $\mathcal{A}$ computed as follows: If $b = 0$, the challenger computes $\mathbf{e}^* \leftarrow \mathsf{MEnc}(\mathbf{PK}, \mathbf{pk}^*, m)$, for $\mathbf{PK} = \{PK_l\}_{l=1,\dots,n}$; If $b = 1$, challenger computes $\mathbf{e}^* \leftarrow \mathsf{MEnc}(\mathbf{PK}', \mathbf{pk}^*, m)$ for $\mathbf{PK}' = \{PK_l'\}_{l=1,\dots,n}$, where each PK_l' is randomly chosen by GInit.

We define the adversary's advantage Adv-SPri($\mathcal{A}$) as the probability that $b' = b$ in the above game. We say that an envelope scheme is $(n, n', T, \epsilon, q_D)$-sender-private if for n groups and n' users, and for any T-time adversary $\mathcal{A}$ with at most q_D decryption queries, $|\mathsf{Adv\text{-}SPri}(\mathcal{A}) - \frac{1}{2}| \leq \epsilon$.

Remarks: This privacy definition is a "real-random" notion of privacy, but it implies a "flexible" privacy notion, where the adversary picks two sets of group keys, V_0 and V_1, $|V_0| = |V_1|$, and the challenger responds with encryption under keys V_b for a random b.

Receiver Privacy. The receiver privacy property states that no one can tell anything about the affiliations of the receiver (except possibly of the group authorities). Formally, the receiver privacy is defined via the following game, parameterized by a tuple $(i^*, \mathcal{G}_0, \mathcal{G}_1)$, where i^* is a index of an attacked user and $\mathcal{G}_0$ and $\mathcal{G}_1$ are two sets of groups s.t. $|\mathcal{G}_0| = |\mathcal{G}_1|$ and $\mathcal{G}_0 \cup \mathcal{G}_1 \subseteq \mathsf{Afl}(U_{i^*})$. This game has five phases: Init, JC-I, Challenge, JC-II, and Guess. Init and Guess are the same as in the security game. JC-I and JC-II are identical, which are the same as JCD-I in the security game except that (1) there is no decryption query, since the answer to an decryption query to an honest receiver leaks the receiver's affiliation completely; and (2) corruption query on user U_{i^*} is rejected. In the Challenge phase, the challenger picks a random bit b, and replies with $(\{pk_{i^*,j}\}_{G_j \in \mathcal{G}_b}, \{cert_{i^*,j}\}_{G_j \in \mathcal{G}_b})$.

We define the adversary's advantage $\mathsf{Adv\text{-}RPri}_{i^*, \mathcal{G}_0, \mathcal{G}_1}(\mathcal{A})$ as the probability that $b' = b$ in the above game. We say that an envelope scheme is (n, n', T, ϵ)-receiver-private if for n groups and n' users, for *any* $(i^*, \mathcal{G}_0, \mathcal{G}_1)$ satisfying the above constraints, any T-time adversary $\mathcal{A}$, $\left| \text{Adv-RPri}_{i^*, \mathcal{G}_0, \mathcal{G}_1}(\mathcal{A}) - \frac{1}{2} \right| \leq \epsilon$.

Remark on Trust in Group Managers: The security and sender-privacy definitions given above are stated for an adversary who is not an authorized recipient, i.e. who is not a valid member of any groups in the sender's authentication policy. In particular, it makes no sense to require such security or privacy against an adversary which colludes with the group managers themselves, since a group manager of any group in the sender's policy can create a virtual user that satisfies this policy, and hence neither the message nor the policy itself can be protected against such adversary. In contrast, the receiver privacy could in principle be achieved even for players colluding with the relevant group managers, but above we only define a simplified notion of receiver privacy where *all* group managers are honest. Indeed, the receiver privacy of our envelope scheme is broken by any dishonest group manager. However, as we show in the full version of the paper [16], this vulnerability can be fixed if the Add protocol we give in Section 5 below for adding a member to a group is changed so that instead of having user U_i send its long-term secret sk_i to GA_j, the two parties compute $(pk_{i,j}, cert_{i,j})$ using an efficient secure computation protocol.

5   The Construction of an Affiliation-Hiding Envelope Scheme

We provide the algorithms that instantiate our affiliation-hiding envelope scheme. The whole operation of the scheme is schematically illustrated in Fig.2.

- Setup(1^λ) sets $\sigma = (g, q)$ where g generates a multiplicative group $\mathbb{G}$ of order q, s.t. DDH and GapDH hold in $\mathbb{G}$ with security parameter λ.
- GInit(σ) executed by GA_j picks $x_j \in \mathbb{Z}_q$, and sets $(SK_j, PK_j) = (x_j, y_j = g^{x_j})$.
- UInit(σ) executed by U_i picks $s_i \in \mathbb{Z}_q$ and sets $sk_i = s_i$.
- Add proceeds as follow: U_i sends s_i to GA_j, who replies with $(pk_{i,j}, cert_{i,j}) = (r_{i,j}, \Pi_{r_{i,j}})$ where $r_{i,j} = (y_j)^{s_i} = g^{x_j \cdot s_i}$ and $\Pi_{r_{i,j}} = \text{NIZK-DL}^{H_1}(r_{i,j})$. Note that $r_{i,j}$ is random in $\mathbb{G}$ for a random s_i, so $r_{i,j}$'s can be publicly sampled. (See also a Remark on Trust in Group Managers above.)
- MEnc$(\mathbf{PK}_{i'}, \mathbf{pk}_i, m)$ picks random $t \in \mathbb{Z}_q$, sets $d_l = (y_l)^t$ for every $y_l \in \mathbf{PK}_{i'}$ and $c_j = \text{SEnc}(K_j, m)$ for every $r_j \in \mathbf{pk}_i$, where $K_j = H_2((r_j)^t)$ and $H_2 : \mathbb{G} \rightarrow \text{KeySpace}(\Pi^{\text{sym}})$ is a hash function modeled as random oracle in the security analysis. The output is $\mathbf{e} = (\mathbf{d}, \mathbf{c})$ where $\mathbf{d} = \{d_l\}_{PK_l \in \mathbf{PK}_{i'}}$ and $\mathbf{c} = \{c_j\}_{pk_j \in \mathbf{pk}_i}$. We define $e_{l,j}$ in Rep$(\mathbf{e})$ as $\langle d_l, c_j \rangle$.
- Dec$(s_i, \langle d, c \rangle)$ outputs $m \leftarrow \text{SDec}(K, c)$ for $K = H_2(d^{s_i})$, if $d \in \mathbb{G}$, and $\perp$ otherwise. Note that even though the resulting algorithm MDec$(s_i, \langle \mathbf{d}, \mathbf{c} \rangle)$ runs in $\Omega(n^2)$ time, it makes only n exponentiation operations, to compute d^{s_i} for every $d \in \mathbf{d}$, and it only needs to make n^2 *symmetric* decryption operations SDec.
- Check$(\sigma, \mathbf{pk}, \mathbf{cert})$ output 1 if Π_{r_j} verifies as NIZK-DL$^{H_1}(r_j)$, for every $r_j \in \mathbf{pk}$ and corresponding $\Pi_{r_j} \in \mathbf{cert}$, and outputs 0 otherwise.

Theorem 1. *If the symmetric key encryption scheme Π^{sym} is complete and ϵ-INT-CTXT, then our envelope scheme is ϵ'-complete, where $\epsilon' = n^2 \cdot \epsilon$.*

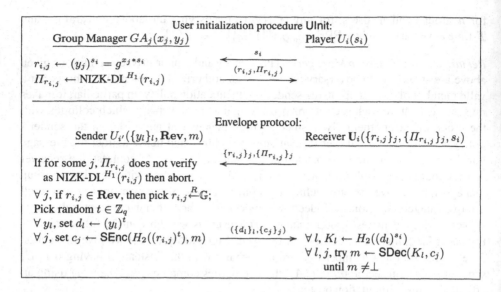

Fig. 2. Our Affiliation-Hiding Envelope Scheme

Theorem 2. *If GapDH is (T_1, ϵ_1, q_1)-hard in $\mathbb{G}$, and the symmetric key encryption scheme Π^{sym} is (T_2, ϵ_2, q_2)-IND-CPA and $(T_3, \epsilon_3, q_E, q_D)$-INT-CTXT, then our construction of the envelope scheme is $(n, n', T', \epsilon', q'_D)$-secure in ROM, where $T' = \min\{T_1/2, T_2, T_3\} - O(nn' + q'_D + q_{H_1} + q_{H_2}) \cdot T_{exp}$, $\epsilon' = n\epsilon_1 + \sqrt{q_{H_1} \cdot \epsilon_1} + n\epsilon_2 + q_D \cdot \epsilon_3 + (2nn' + 1)q_{H_1}/q + 2nq_D/q + n/q$, $(q'_D + n) \cdot q_{H_2} \leq q_1$, where T_{exp} is the time for an exponentiation operation and q'_{H_i} is the number of queries the adversary makes to hash function H_i, for $i = 1, 2$.*

For lack of space we relegate the security proof to the full version of the paper [16], but we sketch it here in the simplified single-policy and single-affiliation setting. In this case the adversary's challenge is $(pk, cert, m_0, m_1)$, where $pk = r$ and $cert = (a, z)$ is a NIZK-DL$^{H_2}(r)$, and the challenger returns a ciphertext (d, c) where $d = y^t$ and $c = \mathsf{SEnc}(H_2(r^t), m_b)$ where $PK = y$ is the single public key in challenger's policy.

Consider first the case when $\mathcal{A}$ never queries the random oracle H_2 on r^t and yet $\mathcal{A}$ has a non-negligible advantage in distinguishing challenger's bit b. In this case we construct a reduction that breaks the IND-CPA security of the symmetric key encryption scheme, since the reduction effectively distinguishes the encryption of m_0 from m_1 without any information about the encryption key $H_2(r^t)$. Without getting into details, the fact that the symmetric encryption scheme has the ciphertext integrity property (INT-CTXT) helps the reduction handle adversary's decryption queries in this case. Now, suppose that $\mathcal{A}$ does query the random oracle on $w = r^t$. We consider two cases depending on the origin of the $(r, (a, z))$ tuple in $\mathcal{A}$'s challenge. If this tuple is a public key and a certificate of some non-revoked user (note that $\mathcal{A}$ learns the public keys and certificates of all honest players), then we construct a reduction that breaks the GapDH assumption by successfully outputting $w = r^t$ given triple (h, h^t, r). This can be achieved by setting the group key y to h, setting d to h^t in the encryption challenge, and simulating the NIZK proof (a, z) of DL between g and r. This reduction uses the DDH oracle to correctly answer the decryption queries, and while not knowing r^t it cannot set the c part of the ciphertext correctly, the only way the adversary can realize that c is incorrect is by querying H_2 on r^t, at which point the reduction solves its challenge. The last case is when $\mathcal{A}$ queries $H_2(r^t)$ but the $(r, (a, z))$ tuple in its challenge is created by $\mathcal{A}$ itself. Note that (a, z) is a NIZK of DL between g and r, so $\mathcal{A}$ can always create such proof if he sets $r = g^\gamma$ for some known γ. The reduction has to extract this value γ from this proof of knowledge, and this is done using a standard forking lemma. (Interestingly, in the full multi-affiliation version of the proof the reduction does not have to extract the witnesses for all (r_j, Π_j) pairs provided by $\mathcal{A}$ since with probability $1/n$ it can guess the index j of r_j for which it needs to extract the discrete logarithm.) If $\mathcal{A}$ queries H_2 on $w = r^t$ in this game, then we construct a reduction which breaks the GapDH assumption by computing g^t given (h, h^t, g). The reduction again sets the group public key y to h and d in the encryption to h^t. Then $g^t = r^{DL(r,g) \cdot t} = (r^t)^{1/\gamma} = w^{1/\gamma}$. Decryption queries are handled as above.

The proofs of the sender/receiver privacy properties of this envelope scheme, stated below, are also given in [16]. We note that the proof of receiver privacy is rather simple, while the proof of sender privacy is similar to the proof of (CCA) security.

Theorem 3. *If GapDH is (T_1, ϵ_1, q_1)-hard and DDH is (T_4, ϵ_4)-hard in $\mathbb{G}$, and the symmetric key encryption scheme Π^{sym} is $(T_3, \epsilon_3, q_E, q_D)$-INT-CTXT, then our envelope scheme is (T', ϵ', q_D')-sender-private, where $T' = \min\{T_1/2, T_3, T_4\} - O(nn' + q_D' + q_{H_1} + q_{H_2}) \cdot T_{exp}$, $\epsilon' = n\epsilon_1 + \sqrt{q_{H_1} \cdot \epsilon_1} + q_D \cdot \epsilon_3 + \epsilon_4 + (3nn' + 1)q_{H_1}/q + nq_D/q + nq_{H_2}/q + n/q$, $(q_D' + n) \cdot q_{H_2} \le q_1$, where q_{H_i} is the number of queries the adversary makes to hash function H_i, for $i = 1, 2$.*

Theorem 4. *If DDH is (T, ϵ)-hard in $\mathbb{G}$, then our envelope scheme is $(n, n', T', 2\epsilon')$-receiver-private where $T' = T - O(nn')T_{exp}$ and $\epsilon' = \epsilon + nn'q_{H_1}/q$.*

6 Affiliation-Hiding Authentication Scheme

An affiliation-hiding authentication scheme consists of algorithms Setup, GInit, UInit and 2-party protocols Add and Handshake. All the algorithms and protocols are as in an envelope scheme except Handshake, which is the actual authentication protocol executed between two players U_i and $U_{i'}$. At the end of this protocol, the two players output the same authentication key if each of them is affiliated with at least one group in the other player's policy. An affiliation-hiding authentication scheme can be constructed from an envelope scheme. We give a simplified description of the construction of the Handshake protocol here and refer the readers to [16] for details. The two players U_i and $U_{i'}$ picks their respective random nonces K_i and $K_{i'}$, and send them to each other using two instances of the envelope scheme. The resulting authentication key is the hash of the two nonces. By a similar argument as in [15], if the envelope scheme is secure and sender and receiver private then this authentication scheme is both secure and affiliation-hiding.

References

1. Balfanz, D., Durfee, G., Shankar, N., Smetters, D.K., Staddon, J., Wong, H.C.: Secret hand-shakes from pairing-based key agreements. In: IEEE Symposium on Security and Privacy (2003)
2. Barth, A., Boneh, D., Waters, B.: Privacy in encrypted content distribution using private broadcast encryption. In: Proceedings of Financial Cryptography 2006 (2006)
3. Bellare, M., Boldyreva, A., Desai, A., Pointcheval, D.: Key-privacy in public-key encryption. In: Boyd, C. (ed.) ASIACRYPT 2001. LNCS, vol. 2248. Springer, Heidelberg (2001)
4. Bellare, M., Boldyreva, A., Staddon, J.: Multi-recipient encryption schemes: Security notions and randomness re-use. In: PKC 2003 (2003)
5. Bellare, M., Kohno, T., Shoup, V.: Stateful public-key cryptosystems: how to encrypt with one 160-bit exponentiation. In: CCS 2006: 13th ACM conference on Computer and communications security (2006)
6. Bellare, M., Namprempre, C.: Authenticated encryption: Relations among notions and analysis of the generic composition paradigm. In: Okamoto, T. (ed.) ASIACRYPT 2000. LNCS, vol. 1976. Springer, Heidelberg (2000)
7. Bradshaw, R., Holt, J., Seamons, K.: Concealing complex policies in hidden credentials. In: CCS 2004: 11th ACM Conference on Computer and Communications Security (2004)
8. Canetti, R., Krawczyk, H.: Universally composable notions of key exchange and secure channels. In: Knudsen, L.R. (ed.) EUROCRYPT 2002. LNCS, vol. 2332. Springer, Heidelberg (2002)
9. Castelluccia, C., Jarecki, S., Tsudik, G.: Secret handshakes from CA-oblivious encryption. In: Lee, P.J. (ed.) ASIACRYPT 2004. LNCS, vol. 3329. Springer, Heidelberg (2004)
10. Chaum, D., van Heyst, E.: Group signatures. In: Davies, D.W. (ed.) EUROCRYPT 1991. LNCS, vol. 547. Springer, Heidelberg (1991)
11. Fujisaki, E., Okamoto, T.: Secure integration of asymmetric and symmetric encryption schemes. In: Wiener, M.J. (ed.) CRYPTO 1999. LNCS, vol. 1666. Springer, Heidelberg (1999)
12. Holt, J., Bradshaw, K., Seamons, K.E., Orman, H.: Hidden credentials. In: 2nd ACM Workshop on Privacy in the Electronic Society (2003)
13. Jarecki, S., Kim, J., Tsudik, G.: Authenticated group key agreement protocols with the privacy property of affiliation-hiding. In: Abe, M. (ed.) CT-RSA 2007. LNCS, vol. 4377. Springer, Heidelberg (2006)
14. Jarecki, S., Kim, J., Tsudik, G.: Beyond secret handshakes: Affiliation-hiding authenticated key exchange. In: Proceedings of CT-RSA (2008)
15. Jarecki, S., Liu, X.: Unlinkable secret handshakes and key-private group key management schemes. In: Katz, J., Yung, M. (eds.) ACNS 2007. LNCS, vol. 4521. Springer, Heidelberg (2007)
16. Jarecki, S., Liu, X.: Affiliation-hiding envelope and authentication schemes with efficient support for multiple credentials (2008), eprint.iacr.org
17. Jarecki, S., Liu, X.: Private conditional oblivious transfer and unlinkable secret handshakes (in submission, 2008)
18. Katz, J., Yung, M.: Unforgeable encryption and chosen ciphertext secure modes of operation. In: Schneier, B. (ed.) FSE 2000. LNCS, vol. 1978. Springer, Heidelberg (2001)
19. Kilian, J., Petrank, E.: Identity escrow. In: Krawczyk, H. (ed.) CRYPTO 1998. LNCS, vol. 1462. Springer, Heidelberg (1998)
20. Tsudik, G., Xu, S.: A flexible framework for secret handshakes. In: Privacy Enhancing Technologies 2006, pp. 295–315 (2006)

Author Index

Lecture Notes in Computer Science

Sublibrary 1: Theoretical Computer Science and General Issues

For information about Vols. 1–4818
please contact your bookseller or Springer